Methods in Molecular Biophysics
Structure, Dynamics, Function

Our knowledge of biological macromolecules and their interactions is based on the application of physical methods, ranging from classical thermodynamics to recently developed techniques for the detection and manipulation of single molecules. These methods, which include mass spectrometry, hydrodynamics, microscopy, diffraction and spectroscopy, electron microscopy, molecular dynamics simulations and nuclear magnetic resonance, are complementary; each has its specific advantages and limitations.

Organised by method, this textbook provides descriptions and examples of applications for the key physical methods in modern biology. It is an invaluable resource for undergraduate and graduate students of molecular biophysics in science and medical schools, as well as research scientists looking for an introduction to techniques beyond their speciality. As appropriate for this interdisciplinary field, the book includes short asides to explain physics aspects to biologists and biology aspects to physicists.

IGOR N. SERDYUK was born in Odessa, Ukraine, and studied physics at Odessa State University. Following research as a postgraduate student at the Department of Polymer Physics, Institute of High Molecular Weight Compounds, Leningrad, he obtained his Ph.D. in 1969. He then studied the physics of proteins as a junior researcher at the Laboratory of Protein Physics, Institute of Protein Research, USSR Academy of Sciences, Pushchino, rising to lead his own research group there and gaining his D.Sc. (Physics and Mathematics) from Moscow State University (1981). Since then he has been head of the Laboratory of Nucleoprotein Physics at the Institute of Protein Research, Pushchino. In 1985 he was awarded the USSR State Prize for Science and Technology and was appointed Full Professor of Molecular Biology in 1993. He has many years of experience teaching molecular biophysics in Moscow University.

NATHAN R. ZACCAI was educated in the French school system and attended Edinburgh University, where he gained a B.Sc. (Physics, 1997) and subsequently completed his D.Phil. at the University of Oxford (Biochemistry, 2001). He has undertaken research in immunology, virology and structural biology, specialising in X-ray crystallography. His initial work was on the structural biology of cell surface receptors involved in the immune system. He is currently based in the department of pharmacology, at the University of Bristol, where he is developing a research programme on the molecular basis of ion-receptor activation.

JOSEPH (GIUSEPPE) ZACCAI was born in Alexandria, Egypt, and educated in English language schools there and in Rome, and at the University of Edinburgh, where he obtained his B.Sc. (1968) and Ph.D. (1972) in Physics. A postdoctoral fellowship at the Brookhaven National Laboratory, New York, allowed him to pursue new interests in the biophysics of biological membranes and the application of neutron scattering methods in biology. He continued to develop these interests upon returning to Europe, first at the Institut Laue Langevin (ILL) in Grenoble, widely recognised as the foremost neutron scattering institute in the world. He holds the position of Directeur de Recherche with the Centre National de la Recherche Scientifique (CNRS) of France, and has been head of the Laboratory of Molecular Biophysics of the Institut de Biologie Structurale (IBS) in Grenoble since it was founded in 1992. Joe Zaccai's current research interests include the exploration of the physical chemical limits for life, especially in relation to the role of water and salt, molecular adaptation, protein dynamics, structure and stabilisation in organisms that live under extreme conditions of salinity, temperature, pressure, and exobiology. He has many years of experience teaching biology to physicists and physics to biologists, and was the first to introduce these courses at the undergraduate and graduate level at the University of Grenoble in the 1980s.

I first asked what methods in molecular biophysics I would expect to use as a biochemist and structural biologist. This text book provides an introduction to the physics of each of [the techniques used by my own group] as well as a review of the applications. . . . [It] will be in demand by third year undergraduates in the many courses run by physicists to introduce them to biological themes. It would also be used by the many post-graduate students doing . . . research degrees as well as post-doctorals in chemical biology, biochemistry, cell biology and structural biology research groups. . . . In summary, this is a valuable contribution to the field. . . . this is an area which has advanced tremendously and the major texts in biophysical methods are now simply out of date. The text covers the methods that young researchers and some undergraduates will wish to learn. I am sure that it will find itself on the shelves of many laboratories throughout the world. There is nothing quite like it at the moment.

Sir Tom Blundell FRS, FMedSci,
Professor and Head, Department of Biochemistry, University of Cambridge

Thank you very much for giving me the opportunity to preview this wonderful text book. It has outstanding breadth while maintaining sufficient depth to follow modern experiments or initiate a deeper understanding of a new subject area. I love the 'Physicist's' and 'Biologist's Boxes' to address specific subjects for researchers with different backgrounds. This is one of the most comprehensive and highly relevant texts on biophysics that I have encountered in the last 10 years, clearly written and up-to-date. It is a must-have for biophysicists working in all lines of research, and certainly for me.

Nikolaus Grigorieff, Professor of Biochemistry, Brandeis University

[This is] a wonderful up-to-date treatise on the many and diverse methods used . . . in the fields of molecular biophysics, physical biochemistry, molecular biology, biological physics and the new and emerging field of quantum nanobiology. The wide range of methods available . . . in these multidisciplinary fields has been overwhelming for most researchers, students and scientists [who fail] to fully appreciate the utility and usefulness of the methods [other than their own]. [In many cases, this has] created disagreements and . . . controversy. The only way to understand and appreciate fully the problems in quantum nanobiology and their complexity is to utilize and fully understand the many diverse methods covered by the authors in this very fine treatise . . . [It] should be in the library of any serious researcher in the many diverse multidisciplinary fields working on problems in quantum nanobiology. . . . They will be greatly rewarded by an ability to see and view the problems and their complexity through different perspectives, aspects and points of view, . . .

Karl J. Jalkanen, Associate Professor of Biophysics,
Quantum Protein Centre, Technological University of Denmark

This most welcome text provides an up-to-date introduction to the vast field of biophysical methods. Written in an accessible style with an eye to a broad audience, it will appeal to biologists who wish to understand how to determine how macromolecules function and to scientists with a physics or physical chemistry background who wish to know how measurement of the physical world can impact our understanding of biological problems. The book succeeds in unifying disparate approaches under the aegis of developing an understanding of how macromolecules work. Importantly, the text also provides the relevant historical background, an invaluable guide that will aid in the appreciation of what has gone before and should serve to orient them towards the future and what may be possible. It is a valuable resource for novice and seasoned biophysicists alike.

Dan Minor, California Institute for Quantitative
Biomedical Research University of California, San Francisco

Methods in Molecular Biophysics is now the book I consult first when faced with an unfamiliar experimental technique. Both classic analytical techniques and the latest single-molecule methods appear in this single comprehensive reference.

Philip Nelson, Department of Physics,
University of Pennsylvania, and author of Biological Physics (2004)

The authors provide an overview of many of the major recent accomplishments in the use of physical tools to investigate biological structure. There are interesting historical and biographical comments that lead the reader into understanding contemporary concepts and results. The book will be valuable both for students and research scientists.

Michael G. Rossmann,
Hanley Professor of Biological Sciences, Purdue University

The melding of physics, chemistry and biology in modern science has changed our view of the natural world and opened avenues for detailed understanding of the origin of biological regulation. *Methods in Molecular Biophysics* provides an up-to-date view of classical biophysics, theory and practice of modern chemical biology and represents an essential text for the interdisciplinary scientist of the 21st Century. A great achievement and presentation awaits the student who reads this book, along with an excellent reference for the seasoned practitioner of biophysical chemistry.

Milton H Werner,
Laboratory of Molecular Biophysics, The Rockefeller University

The methods, concepts, and discoveries of molecular biophysics have penetrated deeply into the fabric of modern biology. Physical methods that were once seemingly arcane are now commonplace in modern cell biology laboratories. This well written, thorough, and elegantly illustrated book provides the connections between molecular biophysics and biology that every aspiring young biologist needs. At the same time, it will serve physical scientists as a guide to the key ideas of modern biology.

Stephen H. White, Professor, Department of
Physiology and Biophysics, University of California at Irvine

Methods in Molecular Biophysics offers a well-written, modern and comprehensive coverage of the properties of biological macromolecules and the techniques used to elucidate these properties. The authors have done a great service to the biophysics community in providing a long-needed update and expansion of previous texts on analysis of biological macromolecules. The choice and organization of material is especially well done. This book will be of considerable value not only to students, but also, due to the scope and breadth of coverage, to experienced researchers. I enthusiastically recommend *Methods in Molecular Biophysics* to anyone who wishes to know more about the techniques by which the properties of biological macromolecules are determined.

David Worcester,
Department of Biological Sciences, University of Missouri – Columbia

Methods in
Molecular Biophysics

Structure, Dynamics, Function

Igor N. Serdyuk

Institute of Protein Research, Pushchino,
Moscow Region

Nathan R. Zaccai

University of Bristol

Joseph Zaccai

Institut de Biologie Structurale and Institut Laue
Langevin, Grenoble

CAMBRIDGE UNIVERSITY PRESS
Cambridge, New York, Melbourne, Madrid, Cape Town, Singapore, São Paulo

Cambridge University Press
The Edinburgh Building, Cambridge CB2 8RU, UK

Published in the United States of America by Cambridge University Press, New York

www.cambridge.org
Information on this title: www.cambridge.org/9780521815246

© Cambridge University Press 2007

First published 2007

Printed in the United Kingdom at the University Press, Cambridge

A catalogue record for this publication is available from the British Library

ISBN-13 978-0-521-81524-6 hardback

To Ol'ga, Brinda, Missy.

Contents

Foreword

D. M. ENGELMAN

It might be thought that this book is about methods, and, as is often supposed, that the field of biophysics is defined as a collection of such methods. But, as is carefully developed in the text, there is a deeper significance – methods define what we (provisionally) know about the world of biological molecules, and biophysics is a field that integrates pieces of information to give substance to our explanations of biology in terms of macromolecular space and time: structure, interactions and dynamics. Two interrelated ideas that biophysicists employ are the structure–function hypothesis and evolution.

The structure–function hypothesis is crisply discussed in the lucid introduction to the book: the idea is that each macromolecule coded by the genome has a function, and that the function can be understood using the chemical structure, interactions and dynamics of the macromolecule. Evolution forms the foundation of this reductionist view, since functionality is the basis of natural selection. Thus, biophysical methods teach us, within the limits of the information they give, about function and evolution. A particular hope, which has been rewarded with a great deal of success, is that understanding particular cases will lead to generalising ideas – base pairing in nucleic acids, oxygen binding by haem, self-association of lipids to create bounded compartments, for example.

A book that teaches the methods well creates the intellectual framework of our understanding, and can guide the field. Earlier efforts by Cohn and Edsall, Tanford, Edsall and Wyman, and Cantor and Schimmel have served this important purpose in the past, but the advance of time and technology has diluted the force of these classic works in contemporary biophysics, both in the teaching and the practice of the field. How welcome, then, a clearly written, thoughtful and modern text that will serve well, both in formal courses and as a reference. The authors have built each method from its fundamental premises and principles, have successfully covered an impressive span of topics, and will be rewarded by attention from an audience that hungers for the next defining text in molecular biophysics.

New Haven

Foreword

PIERRE JOLIOT

As the authors of this book have written in the Introduction the ideal biophysical method would have the capability of observing atomic level structures and dynamics of biological molecules in their physiological environment, i.e. in vivo. Such a method does not exist, of course, and it will probably never exist because of insurmountable technical constraints. Characterisation of structural and functional properties of biological molecules requires the concerted application of an arsenal of complementary techniques. We note that in practice, however, many highly productive molecular biophysics groups are concerned by a single technique that they 'push' to its extreme limits. Such groups develop an essentially methodological approach, in which they seek to characterise by their technique as many biological molecules as they can. High-throughput crystallography or structural genomics is an example of this type of biophysics. Its aim is to provide a precious data base of information on three-dimensional protein structures, analogous to that on primary structures from genome sequence – a data base that will be used intensively by all biologists.

A different approach consists in tackling a biological problem with a multi-disciplinary approach, in which molecular biophysics plays a dominant role. The aim of this approach is to define as finely as possible the functional, structural, and dynamic properties of the molecules implicated in the physiological process as well as their interactions. It is this type of approach that is implicitly defended by the book, which provides an important and exhaustive overview of biophysical techniques currently available, and discusses their strengths and limitations. The usual first step is to study each molecule in purified form. Most biophysical techniques require ordered or disordered samples made up of large numbers of identical molecules (there are 10^{16} molecules in 1 mg of a 60 000 molecular-weight protein!). The large number of molecules makes it possible to attain the required measurement sensitivity while minimising the damage induced by the experiment itself (the probing radiation, for example). These molecules are, therefore, studied in conditions that are quite different from their physiological environment. The next step is to look at associations between the molecules, and, in particular, at the complex supramolecular structures that are now believed to be present in the cell. Where it is not possible to organise these complexes in ordered two- or three-dimensional structures, their structures can only be observed to low resolution. Higher-resolution models can be obtained, however, from a 'theoretical' approach based on the ensemble of structural and dynamics data, on the complexes themselves and their components. This book also provides chapters on promising new developments, for example in single-molecule detection and manipulation. The final step, which is still difficult to climb, concerns the study of molecules and complexes in vivo. In this context, new technical approaches but also new ways of thinking must be explored, even if a few biophysical methods are already able to provide information on molecules in their cellular environment. By applying a function-to-structure approach in addition to the more traditional structure-to-function approach, it is possible to explore what are the structural organization models compatible with the function properties of an ensemble of molecules. For example, it was possible to demonstrate that diffusion of mobile carriers belonging to the photosynthetic electron transfer chain is restricted to small domains, on the basis of a thermodynamic and kinetics analysis of electron transfer reactions in the photosynthetic apparatus. These domains could be small membrane compartments isolated from one another or super complexes formed by the association of several large membrane proteins in which mobile carriers are trapped. In many cases, membranes as well as the cytosol appear to be highly compartmentalised systems. The determination of supramolecular organisation within these compartments will certainly be one of the major goals in modern biophysics.

Preface

André Guinier, whose fundamental discoveries contributed to the X-ray diffraction methods that are the basis of modern structural molecular biology, died in Paris at the beginning of July 2000, only a few weeks after it was announced in the press that a human genome had been sequenced. The sad coincidence serves as a reminder of the intimate connection between physical methods and progress in biology. Not long after, Max Perutz, Francis Crick and then David Blow, the youngest of the early protein crystallographers, passed away. The period marked the gradual closing of the era in which molecular biology was born and the opening of a new era. In what has been called the post-genome sequencing era, physical methods are now increasingly being called upon to play an essential role in the understanding of biological function at the molecular and cellular levels.

Classical molecular biophysics textbooks published in the previous decades have been overtaken not only by significant developments in existing methods such as those brought about by the advent of synchrotrons for X-ray crystallography or higher magnetic fields in NMR, but also by totally new methods with respect to biological applications, such as mass spectrometry and single-molecule detection and manipulation. Our ambition in this book was to be as up-to-date and exhaustive as possible. In their respective parts, we covered classical and advanced methods based on mass spectrometry, thermodynamics, hydrodynamics, spectroscopy, microscopy, radiation scattering, electron microscopy, molecular dynamics and NMR. But rapid progress in the field (we couldn't very well ask the biophysics community to stop working during the few years it takes to write and prepare a book!) and the requirement to keep the book to a manageable size meant that certain methods are either omitted or not perfectly up-to-date.

The key-word in molecular biophysics is *complementarity*. The Indian story of the six blind men and the elephant is an appropriate metaphor for the field. Each of the blind men touched a different part of the elephant, and concluded on its nature: a big snake said the man who touched the trunk, the tusks were spears, its side a great wall, the tail a paint brush, the ears huge fans, the legs were tree trunks. We could add a seventh very short-sighted man to the story who can see the whole elephant but as a blurred grey cloud to illustrate diffraction methods. As we wrote in the Introduction, the ideal molecular biophysics method does not exist. It would be capable of observing not only the positions of atoms in molecules in vivo, but also the atomic motions and conformational changes that occur as the molecules are involved in the chemical and physical reactions associated with their biological function, regardless of the time scale involved. No single experimental technique is capable of yielding this information. Each provides us with a partial field of view with its clear regions and areas in deep shadow. In the twenty-first century, physical methods have to cope with very complicated biological problems, whose solution will depend, on the ability to transfer structural and functional knowledge from the operation of a single molecule to the cellular level, and then to the whole organism. The splendour and complexity of the task is humbling, but the challenge will be met.

We are deeply obliged to Professor Don Engelman of Yale University, USA, and Professor Pierre Joliot of the Institut de Biologie Physico Chimique, France, who agreed to write forewords for the book. Outstanding scientists and teachers, each is both major actor and observer in biophysical research and the development of modern biology. Grateful thanks also to expert colleagues for critical discussions on the different methods: Martin Blackledge and the members of the NMR laboratory, Christine Ebel, Dick Wade, Hugues Lortat-Jacob, Patricia Amara, Jean Vicat the members of the laboratory of mass spectrometry, all of the Institut de Biologie Structurale, Ingrid Parrot and Trevor Forsyth of the Institut Laue Langevin, France, Regine Willumeit of the GKSS, Forschungszentrum Geesthacht, Germany, Victor Aksenov of the Joint Institute of Nuclear Research, Russia, Lesley Greene, Christina Redfield, Guillaume Stewart-Jones, Yvonne Jones and David Stuart of the University of Oxford, UK, Jonathan Ruprecht and Richard Henderson of the Laboratory of Molecular Biology, UK, Simon Hanslip and Robert Falconer of Cambridge University, UK. Hugh Montgomery of Edinburgh University, Rebecca Sitsapesan, Bristol University, David Worcester, University of Missouri, Philip Nelson, University of Pennsylvania, Georg Bueldt and Joachim Heberle of the Juelich Research Centre. We gratefully acknowledge support from the Radulf Oberthuer Foundation, Germany, the Institut de Biologie Structurale and the Institut Laue Langevin, France, and the Cyril Serdyuk Company, Ukraine. We are indebted to Gennadiy Yenin of the Institute of Protein Research, Russia for drawing figures and scientific illustrations, and to Aleksandr Timchenko, Margarita Shelestova, Margarita Ivanova, Tatyana Kuvshinkina, and Albina Ovchinnikova (Institute of Protein Research, Russia) for technical assistance. And finally, we would like to thank all our colleagues, friends and families, and the staff of Cambridge University Press, who supported us with much patience, understanding and encouragement.

Introduction
Molecular biophysics at the beginning of the twenty-first century: from ensemble measurements to single-molecule detection

The ideal biophysical method would be capable of measuring atomic positions in molecules in vivo. It would also permit visualisation of the structures that form throughout the course of conformational changes or chemical reactions, regardless of the time scale involved. At present there is no single experimental technique that can yield this information.

A brief history and perspectives

Molecular biology was born with the double-helix model for DNA, which provided a superbly elegant explanation for the storage and transmission mechanisms of genetic information (Fig. 1). The model by **J. D. Watson** and **F. H. C. Crick** and supporting fibre diffraction studies by **M. H. F Wilkins, A. R. Stokes, and H. R Wilson**, and **R. Franklin and R. G. Gosling**, published in a series of papers in the **25 April, 1953** issue of *Nature*, marked a major triumph of the physical approach to biology.

The Watson and Crick model was based only in part on data from X-ray fibre diffraction diagrams. The patterns, which demonstrated the presence of a helical structure of constant pitch and diameter, could not provide unequivocal proof for a more precise structural model. One of the 'genius' aspects of the discovery was the realisation that A–T and G–C base pairs have identical dimensions; as the rungs of the double-helix ladder, they give rise to a constant diameter and pitch. From a purely 'diffraction physics' point of view, a variety of helical models was compatible with the fibre diffraction diagram, and other authors proposed an alternative model for DNA, the so-called 'side-by-side model', coupling two single DNA helices. This shows that if molecular biology were to be established, it was important to obtain the structure of biological molecules in more detail than was possible from fibre diffraction. Considering the dimensions involved, about 1 Å (0.1 nm) for the distance between atoms, X-ray crystallography appeared to be the only suitable method. Major obstacles remained to be overcome such as obtaining suitable crystals, coping with the large quantity of data required to describe the positions of all the atoms in a macromolecule, and solving the phase problem.

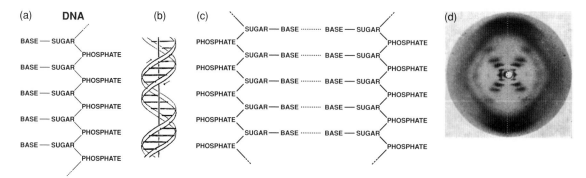

Fig. 1 (a) Chemical organisation of a single chain of DNA. (b) This figure is purely diagrammatic. The two ribbons symbolise the two phosphate–sugar chains, and the horizontal rods the pairs of bases holding the chains together. The vertical line marks the helix axis. (c) Chemical organisation of a pair of DNA chains. The hydrogen bonding is symbolised by dotted lines. (d) X-ray fibre diffraction of the B-form of DNA. The figures are facsimiles from the original papers of Watson and Crick (1953) and Franklin and Gosling (1953).

Protein crystals had already been obtained in the 1930s. It was not until 1957, however, that **Max Perutz** and **John Kendrew** found a way to solve the crystallographic phase problem by isomorphic substitution using heavy-atom derivatives. This permitted the structure of myoglobin to be solved in sufficient detail to describe how the molecule was folded. The difficulties encountered with protein crystallisation, and the labour intensive nature of the crystallographic study itself (this was before powerful computers and long calculations were essentially performed by 'post-doctoral hands') appeared to doom protein crystallography to providing rare, unique information on the three-dimensional structure of a very few biological macromolecules. Structural molecular biologists, therefore, continued the development and improvement of methods that do not provide atomic resolution but have complementary advantages for the study of macromolecular structures. These methods, at the boundary between thermodynamics and structure, had already played crucial roles in the century before the discovery of the double helix. The discovery of biological macromolecules is itself tightly interwoven with the application of physical concepts and methods to biology (biophysics).

The application of physics to tackle problems in biology is certainly older than its definition as biophysics. The *Encyclopædia Britannica* suggests that the study of bioluminescence by **Athanasius Kircher** in the **seventeenth century** might be considered as one of the first biophysical investigations. Kircher showed that an extract made from fireflies could not be used to light houses. The relation between biology and what would become known as electricity has preoccupied physicists for centuries. **Isaac Newton**, in the concluding paragraph of his *Principia* **(1687)**, reflected that ' . . . all sensation is excited, and the members of animal bodies move at the command of the will, namely, by the vibrations of this Spirit, mutually propagated along the solid filaments of the nerves, from the outer organs of sense to the brain, and from the brain into the muscles. But these are things that cannot be explained in few words, nor are we furnished with that sufficiency of experiments which is required to an accurate determination and demonstration of the laws by which this electric and elastic Spirit operates.' One hundred years later, **Luigi Galvani** and **Alessandro Volta** performed the experiments on frogs' legs that would lead to the invention of the electric battery. They also laid the foundations of the science of electrophysiology, even though, because of the excitement caused by the electric battery it was well into the **nineteenth century** before the study of animal electricity was developed further, notably by **Emil Heinrich Du Bois-Reymond**. Another nineteenth century branch of biophysics, however, that dealing with diffusion and osmotic pressure in solutions, would later overlap with physical chemistry, and is more directly relevant to the discovery and study of biological macromolecules. The first papers published in *Zeitschrift für Physikalische Chemie* **(1887)** were concerned with reactions in solution, because biological processes essentially take place in the aqueous environment inside living cells.

The thermal motion of particles in solution ('Brownian' motion) was discovered by **Robert Brown (1827).** The **Abbé Nollet**, a professor of experimental physics, first described osmotic pressure in the early **nineteenth century** from experiments using animal bladder membranes to separate alcohol and water. The further study and naming of the phenomenon is credited to the medical doctor and physiologist **René J. H. Dutrochet (1828)**, who recognised the important implication of osmotic phenomena in living systems and firmly believed that basic biological processes could be explained in terms of physics and chemistry. The theory of osmotic pressure was developed by **J. Van't Hoff (1880). George Gabriel Stokes** (middle of the **nineteenth century**) is best known for his fundamental contributions to the understanding of the laws governing particle motion in a viscous medium, but he also named and worked on the phenomenon of fluorescence. The laws of diffusion under concentration gradients were written down by **Adolf Fick (1856)**, by analogy with the laws governing heat flow. The second half of the **nineteenth century** also saw the discoveries of flow birefringence by **James Clerk Maxwell** and of electric birefringence in solutions by **John Kerr**. Both phenomena depend on the existence of large asymmetric solute particles.

Macromolecules, although large as molecules, are still much smaller than the wavelength of light. They could not be seen through direct observation by using microscopes, which had already shown the existence of cells in biological tissue and of structures within the cells such as the chromosomes (from the Greek, meaning 'coloured bodies'). From the knowledge gained from experiments on solutions it gradually became apparent that the biochemical activity of proteins, studied by **Emil Fischer (1882)**, is due to discrete macromolecules. In **1908, Jean Perrin** applied a theory proposed by **Albert Einstein (1905)** to determine Avogadro's number from Brownian motion. The theory of macromolecules is due to **Werner Kuhn (1930)** after **Hermann Staudinger (1920)** proposed the concept of macromolecules as discrete entities, rather than colloidal structures made up of smaller molecules. The discovery of X-rays by **Wilhelm Conrad Röntgen (1895)**, and its application to atomic crystallography in the **1910s** through the work of **Peter Ewald, Max von Laue** and **William H.** and **W. Laurence Bragg** laid the ground work for the observation of atomic structural organisation within macromolecules almost half a century later. **Theodor Svedberg (1925)** made the first direct 'observation' of a protein as a macromolecule of well-defined molar mass by using the analytical centrifuge he had invented. In parallel, the atomic theory of matter became accepted as fact. There was rapid progress in X-ray diffraction and crystallography, electron microscopy and atomic spectroscopy. The novel experimental tools, provided by the new understanding of the interactions between radiation and matter, were carefully honed to meet the challenge of biological structure at the molecular and atomic levels. Physicists, encouraged by the example of **Max Delbrück**, who chose to study the genetics of a bacteriophage (a bacterial virus) as a tractable model in the **1940s**, and **Erwin Schrödinger's** influential book *What is Life?* **(1944)**, which discussed whether or not biological processes could be accounted for by the known laws of physics, turned to biological problems in a strongly active way.

At the beginning of the **twenty-first century**, biophysics is dominated by two methods, X-ray crystallography and NMR, which play the key role in determining three-dimensional structures of biological macromolecules to high resolution. But even if all the protein structures in different genomes were solved, crucial questions would still remain. What is the structure and dynamics of each macromolecule in the crowded environment of a living cell? How does macromolecular structure change during biological activity? How do macromolecules interact with each other in space and time? These questions can be addressed only by the combined and complementary use of practically all biophysical methods. Mass spectrometry can determine macromolecular masses with astonishing accuracy. Highly sensitive scanning and titration microcalorimetry are applied to determine the thermodynamics of macromolecular folding and stability, and are joined by biosensor techniques in the study of binding interactions. There has been a rebirth of analytical ultracentrifugation, with the advent of new, highly precise and automated instrumentation, and it has joined small-angle X-ray and neutron scattering in the study of macromolecular structure and interactions in solution and the role of hydration. A femtosecond time resolution has been achieved for the probing of fast kinetics by optical spectroscopy. Light microscopy combined with fluorescence probes can locate single molecules inside cells. Scanning force microscopy is determining the profile of macromolecular surfaces and their time-resolved changes. Electron microscopy is approaching close to atomic resolution and is most likely to bridge the gap between single-macromolecule and cellular studies. Neutron spectroscopy is providing information on functional dynamics of proteins within living cells. Synchrotron radiation circular dichroïsm can access a wider wavelength range vacuum ultraviolet for the study of electronic transitions in the polypeptide backbone.

Up to the **late 1970s**, biophysics and biochemistry had only dealt with large molecular ensembles for which the laws of thermodynamics are readily applicable. One hundred microlitres of a 1 mg/ml solution of haemoglobin, for example, contains 10^{18} protein molecules; a typical protein crystal contains of the order of 10^{15} macromolecules. In their natural environment, however, far fewer molecules are involved in any interaction and exciting new methods have been developed that allow the study of single molecules. Single molecules can now be detected and manipulated with hypersensitive spectroscopic and even mechanical probes such as atomic force microscopy, with which a single macromolecule can even be stretched into novel conformations. Conformational dynamics can be measured by single-molecule fluorescence spectroscopy. Fluorescence resonance energy transfer can measure distances between donor and acceptor pairs in single molecules, in vitro or in living cells. Near-field scanning optical microscopy can identify and provide dynamics information on single molecules in the condensed phase.

The historical development of each of the biophysical methods outlined above is discussed in more detail in the corresponding section of the book.

Languages and tools

Physike in Greek is the feminine of *physikos* meaning *natural*. Physics is the science of observing and describing Nature. When one of the authors (J. Z.) was a student at Edinburgh University, physics was taught in the department of Natural Philosophy. The word *philosophy*, love of wisdom, conveyed quite accurately how the wisdom of the observer is brought to bear in science. The observer plays his role through the tools he uses in his experiments and the language he uses to describe his results. Modern science covers so many diverse areas that it is impossible to master an understanding of all the tools and languages involved. Biophysics students are familiar with the language difficulties of trying to communicate with 'pure' physicists, on the one hand, and 'pure' biologists, on the other, despite decades of interdisciplinary teaching and research in universities. Rather than bemoaning this fact, we should recognise that it reflects the richness and depth of each discipline, expressed in its own sophisticated language, and developed in its own set of observational tools. Clearly, physics and biology have different languages, but it is important to appreciate that within each discipline also there are different languages. Language influences tool development, which in turn contributes to refine the concepts described by language. Biophysicists have to be fluent in the various languages of physics and biology and be able to translate between them accurately. This is a very difficult and sometimes impossible task, as any good language interpreter can testify, each language having its own specificity and view point.

Biophysics deals, to a large extent, with the structure, dynamics and interactions of biological macromolecules. What are biological macromolecules? Their biological activity is described in the language of biochemistry and molecular spectroscopy; they were discovered through their hydrodynamic and thermodynamic properties; they are visualised by their radiation scattering properties, and their pictures are drawn in beautiful colour as physical particles. To each language there corresponds a set of tools, the instruments and methods of experimental observation. Progress in probing and understanding biological macromolecules has undoubtedly been based on advances in the methods used. Physical tools capable of ever increasing accuracy and precision require a parallel development in biochemical tools (often themselves of physical basis, like electrophoresis or chromatography, for example) to provide meaningful samples for study. The word *meaningful* is a key word in the previous sentence. It refers to the relevance of the study with respect to *biology* (from the Greek *bios*, life, and *logos*, word or reason), i.e. biophysics has the goal of increasing our understanding of life processes. It should be distinguished from *biological physics*, which deals with the properties of biological matter, for example to design nanomachines based on DNA.

Length and time scales in biology

Biological events occur on a wide range of length and time scales – from the distance between atoms on the ångström scale to the size of the earth as an ecosystem, from the femtosecond of electronic rearrangements when retinal absorbs a photon in the first step of vision to the 10^9 years of evolution. Observation tools have been developed that are adapted to the different parts of the length and time scales. The cell represents a central threshold for biological studies (Fig. 2). With a usual size of the order of 1–10 μm, cells can be seen in the light

Fig. 2 A 'realistic' drawing of the bacterium *Escherichia coli*, based on available experimental data. A flagellum, the double cell membrane and its associated proteins and glycoproteins are shown in hues of green; ribosomes and other protein and nucleic acid cytoplasmic components are in violet and blue; nascent polypeptide chains are in white; DNA and its associated proteins are in yellow and orange. The scale is given by the size of the bacterium of about 1 μm, or the double membrane thickness of about 10 nm. (http://www.scripps.edu/mb/goodsell/)

microscope. Also, the durations of cellular processes, which are of the order of seconds to minutes can be observed and measured with relative ease. If we imagined diving into a eukaryotic cell through its plasma membrane, we would see other membrane structures that separate distinct compartments like the nucleus or mitochondria, large macromolecular assemblies such as chromatin, ribosomes, chaperone molecular machines or multienzyme complexes. Looking for progressively smaller structures we would find RNA and protein molecules, then peptides and other small molecules, water molecules and ions, and finally the atoms that make them up (Fig. 3).

The smaller the length, the shorter the time, the heavier is the implication of sophisticated physical instrumentation and methods for their experimental observation.

The femtosecond (10^{-15} s) is the shortest time of interest in molecular biology; it corresponds to the time taken by electronic reorganization in the light sensitive molecule, retinal, upon absorption of a photon, in the first step in vision. Time intervals of this order can be measured by laser spectroscopy (the distance covered by light in 1 fs is 3×10^{-7} m, or 300 nm, about one half the wavelength of visible light). Thermal fluctuations are in the picosecond (10^{-12} s) range; DNA unfolds in microseconds; enzyme catalysis rates are of the order of 1000 reactions per second; protein synthesis takes place in seconds etc. The longest time of interest in molecular biology is, in fact, geological time, corresponding to the more than one thousand million years of molecular evolution (Fig. 4).

The structure–function hypothesis

This book describes the application of classical and advanced physical methods to observe biological structure, dynamics and interactions at the molecular level. Intensive research since the 1950s has emphasized the fundamental importance of biological activity at this level. The *structure–function* hypothesis is the foundation of molecular biology. One of its implications is that if a protein exists today in an organism it is because it fulfils a certain biological function and its 'structure' has been selected by evolution. The discovery and study of nucleic acids and proteins as macromolecules with well-defined structures has allowed an unprecedented understanding of processes such as the storage and transmission of genetic information, the regulation of gene expression, enzyme catalysis, immune response or signal transduction. In parallel, it became apparent that we could act on biological processes by acting on macromolecular structures and powerful tools were developed not only to further fundamental scientific understanding but also to apply this knowledge in biotechnology or in drug design pharmacology.

The concept of 'structure' should be understood in the broadest sense. The three-dimensional organisation of a protein is not rigid but can adapt to its ligands according to the hypothesis of 'configurational adaptivity' or 'induced fit'. Also,

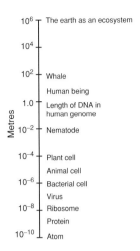

Fig. 3 Length scales in biology.

Fig. 4 Time scales in biology.

many proteins have been found that display a highly flexible random-coil conformation under physiological conditions. An intrinsically disordered protein could adopt a permanent structure through binding, but there are cases of proteins with intrinsic disorder that are biologically active while remaining disordered. A large proportion of gene sequences appear to code for long amino acid stretches that are likely either to be unfolded in solution or to adopt non-globular structures of unknown conformation.

Events taking place on the ångström and picosecond scales have profound consequences for life processes over the entire range of length and time scales – from the length and time associated with a cell, via those associated with an organism to those associated with the relation between an organism and its environment. The development of high-throughput techniques for whole genome sequencing, for the analysis of genomic information (bioinformatics), for the identification of all the proteins present in a cell (functional proteomics), for determining how this population responds to external conditions (dynamic proteomics) and for protein structure determination (structural genomics) has opened up a new era in molecular biology whose revolutionary impact still remains to be assessed.

Biological macromolecule structures usually appear in pictures as static structures. A more precise definition would be 'ensemble and time-averaged' structures. The atoms in a macromolecular structure are maintained at their average positions by a balance of forces. Under the influence of thermal energy, the atoms move about these positions. Dynamics, from the Greek *dynamis* meaning strength, pertains to forces. Structure *and* motions result from forces. It is common usage in biophysics, however, to separate *structure* from *dynamics*, considering the first as referring to the length scale (i.e. to the time-averaged configuration) and the second as referring to the time scale (i.e. to energy and fluctuations). The separation into two separate concepts is validated by the fact that the methods used to study structure and dynamics are usually quite separate and specialised. Modern experiments, however, often address both an average structure and how it changes with time.

Complementarity of physical methods

We know of the existence of macromolecules only through the methods with which they are observed. No single method, however, provides all the information required on a macromolecule and its interactions. Each method gives a different view of the system in space and time: the methods are *complementary*.

Biological macromolecules take up their active structures only in a suitable solvent environment. The forces that stabilise them are weak forces (of the order of kT, where k is Boltzmann's constant and T is absolute temperature), which arise in part from interactions with the solvent. The study of biological macromolecules, therefore, cannot be separated from the study of their aqueous solutions. The

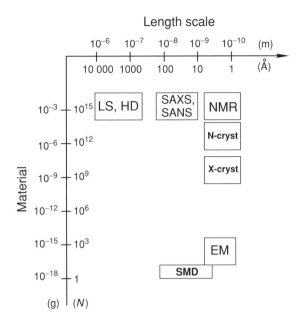

Length scale

Fig. 5 Length resolution achieved and amount of material required for the sample for experiments using different physical methods to determine structure. Abbrevations: g, grams; *N*, number of molecules (assuming a molecular weight of the order of 100 000); LS, light scattering; HD, hydrodynamics; SAXS, SANS, X-ray and neutron small-angle scattering, respectively; NMR, neutron magnetic resonance in solution; N-cryst, neutron crystallography; X-cryst, X-ray crystallography; EM is electron microscopy; SMD is single-molecule detection methods.

macromolecules are usually studied in dilute or concentrated solutions, in the lipid environment of membranes, or in crystals. Protein molecules or nucleic acid molecules in the unit cell of a crystal are themselves surrounded by an appreciable number of solvent molecules, and there are aqueous layers on either side of membranes. According to the experimental method used, we shall consider biological macromolecules in solution as 'physical particles' (mass spectrometry, single-molecule detection . . .), 'thermodynamics particles' (osmotic pressure measurements, calorimetry . . .), 'hydrodynamics particles' (viscosity, diffusion, sedimentation . . .) or 'radiation interaction particles' (spectroscopy, diffraction and microscopy).

The length resolution scale achieved, the techniques involved and the sample mass required for some biophysical methods are illustrated in Fig. 5.

Thermodynamics

It is a result of classical thermodynamics that many properties of solutions, such as an increase in boiling point, freezing point depression, and osmotic pressure, depend on the number concentration of the solute. At constant mass concentration, therefore, these thermodynamics parameters vary sensitively with the molecular mass of the solute. Thus, for example, macromolecular masses and interactions have been determined from osmotic pressure measurements.

Macromolecular folding itself and the stabilisation of active biological structures follow strict thermodynamics rules in which interactions with solvent play a determinant role. Sensitive calorimetric measurements of heat capacity as a function of temperature showed very clearly that stabilisation free energy presents a maximum at a temperature close to the physiological temperature, the stability of the folded particle decreasing at lower as well as higher temperatures. The interpretation is the following. The behaviour of the chain surrounded by solvent is much more complex than if it were in a vacuum. Enthalpy may rise, decrease or even not change upon folding, because bonds can be made equally well within the macromolecule and between the chain and solvent components. Similarly for entropy, the loss of chain configuration freedom upon folding may be more than compensated for by a loss of degrees of freedom for the solvent molecules around the unfolded chain, for example through the exposure of apolar groups to water molecules. A water molecule in bulk has the freedom to form hydrogen bonds with partners in all directions. Apolar groups cannot form hydrogen bonds, so that water molecules in their vicinity lose some of their bonding possibilities; their entropy is decreased.

In a protein solution, the heat capacity is strongly dominated by the water, and that of the macromolecules represents a very small part of the measured total. High-precision microcalorimeters were built to allow experiments on protein solutions to be performed. Nevertheless, early calorimetric studies on biological macromolecules concentrated on relatively large effects such as sharp transitions as a function of temperature. They led to a fundamental understanding of the energetics of protein folding. There are now important modern developments in the field. Very sensitive nanocalorimeters have been developed as well as analysis programs to treat the thermodynamics information and relate it to structural data. The energetics of intramolecular conformational changes, of complex formation and of interactions between partner molecules can now be explored in detail for proteins and nucleic acids. We should recall, however, that calorimetry (like all thermodynamics-based methods) provides measurements of an ensemble average over a very large number of particles (typically of the order of 10^{15}), even if results are usually illustrated in a simple way in terms of changes occurring in one particle.

Hydrodynamics

The first hints of the existence of biological macromolecules as discrete particles came from observations of their hydrodynamic behaviour. The language of macromolecular hydrodynamics is the language of fluid dynamics in the special regime of low Reynolds' numbers. The Reynolds' number in hydrodynamics is a dimensionless parameter that expresses the relative magnitudes of inertial and viscous forces on a body moving through a fluid. Bodies with the same Reynolds, number display the same hydrodynamics behaviour. Because of this, it is possible, for example, to determine the behaviour of an airplane wing from wind-tunnel studies on a small-scale model. The Reynolds' numbers of a small fish and a whale are 10^5 and 10^9, respectively.

Reynolds' numbers in aqueous solutions for biological macromolecules and their complexes, from small proteins to large virus particles and even bacteria, are very small. For example, it is 10^{-5} for a bacterium swimming with a velocity of about 10^{-3} cm/s. Inertial forces are negligible under such conditions, so that the motion of a particle through the fluid depends only on the forces acting upon it at the given instant; it has no inertial memory. Particle diffusion through a fluid under the effects of thermal or electrical energy, and sedimentation behaviour in a centrifugal field can be predicted by relatively simple equations in terms of macro-molecular mass and frictional coefficients that depend on shape. The *resolution* defines the detail with which a particle structure is described. Hydrodynamics provides a *low-resolution* view of a biological macromolecule, for example as a two- or three-axis ellipsoid, but it is also very sensitive to particle flexibility and particle–particle interactions. Modern hydrodynamics includes a number of novel experimental methods. In addition to the classical approaches of analytical ultracentrifugation to measure sedimentation coefficients and dynamic light scattering to measure diffusion coefficients, we now have free electrophoresis to measure transport properties in solution, fluorescence photobleaching recovery to monitor the mobility of individual molecules within living cells, time-dependent fluorescence polarisation anisotropy and electric birefringence to calculate rotational diffusion coefficients, fluorescent correlation spectroscopy and localised dynamic light scattering to measure macromolecular dynamics.

Radiation scattering

We see the world around us because it scatters light, which is detected by our eyes and analysed in our brains. In a diffraction experiment, waves of radiation scattered by different objects interfere to give rise to an observable pattern, from which the relative arrangement (or structure) of the objects can be deduced. The interference pattern arises when the wavelength of the radiation is similar to or smaller than the distances separating the objects. In some cases, the waves forming the pattern can be recombined by a lens to provide a direct image of the object. Atomic bond lengths are close to one ångström unit (10^{-10} m or 0.1 nm), and three types of radiation are used, in practice, to probe the atomic structure of macromolecules by diffraction experiments: X-rays of wavelength about 1 Å, electrons of wavelength about 0.01 Å, and neutrons of wavelength about 0.5–10 Å. Visible light scattering, with wavelengths in the 400–800 nm range, provides information on large macromolecular assemblies and their dynamics. X-rays, however, because they permit studies to atomic detail, provided the foundation on which structural biology has been built and is developing.

Neutron diffraction studies of biological membranes, fibres and macro-molecules and their complexes in crystals and in solution became possible in the 1970s with the development of methods that make full use of the special properties of the neutron.

Following the limitations of staining techniques, cryoelectron microscopy was developed to visualise subcellular and macromolecular structures to increasing resolution.

In the last decade of the twentieth century, the availability of intense synchrotron sources caused a revolution in macromolecular crystallography by greatly increasing the rate at which structures could be solved. Efficient protein modification, crystallisation, data collection and analysis approaches were developed for macromolecular crystallography. Extremely fast data-collection times made it possible to use time-resolved crystallography to study kinetic intermediates in enzymes. In parallel, field emission gun electron microscopes were applied and new methods developed to solve single-particle structures. Spallation sources for neutron scattering promise highly improved data-collection rates.

Light, X-rays and neutrons are scattered weakly by matter and require samples containing very large numbers of particles in order to obtain good signal-to-noise ratios. Experiments provide ensemble-averaged structures. Modern electron microscopy methods, on the other hand, allow single macromolecular particles to be visualised.

Spectroscopy

In spectroscopy, the radiation has exchanged part of its energy with the sample, through absorption effects or excitations due to particle internal or global dynamics, resulting in a change in the wavelength (frequency or colour) of the outgoing beam with respect to the incident beam. Since absorption depends on the location of an atom in a structure, certain types of spectroscopic experiment may also be used to study structure. Nuclear magnetic resonance (NMR) spectroscopy is sensitive to close to atomic resolution. The frequency of absorbed radiation can be measured as a function of time with an accuracy better than one part in a million. The precise nature of the signal depends on the chemical environment of the nucleus; hence structural information is obtained. In magnetic resonance imaging (MRI), millimetre resolution is obtained with metre wavelength probes by placing the body to be observed in magnetic field gradients and by focusing on nuclei in a given chemical environment; an absorption resonance then corresponds to a given field value and therefore to a precise location. As with diffraction, for which the wavelength matches the structural resolution required, the beam energy in spectroscopy is chosen so that differences due to sample excitations or absorption can be measured readily. In general, therefore, radiation of different wavelengths is used for diffraction and for spectroscopic experiments.

Coherent spectroscopy, in which radiation fields of well-defined phase are used, created unprecedented opportunities to study dynamics and time-evolving structures. The 'spin echo' method, applied to NMR and neutron spectroscopy, was extended by the 'photon echo' method when coherent lasers became available. Two-dimensional spectroscopy, first developed for NMR, measures the

Electromagnetic radiation

λ (m)	10^{-15}	10^{-12}	10^{-9}	10^{-6}	10^{-3}	1	10^3
ν (s^{-1})	10^{24}	10^{21}	10^{18}	10^{15}	10^{12}	10^9	10^6

Cosmic rays Gamma rays X-rays Ultraviolet Visible Infrared Radio

E (eV)	10^9	10^6	10^3	1	10^{-3}	10^{-6}	10^{-9}
E/k (K)	10^{13}	10^{10}	10^7	10^4	10	10^{-2}	10^{-5}

Neutrons

λ (m)		10^{-10}	10^{-9}
v (m s^{-1})		4000	400
E (eV)		3×10^{-2}	3×10^{-4}
E/k (K)		400	4

Fig. 6 Wavelength, energy and frequency for electromagnetic and neutron radiation. The scales in the figure give approximate orders of magnitude. The precise values for the constants are obtained from: $\nu\lambda = c$ where $\nu\lambda$ are the frequency and wavelength, respectively, of electromagnetic radiation and c is the speed of light (3×10^8 m/s); $E = h\nu$ (where E is energy and h is Planck's constant ($6.626 = 10^{-34}$ Js $= 4.136 \times 10^{-15}$ eV s); the temperature equivalent of energy, 1 eV/k = 11604.5 K, where k is Boltzmann's constant. In the neutron case, $\lambda = h/mv$ (where v m/s is neutron speed), and $E = \frac{1}{2}mv^2$, where m is neutron mass (1.6726 $\times 10^{-27}$ kg).

coupling within networks of vibrational modes. It has been applied to the infrared region to determine the structure of small molecules. The most exciting aspect of two-dimensional infrared spectroscopy is the combination of its sensitivity to structure and time resolution down to the femtosecond.

Taking electromagnetic radiation as an example, atomic diffraction requires X-ray wavelengths, while intramolecular vibrations correspond to infrared energies (Fig. 6). In NMR spectroscopy, the probing electromagnetic radiation is in the radio-frequency range, corresponding to metre wavelengths. Note that with neutron radiation, wavelengths of about 1 Å (corresponding to interatomic distances and fluctuation amplitudes) have associated energies of about 1 meV (corresponding to the energies of atomic fluctuations), so that diffraction and spectroscopy experiments can be performed simultaneously to measure atomic amplitudes and frequencies of motion in macromolecules. Molecular time scales, corresponding energies and temperatures are shown in Fig. 7 for different biophysical methods.

Single-molecule detection

Until the 1980s, biochemical and biophysical studies of biological macromolecules suffered the fundamental disadvantage of always having to deal with

Fig. 7 Molecular time scales, associated energies and temperatures of various biophysical methods. The range follows the dashed black diagonal but the arrows have been displaced horizontally for clarity. Abbreviations: DLS, dynamic light scattering; NMR, nuclear magnetic resonance; EB, electric birefringence; NS, neutron spectroscopy; FTIR, Fourier transform infrared spectroscopy; LS, laser spectroscopy; 2D-IR, two-dimensional infrared spectroscopy; FB, flow birefringence; FD, fluorescence depolarisation.

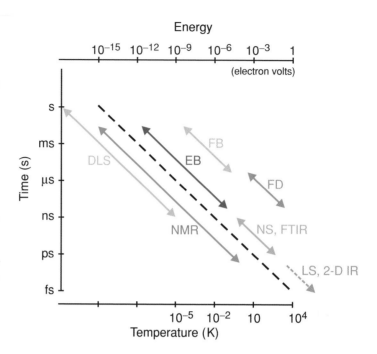

very large numbers of particles, whereas under in-vivo conditions they function as single particles in a dynamic heterogeneous environment. Structures, dynamics and interactions were (and predominantly still are) observed and measured as ensemble averages. Furthermore, enzymatic, binding or signalling reactions are in general stochastic, so that the kinetics of a protein activity measurement, for example, is also hidden in an ensemble average when measured in a large molecular population, even if the reaction is triggered contemporaneously for the entire sample.

Single macromolecules had been visualised by electron microscopy, but only in the last decade have methods become available to observe them while they were active. The development of single-molecule detection (SMD) techniques now permits allows the observation as well as the manipulation of single macromolecules in action. SMD is based on the two key technologies of *single-molecule imaging under active conditions* and *nanomanipulation*. Single-molecule signals that are detectable with good signal-to-noise ratios are given by fluorescent labels, which are observed using fluorescent optical microscopy. Applying total reflection and evanescent field techniques, the resolution of the method is several fold better than the diffraction limit given by the wavelength of light. Single-molecule nanomanipulation techniques include capturing biomolecules using a glass needle or beads trapped by the force exerted by a focused laser beam (optical tweezers), and probing molecular forces with atomic force or scanning probe near-field microscopy. The forces involved are in the piconewton range, comparable to the thermal forces stabilising the active macromolecular structures.

Table 1. *The range of forces in macromolecules*

Tensile strength of a covalent bond	1000–2000 pN
Deformation of a sugar ring	700 pN
Breaking of double-stranded DNA	400–580 pN
Unfolding the β-fold immunoglobulin domain of the muscle protein titin	180–320 pN
Adhesive force between avidin and biotin	140–180 pN
Structural transition of uncoiling double-stranded DNA upon stretching	60–80 pN
Structural transition of double-stranded DNA upon torsional stress	~20 pN
Individual nucleosome disruption	20–40 pN
Unfolding triple helical coiled-coil repeating units in spectrin	25–35 pN
RNA–polymerase motor	14–27 pN
Structural transition of RNA hairpin in ribozyme under stretching (folding–unfolding)	~14 pN
Separation of complementary DNA strands (room temperature, 150 mM NaCl, sequence-specific)	10–15 pN
Stall force of the myosin motor	3–6 pN
Force generated by protein polymerisation in growing microtubules	3–4 pN

Erwin Schrödinger wrote in 1952 that we would never be able to perform experiments on just one electron, one atom, or one molecule. In the early 1980s, however, scanning tunnelling microscopy was invented by G. Binning and H. Rohrer and radically changed the ways scientists view matter. Mechanical experiments to measure the piconewton forces that structure a single macromolecule became possible (Comment 1).

In optical tweezer instruments (Fig. 8(a)) one or two laser beams are focused to a small spot, creating an optical trap for polystyrene beads. One end of a single molecule (DNA, for example) is attached to a bead, while the other end is attached to a moveable surface, which, in this example, is another bead on a glass micropipette. The opposing force is measured, as the molecule is stretched by moving the micropipette.

In magnetic tweezer instruments (Fig. 8(b)), one end of the single molecule is attached to a glass fibre, while the other end is attached to a magnetic bead. A magnetic field exerts a constant force on the bead. The extension and rotation of the molecule as a function of the applied force is then measured.

In an atomic force microscopy experiment (Fig. 8(c)), one end of the molecule is attached to a surface, and the other to a cantilever. As the surface is pulled away, the deflection of the cantilever is monitored from the position of a reflected laser beam.

Comment 1
Entropic force

The typical energy scale for a macromolecule is thermal energy: $k_B T = 4 \times 10^{-21}$ J. Since the length scale of biomacromolecules is of order of 1 nm, the force scale is on the order of the piconewton (10^{-12} N). Therefore an entropic force can be calculated as $k_B T/(1$ nm$)$, which is equal to 4 pN at 300 K.

Fig. 8 A schematic view of three main techniques used in single-molecule force studies: (*a*) optical tweezer, (*b*) magnetic tweezer and (*c*) atomic force microscopy (Carrion-Vazquez *et al.*, 2000).

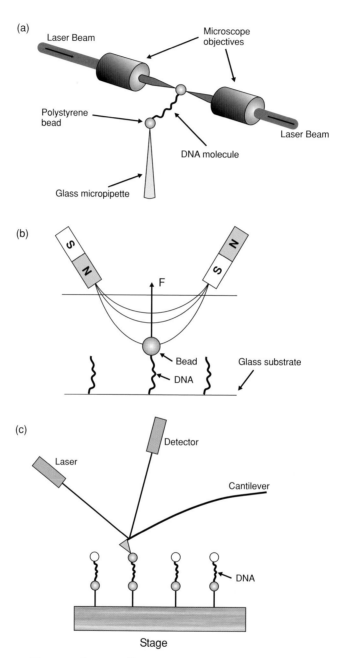

The experiments allow a new structural parameter to be accessed within a single molecule: *force* (Table 1). The upper boundary for force measurements in micromanipulation experiments is the tensile strength of a covalent bond (in the eV/Å range or about 1000–2000 pN). The smallest measurable force limit is set by the Langevin force (about 1 fN), which is responsible for the Brownian motion of the sensor (size of the order of 1 μm).

Note that the total range of forces in Table 1 covers only three orders of magnitude. Until single-molecule techniques became available, information on protein stability could only be obtained by measuring the loss of structure under denaturing conditions (by using temperature, chemical agents or pH) from which folding free energy could be calculated for an ensemble average of molecules. Free energy, however, does not provide direct information on mechanical stability. For mechanical stability, it is important to know how the total energy varies as a function of spatial coordinates. Several proteins were studied to measure the force required to unfold a single molecule. These studies revealed very large differences in magnitude (which can reach the order of a factor of 10) between the unfolding forces for different protein domains whose melting temperatures are very similar. These results demonstrated that the mechanical stability of a protein fold is not directly correlated with its thermodynamic stability. We expect the analysis of the mechanical properties of macromolecules to set the foundation of a new field of study, *mechanochemical biochemistry*.

Part A
Biological macromolecules and physical tools

Chapter A1
Macromolecules in their environment

A1.1 Historical review

The discovery of biological macromolecules is tightly interwoven with the history of physical chemistry, which formally emerged as a discipline in **1887**, when the journal founded by **Jacobus Van't Hoff** and **Wilhelm Ostwald**, *Zeitschrift für Physikalische Chemie*, was first published. Interestingly, the first papers were concerned with reactions in solution, because biological processes essentially take place in the aqueous environment inside living cells.

The **nineteenth century** discoveries of solution properties that led to our knowledge about biological macromolecules are described briefly in the Introduction. We must also mention the Grenoble chemist **François-Marie Raoult (1886)**, who formulated the freezing-point depression law that made it possible to determine the molecular weight of dissolved substances, and **Hans Hofmeister (1895)**, a medical doctor and physiologist, who was interested in the diuretic and laxative effects of salts, and classified them according to how they modified the solubility of protein in aqueous solutions. The *Hofmeister series* was later established as a ranking order of the 'salting-out', or precipitating, efficiency of ions. **Gilbert Newton Lewis** introduced the concepts of *activity* in **1908**, and of *ionic strength*, with **Merle Randall in 1921**. In **1911, Frederick George Donnan** published a paper on the membrane potential developed during dialysis of a non-permeating electrolyte. **Peter Debye** and **Erich Hückel (1923)** proposed a theory for electrolyte solutions. In recent decades, modern methods, such as dynamic light scattering and small-angle neutron scattering, developed for the characterisation of polymers, and especially polyelectrolytes, have contributed significantly to our current understanding of biological macromolecules in solution.

There is now a growing interest in the behaviour of proteins in non-aqueous solvents and even in vacuum, mainly in the context of biotechnology, but also with respect to whether or not water is essential to life. Proteins, as active biological particles, have evolved in the presence of water, however, and, to a large extent, cannot be considered separately from their aqueous environment. We recall that even crystals of biological macromolecules contain an appreciable amount of solvent and should be considered as organised macromolecular solutions.

Comment A1.1

Biologist's box: Molecular mass units

Molar mass is in g mol^{-1} or SI units of kg mol^{-1}.

Relative molecular mass or *molecular weight* is a dimensionless quantity, defined as the ratio of the mass of a molecule relative to 1/12 the mass of the carbon isotope ^{12}C. The molar mass of ^{12}C is very close to 12 g mol^{-1}. Molar mass (in kg mol^{-1}) can, therefore, be converted to molecular weight by dividing by 10^{-3} g mol^{-1} (the equivalent of multiplying by 1000 and cancelling units).

Biochemists use *molecular mass* expressed in *daltons* (Da) (1 Da = 1 atomic mass unit = 1/12 of the mass of 1 atom of ^{12}C).

A1.2 Macromolecular solutions

A solution is a homogenous mixture at the molecular level of two or more components. The majority component is the *solvent*; the others are the *solutes*. We shall deal mainly with macromolecular aqueous solutions, in which the solvent is water, and the solutes are macromolecules and other small molecules, such as simple salts.

A1.2.1 Concentration

The solute concentration can be defined in various ways.

The *weight or mass fraction* is the mass of solute per unit weight of solution (or per 100 weight units of solution if it is expressed as a *weight or mass percentage*).

The usual unit of molecular mass in biochemistry is the dalton (Comment A1.1).

The *molarity* is the number of moles of solute per litre of solution. Expressing a concentration in molar terms has the advantage of being more relevant than using weight fraction with respect to the *colligative* properties of the solution (i.e. properties that depend only on the number of solute particles rather than on the mass of solute or its specific properties; see also Section A1.2.3).

The *molality* is the number of moles of solute per kilogram of solvent. Expressing solute concentration in these terms has the advantage that molality is obtained by weighing masses of solute and solvent, whereas molarity depends also on measuring a solution volume. Masses are invariant, while the volume of the solution is a function of temperature and pressure.

The *mole fraction* and *volume fraction* definitions are similar to that of *weight fraction* but refer to moles or volume of solute per total moles or total volume, respectively.

The usual ways of measuring concentrations in protein and nucleic acid solutions are discussed in Comment A1.2.

A1.2.2 Partial volume

The partial volume of a solute is equal to the volume change of the solution upon addition of the solute, under given conditions. The partial volume is not simply the volume occupied by the added solute, because its presence may lead to a volume change in the solvent. The partial volumes of charged molecules in aqueous solution are an interesting illustration of solvent effects. The water molecule can be represented by a small electric dipole. Liquid water has a rather loose hydrogen bonded structure (see Section A1.3.4); when in the presence of ions, water molecules orient around the charges effectively taking up a smaller volume than in the bulk liquid. The effect is called *electrostriction*. The partial volume of a charged ionic solute may be negative, therefore, as is the case for the

---

Comment A1.2 Physicist's box: Measuring protein and nucleic acid concentrations (see also Section E1.1, Comment E1.3)

The concentration of protein or nucleic acid solutions cannot be measured simply by weighing an amount of material into the solvent. Protein and nucleic acid powders obtained from lyophilisation or precipitation always contain an unknown quantity of hydration water and salt ions, which are necessary for the maintenance of an active conformation. Furthermore, extremely low protein and nucleic acid concentrations are sufficient for many experimental methods, and it is not possible to weigh micrograms or less of material with precision. Protein concentrations are measured by colorimetric assays (e.g. the Bradford assay), in which an indicator interacts chemically with the polypeptide, or by spectrophotometry, in which the absorption of light at a given wavelength is proportional to the amount of material present. The absorbance at 280 nm is particularly sensitive to the presence of tryptophan, tyrosine and cysteine amino acid residues, and for most proteins it is of the order of 1 for a 1 mg ml^{-1} solution and a path length of 1.0 cm. The exact value, however, varies with the number of sensitive residues in the protein. Nucleic acids show a strong absorbance at 260 nm (1 absorbance unit corresponds to about 40 micrograms per millilitre), which is used to measure their concentration in solution. The colorimetric and spectrophotometric measurements yield *relative* values with respect to a calibration series. When absolute concentration values are required, e.g. as they are for the interpretation of small-angle scattering data (Chapter G2), the colorimetric or spectrophotometric values have to be calibrated on an absolute scale for the specific macromolecule, e.g. by precise amino acid analysis of the sample.

Na$^+$ and Mg^{2+} ions, for example. The decrease in volume due to electrostriction for these ions is greater than the volume they effectively occupy in the solution. The partial volume of the K$^+$ ion is slightly positive; electrostriction does not quite compensate for the volume occupied by the ion. Note that ions cannot be added separately to a solution, so that partial ionic volumes were obtained by interpolation from data on different neutral salts. It is also a consequence of electrostriction that ionic partial volume values are solvent composition-dependent; in general they increase with salt concentration, for example, because the water has already suffered some electrostriction.

The *partial specific volume* of a solute is the volume change of the solution per gram of added solute.

The *partial molal volume* of a solute is the volume change of the solution per mole of added solute.

The partial volumes of ions and biological macromolecules, in usual units, are given in Comment A1.3. Nucleic acids and to some extent carbohydrates, depending on their specific chemical natures, are polyelectrolytes and their partial volume values are strongly solvent salt-concentration-dependent. Interestingly,

Comment A1.3 Partial molal volumes of ions and partial specific volumes of biological macromolecules (see also Section D4.8)

Ion	Partial molal volume (ml mole^{-1})
Na$^+$ in water	−5.7
Na$^+$ in sea-water (~0.725 molal NaCl)	−4.4
K$^+$ in water	4.5
K$^+$ in sea-water (~0.725 molal NaCl)	5.9
Mg^{2+} in water	−30.1
Mg^{2+} in sea-water (~0.725 molal NaCl)	−27.0
Cl$^-$ in water	22.3
Cl$^-$ in sea-water (~0.725 molal NaCl)	23.3

From Millero F. J. (1969) *Limnol. Oceanogr.* **14**, 376–385.

Macromolecule	Partial specific volume (ml g^{-1})
Proteins	0.73 (0.70–0.75)
Carbohydrates	0.61 (0.59–0.65)
RNA	0.54 (0.47–0.55)
DNA	0.57 (0.55–0.59)

Note that the spread in values for proteins corresponds to different protein molecules, while that for the other macromolecules also takes into account variations with solvent salt concentration. For example, a given RNA molecule has a partial specific volume of 0.50 ml g^{-1} in water and 0.55 ml g^{-1} in high salt concentration.

the partial volumes of proteins correspond well to the sum of their amino acid component volumes and they are not salt-concentration-dependent despite the fact that protein surfaces may show significant charge. This is because there are compensating effects on the volume from electrostrictive protein–solvent interactions on the one hand, and a looser surface packing of amino acid residues on the other (see also Chapter D4).

A1.2.3 Colligative properties

Colligative properties of solutions (from the Latin *ligare*, to bind) are properties that depend only on the number of solute molecules per volume and not on the mass or the nature of the molecules. The discovery of the colligative properties of solutions played an essential role in early physical chemistry, by allowing accurate measurements of molecular weight, which in turn provided evidence for the very existence of atoms and molecules. *Raoult's law* states that in an *ideal solution* at constant temperature the partial pressure of a component in a liquid mixture is proportional to its mole fraction. Colligative properties related to Raoult's law and applicable to dilute solutions of non-volatile molecules are the rise in boiling point and decrease in freezing point that result when a solute is dissolved in a solvent. The temperature differences between the values for the ideal solution and those for the pure solvent are proportional in each case to the number of solute molecules present. These laws can still be applied to non-ideal solutions by applying the concepts of *chemical potential* and *activity*.

A1.2.4 Chemical potential and activity

Consider the box shown in Fig. A1.1, with a barrier separating solutions of different molar concentration, C_A, C_B, on either side, respectively. If we open a breach in the barrier, there is a net flow of solute molecules from the high- to the low-concentration side, similar to water flowing down a gravity potential gradient. A potential can, therefore, be associated with the solute concentration. The chemical potential, μ, of a solute is the *free energy* gain upon addition of one mole into the solution (see also Chapter C1).

$$\mu = \Delta G / \Delta C$$

$\mu C_A > \mu C_B$

Fig. A1.1 Chemical potential (see text).

In solution thermodynamics, the free energy difference between two solutions of concentrations C_A, C_B, (in number of moles per volume of solution), respectively, is given by:

$$G_A - G_B = \Delta G = -RT \ln \frac{C_A}{C_B} \tag{A1.1}$$

where R and T are the gas constant and absolute temperature, respectively. The expression results from integration of the Boltzmann equation, which describes how molecules in a perfect gas at constant temperature distribute according to their free energy:

$$\frac{p_A}{p_B} = \exp\left(-\frac{G_A - G_B}{RT}\right) \tag{A1.2}$$

where p_A, p_B are pressures at constant volume (proportional to molar concentration) associated with states of free energy G_A, G_B, respectively. In a trivial rewriting Eq. (A1.1) becomes

$$\Delta G = \mu(C_A - C_B) = -RT \ln \frac{C_A}{C_B} \tag{A1.3}$$

Expressing the relation in terms of free energy or chemical potential *differences*, rather than absolute values, avoids having to define a standard free energy or chemical potential (e.g. that associated with an ideal solution at a given concentration, temperature and pressure). The equations apply to ideal solutions,

Comment A1.4
Mole fraction and molarity values of usual solutions of biological macromolecules

The mole fraction in a $300\ \mathrm{g\,l^{-1}}$ aqueous solution of 30 kD macromolecules is 1/4000; the molarity of the solution is 10 mM. Corresponding to a high concentration for most biophysical experiments, this is similar to the protein concentration in cytoplasm. The mole fraction in a $3\ \mathrm{mg\,ml^{-1}}$ solution, which is usual for many types of biophysics experiment, is 1/400 000; the molarity of the solution is 100 μM.

i.e. solutions in which the solute molecules behave as point particles and do not interact in any way with each other (or with the vessel).

G. Lewis introduced the concept of *activity* in 1908 to account for deviations from ideal behaviour in solutions. Writing Eq. (A1.3) in terms of activity, rather than concentration:

$$\Delta G = -RT\,\ln\frac{a_A}{a_B} \qquad (A1.4)$$

The activity of solute A is given by $a_A = \gamma_A C_A$ and γ_A is an activity coefficient in the appropriate units and is equal to 1 for an ideal solution. Activity coefficients are obtained experimentally, for example, from deviations from Raoult's law. By replacing *concentration* with *activity*, equations that were derived in the ideal case could be applied in practice to *real* solutions.

A1.2.5 Temperature

The rise in boiling point and decrease in freezing point of solutions due to the presence of solute are not very useful properties when dealing with biological macromolecules such as proteins or nucleic acids – firstly, because the mole fraction of macromolecule in even highly concentrated solutions is very low (Comment A1.4); secondly, because biological macromolecules are usually not stable in pure water solvent and the molar concentration of buffer solutes and ions strongly dominates the effect; and, finally, because proteins and nucleic acids are usually not stable at the boiling and freezing points of water. Proteins and nucleic acids have evolved to be able to fold into their stable and active conformations in limited solvent conditions, and in limited ranges of the thermodynamic parameters of temperature and pressure (see Part C). It is interesting to note, however, that there exist organisms, called *extremophiles* (lovers of the extreme), which have adapted to various extreme environmental conditions, including temperature (Comment A1.5).

A1.2.6 Osmotic pressure

Osmotic pressure is a colligative property that applies equally usefully to small molecule solutes and macromolecules (Fig. A1.2). Its importance for biological processes has been appreciated since its discovery. Named from the Greek *osmos*, impulse, osmosis is the phenomenon that occurs when two solutions of different concentrations are separated by a semipermeable membrane. A semipermeable membrane is a membrane that allows solvent molecules (water) to cross it unhindered, but not the solute. Osmotic pressure is *a hydrostatic pressure* due to the solvent that develops on the more concentrated side of the membrane; it is due to water trying to move from the dilute to the concentrated side of the membrane in an attempt to equalise its chemical potential. The pressure builds up on the high

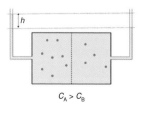

$c_A > c_B$

Fig. A1.2 Definition of osmotic pressure (see text).

Comment A1.5   Extremophiles

Microbial life has adapted to various extreme environments, including very high or very low pH (alkalophiles and acidophiles, respectively), high salt concentration (halophiles, see Section A1.3), high pressure in deep ocean trenches (barophiles), temperatures close to 0 °C in polar or glacier waters (psychrophiles, from the Greek *psychro* meaning cool; *cryo* meaning cold is usually applied to subzero temperatures), temperatures of about 70 °C, in thermal springs, shown in the picture above (thermophiles) and even temperatures as high as 113 °C (the highest known for a living organism) for *Methanocaldococcus jannaschi*, which lives in deep marine hydrothermal vents (hyperthermophiles). The proteins of extremophile organisms have very similar folds to those from the *mesophiles* (defined in an anthropomorphic way as living at around 37 °C). Their shifted stability and activity optima with respect to environmental conditions, arise from amino acid substitutions that modify the internal forces appropriately (see Chapter C2).

concentration side, because the compartment volumes are constant. The osmotic pressure Π is given by

$$\Pi V = NRT \qquad\qquad (A1.5)$$

where R, T are the gas constant and absolute temperature, respectively, and N is the number of moles of solute in the volume V. The mass concentration C and the molar mass M of solute are related to N and V by

$$\frac{N}{V} = \frac{C}{M}$$

The difference between the osmotic pressure in the two compartments in Fig. A1.2 is given by $\rho g h$, where ρ is the density of the solvent and g is the acceleration due to gravity:

$$(\Pi_A - \Pi_B) = \frac{C_A - C_B}{M} RT \qquad\qquad (A1.5a)$$

When only one compartment contains the solute (i.e. C_B is zero), (A1.5a) provides an especially sensitive measure of molar mass, which has been used extensively in the field of polymers.

Note that the apparent mathematical identity of Eq. (A1.5) with the perfect gas equation (expressed in molar units) has no physical basis. As we pointed out above, osmotic pressure, Π, is a hydrostatic pressure due to the solvent and does not arise from solute molecules 'pushing' like gas molecules against the membrane on the high concentration side.

When there is more than one solute in the solution, Eq. (A1.5) becomes

$$\Pi V = \sum_j N_j R T \qquad (A1.6)$$

where the sum is over all non-diffusible solutes, i.e. solutes for which the membrane is impermeable. The effective osmotic pressure across a membrane, therefore, depends also on the quality of the membrane with respect to the different solutes. For example, in order to calculate the osmotic pressure across a dialysis membrane with a 'cut-off' of 10 kD, we apply Eq. (A1.5), but we only count molecules with a higher molecular mass than 10 kD.

A1.2.7 Virial coefficients

Equations (A1.5) and (A1.6) assume we are dealing with ideal solutions, in which the solute particles do not interact with each other. They can be considered as good approximations when the solutions are very dilute. Although interparticle interactions can be taken into account by a chemical potential analysis, using *activity coefficients*, an approach in terms of *virial coefficients* (from the Latin *vires*, forces) has been more widely used for macromolecules in solution.

We now rewrite the single solute equation (Eq. (A1.5) with $N/V = C/M$) for the non-ideal solution; the osmotic pressure of a non-ideal solution is given by an expansion in powers of concentration

$$\frac{\Pi}{CRT} = \frac{1}{M} + A_2 C + A_3 C^2 + \cdots \qquad (A1.7)$$

where A_2, A_3 . . . are the second, third . . . virial coefficients, respectively. Virial coeffciicients are determined from experimental measurements and provide valuable quantitative data on particle–particle interactions. The coefficients may be positive (corresponding to repulsive interactions) or negative (corresponding to attractive interactions).

A1.3 Macromolecules, water and salt

Salt solutions have important specific and non-specific effects on the conformational stability of macromolecules – so much so that proteins from the extreme

Comment A1.6   Extreme halophiles

Extreme halophiles are archaeal and bacterial organisms (see Chapter A2) that can live only in high-salt environments, such as salt lakes like the Dead Sea, the Great Salt Lake or the lac Rose in Senegal (shown in the picture), and in natural or artificial salt flats. They contain carotenoids, and are responsible for the pink colour of these environments. The carotenoids enter the food chain and give their colour to flamingos and salmon flesh, for example. Extreme halophiles are distinguished from moderate halophiles and halotolerant organisms by the fact that they compensate the high osmotic pressure of the environment, due to close-to-saturated NaCl, by a correspondingly high concentration of KCl in their cytoplasm, instead of so-called compatible solutes like glycerol. All their biochemical machinery, therefore, functions in an environment that is usually deleterious to protein stability and solubility. Halophilic proteins have adapted to this environment in interesting ways. Instead of 'protecting' their structure from the salt, as might have been expected, for example by surrounding themselves with a strongly bound water shell, they actually incorporate large numbers of salt ions and water molecules to stabilise their fold and maintain their solubility.

halophiles (Comment A1.6) have evolved specific adaptation mechanisms in order to be stable, soluble and active in the high salt concentrations found in the cytoplasm of these organisms. Specific salt effects on macromolecules are due to ion association or binding. They depend on both salt and macromolecule type. A small solvent concentration of sodium or potassium and magnesium ions, for example, is necessary for transfer RNA (tRNA) to achieve its folded conformation in solution. At solvent concentrations of about 0.1 M NaCl and 1 mM $MgCl_2$ the Na^+ and Mg^{2+} counter-ions neutralise the repulsion between the phosphate groups in the main chain of the nucleic acid so that it can take up a compact conformation. The effect is not purely electrostatic but contains a steric component. If the salt is $N(CH_3)_4Cl$, the $N(CH_3)_4^+$ counter-ion does not allow the tRNA to fold correctly, presumably because the $N(CH_3)_4^+$ is too large. Specific ion binding also plays an essential role in stabilisation and switching mechanisms in certain proteins, such as calmodulin, the amylases, the parvalbumins, for example, which bind Ca^{2+}. Salt effects that are similar for all macromolecules have been named non-specific, even though they could depend strongly on ion type. They include ionic strength effects at low salt concentrations. Salts also have an important action on macromolecules through modifications of the water structure. Such an effect is usually evident at high salt concentrations (in the molar range). Ammonium sulphate is a salt that is frequently used at high concentrations in biochemistry to precipitate or to crystallise proteins. The sulphate ion acts on the water structure to reduce the solubility of apolar (hydrophobic) groups. In fact, in the early days of biochemistry proteins were classified as globulins or albumins according to the ammonium sulphate concentration required for their precipitation.

A1.3.1 Ionic strength and Debye–Hückel theory

The discovery of the dissociation of strong electrolytes into separate ions in aqueous solution caused great excitement in physical chemistry. Ionic solutions are very far from *ideal* even when highly dilute, because of the electrostatic interaction between charges. The concept of *activity* (see above) was introduced for strong electrolytes. Lewis and Randall, in order to take into account the effects of ions of different valency, suggested that the mean activity of a completely dissociated electrolyte in dilute solution depends only on the ionic strength, i, of the solution,

$$i = \frac{1}{2} \sum_j C_j z_j^2 \tag{A1.8}$$

where C_j, z_j are, respectively, the molar concentration and charge of ion j. The value of i calculated for a 1 mM solution of NaCl is 1 mM. The value of i calculated for a 1 mM solution of $MgCl_2$ is 2.5 mM. Of course, the expression for the ionic strength can be calculated for solutions of any concentration, and unfortunately workers in the field have written of ionic strength values for solution concentrations of 1 M or even higher. We wrote 'unfortunately' because it should not be forgotten that the concept of ionic strength is applicable only to (very) dilute solutions, in which the 'nature' of the ions and their interactions with water are neglected so that they can be considered to act as point charges. In a strict sense, ionic strength considerations should be limited to concentrations in the millimolar range.

The Debye–Hückel theory was essential for the understanding of the behaviour of dilute electrolyte solutions. The theory combines Poisson's equation, the general form of Coulomb's law of electrostatics (Comment A1.7), with statistical mechanics to calculate electrostatic potential at a point in the solution, in terms of the concentration and distribution of ionic charges and the dielectric constant of the solvent. The theory established a relation between the ionic strength and the activity coefficient of an electrolyte solution for low-ionic-strength values (for which the concept is valid). It predicts that the activity coefficient decreases with rising ionic strength, in agreement with observations. At higher ionic strength values (in which the concept is no longer valid), the activity coefficient rises and in some cases may become greater than 1. Attempts to extend an approach similar to the Debye–Hückel theory to the high-concentration domain take into account specific ionic properties such as *solvation* (specific solute–water interactions).

A1.3.2 Polyelectrolytes and the Donnan effect

Polyelectrolytes are macromolecular ions. DNA, for example, exists as a neutral salt in powder form; the negatively charged phosphate groups that are covalently bound in the macromolecule are neutralised by positive counter-ions, for example Na^+. The counter-ions dissociate from the macromolecule in solution, giving negatively charged polyions (or macroions) and 'free' Na^+ ions. Interestingly, the distribution of counter-ions in the electrostatic field of the polyion can be calculated in low-ionic-strength conditions from Debye–Hückel theory.

Consider the dialysis experiment in Fig. A1.3. The figure shows a dilute solution of NaCl in a vessel separated into two compartments by a membrane that is perfectly permeable to water and to the Na^+ and Cl^- ions. A 70-nucleotide Na·tRNA salt is dissolved in the left-hand compartment of the vessel (the red ellipsoid); the macromolecule dissociates into a tRNA polyion with 70 negative charges and 70 Na^+ ions (red in the figure). The polyions cannot cross the membrane. Three electroneutral components were used to make up the solution: component 1 is water; component 2 is the Na·tRNA salt macromolecule; and component 3 is NaCl. The negative charges on component 2 are restricted to the left-hand compartment, and, in order to maintain electroneutrality on either side of the membrane, there will be more small positive ions (Na^+) than unbound negative ions (Cl^-) in that compartment. Since there are also 70 positive counter-ions added per macromolecule, the result is an outflow of component 3 from the left-hand compartment to the right-hand compartment. The phenomenon is called the Donnan effect. The distribution of ions across the membrane to establish chemical potential equilibrium between the left- and right-hand compartments is termed the Donnan distribution.

Fig. A1.3 The Donnan effect (see text).

A1.3.3 Macromolecule–solvent interactions

Again we consider two solutions separated by a dialysis membrane (Fig. A1.4). On the left-hand side, the solution is made up of the three components: (1) water,

Fig. A1.4 Density increments. The solvent made up of components 1 (water) and 3 (salt) is light blue. The macromolecule (component 2) is shown as a red ellipsoid. Its solvation shell made up of bound water and salt is shown in darker blue.

(2) macromolecule (which is too large to diffuse across the dialysis membrane) and (3) a small solute, e.g. salt (which can diffuse across the membrane). The solution on the right-hand side does not contain the macromolecule. The water and diffusible solute move between the compartments to equalise their chemical potentials, μ_1, μ_3. In general, the macromolecule interacts specifically with water and the small solute – through hydration, ion binding or the Donnan effect, for example. How do these interactions affect the distribution of components 1 and 3 across the membrane?

Equation (A1.9) defines the *density increment* of the solution on the left-hand side due to the presence of a concentration c_2 of the macromolecule:

$$\left(\frac{\partial \rho}{\partial c_2}\right)_{\mu_1, \mu_3} = \frac{\rho' - \rho}{c_2} \tag{A1.9}$$

where ρ, ρ' are the mass densities of the solutions in the left- and right-hand compartments, respectively.

The density increment is a bulk property of the solution. It can be measured precisely by weighing a given volume of solution at constant temperature in a densimeter. The presence of the macromolecule perturbs the solvent around it. If, for example, it binds salt and water in a ratio that is different from their ratio in the solvent, then this results in salt or water flowing in or out of the dialysis compartment to compensate and maintain constant chemical potential of diffusible components across the membrane. The mass density increment expresses the increase in mass density of the solution per unit macromolecular concentration. Not only does it account for the presence of the macromolecule itself but also for its interactions with diffusible solvent components. It can be written

$$(\partial \rho / \partial c_2)_\mu = (1 + \xi_1) - \rho^\circ \overline{v}_2 + \xi_1 \overline{v}_1 \tag{A1.10}$$

where the subscript μ is shorthand for constant μ_1, μ_3 and ξ_1 is an interaction parameter in grams of water per gram of macromolecule, and $\overline{v}_x$ is the partial specific volume (in millilitres per gram) of component x.

The parameter ξ_1 does not represent water 'bound' to the macromolecules; it represents the water that flows into a dialysis bag to compensate for the change in solvent composition caused by the association (or repulsion) of both water and small solute components with the macromolecule. The mass density increment is the difference in mass between the volume of 1 g of macromolecules and ξ_1 g of water, on the one hand, and the same volume of bulk solvent, on the other.

An equation similar to Eq. (A1.10) can be derived in terms of ξ_3, representing grams of the small solute per gram of macromolecule; ξ_1 and ξ_3 are related by $\xi_1 = -\xi_3 / w_3$, where w_3 is the molality of the solvent in grams of component 3 per gram of water. The *preferential interaction parameters* ξ_1 and ξ_3 are thermodynamic representations of the solvent interactions of the macromolecule in the given solution conditions.

(a)

(b)

(c)

(d)

Fig. A1.5 Water structures: (a) a single molecule, the large dark atom is oxygen, the two light atoms are hydrogen; (b) directional tetrahedral hydrogen bonding around one molecule; (c) a two-dimensional model of an ordered water or ice lattice; (d) a two-dimensional model of a disordered water lattice. (The figures were kindly provided by Professor John Finney.)

A1.3.4 Water, salt and the hydrophobic effect

Whereas at very low salt concentrations (neglecting specific effects related to the properties of different ions) the ionic strength concept appears to be justified, it is certainly not the case at high salt concentrations. The structure of liquid water and its specific interactions with solutes play an essential role in defining the behaviour of macromolecules in solution.

Despite an intensive effort in molecular modelling to account for the extensive thermodynamics data on pure water and solutions, water structure remains incompletely understood at the molecular level. We describe its main features in qualitative terms, bearing in mind that they are based on carefully measured experimental thermodynamic quantities. Liquid water is a highly dynamic system of directional hydrogen bonds. Each molecule can participate in two hydrogen bonds and we can regard it as a small body, the oxygen atom, with two arms stretching out at a fixed angle to each other, the two hydrogen atoms. (Fig. A1.5).

The water molecules in the liquid phase are constantly rearranging in order to form hydrogen bonds with different partners, like a person in the middle of a crowd briefly touching the bodies of different pairs of the surrounding people people, while being touched by others (Fig. A1.6). In thermodynamics terms, forming a hydrogen bond leads to a decrease in enthalpy (negative ΔH) while the configurational freedom of having different partners with which to form bonds leads to an increase in entropy (positive ΔS). Both features contribute to a decrease in free energy ($\Delta G = \Delta H - T\Delta S$) and are, therefore, favoured thermodynamically (see Part C).

Fig. A1.6 Three-dimensional model of water. The structure is highly dynamic with hydrogen bonds forming, breaking and reforming in different patterns. In bulk water, each molecule has the configurational freedom of forming such bonds in all directions around itself. This is not the case in the presence of a solute. (The figure was kindly provided by Professor John Finney.)

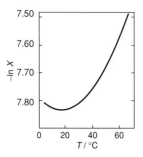

Fig. A1.7 Solubility of an apolar solute in water. The curve shown is that for benzene in water. X is solubility expressed as the mole fraction. The Gibbs free energy of transfer is given by $-RT \ln X$ (Franks *et al.*, 1963). (Figure reproduced with permission.)

What happens to the liquid water picture in the presence of solute? A solute molecule, in general, perturbs the dynamics of water molecules with which it is in contact. If the solute is polar (i.e. ionically charged, or neutral but capable of forming hydrogen bonds) it may orient the surrounding water molecules in a preferential way compared to when they are in the bulk liquid. We saw in Section A1.2.2 how a small charged ion (Li^+ or Na^+, for example) 'pulls' in the water molecules towards itself and causes electrostriction. If the solute is apolar (i.e. non-ionic and incapable of forming hydrogen bonds), it still perturbs the water structure because water molecules in its vicinity have fewer hydrogen bonding possibilities. They lose the configurational freedom of being able to make hydrogen bonds in any direction and their entropy decreases. Apolar solutes, therefore, are poorly soluble in water. Furthermore, in a temperature range close to room temperature their solubility curve is anomalous: the solubility decreases with rising temperature (Fig. A1.7). The temperature range concerned is where the effects of the solute on the entropy of the solvent (inducing a reduction in conformational freedom or in the number of hydrogen bonding possibilities available) dominate the thermodynamics of the system.

A phenomenon named the *hydrophobic effect* results from the low solubility of apolar solutes in water. Because contact between the solute and water is entropically unfavourable (see Part C), apolar solutes tend to aggregate to minimise their surface of interaction with the surrounding water molecules leading to an apparent hydrophobic (water fearing) force bringing them together.

Salt solutes perturb the water structure around them according to specific properties of the ions, especially their charge and volume. At high salt concentrations, a large portion of the water solvent is affected (Comments A1.8, A1.9), which, in turn, affects the solubility of other solutes. So-called 'salting-out' ions reduce the solubility of apolar solutes in water even further; adding such ions to a solution

Comment A1.8 Salt and water molalities

A kilogram of liquid water contains about 55 moles of the water. A 5 molal solution of NaCl is, therefore, made up of 10 moles of ions and 55 moles of water. Assuming an average six-fold coordination of the ions by water molecules, there clearly is not enough water to go around! The solution does not contain water molecules that are free of any ionic contact.

Comment A1.9 Chaotropes and kosmotropes

The structure of liquid water is a highly dynamic organisation of directional hydrogen bonds (see text). Ions have been classified into *kosmotropes* and *chaotropes* according to their effect on water structure. Kosmotropes are 'water structure makers'. They mainly include small ions of high charge density (e.g. Na^+, Li^+, F^-). *Chaotropes* are large ions that are 'water structure breakers'. K^+ and Cl^- are just in this category; Rb^+, Br^-, then Cs^+, I^-, are progressively more chaotropic. There is a complex relationship between the water ordering character of ions and their salting-out, salting-in behaviour. We recall that the Hofmeister series is phenomenological. The kosmotropic character of anions appears to be correlated with their salting-out effects. The situation is less straightforward for cations, for which complex binding behaviour to the folded and unfolded macromolecule complicates the issue.

Table A1.1. *The Hofmeister series expressed separately for cations and anions. Of the anions, phosphate is the most salting-out, while chloride is neutral and thiocyanide the most salting-in. Of the cations, ammonium, potassium and sodium are neutral in their salting behaviour, while calcium and magnesium are more salting-in than lithium*

Cations:

phosphate > sulphate > acetate > chloride < bromide < chlorate < thiocyanide

Anions:

ammonium, potassium, sodium < lithium < calcium, magnesium

savours precipitation ('salting out') of apolar solutes. Biological macromolecules (proteins and nucleic acids) display heterogeneous surfaces, with various charged as well as apolar patches, leading to complex hydration interactions, folding and solubility behaviour (see also Part C). The *Hofmeister series* is a classification of different ions according to how, at high solvent concentrations, they stabilize the native fold of and reduce the solubility of proteins in aqueous solutions (Table A1.1). The series has been established from phenomenological observations and appears to be valid for a wide variety of processes, from protein solubility to helix coil transitions in certain polymers and the stabilization of the folded or unfolded form of macromolecules in general. *Salting-in* is the process in which the solubility of the macromolecule is increased by adding salt. Strong salting-in ions also destabilise the folded structure at high concentration, presumably because the unfolded chain offers a larger number of binding sites. The exact order of ions in the Hofmeister series is not strict and may vary for different proteins, for example. The classification of anions in terms of stabilizing (salting-out) and destabilising (salting-in) ions, nevertheless, appears to have general validity (Comment A1.9).

Specific salt effects on biological macromolecules are used extensively in biochemistry and biophysics, as an important part of the battery of methods for fractionation, solubilization, crystallization etc. The essential role of the solvent in maintaining the integrity of an active biological macromolecule cannot be overestimated and in biochemistry it is vital to work in well-defined, buffered solvent conditions (Comment A1.10).

A1.4 Checklist of key ideas

- Biological macromolecules as active biological particles cannot be considered separately from their aqueous environment.
- The *partial volume* of a solute is the volume change of the solution due to the presence of the solute.

Comment A1.10
Physicist's box: Buffers

In all experiments on proteins or nucleic acids, the starting point is a buffered macromolecule solution, emphasising the fact that a macromolecule is maintained in an active conformation only in the 'correct' solvent environment. The 'buffer' is a chemical compound that has the property of regulating pH within narrow limits (e.g. 50 mM Tris-HCl, pH 7.5). Other solvent components may be a given salt at a given concentration (e.g. 0.1 M NaCl) or a reducing agent to protect SH groups (e.g. dithioerythritol, DTT), or mercaptoethanol.

- The solute concentration in a solution can be expressed in terms of mass per volume or a value that depends on the number concentration of molecules per volume (*molarity* or *molality*).
- *Colligative* properties of solutions depend on the number concentration of solute molecules.
- *Raoult's law* states that in an ideal solution at constant temperature the partial pressure of a component in a liquid mixture is proportional to its mole fraction.
- A consequence of Raoult's law is that the freezing point depression (or rise in boiling point) of a solution with respect to the pure solvent is proportional to the mole concentration of solute.
- The *chemical potential* of a solute is the free energy created by its presence in the solution.
- The concept of *activity* was introduced to replace *concentration* and account for deviations from ideal behaviour in solutions.
- Laws derived for ideal solutions in terms of concentration remain valid in non-ideal circumstances if concentration is replaced by activity.
- The activity of a solute is equal to its concentration multiplied by an *activity coefficient* that is equal to 1 for an ideal solution.
- Biological macromolecules are folded and active in limited ranges of solvent conditions, temperature and pressure.
- *Extremophiles* are organisms that live under extreme conditions of solvent, temperature or pressure.
- Macromolecules from extremophiles have adapted to be stable and active in the solvent, temperature and pressure conditions of their environment.
- A *semipermeable membrane* is permeable to water but not to solutes.
- *Osmotic pressure* is a hydrostatic pressure that develops on the high-concentration side of a semipermeable membrane separating solutions of different molar concentration.
- Osmotic pressure is a colligative property; at constant temperature, it is proportional to the difference in molar concentration of solute on either side of the membrane.
- In non-ideal solutions, osmotic pressure can be expressed as a concentration power series; the coefficients of the different terms are called the *virial coefficients*.
- Salts have important effects on the stability and solubility of macromolecules in solution, according to chemical type and concentration.
- Salt effects act through salt–water interactions or direct binding to the macromolecules.
- *Ionic strength* is a concept that is valid only at very low (millimolar) salt concentrations, in which only the concentration and valency of ions are taken into account, and not their specific character.
- *Debye–Hückel theory* provides a means of calculating the electrostatic potential at a point in an electrolyte solution in terms of the concentration and distribution of ionic charges and the dielectric constant.
- Debye–Hückel theory is applicable in low ionic concentrations, in which the concept of ionic strength is valid.

- The distribution of ions around a polyelectrolyte, such as DNA, can be calculated, at low ionic strength, by Debye–Hückel theory.
- The conventional numbering of components in a macromolecular solution containing a further small solute, such as salt, is: component 1 is water, component 2 is the macromolecule and component 3 is salt.
- Macromolecule–water–small solute interactions can be calculated from *density increment* measurements under dialysis conditions, in terms of *preferential interaction parameters*.
- The structure of liquid water and its interactions with small solutes play an essential role in defining the behaviour of macromolecules in solution.
- Liquid water is a highly dynamic system of directional hydrogen bonds.
- The *hydrophobic effect* results from the low solubility of apolar groups in water because they cannot form hydrogen bonds.
- The hydrophobic effect results in an apparent *hydrophobic force* bringing together apolar groups in aqueous solution.
- Dissolved salts modify the hydrophobic effect according to their type and concentration.
- *Kosmotropic* ions, 'water structure makers', are small, high-charge-density ions that increase the solubility of apolar groups in water.
- *Chaotropic* ions, 'water structure breakers', favour the hydrophobic effect.
- The *Hofmeister series* is a phenomenological classification of ions according to their ability to stabilise the native fold and precipitate proteins and nucleic acids in solution.

Suggestions for further reading

Von Hippel, P., and Schleich, T. (1969). The effects of neutral salts on the structure and conformational stability of macromolecules in solution. In: S. N. Timasheff, G. D. Fasman, eds. *Structure of Biological Macromolecules*. NY: Marcel Dekker Inc., pp. 417–575. This chapter is still an excellent reference on the effects of salts on macromolecules, even though it is more than 30 years old.

Collins, K. D. (1997). Charge density-dependent strength of hydration and biological structure. *Biophys. J.*, **72**, 65–76.

Madern, D., Ebel, C., and Zaccai, G. (2000). Halophilic adaptation of enzyme. *J. Extremophiles*, **4**, 95–98.

Price, P. B. (2000). A habitat for psychrophiles in deep Antarctic ice. *Proc. Natl. Acad Sci. USA*, **97**, 1247–1251.

Jaenicke, R. (2000). Do ultrastable proteins from hyperthermophiles have high or low conformational rigidity? *Proc. Natl. Acad Sci. USA*, **97**, 2912–2940.

Papers presented at a Royal Society (UK) meeting on water and life are published in *Phil. Trans. R. Soc. Lond. B.* **359** (2004).

Chapter A2
Macromolecules as physical particles

A2.1 Historical review and biological applications

Late 1600s

R. Boyle questioned the extremely practically oriented chemical theory of his day and taught that the proper task of chemistry was to determine the composition of substances.

1750

A.-L. Lavoisier studied oxidation, and correctly understood the process. He demonstrated the quantitative similarity between chemical oxidation and respiration in animals. He is considered the father of modern chemistry.

Late 1700s

The work of **J. Priestley**, **J. Ingenhousz** and **J. Senebier** established that photosynthesis is essentially the reverse of respiration.

1800s

The development of organic chemistry, despite the strong opposition of *vitalists* (who believed that transformations of substances in living organisms did not obey the rules of chemistry or physics but those of a *vital* force), led to the birth of biochemistry. In **1828**, **F. Wöhler** performed the first laboratory synthesis of an organic molecule, urea. During the **1840s J. V. Liebig** established a firm basis for the study of organic chemistry and described the great chemical cycles in Nature. In **1869** a substance isolated from the nuclei of pus cells was named nucleic acid. **O. Avery**'s experiments of **1944** on the transformation of *pneumococcus* strongly suggested that nucleic acid was the support of genetic information.

1860s

L. Pasteur is considered the father of bacteriology. He proved that microorganisms caused fermentation, putrefaction and infectious disease, and developed chemical methods for their study. In **1877** Pasteur's 'ferments' were named *enzymes* (from the Greek *en* 'in' and *zyme* 'leaven or yeast'. In **1897 E. Buchner** showed that fermentation could occur in a yeast preparation devoid of living cells.

1882

E. Fischer showed that proteins were very large molecules built of amino acid units; he also discovered the phenomenon of *stereoisomerism* in carbohydrates.

1926

Urease, the first enzyme to be crystallised in a pure form was shown to be a protein by **J. B. Sumner**. In the **1960s M. Perutz** and **J. Kendrew** solved the molecular structures of haemoglobin and myoglobin, respectively, by crystallography, thus proving that proteins had well-defined structures. They were awarded the Nobel prize.

1940

F. A. Lipmann proposed that adenosine triphosphate (ATP), which had been isolated from muscle in **1929**, was the energy exchange molecule in many cell types.

1953

J. D. Watson and **F. Crick** published the double-helix structure of DNA, based on chemical intuition and the fibre diffraction data of **R. Franklin** and **M. Wilkins.**

1955

F. Sanger determined the amino acid sequence of insulin, thus proving proteins are well-defined linear polymers of amino acids, an achievement for which he was awarded the Nobel prize. He went on to develop powerful methods for the determination of nucleotide sequences in DNA and RNA, which earned him a share (with **P. Berg** and **W. Gilbert)** in a second Nobel prize.

1970s

C. Woese classified living organisms in three kingdoms, *Bacteria, Archaea* and *Eukarya*, following sequence analysis of ribosomal RNA (Fig. A2.1). The classification was confirmed and refined when whole genome sequences became available.

1980s

S. Altman isolated ribonuclease P, an enzyme that contains both protein and RNA active components, and opened the way for the discovery of many other types of catalytic RNA. He was awarded the Nobel prize. In **1986, W. Gilbert** coined the phrase *RNA World* to denote a hypothetical stage in the evolution of life, already suggested to have existed by **C. Woese, F. H. C. Crick** and **L. E. Orgel**, in which RNA combined the role of a genetic information storage molecule with the catalytic properties necessary for a form of self-replication. The RNA World hypothesis led to a strong revival of interest in origin of life studies.

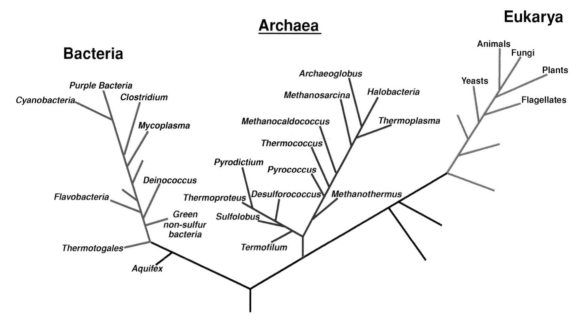

Fig. A2.1   An 'unrooted' philogenetic tree of life showing the three kingdoms resulting from ribosomal RNA sequence analysis. Further work showed that the likely root of the tree lies between the Bacteria and the Archaea. Bacteria and Archaea are *prokaryotes*, unicelluar organisms, whose genetic material is not contained in a nucleus. Archaea were so named because they were thought to be closer to the root of early life forms. They have specific characteristics as well as characteristics in common with either Bacteria or Eukarya. Most of the known extremophile organisms (see Chapter A1) are found in the Archaea. Some of them, like the anaerobic methanogens found in cow gut and marshland, are extremely common.

The spectacular development of molecular biology was accompanied by equally spectacular progress in methods for solving the structures of biological macromolecules and for studying their interactions. These will not be reviewed here as they constitute the subject of the book and are treated in the other chapters. In this chapter we describe the structural components and organisation of biological macromolecules.

A2.2 Biological molecules and the flow of genetic information

A gene is represented by a well-defined nucleotide sequence in the cellular DNA. There are four different nucleotides. The genetic code relates the 64 possible nucleotide triplets (*codons*) to the 20 amino acids found in proteins. Several triplets correspond to the same amino acid, allowing for stability at the protein level with respect to mutation at the DNA level, and there are codons corresponding to a 'stop' signalling the end of the chain. In protein synthesis, the gene is transcribed into messenger RNA and subsequently translated into the corresponding protein on the ribosome, according to the fundamental dogma of molecular biology:

$$DNA \leftrightarrow DNA \leftrightarrow RNA \rightarrow protein$$

The first double arrow represents the replication of a DNA mother molecule into identical daughter molecules, the second double arrow represents DNA to RNA transcription and RNA to DNA reverse transcription (observed in *retroviruses*), and the last arrow stands for translation.

The scheme represents a flow of information and not a sequence with respect to the evolution of macromolecules, since the final product, protein, is required in each and every step. The processes are catalysed and highly regulated by protein transcription factors, protein enzymes such as the polymerases, protein translation factors and ribosomes, themselves made up of protein and RNA. The end result is a protein with an amino acid sequence (*primary structure*) that has a one-to-one correspondence with the gene that initially coded for it. The protein has to undergo a folding process through solvent and intrachain interactions before it gains biological activity. Its *secondary structure* represents local structures, such as helices or extended chain conformations stabilised by intrachain or interchain hydrogen bonding; its *tertiary structure* is the structure achieved when the secondary structure elements pack to form a compact three-dimensional fold; its *quaternary structure* is the structure obtained when different proteins or copies of the same protein molecule associate to form a complex.

DNA has often been discussed using polymer terminology. It is composed of a linear chain of repeating structural units, the nucleotides. Each nucleotide is made up of a constant phosphate–ribose group, to which is bound a variable base taken from a group of four: adenine (A), guanine (G), cytosine (C) and thymine (T). DNA differs from usual synthetic polymers, however, in that it forms specific structures, according to its environment and interactions with proteins, which are intimately related to its biological function.

It is often useful to define a physical object by writing what it is not. A protein, perhaps even more so than DNA, is not a *classical* polymer (Fig. A2.2). This statement might appear to be contradictory, since a protein is often described as a polymer of amino acid residues. Its *biological* activity, however, conveys special properties to a protein molecule – properties that are different from those of the classical polymers, whose study led to the laws of polymer science. (Chemists have moved, however, towards the synthesis of polymers with some of the properties that are specific to proteins.)

First, with respect to composition and molar mass – a classical polymer chain is composed of a small number of different units repeated a large number of times. A sample is made up of chains of different length, and the molar mass is

Unique 3-D structure
in physiological
conditions

Disordered in
denaturing
conditions

Gaussian coil

Fig. A2.2 A protein is not a classical polymer.

defined as an average over a statistical distribution of chain lengths. A protein chain has a well-defined beginning (the N-terminal) and a well-defined end (the C-terminal); its primary structure (the amino acid sequence) and molar mass are perfectly defined (e.g. all the 10^{15} molecular chains in a tenth of a milligram of a given protein have an *identical* mass, to a precision restricted only by the natural isotopic variability); these properties result strictly from the structure of the gene that codes for the protein, and have been confirmed experimentally by mass spectrometry.

Second with respect to solution conformation – the classical polymer chain takes up a solvent-dependent conformation in solution, which again can be described by an average over a distribution of different conformations. Water-soluble and membrane protein chains fold in the solvent environment in which they are active to form well-defined compact structures. These structures are *essentially* identical for all the molecules in a pure protein sample. Protein crystallography and NMR studies of proteins (also on samples containing about 10^{15} molecules) yield structures where atomic positions may be defined with a precision better than 1 ångström unit.

The polypeptide chain that makes a protein has a precise length defined by its gene, with an N-terminal beginning and a C-terminal end. The chain is composed of a number of amino acid residues. There are 20 naturally occurring amino acids. In physiological conditions it takes up well-defined tertiary and quaternary structures, which unfold under denaturing conditions (heat or in certain solvents such as urea). Denatured protein attains conformations that are similar to homopolymers (Gaussian coils). Disordered protein structures, however, are not necessarily biologically inactive. The complete genome sequences now available have shown that a large proportion of gene sequences appear to code long amino acid stretches that are unfolded in physiological conditions.

In addition to DNA, RNA and protein, a cell contains two other major classes of molecule, carbohydrates and lipids, which play essential structural and functional roles. Carbohydrates and lipids are not coded directly in the genome but are the substrates and products of protein enzymes in complex metabolic pathways.

A2.3 Proteins

Essentially all the molecules in a living organism are either proteins or products of protein action. Except for the ribozymes, which constitute a small class of RNA molecules, all enzymes (the biochemical catalysts of metabolic reactions) are proteins. Several hormones (molecules with regulatory functions through their interaction with protein receptors) are themselves proteins, while others are protein components such as single amino acids or small peptides (the only other major hormone classes are steroids and amines). Proteins execute transport functions (e.g. haemoglobin, the oxygen carrier in red blood cells). They constitute the basis of muscle, of the conjunctive tissue that maintains the body structure in animals, and of skin and hair. Proteins are the components of elaborate recognition and signalling pathways (e.g. in the immune system as well as in a variety of cellular responses to external stimuli that can be chemical or physical, as in the case of photons in the processes of vision and photosynthesis). Proteins constitute the pumps and channels that establish the electrochemical gradients across membranes, which are essential to cellular bioenergetics, and play important roles in processes such as nerve transmission. And this is certainly not an exhaustive list of the biological functions accomplished by proteins.

Proteins are mainly made up of 20 amino acid types, linked in a linear chain by peptide bonds. Their biological activity results from the higher levels of organisation achieved by the chain in its physiological solvent environment. The *protein* is defined as the object with biological activity, while *polypeptide* refers to its chemical composition. Non-amino acid *prosthetic groups* may be associated with a protein and be essential for its activity. Examples are the oxygen-binding haem groups in haemoglobin, myoglobin and certain respiratory chain proteins, retinal, which provides the light sensitivity in rhodopsin, the protein associated with vision, and chlorophyll in plant proteins associated with photosynthesis. In fact, since the natural amino acids are colourless (their main absorption is in the UV), the colour of a protein is always associated with a prosthetic group: red for the haem, purple for retinal, green for chlorophyll. Such light-absorbing groups allow particularly useful spectroscopic approaches to the study of their associated proteins (see Part E).

A2.3.1 Chemical composition and primary structure

The general structure of an amino acid and its modification when it forms peptide bonds and enters a polypeptide chain are shown in Fig. A2.3. The *primary structure* of a protein is the sequence of amino acids in its polypeptide. The 20 main naturally occurring amino acid side-chains are shown with their properties in Fig. A2.4.

Except for glycine, amino acids can exist in two *enantiomeric* forms depending on the side of the α-carbon to which the side-chain binds. Natural amino acids are in the D (*dextro*, or right-handed) form (according to the direction of rotation of the plane of polarised light).

A2.3.2 Structures of higher order

A protein structure can be depicted in different ways. The pictures in Fig. A2.5 were drawn from the same structural model, resulting from the analysis of crystallographic data. Each emphasises a different aspect. The protein in our example is the metabolic enzyme, malate dehydrogenase, from the 'halophilic' (which lives in a high-salt environment) organism *Haloarcula marismortui*.

The secondary, tertiary and quaternary structures of the protein are all seen in the top part of Fig. A2.5(a), which traces the polypeptide fold in a *ribbon representation. Secondary structures* are favoured local chain conformations arising from chemical and steric constraints. The most common secondary structures found in proteins are α-helices and β-strands (see below); they are depicted, respectively, as ribbon helices and arrows. The protein is a tetramer. Its *tertiary structure* is the three-dimensional conformation of the subunit (given by the coordinates of the constituent atoms). Its *quaternary structure* is the organisation of subunits in the tetramer. A high-resolution zoom into a section of the structure is shown in *ball and stick* representation in the bottom part of Fig. A2.5(a). It illustrates the detailed relationship between the different chemical groups. Such an illustration is often used to show the active site interactions in an enzyme, for example. In this case, the picture is of complex salt-bridges (ionic bonding between charged amino acids) and solvent salt ion binding, which are related to protein stability in high salt concentrations.

(a)

(b)

(c)

N-terminal

C-terminal

Fig. A2.3 (a) An amino acid in its neutral form, and in the doubly charged ion (zwitterion) form it can have in solution. R is a variable side-chain. (b) Amino acids polymerise to form a polypeptide chain, with the loss of a water molecule for each peptide bond formed. (c) The chain has an N-terminal end and a C-terminal end. The N-to-C-terminal direction corresponds to the direction in which the gene is translated and the chain is synthesised on the ribosome.

Fig. A2.4 The 20 main natural amino acids. The three-letter and one-letter codes are given under each. The main chain is in crimson, apolar side-chains are in black, neutral polar side-chains in green, positively charged (basic) chains in blue, and negatively charged (acidic) chains are in red. Histidine is shown in its charged form (below pH 6.0).

In the *space-filling model* of Fig. A2.5(b) a Van der Waals sphere is drawn around each atom to provide a picture of the protein surface. Negatively and positively charged atoms are in red and blue, respectively. The surface has a net negative charge, which appears to be a general property of proteins from organisms adapted to high salt concentrations.

Secondary structure

The atoms in the main chain of a polypeptide ($N–C_\alpha–C–N–\cdots$) cannot lie in a straight line or rotate freely because of the properties of the bonds between them.

Fig. A2.5 Different ways
of drawing the same
protein structure in order
to emphasise different
features. (a) Top: a *ribbon
diagram* showing how the
secondary structure
elements (ribbon helices
for α-helices and arrows
for β-sheets) are
organised to form the
tertiary structure of each
subunit and how the four
subunits are organised in
the tetrameric quaternary
structure; bottom:
ball-and-stick atomic
'zooms' into the
interdimer interfaces to
show the complex
ion-binding salt-bridges
in these regions of the
structure. (b) An atomic
space-filling
representation showing
the protein surface;
negatively charged atoms
are red, positively
charged atoms are blue.

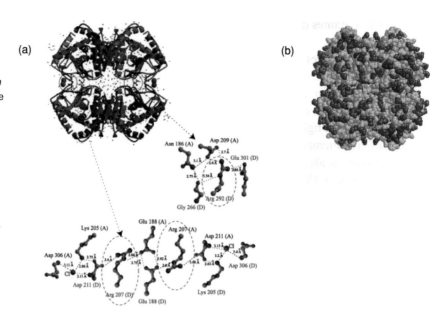

In particular, the peptide bond restricts the atoms in the

$$\underset{-\text{C}-\text{N}-}{\overset{\text{O H}}{\overset{|| \ |}{}}}$$

group to lie in the same plane so that the chain has limited flexibility
(Fig. A2.6).

The chain conformation can be described by the angles ϕ and ψ, formed by
the peptide plane and the C_α atoms on either side. The ϕ rotation is clockwise
facing the NH from the C_α atom, and the ψ is clockwise facing the CO from the
C_α atom. Constant ϕ and ψ values result in the chain assuming a regular helical conformation. A *Ramachandran plot* is a two-dimensional representation of
$\phi-\psi$ pairs and the corresponding secondary structures (see Chapter G3). Some of
these structures, like the α-helix, for example, are stabilised by internal hydrogen
bonds and favoured energetically (Fig. A2.7). The 'straight' chain with side-
chains projecting on alternate sides is called a β-strand; β-sheets are made up of
β-strands joined together by hydrogen bonds. They can be parallel or antiparallel
(Fig. A2.8). α-helices and β-sheets are the most common secondary structures

Fig. A2.6 The planar
peptide bonds in a
polypeptide chain.

found in the structures deposited in protein data banks, which can be accessed on the web.

Secondary structures are asymmetric, not only because of the directionality of the main chain but also because of the asymmetric position of amino acid side chains on the α-carbon (except for glycine). We recall that natural amino acids are in the D (right-handed) enantiomeric configuration. The side-chains therefore all point outwards in the right-handed α-helix configuration. Side-chains in β-sheets project outwards alternately above and below the sheet. A β-sheet in which polar and apolar amino acids alternate, for example, displays a polar face on one side and an apolar face on the other.

Tertiary structure

The tertiary structure results from weak (non-covalent, except for the disulphide bond between two cysteines) interactions between the amino acids in the polypeptide chain. The structure function hypothesis is based mainly on tertiary structure, and it is interesting to note how its general features for a particular protein family can be strongly conserved by evolution, even for molecules with widely varying primary structures.

Most proteins have structural similarities with other proteins. The study of these similarities is important not only in order to understand the evolutionary and functional relationships involved but also because it can assist in the

Fig. A2.7 The α-helix. The main chain carboxyl group of amino acid j is hydrogen bonded to the NH group of amino acid $j + 4$.

Fig. A2.8 Parallel (top) and antiparallel (bottom) β-sheets.

Fig. A2.9 (a) Ribbon model of the diphtheria toxin (DT) dimer, showing the three domain structure, C, T, R. One monomer is in grey and the other in white. The dimer illustrates the phenomenon of *domain swapping*, in which monomers exchange domains R. Each domain is associated with a specific function of the toxin. The beta protein domain R (similar in tertiary structure to immunoglobulin domains) is the receptor recognition domain in the first step of the toxin's entry into a cell. In the process of endocytosis the DT is carried to an endosome within the cell. The alpha protein domain T (similar in tertiary structure to trans-membrane proteins) allows the toxin to cross the endosome membrane in order to release the catalytic domain C into the cytoplasm, after cleavage of the polypeptide chain between C and T. The alpha–beta domain C is an enzyme that catalyses a reaction that is lethal to the cell. Note that the active site of C is hidden by R so that the intact protein is inactive. Similarly, the dimer is less active in cell entry because the R domains shield each other. (b) Ribbon drawing of an 'open' monomer. (Bennet *et al.*, 1994.) (Figure reproduced with permission from *Protein Science*.)

analysis in structural terms of the huge amount of sequence data produced by the genome projects. The tertiary structures of protein subunits can be divided into *single domain* and *multidomain* proteins according to distinct features. For example, the metabolic enzymes called dehydrogenases share a common dinucleotide cofactor domain, called the *Rossman fold*, after the scientist who discovered it; the diphtheria toxin protein subunit is strikingly divided into three recognisable domains, each with a distinct organisation associated with its function (Fig. A2.9).

Domain structures have been classified in groups or families, following the analysis of the entire solved structure database. These structures are accessible at the protein data bank web page.

The CATH protein structure classification is a hierarchical domain classification program in five levels: class (C-level), architecture (A-level), topology or fold family (T-level), homologous superfamily (H-level) and sequence family (S-level). The last two levels are for proteins that share a common ancestor and are based on sequence as well as structural comparisons. Class is determined according to secondary structure composition and packing within the domain. There are four major classes: mainly-alpha, which comprises domains whose secondary structure is mainly α-helix; mainly-beta, which includes domains organised in β-sheets; alpha–beta, which includes both alternating alpha–beta secondary structures and alpha plus beta secondary structures; the fourth class includes domains with low secondary structure content. Architecture describes the overall shape of the domain, determined by the orientations of the secondary structure elements. Topology depends on both the overall shape and connectivity of the secondary structures.

SCOP (structural classification of proteins) provides another domain organisation database, which is based on all solved structures. The levels of the SCOP hierarchy are *species, proteins, families, folds* and *classes.*

Quaternary structure

The organisation of protein subunits at the quaternary structure level has been selected by evolution because of functional advantages that range from simple stability considerations as in the case of the oligomerisation of enzymes from hyperthermophilic organisms that have to be stable and active at temperatures close to 100 °C (see Chapter A1), via the subtle tuning of activity as in the case of *allostery* (from the Greek 'other structure', Comment A2.1), to extremely coordinated complex reaction mechanisms as in the case of the large molecular machines in the cell, such as ribosomes (Fig. A2.14, see also Fig. D4.30, Fig. H2.10), proteasomes, thermosomes (see Chapter G2) and chaperone complexes. Large organised structures made up of different protein molecules as found in muscle or microtubules (see Chapter H2), for example, can also be considered as quaternary structures.

Comment A2.1 Physicist's box: Allostery

Allostery is a change in structure caused by an *effector*, which facilitates a certain activity, for example the binding of a substrate to an enzyme. Haemoglobin is the oxygen carrier in blood. The protein is a tetramer with four bound haem groups, one to each subunit. The subunit conformations can be in either a relaxed (R) or tense (T) form. Oxygen acts as both the allosteric effector and substrate of the protein. When it binds to one of the subunits it causes all of the subunits to go into the R form, which has a high affinity for oxygen binding.

A2.4 Nucleic acids

There are two main classes of nucleic acid: deoxyribonucleic acid (DNA) and ribonucleic acid (RNA). Similarly to proteins, which have a polypeptide primary structure, nucleic acids are polynucleotides, unbranched polymers of a certain chemical type of subunit, the nucleotide. DNA is the depository of genetic information. It is found associated with proteins in the chromosomes of cells. Its structure is predominantly double-stranded, with a helical organisation. RNA displays a range of biological functions (and new ones are still being discovered), and presents an even greater variety of structural organisation. It is usually made up of a single strand, which nevertheless turns around and interacts with itself to form intricate secondary structures. Messenger RNA (mRNA) is transcribed from DNA and carries the genetic message to the ribosome. Transfer RNA (tRNA) is a family of small RNA molecules (of about 70 nucleotides); each type of tRNA carries a specific amino acid to the ribosome and acts as an adapter between the mRNA and the growing polypeptide chain. Ribosomes are themselves organised by a number of small and large RNA molecules, ribosomal RNA (rRNA). Since the purification of the enzyme RNase P, which contains both RNA and protein active components, many types of catalytic RNA molecules have been and are being discovered. These include the so-called ribozymes, and portions of rRNA that play important catalytic roles in translation, and siRNA, small interfering RNA molecules involved in gene-silencing by RNA interference (RNAi) with mRNA. RNAi is currently a hot topic for study because of its potential applications in cancer and viral infection therapies. tmRNA is a remarkable molecule, which combines transfer RNA and messenger RNA functions. It codes for and adds a C-terminal peptide tag to an unfinished protein on a stalled ribosome, which directs the aborted protein for proteolysis. Small nucleolar RNAs (snoRNA) are localised in the eukaryotic cell nucleolus and are involved in rRNA biogenesis by defining nucleotide modification sites. The genome of certain viruses is based on double-stranded viral RNA rather than DNA. This is the case for retro-viruses, like the HIV virus, which also provide the machinery for retro-transcription that transcribes the viral RNA into DNA so that it can incorporate the infected cell's genetic material.

A2.4.1 Chemical composition and primary structure

Like the amino acids, nucleotides are made up of constant and variable parts. The constant part forms the main chain of the polynucleotide and is made up of a *nucleoside*, which is constituted of two covalently linked groups: a deoxyribose group (in DNA) or a ribose group (in RNA) and a phosphate group. The variable part is a nitrogen-containing base bound to the sugar group. There are four main naturally occurring bases in DNA: adenine, thymine, guanine and cytosine. The same bases occur in RNA, except for thymine, which is replaced by uracyl. In certain molecules the bases can be found in modified chemical forms (e.g. via

Fig. A2.10 (a) Chemical structures of the nucleotides, with the constant part shown in blue, made up of a deoxyribose (in DNA) and ribose (in RNA) and a phosphate group bound to the 5′ carbon of the sugar (sugar carbons are numbered clockwise 1′, 2′ etc., starting to the right of the ring oxygen atom), and the four natural bases in black. The base and sugar group together are called a nucleoside. (b) Chain elongation is via the loss of a water molecule and association of the oxygen on the 3′ carbon with the phosphate group of the next nucleotide in the chain. The polynucleotide chain displays a 5′ and a 3′ end. A gene is read in the 5′ to 3′ direction, which is also the direction of chain growth during replication and transcription.

methylation of DNA, or in more complex ways in tRNA). Chemical modification usually has important biological significance with respect to the interactions of the nucleic acid with other molecules. We recall that cells of different types within the same organism contain identical DNA molecules. The different nature of the cells is due to the fact that different sets of genes may be transcribed

Fig. A2.11 Watson–Crick base-pairs, A–T and G–C. Note the equal widths of their structures, which permits the formation of a regular double helix.

in each type. *DNA methylation*, i.e. the addition of methyl groups to specific cytosine residues in a DNA chain, appears to be involved with transcription inhibition in vertebrate cells, with distinct methylation patterns associated with each cell type. Nucleotides and polynucleotide chain elongation are shown in Fig. A2.10.

A2.4.2 Structures of higher order

DNA

DNA is predominantly double-stranded, and its secondary and tertiary structures are based on base pairing. The classical Watson–Crick pairs are A–T, G–C, leading to the rule A + C = T + G obeyed by DNA composition. An important stereochemical consideration that contributed to the discovery of the double helix is that the A–T pair, joined as in Fig. A2.11 by three hydrogen bonds, has the same width as the G–C pair joined by three hydrogen bonds, so that either pair fits into a regular double helix structure. The double-helical tertiary structure allows for sequence specificity through access to the major and minor grooves (Fig. A2.12). DNA can take up a variety of double-helical tertiary structures,

Fig. A2.12 The DNA double helix showing how the bases (in grey) are accessible for interaction in major and minor grooves.

A-DNA	B-DNA	Z-DNA
RH	RH	LH
11 bp/turn	10 bp/turn	12 bp/turn
pitch=28.2Å	pitch=34Å	pitch=43Å

Fig. A2.13 Different DNA structures and the corresponding fibre diffraction diagrams. Right-handed A-DNA changes to right-handed B-DNA when the relative humidity is increased from 75% to 92%; Z-DNA is a *left-handed* helix observed for poly GC sequences in high salt (Fuller *et al.*, 2004). (Figure reproduced with permission from Royal Society.)

Fig. A2.14 Secondary and tertiary structures of 23 S rRNA from the 50 S subunit (Ramakrishnan and Moore, 2001). (Figure reproduced with permission from Elsevier.)

depending on composition and hydration (Fig. A2.13). This structural variability of DNA is likely to be exploited in biological functions. There is, for example, clear evidence of the existence of proteins that bind to the B, A and Z double-helical confirmations. DNA can also fold into structures that involve more than two strands, such as triplexes and quadruplexes. G-quadruplex (guanine tetrad motif) structures have been found in-vitro for telomeric sequences (sequences at the end of eukaryotic chromosomes). The verification of their occurrence in-vitro is at present a hot topic in the field.

RNA

The genome of some RNA viruses is double-stranded. In general, however, the rule $A + C = U + G$ does not hold for RNA because the structures are predominantly single-stranded. The polynucleotide, nevertheless, bends back on itself in hairpin loops to form double-helical stem regions of A–U and G–C base pairs, leading to a much richer variety of secondary and tertiary structures than for DNA.

The family of tRNA molecules displays typical *clover leaf* secondary structures and *L-shaped* tertiary structures. In general, however, RNA secondary and tertiary structures can be extremely complex as is the case for 23 S ribosomal RNA for example (Fig. A2.14). Secondary structure predictions for the larger molecules result in a number of possible patterns. RNA tertiary structures have proven difficult to study because of their relatively labile character.

A2.5 Carbohydrates

Carbohydrate means 'watered carbon' and members of this molecular family can be represented by the general formula $C_x(H_2O)_x$ with more or less minor modifications. Carbohydrates serve living organisms as structural components and as energy and carbon sources, as signalling molecules and as mediators of cell–cell interactions as well as of interactions between different organisms. *Saccharide* (from the Latin '*saccharum*' through the Greek '*sakcharon*' meaning 'sugar') entities are the basic components of the highly complex carbohydrate molecules encountered in *glycobiology* (a term coined in 1988 to recognise the combination of the traditional discipline of carbohydrate biochemistry with the modern understanding of the role of complex sugars in cellular and molecular biology).

Biological carbohydrates are divided into *monosaccharides* (single sugar subunits, e.g. *glucose*, from the Greek '*glycys*', 'sweet'; *-ose* is the generic nomenclature ending chosen for the monosaccharides), *disaccharides* (two covalently bound sugar subunits, e.g. *sucrose*), *oligosaccharides* ('a few' or several covalently bound sugar subunits) and *polysaccharides* ('many' covalently bound sugar subunits) – the division between the oligo and polysaccharide groups being fairly loose.

Cellulose, a large linear polymer of glucose subunits, is the principal structural component in plants and the most common natural polysaccharide. *Glycogen* (from 'sweet' and the root of '*gennaein*', 'to produce' or 'give birth') is a complex branched polysaccharide of glucose subunits, stored as an energy source in liver and muscle cells. Linear and branched polysaccharide mixtures compose the *starch* (from the old English '*stercan*' meaning 'to stiffen') granules, which represent the main energy storage mechanism in plants.

All cell types and many biological macromolecules carry complex arrays of oligosaccharide chains called *glycans*, which also occur as free-standing entities. Most glycans are bound to excreted macromolecules or to protein or lipid molecules on the outer surfaces of cells, which are surrounded by a specific carbohydrate-rich 'shell' called the *glycocalyx* (from the Greek '*kalyx*', a covering, *not* from the Latin '*calix*', a cup). Glycans participate in specific cell–cell and cell–macromolecule recognition events, cell matrix interactions in higher organisms and interactions between different organisms, such as the ones between a parasite and its host. They are topics of intense study because of their implications in disease through immune response or tumour proliferation, for example. Glycans are immunogenic and are recognised and bound specifically by *antibody* molecules. *Lectins* are non-antibody proteins that bind carbohydrates without modifying them.

Polypeptides and polynucleotides are linear chains, each containing one type of bond between their constituent amino acids or nucleotides, respectively. Glycans, on the other hand, result from a very large number of subunit bonding possibilities that can create infinitely complex branched polymeric structures. There are two kinds of linkage (called α and β) between any pair of several positions to form a disaccharide from two monosaccharide subunits. We can easily imagine how the structural possibilities increase dramatically as the number of subunits increases. Fortunately, however, naturally occurring glycans contain relatively few monosaccharide subunit types in a limited number of combinations (if it were otherwise, structural studies of glycans, which are already very difficult, might close to impossible).

A2.5.1 Chemical composition and primary structure

The chemical structures of monosaccharides were first described by E. Fischer, who also discovered their stereoisomerism, at the end of the nineteenth century. Glucose, fructose and galactose (we recall the -ose suffix is typical in carbohydrate nomenclature; five- and six-carbon carbohydrates, for example, are named *pentoses* and *hexoses*, respectively), for example, are stereoisomers; they have the same chemical composition ($C_6H_{12}O_6$) yet differ by the structural arrangement of the atoms, which gives them distinct properties and characteristics. Biological systems are very sensitive to the stereoisomer state (we have already mentioned that natural amino acids are in the D enantiomeric form), since active site recognition by an enzyme, for example, depends on a precise atomic structural arrangement.

Monosaccharides are divided into two classes, *aldoses* and *ketoses*, according to whether they contain a functional *aldehyde* group (–CH=O) or *ketone* group (>C=O). They are then divided into subclasses according to the number of carbon atoms: aldotriose (three carbons), aldotetrose (four carbons), ketotriose, ketotetrose etc. The open and closed forms of glucose, mannose and galactose are shown in Fig. A2.15. The Fisher and Haworth projections are accurate with respect to the orientation of the groups, but misleading in that the ring is not planar but can take various 'chair'-like conformations, as shown.

Common monosaccharide subunits found in glycans of higher animals are given in Table A2.1. The molecules in Table A2.1 are dominant in higher animal (and of course human) glycobiology; several other types are found in the lower animals, plants and microorganisms. Note how chemical modifications involving the monosaccharide hydroxyl, amino and carboxyl groups further increase the variety of structures and enrich polysaccharide biological functionality.

In solution, most sugar molecules form cyclic five- or six-member ring structures that do not display free aldehyde or ketone groups. Ring closure creates a further asymmetric centre at the original carbonyl carbon, termed the *anomeric* carbon, so that D-glucose, for example, can exist in two forms called α and β.

Fig. A2.15 Row 1. The open form of three aldohexoses. The closed forms are shown in different projections in the following rows: row 2, Fisher projection; row 3, Haworth projection; row 4, the 'chair' projection; row 5, the stereo projection.

Glycans are often found associated with a non-carbohydrate moiety, usually protein or lipid, to form a *glycoconjugate*, and are defined according to the linkage between them. When a protein or a lipid forms a glycoconjugate it is said to be *glycosylated*.

Microheterogeneity renders the study of protein glycosylation particularly difficult. It describes the observation that a range of different glycans are found on the same glycosylation site of a given protein.

As carbohydrate primary structures are solved they are deposited in a database that can be consulted by computer. The existence of branching significantly increases the complexity of these primary structures and special software has been developed to access and analyse them.

Table A2.1. *Monosaccharide subunits in glycans of higher animals*

Pentose: five carbon sugar xylose (Xyl)
Hexoses: neutral sugars including glucose (Glc), galactose (Gal), and mannose (Man) (Fig. A2.15)
Hexosamines: hexose with either a free or N-acetylated amino group on carbon 2,
    N-acetylglucosamine (GlcNAc), N-acetylgalactosamine (GalNAc).
Deoxyhexoses: hexose without the hydroxyl group on carbon 6, fucose (Fuc).
Uronic acids: hexose with a carboxylate on carbon 6, glucuronic acid (GlcA), iduronic acid (IdA).
Sialic acids (Sia): nine carbon acidic sugars, mainly N-acetyl neuraminic acid (Neu5Ac, NeuNAc,
    NeuAc, or NANA).

A2.5.2 Higher-order structures

Cellulose is a homopolymer of glucose subunits, organised in a well-defined structure that has been studied by electron microscopy and diffraction (Fig. A2.16). Homo- and hetero-polysaccharides associated with protein or lipid also form ordered structures that fulfil structural roles in connective tissue or cell walls.

Fig. A2.16 Cellulose structure and hydrogen bonding pattern. Carbon, oxygen, hydrogen and deuterium atoms are coloured black, red, white and green, respectively. Hydrogen bonds are represented by dotted lines. Four alternative hydrogen bonding patterns have been identified, and are shown on the four panels (Nishiyama *et al.*, 2002). (Figure reproduced with permission from the American Chemical Society).

Fig. A2.17 A comparison
of the receptor-binding
sites of the cholera toxin
CT and Shiga toxin SHT
families (bound sugars
are shown in stick
representation) (a) CT B
pentamer showing five
binding sites in the
ganglioside oligo
Saccharide.
(b) SHT family pentamer
showing 15 binding sites
for the ganglioside (three
per monomer labelled
I,II,II). (Fan *et al.*, 2000).
(Figure reproduced with
permission from Elsevier.)

(a)   (b)

High-resolution structures of protein glycoconjugates, showing the nature of
the linkage and three-dimensional atomic organisation, have been solved by
X-ray crystallography and NMR. Solution NMR is the preferred technique
because even though mono and disaccharide subunits are relatively rigid
molecules, oligosaccharides usually show a degree of flexibility that makes crystallization difficult (see also Section J3.2.3).

Parasite, bacterial and viral infections often involve specific binding between
adhesion microbial proteins and glycans presented on the host cell surface.
Cholera toxin is an adhesion protein that binds to a cellular oligosaccharide.
The structure of the cholera toxin adhesion protein with a bound sugar has been
solved by X-ray crystallography to high resolution, and provides a good picture
of interactions between carbohydrate and protein atoms (Fig. A2.17).

A2.6 Lipids

Lipids may be defined loosely as a class of molecules whose properties are dominated by a water-insoluble moiety. They constitute a diverse group of compounds
involved in many aspects of cellular function. Lipid biochemistry is correspondingly rich; lipid molecules serve as sources of energy, they participate in complex
pathways such as blood-clot formation, and are necessary for the normal function
of the nervous system. With respect to macromolecular structure and dynamics,
however, we are concerned mainly with the structural roles of lipids in biological
membranes and their interaction with membrane-associated proteins.

A2.6.1 Chemical composition

The main lipid components of biological membranes can be presented schematically as being made up of a water-soluble charged or polar headgroup and water-
insoluble apolar hydrocarbon chains. The general composition of phosphoglyc-
erides and phospholipids is given in Fig. A2.18. The chains are composed of

Fig. A2.18
(a) Phosphatidyl ethanolamine, (b) phosphatidyl choline and (c) cardiolipin. For (a) and (b) the name describes the headgroup and the fatty acid chains may have different compositions (see text).

(a)   (b)   (c)

fatty acids (black) that can be *saturated* (containing only single carbon–carbon bonds) or *unsaturated* (containing one or more double bonds). They are associated by ester linkages to the glycerol group (red). The head groups are shown in blue.

Biological membranes contain many different types of lipid. In addition to phosphatidyl ethanolamine and phosphatidyl choline, higher animal membranes, for example, contain significant amounts of sphingolipids and cholesterol (Fig. A2.19). Cholesterol is a *steroid*, which fits between the fatty acid chains and changes the fluidity of the membrane structure. The apolar chains in lipids are not necessarily fatty acid derivatives. Lipid hydrocarbon chains in the membranes of Archaea, for example, are isoprenoid derivatives, similar to the *phytol* chains attached to chlorophyll in plants, containing alternate double and single carbon–carbon bonds and methyl branches along the length of the chain. Another chemical difference between Archaeal membrane lipids and membrane lipids from Eukarya and Bacteria is that the chains are ether-linked rather than ester-linked to the glycerol backbone. Interestingly, certain Archaea contain a 'double' lipid molecule, with two headgroups one on either side of the hydrocarbon chains. These

Fig. A2.19 Cholesterol.

glycerol dialcyl glycerol tetraethers were first discovered in hyperthermophilic organisms, in which they presumably help to increase membrane stability at very high temperatures. More recently, however, similar structures have been found in Archaea living in cool environments, below 20 °C (see Chapter A1).

A2.6.2 Higher-order structures

Lipid molecules in aqueous solution spontaneously associate in order to sequester their apolar, hydrophobic moieties from thermodynamically unfavourable contacts with water molecules. The process leads to the existence of a number of interesting structural phases for lipid–water mixtures, as a function of lipid composition, water-to-lipid ratio, temperature and pressure (Fig. A2.20).

Fig. A2.20 Structural phases (presented schematically on the left) and the phase diagram for monoolein–water mixtures. Monoolein is a single hydrocarbon lipid that is not found in natural membranes but is commonly used in membrane protein crystallization. The fluid lamellar bilayer phase (L_α) corresponds to the state of most natural membranes (Caffrey, 2000). (Figure reproduced with permission from Elsevier.)

Lamellar bilayer structures in the fluid phase (L_α), in which the hydrocarbon chains are disordered, are the most likely phases in biological membranes, providing an extremely efficient, essentially planar, permeability barrier, which still allows functional flexibility and lateral diffusion motions of associated membrane proteins. With natural membrane lipids, the L_α phase is favoured by lipid mixtures containing unsaturated hydrocarbon chains, higher temperatures and water content. It is interesting to note that in organisms exposed to low temperatures, the membrane lipid composition adapts in order to maintain a fluid bilayer structure. In hyperthermophilic Archaea, a 'bilayer' structure is obtained with a monolayer of the double-polar head lipid found in these organisms. The cubic phase has been suggested but has not been proven to occur in certain natural membranes.

A2.6.3 Lipids and membrane proteins

Membrane proteins are intimately associated with lipids. They have a hydrophobic character, which makes their study by the usual biochemical methods very difficult. Membrane protein studies have, therefore, lagged far behind those of water-soluble proteins. The development of detergent methods allowed active membrane proteins to be purified out of their lipid environment and in some cases to be crystallised. A Nobel prize was awarded for the first high-resolution membrane protein structure obtained by X-ray crystallography from crystals containing detergent. Obtaining membrane protein crystals, however, remains a formidable task. The phase diagram in Fig. A2.20 was studied in the context of membrane protein biophysics. The cubic phase of monoolein, in particular, was found to favour membrane protein crystallization for high-resolution structural studies (see Part G).

A2.7 Checklist of key ideas

- Living organisms are classified, by their genome sequences, into three kingdoms, *Bacteria, Archaea* and *Eukarya*.
- The flow of genetic information in all known organisms is from DNA to RNA to protein.
- The scheme is a product of a high degree of evolution since sophisticated regulated catalytic mechanisms by the final product, protein, are required for all the other steps: from DNA replication, RNA transcription and reverse transcription to genetic translation to protein biosynthesis itself.
- Biological macromolecules, although they are made up of a concatenation of subunits, have evolved to fulfil specific functions and have specific properties that are very different from those of classical polymers.
- All the molecules in a living organism either are proteins or can be considered as products of protein action.
- Proteins are made up of properly folded polypeptides of amino acid residues, and may include *prosthetic* groups with specific properties (such as the haem group, which binds oxygen).

- Colour in proteins (e.g. the red in haemoglobin, or green in chlorophyll binding proteins) is always due to a prosthetic group, because amino acids absorb in the UV region.
- There are 20 main amino acids in natural proteins, with a variety of chemical characteristics: acid and base, polar and non-polar, aliphatic and aromatic.
- *Primary structure* is the subunit sequence in the macromolecule; *secondary structures* are favoured local chain conformations arising from chemical and steric constraints; *tertiary structure* is the three-dimensional conformation of the macromolecular chain (given by the coordinates of the constituent atoms); *quaternary structure* is the organisation of different or similar chains in a macromolecular complex.
- The secondary structures of proteins can be expressed on a *Ramachandran plot* in terms of angles of rotation of the peptide planes in the chain around the so-called alpha-carbons, to which the amino acid side-chains are bound.
- α-helices and β-sheets are the main secondary structures found in proteins.
- The tertiary structure results from weak (non-covalent, except for the disulphide bond between two cysteines) interactions between the amino acids in the polypeptide chain.
- Protein domains with distinct features have been identified in the solved tertiary structures.
- Protein domains have been classified in groups or families, in levels, according to their architecture, topology, homology and sequence.
- The organisation of proteins in quaternary structures has been selected by evolution because of functional advantages with respect to stability, allosteric control, and the coordination of complex reaction mechanisms by large molecular machines in the cell, such as ribosomes, proteasomes, thermosomes and chaperone complexes.
- DNA is the repository of genetic information; its structure is predominantly double-stranded, with a helical organisation depending on its environment.
- RNA displays a range of biological functions and presents a great variety of structural organisation.
- RNA is usually made up of a single strand which, nevertheless, turns around and interacts with itself to form intricate secondary structures.
- Ribozymes are RNA molecules with catalytic activity; their discovery strengthened the *RNA World* hypothesis of primitive life forms based on RNA, both for genetic storage and catalysis of biological reactions.
- DNA is composed of a polynucleotide chain of four main nucleotides, denoted by their bases: adenine, thymine, guanine and cytosine.
- In RNA, thymine is replaced by uracyl.
- Bases may present chemical modification relevant to biological activity such as methylation in DNA or the more complex modifications observed for tRNA.
- Carbohydrates serve living organisms as structural components, energy and carbon sources, signalling molecules and mediators of cell–cell interactions as well as of interactions between different organisms.
- Polysaccharides can be simple linear chains or more complex branched chains composed of the same or different covalently bound monosaccharide units.

- All cell types and many biological macromolecules are *glycosylated* by binding complex arrays of oligosaccharide chains called *glycans*, to form glycoconjugates.
- Glycans are immunogenic and are recognised and bound specifically by *antibody* molecules.
- *Lectins* are non-antibody proteins that bind glycans without modifying them.
- Monosaccharides can exist in different structural forms called *stereoisomers*.
- *Microheterogeneity* renders the study of protein glycosylation particularly difficult. It describes the observation that a range of different glycans are found on the same glycosylation site of a given protein.
- Parasite, bacterial and viral infections often involve specific binding between adhesion microbial proteins and glycans presented on the host cell surface.
- Lipids are the main components leading to the passive permeability barrier in biological membranes.
- The main lipid components of biological membranes can be presented schematically as being made up of a water-soluble charged or polar headgroup, and water-insoluble apolar hydrocarbon chains.
- Lipid molecules in aqueous solution spontaneously associate in order to sequester their apolar, hydrophobic moieties from thermodynamically unfavourable contacts with water molecules.
- Lipid–water mixtures display various interesting structural phases as a function of lipid composition, water-to-lipid ratio, temperature and pressure.
- The so-called L_α phase, in which the lipid hydrocarbon chains are in a fluid, liquid crystalline state, is the most relevant for natural biological membranes.
- The so-called cubic phases obtained under artificial conditions have been useful for the crystallization of membrane proteins.

Suggestions for further reading

Woese, C. (1967). *The Genetic Code*. New York: Harper and Row.

Crick, F. H. C. (1968). The origin of the genetic code. *J. Mol. Biol.*, **38**, 367–379.

Orgel, L. E. (1968). Evolution of the genetic apparatus. *J. Mol. Biol.*, **38**, 381–393.

Gilbert, W. (1986). The RNA World. *Nature*, **319**, 618.

Hirao, I., and Ellington, A. D. (1995). Re-creating the RNA World. *Curr. Biol.*, **5**, 1017–1022.

Dykxhoorn, D. M., and Novina, C. D. (2003). Killing the messenger: short RNAs that silence gene expression. *Nat. Rev. Mol. Cell Biol.*, **4**, 457–467.

Varki, A., Cummings, R. *et al.* (eds) (1999). *Essentials of Glycobiology*. New York: CSHL Press.

Langan, P., Nishiyama, Y., and Chanzy, H. (1999). A revised structure and hydrogen bonding system in cellulose II from a neutron fibre diffraction analysis. *J. Am. Chem Soc.*, **121**, 9940–9946.

Merrit, E. A., Sarfaty, S. *et al.* (1994). Crystal structure of cholera toxin B-pentamer bound to receptor G_{M1} pentasaccharide. *Protein Sci.* **3**, 166–175.

Caffrey, M. (2000). A lipid's eye view of membrane protein crystallization in mesophases. *Curr. Opin. Struct. Biol.*, **10**, 486–497.

Schouten, S., Hopmans, E. C., Pancost, R. D., and Damste, J. S. S. (2000). Widespread occurrence of structurally diverse tetraether membrane lipids: Evidence for the ubiquitous presence of low-temperature relatives of hyperthermophiles. *Proc. Natl. Acad. Sci. USA*, **97**, 14421–14426.

Dickerson, R., and Geiss, I. (1969). *The Structure and Action of Proteins*. Menlo Park: Benjamin Cummings.

Branden, C., and Tooze, J. (1999). *Introduction to Protein Structure*. New York: Garland Science.

Chapter A3
Understanding macromolecular structures

A3.1 Historical review

In **1963 Richard Feynman** traced to **Pythagoras** (*c*. **500** BC) the first example, outside geometry, of the discovery of a numerical relationship in Nature. Pythagoras' discovery, which led to the foundation of a school of thought with mystic beliefs in the power of numbers, was that two strings under the same tension but of different lengths give a pleasant sound when plucked together if the ratio of their lengths is that of two small integers. We now say that a ratio of 1:2 corresponds to an octave, a ratio of 2:3 to a fifth and so on, which are all harmonic sounding chords. Feynman analysed this discovery in terms of three characteristics: its basis in experimental observation, the use of mathematics as a tool for understanding Nature, and its concern with aesthetics (the 'pleasant' quality of the sound). With easy hindsight (as he readily admits!) Feynman wrote that if Pythagoras had been more impressed by the first point on the importance of experimental observation the science of physics might have had an earlier start.

We discovered the existence and seek our biophysical understanding of macromolecules through experiment, and we use mathematical tools not only to set up experiments and analyse the results, but also for the description of the studied 'objects' themselves. The basis of aesthetics may still remain as mysterious as in Pythagoras' time, but there is no doubt that the 'beauty' of the DNA double helix and the satisfyingly elegant way in which it provided an explanation for how genetic information is stored and transmitted was an essential inspiration in the development of modern molecular biology. Similarly, the usual, colourful illustrations of protein structural models are undoubtedly aesthetic and certainly play a role in the acceptance and understanding of these models by biologists, who might have been put off by a purely mathematical interpretation of the data, even if it were more accurate.

Experiments on biological macromolecules are difficult to perform and interpret. Sample material is fragile and a good biophysical experiment must rest on a firm biochemical foundation. Biological macromolecules are much smaller than the wavelength of light and cannot be 'seen'. We gain information on their structures and dynamics by shining suitable electromagnetic or other types of radiation on a sample and analysing what emerges. The sample may absorb or scatter the radiation. Energy differences between the incident and emergent radiation are analysed by *spectroscopy*, whereas in *diffraction* experiments there is no energy difference and the information is obtained from the intensity of the scattered beam as a function of scattering angle. The interpretation of such experiments relies entirely on the physics and mathematics of *waves* and their interactions, be it in the classical description of crystallography in terms of interference of scattered waves, or in the quantum chemical description necessary for the interpretation of electromagnetic spectroscopy. The work of **Joseph Fourier (1768–1830)** on heat transmission in terms of waves has led to an extraordinary set of mathematical tools for the interpretation and understanding of the interactions between radiation and matter.

A useful approach to understanding macromolecular structures is based on an understanding of the forces underlying them – the forces acting to maintain the atoms in their positions in the correctly folded active conformation of the molecule. If we achieve an understanding of these forces, then we shall also understand how the atoms in the structure move about their mean locations. Expressed in the language of physics: our aim is to reach a sufficient understanding of the *force field* or *potential energy function* around each atom in a structure, in order to simulate its molecular dynamics.

In **1975**, the first molecular dynamics simulation of the time course of atomic motions in a biological macromolecule, the small stable protein bovine pancreas trypsin inhibitor (BPTI) for which an accurate X-ray crystallographic structure was available, was published by **J. A. McCammon**, **B. R. Gelin** and **M. Karplus**. Despite it being limited to a duration of less than 10 ps by the computing power available then, the BPTI study was effectively instrumental, together with the earlier hydrogen exchange experiments of **K. Linderstrom-Lang** and his collaborators **(1955)**, in establishing the view that proteins are not rigid bodies but dynamic entities whose internal motions must play a role in their biological activity.

The decades following the publication of the BPTI simulation, the **1980s** and **1990s**, saw a significant extension of the calculations to larger proteins and longer durations, as higher-resolution structures and more powerful computers became available. Parallel developments in experimental approaches provided information on macromolecular energies and internal motions and were essential in order to validate the simulation methods. These included microcalorimetry (see Part C), the analysis of temperature factors in crystallography (see Part G), fluorescence depolarisation (see Part D), NMR (see Part J), inelastic neutron scattering (see Part I) and Mössbauer spectroscopy, Fourier transform infrared spectroscopy (see Part E), and various fast kinetics measurements using laser flashes to trigger reactions. Synergistic relationships were established between molecular dynamics calculations and NMR and X-ray crystallography, starting with the incorporation of energy minimisation in structural refinement. The dynamical transition in proteins was discovered by Mössbauer spectroscopy (**F. Parak** and collaborators) and inelastic neutron scattering (**W. Doster** and collaborators). **H. Frauenfelder** and collaborators published the conformational substate (CS) model for proteins, providing a physical framework for the understanding of macromolecular structure and dynamics.

Since the year **2000**, molecular dynamics simulations have been applied on the supramolecular and even cellular scale. Experimental techniques are being refined continuously and new ones are being developed such as real time crystallography, and kinetic cryocrystallography, in which intermediate structures in an enzyme catalytic cycle are 'frozen' in and examined separately.

A3.2 Basic physics and mathematical tools

Practically all the experiments designed to analyse biological macromolecules (with the exception of calorimetry and classical solution physical chemistry methods such as the ones used to determine osmotic pressure, viscosity etc.) rely completely or in part on the observation of their interaction with radiation. This is patently obvious for diffraction- and spectroscopy-based methods, like crystallography or NMR, but it is also true for hydrodynamics-based methods, such as analytical ultracentrifugation or dynamic light scattering, in which the presence of the macromolecule in the solution is detected via its absorption or scattering of radiation. For most purposes, the sample is a *black box* into which we shine radiation of known properties and analyse what comes out (Fig. A3.1). In order to be able to deal with the interactions between radiation and matter, we require some knowledge of the mathematical tools that describe waves and of quantum mechanics, in which it is useful to consider the particle-like properties of radiation beams and radiation-like properties of moving particles.

A3.2.1 Waves

Sines and cosines

The sine or cosine function is the simplest way to describe the periodic rise and fall of a wave (Fig. A3.2). A wave is characterised by various parameters. Its *amplitude, A*, represents the maximum rise above the average level; the *intensity* of the wave is equal to the square of its amplitude; its *wavelength*, λ, is the repeat distance of a cycle in space; its *frequency, f*, is the number of cycles in unit time. The propagation velocity of the wave, *v* (the distance covered by a point in the cycle per unit time) is, therefore, given simply by its frequency multiplied by its wavelength:

$$v = f\lambda \qquad \text{(A3.1)}$$

(a)

(b)

Fig. A3.1 The sample is a 'black box' into which radiation of known wavelength is shone: (a) a diffraction experiment: the scattering intensity is measured as a function of angle; (b) spectroscopy experiment: the energy of the emerging beam is analysed.

Fig. A3.2 Waves
represented by functions:
(a) $Y = A\sin X$;
(b) $Y = A\cos X$; (c) $Y = (A/\sqrt{2})(\cos X - \sin X)$;
(d) $Y = (A/2)(\cos X - \sqrt{3}\sin X)$; (e) $Y = (A/2)(\sqrt{3}\cos X - \sin X)$
(see text).

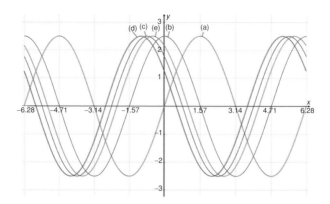

The $\sin X$ and $\cos X$ functions cycle with a period of 2π (Fig. A3.2). The waves with which we are concerned propagate in space or in time, or in both.

We first consider X in space, as in the case of a standing wave or a propagating wave observed at a specified time. X can then be written in terms of the wavelength as $kx = (2\pi/\lambda)x$, where x is distance, so that the wave is described by

$$Y = \cos kx \text{ or } Y = \sin kx \tag{A3.2}$$

where we have assumed unit amplitude for simplicity.

The *wavevector*, $\mathbf{k}$, of magnitude $k = (2\pi/\lambda)$, is defined as pointing in the direction of propagation.

We now consider the parameter X as describing time; we are observing the rise and fall of the wave at one particular position in space, as a function of time. As above, taking into account the cycling with a period of 2π, we write $X = \omega t = 2\pi ft$, where t is time, and (again assuming unit amplitude)

$$Y = \cos \omega t \text{ or } Y = \sin \omega t \tag{A3.3}$$

$\omega = 2\pi f$ is called the angular frequency of the wave.

The phase difference, δ, describes the extent to which two waves of the same wavelength and frequency are out of step with each other (Fig. A3.2). If one wave is described by $\cos \omega t$, for example, then the other will be described by $\cos(\omega t + \delta)$. Waves with different phases are described by the functions $\cos(\omega t + \delta_1)$, $\cos(\omega t + \delta_2)$, $\cos(\omega t + \delta_3)$, etc. We note that the cosine and sine functions are related by a phase difference $\delta = \pi/2$ (Fig. A3.2 (a), (b)).

We recall from the mathematics of sines and cosines that

$$\cos(\omega t + \delta_1) = \cos \delta_1 \cos \omega t - \sin \delta_1 \sin \omega t \tag{A3.4}$$

Since $\cos \delta_1$ and $\sin \delta_1$ are constants we can write

$$\cos(\omega t + \delta_1) = a_1 \cos \omega t + b_1 \sin \omega t \tag{A3.5}$$

In other words, a wave of given frequency and phase can be represented by the sum of a cosine and a sine with appropriate coefficients (a_1, b_1, respectively, in Eq. (A3.5)).

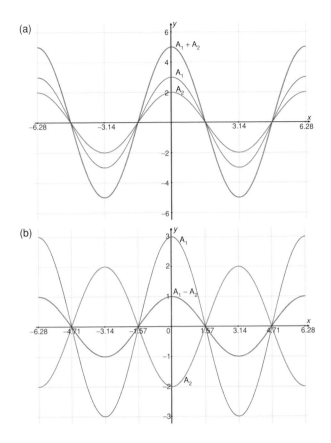

Fig. A3.3 (a) Constructive interference of two waves *in phase* (or out of phase by 2π, 4π, 6π, etc.)

$$A_1 \cos X + A_2 \cos(X + n\pi) = (A_1 + A_2) \cos X$$

where n is zero or an even integer. (b) Destructive interference of two waves, which are perfectly out of phase (phase difference $n'\pi$, where n' is an odd integer)

$$A_1 \cos X + A_2 \cos(X + n'\pi) = (A_1 - A_2) \cos X$$

The wave described by the function $(\cos X - \sin X)$ is out of phase by $\delta = \pi/4$ with the wave described by $\cos X$ (Fig. A3.2 (b), (c)). (This can also be calculated from Eq. (A3.4); $\sin \delta = \cos \delta = 1/\sqrt{2}$ for $\delta = \pi/4$ (45°).) Similarly, waves that are out of phase with $\cos X$ by $\pi/3$ and $\pi/6$ are described by $(\cos X - \sqrt{3} \sin X)$ and $(\sqrt{3} \cos X - \sin X)$, respectively (Fig. A3.2 (d), (e)).

A sum of waves of the same wavelength and frequency but with different amplitudes and different phases can be expressed mathematically, therefore, as a sum of sine and cosine functions with appropriate coefficients. Two extreme cases are illustrated in Fig. A3.3: the sum of two waves with a phase difference of $n\pi$, where n is zero or an even integer, leading to *constructive interference*; the sum of two waves with a phase difference of $n'\pi$, where n is an odd integer, leading to *destructive interference*.

The interference of waves of different frequency leads to more complicated pictures. Consider two waves of equal amplitude and angular frequencies, ω_1, ω_2, respectively. Applying the sine and cosine rules, their sum is given by

$$\cos \omega_1 t + \cos \omega_2 t = 2 \cos \tfrac{1}{2}(\omega_1 + \omega_2)t \cos \tfrac{1}{2}(\omega_1 - \omega_2)t \qquad (A3.6)$$

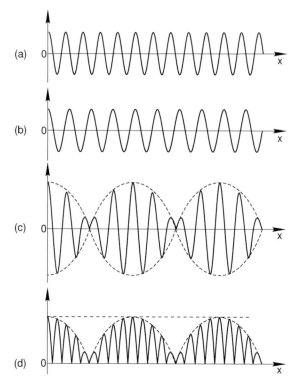

The sum of waves of different frequency and different amplitude, A_1, A_2, respectively, is given by

$$A_1 \cos \omega_1 t + A_2 \cos \omega_2 t = \cos \tfrac{1}{2}(\omega_1 + \omega_2)t \left[A_1 \cos \tfrac{1}{2}(\omega_1 - \omega_2)t \right.$$
$$\left. + A_2 \cos \tfrac{1}{2}(\omega_1 - \omega_2)t \right] \tag{A3.7}$$

The wave sum for an equal-amplitude case is illustrated in Fig. A3.4. To a very good approximation, the resulting wave can be considered as having an average frequency $\tfrac{1}{2}(\omega_1 + \omega_2)$, and an amplitude that oscillates with a lower frequency, $\tfrac{1}{2}(\omega_1 - \omega_2)$. The resulting wave is said to be *modulated* by the lower-frequency oscillation.

The phenomenon of *beats* occurs when ω_1 and ω_2 are close to each other ($\sim\omega$), as when we listen to the vibration of two strings of slightly different tension during the tuning of a guitar. The frequency of the *note* corresponds to $\sim\omega$, but its intensity *pulsates* or beats with a much lower frequency ($\omega_1 - \omega_2$); this frequency is twice the frequency of the amplitude oscillation, because the intensity is equal to the square of the amplitude (Fig. A3.4). The phenomenon of beats is observed in dynamic light-scattering experiments, in which light waves of slightly different frequency interfere after being scattered by molecules moving with different velocities (Doppler effect, see Chapter D10).

The velocity of a wave, as defined in Eq. (A3.1), corresponds to the velocity of propagation of a point of constant phase (e.g. a crest in the wave); it is called the *phase velocity* of the wave. Using the definitions of wavevector and angular frequency, given above, we rewrite Eq. (A3.1) as

$$v = \omega/k \qquad (A3.8)$$

which is the usual definition of phase velocity.

When we have a superposition of waves of different ω and k values travelling in the same direction, as in Fig. A3.4, the *group velocity*, v_g, of the resulting wave is the velocity of propagation of the modulation. It can be shown that

$$v_g = \frac{d\omega}{dk} \qquad (A3.9)$$

The relation between the ω and k values of a wave is called the *dispersion relation*. The phase and group velocities are clearly equal for waves described by a dispersion relation ω proportional to k.

In order to calculate the group velocity of the resultant wave in Fig. A3.4, i.e. the velocity of the modulation, we first add to Eq. (A3.7) the oscillation along the length dimension, x:

$$\begin{aligned}
A_1 \cos(\omega_1 t - k_1 x) + A_2 \cos(\omega_2 t - k_2 x) = \; &\cos \tfrac{1}{2}[(\omega_1 + \omega_2)t - (k_1 + k_2)x] \\
&\times \{A_1 \cos \tfrac{1}{2}[(\omega_1 - \omega_2)t - (k_1 - k_2)x] \\
&+ A_2 \cos \tfrac{1}{2}[(\omega_1 - \omega_2)t - (k_1 - k_2)x]\}
\end{aligned} \qquad (A3.10)$$

The time and length terms have opposite signs because, as the wave moves forward, a crest at a certain point, for example, results from the arrival of a crest that was one wavelength *behind* in the *previous* time period of the wave. The modulation arises from the cosine terms in the curly brackets on the right-hand side of Eq. (A3. 10), and its speed is equal to $v_{mod} = (\omega_1 - \omega_2)/(k_1 - k_2)$. In the limit of very small frequency and wave vector differences v_{mod} is equal to the group velocity defined in Eq. (A3.9).

Complex exponentials

Complex numbers were invented in order to solve equations such as $x^2 = -1$, but because of their properties they became powerful mathematical tools for a large range of problems in physics.

A complex number a is written $x + iy$, where i is the square root of -1; x is called the 'real' part of the complex number and y the 'imaginary' part. Note that in mathematical terms there is nothing imaginary about that part!

Sums and products of complex numbers are themselves complex numbers:

$$\begin{aligned}
(x + iy) + (x' + iy') &= (x + x') + i(y + y') \\
(x + iy)(x' + iy') &= (xx' - yy') + i(xy' + x'y) \qquad (A3.11) \\
a^* &= x - iy \\
aa^* &= x^2 + y^2
\end{aligned}$$

where a^* is called the complex conjugate of a.

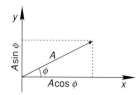

Fig. A3.5 The complex number $a = x + iy$ expressed geometrically on an xy plane. It can also be written $a = A \exp i\phi$, where A is called the amplitude and ϕ, the phase.

$$x = A \cos \phi$$
$$y = A \sin \phi$$
$$A = \sqrt{(x^2 + y^2)}$$

The complex number a can be represented on an xy plot as a radial line of length A (called the amplitude) making an angle ϕ (called the phase) with the x-axis, the 'real' axis; the y-axis is the 'imaginary' axis (Fig. A3.5). Such a plot is called an *Argand diagram*. We see from the figure that $A = \sqrt{(x^2 + y^2)}$, and the complex number can now be written as $a = A(\cos \phi + i \sin \phi)$. We know from mathematics, however, that the *complex exponential*

$$\exp i\phi = \cos \phi + i \sin \phi \tag{A3.12}$$

so that $a = A \exp i\phi$.

Feynman called Eq. (A3.12) 'the most remarkable formula in mathematics . . . our jewel'! The formula embodies the connection between algebra (complex numbers) and geometry (sines and cosines defined as ratios of sides in right-angle triangles), and it is in fact amazing how its properties greatly simplify the mathematical operations in which it is involved. The main one of these properties is

$$\exp[i(\phi_1 + \phi_2)] = \exp(i\phi_1) \exp(i\phi_2) \tag{A3.13}$$

multiplying the exponential by its complex conjugate, we have

$$\exp(i\phi) \exp(-i\phi) = \exp 0 = 1$$

A complex number can be written, therefore, in terms of real and imaginary parts as $x + iy$ or in terms of an amplitude value and a phase angle in a complex exponential as $A \exp i\phi$. Note that, similarly to Eq. (A3.11), the square of the amplitude of the complex exponential is not obtained by squaring the exponential but by multiplying it by its complex conjugate

$$aa^* = A \exp(i\phi A) \exp(-i\phi) = A^2 \tag{A3.14}$$

There are considerable advantages in describing waves in terms of complex exponentials, because they are mathematically easier to work with than sums of sines and cosines (Comment A3.1).

Comment A3.1   Summing complex exponentials

We can derive Eq. (A3.7) in just a few lines by using complex exponentials.

$$A_1 \exp i\phi_1 + A_2 \exp i\phi_2 = A_1 \exp i \tfrac{1}{2} (\phi_1 + \phi_2 + \phi_1 - \phi_2)$$
$$+ A_2 \exp i \tfrac{1}{2} (\phi_1 + \phi_2 - \phi_1 + \phi_2)$$
$$= [\exp i \tfrac{1}{2} (\phi_1 + \phi_2)] [A_1 \exp i \tfrac{1}{2} (\phi_1 - \phi_2)$$
$$+ A_2 \exp -i \tfrac{1}{2} (\phi_1 - \phi_2)]$$

Equation (A3.7) is the 'real' part of the line above. The derivation using sine and cosine rules is much more complicated.

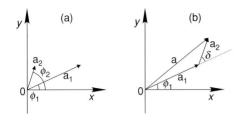

Fig. A3.6 (a) Two waves represented as complex numbers, a_1, a_2, on an Argand diagram and (b) their sum, a, expressed as a vector sum of a_1 and a_2. Note that a etc. denote the complex number and not the length of the line, which is A etc. as in Fig. A3.5. The phases of a_1, a_2 are ϕ_1, ϕ_2, respectively, and $\phi_2 - \phi_1 = \delta$.

A wave in space of amplitude A and wavevector magnitude Q is written $A \exp iQx$. A wave in time of amplitude A and angular frequency ω is written $A \exp i\omega t$. A wave oscillating in both space and time is represented by $A \exp [i(kx - \omega t)]$. As in the previous section, the time term has a negative sign because, as the wave propagates, a crest at a certain point, for example, results from the arrival of a crest that was one wavelength *behind* in the *previous* time period of the wave.

Two waves of amplitudes A_1, A_2, respectively, and phases, ϕ_1, ϕ_2, respectively, are represented on an Argand diagram by $a_1 = A_1 \exp i\phi_1$ and $a_2 = A_2 \exp i\phi_2$ and it is then straightforward to calculate their sum, $A \exp i\phi$, from either algebra or geometry, by treating a_1 and a_2 as vectors in the xy plane (Fig. A3.6, and Eq. (A3.14)):

$$A \exp i\phi = A_1 \exp i\phi_1 + A_2 \exp i\phi_2$$

Equating the real parts we have

$$A \cos \phi = A_1 \cos \phi_1 + A_2 \cos \phi_2 \tag{A3.15}$$

and equating the imaginary parts we have

$$A \sin \phi = A_1 \sin \phi_1 + A_2 \sin \phi_2$$

The properties of the resulting wave are derived following an analysis as in Comment A3.1.

We now write the two waves in terms of the phase difference between them, $\delta = \phi_2 - \phi_1$, and calculate the amplitude of the resulting wave

$$
\begin{aligned}
a = a_1 + a_2 &= A_1 \exp i\phi_1 + A_2 \exp [i(\phi_1 + \delta)] \\
aa^* &= \{[A_1 \exp i\phi_1 + A_2 \exp i(\phi_1 + \delta)]\}\{[A_1 \exp -i\phi_1 + A_2 \exp - [i(\phi_1 + \delta)]\} \\
&= \exp(i\phi_1)\exp(-i\phi_1)[A_1 + A_2 \exp i\delta][A_1 + A_2 \exp(-i\delta)] \\
&= A_1^2 + A_2^2 + A_1 A_2[\exp i\delta + \exp(-i\delta)] \\
&= A_1^2 + A_2^2 + 2A_1 A_2 \cos \delta \tag{A3.16}
\end{aligned}
$$

The amplitude of the resulting wave is equal to $(A_1^2 + A_2^2 + 2A_1 A_2 \cos \delta)$. And, as another illustration of the power of using complex exponentials, we note that the last line of Eq. (A3.16) is simply the *cosine rule* applied to the triangle with sides formed by a_1, a_2, a, in Fig. A3.6.

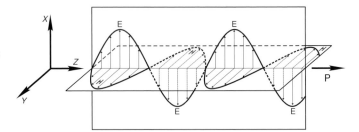

Polarization

The electric and magnetic fields in an electromagnetic wave oscillate along directions perpendicular to the propagation direction of the wave (Fig. A3.7). The light illustrated is said to be *linearly or plane polarised*, because the electric and magnetic fields oscillate along straight lines (the *x*- and *y*-axes, respectively). However, this need not be the case. The electric field (or magnetic field) that describes the light is described by an oscillating vector, which can lie in any direction provided this is perpendicular to the propagation direction. In other words, if the propagation direction is along the *z*-axis, the electric field vector, $\mathbf{E}$, may have components, E_x, E_y, along the *x*- and *y*-axes, respectively.

The light is linearly or plane polarised when E_x, E_y oscillate *in phase*. Cases for different amplitude values for each of the components are illustrated in Fig. A3.8.

When E_x, E_y, oscillate with a phase difference, δ, the light is said to be *elliptically polarised* (*circularly polarised* in the special cases of $\delta = +\pi/2$ or $-\pi/2$), because the tip of the electric field vector traces an ellipse (or a circle) in time (Fig. A3.9). Using the cosine or complex exponential notation introduced above, and assuming unit amplitude,

$$E_x = \cos \omega t \text{ or } 1$$
$$E_y = \cos(\omega t + \delta) \text{ or } \exp i\delta$$

The tip of the electric field vector turns in a clockwise direction for phase angles $0 < \delta < \pi$, and anticlockwise for $\pi < \delta < 2\pi$. The light is linearly or plane polarised for $\delta = 0$ or π.

Fig. A3.8 Linearly polarised light: electric field components along *x*- and *y*-axes of different amplitude oscillating in phase: (a) $E_y=1$; $E_x=0$; (b) $E_y=1$; $E_x=1/2$; (c) $E_y=1$; $E_x=1$; (d) $E_y=0$; $E_x=1$; (e) $E_y=1$; $E_x=-1$; (f) $E_y=-1$; $E_x=1$.

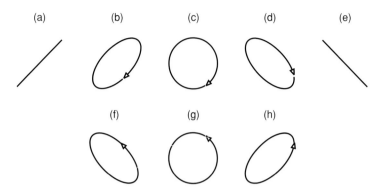

Fig. A3.9 Linearly, elliptically or circularly polarised light: E_x and E_y are assumed to have the same amplitude, and to oscillate with phase ωt and $\omega t + \delta$, respectively: (a) $\delta = 0$; (b) $\delta = \pi/4$; (c) $\delta = \pi/2$; (d) $\delta = 3\pi/4$; (e) $\delta = \pi$; (f) $\delta = 5\pi/4$; (g) $\delta = 3\pi/2$; (h) $\delta = 7\pi/4$. The light in cases (a) and (e) is linearly polarised, it is circularly polarised in (c) and (g), and elliptically polarised in the other cases.

A3.2.2 Simple harmonic motion

A simple harmonic oscillator

A swinging pendulum or a mass on a spring moving periodically up and down are mechanical systems that provide us with good examples of *simple harmonic motion*. We consider the motion of a mass, m, attached to a spring (Fig. A3.10).

The mass undergoes simple harmonic motion in one dimension, described by an equation of motion given by

$$F = -ky = m\frac{d^2 y}{dt^2} \qquad (A3.17)$$

where F is the force exerted on the particle (e.g. by the stretched spring), y is the mass displacement from an equilibrium position and k is a force constant, which depends upon the stiffness of the spring. The negative sign indicates that F is a *restoring force*. The force is proportional to the displacement and acts in the opposite direction. Thus, it tends to restore the mass to its original position. Equation (A3.17) is an expression of Hooke's law, which states that in an elastic

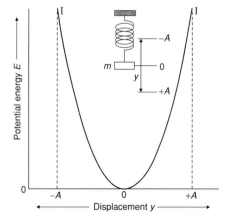

Fig. A3.10 Potential energy diagram for a simple harmonic oscillator.

body the strain (deformation) is proportional to the stress (force per area). The solution of Eq. (A3.17) is in the form of a wave

$$y = A \cos \omega_m t \qquad (A3.18)$$

which describes a periodic motion, where ω_m is the natural vibrational angular frequency of the mass, and A is the maximum displacement from equilibrium, the *amplitude* of the motion. Of course, we could also have written a wave solution in terms of complex exponentials.

In order to find the relation between ω_m and the force constant k, we calculate the second time derivative of y (the acceleration) in Eq. (A3.18), substitute in the equation of motion (Eq. (A3.17)) and rearrange

$$\omega_m = \sqrt{k/m} \qquad (A3.19)$$

We arbitrarily set the potential energy, E, of the system as equal to zero, when the mass is in its equilibrium position. As the spring is compressed or stretched, E increases by an amount equal to the work required to displace the mass:

$$dE = -F\,dy \qquad (A3.20)$$

Combining Eq. (A3.20) with Eq. (A3.17) and integrating, we derive the following relation for the potential energy of the oscillator as a function of displacement:

$$E = \tfrac{1}{2} k y^2 \qquad (A3.21)$$

which describes a parabola (Fig. A3.10). The potential energy is a maximum when the spring is stretched or compressed to the amplitude, A, and it decreases to zero at the equilibrium position.

The total energy of the system is equal to its potential energy plus the kinetic energy of the moving mass. Note that contrary to the potential energy dependence, the kinetic energy is a maximum as the mass passes through the equilibrium position, and decreases to zero at $y = +A$ and $y = -A$, where the mass stops before reversing the direction of its motion. The total energy of the oscillator, E_{total}, is exchanged back and forth between potential and kinetic energy, and it can be shown that it is a constant during all phases of the motion:

$$E_{\text{total}} = \tfrac{1}{2} k A^2 \qquad (A3.22)$$

The above equations may be modified to describe the behaviour of a system consisting of two masses m_1 and m_2 connected by a spring (in the absence of gravity) (Fig. A3.11), simply by substituting the *reduced mass* $m_{1,2}$ for m, where

$$m_{1,2} = \frac{m_1 m_2}{m_1 + m_2} \qquad (A3.23)$$

The vibrational angular frequency of the system is then given by

$$\omega_m = \sqrt{\frac{k}{m_{1,2}}} = \sqrt{\frac{k(m_1 + m_2)}{m_1 m_2}} \qquad (A3.24)$$

Fig. A3.11 Two masses connected by a spring of force constant, k.

Fig. A3.12 A linear chain of $N + 1$ atoms. The boundary conditions are that atoms, $s = 0$ and $s = N$ are fixed; u_s is the displacement of atom s and a is the atom spacing.

Normal modes in one dimension

The motions of a set of masses coupled by Hooke's law springs, or simple harmonic potentials such as the one illustrated in Fig. A3.10, can be seen as resulting from a superposition of fundamental vibrations called *normal modes*.

An analysis of the normal modes in a linear chain of atoms under specific boundary conditions is illustrated in Fig. A3.12. The chosen boundary conditions are that the end atoms do not move, like the ends of a guitar string. And, similarly to a plucked guitar string, the fundamental modes correspond to a set of waves in the line, while leaving the end atoms fixed.

The longitudinal (parallel to the line) or transverse (perpendicular to the line) wave displacement of atom s is given by

$$u_s \propto \sin s\, Q a \qquad (A3.25)$$

where Q is called a wavevector. (It is not written as a vector because our example is in one dimension; the wavelength of the atomic displacements is given by $\lambda = 2\pi / Q$.) The first boundary condition, $u_s = 0$ for $s = 0$, is automatically satisfied by Eq. (A3.5). The second boundary condition, $u_s = 0$ for $s = N$, sets limits on the values of Q. It is satisfied by choosing

$$Q = \frac{\pi}{Na}, \frac{2\pi}{Na}, \frac{3\pi}{Na}, \dots, \frac{N\pi}{Na} \qquad (A3.26)$$

Note that the solution for the maximum value of Q ($Q = \pi/a$) results in $u_s = 0$ for all atoms. There are, consequently, $N - 1$ normal modes for the line of atoms with the given boundary conditions. Each normal mode is defined by its Q vector and corresponds to a standing wave described by the equation

$$u_s = u(0) \exp(-i\omega_Q t) \sin s\, Q a \qquad (A3.27)$$

where $u(0)$ is an initial amplitude, ω_Q is the angular frequency of the mode and t is time.

We recall that the mathematical relation between ω_Q and Q is called a *dispersion relation*. For the above example, the dispersion relation is given by

$$\omega_Q = \begin{cases} \frac{Na}{2\pi}, & \text{for } -\frac{\pi}{a} \le Q \le \frac{\pi}{a} \\ 0, & \text{otherwise} \end{cases} \qquad (A3.28)$$

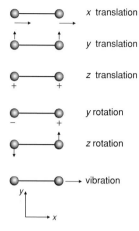

Fig. A3.13 Intrinsic degrees of freedom of a diatomic molecule. The z-axis points out of the page. Because the atoms are considered as point masses, there is no *x* rotation.

Normal modes in three dimensions

The equation of motion of a mass, m, undergoing simple harmonic motion in one dimension is given in Eq. (A3.17). Extending the analysis to a body of N harmonically coupled masses moving in three dimensions, the m and k terms are replaced by two $3N \times 3N$ matrices, respectively, of effective masses and force constants between all mass pairs, and y is replaced by a $3N$-dimensional coordinate vector. By analogy with Eq. (A3.18), the solutions for the three-dimensional case form a set of $3N$ periodic functions. Six of these, however, correspond to the translational and rotational degrees of freedom of the body as a whole, for which the masses do not move relative to each other, so that the system has $3N - 6$ normal modes. It can be shown that if the masses are in a line (as in the example discussed below) the number of normal modes corresponds to $3N - 5$.

The $3N = 6$ degrees of freedom of a diatomic molecule are shown in Fig A3.13; three correspond to translational displacements (along the x-, y-, and z-axes); two are rotational modes (about the y- and z-axes); and there is only one vibrational (bond stretching) mode as expected from the $(3N - 5)$ relation for masses on a line.

In the harmonic approximation, the vibrations of atoms in a macromolecule result from a superposition of the normal modes. Atomic collective motions calculated, using harmonic potentials, for two low-frequency normal modes in the small protein BPTI are shown in Fig. A3.14. Note the relatively large displacements of some of the atoms. In each normal mode, all atoms move with the same phase, i.e. they achieve maximum and minimum displacements and pass through their equilibrium positions simultaneously.

Fig. A3.14 Collective atomic displacements for two low-frequency normal modes from a calculation for BPTI. The mode frequencies are 3.56 ps^{-1}, 0.21 ps^{-1}, for (a), (b), respectively (Gō *et al.*, 1983).

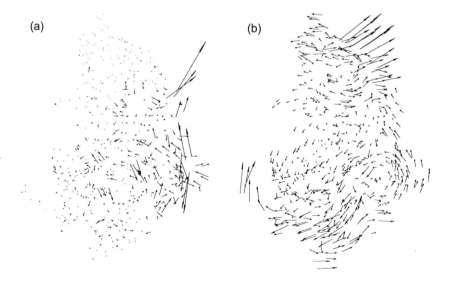

A3.2.3 Fourier analysis

Periodic functions, Fourier series and Fourier transforms

We showed above that a periodic function could be expressed as a sum of sines and cosines. In fact, according to a theorem due to Fourier, *any periodic function may be represented by a sum of cosines and sines*, known as a *Fourier series* (Eq. A3.29):

$$
\begin{aligned}
y(t) = {} & a_0 + a_1 \cos \omega t + b_1 \sin \omega t \\
& + a_2 \cos 2\omega t + b_2 \sin 2\omega t \\
& + a_3 \cos 3\omega t + b_3 \sin 3\omega t \\
& + \cdots \\
= {} & \sum_{n=0}^{\infty} (a_n \cos n\omega t + b_n \sin n\omega t)
\end{aligned}
\tag{A3.29}
$$

The cosine and sine terms describe waves of increasing frequency in integer multiples, $n\omega$, of a fundamental frequency, ω, with amplitudes, a_n, b_n, respectively. We can also write the Fourier series in terms of complex exponentials:

$$
y(t) = \sum_{n=-\infty}^{+\infty} A_n \exp i n\omega t
\tag{A3.30}
$$

where A_n is the amplitude of the wave of frequency $n\omega$, and the sum is over negative as well as positive integers because they are all required to establish the correspondence between Eqs. (A3.29) and (A3.30). Consider the simple cosine wave as an illustration of the relationship between the complex exponential and sine cosine series (see also Fig. A3.15):

$$
y(t) = a \cos \omega t = \sum_{n=0}^{\infty} (a_n \cos n\omega t + b_n \sin n\omega t)
$$

clearly $a_1 = a$, and all the other a_n, b_n are zero. In terms of complex exponentials,

$$
y(t) = a \cos \omega t = \sum_{n=-\infty}^{\infty} A_n \exp n \, i\omega t
$$

has two terms, for $n = \pm 1$ so that the sine terms making up the imaginary part cancel out,

$$
y(t) = a \cos \omega t = A_{-1} \exp(-i\omega t) + A_1 \exp(+i\omega t)
$$

$$
= 2A \cos \omega t
$$

where $A_{-1} = A_1 = A$, and $a = 2A$.

Another way to describe the series expansion for $y(t)$ is to consider its wave components separately, each defined by an amplitude A_n and corresponding frequency value $n\omega$. This yields a set of lines called *Fourier components*; the amplitude of each component is called a *Fourier coefficient*; and the function $A_n(n\omega)$ is called the *Fourier transform* of $y(t)$.

Fig. A3.15 Waves and
their corresponding
Fourier transfroms: (a) a
wave described by
$2A\cos \omega_1 t$ oscillating
about zero; (b) a wave
described by
$B + 2A\cos \omega_1 t$ oscillating
about the line at plus B;
(c) a wave described by
$2A_1\cos \omega_1 t + 2A_2\cos \omega_2 t$,
oscillating about zero.

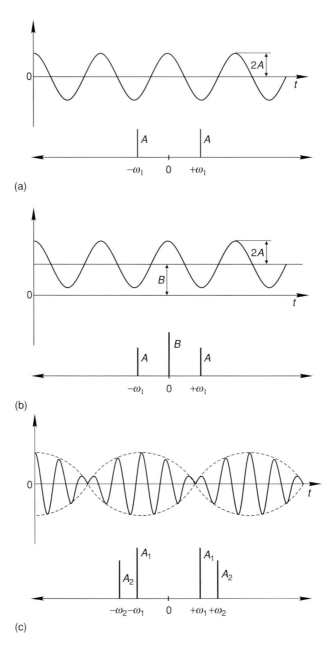

The Fourier transform of a wave form, which is described perfectly by $A\cos \omega_1 t$, for example, has Fourier components of amplitude, A, at $n\omega = \pm\omega_1$. A wave resulting from the sum of two cosine functions as in Eq. (A3.7), has Fourier components on $n\omega = \pm\omega_1, \pm\omega_2$, of height, A_1, A_2, respectively, and so on (Fig. A3.15). Note from Fig. A3.15(b) that the Fourier transform of a constant function, $y(t) = B$, is a single line at the origin ($n\omega = 0$); i.e. a constant level can be described by a wave of zero frequency or infinite wavelength.

Fourier analysis of a periodic function is equivalent to *spectral analysis*, in which light is separated by a prism or diffraction grating into its component colours, or to analysis of a musical sound in terms of its harmonics. Fourier analysis is also equivalent to normal mode analysis of the complex vibration pattern of a set of masses coupled by Hook's law springs (see above).

Fourier devised a mathematical method to derive the values of the coefficients A_n in the series, and calculated that

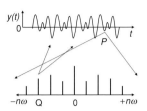

Fig. A3.16 The periodic function $y(t)$ and its Fourier transform $A_n(n\omega)$. Each point P on $y(t)$ is fully defined by the entire set of $A_n(n\omega)$ components and each component Q is completely defined by the full periodic pattern in $y(t)$.

$$A_n = \frac{2}{T} \int_0^T y(t) \exp(-\,\mathrm{i}\,n\omega t)\,\mathrm{d}t \qquad (A3.31)$$

where $T = 2\pi/\omega$ represents one *period* of the fundamental oscillation of frequency ω, i.e. the periodic repeat value of the variable t.

Since integration is equivalent to taking the sum of the function at very small intervals, the symmetry apparent in Eqs. (A3.30) and (A3.31) is quite remarkable. It essentially tells us that, on the one hand, each Fourier component is fully defined by the fundamental pattern of the function y, while, on the other hand, the value of y at each point t is, itself, defined by the entire set of Fourier components (Fig. A3.16).

We note another important property of Fourier transforms: if y is defined as a function of t, then its Fourier transform is defined as a function of frequency, i.e. a parameter that is proportional to $1/t$. The application of Fourier analysis and this reciprocal relation is quite general. In the case of a wave in space, if y is a function of x, then its Fourier transform is a function of the wavevector magnitude Q, which is proportional to $1/x$. The analogous equation to Eq. (A3.30) is written

$$y(x) = \sum_{n=-\infty}^{+\infty} A_n \exp\,\mathrm{i}\,n Q x \qquad (A3.32)$$

Because of this reciprocal relation, the space of x is sometimes called *real space* and the space of Q, in which the Fourier transform is defined, is called *reciprocal space*.

The Dirac delta function

The lines drawn for the Fourier coefficients in Fig. A3.15 were given heights proportional to their values. Although this is quite illustrative and pleasing to the eye, it is not mathematically correct. In fact, each line corresponds mathematically to an infinitely narrow curve, which has, nevertheless, a surface area equal to the value of the Fourier coefficient. The mathematical function called the *Dirac delta function* describes such a curve of unit surface area. A Dirac delta function at position $x = x'$ in one dimension is written $\delta(x - x')$. The function has the

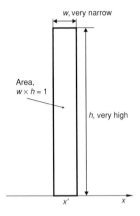

w, very narrow

Area,
$w \times h = 1$

h, very high

x' x

Fig. A3.17 The Dirac delta function.

following properties:

$$\text{if } x = x', \text{ then } \delta(x - x') = \infty;$$
$$\text{if } x \neq x', \text{ then } \delta(x - x') = 0;$$

the surface area of the curve is equal to unity, so that $\int \delta(x - x')\mathrm{d}x = 1$. Since the function is extremely narrow, the integral need not be from $-\infty$ to $+\infty$, but only over a small range of x around x' (Fig. A3.17). It can be shown that

$$\delta(x - x') = \int_{-\infty}^{+\infty} \exp\left[\mathrm{i}\, Q(x - x')\right]\mathrm{d}Q \tag{A3.33}$$

where Q is a wavevector magnitude as described above. There is a striking similarity between Eqs. (A3.32) and (A3.33). Equation (A3.33) also appears to be a Fourier series. In this case, however, the terms of the series are not separated by integer increments in n but by very small $\mathrm{d}Q$ intervals.

The Fourier integral and continuous Fourier transform

Equation (A3.33) describes a *Fourier integral*. It shows that Fourier analysis is not limited to functions of a periodic character. We recall that any periodic function can be represented by a sum of waves with frequencies increasing in integer multiples. We now generalise this statement and write that *any function $y(x)$ can be represented as a Fourier integral in terms of a sum of waves of continuously decreasing wavelength* (or of continuously increasing frequency if x is replaced by t):

$$y(x) = \frac{1}{2\pi} \int_{-\infty}^{+\infty} F(Q)\exp\left(\mathrm{i}\, Qx\right)\mathrm{d}Q$$

$$\tag{A3.34}$$

$$y(t) = \frac{1}{2\pi} \int_{-\infty}^{+\infty} F(\omega)\exp\left(\mathrm{i}\,\omega t\right)\mathrm{d}\omega$$

where the $1/2\pi$ factor results from replacing a sum by an integral.

In the corresponding Fourier transform, the A_n coefficients are now replaced by a continuous function $F(Q)$ (or $F(\omega)$).

$$F(Q) = \int y(x)\exp(-\mathrm{i}\, Qx)\,\mathrm{d}x$$

$$\tag{A3.35}$$

$$F(\omega) = \int y(t)\exp(-\mathrm{i}\,\omega t)\,\mathrm{d}t$$

Equation (A3.34) is called the *reverse Fourier transform* of $F(Q)$. Fourier transforms of useful non-periodic functions are given in Fig. A3.18.

The bell-shaped *Gaussian* and *Lorentzian* functions often occur in molecular biophysics. The statistical distribution of data measured on a large number of molecules (ensemble) follows a Gaussian curve (normal distribution). Practically all time-dependent phenomena can be described by an exponential decay

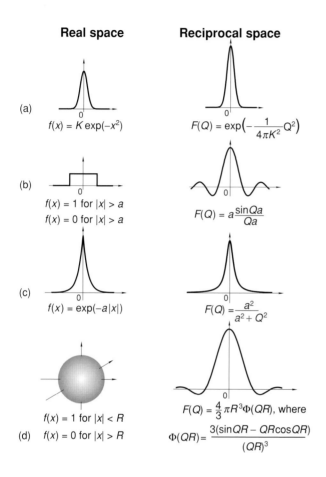

Real space | **Reciprocal space**

(a) $f(x) = K \exp(-x^2)$ | $F(Q) = \exp\left(-\dfrac{1}{4\pi K^2}Q^2\right)$

(b) $f(x) = 1$ for $|x| > a$ | $F(Q) = a\dfrac{\sin Qa}{Qa}$
$f(x) = 0$ for $|x| > a$

(c) $f(x) = \exp(-a|x|)$ | $F(Q) = \dfrac{a^2}{a^2 + Q^2}$

(d) $f(x) = 1$ for $|x| < R$
$f(x) = 0$ for $|x| > R$ | $F(Q) = \dfrac{4}{3}\pi R^3 \Phi(QR)$, where
$\Phi(QR) = \dfrac{3(\sin QR - QR\cos QR)}{(QR)^3}$

Fig. A3.18 Some non-periodic functions and their Fourier transforms. (a) The Fourier transform of a Gaussian of width $1/K$ is itself a Gaussian of width K. (b) The Fourier transform of a slit of width $2a$ is an oscillating function with a large central maximum of Q width $2\pi/a$ and subsidiary maxima that rapidly fade away. (c) The Fourier transform of an exponential decay is a bell curve with longer wings than a Gaussian called a Lorentzian function. (d) The Fourier transform of a sphere of radius R is a function with spherical symmetry called a Bessel function with a large central maximum of Q width about $2\pi/R$ and weak oscillations on either side of it.

in the time domain (e.g. the signal in dynamic light scattering, Chapter D10, and quasi-elastic neutron scattering, Chapter I2, related to translational diffusion; fluorescence depolarisation, Chapter D8 or electric birefringence, Chapter D6, related to rotational diffusion, etc.). The Fourier transform of an exponential function is a Lorentzian curve, so that the time-dependent phenomenon can be described by a Lorentzian in the frequency or energy domain.

In three dimensions, the Fourier transform is written in terms of vectors

$$F(\mathbf{Q}) = \int y(\mathbf{r})\exp(-i\,\mathbf{Q}\cdot\mathbf{r})\,d\mathbf{r} \qquad (A3.36)$$

Convolution
Consider the functions in Fig. A3.19. Clearly, the function, $C(x)$, in line (c) must be related mathematically to those in lines (a) and (b), $A(x)$, $B(x)$, respectively. It results, in fact, from an operation called convolution. $C(x)$ is equal to the convolution of $A(x)$ and $B(x)$.

(a)

(b)

(c)

Mathematically, the convolution of two one-dimensional functions $f(x)$, $g(x)$
is written

$$f(x) \otimes g(x) = \int_{-\infty}^{\infty} f(u)g(x - u)\,\mathrm{d}u \qquad\qquad (A3.37)$$

where $\otimes$ is the convolution symbol. The convolution operation is illustrated in
Fig. A3.20(a) and (b) for two simple functions, centred on $x = 0$ and x', respectively. We see in Fig. A3.20(c) the situation of one of the terms in the integral,
for $x = x'$ and $u = u'$. The function $g(x - u)$ has a finite value since it has been
displaced and it is now centred on u'; we see, however, that $f(u')$ is equal to
zero at this point, so that the term in the integral is zero. Clearly, the term is
non-zero, only if both $g(x - u)$ and $f(u)$ are non-zero. This only happens for x
close to x', otherwise g is zero, and for values of u that shift g so that it *overlaps*
with $f(u)$, i.e. u'' close to zero (Fig. A3.20(d)). The result of the convolution is a
function in which each point in g has been multiplied by f (Fig. A3.20(e)). If
$f(x)$ is not centred on zero but on $x + x''$, the convolution is centred on
$x' + x''$. It can be shown that $f(x) \otimes g(x) = g(x) \otimes f(x)$, so that we
should obtain the same result by scanning either of the functions over the
other.

   The convolution product plays a very important role in physical measurement. Consider a sample that absorbs only in a very narrow wavelength band
of radiation in a spectroscopy experiment. It is likely that the instrument used
for the measurement itself has an insensitivity to wavelength that is broader
than the band to be observed. The result of the measurement is the convolution product of the band shape function and instrument *wavelength resolution*
function.

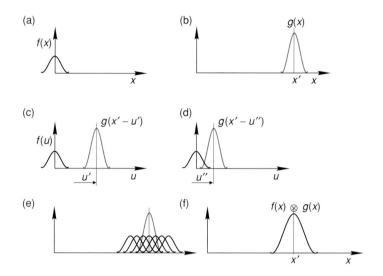

Fig. A3.20 An illustration of *convolution*: in (a) and (b) are the two functions whose convolution product is to be calculated; (c) and (d) show the terms in the integral of Eq. (A3.37) for $x = x'$ and $u = u'$ and u'', respectively; (e) shows how the result can be seen as each point in g being multiplied by f; (f) shows the result of the convolution.

The convolution product in three dimensions is written in terms of vectors

$$f(\mathbf{r}) \otimes g(\mathbf{r}) = \int_{-\infty}^{\infty} f(\mathbf{u})g(\mathbf{r} - \mathbf{u})\,d\mathbf{u} \qquad (A3.38)$$

Calculating the Fourier transform of a convolution product leads to an extremely useful result (shown in one dimension for simplicity)

$$FT\,[f(x) \otimes g(x)] = FT\,[f(x)] \times FT\,[g(x)] \qquad (A3.39)$$

where FT stands for the Fourier transform operation. Equation (A3.39) states that in order to obtain the Fourier transform of the *convolution product* of two functions, we simply *multiply* together the Fourier transforms calculated separately for each. By using the terms *real* and *reciprocal space*, we can write that convolution in real space leads to multiplication in reciprocal space (and vice versa).

We illustrate the result in Eq. (A3.39) with an example taken from crystallography (Fig. A3.21). A lattice is described by a set of Dirac delta functions. Placing any function on each node of the lattice can be seen as the result of a convolution product of the lattice and the function (left-hand panel in the figure). The right-hand panel shows the corresponding Fourier transforms. The Fourier transform of the repeating function is obtained by multiplying the Fourier transform of the function (the broad red bell curve) with the Fourier transform of the lattice (a set of delta functions separated at nQ_d, where n is an integer and $Q_d = 2\pi/d$, where d is the lattice repeat in this one-dimensional example). The resulting transform is the set of red lines (delta functions of decreasing amplitude) in the bottom of the right-hand panel.

Fig. A3.21 Left panel: The
convolution of a function
(red curve) with a lattice
represented by a set of
delta functions. Right
panel: the Fourier
transform of the
repeating function is
equal to the Fourier
transform of the red curve
multiplied by the Fourier
transform of lattice.

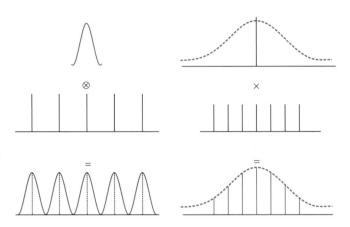

A3.2.4 Quantum mechanics

The concepts of quantum mechanics are necessary for the description of the
interaction between radiation and matter on an atomic scale. Quantum mechanics
was largely developed to provide an interpretation of spectroscopic observations,
in particular of the discrete lines in atomic spectra, and it is essential for many
of the methods described in this book. Even a succinct description of quantum
mechanics, however, would be too long to be included here, and only a few key
concepts are summarised below.

Planck's constant, energy quanta and photons

In quantum mechanics, energy may be transmitted during a given time from
one system to another only in discrete multiples of a universal minimum limiting
value. This value, given by Planck's constant ($h = 6.626 \times 10^{-34}$ J s), can be seen
as an indivisible packet of *energy multiplied by time*. In a quantum mechanical
description, a wave of frequency, ν, which, therefore, oscillates ν times a second,
transmits well-defined packets of energy, $h\nu$, which are called *quanta* (the singular
is *quantum*). The energy of a wave in quantum mechanics is described by two
components. The first component, which is fully defined by its frequency, is its
quantum energy. The second component is given by the number of quanta carried
per unit time, which is proportional to the intensity of the wave.

The quantum of a wave of electromagnetic radiation is called a *photon*. The
quantum energy of visible light, for example, depends on its colour. Quanta of
red light are of lower energy than quanta of blue light, but a more intense beam of
red light may still carry more energy. The wavelength associated with a photon is
inversely proportional to the frequency, $\lambda = c/\nu$, where c is the speed of light. The
quantum energy of a beam of electromagnetic radiation is fully defined, therefore,
by either the frequency ($h\nu$) or the wavelength (hc/λ).

Table A3.1. *Particles and waves*

	Particle	Wave
	$\bullet\ \bullet\ \bullet\ \bullet\ \bullet\ \bullet \longrightarrow$	
Intensity	rate = number of particles per unit time	square of the amplitude = A^2
Momentum	neutrons: $\mathbf{p} = m_n v$	neutrons: $\mathbf{p} = \hbar\mathbf{k} = (h/\lambda)\mathbf{u}$
	electrons: $\mathbf{p} = \dfrac{m_e v}{\sqrt{1 - v^2/c^2}}$	electrons: $\mathbf{p} = \hbar\mathbf{k} = (h/\lambda)(\mathbf{u})$
	photons: $\mathbf{p} = \hbar\mathbf{k} = (h/\lambda)\mathbf{u}$	photons: $\mathbf{p} = \hbar\mathbf{k} = (h/\lambda)(\mathbf{u})$
Energy	neutrons: $E = \frac{1}{2}m_n v^2 = \frac{1}{2}p^2/m_n v^2$	neutrons: $E = h\nu = \hbar\omega = \dfrac{h^2}{2\lambda^2 m_n} = \dfrac{\hbar^2 k^2}{2m_n}$
	electrons: $E = \frac{1}{2}p^2/m_e v^2 = \dfrac{m_e v^2}{2(1 - v^2/c^2)}$	electrons: $E = h\nu = \hbar\omega = \dfrac{h^2}{2\lambda^2 m_e} = \dfrac{\hbar^2 k^2}{2m_e}$
	photons: $E = h\nu = \hbar\omega = hc/\lambda$	photons: $E = h\nu = \hbar\omega = hc/\lambda$

Equations relating intensity, momentum and energy in the wave and particle pictures for neutrons ('classical' particles), electrons ('relativistic' particles) and photons (particles of zero rest mass). The De Broglie equation relates wavelength, λ, and momentum, $\mathbf{p}$; $\mathbf{u}$ is a unit vector in the direction of wave propagation; $\mathbf{k}$ is the wave vector, $k = 2\pi/\lambda$; $\hbar$ is Planck's constant divided by 2π; m_n, m_e, are the mass of the neutron and electron, respectively. Other symbols have their usual meanings.

The wave-particle duality

It is convenient for the analysis of its behaviour to picture a beam of radiation either as a stream of particles or as a wave. Since the events involving the beam happen in times and over distances that are very much smaller than can be grasped by our common-sense experience, the question 'Is the beam made up of particles or waves?' is meaningless. Suffice it to say that the beam can be described in either way. And, since the two descriptions must be self-consistent, there are rules linking them. Thus, the momentum and energy in terms of mass and velocity of the particle description are related by equations involving Planck's constant to the wavevector and frequency of the wave description. The relations are given in Table A3.1 for three cases: a beam of neutrons (particles of rest mass m_n moving 'slowly' with velocity u), a beam of electrons (particles of rest mass m_e, moving at a velocity close to the velocity of light), and a beam of photons (particles of rest mass zero, moving at the speed of light).

De Broglie's relation states that the wavelength associated with a moving particle is inversely proportional to its momentum, with the proportionality constant equal to Planck's constant, $\lambda = h/p$. Faster particles are therefore associated with shorter wavelengths and vice versa. Note that the neutron, because of its large mass, is a particle with unique properties for the study of matter. If we express the kinetic energy of a neutron in terms of temperature, room temperature neutrons (*thermal* neutrons) have an associated wavelength between 1 and 2 Å; *cold* neutrons (about 10 K) have wavelengths close to 10 Å, while *hot* neutrons have wavelengths of a fraction of an ångström.

Heisenberg's uncertainty principle

The only way that we can measure the position, momentum or energy of a particle or wave is by interfering with it in some way. The perturbation due to the measurement cannot be infinitely small but is itself limited by the minimum energy–time packet represented by Planck's constant, which, of course, is not negligible for measurements on an atomic scale. Heisenberg's uncertainty principle is a result of this. Expressed in two inequalities, it relates momentum–spatial coordinate and energy–time, respectively:

$$\Delta p \Delta x \geq h$$
$$\Delta E \Delta t \geq h$$

(A3.40)

where x is a spatial coordinate. The uncertainty principle states that in order to increase the precision of a position measurement (i.e. make Δx as small as possible) at the atomic level, for example, we have to sacrifice knowledge of its momentum, and vice versa. In other words, we can hope to know *either* how fast a particle is moving *or* where it is, but we cannot hope to have both bits of information at once.

An illustration of the uncertainty principle is provided by the wave description of matter. We recall from the section on Fourier analysis above that a non-periodic function can be represented by a sum of waves of different wavelengths. A particle that is localised in space can, therefore, be described by such a sum of waves. Since the waves have a range of different wavelengths we can only describe its momentum with a large uncertainty. Consider now a particle whose momentum is known precisely. It is represented by a wave of well-defined wavelength (e.g. a single sine wave). But such a wave extends infinitely in space, so that we have no idea where the particle might be! Similar arguments can be developed for energy and time measurements.

The uncertainty principle is also illustrated rather nicely by the resolution condition in crystallography (see Chapter G1, and Fig. A3.22). It is a consequence of the reciprocal relation between a crystal structure and the waves it scatters (which in effect are represented by its Fourier transform) that the smallest distance, d_{min}, that can be resolved in a crystallography measurement is related to the maximum observed value of scattering vector magnitude, Q_{max}, by

$$d_{min} = \frac{2\pi}{Q_{max}}$$

Fig. A3.22 We 'see' the atomic arrangement in a crystal by the way it scatters radiation. The crystal (depicted as a rectangle) is put in a beam described by a wave of wavevector k_0 and the distance between atomic planes is measured by observing the intensity of scattered waves as a function of wavevector k_1. The scattering vector Q is defined as the difference between k_1 and k_0.

where

$$\mathbf{Q}_{max} = \mathbf{k}_{max} - \mathbf{k}_0 \qquad\qquad (A3.41)$$

and $\mathbf{k}_0, \mathbf{k}_{max}$ are the wavevectors of the incident and diffracted beams, respectively. Now, we know from Table A3.1 that the momentum of a wave is expressed in terms of its wavevector, so that Q_{max} can be seen as proportional to the difference in momentum between the diffracted and incident waves. The spatial resolution, d_{min}, can be seen as Δx, the minimum spacing that can be determined in the experiment:

$$\Delta p = \frac{h}{2\pi} Q_{max}$$
$$\Delta x = \frac{2\pi}{Q_{max}}$$

so that

$$\Delta p \Delta x = h \qquad\qquad (A3.42)$$

The Schrödinger and Dirac approaches

It is a consequence of the uncertainty principle that the results of quantum mechanical calculations (e.g. the solution of the equations of motion of a system of atoms) are not in terms of determined values but in terms of probabilities. Thus, trajectories of varying probability are calculated for a beam of particles hitting a screen in which there is a small slit; the probability pattern found corresponds to the diffraction fringes observed experimentally. In the Schrödinger *wave mechanics* approach the probability density in the beam of particles before and after they cross the slit is depicted as the square of the amplitude of a wave function. The Dirac approach is in terms of the probabilities of different final *states*, given an initial *state*, expressed as discrete elements in a mathematical matrix.

Energy levels

A fundamental concept of quantum mechanics is that the energy of systems such as atomic nuclei or electrons is quantised in levels that are occupied according to specific statistical laws. It is the basis of spectroscopy experiments (see Part E) that transitions to higher levels can be stimulated by the absorption of radiation and that radiation is emitted when the system relaxes to a lower level. *The energy quanta, hv, absorbed or emitted correspond exactly to the differences between levels.*

The energy levels of an atomic electron are characterised by *orbital* and *spin* quantum states. In a classical analogy, these may be described as discrete energy states corresponding to the electron orbiting around the nucleus and spinning on its axis, respectively. The analogy should not be taken too far, however, because it is the very essence of quantum mechanics that it was invented because events

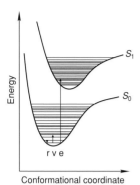

Fig. A3.23 Potential energy diagram of the lowest (S_0) and first excited (S_1) electronic states in a molecule: r, v, and e are rotational, vibrational and electronic transitions, respectively. (After Cantor and Schimmel, 1980.)

could not be described in a classical manner. The occupation of electronic states is governed by strict rules; e.g. a state of given orbital and spin quantum numbers may only be occupied by one electron. The lowest-energy state is called the *ground state*, and states of higher quantum number are called the *first excited state, second excited state* etc.

Atoms that are bonded together to form a molecule vibrate and rotate like the balls connected by springs we discussed in the section on simple harmonic motion. In quantum mechanics, these vibrations and rotations also correspond to discrete energy levels characterised by quantum numbers. The 'springs' of the atomic bonds comprise shared electrons whose energy levels are themselves affected by the molecular vibrational and rotational energies (Fig. A3.23). The lowest (ground) state and first excited electronic states in a molecule are each described by a potential energy well with respect to a conformational coordinate, in which the electrons may be seen as vibrating and rotating with the molecule. Contrary to the classical case, in quantum mechanics only discrete energy levels may be occupied in these potential wells. Note the large separation between electronic levels compared with the vibrational and rotational transitions, and that difference between vibrational energy levels is larger than that between rotational levels.

Atomic vibrations in a molecule may be discussed, to a first approximation, in terms of simple harmonic motion. Above, we derived the potential energy function of a classical simple harmonic oscillator. The quantum mechanical potential energy of a simple harmonic oscillator made up of two bonded atoms of reduced mass $m_{1,2}$ is of the form

$$E = \left(n + \frac{1}{2} \right) \hbar \omega$$

where

$$\omega = \sqrt{\frac{k}{m_{1,2}}} \tag{A3.43}$$

and n is a vibrational quantum number, which can take only positive integer values including zero, $\hbar$ is Planck's constant divided by 2π, k is a force constant and $m_{1,2}$ is reduced mass (compare Eq. (A3.24)). In contrast to classical mechanics where vibrators can assume any potential energy, quantum vibrators can take only certain discrete energies (Fig. A3.23). Transitions in energy levels can be brought about by absorption of radiation, provided the energy of the radiation exactly matches the difference in energy levels ΔE between the quantum states.

The potential energy of electronic vibrations in a bond between two atoms, however, is not perfectly described by a harmonic model. For example, as the two atoms approach one another, Coulombic repulsion between the two nuclei

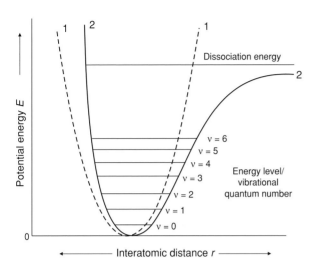

Fig. A3.24 Potential energy diagram for electronic vibrations in a bond between two atoms: curve 1, harmonic oscillator; curve 2, anharmonic oscillator. Note, that the harmonic and anharmonic curves nearly coincide at low potential energies. The dissociation energy is that required to 'break the spring' between the atoms.

produces a force that acts in the same direction as the restoring force on the bond; thus the potential energy can be expected to rise more rapidly than predicted by the harmonic approximation. Qualitatively, the potential energy curves takes the anharmonic form shown in Fig. A3.24 (and in Fig. A3.23). Such curves deviate by varying degrees from harmonic behaviour, depending upon the nature of the bond and the atoms involved.

A3.2.5 Measurement space, mathematical functions and straight lines

Practically all experimental measurements in biophysics are performed in a parameter space that is related to real space by a mathematical transformation. In crystallography, for example, the *measurement* space is *reciprocal* space (the space in which diffraction is measured), which is related to real space (the space of atomic coordinates) by Fourier transformation. In hydrodynamics, the variation of a parameter (translational or rotational diffusion, for example) is plotted as a function of time in the measurement space of the experiment. A model provides the means to transpose this information to real space (for example, if a sphere is assumed its radius will be determined).

It is important to emphasise that the experiment always takes place in measurement space; the information in real space is usually model dependent and will be obtained with a limited accuracy.

In order to simplify the mathematical interpretation of experimental data it is often possible to express complicated equations in linear form (see Comment A3.2).

Comment A3.2   Biologist's box: Mathematical functions and straight lines

Parameters measured in an experiment often occur in complicated forms in mathematical equations. Consider, for example, the angular dependence of scattered intensity from a solution of macromolecules:

$$I(Q) = I(0) \exp\left(-\frac{1}{3} R_G^2 Q^2\right)$$

where Q is related to the scattering angle. The formula is known as the Guinier approximation (see Chapter G2). Two experimental parameters can be derived from $I(Q)$: $I(0)$, which is related to the molar mass of the particle in solution, and R_G, the radius of gyration, which is a measure of the particle shape and dimensions. The easiest way to obtain these parameters is to *linearise* the equation:

$$\ln I(Q) = \ln I(0) - \frac{1}{3} R_G^2 Q^2$$

In this form a plot of $\ln I(Q)$ against Q^2 will yield a straight line of intercept $\ln I(0)$ and slope $-1/3R_G^2$.

The application of linearization procedures in order to express a mathematical function as a straight line is generally useful in biophysical studies. Another example of the application is in hydrodynamics. The translational friction suffered by particles of *the same shape* in solution is proportional to the cube root of the molar mass (linear dimension)

$$f = kM^{\frac{1}{3}}$$

where k is a constant. We can linearise the equation in the following way:

$$\log f = \log k + 1/3 \log M$$

and plot the f *versus* M values on a double logarithmic scale. If the particles are indeed of the same shape, the equation is obeyed and the data fall on a straight line with slope $1/3$ (see Chapter D2).

A3.3 Dynamics and structure, kinetics, kinematics, relaxation

The concept of *dynamics* is often used to describe time dependence (or motions), as opposed to *structure*, which describes a static or (more accurately) time-average view. The Greek origin of the word, *dynamis*, means strength, however, and refers to *forces*, and *dynamics* (a singular noun) is the branch of *mechanics* in physics that deals with the motion of objects and the forces that act to produce such motion.

In physics, dynamics is divided into *kinetics* (from the Greek *kineein*, to move), which is concerned with the relationship between moving objects, their masses and the forces acting upon them, and *kinematics*, which is concerned only with the motion of objects, without consideration of forces. Kinematics and the word *cinema* share the same Greek origin (*kinema*, movement). In chemistry, *kinetics* refers to the study of the *rates of chemical reactions*.

Relaxation refers to the return to equilibrium of a disturbed system. The simplest relaxation can be described by an exponential decay:

$$A(t) = A(0) \exp(-t/\tau) \qquad (A3.44)$$

The parameter $A(t)$ represents the deviation of a property of the system from equilibrium at time t. It is $A(0)$ at time 0 and decays to zero at infinite time. The relaxation time, τ, is the time at which A has decayed to $1/e$ (about $1/2.7$) of the value of $A(0)$ (see Chapters D6 and D8).

In this chapter, we discuss the present understanding of biological macromolecular structure in terms of the acting forces. We, therefore, concentrate on the internal dynamics that define the macromolecule as a physical particle (i.e. the dynamics of the atoms in the correctly folded macromolecule) rather than on the dynamics of the particle moving as a rigid body, which is treated in Part D on hydrodynamics.

A3.3.1 Macromolecular stabilisation forces

The forces that maintain atoms in position in a correctly folded, biologically active macromolecular structure are known (see Section C2.4.9). They arise from van der Waals interactions, hydrogen bonds, electrostatic interactions, hydrophobic interactions, and S–S bonds between cysteine residues (Fig. A3.25). The *stabilisation energy* of a macromolecular structure is the difference in free energy between

Fig. A3.25 Interactions stabilising the folded form of the malate dehydrogenase tetramer from the halophilic archaeon *Halobacterium marismortui*. The protein does not contain cysteine residues. Its physiological environment is close to saturated in KCl and it requires high salt concentration to be folded and stable. In fact, weak electrostatic interactions lead to specific binding of solvent ions that participate in the stabilisation. The orange balls between the subunits in the picture are chloride ions (Richard *et al.*, 2000).

H-bonds in α-helices and β-sheets

Salt bridges between Asp and Arg residues in adjacent subunits

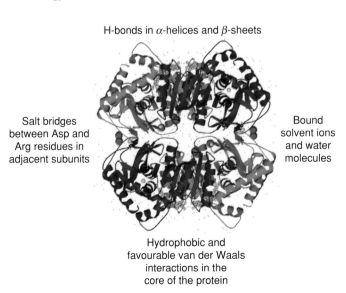

Bound solvent ions and water molecules

Hydrophobic and favourable van der Waals interactions in the core of the protein the folded and unfolded forms. Covalent bonds, which are very strong with respect to temperature, are not usually broken when a macromolecule unfolds, so that (except for S–S bonds) they do not contribute to the stabilisation energy of macromolecular tertiary or quaternary structure. Biological macromolecules are *soft* because the energies associated with the stabilisation forces are *weak* in the sense that they are of the order of thermal energy at physiological temperatures. This softness may at first appear surprising because proteins, for example, are known from crystallography and calorimetry to form compact structures in the core of which the atoms are tightly packed; also, the electrostatic (Coulombic) interaction between two charges is long-range and very intense, and S–S bonds are essentially covalent bonds. Stabilisation forces, however, are all strongly environment-dependent. The electrostatic interaction in biological macromolecules is shielded by the dielectric properties of water and by solvent counter-ions. S–S bonds are easily broken in a reducing environment to form two separate SH groups. Hydrogen bonds of different strengths are formed between donor and acceptor groups in the macromolecule and between the macromolecule and solvent ions. The hydrophobic interaction is environment-dependent by definition, since it arises from entropic effects due to the low solubility in water of certain chemical groups (see also Sections A1.3.4, C2.4.9). Even the van der Waals interaction, which arises from the close packing of atoms, is environment-dependent, because certain atoms pack closely better than others, and it has been suggested as an important driving force of protein folding.

Molecular dynamics simulations represent our theoretical understanding of macromolecules as *physical particles*. Despite the fact that the types of macromolecular stabilisation forces are well known, our quantitative knowledge of the determinants involved is very far from complete, mainly because of complex environment effects. Such calculations, therefore, still have to remain firmly anchored by experimental results.

A3.3.2 Length and time scales in macromolecular dynamics

The amplitudes of atomic motions in macromolecules at ambient temperature (300 K) range from 0.01 Å to >5 Å for time periods from 10^{-15} s to 10^3 s (a femtosecond for electronic rearrangements to about 20 min for protein folding or local denaturation) (Table A3.2). Different biophysical methods are adapted to each time and length scale (see Introduction). Laser triggered optical spectroscopy can now reach subfemtosecond resolution and spans the entire 18 orders of magnitude of the time scale to kinetic measurements taking minutes. NMR is sensitive to dynamics from the picosecond upwards. Neutron scattering can resolve frequencies in the 10^9–10^{12} s^{-1} range and amplitudes in the 1–5 Å range.

Table A3.2. *Length and time scales in macromolecular dynamics*

Type	Amplitude (Å)	Time (s)
Electronic changes	0.01	10^{-15}
Atomic bond fluctuations	0.01–0.1	$10^{-14}-10^{-13}$
Side-chain motions	1	$10^{-12}-10^{-9}$
Side-chain rotations in protein interior	5	$10^{-4}-1$
domain motions	1–5	$10^{-9}-1$
Protein folding, complex formation,	1 to >5	$10^{-6}-10^{3}$
conformational changes . . .		

A3.3.3 A physical model for protein dynamics

Our current understanding of protein dynamics (see Chapter I1) is based on experimental results from a wide range of biophysical methods, in particular, time-resolved optical, and neutron spectroscopies, NMR and crystallography.

A protein is a densely packed object with the atoms occupying definite average positions corresponding to the native structure. Substantial restoring forces act on the atoms so that protein molecular dynamics resembles that of an amorphous solid. Only to a first approximation, can the protein be seen as forming a continuous elastic medium, with local side-chain motions and global domain motions displaying simple Hooke's law character (see Section A3.2.2).

Conformational substates in the energy landscape and protein specific motions

The *conformational substate* (CS) model has been put forward for proteins, following experiments on carbon monoxide binding to the haem group in myoglobin (Comment A3.3). An analysis of the rebinding kinetics indicates a heterogeneous protein population at low temperature (below about 200 K). At higher temperatures, the ligand overcomes a number of energy barriers on its path to rebinding in a way that suggests the protein fluctuates between subtly different conformations. A schematic energy landscape characteristic of the CS model is shown in Fig. A3.26.

Below about 200 K, the protein is trapped in one of a number of energy minima – one of the CS (bottom panel in Fig. A3.26).

The structural difference between two CS may be very small, e.g. due to a different orientation of just one amino acid side-chain. With increasing temperature, there comes a point where the macromolecule has sufficient thermal energy to 'jump' the barrier from one CS to another. The resulting motions have been called *protein-specific motions*. In other words, the CS represent a 'static' disorder in protein structure at low temperature – each molecule in a population

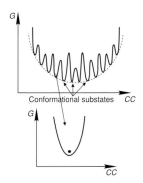

Fig. A3.26 One-dimensional schematic diagram of the CS model and potential free energy (*G*) landscape of a protein. *CC* is a conformational coordinate (from Frauenfelder *et al.*, 1988). (Figure reproduced with permission from *Annual Reviews.*)

Comment A3.3   Flash photolysis experiments on myoglobin at low temperatures (see also Section E1.3.3)

Myoglobin is a small monomeric oxygen binding protein (molecular weight about 16 K) found in muscle. It was one of the first (with haemoglobin, the related oxygen carrier protein in red blood cells, which can be seen as a tetramer of myoglobin-like subunits) whose structure was solved by X-ray crystallography. The Nobel prize was awarded to Max Perutz and John Kendrew for these studies. Haemoglobin and myoglobin have been studied extensively by crystallography and functionally important structural differences have been elucidated between the unbound (deoxy-)states and various ligand bound states. Myoglobin contains a prosthetic haem group with an iron atom, which binds the oxygen molecule. Myoglobin has a higher affinity for carbon monoxide than for oxygen, which is the reason why CO is a poison (it blocks the oxygen binding site). Its CO-binding properties, however, have provided important tools for the study of the protein. The binding of CO to the haem group in myoglobin can be followed readily by its effect on the protein's absorption spectrum for visible light. In a flash photolysis experiment, rebinding kinetics is observed after the CO bond to the haem iron is broken by a laser flash. At low temperatures (below 200 K), the CO does not diffuse out of the protein. In position A when it is bound, it released by flash photolysis to position B in the haem pocket of the protein.

Contrary to expectations for a simple relaxation model (Eq. (A3.34)), it has been observed that rebinding kinetics did not fit a single exponential in time, but is constituted of a mixture of fast and slow processes. This can be explained in one of two ways: either the proteins make up a homogeneous population with several B sites in each (e.g. a 'fast' and a 'slow' rebinding site)

or the proteins make up a heterogeneous population, with different molecules displaying different sites

It was possible to distinguish unambiguously between the two models by *multiple flash* experiments, using different flash rates. When the flash rate is intermediate between the fast and slow rates, very different behaviour is expected for the homogeneous and inhomogeneous protein models. It has been established that the inhomogeneous protein model is the correct one. The *pure protein* population at low temperatures, therefore, is made up of molecules with slightly different structures, represented by the *conformational substates*. Direct rebinding from the haem pocket becomes exponential at higher temperatures, with a homogeneous protein population fluctuating between the conformational substates (Frauenfelder *et al.*, 1988.) (Figures reproduced with permission from *Annual Reviews.*)

having a slightly different structure (e.g. a side-chain in conformation A in one molecule and in conformation B in another); and they represent a 'dynamic' disorder at higher temperature (e.g. the side-chain moving, within the same molecule, between conformations A and B). In fact, careful analysis of the CO rebinding showed that the *energy landscape* of the protein presents a complex hierarchy with each CS potential well, itself divided into further CS. The CS model has received support from various other experimental methods, including analysis of the temperature dependence of Debye–Waller factors in X-ray crystallography, which clearly demonstrated the static disorder in the crystal population.

The dynamical transition and mean effective force constants

Neutron scattering experiments showed that the mean square fluctuation, $\langle u^2 \rangle$, of atoms in myoglobin has a temperature dependence as shown in Fig. A3.27. Note that the magnitude of the fluctuations is of the order of an ångström. The break in slope at 180 K was called a *dynamical transition*. Since the atomic fluctuations in the protein are due to thermal energy, the temperature axis actually represents the energy of the system. Below about 180 K ($-93\ °C$) $\langle u^2 \rangle$ increases linearly with energy, with a backward extrapolation to zero at zero absolute temperature (Comment A3.4). The straight-line dependence can be accounted for by a model in which the atoms move in simple harmonic potentials, for which the restoring force is proportional to the displacement from the mean position. The energy of a set of harmonic oscillators varies with the mean square displacement with a proportionality constant, $\langle k \rangle$, equal to the mean elastic force constant or 'spring' constant (Section A3.2.2). The value of $\langle k \rangle$ can, therefore, be calculated from the inverse of the slope of $\langle x^2 \rangle$ versus T (Fig. A3.27). Converting T to energy units by multiplying by Boltzmann's constant, the mean force constant maintaining

Comment A3.4
Zero point energy

In simple harmonic motion, the mean square fluctuation is linear with temperature and extrapolates to zero at absolute zero temperature. Because of quantum effects, however, the measured data points are expected to deviate from the straight line and approach a non-zero value (zero point motion) as T approaches absolute zero. This is a consequence of Heisenberg's uncertainty principle (see Section A3.2.4), since any measurement of a system at absolute zero perturbs the system so that it is no longer at absolute zero.

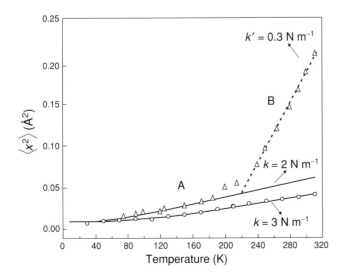

Fig. A3.27 Mean square fluctuations as a function of absolute temperature, measured by neutron scattering (see Chapter I2) in myoglobin surrounded by water ($\triangle$) and in a trehalose glass (o). The break in slope at about 180 K is called a dynamical transition. Effective force constants, k and k' were calculated for different parts of the curves as explained in the text.

Comment A3.5
Force constants on atoms and in elastic bands

An effective force constant of about 2 N/m holds an atom in the myoglobin structure in place at low temperature. An elastic band of force constant 2 N/m stretches by 1 cm under a weight of 2 g. The force exerted by the weight is 0.002×10 N.

the atoms in the myoglobin structure at low temperature was calculated to be 2 N/m, corresponding to the macroscopic force constant for a fairly stiff elastic band (Comment A3.5). Trehalose is a natural disaccharide, which is synthesised by desert and other organisms that have to survive extreme drought. It coats and protects their cellular structures while their metabolism is arrested and they are waiting for better, wetter, times. Myoglobin encased in a trehalose glass (circle data in Fig. A3.27) is stiffer than the protein in water with a $\langle k \rangle$ value of 3 N/m.

In order to avoid terms like rigidity or stiffness, which are often used qualitatively, the mean effective force constant derived from the temperature dependence of the mean square fluctuation was called the *resilience* of the structure. Note that there is no dynamical transition in trehalose, leading to the suggestion that the protective effect of the sugar is due to its trapping the macromolecule in a stiff harmonic state to relatively high temperatures, which precludes the motions that would lead to unfolding.

In terms of the CS model, the protein atoms, below the dynamical transition, oscillate in harmonic potential wells. The dynamical transition then corresponds to their being able to fluctuate between different CS wells in the potential energy landscape. The data in Fig. A3.27 have been modelled successfully by using a two-well potential model (two CS) for all the atoms in the protein surrounded by water (Fig. A3.28). This is clearly an oversimplification because different protein atoms move in different potentials, but it does provide an illustration of the link between the dynamical transition and the CS model. Even though the motions are not simple harmonic above the dynamical transition, it is still possible to consider a quasi-harmonic approximation in order to calculate an effective phenomenological force constant from the inverse slope of $\langle u^2 \rangle$ versus *T*. The value for myoglobin is 0.3 N/m, about ten times 'softer' than for the protein at low temperature.

Dynamical transitions have been observed in various other protein samples, as well as in myoglobin, and they may well represent a general feature of protein molecular dynamics. The values of the mean square fluctuations and effective force constants appear to be specific to each case and related to protein stability function and activity. It has been shown in cryocrystallography experiments (see Part G), for example, that the enzyme ribonuclease A cannot bind its substrate at temperatures below the dynamical transition. If the substrate is bound at a higher temperature and the protein cooled to below the transition, then it cannot be released. The additional dynamical flexibility of the protein, permitting it to sample different CS appears, therefore, to be necessary for this enzyme's activity. Correlations have also been established between the dynamic transition and activity for bacteriorhodopsin (BR), a membrane protein, which functions as a light-driven proton pump (Comment A3.6, Fig. A3.29).

There are similarities between the dynamical transition in proteins and the glass transition in amorphous materials, but they cannot be considered as completely

Comment A3.6   Purple membrane and bacteriorhodopsin

Purple membrane forms patches on the cell membrane of the archaeal extreme halophile (see Chapter A1) *Halobacterium salinarum*. It is an extraordinary membrane, made up of a single membrane protein type, organised with specific lipids on a highly ordered two-dimensional lattice. The existence of the lattice permitted high-resolution crystallographic studies by electron microscopy of the natural membrane (see Part G), so that we know a lot about its structure. The protein was called *bacteriorhodopsin* (BR) because it binds a molecule of the chromophore, retinal, which gives the purple colour to the membrane. Before this discovery, retinal had been found only in the rhodopsins, the vision proteins in animals. In *H. salinarum,* BR functions as a light-activated proton pump, transferring one proton out of the cell (against its potential gradient) for each photon absorbed by the retinal. A millisecond photocycle of colour changes, associated with the proton pump activity, provide a powerful tool for the study of structure function relations in Purple Membrane by spectroscopic methods (see Section E1.3.3).

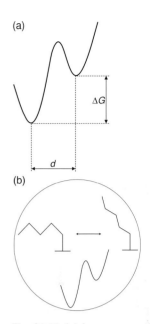

Fig. A3.28 (a) A double-well free energy (*G*) potential for protein atoms. Such a potential would be illustrated in practice by side-chains being able to sample two different local energy minima in the structure. (b) By using a quasi-harmonic approximation, an approximate effective force constant can be calculated as $\langle K \rangle \sim \delta G/2d^2$

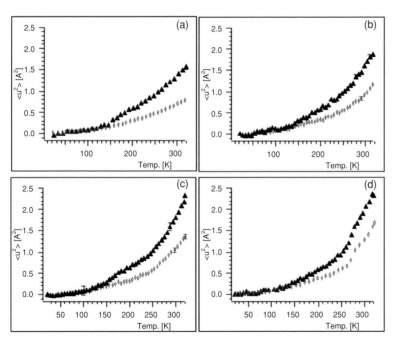

Fig. A3.29   BR dynamics measured by neutron scattering (see Part H). Mean square fluctuations are plotted versus temperature for the native membrane ($\triangle$) and a labelled membrane ($\diamond$) in which the data correspond to the motions of specific amino acids and retinal in the core of the protein. Parts (a)–(d) correspond to progressively higher hydration; 0%, 75%, 86%, 93% relative humidity, respectively (Lehnert, thesis 2002).

Fig. A3.30 A schematic diagram of purple membrane, showing a BR molecule as a ribbon diagram and lipids in blue. Two lipid molecules and the labelled groups discussed in Fig. A3.29 are shown in ball and stick representation. Converting light energy to chemical energy, BR pumps a proton out of the cell for each photon absorbed. Experiments demonstrated that the core and extracellular half of BR (within the red line), which act as the valve of the pump, are more resilient (stiffer) than the membrane as a whole. (From Zaccai, 2000.)

analogous. Proteins are complex, heterogeneous structures, whose dynamics has evolved to fulfil specific functions. Figure A3.29 shows that in BR, for example, the mean square fluctuations and effective force constants are different for different amino acid groups within the protein structure. The active core of the proton pump mechanism of the protein lies around the retinal, which remains bound to a lysine residue, via a so-called Schiff base linkage, about half way down the membrane thickness. During the BR photocycle (see Comment A3.6), the Schiff base proton changes its orientation from being accessible to the extracellular side of the membrane to being accessible to the intracellular side. The retinal linkage to the protein, therefore, acts as the valve of the pump, allowing the unidirectional transmission of the proton. Experiments have been designed in this context to examine the dynamics of different parts of the BR structure. The active core of the protein presents lower mean square fluctuations and higher effective force constants than the protein average (Fig. A3.30), as expected for a stiffer environment required to regulate the valve function of the retinal binding site.

Solvent effects, membrane and protein hydration

From a dynamics standpoint, a protein structure cannot be considered separately from its environment. The effect of trehalose on myoglobin dynamics is striking (Fig A3.27). Similarly, we note in Fig. A3.29 that the mean square fluctuations in BR, are strongly hydration-dependent, above a transition temperature at about 150 K. Hydration was defined, in the experiments, by putting the sample in a controlled relative humidity environment (Comment A3.7). Above 150 K, both the labelled and native sample have higher fluctuations and become softer with increasing relative humidity, illustrating the effects of hydration on the membrane and protein dynamics. It is interesting to recall that purple membrane is obtained

(a)

(b)

Comment A3.7 Salt and relative humidity

The partial pressure of water above a saturated salt solution is a constant for a particular salt type, at a given temperature and pressure. Relative humidity (RH) can, therefore, be regulated under experimental conditions by putting the sample in a closed temperature-controlled vessel, in the presence of an appropriate saturated salt solution. At 25 °C and normal pressure, for example, the RH above a saturated LiCl solution is 15%; for the same conditions it is 75% for NaCl, 86% for KCl, 93% for KNO_3. An advantage of these salts is that the corresponding RH values are not very sensitive to small temperature variations. 0% ('dry'), and 100% ('wet') relative humidity values are more difficult to define precisely. A 'dry' atmosphere may be obtained with silica gel, but 'even drier' conditions are achieved by pumping a vacuum over P_2O_5. 100% RH is obtained, in principle, above pure water, but even small temperature gradients in the walls lead to condensation and fluctuating RH values within the vessel.

Fig. A3.31 (a) Purple membrane samples for neutron scattering are made up of a regular stack of membrane and water layers, of periodicity, *d*. The membrane thickness, d_m, is constant and equal to about 50 Å; the water layer thickness, d_w, varies with RH from the thickness of one layer of water (about 3 Å) at 75% RH to the thickness of 3 layers at 93% RH. (b) Schematic representation of the hydration shell (light blue) around a lysozyme (red). The shell thickness corresponds to approximately one molecular layer and its density is about 10% higher than bulk water. (Weik *et al.*, 2004); Svergun *et al.*, 2001; Merzel and Smith, 2003.) (See also Section D1.3.)

from an extreme halophile that lives in a close to saturated salt environment (see Chapter A1). Saturated NaCl and KCl solutions correspond, respectively, to relative humidity values of 75%, and 86%. Parts (b) and (c) of Fig. A3.29 might, therefore, be the closest representations of purple membrane physiological conditions. Purple membrane samples in the neutron scattering experiments were made up of stacks of alternating membrane and water layers (Fig. A3.31). The proton pump photocycle of purple membrane (also measured in stack samples) is inhibited below about 75% RH, a condition in which there is one layer of water between the membranes in the stack-a hydration level that appears to be essential to ensure the minimum dynamic level for protein activity in the membrane. Careful measurements of activity as a function of hydration in lysozyme and other soluble enzymes have produced a similar conclusion. It is now generally accepted that at least one hydration layer around globular soluble proteins is required for activity. Depending on the size of the protein, this corresponds to a minimum hydration of between about 0.2 and 0.4 g of water per gram of protein. It has also been observed experimentally and successfully simulated in a molecular dynamics calculation that the density of this first hydration shell is about 10% higher than that of bulk liquid water (Fig. A3.31).

Motion pictures of intermediate structures

Kinetic crystallography provides the closest approximation we currently have to motion picture (cinema) images of macromolecules. (The method should really be called kinematic crystallography.) Macromolecular crystallography produces beautiful pictures of structures to atomic detail, and it was a logical consequence to explore its possibilities for obtaining images of a structure as a function of time.

The extremely high intensity of broad wavelength range X-ray beams produced in synchrotrons and the possibility of obtaining them in periodic pulses as short as a few picoseconds made time-resolved crystallography appear to be an attainable goal. But a crystal contains about 10^{15} molecules, and the information on atomic positions is averaged not only over time, but also over all the molecules in the crystal. In other words, in order to observe structural changes with time, all the molecules should be in phase, undergoing the same changes at the same time. Photoactivation is extremely fast and reaction synchrony has been achieved by using a laser flash to trigger the conformational changes simultaneously in the entire crystal, either if the protein itself naturally binds a light sensitive molecule (e.g. the haem in haemoglobin and myoglobin (Comment A3.3) or the retinal in rhodopsin and BR (Comment A3.6) or by the use of *caged compounds*. Binding a *blocking* chemical group to a substrate or cofactor can effectively cage it in and greatly reduce its affinity for the enzyme, or, if it binds to the enzyme, effectively stop the reaction from proceeding. By choosing a blocking group bond that is photolabile (i.e. sensitive to light), after diffusion into the crystal, the caged compound can be released by a laser flash of appropriate wavelength very rapidly and homogeneously to initiate the reaction.

Real-time crystallography provides motion pictures of working photosensitive proteins down to nanosecond resolution. For various technical reasons, however, in the middle of 2004, most published real-time crystallography studies of enzyme reactions using caged compounds related to second or even minute time resolution, and there was only one study reporting millisecond resolution. A complementary and perhaps more promising approach to obtaining motion pictures of proteins than real-time crystallography is based on trapping transient structural intermediates at *cryotemperatures* (from the Greek *cryo*, very cold). This approach has been very fruitful in studies of naturally photosensitive proteins and in studies using caged compounds. The effective time resolution is obtained from the choice of temperature. The reaction is stopped if it reaches an energy barrier that cannot be crossed because of insufficient thermal energy.

Various authors have investigated the structural intermediates formed when CO bound to myoglobin dissociates and migrates out of the protein molecule, by using real-time crystallography at room temperature or cryocrystallography. The results of these experiments were described in terms of motion pictures. A careful examination of these movies is particularly instructive about some of the general features of protein dynamics involved in biological function. The first frame of the movie corresponds to CO-ligated myoglobin, the last frame to the structure of deoxymyoglobin (Comment A3.3). An X-ray crystallography study provides information on the electron density distribution in a structure. Structural changes in real time appear as positive electron density peaks, indicating the new presence of atoms at these locations, or negative peaks, indicating that atoms have moved away from these sites. A short laser pulse triggers CO dissociation. The very first changes, a movement of the iron atom and 'doming' of the haem plane, take place

(a)

(b)

Fig. A3.32 (a) The time courses of the integrated electron content of the positive difference density at the Xe 1 binding site and of the integrated electron content of the negative Leu89 feature. The time courses of the other three Xe binding sites (Xe 2, Xe 3, and Xe 4) are shown for comparison; they indicate no occupation by CO on the time scale. The solid line represents a fit of the time course of the Xe 1 density by two exponential phases and a bimolecular phase, fixed to that of ligand rebinding. (b) Difference electron density map of the Xe 1 region at 362 ns. Positive density at the Xe 1 site is labelled X, while positive and negative densities indicating rearrangement of the Leu89 side-chain are labelled L1 and L2, respectively (corresponding to two conformational substates). The CO- and deoxy-myoglobin structures are in red and blue ball-and-stick representation, respectively. Positive and negative difference electron density appears as blue and red 'nets', respectively. The heme with the iron atom (Fe) at its centre and the position of the bound CO are seen at the top of the structure (Srajer et al., 2001). (Figure reproduced with permission from the American Chemical Society.)

Fig. A3.33 Movie of the radiation damage to a glutamic acid side-chain (top) and a cysteine–cysteine bond (bottom) in acetylcholinesterase at 100 K (from Weik *et al.* (2000)). (Figure reproduced with permission from *Proceedings of the National Academy of Sciences* (USA).)

too rapidly to be resolved in the time frames of the movie, which follow changes in the crystal structure occurring between 1 ns and 1 ms. On this time scale, structural fluctuations in the protein structure open up a diffusion path for the CO, which proceeds outwards from its binding pocket by pausing at a number of specific 'docking' sites, where it interacts favourably with protein side-chains. Four xenon binding sites, labelled Xe 1 to Xe 4, have been identified as potential such docking sites. Meanwhile the protein structure changes towards the conformation of the unligated deoxymyoglobin. In particular, there is a rearrangement of the side-chain of Leu89 to accommodate CO in the Xe 1 site. The time courses of the integrated electron content of the positive difference density at the Xe 1 binding site and of the integrated electron content of the negative Leu89 feature are shown in Fig. A3.32(a). The map showing the corresponding difference electron density peaks at 362 ns is in Fig. A3.32(b).

We recall, the structure observed in a crystallographic experiment corresponds to that adopted by an appreciable number of molecules in the crystal. The frames in the movie, therefore, illustrate specific intermediates (or energetically favoured CS) in the relaxation between CO-ligated and deoxymyoglobin structures. The crystallographic movies do not show continuous changes in atomic positions between one intermediate and the next, indicating that individual molecules in the crystal take different paths across the energy barriers. The picture fits well with the hypothesis that a large number of fluctuation-like motions on the nanosecond

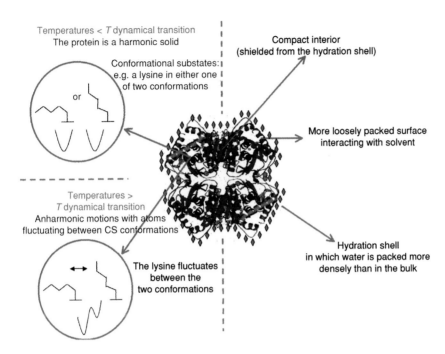

Temperatures < T dynamical transition
The protein is a harmonic solid

Conformational substates:
e.g. a lysine in either one
of two conformations

or

Temperatures >
T dynamical transition
Anharmonic motions with atoms
fluctuating between CS conformations

The lysine fluctuates
between the
two conformations

Compact interior
(shielded from the hydration shell)

More loosely packed surface
interacting with solvent

Hydration shell
in which water is packed more
densely than in the bulk

Fig. A3.34 Schematic
diagram illustrating the
main features of the
physical model for
protein dynamics.

time scale result in the much slower conformational changes between intermediates and the opening of a way out for the CO molecules.

Figure A3.33 illustrates another type of crystallographic motion picture. It shows the time course under irradiation of two locations in the structure of an enzyme called acetylcholinesterase, which is involved in nerve transmission. The amino acid side-chains in the initial structure are shown as ball-and-stick and the observed electron density is shown as a blue 'net'. The top panel shows the progressive decarboxylation of a glutamic acid side-chain due to radiation damage. The bottom panel shows the breaking of a S–S bond between two cysteines. The nine data sets, A–I, were collected at 100 K. At 150 K the damage affected more residues including those in the active site, which indicates that the side-chains can sample different conformations at the higher temperature.

The main features of the current physical model for protein dynamics are summarised and illustrated in Figure A3.34.

A3.4 Checklist of key ideas

- Molecular biophysics is a predominantly experimental science, in which even *theoretical* approaches such as *molecular dynamics simulations* have to be based firmly on observation.
- Practically all the experiments in biophysics (with the exception of calorimetry and classical solution physical chemistry methods such as the ones used to determine osmotic

pressure, viscosity, etc.) rely completely or in part on the observation of the interaction between macromolecules and radiation.

- The mathematical tools required include those dealing with *the general properties of waves, complex exponentials, simple harmonic motion and normal modes* and *Fourier analysis*.

- Experiments are performed in a *measurement space* related to real space by a mathematical transformation, such as *reciprocal space* in crystallography, which is related to real space by Fourier transformation.

- The concepts of quantum mechanics are essential for the description of the interaction between radiation and matter on an atomic scale.

- *Dynamics* (a singular noun) is the branch of *mechanics* in physics that deals with the motion of objects and the forces that act to produce such motion. Dynamics is divided into *kinetics*, which is concerned with the relationship between moving objects, their masses and the forces acting upon them, and *kinematics*, which is concerned only with the motion of objects, without consideration of forces.

- In chemistry, *kinetics* refers to the study of the *rates of chemical reactions*.

- *Relaxation* refers to the return to equilibrium of a disturbed system.

- The folded native structures of biological macromolecules are maintained by forces arising from hydrogen bonds, salt bridges or screened electrostatic interactions, so-called hydrophobic interactions, and van der Waals interactions.

- The amplitudes of atomic motions in macromolecules at ambient temperature range from 0.01 Å to >5 Å for time periods from 10^{-15} s to 10^{3} s (a femtosecond for electronic rearrangements to about 20 min for protein folding or local denaturation).

- Our understanding of protein dynamics is based on results from various spectroscopic experiments, and crystallographic studies of time-averaged structures and transient intermediates.

- At low temperature, below about 200 K, a protein structure can be represented by one of many slightly different conformational substates, CS; above about 200 K, the protein can sample several CS, by fluctuating between them.

- The change between the low-temperature harmonic vibration dynamics regime, where the protein is trapped in one CS, and the higher temperature regime, where it fluctuates between CS, is called a *dynamical transition*.

- Below the dynamical transition, mean square atomic fluctuations measured by neutron scattering are found to increase linearly with temperature, as expected for a set of simple harmonic oscillators, and a mean force constant can be calculated from the inverse slope of the dependence.

- Above the dynamical transition, an effective force constant can be calculated from neutron scattering, mean square fluctuation, temperature dependence data, by applying a quasi-harmonic approximation.

- Harmonic force constants below the dynamical transition for proteins are of the order of the newtons per metre: above the dynamical transition, effective force constants are ten times softer.

- Protein dynamics is sensitively solvent-dependent, and measured mean square fluctuations and force constants indicate greater flexibility and lower *resilience* with increasing relative humidity (water partial pressure) of the macromolecular environment.
- One hydration layer appears to be necessary for a protein to achieve the dynamic level required for biological activity.
- A trehalose glass maintains proteins in a high-resilience harmonic state to high temperatures, protecting them from unfolding.

Part B
Mass spectrometry

Chapter B1
Mass and charge

B1.1 Historical review

1897
J. J. Thomson made the first measurement of the mass-to-charge ratio of elementary particle 'corpuscles', which later became known as electrons. This can fairly be considered as the birth of mass spectrometry.

1918–1919
A. Dempster and **F. Aston** developed the first mass spectrographs. Photographic plate was used as the array detector. The instruments were used for isotopic relative abundance measurements.

1951
W. Pauli and **H. Steinwedel** described the development of a quadrupole mass spectrometer. The application of superimposed radio-frequency and constant potentials between four parallel rods acted as a mass separator in which only ions within a particular mass range perform oscillations of constant amplitude and are collected at the far end of the analyser.

1959
K. Biemann was the first to apply electron ionisation mass spectrometry to the analysis of peptides. Later it was shown that for sequence determination, peptides had to be derivatized prior to analysis by a direct probe.

1968–1970
M. Dole was the first to bring synthetic and natural polymers into the gas phase at atmospheric pressure. This was done by spraying a sample solution from a small tube into a strong electric field in the presence of a flow of warm nitrogen, to assist desolvation. First experiments on lysozyme demonstrated the phenomenon of multiple charging.

1974

D. Torgerson introduced plasma desorption mass spectrometry. This technique uses ^{252}Cf fission fragments to desorb large molecules from a target. It was the first of the particle-induced desorption methods to demonstrate that gas-phase molecular ions of proteins could be produced from a solid matrix.

1974

B. Mamyrin made the most important contribution to the development of time-of-flight (TOF) mass spectrometry. He constructed the so-called reflectron device, which had been proposed by **S. Alikanov** in **1957**. The reflectron essentially improves mass resolution in the TOF mass spectrometer.

1978

N. Commisarow and **A. Marshall** adapted Fourier transform methods to ion cyclotron resonance spectrometry and built the first Fourier transform mass instrument. Since that time, interest in this technique increased exponentially, as has the number of instruments.

1981

M. Barber discovered fast atom bombardment (FAB), a new ion source for mass spectrometry. The mass spectrum of an underivatised undecapeptide, Met-Lys-bradykinin of $M = 1318$ was obtained by bombarding a small drop of glycerol containing a few micrograms of the peptide with a beam of argon atoms of a few kiloelectron-volts. The technique revolutionised mass spectrometry and opened it to the biologist.

1984

R. Willoughby and, independently, **M. Aleksandrov** proposed the coupling of liquid chromatography and mass spectrometry for analysing high-molecular-weight substances delivered by a liquid phase.

1988

J. Fenn and, independently, **M. Yamashita** were able to bring biological macro-molecules into the gas phase at atmospheric pressure. They proposed a new type of ionisation technique called electrospray ionisation (ESI) to generate intact biological molecular ions, by spraying a very dilute solution from the tip of a needle across an electrostatic field gradient of a few kV. **M. Karas** and **F. Hillencamp** and, independently, **K. Tanaka** developed a new ionisation technique called matrix-assisted laser desorption–ionisation (MALDI). It was shown that proteins up to a molecular weight of 60 000 could be ionised if embedded in a large molar excess of a UV-absorbing matrix and irradiated with a laser beam. Taking advantage of high resolution, mass measurement accuracy, and ion-trapping capabilities, MALDI provides not only molecular mass information but also structural information for various peptides and oligonucleotides.

K. Tanaka received the 2002 Nobel prize in Chemistry for his contribution to mass spectrometry.

1992–1999

The molecular specificity and sensitivity of MALDI-MS gave rise to a new technology for direct mapping and imaging of biological macromolecule distributions present in a single cell or in mammalian tissue. By rastering the ion beam across a sample, and collecting a mass spectrum for each point from which ions are desorbed, it is possible to create mass-resolved images of molecular species across a cell surface or in a piece of tissue.

2000 to present

Mass spectrometry has developed into an important analytical tool in the life sciences. Soft-ionisation techniques, such as FAB, ESI and MALDI, allow routine mass measurements of proteins and nucleic acids with high resolution and accuracy. Mass spectrometry has become one of the most powerful experimental tools for the direct observation of gas-phase biological complexes, their assembly and their disassembly in real time. Developments include the combination of mass spectrometry with isotopic labelling, affinity labelling and genomic information. It is clear that the rapid growth phase of bioanalytical mass spectrometry has not yet reached its peak. There is no doubt that in the next decade mass spectrometry will move at an extraordinary pace, extending from the world of structural biology to that of medicine and therapeutics.

B1.2 Introduction to biological applications

Since the 1930s, mass spectrometry (Comment B1.1) has become an important analytical tool in structural biology. This is a result of the ability to produce intact, high-molecular-mass gas-phase ions of various biological macromolecules. Several ionisation techniques such as FAB, MALDI, and ESI revolutionised mass spectrometry and opened it up to biology. New methods for ultrasensitive protein characterisation based upon Fourier transform ion cyclotron resonance mass spectrometry (FTIR-MS) have been developed, providing a detection limit of approximately 30 zmol (30×10^{-21} mole) for proteins with molecular mass ranging from 8 to 20 kDa. Using this technique individual ions from polyethylene glycol to DNA, with masses in excess of 10^8 Da can be isolated (Comments B1.2 and B1.3).

Comment B1.1
The term 'mass spectroscopy'

We would like to warn the reader against the term 'mass spectroscopy'. The term 'mass spectroscopy' is not correct because it bears no relation to real spectroscopic techniques described in Parts E, I and J. The mass spectrum depends mainly on the stability of ions produced and collected during the experiment. The stability of ions strongly depends on experimental conditions and therefore predicting of a mass spectrum is practically impossible.

Comment B1.2
Absolute and relative masses

A mass spectrometer does not measure absolute mass, M. The instrument needs to be calibrated with standard compounds, whose M values are known very accurately. The carbon scale is used most frequently with $^{12}C = 12.000\ 000$.

Comment B1.3   Molecular mass and molecular weight

Some confusion may arise when M_r is used to denote relative molecular mass. M_r is a relative measure and has no units. However, M_r is equivalent in magnitude to M and the latter does have units and for high-mass biological macromolecules the dalton is usually used. Note that molecular weight (which is a force and not a mass) is an incorrect term in this case.

Mass spectrometry has now become the method of choice for a number of important aspects of protein structure: (1) precise protein and nucleic acid mass determination in a very wide mass range, (2) peptide and nucleotide sequencing, (3) identification of protein post-translational modifications, (4) protein structural changes, folding and dynamics, (5) identification of subpicomole quantities of proteins from two-dimensional electrophoresis, (6) identification of isotope labelling.

Mass spectrometry allowed the characterisation of fully functional biological subcellular complexes and organelles as well as intact bacteria. Large perspectives have been opened for mass spectrometry applications in functional proteomics, bacterial taxonomy and medicine.

B1.3 Ions in electric and magnetic fields

An ion that is accelerated out of a source acquires kinetic energy

$$E_{kin} = z V_{acc} = \frac{mv^2}{2} \tag{B1.1}$$

where z is the charge of the ion, V_{acc} represents the potential difference that defines the acceleration region, m is the ion mass, and v is the ion velocity. When entering a homogeneous magnetic field $\mathbf{B}$ perpendicular to its trajectory, the ion experiences the Lorentz force $\mathbf{F}$ (Comment B1.4), which is perpendicular to both $\mathbf{B}$ and $\mathbf{v}$ (Fig. B1.1). The resulting trajectory of the ion in a magnetic field is a circle with a radius r, because the Lorentz force just balances the centrifugal force

$$F = zvB = \frac{mv^2}{r} \tag{B1.2}$$

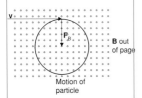

Comment B1.4
Lorentz force

The Lorentz force ($\mathbf{F}$) experienced by an anion having a charge (z) moving in an electromagnetic field. $\mathbf{F} = z(\mathbf{E} + \mathbf{v} \times \mathbf{B})$, where $\mathbf{E}$ represents the electric field strength and $\mathbf{v} \times \mathbf{B}$ is the vector product of the magnetic field strength, $\mathbf{B}$, and the ion's velocity, $\mathbf{v}$.

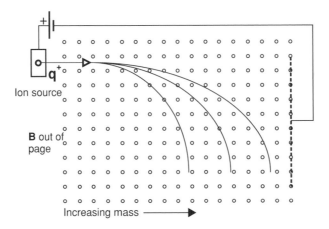

Fig. B1.1 Charged particles in electric and magnetic fields.

The mass to charge ratio m/z is given by

$$\frac{m}{z} = \frac{B^2 r^2}{2V_{acc}} \tag{B1.3}$$

As seen from Eq. (B1.3) the lightest ions have the smallest radius of curvature. The radius increases as the mass of the ions and the strength of the electric field grow.

B1.4 Mass resolution and mass accuracy

B1.4.1 Mass resolution

The ability to separate mass signals is affected by the resolving power of the mass spectrometer. The resolution R in mass spectrometry is defined as $R = m/\Delta m$, where Δm is the mass difference of two neighbouring masses, m and $m + \Delta m$, of equal intensity, with signal overlap of 10%. A resolution of 100 000 makes it possible to distinguish an ion of mass 100 000 Da from one of mass 100 001, i.e. to 10 parts per million.

Because peaks in a mass spectrum have width and shape, it is necessary to define the extent of overlap between adjacent peaks when determining the resolution. There are two definitions in widespread use, and it is essential to know which is being used when resolution figures are quoted (Fig. B1.2). The first one is the so-called 10% valley definition in which the two adjacent peaks each contribute 5% to the valley in between them. The second one is the 'full-width, half-maximum' (FWHM) definition. The resolution of a peak using this definition is the mass of the peak (in daltons) divided by the width (in daltons) measured at the half-height of the peak. A useful rule of thumb is that the value for the resolution determined using the FWHM definition is approximately twice that obtained using the 10% valley definition. A resolution of 1000 using the 10% valley definition is approximately equivalent to a resolution of 2000 using the FWHM definition. Note that, for proteins, resolving the isotopes in the protonated molecular ion envelope is possible in the case of very high resolution using FTICR mass spectrometers (Section B1.6.6).

Fig. B1.2 Two definitions of mass resolution in mass spectrometry (see text for details. (After Carr and Burlingame, 1996.)

B1.4.2 Molecular mass accuracy

The molecular mass accuracy of a measurement is defined as the difference between the measured and calculated masses for a certain ion. The accuracy is stated as a percentage of the measured mass (e.g. molecular mass = 10 000 ± 0.01%) or as parts per million (e.g. molecular mass = 10 000 ± 100 ppm). As the mass considered increases, the absolute mass error corresponding to the percentage or ppm error also increases proportionally.

Usually, accurate mass measurements do not require highest mass resolution, provided that for the observed signal there is only one species at that mass.

Comment B1.5   Monoisotopic mass

Most chemical elements have a variety of naturally occurring isotopes, each with a unique mass and natural abundance. The monoisotopic mass of an element refers specifically to the lightest stable isotope of the element. For example, there are two principal isotopes of carbon, ^{12}C and ^{13}C, with masses of 12.000 000 and 13.003 355 and natural abundances of 98.9 % and 1.1%, respectively. Similarly, there are two naturally occurring isotopes for nitrogen, ^{14}N and ^{15}N, with masses of 14.003 074 (monoisotopic mass) and 15.000 109 and natural abundance of 99.6% and 0.4%, respectively. A monoisotopic peak means that all the carbon atoms in the molecule are ^{12}C, all the nitrogen atoms are ^{14}N, all the oxygen atoms are ^{16}O, etc. The monoisotopic mass of the molecule is thus obtained by summing the monoisotopic masses of each element present.

Comment B1.6   Biologist's box: Measured mass

Measurements are made on a large, statistical ensemble of molecules and consist not only of species having just the lightest isotopes of the element present, but also of some percentage of species having one or more atoms of one of the heavier isotopes. The contribution of these heavier isotope peaks in the molecular ion cluster depends on the abundance-weighted sum of each element present. The theoretical probability of occurrence of these isotope clusters may be precisely calculated by solving the polynomial expression shown below:

$$(a + b)^m$$

where a is the percentage natural abundance of the light isotope, b is the percentage natural abundance of the heavy isotope, and m is the number of atoms of the element concerned in the molecule.

Calculations show that for small molecules such as n-butane (C_4H_{10}) there is a small but significant probability ($\sim$4%) that natural n-butane will have a molecule containing a ^{13}C atom. The probability of there being two or three ^{13}C atoms is negligible. For biological macromolecules containing several hundred carbon and nitrogen atoms the isotopic distribution pattern becomes extremely complicated. It can be calculated, however, with commercially available programs.

The main factor limiting accurate molecular mass determination for high-mass biological macromolecules is peak overlap. For MALDI the peaks correspond to $[M + H]^+$, $[M + Na]^+$, and $[M + matrix]^+$.

High mass resolution is usually deemed to be a requirement for accurate mass measurements, but under appropriate circumstances (sample ion completely separated from background ions), measurements with comparable accuracy may be made at low resolution (see Comments B1.5–B1.7).

Comment B1.7   Average mass

The chemical average mass of an element is simply the sum of the abundance-weighted masses of all of its stable isotopes (e.g., 98.9% for ^{12}C and 1.1% for ^{13}C, to give the isotope weighted average mass of 12.011 for carbon). The average mass of the molecule is then the sum of the chemical average masses of the elements present.

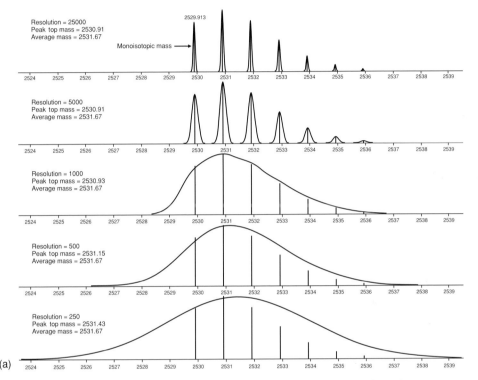

Comment Fig. B1.7(a) The molecular ion cluster for the oxidised β-chain of insulin (formula $C_{97}H_{151}N_{25}O_{46}S_4$) is shown at various resolutions. The asymmetry of the cluster becomes less apparent as resolution is decreased and the peak top mass and the average mass become almost identical. (Carr and Burlingame, 1996.)

The relationship between the monoisotopic mass, average mass, and peak top mass is shown in Comment Figs. B1.7(a), B1.7(b) for a protein of mass 2.5 kDa and a protein with mass of 25 kDa respectively. The important consequence of the contribution of the heavier isotope peaks is that for peptides with masses greater than 2000 Da, the peak corresponding to the monoisotopic mass is no longer the most abundant in the isotopic cluster (Comment Fig. B1.7(b)). With increasing molecular mass, the peak top mass continues to shift upward relative to the monoisotopic mass. Above masses of 8000 Da, the monoisotopic mass has an insignificant contribution to the isotopic envelope (Fig. B1.7(b)). Whether the monoisotopic mass or the average mass should be used when measuring and reporting molecular masses depends on the mass of the substance and the resolving power of the mass spectrometer. Another very interesting point is that at very high resolution, the satellite peaks become visible.

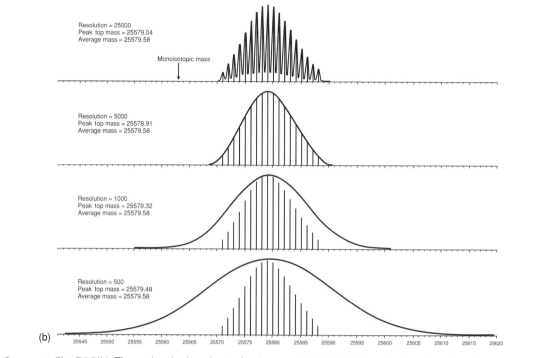

(b)

Comment Fig. B1.7(b) The molecular ion cluster for the protein HIV-p24 (formula $C_{1129}H_{1802}N_{316}S_{13}$) is shown at various resolutions. The position of the monoisotopic mass is indicated by the arrow. (Carr and Burlingame, 1996.)

B1.5 Ionisation technique

B1.5.1 From ions in solution to ions in the gas phase

The transfer of ions from the gas phase to solution is a natural process. In the presence of solvent molecules such as H_2O, naked gas-phase ions such as Na^+ spontaneously form ion–solvent molecule clusters, $Na^+(H_2O)_n$. If the pressure of the solvent vapour is somewhat above the saturation vapour pressure, these clusters grow to small droplets.

The transfer of ions from solution to the gas phase is a desolvation process, which requires energy (it is endoergic) and hence does not occur spontaneously. For example, the free energy required when Na^+ ions are transferred from aqueous solution to the gas phase is very large, about 98 kcal/mol. The energy required for ion transfer to the gas phase in most analytical mass spectrometry methods is very high and is supplied by complex high-energy collision cascades or highly localised heating.

Ions may be produced in three different ways. First, by removing an electron from the molecule to produce a positively charged cation, which can be accelerated in either an increasing negative gradient field or decreasing positive gradient field. Secondly, by adding an electron to form an anion. In this case the accelerating fields are exactly the opposite to what they were for cations. Thirdly, by removal or addition of protons. In this case, the mass of the resulting ion differs by ± 1 from the mass of the original neutral one. Below we describe the more common ways of producing ions in a mass spectrometer.

B1.5.2 Electron ionisation (EI)

EI is the most widely used ionisation technique in mass spectrometry. EI is a relatively simple illustration of the general principles of ionisation under electron bombardment. The electron energy generated by a heated filament in the ion source is usually set to 70 eV (Comment B1.8). Upon impact with 70-eV electrons, the gaseous molecule may lose or capture one electron. The possible events that may occur are described below.

Covalent bonds are formed by the pairing of electrons. Ionisation resulting in a cation requires loss of an electron from one of these bonds, leaving a bond with a single unpaired electron. In this case events are

$$M \text{ (neutral)} + e^- \rightarrow M^{*+} + 2e^-$$

where M^{*+} means positively charged molecular ion. In the case of electron capture, an anion is formed by the addition of an unpaired electron and therefore

$$M \text{ (neutral)} + e^- \rightarrow M^{*-}$$

where M^{*-} denotes a negatively charged molecular ion. Such ions are relatively unstable under conditions of electron bombardment. They give a series of daughter ions, which are recorded as the mass.

B1.5.3 Field ionisation (FI)

FI requires the sample to be introduced in the vapour state. The molecules are subjected to a high intense electric field, of the order of 10^7–10^8 V/cm. The electric field strength required in FI is achieved by using a metal (Pt, W) tip emitter with a tip radius of 100–1000 nm, to which a voltage of about 5 kV is applied. Under such conditions the outer shell electrons are subject to large forces, sufficient to generate molecular cations.

B1.5.4 Fast atom bombardment (FAB)

In FAB mass spectrometry (FAB-MS) ionisation is produced by bombarding the sample surface with an atomic beam of Ar or Xe, accelerated to an energy of a few kiloelectron-volts. It is supposed that, in this case, a primary particle induces a collision cascade in a small volume of the sample.

Comment B1.8
Electron-volt (eV) and joule (J)

The electron-volt (eV) is a unit of energy equal to the kinetic energy a single electron acquires accelerating through a potential difference of 1 V.

$$1 \text{ eV} = 1.6 \times 10^{-19} \text{ J}$$

The joule (J) ist the standard unit of energy in SI units. 1 J = 1 (McLafferty, 1993).

Fig. B1.3 Desorption
mechanism in FAB
ionisation. A⁻ represents
a negatively charged ion
and C⁺ a positively
charged ion. The droplet
is bombarded with
energetic Xe or Ar atoms.
(After Caprioli and Suter,
1995.)

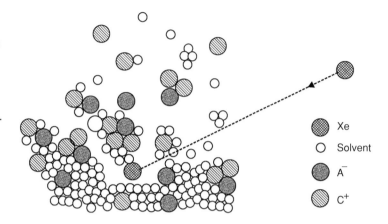

Fig. B1.3 Desorption mechanism in FAB ionisation. A^- represents a negatively charged ion and C^+ a positively charged ion. The droplet is bombarded with energetic Xe or Ar atoms. (After Caprioli and Suter, 1995.)

The sample is placed in a liquid matrix (usually glycerol) to maintain a relatively constant concentration of molecules in the surface layer, vaporised molecules being replaced by molecules coming up from the bulk. This procedure permits a stable mass spectrum to be observed during a considerable period of time (about 1 h).

In FAB ionisation with Ar or Xe atoms, the sample droplet is bombarded with energetic atoms of 8–10 keV kinetic energy. At this energy, initially charged sample molecules are desorbed as such, whereas the formation of protonated molecular ions $(M+H)^+$ in positive mode, and of $(M-H)^-$ in negative ion mode, arises from both the gas-phase transition and solution chemistry.

The main disadvantage of the FAB technique (i.e. a very high concentration of the organic liquid matrix) has been overcome by the development of continuous-flow FAB. In this, a sample solution is continuously delivered to the target at slow flow rates up to 10 μl/min. As a consequence, less organic matrix is necessary to maintain the droplet, which results in an increase in signal-to-noise ratio. Figure B1.3 illustrates the main components of the FAB technique. The major advantage of FAB is its simplicity. The spectra are easy to interpret. Currently, FAB-MS is widely used for the analysis of compounds having molecular mass below about 5 kDa.

B1.5.5 Plasma desorption (PD)

In plasma-desorption mass spectrometry, a sample is deposited on a metal surface and bombarded with fission fragments of a radioactive isotope. Plasma, in this context, comprises atomic nuclei stripped of electrons. Spontaneous fission of ^{252}Cf produces two atomic particles of high kinetic energy (80 MeV), ^{144}Cs and ^{108}Tc, travelling in opposite directions. Owing to their high energy, these particles can pass through a thin metal foil and ionise biological material deposited on the other side of the foil. Fission is a discrete event occurring at a rate of $(1-5)\times 10^3$ s^{-1} and resulting in an ion beam pulsed at that frequency. For this reason, a

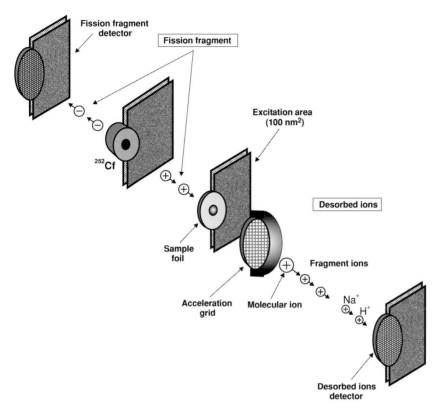

Fig. B1.4 The main features of the PD-MS technique with a ^{252}Cf source and TOF mass analyser. (After Caprioli and Suter, 1995.)

time-of-flight (TOF) mass spectrometer is used (Section B1.6.5). By measuring the flight time of the secondary ions and knowing their energy and drift path, it is straightforward to transform the ion TOF spectrum into a mass spectrum. Figure B1.4 shows the main features of the PD-MS technique with a TOF mass analyser.

PD-MS has a reasonably good sensitivity with peptides and relatively small proteins (7–20 kDa). Typically, about 10 pmol material is necessary for a molecular mass determination. Mass resolution is about 1000.

B1.5.6 Laser desorption and matrix-assisted laser desorption ionisation

In laser desorption ionisation (LDI), laser radiation is focused onto a small spot with a very high power density that gives an extremely high rate of heating. This leads to the formation of a localised laser 'plume' of evaporated molecular species, either from adsorbed material or from the solid substrate itself. Direct LDI of intact biological molecules without using the matrix is limited to molecular masses of about 1 kDa. The mass range limitation gave rise to the development of MALDI.

Fig. B1.5 Schematic
mechanism for MALDI
using lasers:
(a) absorption of radiation
by the matrix;
(b) dissociation of the
matrix, phase change to
supercompressed gas,
and transfer of charges to
sample molecules;
(c) expansion of the
matrix at supersonic
velocity, entrainment of
sample molecules in
expanding matrix plume,
and transfer of charge to
molecule.

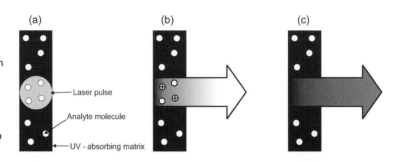

The MALDI process differs from direct laser desorption because it utilises a specific matrix material mixed with the sample. From this point of view, MALDI is similar to FAB; the latter using liquid matrices to provide soft ionisation. However, MALDI provides much softer ionisation than FAB, which allows the analysis of large molecules up to 1000 kDa with minimum fragmentation.

The details of energy conversion and sample desorption and ionisation are still not fully known. A general outline of the mechanism is presented in Fig. B1.5. Energy from the laser beam is absorbed by the chromophor(ic) matrix, which rapidly expands into the gas phase, carrying with it sample molecules. Ionisation occurs by proton transfer between excited matrix molecules and sample molecules, presumably in the solid phase, and also by collisions in the expanding plume.

The matrix is the key component in the MALDI technique. The matrix functions as an energy 'sink' resulting in longer sample life. The material to be analysed is mixed with an excess of matrix, which preferentially absorbs the laser radiation. Commonly used matrix materials are aromatic compounds that contain carboxylic acid functional groups. The aromatic ring of the matrix acts as a chromophore for the absorption of laser irradiation leading to the desorption of matrix and sample molecules into the gas phase. The matrix not only increases sample ion yield, but also prevents its extensive fragmentation.

Two types of laser are most useful for laser desorption of biological materials: the IR laser, which can couple efficiently with molecular vibrational modes, and the UV laser, which can excite electronic modes in aromatic molecules. Pulses of 100 ns or less duration are used in both wavelength ranges, because longer exposure times would lead to thermal heating resulting in the pyrolytic decomposition of biological molecules.

Because most laser sources are pulsed, TOF and Fourier transform ion cyclotron resonance (FTR-ICR) mass spectrometers have been most widely used with MALDI (Section B1.6). A mass accuracy of $\pm 0.01\%$ (± 1 Da at a molecular mass of 10 kDa) can be achieved under favourable conditions. If high-resolution conditions are available, it is possible to resolve individual carbon isotope peaks, for example (see Section B2.1). The MALDI technique is still under active development and improvements are occurring at rapid rate.

Nanoflow electrospray ionisation	Free jet expansion in the ion source	Disassembly in the collision cell	Mass analysis

Atmospheric pressure		$1e^{-7}$ - $1e^{-10}$ mbar

Fig. B1.6 Schematic representation of the passage of ions from the nanoflow electrospray needle to the detector of the mass spectrometer. Protein solution, typically 1–2 μl of 5 μM concentration, is placed in a fine-drawn capillary of internal diameter approximately 10 μm. A voltage of several kilovolts is applied to the gold-plated needle, causing an electrospray of fine droplets. The positively charged droplets are electrostatically attracted, dissolvated and focused in the mass spectrometer for detection. (After Rostom, 1999.)

B1.5.7 Electrospray ionisation (ESI)

ESI produces intact ions from sample molecules directly from solutions at atmospheric pressure. Ions are formed by applying a 1–5 kV voltage to a sample solution emerging from a capillary tube, at a low flow rate (1–20 nl/min). The high electric potential, which is applied between the tip of the capillary tube and a counter-electrode located a short distance away causes the liquid at the tip of the tube to be dispersed into a fine spray of charged droplets (Fig. B1.6). The solvent evaporates from the droplets as they move from the atmospheric pressure of the ionisation region into the vacuum chamber containing the mass analyser. The evaporation of the solvent is aided either by a counter-current flow of drying gas or by heating the tube that transports the droplets from the ion source into the vacuum of the mass analyser. The production of positive or negative ions is determined by the polarity of the voltage applied to the capillary.

Comment B1.9 Number of attached protons

In general, the maximum number of protons that attach to a peptide or protein under ESI conditions correlates well with the total number of basic amino acids (Arg, Lys, His) plus the N-terminal amino group, unless it is acylated. However, the accessibility of these basic sites is an important factor. The distribution of charge states thus depends on pH, temperature and any denaturating agent present in the solution. This information can be used to probe conformational changes in the protein.

For example, for bovine cytochrome c the most abundant ion has 10 positive charges when electrospraying a solution at pH 5.2, but 16 charges at pH 2.6. A similar effect is observed upon reduction of disulphide bonds. Hen egg white lysozyme with four disulphide bonds shows a charge distribution centred at 12^+, but upon reduction with DTT (dithiothreitol), a new cluster centred around 15^+ appears (see also Comments B2.5, B2.6 and B2.7).

An attractive feature of the electrospray process is the formation of multiply-charged molecular species, if the sample molecule can accept more than one charge (Comment B1.9). Because of multiple charging, high-mass ions can be detected within a low m/z range. The shifted scale makes the high-resolution detection of large mass ions possible, because the mass-resolving power is inversely proportional to m/z (see examples in Section B2.7). The attainable mass accuracy for measuring molecular masses with the ESI technique in conjunction with Fourier transform mass spectrometry is about 0.001–0.005%. If high-resolution conditions are available, the individual carbon isotope peaks can be resolved (Section B1.8).

Finally, it should be pointed out that ESI is one of the most gentle ionisation methods available, yielding no molecular fragmentation in practice. ESI is still under active development.

B1.6 Instrumentation and innovative techniques

The first mass spectrometer was built in 1913 when J. J. Thomson proposed using fixed magnetic and electric fields to separate two different isotopes of the noble gas Ne, by making use of the different behaviours of charged particles of differing momentum and energy in an electromagnetic field. The essential requirements to obtain a mass spectrum are to produce ions in the gas phase, to accelerate them to a specific velocity using electric fields, to introduce them into a suitable mass analyser for separation, and finally to detect each charged entity of a particular mass sequentially in time. All mass spectrometers are made up of five main parts: (1) a sample injector, (2) an ionisation chamber, (3) a mass analyser, (4) an ion detector and (5) a data handling facility (Fig. B1.7).

Sample introduction systems consist of controlled leak devices, through which sample vapour is introduced from a reservoir, various direct insertion probes for the injection of low-volatility liquids, and combinations with various chromatographic techniques. The ions that are produced in a number of ways in the ionisation chamber (Section B1.5) are analysed according to their mass-to-charge ratio in the mass analyser.

Five types of mass analyser are currently available: magnetic sector, quadrupole mass filter, quadrupole ion trap, TOF and ion cyclotron resonance devices. The detection of ions after mass analysis can be performed by destructive or non-destructive techniques (see below). Modern mass spectrometers have

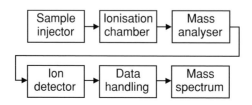

Fig. B1.7 Schematic diagram of a mass spectrometer.

almost total computer control over the various parts of a spectrometer, and advanced computer programs are available for data handling and interpretation.

B1.6.1 Single- and double-focusing mass spectrometers

In a single-focusing-sector instrument, ions with mass m, charge z and a particular kinetic energy are introduced into a magnetic field $\mathbf{B}$. Equation (B1.3) shows that various values m/z are obtained if either $\mathbf{B}$ or V_{acc} is changed. As follows from Eq. (B1.3), for given values of $\mathbf{B}$, V and z, ions of different masses follow paths of different radius r.

For a mixture of ions covering a wide range of masses it would be impractical to have separate detectors positioned to accommodate all possible radii. The problem is solved by making all ions follow the same radius. There are two ways to achieve this. In the electromagnet analyser (EMA) technique, electromagnets are used to vary $\mathbf{B}$ such that ions of different mass (but the same velocity) are forced to follow the given radius (Fig. B1.8).

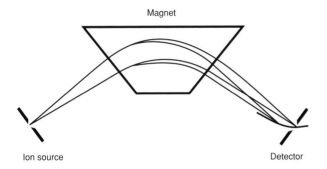

Magnet

Ion source   Detector

Fig. B1.8 Principal scheme of the single-focusing mass spectrometer based on EMA. (After Caprioli and Suter, 1995.)

In the electrostatic analyser (ESA) technique, the accelerating voltage V can be varied so as to accelerate ions of different mass to different terminal velocities (Fig. B1.9). The single-focusing mass spectrometer has a limited mass range and low resolution.

The combination of the ESA and EMA modes results in the double-focusing mass spectrometer (Fig. B1.10). In this instrument, both direction (line A in

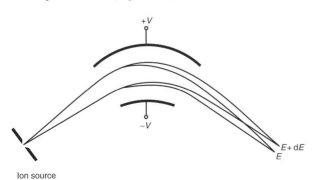

$+V$

$-V$

$E+dE$
E

Ion source

Fig. B1.9 Principal scheme the of single-focusing mass spectrometer based on ESA. (After Caprioli and Suter, 1995.)

Fig. B1.10 Principal
scheme of the
double-focusing mass
spectrometer. (After
Caprioli and Suter, 1995.)

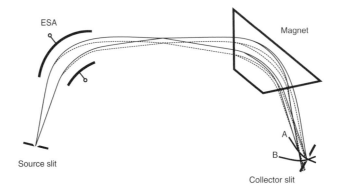

Fig. B1.10) and energy (line B in Fig. B1.10) focusing occurs at the intercept of
lines A and B. The double-focusing instrument has much higher resolution than
the single-focusing one.

B1.6.2 Quadrupole mass filter

Mass separation in a quadrupole mass filter is based on achieving a stable
trajectory for ions of specific m/z values in a rapidly changing electric field.
An idealised quadrupole mass filter consists of four parallel cylindrical rods of
circular cross-section as shown in Fig. B1.11. To one pair of diagonally opposite
rods a negative direct current ($-$dc) voltage and an alternating radio-frequency
(rf) voltage are applied. To the other pair of rods, a positive dc voltage of opposite
polarity and the inverse ($180°$ out of phase) rf voltage are applied. Mass filter-
ing occurs as these voltages are scanned but the ratio of dc to rf voltage is kept

Fig. B1.11 Quadrupole
mass filter. In the
quadrupole mass filter
instrument one pair of
rods has a negative dc
voltage, $-V_{dc}$, applied and
the other pair a positive
dc voltage $+V_{dc}$. There is
also a superimposed
radio-frequency (rf)
voltage, $V_{rf}\cos\omega t$, which
is $180°$ out of phase
between rod pairs. In an
ideal situation, rods with
hyperbolic cross-section
would be used. In order to
scan between $m/z = 1$
and 500, the dc voltage is
varied between 0 and 300
V and the ac voltage
between 0 and
1500 V. The ac frequency
is in the megahertz range.
(After Gordon, 2000.)

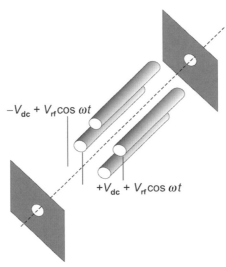

constant. For a given set of field conditions, only certain trajectories are stable, allowing ions of specific mass to be transmitted in the direction of the detector. Ions that have unstable trajectories come in contact with the rods and are not transmitted, hence the term filter.

One of the advantages of a quadrupole mass filter over an ESA instrument is the low voltage applied to the ion source (5–20 V) compared to several kiloelectron-volts for ESA instruments. The low voltage makes interfacing to liquid chromatography easier.

B1.6.3 Quadrupole ion trap

The quadrople ion trap was originally developed by physicists who were interested in increasing the observation time available for spectroscopic measurements on elementary particles. The quadrupole ion trap is based on the same principle as the quadrupole mass filter, except that the quadrupole field is generated within a three-dimensional device consisting of a ring electrode and two end caps, as shown in Fig. B1.12. The ring electrode is a hyperboloid of one sheet. It is similar to a torus except that the cross-section of the ring is hyperbolic.

Ions that are produced in the trap itself or in an external ion source are stored in the trap. By raising the rf potential, the trajectories of ions of successive m/z values are made unstable and these ions are ejected out of the trap where they are detected by means of an electron multiplier. In contrast to the quadrupole filter, where the ions with stable trajectories are detected, ions with unstable trajectories are detected in the ion trap.

A schematic view of an atmospheric pressure MALDI ion-trap mass spectrometer is shown in Fig. B1.13.

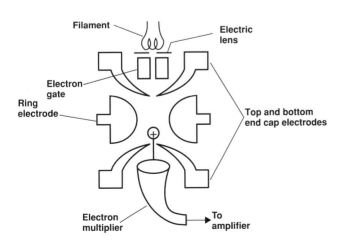

Fig. B1.12 A longitudinal cross-section of a quadrupole ion trap. Ions are created within the trap by radial injection of a pulse of electrons through holes in the ring electrode or axial injection through an end cap. For given values of m/z the ions are held in stable orbits, provided the correct amplitudes and frequency of dc and rf potentials are applied between the end caps and the ring. (After Gordon, 2000.)

Fig. B1.13 Schematic view of a MALDI mass spectrometer based on an ion trap.

Laser

To Detector

Ion Trap

Octopole ion guide

Skimmer   Quadrupole ion guide

Heated capillary tube and lens

Sample plate

B1.6.4 Ion cyclotron resonance mass spectrometry (ICR-MS)

As an ion-trapping technique, ion cyclotron resonance mass spectrometry (ICR-MS) differs substantially from mass spectrometry that uses ion transmission to separate masses (Comment B1.10). In ICR, ions trapped in magnetic and dc electric fields are detected when the frequency of an applied rf field comes into resonance with the cyclotron frequency (Comment B1.11). The resonance frequency ω_c is directly proportional to the strength of the magnetic field (typically 3–7 T) and inversely proportional to the mass-to-charge ratio, m/z, of the ions

$$\omega_c = \frac{Bz}{m}$$

(B1.4)

Comment B1.10   Ion cyclotron principle

In 1932, E. Lawrence and S. Livingstone demonstrated that a charged particle moving perpendicular to a uniform magnetic field is constrained to a circular orbit in which the angular frequency of its motion is independent of the particle's orbital radius and is given by the cyclotron equation (Eq. (B1.4)). Lawrence showed that cyclotron motion of a particle could be excited to a larger orbital radius by applying a transverse alternating electric field whose frequency matched the cyclotron frequency of the particle. The significance of Lawrence's discovery was that a particle could be excited to very large kinetic energy by use of only modest electric field strength. An alternating voltage of 1 kV would, after 1000 cyclotron cycles, accelerate the particle to a kinetic energy of 1 MeV.

Comment B1.11   Ion cyclotron frequencies

It follows from Eq. (B1.4), that ions of different m/z have unique cyclotron frequencies. At a magnetic field strength of 6 T, an ion of $m/z = 36$ has a cyclotron frequency of 2.6 MHz, whereas an ion of m/z 3600 has a cyclotron frequency of 26 kHz. Equation (B1.5) also shows that increasing the magnetic field linearly increases the cyclotron frequencies of the ions, making high-mass ions easier to detect over the environmental noise in the low-kilohertz region. Additional benefits of increasing the magnetic field include an improvement in mass-resolving power and the extension of the upper mass limit.

It should be noted that Eq. (B1.4) does not account for the presence of the electric field produced by two trapping plates and can be considered as a first approximation.

Tesla (T)

The standard unit of magnetic flux density in the SI system

$$1\ \mathrm{T} = 1\ \mathrm{kg\,s^{-2}A^{-1}}$$

Torr

A unit of pressure, being that necessary to support a column of mercury 1 mm high at $0\,^{\circ}\mathrm{C}$ at standard gravity

$$1\ \mathrm{Torr} = 133.322\ \mathrm{Pa}$$

Pascal (Pa)

The standard unit of pressure in the SI system.

$$1\ \mathrm{Pa} = 1\ \mathrm{kg\ m^{-1}\ s^{-2}}$$

Once formed, ions in the ICR-MS analyser cell are constrained to move in circular orbits of radius r

$$r = \frac{mv}{zB} \tag{B1.5}$$

with the motion confined perpendicular to the magnetic field (xy plane) but not restricted parallel to the magnetic field (z-axis) (Fig. B1.14). Ion trapping along the z-axis is accomplished by applying an electrostatic potential to the two plates on the ends of the cell. The trapped ions can be in the cell for up to several hours, provided that a high vacuum (10^{-8}–10^{-9} Torr) is maintained to reduce the number of destabilising collisions between the ions and residual neutral molecules.

After formation by an ionisation event, trapped ions of a given m/z have the same cyclotron frequency but a random position in the cell. The net motion of the ions under these conditions does not generate a signal on the receiver plates of the ICR-MS cell because of their random location. To detect cyclotron motion,

Fig. B1.14 FTICR cell,
identifying the trapping
plates, the transmitter
plates, the receiver plates
and direction of the
magnetic field (**B**). (After
Buchanan and Hettich,
1993.)

Fig. B1.14 FTICR cell, identifying the trapping plates, the transmitter plates, the receiver plates and direction of the magnetic field (**B**). (After Buchanan and Hettich, 1993.)

an excitation pulse must be applied to the ICR-MS cell so that the ions spatially 'bunch' together into a coherently orbiting ion packet. As a result, the net coherent ion motion produces a time-dependent signal on the receiver plates.

Fourier transform ion cyclotron resonance mass spectrometry (FTICR-MS) is a further development of the ICR technique. Time-domain signals are digitised and subjected to Fourier transformation to generate an ICR frequency-domain signal which can subsequently be converted into a mass spectrum (Section B1.6.6).

The essential advantages of FTICR-MS are: (1) an extremely high mass resolution, (2) a wide range of m/z values detected simultaneously, (3) the ability to study ion–molecule reactions at low pressure.

Finally, we would like to point out the similarities between ICR and NMR (Comment B1.12).

Comment B1.12   Similarities and differences between NMR and ICR

NMR spectroscopy is a technique that directly measures the Zeeman splitting of the otherwise degenerate energy levels of a nucleus possessing a magnetic moment (see Chapter J1). In NMR, the resonant frequency of a nucleus is given by

$$\omega = \gamma B (1 - \sigma)$$

where ω is the resonant frequency in radians per second, γ is the gyromagnetic ratio characteristic of a particular magnetic nucleide, **B** is the applied magnetic field, and σ is the chemical shift parameter that is characteristic of the chemical environment of the magnetic nucleus.

Note that in both Eq. (J1.17) and Eq. (B1.4), the angular frequency is proportional to the magnitude of the applied longitudinal magnetic field, **B**. However, bandwidths in NMR and ICR techniques are very different. For example, for high-resolution liquid ^{1}H NMR at a magnetic field of 2.3 T, all ^{1}H NMR frequencies fall within a narrow frequency band of 100 MHz $\pm$ 5 kHz. In contrast, at 2.3 T a mass range of 15–1500 Da covers an ICR frequency band of 20 kHz–2 MHz for singly charged ions.

B1.6.5 TOF mass spectrometer

Mass analysis in a TOF mass spectrometer is based on the principle that ions of different m/z values have the same energy, but different velocities, after acceleration out of the ion source. It follows that the time required for each ion to pass the drift tube is different for different ions: low-mass ions are quicker to reach the detector than high-mass ions. From Eq. (B1.1) we derive the expressions for the velocity u of an ion of mass m and charge z

$$u = \left(\frac{2z V_{acc}}{m} \right)^{1/2} \tag{B1.6}$$

and for the time t, spent to cover a length L

$$t = \left(\frac{m}{2z V_{acc}} \right)^{1/2} L \tag{B1.7}$$

Equation (B1.7) shows that with an accelerating voltage of 20 kV and L of 1 m, a singly charged ion of mass 1 kDa has a velocity of about 6×10^4 m/s and the time spent traversing the drift tube is 1.4×10^{-5} s.

It is evident that for a TOF mass analyser the suitable ionisation techniques are those by which ions are generated in a pulsed regime: using ^{252}Cf fission particles, a laser pulse, and introduction of ions from continuous ionization sources (EI, ES, FAB and so on) with pulsed deflection of an ion beam or pulsed extraction from an ion source. The pulse gives the start signal for data acquisition.

The TOF method can be advantageous compared with scanning technologies because of its 'unlimited' mass range, high transmission (most of the ions injected into the analyser are detected), high speed (the experiment involves nearly simultaneous detection of the mass spectrum on the microsecond time scale), and the potential for high duty factors (percentage of ions formed that are detected). A major drawback is the low mass resolving power. From Eq. (B1.7) it follows that m/z is proportional to t^2, which leads to the formula for resolution

$$R = \frac{m}{\Delta m} = \frac{1}{2} \frac{t}{\Delta t}$$

Standard linear TOF instruments typically have a resolution no greater than 1000.

A significant improvement of the resolution in the TOF method can be obtained by using an electrostatic mirror or 'reflectron' and the orthogonal TOF mass spectrometer (o-TOF-MS).

A reflectron TOF mass spectrometer is based on the fact that high-energy ions penetrate deeper into the reflection electric field and, therefore, spend more time there than low-energy ions. Because they must traverse a greater distance, the more energetic ions arrive at the detector at the same time as the less energetic one. With the reflectron, the resolution of the TOF mass spectrometer increases up to 6000.

The main feature of o-TOF-MS is its use of orthogonal dimensions, x and y, respectively, for the continuous ion beam and distance over which the TOF is measured. Ions are sampled from a nearly parallel ion beam from a continuous ion source. The electric fields are designed to apply a force that is strictly and exclusively at right angles to the axis of the ion beam. The decoupling of the ion beam velocity spread from the TOF axis leads to the resolving power advantage of orthogonal acceleration. The resolving power of such instruments is about 4000. o-TOF-MS is highly compatible with the reflectron geometry.

B1.6.6 Fourier transform mass spectrometry (FT-MS)

In most mass spectrometers, ions are detected by electrical current when they hit the surface of a device such as an electron multiplier (destructive method). Although this method is widely used and very sensitive, it has the disadvantage that the ions must be destroyed in order to be detected. In other words an ion signal can be measured only once. In FT-MS the detection method is fundamentally different: a strong magnetic field traps ions inside an analyser cell and electrical signals produced by their cyclotron motion are detected by a pair of metal electrodes connected to a high-impedance amplifier (Fig. B1.15).

The image current detection method 'senses' the number of ions without removing them from the analyser cell and without destroying them (non-destructive method). A signal from the same ion can be measured repeatedly. The detection problem is similar to that in NMR (Comment B1.9, and Part J). With FT-MS, detection of just a few hundred ions produces a signal that contains complete information on the frequencies and abundances of all the ions trapped

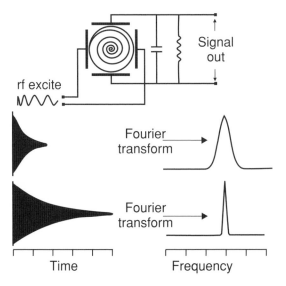

Fig. B1.15 General scheme of the cyclotron motion of excited ions in the FTICR cell. The resulting time-domain signal is then Fourier transformed to the frequency domain, from which the mass spectrum is obtained. (Carr and Burlingame, 1996.)

in the cell. Because frequency can be measured precisely, the mass of an ion can be determined to 1 part in 10^9 or better. It should be noted that resolution in FTICR-MS is mass-dependent; ultrahigh resolution can be obtained at low mass. The sensitivity of FTICR is so high that the method has been successfully applied to study individual multiply charged macro-ions.

B1.6.7 Tandem mass spectrometry (MS-MS)

To obtain structural information by mass spectrometry the molecule must undergo fragmentation of one or more bonds in such a manner that ions are formed, the m/z ratio of which can be related to the structure. We recall that 'soft' ionisation methods, such as FAB, MALDI and ESI, generate single molecular ions that contain insufficient excess energy to fragment. However, by converting the kinetic energy of the ion into vibrational energy, fragmentation can be achieved. This can be done in MS-MS using a special collision cell.

The most common MS-MS experiment is the product ion scan. In the experiment, ions of a given m/z value are selected with the first mass spectrometer (MS 1, Fig. B1.16). The selected ions are passed into the collision cell (CC), typically filled with helium, argon or xenon. The ions are activated by collision, and induced to fragment. The product ions are then analysed with the second mass spectrometer (MS 2), which is set to scan over an appropriate mass range. Since it takes only 1–2 min to record the spectrum, one can then set MS 1 for the next precursor ion and obtain its collision spectrum, and so on.

There are two main types of instrument that allow MS-MS experiments. The first is made of two mass spectrometers assembled in tandem. Two mass analysing quadrupoles, or two magnetic analyser instruments or hybrids containing one magnetic and one quadrupole spectrometer are representative cases. From this standpoint coupling of a magnetic and an electric sector can be considered as MS-MS (double-focusing MS, Section B1.6.1). The second type of MS-MS instrument consists of analysers capable of storing ions: the ICR (Section B1.6.5) and the quadrupole ion trap (Section B1.6.3) mass spectrometers. These devices allow the selection of particular ions by ejection of all others from the trap. The selected ions are then excited and caused to fragment during a selected time period, and the ion fragments can be observed with a mass spectrometer. The process may be

Fig. B1.16 Principle of tandem mass spectrometry. MS 1 and MS 2 are the first and the second mass spectrometer respectively. CC is the collision cell. A mixture of five peptides is scanned to produce the spectrum of the five $(M + H)^+$ ions $(P_1–P_5)$. After the scan only one selected ion (P_4) passes into collision cell. The fragments $(F_1–F_6)$ produced upon collision-induced decomposition of the precursor ion (part of which remains intact) are then mass analysed by scanning MS 2 to record the product ion spectrum. (After Biemann, 1992.)

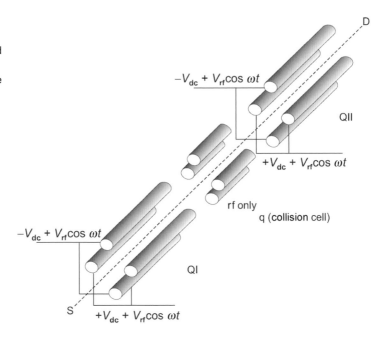

repeated to observe fragments of fragments, over several generations. The instruments exploit a sequence of events in time.

An alternative approach is to use the triple quadrupole design (Fig. B1.17, and Section B1.5.2), which, although much cheaper, suffers from poor sensitivity and mass limitation. The first quadrupole, QI, is used as a mass spectrometer, a selected peak being injected into the collision cell (CC), and the decomposition products are analysed in the second quadrupole QII.

Finally, there are also 'hybrid' instruments, which are so-named because they combine the use of magnetic sectors, quadrupoles and TOF instruments in linear and orthogonal projections.

B1.7 Checklist of key ideas

- A mass spectrometer does not measure absolute mass. The instrument needs to be calibrated with standard compounds, whose mass values are known very accurately.
- The ESI technique produces intact ions from samples directly from solutions at atmospheric pressure by spraying a very dilute solution from the tip of a needle across an electrostatic field gradient of a few kilovolts.
- A unique feature of ESI process is the formation of multiply-charged molecular species. ESI is the most gentle ionisation method yielding no molecular fragmentation in practice.
- The MALDI technique produces intact ions from the sample mixed with specific matrix material, which preferentially absorbs the laser radiation.

Suggestions for further reading

Historical review

Griffiths, I. W. (1997). J. J. Thomson – the centenary of his discovery of the electron and his invention of mass spectrometry. *Rapid Commun. Mass Spectr.*, **11**, 2–16.

Comisarow, M. B., and Marshall, A. G. (1996). The early development of Fourier transform ion cyclotron resonance (FT-ICR) spectroscopy. *J. Mass Spectr.*, **31**, 581–5.

Ionisation techniques

Smith, D. R., Loo, J. A., Loo, R. R. O., Busman, M., and Udseth, H. R. (1991). Principles and practice of electrospray ionization – mass spectrometry for large polypeptides and proteins. *Mass Spectr. Rev.*, **10**, 359–451.

Muddiman, D. C., Gusev, A. I., and Hercules, D. M. (1995). Application of secondary ion and matrix-assisted laser desorption-ionization time-of-flight mass spectrometry for the quantitative analysis of biological molecules. *Mass Spectr. Rev.*, **14**, 383–429.

Gordon, D. B. (2000). Mass spectrometric techniques. In *Principles and Techniques of Practical Biochemistry*, Chapter 11, eds. K. Wilson and J. Walker. Cambridge: Cambridge University Press.

Instrumentation and innovative techniques

Caprioli, R. M., and Suter, M. J.-F. Mass spectrometry. Chapter 4 in *Introduction to Biophysical Methods for Protein and Nucleic Research*, Academic Press.

Amster, I. J. (1996). Fourier transform mass spectrometry. *J. Mass Spectr.*, **31**, 1325–1337.

Hofmann, E. (1996). Tandem mass spectrometry: a primer. *J. Mass Spectr.*, **31**, 129–37.

Dienes, T., Pastor, J. S., *et al.* (1996). Fourier transform mass spectrometry – advancing years (1992–mid 1996). *Mass Spectr. Rev.*, **15**, 163–211.

Guilhaus, M., Mlynski, V. and Selbi, D. (1997). Perfect timing: time-of-flight mass spectrometry. *Rapid Commun. Mass Spectr.*, **11**, 951–962.

Belov, M. E., Gorshkov, M. V., Udeseth, H. R., Anderson, G. A. and Smith, R. D. (2000). Zeptomole-sensititivity electrospray ionization – Fourier transform ion cyclotron resonance mass spectrometry proteins. *Anal. Chem.*, **72**, 2271–2279.

Chapter B2
Structure function studies

B2.1 Protein structure and function

ESI and MALDI have become increasingly useful for the mass spectrometric analysis of proteins. The two ionisation techniques have been exploited to study protein folding, to characterise non-covalent protein complexes, to map protein function and for many other applications (Comment B2.1).

Comment B2.1 Mass spectrometry and X-ray crystallography

It is very useful to use mass spectrometry prior to X-ray crystallography. Indeed, the determination of a protein structure at atomic resolution takes a considerable investment of time and effort (see Chapter G3). Mass spectrometry permits a rapid check of the correctness of the accepted primary structure of a protein, and high-resolution determination of the purity of protein preparations. The information is particularly important for proteins obtained by recombinant techniques, which are subject to a number a special sources of error, including unanticipated mutations, modifications, termination and proteolitic degradation. A simple molecular mass measurement is so informative and time saving that there is a reason to obtain the mass spectrum of virtually every protein before it is subject to X-ray crystallography. It is especially important for proteins produced with special amino acid residues (e.g. selenomethionine), for NMR with ^{13}C and/or ^{15}N enrichment and for neutron small-angle scattering with specific deuteration. In the last case the incorporation of deuterium into biological macromolecules by biosynthetic methods is routinely determined by mass spectrometry measurements.

B2.1.1 Mass determination

ESI and MALDI have made the mass analysis of proteins a routine procedure. For many reasons, peptides and proteins are particularly suited to these ionisation techniques, which were developed and first demonstrated with molecules of this type.

(a)

[M + 15H]$^{15+}$

(b)

[M + 17H]$^{17+}$

813.85   815.14   816.44   817.73   819.02   983.81   985.97   988.14   990.3   992.47

m/z   *m/z*

Fig. B2.1 A portion of the mass spectra of (a) horse cytochrome *c* and (b) horse myoglobin obtained by ESI-FTICR mass spectrometry. (Belov *et al.*, 2000.)

Figure B2.1 shows a portion of the mass spectra of horse cytochrome *c* and horse myoglobin obtained by ESI-FTIR mass spectrometry. Sample concentrations were 0.4 nM (Comment B2.2). The total consumed amount for each protein was 135 zmol (about 80 000 molecules).

Figure B2.2 depicts a high-resolution MALDI mass spectrum [Arg^8]-vasopressin. The base peak at *m/z* 1084.446 is the 'monoisotopic' peak for the intact protonated peptide. The three peaks at higher mass, each separated by 1 Da, result from the incorporation of one or more of the less abundant carbon

Comment B2.2
Unit prefixes

10^{-9}	nano	n
10^{-12}	pico	p
10^{-15}	femto	f
10^{-18}	atto	a
10^{-21}	zepto	z

(a)

100 ─

1084.4

50 ─

1085.4

1086.4

1087.4

0 ─

1083   1084   1085   1086   1087   1088   1089
m/z

(b)

100 ─

1084.4

50 ─

→| |← Δm = .001 u

0 ─

1084.39   1084.41   1084.43   1084.45   1084.47
m/z

Fig. B2.2 High-resolution MALDI mass spectrum for [Arg^8]-vasopressin: (a) narrow band acquisition from *m/z* 1080 to 1090 and (b) an expanded mass axis to show a mass resolution of 1 100 000. (Li *et al.*, 1994.)

Comment B2.3   Biologist's box: Computation of protein molecular mass

A mass spectrum is a plot of the intensity as a function of mass-to-charge ratio. The peak in the spectrum with highest intensity is called the base peak. Generally, the spectrum is normalised to the intensity of the base peak, resulting in relative intensities.

Comment Fig. B2.3 Electrospray mass spectrum of multiply charged cytochrome *c* (16951.5 Da) at low resolution from +12 to 18. (After Gordon, 2000.)

The figure shows the electrospray mass spectrum of multiply charged cytochrome *c*. The molecular mass of the protein can be calculated easily according to the mathematical formalism presented below, remembering that *z* values are integers. It is assumed that the ions are adducts of neutral molecule and protons.

The molecular mass, M, of the neutral molecule can be found from recorded masses m_1 and m_2 (equivalent to the m/z values) and the number of charges or protons added n_1 and n_2. Such that

$$M = n_2(m_2 - 1) \tag{B2.1}$$

where

$$n_1 = n_2 + 1 \text{ and } n_2 = \frac{m_1 - 1}{m_2 - m_1}$$

By taking peaks in pairs, n_2, and hence M, can be calculated from the recorded masses.

Applying Eq. (B2.1) to calculate the molecular mass of cytochrome c (see figure) using two peaks, $m_1 = 952.3$ and $m_2 = 1031.3$,

$$n_2 = \frac{m_1 - 1}{m_2 - m_1} = \frac{951.3}{1031.3 - 952.3} \approx 12.04$$

or $Z = 12$. This means that 12 positive charges are associated with a relative mass 1031.3. The molecular mass calculated from this peak is given by $M = n_2(m_2 - 1)$ is 12 363.6.

Taking the next two peaks with relative masses $m_1 = 884.3$ and $m_2 = 952.3$, we have:

$$n_2 = \frac{m_1 - 1}{m_2 - m_1} = \frac{883.3}{952.3 - 884.3} \approx 12.989$$

or $Z = 13$, i.e. positive charges associated with a relative mass 952.3. The molecular mass calculated from this peak is 12 366.9.

Continuing this procedure for other pairs of peaks we have:

- from the two peaks with relative masses $m_1 = 825.5$ and $m_2 = 884.3$, a molecular mass of 12 366.2;
- from the two peaks with relative masses $m_1 = 773.9$ and $m_2 = 825.5$, a molecular mass of 12 367.5;
- from the two peaks with relative masses $m_1 = 825.5$ and $m_2 = 884.3$, molecular mass of 12 366.4.

So we can conclude that the observed molecular mass average, calculated from the five peaks is 12 366.1. The theoretical mass of cytochrome c is 12 366.

isotopes into the molecule, with ^{13}C at a natural abundance of 1.108% being the main contribution. For an example of the computation of protein molecular mass the reader is referred to Comment B2.3.

B2.1.2 Non-covalent complexes

Non-covalent interactions, which are highly specific in biological macro-molecules, are of fundamental interest. The physiologically active forms of many proteins are multimeric, with active sites often at subunit interfaces. The strength of non-covalent interactions generally arises from a multitude of relatively weak bonds and can vary widely. It is reflected in the dissociation constants (K_D) typically determined for a specific set of solution conditions. Clearly, the detection of weakly bound, thermally sensitive complexes requires gentle ESI inter-face conditions (Comment B2.4). Several non-covalent complexes have been reported to remain intact in ESI-MS experiments. They include metals, haem

Fig. B2.3 (a) ESI mass spectrum of human cytoplasmic receptor binding protein (FKBP) for cyclosporin at pH 7.5. The molecular mass, M, of FKBP is 11 812 Da. The envelope of multiply charged ions ranges from the $(M + 6H)^{6+}$ to the $(M + 8H)^{8+}$ charge state of FKBP. (b) ESI of FKBP with small substance FK506 ($M = 804$ Da). (c) Competitive binding of FKBP with FK506 and rapamycin (RM). The molecular mass of RM is 912 Da. (Ganem et al., 1991.)

Comment B2.4 Detection of weakly bound complexes

In most cases, the solution conditions that are needed to maintain an intact complex are not optimal for normal ESI operation. Thus, for maximum sensitivity, solutions of pH 2–4 for positive ionisation and pH 8–10 for negative ionisation are typical for polypeptide analysis. Many protein complexes are denatured in solution at pH values outside the pH 6–8 range. It is clear therefore that protein solutions for analysis in ESI-MS should be maintained close to physiological conditions of neutral pH and ambient temperature such that the protein remains close to its native state. However, ESI-MS with neutral pH solutions of proteins generally demands a more extended m/z mass spectrometer than typically required when using conventional conditions.

It is not yet fully understood which weakly bound complexes known to exist in solution are observable by ESI-MS, or what minimum binding strength may be required for an ESI-MS observation. Evidence from a growing body of literature suggests that the ESI-MS observation for these weakly bound systems reflects, to some extent, the nature of the interaction found in the condensed phase. However, the results of all ESI-MS experiments show that each biomolecular system has its inherent experimental features and experience obtained from studying one protein complex may not be the proper preparation for investigating the properties of another one.

---

groups, and peptides bound to protein as well as multimeric protein complexes, oligonucleotides, enzyme–substrate and receptor–ligand complexes and large RNA–protein complexes.

One of the most impressive illustrations of ESI-MS was the observation of very tight complexes between proteins and other molecules. The ESI spectrum of human cytoplasmic receptor for cyclosporin exhibits an abundant $(M + 7H)^{7+}$ ion at m/z 1688.7 (Fig. B2.3(a)). Upon addition of the immunosuppressive drug with molecular mass $M = 804$ Da a new peak appears at m/z 1803.1, corresponding to the FKBP-FK506 complex (1:1) in the 7^+ charge state (Fig. B2.3(b)). The same effect is observed with rapamycin which has $M = 913$ Da, and it is even possible to estimate the relative ratio of their binding constants from the relative peak height when adding a 1:1 mixture of the rapamycin (Fig. B2.3(c)). Nevertheless, a substantial amount of free FKBP was also detected in the mass spectra in both cases.

Using this methodology it is possible to monitor the hydrolysis of hexa-N-acetylglucosamine by hen egg white lysozyme.

B2.1.3 Protein folding and dynamics

The speed, accuracy and sensitivity of ESI and MALDI have been exploited in the development of several different mass-spectrometry-based approaches for studying protein folding and dynamics and mapping protein function. In many cases, mass spectrometry has provided data on the functional properties of a protein complementary to data obtained from traditional techniques such as circular dichroism, NMR and fluorescence spectroscopy. Also, the relative speed and ease with which ESI and MALDI can be used to acquire very accurate molecular mass information on limited amounts of sample has made possible the acquisition of data that are not readily obtainable by other techniques.

Folded and unfolded states

In the early 1990s it was shown that folded and unfolded proteins produced different distributions of charge states in their ESI spectra. Proteins electrosprayed from solution conditions that preserve their native conformation tend to have a narrow distribution with a low net charge, whereas proteins electrosprayed from denaturing solutions produce a broad distribution centred on a much higher charge (Fig. B2.4). The difference in the distribution of charge states is believed to be related to the accessibility of ionisable groups. It is likely, for example, that in an unfolded state the basic amino acids (Arg, Lys, His residues) are more accessible to accumulating charge than when they are in the native state (Fig. B2.4).

The folding state of a protein in solution can be monitored by the charge state distribution produced during ESI. Thus, in the case of acid-induced unfolding of cytochrome c ($M = 12\,360$ Da) the observed changes clearly indicate a highly cooperative unfolding behaviour (see Fig. B2.5(a)). In contrast, the unfolding of horse heart apo-myoglobin (myoglobin after loss of the haem group $M = 16\,951.5$ Da) is accompanied by gradual shifts in the maximum of the observed charge state distribution (Fig. B2.5(b)). The observations suggested that ESI-MS can be considered as a general experimental method for assessing the cooperativity of protein folding transitions.

It is interesting to note that processes resembling folding and unfolding of equine cytochrome c ions *in vacuo* were observed by ESI-MS, raising the question of the role of water in protein folding.

Fig. B2.4 The multiple-charging characteristics of folded and unfolded proteins in ESI. (Winston and Fitzgerald, 1997.)

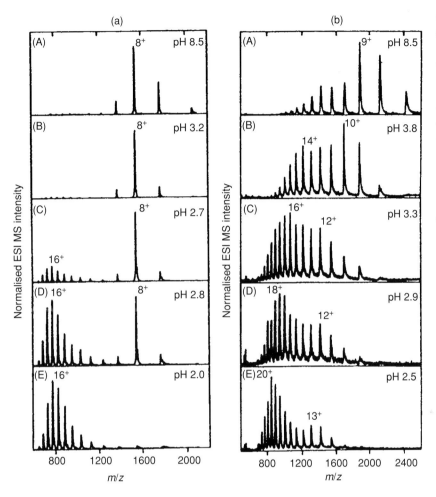

Fig. B2.5 (a) ESI mass spectra of cytochrome c recorded at different pH: (A) pH 8.5, (B) pH B2.2, (C) pH 2.7, (D) pH 2.6 and (E) pH 2.0. The pH was adjusted by addition of ammonium hydroxide and/or acetic acid. (b) ESI mass spectra of apo-myoglobin recorded at different pH: (A) pH 8.5, (B) pH B2.8, (C) pH B2.3, (D) pH 2.9 and (E) pH 2.5. The pH was adjusted by addition of ammonium hydroxide and/or acetic acid. (Konermann and Douglas, 1998.)

Protein folding intermediates

Taking advantage of the fact that charge distribution depends on the folded state of a protein, a novel approach for studying protein folding using 'time-resolved ESI' has been proposed. With a time resolution of 0.1 s, ESI has been used to monitor folding processes for cytochrome c and myoglobin. In the case of the first protein, no conformational intermediates between folded and unfolded states were detected. In contrast, a similar experiment with myoglobin revealed the presence of intermediates during its acid-induced denaturation (Fig. B2.6). The initial experiments produced only qualitative information. The extraction of quantitative information from ESI mass spectra became possible after the procedure of deconvolution of the charge-state distribution was introduced.

The ESI mass spectrum of any protein can be represented by a linear combination of charge-state distributions, called 'basis functions', which can be approximated by a Gaussian distribution. The intensity changes are represented by

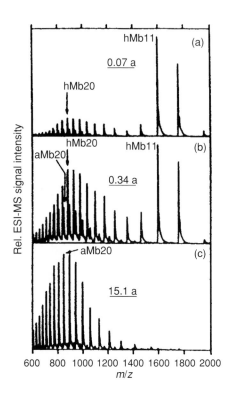

a weighting factor, which accounts for the relative contribution to the overall charge-state distribution. In this way, an observed ESI mass spectrum can be considered as a sum of the contributions from each protein conformation (conformer). Figure B2.7 shows ESI mass spectra of holo-myoglobin (hMb) and apo-myoglobin (aMb) over a wide pH range (2.5–8.6). The hMb spectra exhibit a very narrow charge-state distribution at pH 4.5 and above. The spectrum contains only two peaks, for charge states +8 and +9, respectively. Further decrease of solution pH (i.e. to pH 4) results in large-scale conformational changes, as manifested by the appearance of the highly protonated (low m/z) protein ions and partial dissociation of the haem group from the protein (Fig. B2.7(c)). Further decrease of solution pH down to 2.5 leads to a continuous increase of the average charge state of protein ions and disappearance of the protein–haem complex ions.

Unlike hMb, the aMb spectrum exhibits a multimodal character even at neutral pH (Fig. B2.7(f)). In addition to +9 and +8 ions, a wide distribution of less abundant ion peaks is seen in the spectrum at charge states ranging from +10 to +23 (Fig. B2.7). At pH 2.5, the aMb spectrum is indistinguishable from that of hMb, fully consistent with the expectation that any interaction between the haem group and the acid-destabilised form of the protein would be minimal.

The results of deconvolution of some of the charge-state distributions using basis functions are shown in Fig. B2.8 (a)–(f). Only one basis function is required to obtain a satisfactory fit to the spectra at pH 4.5 and above for hMb, while three

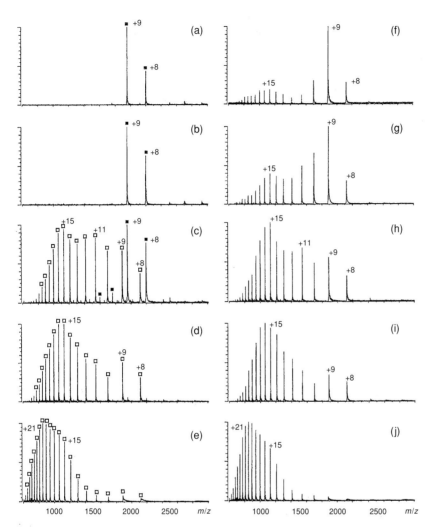

Fig. B2.7 Positive ion ESI mass spectra of (a)–(e) holo-myoglobin (*h*Mb) and (f)–(j) apo-myoglobin (*a*Mb) acquired at pH 7.4 ((a), (f)), 4.5 ((b), (g)) 4.0 ((c), (h)), 2.5 ((d), (i)) and 2.5 ((e), (j)). Intact *h*Mb ion peaks are indicated with filled squares. (Dobo and Kaltashov, 2001.)

basis functions are needed for data fitting in the pH range 3.5–4.5. Finally, a fourth basis function has to be added to the set in order to fit the data at low-pH levels both for *h*Mb and the *a*Mb. The four basis functions were assigned to four different conformational states of the protein (in order of decreased folding): native (*N*), so-called 'pH 4 intermediate' (*I*), extended conformation (*E*) and unfolded state (*U*). All the ESI spectrum features, interpreted as a linear combination of ionic contributions from *N*, *I*, *E* and *U*, are fully consistent with the existing picture of the acid unfolding of the protein using a wide variety of other experimental techniques.

The experiments give an excellent illustration of the unique ability of ESI to monitor distinct populations of folding intermediates.

Fig. B2.8 Curve fitting of positive ion charge-state distributions in ESI mass spectra of (a)–(c) *h*Mb and (d)–(f) *a*Mb acquired at pH 7.4 ((a), (d)), 4.0 ((b), (e)) and 2.5 ((c), (f)). Experimental data points are shown with squares (■ for intact *h*Mb ions and □ for *a*Mb ions). The Gaussian curves represent the weighted basis functions used for curve fitting (shaded for *h*Mb). The thick solid lines represent the summation of weighted basis functions. (Dobo and Kaltashov, 2001.)

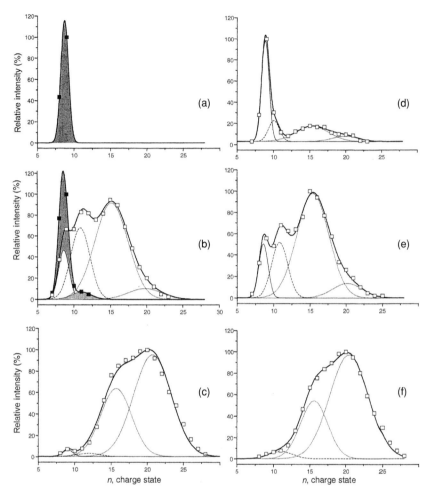

B2.1.4 Protein sequencing

The lack of fragmentation of proteins under soft ionisation is very useful for the unambiguous determination of their molecular mass, but the resulting absence of structural information is an essential drawback. This was firstly overcome by using the FAB technique, which unfortunately is limited to relatively short peptides and requires large amount of sample (2–50 nmol).

MS-MS (Section B2.6.7) is the most suitable technique for protein sequencing. MS-MS can sequence not only the 20 common amino acids but also known or unknown modified amino acids according to their mass. MS-MS is fast, sensitive and can analyse peptide mixtures directly.

A number of mass spectrometric approaches have been devised for sequencing large peptides and proteins. These include: (1) MS-MS approaches combined with enzymatic or chemical degradation to form oligopeptides (<3 kDa) as a first step, with MS of the resulting products, as a second step, (2) using ESI-FTMS as the

Comment B2.5   Edman degradation

In 1950 P. Edman proposed a chemical method for the stepwise removal of amino acid residues from the N-terminus of a polypeptide or protein. The series of reactions has come to be known as the Edman degradation, and the method remains the most effective chemical means for polypeptide sequencing. The Edman method requires a free amino group at the N-terminus. The peptides in which the terminal amino group is blocked (e.g. by a formyl, acetyl or acyl group) have to be cleaved by either chemical or enzymatic degradation. Edman degradation is carried out in an automated analyser. Sequence data can now be obtained from as little as 10–100 ng of protein.

first step of degradation, (3) combining classical Edman degradation (Comment B2.5) with MS-MS.

Computation of protein sequence

To obtain structural information by mass spectrometry, the molecule to be studied must undergo fragmentation of one or more bonds in such a manner that ions are formed. Protein-derived peptides are linear molecules whose structures have two constraints: the linear backbone consists of only repeating α-amino acid–amide bonds, and the substituent groups of each α-carbon are side-chains made of one of the 20 naturally occurring amino acids. 'Soft' ionisation methods generate singly or poly-protonated molecular ions that contain insufficient excess energy to fragment. However, using a special collision cell in tandem mass spectrometry (Section B2.6.7) or using Ar (Ne) atoms in fast bombardment mass spectrometry (Section B2.5.4) fragmentation can be achieved.

Although different types of fragmentation may occur, there is a predominance of peptide bond cleavages. The loss of one amino acid residue at a time gives rise to peaks in the spectrum that differ sequentially by the mass of the amino acid minus H_2O (the so-called residue mass, Table B2.1). The sequential mass differences represent exactly the primary structure of the peptide.

The fragmentation of an idealised peptide is shown in the figure below. The bond cleavages indicated at the top of the figure lead to fragments of types a_n, b_n, and c_n, if the charge is retained on the N-terminal fragment. The exception is that for a c_n ion two daltons have to be added because it retains the original protonating hydrogen and picks up one more from the other side of the peptide bond.

The x_n, y_n, and z_n ions are formed when the charge is retained on the C-terminal, with the y_n ions having added two hydrogen atoms analogous to the c_n ions. From a single, complete ion series we can deduce the amino acid sequence of a peptide, although leucine and isoleucine, and lysine and glutamine, cannot be differentiated in this manner. A single, complete ion series is not generally observed, and overlapping N- and C-terminal ions are used to determine the full sequence

Table B2.1. *Residue mass [-NH-CHR-CO-] of amino acids*

Amino acid	Letter code		Mass
	three	single	
Glycine	Gly	G	57
Alanine	Ala	A	71
Serine	Ser	S	87
Proline	Pro	P	97
Valine	Val	V	99
Threonine	Thr	T	101
Cysteine	Cys	C	103
Isoleucine	Ile	I	113
Leucine	Leu	L	113
Aspargine	Asn	N	114
Aspartic acid	Asp	D	115
Glutamine	Gln	Q	128
Lysine	Lys	K	128
Glutamic acid	Glu	E	129
Methionine	Met	M	131
Histidine	His	H	137
Phenylalanine	Phe	F	147
Arginine	Arg	R	156
Tyrosine	Tyr	Y	163
Tryptophan	Trp	W	186

The fragmentation processes of protonated peptides are now so well understood and reproducible that a set of empirical rules has been devised.

(1) The highest mass peak in the FAB spectrum represents $(M + H)^+$.
(2) The mass difference (Δ) between sequence ions represents the amino acid (aa) residual mass.
(3) The sum of the amino acid residues should give the total molecular mass.

Fig. B2.9 Fragmentation of an idealised protonated peptide. (After Biemann, 1992.)

The application of these rules in the case of FAB-MS of an idealized protonated peptide from the following m/z data 128; 185; 299; 396; 528 is presented below. The highest mass peak in the FAB spectrum is $m/z = 528$, hence $M = 527$. The set of m/z data can be converted to sequence of peptides in the following manner:

m/z	128	185	299	396	527
Δ		57	114	97	131
aa		Gly	Asn	Pro	Met

The lowest mass sequence ion, $m/z = 128$ is Lys, hence the sequence from the N-terminal end is Met-Pro-Asn-Gly-Lys.

The sum of the aa residues = 527 Da. This agrees with the mass spectrometry value for M.

B2.1.5 Protein identification from two-dimensional electrophoresis

Two-dimensional gel electrophoresis (2-DE) allows the separation of proteins according to their chemical structure (Chapter D5). In the 1970s, identification of proteins from gels was achieved by coelectrophoresis of known purified proteins or by immunostaining. In later innovations the power of 2-DE was increased

Fig. B2.10 The 'End sequencing' principle. Isolated peptides are fragmented. The masses of the low-mass fragments such as y_1, y_2, b_2 and b_3 fragment ions plus the accurate molecular mass of the intact peptide are used to 'end sequence' the peptide. The pattern and the molecular mass uniquely identify the peptide in the database. (After Nielsen *et al.*, 2002.)

Fig. B2.11 Analysis by o-MALDI of a silver-stained gel fragment excised from a two-dimensional gel. (a) Magnification of a region of the two-dimensional gel. The excised and digested sample spot is marked by an arrow. (b) Mass spectrum of peptides derived from the gel spot. (c) Tandem mass spectrum of the low-mass region from the most intense precursor ion at m/z 1298.762. Calculated tryptic peptide masses from this protein are labelled with an asterisk. (After Nielsen *et al.*, 2002.)

to resolve up to 10 000 protein spots on one gel by using silver staining and by increasing the gel size (see Chapter D5). The identification of the 2-DE-separated spots in the femtomole range has been achieved by MALDI fingerprinting with TOF mass spectrometry or by ESI MS-MS.

A high-sensitivity and high-throughput system based on orthogonal MALDI MS-MS (Section B1.6.7) provides automated recognition of fragments corresponding to the N- and C-terminal amino acid residues. The pulsed feature of the instruments enhances the low-mass region of the spectra by approximately 1 order of magnitude. The low-mass range of the spectrum is very simple to interpret, given the fact that only a few amino acid combinations lead to the observed low fragment masses. Typically, the sum of the masses of the second N-terminal ion (the b_2 ion), the third N-terminal ion (the b_3 ion), and the two C-terminal fragments of the peptide (y_1 and y_2) can be determined. Given the mass accuracy in the low parts per million range, peptide end sequencing on one or two tryptic peptides is sufficient to uniquely identify a protein from gel samples in the low silver-stained range (a few nanograms of protein per band). Figure B2.10 shows the 'end sequencing' principle. Figure B2.11 presents the analysis by o-MALDI of silver-stained fragments excised from a two-dimensional gel. Analysis of the mass spectrum obtained from the gel spot (Fig. B2.11(b)) shows that the peptide map does not yield a distinct protein match when searching the protein database. The precursor ion at m/z 1298.762 was isolated and a tandem mass spectrum was obtained (Fig. B2.11(c)).

The combination of N- and C-termini sequencing unambiguously identified the peptide as VHLVGIDIFTGK (Val–His–Leu–Val–Gly–Ile–Phe–Thr–Gly–Lys) and the protein as an eucaryotic translation initiation factor protein 5A.

B2.2 Functional proteomics

In contrast to the static genome, the proteome is dynamic and highly dependent not only on the type of the cell, but also on the state of the cell. The proteome is defined as the full complement of proteins in a cell, representing the products of expression of the genome together with the influence of post-translational modifications.

In the post-genome-sequencing era, the formidable challenge is to reach an understanding of living systems at the genomic and proteomic levels.

B2.2.1 The role of mass spectrometry

In contrast to genomics, where the contribution of mass spectrometry has been rather limited, the study of the proteome involves mass spectrometry in a cardinal role.

Two general strategies are commonly used for high-throughput proteomics analysis (Fig. B2.12). In the first, 2-DE constitutes the display and quantification methodology. The identification of spots on the two-dimensional gels is generally performed by excising them from the gel, fragmenting the protein with trypsin, extracting the peptides, and identifying the fragments by MALDI-TOF and/or ESI mass spectrometry.

In the second strategy, fragments from protein mixtures treated with trypsin are separated by liquid chromatography. The resulting peptides pass directly from the liquid chromatography column to the mass spectrometer. One clear disadvantage of two-dimensional gels is that they involve denaturation of the protein, with

Fig. B2.12 There are two strategies for proteome analysis of protein samples prepared from the cells or tissues. (1) The two-dimensional gel strategy in which the proteins are separated and displayed on two-dimensional gels. Individual proteins are recovered from the gels, subjected to fragmentation by trypsin, and analysed by mass spectroscopy. (2) The liquid chromatography strategy in which all the proteins in a mixture are fragmented by trypsin, separated by liquid chromatography, and analysed by mass spectrometry. (After Godovach-Zimmermann and Brown, 2001.)

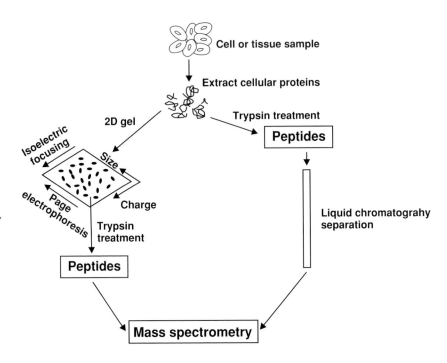

concomitant destruction of protein–protein interactions. A closer examination of the proteins routinely identified by proteome studies also suggested that 2-DE coupled with MS does not represent a truly global technique. Some classes of proteins, such as very acidic or basic proteins, excessively large or small proteins, are absent in the two-dimensional gel pattern and low-abundance proteins are not detected.

A new strategy for the visualisation and analysis of proteomes by mass spectrometry was developed in the beginning of the twenty-first century. The strategy is based on mass spectrometric surface analysis of isoelectric focusing gels and named 'virtual two-dimensional gel'. The 'virtual two-dimensional gel' approach, in which mass spectrometry is substituted for size-based gel separation, provides advantages in mass resolution and accuracy over classical two-dimensional gels and can be readily automated.

Classical two-dimensional polyacrylamide gel (PAGE) employs isoelectric focusing (IEF) in the first dimension and sodium dodecyl sulphate in size analysis as the second dimension (Chapter D5). Analogously, in the MS-based surface scanning method, proteins are desorbed *directly* from one-dimensional IEF gels to create a virtual two-dimensional gel. Analytical data are present as an image, similar to that obtained from a stained two-dimensional gel.

Figure B2.13 demonstrates protein detection capabilities by surface scanning from an IEF gel of *E.coli* whole cell lysate and MALDI mass spectra corresponding to the positions indicated. Consistent with the complexity of classical

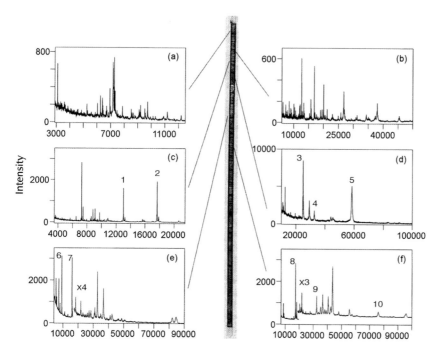

Fig. B2.13 Silver-stained IPG gel of *E.coli* whole cell lysate (centre) and MALDI mass spectra obtained directly from dried, ultrathin polyacrylamide gel corresponding to the positions indicated (but acquired from duplicate, unstained gels prepared for mass spectrometry). (a)–(f) Singly and doubly charged ions observed at various positions (pl) from duplicate, unstained gels. The *y*-axes indicate relative ion intensity. The *x*-axes indicate molecular mass. Mass spectra acquired at: (a) 3 mm from the basic edge of gel (pl 6.6), (b) 15 mm (pl 6.4), (c) 22 mm (pl 6.3), (d) 30 mm (pl 6.2), (e) 55 mm (pl 5.8). (beyond 20 kDa, the *y*-axis has been expanded by factor of 4), (f) 71 mm (pl 5.6), (beyond 20 kDa, the *y*-axis has been expanded by factor 3). Protein identities for labelled peaks postulated from pl (±0.3 pH unit) and molecular mass (±0.1– 0.2%); identities of 1, 3, 5–9 confirmed by Edman degradation; 1 – *yifE*, 2 – *purE*, 3 – *artJ*, 4 – *yeaD* or *gutQ*, 5 – *oppA*, 6 – *ptsH*, 7 – *ygaU*, 8 – *ppiB*, 9 – *mdh*, 10 – *metG* or *yjcQ*. (Ogorzalek Loo *et al.*, 2001.)

two-dimensional gels, proteins spanning a range of masses are observed at each pH value; e.g., 33 proteins are observed in the mass spectrum shown in Fig. B2.13(b). Low-molecular-mass proteins, which are crucial to confirming the expression of small open reading frames or protein processing, are often lost in the second dimensions of classical two-dimensional gels but are readily detected by MS surface scanning of the IPG. A mass resolution and accuracy superior to SDS-PAGE are apparent in Fig. B2.13(e) and (f), where 82.1- and 84.6-kDa proteins are resolved as well as eight proteins of from 30 to 50 kDa. Proteins over 100 kDa in size can be analysed by this method, as well as small polypeptides (<8 kDa) which are often lost in the second-dimension SDS gel without special electrophoresis conditions.

Finally, virtual two-dimensional gels are also amenable to protein cleavage carried out on a gel-wide scale with analysis directly from the gel. The technologies of two-dimensional gel and MS are the driving force in contemporary proteomics and are likely to remain so in the immediate future. However, in the long term, new approaches, including antibody-based techniques and protein chips, may challenge their pre-eminence.

B2.3 Nucleic acids

The analysis of nucleic acids by MS lags behind that of proteins. The main reason for this is the high affinity of nucleic acids for sodium ions, which can greatly reduce the ionisation efficiency. Moreover, the generation of intact molecular ions of oligomers of more than two nucleotides turned out to be difficult when using classical ionisation techniques such as electron impact and chemical ionisation, due to the high polarity of nucleic acids and a high tendency for their molecular ions to fragment. In 1982, the capability of the FAB technique to generate intact molecular ions of nucleotides was demonstrated. Ultrafast sequencing of oligonucleotides by FAB mass spectrometry was proposed in the following year. The introduction of ESI and MALDI made way for a significant improvement of the accessible mass range for the analysis of nucleic acids. Notable advances have followed in oligonucleotide sequencing, mixture analysis, studies of non-covalent complexes and microscale sample handling.

Computation of nucleotide sequence

The figure below shows the shorthand structure of the oligodeoxyribonucleotide d(A-C-T-C-G-A-T-G) with bonds marked that on breakage give rise to the main fragment ions. In the same way as for a peptide sequence, a set of empirical rules has been devised for a nucleotide (nt) sequence. The four possible phosphodiester bound cleavages are labelled a, b, c and d for the fragments containing the 5′-OH group and w, x, y and z for fragments containing the 3′-OH group. The subscript indicates the position of the cleavage, counting the number of bases from the appropriate terminal group. An example of the application of these rules in the case of FAB-MS is presented in Fig. B2.14.

The sequence of an oligodeoxyribonucleotide can be determined from the following set of m/z data

m/z (5′–P):   2407, 2174, 1885, 1581, 1292, 963,   650,   346
m/z (3′–P):   330,   619,   923,   1212, 1541, 1854, 2158, 2407

The necessary rules are the following:

(1) The mass differences (Δ) in m/z values between sequence ions represent the nucleotide (nt) residual mass. This means that the set of m/z data can be converted to sequence of oligonucleotides in the following manner:

(5′-P) Δ		233*	289	304	289	329	313	304
m/z 2407		2174	1885	1581	1292	963	650	346*
nt	A	C	T	C	G	A	T	G

Fig. B2.14 The shorthand structure of the octatomic oligodeoxyribonucleotide d(A–C–T–C–G–A–T–G) with bonds marked that on breakage give rise to the main fragment ions. The dashed lines show the main fragmentations with attached mass, those above corresponding to the ions with 5′-phosphate ends (5′-P) and those below to the ions with 3′-phosphate ends (3′–P). (Grotjahn et al., 1982.)

The first nucleotide is A (molecular mass $= 233 + 96(PO_4H) - 17(OH) + 1(H) = 31B2$). The last nucleotide is G (molecular mass $= 346 - 17(OH) = 329$). So, the sequence is

$$d(A\text{-}C\text{-}T\text{-}C\text{-}G\text{-}A\text{-}T\text{-}G)$$

The reader can check the sequence by applying the same rules to a set of m/z values for the (3′-P) end.

(2) The highest mass peak in the FAB spectrum is $m/z = 2407$ which represents $(M - H)^-$. Hence $M = 2408 + 96(PO_4H) - 34(2OH) = 2470$.

The summation of the nt residues $= 2470$ Da. This agrees with the mass spectrometry value for M.

B2.3.1 Oligonucleotide mixture analysis

The direct combination of liquid chromatography and electrospray ionisation techniques in mass spectrometry should have been highly beneficial in the analysis of a nucleotide mixture. However, developments in this field have been relatively limited owing to the incompatible demands of the two methods. Anion exchange and reversed-phase liquid chromatography utilise a salt-containing mobile phase for efficient separation, which reduces the quality of mass spectra. This problem has been solved by employing a new mobile phase additive (hexafluro-2-propanol). The broad potential of this new method was demonstrated for synthetic homopolymers of thymidine up to 75 bases, fragments based on the pBR322 plasmid sequence, and phosphorothioate ester antisense oligonucleotides. This approach is particularly useful for the characterisation of DNA probes, polymerase chain reaction (PCR) primers, as well as for materials used in clinical trials on a level below 10 pmol.

B2.3.2 Non-covalent complexes

Due to the gentle nature of the ESI process, complementary oligonucleotide duplexes can be transferred with fidelity to the gas phase. Highly accurate mass measurements of the complexes formed provide strict evidence for the identity and stoichiometry of the constituent subunits. A number of successful measurements for protein–oligonucleotide complexes including a small protein called gene V protein (molecular mass 9688 Da) and bovine serum albumin (molecular mass 66 497 Da) have been reported. Such experiments are highly dependent on experimental conditions, however, and the question remains as to whether or not the measurements reflect real solution structures.

B2.3.3 Large and very large nucleic acids

Attempts to analyse large nucleic acids by MALDI-MS with lasers emitting in the UV region have been only moderately successful. The state-of-the-art mass upper limit is $\sim$ 90 kDa for DNA and $\sim$150 kDa for RNA, and is restricted mainly by ion fragmentation.

Using an IR laser and a glycerol matrix was found to be the most gentle combination for the intact desorption and ionisation of nucleic acids in a broad range of the mass, from small oligonucleotides to molecules of more than 2000 nucleotides. Thus, for a synthetic 21-nucleotide DNA (molecular mass 6398 Da) a mass resolution of 1200 has been obtained, similar to that for proteins of comparable mass.

Detection of very large DNA molecules was accomplished in the original FTICR experiment when a single pass of the ions permitted remeasurement on the same ions more than 450 times. This greatly improved the signal-to-noise ratio and the resulting precision of mass and charge measurement for very large ions. Figure B2.15 shows the mass spectrum of 2.88-MDa ions of DNA carrying more than 250 charges obtained with low mass resolution (about 25).

At first sight, ESI appears to be a superior method to MALDI for the intact desorption of large nucleic acids. However, mass assignment is usually very poor (with an uncertainty of around 10%). The largest nucleic acids whose masses have been accurately determined by ESI-MS have a mass of about 40 kDa.

Coupling of ESI techniques with FTICR-MS allows the analysis of individual multiply-charged macroions. Figure B2.16 shows FTICR spectra obtained for Coliphage T4 DNA (expected molecular mass 111.5 MDa). The signal from individual Coliphage T4 DNA ions at a pressure of $\leq 1.0 \times 10^{-9}$ Torr was detected more than 1.5 h after the initial excitation.

Fig. B2.15 Histogram of the average mass obtained from a DNA sample. The dominant peak at 2.88 MDa corresponds to the sodium adducted bacterial plasmid (pBR322) and the much smaller peak at 5.85 MDa corresponds to pBR322 dimer ions. (After Benner, 1997.)

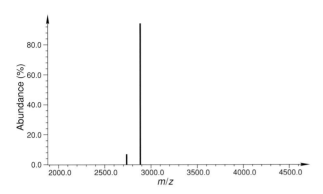

Fig. B2.16 FTICR spectrum of the ion at m/z 2883 obtained for Coliphage T4 DNA. The calculated molecular mass is about 90.9 ($\pm$ 9.1) 10^6 Da. (Smith *et al.*, 1996.)

B2.3.4 DNA sequencing

For mass spectrometry to be a useful method for sequencing DNA, the ionisation method must be capable of generating ions of DNA containing several hundred bases with sufficient mass resolution and mass accuracy to identify peaks corresponding to a difference of one base or better. From the methodological point of view, mass spectroscopy approaches can be subdivided into two groups as sequence information can be derived either directly from gas-phase fragmentation, or by mass measurements of chain cleavage products formed in the condensed phase (mass analysis of Sanger sequencing reaction products, or ladder sequencing).

The most straightforward approach to using mass spectrometry for DNA sequencing is to replace the polyacrylamide gel separation of the Sanger sequencing (Comment B2.6) products by mass analysis. Instead of having one lane for

---

Comment B2.6   Sanger sequencing

There are several common non-MS-based sequencing procedures used in molecular biology. The preferred method for large-scale DNA sequencing is the Sanger method. DNA complementary to the chain of interest is synthesised with a primer labeled on the 3′-end of each nucleotide. The reaction mixture is divided into four aliquots, and 2′,3′-dideoxynucleoside triphosphates (ddNTP) are added to the reaction mixtures. When a ddNTP is added to the chain in the place of its deoxy analogue, the lack of a free hydroxyl group at the 3′-end of the molecule terminates the chain. Each of the oligonucleotides is labelled at the common end with either a radioactive or a fluorescent tag. Identification is made by the separation of various fragments, using polyacrylamide gel electrophoresis and visualisation either by exposure to a film in the case of the radioactively labelled primers or by detection of the fluorescence of the primer when such a tag is used. The Sanger method generally requires 1.0 pmol of template DNA, and can routinely determine the sequence of fragments up to the 1000-mer level.

Comment B2.7
Human Genome Project (HGP)

In the HGP, DNA sequencing was performed by electrophoretic separation of fragments generated by the Sanger method. The four reaction product mixtures were loaded into lanes and separated according to their electrophoretic mobility, an intrinsic property that corresponds roughly to the length of the DNA fragment. Depending on the length of the gel, the separation can take several hours. This extended period was compensated for by performing several analyses in parallel. Automated sequencers running 96 lanes in parallel worked day and night and a total of about 3 billion bases were identified directly from the gels.

each of the four dideoxy termination reactions, four mass spectra are obtained, one for each of the A, C, T, G bases. Each sequencing solution contains a mixture of DNA strands several hundred bases in length. The mass spectrometer must be able to generate ions from this mixture at a resolution sufficient to distinguish between two bases in a strand up to 500 bases in length. This corresponds to a mass resolution of 500 for a mass range in excess of 100 kDa.

Mass spectrometric ladder sequencing involves cleaving successive nucleotides from the strand with an enzyme and measuring the resulting change in mass. The mass resolution and mass measurement accuracy requirements are more stringent for ladder sequencing than for Sanger sequencing. A mass difference of 9 Da must be measured to distinguish A from T, whereas for Sanger sequencing the ability to determine a mass difference of ~300 Da is sufficient to define the sequence.

In the gas-phase fragmentation methods all of the sequencing process is done in the mass spectrometer by identification of fragment ions resulting from collision-induced dissociation. Sequencing by gas-phase fragmentation has the potential of being much faster than methods that rely on solution-phase chemistry.

Each approach has advantages and drawbacks. Thus, direct sequencing by gas-phase fragmentation is simple from the experimental standpoint but is limited to chain lengths of ~80 nucleotides. This is because the mass resolution of MALDI degrades rapidly and it is difficult to separate two oligonucleotides that differ by one base if the DNA size is larger than 80 bases. On the other hand, indirect sequencing methods, although requiring pre-mass-spectrometry sample treatment of varying complexity, are capable of reaching chain lengths of ~100 nucleotides. However, in some cases the quality of the results is improved by lack of enzyme cleavage peculiarity.

At the beginning of the Human Genome Project (HGP), mass spectrometrists surmised that mass spectrometry would eventually allow DNA sequencing. When the HGP was completed, a total of about 3 billion bases had been identified directly from electrophoretic gel sequencing (see Comment B2.7) without using mass spectrometry. The real advantage of electrophoretic gel sequencing was that DNA with up to 1200 nucleotides could be analysed. Such lengths were inaccessible to mass spectrometry, which is limited to <100 bases. Mass spectrometry is best suited for many genotyping applications in which fast and accurate mass measurements of short DNA lengths are required. Many mass spectrometrists continue to share the opinion that the potential advantages of mass spectrometry DNA sequencing are quite substantial, and that in the long run mass spectrometry will replace electrophoretic sequencing.

B2.4 Carbohydrates

Biological carbohydrates constitute a diverse group of polymers playing various roles in cellular processes, ranging from energy storage (glycogen) and structural support (cellulose, chitin) to signalling, adhesion and protein modifications. Biological carbohydrates range from simple monosaccharides to megadalton homopolymers, from precisely structured signalling oligosaccharides to complicated mixtures of carbohydrate-modified proteins and lipids.

The main difficulty for the structural identification of carbohydrate oligomers is the large number of isomers arising both from variation in linkage positions between monomer residues and from the possibility of multiple linkages to a single residue (branching). Hence, for a straight-chain oligomer, structural identification requires the sequence of glycosyl linkage types, i.e. the linkage position and anomeric configuration, as well as the sequence of monomer residues.

Many of the structurally interesting biological carbohydrates are glucoconjugates, e.g. glycoproteins, glycolipids and glycolipoproteins. Conjugation also presents an additional dimension of structural isomers in terms of the specific location of the carbohydrate on a glycoprotein. The combination of conjugation and isomeric structural complexity has prevented the development of standard sequencing methodologies for carbohydrates, which have long been available for polypeptide or nucleic acid analysis.

Compared with the methodologies currently in use for protein and nucleic acid sequencing, the methods used for glycan analysis are far more diverse and are still at an early stage of development.

Below we give two simple examples of studies of carbohydrates by mass spectrometry.

B2.4.1 Oligosaccharide

Compared with peptides, native oligosaccharide substrates are difficult to ionise. However, various derivatisation strategies have been reported, including permethylation. Permethylated oligosaccharides undergo well-characterised crossring fragmentation, which enables the assignment of particular linkage types to the carbohydrate structure. Cleavage of glycosidic bonds produces fragments that are easily distinguished by differences in the mass arising from methylation.

Linear oligosaccharides and N-linked protein oligosaccharide mixtures (Fig. B2.17) have been studied successfully using ESI-quadrupole ion trap mass spectrometry. MS-MS experiments with orders higher than two offer a number of ways to enhance MS-MS spectra and to derive information not present in spectra from mass-spectrometer and MS-MS experiments (Fig. B2.18). Collisional activation of permethylated oligosaccharide molecular ions (MS-MS (or MS^2), as illustrated with maltoheptaose (Fig. B2.18), produces abundant fragments from

Glcβ1—4Glcβ1—4Glcβ1—4Glcβ1—4Glcβ1—4Glcβ1—4Glc

Fig. B2.17 Structure of maltoheptaose, Examples of monosaccharide abbreviations are given in Table A2.1.

(a)

MS^2 of 1497.8, [M+Na]$^+$ >

(b)

MS^3 of 1279.6, [M+Na]$^+$ > Y$_6$ >

(c)

MS^4 of 1075.5, [M+Na]$^+$ > Y$_6$ > Y$_5$ >

(d)

Fig. B2.18 (a) MS^2 spectrum of permethylated maltoheptaose, [M + Na] = m/z 1497.8. Scheme of the ion fragments produced upon (b) MS^2, (c) MS^3 and (d) MS^4. The nomenclature system used is based upon that introduced by Domon and Castello in 1987. (After Weiskopf et al., 1997.)

glycosidic bond cleavages. This indicates composition and sequence, and weak cross-ring cleavage products, which denote specific linkages within the oligosaccaride. Through the trapping and further dissociation of these fragments (MS^n), cross-ring cleavage products can be confirmed and their relative abundances increased to facilitate interpretation.

Methylation also enables one to establish the branching motif of a given sugar; since only unoccupied linkage sites are derivatised, the extent of methylation is inversely related to the degree of branching at a given monosaccaride.

B2.4.2 Glycopeptides

Numerous nuclear and cytoplasmic proteins are modified by single N-acetylglucosamine residues attaching to the hydroxyl side-chain of serine or threonine (O-GlcNAc). This post-translational modification is believed to be a regulatory modification similar to phosphorylation since it is dynamic, rapidly changing in response to external cellular stimuli and fluctuating in a cell-cycle-dependent manner. The current method of site mapping, which involves galactosyltransferase labelling, generation of glycopeptides by proteolysis, purification by several rounds of HPLC and gas-phase and manual Edman sequencing, is very tedious and requires about 10 pmol of pure, labelled glycopeptide.

A very sensitive ESI-MS method was developed for selective detection of glycopeptides and identification of the exact sites of glycosylation at the low picomole level. Synthetic glycopeptides have been generated (Fig. B2.19) and

(a)

Tyr-Ser-Pro-Thr-Ser-Pro-Ser-Lys

(b)

Tyr-Ser-Pro-Thr-Ser-Pro-Ser-Lys

(c)

Tyr-Ser-Pro-Thr-Ser-Pro-Ser-Lys

Fig. B2.19 Structure of three synthetic glycopeptides:
(a) O-GlcNAc peptide,
(b) O-GalNAc peptide,
(c) LacNAc peptide.
(Greis et al., 1996.)

Fig. B2.20 ESI mass spectra of the HPLC-purified glycopeptides obtained by triple quadrupole mass spectrometry: (a) O-GlcNAc peptide, (b) O-GalNAc peptide, (c) LacNAc peptide. The insets represent the predicted b- and y-ion mass fragments for each glycopeptides. The nomenclature system used is the same as for proteins. The calculated average masses for the glycopeptides are 1069 for the O-GlcNAc and O-GalNAc glycopeptides and 1231 for O-LacNAc peptide. The mass spectra of both the O-GlcNAc (Fig. B2.19(a)) and O-GalNAc (Fig. B2.19(b) glycopeptides were consistent with the expected mass as demonstrated by the major peaks at mass charge ratios (m/z) of 535 and 1070. These peaks correspond to the doubly protonated and singly protonated forms, respectively, of the glycopeptides. Other prominent peaks at 866 and 204 (corresponding to the deglycosylated peptides and O-GlcNAc or O-GalNAc) as well as those at 819, 616 and 251 were generated by fragmentation of the glycosidic and peptide bonds (see inset for predicted fragmentation patterns). The mass spectrum of the O-LacNAc peptide displays the same characteristics. (Greis *et al.*, 1996.)

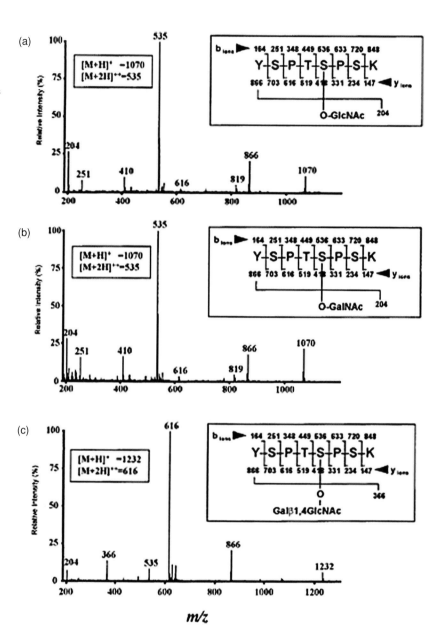

used to demonstrate that O-GlcNAc-modified peptides can be rapidly identified in complex mixtures by HPLC-coupled ESI-MS due to partial loss of linked O-glycan ($M = 204$ Da) at modest orifice potential. The resulting mass spectra are presented in Fig. B2.20. The site of the glycosylation was directly identified by collision-induced dissociation (CID, Section B2.6.7) of the glycopeptide after removal of O-GlcNAc by alkaline β-elimination.

The conversion of glycosylserine to 2-aminopropeonic acid (see the structures in Fig. B2.19) by β-elimination both decreased the mass of the glycopeptide by 222 and resulted in a CID fragmentation representing the loss of 69 units of molecular mass instead of 87 (Ser) at the position of the glycosylserine.

The results show that this method may be a powerful tool for determining the sites of O-GlcNAc modification on proteins of low abundance such as transcription factors and oncogenes (see Comment B2.8).

B2.5 Subcellular complexes and organelles

B2.5.1 Ribosomes, ribosomal subunits and ribosomal proteins

The work done on ribosomes is a good example of the possibilities of mass spectrometry. Ribosomes 70 S *E.coli* in a buffer containing 5 mM Mg^{2+} were projected into the gas phase of a mass spectrometer by means of nanoflow ESI techniques (Fig. B2.21(a)). By lowering the Mg^{2+} concentration in the solution, the particles were found to dissociate into 30 S and 50 S subunits (Fig. B2.21(b)). The resolution of the charge states in the spectrum of the 30 S subunit enabled its mass to be determined as 852 187 ± 3918 Da, a value within 0.6% of that calculated from its constituent proteins and 16S RNA. Further dissociation into smaller macromolecular complexes and then individual proteins can be induced by subjecting the particles to increasingly energetic gas-phase collisions. Of the 56 proteins in this complex 55 were observed as complete proteins (without fragmentation) and a wide variety of post-translational modifications was observed. It is interesting to note that the ribosomal protein of highest molecular mass (S1, 61 kDa) was not detected.

A similar study using another approach has been carried out on the yeast ribosome. All the proteins of this complex were isolated, denatured, subjected to trypsin digestion and desalted. The peptide mixture was then separated by chromatography and was observed by MALDI-TOF mass spectrometry. Of the predicted 78 proteins in the ribosomal complex, only three were not identified.

Comment B2.8
Alkaline-induced β-elimination

Alkaline-induced β-elimination is one of the most common methods used to liberate O-linked glycans from proteins and peptides. The method converts glycosylated serine and threonine residues to alanine and 2-aminobutyric acid, respectively.

Fig. B2.21 Nanoflow ESI mass spectra of 70S ribosome particles: (a) in the presence of 5 mM Mg^{2+} and (b) after a three-fold dilution of the solution. (Rostom *et al.*, 2000.)

Fig. B2.22 MALDI-TOF
mass spectra from
cellular suspensions of
four strains of *E.coli*: (a)
BLR, (b) XL1 blue, (c)
RZ1032, and (d) CSH23,
that show the typical
'whole cell' pattern of
peaks in the interval from
m/z- 3000 to 10 000. (Lay,
2001.)

B2.5.2 Bacteria and bacterial taxonomy

The taxonomic identification of bacteria based on constituent proteins represents a logical extension of the methods developed for bacterial proteomic studies. Ideally, this approach should be rapid and yet be based on a sufficiently large group of proteins so that a unique mass spectrum fingerprint for the organism or strain can be obtained. Figure B2.22 shows spectra obtained from four strains of *E. coli*. They are typical of 'whole cell' or 'intact cell' spectra that show a large number of peaks between m/z 3000 and m/z 10 000. The different *E.coli* strains share many peaks in common, but also display unique features. It is important to note that such spectra depend on the growth stage of the cells. Bacteria respond rapidly to environmental changes, and the production of stress proteins when cells are stored, handled, or cultured over different time periods results in non-identical samples. Such considerations must be taken into account when MALDI-TOF-MS is applied to identify bacteria.

B2.6 Mass spectrometry in medicine

A variety of problems in clinical chemistry involving nucleic acid and protein analysis can be addressed by mass spectrometry. It has already replaced electrophoresis for analysing products from routine molecular biological procedures. It allows the sizing of DNA amplified from the polymerase chain reaction (Comment B2.9), the sequencing of short oligonucleotides, the detection of genetic changes and mutations and many other applications. Technology platforms for full comparative DNA analysis based on the generation of short diagnostic products and precise measurements of their masses by MALDI-TOF have been developed. They are the first steps in preclinical and clinical drug trials moving from a statistics-based approach to a non-statistics based approach, enabling individual patient-specific response profiles for drug candidates.

Another potential application of MS is for the early detection of certain cancers. Cancer can arise from the gradual accumulation of genetic changes in a single cell and many important mutations such as the point mutations in *K-ras* and *p53* genes have been identified. It is common that cancer samples contain only a few mutated cells in a large background of normal cells (Comment B2.10). Hence, great efforts are being made to develop a mass spectrometry method for detecting the presence of these cells at an early stage.

B2.6.1 Clinical oncology

MALDI-TOF-MS offers the opportunity to detect rapidly and to monitor abnormalities in oncoprotein expression (Comment B2.10). For example, mass spectrometry has been used to study the protein profile of human cells as they

Fig. B2.23 MALDI-TOF
mass spectra of: (a)
malignant and (b) normal
breast cell lysates from
the MCF-10 cell line.
Abbreviations are the
following: c-H-Ras is
transforming protein
H-ras-1, int-2 is Int-2 proto
oncogene protein
precursor, p53 is nuclear
phosphoprotein, c-myc is
MYC proto-oncogene
protein, THR beta-2 is
Thyroid hormone
receptor beta-2. (Chong
et al., 1998.)

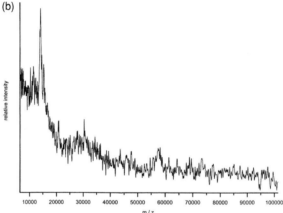

progressed from normal breast epithelium to fully malignant cells. Figure
B2.23(a) and (b) shows the MALDI-TOF mass spectra of malignant and normal whole cell lysates. The mass spectrum of the malignant breast cell lysate is
dominated by several protein peaks identified as specifically associated with car-
cinogenesis (Fig. B2.23(a)). In contrast, the mass spectrum of the cell lysate of a
normal cell line shows very few protein peaks standing out above the background
(Fig. B2.23(b)).

B2.7 Imaging mass spectrometry

While conventional methods such as transmission and scanning electron
microscopy are routinely used to create images of biological samples, they are
generally limited to providing mainly morphological information (see Part H).
The molecular specificity and sensitivity of MALDI-MS is being employed in the
development of a new technology for direct mapping and imaging of biomacromolecules present in single-cell or tissue sections.

B2.7.1 Single-cell level

Frozen-hydrated freeze-fractured biological samples provide the best targets for direct molecular characterisation by imaging MALDI-TOF-MS. The potential of the technique has been demonstrated on concentrated *Paramecium* cells that were freeze-fractured and introduced into a TOF secondary ion mass spectrometer. The single-cell organisms were ideal for study because their large size (~180–310 μm long) allowed the unambiguous positioning of individual cells directly in the path of the primary ion beam. Images of the distribution of molecular species across a cell surface with a submicrometre ion probe beam were obtained. The distribution of dimethylsulphoxide and cocaine added to the *Paramecium* cells was also imaged.

B2.7.2 Mammalian tissue level

Tissue samples for molecular distribution image analysis can be prepared using several protocols. In a typical preparation procedure, a frozen section of tissue is mounted on a stainless steel target plate, coated with a matrix solution, then dried and introduced into the vacuum inlet of the mass spectrometer. Molecular images from a raster over the surface of the sample with consecutive laser spots (~25 μm in diameter) are created (Fig. B2.24). With a laser frequency of 20 Hz, the time cycle is about 2.5 s per data point, including acquisition, data download to the computer, data processing and repositioning of the sample stage. A typical data array is made up of 1000–30 000 spots depending on the desired image resolution. The image is made up of the intensity of ions desorbed at each spot in a molecular mass range of 500 Da to over 80 kDa.

   Figure B2.25 shows mass spectrometric images of a mouse brain section. Although many of the protein signals were common to all areas of the brain,

Fig. B2.24 Schematic demonstration of methodology developed for the spatial analysis of tissue by MALDI-MS. Frozen sections are mounted on a metal plate, coated with a UV-absorbing matrix and placed in the mass spectrometer. A pulsed UV laser desorbes and ionises molecules in the tissue and the *m/z* values are determined using a TOF analyser. From a raster over the tissue and measurement of the peak intensities over thousands of spots, mass spectrometric images are generated of specific molecular mass value distributions. (Stoeckli *et al.*, 2001.)

Fig. B2.25 Mass spectrometric images of a mouse brain section: (a) optical image of a frozen section mounted on a gold-coated plate; (b) m/z 8528 in the regions of the celebral cortex and the hippocampus; (c) m/z 6716 in the regions of the substantia nigra and medial geniculate nucleus; (d) m/z 2564 in the midbrain. (Stoeckli et al., 2001.)

some were found to be highly specific to a given brain region. For example, the protein at m/z 8528 (Fig. B2.25(b)) was present in the regions of the cerebral cortex and the hippocampus; the protein at m/z 6716 was localised in the regions of the substantia nigra and medial geniculate nucleus; and the peptide at m/z 2564 was in the midbrain. Identification of the proteins can be done through extraction, high-performance liquid chromatography fractionation, proteolysis, mass spectrometric sequencing of one or more of the fragments and protein database searching. The potential of the new technology has been illustrated in cancer research by identification of specific tumor markers in proliferating tissue.

B2.8 Checklist of key ideas

- MALDI provides not only molecular mass information but also structural information for various peptides and oligonucleotides.
- The speed, accuracy and sensitivity of ESI and MALDI have been exploited in the development of several different mass-spectrometry-based approaches for studying protein function, structure and dynamics.
- Subcellular complexes and organelles (ribosomes and bacteria) can be projected into the gas phase without fragmentation.
- Mass spectrometry may be best suited for many genotyping applications for which fast and accurate mass spectrometric measurements of short DNA lengths are required.
- In contrast to genomics, where the contribution of mass spectrometry has been only limited, the study of the proteome involves mass spectrometry in a cardinal role.
- Coupling the modern technologies of 2-DE gel and tandem MS is the driving force in contemporary proteomics at sub-pmol level.

- Particularly large perspectives have been opened for mass spectrometry in bacterial taxonomy and medicine.
- MALDI-MS provides direct mapping and imaging of biological macromolecule distributions present in the single cell or in mammalian tissues.
- In the next decade mass spectrometry is expected to move at an extraordinary pace, extending from the world of structural biology at the molecular level to the worlds of medicine and therapeutics.

Suggestions for further reading

Protein structure and function

Chait, B. T. (1994). Mass spectrometry – useful tool for protein X-ray crystallographer and NMR spectroscopist. *Structure*, **2**, 465–467.

Winston, R. L., and Fitzgerald, M. C. (1997). Mass spectrometry as a readout of protein structure and function. *Mass Spectr. Rev.*, **16**, 165–179.

Rostom, A. A., and Robinson, C. V. (1999). Disassembly of protein complexes in the gas phase. *Curr. Opin. Struct. Biol.*, **9**, 135–141.

Miranker, A. D. (2000). Protein complexes and analysis of their assembly by mass spectrometry. *Curr. Opin. Struct. Biol.*, **10**, 601–606.

Dobo, A., and Kaltashov, I. A.(2001). Detection of multiple protein conformational ensembles in solution via deconvolution of charge-state distribution in ESI MS. *Anal. Chem.*, **73**, 4763–4772.

Nielsen, M. L., Bennet, K. L., Larsen, B., Monniate, M., and Mann, M. (2002). Peptide end sequencing by orthogonal MALDI tandem mass spectroscopy. *J. Proteome Res.*, **1**, 63–71.

Link, A. J., Eng, J., *et al.* (1999). Direct analysis of protein complexes using mass spectrometry. *Nature Biotechnol.*, **17**, 676–682.

Functional proteomics

Jungblut, P., and Thiede, B. (1997). Protein identification from 2-DE gels by MALDI mass spectrometry. *Mass Spectr. Rev.*, **16**, 145–162.

Godovach-Zimmermann, J., and Brown, L. R. (2001). Perspectives for mass spectrometry and functional proteomics. *Mass Spectr. Rev.*, **20**, 1–57.

Ogorzalek Loo, R. R., Cavalcoli, J. D., *et al.* (2001). Virtual 2-D gel electrophoresis: visualization and analysis of the *E. coli* proteome by mass spectrometry. *Anal. Chem.*, **73**, 4063–4070.

Nucleic acids

Nordhoff, E., Kirpekar, F., and Roepstorff, P. (1996). Mass spectrometry of nucleic acids. *Mass Spectr. Rev.*, **15**, 67–138.

Murray, K. K. (1996). DNA sequencing by mass spectrometry. *J. Mass Spectr.*, **31**, 1203–1215.

Berkenkamp, S., Kirpekar, F., and Hillenkamp, F. (1998). Infrared MALDI mass spectrometry of large nucleic acids. *Science*, **281**, 260–262.

Crain, P. F., and McCloskay, J. A. (1998). Application of mass spectrometry to the characterization of oligonucleotides and nucleic acids. *Curr. Opin. Biotechnol.*, **9**, 25–34.

Carbohydrates

Karlsson, K. A. (1989). Animal glycosphingolipids as membrane attachment sites for bacteria. *Ann. Rev. Biochem.*, **58**, 309–350.
Solouki, T., Reinhold, B. B., *et al.* (1998). Electrospray ionization and matrix-assisted laser desorption/ionisation Fourier transform ion cyclotron resonance mass spectrometry of permethylated oligosaccarides. *Anal. Chem.*, **70**, 857–864.

Subcellular complex and organelles

Rostom, A., Fucini, P., *et al.* (2000). Detection and selective dissociation of intact ribosomes in a mass spectrometer. *Proc. Natl. Acad. Sci. USA*, **97**, 5185–5190.
Lay, J. O. (2001). MALDI-TOF mass spectrometry of bacteria. *Mass Spectr. Rev.*, **20**, 172–194.

Mass spectrometry in medicine

Guo, B. (1999). Mass spectrometry in DNA analysis. *Anal. Chem.*, **71**, 333–337R.

Imaging mass spectrometry

Pacholski, M. L., and Winograd, N. (1999). Imaging with mass spectrometry. *Chem. Rev.*, **99**, 2977–3005.

Part C
Thermodynamics

Chapter C1
Thermodynamic stability and interactions

C1.1 Historical overview and biological applications

The term thermodynamics is derived from the Greek *therme* meaning heat and *dynamis* meaning strength or force. Thermodynamics, as a science, has its beginnings in the **nineteenth century**, with the first experiments exploring the relationship between heat and work, the definitions of the concepts of temperature and energy and the clear enunciation of the first and second laws of thermodynamics. Classical thermodynamics uses a phenomenological approach based upon these laws, in contrast to statistical thermodynamics, which tries to establish a more fundamental understanding of heat in terms of the kinetics of large assemblies of atoms or molecules. The concept of work was generalised beyond mechanical work to include all forms of energy such as electric, magnetic, chemical, and radiation energy. The validity of thermodynamics was established for all types of system, from a volume of perfect gas at a given pressure to thermonuclear plasma, encompassing magnetic systems, liquid–vapour systems, macromolecular solutions, chemical reactions etc. In this chapter, we are concerned, in particular, with the applications of both classical and statistical equilibrium thermodynamics to biological macromolecules, their solutions and interactions.

Early in the **seventeenth century, Galileo Galilei** and his followers constructed the first thermometers, allowing reproducible measurements of temperature. **Joseph Black (1728–1799)** built the first calorimeter; he is considered the founder of calorimetry. His measurements led to the theory of *caloric*, a conserved elastic fluid that had weight and was associated with temperature. **Antoine-Laurent Lavoisier (1780)** performed the first calorimetric measurements in biology. He studied energy exchange in animals (even though the concept of energy had not yet been enunciated) by placing a guinea pig in a more elaborate ice calorimeter (developed with **P. S. Laplace**) than **Black**'s. The heat output was measured by the rate at which the ice melted; heat input was calculated by subtracting measurements on faeces and urine from measurements on food intake. **Thomas Young (1807)** introduced the term *energy* (from the Greek *en* meaning 'in' and *ergon* meaning 'work') but the distinction between force and energy was not made clearly at the time. **Sadi Carnot**, a military engineer, introduced the concept of the heat engine cycle, published in his memoir '*Sur la puissance motrice du feu*' **(1824)** ('*On the motor power of fire*') and interpreted his results in terms of caloric theory. **Julius Robert von Mayer (1842)**, a medical doctor, announced the equivalence of heat and mechanical energy from physiological considerations. In **1845**, he published a numerical value for the mechanical equivalent of heat. **James Prescott Joule**, a Manchester brewer and amateur scientist, provided convincing experimental evidence that heat and work could be converted one into the other **(1840s)**. **Lord Kelvin (William Thomson) (1849)** pointed out the incompatibility between Joule's results and the caloric theory used by Carnot to interpret his experiments. The problem was solved by **Rudolf Julius Clausius (1850)**, who explicitly stated the first and second laws of thermodynamics; his paper marked the birth of the new science of thermodynamics. Clausius defined the property *entropy* (from the Greek *en* meaning 'in' and *tropos* meaning 'turn' and used in the sense of transformation, so that entropy is transformation content). The property of *energy* was defined by Kelvin, who also discussed various temperature scales. **James Clerk Maxwell** derived mathematical relations between thermodynamics parameters that now bear his name. **J. Willard Gibbs (1873, 1874)** published three definitive papers discussing the conditions of equilibrium in mathematical terms so elegant that they became the foundations of classical thermodynamics and physical chemistry. In **1918, Walther Nernst** stated the third law, essentially completing the exposition of classical thermodynamics.

Mechanical theories of heat received great impetus from Joule's experiments. The science of statistical thermodynamics is based on the relation between heat and the motions of atoms and molecules. It developed in parallel with classical thermodynamics, mainly through the work of **Hermann Ludwig Ferdinand von Hemholtz, Clausius, Maxwell, Ludwig Boltzmann** and **Gibbs**.

Calorimetry was developed in the **twentieth century** for the careful study of the physical chemistry of solutions. **John T. Edsall (1930s)** measured thermodynamics parameters for the dissolution of amino acids and other organic compounds in water. With respect to the application of thermodynamics ideas to protein folding, **Christian Anfinsen** proposed the 'thermodynamic hypothesis' in the **1950s**, following experiments on ribonuclease. The hypothesis states that all the information required for a protein to fold into its active structure under physiological conditions is contained in its sequence. The first highly sensitive calorimeter for protein solution studies was built in the **1960s** by **Peter Privalov** and his collaborators, and in the following decades differential scanning calorimetry and, later, isothermal titration calorimetry were used to study protein folding and binding interactions, respectively. Experiments allowing the measurement of thermodynamic parameters with great precision became possible with the development of ever more sensitive calorimeters – microcalorimeters and with the approach of the year **2000**, nanocalorimeters (the micro and nano prefixes referring to the precision with which heat can be measured in these devices).

Important microcalorimetric work on proteins was undertaken by **Privalov**, **John Sturtevant**, **Ernesto Freire** and their collaborators.

Sensitive biosensor devices that measure the binding between molecules in solution and immobilised ligands became available in the **1990s**, and entered routine operation in many biochemistry laboratories.

Non-equilibrium thermodynamics is beyond the scope of this book, but it should be mentioned in this brief historical survey, because of its important applications in many disciplines including experimental physics, reaction chemistry, biology, ecology and even sociology. The field of irreversible and non-equilibrium thermodynamics has known a period of strong development since the **late 1950s**, mainly through the work of **Ilya Prigogine** and his collaborators. They introduced the term *dissipative structures* to define open systems that could be maintained in far-from-equilibrium states by a flow of energy or matter. It is remarkable that spectacular examples of self-organisation in biology, such as cell division, differentiation and morphogenesis, can be modelled in terms of dissipative structures.

There is great scope in biophysics for applying modern thermodynamics, based on high-resolution structural information (see Part G) and the tools of molecular biology. The stability of a given protein under different conditions, for example, can be understood and tuned to a fine degree, by studying the folding–unfolding transition of various especially designed mutants. For thermodynamics as for other biophysical methods, a high-resolution structure acts as a starting point for understanding rather than as an end in itself.

C1.2 The laws of thermodynamics

The language of classical and statistical thermodynamics is a difficult but extremely useful one (in fact it cannot be avoided). It is essential for the organisation and understanding of experimental data on the energetics of defined systems, from very large assemblies of particles to single molecules. In writing this chapter we have not attempted more than a brief description of the principles governing the language. There are many good books on the subject and the reader is strongly encouraged to learn the language as thoroughly as possible, because thermodynamics is fundamental to the development and applications of biophysical methods.

Classical thermodynamics is a phenomenological science, concerned with precise observations on defined systems (e.g. a macromolecular solution). Statistical thermodynamics attempts an explanation of the observations in terms of the motions of the particle components of the system (e.g. the solute and solvent molecules that make up a solution).

Only a brief overview of the laws of classical thermodynamics is given in this chapter. We discuss the most useful parameters and the mathematical relations between them required mainly for the understanding of calorimetric and binding experiments on the folding and interactions of biological macromolecules

> **Comment C1.1　Biologist's box: Thermodynamics and mathematics**
>
> It would be gravely misleading to attempt an explanation of *thermodynamics* without using mathematics. The triumph and usefulness of thermodynamics are due to its formulation of a few laws in terms of mathematical equations, which establish precise relations between measurable quantities. *Partial differential equations*, in particular, which themselves follow definite mathematical rules, provide a simple and elegant way of expressing the changes expected in an experimental parameter (e.g. the pressure of a gas), when another one is changed (e.g. the temperature), while a third parameter is kept constant (e.g. the volume). These equations not only tell us that the pressure will rise, as we might expect intuitively, but also by exactly how much for a given temperature change. This example is a simple illustration of the application of thermodynamics, but similar strict predictions can be made for much more complex situations such as chemical reactions or protein folding.

(Comment C1.1). The basic principles presented are also valid and useful for the understanding of macromolecules in solution (Part A), hydrodynamics experiments (Part D), scattering experiments (Part G), spectroscopy (Part E) and single molecule studies (Part F). The treatment is limited to equilibrium thermodynamics, which provides an adequate basis for the understanding of biological macromolecules and their (mainly in vitro) interactions. Non-equilibrium thermodynamics, which is essential for the analysis of higher orders of biological complexity, is beyond the scope of this book.

C1.2.1 Fundamental definitions and the zeroth law

Adiabatic, isothermal systems and the concept of equilibrium

We consider a system to be a collection of matter contained by walls with which it does not react chemically. An *adiabatic* (from the Greek for 'does not allow passage') wall isolates the system perfectly from external effects. Walls that are not adiabatic are said to be *diathermal* (from the Greek, *dia* meaning 'through' and *therme* meaning 'heat'). Two systems separated by a diathermal wall are *isothermal*. The condition of the system is called its *state*. The properties of a given state can be determined by experimental measurements. Two states are identical when all the values measured for their properties are identical. A system enclosed in an adiabatic vessel will, in time, reach a state in which its properties no longer evolve; the system is then said to be in *equilibrium*. Note that equilibrium denotes a static (unchanging) condition only with respect to macroscopic properties. A gas or solution at equilibrium, for example, would be seen to be a highly dynamic system if we observed the Brownian motion of the molecules or solute particles, respectively.

The zeroth law

The zeroth law of thermodynamics was stated after the first and second laws. It essentially defines the concept of temperature. As this concept is useful for the understanding of the other laws, it was called the zeroth law.

The zeroth law states that if two bodies, A and B, are separately in equlibrium with a third body, C, then each of A and B is in equilibrium with the other.

A consequence of the law is that there exists an empirical function of the state of a system, the temperature, which takes the same value for different systems in equilibrium.

Functions of state and conjugate variables

An observable parameter, whose value depends only on the condition of a system and not on the means by which the condition was attained, is called a function of state. Temperature is a function of state. Internal energy, entropy, pressure and volume are other examples of functions of state. A parameter that depends on the size of the system is called an extensive variable, whereas a parameter that depends on the condition but not on the size of the system is called an intensive variable. Volume, internal energy and entropy are all extensive variables; pressure, temperature and the pH value of a solution, for example, are intensive variables.

In the thermodynamic equations, which describe relations between different functions of state, the terms appear in units of energy. When an energy term contains two parameters, the pair is made up of so-called *conjugate variables*. Pressure and volume are conjugate variables, as are entropy and temperature (see Eq. (C1.5), for example).

C1.2.2 The first law and energy

The first law of thermodynamics has been worded in various ways. In essence, it states that 'The amount of work needed to change the state of a system from an initial to a final state depends solely on the change accomplished, and not on the means by which the work is performed, nor on the intermediate states by which the system may pass.'

The first law defines *internal energy (U)* as a property of the initial and final states whose value changes by the amount of work done on the system.

Energy conservation

Conservation of energy (with no distinction between whether it is mechanical, chemical, heat or any other type of energy) is a consequence of the first law of thermodynamics. It is interesting to note, however, that the first law cannot be derived from a statement of conservation of energy.

According to the first law, the change, ΔU, in the internal energy of a system is given by

$$\Delta U = Q + W \qquad\qquad (C1.1)$$

Comment C1.2   Definition of heat

Equation (C1.1) defines heat. For a system enclosed by adiabatic walls, the change in internal energy is given by $\Delta U = W$, so that the value of the quantity Q is a measure of the extent to which the walls are not adiabatic and *heat* can flow into the system from the outside. It can be shown that Q has the properties we usually associate with heat:

The addition of heat to a system changes its state.
Heat is conveyed from one system to another by conduction, convection or radiation.
In a calorimetric experiment on an adiabatically enclosed system, heat is conserved.

---

where W and Q are, respectively, the amount of mechanical work (or chemical work, or any work other than in the form of heat) done on the system and the heat given to the system (Comment C1.2).

Chemical potential

In the case of a solution, solute molecules are brought together into a volume of solvent. The chemical potential, μ, of a solute is the energy required to add one molecule (or one mole of molecules, according to definition) into the solution. We can split the work, W, in Eq. (C1.1) into a chemical potential term and a mechanical work term

$$\Delta U = Q + \mu \Delta n - P \Delta V \qquad (C1.2)$$

where Δn is the change in number of solute molecules between the initial and final states (depending on the scale used, this can also be expressed in terms of moles). We have expressed the mechanical work in terms of pressure, P, and volume change, ΔV. Chemical potential is an intensive variable. Concentration is the corresponding, conjugate, extensive variable. For a change at constant pressure, the chemical potential term is related to a function called the Gibbs free energy (Section C1.3.1).

C1.2.3 The second law and entropy

A direct implication of the concept of equilibrium is that many types of change that would satisfy the first law simply do not occur in practice. The observed fact that an isolated system tends towards equilibrium (for example, solute molecules moving to fill a vessel homogeneously after the removal of a wall that confined them to part of the vessel) excludes the occurrence of the reverse path (the molecules moving back to one corner of the vessel, in the example) from taking place, even though it would not violate the first law.

The second law

We can express the directionality of certain thermodynamic changes in terms of the subjective notions of hotter and colder. There is a clear tendency for heat to flow from a hotter to a colder body. Clausius' formulation of the second law includes these concepts:

> It is impossible to devise an engine which, working in a cycle, shall produce no effect other than the transfer of heat from a colder to a hotter body.

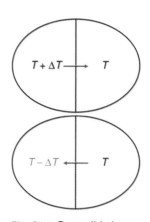

Fig. C1.1 Reversible heat transfer between two bodies.

Reversibility

Carnot demonstrated, following his experiments with heat engines, that the closest we can get to converting heat into work is by using an engine in which every process is reversible. Heat flow in such an ideal engine would be between bodies of essentially the same temperature (Fig. C1.1). If we cool one body just a little, heat will flow to it from the other; if now we heat the same body just a little, heat will flow the other way. The process is reversible. Similarly, we can imagine exerting work in a reversible manner on a volume of gas that is isothermal with its surroundings at a temperature T by moving a frictionless piston very slowly in one direction or the other so that heat Q flows in or out of the gas, while its temperature stays very close to T (Fig. C1.2).

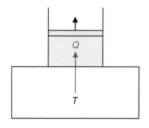

Fig. C1.2 Exerting work in a reversible manner on a volume of gas.

Entropy

Clausius expressed the tendency of an isolated system (one which cannot exchange energy or matter with its surroundings) to reach equilibrium by defining a function of state, the entropy S, which increases with time until it reaches a maximum at equilibrium

$$\Delta S \geq 0 \qquad (C1.3)$$

The equality (zero change in entropy) is valid in the limit of a reversible change, during which, in any case, the system is considered as being very close to equilibrium.

Exploring the consequences of the second law, it was found that entropy is related to the heat term in Eq. (C1.1) by

$$\Delta S = Q/T \qquad (C1.4)$$

Equation (C1.4) defines the units of entropy in terms of energy and temperature. We can now rewrite Eq. (C1.2) as

$$\Delta U = T\Delta S + \mu\Delta n - P\Delta V \qquad (C1.5)$$

Entropy and disorder

The example we gave above of molecules moving to fill the entire volume available to them as the system evolves towards equilibrium is a representation of an intuitive concept that the equilibrium state is a disorganised state. Its components

---

Comment C1.3   Boltzmann's constant and the definition of a temperature scale

The constant k_B has the same dimensions as entropy

$$k_B = 1.380\,650\,3 \times 10^{-23} \text{ J K}^{-1}$$

The energy associated per degree of freedom at the molecular level with a temperature T (in kelvins) is $(1/2)k_B T$.

On a mole scale Avogadro's number (the number of particles per mole)

$$N_A = 6.022\,141\,99 \times 10^{23} \text{ mol}^{-1}$$

multiplied by Boltzmann's constant is equal to the gas constant R

$$R = N_A k_B$$
$$R = 8.346\,24 \text{ J K}^{-1} \text{ mol}^{-1}$$

---

sample all possible configurations available to them. Boltzmann, in his development of statistical thermodynamics, expressed this concept quantitatively and related entropy, S, to the number of configurations, p, available to the system:

$$S = k_B \ln p \tag{C1.6}$$

where k_B is Boltzmann's universal constant (Comment C1.3).

Equation (C1.6) combined with the second law of thermodynamics (Eq. (C1.3)) indicates that at equilibrium the number of configurations sampled by the system is a maximum. Reconsidering Eq. (C1.4) in the light of Eq. (C1.6), we conclude that TS is the heat energy invested by the system in order to sample a number p of configurations at temperature T (in kelvins). We shall see in Section C1.3.1 that this energy is not 'free energy', i.e. it is not available to perform work.

C1.2.4 The third law and absolute zero

The Nernst heat theorem also called the third law of thermodynamics states that

> The entropy of any body is zero at the zero of the absolute temperature scale.

The classical formulation did not have a zero point for entropy (or temperature) and only *differences* in these quantities between one state and another were defined. The third law provided the zero point of an absolute entropy scale.

Since it has now been shown that there is no process in which entropy can become zero, alternative statements of the third law are:

By no finite series of processes is the absolute zero attainable.

and

As the temperature tends to zero, the magnitude of the entropy change in any reversible process tends to zero.

Absolute temperature

The law defines the absolute zero of the temperature scale. Absolute temperature is now expressed in kelvins. It is zero at absolute zero and rises in degree steps, where the degree is defined as one hundredth of the temperature difference between the freezing and boiling points of water at atmospheric pressure.

The third law plays a valuable role in physical chemistry. By setting the zero point for entropy and an absolute temperature scale it allowed the calculation of *equilibrium constants* from the thermal properties of reactants (Section C1.3.2), and of *total internal energy* in calorimetry (Section C1.3.3).

C1.3 Useful concepts and equations

Classical thermodynamics is built on a set of mathematically simple equations, which establish how different variables are related to each other. The equations can also be written in terms of variable changes by applying the rules of differential geometry. Maxwell's relations, for example, relate changes in the pressure P, volume V, entropy S and temperature T of a fluid system (Comment C1.4). In what follows, we shall concentrate on a few concepts from chemical applications of classical and statistical thermodynamics that are useful for the understanding of biological macromolecules and their interactions.

C1.3.1 Free energy and allied concepts

Enthalpy

The enthalpy H of a system is defined as

$$H = U + PV \qquad \text{(C1.7)}$$

The term enthalpy is from the Greek meaning 'warmth within'. Under constant pressure (a usual condition for solutions of biological molecules) the enthalpy of a system at temperature T represents the total energy required to heat the system from absolute zero to T.

The change in enthalpy, ΔH, during a chemical reaction is the heat absorbed or released in the breaking and formation of bonds. ΔH represents the experimentally measured quantity in calorimetry.

Comment C1.4   Maxwell's relations in thermodynamics

The following equations of thermodynamics are known as Maxwell's relations:

$$\left(\frac{\partial T}{\partial V}\right)_S = \left(\frac{\partial P}{\partial S}\right)_V$$

$$\left(\frac{\partial T}{\partial P}\right)_S = \left(\frac{\partial V}{\partial S}\right)_P$$

$$\left(\frac{\partial V}{\partial T}\right)_P = \left(\frac{\partial S}{\partial P}\right)_T$$

$$\left(\frac{\partial P}{\partial T}\right)_V = \left(\frac{\partial S}{\partial V}\right)_T$$

The first equality follows directly from energy conservation in a closed simple system

$$\Delta U = T\Delta S - P\Delta V$$

The others may be derived from the definitions of enthalpy and free energy. The equations relate entropy, a thermodynamic quantity that cannot be measured directly, to pressure, volume and temperature, all of which are measurable properties of a system.

Gibbs free energy

The Gibbs free energy G is defined as

$$G = U + PV - TS = H - TS \tag{C1.8}$$

G represents the part of the energy in a system that is not used for the population of the different entropy configurations. The next step in the reasoning is as follows: because the Gibbs energy is not involved in maintaining thermal agitation, it can be transformed into useful work, hence its name of *free energy*.

A reversible chemical reaction (Comment C1.5) can be written as

$$A \rightleftharpoons B \tag{C1.9}$$

At constant temperature (the system is maintained in isothermal contact with a large constant temperature bath, as is the case for molecules in dilute solution for example) the free energy released by the reaction is given by

$$\Delta G = \Delta H - T\Delta S \tag{C1.10}$$

The second law postulates that an isolated system evolves in order to maximise its entropy. Analogously, a system that exchanges energy with its surroundings

---

Comment C1.5 Chemical reactions involving biological macromolecules

A chemical reaction involves bonds of various types being made or broken between atoms. Folding and ligand binding interactions in biological macromolecules are chemical reactions in this general sense. Covalent bonds have associated energies of the order of $10^4 \, k_B T$ van der Waals bonds, hydrogen bonds, electrostatic bonds in aqueous solution, so-called hydration and hydrophobic bonds are all called weak bonds, because their associated energies are of the order of $k_B T$ for T close to room temperature, 300 K. Folding (a) and ligand (b) interactions may involve the formation and breakage of both covalent and weak bonds:

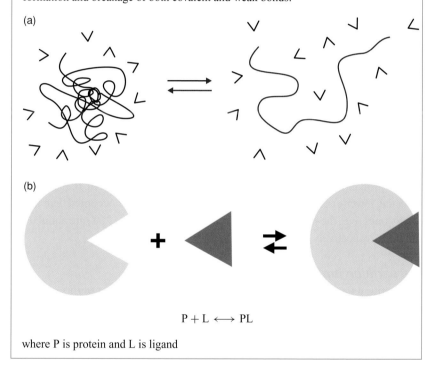

(a)

(b)

$$P + L \longleftrightarrow PL$$

where P is protein and L is ligand

---

evolves in order to maximise its free energy. Then, the spontaneous sense of the reaction in Eq. (C1.8) is in the direction for which $\Delta G < 0$. When two states of a system are at equilibrium, their free energy difference is zero, $\Delta G = 0$.

The Gibbs free energy is a very useful parameter in the analysis of binding events. It is dependent on the changes in enthalpy (ΔH) and entropy (ΔS) of the ligands upon binding. The enthalpy term reflects internal energy changes associated with bond formation and breakage, the entropy term the number of configurations available to the system.

Fig. C1.3 A free energy
diagram of a system as a
function of its
conformation. Two states
separated by a free
energy difference, ΔG, are
shown in the example.
The excess free energy
barrier, E^*, is called the
activation energy
between the two states.

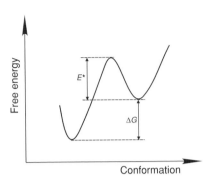

The free energy landscape and Boltzmann's distribution

When there are states of different energies available to a system, they are populated to different extents as a function of temperature (Fig. C1.3).

The probability that a state of free energy ΔG_j at temperature T is populated is given by Boltzmann statistics:

$$C_j = \exp(-\Delta G_j / RT) \tag{C1.11a}$$

so that

$$\Delta G_j = -RT \ln C_j \tag{C1.11b}$$

The probability of a state of energy RT is $\exp(-1)$ or about 0.36, and, as expected, lower-energy states have a greater likelihood to be occupied than high-energy ones (e.g. the probabilities for free energies, $RT/2$ and $2RT$, are about 0.60 and 0.13, respectively).

The partition function

The partition function Q is defined as the sum of the statistical weights of all the states accessible to the system

$$Q = \sum_{j=0}^{N} C_j = \sum_{j=0}^{N} \exp(-\Delta G_j / RT) \tag{C1.12}$$

All the thermodynamics variables of a system can be written in terms of Q. Equation (C1.12) defines the *canonical ensemble partition function* for a system containing a constant number of particles and a constant average energy.

Defining state $j = 0$ as the reference state with respect to which the thermodynamics variables are expressed as *excess* quantities, Eq. (C1.12) can be re-written as

$$Q = 1 + \sum_{j=1}^{N} \exp(-\Delta G_j / RT) \tag{C1.12a}$$

The population of state j, P_j, is given by the ratio of the statistical weight of that state over the sum of the statistical weights of all states accessible to the system

$$P_j = \exp\left(-\Delta G_j/RT\right)/Q \qquad \text{(C1.12b)}$$

The average free energy of the system, $\langle \Delta G \rangle$, is written

$$\langle \Delta G \rangle = \sum_{j=1}^{N} P_j \Delta G_j = -RT \ln Q \qquad \text{(C1.12c)}$$

The average excess enthalpy is

$$\langle \Delta H \rangle = \sum_{j=1}^{N} P_j \Delta H_j = RT^2 \left(\partial \ln Q / \partial T \right) \qquad \text{(C1.12d)}$$

and the average excess entropy is

$$\langle \Delta S \rangle = \sum_{j=1}^{N} P_j \Delta S_j = RT \left(\partial \ln Q / \partial T \right) + R \ln Q \qquad \text{(C1.12e)}$$

Equations similar to Eqs. (C1.12c)–(C1.12e) apply to any property of the system. In general, the experimental average value of a property α of the system given by

$$\langle \alpha \rangle = \sum_{j=1}^{N} P_j \alpha_j \qquad \text{(C1.12f)}$$

where α_j is the value of α associated with state j.

Equation (C1.12f) provides the basis for theoretical or model calculations of $\langle \alpha \rangle$ to compare with experimental observations.

Chemical potential

We recall from the discussion of Eq. (C1.2) that the chemical potential, μ, of a solute is the *energy* required to add one mole of molecules to the solution. This *energy* is in fact *free energy*, so that chemical potential can be equated to the free energy change resulting from the addition of one mole of solute to the solution. The chemical potential or change in free energy of a solution due to the dissolution of a solute to a concentration N_j (moles per unit volume) is given by

$$N_j \mu_j = \Delta G_j = -RT \ln N_j \qquad \text{(C1.13)}$$

Note that by comparing Eqs. (C1.11b) and (C1.13) we can equate N_j and C_j; the solution defined by the concentration N_j represents a state C_j of the system.

For a solution containing several different solutes

$$\Delta G = -RT \sum \ln N_j = -RT \ln N_1 N_2 N_3 \cdots = -RT \ln \prod N_j \qquad \text{(C1.13a)}$$

The biological standard state

Since we are dealing with differences in potentials or between energy levels, it is useful to define *standard states* with respect to which these differences are measured. The temperature T is taken as 298.15 K (25 °C) unless otherwise defined. Gases are at 1 atm partial pressure. For aqueous solutions, solutes have 1 M concentrations. When protons are involved in a reaction, the *biological standard state* is defined as pH 7.

Chemical reactions

We rewrite the reaction in Eq. (C1.9)

$$A_1 + A_2 + A_3 + \cdots \leftrightarrow B_1 + B_2 + B_3 + \cdots$$

where A_j are reactants and B_j are products. By using Eq. (C1.13a) we calculate the free energy change in the reaction as it proceeds from left to right, in terms of reactant and product concentrations.

$$\Delta G = -RT \ln \Pi[B] - RT \ln \Pi[A] = -RT \ln (\Pi[B]/\Pi[A]) = -RT \ln K \qquad \text{(C1.14)}$$

(where $K = [B_1][B_2][B_3]\cdots/[A_1][A_2][A_3]\cdots$).

It is a consequence of the definition that $K = 1$ for the standard conditions. The standard free energy ΔG° is defined as the free energy change when a reacting system starts at $K = 1$ and ends at equilibrium, $K = K_{eq}$, where there is no difference between the free energy of the products and reactants

$$\Delta G^\circ = -RT \ln K_{eq} \qquad \text{(C1.15)}$$

The free energy change for the reaction starting from any value of K is now written

$$\Delta G = \Delta G^\circ - (-RT \ln K) = \Delta G^\circ + RT \ln K \qquad \text{(C1.16a)}$$

or

$$\Delta G = -RT \ln K_{eq}/K \qquad \text{(C1.16b)}$$

Note that ΔG° is equal to zero only if $K_{eq} = 1$ (i.e. when concentrations are such that $[B_1][B_2][B_3]\cdots = [A_1][A_2][A_3]\cdots$).

If $K > 1$, ΔG is negative and the reaction occurs spontaneously in standard conditions (25 °C etc.). If $K < 1$, ΔG is positive and the reaction requires energy input in order to occur, under standard conditions.

Van't Hoff analysis

Equation (C1.15) provides the basis of a van't Hoff analysis for a two-state system ($A \rightleftharpoons B$). In such an analysis, standard free energy, and consequently enthalpy and entropy, are calculated from the measured equilibrium concentrations of the two states.

Standard free energy, enthalpy and entropy are related by Eq. (C1.10)

$$\Delta G^\circ = \Delta H^\circ - T\Delta S^\circ \qquad \text{(C1.17)}$$

Combining with Eq. (C1.15), we obtain $\ln K_{eq} = -\Delta H^\circ/RT + \Delta S^\circ/R$, from which we derive the van't Hoff equation

$$\Delta H_{v'tH} = RT^2 \partial \ln K_{eq} / \partial T \qquad \text{(C1.17a)}$$

C1.3.2 Binding studies

Binding constants and affinity

If molecule A binds to molecule B to form the complex A·B, the thermodynamic equilibrium dissociation constant, K_d, is defined in terms of the different species' concentrations, [A], [B] and [AB], at equilibrium

$$K_d = \frac{[A][B]}{[AB]} \qquad \text{(C1.18)}$$

Here, we have followed the binding experiment convention and written the equilibrium constant as the dissociation constant. Referring to Eq. (C1.14) and following $K_d = 1/K_{eq}$, *where the association constant, K_{eq}, is a measure of the binding 'strength' or affinity of the interaction.* The constant K_d is related to the chemical on-rate constant k_{on} for the association of A and B, and the chemical off-rate constant k_{off} for the dissociation of the complex A·B by (Comment C1.6):

$$K_d = k_{off} / k_{on} \qquad \text{(C1.19)}$$

Binding to many sites

If A has more than one site for the binding of B, different conditions may apply. The binding is said to be *cooperative* if the occupation of one site favours the occupation of another. The binding is *anticooperative* if, on the contrary, the occupation of a site causes a decrease in the binding affinity of another. These

---

Comment C1.6   Equilibrium and rate constants

Consider the single step binding reaction

$$A + B \leftrightarrow AB$$

The rate of the forward reaction, which converts A and B to AB is

$$k_{on} \times [A][B]$$

The rate of the reaction converting AB back to A and B is

$$k_{off} \times [AB]$$

Equilibrium is reached when these two rates are equal:

$$k_{on} \times [A][B] = k_{off} \times [AB]$$

Rearranging gives

$$k_{off} / k_{on} = [A][B]/[AB] = K_d$$

types of complex behaviour can be identified and analysed in binding experiments. In the case in which the sites are *identical and independent*, the dissociation constants for each site are identical and do not change as a function of whether or not other sites are occupied. If molecule A has n identical and independent binding sites for a *ligand* B, the *average number*, $\bar{n}$, of ligands actually bound per molecule A is given by

$$\bar{n} = \frac{n[B]}{K_d + [B]}$$ (C1.20)

where [B] is the free ligand equilibrium concentration. *Scatchard* analysis recasts this equation as

$$\frac{\bar{n}}{[B]} = \frac{n}{K_d} - \frac{\bar{n}}{K_d}$$ (C1.21)

Both the average number bound, $\bar{n}$, and the free ligand concentration, [B], are obtained from a binding experiment at equilibrium. Plotting $\bar{n}/[B]$ versus $\bar{n}$ then yields a straight line of slope $-1/K_d$ and the intercept on the *x*-axis is n, the number of sites. A Scatchard plot is shown in Chapter C4.

Binding free energy

Knowledge of K_d enables the calculation of the energetics involved in the interaction between a ligand and its receptor. The difference in Gibbs free energy between bound and unbound states can be attributed to various chemical and physical factors. We recall Eq. (C1.10) ($\Delta G = \Delta H - T \Delta S$) and that the reaction is spontaneous in the direction of decreasing free energy, i.e. $\Delta G < 0$. The enthalpy term, ΔH, reflects internal energy changes associated with the intermolecular bonding interactions. The entropy term, ΔS, provides a description of the degrees of freedom of the components involved in the binding reaction. Since, whether or not the interaction takes place depends on ΔG and not on the other terms separately, large negative (unfavourable) entropy changes can occur on binding if there are compensating enthalpy differences. These may reflect the formation of favourable atomic interactions, like salt bridges and hydrogen bonds.

Calculating standard enthalpy and entropy changes upon binding

We recall Eq. (C1.15) for the standard free energy of the reaction:

$$\Delta G^\circ = -RT \ln K_{eq}$$

in terms of K_d, is

$$\Delta G^\circ = RT \ln K_d$$ (C1.22)

Using Eq. (C1.10) and rearranging, we obtain

$$\ln K_d = -\frac{\Delta H^\circ}{RT} + \frac{\Delta S^\circ}{R}$$ (C1.23)

Measuring affinities as a function of temperature we obtain

$$\frac{\partial}{\partial T} (\ln K_{\mathrm{d}}) = -\frac{\partial}{\partial T} \left(\frac{\Delta H^{\circ}}{RT} \right) + \frac{\partial}{\partial T} \left(\frac{\Delta S^{\circ}}{R} \right) \tag{C1.24}$$

Since ΔH° and ΔS° are temperature-independent, Eq. (C1.24) reduces to

$$\frac{\partial}{\partial T} (\ln K_{\mathrm{d}}) = \frac{\Delta H^{\circ}}{R} \times \frac{1}{T^2} \tag{C1.25a}$$

which can be rewritten as

$$\frac{\partial (\ln K_{\mathrm{d}})}{\partial (1/T)} = -\frac{\Delta H^{\circ}}{R} \tag{C1.25b}$$

A plot of $\ln K_{\mathrm{d}}$ against $1/T$ (an Arrhenius plot) therefore gives a straight line with a slope proportional to the change in standard enthalpy upon binding. The associated change in standard entropy can then be determined from Eqs. (C1.22) and (C1.10).

Avidity

When the binding event is due to a network of smaller events occurring at the interface between the molecules, the overall binding strength is called the *avidity* and, analogously to affinity, is measured through the overall equilibrium constant. The avidity then results from a set of affinities for the different sites involved. The avidity, however, is larger than the sum of affinities (Comment C1.7).

An important consequence of the relationship between the standard ΔG and avidity is that a small change in one of the sites in the network can lead to a very large change in the overall equilibrium constant.

Comment C1.7 Avidity

The standard free energy of binding of a network of weak independent sites is given by

$$\Delta G = \sum_{n=1}^{k} \Delta G_n = \Delta G_1 + \Delta G_2 + \Delta G_3 + \cdots$$

and

$$\Delta G = \sum_{n=1}^{k} (-RT \ln K_{\mathrm{d},n}) = -RT \ln K_{\mathrm{d},1} - RT \ln K_{\mathrm{d},2} - RT \ln K_{\mathrm{d},3} - \cdots$$

where $K_{\mathrm{d},j}$ is the dissociation equilibrium constant of binding event j.

The overall K_{d} is called the avidity of the reaction. It is given by the product (and not the sum) of the individual affinities

$$K_{\mathrm{d}} = K_{\mathrm{d},1} \times K_{\mathrm{d},2} \times K_{\mathrm{d},3} \times \cdots$$

C1.3.3 Calorimetry and binding

The complete set of thermodynamics equations describing a system, in a given temperature range, can be calculated from the functional dependence of enthalpy on temperature, which can be obtained, in principle, from careful calorimetric measurements.

Calorimetry measures heat exchange during temperature-induced changes in a system. Temperature (T) and enthalpy (H) are conjugate extensive and intensive variables (Section C1.2.1). When a change in the state of a system (e.g. protein unfolding, or ligand binding) results from a change in temperature, there is a corresponding change in enthalpy. Heat is absorbed ($\Delta H > 0$) if the change results from an increase in temperature or heat is released ($\Delta H < 0$) if the change results from a decrease in temperature.

Heat capacity

The quantity measured in a calorimetry experiment on a macromolecular solution is the heat capacity at constant pressure, C_p

$$C_p = (\partial H / \partial T)_p \tag{C1.26}$$

The corresponding enthalpy and entropy functions, *H(T)* and *S(T)*, are obtained by integrating C_p in the temperature range:

$$\Delta H = H(T_0) + \int_{T_0}^{T} \Delta C_p(T) \mathrm{d}T \tag{C1.27a}$$

$$\Delta S = S(T_0) + \int_{T_0}^{T} (\Delta C_p(T)/T) \mathrm{d}T \tag{C1.27b}$$

We recall from Section C1.3.1 that the reference temperature, T_0, is equal to absolute zero in physics, so that under constant pressure the enthalpy of a system at temperature T represents the total energy required to heat the system from absolute zero to T. In biophysics, however, it is usual to take the reference temperature T_0 to be equal to room temperature (298.15K).

In the case of a reaction, the H and S terms in the equations above are replaced by ΔH and ΔS, corresponding to the respective *changes* during the reaction. The heat capacity change during a reaction at temperature, T, is given by

$$\Delta C_T = \Delta C_{p,T_0} + \int_{T_0}^{T} \left(\mathrm{d}\Delta C_{P,T} / \mathrm{d}T \right) \mathrm{d}T \tag{C1.28}$$

where the subscripts define temperature and pressure conditions.

The effects of temperature on binding

Subsuming the equations above and Eq. (C1.15), which relates the dissociation constant to standard free energy change, the standard free energy change can be written

$$\Delta G_T^\circ = \Delta H_{T_0}^\circ + T\Delta S_{T_0}^\circ + T\Delta C_{T_0}^\circ \left[1 - \frac{T_0}{T} - \ln\left(\frac{T_0}{T}\right)\right]$$

$$+ \tfrac{1}{2}TT_0\left(\frac{\mathrm{d}\Delta C_{T_0}^\circ}{\mathrm{d}T}\right)\left[\frac{T_0}{T} - \frac{T}{T_0} + 2\ln\left(\frac{T}{T_0}\right)\right]$$

(C1.29)

If the heat capacity is the constant in the temperature range the equation simplifies to

$$\Delta G_T^\circ = \Delta H_{T_0}^\circ + T\Delta S_{T_0}^\circ + T\Delta C_{T_0}^\circ \left[1 - \frac{T_0}{T} - \ln\left(\frac{T_0}{T}\right)\right]$$

(C1.30)

which is a form of the *van't Hoff equation*. $\Delta H_{T_0}^\circ$, $\Delta S_{T_0}^\circ$ and $\Delta C_{T_0}^\circ$ are estimated from measurements of a non-linear fit of ΔG_T° measured values at different temperatures.

C1.3.4 Activation thermodynamics

The thermodynamic path of a reaction or conformational change often includes an energy barrier that must be overcome. Whereas the reaction or change will occur if the equilibrium ΔG is favourable ($\Delta G < 0$), the rate at which it occurs depends on the height of the intervening energy barrier (Fig. C1.4). The time constants of a reaction or change observed at different temperatures provide a measure of the thermodynamics parameters involved.

The temperature dependence of the rate constant, k, is given by the Arrhenius equation

$$k = A\exp\left(\frac{-E^*}{RT}\right)$$

(C1.31)

where A is a constant. In *statistical thermodynamics*, A includes the probability of the change taking place (e.g. the probability of collision between two interacting partners), and E^* is activation energy. In *transition state theory*, A depends on the *activation entropy* and E^* is the *activation enthalpy*. A is given in the quantum mechanical *Eyring transition state model* as

$$A = \frac{k_B T}{h}\exp\left(\frac{\Delta S^*}{R}\right)$$

(C1.32)

where k_B is Boltzmann's constant, h is Planck's constant and ΔS^* is activation entropy. Combining Eqs. (C1.31) and (C1.32), and putting $E^* = \Delta H^*$, the activation enthalpy, yields

$$\ln\left(\frac{k}{T}\right) = \ln\left(\frac{k_B}{h}\right) + \frac{\Delta S^*}{R} - \frac{\Delta H^*}{RT}$$

(C1.33)

Fig. C1.4 Free energy profile of a reaction or conformational change. E^* is the activation energy or activation enthalpy depending on the theory used (see text).

Activation enthalpy and activation entropy can be obtained directly from the slope and intercept of the line $\ln(k/T)$ versus $1/T$ (called an *Eyring plot*).

C1.4 Checklist of key ideas

- Classical thermodynamics is a phenomenological science, concerned with precise observations on defined systems.
- Macromolecular solutions can be considered as thermodynamic systems.
- An *adiabatic* system is perfectly isolated from external effects; an *isothermal* system is in perfect thermal contact with its surroundings.
- Two systems are in identical *states* when all the values measured for their properties are identical.
- A system enclosed in an adiabatic vessel will, in time, reach a state in which its properties no longer evolve; it is then said to be in *equilibrium*.
- The zeroth law of thermodynamics states that if two bodies, A and B, are separately in equilibrium with a third body, C, then each of A and B is in equilibrium with the other.
- The first law of thermodynamics states that the amount of work needed to change the state of a system from an initial to a final state, depends solely on the change accomplished, and not on the means by which the work is performed, nor on the intermediate states by which the system may pass.
- The second law of thermodynamics states that it is impossible to devise an engine which, working in a cycle, produces no effect other than the transfer of heat from a colder to a hotter body.
- The *entropy* of a system is a function of state, which increases with time until it reaches a maximum at equilibrium.
- The entropy is a direct function of the number of different configurations (disorder) available to the system.
- The heat energy invested by a system in order to sample all the configurations available to it at a given temperature is equal to its entropy multiplied by the temperature.
- The third law of thermodynamics states that the entropy of any body is zero at the zero of the absolute temperature scale.
- The third law, by setting the zero point for entropy and an absolute temperature scale (the Kelvin scale), allows the calculation of *equilibrium constants* from the thermal properties of reactants, and of *total internal energy* in *calorimetry*.
- The *enthalpy* of a system at temperature T represents the total energy required to heat the system from absolute zero to T.
- The change in enthalpy during a chemical reaction is the heat absorbed or released in the breaking and formation of bonds.
- The *Gibbs free energy* represents the part of the energy in a system that is not used for the population of the different entropy configurations.
- A system that exchanges energy with its surroundings evolves in order to maximise its free energy.
- When two states of a system are in equilibrium, their free energy difference is zero.

- All the thermodynamic variables of a system can be written in terms of its partition function, which is defined as the sum of the statistical weights of all the states accessible to the system.
- The *chemical potential* of a solute is the *free* energy required to add one mole of molecules to the solution.
- The *biological standard state* is defined at a temperature of 298.15 K (25 °C), 1 atm partial pressure for gases, 1 M concentration for solutes in aqueous solution and pH 7.
- The equilibrium constant of a chemical reaction is equal to the mathematical product of reaction product concentrations divided by the mathematical product of reactant concentrations, when the free energies of products and reactants are equal.
- The van't Hoff equation relates the enthalpy change to the equilibrium constant of a chemical reaction.
- The equilibrium constant of an association reaction is a measure of the binding strength or *affinity* of the interaction.
- Binding is said to be *cooperative* if the occupation of one site favours the occupation of another; binding is *anticooperative* if, on the contrary, the occupation of a site causes a decrease in the binding affinity of another; sites are *identical and independent*, if the dissociation constants for each are identical and do not change as a function of whether or not other sites are occupied.
- A Scatchard plot relates the number of binding sites to the dissociation constant and free ligand concentration.
- Equilibrium thermodynamic parameters can be obtained from the Arrhenius plot relating dissociation constants to temperature.
- For a binding event due to a network of smaller events occurring at the interface between the molecules, the overall binding strength is called the *avidity* and, analogously to affinity, is measured through the overall equilibrium constant.
- A calorimetry experiment on a macromolecular solution measures the heat capacity at constant pressure.
- The corresponding enthalpy and entropy functions in calorimetry are obtained by integrating the heat capacity and the heat capacity divided by the temperature, respectively, in the temperature range of the experiment.
- The rate of a reaction or conformational change depends upon the height of the activation energy barrier in the path between the higher and lower free energy states; activation thermodynamics parameters can be obtained from the Eyring plot relating rate constants to temperature.

Suggestions for further reading

Feynman, R. P., Leighton, R. B., and Sands, M. (1963). *The Feynman Lectures on Physic*, Volume 1. Reading, MA: Addison Wesley.

Price, N. C., Dwek, R. A., Ratcliffe, R. G., and Wormald, M. R. (2001). *Principles and Problems in Physical Chemistry for Biochemists*. Third edition. Oxford: Oxford University Press.

Chapter C2
Differential scanning calorimetry

C2.1 Historical overview

The application of differential scanning calorimetry (DSC) to protein studies dates from the **mid-1960s**, when the first instruments with sufficient sensitivity were developed independently by **P. L. Privalov, S. J. Gill** and **J. M. Sturtevant**. Effort was put in this development, because a number of problems in biophysics had arisen that could not be solved without measuring small heat effects in relatively dilute samples. In the following decades, experiments allowing the measurement of thermodynamics variables with progressively greater precision became possible with the development of ever more sensitive calorimeters. Microcalorimeters and, with the approach of the year **2000**, nanocalorimeters became available (the micro and nano prefixes referring to the precision with which heat can be measured in these devices).

C2.2 Basic theory

A full quantitative thermodynamics description of a macromolecule in solution for a given temperature range may be obtained from the DSC measurement of its partial specific heat capacity at constant pressure, C_p, as a function of temperature. We recall from Chapter C1

$$\Delta H = \int \Delta C_p(T)\,dT \qquad (C2.1a)$$

$$\Delta S = \int \frac{\Delta C_p(T)}{T}\,dT \qquad (C2.1b)$$

where ΔC_p is the heat capacity change due to a given macromolecular process (e.g. protein unfolding) and ΔH, and ΔS are the associated enthalpy and entropy changes, respectively.

C2.3 Experimental considerations

C2.3.1 Instrument specifications

The general specifications of a DSC instrument for studies of biological macromolecules can be summarised in three points:

(1) very high sensitivity and precision;

(2) an operational range from negative to positive temperatures;

(3) the possibility of scanning down as well as up in temperature.

Modern instruments of different makes are available commercially. Typical specifications include macromolecule requirements of less than 50 μg per test, a temperature range from −10 °C to 130 °C, ∼10 μJ/deg sensitivity, and, of course, powerful and user-friendly computer control to pilot the experiments and for on-line analysis.

The reversibility of processes should be checked by scanning down as well as up in temperature, with precautions taken to avoid irreversibility due to secondary effects such as aggregation or chemical degradation at high temperatures.

Calorimetric measurements down to −10 °C (in supercooled solutions) are not trivial but they are of special interest because they allow certain processes, such as cold denaturation, the folding of an α-helix or of a single-stranded oligonucleotide, to be followed to completion. Measurements below −10 °C are not possible in practice because of the high probability of spontaneous freezing in the aqueous solutions at these temperatures.

Measurements above 100 °C are equally difficult to perform; they require pressure cells and special precautions in sample preparation and experimental design. The heating rate could be increased, for example, to minimise exposure time to high temperatures. High-temperature measurements, up to about 130 °C (above this temperature the chemical degradation of biological molecules is too fast), are important in order to study the proteins and complexes of hyperthermophile organisms, which are stable and active beyond 100 °C (see Chapter A1).

C2.3.2 Sensitivity of heat capacity measurements

The variable, $C_p(T)_{obs}$, measured in DSC is *the heat capacity difference at constant pressure as a function of temperature between a macromolecular solution and the same volume of solvent alone*. The heat capacity of the solvent being greater than that of an equal volume of solution, the observed heat capacity difference is negative. It is related to the partial specific heat capacity of the macromolecule by

$$C_p(T)_{obs} = C_p(T)_{macrom.} m(T)_{macrom.} - C_p(T)_{solv.} \Delta m(T)_{solv.} \qquad (C2.2)$$

where $C_p(T)_{macrom.}$ is the *partial specific heat capacity* (i.e. the heat capacity per mass of substance) of the macromolecular solute at temperature T, $m(T)_{macrom.}$ is the mass of macromolecule in the measuring cell at temperature T, $C_p(T)_{solv.}$ is the partial heat capacity of the solvent at temperature T, and $\Delta m(T)_{solv.}$ is the mass of solvent displaced by the presence of mass m of macromolecules. The mass of displaced solvent can be calculated from the partial specific volumes of

---

Comment C2.1 Partial heat capacity contributions

The signal is very small in DSC experiments on protein solutions, and highly sensitive and precise instrumentation is required for its determination. We calculated values for the protein and solvent terms in Eq. (C2.2) based on a volume of 1 ml of aqueous solution containing 1 mg of macromolecule. We assumed a value of 2 J K^{-1} g^{-1} for the protein partial specific heat capacity (protein partial specific heat capacities lie between 1.3 and 2.3 J K^{-1} g^{-1}) and a value of 0.73 cm^3 g^{-1} for the protein specific volume; the partial specific heat capacity of water is of course 4.2 J K^{-1} g^{-1} (1 calorie per degree per gram) and its partial specific volume is 1 cm^3 g^{-1}.

$$C_p(T)_{macrom.} m(T)_{macrom.} = 2 \times 10^{-3} \text{ J K}^{-1}$$

$$C_p(T)_{solv.} \Delta_m(T)_{solv.} = 4.2 \times 10^3 \times 0.73/1 \text{ J K}^{-1}$$

$$C_p(T)_{obs} = -1 \times 10^{-3} \text{J K}^{-1}$$

The heat capacity measured for 1 ml of solution is $\sim$4 J K^{-1} g, so that the heat capacity difference due to the macromolecules is $\sim 10^{-3}$ that of the solvent alone. An instrumental sensitivity of 10^{-6} is required in order to have a measurement with 0.1% precision.

---

the macromolecule and solvent, $\overline{v}(T)_{macrom.}$, $\overline{v}(T)_{solv.}$, respectively:

$$\Delta m(T)_{solv.} = m(T)\overline{v}(T)_{macrom.}/\overline{v}(T)_{solv.} \tag{C2.3}$$

By combining Eqs. (C2.2) and (C2.3), and putting in typical values for the parameters, we see that the observed difference in heat capacity is very small compared with the heat capacity measured for the solution (Comment C2.1).

C2.3.3 Sample requirements

Macromolecular sample solutions for calorimetry experiments should be well-characterised, very pure and homogeneous. It has been estimated that no more than 3% of contaminants can be tolerated for quantitative studies. Similarly, the absolute concentration should be known with a precision of a few percent. The solvent should be chosen such that the macromolecules remain soluble in the temperature range of the study, the macromolecular states to be studied are attainable and transitions between these states are reversible. The solution should be sufficiently dilute to reduce the risk of aggregation.

In the decades when the fundamental thermodynamics of protein folding was being established, DSC sample cells had volumes of the order of 1 ml; sample macromolecular concentrations were of the order of 0.1–1 mg ml^{-1}, depending on the sensitivity required. Since then, instruments have attained greatly improved characteristics by integrating hardware and software innovations, and measurements on samples of as little as 50 μg of protein have become possible.

C2.4 The heat capacity of proteins

C2.4.1 The heat capacity versus temperature curve

A molar heat capacity as a function of temperature curve, obtained for a protein in a DSC experiment, is illustrated in Fig. C2.1. The native state is assigned as the reference relative to which the thermodynamic parameters of the system are defined (see Chapter C1), so that the excess heat capacity, ΔC_p, becomes the significant parameter measured from the curve in Fig. C2.1. Contributions to ΔC_p can be split into two terms

$$\Delta C_p(T) = \Delta C_p(T)_{tr.} + \Delta C_p(T)_{bl.} \qquad (C2.4)$$

The first term refers to the transition peaks in the scan, while the second describes the baseline shift associated with the transition (the sigmoid shift between the two dashed lines in Fig. C2.1).

C2.4.2 Partition function analysis of the heat capacity curve

In the analytical approach described in Chapter C1, the value of a thermodynamic variable, $\langle \alpha \rangle$, is expressed as a sum of contributions from the different states of the system, which are themselves expressed in terms of a partition function (Eqs. (C1.12f), (C1.12b))

$$\langle \alpha \rangle = \sum_{j=0}^{N} P_j \alpha_j$$

$$P_j = \exp(-\Delta G_j / RT)/Q$$

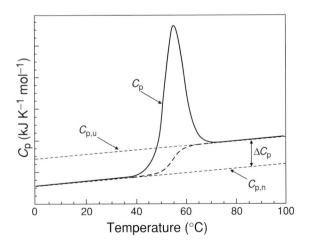

Fig. **C2.1** Molar heat capacity, C_p, of a protein observed as a function of temperature by DSC $C_{p,u}$ and $C_{p,n}$ are the molar heat capacities of the unfolded and native states, respectively. ΔC_p is the heat capacity change associated with the transition. When there are many states above the main transition temperature, the excess heat capacity is expressed as an average over all these states, $\langle \Delta C_p \rangle$.

where Q is the partition function and P_j, α_j are, respectively, the values of the relative population and thermodynamic variable associated with state j. If state $j = 0$ is taken as the reference native state, then the sum is from $j = 1, \ldots, N$; and $\langle \alpha \rangle = \langle \Delta C_p \rangle$ represents the excess heat capacity associated with the population of states other than the native state. Note that in a rigorous analysis the $\langle \rangle$ brackets should have been included for the terms in Fig. C2.1 and Eq. (C2.4), since in general there could be an ensemble of denatured states for the protein.

The excess heat capacity is given by the gradient of the excess enthalpy versus temperature curve (the inverse of Eq. (C2.1a))

$$\langle \Delta C_p \rangle = (\partial \langle \Delta H \rangle / \partial T)_P \qquad \text{(C2.5)}$$

Splitting the right-hand side of Eq. (C2.5), in a similar way to Eq. (C2.4) and expressing the terms as sums over the different states, we obtain

$$\langle \Delta C_p \rangle = \sum_{j=1}^{N} \Delta H_j (\partial P_j / \partial T)_P + \sum_{j=1}^{N} P_j \Delta C_{pj} \qquad \text{(C2.6)}$$

The first term (by analogy with the transition term, subscript tr, in Eq. (C2.4) is the heat capacity associated with the transitions from the native state $j = 0$ to the states, $j = 1, \ldots, N$. In a modelling approach (see below), this term may correspond to a calculated value, which would be compared with the experimentally measured $\Delta C_{p, \text{tr.}}$. The second term (by analogy with the baseline term, subscript bl, in Eq. (C2.4)) is the sum of the heat capacities of the states other than the native state, each weighted by its population, P_j.

The partition function analysis of the curve is quite general and applies to any monomeric (so that the number of moles present in the solution does not change upon unfolding) system at equilibrium, independently of transition mechanisms and of the number of intermediate states involved.

C2.4.3 Two-state transition: calorimetric and van't Hoff enthalpies are equal

The first DSC studies of small single-domain proteins established that their temperature unfolding could be modelled adequately as a cooperative two-state transition. The native structure unfolds very quickly so that only native and unfolded states coexist in the temperature range of the transition (see the figure in Comment C1.4). Any transient intermediates involved would be too short-lived to be observed through their contribution to the heat capacity curve.

The modelling approach is based on a comparison of the calorimetric transition enthalpy, integrated from the DSC curve, with a value calculated using the van't Hoff equation (Chapter C1) assuming a two-state transition. If the two values are found to be the same then the transition can be considered to be between two states with negligible intermediates.

The experimental calorimetric transition enthalpy is given by

$$\Delta H_{\text{tr.,exp.}} = \int \Delta C_{p,\text{tr.,exp.}}(T) \, dT \qquad \text{(C2.7)}$$

Comment C2.2   Equilibrium constants and thermodynamics variables

We recall from Section C1.3.1 that

$$\Delta G = -RT \ln K$$

Combining with the fundamental relation

$$\Delta G = \Delta H - T\Delta S$$

we obtain

$$\ln K = -\Delta H/RT + \Delta S/R$$

from which we derive the van't Hoff equation

$$\Delta H_{v'tH} = RT^2 \partial \ln K / \partial T$$

(ΔH is the slope of $\ln K$ versus $1/RT$).

---

where $\Delta C_{p,tr,exp}(T)$ (Eq. (C2.4) and Fig. C2.1) is the value in the transition peak above the $\Delta C_{p,bl,exp}(T)$ experimental baseline value.

The van't Hoff equation relates enthalpy and the temperature-dependent equilibrium constant between the states (Comment C2.2):

$$\Delta H_{v'tH} = RT^2 \partial \ln K_{eq}/\partial T \qquad (C2.8)$$

which is the slope of $\ln K$ versus $1/RT$. We recall from Chapter C1 that for a two-state transition

$$A \rightleftharpoons B$$

the equilibrium constant at temperature T is given by

$$K(T) = \frac{[B(T)]}{[A(T)]} \qquad (C2.9)$$

where $[B(T)]$, $[A(T)]$ are equilibrium concentrations. The equilibrium constant, $K(T)$, can therefore be measured from whichever experiment will give the ratio $[B(T)]/[A(T)]$, as a function of temperature. For example, $K(T)$ can be calculated from circular dichroism spectroscopy, if the signals from states A and B are significantly different as is the case for the unfolding transition in proteins (see Chapter E4).

It is interesting to note that within the two-state assumption the population ratio between the states can be obtained from the shape of the DSC curve itself (Comment C2.3).

If the calorimetric and van't Hoff transition enthalpies calculated from Eqs. (C2.7) and (C2.8), respectively, are equal, then the transition can be considered

---

Comment C2.3   Van't Hoff enthalpy from the DSC curve

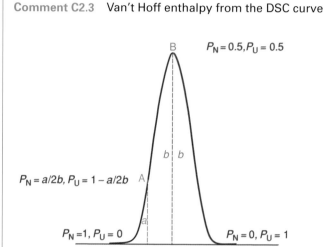

The figure represents a schematic heat capacity curve after baseline subtraction of the pure native and unfolded state contributions. The heat capacity represents an equilibrium measurement, i.e. the different states contributing to each point on the curve are at equilibrium with each other. The population ratio of states N and U, P_N/P_U, can be calculated for each temperature point on the curve. At the left-hand baseline level P_N and P_U are 1 and 0, respectively; at the right-hand baseline level they are 0 and 1, respectively. At the peak, B, of the curve, $P_N = P_U = 0.5$. At point A, for example, the ratio can be calculated from the ratio of the areas $a/2b$, where a is the area under the curve up to the blue dashed line, and $2b$ is the total area. By applying these geometrical considerations to the curve, it was shown that the van't Hoff enthalpy is given by

$$\Delta H_{v'tH} = 4RT^2 \frac{\Delta C_p(T_{tr})}{\Delta H_{tr}}$$

where $\Delta C_p(T_{tr})$ is the heat capacity peak value at the transition temperature T_{tr} and ΔH_{tr} is the total heat absorbed in the transition peak.

---

to be cooperative between two states. In practice, this means that any intermediate state would be too short-lived to be detected.

C2.4.4 Calorimetric and van't Hoff enthalpies are not equal: cooperative domains

Useful information on the transition is also obtained when the van't Hoff and calorimetric enthalpies are not equal. The calorimetric transition enthalpy measured by DSC represents the actual heat change associated with the transition

Comment C2.4   Calorimetric and van't Hoff enthalpies for
monodomain, two-domain macromolecules, and dimers

[t]

$$\Delta H_{vt'H} = \Delta H_{tr}$$

$$\Delta H_{vt'H} = 0.5\,\Delta H_{tr}$$

$$\Delta H_{vt'H} = 2\,\Delta H_{tr}$$

per mole of the system. The validity of the van't Hoff analysis, on the other hand, depends on the calculated populations of states corresponding in fact to the true populations. The van't Hoff enthalpy reflects the heat associated with the transformation of one mole of cooperative unit within the system.

When the cooperative domain extends over the whole molecule the van't Hoff and calorimetric enthalpies are equal. When the cooperative domain is smaller than the molecule, as would be the case for a multidomain protein, the van't Hoff enthalpy is smaller than the calorimetric enthalpy.

A van't Hoff enthalpy that is larger than the calorimetric enthalpy is an indication that the cooperative domain is larger than the molecule, i.e. it provides evidence of intermolecular interactions such as oligomerisation.

In all cases where the van't Hoff and calorimetric enthalpies are not equal the transition proceeds *via* intermediate states (Comment C2.4).

C2.4.5 Folding intermediates and effects of mutations

Multidomain proteins

The partition function equations developed in Section C2.4.2 have been used as a basis to calculate computer-simulated heat capacity functions for transitions that involve more than two states.

Fitting simulated functions to experimental curves provides an efficient approach to the thermodynamics analysis of DSC data. The precision of DSC nanocalorimeters is such that very small deviations between simulated and

Fig. C2.2 Experimental
and simulated DSC scans
of hen egg white
lysozyme. The solid curve
is the experimental scan
at pH 2.5, in both panels.
The dashed lines are
computer simulated
curves assuming a
two-state transition (top)
and a three-state
transition (bottom).
(Privalov *et al.*, 1995.)
(Figure reproduced with
permission from Elsevier.)

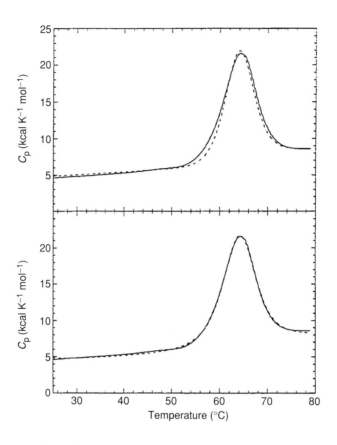

Comment C2.5
The Levinthal
paradox

The rate of polypeptide
folding to form unique
three-dimensional
protein structures
defined by the
sequence information
rules out the
possibility that folding
follows a stochastic
process in which the
huge number of
possible conformations
is sampled. This
statement is known as
the *Levinthal paradox*.

experimental curves can be observed and interpreted (Fig. C2.2). Hen egg white lysozyme is a two-domain protein and the data in Fig. C2.2 established that its temperature-induced unfolding is not a cooperative transition over both domains. The independent unfolding of the two domains was more pronounced in the related protein equine lysozyme, discussed in Section C2.4.7 (Fig. C2.5).

Are there folding intermediates in monodomain proteins?

Because it would take an essentially infinite time to sample all possible configurations, the folding process of even a small monodomain protein must be directed in some way by proceeding through a series of intermediate states of decreasing free energy (Comment C2.5). The *molten globule* has been suggested to be such an intermediate that can be stabilised under certain conditions. It represents the fast collapse of the chain into a compact liquid-like structure, from which it later reaches the correct native structure (see Chapter E4). The molten globule state would solve the Levinthal paradox by greatly reducing the configurational space search. The properties of the molten globule and its very existence, however, are controversial. Various authors have defined it in different ways, especially with respect to its dynamics and degree of structure. P. L. Privalov reviewed calorimetric data in the search for protein folding intermediates. He concluded that observed intermediate states were either misfolded forms obtained under conditions inappropriate for folding or partially unfolded forms that retained a correctly folded subpart of the molecule. These subparts have all the characteristics of cooperative folding domains. Small monodomain proteins were found to fold very rapidly with any transient intermediates being too unstable to be observed.

Mutant stability

Changes in protein structure resulting from the mutation of as few as one or two amino acid residues may result in significantly different experimental DSC curves (Fig. C2.3). It is tempting to assign a change in protein stability due to a mutation to enthalpic or entropic contributions of the amino acid replacement itself. Such an assignment should be done with great care, however, because the stability change may very well be due to a change in cooperativity of the unfolding process. Unlike other measures of protein stability, the DSC curve is extremely sensitive to changes in cooperativity. It can indicate, for example, that the amino acid change has significantly affected domain–domain interactions within the protein. Note how the serine to lysine replacement in T_4 lysozyme has not only destabilised the protein (it unfolds at a lower temperature than the wild type)

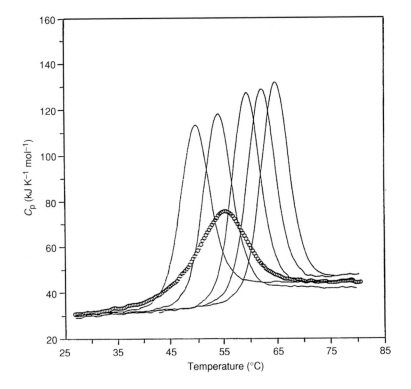

Fig. C2.3 DSC scans of T_4 lysozyme. The solid lines are the data for pH 2.8, 3.0, 3.3, 3.5, 3.7 (left to right). The circles are the data points for the S44A mutant of the proteinat pH 4.0. (Carra et al., 1996.) (Figure reproduced with permission from Biophysical Journal.)

but also broadened the temperature range of the unfolding showing it to be less cooperative (Fig. C2.3).

C2.4.6 Complex proteins

The heat capacity versus temperature curve of a complex protein structure can be separated into component curves corresponding to the transitions of the different states of the system. This is done without any a priori assumptions, by using computer algorithms based on recurrent procedures. The reliability of the curve fitting process (authors in the field call it 'deconvolution' but it is not a

Fig. C2.4 Calorimetric analysis of myosin. Top panel: Schematic view of myosin. The hatched parts are susceptible to complete proteolysis. LMM and HMM (light and heavy meromyosin, respectively) are the main fragments obtained by trypsin or pepsin treatment. The HMM fragment is further split by papain as shown. LMM is further split by trypsin into the fragments shown. The numbered curly brackets denote the cooperative blocks revealed by the calorimetric analysis shown in the bottom panel. Bottom panel: The partial heat capacity functions of myosin fragments in 0.5 M KCl, 25 mM potassium phosphate buffer, pH 6.5: (a) trypsin fragment LMM; (b) pepsin fragment LMM; (c) LF-3; (d) TR, the total rod. The two-state transition fitted peaks have been numbered according to increasing stability and assigned to the structure as shown in the top panel. (Privalov, 1982.) (Figure reproduced with permission from Elsevier.)

deconvolution in the strict sense) depends mainly on the precision of the experimental heat capacity curve, and, therefore, on the microcalorimetry instrumentation. Ideally, each component curve corresponds to a two state transition, for which it defines a transition temperature and transition enthalpy. Component curves may be assigned to structural units in the complex from model compound information, or by studying fragments, for example. Observing whether or not component transition temperatures and enthalpies vary independently with conditions such as pH provides information on the relationships between the structural units. The method has been applied successfully to multidomain proteins, protein–protein and protein–nucleic acid complexes and fibrous protein structures such as the muscle proteins and collagens. The calorimetric analysis of the melting process of the coil–coil structure of the myosin rod is shown as an example in Fig. C2.4. The complexity of the melting profile increases with increasing fragment size. All the profiles could be fitted by different combinations of the same six component peaks, each corresponding to a two-state transition. Each component was then assigned to a discrete cooperative unit in the structure, by estimating the approximate size of the cooperative block from the transition enthalpy (assuming a constant specific melting enthalpy along the rod). The analysis yielded very interesting results that could not be obtained by other methods. The least stable cooperative block (component 1) was close to the middle of the rod, precluding the rod melting by an 'unzippering' process. It is also noteworthy that the block corresponding to component 4 was apart from the others and located between the easily proteolised flexible regions (hatched in Fig. C2.4), which do not take part in the melting process.

C2.4.7 Solvent effects on the transition and the absolute partial heat capacity difference between folded and unfolded states of a macromolecule

The solvent plays an important role in the stabilisation of both the folded and native states of a protein (see Chapter C1), and in defining the nature of the transition between them. Protein–solvent interactions may be specific, as in the case of ion binding to specific sites, or more general through, for example, pH, ionic, or small solute effects on water structure affecting the hydration and solubility of protein polar and apolar groups.

Ion-binding stabilisation
Equine lysozyme is made up of two domains that unfold independently, giving two peaks in the DSC scan. One of the domains is stabilised significantly by calcium ion binding. The peak positions in temperature, their relative heights and widths all depend significantly on the $CaCl_2$ concentration in the solvent (Fig. C2.5).

Fig. C2.5 Ion binding to equine lysozyme: DSC temperature scans of equine lysozyme at pH 4.5, in the presence of various concentrations of CaCl$_2$: curve (1) 0.00 mM; (2) 0.10 mM; (3) 0.75 mM; (4) 1.50 mM. The dashed line is the heat capacity function calculated for the unfolded peptide, and the dot-dash line is that of the native state obtained by linear extrapolation, using the slope specific for native globular proteins. (See the Section C2.4.8.) (Griko *et al.*, 1995.) (Figure reproduced with permission from Elsevier.)

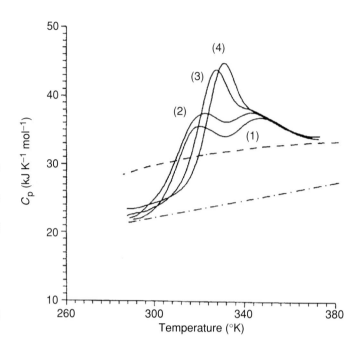

Fig. C2.6 Effects of pH: DSC temperature scans of barnase in solutions of different pH (indicated by each curve). The dashed and dot-dash lines, respectively, are the partial molar heat capacities of the native protein and unfolded proteins.
Inset: the solid line is the heat capacity versus temperature of irreversibly denatured barnase (obtained by heating the protein to 100 °C); the dashed line is the calculation for the heat capacity of the unfolded peptide obtained by summing the heat capacities of individual amino acid residues. (Privalov and Makhatadze, 1992.) (Figure reproduced with permission from Elsevier.)

Effects of pH

The unfolding temperature of barnase varies from below 25 °C at pH 1.8 to above 50 °C at pH 5.5 (Fig. C2.6). We recall that knowledge of $\langle \Delta C_p(T) \rangle$, the heat capacity difference between the native and unfolded states, provides a full characterisation of the thermodynamics of the transition. The pure native and unfolded states, however, are usually realised at different temperatures, so that $\langle \Delta C_p(T) \rangle$ can only be estimated by extrapolation or model calculation (see, for example, the dashed and dash-dot lines in Figs. C2.5 and C2.6). Solvent effects,

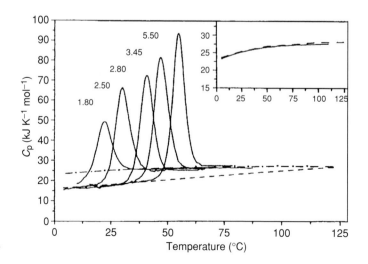

if they do not affect the heat capacity of the protein, could be very useful in this context. The pH variation illustrated in Fig. C2.6, by shifting the unfolding transition, allows the measurement of $\langle \Delta C_p(T) \rangle$ values at the same temperature.

C2.4.8 Heat capacity calculations from structural data

The heat capacity of macromolecules is due mainly to internal vibrational and rotational modes and solvent interactions. Reliable free energy predictions are an essential component of approaches to protein and ligand design. It is not possible, however, to predict the heat capacity of a complex system from *ab initio* considerations. Nevertheless, methods to calculate heat capacities of proteins in their various states have been developed, based on DSC information and data from model compounds. In turn, these methods have greatly enriched the analysis of DSC data, through the interplay between simulated and experimental curves.

Calculated heat capacities of native and unfolded proteins
The heat capacity at constant pressure of an unfolded peptide can be calculated accurately and reliably by summing the heat capacities of the individual amino acids (e.g. see the dashed line in Fig. C2.5 and the inset in Fig. C2.6). We would expect a more complicated situation, however, with respect to the native state.

The heat capacity of a native or unfolded protein in aqueous solution can be considered as being made up of a set of contributions each referring either to the intrinsic chemical properties of the macromolecule or to solvent effects. Extensive work on model organic compounds has confirmed that these contributions to the heat capacity are additive. The heat capacity of a protein can be written

$$C_p = C_{p,a} + C_{p,b} + C_{p,c} + C_{p,other} \qquad (C2.10)$$

$C_{p,a}$ is called the primary heat capacity and contains the atomic and covalent bond contributions. It depends only on the amino acid composition of the macromolecule. $C_{p,b}$ contains the contributions from non-covalent bonds in the folded protein. $C_{p,c}$ contains the contributions due to hydration or protein–solvent interactions, in general. $C_{p,other}$ contains other contributions, for example, those due to the protonation state of groups with ionisable side-chains such as histidine, aspartic acid, glutamic acid, arginine and lysine. It is usually small compared with the other terms in Eq. (C2.10), or can be corrected for quite easily, so that we shall neglect it in the following discussion.

Polypeptide chain energies: the primary heat capacity
The primary heat capacity, $C_{p,a}$, is the heat capacity of the anhydrous unfolded polypeptide. It can be calculated accurately by summing contributions from individual amino acids (Table C2.1) and from peptide bond formation (Comment C2.6).

Comment C2.6
Heat capacity of peptide bond formation

The average heat capacity contribution of peptide bond formation with the release of a water molecule was calculated from the difference between experimental heat capacities of dipeptides and the corresponding individual amino acids, all in the anhydrous state. It is equal to -39.2 ± 0.74 $J\,K^{-1}\,mol^{-1}$.

Table C2.1. *Experimental heat capacities of amino acids*

Amino acid	C_p[a] (J/Kmol)
Ala	122.7
Arg	234.4
Asn	160.9
Asp	155.8
Cys	163.0
Gln	184.9
Glu	175.7
Gly	99.5
His	216.2
Ile	189.0
Le	201.7
Lys	205.4
Met	290.6
Phe	203.7
Ser	136.1
Thr	147.8
Trp	239.0
Tyr	217.1
Val	169.5

[a] Values for anhydrous amino acids at 25 °C are from Hutchens (1970).

   Heat capacities of amino acid residues in solution are tabulated in Makhatadze and Privalov (1990), Privalov and Makhatadze (1990, 1996).

Internal protein interactions: the heat capacity of an anhydrous folded protein

The heat capacity of an anhydrous folded protein is equal to $C_{p,a} + C_{p,b}$. When expressed in units per mass, the heat capacities of different soluble proteins turn out to be quite similar, with the following values: 1.17 kJ K^{-1} g^{-1} for the heat capacity calculated from the primary structure; 1.21 kJ K^{-1} g^{-1} for the experimental value for the anhydrous state at 25 °C; 1.46 kJ K^{-1} g^{-1} for the folded state in solution; and 1.96 kJ K^{-1} g^{-1} for the unfolded state in solution. Note also that the measured heat capacities of anhydrous proteins are very close to the calculated primary heat capacity values, showing that most of the absolute heat capacity originates in the covalent structure. In the case of the anhydrous protein,

$C_{p,a}$ accounts for 97.5% of the experimental heat capacity (see also Comment C2.6).

The heat capacity of anhydrous proteins has been shown to increase linearly with temperature, with a slope, 4.10×10^{-3} J K^{-2} g^{-1}, which is close to constant for all the proteins studied.

Heat capacity of a native protein in solution

According to Eq. (C2.10), the heat capacity of a native protein in solution is equal to $C_{p,a} + C_{p,b} + C_{p,c,\text{native}}$. It is larger than that of the anhydrous protein, revealing mainly the positive magnitude of the hydration term, $C_{p,c}$. At 25 °C, hydration accounts for about 15% of the total heat capacity of a native protein in solution. It is also responsible for the larger (linear) temperature dependence observed for that heat capacity (6.80×10^{-3} cal K^{-2} g^{-1}), compared to that of the anhydrous folded state (4.1×10^{-3} J K^{-2} g^{-1}).

The hydration term varies according to the protein. It reflects the role of the macromolecular surface composition in the solvent interactions, especially with respect to the accessible surface area of polar and apolar groups. The hydration contribution of each group has been established from model compound studies to be proportional to its solvent accessible area. The heat capacity of polar group hydration is negative, while that of apolar group hydration is positive. Qualitatively, therefore, the larger heat capacity of native proteins in solution compared to the anhydrous folded state is as expected from the fact that, on average, 55% of the surface of a soluble protein is made up of apolar groups.

The heat capacity of an unfolded peptide in solution

The heat capacity of an unfolded peptide in solution is equal to $C_{p,a} + C_{p,c,\text{unfolded}}$, where the hydration term $C_{p,c,\text{unfolded}}$ is the sum of the appropriate values for the individual amino acids in the protein sequence, N and C terminal groups. The partial heat capacities of amino acid residues in solution have been measured accurately and are well known over the temperature range 5–125 °C (see Table C2.1 for references).

The heat capacity of the unfolded state in solution is about 0.4 J K^{-1} g^{-1} larger than that of the native state (Comment C2.7). It is dominated by the hydration terms arising from the higher solvent exposure of apolar groups, rather than the lack of non-covalent interactions within the chain, in the unfolded state. Unlike that of the native state in solution, however, the heat capacity of the unfolded state in solution does not vary linearly with temperature. In the range 0–100 °C it is approximated well by a second order temperature polynomial.

Global fit of the heat capacity by a single mathematical function

A single mathematical function, based on Eq. (C2.10), has been derived for the heat capacity as a function of temperature of proteins in their various states

**Comment C2.7
Heat capacity values**

Expressed in units per gram, the average experimental heat capacity of the anhydrous folded proteins is 1.22 J K^{-1} g^{-1}, their average primary heat capacity is equal to 1.19 J K^{-1} g^{-1}, so that the non-covalent interaction term is estimated to contribute only about 0.03 J K^{-1} g^{-1}.

The heat capacities of the native and unfolded states in solution are larger than the values corresponding to the anhydrous states. Because of the positive contribution of apolar capacity, values vary from protein to protein according to surface composition (for the native states) and amino acid composition (for the unfolded states). The difference between the heat capacities of the unfolded and native states in solution ranges between 0.4 and 0.7 J K^{-1} g^{-1}.

Comment C2.8
Protein heat
capacity
calculations

Heat capacity
calculations for
macromolecules based
on composition,
hydration values and
structural parameters
such as the solvent
accessible surfaces of
different groups have
been developed
essentially by the
groups of Privalov and
collaborators and
Freire and
collaborators
(reviewed in Privalov
and Privalov (2000)
and Freire (1995)).

(Comment C2.8). We recall Eq. (C2.10), neglecting the last term,

$$C_p = C_{p,a} + C_{p,b} + C_{p,c} \tag{C2.11}$$

and express each of the heat capacities on the right-hand side in terms of a number of parameters (v, w, p, q, a, b) and protein structural properties (M_W, molecular weight; A_{buried}, the total buried area inaccessible to the solvent; A_{apolar} and A_{polar}, the apolar and polar solvent-accessible areas, respectively):

$$C_{p,a} = [v + w(T - 25)]M_W \tag{C2.12}$$

$$C_{p,b} = [p + q(T - 25)]A_{buried} \tag{C2.13}$$

$$C_{p,c} = a(T)A_{apolar} + b(T)_{Apolar} \tag{C2.14}$$

where

$$a(T) = a_1 + a_2(T - 25) + a_3(T - 25)^2 \tag{C2.14a}$$

$$b(T) = b_1 + b_2(T - 25) + b_3(T - 25)^2 \tag{C2.14b}$$

The equations above were fitted to the entire protein thermodynamics database available. The best values obtained for the parameters are given in Table C2.2.

The resulting unique set of parameters gives remarkably good predictions, in most cases, not only of the measured heat capacity of a given protein state, but also of its temperature dependence – in particular, the almost linear and quadratic temperature curves of the native state and unfolded states in solution, respectively.

Table C2.2. *Parameters representing the best fit to the heat capacities of the protein data base*

Parameter	Value (units)
v	1.18 (J K^{-1} g^{-1})
w	41.0 × 10^{-4} (J K^{-2} g^{-1})
p	36.5 × 10^{-3} (J K^{-1} mol^{-1} Å^{-2})
q	27.0 × 10^{-4} (J K^{-1} mol^{-1} Å^{-2})
a_1	1.89 (J K^{-1} mol^{-1} Å^{-2})
a_2	11.0 × 10^{-4} (J K^{-2} mol^{-1} Å^{-2})
a_3	−17.6 × 10^{-5} (J K^{-3} mol^{-1} Å^{-2})
b_1	−1.11 (J K^{-1} mol^{-1} Å^{-2})
b_2	12.0 × 10^{-4} (J K^{-2} mol^{-1} Å^{-2})
b_3	18.1 × 10^{-5} (J K^{-2} mol^{-1} Å^{-2})

The heat capacity of proteins in different states is expressed as a single mathematical function in terms of parameters that could be fitted from experimental data (Eqs. (C2.12)–(C2.14)). (Gomez *et al.*, 1995.)

Although the average parameter values given in Table C2.2 appear to be good approximations in most cases, it should be pointed out that hydration heat capacities do vary considerably according to the chemical group. A better approximation is obtained by dividing apolar groups further into aliphatic and aromatic groups; the latter, in fact, have a slightly polar character. The corresponding average parameters at 25 °C are 2.14 J K^{-1} mol^{-1} Å^{-2} for aliphatic groups and 1.55 J K^{-1} mol^{-1} Å^{-2} for aromatic groups, so that the parameter a_1 (given above as an average for aliphatic and aromatic groups) can itself be divided accordingly into two terms.

Heat capacity changes upon protein folding

The heat capacity difference between the unfolded and native states in solution is dominated by the difference in hydration heat capacity:

$$\Delta C_{p,\text{unfolded-native}} = \Delta C_{p,b,\text{unfolded-native}} + \Delta C_{p,c,\text{unfolded-native}} \qquad (C2.15)$$

where $\Delta C_{p,\text{unfolded-native}}$ is the difference between the heat capacities of the unfolded and native states in solution.

The $\Delta C_{p,c}$ term accounts for more than 90% of ΔC_p at 25 °C. On average, the contribution of apolar hydration decreases by a half as the temperature increased from 25 °C to 100 °C, from 1.89 J K^{-1} mol Å^2 at 25 °C to about 0.96 J K^{-1} mol Å^2 at 100 °C, while that of polar hydration goes from -1.11 J K^{-1} mol Å^2 at 25 °C to practically zero at 100 °C. The $\Delta C_{p,b}$ term is negative because non-covalent interactions are disrupted in going from the native to the unfolded state. Its magnitude is smaller than the hydration term; it increases with temperature to an average value of -0.25 J K^{-1} (mol buried)$^{-1}$ Å^{-2} at 100 °C. The net effect of hydration and non-covalent interactions on the temperature dependence of the ΔC_p of unfolding is that its extrapolated value tends to vanish as the temperature approaches 120 °C (Fig. C2.7).

Reliability of the calculations

Because of the importance of free energy predictions in protein and ligand design, accurate, structure-based, heat capacity calculations are extremely useful. Such calculations can be performed reliably for a completely unfolded protein in solution (so that all the amino acids are solvent accessible), based on its sequence and tabulated values for the individual amino acids. The heat capacity of a *native* protein in solution cannot be calculated from its sequence, because it depends on the macromolecular surface and dynamics of the folded structure. When the native structure is known, however, a parametrised approach, as described above, could be a very good approximation (Fig. C2.7), when the protein belongs to the same structural family (e.g. small monodomain) as the proteins in the database utilised to derive the parameters. In principle, however, such calculations should be treated with care, because the heat capacities of individual chemical groups may vary significantly from best-fit values obtained for all the proteins in a given

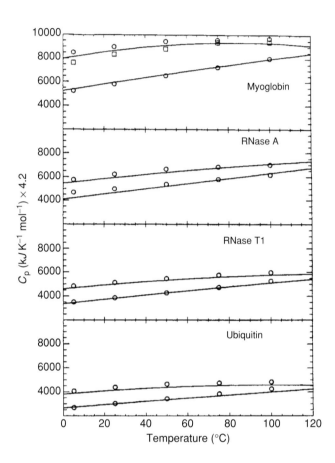

Fig. C2.7 Calculated (lines) and measured heat capacities (small circles) of the unfolded (top curve in each panel) and native states (lower curve) of different soluble proteins. The calculations were peformed by using the parameters in Table C2.2. Data points for the native state above 50 °C were extrapolated. In the case of unfolded myoglobin the squares are calculated values taken from Privalov and Makhatadze (1990). (From Freire, 1995.) (Figure reproduced with permission from Elsevier.)

database. The difference in the chain dynamics heat capacity is expected to be small between the native and unfolded states, so that calculations of the heat capacity *difference* upon unfolding are more reliable than calculations on the native state, because they are dominated by hydration terms (Comment C2.7). The hydration heat capacity contribution of a given group is proportional to its solvent accessible area. In the unfolded state, the group is assumed to be totally accessible to solvent, while its solvent accessible area in the native protein can be determined accurately from the structure coordinates.

C2.4.9 Protein stabilisation forces

Protein stabilisation forces arise from the competition between various protein–protein and protein–solvent interactions. They favour the folded structure in a given temperature interval. Calorimetric studies have made fundamental contributions to our understanding of these forces. As we saw above, the ΔC_p between the unfolded and folded states is accounted for primarily by hydration interactions. Figure C2.7 shows its temperature dependence for small monodomain proteins, and how it tends to very small values at the highest temperatures. What does a full thermodynamic analysis of the experimental heat capacity in terms of entropy and enthalpy terms yield about stabilisation forces in protein folding?

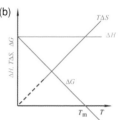

Chain folding in a vacuum

A straightforward thermodynamic analysis of protein folding in a vacuum (or for conditions in which solvent effects can be neglected) is shown in Fig. C2.8. The chain is maintained in its folded configuration by internal bonds such as hydrogen bonds, van der Waals' or electrostatic interactions. Breaking these bonds leads to a temperature-independent increase in internal energy, so that, upon unfolding, ΔH is positive and constant with temperature. Now, considering entropy, the unfolded state clearly has many more degrees of freedom than the folded one, so that ΔS is also positive upon unfolding. The free energy change, $\Delta G = \Delta H - T\Delta S$, upon unfolding therefore decreases with temperature and crosses zero at T_m, as shown in Fig. C2.8(b); the folded state is favoured above T_m, and the unfolded state is favoured below T_m, so that T_m can be seen as a chain melting temperature.

Fig. C2.8
Thermodynamics of protein folding in a vacuum: (a) schematic representation of the equilibrium, upon unfolding ΔH and ΔS are both positive; (b) the changes in enthalpy (ΔH), entropy (ΔS) and free energy (ΔG) upon unfolding. T_m is the melting temperature of the chain.

Solvent effects in protein folding

It has been clearly established, however, from careful calorimetric measurements that the simple picture shown in Fig. C2.8 is not applicable to proteins. The ΔH, ΔS and therefore ΔG upon unfolding calculated from the calorimetric heat capacity curve, all vary with temperature in complex ways establishing that solvent effects in protein folding are far from negligible.

The general shapes of thermodynamic functions derived from calorimetric data for myoglobin (Fig. C2.9 and Comment C2.9) are fairly typical of small globular proteins, although they may be shifted in temperature or magnitude according to the specific nature of each protein.

The native state is stabilised in a narrow temperature range by a positive unfolding free energy that peaks close to room temperature, falling off at both higher and lower temperatures. The maximum stabilisation free energy is interestingly small. At about 50 kJ mol^{-1} it corresponds, for example, to the enthalpy gained by breaking two or three hydrogen bonds in the protein interior (Table C2.3), when hundreds are involved in internal protein and protein–solvent interactions (Fig. C2.10 and Table C2.3). For comparison, the covalent bond has an associated energy of about 300 kJ mole^{-1}. With the exception of S–S bonds that can be formed between cysteine side-chains in non-reducing conditions, covalent bonds do not intervene in native fold stabilisation. Bonds involved in stabilisation of the native protein fold are weak bonds, with associated energies corresponding to a few kT.

Fig. C2.9 The temperature dependence of thermodynamic functions for myoglobin unfolding. The enthalpy, entropy and free energy differences upon unfolding calculated from the heat capacity as a function of temperature. The dashed lines correspond to the functions that would have been obtained in the absence of solvent, so that $\delta\Delta G$ is due to protein–solvent interactions. (Privalov and Khechinashvili, 1974.) (Figure reproduced with permission from Elsevier.)

Fig. C2.10 Protein folding in a solvent. Internal protein bonds, broken upon unfolding, could reform with solvent molecules. The unfolding enthalpy, therefore, could equally well be positive, negative or zero. The same is true for the unfolding entropy. The gain in chain configurational entropy could be more than compensated for by a loss in the configurational freedom of solvent molecules in contact with protein groups.

Comment C2.9 Relating ΔH, ΔS and ΔC_p

We recall that the enthalpy and entropy terms are calculated from the heat capacity measurements according to Eqs. (C2.1):

$$\Delta H = \int \Delta C_p(T)\, dT \qquad (C2.1a)$$

$$\Delta S = \int \frac{\Delta C_p(T)}{T}\, dT \qquad (C2.1b)$$

The converse of the equations is that ΔC_p is related to the slopes of the ΔH, and ΔS temperature dependence.

The hydrophobic effect

The stabilisation of the protein native state could be discussed in solubility terms. The protein folds in order to 'hide' as many as possible of the 'insoluble' apolar groups in the chain from solvent contact. In this sense, the observation that the unfolding free energy has a maximum close to 30 °C is reminiscent of the anomalous temperature dependence of the solubility of apolar compounds (see Fig. A1.7).

The solubility of apolar compounds in water is minimum close to room temperature, corresponding to a maximum transfer free energy. The dominant term at this temperature is a large negative entropy change, which has been interpreted as arising from an ordering of water molecules around the apolar groups. The

Table C2.3. *Energies associated with protein stabilisation forces*

'Bond'	Energy (kJ mole^{-1})	Equivalent in *RT* at 300 K (~2.5 kJ mole^{-1})
Electrostatic in water at ~1 Å[a]	~20	~10 ×
Hydrogen bond[b]	~4 to 25	~2 × to 10 ×
Hydrophobic[c]	~4	~2 ×
Van der Waals	~0.5	~0.1 ×

[a] The Coulombic interaction between charged groups depends on the charge magnitudes, their separation and the dielectric constant of the medium; the energy given is for unit charges separated by 1 Å in water.
[b] The hydrogen bond energy depends strongly on the environment; it is minimum in water and maximum in a vacuum or in an otherwise apolar environment.
[c] The energy given corresponds to the approximate free energy change when a CH, CH_2, or CH_3 group is shielded from water contact. Note that the precise free energy changes are different for the three apolar group types.

structure of liquid water results in a large entropy due to the configurational freedom of each water molecule to make two hydrogen bonds with the nearest neighbours in all directions (see Chapter A2); this freedom is clearly reduced when the water molecule finds itself close to an apolar group that does not offer hydrogen bonding possibilities. This phenomenon is called the *hydrophobic effect*.

A system evolves in order to avoid the thermodynamically unfavourable exposure of apolar groups to water. The hydrophobic effect then results in an effective *hydrophobic force,* bringing together apolar molecules in aqueous solution, in order to minimise their water contact surfaces. A free energy loss of about 4 kJ mole^{-1} results from the burying of a CH, CH_2 or CH_3 group away from contact with water (Table C2.3). Although the hydrophobic effect is dominated by the negative entropy change of apolar group hydration, an enthalpic contribution favouring the van der Waals' packing of the apolar molecules in their pure liquid state is not negligible.

The hydrophobic effect has been suggested to be the main driving force behind protein stabilisation. Careful analysis of calorimetric data, such as those shown in Fig. C2.9, however, introduced considerable qualifications on its role in the stabilisation of the native protein fold.

The enthalpy and entropy of native protein stabilisation: apolar and polar group hydration and the compactness of the protein interior
We can see from the curves in Fig. C2.9 that, close to room temperature, the respective enthalpy and entropy changes due to protein–solvent interactions compensate to a large extent the enthalpy changes due to internal protein bonds and

Comment C2.10
Destabilisation of protein complexes at low temperature

The mechanisms of cold denaturation also act on protein complexes. It is well known that certain protein complexes dissociate if their sample solutions are put in a cold room. F_1-ATPase, the soluble part of the membrane-associated F_1F_0-ATPase, is an example of such a cold-sensitive structure. It can be concluded that hydrophobic forces dominate the protein–protein interactions within the complex.

entropy changes associated with chain configurational freedom. At the maximum value of unfolding ΔG, ΔS is zero, so that the native protein is stabilised by a small enthalpy term. At lower temperatures, ΔS is negative and the stabilisation is entropy driven, while at higher temperatures enthalpy terms dominate. Note how, in Fig. C2.9, $\Delta G'$, the stabilisation free energy in the absence of solvent interactions is everywhere higher than ΔG and that the difference between them tends to vanish with increasing temperature. The hydrophobic effect and solvent interactions, in general, therefore, make the native state *less* stable with respect to the vacuum condition.

The entropy of apolar group solvation dominates towards the lower temperature end of the ΔG curve. It leads to a destabilisation of the native protein fold as the temperature decreases, which has been called cold denaturation (Comment C2.10). Total cold denaturation of a protein (ΔG crossing zero) is not observed in the usual physiological solvents, because it would occur at subzero temperatures. It has been observed in denaturing solvents, however, such as concentrated guanidinium hydrochloride solutions, which shift the thermodynamics variable curves towards higher temperature.

Further analysis has emphasised the non-negligible role of polar group hydration. We recall from Section C2.4.8 that, in contrast to the positive contribution of apolar group hydration, polar group hydration provides a negative contribution to ΔC_p, which vanishes at very high temperature. We recall (Comment C2.9) that ΔC_p is related to the slope of the entropy temperature dependence. The exposure of apolar and polar groups has an ordering effect on water molecules, be it in different ways – apolar groups, because they cannot form hydrogen bonds, and polar groups, because they can form hydrogen bonds with preferential orientations. Close to room temperature, the hydration entropies of apolar and polar groups are negative and similar. The hydration entropy of polar groups has a negative slope and decreases even further with temperature, however, while that for apolar groups has a positive slope and approaches zero above 100 °C. It must be emphasised that the 'general' values assumed for the heat capacities of apolar and polar group hydration in the calculations of Section C2.4.8 are, at best, good approximations. The value and temperature dependence of hydration heat capacity depend on the specific nature of the group, and within the apolar family, they are significantly different for aliphatic and aromatic groups (aromatic groups having a degree of polar character).

Analysis of calorimetric data after hydration effects have been accounted for has shown that the tight van der Waals' packing of the protein interior plays a significant part in native fold stabilisation. From a consideration of the hydrophobic effect, we would expect as many as possible of the apolar groups to gather together away from water exposure. The hydrophobic force, however, is due mainly to an undirected entropic effect, i.e. which cannot provide a mechanism for specific interactions. In the case of hydrocarbons in water, the aggregation of apolar molecules forms a liquid phase. This is *not* the case for the protein interior,

---

Comment C2.11   Thermodynamics and protein structure

Comment Fig. C2.11 Space filling model of a protein structure solved by X-ray crystallography.

Note the compact packing of atoms. The red and blue spheres on the protein surface represent negatively and positively charged side-chains, respectively. The main processes involved in the difference in stability between the native and unfolded states are:

Surface group hydration: this is proportional to the solvent accessible area for each group, which can be calculated from the structure in the native state and assumed to be total in the unfolded state.

Detailed packing of the native protein interior (which can be analysed in the structure).

---

which has been shown by inspection of thousands of protein crystal structures to be more like a compact *amorphous solid*. Individual van der Waals' interactions are very weak (Table C2.3). The sum of van der Waals' energies from the specific packing of all the groups in the protein interior is quite significant, however, and could well be an important driving force behind the cooperativity of the native fold.

Detailed structural analyses of the macromolecular surface and internal packing density are required in order to describe the energetics of protein stabilisation and to make the most efficient use of calorimetric data. Considerable effort is being devoted in the development of methods to perform such analyses on protein structures that have been solved by crystallography or NMR (Comment C2.11).

C2.5 Nucleic acids and lipids

The arguments and analytical approach in DSC developed above for protein fold stabilisation are readily extended to other macromolecules such as nucleic acids (Fig. C2.11), as well as to the interactions involved in complex formation between macromolecules and between macromolecules and smaller ligands.

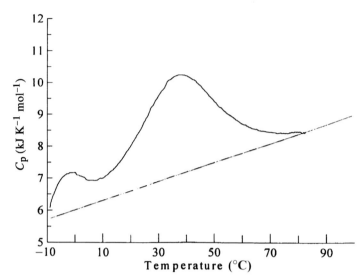

Partial heat capacity as a function of temperature of the single strand deoxyoligonucleotide 5'-GCGAACAATCGG-3', at a concentration of 1.88 mg ml^{-1}. The total unfolding enthalpy from $-10\,°C$ to $80\,°C$ is equal to 77 kJ mol^{-1}. (Privalov and Privalov, 2000.) (Figure reproduced with permission from Elsevier.)

The energetics of DNA double-helix formation has been examined through the study of oligonucleotides. The residual structure of oligonucleotides appears as the temperature decreases below 80 °C. The structure is formed in a diffuse process with quite a low enthalpy of formation, which ends below 0 °C, requiring supercooled conditions for its observation.

DSC is also an important tool for the study of membrane formation when lipids are mixed with water. Complex phase diagrams are obtained as a function of temperature and water/lipid ratio. The energetics of such systems is examined by measuring the temperature dependence of the heat capacity for each phase and the peaks associated with the transitions. Review references are given in the suggestions for further reading.

C2.6 Checklist of key ideas

- A full quantitative thermodynamics description of a macromolecule in solution for a given temperature range may be obtained from the measurement of its partial specific heat capacity at constant pressure, as function of temperature, by DSC.
- Specifications for a modern differential scanning calorimeter, include macromolecule requirements of less than 50 µg per test, a temperature range from -10 °C to 130 °C, $\sim$10 µJ deg^{-1} sensitivity, and the possibility of scanning down as well as up in temperature to check the reversibility of processes.
- In a heat capacity versus temperature scan of a macromolecule, the native state is assigned as the reference relative to which the thermodynamics parameters of the system are defined, so that the excess heat capacity, with respect to the native state becomes the significant parameter.

- A partition function analysis of the heat capacity curve is quite general and applies to any monomeric (so that the number of moles present in the solution does not change upon unfolding) system at equilibrium, independently of the transition mechanisms and of the number of intermediate states involved.
- When the cooperative domain extends over the whole molecule the van't Hoff and calorimetric enthalpies are equal, and a two-state transition may be assumed.
- In all cases where the van't Hoff and calorimetric enthalpies are not equal the transition proceeds via intermediate states.
- When the cooperative domain is smaller than the molecule, as would be the case for a multidomain protein, the van't Hoff enthalpy is smaller than the calorimetric enthalpy.
- A van't Hoff enthalpy that is larger than the calorimetric enthalpy is an indication that the cooperative domain is larger than the molecule, i.e. it provides evidence of intermolecular interactions such as oligomerisation.
- There is no calorimetric evidence for a *molten globule* intermediate for small monodomain proteins.
- The difference in stability measured by calorimetry between a native protein and a mutant may very well be due to a change in cooperativity of the unfolding process, due to the mutation, and not to enthalpic or entropic contributions due to the amino acid replacement itself.
- The heat capacity versus temperature curve of a complex protein structure can be separated into component curves, corresponding to the transitions of the different states of the system without any a priori assumptions, by using computer algorithms based on recurrent procedures.
- The solvent plays an important role in the stabilisation of both the folded and native states of a protein and protein–solvent interactions may be specific, as in the case of ion binding to specific sites, or more general through, for example, pH, ionic or small solute effects on water structure affecting the hydration and solubility of protein polar and apolar groups.
- Methods to calculate heat capacities of proteins in their various states have been developed based on DSC information and data from model compounds.
- The primary heat capacity, the heat capacity of an anhydrous unfolded polypeptide, can be calculated accurately by summing contributions from individual amino acids and peptide bond formation.
- The heat capacities of anhydrous folded proteins are very similar.
- The heat capacity of a native protein in solution is larger than that of the anhydrous folded protein revealing mainly the positive magnitude of the hydration term.
- The hydration term varies according to the protein, reflecting the role of the macromolecular surface composition in the solvent interactions, especially with respect to the accessible surface area of polar and apolar groups.
- The heat capacity of an unfolded peptide in solution is equal to the primary heat capacity and a term equal to the sum of the hydration values for the individual amino acids in the protein sequence, and the N and C terminal groups.

- The heat capacity of the native state in solution varies linearly with temperature; the heat capacity of the unfolded state in solution is approximated well by a second order temperature polynomial in the range 0–100 °C.
- Because of the importance of free energy predictions in protein and ligand design, accurate structures based on heat capacity calculations are extremely useful.
- The heat capacity of a *native* protein in solution cannot be calculated from its sequence, because it depends on the macromolecular surface and dynamics of the folded structure, but parametrised approaches have been developed that provide very good approximations.
- Calorimetric studies have made fundamental contributions to our understanding of protein stabilisation forces.
- Detailed structural analyses of the macromolecular surface and internal packing density are required in order to describe the energetics of protein stabilisation and to make the most efficient use of calorimetric data.
- The *hydropobic effect* results in an effective *hydrophobic force*, bringing together apolar molecules in aqueous solution, in order to minimise their water contact surfaces.
- The entropy of apolar group solvation leads to a destabilisation of the native protein fold as the temperature decreases, which has been called cold denaturation.
- Analysis of calorimetric data after hydration effects have been accounted for has shown that the tight van der Waals' packing of the protein interior plays a significant part in native fold stabilisation.
- Subzero temperatures and supercooled conditions are required to observe the full heat capacity curve of a nucleic acid corresponding to the appearance of oligonucleotide structure.

Suggestions for further reading

Tristram-Nagle, S., and Nagle, J. F. (2004). Lipid bilayers: thermodynamics, structure, fluctuations, and interactions. *Chem. Phys. Lipids*, **127**, 3–14.

Weber, P. C., and Salemme, F. R. (2003). Applications of calorimetric methods to drug discovery and the study of protein–protein interactions. *Curr. Opin. Struct. Biol.*, **13**, 115–121.

Privalov, G. P., and Privalov, P. L. (2000). Problems and prospects in microcalorimetry of biological macromolecules in 'Energetics of biological macromolecules', *Meth. Enzymol.*, **323**, 31–62.

Privalov, P. L. (1980). Scanning microcalorimeters for studying macromolecules. *Pure & Appl. Chem.*, **52**, 479–497.

Chapter C3
Isothermal titration calorimetry

C3.1 Historical review

T. Wiseman, S. Williston, J. F. Brandts and **L. N. Lin** published a paper in **1989** with the title: 'Rapid measurement of binding constants and heats of binding using a new titration calorimeter'. The term isothermal titration calorimetry (ITC) was introduced by **E. Freire** and colleagues, in **1990**. The method is unique in providing not only the magnitude of the enthalpy change upon binding but also, in favourable experimental conditions, values for the binding affinity and entropy changes. Because these parameters fully define the *energetics* of the binding process, ITC is playing an increasingly important role in the detailed study of protein–ligand interactions and the associated molecular design approaches, in particular with respect to drug design. In the **1990s**, a number of critical reviews of ITC results and analytical developments has been published.

C3.2 Experimental aspects and equations

C3.2.1 Measuring protocol and samples

The reaction cell in ITC has a volume close to 1 ml and contains one of the reactants. The other reactant is added to it by injection in small volumes (close to 10 ml), and stirred in. The amount of power (in millijoules per second or in watts) required to maintain a constant temperature difference between the reaction bath and a reference cell is measured by the calorimeter. The heat absorbed or released by the chemical reaction is determined from the integral of the power curve over the appropriate time (Fig. C3.1).

In the ITC experiment of Fig. C3.1, the 0.6 ml calorimeter cell was filled with protein solution (68.8 mM pancreatic RNase A), and the ligand solution (0.5 mM cyclic monophosphate) was added in precise 5 ml volumes through an injection syringe that also stirred the resulting solution to ensure proper mixing. The quantity plotted on the *y*-axis is the time dependence of the power needed to maintain a constant temperature difference between the measuring

Fig. C3.1 ITC
experimental curves of
cyclic monophosphate
binding in RNase A: (a)
the power required to
maintain a constant
temperature difference
between the measuring
and the reference cell, as
a function of time; each
peak corresponds to one
injection; (b) the integral
heat effect due to the
binding as a function of
injection number.
(Privalov and Privalov,
2000.) (Figure reproduced
with permission from
Elsevier.)

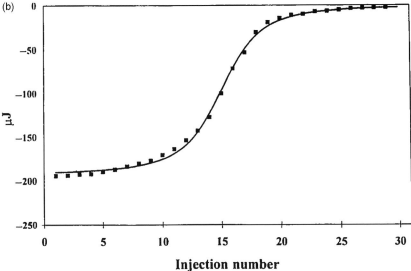

and a reference cell. Each peak corresponds to one injection, and its integral
(after subtraction of the background due to ligand dilution, mechanical mixing effects etc.) to the heat associated with the amount of binding (q_j in
Eq. (C3.1)).

The peaks are progressively smaller with each injection because there is less
free protein available for binding. When the protein is saturated, the peak height
represents the background level. The integral heat effect of the binding as a
function of injection number is plotted in (b).

C3.2.2 Binding enthalpy and heat capacity

The amount of heat q_j absorbed or released in the injection j (Fig. C3.1) is equal to

$$q_j = v\Delta H \Delta \left[L_{j,\text{bound}}\right] \qquad (C3.1)$$

where v is the volume of the reaction cell and $\Delta[L_{j,\text{bound}}]$ is the change in concentration of bound ligand after the jth injection, so that q_j is proportional to the amount of ligand, $v\Delta[L_{j,\text{bound}}]$, that binds to the protein. The constant of proportionality is the enthalpy change per mole of bound ligand, ΔH. The change in binding heat capacity can be determined by repeating the titration at different temperatures (Eq. (C3.2) and Fig. C3.2)

$$\Delta C_P = \partial \Delta H / \partial T \qquad (C3.2)$$

If the amount of added ligand is small with respect to the affinity of the reaction, we can assume that all the added ligand is bound and hence derive the corresponding enthalpy change directly by using Eq. (C3.1). If, furthermore, the association constant, K_a, of the reaction is known from other measurements (such as fluorescence titration or surface plasmon resonance, see Chapters C4, D8), the free energy change, ΔG and hence the entropy change, ΔS, of the reaction can be calculated from ΔG and ΔH by using the classical Eq. (C3.3), and the reaction thermodynamics is fully defined:

$$\Delta G = -RT \ln K_a = \Delta H - T\Delta S \qquad (C3.3)$$

Depending on the number of sites and other aspects of the binding model, the mathematical expression relating affinity constants to Eq. (C3.1) may be quite complicated.

Fig. C3.2 Binding enthalpy and heat capacity. Plot of the binding or association enthalpy, ΔH^{ass}, as a function of temperature for a protein–DNA interaction (the HMG box of Sox-5 protein with a 12-base-pair DNA duplex containing the recognition sequence). The binding heat capacity is given by the initial slope of the line ($-6\text{kJ mol}^{-1}\text{ K}^{-1}$) at the lower temperatures. The break in slope at the higher temperatures is due to temperature-induced alterations in the components of the complex (see also Section C3.3.3).

C3.2.3 Affinity constants

In the simple case of a protein, P, having one binding site for a ligand, L, we write

$$P + L \xrightarrow{K_a} PL$$

$$K_a = \frac{[PL]}{[P][L]}$$

(C3.4)

$$P_{total} = [P] + [PL]$$

$$L_{total} = [L] + [PL]$$

where P_{total}, L_{total}, are the total concentrations of protein and ligand, respectively, and the square brackets indicate molar concentrations. Combining Eqs. (C3.1), (C3.4)

$$q_j = v(\Delta H)P_{total}\left(\frac{K_a[L]_j}{1 + K_a[L]_j} - \frac{K_a[L]_{j-1}}{1 + K_a[L]_{j-1}}\right)$$

(C3.5)

where $[L]_k$ is free ligand concentration after the kth titration. Equation (C3.5) can also be written in a more complicated form in terms of total ligand concentration, by solving Eqs. (C3.4).

By using the appropriate set of binding equations, the analysis can be extended to reactions with more than one binding site or involving equilibrium constants between many different states. Complementary experiments, with other methods, are necessary to fully solve such problems, however, because of the larger number of parameters.

Fig. C3.3 ITC curves for affinity constants of different strength. Simulated ITC curves are presented in the three panels to illustrate the effects of increasing binding affinity. The following parameters were used for the simulation: $[p] = 0.05$ mM; $[l] = 0.6$ mM; $\Delta H = -42$ kJ mol^{-1}; K_a and c are shown in the figure, where $c = K_a[p]$. The binding enthalpy can always be measured accurately. The binding affinity, however, can only be determined in cases illustrated by the first panel ($c < 1000$), in which the peaks are progressively smaller. If the affinity is too high with respect to experimental conditions ($c > 1000$), only a lower limit can be set on its value. Microcalories per second can be converted to microwatts by multiplying by 4.2. (Leavitt and Freire, 2001.) (Figure reproduced with permission from Elsevier.)

In cases for which the reaction enthalpy is large enough to be measurable over a range of ligand concentrations with respect to the affinity constant, the titration series also provides a measure of the affinity constant itself (Fig. C3.3). A parameter, c, was defined to relate binding affinity and experimental conditions

$$c = K_a[P] \qquad (C3.6)$$

The value of c must be <1000 for the K_a of the reaction to be determined by ITC, which, in practice, sets an upper limit of 10^8-10^9 M^{-1} on the binding constant (Fig. C3.3).

Reactions with high binding affinities

For values of c beyond 1000, ITC curves lose their characteristic shape and no longer contain information concerning the binding constant (Fig. C3.3). This has been an important limitation in the application of ITC to drug design, because lead compounds have to be optimised to nanomolar or even higher affinities (see section on drug design, below). A protocol has been developed, however, that allows high affinity binding constants to be determined in ITC experiments, by using ligand competition. Three titrations are performed on the protein solution. First, the binding thermodynamics parameters of a weak inhibitor are determined. Second, the protein is titrated by the high-affinity ligand in order to determine its binding enthalpy. In the third titration, *the displacement titration*, the high-affinity ligand is titrated into a solution, in which the protein has been prebound to the weaker inhibitor. The method is illustrated in Fig. C3.4 by a study of inhibitors of HIV-1 protease, an important drug target in the fight against AIDS (Comment C3.1). The binding enthalpy of the inhibitor-HIV-1 protease interaction was measured accurately by ITC (Fig. C3.4, top right-hand panel, the y-axes in these panels are in mcal/M) but not its association constant K_a, which is too high to be determined by ITC. The displacement titration (binding constant K_{app}, determined in Fig. C3.4, bottom panel) was performed in the presence of the weak inhibitor acetyl-pepstatin (binding constant, $K_{a,w}$, determined in the top left-hand panel). The selection of a weak competitive inhibitor with a binding enthalpy of opposite sign increases the sensitivity of the experiment. K_a is determined from K_{app} by using the equation:

$$K_{app} = \frac{K_a}{1 + K_{a,w}[X]}$$

where [X] is the concentration of the weak inhibitor, which can be adjusted to give a value of K_{app} in the desired range. The values found for the affinity constants were $K_{a,w} = 2.4 \times 10^6$ M^{-1} and $K_a = 3.1 \times 10^{10}$ M^{-1}.

Drug design

An important medical application of three-dimensional structure determination of enzymes involved in disease is the design of high-affinity inhibitors that could be potential pharmaceutical agents. Affinities in the nanomolar range (dissociation

Comment C3.1

Physicist's box: HIV infection initiation

Infection by HIV (human immunodeficiency virus), the AIDS (acquired immune deficiency syndrome) virus, is initiated by the specific interaction between gp120, the external envelope glycoprotein of the virus, and membrane bound cellular receptor, CD4, and obligatory chemokine receptors, CCR5 or CXCR4). The interaction with CD4 triggers conformational changes in the virus envelope that lead to recognition of the chemokine receptors and subsequently to fusion between the viral and cell membranes and internalisation of the virus.

Fig. C3.4 Displacement
titration to measure high
affinity binding constants
(see text). The
experiments on the
inhibitor-HIV-1 protease
interaction were
performed at 25 °C in 10
mM acetate, pH 5.0, 2%
DMSO.
(Velazquez-Campoy et al.,
2001.) (Figure reproduced
with permission from
Elsevier.)

constants of 10^{-9} M) are required for a drug to be useful. Briefly, the drug design approach is divided into a number of steps. After having chosen the target protein a very large combinatorial search is effected for lead compounds, which could interact with its active site. Hundreds of thousands of such compounds are found in 'libraries'. Many of these compounds are natural molecules obtained from plants, for example, with structures that are so complex they could never have been invented. Taxol, an antimitotic drug, purified from yew bark is an example of such a structure. Promising lead compound–protein complexes are characterised by different methods including ITC, which can provide essential information on binding energetics. Chemical modelling and simulation approaches are then applied to propose structural variations that would increase the affinity of the compound for the protein. Modified inhibitors are synthesised and characterised and the process is repeated until a satisfactory compound is obtained.

C3.3 Applications

C3.3.1 Entropic versus enthalpic optimisation

We see from the classic Eqs. (C3.3) that the same binding constant could be achieved from different combinations of entropic and enthalpic contributions. The binding enthalpy reflects the competition between protein–ligand bonding interactions (e.g. van der Waals and hydrogen bonds), on the one hand, and protein–solvent and ligand–solvent bonding interactions, on the other. The binding entropy essentially reflects changes in solvation entropy (e.g. through the shielding of hydrophobic groups from solvent contact) and changes in the conformational entropies of the protein and ligand.

Optimisation of binding affinity invokes different tactical approaches depending on if it is achieved predominantly via the enthalpic or the entropic terms. Superhigh binding affinity requires a synergy of favourable binding enthalpy and binding entropy contributions. Entropic binding is favoured by the design of conformationally constrained ligands with a high degree of hydrophobicity. HIV-1 protease inhibitors (Fig. C3.4) displaying these thermodynamics properties have been approved for clinical use. Such ligands exhibit very high specificity towards their targets. This is an important contribution to high binding affinity when dealing with an immutable target, but it also makes the ligand more susceptible to drug-resistant mutations or ineffectual against related but not identical target molecules. A certain degree of ligand flexibility may be desirable, in which case the resulting lower binding affinity due to the decrease in entropy upon binding has to be compensated by other favourable interactions. Note also that the hydrophobicity of a ligand affects its solubility. The selection of the entropy/enthalpy balance, therefore, influences not only the binding affinity but also the general properties of the ligand itself.

C3.3.2 Relating binding energy and structure

Accurate structural interpretations of the binding enthalpy would obviously be of great help in ligand design. They are far from straightforward, however, because different processes of roughly equal magnitude contribute to the binding enthalpy. A structural survey of protein complexes with low-molecular-weight ligands has shown that the binding sites in the unbound proteins are characterised by low structural stability. The stabilisation by ligand binding of these protein regions appears in the thermodynamics as a conformational binding enthalpy term. Despite the inherent difficulties of the problem, there are attempts to parametrise the binding enthalpy in terms of structural features, in a similar way to that developed for DSC (see Chapter C2).

C3.3.3 Combining ITC and other biophysical methods

ITC, Mass Spectrometry, analytical ultracentrifugation (AUC) and surface plasmon resonance (SPR)

The energetics of the HIV gp120-CD4 binding reaction (Comment C3.1) has been studied by a series of biophysical experiments: MALDI-MS and AUC to characterise the masses of the proteins and complexes, ITC to characterise the binding affinites and energetics, SPR to characterise binding kinetics and circular dichroism (CD) to characterise secondary structure rearrangements upon binding (see also Chapters B1, D4, C4, E4).

Fig. C3.5 Sedimentation equilibrium analysis of the gp120–CD4 interaction (see text). (Myszka *et al.*, 2000.) (Figure reproduced with permission from *Proceedings of the National Academy of Sciences (USA)*.)

Soluble forms of the two proteins were obtained by expression in Chinese hamster ovarian cells and Drosophila cell lines, respectively. gp120 was prepared as the full-length glycosylated protein and as a 'core gp120', which had certain amino acid deletions and the removal of most of the sugar groups. The masses of the expressed proteins were checked by MALDI and AUC, and their assembly state in the complex was determined by AUC (Fig. C3.5).

The data shown in Fig. C3.5 (squares) are for a mixture of CD4 and core gp120. The solid line represents the best fit (the differences with the data are expressed as residuals in the top panel). The masses of CD4 and core gp120 were 45 kDa and 38 kDa, respectively. The complex was measured at 83 kDa by AUC, showing it to be a one-to-one assembly of the two proteins. The component curves of the fit are also shown in the figure: the long dashed line is for the one-to-one complex, the dotted line is for excess CD4 and the baseline is the dashed-dotted line.

The binding thermodynamics of the interaction was determined by ITC (Fig. C3.6). Figure C3.6 shows the power data and integrated enthalpy for the titration of full length (part (a)) and core (part (b)) gp120 by CD4. Best-fit lines are for binding enthalpies of -63 and -62 kcal (mol of CD4)$^{-1}$ for full length and core gp120, respectively. Equilibrium dissociation constants K_D were determined

Fig. C3.6 ITC analysis of the gp120-CD4 interaction: (a) power data and integrated enthalpy for the titration of the full length gp120 by CD4; (b) power data and integrated enthalpy for the titration of the gp120 core by CD4; (c) temperature dependence of the CD4 binding enthalpy to full length (open circles) and core (filled circles) gp120. From Myszka *et al.* (2000). (Figure reproduced with permission from *Proceedings of the National Academy of Sciences (USA).*)

as 5 and 190 nM, for the full length and core gp120, respectively. Figure C3.6(c) shows the temperature dependence of the CD4 binding enthalpy for full length (open circles) and core (filled circles) gp120. The slopes of the best-fit lines gave binding heat capacities of -1.2 and -1.8 kcal mol^{-1} K^{-1}, respectively.

The thermodynamic parameters in Table C3.1 were measured or calculated from the ITC data in Fig. C3.6. We note that these parameters are essentially similar for the full length and core gp120, indicating common binding mechanisms. The apparently large difference in dissociation constant values (5 nM and 190 nM for the full length and core gp120, respectively) in fact corresponds to just over 2 kcal mol^{-1} in binding free energy, less than the energy of formation of one hydrogen bond.

Table C3.1. *Thermodynamics parameters of the CD4-gp120 interactions at 37 °C*

	ΔG° (kcal mol^{-1})	ΔH° (kcal mol^{-1})	$-T\Delta S^\circ$ (kcal mol^{-1})	ΔC° (kcal K^{-1})	K_D (nM)
Full length gp120	-11.8 ± 0.3	-63 ± 3	51.2 ± 3	-1.2 ± 0.2	5 ± 3
Core gp120	-9.5 ± 0.1	-62 ± 3	52.5 ± 3	-1.8 ± 0.4	190 ± 30

The thermodynamic parameters for CD4-gp120 binding were compared to those of other protein–protein interactions. Thirty protein–protein complexes were sampled, as well as thirteen antibody–antigen interactions, and two T cell receptor–major histocompatibility complex peptide interactions (see also Fig. C4.7).

The thermodynamics of the CD4/gp120 interaction is qualitatively different from that of the thirty protein–protein complexes sampled, which had small positive ΔG, ΔH and $-T\Delta S$. It is similar, however, to that of antibody–antigen and T cell receptor–MHC peptide interactions. The favourable binding enthalpy of the CD4–gp120 interaction ($\Delta H^\circ \sim -63$ kcal mol^{-1}) is strikingly large, while the binding entropy $-T\Delta S^\circ \sim 52$ kcal mol^{-1}) is also very large but unfavourable. CD4–gp120 complex formation, therefore, involves a large number of bonding interactions, such as hydrogen bonds or van der Waals interactions, as well as substantial loss of degrees of freedom. Analysis of the high-resolution crystal structure suggests that the entropy decrease is not accounted for by trapped water molecules but must involve major conformational rearrangements of the proteins. The large binding enthalpy and binding entropy values compensate to give low values for the dissociation constants and binding free energy, similar to those for other complexes.

The binding heat capacities (Table C3.1) are significantly greater than the values (-0.2 to -0.7 kcal mol^{-1} K^{-1}) usually observed for protein–protein complexes. Recalling the structural interpretation of heat capacity changes discussed in Section C3.3, this suggests extensive apolar surface area is buried upon CD4–gp120 complex formation. This is not observed at the binding site in the crystal structure of the complex, suggesting that this apolar surface area is buried elsewhere in the structure through major conformational rearrangements. In another modelling approach (Comment C3.2), binding heat capacity and entropy were interpreted in terms of a number of residues that reorganise during binding. Application of the method to the CD4–gp120 complex gave values for the number of residues that were close to 100, among the largest reported for a protein–protein interaction.

Given the strong indications from the thermodynamic data of major conformational changes in the proteins upon binding, the study continued with a CD

Comment C3.2
Structural modelling of thermodynamics parameters

In an independent approach to that of Freire and his collaborators (Chapter C2), Spolar and Record (1994) have related thermodynamic parameters from calorimetry to the number of residues that reorganise during binding, by analysing high-resolution structures.

Fig. C3.7 SPR analysis of CD4–gp120 binding kinetics. (a) Normalised binding responses of CD4 (1 μM) over full length gp120 (black curve), glycosylated core (red curve), core gp120 (blue curve) and unmodified sensor (grey curve). Responses were normalised to 100% bound at the beginning of the dissociation phase. (b) Kinetic data for core gp120 binding to a CD4 surface, injected at varying concentrations, from 975 (top curve) to 36 nM, the baseline is for 0 nM added. The data were globally fitted (lines) by a single site interaction model. (Myszka et al., 2000.) (Figure reproduced with permission from *Proceedings of the National Academy of Sciences (USA).*)

investigation. A substantial change in ellipticity was observed during complex formation, consistent with a reduction in random coil content. These data suggest that binding induces structural rigidity.

The kinetics of CD4–gp120 binding has been analysed by SPR (Fig. C3.7, see also Section C4.4.3.). The similar and relatively slow association rate constants measured for both gp120 (6.72×104 M^{-1} s^{-1} and 6.27×104 M^{-1} s^{-1}, respectively) are consistent with major structural rearrangements during binding.

The different equilibrium dissociation constants found for the full length and core gp120 are consistent with the values measured by ITC. They arise from different dissociation rate constants probably because of a loss of favourable contacts in the core gp120. Glycosylation had little effect on binding kinetics.

The flexibility of gp120 and the masking of its receptor and chemokine binding sites contribute to the fact that monomers of this protein, shed by the virus, elicit mainly non-neutralising antibodies and are very inefficient in eliciting broadly neutralising antibodies – a decoy strategy that may help the virus evade humoral immune response.

The gp120 protein is organised on the virus in a trimeric complex with another protein called gp41. The complex is stable and inert. Binding to CD4 and the resulting large conformational change in gp120 induces a properly timed metastable state in gp41 that triggers fusion between the viral envelope and the cell membrane.

The results from hydrodynamics, thermodynamics and spectroscopy when combined with the crystallographic structure of the CD4–gp120 complex present an intriguing picture of the role of gp120 in the biology of HIV infection. As the authors of the study cited in Capitons to Figs. C3.5–C3.7 wrote: 'conformational flexibility in gp120 provieds an attractive hypothesis for understanding of both immune evasion and cell invasion.'

Fig. C3.8 Component partial molar heat capacities for the HMG box-DNA interaction. The dash-dot line is for 12-base-pair DNA duplex; the dotted line is for Sox-5 HMG box protein; the solid line is for their complex. (Privalov and Privalov, 2000.) (Figure reproduced with permission from Elsevier.)

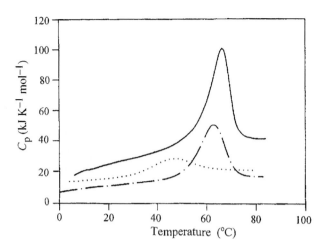

ITC and DSC

The gradient of the binding enthalpy temperature can be interpreted as a binding heat capacity only if the temperature change does not induce alterations in any of the reaction components. If the apparent heat capacity changes with temperature, as it does in the reaction shown in Fig. C3.2, it is an indication that such alterations with temperature do occur. In these cases, the correct approach is to combine ITC with DSC to study the heat capacity of each of the reaction components in the appropriate temperature range.

We treat here the example of the HMG box–DNA. The temperature dependence of the binding enthalpy has already been shown in Fig. C3.2. The molar partial heat capacity functions of the reaction components, measured by DSC, are shown in Fig. C3.8. Clearly, the interpretation of the binding heat capacity of the interaction is considerably complicated by the unfolding of the protein (dotted line in Fig. C3.8) and partial dissociation of the complex (solid line) in the temperature range of the study. The HMG box–DNA reactions were expressed as a set of equilibrium equations, similar to Eq. (C3.4). By fitting the observed heat capacity functions of the complex and its free components, all the thermodynamic parameters of the system were evaluated.

The analysis revealed that there are two stages in the temperature-induced changes of the complex (solid line in Fig. C3.8). In the first, there is a gradual increase in heat capacity reflecting accumulation of thermal energy. In the second, there is a highly cooperative dissociation of the complex with concomitant unfolding of its protein and DNA duplex components. The second stage proceeds with a significant peak in heat capacity, which drops sharply with temperature, due to the decrease in complex population.

C3.4 Checklist of key ideas

• In ITC, the amount of power required to maintain a constant temperature difference between a reaction bath and a reference cell during a titration is measured by the calorimeter; the heat absorbed or released by the chemical reaction is determined from the integral of the power curve over the appropriate time.

- The change in binding heat capacity during the reaction can be determined by repeating the titration at different temperatures.
- In cases for which the reaction enthalpy is large enough to be measurable over a range of ligand concentrations with respect to the affinity constant, the titration series also provides a measure of the affinity constant itself.
- Promising lead compound–protein complexes in *drug design* can be characterised by ITC, which provides essential information on binding energetics.
- The same binding constant can be achieved from different combinations of entropic and enthalpic contributions; binding enthalpy reflects the competition between protein–ligand bonding interactions, on the one hand, and protein–solvent and ligand–solvent bonding interactions, on the other; binding entropy essentially reflects changes in solvation entropy and changes in the conformational entropies of the protein and ligand.
- ITC has been combined with mass spectrometry, AUC and SPR to provide as full a description as possible of the energetics and structural changes during a binding reaction.
- In reactions for which the heat capacity of the components may change with temperature, ITC is combined with DSC, which is used to measure the heat capacity of each of the reaction components in the appropriate temperature range.

Suggestions for further reading

Seelig, J. (2004). Thermodynamics of lipid–peptide interactions. *Biochim. Biophys. Acta,* **1666**, 40–50.

Velazquez-Campoy, A., Leavitt, S. A., and Freire, E. (2004). Characterisation of protein–protein interactions by isothermal titration calorimetry. *Methods Mol. Biol,* **261**, 35–54.

Chapter C4
Surface plasmon resonance and interferometry-based biosensors

C4.1 Historical overview and introduction to biological problems

1801

Thomas Young observed that two slits placed in front of a light source created a pattern of intensity fringes on a screen. This phenomenon is the basis of modern interferometers, which have now been adapted to measure ligand–receptor interactions.

1902

R. W. Wood observed for the first time the phenomenon of surface plasmon resonance (SPR), which provided a simple and direct sensing technique for probing refractive index changes that occur in the very close vicinity of a thin metal film surface.

1968

E. Kretschmann and **H. Z. Raether** proposed the basic configuration of an SPR sensor, with an optical system containing a prism coupled to a reaction cell.

1982–1983

B. Liedberg and colleagues realised the potential of SPR for macromolecular binding studies since the change in refractive index is dependent on the molecules accumulating at the metal surface. They adsorbed an antibody specific to immunoglobulin G onto a gold sensing film, resulting in the selective binding and detection of the protein.

1980s, 1990s

Several biosensors, based on either interferometry or SPR, that allowed the simple, rapid and non-labelled assay of various biochemical analytes such as proteins, DNA and small compounds, became commercially available.

1993

The potential to analyse weak affinities by SPR instrumentation was demonstrated by **S. Davis** and coworkers in their studies of cell surface receptor interactions.

2000 and after

Optical arrays were developed to analyse the interactions of thousands of different molecules simultaneously, with high selectivity and sensitivity. Biosensors are used in the study of intermolecular interactions with a broad range of affinities under a variety of chemical conditions and temperatures. The configuration of the biosensors makes them particularly advantageous in the rapid screening of putative ligands. These instruments have become essential tools for proteomics. In conjunction with other biophysical techniques, such as isothermal titration calorimetry, analytical centrifugation, and mass spectrometry, they provide very precise thermodynamic analysis, characterisation and identification of binding partners and binding events. The advantages of biosensor experiments compared with other binding studies include the small amounts of material required (in the nanogram range), the wide variety of systems that can be studied (ranging from 200 Da molecules to unicellular organisms), the facts that labelling is not required and that a high degree of purity is not essential so that the binding properties of whole cell extracts can be scrutinised rapidly, and finally that kinetic information is obtained in real time.

C4.2 Measuring surface binding

C4.2.1 Lay-out of a biosensor

Different types of biosensors have been developed for the study of biological macromolecule interactions. Surface binding experiments typically involve immobilising a molecule (*the ligand*) on a surface and monitoring its interaction with a second molecule (*the analyte*) in solution. The set-up has the advantage of mimicking many biological recognition events, such as those between cell-bound receptors and soluble signalling proteins.

In the biosensor method, the build up of bound analyte concentration leads to a change in refractive index (Comment C4.1) near the surface. Two types of instrument have been developed to measure this change with high sensitivity. One is an interferometer and the other is based on the phenomenon of SPR. The biochemical binding constants, such as the equilibrium dissociation constant and the on and off rates of a reaction (see Chapter C1) can be calculated directly from the change in refractive index of the solution over time.

A biosensor comprises a flow cell attached to an optical device. Buffer and analytes enter and exit the cell, and biochemical interactions are monitored through changes in refractive index.

> **Comment C4.1**
> **Biologist's box:**
> **Refractive index**
>
> A change in the protein concentration of a solution is associated with a change in the refractive index of the solution. Molecular interactions occurring in a solution can therefore be monitored by measuring the local variations of the refractive index. Biochemical binding constants can be calculated directly from the change in refractive index as a function of time. The power of the biosensor SPR- and interferometry-based techniques is that they can measure refractive index changes arising from ligand binding to a monolayer of macromolecules on a surface.

(a)

(b)

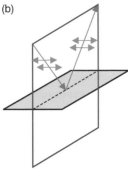

Fig. C4.1 (a) p-polarised light: the electric field vector (blue arrows) lies parallel to the plane (blue square) formed by the incident and reflected beams (red lines). (b) s-polarised light: the electric field vector (green arrows) lies perpendicular to the plane formed by the incident and reflected beams and is, therefore, parallel to the surface (green plane).

C4.2.2 SPR biosensor

Several biosensors have been designed around the phenomenon of SPR. Due in large part to its relative simplicity and high sensitivity, this method has become very popular.

Total internal reflection occurs at an interface between two non-absorbing media (see Part F). When light is shone upon the interface between two transparent media of different refractive index above a *critical angle* of incidence most of the light coming from the side with the higher refractive index is reflected back. Some of the incident light, however, leaks into the lower refractive index medium as an *evanescent field wave*. The amplitude of the evanescent wave decreases exponentially with distance from the interface, with a decay constant of about one light wavelength.

If the interface between the two media is coated with a thin layer of a suitable conducting material, such as a metal, the p-polarized component of the evanescent field wave penetrates the metal layer (Fig. C4.1). In fact, at a specific angle of incidence, plane polarised light excites the delocalised surface electrons (or plasmons) of the metal, which results in a larger evanescent wave. As a consequences, at this angle of resonance the intensity of the reflected light decreases *drastically* due to the energy transferred to the plasmons.

The two media in the SPR biosensor are, respectively, a glass prism and a flow cell in which the biochemical reaction occurs (Fig. C4.2). A thin film of metal separates the two. The conditions for *surface plasmon excitation* at the interface between the metal and the biochemical solution are achieved by matching the projection of the wave vector of the incident light in the direction of the interface (k_x) and wave vector (k_{sp}) of the surface plasmon oscillation and are

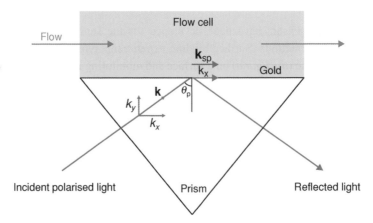

Fig. C4.2 General principle of biomolecular interaction analysis using SPR. The polarised light at incident angle θ_p excites the surface plasmons at the gold interface. $\mathbf{k}_{sp}$ is the surface plasmon wave vector, and $\mathbf{k}$ is the wave vector of the incident beam with components k_x and k_y parallel and perpendicular to the surface, respectively.

Fig. C4.3 General
principles of biomolecular
interaction analysis using
SPR. An increased sample
concentration on the
sensor chip surface
causes a corresponding
increase in the refractive
index, which alters the
SPR angle. The angle
change is monitored as a
change in the detector
position for the reflected
intensity dip (from I to II).

given by

$$k_x = k_{sp} \tag{C4.1}$$

with

$$k_x = \frac{\omega}{c}\sqrt{\varepsilon_{\text{prism}}} \times \sin\theta, \qquad k_{sp} = \frac{\omega}{c}\sqrt{\frac{\varepsilon_{\text{metal}} \times \varepsilon_{\text{cell}}}{\varepsilon_{\text{metal}} + \varepsilon_{\text{cell}}}} \tag{C4.2}$$

$$n_{\text{glass}}\sin\theta_p = \sqrt{\frac{\varepsilon_{\text{metal}} \times \varepsilon_{\text{cell}}}{\varepsilon_{\text{metal}} + \varepsilon_{\text{cell}}}}$$

where ω is the angular frequency of the incident wave, c is the speed of light, and ε is the wavelength-dependent complex dielectric permittivity. The incidence angle θ_p at which SPR conditions are satisfied therefore depends on the refractive index of the material on the non-illuminated side of the metal (the flow cell in this case). The prism enables a range of incidence angles to be observed simultaneously in a wedge of light beams. When the resonance condition described in Eq. (C4.2) is satisfied, there is a strong absorption dip within the angular dependence of the wedge of reflected light (Fig. C4.3). At optical wavelengths, the SPR condition is fulfilled by several metals, of which gold and silver are the most commonly used (Comment C4.2). The resonance is influenced by the

---

Comment C4.2 Surface plasmon propagation

Once light has been converted into a surface plasmon mode on a flat metal surface it propagates but gradually attenuates owing to losses arising from absorption in the metal. This attenuation depends on the dielectric function of the metal at the oscillation frequency of the surface plasmon. Gold and silver are the best candidates for the metal film in the visible light region.

Although, silver films yield a more distinct SPR spectrum than gold ones, this metal tends to be unstable chemically. Thin gold films are therefore a better choice for SPR biosensing applications.

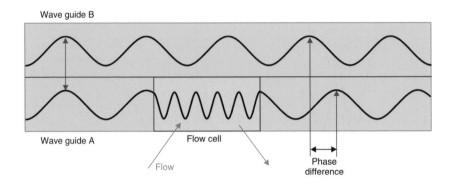

refractive index in the evanescent wave path. The refractive index beyond the penetration distance of about 600 nm (the wavelength of the light) therefore does not affect the experimental outcome. The signal, measured in resonance units (RU), directly correlates with the amount of protein interacting near the surface (typically 1000 RU = 1 ng bound protein mm^{-2}). Results are plotted as a *sensorgram*, which represents changes in resonance signal as a function of time (Fig. C4.3).

C4.2.3 Interferometers as biosensors

The speed of light in a medium of refractive index n is given by

$$c = \frac{\omega}{n}\sqrt{\varepsilon} \qquad \text{(C4.3)}$$

where ω is the light wave's angular frequency, and ε is the dielectric permittivity of the medium.

In the dual wave guide interferometer, a laser beam travels through two parallel optical wave guides, A and B, one of which (A) contains the sample. The sample, because it has a different refractive index, introduces a velocity change, dephasing the beam emerging from A with respect to that emerging from B. The two waves interfere to give a classical Young's pattern of fringes related to the refractive index of the sample. Any changes in the refractive index of the sample alters the speed of propagation of the laser beam and causes a shift in the Young's fringes. The magnitude of the shift is calibrated by using known refractive index changes. The optical configuration is shown in Fig C4.4.

In the biosensor interferometer, the sample is the flow cell in which molecular reactions take place. Receptors are attached to the cell surfaces and potential analytes flow past. In contrast to SPR biosensors, the experiment is sensitive to refractive index changes throughout the volume of the cell.

C4.2.4 Other types of biosensor

A variety of physical phenomena can form the basis of biosensor devices. Coupled plasmon-wave guide resonance spectroscopy (CPWR), for example, combines features of both SPR and wave guide instruments. Optical diffraction gratings have also been incorporated into SPR instruments. Instruments are designed to reproduce as closely as possible natural biological environments in which to monitor intermolecular interactions.

C4.2.5 Coupling ligands to a surface

The preparation of the ligand surface in the flow cell poses special challenges, which have been met by developments in the field of surface nanoscience. Ligands in biosensor experiments are usually proteins but they can also be small molecules, DNA or putative drugs tethered to a surface in the flow cell. Long-lived, highly stable associations prevent signal drift, and enable efficient measurement and control of the amount of analyte binding.

Proteins can be bound via their hydrophobic surface, or covalently coupled to the chip via their amino, carboxyl or sulphydryl groups. With these types of bonding, the ligands are likely to adopt a random orientation on the chip surface, and the analyte-binding interface is not always accessible. This affects the maximum binding capacity but not the measured rates and dissociation constants (next section).

Tags are often used in coupling. Chosen wisely they can orient a receptor protein as it is naturally found at the cell membrane's surface. Chips have been specifically designed to bind biotin, GST (glutathione S-transferase) and poly-histidine tags. More realistic systems, involving lipid-embedded ligands with lateral fluidity, have been developed.

A biosensor chip can be used over and over again, because after each experiment the ligand surface is *regenerated* so as to remove any remaining bound analyte. A new binding analysis can then be performed.

C4.3 Binding between a soluble molecule and a surface

C4.3.1 Thermodynamics of surface interactions

Recall from Chapter C1 that the thermodynamic equilibrium dissociation constant for the association of A and B is given by

$$K_d = \frac{[A] \cdot [B]}{[AB]} = \frac{k_{on}}{k_{off}}$$

In biosensor instruments, k_{on} is the rate of analyte binding to the surface and k_{off} is the rate at which the analyte is removed from the surface. Typical ranges of values are $K_d = 10^{-7}-10^{-9}$ M, $k_{on} = 10^3-10^8$ M^{-1} s^{-1} and $k_{off} = 10^0-10^{-6}$ s^{-1}.

A typical sensorgram is shown in Fig. C4.5. The biosensor is calibrated such that the measured signal, R, is proportional to the concentration of bound analyte. The signal varies over time according to the molecular concentration in solution

Fig. C4.5 A sensorgram is a plot of the SPR signal versus time. The increase in resonance units from the initial baseline represents the association of molecules. The plateau represents the steady-state phase of the interaction where binding is balanced by dissociation from the complex. The decrease in resonance units from the plateau represents the dissociation from the surface after removing the analyte from the running buffer.

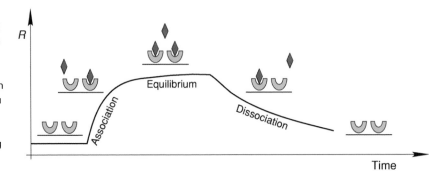

and to how fast the molecule dissociates and associates from the flow cell-bound receptor.

C4.3.2 Measurement of the equilibrium constant

The equilibrium dissociation constant K_d of an interaction is determined from the equilibrium response. The standard equation for the change in signal over time, for a one-to-one interaction, is

$$dR(t)/dt = k_{on}C(t)[(R_{max} - R(t))] - k_{off}R(t) \tag{C4.4}$$

for an analyte at concentration C. Here R_{max} is the maximal response. The equilibrium signal R_{equ} is reached when $dR(t)/dt = 0$, and the previous equation can be rewritten as,

$$R_{equ} = [1/(k_{off}/k_{on} + C)]\,C\,R_{max} \tag{C4.5a}$$

Or, by using the equilibrium constant $K_d = k_{off}/k_{on}$,

$$R_{equ} = [1/(K_d + C)]\,C\,R_{max} \tag{C4.5b}$$

which is similar to a Langmuir isotherm (Comment C4.3). The equilibrium dissociation constant K_d is calculated from the dependence of the equilibrium plateau signal on the concentration of the analyte. The standard method of obtaining an estimate of K_d is to measure the resultant equilibrium response of R_{equ} for various concentrations C_k of ligand. Recasting the previous equation into linear form gives the following,

$$(1/C_k)\,R_{equ}(C_k) = (-1/K_d)R_{equ}(C_k) + (1/K_d)R_{max} \tag{C4.6}$$

The *Scatchard* analysis is based on the binding responses when at equilibrium and the corresponding analyte concentrations (see Chapter C1). The slope of R_{equ}/C_k plotted against R_{equ} is $-1/K_d$ and its x-axis intercept is R_{max}/K_d (see the example plot in Fig. C4.6). Note that Eq. (C4.6) allows K_d and R_{max} to be calculated from the equilibrium responses at different concentrations without actually reaching the R_{max} plateau (the plateau corresponds to saturating the ligand

**Comment C4.3
The Langmuir isotherm**

The Langmuir isotherm was initially proposed for an equilibrium in which a gas is in contact with a solid:

$$X/M = abC_e/(1 + bC_e)$$

where X is the mass of gas adsorbed, M is the mass of the solid, C_e is the equilibrium concentration remaining in solution and a and b are Langmuir constants. The relation is called an *isotherm* because it is only valid at constant temperature.

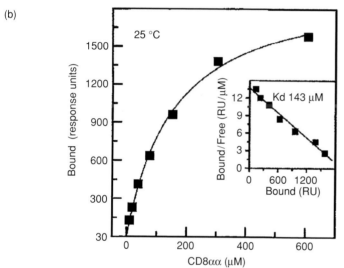

Fig. C4.6 Analysis of the binding between the ligand protein CD8αα and a HLA-A2 molecule (analyte). (a) Serial dilutions show a decrease in signal above the control. (b) The binding response and Scatchard plot (inset). (From Wyer et al., 1999.) (Figure reproduced with permission from *Immunity*.)

binding surface with analyte). In fact, with weak affinity interactions (K_d in the micromolar range ($\sim$0.1–500 µM)), saturation of the sensor surface is seldom achieved.

During an experiment, the analyte is usually diluted serially to achieve ideally a 100-fold range of concentrations. The initial concentration value should be of the order of the dissociation constant (Fig. C4.6).

The standard free energy change upon binding, ΔG°, can be calculated from the dissociation constant as outlined in Chapter C1. The standard entropy and enthalpy contributions to the binding can be calculated from the temperature dependence of ΔG°, via a non-linear fit of Eq. (C1.28).

C4.3.3 The determination of the k_{off} and k_{on} of an interaction

When the analyte concentration in the flow is reduced to zero, bound analytes that detach from the ligand are not replaced and the signal progressively returns to its initial value, at a rate that depends on the binding strength (Fig. C4.5). The dissociation rate k_{off} is deduced from the decrease of the response (R) over time (t):

$$R(t) = R(t_0) \exp[-k_{\mathrm{off}}(t - t_0)] \qquad \text{(C4.7)}$$

The time t_0 denotes the instant the signal started decreasing, i.e. when more molecules detach from the ligand surface than bind to it.

In the association phase, the measured signal increases rapidly, depending on the concentration C of analyte. Equation (C4.4) can be reformulated as

$$\frac{\mathrm{d}R(t)}{\mathrm{d}t} = -(k_{\mathrm{on}} \times C_{\mathrm{k}} + k_{\mathrm{off}})\, R(t) + k_{\mathrm{on}} C_{\mathrm{k}} R_{\mathrm{max}} \qquad \text{(C4.8)}$$

and the k_{on} rate determined from the linear regression plot of d$R(t)$/dt against $R(t)$ in the association phase (Fig. C4.4). Fast association rates make this estimate quite error prone, and k_{on} is better established from

$$k_{\mathrm{on}} = K_{\mathrm{d}} \times k_{\mathrm{off}}$$

Activation processes

The rate of a reaction measured as a function of temperature provides information on the activation thermodynamics parameters. Recall Eq. (C1.31)

$$\ln\left(\frac{k}{T}\right) = \ln\left(\frac{k_{\mathrm{B}}}{h}\right) + \frac{\Delta S^*}{R} - \frac{\Delta H^*}{RT}$$

In the binding experiment, the rate constant, k, in the equation is equal to k_{off} for the dissociation interaction and $C^\circ k_{\mathrm{on}}$ for the association reaction, where C° is the standard concentration (1 M). The activation enthalpy and activation entropy are then obtained directly from the slope and intercept of the *Eyring plot*, $\ln(k/T)$ versus $1/T$.

C4.4 Experimental analysis

C4.4.1 Scope of analytes

In early experiments, biosensors were used mainly for analysis of large molecules ($>5\,\mathrm{kDa}$), because the refractive index signal is mass-dependent. Biosensors were able to identify *E. Coli* strains from which ligands they bound, and to measure binding affinities between viral particles. Current biosensors are sufficiently sensitive to detect directly analytes as small as 200 Da. Even smaller molecules have been characterised indirectly through *inhibition or competition assays*, in which the strength of binding is measured by mixing the molecule with another much bulkier molecule that also binds the ligand. The affinity of the smaller molecule for the ligand can be calculated from the assay, which determines how the extent to which the bulkier analyte is prevented from binding by the addition of the smaller molecule. A *caveat* of the method is that the two analytes must bind at the same site on the ligand.

C4.4.2 Experimental controls and pitfalls

A number of experimental controls must be performed to assess the reliability of the analysis and avoid potential pitfalls. Non-specific binding is assessed by using blank cells that do not contain ligand. The Scatchard-like plot assumes that the maximum binding capacity of the receptor to the ligand is invariant in time. This is often not the case and experimental procedures must be adapted accordingly. The ligand may lose its binding capacity, through, for example, denaturation or degradation, since experiments are usually performed at room temperature. Repeating the experiment at similar analyte concentrations is a potential way of assessing ligand decay. Concentrations of either ligand or analyte that are too high may cause steric hindrance and prevent molecules from binding due to overcrowding at the flow cell surface. The diffusion of analytes from bulk solvent to the ligand surface may be impeded. This would lead to wrong estimates for the on and off rates. Such *mass transport* effects can be identified by varying the flow rate of the analyte. Very fast on rates lead to ligand competition for a limited number of analytes and rebinding, and to an underestimate of the dissociation rate. Difficulty in measuring high-affinity interactions can also originate from a slow dissociation rate. For example, for an analyte whose k_{off} is in the order of 10^{-6} s^{-1} it would take almost three hours (10^4 s) for only 1% of bound material to dissociate from the ligand. The observed affinity should be confirmed by swapping the ligand and the analyte with each other. The new measurement series then indicates if the binding is influenced by the coupling method to the cell, by the presence of aggregates or by errors in determining concentrations of active analyte.

Biosensor instruments should not be used for certain types of molecular interactions. Attaching the ligand to a surface may alter its binding properties. Binding may also be precluded if large ligand conformational changes or oligomerisation are involved.

C4.4.3 Cell–cell interactions

Biosensors have been applied successfully to the study of cell–cell interactions, since the immobilisation of a normally membrane-bound ligand to a flow cell mimics to a certain extent the in vivo situation.

Antigen recognition by T cells is the key event controlling the adaptive immune responses and has been studied extensively with SPR biosensors. The T cell receptor (TCR) recognises antigens presented by major histocompatibility complex (MHC) molecules. The antigen is usually a peptide derived from proteins synthesised by the cell. The T cell checks for the presence of unusual peptide

Fig. C4.7 Comparison of the average thermodynamic parameters of thirty protein–protein interactions and of the TCR–peptide–MHC interaction. (From Willcox *et al.*, 1999.) (Figure reproduced with permission from *Immunity*.)

antigens as these indicate that something is wrong. For example, T cells recognise and destroy cells displaying viral antigens during a viral infection. The interaction between the MHC–peptide complex and the TCR controls the fate of the MHC-presenting cell. The measured affinity is usually low ($K_d \sim 0.1$–500 μM) as it is due to slow association and fast dissociation reactions. The large k_{off} (~ 0.01–5 s^{-1}) corresponds to half-lives of 70–0.1s and is highly significant because it indicates that, once formed, the TCR–peptide–MHC complex is more stable than other cell–cell recognition molecule interactions. SPR measurements of the binding at different temperatures further showed that the binding is characterized by unusually favourable enthalpic changes and highly unfavourable entropic changes (Figure C4.7 (see also Section C3.3.3, Fig. C3.7)).

A number of different MHC–peptide complexes and TCRs have been studied using biosensors, and a broad correlation between affinity/half-life and functional effect has been observed. This suggests that the duration of binding determines the outcome of TCR–peptide–MHC interactions; the longer interaction offers the opportunity for other T cell molecules to assemble at the contact point between the two cells, and to initiate the death of the antigen-presenting cell.

C4.4.4 SPR and mass spectrometry

Combining a biosensor with mass spectrometry makes it possible to link analyte binding and kinetic analysis with its identification.

Fig. C4.8 Binding of saquinavir and indinavir to HIV-1 protease: (a) MALDI-TOF spectrum showing saquinavir (670 Da) and indinavir (613 Da) eluted from the surface; (b) Simulated sensorgram from the mass spectrometry data, for the dissociation of saquinavir and indinavir to the sensor chip. The vertical line indicates the time at which the ratio between the two inhibitors should be 0.4, corresponding to the concentration ratio between species in (a). Dashed line, saquinavir plus indinavir; long dashed line, saquinavir; dotted and dashed line, indinavir. (From Mattei *et al.*, 2004.)

The biosensor can isolate binding partners for a receptor by analysing whole cell lysate. Molecules of interest are separated from the crude extract as they are retained by their ligand in the flow cell. The analyte is then eluted and subsequently identified by mass spectrometry. For proteins, a proteolytic step can be included in the mass spectrometric analysis, in order to identify post-translational modifications.

In another application, mass spectrometry and SPR can be combined to analyse the competition of two analytes for the same ligand. Figure C4.8 shows results on the competition between two drugs (saquinavir and indinavir) for a sensor surface containing immobilised HIV-1 protease. Mass spectrometry analysis identified the contribution of each inhibitor to the dissociation phase.

C4.5 Checklist of key ideas

- The phenomenon of SPR occurs when monochromatic, p-polarised light is reflected on a metal-coated interface between two media of different refractive index.
- A biosensor binding experiment involves immobilising a molecule on a surface (the ligand) and monitoring its interaction with a second molecule in solution (the analyte).
- A change in local macromolecular concentration is associated with a change in refractive index.
- If light is shone upon the interface between two transparent media of different refractive index, above a critical angle of incidence, total internal reflection occurs, i.e. the light coming from the side with higher refractive index reflects back.
- Refractive index changes in a biosensor experiment can be measured by SPR or interferometry.
- The binding progress curve is called a sensorgram.
- A number of tags have been developed to secure the ligand to the sensor chip.
- If the diffusion of the analytes from the bulk solvent to the chip's surface is impeded, this mass transport effect gives misleading binding data.
- Serial dilutions of analytes over a 100-fold range of concentrations are often used in biosensor experiments in order to provide a precise measure of the thermodynamics parameters.
- The dissociation rate k_{off} is deduced from the decrease of the binding response over time.
- The association constant k_{on} rate can be determined from the association phase using different concentrations of analyte.
- The *Eyring plot*, k/T against $1/T$, where k is k_{off} or $C^{o}k_{on}$, provides the values of the activation entropy and enthalpy of the reaction.
- Analytes can range in size from molecules with a mass of a few hundred daltans to whole cells.
- Cell–cell interactions have been mimicked in biosensor experiments.
- Combining SPR with mass spectrometry makes it possible to link binding and kinetic analysis with analyte identification.

Suggestions for further reading

Mullet, W. M., Lai, E. O. C. and Yeung, J. M. (2000). Surface plasmon resonance-based immunoassays. *Methods*, **22**, 77–91.

Schultz, D. A. (2003). Plasmon resonant particles for biological detection. *Curr. Opin. Biotech.*, **14**, 13–22

Barnes, W. L., Dereux, A. and Ebbesen, T. W. (2003), Surface plasmon subwavelength optics. *Nature*, **424**, 824–830.

Part D
Hydrodynamics

Chapter D1
Biological macromolecules as hydrodynamic particles

D1.1 History and introduction to bioliogical problems

Traditionally, hydrodynamics deals with the behaviour of bodies in fluids and, in particular, with phenomena in which a force acts on a particle in a viscous solution. Very eminent scientists, such as **Isaac Newton, James Clerk Maxwell, Lord Rayleigh (J. W. Strutt)** and **Albert Einstein**, started their careers with major contributions to the science of hydrodynamics. Note that not only are the discoveries from more than 100 years ago still highly relevant today, but also that they continue to stimulate important new developments in the field.

1731
The science of hydrodynamics arose from the classical book *Hydrodynamics* by **Daniel Bernoulli**, which contained the 'Bernoulli law' relating pressure and velocity in an incompressible fluid, as well as a number of its consequences. The next fundamental contribution to the field was in **1879** when **Sir Horace Lamb** published another classical book also named *Hydrodynamics*.

1821
Botanist **Robert Brown** described the random, thermal motions of small plant particles suspended in water, a phenomenon that was later named Brownian motion. In **1855 Adolf E. Fick** published a phenomenological description of translational diffusion and deduced the fundamental laws governing transport phenomena in solutions. In the **1990s**, the method of video-enhanced microscopy was proposed for the direct observation of Brownian motion of labelled macro-molecules in a membrane.

1846
J. L. M. Poiseuille produced a theory of liquid flow in a capillary. Based on this theory, **Wilhelm Ostwald** invented the viscometer and introduced its use in physical and chemical experiments. The instrument was later named after him. In **1962**, **Bruno Zimm** proposed an original design for a rotational viscometer, which operates at very low velocity gradients, and has been very useful for

measurements on asymmetric structures such as DNA, fibrous proteins and rod-like viruses.

1856

Sir George Stokes demonstrated that the coefficient of translational friction of a particle depends on its linear dimensions. He described the particle in terms of the radius of an equivalent sphere (the Stokes radius). The way was then clear for a direct determination of particle dimensions from hydrodynamic measurements. In **1880**, **Stokes** analysed rotational friction and deduced an expression to relate the rotational friction coefficient of a particle to its volume. In the **1930s**, **Francis Perrin** extended Stokes' formula to ellipsoids of revolution. He also presented equations that give the three translational friction coefficients as functions of the dimensions of a general ellipsoid.

1856

James Clerk Maxwell discovered that certain liquids became birefringent when they flowed. In the **1960s** and **1970s**, **Victor N. Tsvetkov** and **Roger Cerf** developed a detailed method to measure flow birefringence for macromolecular solutions. The method has proven to be effective in studying the flexibility and optical features of polymer and biological macromolecules that have no fixed, rigid structure.

1887

Osborne Reynolds pointed out that the ratio of inertial and viscous forces is a key feature for the characterisation of any fluid movement. In the **1970s**, **Howard Berg** and **Edward Purcell** applied this idea to describe the movement of different objects (from molecules to animals) in solution. It was shown that the movement of particles with molecular dimensions ($10-10\,000$ Å) was described in terms of so-called low Reynolds numbers. This means that biological macromolecules 'live' in a world without inertia.

1881

Albert Wiedemann proposed the term 'luminescence' to emphasise the difference between thermal equilibrium and non-equilibrium radiation emission. Among non-equilibrium processes, he studied light emission by molecules at room temperature caused by incident radiation. In **1945**, **Serguei Vavilov** proposed a way to determine the hydrodynamic volume of a particle by using experimental data on fluorescence polarization, a form of luminescence.

1896

John Kerr found that some solutions become birefringent under the influence of an electric field. Electric birefringence measurements became a way of obtaining information on the nature of dipole–dipole interactions and on the flexibility of macromolecules with either a natural or an induced dipole moment. The introduction of pulsed voltage in electric birefringence experiments and the development of the basic theory of transient phenomena for macromolecules, by **Henri Benoit** in **1950**, opened the way to studying biological macromolecules in solution by electric birefringence.

1905

Albert Einstein created the theory of Brownian motion. He characterised molecular motions in simple solutions and gases quantitatively. A year later he derived an equation relating the diffusion coefficient of a macromolecule in solution to its coefficient of translational friction and demonstrated that the specific viscosity of a suspension of rigid spheres is proportional to their volume fraction, and is independent of their radius. In **1940 Robert Simha** obtained the equation for the viscosity of a solution of ellipsoids of revolution and in **1981 Stephen Harding** and **Arthur Rowe** solved the viscosity equation for a three-axis ellipsoid.

1920

Theodor Svedberg invented the high-speed centrifuge, opening the epoch of analytical and preparative ultracentrifugation. He proposed the combined use of sedimentation and diffusion coefficients to obtain a direct estimate of particle molecular mass. In **1929 Otto Lamm** deduced a general equation describing the behaviour of the moving boundary in the ultracentrifuge field that was later used to propose several ways for determining diffusion coefficients of macromolecules during centrifugation. A new generation of ultracentrifuges, highly automated for data collection and analysis, appeared in the **1990s** and provided direct methods for the precise molecular mass determination of biological macromolecules in solution, from several hundreds to tens of millions of daltons.

1937

Albert Tiselius used the difference in their charge to separate macromolecules during a fractionation process, thus introducing electrophoresis into wide biochemical use. In the **1950s Oliver Smithies** showed that electrophoresis with molecular sieving of a gel can give much higher resolution than electrophoresis in free solution. In **1975 Patrick O'Farrell** developed the method of two-dimensional electrophoresis, which revolutionised protein biochemistry and is now used extensively in proteomics.

1964

Herman Z. Cummings, following the theory of **Robert Pekora**, published the first experimental paper on dynamic light scattering and demonstrated that the diffusion coefficients of latex particles in solution can be extracted by this method. This work confirmed the theoretical predictions of **Leonid Mandelstamm**, in **1923**, concerning the modulation of scattered light intensity by Brownian motion.

It marked the beginning of a new trend in structural biology for the rapid and accurate determination of macromolecular diffusion coefficients from dynamic light scattering.

1967—1985

A new theoretical formalism was developed to calculate hydrodynamic properties of biological macromolecules of different shapes **(Victor Bloomfield, J. Garcia de la Torre, Stuart Allison)**. The hydrodynamic properties of a particle can be calculated by modelling it by a set of spheres (beads) of different radius. In alternative approaches, the particle surface is modelled as a set of small equal-sized spheres, or as a set of panel elements. The formalism opened a way to calculate hydrodynamic characteristics for biological macromolecules of arbitrary shape.

1993

Joseph Hubbard and **Jack Douglas** proposed to using the mathematical similarity of the equations of hydrodynamics and electrostatics to perform model calculations of hydrodynamic parameters by using their electrostatic counterparts. This opened up new possibilities for accurate hydrodynamic calculations **(Huan-Xiang Zhou)**. The main point is that electrostatic calculations are much easier to perform than hydrodynamic ones.

1994—2000

The spectacular progress in solving protein and nucleic acid structures to high resolution by X-ray crystallography and NMR stimulated the development of novel approaches to calculate hydrodynamic parameters from atomic-level structural details. It was shown that the frictional parameters of a protein can be calculated with an accuracy of about $1-3\%$ from its atomic structure by including a hydration shell.

2000 and now

Modern hydrodynamics is undergoing a renaissance, and is one of the recognised approaches for determining the size, shape, flexibility and dynamics of biological macromolecules. Modern hydrodynamics includes many novel experimental physical methods: fluorescence photobleaching recovery to monitor the mobility of individual molecules within living cells; time-dependent fluorescence polarization anisotropy to calculate Brownian rotational diffusion coefficients for macromolecules; fluorescent correlation spectroscopy and localised dynamic light scattering to study the dynamical properties of macromolecules. But in spite of all these achievements we must remember that *hydrodynamics is a low-resolution method* (Comment D1.1). It operates on a few parameters only. The

---

Comment D1.1   Biologist's box: The units of force and viscosity in hydrodynamics

Because hydrodynamics is a technique developed decades ago, calculations have traditionally been performed in cgs units.

The dyne is the unit of force in the cgs system. 1 dyne is the force necessary to accelerate a one-gram mass by one centimetre per second per second:

$$\text{dyne} = \text{g cm s}^{-2}$$

All fluids possess a definite resistance to change of form. This property, a sort of inertial friction, is called viscosity. The unit of viscosity, defined as the tangential force per unit area (dyne cm^{-2}) required to maintain unit difference in velocity (1 cm s^{-1}) between two parallel planes separated by 1 cm of fluid, is the poise:

$$1 \text{ poise} = 1 \text{ dyne s cm}^{-2} = 1 \text{ g cm s}^{-1}$$

Kinematic viscosity is the ratio of viscosity to density. The cgs unit of kinematic viscosity is the stoke:

$$1 \text{ stoke} = 1 \text{ cm}^2 \text{ s}^{-1}$$

---

highest level of data interpretation that can be achieved by using direct methods is to define a particle as a three-axis body (Section D2.5.3).

D1.2 Hydrodynamics at a low Reynolds number

In order to construct reasonable physical models for flow systems involving biological particles, it is necessary to make a number of simplifications. In this section it is assumed that the flow is laminar and, further, that it is sufficiently 'slow' that inertial effects need not be considered in the equations of motion, which describe the movement of particles relative to fluid (solvent). The approximation is justified, since biological systems of interest consist of very small particles, and even though the particles move rapidly with respect to the container wall, in, for example, the viscosity and flow birefringent methods, they still move slowly with respect to the fluid surrounding them.

D1.2.1 Reynolds number

We consider an object moving with some velocity through a fluid of specific density and viscosity. The Reynolds number is a dimensionless parameter, which determines the relative importance of inertial and viscous effects

$$\text{Reynolds number} = \frac{\text{fluid density} \times \text{speed} \times \text{particle size}}{\text{viscosity}} = R = \frac{\rho u l}{\eta} \quad \text{(D1.1)}$$

The ratio was proposed as a significant intrinsic number to characterise a system more than 100 years ago by Reynolds. When the Reynolds number is low, viscous forces dominate. If it is high, inertial forces dominate.

D1.2.2 Movement at low Reynolds number

We calculate the Reynolds number for a virus 500 Å (5×10^{-6} cm) in diameter moving in water with a speed of order 10^{-3} cm s^{-1}. Taking $\rho = 1$ g cm^{-3} and $\eta = 10^{-2}$ g cm^{-1} s^{-1} we obtained a Reynolds number of 5×10^{-7}, i.e. the Reynolds number for the virus is negligibly small. A small Reynolds number means that the virus molecule will stop moving immediately when the force disappears. Of course, the virus is still subject to Brownian motion, so in reality it does not stop.

Calculations show that for large biological complexes including bacteria in water the Reynolds number is also very small (Comment D1.2). So all biological macromolecules from small proteins to bacteria live in a world without inertia where viscous forces predominate (Comment D1.3).

Comment D1.3
Definition of movement at very low Reynolds number

The best definition of movement at very low Reynolds number is by E. Purcell: 'What You are doing at the moment is entirely determined by the forces that are exerted on You *at the moment*, and by *nothing in the past*.' (Purcell, 1977).

Comment D1.2   Different objects in water

Bacteria

E. Purcell was the first to calculate the Reynolds numbers for bacteria and fish (Purcell, 1977). He considered that a bacterium is 10^{-4} cm in diameter and swims with a velocity of the order of 2×10^{-3} cm s^{-1}. Taking $\rho = 1$ g cm^{-3} and $\eta_0 = 10^{-2}$ g cm^{-1} s^{-1}, he obtained a Reynolds number of 10^{-5}, i.e. very small. The bacterium therefore lives in a world *without inertia*.

Fish

The same calculation for a fish of length $l = 10$ cm, moving with velocity ~ 100 cm s^{-1} in water yields a Reynolds number of about 10^5. This is an example of hydrodynamics at high Reynolds number. The fish lives in a water medium *with inertia*.

Whale

For a whale $l = 10$ m (1000 cm) moving with velocity 36 km h^{-1} (1000 cm s^{-1}) in water the Reynolds number is about 10^8. The whale swims in a water medium with *very large inertia*.

D1.3 Hydration

In hydrodynamic experiments, a biological macromolecule moves with a certain amount of bound solvent, thus defining the concept of a hydrated particle as a core of particle material and an envelope of bound water (see Section A3.3.3). Figure D1.1 shows that hydration is manifested as increased size or volume of the core particle. The hydrated volume, V_{hyd}, is larger than the 'dry' volume, V_{anh}, which can be obtained from molecular mass, M, and the partial specific volume, $\bar{v}$ of the protein:

$$V_{anh} = M\bar{v}/N_A \qquad (D1.2)$$

Protein hydration, δ (g g^{-1}), expresses the ratio of the mass of the bound water to that of the protein

$$\delta = \frac{\text{grams (water)}}{\text{grams (protein)}} \qquad (D1.3)$$

If ρ is the density of the solvent, then we have

$$\delta = (V_{hyd}/V_{anh} - 1)\rho\bar{v} \qquad (D1.4)$$

There are two interpretations of the δ value. The first is based on the uniform expansion hypothesis (Fig. D1.2). It originates in the classical representation of globular proteins as ellipsoidal particles. For a particle of arbitrary shape, uniform expansion assumes that the linear dimension, l, of the particle is expanded by constant factor, h,

$$h = u_{hyd}/u_{anh} \qquad (D1.5)$$

such that

$$h^3 = V_{hyd}/V_{anh} \qquad (D1.6)$$

It follows that in this representation h is related to δ by

$$h = (1 + \delta/\bar{v}\rho) \qquad (D1.7)$$

The uniform expansion is applicable for compact particles, but is not realistic for very elongated or rod-like particles (Comment D1.4).

The second interpretation of the δ value is based on the assumption that the anhydrous core is coated by a bound water shell which has a constant thickness t_h, measured in the direction normal to the protein surface (Fig. D1.3). The hydration shell is considered as an intrinsic property of all proteins (Chapter D3).

Early interpretations of hydrodynamic data led to hydration levels that varied widely from protein to protein. For example, hydration values deduced from diffusion coefficients and intrinsic viscosity are in a broad range, from 0.1 to more than 1 g of water per gram of protein. These results were obtained by modelling proteins as spheres or ellipsoids, for which the translation, friction and intrinsic viscosity are known analytically.

Fig. D1.1 Schematic presentation of a rigid hydrated particle. Hydration influences the overall shape of the protein in the sense of smoothing out some structural details such as pockets or cavities. (After Garcia de La Torre, 2001.)

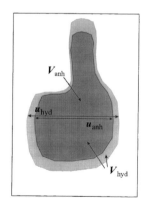

Fig. D1.2 Illustration of the δ value as a uniform expansion in globular proteins. (After Garcia de La Torre, 2001.)

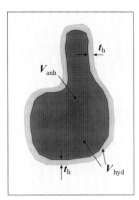

Fig. D1.3 Illustration of the interpretation of the δ value as a thick uniform hydration shell; t_h is the average thickness of the hydration shell. (After Garcia de La Torre, 2001.)

Comment D1.4   Uniform expansion for compact and elongated particles

For a typical globular protein in water if $\rho = 1$ cm^3 g^{-1}, $\bar{v} = 0.73$ g cm^{-3} and $\delta = 0.3$ g g^{-1}, then $h = 0.12$, which corresponds to 12% in linear dimensions and 41% in volume.

For a very elongated particle that is 20 Å in diameter and 200 Å in length hydration leads to an increase in diameter to approximately 22 Å. The same hydration applied to the particle length leads to an increase to 280 Å, i.e. 40 Å at each end. Evidently this result is not realistic because it leads to abnormal hydrodynamic solution properties.

The use of models obtained from detailed protein structures for the calculation of translation, rotational friction and intrinsic viscosity leads to a much smaller hydration range, from $0.3-0.4$ g of water per gram of protein, corresponding to less than a single molecular layer in the hydration shell. The development of this type of calculation (Chapter D3) allows hydrodynamic measurements to join other techniques such as NMR, IR spectroscopy, calorimetry, and small-angle X-ray and neutron scattering to provide a unified picture of protein hydration.

The following picture of protein hydration has emerged (Fig. D1.3). A protein is hydrated at a definite level, corresponding to a 1.2-Å-thick hydration shell on average. The local density of water in the hydration shell is about 10% higher than that of bulk water. If in a hydrodynamic experiment a hydration value that differs greatly from the usual levels were required to fit the data, this is an indication that the hydrodynamic equivalent model used is probably wrong (Comment D1.5).

Because of their net negative charge in solution, more water is associated with RNA or DNA molecules, leading to hydration values of about 0.6 ± 0.2 g g^{-1}. For glycosylated proteins and for carbohydrates, hydration values are also larger (0.5 g g^{-1}) owing to the generally higher affinity for water of these glycopolymers.

Comment D1.5   Estimation of hydration

The estimation of hydration from hydrodynamic properties of a protein is sensitive to several types of error because the extent of hydration is determined as 'a small difference of two large values' between hydrated and dry volumes (Eq. (D1.4)). The main source of uncertainty in the estimation of hydration from hydrodynamic parameters is the experimental errors in the data of hydrodynamic and other solution properties. Many of the tabulated data for common proteins are up to 50 years old, and it is evident that for a quantitative, more accurate evaluation of hydration more precise data are required.

D1.4 Determination of particle friction properties

D1.4.1 'Stick' and 'slip' boundary conditions

In classical hydrodynamic theory two extreme cases of solvent–particle inter-action called 'slip' and 'stick' boundary conditions are usually considered (Comment D1.6). In the 'slip' approximation, there is no interaction between solvent and particle and the solvent slips over the particle surface (Fig. D1.4(a)). The other extreme is represented by the 'stick' approximation, in which the first solvent layer sticks to the particle surface and moves with it (Fig. D1.4(b)).

It is important to note that the value of the coefficient in the equation con-necting measured and calculated hydrodynamic values depends on the boundary conditions (see Eqs. (D1.11) and (D1.12), (D1.14), and (D1.15) and (D1.16)). It is generally accepted that for small protein molecules (<2000 Da) 'slip' condi-tions are more applicable, whereas for the large Brownian particles (>5000 Da) 'stick' boundary conditions hold (see discussion in Chapter D3).

Comment D1.6
Mathematical definitions of stick and slip boundary conditions

Mathematical definitions of stick and slip boundary conditions are particularly complex. Interested readers can find them in the specialised literature (Hu and Zwanzig, 1974).

D1.4.2 Hydrodynamic experiments

Experiments in hydrodynamics can be divided into four groups. In the first, we find experiments that measure the *equilibrium velocity* of the particles. Translational diffusion is observed when the effective force arises from particle concentration gradients in the solution. When the acting force is gravitational (either under natural gravity or through ultracentrifugation), the phenomenon is called sedi-mentation. If the acting force is electrical in nature, the phenomenon is called electrophoresis.

The second group includes experiments in which the *rate of particle rotation* under the action of a pair of forces (a torque) is determined. If a velocity gradient in the solvent plays the role of an orienting force, the phenomenon is known as the Maxwell effect or flow birefringence. If the force is of an electrical nature, the phenomenon is called the Kerr effect or electric birefringence.

(a) Hydrodynamic (slip)

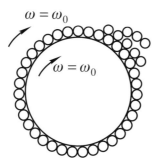

(b) Hydrodynamic (stick)

Fig. D1.4 Hydrodynamic slip (a) and stick (b) boundary conditions: ω is the frequency of rotation.

The third group is represented by experiments that measure *energy loss* due to friction of the molecule in the solution. The phenomenon is called viscosity.

In the fourth group are phenomena in which no external force acts on the particles and their displacements and rotations occur only under the action of *thermal agitation or Brownian motion*. The behaviour of a particle in fluorescence and dynamic light scattering experiments is of this type.

Table D1.1 presents a summary of hydrodynamic methods currently in use. Note, that in spite of the variety of experimental methods presented in Table D1.1, in the end only three parameters can be calculated: translational and rotational friction coefficients and intrinsic viscosity.

Table D1.1. *Hydrodynamic methods currently used by research in structural biology*

Experimental method	Measured parameter	Calculated parameters (depending on)
Translational diffusion (Chapter D3)	Translational diffusion coefficient	Translational friction coefficient (solvent viscosity, linear dimension, shape)
Sedimentation velocity (Chapter D4)	Sedimentation coefficient	Translational friction coefficient (solvent viscosity, molar mass, buoyancy, linear dimension, shape)
Electrophoretic mobility (Chapter D5)	Electrophoretic mobility	Translational friction coefficient (solvent viscosity, molar mass, linear dimension, shape)
Fluorescence correlation spectroscopy (Chapter D11)	Diffusion time	Translational friction coefficient (solvent viscosity, molar mass, linear dimension, shape)
Recovery of fluorophore after photobleaching (Chapter D3)	Diffusion time	Translational friction coefficient (solvent viscosity, molar mass, linear dimension, shape)
Direct hydrodynamic modelling experiments (Chapter D2)	Translational friction coefficient	
Electric and flow birefringence (Chapters D6 and D7, respectively)	Orientation angle and birefringence value	Rotational friction coefficient, molecule anisotropy (solvent viscosity, volume, shape)
Fluorescence depolarisation (Chapter D8)	Depolarisation coefficient	Rotational friction coefficient (solvent viscosity, volume, shape)
Dynamic light scattering (Chapter D10)	Correlation function and number fluctuation	Translational friction and/or rotational friction coefficients (solvent viscosity, linear dimension and/or volume, shape)
Viscosity (Chapter D9)	Specific viscosity	Intrinsic viscosity (solvent viscosity, volume, shape)

Table D1.1 also includes the method, which we called hydrodynamic modelling. In hydrodynamic modelling experiments, the actual translational friction coefficient of a rigid object is determined by observing the settling rates of a macroscopic model in a high-viscosity fluid at low Reynolds number. Note that in such experiments, the translational friction coefficients of the model are determined directly without any intermediate equations or calculations. A more detailed description of the method and a few results are given in Chapter D2.

D1.4.3 Hydrodynamic quantities

The hydrodynamic methods presented in Table D1.1 allow the *translational and rotational friction coefficients, and the intrinsic viscosity* of biological macromolecules to be determined. These, in turn, depend on the viscosity of the solvent and also on particle properties, which are of significant interest in the characterisation of macromolecular structures and interactions.

Fig. D1.5 Stream lines for flow around a sphere. The sphere is moving to the right at a constant velocity in a stream of viscous liquid. Solvent drag creates a force on the particle opposite to the velocity direction.

Translational friction coefficient

The translational friction influences translational diffusion, high-speed sedimentation and electrophoretic mobility. In each case, a force **F** acts on the particles and causes them to accelerate. The movement of a particle of mass m due to this force is described by the fundamental relation in mechanics: force is equal to mass times acceleration or rate of change in velocity

$$\mathbf{F} = m\,\mathrm{d}\mathbf{u}/\mathrm{d}t \tag{D1.8}$$

where **u** is the particle velocity and t is time.

In a viscous solution, the motion is opposed by solvent drag. The force, $\mathbf{F}_{\mathrm{frict}}$, due to this friction is proportional to the velocity of the particle and is in the opposite direction (Fig. D1.5). The proportionality constant is defined as the friction coefficient f:

$$\mathbf{F}_{\mathrm{frict}} = -f\mathbf{u} \tag{D1.9}$$

The negative sign denotes that the force is in the direction opposite to the velocity (see Comment D1.7).

When the two opposing forces are equal in magnitude, acceleration goes to zero, and the particle moves with a constant velocity, **u**, given by

$$\mathbf{u} = \mathbf{F}/f \tag{D1.10}$$

where **F** is the magnitude of each force, when $\mathbf{F} = -\mathbf{F}_{\mathrm{frict}}$.

Equation (D1.10) relates this constant velocity with the frictional coefficient and the magnitude of the two opposing forces. It provides the basis for the

Comment D1.7

In usual cgs system units the dimensions of the translational frictional coefficient are grams per second.

In usual cgs system the unit of the rotational frictional coefficient is $(\mathrm{second})^{-1}$.

Specific viscosity is dimensionless (Chapter D9).

Comment D1.8
Time required to reach constant velocity for macromolecules

We can estimate from Eq. (D1.10) that the time required to reach constant velocity in a macromolecular solution is very small. The molar mass of a 300 amino-acid residue protein is about 33 000 g mol^{-1}; its f value is 5×10^{-8} g s^{-1}, and m is about 5×10^{-20} g. The final velocity is achieved in no longer than 10^{-12} s (1 ps). This is very fast, and is close to the relaxation time of thermal vibrations in the molecule.

determination of the translational friction coefficient, by the experimental methods of group 1 in Table D1.1.

It follows from (D1.9) that the greater the particle friction, the greater the force that needs to be applied to make it move with constant velocity. If the ratio of the applied force to the coefficient of friction ($\mathbf{F}/f$) is large, the stable velocity may be too high to be reached in the experiment. However, if the ratio, $\mathbf{F}/f$, is sufficiently small, the final velocity is achieved almost instantaneously after application of the force (Comment D1.8).

Stokes derived the relation for a spherical particle in the two extreme solvent interaction cases. In the 'stick' approximation, the relation between the translational friction coefficient and solvent viscosity for a sphere of radius r is

$$f = 6\pi \eta_0 R_0 \qquad (D1.11)$$

The Stokes relation for 'slip' conditions is

$$f = 4\pi \eta_0 R_0 \qquad (D1.12)$$

Thus the friction is reduced by one third when the boundary conditions are changed.

Rotational friction coefficient

Rotational friction influences rotational diffusion, the Kerr effect, the Maxwell effect and fluorescence polarization. In analogy with translational motion we may define a rotation frictional coefficient. If a constant torque $\mathbf{F}_{rot}$ is placed on a particle in a fluid, the particle will reach a constant angular velocity, ω, after a transient period (Fig. D1.6). The parameter relating the angular velocity to torque is the rotational friction coefficient θ:

$$\omega = \mathbf{F}_{rot}/\theta \qquad (D1.13)$$

This equation is the analogue of Eq. (D1.4). It provides a basis for the determination of the rotational friction coefficient by many of the experimental methods presented in Table D1.1.

According to Stokes, in the 'stick' approximation the relation between the rotational friction coefficient and the solvent viscosity for a sphere of volume V is given by:

$$\theta = 8\pi \eta_0 V \qquad (D1.14)$$

In the case of 'slip' boundary conditions, the rotational friction coefficient is zero.

Fig. D1.6 Stream lines for flow around a sphere. The sphere rotates about an axis normal to the plane of the page with constant angular velocity, ω.

Intrinsic viscosity

In viscosity measurements we determine the local energy dissipation produced when large particles are introduced into a solvent (Fig. D1.7). Let η_0 denote the solvent viscosity in the absence of particles and η the viscosity when the number concentration of particles is C. It is assumed that the solution contains a monodisperse suspension of spherical particles. Einstein's relation for specific viscosity, in the case of 'stick' conditions, is

$$(\eta - \eta_0)/\eta_0 = 2.5\Omega \qquad \text{(D1.15)}$$

where Ω is the volume fraction of the solution occupied by spheres.

Under slip conditions the relation for the specific viscosity is

$$(\eta - \eta_0)/\eta_0 = 1\Omega \qquad \text{(D1.16)}$$

It should be noted that radius of the spheres does not enter into Eqs. (D1.15)–(D1.16).

Fig. D1.7 Stream lines for flow around a sphere in a liquid.

D1.4.4 Hydrodynamic equivalent bodies

Equations (D1.11), (D1.14) and (D1.15) are very useful for work on large biological macromolecules (more than several kilodaltons, see discussion in Chapter D3). They open the way to express a measured value for the friction coefficients and viscosity in terms of the radius, R_0, of a hydrodynamic equivalent sphere. R_0 has been named the Stokes radius. The concept can be extended to other hydrodynamic equivalent shapes (Fig. D1.8).

It should be clearly understood that the description of the particle in terms of a hydrodynamic equivalent does not mean that it has this shape, only that it has the same hydrodynamic properties. In spite of the fact that this appears to be a rough approximation, in many cases the hydrodynamic equivalent body describes the behaviour of the particle in a real experiment quite reliably.

In the case of a rigid sphere, only one parameter (the sphere volume or its radius) is sufficient to describe it fully. To describe an ellipsoid of revolution (the shape of a rugby football), it is necessary to determine two parameters, characterising its volume and axial ratio. To describe a three-axis ellipsoid of rotation, three parameters are required, characterising its volume and two axial ratios (Chapter D2).

The groups of methods presented in Table D1.1 differ in their comparative sensitivity relative to these parameters. Thus, all methods based on translational friction are sensitive to the linear dimensions of the particle, whereas methods based on rotational friction and viscosity are sensitive to its volume (the cube of linear dimensions).

(a)

Hemerythrin

(b)

Arabinose-binding protein

Fig. D1.8 (a) A sphere as the hydrodynamic equivalent body for hemerythrin; (b) an ellipsoid of revolution with axial ratio 2:1 as the hydrodynamic equivalent body for arabinose-binding protein.

D1.5 Prediction of particle friction properties

Modern hydrodynamics allows the prediction of the frictional properties of biological macromolecules of any shape. The computing procedure used essentially depends on the particle shape.

(a)

(b)

Fig. D1.10 Particles with a 'broken' shape have sharp edges. They include (a) the cube and (b) the circular cylinder. Their frictional properties can be calculated only as an approximation.

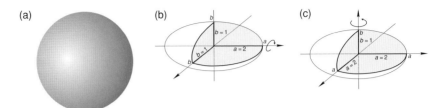

Fig. D1.9 Sphere (a) and two ellipsoids of revolution ((b), (c)), with equal volumes. A prolate ellipsoid (b) is a rod-like shape, generated by rotating an ellipse around its long semiaxis a; the two short semiaxes, b, are identical. The oblate ellipsoid (c) is a disc shape, generated by rotating an ellipse around its short semiaxis b; the two long semiaxes, a, are identical. For either kind of ellipsoid, the axial ratio (p) is defined as a/b, the ratio of the long to the short semiaxes. The near right octant of each ellipsoid is cut away to show the long (a) and short (b) axes.

D1.5.1 Particles of 'round' shape

Biological macromolecules of a 'round' shape can be approximated reasonably well by a sphere or a two-axis ellipsoid of revolution as hydrodynamic equivalents (Fig. D1.9). Hydrodynamic properties for such particles can be calculated analytically (Chapter D2).

D1.5.2 Particles with a 'broken' shape

As a first approximation, many biological macromolecules can be modelled as rigid rods. These include small, rigid, rod-like monomers, oligomers, duplex oligonucleotides, α-helical polypeptides, and rod-like proteins and viruses. The difficulty for all hydrodynamic theories that attempt to model the translation and rotation of rod-like molecules arises from the effects of the sharp ends of the rod ('end correction'). The friction of a *cube* which has sharp edges and corners is one of the major unsolved problems of hydrodynamics (Fig. D1.10).

At present there are several different approaches to calculating the frictional properties of particles with a 'broken' shape: modelling the *entire* particle with a set of spheres, modelling the *surface* of a particle with a set of small equal spheres, modelling the *surface* of a particle with a set of panel elements, and using their electrostatic counterparts. These approaches allow only an approximation of the friction properties of particles with a 'broken" shape (Chapter D2).

Fig. D1.11 Examples of particles of arbitrary shape: t-RNA and immunoglobulin. Their frictional properties can be calculated only as an approximation.

D1.5.3 Particles of arbitrary shape

The detailed description of a particle with a more complex shape (Fig. D1.11) might require the determination of a far greater number of parameters than can be obtained from hydrodynamic measurements alone. The approaches above are used to calculate hydrodynamic characteristics of such biological macromolecules with good accuracy (Chapter D2).

D1.5.4 Particles with a known three-dimensional structure

The impressive progress in solving protein and nucleic acid structures to high resolution by X-ray crystallography and NMR stimulated the development of novel approaches to calculating hydrodynamic parameters from atomic-level structural details. Figure D1.12 shows the structure of lysozyme as an example. Several algorithms have been proposed to calculate its hydrodynamic properties including: modelling the particle with a set of beads, modelling the protein with a set of beads in C-α carbon positions, and modelling the surface of the protein with a set of panel elements (Chapter D2). Comparison with experiment shows that the algorithms predicted diffusion coefficients with an accuracy about of 1% (Chapter D3) and intrinsic viscosity with an accuracy of about 2–3% (Section D2.4). Special approaches have been proposed for DNA molecules (Chapter D2).

Fig. D1.12 The structure of the lysozyme molecule with atomic resolution. The frictional properties of the molecules can be predicted with accuracy of about a few per cent.

D1.6 Checklist of key ideas

- A dimensionless parameter, the Reynolds number, characterises the ratio of inertial and viscous forces acting on a particle in a viscous medium.
- The movement of biological particles with molecular dimensions (10–10 000 Å) is described in terms of low Reynolds number hydrodynamics, where viscous forces dominate.
- Numerical coefficients in equations relating hydrodynamics parameters with solvent viscosity and particle dimensions depend on 'slip' and 'stick' boundary conditions.
- In translational diffusion, velocity centrifugation and electrophoretic mobility experiments the *equilibrium velocity* of the particle under the action of force is measured.
- Using the Maxwell effect (flow birefringence) or the Kerr effect (electric birefringence) the *rate of particle rotation* under the action of a pair of forces (a torque) is determined.
- In viscosity experiments *energy loss* due to friction as the particle moves through the viscous environment is determined.
- Fluorescence and dynamic light scattering experiments measure *displacements and rotations* of the particles under only the action of Brownian motion (no external force).
- The results from all hydrodynamics methods can be described by only three hydrodynamic parameters (translational and rotation frictional coefficients and intrinsic viscosity).
- Hydrodynamic properties for particles with a 'round' shape (spheres, two-, three-axis ellipsoids) can be calculated analytically.

- Hydrodynamic properties for particles with a 'broken' shape (cubes, circular cylinders) can only be approximated.
- Hydrodynamic properties for globular proteins whose detailed atomic structures are known can be calculated by using one of several algorithms with an accuracy of about 1–3%, assuming a 1-Å hydration shell (for all proteins).

Suggestions for further reading

Hydrodynamics at low Reynolds number

Purcell, E. M. (1977). Life at low Reynolds number. *Am. J. Phys.*, **45**, 3–11.

Hydration

Kuntz, I. D., and Kauzmann, W. (1974). Hydration of proteins and polypeptides. *Adv. Prot. Chem.*, **28**, 239–345.

Finny, J. L. (1996). Overview lecture. Hydration processes in biological and macromolecular systems. *Faraday Discuss. Chem. Soc.*, **103**, 1–395.

Wuthrich, K., Billeter, M., *et al.* (1996). NMR studies of the hydration of biological macromolecules. *Faraday Discuss. Chem. Soc.*, **103**, 245–253.

Garcia de la Torre, J. (2001). Hydration from hydrodynamics. General consideration and applications to bead modelling to globular proteins. *Biophys. Chem.*, **93**, 159–170.

Zhou, H.-X. (2001). A unified picture of protein hydration: prediction of hydrodynamic properties from known structures. *Biophys. Chem.*, **93**, 171–179.

Perkins, S. J. (2001). X-ray and neutron scattering analyses of hydration shells: a molecular interpretation based on sequence predictions and modelling fits. *Biophys. Chem.*, **93**, 129–139.

Engelsen, S. B., Monteiro, C., Herve de Penhoat, C., and Perez, S. (2001). The dilute aqueous solvation of carbohydrates as inferred from molecular dynamics simulations and NMR spectroscopy. *Biophys. Chem.*, **93**, 103–127.

Determination of particle friction properties

Happel, J., and Brenner, H. (1973). *Low Reynolds Number Hydrodynamics*. Second edn. Groningen: Noordhoff Int.

Harding, S. E. (1998). The intrinsic viscosity of biological macromolecules. Progress in measurement, interpretation and application to structures in dilute solution. *Prog. Biophys. Mol. Biol.*, **68**, 207–262.

Hu, C.-M., and Zwanzig, R. (1974). Rotational friction coefficients for spheroids with the slipping boundary conditions. *J. Chem. Phys.*, **60**, 4354–4357.

Prediction of particle friction properties

Brune, D., and Kim, S. (1994). Predicting protein diffusion coefficients. *Proc. Natl. Acad. Sci. USA*, **90**, 3835–3839.

Zhou, H-X. (1995). Calculation of translational friction and intrinsic viscosity. II. Application to globular proteins. *Biophys. J.*, **69**, 2298–2303.

Garcia de la Torre, J., Huertas, M. L., and Carrasco, B. (2000). Calculation of hydrodynamic properties of globular proteins from their atomic-level structure. *Biophys. J.*, **78**, 719–730.

Allison, S. A. (2001). Boundary element modelling of biomolecular transport. *Biophys. Chem.*, **93**, 197–213.

Chapter D2
Fundamental theory

D2.1 Historical review

1845

G. Stokes showed that the translational friction for a sphere is proportional to its radius, and to the viscosity of its surrounding solvent. In **1856** he demonstrated that for small angular velocity the rotation of the sphere may be characterised by a single parameter, which is proportional to the linear dimensions cubed.

1893

D. Edwardes calculated two frictional coefficients of the rotation for an ellipsoid of revolution: one for rotation around the axis of revolution and another for rotation around a direction normal to the first. In **1906 A. Einstein** showed that rotation of the sphere in Stokes' approximation may be characterised by a single constant which has the dimensions of time. In **1928 R. Gans** used the Edwardes frictional coefficients for an ellipsoid of revolution to calculate the ratios of the principal relaxation times to the relaxation time of a sphere of equal volume. In **1936 F. (Francis) Perrin** presented equations that give the three rotational coefficients and three rotational relaxation times as functions of the dimensions of a three-axis ellipsoid. These equations could not be expressed in terms of elementary functions. In **1960 L. D. Favro** showed that diffusion coefficients related to the rotational motion of a general particle involve five relaxation times; when two of the diffusion coefficients are equal the number of relaxation times is reduced to three. In **1977 E. Small** and **I. Isenberg** solved Perrin's equations for the rotational diffusion of a general ellipsoid using a numerical integration procedure. Rotational friction coefficients, rotational relaxation times and the five exponential terms in the fluorescence anisotropy can be expressed as functions of the axial ratios of the ellipsoid. In **1981 P. J. Hagerman** and **B. H. Zimm** presented a Monte Carlo approach to the analysis of the rotational diffusion of worm-like chains. A quantitative relationship between the longitudinal (longest) rotational relaxation time of a linear polymer of a given axial length and the flexibility (persistence length) was established.

1906

A. Einstein was the first to treat the viscosity of suspensions of rigid spherical particles that are large relative to the size of the solvent molecules. He showed that the specific viscosity of the solution is proportional to the volume fraction of the solution occupied by the particles and does not depend on the absolute size of the spheres. In **1945–1951 W. Kun** and **H. Kun, J. J. Kirkwood** and **P. L. Auer** completed the theory of intrinsic viscosity for long rods and ellipsoids. In **1967 V. A. Bloomfield, W. O. Dalton** and **K. E. Van Holde** presented the shell model which in theory should be the ideal solution for the rigorous calculation of the viscosity of macromolecules. However, this approach turned out not to be practical owing to the large number of beads necessary to model the structure and the consequent extraordinarily lengthy computation time. In **1970–1977** several groups independently (**K. Tsuda, J. A. McCammon** and **J. M. Deutch, H. Nakajima and Y. Wada, J. Garcia de la Torre and V. A. Bloomfield**) developed a general theory for the calculation of the intrinsic viscosity of complex rigid macromolecules. In all these studies the shape of the macromolecule was modelled as an array of a relatively small quantity of spherical beads. In **1981 S. E. Harding, M. Dampier** and **A. J. Rove** obtained a solution for the viscosity for a three-axis ellipsoid assuming that the particles rotate on average with the local undisturbed angular velocity of the fluid.

1927

C. Oseen derived a hydrodynamic friction tensor describing the hydrodynamic interaction between point sources of friction fixed in space. In **1932 J. Burgers** applied Oseen's tensor to objects of various geometry. In **1953 J. Kirkwood** introduced the hydrodynamic description of a chain macromolecule and proposed a computationally simple formula, based on the Oseen's original tensor, for the orientation-averaged translation friction of any assembly of equal-size spheres. In **1963 J. E. Hearst** applied Kirkwood's formalism, which was developed for the case of translational friction of macromolecular complexes composed from identical subunits, to the rotational motion of flexible and semiflexible macromolecules. It opened the way to studying DNA by different hydrodynamic methods. In **1967 V. Bloomfield** pioneered the calculation of the hydrodynamic properties of arbitrarily shaped particles and introduced modelling solid objects by a shell of small beads describing the surface. In the **late 1960s J. Rotne and S. Prager** and, independently, **H. Yamakawa** proposed new approaches to hydrodynamic interactions.

1932

H. Staudinger proposed using intrinsic viscosity for molecular mass determination of polymers. He found that the dependence of the intrinsic viscosity $[\eta]$ on the molecular mass M for a homologous series of polymers can be expressed by a simple formula of the type $[\eta] \approx M^{\alpha}$. It was understood later (**H. Mark,**

R. Houwink, **W. Kuhn** and **H. Kuhn**) that the constant α is related to the molecular conformation of macromolecules in solution.

1934

F. Perrin obtained the analytical solution to the problem of the hydrodynamic motion of a solid ellipsoid of revolution. He presented equations that give the three translational friction coefficients as a function of the dimensions of a three-axis ellipsoid. These equations involve a set of elliptic integrals and cannot be expressed in terms of elementary functions. In **1977 E. Small** and **I. Isenberg** performed a numerical integration of the Perrin elliptical integrals for the translational friction coefficients of a general ellipsoid and showed that the restriction to ellipsoids of revolution is not necessary. In the **early 1980s**, **S. Harding** and **A. Rowe** introduced a three-axis hydrodynamic ellipsoid as a more realistic model for description of the structure of biological macromolecules.

1936

A. Peterlin introduced a rotational diffusion coefficient by means of which the intensity of Brownian motion can be characterised. He computed the distribution function for the orientation of the major axes of the ellipsoid at any time at a given value of the gradient velocity taking into account only the energy dissipation due to the rotation of the particle in the hydrodynamic field. In **1940 R. Simha** extended Peterlin's result and also took into account the energy dissipation arising from Brownian motion. He obtained the equation for the viscosity of a solution of ellipsoids of revolution in the limiting case where the particles have a random orientation due to the Brownian motion and can be considered to be rotating with uniform angular velocity. Simha's result was rigorously reproduced by **N. Saito** using different boundary conditions.

1941

J. L. Oncley presented the first attempt to solve the hydration problem for the proteins. He demonstrated that a choice always exists between a prolate and an oblate ellipsoid, each of specified axial ratio, even when a reasonable degree of hydration has been assumed. In 1953 in an attempt to overcome ambiguity in solving the problem of the volume–axial ratio, **H. A. Sheraga** and **L. Mandelkern** proposed combining frictional coefficients and intrinsic viscosity. The first volume-independent β-function obtained was almost completely insensitive to the axial ratio of the oblate ellipsoids and can only be used to distinguish between prolate and oblate ellipsoids for highly asymmetrical particles. A whole series of combinations of different hydrodynamic parameters have been proposed: in **1954**, the ratio of the sedimentation regression coefficient to the intrinsic viscosity (**M. Wales** and **K. E. Van Holde**); in **1970**, the combination of the translational frictional coefficient with the harmonic mean rotational relaxation time (**P. G. Squire**); in **1977**, the ratio of the sedimentation regression to the intrinsic viscosity (**A. J. Rowe**); also in **1977**, the combination of a frictional coefficient and an operationally defined molecular covolume (**P. D. Jeffrey** with colleagues); in **1980–81**, the combination of the harmonic mean rotational relaxation time with the intrinsic viscosity and the intrinsic viscosity with the molecular covolume (**S. E. Harding**). Some of these combinations lead to functions that are potentially capable of distinguishing between an oblate and a prolate ellipsoid revolution except as the limiting shape of a sphere is approached.

1960

S. Broersma was the first to examine the frictional properties of a circular cylinder. He obtained that the transverse (smallest) rotational diffusion coefficient is roughly proportional to the inverse cube of the length. It was later found that his formula is correct for a long cylinder only. In **1965 H. Brenner** made an essential contribution to the hydrodynamics of rigid particles. He obtained the analytical solution for intrinsic viscosity for ellipsoids, long cylinders and dumbbells. He also showed that the translational and rotational motions of a rigid macromolecule immersed in a viscous fluid are intrinsically coupled if the particle has a complicated shape.

1975

G. Youngren and **A. Acrivos** introduced the boundary element technique (BE) to describe the surface of the particle with a set of panel elements ('platelets'). In **1992–1995 D. Brune, S. Kim** and **H.-X. Zhou** applied the BE technique to calculations of the frictional properties of proteins. In **1996 S. Allison** used the BE technique to model the free solution electrophoretic mobility of short DNA fragments.

1978

D. Teller made a first attempt to calculate the friction coefficient of proteins from atomic coordinates. **R. Venable** and **R. Pastor** calculated the frictional properties of proteins using a detailed picture of the distribution of different amino acids in proteins. **J. Garcia de la Torre** extended the calculations to nucleic acids. In **1995–1999** several approaches were proposed for constructing hydrodynamic models of the biological macromolecules on the basis of their atomic coordinates. Special approaches were put forward to try to model the frictional properties of short DNA fragments and closed circular DNA.

1982

S. E. Harding and **A. J. Rowe** proposed a new approach to modelling biological macromolecules using a three-axis ellipsoid. The new method involved the graphical intersection of two three-axial hydrodynamic functions involving intrinsic viscosity, sedimentation and electric birefringence decay. A year later the same authors proposed an alternative method including intrinsic viscosity, sedimentation and the harmonic mean relaxation time. In **1997** the new concept for the description of biological macromolecules was introduced, based on an ellipsoidal representation of macromolecular shape in solution using *universal shape functions*. Several computer programs *ELLIPS* (**Harding** and colleagues) and *SOLPRO* (**Garcia de la Torre** and colleagues) that provide a method for the unique evaluation of the three-axial dimensions of biological macromolecules without having to guess the value of hydration and volume are now available to users.

1993

J. Douglas and **J. Hubbard** discovered a relationship between hydrodynamics and electrostatics and opened up new possibilities for hydrodynamic calculations. In **1995 H.-X. Zhou** calculated the frictional coefficients of several proteins using the relationship between hydrodynamics and electrostatics. He also proposed an algorithm for the calculation of intrinsic viscosity of arbitrarily shaped particles and globular proteins.

2000 to now

Hydrodynamic properties of different rigid biological macromolecules can be calculated with good accuracy by covering the surface of the macromolecules with a shell of small beads or by dividing the surface into small elements. Automated methods of converting crystallographic data into a bead model have been proposed for nucleic acids and proteins. A method for the evaluation of the three-axial dimensions of biological macromolecules without having to guess the value of hydration and volume now are available.

D2.2 Translational friction

D2.2.1 Regularly shaped particles

Sphere

In the simplest view, a frictional force arises in a fluid because of the attraction between the molecules in the fluid. In order to move a solid object through the fluid, it is necessary to also move some solvent molecules. This displacement is shown as stream lines in Fig. D2.1. The solvent molecules closest to the moving particle are the most perturbed. The perturbation caused by the particle decreases to zero as the distance from it increases. To compute the frictional force, we must calculate the force required to maintain the perturbed velocity distribution of the solvent molecules. This force is related to a property of the fluid called the viscosity. For the motion of a spherical particle, we need to find an equation that relates the particle's coefficient of translational friction, f, to the fluid viscosity, η_0. The actual derivation is extremely difficult because one must calculate explicitly how the sphere's motion induces velocity gradients in the fluid. According to

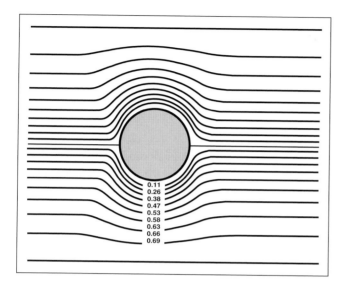

Fig. D2.1 The flow lines around a sphere, radius R_0, moving to the right through an incompressible viscous fluid at constant velocity v_d. The numbers on the flow lines at $\theta = -90°$ indicate the magnitudes of v_θ at these points in units of v_d. Even at the outermost flow lines shown in the figure, which are at a distance from the sphere that is equal to the sphere's diameter, the fluid moves approximately 30% faster than the sphere. Individual erratic motions of water molecules due to Brownian motion have been neglected.

Stokes, the translational friction of a sphere is proportional to its radius, R_0, and the viscosity η_0 of the solvent through which the particle moves:

$$f_0 = 6\pi \eta_0 R_0 \qquad\qquad (D2.1)$$

There are three important remarks concerning Stokes' equation. First, the coefficient (i.e. 6π in Eq. (D2.1)) is determined by the boundary conditions of fluid flow at the surface of the particle ('stick' or 'slip' conditions). The number 6 in Eq. (D2.1) indicates the use of stick conditions (Chapter D1). Second, from a mathematical point of view, Stokes' approach is correct when the solvent is considered as an unstructured medium. It is evident that this condition holds if the molecules under consideration are much larger than the solvent molecules. It is generally accepted that for proteins with a molecular mass in excess of 5000 Da (so-called 'large' molecules), the motion of molecules in solution follows a continuous flow pattern. However, the low limit for the correct application of Stokes' law (for so-called 'small' particles) is still under discussion (see Chapter D3). Third, the equation was derived with the assumption that inter-particle interactions are absent. This condition is realised upon extrapolation to infinite dilution of the solution.

Since the frictional force is directly connected with the surface area of the particle studied and a sphere has the smallest surface area of all geometrical objects, it can be concluded that in general frictional coefficients for a spherical molecule are smaller than for any non-spherical molecule of the same volume.

Ellipsoid of revolution

Theoretically, the influence of the asymmetrical shape of the particle on its hydrodynamic properties can be strictly taken into account if it is modelled by an

Fig. D2.2 Perrin function $F(p)$ for oblate and prolate ellipsoids revolution. NB. Now there is no need to follow the customary practice of quoting extensive tables of the Perrin function. Useful hydrodynamic shape functions are available from an easy to use computer program. (Harding *et al.*, 1977.)

Comment D2.1

Three-axis hydrodynamic ellipsoid

At the beginning of the 1980s, Harding and Rowe proposed using a three-axis hydrodynamic ellipsoid as a model for the structure of biological macromolecules. This is a more realistic model than one based on a two-axis ellipsoid. An explicit expression for the translational friction of a three-axis ellipsoid can be found in literature (Harding and Rowe, 1982).

ellipsoid of revolution (either elongate or oblate). From general considerations it is clear that the frictional force of asymmetrical particles depends on their orientation relative to the flow direction. Thus, an elongated particle has a smaller friction when it is oriented along the flow as compared to that when it is oriented perpendicular to the flow. In the averaged case the solution is

$$f_0 = 6\pi \eta_0 R_0 F(p) \tag{D2.2}$$

Theoretical values of the Perrin function $F(p)$ for prolate and oblate ellipsoids revolution are presented in Fig. D2.2. Three important conclusions can be drawn from this figure. First, as the axial ratio increases, the numerical values of the Perrin function increase relatively slowly. Thus, the Perrin function changes by only 4% between a sphere and an ellipsoid of the same volume with an axial ratio of 2. At the same time the analogous function describing rotational friction and intrinsic viscosity changes by about 12% (see below). Second, at the same axial ratio, the coefficient of translational friction for a prolate ellipsoid is always larger than that for an oblate one. Third, for an asymmetrical particle, the coefficient of its translational friction depends on two parameters: its volume, given as the radius of an equivalent sphere, and the axial ratio. Therefore the calculated coefficient of translational friction can always be explained by either a spherical particle of a definite volume, or an elongated particle of a smaller volume. The three-axis hydrodynamic ellipsoid is discussed in Comment D2.1.

Circular cylinder

A variety of biological macromolecules, ranging from short DNA fragments to filamentary viruses like the tobacco mosaic virus, exist as elongated shapes of uniform thickness. The proper hydrodynamic model in this case is the circular cylinder with flat ends.

For cylinders of length L, radius r and ratio $L/r = p$ the average friction coefficient f is

$$f = 6\pi \eta_0 L / (2 \ln p + \gamma) \tag{D2.3}$$

where γ is a function of p and depends on how we take into account the end effects.

For relatively short cylinders $\gamma(p)$ is

$$\gamma(p) = 1.65 - 4(1/\ln p - 0.43)^2 - 8(1/\ln p - 0.30)^2 \tag{D2.4}$$

Equation (D2.4) has a defect. It is only correct when $p \to \infty$. Now, it is accepted that the limiting value of γ is

$$\gamma = 0.386 \quad \text{as } p \to \infty. \tag{D2.5}$$

A good approximation of $\gamma(p)$ is

$$\gamma(p) = 0.312 + 0.561/p + 0.100/p^2 \quad 2 < p < 20 \tag{D2.6}$$

D2.2.2 Arbitrary shaped rigid particles

Figure D2.3 shows the shapes of a number of biological macromolecules determined from X-ray crystallography data. A sphere or an ellipsoid of revolution can quite satisfactorily describe the hydrodynamic features of several biological molecules. A three-axis spheroid, in particular, describes the shape of most globular proteins. However, many biological macromolecules, e.g. tRNA, immunoglobulins and phosphoglyceratemutase, whose shapes are far from being spherical or ellipsoidal (see Fig. D2.3) require more complex models for the calculation of their hydrodynamic properties. Over the last three decades, three distinct but related modelling techniques have been proposed to describe particles of complex form.

In the 'bead' model approach, which has its roots in the pioneering work of Kirkwood and Riseman, the particle is modelled as a set of spheres (beads) of equal or unequal size. In the 'shell' model approach, which has its roots in Bloomfield's work, the surface of a particle is modelled as a set of small equal-sized spheres. In a formal sense, the 'bead' and 'shell' model approaches are closely related with the fundamental hydrodynamic units being *spheres*. The 'boundary element' (BE) approach has its roots in work of Youngren and Acrivos. In this approach the surface of a particle is modelled via a set of small panel elements (platelets). Here the fundamental hydrodynamic units are *platelets*.

Modelling the entire particle with a set of spheres ('beads' method)
In a general sense, a 'bead' model represents a particle as an array of spherical frictional elements (beads) to each of which individual Stokes' law frictional coefficients are assigned.

The first key problem is *how to treat* hydrodynamic interactions between the different spherical elements. There are two approaches: the rigorous tensor method and the double sum approximation method (Comment D2.2).

Comment D2.2   Classical and rigorous bead model treatments

In rigorous bead model treatments, a tensor equation must be used because the velocity perturbation is not necessarily parallel to the force that causes it. The frictional forces are obtained as the solution of a system of $3N$ linear equations. Accordingly, the required computer time for the calculations is proportional to N^3, and increases greatly with N. This conflicts with the need to use a large number of beads to reproduce fine structural details.

In classical Kirkwood–Riseman theory, approximations in the treatment of hydrodynamic interactions lead to simple equations in which hydrodynamic properties are computed from double sums over the elements. This requires a number of operations of the order of N^2, so that the computer time needed for high N is much smaller than for rigorous treatments.

Fig. D2.3 Shapes of different biological molecules determined from X-ray data (magnification 4×10^6). The scale is such that individual atoms have a diameter of 1–2 mm. The orientation of every molecule is chosen so that specific features of its shape are defined. (After Goodsell and Olson, 1993.)

Fig. D2.3   (*cont.*)

Adenylate
Kinase

Carbonic Anhydrase

Aconitase

Phosphoglycerate
Kinase

Dihydrofolate
Reductase

Cytochrome P-450$_{cam}$

Chloramphenicol
Acetyltransferase

Phosphofructokinase

Alcohol Dehydrogenase

Citrate Synthase

Phosphoglycerate
Mutase

D-Glyceraldehyde-3-phosphate
Dehydrogenase

Glycolate Oxidase

Malate
Dehydrogenase

Aspartate Carbamoyltransferase

Tryptophan
Synthetase

Tyrosyl-tRNA
Synthetase

Glutamine Synthetase

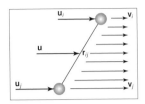

Fig. D2.4 Hydrodynamic interaction between spheres moving with velocities $\mathbf{u}_i$ and $\mathbf{u}_j$. The fluid is moving at velocities $\mathbf{v}_i$ and $\mathbf{v}_j$. The centre of mass is moving with velocity $\mathbf{u}$ and r_{ij} is the distance between two spheres. As it moves through the fluid, each sphere perturbs the velocity distribution of the fluid nearby. (After Cantor and Schimmel, 1980.)

The second problem is *how to build* the model so that the model's size and shape reflect that of the given particle. There is always a conflict between offering a high-resolution description and the required computing time. Below we shall briefly describe potential solutions to these two problems.

In the 1950s, J. Kirkwood proposed a simple method for describing the hydrodynamic features of flexible macromolecules: a linear or coiled polymer is approximated as a string of beads of the same radius each with identical hydrodynamic properties.

The hydrodynamic interaction between the beads endowed with the friction of Stokes' spheres was treated by the Ozeen–Burgers method. Figure D2.4 schematically shows the hydrodynamic interaction between two spheres of a macromolecule fixed in the space. According to the Ozeen–Burgers approach, the solvent velocity ($\mathbf{v}_i$) at the position of subunit i is the sum of the unperturbed velocity $\mathbf{u}$ and of a perturbation term that arises from hydrodynamic interactions with the other spheres of the macromolecule. The perturbation of sphere i depends on the frictional forces $\mathbf{F}_j$ exerted by all the other subunits j through the hydrodynamic interaction tensors $\mathbf{T}_{ij}$, so that

$$\mathbf{v}_i = \mathbf{u} - \mathbf{T}_{ij} \cdot \mathbf{F}_j \tag{D2.7}$$

The Oseen–Burgers hydrodynamic interaction tensor $\mathbf{T}_{ij}$ is

$$\mathbf{T}_{ij} = (1/8\pi \eta_0 r_{ij})\left(\mathbf{I} + \mathbf{r}_{ij}\mathbf{r}_{ij}/r_{ij}^2\right) \tag{D2.8}$$

where $\mathbf{I}$ is the unit tensor and $\mathbf{r}_{ij}$ is the distance vector connecting spheres i and j.

The expression for the coefficient of translational friction of a particle consisting of N spheres, each of which has the friction coefficient ξ, has a very simple form (Comment D2.3):

$$f = \frac{N\xi}{\left[1 + \frac{\xi}{6\pi\eta_0 N}\sum_i \sum_{i\neq j} r_{ij}^{-1}\right]} \tag{D2.9}$$

Comment D2.3

Oseen–Burgers interactions between beads

The main features of the hydrodynamic interaction between the beads according to the Oseen–Burgers method are: (1) each bead is a point source of friction; (2) only pairwise interactions between point sources are taken into account. As a result Eq. (D2.9) always has a double sum of reciprocal distances.

Equation (D2.9) can be readily interpreted from a physical point of view. In the absence of hydrodynamic interaction between the spheres the frictional coefficient of the particle is a sum of frictional coefficients of spheres $N\xi$, whereas in the presence of hydrodynamic interaction between the spheres the frictional coefficient of the particle is less than $N\xi$. This happens because, on average, each sphere is subject to interaction with the solution flow and, consequently, has lower friction. In this case the hydrodynamic interaction should be taken into account twice.

The simplicity of Eq. (D2.9) is very attractive. It allows the required calculations to be done manually for a small number of spherical subunits. But we must note that the attraction of Eq. (D2.9) is very illusory (see Comment D2.4).

The Oseen–Burgers hydrodynamic interaction tensor has been modified to eliminate two basic disadvantages of the Kirkwood approach. Subunits were considered to be point sources of friction in spite of their final dimensions.

Comment D2.4 Correctness of Eq. (D2.9)

In the derivation of Eq. (D2.9), the interaction spheres are considered as point sources of friction. This is only correct if the dimensions of that spheres are much smaller than the distance between them, r_{ij}. This is not the case for many biological structures of interest. It is also not the case in a classical presentation of Eq. (D2.9) for four beads, where the dimension of each bead is of the order of the distances between them.

Three likely packings of four identical subunits are linear, square planar, and tetrahedral. Below we presented calculations of two of them.

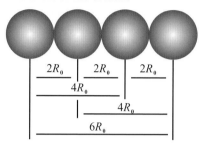

The different distances are $2R_0$, $4R_0$, $6R_0$. For the case

$$f = \frac{4 \times 6\pi \eta_0 R_0}{\left[1 + \frac{6\pi \eta_0 R_0}{4 \times 6\pi \eta_0} \left(\frac{6}{2R_0} + \frac{4}{4R_0} + \frac{2}{6R_0} \right) \right]}$$

$$f_0 = 6\pi \eta_0 R_0$$

$$f/f_0 = \frac{4}{1 + \frac{1}{4} \left(3 + 1 + \frac{1}{3} \right)} = \frac{4}{1 + \frac{13}{12}} = 1.92$$

Linear tetramer

The results of computations of translational friction for a linear tetramer using the shell model (see below) with extrapolation to an infinite number of beads of infinitesimal size show that computation according the Kirkwood formula gives the error about 6%.

All the distances are $2R_0$. For the case

$$f = \frac{4 \times 6\pi \eta_0 R_0}{1 + \frac{6\pi R_0}{4 \times 6\pi \eta_0} \left(\frac{12}{2R_0} \right)} = 24\pi \eta_0 R_0 (1 + 3/2)$$

$$f_0 = 6\pi \eta_0 R_0$$

$$\frac{f}{f_0} = \frac{4}{1 + \frac{1}{4} \times 6} = \frac{4}{1 + \frac{3}{2}} = 1.60$$

Tetrahedral tetramer

The results of computations of the translational friction of a tetrahedral tetramer using shell model computations with extrapolation to an infinite number of beads of infinitesimal size show that computation according the Kirkwood formula gives an error of about 12%.

These results show that calculations according Kirkwood's equation provide more inexact results for close contacts (tetrahedral packing). As distances between the particles increase, we can expect the accuracy of the results from direct application of the Kirkwood formula to increase.

A positive judgement on the type of the subunit packing can be made in two cases: in the case of the most compact tetrahedral packing, when the coefficient of friction is the lowest, and in the case of a more elongated linear packing, when the coefficient of friction is the highest. In all other cases, the square in the plane is included, and it is impossible to judge the type of subunit packing.

Note that in these calculations the numbers are the ratio of the friction of the particle to the friction of a sphere of equal volume. Therefore f/f_0 should be divided by $4^{1/3} = 1.5873$.

Comment D2.6
Translational diffusion tensor

The translational diffusion tensor is only meaningful when referred to a specific point, the so-called centre of diffusion, which coincides with the symmetry centre for a centrosymmetric particle.

Theoretical foundations of the hydrodynamic calculation necessary for rigorous bead modeling can be found in the specialist literature (e.g. Navarro et al., 1995).

Comment D2.5   Particle of arbitrary shape

For a particle of arbitrary shape, the hydrodynamic resistance is expressed by means of a 6×6 friction tensor, Ξ. Similarly, the Brownian diffusivity is expressed by a 6×6 diffusion matrix D, which is related to Ξ through the generalised Einshtein relationship $\mathbf{D} = kT\,\Xi^{-1}$. Both Ξ and $\mathbf{D}$ can be partitioned in 3×3 blocks, which correspond to translation, rotation and translation–rotation coupling (Navarro et al., 1995).

Hydrodynamic interactions were also considered as first-order amendments for unperturbed forces. In the case of spheres of equal size with a radius R_0, the hydrodynamic interaction tensor $\mathbf{T}_{ij}$ becomes

$$T_{ij} = \frac{1}{8\pi\eta_0 r_{ij}}\left[\left(I + \frac{r_{ij}r_{ij}}{r_{ij}^2}\right) + \frac{2R_0^2}{r_{ij}^2}\left(\frac{1}{3}I - \frac{r_{ij}r_{ij}}{r_{ij}^2}\right)\right] \tag{D2.10}$$

The Rotne–Prager–Yamakawa tensor is usually called the modified Oseen–Burgers tensor.

Equation (D2.10) has been generalised for spheres of different radii, σ_i and σ_j, such that

$$T_{ij} = \frac{1}{8\pi\eta_0 r_{ij}}\left[\left(I + \frac{r_{ij}r_{ij}}{r_{ij}^2}\right) + \frac{(\sigma_i^2 + \sigma_i^2)}{r_{ij}^2}\left(\frac{1}{3}I - \frac{r_{ij}r_{ij}}{r_{ij}^2}\right)\right] \tag{D2.11}$$

Note that Eq. (D2.11) is only valid if the interparticle distance is larger than the sum of the radii $r_{ij} \geq \sigma_i + \sigma_j$.

Now we shall present a short summary of the hydrodynamic theory focusing only on translational friction. Rotational friction and viscosity will be discussed later.

When a particle of arbitrary shape moves with velocity $\mathbf{u}$ through a fluid, the frictional force exerted on the fluid depends on $\mathbf{u}$ through the translational friction tensor Ξ_t, as

$$\mathbf{F} = \Xi_t \mathbf{u} \tag{D2.12}$$

The translational dynamics of the particle is described by means of the translational diffusion tensor $\mathbf{D}_t$, given by

$$\mathbf{D}_t = kT/\Xi_t \tag{D2.13}$$

where k is Boltzmann's constant and T is the temperature in kelvins. Equation (D2.13) is only rigorously correct when there are no translation–rotation coupling effects (Comment D2.5). Ξ_t and $\mathbf{D}_t$ are symmetric tensors whose components depend on the orientation of the particle with respect to the reference axes (Comment D2.6). The trace of $\mathbf{D}_t$, however, is an invariant, and so is the translational diffusion coefficient, defined as

$$D_t = \frac{1}{3}tr(\mathbf{D}_t) \tag{D2.14}$$

(a)

(b)

Fig. D2.5 Schematic two-dimensional presentation of a particle: (a) a bead model composed of equal spherical elements; (b) a bead model composed of unequal spherical elements. (After Carrasco and Garcia de la Torre, 1999.)

The orientationally averaged resistance to translational motion can be expressed in terms of a scalar, the translational friction coefficient, f, which is given by the Stokes–Einstein relationship.

The second problem associated with bead modelling is how to build the hydrodynamic model. It is evident that we should have two aims. First, the volume of the particle, represented by beads, which can identical or different, should be as close as possible to reality. Second, the array of beads should have an envelope that resembles the shape of the particle. A schematic illustration of a particle model in two dimensions is shown in Fig. D2.5.

When modelling an elongated structure, such as an ellipsoid or a rod, the essential criterion is that the model has the same length and volume as the particle.

Modelling the surface of a particle with a set of small equal spheres (the 'shell' method)

For a compact solid particle, hydrodynamic friction actually occurs on its *surface*. In the case of a real macromolecule, such as a globular protein, this is indeed the case, because the interior of the protein is inaccessible to solvent. Even if the macromolecule is somewhat porous or permeable to the solvent, the fluid inside is trapped, and moves along with it. It is hence a part of the hydrodynamic particle.

In the method, known as the *shell method*, a macromolecule is modelled by n specific subunits, each of which is surrounded by a shell consisting of small spheres of the radius r (Fig. D2.6). In this procedure, a large number of small spheres or beads, are placed on the surface of the particle and packed tightly. The translational friction of the assembly of beads is calculated using the modified Oseen–Burgers tensor. The modelling process is repeated using smaller bead sizes, and finally the friction in the limit of an infinite number of beads of infinitesimal size is estimated by extrapolation (Fig. D2.7).

The modelling procedure is particularly easy for revolution bodies, like ellipsoids or cylinders, for which beads can be placed at the parallel circumferences defined by planes perpendicular to the main symmetry axis. Thus, by stacking rings of beads of varying ring radius, we can build a smooth shell model. An example is presented in Fig. D2.8. The procedure is also very suitable for the prediction of the hydrodynamic properties of rigid macromolecular structures obtained from electron microscopy images.

Modelling the surface of a particle with a set of small panel elements (the 'platelets' method)

In the 'platelets method', the surface of the molecule is described as an interconnected set of N small panel triangular platelets. Representative examples for a sphere and a capped cylinder are shown in Fig. D2.9(a) and (b), respectively. The frictional properties of the corresponding 'smooth' particles are estimated by

Fig. D2.6 Schematic two-dimensional presentation of a particle: (a) ideal shell model; (b) rough-shell model. (After Carrasco and Garcia de la Torre, 1999.)

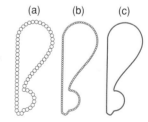

Fig. D2.7 Three logical models in the calculation of translational friction of the particle by the shell-model approach. Models (a), (b), and (c) have different size beads. Extrapolation to beads of an infinitesimally small size leads to correct the translational frictional coefficient for the particle. (Carrasco and Garcia de la Torre, 1999.)

Fig. D2.8 Smooth shell model for an ellipsoid with axial ratio $p = 2$. (After Carrasco and Garcia de la Torre, 1999.)

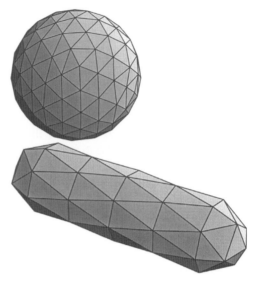

Fig. D2.9 (a) A 20 Å sphere modelled by 256 platelets and (b) a capped cylinder as a model of 20-basepair DNA represented by 96 platelets. (After Allison, 2001.)

Fig. D2.10 An example of the discretisation of tobacco mosaic virus. The roughly cylindrical shape of the tobacco mosaic virus was divided into an n-sided prism, and n was increased. Consistent results occurred at about $n = 16$. Computed results for the tobacco mosaic virus are in good agreement with experimental values. (Brune and Kim, 1993.)

carrying out numerical calculations of several model structures and extrapolating to a model in which the number of plates is infinite. This technique also works in the same way for both 'stick' and 'slip' conditions (Chapter D1).

For proteins, whose shape is quite irregular, the procedure of discretisation is not so simple. A smoothed surface of the protein of interest should be obtained from atomic coordinates (see below).

For a large molecule, it is not necessary to know the exact position of each atom in order to discretise the surface. For instance, tobacco mosaic virus is large enough that the surface irregularities are of little hydrodynamic significance. The particle can therefore be modelled as a cylinder, divided into ever-smaller panels (Fig. D2.10).

D2.2.3 Translational friction coefficient and electrostatic capacitance

Traditionally the frictional coefficient of an arbitrarily shaped particle is calculated by the methods described in the previous section. These methods require delicate tailoring to give the best fit for individual cases and are time consuming. Electrostatics calculations are much simpler, and have a long and rich tradition. The electrostatic capacitances of many geometrical shapes, including the touchstone of hydrodynamics – the *cube*, are well known (Comment D2.7). In 1993

Comment D2.7 Capacitance

An electric charge q on a body generates an electrostatic potential, U, proportional to q

$$U = q/C$$

The proportionality constant is $1/C$, where C is defined as the capacitance of the body.

C has the dimensions of length. For the mathematically minded: C is equal to the charge required to maintain a body at unit electrostatic potential with respect to infinity or C is equivalent to the electrostatic capacitance of the particle in units in which the capacity of a sphere equals its radius.

---

the connection between hydrodynamics and electrostatics was recognised and a simple and accurate method of calculating the translational hydrodynamic friction for rigid particles of arbitrary shape was proposed. The translational friction coefficient f of a particle is related to its capacitance C by

$$f = 6\pi \eta_0 C \qquad\qquad (D2.15)$$

This relation is exact for a sphere, and two- and three-axis ellipsoids (Comment D2.8). It is accurate to within 1% for many cases where analytical results are known.

Comment D2.8 On the similarity of equations describing the properties of a sphere in reaction kinetics, electrostatics and hydrodynamics

It is interesting to note that mathematical equations describing the properties of a sphere (with radius R_0) in reaction kinetics, electrostatics and hydrodynamics are similar. Indeed, the diffusion-controlled reaction rate of particles (with diffusion constant, D) wandering towards an absorbing sphere (Chapter D3) is given by.

$$k = 4\pi D R_0$$

In electrostatics, the capacitance, C, of a conducting sphere is

$$C = R_0$$

The translational friction coefficient, f of a sphere in a solvent with viscosity η_0 under 'stick' boundary condition, is given by Stokes' law (Eq. (D2.1))

$$f = 6\pi \eta_0 R_0$$

All three quantities are proportional to the radius of a sphere (they scale as the radius of the sphere).

Equation D2.15 offers a simple way to calculate the hydrodynamic friction of arbitrarily shaped particles via the electrostatic capacitance C. The main point of this approach is that C is generally much easier to calculate than the components of the friction tensor for bodies of arbitrary shape.

D2.2.4 Particles with known structure

Modelling proteins from atomic coordinates

The availability of a database that contains the atomic coordinates of a vast number of proteins allows the construction of hydrodynamic models on the basis of these coordinates. Modelling each atom of large molecules by an individual bead would result in too many frictional centres, and hence make the computation impractical. A few approaches have been proposed to decrease the quantity of beads. They include modelling the proteins with a set of beads (the cubes method) and modelling the surface of a protein with a set of panel elements.

Modelling proteins with a set of beads (the cubes method)

The 'A to B' (from Atoms to Beads) program can be used to construct an intact bead model with various three-dimensional resolutions from the atomic coordinates. The space occupied by the molecule is divided into small cubes. In each of these there is a bead, the size and position of which are determined by the positions of all the atoms in the cube. Such an approach allows an easy adjustment of the model 'resolution' (cube size) to the molecule size. The real shape of a macromolecule dictates the required resolution. For more rounded globular particles, model representation can be done using beads of a larger size. Figure D2.11 shows an example of such a construction for aldolase. The model was generated at resolution of 5–30 Å by the AtoB program from the atomic coordinates. The three-dimensional resolution of bead models shows the size of the lattice in which the molecule was divided.

As observed, a resolution of between 10 and 20 Å is sufficient to create a model that retains a reasonable similarity to the original crystal structure; at lower resolutions important structural details are lost.

Modelling the surface of a protein with a set of panel elements (the 'platelet' method)

In the platelet approach, the set of atomic coordinates is used to construct the surface of the molecules. The surface of lysozyme defined by the atomic radii shows a large number of small irregularities (Fig. D2.12(a)). On the level of continuum hydrodynamics, the inclusion of small irregularities in the protein surface has only a small effect on the results, but they significantly increase the computational time required to get the results. To produce a surface amenable to hydrodynamic calculations, small-scale irregularities are smoothed by rolling a probe sphere. (See Fig. D2.12(b) which shows the results of using a probe of

Resolution (Å)	No of beads	Plane of view		
		xy	yz	zx
30	26			
20	67			
15	125			
10	319			
5	1103			

Fig. D2.11 Example of a representation of the protein aldolase generated by the AtoB program from its atomic coordinates with a resolution of 5–30 Å. (After Byron, 1997.)

radius 3Å.) The surface of the protein is then discretised into boundary elements. (See the triangulated surface shown in Fig. D2.12(c).) The diffusion coefficient can then be calculated using the boundary element technique.

Modelling DNA structure

Most dynamic properties of long DNAs are dominated by the contour length and the bending flexibility of macromolecules, whereas the thickness and, particularly,

(a)   (b)   (c)

Fig. D2.12 Surface of lysozyme molecules: (a) at atomic resolution, (b) with probe radius of 3 Å, (c) after discretisation into boundary elements. (After Brune and Kim, 1993.)

Fig. D2.13 Approximation of DNA by double-helical bead model in which each nucleotide is represented by one bead. (After Garcia de la Torre *et al.*, 1994.)

the detailed cross-sectional structure are less important. However, for short DNA fragments cross-sectional structures are more important.

DNA as a helical structure

In this approximation, DNA is modelled as a double helix in which each nucleotide is represented by one bead (Comment D2.9). The radius of the helix is regarded as an adjustable parameter. Figure D2.13 shows a double in which each helix has 15 beads per turn, a radius $A = 10$ Å and pitch $P = 34$ Å. One of the helices is symmetric ($\varphi = 180°$), with two equal grooves, whereas the other ($\varphi = 120°$) shows two grooves of different widths. The set of data with $\varphi = 120°$ is roughly equal to those for the B form of DNA.

Comment D2.9   DNA geometry

The geometry of a single helix can be described by the following set of quantities: A, radius of the helix; P, pitch; n, number of turns (not necessarily an integer). If the helical axis is z, then the parametric equations of the helix can be written as

$$X = A\cos(t + \varphi)$$
$$Y = A\sin(t + \varphi)$$
$$Z = Pt/2\pi$$

where the value of the phase angle, φ, can be arbitrary and t is a continuous parameter that goes from $t = 0$ to $t = 2\pi n_t$. The two strands of a double helix can also be described by an equation using different values for the phase angle of each strand. In the bead model of the helix, beads of radius R_0 are placed along the contour of the helical line and are chosen such that the beads touch. Two approaches for modelling the DNA structure are described in the text.

Beads modelling DNA from atomic coordinates

In this approach the beads are placed at the geometrical centres of the groups of atoms (10–30) to maintain the actual shape of the molecule surface. A schematic view of DNA 20-mer surrounded by its bead model is shown in Fig. D2.14(a). The best agreement with experiments was found when the bead radius was 7.3 Å. In the 'double-bead' model (Fig. D2.14(b)), in which the nucleotides were divided into two groups of atoms (one containing the base and other one containing the sugar and the phosphate group), the best agreement with experiment was found when the bead radius was 5.6 Å. The use of the second sphere is important for small fragments of DNA only.

Platelets modelling of DNA from atomic coordinates

An example of the partitioning of a hydrodynamic surface for a 20-base-pair DNA fragment into 352 triangular platelets is shown in Fig. D2.15(a). The 'platelets'

Fig. D2.14 (a) A schematic view of a B-DNA 20-mer in a ball-and-stick representation immersed in its single-bead model. (b) A schematic view of a B-DNA 20-mer in a ball-and-stick representation immersed in its double-bead model. (Banachowicz et al., 2000.)

Fig. D2.15 (a) Platelets model of 20-base-pair DNA $(pd(A)_{20} pd(T)_{20})$ represented as 352 flat triangular plates. (b) Platelets model of a 375-base pair closed circular DNA with linking number $= -4$. The structure, represented by 640 interconnected platelets, is flexurally and torsionally flexible. (Allison, 2001.)

method is not restricted to relatively small macromolecules such as the 20-base-pair DNA fragment considered above. Figure D2.15(b) shows a torsionally stressed supercoiled 375-base-pair closed circular DNA with a linking number of -4. The resultant structure is sheathed in a closed circular cable consisting of 640 platelets with radius 10 Å (contour length $=$ 1275 Å).

D2.2.5 Rigid particles with segmental mobility

The hydrodynamic description of flexible macromolecules requires the amount of both the solvent within and travelling together with the coil and the solvent flowing freely through the coil to be taken into account. It must also take account of the static distribution of friction elements of the coil. Calculations for such particles on the basis of the Kirkwood–Riseman formalism, Monte-Carlo methods and molecular dynamics can be found in the literature and will not be analysed here.

However, intermediate structures can be found in biology in which a molecule is considered as a rigid structure with definite segmental mobility. Such molecules include immunoglobulins, myosin and some of the globular proteins, e.g. calmodulin. Since the structural mobility of such molecules is directly connected with their function, studies of their hydrodynamic behaviour are of interest.

The simplest and most studied model imitating segmental mobility is a rigid rod consisting of two shoots linked by a hinge (Fig. D2.16). Calculations, based on both the simple averaging of various conformations and the hydrodynamic interactions, have shown that the coefficient of translational friction of such a loosely joined rod varies by only 3% from that of a rigid rod of the same length, and depends little on the location of the hinge. This means that an experimental study of segmental mobility of a rod model, using such methods as sedimentation or translational diffusion, is not justified, because the experimental error of these methods is about 1–2%. However, in this connection an important result has been found: flexibility influences the dynamic features of the particle. It appears that the coefficient of translational diffusion of a broken rod calculated from dynamic experiments, e.g. dynamic light scattering, depends on time. In other words, the function of autocorrelation that describes the time dependence of fluctuations in the intensity of light dispersed by an ensemble of such rods consists of at least two exponents. Similar time effects were revealed in the phenomena of polarised fluorescence and electric birefringence. The reader can find a discussion of experimental studies employing these methods in Chapter D7.

The model of a worm-like chain (Fig. D2.17, see also Fig. D6.11), in which flexibility is distributed along its contour length, is particularly interesting from the hydrodynamic point of view because it describes well the translational friction of DNA molecules. Comparison of this model with the broken rod model has shown that their equilibrium averaged conformations are quite close, but their dynamic properties expressed via a spectrum of relaxation times differ. We will discuss the dynamic properties of DNAs in Section D6.7.1.

Fig. D2.16 A broken rod with two small shoots with an angle α between them simulated with a set of beads.

Fig. D2.17 A worm-like chain. Because relatively stiff chains may be envisioned to bend only gradually and smoothly in solution, somewhat like a worm, they are often called worm-like chains.

D2.2.6 Experimental methods for measurement of the translational frictional coefficient

Table D2.1 shows experimental methods currently used to determine translational friction coefficients, f, of biological macromolecules, their measured values, and the range of measured values. It also includes the range of applicability (in dimensions and shape) of each from the methods and gives the chapters where each experimental technique is treated in detail and where applications and the applicability range of each method are discussed.

Table D2.1. *Methods currently used to determine translational friction coefficients, f,*
of biological macromolecules

Experimental method	Measured value	The range of measured values	The applicability range
Translational diffusion (Chapter D3)	Translational diffusion coefficient D	5×10^{-5}–5×10^{-12} ($cm^2 \ s^{-1}$)	Applicable to biological macromolecules of different shapes
Sedimentation velocity (Chapter D4)	Sedimentation coefficients (in Svedberg units)	0.5–$200\,000$ (10^{-13}s)	Not applicable to very small or very large molecules
Electrophoretic mobility (Chapter D5)	Electrophoretic mobility μ	1×10^{-4}–5×10^{-4} ($cm^2 \ V^{-1} \ s^{-1}$)	Applicable mainly to DNA molecules
Fluorescence correlation spectroscopy (Chapter D11)	Translational diffusion coefficient D	5×10^{-5}–5×10^{-12} ($cm^2 \ s^{-1}$)	Applicable to fluorescent labelled biological macromolecules of different shapes
Recovery of fluorophore after photobleaching (Chapter D3)	Translational diffusion coefficient D	5×10^{-5}–5×10^{-12} ($cm^2 \ s^{-1}$)	Applicable to fluorescent labelled biological macromolecules of different shapes
Hydrodynamic modelling experiment (Chapter D2)	Direct determination of translational friction coefficient f		Applicable to biological macromolecules of quasi-spherical shape

---

Comment D2.10   Direct determination of the translational friction coefficient

The standard apparatus consists of a glass cylinder (usually ~10 cm in internal diameter, 100 cm high), immersed in a water bath at constant temperature. Two circular marks approximately 7–10 cm from each end of the cylinder indicate the distance over which the settling particle is timed. Rigid models are constructed from balls using different materials. The diameters of the balls range from 0.25 cm to 1 cm. Multisphere particles are constructed by gluing matched balls together with a minimum amount of epoxy cement.

Settling rates for non-spherical particles (cylinders, dimers and tetramers) are determined in two orientations for different particle sizes and different fluid viscosities, covering Reynolds number in the range 0.0001–0.01. For each multisphere particle, the quantity of interest is the ratio of the settling rate of a single sphere to that of the multisphere particle. The ratio should be free from any influence of the experimental set-up, especially wall effects, and should not be dependent on particle size.

Such simple experiments on macroscopic models allow the determination of the translational friction properties of microscopic particles of arbitrary shape. Theoretical values coincide with experimental ones in the limit of experimental error (0.1%) in all cases for which analytical results are known.

Comment D2.11

Impressive progress has been made in the last 10 years that allows the detection of isolated single molecules in solution and on surfaces (see Chapter F4).

Actual translational frictions of rigid microscopic particles can be determined by observing the settling rates of their macroscopic models in a high-viscosity fluid at low Reynolds number. A typical experiment for rigid assemblies of spheres is described in Comment D2.10.

D2.3 Rotational friction

D2.3.1 Rotational motion in one dimension

Brownian motion involves not only the translation of the molecules but also their rotational movements. Except when the particles are so large that they are visible in a microscope the Brownian motion itself is not observed but can be deduced from a study of the physical properties of the solution (Comment D2.11). We begin our discussion with rotational diffusion in one dimension because this allows us to make important analogies with translational diffusion.

The rotational motion of a particle may be described as one-dimensional if the motion of every point of the particle is a simple rotation about a single axis, which passes through the centre of gravity of the particle (Fig. D2.18). Motion of this type can be described in terms of a single angle θ between a suitable reference axis of the particle (in our case it is a line passing through long axis of the rod) and any suitable reference axis in space (in our case it is a horizontal line). The change in θ with time can be expressed in terms of the angular velocity, $\omega = d\theta/dt$.

Now we introduce a variable $\rho(\theta)$, such that the number of particles, per cubic centimetre of solution with orientation between θ and $\theta + d\theta$ is $\rho(\theta)\, d\theta$. The physical sense of the variable $\rho(\theta)$ is simple. At equilibrium all values of θ are equally probable and $\rho(\theta)$ is a constant. In non-equilibrium states of the ensemble of molecules, under the influence of a torque, some orientations become more preferable and $\rho(\theta)$ becomes dependent on θ. If the torque producing an

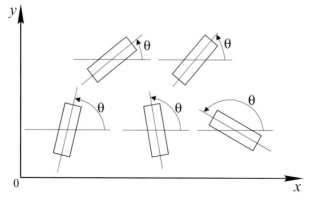

Fig. D2.18
One-dimensional rotation. All rods lie in the *xy* plane. Rotation takes place about the *z*-axis which is perpendicular to the *xy* plane. The rotational motion of the rod can be described by one parameter, the angle θ.

orientation is removed the system gradually loses orientation and in the end $\rho(\theta)$ again becomes constant. This process is called *rotational diffusion.*

The treatment of one-dimensional rotational diffusion is similar to the treatment of one-dimensional translational diffusion (Chapter D3). If we define $J(\theta)$ dt as the net of number of molecules per cubic centimetre that in time dt traverse the orientation angle θ in the direction of positive θ, then a phenomenological law analogous to Fick's first law (Eq. (D3.7)) may be written as

$$J(\theta) = -\Theta[d\rho(\theta)/d\theta]t \qquad (D2.16)$$

In this equation Θ is called the rotational diffusion coefficient.

Fick's second law can be written by an analogy with Eq. (D3.9)

$$d\rho(\theta)/dt = \Theta d^2\rho(\theta)/d\theta^2 \qquad (D2.17)$$

Continuing the analogy with translational motion we may define the rotational frictional coefficient ζ. A torque F_{rot} acting in the xy plane (Fig. D2.18) in the direction of positive θ, of any macromolecule, leads to an angular acceleration. In a viscous solution, the motion is opposed by solvent friction. The force, F_{frict}, due to this friction is proportional to the angular velocity ω. The proportionality constant is defined as the rotational friction coefficient ζ:

$$F_{frict} = \zeta\omega \qquad (D2.18)$$

In steady-state conditions the two opposing forces are equal in magnitude, the acceleration goes to zero and the macromolecule rotates with constant angular velocity

$$\omega = F/\zeta \qquad (D2.19)$$

This equation is analogous to Eq. (D1.10). It provides the basis for the determination of the rotational friction coefficient by many of the experimental methods presented in Table D2.1.

The relation between the rotation frictional and diffusion coefficients is similar to Eq. (D3.24):

$$\Theta = kT/\zeta \qquad (D2.20)$$

D2.3.2 Rotational motion in three dimensions

For a three-axis ellipsoid there are three rotational friction coefficients, each characterising the resistance to motion of the ellipsoid parallel to one of its principal axes. The experimentally determined rotational diffusion and frictional coefficients are given by the relationship

$$\Theta = (\Theta_1 + \Theta_2 + \Theta_3)/3 \qquad (D2.21a)$$

or

$$\Theta = kT(1/\zeta_1 + 1/\zeta_2 + 1/\zeta_3)/3 \qquad \text{(D2.21b)}$$

For an ellipsoid of revolution, two of parameters are identical, and

$$\Theta = (\Theta_1 + 2\Theta_2)/3 \qquad (\Theta_2 = \Theta_3) \qquad \text{(D2.22a)}$$

or

$$\Theta = kT(1/\zeta_1 + 2/\zeta_2)/3 \qquad \text{(D2.22b)}$$

D2.3.3 Rotational motion and relaxation times

Let us assume that we have a set of macromolecules in solution. Because the molecules are randomly distributed we can consider that all possible directions are equally represented (Fig. D2.19(a)). Now assume at any given time we can orient macromolecules in a given direction using electric forces (Fig. D2.19(b)). Once the orienting forces are removed the molecules very rapidly adopt partially randomised orientations as shown in Fig. D2.19(c). To characterise stages (b) and (c) we introduce the angle θ which is defined by the molecule's original direction in (b) and its new direction in (c). The mean value of its cosine $\langle \cos\theta \rangle$ can serve as a measure of orientation; $\langle \cos\theta \rangle$ is unity when the molecules are aligned and goes to zero when the orientation is random once more (Fig. D2.19(d)). The time necessary to pass from state (b) to state (d) is the called *relaxation time* of the system. So the relaxation time τ, which is defined by

$$\frac{\langle \cos\theta \rangle_t}{\langle \cos\theta \rangle_{t=0}} = \exp(-t/\tau) \qquad \text{(D2.23)}$$

Fig. D2.19 Different states of macromolecules in solution. (a) All molecules are randomly disposed; (b) all molecules are oriented in the vertical direction; (c) some of molecules are randomly disposed, but randomisation is not complete; (d) randomisation is complete. Note: the time interval between states (b) and (c) is short, the time interval between states (c) and (d) is long in comparison with the rotational properties of macromolecules.

is the time required for $\langle \cos\theta \rangle$ to fall to $1/e$ of its initial value.

As specified in the chapters devoted to different experimental techniques, such as electrical birefringence (Chapter D6) and depolarised fluorescence (Chapter D8) observable properties are always multiexponential functions containing up to five relaxation times (Comment D2.12). But, the extraction of the five relaxation times and their corresponding amplitudes from an experimentally measured decay

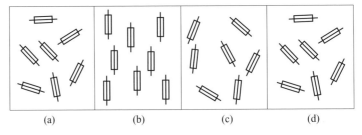

(a)   (b)   (c)   (d)

Comment D2.12   Shape of the body and relaxation times

For three-axis ellipsoids there are three relaxation times, whereas for so-called 'rigid general particles' there are five relaxation times. There are two approximate equalities between the five decay times. For this reason at most only three decays are observed. However, in addition, the inverses of these decay times are linearly related to a high degree of approximation. Therefore, the number of independent decay times, in practice, is reduced to at most only two.

---

is impossible owing to the well-known ill-conditioned nature of multiexponential fitting. Simplification of this problem leads to a 'mean relaxation time', which is defined as a weighted arithmetic mean of the five relaxation times and an 'initial relaxation time', which is defined as a weighted harmonic mean of the relaxation times (see Section D7.5.2).

D2.3.4 Regularly shaped rigid particles

In order to rotate a solid object in a viscous fluid we must rotate some solvent molecules with respect to others. The solvent molecules nearest the rotating particle are the most perturbed. The perturbation caused by the particle vanishes as we move away from the particle. Figure D2.20 shows the flow lines (dashed lines) around a sphere, radius R_0, rotating with a constant angular velocity ω. The

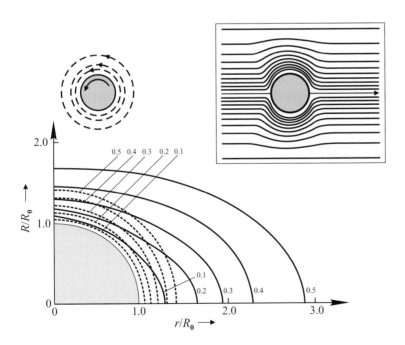

Fig. D2.20 Comparison of rotational and translational flows. The figure shows the calculated fluid velocity profile generated by spherical macromolecules obeying stick boundary conditions. The dashed lines represent the flow lines for a sphere rotating around an axis normal to the plane of the page. The solid lines represent the flow lines for a sphere moving to the right in the plane of the page. In each case, a frame of reference has been chosen to make the sphere stationary. A cross-section of the fluid, coincident with the centre of the sphere, is shown at the top; at the bottom, just one quadrant is shown in an expanded view. The radius of the sphere is R_0, and the distance is expressed in units of R_0. (After Kuntz and Kauzman, 1974.)

flow lines in this case are circles. At the surface of the sphere the fluid moves with the sphere, as required by the stick boundary conditions. Figure D2.20 shows also the flow lines (solid lines) around a sphere when it moves in a stream to the right in the plane of the page. Evidently the disturbance for translational motion drops off with distance considerably more slowly than that for rotational motion.

Sphere

The rotational friction coefficient ζ of a sphere of radius R_0 in stick boundary conditions is given by

$$\zeta_0 = 8\pi \eta_0 R_0^3 \tag{D2.24}$$

or

$$\zeta_0 = 6\eta_0 V \tag{D2.25}$$

For a rotational diffusion coefficient Θ and rotational relaxation time τ_0 Stokes' law is

$$\Theta_0 = kT/8\pi \eta_0 R_0^3 = kT/6\eta_0 V \tag{D2.26}$$

$$\tau_0 = \frac{4\pi \eta_0 R_0^3}{3kT} = \frac{\eta_0 V}{kT} \tag{D2.27}$$

Rotational friction is much more sensitive to the size of the particle than translational friction. The reason is that the value of Θ is approximately inversely proportional to the cube of the dimension of the molecule, but the value of f depends on the dimension to the first power.

The rotational diffusion coefficients for some macromolecules are given in Table D2.2. The range of values, a factor of 10^8, is really impressive compared with the range of values for the translational diffusion or sedimentation coefficients (Tables D3.3 and D4.3).

Table D2.2. *Some typical values of rotational diffusion constants of macromolecules in water*

Macromolecule	Θ (s^{-1})
Gramicidin (dimer)	60 000 000
Lysozyme	16 700 000
Kinesin(349)	5 000 000
DNA fragment (104 bp)	172 000
T7 bacteriophage	5290
Tobacco mosaic virus	330
DNA of T7 bacteriophage	5.2
DNA of T4 bacteriophage	0.41

Ellipsoid of revolution

In the case of an ellipsoid of revolution we cannot speak of just a single rotational friction coefficient. There are two coefficients for ellipsoids of revolution with semi-axes a, $b = c$: one ζ_a for rotation about semi-axis a, and second, ζ_b for rotation about semi-axis b.

In practice it is very difficult to measure ζ_a and ζ_b independently. The average rotational friction coefficient ζ can calculated according Eqs. (D2.25) and (D2.26):

$$\zeta = 6\eta_0 V / J(p) \tag{D2.28}$$

$$\Theta = kT J(p)/6\eta_0 V \tag{D2.29}$$

where $1/J(p)$ is Perrin's function which depends on the axial ratio of the ellipsoid of revolution. Theoretical values of $1/J(p)$ are plotted in Fig. D2.21. Comparison of the dependence in Fig. D2.21 with the analogous one for translational friction in Fig. D2.3 clearly shows that rotational friction is much more sensitive to the shape of macromolecules than translational friction (Comment D2.13).

Fig. D2.21 Dependence of the orientationally averaged rotational friction coefficient ζ/ζ_0 on the axial ratio for prolate and oblate ellipsoids. This hydrodynamic shape function is available from a computer program. (Harding *et al.*, 1997.)

Comment D2.13 Rotation frictional coefficients and boundary conditions

Rotation frictional coefficients are more sensitive to the choice of boundary conditions. When slip boundary conditions change to stick ones the translational friction is only moderately altered. But in rotation phenomena the choice of boundary conditions is a crucial point (we mentioned this before, in the case of a sphere). For example, with slip boundary conditions, there is no resistance for rotation about the long axis of a prolate ellipsoid or the short axis of an oblate ellipsoid. Calculating the rotational properties of macromolecules of different shapes at stick and slip boundary conditions requires their special properties to be taken into account (Allison, 1999).

For a long prolate ellipsoid of revolution with semi-axes of length a and b ($a > 5b$), the rotation about one of the b axes is given by

$$\Theta = \frac{3kT[(2\ln(2a/b)) - 1]}{16\pi \eta_0 a^3} \tag{D2.30}$$

In Eq. (D2.30) the small dimension b appears only in the logarithm and so a large change in it produces only a small change in Θ. Thus the rotational diffusion constant may be used to measure the *length* of a molecule.

Circular cylinder

The rotational diffusion constant of right circular cylinders (rods) is given by

$$\Theta = \frac{3kT}{\pi \eta_0 L^3}\left[\ln\left(\frac{L}{d}\right) - \gamma\right] \tag{D2.31}$$

where η_0 is the solvent viscosity, kT is the thermal energy, L is the rod length, d is the rod diameter and γ is a frictional factor which depends on the exact model. In Eq. (D2.31) the rod diameter d appears only in the logarithm and so just as for an ellipsoid of revolution a large change in the diameter of the rod produces only a small change in Θ. Thus the rotational diffusion constant may be used to measure the *length* of a right circular cylinder.

All the theories agree on the functional form of Eq. (D2.31) and differ in the dependence of the end-effect corrections on the parameter p $(= L/d)$. The exact form of the end-effect correction, γ, depends on the model and the manner in which the hydrodynamic interactions are treated. In general, it is only a function of the length and diameter of the cylinder (Comment D2.14).

For very long cylinders $(p = L/d > 30)$ the improved Broersma formula can be used:

$$\gamma = 0.757 - 7\left(\frac{1}{\ln 2p} - 0.27\right)^2 \qquad (D2.32)$$

For short cylinders $(2 \leq p \leq 30)$ the most suitable formula is that obtained by Tirado and Garcia de la Torre:

$$\gamma = 0.667 - \frac{0.917}{p} + \frac{0.05}{p^2} \qquad (D2.33)$$

Finally it should be mentioned that there are two formulae for describing long (in the framework of a worm-like model) and short (in the framework of a weakly bending rod model) DNA molecules. These formulae are too lengthy to be reproduced here.

Short rods

For short rods $(p \sim 2)$ a practical way to determine the cylinder dimensions from D_t and Θ has been proposed. We can combine Eqs. (D2.3) and (D2.31), and define a function of the axial ratio, $\mu(p)$, as

$$\mu(p) \equiv \left(\frac{9\pi\eta_0}{kT}\right)^{2/3}\frac{D_t}{\Theta^{1/3}} = \frac{\ln p + \varphi}{(\ln p + \gamma)^{1/3}} \qquad (D2.34)$$

The function $\mu(p)$ against p is plotted in Fig. D2.22 from which p can be interpolated. Equation (D2.34) opens the way for the direct determination of the shape of a short rod particle, if D_t and Θ coefficients are determined from one experimental approach, for example, from dynamic light scattering (Ch. D10).

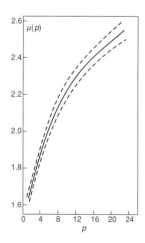

Fig. D2.22 Plot of $\mu(p)$ versus p. Solid line is calculated from Eq. (D2.27); the dashed lines are the limits for the $\mu(p)$ function with a relative error of $\pm 1\%$ in D_t and $\pm 3\%$ in Θ. (After Garcia de la Torre and Martinez, 1984.)

D2.3.5 Arbitrarily shaped particles

Modelling techniques for calculations of the rotational friction of arbitrarily shaped particles using a set of beads are the same as those used for calculation of translational friction (Sections D2.3.1–D2.3.3). However, to calculate the rotational friction for structures of arbitrary shapes it is necessary to know the actual axis of rotation. This axis must contain the centre of frictional resistance. In some cases, such as spheres, rods or rings, the centre of frictional resistance coincides with both the geometrical centre and the centre of mass. For structures of lower symmetry, such as bacteriophages, this is not generally true. It is obvious that the molecule rotates in such a way as to dissipate the minimum amount of energy in friction with the solvent. Thus, the actual axis of rotation is that giving a minimum for ζ or a maximum for D.

Another important distinction between rotational and translational diffusion pertains to the magnitude of the direct fractional contribution of the bound element. For centrally placed elements rotational diffusion is relatively insensitive to the frictional surface of the added protein since such elements lie close to the centre of rotation (Comment D2.15).

D2.3.6 Experimental methods for the measurement of rotational frictional coefficients

Experimental methods for the measurement of rotational frictional coefficients can be divided into two main groups (see also Section D1.4.2). In the first, we find experiments in which the rate of particle rotation under the *action of a pair* of forces (a torque) is determined. If a velocity gradient in the solvent takes the role of an orienting force, the phenomenon is known as birefringence in flow or flow birefringence (the Maxwell effect). If the force is electrical in nature, the phenomenon is called electric birefringence (the Kerr effect). In the second group, we place phenomena in which no external force acts on the particles and their rotations occur only *under Brownian motion*. The behaviour of a particle in fluorescence and dynamic light scattering experiments is of this type.

Table D2.3 contains a summary of five methods currently in use. A summary of five methods currently used to determine rotational friction coefficients is given in Table D2.3. The table lists the experimental methods used to measure the experimental parameters, the range of experimental values of rotational coefficients and the relaxation times accessible to each method and its the range of applicability (in dimensions and shape). It also gives the chapter in which each experimental technique is treated in detail and where applications are discussed.

D2.4 Viscosity

D2.4.1 Viscosity as a local energy dissipation effect

The internal friction or viscosity of a liquid appears during flow when a non-zero velocity gradient is set up. The simplest example is a laminar flow with a constant velocity gradient $G = \mathrm{d}u_x/\mathrm{d}y$ at a right angle to the direction of flow (Fig. D2.23

Comment D2.15 Boundary element technique for the calculation of rotation friction

An alternative approach is based on the boundary element technique. A numerical boundary element algorithm for calculating the rotational diffusion constant of an arbitrarily shaped particle is now available (Allison, 1999). In cases in which the rotational properties are already known, the algorithm is accurate to within a small percentage.

Fig. D2.23 Velocity and gradient distribution in a cylinder (infinitely narrow gap).

Table D2.3. *Methods currently used for the determination of rotational friction coefficients, Θ, of biological macromolecules*[a]

Experimental method	Measured parameter	The range of measured Θ in s^{-1}	The range of measured τ values	The range of applicability of the method
Electric birefringence (Chapter D6)	Relaxation time τ	1.7×10^6–0.3	100 ns–500 ms	Not applicable to very small spherical molecules
Flow birefringence (Chapter D7)	Orientation angle χ	5×10^4–1.7	3 μs–100 ms	Applicable to highly elongated molecules
Fluorescence and phosphorescence depolarization (Chapter D8)	Relaxation time τ	50×10^6–1.7×10^6	3–100 ns	Applicable to small spherical molecules
Dynamic Light scattering (Chapter D10)	Decay of correlation function	50×10^6–0.17	1 ns–1 s	Applicable to biological macro-molecules of different shape
Nuclear Magnetic Resonance (Chapter J3)	Relaxation time τ	50×10^6–1.7×10^6	3–100 ns	Applicable to small spherical or quasi-spherical molecules

[a] It should be noted that workers in different experimental methods frequently use different definitions of the relaxation time. Workers in electric and flow birefringence usually use the term *rotational relaxation time*, denoted as ρ, whereas the workers in magnetic resonance and fluorescence polarisation spectroscopy customarily use *rotational correlation time*, denoted as ϕ. They are related by $\phi = \rho/3$, $\phi = 1/6D$, $\rho = 1/2D$. In this book τ_r means rotational relaxation time, τ_c means rotational correlation time.

and Comment D2.16). The liquid velocity u is then given by the expression

$$u = u_x = Gy$$
$$u_y = u_z = 0 \tag{D2.35}$$

The viscosity of the liquid is a measure of the internal friction which determines the value of the tangential force F required to maintain the velocity gradient G between the planes. The greater the internal friction, the greater the shear that has to be applied to maintain flow with a given velocity gradient G. These are related by Newton's formula:

$$F = F_x = \eta_0 \frac{du_x}{dy} = \eta_0 G \tag{D2.36}$$

The constant of proportionality η_0 is called the coefficient of viscosity or simply the viscosity of the liquid. The dimension of viscosity is dynes second per square centimetre (or erg cm^{-3} s^{-1}). Thus viscosity can be defined as the energy

Fig. D2.24 Laminar flow (see text).

dissipation per unit volume per unit time in a fluid deformed at unit rate of shear.

To maintain a constant gradient G in the liquid, energy must be expended, and to calculate the amount consider two layers 1 and 2, separated by a third of thickness dy and area 1 cm^2 (Fig. D2.24). In time t the shift in the x-direction of layer 2 relative to layer 1 is

$$dx = t(du_x/dy) = tgdy$$

Thus in the central layers of thickness dy and area 1 cm^2 the work done is $dA = Fdx = F\,Gt\,dy$ for a volume of liquid dy cm^3. Thus the amount of work done in overcoming the internal friction per unit volume is $A = FGt$. Substituting for F from Eq. (D2.30), the work done in unit time per unit volume due to the directional flow is

$$E = \frac{dA}{dt} = \eta_0 G^2 \qquad (D2.37)$$

This expression is central in discussions of the viscosity of macromolecules.

D2.4.2 Relative, specific (reduced) and intrinsic viscosity

The viscosity of a pure solvent is denoted by η_0. Addition of macromolecules should raise the viscosity to a new value η. This occurs because the large macromolecules, which extend across the streaming lines, greatly enhance resistance to flow. As a result the viscosity of a macromolecular solution is always greater than that of pure solvent (Comment D2.17).

The fractional increase in viscosity is called the specific viscosity and is given by

$$\eta_{sp} = \eta/\eta_0 - 1 = \eta_{rel} - 1 \qquad (D2.38)$$

where the relative viscosity is

$$\eta_{rel} = \eta/\eta_0 \qquad (D2.39)$$

Comment D2.17   Negative value of intrinsic viscosity

In practice, the viscosity of a solution can be below than that of the pure solvent. In fact, negative intrinsic viscosity was discovered a long time ago in a variety of binary simple liquid mixtures, e.g. for benzene in ethanol and oligoisobutilenes in benzene. It may be regarded as arising from specific interactions between solute and solvent molecules such that a liquid structure of some kind in the solvent is destroyed in the vicinity of a solute molecule. Such effects cannot be treated within the framework of classical hydrodynamics.

Comment D2.16 Laminar and turbulent flow

When a fluid flows slowly along a tube or space between two parallel planes, the flow pattern observed experimentally is a *laminar* flow. In this case η_0 is independent of G and the liquid is said to be Newtonian in its viscous behaviour. At sufficiently high velocity this pattern of flow is disturbed. This disturbed flow is called *turbulent* flow. In this case η_0 depends on G and it is said to be non-Newtonian. In cylindrical tubes the transition from laminar to turbulent flow generally occurs when the Reynolds number exceeds 2000. Fluid flow in narrow capillaries or between slowly rotating cylinders placed closed together is laminar at all reasonable flow speeds, and these form the basis for the measurement of viscosity.

Comment D2.18   The term intrinsic viscosity

The term intrinsic viscosity was proposed by Kraemer in 1938. Strictly speaking, the use of the word 'viscosity' for η_{rel} η_{sp}, and $[\eta]$ is incorrect because these quantities do not have the dimensions of viscosity. In 1957 the International Union of Pure and Applied Chemistry proposed that intrinsic viscosity $[\eta]$ should be renamed 'limiting viscosity number'. However, the proposed terminology appears not to have gained acceptance because the new term is rather awkward in practice.

In the limit of low concentration, C, η_{sp} is proportional to C. So, we can define the intrinsic viscosity, $[\eta]$, as the fractional increment in viscosity of the solution due to addition of a grams of macromolecule (see Comment 2.18). The absolute value of intrinsic viscosity depends on the concentration units. So, if we take concentration in milligram per millilitre then the intrinsic viscosity of standard globular proteins is about 3 ml mg^{-1}. If we take the concentration in milligrams per decilitre analogous values are 0.03 dl mg^{-1}:

$$[\eta] = \lim_{C \to 0} \frac{\eta - \eta_0}{\eta_0 C} \qquad (D2.40)$$

D2.4.3 Regularly shaped rigid particles

Spherical particles

If the solution is undergoing laminar flow defined by Eq. (D2.35) then different parts of a spherical particle are in liquid layers moving with different velocities. Thus frictional forces due to the solvent which is flowing past the particle give rise to rotational as well as translational motion. When the sum of all turning moments is zero the angular velocity of the spherical particle is constant and is given by

$$\omega = \frac{1}{2} G \qquad (D2.41)$$

Fig. D2.25 A spherical particle in laminar flow. (Adapted from Tsvetkov, 1989.)

Figure D2.25 shows undisturbed laminar flow before the introduction of the particle in a coordinate system moving with the translational motion of the particle and having its origin at the centre of the spherical particle. The linear velocities of points a, b, c and d on the surface of the particle are $u_0 = \omega R_0 = \frac{1}{2} G R_0$, where R_0 is the radius of the spherical particle. The undisturbed flow velocities u_0 in this coordinate system are different at different points. At c $du = 0$ whereas at a and b the absolute magnitude of u_0 is $G R_0$, i.e. twice the surface velocity of the particle in the same direction. At other points on the surface of the particle the magnitude of u_0 has intermediate values, but is proportional to ω.

Fig. D2.26 The velocity directions of unperturbed solvent relative to the surface of a spherical particle in laminar flow. The flow lines indicate the magnitude of $u - u_0$ in the *xy* plane. (Adapted from Tsvetkov, 1998.)

In the presence of the particle the true velocity distribution is somewhat different from that shown in Fig. D2.26. The velocity change at points a, b, c and d is not a sharp discontinuity at the particle–solvent surface, but occurs over a specific layer of liquid surrounding the particle. Figure D2.26 shows the undisturbed flow of the solvent in the xy plane relative to the surface of the particle. It can be seen from the diagram that the sum of the turning moments due to the viscous forces is zero. However, the work done by these forces results in an additional energy loss and therefore in an increase in the solution viscosity. A qualitative assessment of the effect shows that the velocity of the solvent relative to the surface of the sphere is always proportional to GR_0, which is equivalent to rotation of the particle in the solvent with a relative angular velocity of

$$\omega_0 = \alpha G \tag{D2.42}$$

Equation (D2.42) is universal. For particles of different shapes the proportionality coefficient has different values, but the proportionality between ω_0 and G is always maintained.

Figure D2.27 shows the true velocity distribution contours for rotational and shear flow patterns. Here we clearly see the main difference between pure rotation and rotation in shear flow: in a certain direction the fluid velocity profile of the shear flow decreases much more rapidly than for the rotational case. In shear flow the molecule is extended along one direction and squeezed along another direction. This effect is very dramatic for flexible chains (see Chapter D7).

For rotation of a solid particle of any form in a viscous solvent the work done in unit time in overcoming friction is

$$W = \omega_0 Q \tag{D2.43}$$

where Q is the turning moment, which is related to the coefficient of rotational friction of the particles, Θ, by the relationship

$$Q = \zeta \omega_0 \tag{D2.44}$$

If 1 cm^3 of solution contains N_0 particles, then the energy loss, ΔE, in unit time due to rotational friction is WN_0. Using Eqs. (D2.42)–(D2.44)

$$\Delta E = \alpha^2 G^2 \zeta N_0 \tag{D2.45}$$

where the ΔE is the difference between energy losses due to the friction of the solution and the solvent. Thus from Eq. (D2.45)

$$\Delta E = \eta G^2 - \eta_0 G^2 \tag{D2.46}$$

where η is viscosity of the solution. Equating Eqs. (D2.41) and (D2.42), the specific viscosity of the solution is

$$\eta_{sp} \equiv \frac{\eta - \eta_0}{\eta_0} = \frac{\alpha^2 N_0 \zeta}{\eta_0} \tag{D2.47}$$

Fig. D2.27 Comparison of rotational and shear flows. The calculated fluid velocity profile was generated for spherical macromolecules obeying stick boundary conditions. The dashed lines represent the flow lines for a sphere rotating around an axis normal to the plane of the page. The solid lines represent the flow lines in shear flow. The radius of the sphere is R_0, and distance is expressed in units of R_0. The disturbance to the fluid due to the presence of the sphere is a function of the velocity components of the fluid along three axes. This function is too lengthy to be reproduced here and we focus our attention on the flow pattern only. Interested readers should refer to Kuntz and Kauzmann (1974) for details.

This equation has general applicability to particles of different shapes. The coefficient α^2 depends on the shape of the particle. Thus, for a spherical particle $6\alpha^2 = 2.5$ and $\Theta = 6\pi \eta_0 V$ and therefore

$$\eta_{sp} = 2.5 V N_0 \tag{D2.48}$$

Equation (D2.48) is the well-known Einstein equation.

Remembering that $N_0 = cN_A/M$, where c is the concentration of the solution in grams per cubic centimetre, N_A is the Avogadro number and M is the molecular weight of the particles, it follows from Eq. (D2.48) that

$$[\eta] \equiv \lim_{c \to 0} \left(\frac{\eta_{sp}}{c} \right) = 2.5 V N_0 \tag{D2.49}$$

where $[\eta]$ is the intrinsic viscosity of the macromolecular solution.

For rigid non-spherical particles, Eq. (D2.47) gives

$$[\eta] = \frac{\alpha^2 N_A \zeta}{\eta_0 M} \tag{D2.50}$$

and using Eq. (D2.20) we have

$$[\eta] = \alpha^2 \frac{RT}{\eta_0 \Theta M} \tag{D2.51}$$

Equation (D2.51) reflects the fact that the intrinsic viscosity of the solution is in all cases (no matter what model is adopted for the macromolecules) a measure of energy loss due to the rotation of the particle in the solvent.

Ellipsoidal particles

For solutions of non-spherical particles the situation is more complicated and the physical picture can be described qualitatively as follows. The evaluation of the energy dissipation is based on the assumption that a solution of ellipsoidal particles in laminar flow is in an equilibrium steady-state dependent upon the magnitude of two opposing forces. One arising from the velocity gradient tends to orient the particles in the direction of the streaming lines, the other due to the Brownian motion tends to produce a random orientation. Thus the total energy losses due to the friction is the sum of two contributions, the first being the purely hydrodynamic losses which were considered in the previous paragraph, and the second being the additional loss due the directional Brownian rotation of the particles. The relative effects of the two contributions depend on the ratio G/Θ.

At low values of G/Θ, the asymmetry of the angular distribution function is completely removed by the opposing Brownian rotation. Under these conditions, even highly asymmetric particles rotate with constant angular velocity $\omega = G/2$.

Simha solved Eq. (D2.49) for the viscosity of a solution of ellipsoids of revolution for the limiting case when $G/\Theta \to 0$. It is common practice to express

Fig. D2.28 Plot of Simha's function ν against axial ratio for prolate and oblate ellipsoids of revolution. For a sphere $\nu = 2.5$.

Simha's equation as

$$[\eta] = \nu(p) N_A V_h / M \tag{D2.52}$$

where N_A is Avogadro's number, M is the molecular mass and V_h is the hydrodynamic volume of the macromolecules (Comment D2.19).

The Simha functions $\nu(p)$ for prolate and oblate ellipsoid revolution are presented in Fig. D2.28. This figure demonstrates that the Simha functions are much more sensitive to shape than the Perrin function (Fig. D2.3). For axial ratios greater than 15, the Simha function for a prolate ellipsoid is given by the asymptotic equation:

$$\nu(p) = \frac{4}{15} \frac{(a/b)^2}{\ln(2a/b)} \tag{D2.53}$$

Equation (D2.53) shows that we need not even know the molecular weight in order to calculate the shape of macromolecules.

Rod-like particles

In 1951 by using a linear array of spheres as a model for rod-like particles Kirkwood and Auer calculated the intrinsic viscosity of these rods as a function of the axial ratio. In this model a rod is treated as a linear array of n spherical monomers with diameter d and length $L = nd$. The asymptotic form of their equation is

$$[\eta] = \frac{\pi N_A L^2 d}{2250 M_0 \ln(L/d)} \tag{D2.54}$$

where N_A is Avogadro's number and M_0 is the molecular weight of the monomer. This equation is essentially identical to the asymptotic behaviour of the Simha equation for a prolate ellipsoid with a very large semi-major axis, $L/2$, if M_0 is taken as $\pi \rho b^2 L/3n$ and it is recognised that (for large a/b) $\ln(2a/b) = \ln 2 +$

Comment D2.20   Three-axis ellipsoids

The viscosity of a tri-axial ellipsoid is given by

$$v = [\eta]/V_s = Y(a, b, c)$$

where $[\eta]$ is the intrinsic viscosity in millilitres per gram, V_s is the swollen specific volume in millilitres per gram and $Y(a, b, c)$ are elliptic integrals which can be solved numerically. A given value of $(a/b, b/c)$ uniquely fixes a value of v, but a given value of v has a line solution corresponding to values $(a/b, b/c)$. Unfortunately, the intersection of two lines of solutions doesn't occur in the interval of low axial ratio of tri-axial ellipsoids (Harding and Rowe, 1982).

Spheroid–cylindrical molecules

Data on the intrinsic viscosity of the spheroid cylinders with oblate, spherical, or prolate hemispheroidal caps at the end (see Fig. D2.9) are available in the literature The end-effect on the intrinsic viscosity was found remarkable, and depends appreciably on the shape of the ends for relatively short cylinders (compared with the analogous effect for the translational diffusion coefficient (Yoshizaki and Yamakawa, 1980).

Dumb bell

Data on the intrinsic viscosity for dumbbells consisting of two equal-radius spheres at various separations are known (Wakia, 1971).

$\ln(a/b) \approx \ln(a/b)$:

$$[\eta] = \frac{24}{9000\rho} \frac{p^2}{\ln p} \tag{D2.60}$$

This gives a method for determining the axial ratio of rod-like particles and, if the molecular weight is known, it is possible to obtain the length of the rod. The intrinsic viscosity of some other types of regularly shaped molecules is discussed in Comment D2.20.

D2.4.4 Arbitrarily shaped particles

The possibilities for the calculation of the intrinsic viscosity of complex particles are more limited than for rotational diffusion and especially translational diffusion. First, because in the limit of zero shear rate, $[\eta]$ must be orientationally averaged. This implies finding the intrinsic viscosity for an infinite number of possible orientations of the particle with respect to the external axes. Second, for non-spherical molecules $[\eta]$ is origin-dependent. The origin to which $[\eta]$ must be referred is the viscosity centre. The velocity field in the viscosity experiment is much more complicated than those characterising translational and rotational dynamics, and therefore the viscosity centre may differ from the centres of diffusion and rotation.

Several approaches are now available for the calculation of the intrinsic viscosity of arbitrarily shaped particles: shell modelling bead modelling, triangular plate modelling and via polarisability (Comment D2.21). In the first of these a rigid macromolecule is modelled using N spherical elements of arbitrary radius, as in translation friction. The size and shape of the model should be as close as possible to that of the macromolecule (see Fig. D2.6(a)). Hydrodynamic interaction between the spheres is taken into account by classical hydrodynamic methods. Unfortunately, all theories for the calculation of intrinsic viscosity operate very well for very elongated particles, and fail in the single-sphere limit or when the model contains only a few spheres.

The second approach is based on modelling of the particle surface as an array of flat triangular plates. The intrinsic viscosity of the corresponding 'smooth' particles is estimated by carrying out numerical studies of several model structures and extrapolating to the limit of a model in which the number of plates goes to infinity. The main advantage of this procedure is that it is general enough to accommodate stick and slip boundary conditions as well as the presence of external forces on the surrounding fluid.

Boundary element calculations for ellipsoids of revolutions are, in the limit of 1%, in good agreement with the exact values. A range of different rod-like structures were modelled with p varying from 2.04 to 17.0. This corresponds to models of DNA fragments with a length ranging from 12 base-pairs ($p = 2.04$) to 100 base-pairs ($p = 17.0$).

The third approach is based on the correlation between the intrinsic viscosity $[\eta]$ and the polarisability α. Based on direct comparison of $[\eta]$ and α in the case of ellipsoids and dumbbells, Douglas and Hubbard proposed the relation

$$[\eta] = \lim_{c_p \to 0} \frac{\eta - \eta_0}{c_p \eta_0} = 0.79\alpha \tag{D2.56}$$

which is accurate to within 5% (Comment D2.22). The orientational averaging of the Oseen tensor leads to the relation

$$[\eta] = \frac{3}{4}\alpha \tag{D2.57}$$

In the case of a sphere its polarisability is $\alpha = 3V_p$ and the intrinsic viscosity predicted by Eq. (D2.57) is thus $[\eta] = \frac{9}{4}V_p$, which differs from the exact Einstein result $[\eta] = 5V_p/2$ by 10%. Adding the constant to the right-hand side of Eq. (D2.57) results in the relation becoming empirically useful for globular proteins. This leads to the empirical relationship

$$[\eta] = \frac{3}{4}\alpha + \tfrac{1}{4}V_p \tag{D2.58}$$

Comment D2.21
Shell model approach in the calculation of intrinsic viscosity

The shell model approach in theory should be the ideal solution for the rigorous calculation of the viscosity of macromolecules. In this procedure, a large number of small spheres are placed on the surface of the object, and packed as tightly as possible (see Fig. D2.7). However, precise calculations of intrinsic viscosity by extrapolation of shell model computations to zero bead size require so much computer time that the method is not generally practical.

Comment D2.22
Calculation of intrinsic viscosity via polarisability

In this calculation it is considered that c_p is measured from the number density, then $[\eta]$ has the unit of volume (Zhou, 1995).

For ellipsoids, cylinders and dumbbells this equation is accurate to within about 3%. The polarisability of arbitrarilily shaped particles can be found in a single calculation using the boundary-element technique.

D2.5 From hydrodynamic equivalent sphere to a whole body approach

D2.5.1 Volume-dependent shape functions

The translation, rotation, and viscosity phenomena described above depend on the axial ratio and the hydrated volume of the particle. Determining this particle volume requires knowledge of the hydration δ (mass in grams of H_2O bound per gram of dry macromolecule). Table D2.4 summarises all volume-dependent shape functions.

D2.5.2 Two-axis ellipsoid: volume-independent shape functions

A first attempt to solve the volume or axial ratio of a two-axis ellipsoid was made by Oncley in 1941. He calculated the numerical values of the axial ratio and the hydration for an ellipsoid of revolution for various values of frictional coefficients and viscosity increments and represented it by very vivid contour maps. This result clearly demonstrated for the first time that a choice always exists between a prolate and an oblate ellipsoid, each of specified axial ratio, even when a reasonable degree of hydration has been assumed.

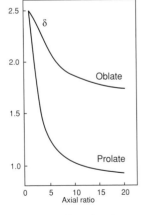

Fig. D2.29 Translational friction coefficient and intrinsic viscosity (the β-function). For a sphere $\beta = 2.12$ (Sheraga and Mandelkern, 1953.)

Translational frictional coefficient and intrinsic viscosity (the β-function)

In 1953 Sheraga and Mandelkern proposed combining the translational friction coefficient and the intrinsic viscosity. The resulting function β (see Fig. D2.29) is very insensitive to the axial ratio, especially for oblate ellipsoids. It can only be used to distinguish between prolate and oblate ellipsoids for highly asymmetrical particles. In practice, the function β has been used in molecular weight determination.

Rotational frictional coefficient and intrinsic viscosity (the δ-function)

The combination of the rotational frictional coefficient (Eq. (D2.24a)) and the intrinsic viscosity (Eq. (D2.45)) results in the δ-function (Fig. D2.30), which is more sensitive to the axial ratio, especially to small ones. The function δ has not been widely used mainly owing to difficulties in obtaining the rotational diffusion coefficient for proteins.

Fig. D2.30 Rotational friction coefficient and intrinsic viscosity (the δ-function). (After Sheraga and Mandelkern, 1953.)

Table D2.4. *Experimentally measured volume-dependent shape functions*

Shape function	Related experimental parameter	Experimental method
Translational friction ratio f/f_0 (for a sphere $f/f_0 = 1$)	$\dfrac{f}{f_0} = \dfrac{M(1 - \bar{v}\rho_0)}{6\pi \eta_0 s N_A} \left[\dfrac{4\pi}{3V_h} \right]^{1/3}$	Sedimentation
	$\dfrac{f}{f_0} = \dfrac{kT}{6\pi \eta_0 D} \left[\dfrac{4\pi}{3V_h} \right]^{1/3}$	Diffusion
Rotational friction ratio Θ_i/Θ_0 (for a sphere $\Theta_i/\Theta_0 = 1$)	$\dfrac{\Theta_i}{\Theta_0} = \dfrac{6\pi \eta_0 V_h}{kT} \Theta_i \; (i = a, \, b)$	Dynamic light scattering Flow birefringence
Intrinsic viscosity, ν (for a sphere $\nu = 2.5$)	$\nu = \dfrac{[\eta]M}{N_A} \dfrac{1}{V_h}$	Viscosity
Reduced excluded volume[a] V_{red} (covolume) (for a sphere $V_{red} = 8$)	$V_{red} = \dfrac{V_{exc}}{V_h} = \dfrac{2A_2 M^2}{N_A} \dfrac{1}{V_h}$ where V_{exc} is macromolecular excluded volume and A_2 is the second virial coefficient	Concentration dependence of osmotic pressure, scattering intensity and sedimentation equilibrium
Harmonic mean relaxation time ratio (for a sphere $\tau_h/\tau_0 = 1$) (Chapter D6)	$\dfrac{\tau_h}{\tau_0} = \dfrac{3}{(\tau_0/\tau_a) + (2\tau_0/\tau_b)}$ $= \dfrac{kT}{3\eta_0} \tau_h \dfrac{1}{V_h}$ where τ_0 is the corresponding time for a spherical macromolecule of the same volume	Steady-state fluorescence depolarisation NMR
Correlation times ratio (for a sphere $\tau_i/\tau_0 = 1$) (Chapter D6)	$\dfrac{\tau_i}{\tau_0} = \dfrac{kT}{3\eta_0} \tau_i \dfrac{1}{V_h}$ ($i=1, \ldots, 5$ for a general body; $i = a, b$ for an ellipsoid revolution) where τ_0 is the corresponding time for a spherical macromolecule of the same volume	Fluorescence depolarisation anisotropy Electric birefringence or electric dichroism
Sedimentation concentration regression coefficient k_s (Chapter D4)	$s_c = s_0(1 - k_s c) \cong \dfrac{s_0}{1 + k_s c}$ where s_c and s_0 are the sedimentation coefficients at concentration c and infinite dilution, respectively	Concentration dependence of the sedimentation coefficient

[a] The excluded volume is a measure of the interaction between two macromolecules. It is a measure of the volume occupied as modified by intermolecular attractions. Hydrodynamic measurements, however, are made in (or extrapolated to) very dilute solutions, in which intermolecular interaction plays no part at all. The hydrodynamic volume is always a measure of the volume occupied in the solution by a single particle. The molar covolume for a system of macromolecules can be obtained from the thermodynamic second virial coefficients A_2 after correction (or suppression) of charge effects.

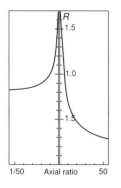

Fig. D2.31 Sedimentation regression coefficient and intrinsic viscosity (the R-function). (Rowe, 1977.)

Fig. D2.32 Translational friction coefficient and harmonic mean rotational relaxation time ratio (the Ψ-function). (Squire, 1970.)

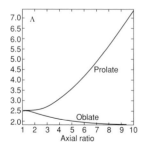

Fig. D2.33 Intrinsic viscosity and harmonic mean rotational relaxation time ratio (the Λ-function). (Harding, 1980.)

Sedimentation regression coefficient and intrinsic viscosity (the R-function)

In 1954 Wales and Van Holde pioneered the theoretical consideration of a combination of the sedimentation regression coefficient k_s (Section D4.5.4) and intrinsic viscosity $[\eta]$ (Eq. (D2.45)). They proposed that the value of the ratio $k_s/[\eta]$ is close to 1.6. Analysis of data from a wide variety of biological macromolecules showed that for globular proteins the ratio $k_s/[\eta]$ in fact lies in the range 1.5–1.7, while more extended particles have a smaller ratio.

On the basis of the general description of the concentration dependence of transport processes a hydrodynamic shape function R that is the ratio of sedimentation regression coefficient to the intrinsic viscosity was proposed. The function R, which depends on the particle asymmetry, is presented in the Fig. D2.31. It is seen that R varies with the axial ratio for ellipsoids, with the variation being very rapid for low values of p. Therefore, this function provides a precise method for calculating the axial ratio of relatively symmetrical particles. The computation of the R-function does not require a value for the absolute concentration since the latter cancels out in the ratio $k_s/[\eta]$.

Harmonic mean rotational relaxation time ratio and translational friction coefficient (the Ψ-function)

The combination of the harmonic mean rotational relaxation time ratio, τ_h/τ_0 and the translational friction coefficient results in the Ψ-function. The dependence of the Ψ-function on the axial ratio for prolate and oblate ellipsoids is presented in Fig. D2.32. It is clearly seen that the Ψ-function is practically insensitive to the axial ratio for small asymmetries ($p < 5$). It can only be used to distinguish between prolate and oblate ellipsoids for highly asymmetrical particles.

Intrinsic viscosity and harmonic mean rotational relaxation time ratio (the Λ-function)

The volume-independent hydrodynamic function Λ can be obtained by combining the intrinsic viscosity with the harmonic mean rotational relaxation time ratio. The Λ-function is presented in Fig. D2.33. It is evident that Λ is much more sensitive to the axial ratio than Ψ. The values of Λ for prolate and oblate ellipsoids diverge from each other. For prolate ellipsoids Λ increases with increasing axial ratio, while for oblate ellipsoids Λ decreases with increasing axial ratio. The main difficulty in applying this approach, especially for particles with a small asymmetry ($p < 3$), is determination of τ_h to a high precision (see Chapter D6).

Translational frictional coefficient and molecular covolume (the ψ-function)

The theoretical dependence of the ψ-function on the axial ratio of prolate and oblate ellipsoids is presented in Fig. D2.34. The figure shows diverging plots which demonstrate a capability to distinguish between prolate and oblate models.

However, inspection of Fig. D2.34 shows that in the limit of experimental error for the ψ-function (about 5%) it is very difficult to distinguish between prolate and oblate models at axial ratios < 3. At larger axial ratios the ψ-function can be used to distinguish between the two models.

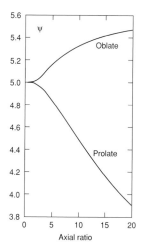

Fig. D2.34 Translational frictional coefficient and molecular covolume (the ψ-function). (Jeffrey *et al.*, 1977.)

Intrinsic viscosity and molecular covolume (Π-function)

A combination of the covolume and the intrinsic viscosity results in the Π-function. Plots of the Π-function versus the axial ratios of prolate and oblate ellipsoids are shown in Fig. D2.35. Comparison of the ψ- and Π-functions shows that the latter should be the more sensitive when determining the axial ratio of a prolate ellipsoid.

Table D2.5 summarises experimentally measured volume-independent shape functions. This table shows that rotational phenomena are more sensitive in general to shape than the corresponding translational ones (Comment D2.23). However, this extra sensitivity comes at a price – in terms of the greater difficulty in both making precise measurements and extracting the parameters. For measuring rotational diffusion coefficients of globular proteins (or equivalently rotational relaxation times) electric birefringence and fluorescence anisotropy depolarisation decay remain the principal probes. Unfortunately, these two

Table D2.5. *Experimentally measured volume-independent shape functions*

Volume-independent shape function	Combination of the hydrodynamic probes	Sensitivity to shape
β-function $(v, f/f_0)$ (Sheraga-Mandelkern, 1953)	Viscosity and translation friction	Very poor sensitivity to axial ratio
δ-function $(v, \Theta/\Theta_0)$ (Sheraga-Mandelkern, 1953)	Viscosity and rotational friction	Sensitive function for small axial ratios
R-function (v, k_s) (Rowe, 1977)	Viscosity and sedimentation regression coefficient	One of the sensitive functions for small axial ratio
Ψ-function $(f/f_0, \tau_h/\tau_0)$ (Squire, 1970)	Translational friction coefficient and harmonic mean rotational relaxation time	Very poor sensitivity to axial ratio
Λ-function $(v, \tau_h/\tau_0)$ (Harding, 1980)	Viscosity and harmonic mean rotational relaxation time	Very sensitive function, except at very low axial ratio $(p < 2)$
ψ-function $(f/f_0, V_{red})$ (Jeffrey *et al.*, 1977)	Translational friction and covolume	Poor sensitivity to axial ratio
Π-function (v, V_{red}) (Harding, 1981)	Viscosity and covolume	Sensitive function, except at very low axial ratio $(p < 3)$

Fig. D2.35 Intrinsic viscosity and molecular covolume (the Π-function). (After Harding, 1981.)

Comment D2.23 Volume-dependent and volume-independent functions

It is necessary to point out that all the volume-independent shape functions presented in Table D2.5 are not as sensitive to shape as their volume-dependent precursors presented in Table D2.4.

techniques have some important practical limitations (see Chapters D7 and D9). Moreover, both techniques are connected with the resolution of multi-exponential decay terms.

Fortunately four volume-independent functions, $\delta\,(v, \Theta), R\,(v, k_s), \Lambda\,(v, \tau_h/\tau_0)$ and $\Pi\,(v, V_{red})$ do appear particularly useful and each of them is largely free from these problems and is a sensitive function of the axial ratio.

D2.5.3 Tri-axial ellipsoids: volume-independent shape functions

In 1936 Perrin provided an explicit expression for the translational friction ratio for a tri-axial ellipsoid and in 1981 Harding, Dampier and Rove obtained an analogous solution of the viscosity for a tri-axial ellipsoid. All the tri-axial ellipsoid shape functions share the common property of having a line solution of possible values for the axial ratios $(a/b, b/c)$ for any given value of the hydrodynamic function (Comment D2.24).

Comment D2.24 Unique solution

It is clear that a given value of $(a/b, b/c)$ uniquely fixes a value of $f/f_0(v)$, but a given value of $f/f_0(v)$ has a line solution corresponding to values $(a/b, b/c)$. A unique solution for these two axial ratios can be found from the intersection of two or more of these 'line solutions'.

Fig. D2.36 Plots of constant values for the Simha function and the Perrin function f/f_0 in the $(a/b, b/c)$ plane corresponding to a hypothetical ellipsoid particle $(a/b, b/c) = (2.0, 2.0)$. (After Harding, 1995.)

Figure D2.36 illustrates two line solutions for two volume-dependent functions, v and f/f_0, for a hypothetical tri-axial ellipsoid of $(a/b, b/c) = (2.0, 2.0)$. It is clear that these two functions form a poor combination of functions, because of the shallowness of the intersection and their dependence on assumed values of hydration.

To use the tri-axial ellipsoid we have to find two volume-independent shape functions that give a reasonable intersection which is as orthogonal as possible. These criteria are quite restrictive. The best result was found to be a combination of two functions (R and Λ) involving viscosity, the sedimentation regressive coefficient and the harmonic mean rotational relaxation. Figure D2.37 shows the dependence of the functions of R and Λ in the $(a/b, b/c)$ plane for neurophysin monomers (a) and dimers (b). This combination of line solutions

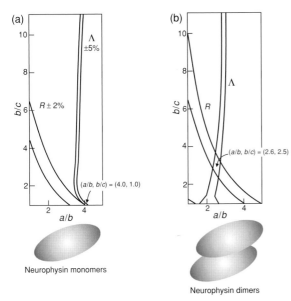

Neurophysin monomers

Neurophysin dimers

Fig. **D2.37** Plots of the functions R and Λ in the $(a/b, b/c)$ plane for neurophysin monomers (a) and dimers (b). The monomer appears as a prolate model in which the two axial ratios are approximately $(a/b, b/c) = (4.0, 1.0)$ while the dimer looks like a more compact body with $(a/b, b/c)$ of about $(2.8, 2.5)$. This indicates that the association process is of a side-by-side rather than an end-to-end type. (Harding, 1995.)

has been used to provide an indication of the likely mode of association of monomers of the neural protein neurophysin into dimers. However, to date this sensitivity and experimentally measurable precision have not been enough to solve the Oncley problem, i.e. make a definite choice between a prolate and an oblate ellipsoid for the low axial ratios typical for the most of globular proteins.

D2.5.4 Whole-body approach and bead model

There are two basic approaches for determining the conformation of biological macromolecules using hydrodynamic techniques. In *bead* modelling a structure is assumed and then its hydrodynamic properties (intrinsic viscosity, translational and rotational diffusion coefficients) are calculated. After that these experimentally determined properties are compared with predicted ones for the unknown structure. The model is then refined until the predicted properties converge to agree with the actual properties. A serious disadvantage in such an approach is that the final model may not be the *only one* that gives these properties. In practice, the original assumed model should be a good starting estimate for the structure, based on, for example, X-ray crystallography.

The second approach is based on calculating a structure *directly* from the known hydrodynamic properties. In this approach, called the whole-body approach, a rigid particle is represented as a tri-axial ellipsoid with three distinct semi-axes. A unique structure can be predicted by the combination of three appropriate measurements to avoid the problem of having to assign a value for the hydrodynamic volume (molecular hydration) and provide a unique pair of axial ratios which define such ellipsoids. The main drawbacks of such an approach are the insufficient sensitivity of volume-independent functions for small axial ratios and insufficient experimentally measurable precision.

In spite of significant advances in the hydrodynamic methods, particularly in terms of hydrodynamic modelling (bead modelling, tri-axial ellipsoid, analysis of flexible macromolecules) hydrodynamic methods will forever be labelled 'low-resolution' methods. Because hydrodynamic methods are generally rapid and non-destructive they can provide either 'low-resolution' information on macro-molecular structure prior-X-ray or NMR structure analysis or the final refinement of a 'high-resolution' model for dilute solution behaviour, especially in terms of intermolecular interaction phenomena.

D2.6 Homologous series of macromolecules

Figure D2.38 shows three typical shapes for homologous series of biological macromolecules. The first is a series of spherical particles that preserve the sim-ilarity of their shape; the second comprises rod-like particles whose long axis increases proportionally to M while the transverse dimensions remain constant; the last comprises Gaussian coils whose asymmetry gradually increase with increasing molecular mass.

The dependence of the translational friction coefficient on the molecular mass M in a homologous series is generally given by the Kuhn–Mark–Houwink equation:

$$f = kM^X \tag{D2.59}$$

where k and X are constants and X is related to the shape of the macromolecules in a homologous series.

Thus, according to Stokes' law for a series of spherical particles their trans-lational friction constant is proportional to their molecular mass to the $1/3$ power (Comment D2.25). For a homologous series of rod-like particles, a theoretical examination shows that the exponent of power, X, in Eq. (D2.59) is about 0.85. For a homologous series of Gaussian coils in an 'ideal' solution, the exponent

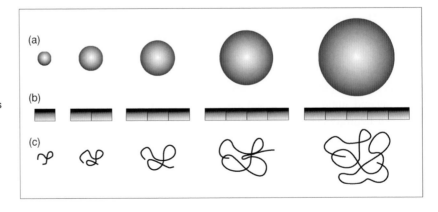

Fig. D2.38 Homologous series of macromolecules with different shapes: (a) spheres, (b) rod-like molecules joined to each other in a 'head to tail' fashion, (c) Gaussian coils.

(a)

(b)

(c)

of power is 0.5. Thus, the greater the asymmetry grows with an increase in the molecular mass, the larger the absolute value of the exponent of power in Eq. (D2.59).

As the coefficient of translational friction of a particle can be determined from the experimental diffusion and sedimentation constants, Eq. (D2.59) predicts varying dependences of the sedimentation and diffusion constants on the molecular mass for different homologous series. Examples of the application of Eq. (D2.59 to determine the shape of different biological macromolecules will be given in Chapters D3 and D4.

By analogy with translational friction the dependence of the rotational friction coefficient on the molecular mass M in a homologoues series can be interpreted as a gradual series of type

$$\Theta = k_\Theta M^c \tag{D2.60}$$

where k_Θ and c are constants related to the shape of the macromolecule. According to Perrins's law (Eq. (D2.2)) for a series of spherical particles their rotational friction constant is proportional to their molecular mass to the power 1.

For a homologous series of impermeable Gaussian coils in an 'ideal' solution, $c = 1.5$. For a homologous series of infinitely thin ($L >> d$) rod-like particles $c = 3.0$.

Thus, the more the asymmetry increases as the molecular mass increases, the larger is the absolute value of c in Eq. (D2.60). Inasmuch as the coefficient of rotational friction of a particle can be determined from the experimental rotational friction constants and rotational relaxation times, Eq. (D2.60) predicts varying dependences of coefficient c on the molecular mass for different homologous series. Examples of the application of Eq. (D2.60) to determine the shape of different biological macromolecules will be given in Chapters D6 and D8.

The dependence of the intrinsic viscosity on the molecular mass M in a homologous series is given by

$$[\eta] = K_{[\eta]} M^\alpha \tag{D2.61}$$

where k and α are constants. It follows directly from Eq. (D2.56) that for rigid ellipsoids or rod-like particles, i.e. those having a constant value of V/M for a homologous series, the form of the dependence of $[\eta]$ on M is determined by the dependence of $v(p)$ on M (Eq. (D2.52)).

Thus, for example, if the mass and therefore the dimensions of the particles increases while the shape remains the same, i.e. while p stays constant, then for a homologous series $v(p) = $ constant, $[\eta] = $ constant and therefore α in Eq. (D2.56) is zero, regardless of the shape of the particles. On the other hand, for rod-like particles (cylindrical), the long axis L of which increases in proportion to M, while the diameter d remains constant, the value of p increases in proportion to M and thus $[\eta] = f(M)$ coincides with the dependence of $v = f(p)$. For

Comment D2.25

On the strict interpretation of power 1/3 in homologous series

To treat a homologous series it is convenient to present the frictional coefficient of a spherical particle in the form

$$f_0 = 6\pi \eta_0 (3V_h/4\pi)^{1/3}$$

It follows that the frictional coefficient for spherical molecules with similar hydration and partial specific volumes is proportional to $M^{1/3}$.

A power 1/3 for a homologous series of the particle does not mean that macromolecules under study really have a spherical shape. For example, in a homological series of ellipsoids with a fixed axial ratio X is also equal to 1/3. A power 1/3 strictly speaking means that the particles in the homologous series preserve similarity of their shape.

such molecules, when p is large, the dependence in Eq. (D2.56) follows from Eq. (D2.51).

For a homologous series of prolate ellipsoids that have the same minor axis Eq. (D2.56) can be approximated as

$$
\begin{aligned}
\nu &= 0.233 p^{1.698} \quad 20 < p < 100 \\
\nu &= 0.207 p^{1.732} \quad 20 < p < 300
\end{aligned}
\tag{D2.62}
$$

Thus, ν should increase with the molecular mass or length to about the 1.7 power. If, however, the mass increase of the particles is accompanied by a growth in a diameter b for constant length L, then $\alpha < 0$.

Thus, the more the asymmetry of the extended particles depends on M, the greater is the magnitude of α in Eq. (D2.56). If the asymmetry, i.e. p, increases with M, then $\alpha > 0$; and if it decreases with M, then $\alpha < 0$. For a homologous series of Gaussian coils in an 'ideal' solvent, α is 0.5; in a 'good' solvent α is 0.8.

D2.7 Checklist of key ideas

- The coefficient of translational friction of a sphere is proportional to its radius and to the viscosity of the solvent within which the particle moves.
- The coefficient of translational friction of a two-axis ellipsoid of revolution depends on its volume and on the axial ratio.
- The translational friction of short particles with a 'broken' shape (different types of circular cylinders, cube . . .) can be calculated only as an approximation.
- Three distinct but related techniques for modelling hydrodynamic friction of arbitrarily shaped rigid particles have emerged.
- In the 'bead' model approach, the particle is modelled with a set of spheres (beads) which can be of equal or unequal size.
- In the 'shell' model approach, the surface of a particle is modelled by a set of small equal-sized spheres.
- In the boundary element approach, the surface of a particle is modelled by a set of small panel elements (platelets).
- The connection between hydrodynamics and electrostatics provides a simple and accurate method of calculating the translational friction coefficient of rigid particles of arbitrary shapes.
- The coefficient of translational friction is not very sensitive to segmental mobility: for a loosely jointed rod it varies by only 3% from that of a rigid rod of the same length, and does not depend significantly on the location of the hinge.
- The actual translational friction of rigid microscopic objects can be determined by observing settling rates of their macroscopic model in a high-viscosity fluid at low Reynolds number.
- Rotation of the sphere in Stokes' approximation may be characterised by a single constant which has the dimensions of time.

- The coefficient of rotational friction of a two-axis ellipsoid of revolution depends on its volume and on the axial ratio.
- The rotational friction of short particles with a 'broken' shape (different types of circular cylinders, cube, . . .) can be calculated as an approximation only.
- Rotational friction is much more sensitive to the dimensions of the particle than translational friction.
- For long elongated particles the rotational diffusion constant may be used to measure the *length* of a molecule.
- The rotational friction coefficient of proteins and short DNA fragments with known structure at the atomic level can be calculated using 'bead' or 'platelet' approaches.
- The viscosity η of a macromolecular solution is always greater than that of pure solvent. The fractional increase in viscosity is called the specific viscosity and denoted $\eta_{sp} = \eta/\eta_0 - 1 = \eta_{rel} - 1$, where η_{rel} is the relative viscosity.
- The intrinsic viscosity $[\eta]$ is defined as the fractional increment in viscosity of the solution due to the addition of 1g of macromolecule; it can be obtained experimentally as $[\eta] = \lim_{c \to 0} \eta_{sp}/c$ and has units of cubic centimetres per gram of decilitres per gram.
- The specific viscosity η_{sp} of a solution of spheres is proportional to the volume fraction occupied by the spheres; η_{sp} is independent of the absolute size of the spheres.
- The intrinsic viscosity of a two-axis ellipsoid of revolution depends on its volume and on the axial ratio. The dependence is described by the Simha equation.
- The intrinsic viscosity of a protein can be predicted with good accuracy from its atomic coordinates, provided hydration contributions are taken into account appropriately.
- The connection between hydrodynamics and electrostatics provides a simple method of calculating the intrinsic viscosity of particles of regular and arbitrary shape.
- The intrinsic viscosity of proteins and short DNA fragments with known structure at the atomic level can be calculated using 'bead' and 'platelet' approaches.
- The hydrodynamic parameters of a macromolecule (intrinsic viscosity, frictional and rotational diffusion coefficients) depend on shape and hydrodynamic volume (including hydration); it follows that it is not possible to determine the shape or volume by measurement of one hydrodynamic parameter.
- Two shape functions can be combined to eliminate the requirement for an estimate of the hydrodynamic volume; as a result for an ellipsoid of revolution seven volume-independent shape functions (β, δ, R, Ψ, Λ, ψ, and Π) can be obtained.
- All volume-independent shape functions are not as sensitive to shape as their volume-dependent precursors.
- Each of these seven volume-independent functions has a different sensitivity to the axial ratio of its approximated ellipsoid.
- Thus, a combination of the translational friction coefficient and the intrinsic viscosity on the translational friction coefficient and the harmonic mean rotational relaxation time ratio results in a function (β and Ψ, respectively), which is very insensitive to the axial ratio, specially for oblate ellipsoids.
- In contrast, combination of the sedimentation regression coefficient and the intrinsic viscosity or the intrinsic viscosity and the harmonic mean rotational relaxation time

ratio results in a function (Λ and R, respectively), which provides a precise method for the calculation of the axial ratio of relatively symmetrical particles.

- For a tri-axial ellipsoid combination of two functions (R and Λ) involving the viscosity, the sedimentation regression coefficient, and the harmonic mean rotational relaxation time ratio leads to the best results for the calculation of its three axes.
- In spite of significant advances in hydrodynamic methods, hydrodynamic methods will always be 'low-resolution' methods.
- The coefficient of translational friction for a homologous series of spherical particles is proportional to their molecular mass to the power of $1/3$.
- The coefficient of translational friction for a homologous series of rod-like particles is proportional to their molecular mass to about the power of 0.8.
- The coefficient of rotational friction for a homologous series of spherical particles is proportional to their molecular mass to the power of 1.
- The coefficient of rotational friction for a homologous series of thin rod-like particles is proportional to their molecular mass to the power of 3.
- The coefficient of rotational friction for a homologous series of Gaussian coils is proportional to their molecular mass to the power of 1.5.
- The intrinsic viscosity for a homologous series of spherical particles does not depend on their molecular mass.
- The intrinsic viscosity for a homologous series of rod-like particles is proportional to their molecular mass to the power of about 1.7–1.8.

Suggestions for further reading

Harding, S. E. (1995). On the hydrodynamic analysis of macromolecular conformation. *Biophys. Chem.*, **55**, 69–93.

Garcia de la Torre, J., Carrasco, B., and Harding, S. E. (1997). SOLPRO: theory and computer program for the prediction of SOLution PROperties of rigid macromolecules and bioparticles. *Eur Biophys J.* **25**, 361–372.

Regularly shaped rigid particles

Garcia de la Torre, J., and Bloomfield, V. A. (1981). Hydrodynamic properties of complex, rigid, biological macromolecules: theory and applications. *Quart. Rev. Biophys.*, **14**, 81–13D2.

Garcia de la Torre, J. (1988). Hydrodynamic properties of macromolecular assemblies. In *Dynamic Properties of Biomolecular Assemblies,* eds. S. E. Harding and A. J. Rowe, pp. 3--31. Nottingham: The Royal Society.

Arbitrarily shaped particles

Swanson, E., Teller, D. C., and de Haen, C. (1978). The low Reynolds number translation friction of ellipsoids, cylinders, dumbbells, and hollow spherical caps. Numerical testing of the validity of the modified Oseen tensor in computing the friction of objects modelled as beads on a shell. *J. Chem. Phys.*, **68**, 5097–5102.

Carrasco, B., and Garcia de la Torre, J. (1999). Hydrodynamic properties of rigid particles: comparison of different modelling and computational procedures. *Biophys. J*, **75**, 3044–3057.

Allison, S. A. (2001). Boundary element modelling of biomolecular transport. *Bioph. Chem.*, **93**, 197–213.

Translational friction and electrostatic capacitance

Hubbard, J. B., and Douglas, J. F. (1993). Hydrodynamic friction of arbitrarily shaped Brownian particles. *Phys. Rev.*, **47**, 2983–2986.

Zhou, H.-X. (1995). Calculation of translational friction and intrinsic viscosity. I. General formulation for arbitrarily shaped particles. *Biophys. J.*, **69**, 2286–2297.

Particles with known structure

Venable, R. M., and Pastor, R. W. (1988). Frictional models for stochastic simulations of proteins. *Biopolymers*, **27**, 1001–1014.

Garcia de la Torre, J., Huertas, M. L., and Carrasco, B. (2000). Calculation of hydrodynamic properties of globular proteins from their atomic level structure. *Biophys. J.*, **78**, 719–730.

Rigid particles with segmental mobility

Garcia de la Torre, J. (1994). Hydrodynamics of segmentally flexible macromolecules. *Eur. Biophys. J.*, **23**, 307–322.

Determination of the shape of macromolecules from translational friction: homologous series

Tsvetkov, V. N. (1989). *Rigid-chain polymers. Hydrodynamic and Optical Properties in Solution.* New York: Consultants Bureau.

Hearst, J. E. (1963). Rotary diffusion constants of stiff-chain macromolecules. *J. Chem. Phys.*, **38**, 1062–1065.

Garcia de la Torre, J., and Bloomfield, V. A. (1981). Hydrodynamic properties of complex, rigid, biological macromolecules: theory and applications. *Quart. Rev. of Biophys.*, **14**, 81–139.

Hagerman, P. J., and Zimm, B. (1981). Monte Carlo approach to the analysis of wormlike chains. *Biopolymers*, **20**, 1481–1502.

Harding, S. E. (1995). On the hydrodynamic analysis of macromolecular conformation. *Biophys. Chem.*, **55**, 69–93.

Garcia de la Torre, J. (1994). Hydrodynamics of segmentally flexible macromolecules. *Eur. Biophys. J.*, **23**, 307–322.

Allison, S. (1999). Low Reynolds number transport properties of axisymmetric particles employing stick and slip boundary conditions. *Macromolecules*, **32**, 5304–5312.

Chapter D3
Macromolecular diffusion

D3.1 Historical review

1850

T. Graham observed that egg albumin diffuses much more slowly than common compounds such as salt or sugar. In **1861** he used dialysis to separate mixtures of slowly and rapidly diffusing solutes, and made quantitative measurements of diffusion on many substances. On the basis of this work he classified matter in terms of *colloids* and *crystalloids*.

1855

A. Fick, in the equations now known as his first and second laws, defined a *diffusion coefficient (D)* to describe the flow of a solute down its concentration gradient. **J. Stefan (1879)** and **L. Boltzmann (1894)** developed mathematical integral forms of the second law, opening the way for the experimental determination of diffusion coefficients by various methods.

1905–1906

A. Einstein, W. Sutherland and **M. von Smoluchowski** determined that steady-state solutions of Fick's equations have a direct analogy in heat flow along a temperature gradient, and established mathematical relations between diffusion and frictional coefficients in Brownian motion. In **1905, W. Sutherland** used a value of D calculated by **J. Stefan** from data collected by **T. Graham**, to obtain a molecular weight of 33 000 for egg albumin (the true value is about 45 000).

1926

L. Mandelshtam recognised that the translational diffusion coefficient of macro-molecules could be obtained from the spectrum of scattered light. However, the lack of spatial coherence and the non-monochromatic nature of conventional light sources rendered such experiments impossible until **1964** when **H. Cummins, N. Knable** and **Y. Yeh** used an optical-mixing technique to resolve spectrally the light scattered from dilute suspensions of polystyrene latex spheres. In **1967, S. Dubin, S. J. Lunacek** and **G. Benedek** measured the translational diffusion coefficients of bovine serum albumin, lysozyme, tobacco mosaic virus and DNA by dynamic light scattering.

1927
T. Svedberg established that the molecular weight of a protein can be calculated from its sedimentation behaviour, partial specific volume and diffusion coefficient. This stimulated the development of new and improved methods and apparatuses for the study of the diffusion process itself.

1938–1960
Various experimental schemes were introduced to measure diffusion coefficients of small and large biological macromolecules (the **Lamm** scale method, the **Philpot** and **Svensson** Schlieren method, the **Jamin, Gouy, Rayleigh** interferometric and **Lebedev** polarisation interferometric methods).

Fig. D3.1 Diffusion as a macroscopic change in concentration.

1972–1974
D. Magde, E. L. Elson and **W. W. Webb** published a rigorous formalism for fluorescence correlation spectroscopy (FCS), highlighting the great potential of the method for the measurement of macromolecular diffusion coefficients. In **1990**, **R. Rigler** and coworkers reached the single-molecule detection limit by combining FCS with confocal fluorescence microscopy.

Mid-1970s
Fluorescence photobleaching recovery (FPR) was developed to measure the diffusion coefficients of fluorescently labelled molecules. FPR provided an ideal tool for the determination of two-dimensional lateral mobility in membranes and individual living cells.

2000 to now
In addition to the experimental methods outlined above, modern theoretical hydrodynamics can predict diffusion coefficients for biological macromolecules of known structure with high accuracy.

D3.2 Translational diffusion coefficients

The translational diffusion coefficient, D_{transl}, can be determined by different types of measurement. Historically, D_{transl} was defined via macroscopic fluctuations in the local concentration (Fig. D3.1) and determined experimentally by the spreading boundary technique (Section D3.5).

The tracer diffusion coefficient, D_{tracer}, is obtained from a *microscopic* approach, in which the average motion of individual particles is considered (Fig. D3.2). Experimentally, D_{tracer} is determined by FPR photobleaching recovery (Section D3.5), nanovid microscopy (Section D3.5) and number-fluctuation techniques in dynamic light scattering under non-Gaussian statistics (Section D10.5) or fluorescence correlation spectroscopy (Chapter D11).

Fig. D3.2 Diffusion as chaotic motion of a single particle.

Table D3.1. *Methods currently used to determine diffusion coefficients of biological macromolecules*

Approach	Experimental method	Diffusion coefficient
Observations of the average microscopic motion of particles in an ensemble	Fluorescence photobleaching recovery (FPR Section D3.5)	Tracer diffusion coefficient, D_{tracer}
Direct observation of the diffusional motion of single fluorescent particles	Confocal fluorescent microscopy (Section F1.3)	Tracer diffusion coefficient, D_{tracer}
Stochastic appearance and disappearance of molecules in a small volume	Number fluctuation in dynamic light scattering (Section D10.5)	Tracer diffusion coefficient, D_{tracer}
Stochastic appearance and disappearance of fluorescent molecules in a very small volume.	Fluorescence correlation spectroscopy (Section D11.3)	Tracer diffusion coefficient, D_{tracer}
Mutual motion of particles in an assembly	Dynamic light scattering (Section D.10.3)	Mutual diffusion coefficient, D_{mutual}
Macroscopic change in concentration	Spreading boundary technique (Section D3.5)	Translational diffusion coefficient, D_{transl}

Fig. D3.3 Diffusion as a mutual motion of particles in an assembly.

The mutual diffusion coefficient, D_{mutual}, is also a *microscopic* diffusion coefficient. It describes the mutual motions of particles in an assembly (Fig. D3.3). Experimentally, D_{mutual} can be determined by dynamic light scattering under Gaussian statistics (Chapter D10).

In the case of non-interacting particles the three types of diffusion measurement lead to the same result, the D_{transl} diffusion coefficient. Table D3.1 presents a summary of hydrodynamics methods currently in use for the determination of the various diffusion coefficients. The methods in Table D3.1 are listed according to the approaches defined above, the experimental techniques used and the type of diffusion coefficient. The chapters and sections in which each experimental technique is treated in detail and applications are discussed, are also given.

As we have already noted, for non-interacting Brownian particles the three diffusion coefficients provide equivalent results, i.e. D_{tracer}, D_{mutual} and D_{transl} are identical.

D3.3 Microscopic theory of diffusion

The diffusion of macromolecules in solution can be considered as a phenomenon connected with Brownian motion. Owing to thermal energy macromolecules in solution are in permanent chaotic motion (Comment D3.1).

It was shown by Einstein that for one-dimensional diffusion the mean square displacement $\langle x^2 \rangle$ is proportional to time t:

$$\langle x^2 \rangle = 2D_1 t \tag{D3.1}$$

and

$$\langle x^2 \rangle^{1/2} = (2D_1 t)^{1/2} \qquad \text{(D3.2)}$$

Equations (D3.1) and (D3.2) show that a knowledge of the mean-square displacement $\langle x^2 \rangle$ is sufficient to determine D_1 and vice versa.

Consider that motions in the x-, y- and z-directions are independent. If $\langle x^2 \rangle = 2D_1 t$, then $\langle y^2 \rangle = 2D_2 t$, $\langle z^2 \rangle = 2D_3 t$. In two dimensions, the square of the distance from the origin to the point (x, y) is $r^2 = x^2 + y^2$; therefore

$$\langle r^2 \rangle = 4Dt \qquad \text{(D3.3)}$$

where D is the average of the diffusion coefficients D_1 and D_2 in a two-dimensional random walk. In three dimensions, $r^2 = x^2 + y^2 + z^2$, and

$$\langle r^2 \rangle = 6Dt \qquad \text{(D3.4)}$$

where D is average of the D_1, D_2 and D_3 diffusion coefficients in a three-dimensional random walk. Again, knowledge of the $\langle r^2 \rangle$ is sufficient to determine D and vice versa.

Diffusion coefficients determined with using Eqs. (D3.1), (D3.3) and (D3.4) should be considered as *tracer* diffusion coefficients, D_{tracer}. Four experimental methods currently used for determining D_{tracer} are presented in Table D3.1. Examples of calculations of the diffusion–velocity and diffusion–time relationships are given in Comments D3.2 and D3.3.

Comment D3.2   Biologist's box: Diffusion and velocity

From Eq. (A) in Comment D3.1 we can calculate the instantaneous velocity of a small particle. Consider two examples.

Sucrose in a vacuum

Sucrose has a molecular mass of 342 Da. This is the mass in grams of one mole or 6.02×10^{23} molecules; the mass of one molecule is $m = 5.7 \times 10^{-23}$ g. The value of kT at room temperature, 293 K, is 4.04×10^{-14} g cm^2 s^{-2}. Therefore, $\langle u_x^2 \rangle^{1/2} = 8.3 \times 10^3$ cm s^{-1}. If collisions were absent the sucrose molecule would cross a typical swimming pool in about 1 s. According to Eq. (B) in Comment D3.2 the velocity of small particle is inversely proporational to the root-square of its molecular weight.

Biological macromolecules

For proteins with a typical molecular mass of of 20 kDa the velocity is 1.09×10^3 cm s^{-1}. For the DNA with a molecular mass of 900 000 kDa the velocity is 33 cm s^{-1}.

Comment D3.1

Mean-square velocity of a particle in Brownian motion

A particle at absolute temperature T has on the average a kinetic energy associated with movement along each axis of $kT/2$, where k is Boltzmann's constant. A particle of mass m and velocity v_x on the x- axis has a kinetic energy $mv_x^2/2$. On the average

$$\langle mv_x^2/2 \rangle = kT/2,$$

where $\langle \, \rangle$ denotes an average over time or over an ensemble of similar particles. From this relationship we can compute the mean-square velocity as

$$\langle v_x^2 \rangle = kT/m \qquad \text{(A)}$$

and the root-mean-square velocity as

$$\langle v_x^2 \rangle^{1/2} = (kT/m)^{1/2} \qquad \text{(B)}$$

Comment D3.3   Biologist's box: Diffusion and time

From Eq. (D3.2) we can calculate the instantaneous velocity of a small particle. Consider a few examples.

Urea in water

The diffusion coefficient of urea in water is 118 cm^2 s^{-1} (Table D3.3). A particle with such a diffusion coefficient diffuses a distance $x = 10^{-4}$ cm in a time $t = x^2/2D = 5 \times 10^{-4}$ s, or about 0.5 ms. It diffuses a distance $x = 1$ cm in a time $t = x^2/2D = 5 \times 10^5$ s, or about 14 h. It is clear that diffusive transport takes a long time when distances are large.

Proteins in water

A particle with a diffusion coefficient D of the order 10^{-6} cm^2 s^{-2} (a small globular protein) diffuses a distance $x = 10^{-4}$ cm in a time $t = x^2/2D = 5 \times 10^{-3}$ s, or about 5 ms. It diffuses a distance $x = 1$ cm in a time $t = x^2/2D = 5 \times 10^5$ s, or about 138 h.

Tobacco mosaic virus in water

The tobacco mosaic virus has a small diffusion coefficient (0.44×10^{-7} cm^2 s^{-2}) and diffuses very slowly: it travels a distance $x = 0.5$ cm in a time $t = x^2/2D = 2.5 \times 10^6$ s, or about 700 h. This explains why the moving boundary method in which the minimal required diffusing distance is about a few millimetres is never used for measurements of diffusion coefficients of such large molecules.

D3.4 Macroscopic theory of diffusion

D3.4.1 Fick's first equation

Fick's first equation states that the net flux J_x is always proportional to the first power of the solute concentration gradient dC/dx with $-D$ as the constant proportionality:

$$J_x = -D[dC/dx] \tag{D3.5}$$

If the particles are uniformly distributed, the slope is 0, i.e. $dC/dx = 0$ and the system is at equilibrium. If the slope is constant, i.e. $dC/dx = $ constant, J_x is also constant. This occurs when C is a linear function of x, as shown in Fig. D3.4.

D3.4.2 Fick's second equation

Fick's second equation follows from the first, provided that the total number of the particles is conserved, i.e. that particles are neither created nor destroyed.

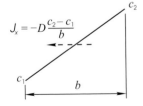

$$J_x = -D\frac{c_2 - c_1}{b}$$

Fig. D3.4 The flux due to a concentration gradient. Molecules move from right to left only because there are more particles on the right than on the left. (After Berg, 1983.)

Consider the box of volume $A\delta$ shown in Fig. D3.5. In the period of time τ, $J_x(x)A\tau$ particles will enter from the left and $J_x(x+\delta)\,A\tau$ will leave from the right. The number of the particles per unit volume in the box therefore increases at the rate

$$\frac{1}{\tau}[C(t+\tau)-C(\tau)] = -\frac{1}{\tau}[J_x(x+\delta)-J_x(x)]\frac{A\tau}{A\delta}$$

$$= -\frac{1}{\delta}[J_x(x+\delta)-J_x(x)]$$

In the limit $\delta\to 0$ and $\tau\to 0$ this means that

$$dC/dt = -dJ_x/dx \qquad (D3.6)$$

or using Fick's first law (Eq. (D3.5)), that

$$dC/dt = D d^2C/dx^2 \qquad (D3.7)$$

This equation states that the time rate of change in concentration, dC/dt, is proportional to the second concentration derivative of the solute, d^2C/dx^2, where D is the proportionality constant.

In three dimensions the concentration changes in time as

$$dC/dt = D\nabla^2 C \qquad (D3.8)$$

where ∇^2 is the three-dimensional Laplacian, $d^2/dx^2 + d^2/dy^2 + d^2/dz^2$.

If the problem is spherically symmetric, the flux is radial

$$J_r = -D dC/dr, \qquad (D3.9)$$

and

$$dC/dt = D(1/r^2)d(r^2 dC/dr)/dr \qquad (D3.10)$$

D3.4.3 Time-dependent solutions of Fick's equations

Consider that we have diffusion from a column of liquid initially containing particles at concentration C_0 into a column of liquid that initially does not contain any particles (Fig. D3.6). In this case, the initial conditions are $C = C_0$ for $x > 0$ and $C = 0$ for $x < 0$ and Eq. (D3.8) has the solution

$$C(x,t) = \frac{C_0}{2}[1 - 2/\pi^{1/2}]\int_0^{(x/2)(Dt)^{1/2}} \exp(-y^2)dy \qquad (D3.11)$$

The integral in Eq. (D3.11) is known as the probability integral; it is a function of $(x/2)(Dt)^{1/2}$ and varies in value from 0 to $\frac{1}{2}$ as $(x/2)(Dt)^{1/2}$ varies from 0 to ∞. A graphical representation of the Eq. (D3.11) is shown in Fig. D3.7(a). By taking derivatives of $C(x,t)$ with respect to x or t, we obtain

$$dC/dx = C_0/(4\pi Dt)^{1/2}\exp(-x^2/4Dt) \qquad (D3.12)$$

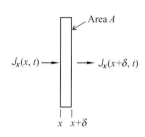

Fig. D3.5 Fluxes through the faces of a thin box extending from position x to position $x+\delta$. The area of each face is A.

Fig. D3.6 Diffusion from a column of liquid initially containing particles at a concentration C_0 (bottom left) into a column of liquid initially containing no particles (top left). Spreading of the boundary after a certain time (right).

Fig. D3.7
(a) Concentration as a function of position at times t_0, t_1, t_2, and t_∞. The initial concentration of particles at time t_0 is C_0. At infinite time, the concentration is uniform throughout the column and equals $C_0/2$.
(b) Derivatives of concentration with respect to x as a function of position at times t_0, t_1, t_2 and t_∞ for the same process as described in (a).

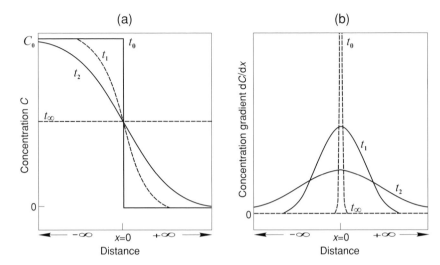

The behaviour of these two equations is shown in Fig. D3.7(b). The Eq. (D3.14) shows that the maximum value of dC/dx will occur at $x = 0$, and its value is $C_0/(4\pi Dt)^{1/2}$.

D3.4.4 Steady-state solutions of Fick's equations

In the steady-state limit $dC/dt = 0$, and Eq. (D3.8) reduces to

$$\nabla^2 C = 0 \tag{D3.13}$$

For the case of spherical symmetry

$$(1/r)d(r^2 dC/dr)dr = 0 \tag{D3.14}$$

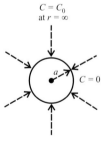

Fig. D3.8 A spherical absorber of radius R_0 in an infinite medium containing particles at initial concentration C_0. The dashed arrows are lines of flux.

One other important case is diffusion to a spherical absorber. Consider a spherical absorber of radius R_0 in an infinite volume, as shown in Fig. D3.8. Each particle reaching the surface of the sphere is captured, so the concentration on the surface at $r = R_0$ is 0. At $r = \infty$ the concentration is C_0. With these boundary conditions Eq. (D3.13) has a solution

$$C(r) = C_0(1 - R_0/r) \tag{D3.15}$$

and the flux, Eq. (D3.6), is

$$J_r(r) = DC_0 R_0/r^2 \tag{D3.16}$$

If we define the diffusion current I as $I = 4\pi r^2 J_r(r)$, then

$$I = 4\pi D R_0 C_0 \tag{D3.17}$$

if C_0 is expressed in particles per cubic centimetre and I is in particles per second. This current is proportional not to the area of the sphere but to its radius.

It comes from the fact that as the radius R_0 increases the area increases as R_0^2, but the concentration gradient to which flux is proportional decreases as $1/R_0$.

It is important to note that Eq. (D3.13) is analogous to Laplace's equation for the electrostatic potential in charge-free space. This implies that the diffusion current to an isolated absorber of any size and shape can be written as

$$I = 4\pi DCC_0 \tag{D3.18}$$

where C is the electrical capacitance (in cgs units c in centimetres, see Comment D2.7) of an isolated conductor of that size and shape. The resemblance between Eqs. (D3.17) and (D3.18) is evident. Since the electrical capacitance of bodies with different shapes is known now we can use Eq. (D3.18) in many practical cases to calculate the frictional properties of molecules.

D3.5 Experimental methods of determination of diffusion coefficients

D3.5.1 Spreading boundary technique

In this technique, a sharp boundary is initially set up between a solution of uniform concentration C_0 and pure solvent. The concentration, C, as a function of distance, x, from the boundary at a given time is then measured, for example, by monitoring the light absorption peak of the diffusing solute (Eq. (D3.11)). A value of D at each time t is calculated from $C(x)$.

An alternative analysis of the experimental concentration dependence is based on Eq. (D3.12), which describes the dependence dC/dx on x (Fig. D3.7(b)). The maximum height of the curve, $H_{\max}$, is equal to $kC_0/(4\pi Dt)^{1/2}$ and the area, A, under curve is equal to kC. The diffusion coefficient is calculated from

$$(A/H_{\max})^2 = 4\pi Dt \tag{D3.19}$$

D3.5.2 Fluorescence photobleaching recovery (FPR)

FPR, also called *fluorescence recovery after photobleaching* (FRAP), is a tracer technique that measures the diffusion of a labelled solute. If the solution contains differently labelled species, their transport coefficients can be obtained separately from the mobility of each tracer.

The method is simple both in concept and in application. A small volume containing mobile fluorescent molecules is exposed to a brief intense pulse of light, which causes irreversible photochemical bleaching of the fluorophore in that region (Fig. D3.9). Fluorescence in the bleached region is then excited by a greatly attenuated beam in order to avoid significant photolysis during the recovery phase. The subsequent exchange between the bleached and non-bleached

Fig. D3.9 Fluorescence photobleaching recovery: the fluorescent species in the small volume (white circle) is bleached by a short, high-intensity laser pulse; the recovery of fluorescence in the volume due to the diffusion of unbleached particles is observed as a function of time. (After Bastiaens and Pepperkok (2000).)

Fig. D3.10 An idealised plot of fluorescence intensity (I) as a function of time shows the parameters of a quantitative FPR experiment. The bleached region is monitored during a prebleach period to determine the initial intensity I_i. The region is then bleached using high-intensity illumination from time t_b to t_0, and recovery is monitored starting at t_0 until I reaches a final I_f, when no further increase can be detected. Some methods calculate the effective diffusion coefficient, D_{eff}, directly from the time ($t_{1/2}$) required to reach half the final intensity ($I_{f/2}$) or from a fit of all the curves (solid line). (Adapted from Bastiaens and Pepperkok, 2000.)

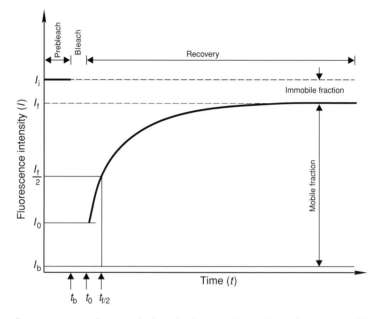

fluorescent species populations in the sampled volume is monitored by quantitative time-lapse microscopy and the fluorescence intensity relative to the prebleach period is plotted as a function of time. Diffusion coefficients are determined from the rate of fluorescence recovery, resulting from transport of fluorophore into the bleached region (Fig. D3.10). The time profile of the FPR pattern can also be used to monitor the nature of the transport, and to distinguish between diffusive and flow motions (Fig. D3.11).

FPR may well be the technique most widely used to study lateral diffusion in a plane – the geometry corresponding to extended regions of cell plasma membranes, reconstituted membranes and thin layers of solution or cytoplasm. It has been shown by FPR that several classes of cell surface proteins and glycoproteins undergo lateral movements within the membrane plane, which have considerable functional significance. Direct experimental measurements have been made of

Fig. D3.11 Theoretical fluorescence recovery curves expressed as the fractional recovery $f_k(t)$ for pure diffusion, flow and a combination of the two. (After Koppel, 1979.)

Comment D3.4 'Nanovid microscopy'

In 1988 M. Sheets devised nanovid microscopy, which enables small colloidal gold particles to be located with a light microscope with nanometre precision. One year later, he applied the method to track the motion of 40-nm diameter gold particles attached to the lectin, concanavalin A, on the surface of living cells.

Diffusion coefficients of macromolecules determined by this experimental method as a rule have *no direct relation* to the shape of molecule and usually demonstrate the character of the mobility of the molecules of interest under specific conditions (Sheetz *et al.*, 1989).

---

the apparent diffusion coefficients of different membrane proteins *in situ*. These have yielded estimates ranging from 5×10^{-9} cm^2 s^{-1} for rhodopsin in photo-receptor membranes to less than 10^{-12} cm^2 s^{-1} for fluorescence in-labelled surface proteins in the human erythrocyte. Values of 10^{-11} cm^2 s^{-1} were recorded for fluorescein-labelled concanavalin A receptor complexes on the plasma membrane of cultured rat myoblasts. Finally, we note that the diffusion coefficients of macromolecules determined by the FPR method often have *no direct relation* to the shape of the molecule nor usually to the character of the mobility of the molecules under specific conditions. The same can be written for the method of nanovid microscopy (Comment D3.4)

D3.6 Prediction of the diffusion coefficients of globular proteins of known structure and comparison with experiment

The frictional behaviour of proteins is determined by several factors: overall dimensions, surface roughness, surface hydration of charged and polar groups. There are two different strategies for the prediction of protein diffusion coefficients from known structures. The first involves a systematic parametrisation of friction constants for all the atoms in the protein; in this approach, frictional resistance is placed explicitly on each atom accessible to the solvent. It has been found, however, that if only protein atoms are included in the calculation, no reasonable range of atomic radii can reproduce the experimental translational diffusion constant. For a satisfactory prediction, water molecules must be explicitly assigned positions on the structural surface and considered as part of the protein. In the second strategy, protein-associated water is represented by a hydration shell of uniform thickness σ (Section D1.3).

Automated methods of converting crystallographic data into bead or other types of model built up of equivalent discrete units are now available Details of the modelling procedures are presented in Chapter D2. A comparison of predicted and experimental diffusion coefficients for various proteins is presented in

Table D3.2. *Predicted and experimental diffusion coefficients*

Object	PDB entry	Molecular mass in kDa	Method of constructing	Predicted D_{20w} (in Fick's units)	Measured D_{20w} (in Fick's units)
BPTI	4pt1	6.4	By double bead[b]	14.4	14.4
RNase A	3rn3	13.6	By double bead[b]	11.1	11.2
RNase A	7rsa	13.7	Via capacitance[c]	11.1	11.2
Lysozyme	6lys	14.3	Via capacitance[c]	11.2	11.2
Lysozyme	11z3	14.0	By double bead[b]	11.3	11.1
Lysozyme	6l1ys	14.0	By discretisation[d]	11.7	11.1–11.2
Profilin	1pne	14.8	By double bead[b]	11.0	10.6
Myoglobin	1mbo	17.2	Via capacitance[c]	10.4	10.3
Cellulase	1eng	22.0	By double bead[b]	9.7	9.8
Chymotrypsinogen	2cga	25.66	Via capacitance[c]	9.2	9.1–9.5
Insulin	1aio	34	By double bead[b]	8.0	7.9
Nitrogenase	Av1	220	By single bead[a]	4.33	4.0
VTM	Circular cylinder (2800Å × 180Å)	40 000	Smooth polygon[d]	0.44	0.44

[a] In the single-bead model, spheres (one for each amino acid with a radius of 6.6 Å) are positioned on the C_α atoms (Banachowicz *et al.*, 2000).
[b] In the double-bead model, a pair of identical, partially overlapping spheres with a total bead radius of 4.5 Å are positioned on the C_α atoms (Banachowicz *et al.*, 2000).
[c] The diffusion coefficients of the proteins were calculated via capacitance, assuming a hydration shell with a uniform thickness of 0.9 Å (Zhou, 2001).
[d] In order to construct a model amenable to hydrodynamic calculations, a 3-Å radius spherical ball was rolled over the particle surface to smooth out small-scale irregularities. The smoothed molecular surface was divided into a closed polygon composed of a number of triangles (Brune and Kim, 1993).

Table D3.2. Current prediction algorithms allow an accuracy of about 1–2%, and it is striking that it is the inclusion of a 1 ± 0.2 Å thick hydration shell (for all globular proteins) that leads to the good agreement between prediction and experiments.

Fig. D3.12 A particle of a mass m subjected to an externally applied force F_x while undergoing a one-dimensional random walk. (After Berg, 1983.)

D3.7 Translational friction and diffusion coefficients

D3.7.1 Einstein–Smoluchowski relation

The kinetic theory of diffusion presented above allows us to estimate the value of the diffusion coefficients by computing the velocity at which a particle drifts through the medium when exposed to an externally applied force. In practice, the velocity at which the particle moves in response to such a force is infinitesimal in comparison with the instantaneous root-mean-square velocity given by Eq. (A) of Comment D3.1. It means that particles diffuse much as they would in the absence of the field, but with a small persistent directional bias, as shown in Fig. D3.12.

Now we deduce the equation connecting the diffusion coefficient of a particle D with its frictional coefficient f using Berg's 'random walk with drift' approach.

Consider a particle of mass m at position x on which acts an externally applied force in the direction $+x$. According to definition, the coefficient of translational friction is a coefficient of proportionality in the equation:

$$F_x = f u_x \qquad (D3.20)$$

where F_x is the active force in the x direction and u_x is the average velocity of the particle. From the other side, in accordance with Newton's second law, the force causes the particle to accelerate uniformly to the right with acceleration $a = F_x/m$. According to the random walk approach a particle steps once every τ seconds to the right with an initial velocity $+v_x$ or to the left with initial velocity $-v_x$. A particle starting at position x with an initial velocity $+u_x$ moves in time τ a distance $\delta_+ = u_x + a\tau^2/2$, while a particle starting at position x with an initial velocity $-u_x$ moves in time τ a distance $\delta_- = -u_x + a\tau^2/2$. Since a step to the right and one to the left are equally probable, the average displacement in time τ is $a\tau^2/2$, and the particle drifts to the right with an average velocity

$$u_d = \tfrac{1}{2}a\tau = \tfrac{1}{2}F_x\tau/m \qquad (D3.21)$$

In the model random walk, $f = 2m/\tau$. Multiplying both the numerator and the denominator of Eq. (D3.21) in the form $u_d = F_x/f$ by (δ/τ^2) and remembering that $u_x = \delta/\tau$ and $D = \delta^2/2$ we find that $f = mu_x^2/D$. But by Eq. (A) of comment D3.1 $mu_x^2 = kT$; therefore $f = kT/D$, or

$$D = kT/f \qquad (D3.22)$$

The result is known as the Einstein–Smoluchowski relation (Comment D3.5), and is very general. It does not depend on any assumptions made about the structure of the particle or the details of its motion. The particle always moves through the medium with a velocity proportional to the externally applied force. In the case of diffusion the constant of proportionality is D/kT. In the case of sedimentation, this constant is $s/m(1-\bar{v}\rho_0)$ (Chapter D4)). In the case of free electrophoresis this constant is μ/Q, where μ is the electrophoretic mobility and Q is the charge of the particle (Chapter D5).

The Einstein–Smoluchowski relation shows that if we can measure D, we can calculate f, the coefficient of friction which gives us information about the dimension and shape of the molecules.

Comment D3.5

Derivation of Eq. (D3.22)

As pointed out in Berg's book in the strict sense such a derivation of Eq. (D3.22) has no good physical basis. Of course, in reality particles do not step at a fixed interval or start each step at a fixed velocity. There are distributions of step intervals, directions, velocities owing to exchange of energy between molecules. But the end result is the same. The essential point is that a particle is accelerated by the externally applied force, but this force is so small (see Section D1.2.2) that it forgets about this acceleration when it exchanges energy with molecules of the fluid. The cycle repeats again and again. This is the reality for molecules at low Reynolds number (Berg, 1983).

D3.7.2 Diffusion coefficients of biological macromolecules

The diffusion coefficient of a macromolecules is a function of the solvent viscosity and the temperature at which the measurement was carried out. It is generally accepted that any measured diffusion coefficient should be corrected to the value it would be if the measurements were performed at 20 °C in pure water, where

Table D3.3. *Diffusion coefficients of some globular proteins in aqueous solution*

Protein	Molecular mass (Da)	Diffusion coefficient $D_{20,w}^0$ (Fick units)
Somatostatin (tetradecapeptide)	31 636	24.5
Gramicidin (dimer)	32 500	18.4
BPTI (4pti)	36 400	14.4
Lipase (milk)	36 669	14.5
Ribonuclease A (bovine pancreas)	12 640	13.1
Ribonuclease A (7rsa)	13 690	11.2
Cytochrome *c* bovine heart	13 370	11.4
Lysozyme (chicken egg white)	13 930	11.2
Lysozyme (6Lyz)	14 320	11.2
Profilin (1PNE)	14 800	10.6
Myoglobin (1mbo)	17 190	10.3
Cellulase (2ENG)	22 000	9.8
Chymotrypsinogen A (bovine pancreas)	23 240	9.5
Chymotrypsinogen A (2cga)	25 660	9.0–9.5
Insulin (1AI0)	34 400	7.9
Carboxypeptidase B	34 280	8.2
α-Lactoglobulin	35 000	7?
Ovalbumin	45 000	7.8
Albumin	68 600	6.4
Haemoglobin (pig heart)	68 000	6.9
Citrate synthase	97 938	5.8
Lactic dehydrogenase (beef heart)	133 000	5.1
GPD	136 800	5.0
Aldolase (rabbit muscle)	156 000	4.8
Nitrogenase (bovine liver)	220 000	4.5
Catalase (MoFe)	250 000	4.1
Apoferritin (horse spleen)	467 000	3.6
Urease (Jeack bean)	482 700	3.5
Glutamate dehydrogenase	1 015 000	2.5

Diffusion coefficients are taken from: Zipper and Durchshlag (2000); Banachowicz *et al.* (2000); Zhou (2001); Hellweg *et al.* (1997); Smith (1970); Garcia de la Torre (2001); Byron (1977); Brune and Kim (1993).

Table D3.4. *Diffusion coefficients of homogeneous double-strand DNA*

Particle	Molecular mass (Da)	Diffusion coefficient
8 bp	35 304	15.26
12 bp	37 956	13.41
20 bp	13 260	10.86
89 bp	59 007	4.27
104 bp	68 952	3.88
124 bp	82 212	3.41
2311 bp	1 532 193	0.46
DNA-pUC19	1 829 880	0.35
DNA-pDS1	2 538 637	0.29
DNA ($M_w/M_n \leq 1.15$)	3 730 000	0.223

Values are taken from: Eimer and Pecora (1991); Tirado *et al.* (1984); Sorlie *et al.* (1988); Seils and Dorfmuller (1991); Jolly and Eisenberg (1976).

η_{20w}, is the viscosity of pure water at 20 °C:

$$D_{20,w} = D^o(293/T)(\eta_{\text{soln},T}/\eta_{\text{w},20}) \qquad \text{(D3.23)}$$

Here $D_{20,w}$ is the diffusion constant in pure water at 20 °C, T is the absolute temperature and, $\eta_{\text{soln},T}$ is is the viscosity of the real solution used at temperature T.

Equation (D3.23) has no solid theoretical background and is based mainly on phenomenological consideration. Note that $D_{20,w}$ is a quantity that can be defined even if the species does not exist in water at 20 °C. Diffusion coefficients of biological macromolecules are normally obtained at finite concentration and should be extrapolated to zero concentration (Comment D3.6).

Table D3.3 lists some representative measured values of $D_{20,w}$ and M of some proteins, starting from small synthetic peptides (somatostatin) and finishing with large oligomeric protein complexes (glutamate dehydrogenase). Table D3.4 gives diffusion coefficients of homogeneous double-stranded DNA.

D3.7.3 Dependence of the diffusion coefficient on the molecular mass of globular proteins

Assume that diffusing molecules are spherical particles. In this case a combination of the Enstein–Smoluchowski relation (Eq. (D3.22)) and Stokes' law (Eq. D2.1) opens the way for the calculation of the radius, R_0, of a sphere from the experimentally measured diffusion coefficient D and the solvent

Comment D3.6
Concentration dependence of diffusion coefficients

The concentration dependence of translational diffusion can be characterised by a coefficient k_D in the equation

$$D_c = D_0(1 - k_D C) \qquad \text{(A)}$$

The coefficient k_D is specified at low solute concentration by the algebraic sum of the effects of the 'excluded volume' effect and hydrodynamic friction retardation and is analogous to coefficients k_s which describes concentration dependence in sedimentation coefficients (Chapter D4). In sedimentation these two effects are of similar magnitude but opposite sign. It follows, as has long been appreciated, that k_D is small in magnitude and of variable sign.

Fig. D3.13 The
dependence of the
experimental values of
diffusion coefficients on
the molecular mass for
globular proteins on a
double logarithmic scale.

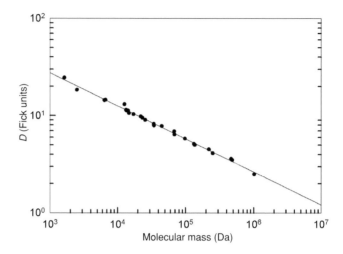

viscosity η_0

$$R_0 = kT/6\pi\eta_0 D \tag{D3.24}$$

Thus, from D and η_0 we can calculate R_0, and if the density of a molecule is ρ, mass is given by

$$m = (4/3)\pi R_0^3 \rho \tag{D3.25}$$

and the molecular mass M is given by

$$M = (4/3)\pi R_0^3 \rho N_A \tag{D3.26}$$

The important point brought out of this sample case is that diffusion is inversely proportional to the cube root of the molecular mass of spherical particles

$$D \sim M^{-1/3} \tag{D3.27}$$

Figure D3.13 shows the dependence of the experimental values of the diffusion coefficients on the molecular mass for globular proteins starting with the tetra-decapeptide (somatostatin) and ending with the large proteins (the oligomeric protein, glutamate dehydrogenase from bovine liver) using the data of Table D3.3. The plot of log D against log$M$ correlates very well with a straight line of slope -0.336. In the limit of experimental error the slope coincides with the expected theoretical value $-1/3$ for spherical particles (Eq. (D3.27)).

D3.7.4 Dependence of the diffusion coefficient on the molecular mass of DNA

Figure D3.14 demonstrates the dependence of diffusion coefficients on molecular mass for double-stranded DNA using the data of Table D3.4. For DNA molecules

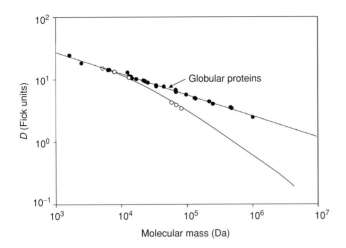

Fig. D3.14 Dependence of log $D_{20,w} \times \bar{v}^{1/3}$ on the logarithm of the molecular mass, for globular proteins (•) and for the DNA molecules (○) in range of molecular mass 8 bp–6 kbp. Data are taken from Tables D3.3 and D3.4, assuming partial specific volumes of DNA and protein of 0.56 cm³/g⁻¹ and 0.73 cm³ g⁻¹, respectively.

ranging from 70 bp (molecular weight ∼ 46 kDa) to 6 kbp (molecular weight ∼ 4 MDa) the dependence is described by one straight line:

$$D = 276M^{-0.72} \tag{D3.28}$$

The value of the exponent is close to the theoretical value for a homological series of rod particles with a finite thickness (see Section D2.6, D4.10.3 and D6.7.1 for further discussion).

It is interesting to note that the frictional properties of short DNA fragments (12–20 bp) are similar to those of globular proteins of corresponding molecular weight (Fig. D3.14). At the same time these properties are very different for large molecular weights. This fact reflects the principal difference between these two homologous series. In the globular protein series, asymmetry remains constant with increasing molecular weight, whereas in the DNA fragment series, asymmetry grows with molecular weight. The reader can find the same phenomenon in Section D4.10.3 in which we discuss the sedimentation behaviour of DNA.

D3.7.5 The limit of application of Stokes' law. 'Large', 'medium' and 'small' molecules

Figure D3.15 shows the dependence of the experimental values of diffusion coefficients on the molecular mass for a wide range of masses (from gases to large macromolecules) in a double logarithmic plot using the data presented in Tables D3.3, D3.5 and D3.6. As seen from the figure the dependence can be subdivided for three different regions.

Region I describes the dependence of the protein diffusion coefficients on molecular masses ranging from somatostatin to glutamate dehydrogenase. In this region the slope of the line of log D against log M is − 0.336. The linear fit is

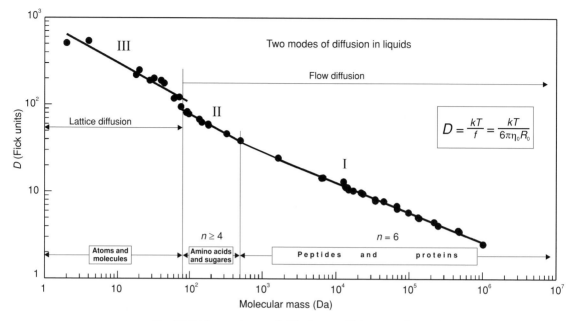

Fig. D3.15 Dependence of diffusion coefficients on the molecular mass M for different substances on a double logarithmic scale.

Region I shows the data for globular proteins. The data are taken from Table D3.3. The equation of the straight line is

$$D = 276\,M^{-0.336} \tag{I}$$

Region II shows the data for small molecules, amino acids and sugars. The data are taken from Table D3.5. The equation of the straight line is

$$D = kM^{-0.47} \tag{II}$$

Region III shows the data for atoms and gases. The data are taken from Table D3.6. The equation of the straight line is

$$D = kM^{-0.42} \tag{III}$$

In the Equations (I)–(II) M in daltons and D in Fick units (1 Fick $= 1 \times 10^{-7}$ cm^2 s^{-1}).

very good. In the limit of experimental error the slope coincides with the expected theoretical value of 1/3 for spherical particles.

Region II describes the dependence for amino acids and sugars, ranging from glycine ($M = 75$ Da) to raffinose ($M = 504$ Da). A linear fit of log D against log M yields a slope of 0.47.

Region III (2–100 Da) describes the motion of gases and small molecules in water, ranging from H$_2$ ($M = 2$ Da) to glycerol ($M = 92$). In this case D is not strictly a monotonically decreasing function of M. The linear fit gives a slope of 0.42.

Table D3.5. *Diffusion coefficients of some amino acids and sugars in aqueous solutions*

Molecule	Molecular mass (Da)	Diffusion coefficient (Fick units)
Glycine	75	94.3[a]
Alanine	89	81.0[a]
Proline	97	78.1[a]
Pentaerythritol	136	67.8[b]
Phenylalanine	147	62.7[a]
Glucose	180	59.8[a]
Mannitol	182	58.0[b]
Sucrose	324	46.3[a]
Raffinose	504	38.6[a]

Data taken from:
[a]Longsworth (1953).
[b]Nir and Stein (1971).

Table D3.6. *Diffusion coefficients of some small molecules*

Molecule	Molecular mass (Da)	Diffusion coefficient (Fick units)
H_2	2	510
He	4	540
H_2O	18	220
Ne	20	250
N_2	28	190
O_2	32	201
Ar	40	190
CO_2	44	177
Cl_2	71	122
Urea	60	118
Glycerol	92	83

Data taken from from Nir and Stein (1971).

The three different regions in Fig D3.15 correspond to three different modes of diffusion in liquids. Region I relates to the 'flow' mode of the 'large' molecules in liquids. In the 'flow' mode the 'large' molecules experience the solvent as a continuum. The molecules correspond to particles moving at very low Reynolds number ($< 10^{-4}$). In this case the movement of the molecules obeys Stokes' law with 'stick' boundary conditions (Eq. (D1.11)). It is possible to assume that the lower limit for the application of Stokes' law in 'stick' boundary conditions is about 1000 Da (Fig. D3.15). Thus we may define 'large' molecules as molecules which have a mass of 1000 Da or more.

Region III relates to the 'lattice' mode of 'small' molecules in liquids. In the 'lattice' mode 'small' molecules move in a medium composed of particles of a similar size to the molecules (i.e., in the self-diffusion of a liquid). The molecules move through the solvent by a 'jump and wait' mechanism for a significant fraction of their net movement, taking advantage of cavities in the liquid. Consequently the molecules do not experience the solvent as a continuum. The diffusional coefficient calculated from the Einstein–Smoluchowski relation is generally greater than the value predicted by Stokes' law. We define 'small' molecules as molecules which have masses of no more than 100 Da.

Region II is the transition region between two modes of diffusion in liquids. Molecules move through the solvent by mixed mechanisms. Region I gradually turns into the region II, whereas there is a jump between regions I and II (Fig. D3.15). We define the molecules in this interval of masses (100–1000 Da) as 'medium-sized' molecules. For 'medium-sized' molecules 'slip' conditions are more applicable; however, Eq. (D3.24) is not used in practice.

D3.8 Checklist of key ideas

- The diffusion coefficient determined from the stochastic motion of a single particle is defined as the *tracer diffusion coefficient, D_{tracer}*.
- The diffusion coefficient determined from the mutual motion of particles in an assembly is defined as the *mutual diffusion coefficient, D_{mutual}*.
- The diffusion coefficient determined from the macroscopic change in concentration is defined as the *translational diffusion coefficient, D_{transl}*.
- For non-interacting Brownian particles, D_{tracer}, D_{mutual} and D_{transl} are identical.
- Fick's first equation states that the net flux is proportional to the slope of concentration gradient with proportionality constant $-D$.
- Fick's second equation states that the time rate of change in concentration is proportional to the concentration gradient with proportionality constant D.
- The steady-state solution Fick's equation is analogous to Laplace's equation for the electrostatic potential in charge-free space; it follows that diffusion properties of macromolecules can be calculated by using approaches developed for electrostatics.
- The spreading boundary method allows the determination of translational diffusion coefficients for biological macromolecules.
- FPR is an ideal tool for the study of lateral diffusion in a plane – the geometry corresponding to extended regions of cell plasma membranes or thin layers of solution.
- Diffusion coefficients of macromolecules are conventionally corrected to 'standard conditions', the standard solvent being pure water at 20 °C.
- Protein diffusion coefficients can be predicted with an accuracy of 1–2% from atomic coordinates, provided hydration contributions are taken into account appropriately.

- The molecular mass dependence of the experimental diffusion coefficients of globular proteins in aqueous solution on a double logarithmic scale is a straight line with slope -0.33_6; the slope coincides with the theoretical value of 1/3 expected for spherical particles.
- Molecules of molecular mass > 1000 Da can be defined as 'large'; they move in solution by a flow mechanism and experience the solvent as a continuum.
- Molecules with a molecular mass < 100 Da are termed 'small' molecules; such molecules move through solvent by a 'jump and wait' mechanism and do not experience the solvent as a continuum.
- Molecules with a molecular mass between 100 and 1000 Da are termed 'medium-sized' molecules; they move through solvent by a mixed flow jump–and-wait mechanism.
- A molecular mass of about 1000 Da is the lower limit of applicability for the Einstein–Smoluchowski–Stokes' relation ($D = kT/6\pi \eta_0 R_0$) for boundary 'stick' conditions.

Suggestions for further reading

Diffusion coefficients

Gosting, L. J. (1956). Measurement and interpretation of diffusion coefficient of proteins. *Adv. Prot. Chem.*, **11**, 429–554.

Microscopic theory of diffusion

Einstein, A. (1956). *Investigations on the Theory of the Brownian Movement*, ed. R. Furth, transl. by A. D. Cowper. New York: Dover Publications, Inc.
Berg, H. C. (1983). *Random Walks in Biology*. Princeton: Princeton University Press.

Macroscopic theory of diffusion

Berg, H. C. (1983). *Random Walks in Biology.* Princeton: Princeton University Press.

Experimental methods of determination diffusion coefficients

Gosting, L. J. (1956). Measurement and interpretation of diffusion coefficient of proteins. *Adv. Prot. Chem.*, **11**, 429–554.
Axelrod, D., Koppel, D. E., Schlessinger, J., Elson, E., and Webb, W. W. (1976). Mobility measurement by analysis of fluorescence photobleaching recovery kinetics. *Biophys. J.*, **16**, 1055–1069.
Kucik, D. F., Elson, E., and Sheetz, M. P. (1989). Forward transport of glycoproteins on leading lamellipoda in locomoting cells. *Nature*, **340**, 315–316.
Lee, G. M., Ishihara, A., and Jacobson, K. A. (1991). Direct observation of Brownian motion of lipids in a membrane. *Proc. Natl. Acad. Sci. USA*, **88**, 6274–6278.

Concentration dependence of diffusion coefficient

Rowe, A. J. (1977). The concentration dependence of transport processes: A general description applicable to the sedimentation, translation diffusion, and viscosity coefficients of macromolecular solutes. *Biopolymers*, **16**, 2595–2611.

Harding, S. E., and Johnson, P. (1985). The concentration dependence of macromolecular parameters. *Biochem. J.*, **231**, 543–547.

Prediction of protein diffusion coefficients and comparison with experiment

Teller, D., Swanson, E., and Haen, C. (1979). The translation friction coefficients of proteins. *Meth. Enzymol.*, **61**, 103–124.

Venable, R. M., and Pastor, R. W. (1988). Frictional models for stochastic simulations of proteins. *Biopolymers*, **27**, 1001–1014.

Garcia de la Torre, J. (2001). Hydration from hydrodynamics. General consideration and applications to bead modelling to globular proteins. *Biophys. Chem.*, **93**, 159–170.

Zhou, H.-X. (2001). A unified picture of protein hydration: prediction of hydrodynamic properties from known structures. *Biophys. Chem.*, **93**, 171–179.

Diffusion coefficients and molecular mass

Gosting, L. J. (1956). Measurement and interpretation of diffusion coefficient of proteins. *Adv. Prot. Chem.*, **11**, 429–554.

Translational friction and diffusion coefficients

Einstein, A. (1956). Investigations on the theory of the Brownian movement, ed. R. Furth, transl. by A. D. Cowper. New York: Dover Publications, Inc.

Chapter D4
Analytical ultracentrifugation

D4.1 Historical review

1913

A. Dumansky proposed the use of ultracentrifugation to determine the dimensions of colloidal particles.

1923

T. Svedberg and **J. B. Nichols** constructed the first centrifuge with an optical system to follow particle behaviour in a centrifugal field. One year later, **Svedberg** noted the decrease in absorbance at the top of the cell during centrifugation of a haemoglobin solution.

1926

Svedberg made the first measurements of protein molecular weights (haemoglobin and ovalbumin) by sedimentation equilibrium and in **1927** he determined the molecular weight of haemoglobin using a combination of sedimentation and diffusion data. These pioneering studies led to the undeniable conclusion that proteins are truly macromolecules, made up of a large number of atoms linked by covalent bonds (Comment D4.1).

1929

O. Lamm deduced a general equation describing the behaviour of the moving boundary in the ultracentrifuge field. The exact solution of the equation is an infinite series of integrals, which can be computed only by numerical integration. In later work, the Lamm equation was solved analytically for specific limiting cases (**H. Faxen, W. J. Archibald, H. Fujita**).

1930s

Schlieren optical systems were designed by **J. St. L. Philpot** and **H. Svenson**, and independently by **L. G. Longsworth**; these allowed a representation of the concentration gradient (or, more precisely, the refractive index increment) as a function of distance in the centrifuge sample cell. Physicists started to use **Perrin's** hydrodynamics theories for ellipsoids of revolution to interpret the frictional

> **Comment D4.1**
>
> It is interesting to note that Theodor Svedberg was awarded the Nobel prize for his work on colloidal systems and not for inventing the analytical centrifuge.

coefficients deduced from sedimentation experiments in terms of the shape and hydration of macromolecules. In this decade important achievements in the field of molecular structure included the prediction of the dimensions and shape of the tobacco mosaic virus, before the rod-like particle was visualised by electron microscopy, and further demonstrations that proteins represent individual molecules of a definite molecular mass.

1940s

The Spinco Model E analytical centrifuge, an extremely reliable electrically driven instrument, was constructed and became commercially available. In **1942**, **W. J. Archibald** proposed new methods for determining molecular mass from AUC data. In **1943**, **E. J. Cohn** and **J. T. Edsall** showed that protein partial specific volumes could be successfully calculated from their composition and the partial specific volumes of component residues. The first textbook devoted to analytical centrifugation, *The Ultracentrifuge* by **T. Svedberg** and **K. O. Pedersen**, appeared in **1942**. This monograph became known as *The Bible* of ultracentrifugation.

1950s

This decade marked the beginning of the widespread use of the sedimentation method. Many brilliant achievements in molecular biology were accomplished with the ultracentrifuge. **M. Meselson** and **F. W. Stahl**, using density gradient ultracentrifugation and isotope labelling, proved the semiconservative mechanism of DNA replication, the experiments of **A. Tissier** and **J. D. Watson (1958)**, **F.-C. Chao** and **H. K. Schachman (1959)** led to the discovery of ribosomes. A second textbook on ultracentrifugation, *Ultracentrifugation in Biochemistry* by **H. K. Schachman**, was published in **1959**.

1960s

The first scanning photoelectric absorption optical system was developed. In this period there was widespread use of the ultracentrifuge to study of proteins, ribosomes, DNA and viruses. Essential contributions to the theory and practice of AUC analysis were made by **J. W. Williams, K. E. van Holde, H. K. Schachman, D. A. Yphantis**, and **H. Fujita**.

1970s

These were the golden years of AUC. By 1980, there were about 2000 analytical ultracentrifuges routinely operating in the world. The incorporation of a monochromator in the absorption optics allowed extremely dilute solutions to be studied. Rayleigh interference optics yielded highly accurate data for non-absorbing solutes, and was used mainly for the determination of macromolecular mass by sedimentation equilibrium. In **1971** the introduction of the differential sedimentation method opened the way for the observation of small differences

---

Comment D4.2 'Analytical ultracentrifugation reborn'

H. Shachman in an article called 'Analytical ultracentrifugation reborn' wrote:

> But by 1980 – despite the almost frenetic activity devoted to the design of new cells, the incorporation of different optical systems, the development of additional treatments and the application of ultracentrifugation to a divers host of important biological problems – the use of the instruments came to a rather abrupt end. What happened? Were better, more reliable and more versatile techniques developed that could be used instead of ultracentrifugal methods? Or were the questions asked by protein chemist and molecular biologist in the 1980s so different that sedimentation techniques had become obsolete?... But molecular biologists in the early 1980s who were interested in molecular interactions between different proteins, or between proteins and nucleic acids, were content with 'yes or no' answers rather than equilibrium constant and free energy changes. For them, filter binding assays, Sephadex columns and polyacrylamide gels provided the desired information. Now, however, with literally hundreds of proteins produced by the technique of site-directed mutagenesis awaiting characterisation, precise techniques for physical chemical investigations are needed. (Schachman, 1989.)

---

between sedimentation coefficients (**M. W. Kirschner** and **H. K Schachman**). In **1978** the development of new data analysis methods, which removed the contribution of diffusion from sedimentation velocity boundaries, yielded integral distributions of sedimentation coefficients (**K. E. van Holde** and **W. O. Weischet**). The third textbook on ultracentrifugation, *Foundation of Ultracentrifugation Analysis*, by **H. Fujita** appeared in **1975**.

1980s

In the early 1980s the sedimentation method started to lose popularity among biochemists. The reasons for this were two-fold: first, because of the development of gel electrophoresis and gel chromatography, which require very small quantities of material; second, in spite of the fact that the ultracentrifuge was linked to a computer the possibilities for fast data treatment were very limited. Data analysis was a tedious process and still based on pattern photography (Comment D4.2).

1990s

The situation changed dramatically with the appearance of a new generation of instruments, which were highly automated for data collection and analysis. The Beckman XL ultracentrifuge includes two different optical detection systems: a UV absorption system that makes it possible to study proteins or nucleic acids at a very low concentration ($\sim$2–3 $\mu g\ ml^{-1}$) and a Rayleigh interference optical system, which can detect macromolecules with low or no light absorbance. A Schlieren optical system is unnecessary, owing to the fact that precise derivative patterns can be obtained from integral curves by computer calculation. New data treatment programs became available to users – for example, to provide numerical solutions of the Lamm equation in the case of complex systems.

2000 to now

AUC is currently undergoing a renaissance. It can be applied to molecular weights from several hundreds to tens of millions. It is now the recognised method for accurate determination of the purity, mass, shape, self-association and other binding properties of macromolecules in solution, and is rapidly becoming, once again, a necessary technique in most biological laboratories.

D4.2 Instrumentation and innovative technique

Below, we describe briefly the main characteristic features of modern analytical ultracentrifuges, paying special attention to the Beckman Instruments Optima XL (Comment D4.3). The introduction in the 1990s of this series of instruments has brought about substantial improvements in the accuracy, precision and range of data that could be obtained. This was made possible by digital data acquisition, microprocessor-controlled experimental parameters, high-quality optical components and by the introduction of a Rayleigh interference system that permits the acquisition of data from macromolecular solutions of low optical absorbance.

The main parts of an analytical ultracentrifuge are a rotor contained in a refrigerated and evacuated protective chamber, an optical system to observe the concentration distribution in the sample during centrifugation and a data acquisition and analysis system.

D4.2.1 Rotors and cells

Rotors must be capable of withstanding large gravitational stresses. At 60 000 rev min^{-1}, a typical ultracentrifugation rotor generates a centrifugal field in the cell of about 300 000g (see Comment D4.4). Under these conditions, a mass of 1 g has an apparent weight of 250 kg (Comment D4.4).

A schematic diagram of a rotor and sedimentation cell is shown in Fig. D4.1. The rotor is solid, with holes to hold the sample cells. The simplest type of rotor incorporates two cells: the analytical cell and the counterpoise cell, which acts as a counterbalance. Cells have upper and lower plane windows of optical grade quartz or synthetic sapphire. A variety of analytical cells are available with volume capacities between 0.02 and 1.0 cm^3.

A sector-shaped cell is essential in velocity experiments since the sedimenting particles move along radial lines. In a sample compartment with parallel sides, sedimenting molecules at the periphery would collide with the walls and cause convection. Sectors that diverge more widely than radial lines would also cause convection.

Fig. D4.1 A typical
analytical centrifuge rotor
and sample double
sector-shaped cell.

Comment D4.4   Biologist's box: Centrifugal acceleration and centrifugal force

The applied centrifugal field (G) is determined by the square of the angular velocity of the rotor (ω, in radians per second) and the radial distance (r, in centimetres) of the particle from the axis of rotation, according to the equation

$$G = \omega^2 r$$

Since one revolution of the rotor is equal to 2π radians, its angular velocity, in radians per second, can be readily expressed in terms of revolutions per minute (rev min^{-1}), the common way of expressing rotor speed being

$$\omega = 2\pi (\text{rev min}^{-1})/60$$

The centrifugal field (G) in terms of rev min^{-1} is then

$$G = 4\pi^2 (\text{rev min}^{-1})^2 r/3600$$

The G value is generally expressed as a multiple of the earth's gravitational field ($g = 981$ cm s^{-2}), i.e. the ratio of the weight of the particle in the centrifugal field to the weight of the same particle when acted on by gravity alone, and is then referred to as the relative centrifugal field or 'number times' g. Hence

$$g = 4\pi^2 (\text{rev min}^{-1})^2 r/3600 \times 981$$

For 60 000 rev min^{-1} $g' = 250\,000$.

Fig. D4.2 Two types of cells used in AUC: (a) double sector-shaped cell; (b) boundary forming cell. Note the two small connecting channels between compartments. Liquid flow across the lower channel occurs only in the centrifugal field. The upper channel permits the return flow of air.

Double-sector cells permit us to account for absorbing components in the solvent, and to correct for the redistribution of solvent components (Fig. D4.2(a)). A sample of the solution is placed in one sector, and a sample of solvent alone acts as a reference in the second compartment. Boundary forming cells (Fig. D4.2(b)) allow the layering of the solvent over a sample of the solution, while the cell is spinning at a moderately low speed. These cells are useful for setting up a sharp boundary in diffusion coefficient determinations, using the boundary spreading

Fig. D4.3 Sedimentation profile of (a) 70 S ribosome *E. coli* and (b) 30 S and 50 S ribosomal subunits, obtained using the Schlieren optical system.

technique (Section D3.5.1), or for the examination of small molecules, for which the rate of sedimentation is insufficient to produce a sharp boundary that clears the meniscus.

D4.2.2 Optical detection systems

Depending on the type of optical system used, sedimentation in an ultracentrifuge has traditionally been observed in one of two main ways: as either the concentration or concentration gradient as a function of radius. The Schlieren optical system displays the boundary in terms of the refractive index gradient as a function of radius; the Rayleigh optical system, in terms of the refractive index as a function of radius; and the absorption optical system, in terms of optical density as a function of radius.

The Schlieren optical system dominated the field for 70 years and has since passed into history (Comment D4.5). Many brilliant achievements in molecular biology have been accomplished using this optics, including the discovery of ribosomes (Fig. D4.3). The Beckman XL ultracentrifuge contains only two optical detection systems: absorption optics in the UV and Rayleigh interference optics. Schlieren-type sedimentation profiles can be obtained by direct differentiation of the centrifugation absorption and interference pattern, using computer analysis.

Comment D4.5   Schlieren optical system

In the Schlieren optical system (named for the German word meaning 'streaks'), light is deflected by passing through a region in the cell where the concentration and hence the refractive index is changing. The optical system converts the radial deviation of the light into a vertical displacement of an image in the camera. The displacement is proportional to the refractive gradient. Thus the Schlieren image provides a measure of the refractive gradient, dn/dr, as a function of radial distance, r.

The absorption optics system is based on the proportional relation between the concentration of the solute in a solution and its optical density. Most biological macromolecules absorb light in the near UV region. Nucleic acids have a very strong absorption in the region 258–260 nm, while proteins are characterised by an absorption peak close to 280 nm (Chapter E1). Figure D4.4 shows a typical sedimentation profile measured using an absorbance optics system.

The most commonly used interference optics is based on the Rayleigh interferometer (Comment D4.6). The system can be applied to macromolecules that do not absorb significantly in the UV–visible range. Figure D4.5(a) shows a fringe pattern produced by a boundary in Rayleigh interference optics and Fig. D4.5(b) illustrates fringe displacement as a function of radius calculated by the software of the XL-I ultracentrifuge.

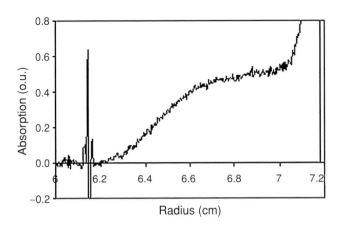

Fig. D4.4 Typical sedimentation profile measured by an absorbance optics system.

Comment D4.6   The Rayleigh interferometer

An interference pattern is produced by splitting a beam of coherent light and passing it through paired sectors of a cell in a spinning rotor. When the two beams are merged after passage through the sectors the waves form an interference pattern. If both sectors contain identical solutions the pattern is of straight interference fringes. However, when one sector contains the solvent and the other contains a macromolecular solution, the fringe pattern is shifted to an extent that corresponds to the concentration difference between the sample and solvent cells.

Fig. D4.5 A typical sedimentation profile measured by interference optics. (a) The fringe pattern produced by a boundary in Rayleigh interference optics. The refractive index difference (Δn) between two points of the cell can be calculated from the number of fringes N crossed in going between these to points from the classical relationship $N = a(\Delta n)/\lambda$, where a is the cell thickness and λ is the wavelength of light used. Since the refractive index is proportional to the concentration, the number of fringes can be used to monitor concentration as a function of radius in the cell. (b) A graphical representation of fringe displacement as a function of radius obtained by the software of the XL-I centrifuge.

(a)

(b)

(c)

Fluorescence
intensity

Radius

Time

Fig. D4.6 Two sedimentation profiles of BSA recorded with a fluorescence detecting system. BSA was labelled by FITC. (a) The highest concentration, 65 ng ml^{-1}; (b) the lowest concentration, 6.5 ng ml^{-1}; (c) axis directions of the fluorescence: intensity (Z), radius (X) and time (Y). A fluorescence detecting system was developed for the Spinco Model E analytical ultracentrifuge and was installed in place of the standard Schlieren optics (Schmidt and Reisner, 1992).

> **Comment D4.7   Prototype fluorescence detector for the XL-I analytical ultracentrifuge**
>
> A prototype fluorescence detector for the XL-I analytical ultracentrifuge has been developed for sedimentation velocity and equilibrium experiments. It is capable of detecting concentrations as low as 300 pM for fluorescein-like labels. The radial resolution of the detector is comparable to that of the absorbance system (MacGregor *et al.*, 2004).

> **Comment D4.8   Gravitational-sweep sedimentation**
>
> In 1972, a light-scattering/turbidity detector was installed in an ultracentrifuge to monitor the concentration changes of polymer molecules in the cell during centrifugation (Scholtan and Lange, 1972). This method was further developed for high-resolution, submicron, particle size distribution analysis of polymers (Machtle, 1999). The main technical feature of this method is gravitational-sweep sedimentation in which the rotor speed is increased reproducibly and exponentially from 0 to 40 000 rpm over a 1 hour period. This innovation allowed extremely broadly distributed polymer dispersions (from 10 to 3000 nm) to be studied without fractionation (see Section D4.5.3 for further discussion).

Significantly higher sensitivity would be expected from an analytical ultracentrifuge with fluorescence detection, because much lower concentrations would be detectable than with UV absorption optics. Furthermore, molecules with different fluorophores could be recorded selectively, even if they were present as a minor component in a mixture. Sedimentation profiles of bovine serum albumin (BSA), recorded for two different concentrations with a fluorescence detection system, are shown in Fig. D4.6. The lowest concentration (6.5 ng ml^{-1}) is 4–5 orders of magnitude lower than would be detectable with UV absorption optics for proteins (Comment D4.7). Finally, we point out that detection of turbidity has also been used and interpreted in AUC (Comment D4.8).

D4.2.3 Data acquisition

With the introduction of the Beckman Instruments Optima XL analytical ultracentrifuge, data acquisition in sedimentation experiments has now become largely automated. Computer programs for the analysis of sedimentation data commonly in use today rely on the graphical transformation of experimental data to obtain integral distributions of the sedimentation velocity boundary. Because of the improved data quality it has now become feasible to extract additional information from the sedimentation velocity boundary, such as diffusion coefficients

---

Comment D4.9 Home pages for AUC

Currently, there are three home pages on the web, where we can find online information that is useful to both the novice and the expert in the ultracentrifugation field.

The Beckman home page (http://www.beckman.com) besides containing information about their commercial products, also has an extensive list of background scientific papers and information about product development and data analysis for the XL-A and XL-I instruments.

The RASMB (reversible association in structural and molecular biology) site was created as a mail server to allow communication between research groups (http://www.eri.harvard.edu).

Bo Demeler's XL-A and RASMB/Analytical Ultracentrifugation page is hyperlinked via RASMB (http://bioc02.uthsca.edu). The site has the same downloadable software available on the RASMB site but in addition presents a mailing list with past communications, collaborations and fee-for-service information.

---

and partial concentrations, and in some cases to determine association constants (Comment D4.9).

D4.3 The Lamm equation

The Lamm equation describes the transport process in the ultracentrifuge. It was derived by introducing drift into Fick's diffusion equation. As discussed in Section D3.4.1, the flux is proportional to the concentration gradient with the constant of proportionality equal to minus the diffusion coefficient, $-D$ (Eq. (D3.5)). Assuming that all the particles in the cell drift in the direction $+x$ with a speed u, then the flux at point x should increase by an amount $uC(x)$, where $C(x)$ is the local concentration. Fick's first equation then becomes

$$J_x = -D[dC/dx] + uC(x) \tag{D4.1}$$

Recalling that $u = s\omega^2 x$ we obtain

$$J_x = -D[dC/dx] + s\omega^2 x C(x) \tag{D4.2}$$

The first and second terms on the right-hand side of Eq. (D4.2) correspond to transport by diffusion and sedimentation, respectively.

Equation (D4.2) was derived for a solution in an ideal infinite cell without walls and is not a strictly correct description of the sedimentation–diffusion process under real sector cell experimental conditions. The cross-section of a sector cell is proportional to r. The appropriate continuity equation is

$$\left(\frac{dC}{dt}\right)r = -\frac{1}{r}\left(\frac{dr\,J}{dr}\right)t \tag{D4.3}$$

Combining Eqs. (D4.2) and (D4.3) we obtain the following partial differential equation, known as the Lamm equation:

$$\left(\frac{dC}{dt}\right)r = -\frac{1}{r}\left\{\frac{d}{dr}\left[\omega^2 r^2 sC - Dr\left(\frac{dc}{dr}\right)t\right]\right\}t \tag{D4.4}$$

The Lamm equation describes diffusion with drift in an AUC sector cell, under real experimental conditions.

D4.4 Solutions of the Lamm equation for different boundary conditions

D4.4.1 Exact solutions

The general analytical solution of Eq. (D4.4) is an infinite series of integrals that can be calculated only by numerical integration. However, exact solutions of Eq. (D4.4) exist in two limiting cases: 'no diffusion' and 'no sedimentation'.

No diffusion

In the absence of diffusion, the Lamm equation has an exact solution for a homogeneous macromolecular solution:

$$C_2(x, t) = \begin{cases} 0 & \text{if } x_m < x < \bar{x} \\ C_0 \exp(-2s\omega^2 t) & \text{if } \bar{x} < x < x_b \end{cases} \tag{D4.5}$$

where C_0 is the initial concentration, and $\bar{x} = x_m \exp(s\omega^2 t)$. A graphical presentation of Eq. (D4.5) is shown in Fig. D4.7. The molecule concentration changes sharply from zero to a value that is independent of x at any time. The step function defines the boundary. The region $x > \bar{x}$ is called the plateau. The plateau level decreases with time owing to sectorial dilution (Fig. D4.7). The position of the step function also changes with time.

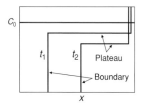

Fig. D4.7 Graphical presentation of Eq. (D4.5). In the absence of diffusion, a sharp boundary is formed and moves down the cell in time. There is a flat plateau below the boundary. The concentration in the plateau decreases with time owing to sectorial dilution.

No sedimentation

In this limiting case, the Lamm equation (D4.4) is given by

$$\left(\frac{dC}{dt}\right)r = -D\left(\frac{d^2C}{dt^2}\right)t \tag{D4.6}$$

The concentration gradient is given by

$$\left(\frac{dC}{dt}\right)r = -C_0(\pi Dt)^{1/2}\exp\left(\frac{-x^2}{4Dt}\right) \tag{D4.7}$$

The solution of this equation was discussed in Section D3.2. The diffusion coefficient is determined by measuring the standard deviation of the Gaussian fit to the boundary function, which is equal to $(2Dt)^{1/2}$. In practice, the concentration gradient curves are measured as a function of time, according to the procedure described in Section D3.4.3 for pure diffusion (Fig. D3.7(a) and (b)). This approach is valid if the term describing sedimentation flow $\omega^2 r^2 sC$ is essentially smaller than the diffusion term. In practice, this condition is fulfilled for small globular proteins (s value about 2 S) at low speed (2000–6000 rpm) with a synthetic boundary cell. The time-dependent spreading of the boundary is measured after overlaying the macromolecular solution with buffer. (See definition of s in Section D4.5.1.)

D4.4.2 Analytical solutions

Analytical solutions of the Lamm equation with very specific boundary conditions date from before 1930 (Comment D4.10). Now, modern computer methods allow approximate solutions of the equation for more realistic experimental situations. Two methods have been developed, each of which enables the simultaneous determination of sedimentation and diffusion coefficients and, as a consequence, of molecular mass. Which method is applicable depends on the relation between the respective contributions of diffusion and sedimentation to the moving boundary.

The van Holde–Weischet method is based on the fact that sedimentation is proportional to the *first power of time*, while diffusion is proportional to *the square root of time* (compare Eqs. (D4.18) and (D3.12)). It follows that extrapolation to infinite time must eliminate the contribution of diffusion to the boundary shape. In a first step, the distance between the baseline and the plateau for each scan is divided into N (usually a few tens) horizontal divisions. N apparent sedimentation coefficients s^* are calculated as

$$s^* = \ln[(r/r_{\mathrm{m}})/\omega^2 t] \tag{D4.8}$$

where r, r_{m} are the radius of the division and the meniscus radius, respectively. The s^* values are then plotted versus the inverse square root of the time of each scan. The y-intercept, which corresponds to infinite time, is equal to the diffusion-corrected s value.

For a homogeneous sample, all lines should converge to the same limit, s. If the sample is not homogeneous, the average of the y-intercepts is used to calculate a weight-average sedimentation coefficient, s_{w}. The results are then plotted as a fraction of total sedimenting material versus s. Such a plot permits the characterisation a priori of sample quality. Figure D4.8 shows sedimentation velocity boundaries (parts (a), (d)) and an analysis by the van Holde–Weischet method of a homogeneous (parts (a)–(c)) and a heterogeneous (parts (d)–(f)) sample (Comment D4.11).

Comment D4.10
Analytical solutions of the Lamm equation with various specific boundary conditions

Faxen-type solutions: the centrifugation cell is considered as an infinite sector, diffusion is small ($D/\omega^2 sx \ll 1$), 3), and only early sedimentation times are considered ($\omega^2 st \ll 1$).

Archibald solutions: s and D are considered to be constant.

Fujita-type solutions: an extended Faxen solution for the case when D is constant, but s depends on concentration as $s = s^0(1 - k_s c)$

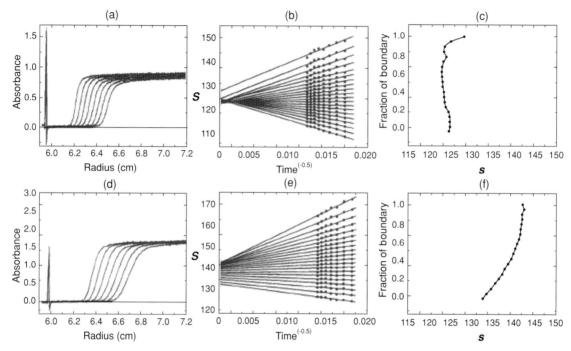

Fig. D4.8 Analysis of sedimentation velocity boundaries by the van Holde–Weischet method: (a)–(c) data and analysis for bacteriophage T7 MLD capsid II; (d)–(f) data and analysis for bacteriophage T7 MLD capsid II, in which the capsids contained measurable amounts of subgenomic, packaged DNA. The speed of the runs was 10 000 rpm. (Hansen *et al.*, 1994.)

Comment D4.11   Method of van Holde–Weischet

In practical applications, the method requires a well-defined plateau region and simulations which can be used to demonstrate the validity of the approach.

The van Holde–Weischet method has been introduced by B. Demeler into the software platform ULTRASCAN (http://www.cauma.uthscsa.edu//), which exists in free or commercial versions, depending on the operating system.

The second approach is the Stafford method in which the sedimentation coefficient distribution is computed from the time derivative of the sedimentation velocity concentration profile. An apparent sedimentation coefficient distribution (i.e. uncorrected for diffusion), $g(s^*)_t$ can be calculated from $(dc/dt)_r$ as

$$g(s^*)_t = (dC/dt)_{\text{corr}}(1/C_0)[\omega^2 t^2 / \ln(r_{\text{m}}/r)(r/r_{\text{m}})^2] \qquad (D4.9)$$

The time derivative sedimentation velocity approach

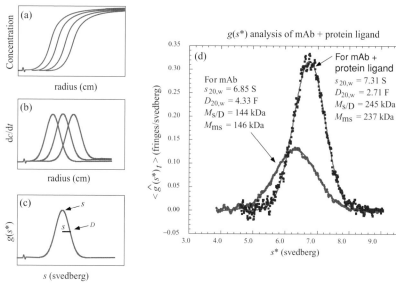

$g(s^*)$ analysis of mAb + protein ligand

(d)

For mAb
$s_{20,w} = 6.85$ S
$D_{20,w} = 4.33$ F
$M_{S/D} = 144$ kDa
$M_{ms} = 146$ kDa

For mAb + protein ligand
$s_{20,w} = 7.31$ S
$D_{20,w} = 2.71$ F
$M_{S/D} = 245$ kDa
$M_{ms} = 237$ kDa

Fig. D4.9 Illustration of the Stafford method ((a)–(c) see text) Typical results are given in (d). Here we see the $g(s^*)$ results for a monoclonal antibody (mAb) and a mAb–protein complex. The molecular mass determined for mAb from s and D was 144 kDa and that from MALDI-MS was 146 kDA, indicating an error of 1.4% for the AUC value. For the mAb--protein complex, a molecular mass of 245 kDa was calculated from s and D, indicating an error of 3.8% when compared to the MALDI-MS of 237 kDA. The very good precision comes from the fact that all the data from 40 boundaries (~1000 points per curve) were used in the analysis. (Hensley, 1996.)

An improved commercial version of the Stafford time derivative method (DCDTPLUS) is available from Philo (http://www.jphilo. mailway.com).

where s is the sedimentation coefficient, ω is the angular velocity of the rotor, C_0 is the initial concentration, r and r_m are the radius and radius of the meniscus, respectively. With this approach, velocity data are obtained in the standard way (see Fig. D4.9(a)). Closely spaced data sets are subtracted, giving time derivative profiles, as shown in Fig. D4.9(b). Each boundary yields a Gaussian-shaped profile. A plot of $g(s^*)$ has essentially the same properties as a Schlieren gradients plot (dC/dr plot). The curves are then normalised for time. As a result the separate Gaussian distributions are superimposed. The y-axis is renormalised so that the areas under the curves are equal to the concentration. The resulting peak is a Gaussian whose position and width give the sedimentation and diffusion coefficients, respectively (Fig. D4.9(c)).

D4.4.3 Numerical solutions

Numerical solutions of the Lamm equation for a continuous particle size distribution are currently available. The problem is stated as an integral equation, and regularisation is used for its numerical inversion (Comment D4.12). The approach requires prior knowledge of the partial specific volumes and shapes of the molecules (in terms of frictional ratios, f/f_0), and of the density and viscosity of the solvent. Sedimentation coefficient or molar mass distributions can be obtained at relatively high resolution from the analysis. An example application is given in Section D4.5.2.

Comment D4.12 Continuous particle size distribution

In the absence of interactions between particles, the experimentally observed sedimentation profiles of a continuous size distribution can be described as a superposition of subpopulations, $C(M)$, of particles of mass between M and $M + \mathrm{d}M$. If $L(M, r, t)$ denotes the sedimentation profile of a monodisperse species of mass M and radius r at time t, the problem is described by a Fredholm integral equation of the first kind,

$$s(r, t) = \int C(M) L(M, r, t)\, \mathrm{d}M + \varepsilon \qquad \text{(A)}$$

where $s(r, t)$ denotes the experimentally observed signal with an error ε.

The equation is encountered in similar form in particle characterisation problems occurring in many other techniques, such as dynamic light scattering (Chapter D10), for example. In the case of sedimentation velocity AUC on a dilute solution, the kernel $L(M, r, t)$ of the Eq. (A) is the solution of the Lamm equation.

The continuous size distribution analysis program SEDIT of P. Schuck is available from http://www.analyticalultracentrifugation.com.

D4.5 Sedimentation velocity

There are two main complementary methods in AUC, with different experimental protocols but using the same instrumentation (Table D4.1). Sedimentation velocity, in which a high rotor speed and a long solution column are used to maximise the resolution, provides information about size, shape and molecular

Table D4.1. *The two types of experiment in AUC*

	Sedimentation velocity	Sedimentation equilibrium
Angular velocity	Large (chosen according to the particles' sedimentation properties)	Small
Analysis	As a function of time (ex: 3–6 hours)	At equilibrium (ex: after 24 hours)
Measurement	Forming a boundary (sedimentation profile as function of time)	The particle distribution in the cell
Calculated parameters	Shape, mass composition	Mass composition

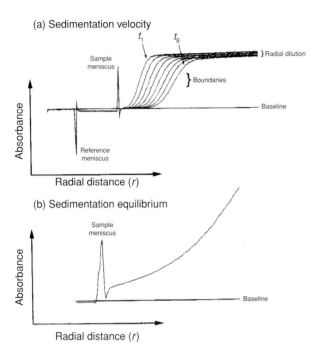

(a) Sedimentation velocity

Absorbance

Radial distance (r)

(b) Sedimentation equilibrium

Absorbance

Radial distance (r)

Fig. D4.10 (a) Boundaries in a sedimentation velocity experiment. Eight successive scans on a sample of reconstituted oligonucleosomes in a low salt buffer are shown. The speed of this run was 20 000 rpm. The direction of sedimentation is from left to right. The scans were collected at 8-min intervals (the earliest recorded scan is labelled 't_1'; the latest recorded scan is labelled 't_8'). The baseline is the absorbance of the solvent alone. The centrifuge cell is composed of two sector-shaped compartments. The radial position of the air–solvent interface in the sample sector is labelled 'sample meniscus'.
(b) Sedimentation equilibrium data on the same oligonucleosome sample, after centrifugation at 1600 rpm for 72 h. The lower rotor speed resulted in the formation of an equilibrium concentration gradient rather than the formation of a moving boundary as in (a). Equilibrium was achieved as judged by the fact that scans measured after 76–90 h were identical with the one after 72 h shown in the figure. (Hansen *et al.*, 1994.)

interactions. Sedimentation equilibrium, in which low rotor speeds and a short solution column are used to minimise the time needed to reach equilibrium, provides information about molecular mass, association constants and stoichiometry. Figure D4.10 shows typical data from a sedimentation velocity experiment (part (a)) and from a sedimentation equilibrium experiment (part (b)).

D4.5.1 Macromolecules in a strong gravitational field

The usual way of determining masses is to observe the movement of particles under the action of known forces, since the operational definition of mass depends on the relative acceleration of an unknown and a standard mass when acted upon by the same force. The most convenient force to use on a macroscopic scale is the force of gravity. For molecules, however, this force is so small that it is outweighed by far by forces generated in collisions from the random motion of surrounding molecules.

A simple calculation shows that, at the molecular level, gravitational potential energy is smaller than thermal energy by a factor of about a few hundred (Comment D4.13). It follows that the experimental detection of macromolecular movement due to natural gravity is impossible. A way to avoid the difficulty is to increase the effective gravitational potential energy to a value greater than kT, by putting the particles in a cell rotating at high speed. Let us consider a particle

---

Comment D4.13   Biologist's box: Comparison of gravitational potential and thermal energies

We shall calculate the difference in gravitational potential energy between two points separated by a vertical distance L for a molecule of molecular mass M. Take $M = 10^5$ Da, and $L = 1$ cm. Then $M = 10^5/(6.03 \times 10^{26}) = 1.6 \times 10^{-22}$ kg (by using the number of molecules in a kilogram mole). The acceleration due to gravity is 10 m s^{-2}, the potential energy difference E_p, is $E_p = mgh = 1.6 \times 10^{-22} \times 10 \times 10^{-2} = 1.6 \times 10^{-23}$ J. The thermal agitation energy is $\sim kT$, and at room temperature $kT = 1.38 \times 10^{-23}$ J K^{-1}, $293 \times K = 400 \times 10^{-23}$ J. We conclude that the gravitational potential energy is smaller than the thermal energy by a factor of 250, and therefore experimental detection of particle sedimentation due to gravity is impossible. An analogous calculation for $M = 10^9$ Da and $L = 10$ cm shows that in this case gravitational potential energy is larger than the thermal energy by a factor of 400 and experimental detection of the particle sedimentation due to gravity is possible.

Comment D4.14
Mechanical and thermodynamic approaches

The discussion in the text is in terms of a simple mechanical model for sedimentation processes. Some of the ambiguities that arise from this type of treatment can be avoided by using a thermodynamic approach.

of mass m and density ρ in a sector-shaped cell that is in a rotor spinning about the z-axis with angular velocity ω moving through a solvent of density ρ_0. The molecule is acted on by centrifugal, F_c, buoyant, F_b, and frictional drag forces, F_d (Comment D4.14, Fig. D4.11).

The centrifugal force is proportional to the product of the particle mass m and the linear acceleration $\omega^2 r$

$$F_c = m\omega^2 r \qquad (D4.10)$$

where r is the distance from the centre of rotation. Under the influence of this force the particle moves through the surrounding medium radially out from the axis of rotation.

The buoyant force (Comment D4.15) is equal in magnitude to the weight of the fluid displaced by the particle

$$F_b = -m_0\omega^2 r \qquad (D4.11)$$

Centrifugal force: $F_c = m\omega^2 r$

Buoyant force: $F_b = -m_0\omega^2 r$

Frictional force: $F_d = -fu$

ω - angular velocity
r - distance from the centre of rotation
u - velocity of molecule

Fig. D4.11 Forces on the particle in a centrifuge cell (see text).

> **Comment D4.15 Archimedes' force: macroscopic and microscopic approaches**
>
> Archimedes' principle states that 'the buoyancy of a body in a liquid equals the weight of the liquid in a volume equal to the volume of the body'. It is well known that the law was formulated for macroscopic bodies under hydrodynamics conditions and, therefore, it would seem that it is not relevant to individual molecules. In this connection the question arises: can a molecule in a liquid be considered as a body immersed in the same liquid? A macromolecule can be considered as a physical body possessing finite volume, and therefore an additional 'buoyancy' acts upon it in the presence of a pressure gradient. There is an entirely rigorous definition for the Archimedes' force acting on individual molecules (Luboshitz and Podgoretskii, 1991): 'on the body immersed in a liquid, there acts a buoyancy equal to the weight of the liquid in that volume which became inaccessible to its molecules.' (See also Section G1.7.)

The resulting net 'gravitational' force is

$$F_c + F_b = (m - m_0)\omega^2 r = m\omega^2 r(1 - \rho_0/\rho) \tag{D4.12}$$

Substituting the density ρ by its reciprocal, the partial specific volume $\bar{v}$ (see below, Section D4.3.6) we obtain

$$F_c + F_b = m\omega^2 r(1 - \bar{v}\rho_0) \tag{D4.13}$$

The frictional force experienced by a molecule depends on its velocity u and frictional coefficient f (Eq. (D1.9))

$$F_d = -fu$$

A constant velocity is reached when the total force on the particle is zero, i.e.

$$F_c + F_b + F_d = 0 \tag{D4.14}$$

$$m\omega^2 r(1 - \bar{v}\rho_0) = fu \tag{D4.15}$$

In order to put things on a mole basis we multiply Eq. (D4.14) by Avogadro's number. Placing experimentally measured parameters on the left-hand side of the equation and molecular parameters on the right-hand side we obtain:

$$s \equiv u/\omega^2 r = M(1 - \bar{v}\rho_0)/N_A f \tag{D4.16}$$

The ratio of the measured velocity of the particle to its centrifugal acceleration is called the *sedimentation coefficient*, s. Its dimension is time in seconds. The sedimentation coefficients of most biological particles are very small, and for convenience the basic unit is taken as 10^{-13} s, a unit that was called 1 svedberg (1 S), in recognition of Teo Svedberg's fundamental pioneering work in AUC.

D4.5.2 Determination of sedimentation and diffusion coefficients from the moving boundary

Routine mode AUC: *s* and *D* are well defined

AUC in routine mode is used to analyse dilute solutions of one type of macro-molecule (a monodisperse solution) or of a mixture of a few different types for which *s* and *D* are well defined. The aqueous solvent typically contains a pH buffer and about 100 mM salt to screen electrostatic interactions between the macromolecules.

In a sedimentation velocity experiment, the centrifuge cell contains an in-itially homogeneous solution (the concentration is constant over the length of the cell). In the process of centrifugation, the particles move along gravitational force lines, the meniscus is cleared of macromolecules, and a moving boundary forms between solvent depleted of macromolecules and the solution of finite concentration (Fig. D4.12). The velocity of the moving boundary is given by the rate of change in its radial coordinate, dr_b/dt (Fig. D4.12). The velocity is expressed as a scalar quantity *u*, since the direction is well defined in the radial direction. We obtain, from Eq. (D4.16)

$$u = dr_b/dt = \omega^2 r s$$

After integrating,

$$\ln r_b(t)/r_b(t_0) = \omega^2 s(t - t_0) \tag{D4.17}$$

The slope of $\ln r_b(t)/r_b(t_0)$ versus $(t - t_0)$ gives $\omega^2 s$ and hence *s* (Fig. D4.12). The method is applicable *only* to 'ideal' solutions, of a single type of macromolecule with no interparticle interactions. For slightly asymmetric or polymodal bound-aries the second moment method is a useful approach (Comment D4.16).

Fig. D4.12 Hypothetical sedimentation boundary showing the meniscus, equivalent boundary position, plateau and baseline. The traditional analysis of routine mode velocity sedimentation data is based on plotting the logarithm of the radial position of the midpoint of the sedimenting boundary versus time.

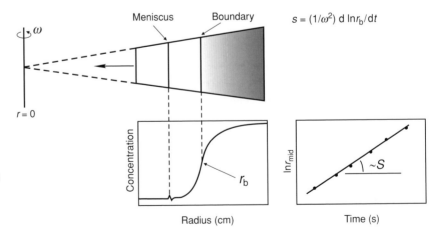

Comment D4.16   The second moment method

The second moment method provides a very useful approach for obtaining accurate estimates of the sedimentation coefficient in cases that exhibit slightly asymmetric or polymodal boundaries. The method permits the calculation of the position of a hypothetical boundary that would have been observed in the absence of diffusion and polydispersity (Fig. D4.13). The rate of movement of the equivalent boundary position gives the weight average sedimentation coefficient.

The second moment approach has generally been applied to data obtained with the Schlieren optical system, from which the equivalent boundary position was calculated from the second moment of the refractive index gradient curve:

$$\langle r^2 \rangle = \frac{\displaystyle\int_{r_m}^{r_p} \frac{\mathrm{d}c}{\mathrm{d}r} r^2 \mathrm{d}r}{\displaystyle\int_{r_m}^{r_p} \frac{\mathrm{d}c}{\mathrm{d}r} \mathrm{d}r}$$

However, when the data are obtained as concentration versus radius instead of concentration gradient versus radius, it is convenient to compute the equivalent position from the following relation:

$$\langle r^2 \rangle = \frac{\displaystyle\int_{r_m}^{r_p} r^2 \mathrm{d}c}{\displaystyle\int_{r_m}^{r_p} \mathrm{d}c}$$

Complex mode AUC: analysis of the boundary in terms of s and D

Sedimentation velocity analysis of complex systems requires a more elaborate approach than the routine mode described above. In the case of very large particles, for which separation is achieved during the time of the experiment, a high-resolution analysis can be performed based on the spatial derivative of the sedimentation profiles, $\mathrm{d}c/\mathrm{d}r$, or by the related method of observing the time course of sedimentation at a single radial position. For smaller particles, however, diffusion broadening of the sedimentation boundary makes it more difficult to resolve subpopulations of the distribution. The Stafford or van Holde–Weischet methods (Section D4.4.2) are very useful in this regime. Both have the virtue of *model-free* analysis.

If a model is available for the sedimentation behaviour of the macromolecules, size-distribution analysis can provide superior information in terms of precision in the derived parameters, although this is computationally more difficult than

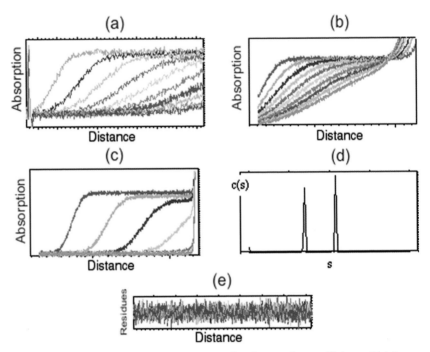

Fig. D4.13 Obtaining a discrete distribution of sedimentation coefficients, C(s), by the SEDFIT program. Part (a) shows experimental files for a solution comprising a 14-kDa fragment of heparin and a protein of 34 kDa (Interferon-γ) in the same proportion. Ten profiles obtained at 42 000 rpm and 10 °C are shown. Parts (b) and (c) depict simulated profiles with $\rho = 1.0105$ g ml^{-1}, $\eta = 1.429$ cp, $\bar{\upsilon} = 0.64$ ml g^{-1}, $f/f_0 = 1.25$. The values of ρ and η are those of the solvent, those of $\bar{\upsilon}$ and f/f_0 are reasonable assumptions corresponding to globular, usually hydrated complexes with about 1 gram heparin per gram protein. Part (b) corresponds to the smallest considered particle with $s = 1$ S and a corresponding calculated D value of 10×10^{-7} cm^2 s^{-1}. Part (c) corresponds to the largest considered particle with $s = 5$ S and a corresponding calculated D value of 4.5×10^{-7} cm^2 s^{-1} (see the text). Comparison of (b) and (c) shows clearly how broadening of the sedimentation profiles is taken into account by the program. SEDFIT was asked to simulate 100 sets of sedimentation profiles corresponding to s values between $s = 1$ S and $s = 5$ S. The best combination of these profiles for the description of the experimental profiles corresponds to the continuous distribution C(s) presented in part (d). Part (e) shows the superposition of the differences between the profiles calculated with the distribution of particles of part (d) and the experimental profiles. The C(s) distribution on part (c) suggests there are two types of species in solution. (Courtesy Chr. Ebel.)

Numerical solutions of the Lamm equations with applications for direct fitting of ultracentrifuge data are available (program SEDFIT of P. Schuck).

the Stafford or van Holde–Weischet methods. Figure D4.13 shows an example depicting the sequential steps in the analysis procedure for experimental sedimentation profiles.

If ρ, η, $\bar{\upsilon}$ and f/f_0 are inadequately chosen, the C(s) distribution is obtained with decreased resolution, while the derived C(M) distribution is definitely wrong

(s is an experimental parameter, M is calculated from the s values and shape and $\bar{v}$ estimates). For heterogeneous systems, and particularly for broad or complex species distributions, a regularisation procedure has to be applied. As a result, the description of the sedimentation profiles in terms of a continuous distribution permits the determination of the level of non-homogeneity in the sample. Furthermore, evidence of interactions between the macromolecules can be provided from the changes in the apparent distribution of sedimentation coefficients with sample concentration.

D4.5.3 Highly heterogeneous systems

It is usual, in routine sedimentation velocity experiments, for a single rotor speed to be chosen for the entire run. Consequently, the range of observable sedimentation coefficients can be severely limited. For example, at 50 000 rpm and a usual acceleration rate of 400 rpm s^{-1}, the largest particle half way down the cell has a sedimentation coefficient of about 280 S. The limitation can be removed and the observable s range expanded by approximately three orders of magnitude, by varying the rotor speed during the run, starting with a relatively low speed so that the largest particles can be easily observed. The speed is gradually increased and the run continued at full speed until the smallest species of interest have cleared the solution. The data from each speed are combined into a single continuous distribution function.

A speed protocol is suggested in Table D4.2 that provides an observable s range from over 200 000 S to 2.0 S in a single experiment. The centrifuge is run for 600

Table D4.2. *Typical speed protocol. Radius of the meniscus = 5.9 cm (Stafford et al., 2004)*

Speed (rpm)	Time at each speed (s)	s^* (at 6.5 cm) (svedbergs)
0–6000	15	~500.000
6000	600	4220
9000	600	1300
13 000	600	550
18 000	600	250
25 000	600	125
35 000	600	62
50 000	3600	31
50 000	3600	7.6
50 000	3600	4.4
50 000	3600	3.0
50 000	3600	2.3

$s^*(6.5 \text{ cm}) = \ln(6.5/5.9)/j(\omega^2 dt)$.

Fig. D4.14 Plot of $s^* g(s^*)$ verses $\ln(s^*)$ for a *Limulus polyphemus* haemocyanin solution (2 mg ml^{-1}). The rotor velocity was ramped through speeds of 12, 18, 25, 35, and 50 krpm, and between 50 and 90 scans were made at each speed ($\sim$20 min) except at the highest speed. The run was continued at 50 krpm until it was obvious that the solution in the lower half of the cell was clear of protein. In this case the run was held at 50 krpm for 3 hours in order to bring most of the 5 S material to the lower quarter of the cell. The points acquired at each speed are shown in a different colour. The areas under the peaks give the concentration for the boundary corresponding to each peak distribution. (Stafford and Braswell, 2004.)

seconds at each speed until the maximum speed is reached. The rotor is then maintained at 50 000 rpm until the smallest material has sedimented more than half way to the bottom of the cell. Figure D4.14 shows the application of the method (Comment D4.17) to a haemocyanin solution. The data were analysed according to the time derivative method of Stafford (Section D4.2.2), and plotted as $s^* g(s^*)$ versus $\ln(s^*)$ because of the large range covered in s^*). The haemocyanin sample resolved into four peaks (s values of $\sim$5, 16, 40 and 60 S), corresponding to monomers, hexamers, 24-mers and 36-mers, demonstrating how an extremely wide range of sedimentation coefficients can be accommodated in a single multi-speed run.

D4.5.4 Sedimentation coefficients of biological macromolecules

Sedimentation coefficients of biological macromolecules are normally obtained in buffered solutions whose viscosity and density may differ from those of water. Sedimentation coefficients can also be measured at different temperatures. It is customary to correct sedimentation data to the 'standard values' that would be observed if the measurements were performed in pure water at 20 °C

$$s_{20,w} = s_{exp} \frac{1 - \bar{\upsilon}\rho_{20,w}}{1 - \bar{\upsilon}\rho_{exp}} \frac{\eta_{exp}}{\eta_{20,w}} \qquad (D4.18)$$

where $s_{20,w}$ is the sedimentation coefficient of the molecule in water at 20 °C, s_{exp} is the experimentally measured sedimentation coefficient of the molecule,

η_{exp} is the viscosity of the solvent at the experimental temperature $T(^{\circ}\text{C})$, $\eta_{20,w}$ is the viscosity of water at 20 °C, $\rho_{20,w}$ is the density of water at 20 °C, ρ_{exp} is the density of the solvent at given temperature $T(^{\circ}\text{C})$ $\bar{\upsilon}$ is the partial specific volume of the molecule. Note that $s_{20,w}$ can be defined even if the species does not exist in water at 20 °C.

Sedimentation coefficients of biological macromolecules are normally obtained at finite concentration and should be extrapolated to zero concentration (Comment D4.18). The sedimentation coefficients of biological macromolecules cover a very large range of values. For small peptides, globular proteins, they range from 0.4 S to 100 S; for ribosomal RNA, from 4 S to 30 S; for ribosomes and polysomes, from 20 S to 100 S; for viruses, from 40 S to 1000 S; for lysosomes, from 4000 S; for membranes, from 100 000 S; for mitochondria, from 20 000 S to 70 000 S; and for nuclei, from 4 000 000 S to 40 000 000 S. Generally, the larger the particle, the larger is its s value. Very large molecules may sediment in the earth's gravitational field.

Table D4.3 lists some representative measured values of $s_{20,w}$, $\bar{\upsilon}$ and M of proteins, starting from small synthetic peptides (peptide I, molecular mass 2 kDa) and finishing with large oligomeric protein complexes (haemocyanin, molecular mass about 9 MDa). Table D4.4 presents measured values of $s_{20,w}$, for native DNA and DNA fragments.

Comment D4.17
Sedimentation with increasing rotor speed

A sedimentation velocity technique with increasing rotor speed was proposed originally by D. Yphantis and coworkers in 1981. In 1984 W. Machtle proposed the term 'gravitational sweep' to describe size distribution studies of large particles by sedimentation velocity in which rotor speed is varied.

Comment D4.18   Concentration-dependence of the sedimentation coefficient

It is generally accepted that the concentration dependence of the sedimentation coefficient s_c, C in grams per millilitre, is given by

$$s = s^0(1 + k'C)^{-1} \qquad\qquad\text{(A)}$$

or

$$s = s^0(1 - k''C) \qquad\qquad\text{(B)}$$

where s^0 is the infinite dilution value and k' and k'' are empirical constants, in milli-litres per gram. Usually, k' and k'' are positive, i.e. s decreases with increasing concentration. These equations are equivalent for small concentrations, but predict very different behaviour for high concentrations. Eq. (A) is preferably used for compact particles whereas Eq. (B) is better for synthetic polymers and highly asymmetric particles.

In some cases an increase in s with increasing C is observed. This is presumptive evidence for reversible association or weakly attractive macromolecular interactions.

Empirical constant k' (k'') depends on the shape of macromolecules. This means that this constant can provide a sensitive measure of molecular asymmetry (Section D2.5).

Table D4.3. *Sedimentation coefficients, partial specific volumes of proteins in aqueous solvent. Bold type indicates molecules that have no globular conformation.*

Peptides and proteins	Molecular mass (Da)	Sedimentation coefficient $s_{20,w}^0$ (Svedberg units)	Partial specific volume $\bar{v}$ (cm^3 g^{-1})
Small synthetic peptide I	2049	0.37	0.706
Small synthetic peptide II	2777	0.45	0.715
Small synthetic peptide III	3505	0.52	0.720
Lipase[a]	6669	1.14	0.7137
Insulin (dimer)	11 466	1.60	0.744
Cytochrome C	12 400	1.80	0707
Ribonuclease A (bovine pancreas)	13 683	1.78	0.696
Lysozyme (chicken egg white)	14 305	1.91	0702
α-Lactalbumin (bovine milk)	14 180	1.92	0.704
Myoglobin	17 836	1.98	0.741
Papain	23 350	2.42	0.723
α-Chymotrypsin	25 000	2.40	0.736
Chymotrypsinogen A (bovine pancreas)	25 767	2.58	0.736
Elastase	25 900	2.60	0.73
Subtilisin BPN	27 537	2.77	0.731
Carbonic anhydrase	30 000	3.0	0.735
Riboflavin-binding protein	32 500	2.76	0.720
Carboxypeptidase	34 472	3.2	0.725
Pepsin	34 160	2.88	0.725
β-Lactoglobulin A, dimer (bovine milk)	36 730	2.87	0.751
β-lactoglobulin	35 000	3.08	0.75
Kinesin motor domain construct K366	41 404	3.25	0.733
Albumin ovum	44 000	3.6	0.74
Bovine serum albumin	66 300	4.50	0.735
Haemoglobin	64 500	4.60	0.749
Anthax protective antigen	85 000	5.01	0.762
Tropomyosin	**93 000**	**2.6**	**0.71**
Lactate dehydrogenase (dogfish)	138 320	7.54	0.741
β-Lactoglobulin A, octamer (bovine milk)	146 940	7.38	0.751
GPD, apo (bakers' yeast)	142 870	7.60	0.737
Aldolase	156 000	7.40	0.742
Malate synthetase	170 000	8.25	0.735

(cont.)

Table D4.3. (*cont.*)

Peptides and proteins	Molecular mass (Da)	Sedimentation coefficient $s^0_{20,w}$ (Svedberg units)	Partial specific volume $\bar{v}$ (cm³ g⁻¹)
Catalase (bovine liver)	248 000	11.3	0.730
Glutamate dehydrogenase (bovine liver)	312 000	11.4	0.749
Fibrinogen	**330 000**	**7.6**	**0.706**
Apoferritin	467 000	17.6	0.750e
Apoferritin (horse spleen)	502 000	18.3	0.728
Urease	482 700	18.6	0.742
Myosin	**570 000**	**6.43**	**0.728**
Glutamate dehydrogenase	1 015 000	26.6	0.73
Haemoglobin	3 500 000	58.9	0.747
Haemocyanin	8 950 000	105.8	0.728

[a] Data are taken from different literature sources

Table D4.4. *Sedimentation coefficients of double stranded DNA*

Particle	Molecular mass (bp)	Sedimentation coefficient $s^0_{20,w}$
Fragments of DNA,	12	1.86[a]
prepared by digestion	20	2.54[a]
of PM2 phage with the	50	3.51
restriction	94	4.58
endo-nuclease Hae III	117	4.92
(Kovacic *et al.*, 1972)	145	5.20
	160	5.41
	263	6.24
	288	6.37
	322	6.70
	498	7.62
	592	7.89
	642	8.03
	794	8.59
	854	8.72
	1310	9.99
	1606	10.67
	1735	10.78
Native DNA:		
ØX 174RF	5386	14.3

(*cont.*)

Table D4.4. (*cont.*)

Particle	Molecular mass (bp)	Sedimentation coefficient $s^0_{20,w}$
T7 DNA	39.937	32.0
λ DNA	48.502	34.4
T4 DNA	168.903	62.1
T2 DNA	172.000	64.5
Highly fractionated	368[a]	7.2
samples by gel	289[a]	6.3
chromatography: ratio	172[a]	5.7
M_w/M_n is about 1.1	136[a]	5.4
	133[a]	5.4
	111[a]	4.7
	68[a]	3.9

[a] Sedimentation coefficients of 12 and 20 bp DNA fragments are calculated from diffusion coefficients according to Svedberg equation (Eq. (D4.20)).

D4.5.5 Differential sedimentation for measuring small changes in sedimentation coefficients

Very small changes in sedimentation coefficient can be measured by the technique of differential sedimentation. The uncertainties due to the rotor speed and temperature are the principal errors in the determination of sedimentation coefficients. They can be eliminated, essentially, by placing the sample solution and solvent in two double-sector cells in the same rotor. Schlieren optics were used in the first experiments with two double-sector cells. One of the cells was assembled using a wedge window, so that its image was displaced on the photographic plate, which then revealed two separate Schlieren patterns.

The technique proposed by Kirschner and Schahman in 1971 is analogous to difference spectroscopy (Chapter E1). Each compartment in one double-sector cell is filled with a solution. If the two solutions are identical, the Rayleigh interference pattern consists of a series of straight, parallel, interference fringes. If the boundary in one solution moves more slowly than in the other, the difference in their positions produces a pattern of curved fringes, similar to that observed with a Schlieren optical system.

The method was tested by measuring differences in sedimentation properties between a solution of bushy stunt virus (BSV) in H_2O buffer and a solution of BSV in an H_2O–D_2O mixture. The results indicated that both the precision and the accuracy of the difference sedimentation technique are within 5% even for values of $\Delta s/s$ as low as 0.002.

The technique has been applied to a variety of proteins in the molecular mass range (30 000–300 000 Da) and it has been shown that changes in sedimentation coefficients of as low as 0.01 S can be measured with an accuracy better than ±5%.

D4.6 Molecular mass from sedimentation and diffusion data

Substituting the Einstein relation for the frictional coefficient in terms of the diffusion coefficient (Eq. (D4.15)) into Eq. (D4.7) we have

$$\frac{s}{D} = \frac{M(1 - \bar{v}\rho_0)}{RT} \qquad (D4.19)$$

Rewriting, we obtain the first Svedberg equation for the determination of macro-molecular mass (Comment D4.19):

$$M = sRT/D(1 - \bar{v}\rho_0) \qquad (D4.20)$$

The values of $\bar{v}$ and ρ_0 need to be determined for the same conditions as for s and D (Comment D4.20).

Comment D4.20   Biologist's box: Calculation of molecular mass

A macromolecule with $\bar{v} = 0.74$ sediments in H_2O at 20 °C; its sedimentation coefficient $S_{20,w}^0$ is 14.2 S and its diffusion coefficient is $D_{20}^0 = 5.82 \times 10^{-7}$ cm^2
    According to Eq. (D4.20)

$$M = \frac{S_{20w}^0}{D_{20}^0} \frac{RT}{1 - \bar{v}\rho_0} = \frac{14.2 \times 10^{-13}(8.31 \times 10^7)293}{5.82 \times 10^{-7}(1 - 0.74 \times 0.998)} = 227 \text{ kDa}$$

In deriving Eq. (D4.20) we assumed that the frictional coefficients affecting diffusion, f_d and sedimentation, f_s, are identical. The assumption is strictly correct only from a theoretical point of view. In reality, f_d and f_s are measured under different experimental conditions: f_d, under the action of zero or weak forces, and f_s, in a high gravitational field. Strong gravitational forces may deform flexible molecules or orient asymmetric ones. Extrapolation of s to zero angular velocity eliminates this effect and we can consider the value of f_s obtained in this way as identical to f_d.

D4.7 Sedimentation equilibrium

D4.7.1 Molecular mass

What happens to a solution of macromolecules if it is centrifuged for a very long time? Large molecular mass solutes form a pellet at the bottom of the centrifugation cell, but macromolecules of medium or small molecular mass are distributed near the bottom of the cell in a way that does not change in time, because of compensation between gravitational and diffusion forces.

Comment D4.19
Molar mass

The use of SI units in the Svedberg equation (D4.20) yields a value of M technically referred to as the *molar mass*, with units of kilogram per mol.
    We recall (see Chapter B1) that the *relative molecular mass* or *molecular weight ratio* is a dimensionless relative quantity, defined as the ratio of the mass of a molecule relative to 1/12 of the mass of the carbon isotope ^{12}C (very close to 1g mol^{-1}).
    Molar mass (in kilogram per mol) can be converted to a molecular weight by dividing by 10^{-3} g mol^{-1} (the equivalent of multiplying by 1000 and cancelling units).
    Biochemists continue to use *molecular mass expressed in daltons* (Da) or atomic mass units (1 dalton = 1 atomic mass unit which equals one-twelfth of the mass of one atom of ^{12}C).

(a)

t: **24 h**

(b)

Absorbtion

5.8   5.9   6.0   6.1

Radius

Fig. D4.15 (a) Schematic representation of sedimentation equilibrium. The flow of macromolecules due to sedimentation increases with radial distance. This process is balanced at equilibrium by the reverse flow from diffusion, which increases with concentration gradient. (b) At equilibrium, the resulting concentration distribution is exponential with the square of the radial distribution. The molecular weight of the macromolecule is directly related to the concentration gradient at equilibrium according to Eq. (D4.23).

> **Comment D4.21   Equilibrium time**
>
> The time required to reach equilibrium depends on the square of the length of the solution column in the radial direction. If a 6-mm solution column attains equilibrium in 72 h, a 3-mm column will reach equilibrium in 24 h.

In a sedimentation equilibrium experiment, a volume of an initially uniform solution of macromolecules is centrifuged at a lower angular velocity than is required for sedimentation velocity experiments (Comment D4.21). As macromolecules begin to sediment, the process of diffusion opposes the gravitational force. After an appropriate period of time, an equilibrium is attained in which the macromolecular concentration increases exponentially towards the bottom of the cell, in the same way that gases in the atmosphere distribute exponentially in the earth's gravitational field (Fig. D4.15). The molecular mass of the sedimenting macromolecules can be calculated from the concentration values at different levels in the cell.

An equation describing the equilibrium sedimentation distribution is obtained by setting the total flux equal to zero in Eq. (D4.13), since, at equilibrium, there are no changes in concentration with time:

$$\left(\frac{dC}{dt}\right)r = -\frac{1}{r}\left\{\frac{d}{dr}\left[\omega^2 r^2 sC - Dr\left(\frac{dc}{dr}\right)t\right]\right\}t = 0 \qquad (D4.21)$$

Consequently

$$D(dC/dr)_t - \omega^2 rCs = 0$$

By rearranging and substituting D from the Svedberg equation (Eq. (D4.20)) we obtain

$$\frac{d\ln(C)}{d(r^2/2)} = \frac{1}{r}\frac{1}{C}\frac{dC}{dr}\frac{\omega^2 s}{D} = \frac{M(1 - \bar{v}\rho)\omega^2}{RT} \qquad (D4.22)$$

The macromolecular concentration distribution between the meniscus at a and point r obeys an exponential law, called the second Svedberg equation:

$$C(r) = C(a)\exp[\omega^2 M(1 - \bar{v}\rho)(r^2 - a^2)/2RT] \qquad (D4.23)$$

A plot of $\ln[C(r)]$ vs $r^2/2$ is presented in Fig. D4.16(a). In the case of a monodisperse solution, the slope of the straight line is proportional to the macromolecular mass. In a polydisperse solution, containing macromolecules of different molecular mass, each will be distributed according to Eq. (D4.23). The higher molecular masses concentrate selectively towards the bottom of the cell, while the lower molecular masses dominate the distribution at the top, and the plot of $\ln[C(r)]$ vs $r^2/2$ shows an upward curvature (Fig. D4.16(b)). The tangential slope

Fig. D4.16 (a) Plot of the logarithm of the equilibrium concentration versus the radius squared for a monodisperse macromolecule solution. The slope of this plot is proportional to the molecular weight of the macromolecule. (b) Plot of the logarithm of the equilibrium concentration versus the radius squared for a polydisperse macromolecule solution.

Comment D4.22   The average molecular mass

The average molecular mass calculated for a polydisperse solution is based on the proportion by number of the various components present. This type of average is known as a *number-average* molecular mass, M:

$$M_n = \sum N_i m_i \Big/ \sum N_i == \sum n_i m_i$$

where n_i is the number fraction of molecules with mass m_i.

The *weight-average* molecular mass, M_w is defined as

$$M_w = \sum N_i m_i^2 \Big/ \sum N_i m_i = \sum w_i m_i$$

where w_i is mass fraction of molecules with masses m_i.

For example, for a mixture of just two molecules with masses a 10 000 Da and 100 000 Da

$$M_n = 55\,000 \text{ Da}$$
$$M_w = 91\,818 \text{ Da}$$

---

at each point on the curve yields the average molecular mass at the corresponding point in the cell (Comment D4.22).

Since the derivation of Eq. (D4.23) did not rely on the molecular shape, the molecular masses obtained in this fashion are independent of this parameter. This is the power of the method. Macromolecular shape affects only the rate at which equilibrium is reached, not the final distribution.

Equation (D4.23) can be used to estimate the rotation speed needed for an equilibrium measurement. With a typical optical system, an accurate measurement can be made when the concentration falls by a factor of 2 across a 1 mm sample height. If the molecular mass of the particle is 50 000 Da, $a = 6$ cm, and $\bar{v} = 0.75$ cm^3 g^{-1}, then $\omega = 1000$ rad s^{-1} or 10 400 rpm.

Figure D4.17 shows optimum speeds for an equilibrium run if either the molecular weight or the sedimentation coefficient of the sample can be estimated.

Fig. D4.17 Optimum
speeds for equilibrium
runs if either the
molecular weight or the
sedimentation coefficient
can be estimated. (After
Chervenka, 1969.)

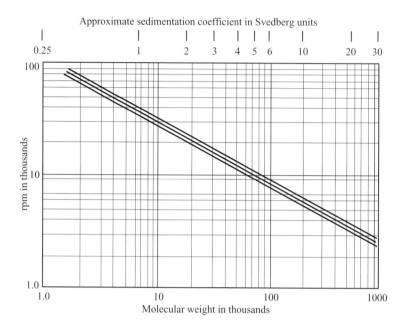

A set of at least three speeds is chosen to yield significantly varied data for diagnostics.

D4.7.2 Binding constants

Measuring the concentration dependence of an effective average molecular mass (for example, to determine the amount of monomers and oligomers) is a highly useful way to describe different kinds of association phenomenon (see also Chapter C1).

For a heterogeneous associating system of the form $A + B \rightleftarrows AB$ the equilibrium sedimentation distribution is described by three exponentials, one for each molecular mass species (the exponentials are written in condensed form as σ_i):

$$C(r) = C_A(r)\sigma_A + C_B(r)\sigma_B + C_{AB}(r)\sigma_{AB} \tag{D4.24}$$

The pre-exponential term for the AB association can be recast in terms of the concentrations of A, B, and the dissociation equilibrium constant, K_{AB}

$$C(r) = C_A(r)\sigma_A + C_B(r)\sigma_B + \frac{C_A C_B}{K_{AB}}\sigma_{AB} \tag{D4.25}$$

and from the law of mass action

$$K_{AB} = \frac{C(r)_A C(r)_B}{C(r)_{AB}} \quad \text{and} \quad C(r)_{AB} = \frac{C(r)_A C(r)_B}{K_{AB}} \tag{D4.26}$$

The dissociation equilibrium constant may therefore be determined directly from the equilibrium sedimentation data by applying Eq. (D4.26).

Fig. D4.18 Equilibrium sedimentation data for a mutant V_L (immunoglobulin variable light chain) and domain of REI (a Bence–Jones protein) analysed in terms of a monomer ↔ dimer ↔ tetramer model (see text). The thickened portions of the curves in (c) indicate the concentration range where the analysis was performed. The thin portions of the curves were obtained by extrapolation (After Hensley, 1996.)

Fig. D4.19 Concentration dependence of molecule weight, M_w, for carbonic anhydrase from *Methanosarcina thermophila*. The solid line describes a trimer–hexamer equilibrium. The data are in agreement with crystallographic data demonstrating that in the crystal two trimers dimerise to form a hexamer by association of the N-termini of six monomers into a six-stranded β-structure with a hydrophobic core. The dashed line was calculated for a dimer–tetramer equilibrium. (After Belhke and Ristau, 1997.)

As an example, Fig. D4.18 shows the data and analysis for two associating proteins forming a monomer⇌dimer⇌tetramer reversibly associating system. The values determined for $K_{1,2}$ and $K_{2,4}$ were 8.3×10^{-6} M and 2.0×10^{-6} M, respectively. Figure D4.18(b) shows the fit as a sum of three exponentials and Fig. D4.18(a) shows the distribution of residuals. The computed relative concentrations of monomer, dimer and tetramer are plotted as a function of total monomer concentration in Fig. D4.18(c). Note the wide concentration range (10^{-3}–10^{-8} M) that can be analysed usefully in a sedimentation equilibrium experiment.

In a second example, Fig. D4.19 shows the concentration dependence of molecular weight, M_w, for carbonic anhydrase in a range covering about three orders of magnitude. The solid line describes a trimer–hexamer equilibrium, and the dashed line a dimer–trimer equilibrium. It is clearly seen that the data preclude a dimer–tetramer equilibrium (dashed line).

D4.8 The partial specific volume

Determination of the molecular mass of a macromolecule by using sedimentation equilibrium (Section D4.6) or by combining sedimentation and diffusion data (Section D4.7) requires an accurate knowledge of its partial specific volume, $\bar{v}$. For a simple two-component system containing one type of solute dissolved in one type of solvent (in our case buffer), $\bar{v}$ is the thermodynamically defined partial specific volume of the solute, which is related to the volume increase of the solution when adding a given quantity of dry macromolecules (see Chapter D1).

The partial specific volume is usually determined from a series of precise density measurements on solutions containing a different weight concentration, w, of solute:

$$\rho = \rho_0 + w(1 - \rho_0\bar{v}) \tag{D4.27}$$

where ρ and ρ_0 are the densities of the solution and solvent, respectively. The values of $\bar{v}$ are obtained from the slope of ρ plotted as a function of w. In the case of a dilute macromolecular solution, $\bar{v}$ is independent of the macromolecular concentration and the plot is a straight line. Partial specific volume, $\bar{v}$, is usually expressed in units of cubic centimetre per gram.

Inspection of Eq. (D4.27) shows that extremely precise density measurements are needed to obtain accurate values of $\bar{v}$. In order to obtain 1% accuracy in $\bar{v}$ (required for 3–4% accuracy in M in the case of protein), density values must be more precise than 1 part in 10^5 at a solute concentration of 1 mg ml^{-1}. It is possible but difficult to obtain the necessary precision with most of the standard techniques, which include pycnometry, magnetic float devices, electrobalance with plummet instruments, vibrating tube densitometers and AUC in solvents with different isotope compositions. We must keep in mind, however, that such measurements are very tedious and difficult. Fortunately, there is a way to overcome the difficulty, at least for proteins and carbohydrates. The partial specific volumes of proteins or carbohydrates can be calculated with adequate precision from the amino acid or monosaccharide composition, respectively, on the assumption of additivity of residue volumes,

$$\bar{v} = \sum_i \bar{v}_i W_i \Big/ \sum_i W_i \tag{D4.28}$$

where W_i and $\bar{v}_i$ are the weight fraction and partial specific volume, respectively, of residue i. Residue volumes are given in Table D4.5 for amino acids, and in Table D4.6 for carbohydrates.

Table D4.5. *Partial specific volumes at 25 °C, $\bar{v}$, and molecular mass, M, of amino acid residues*[a]

Amino acid	M	$\bar{v}$ (cm^3 g^{-1})
Glycine	57	0.64
Alanine	71	0.74
Serine	87	0.63
Threonine	101	0.70
Valine	99	0.86
Leucine	113	0.90
Isoleucine	113	0.90
Proline	97	0.76
Methionine	131	0.75
Phenylalanine	147	0.77
Cystine	103	0.61
Tryptophan	186	0.74
Tyrosine	163	0.71
Histidine	137	0.67
Arginine	156	0.70
Lysine	128	0.82
Aspartic acid	115	0.60
Glutamic acid	129	0.66
Glutamine	128	0.67
Asparagine	115	0.60

[a]Data are taken from Cohn and Edsall (1965).

Table D4.6. *Partial specific volumes at 20 °C, $\bar{v}$, and molecular mass, M, of carbohydrates*[a]

Carbohydrate	M	$\bar{v}$ (cm^3 g^{-1})
Fructose	180	0.614
Fucose	164	0.671
Galactose	180	0.622
Glucose	180	0.622
Hexose	180	0.613
Sucrose (0.05 M)	342	0.613
Sucrose (0.15 M)	342	0.616
Sucrose (1 M)	342	0.620
Raffinose (25 °C)	486	0.608

[a] Data are taken from Chervenka (1969).

Comment D4.23
Prediction of $\bar{v}$

The partial specific
volume is now
routinely calculated
from the chemical
composition of
proteins with the
program SEDNTEP
(Harpaz *et al.*, 1994).

Table D4.7. *Approximate values of partial specific
volume $\bar{v}$ for the main types of biological macromolecule*

Macromolecules	$\bar{v}$	(cm^3 g^{-1})
Proteins	0.73	0.70–0.75
Carbohydrates	0.61	0.59–0.65
RNA	0.54	0.47–0.55
DNA	0.57	0.55–0.59

Careful studies of the problem have shown that the initial approach gives the closest agreement with experiment, while later attempts to refine the method, for example by attributing different volumes to amino acid residues according to whether they are found in the core or on the surface of the molecule only made matters worse (Comment D4.23). Typical $\bar{v}$ values for the main types of biological macromolecule are listed in Table D4.7.

D4.9 Density gradient sedimentation

Density gradient sedimentation in AUC has a long history in the study of nucleic acids and proteins, and is still applied in a variety of studies, with topics ranging from the composition of genomes to the characterisation of protein–detergent, protein–nucleic acids, and protein–lipid complexes.

Below, we describe the classical approaches based on *analytical zonal sedimentation velocity* and *density gradient sedimentation equilibrium*, in which macromolecules move through concentration density gradients of smaller cosolutes (Comment D4.24).

D4.9.1 Velocity zonal method

A sample solution containing a mixture of particles is layered on a preformed gradient of increasing density (Fig. D4.20(a)). Upon centrifugation, the particles sediment through the gradient to separate zones based on their sedimentation velocity. The centrifugation is terminated before particles in the separate zones reach the bottom of the tube (Fig. D4.20(b)). The maximum density of the gradient should be chosen not to exceed that of the densest particle to be separated. The density and viscosity of the gradient medium and the velocity of centrifugation also affect the separation. Sucrose and glycerol solutions are the most frequently used as a non-ionic gradient material for velocity zonal centrifugation. Linear 5–20% sucrose gradients are a traditional choice for the determination of the sedimentation coefficients of protein and nucleoprotein particles. Velocity zonal centrifugation separates molecules in the mixture according to their sedimentation coefficients, which are determined by their size, shape and buoyant

Comment D4.24
Dynamic density gradient

A new *dynamic density gradient from sedimenting cosolute* method was proposed recently. It is based on a numerical solution of the Lamm equation with spatial and temporal variations of the local solvent density and viscosity (Schuck, 2004).

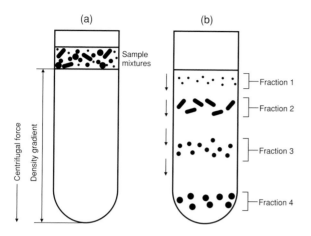

Fig. D4.20 Velocity zonal separation of a mixture of particles using a density gradient. (a) A sample containing mixtures of particles of varying size, shape and density is added on the top of a preformed density gradient. The gradient is higher in density toward the bottom of the tube. (b) Centrifugation results in separation of the particles depending on their size, shape and buoyant density. For globular proteins, those with larger molecular mass are likely to sediment to positions toward the bottom of the gradient, compared to those with smaller molecular mass that remain near the top of the gradient. Fractions of defined volume are collected from the gradient.

Comment D4.25   Sedimentation coefficients from velocity zonal centrifugation

Sedimentation coefficients obtained by velocity zonal centrifugation in a sucrose gradient may be erroneous for any protein undergoing reversible self-association or interaction with other macromolecules. The source of error comes from the molecular crowding effect due to high sucrose concentration that shifts an equilibrium to favour polymeric states of the macromolecule. In this case the use of classical AUC analytical ultracentrifugation methods is strongly recommended for the determination of sedimentation coefficients.

---

density. The method also allows the estimation of relative molecular masses (Comment D4.25).

To illustrate the method, Fig. D4.21 shows the separation of 22 S RNP particles by centrifugation in a sucrose concentration gradient. Peak I corresponds to the 22 S RNP.

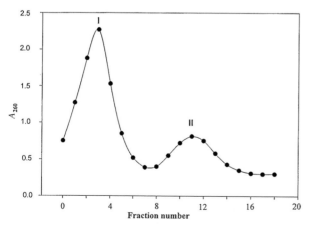

Fig. D4.21 Isolation of the 22 S RNP particles by centrifugation in a 5–20% sucrose gradient at 50 000 rpm for 3 h at 20 °C in an SW 55 rotor. (After Ulitin *et al.*, 1997.)

D4.9.2 Equilibrium sedimentation in a density gradient

The large forces that can be generated in AUC have been used to great advantage in a sedimentation–equilibrium method in which the macromolecules are spun through a small-solute density gradient, which is itself formed by the centrifugal field. Such a gradient may be obtained by spinning a solution of a heavy salt, such as CsCl or RbCl, until equilibrium is reached.

Consider a three-component molecular solution: solvent molecules (component 1), the macromolecules under study (component 2), and small solutes (component 3). According to Eq. (D4.23) the small solute is distributed in the cell in just the same way as a large molecule. Because of the small molecular mass of component 3, we can expand the exponential in Eq. (D4.23) and only keep the first terms

$$C_3(r)/C_3(a) = 1 + M_3(1 - \bar{\upsilon}\rho)\omega^2(r^2 - a^2)/2RT \qquad \text{(D4.29)}$$

Equation (D4.29) shows that a parabolic concentration distribution of component 3 is produced in the sample cell. The density of the solvent is an approximately linear function of the concentration of component 3 so that, at equilibrium, a parabolic density gradient is established (Fig. D4.22). If the density, ρ, of component 2 falls between the two extreme density values of the solvent, the macromolecules tend to gather in the region of the cell where their buoyant density is zero, i.e. $\rho = \rho_0$. The width of the component 2 band is a function of the macromolecular mass and of the steepness of the local density gradient. The bands may be very narrow for large masses, because of their small diffusion coefficients. Native DNA, for example, produces such a narrow band that DNA containing ^{15}N ('heavy nitrogen') is clearly separated from the natural abundance, ^{14}N ('light' nitrogen), macromolecule by equilibrium centrifugation in a caesium chloride

Fig. D4.22 Equilibrium ultracentrifugation of DNA in a CsCl gradient: (a) the density gradient distribution at the beginning of the experiment; (b) the density gradient at equilibrium. (After Cantor and Shimmel, 1980.)

 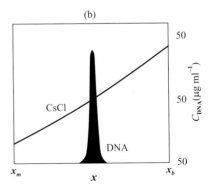

density gradient (Comment D4.26). Based on this fact one of the *gold medal experiments* in structural molecular biology was performed on the mechanism of DNA replication by Messelson and Stahl in 1957.

The most striking feature of the Watson–Crick double-helix model was its postulate that the two DNA strands are complementary, and that replication is based on new complementary strands forming on each strand. The result is two daughter duplex DNA molecules, each of which contains one strand from the parental DNA. The process is called *semiconsevative* replication and is shown in Fig. D4.23.

Messelson and Stahl grew *E.coli* cells in a medium in which the sole nitrogen source was ^{15}N-labelled ammonium chloride. The ^{15}N-containing *E.coli* cell culture was then transferred to a light ^{14}N medium and allowed to continue growing. Samples were harvested at regular intervals. The DNA was extracted and its buoyant density determined by centrifugation in CsCl density gradients. The isolated DNA showed a single band in the density gradient, midway between the light ^{14}N-DNA and the heavy ^{15}N-DNA bands (Fig. D4.24(c)). After two generations in the ^{14}N medium the isolated DNA exhibited two bands, one with a density equal to light DNA and the other with a density equal to that of the hybrid DNA observed after one generation (Fig. D4.24(d)). After three generations in the ^{14}N medium the DNA still has two bands, similar to those observed after two generations (Fig. D4.24(e)). The results were exactly those expected from the semiconservative replication hypothesis.

Comment D4.26
^{15}N ('heavy') and ^{14}N ('light') isotopes

At sedimentation equilibrium in CsCl, the difference in density of DNA molecules containing the ^{15}N isotope and the ^{14}N isotope is equal to 0.014 g cm^{-3}. As a result the difference between bands is equal to 0.5 mm in a standard ultracentrifuge cell at 40 000 rpm.

Parent

F$_1$

F$_2$

Fig. D4.23 The semiconservative mechanism of DNA replication. Each F$_1$ duplex contains one parent strand. The F$_2$ generation consists of two hybrid DNAs and two totally new DNAs. Newly replicated strands are shown in grey; parental strands in black.

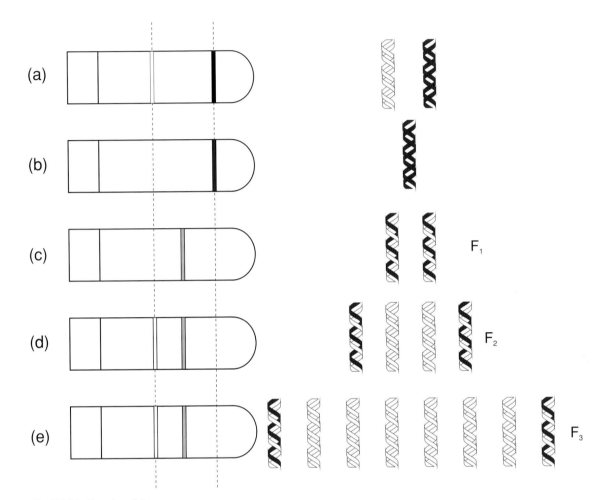

Results of the
Messelson–Stahl
experiment: (a) control
experiment, a mixture of
heavy (^{15}N) and light
(^{14}N) DNAs; (b) parent
DNA is the heavy DNA; (c)
DNA after one division in
the ^{14}N medium; (d) DNA
after two divisions in ^{14}N;
(e) DNA after three
divisions in ^{14}N.

D4.10 Molecular shape from sedimentation data

We recall that Eq. (D4.19) relates the sedimentation coefficient, s, to the molecular
properties, M, D and $\bar{v}$:

$$s = \frac{MD(1 - \bar{v}\rho_0)}{RT} = \frac{MD(1 - \bar{v}\rho_0)}{s} \tag{D4.30}$$

The sedimentation coefficient (itself related to the translational friction coefficient) depends on particle volume and shape. It is usually expressed in terms
of the radius of an equivalent sphere (the Stokes radius) but can equally well be
accounted for by different volume–shape combinations – for example, an elongated particle of smaller volume than the Stokes radius sphere (Section D2.2.1).
It follows from Eq. (D2.2) that the direct determination of molecular shape from
the sedimentation coefficient is not possible. The sedimentation coefficient can

---

Comment D4.27   Kuhn–Mark–Houwink relations (see Section D2.6)

For a homologous series of particles the following *KMX* relations are valid:

$$[\eta] = K_{[\eta]} M^{\alpha} \tag{A}$$
$$s = K_s M^a \tag{B}$$
$$D = K_D M^{-b} \tag{C}$$

The relations between constants α, a and b are:

$$b + a = 1 \tag{D}$$
$$b = (1 + \alpha)/3 \tag{E}$$

Thus, for a homologous series of rigid rod-like particles:

$$\alpha = 1.7, a = 0.15, b = 0.85$$

for a homologous series of spherical particles:

$$\alpha = 0, a = 2/3, b = 1/3$$

for homologous series of Gaussian coils in an 'ideal' solvent:

$$\alpha = 0.5, a = 0.5, b = 0.5$$

and for homologous series of Gaussian coils in 'good' solvent:

$$\alpha = 0.8, a = 0.4, b = 0.6$$

---

be explained, for example, in terms of either a spherical particle of definite $\bar{v}$ and total volume, or an elongated particle of the same $\bar{v}$ and smaller total volume.

Molecules of the same shape but different molecular mass form a so-called homologous series (see Section D2.6). In this case, the relation between M and s follows from Eq. (D2.64)

$$s = K_s M^a \tag{D4.31}$$

where K_s is a coefficient that depends on the partial specific volume. The value of the exponent, a, depends on the *shape* of the molecules in the series (Section D2.6). The usefulness of such an analysis is illustrated below for different homologous series (see also Comment D4.27).

D4.10.1 Homologous series of quasi-spherical particles: globular proteins in water

We recall (Section D2.2.1) that the frictional coefficient of a sphere of radius R_0 under slip boundary conditions in a solvent of viscosity η_0 is given by Stokes',

Fig. D4.25 The log–log
plot of the dependence
$s^0_{20,w}\bar{v}^{1/3}/(1-\bar{v}\rho_0)$ on M
for proteins in a wide
mass interval using the
data presented in
Table D4.3. For globular
proteins plot the slope of
this is 2/3, which is
typical for quasi-spherical
particles. Exceptions to
the rule are: synthetic
peptides (unfolded
particles), tropomyosin,
fibrinogen, myosin
(rod-like particles).

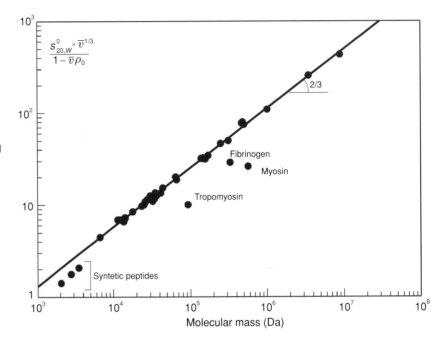

Fig. D4.25 The log–log plot of the dependence $s^0_{20,w}\bar{v}^{1/3}/(1-\bar{v}\rho_0)$ on M for proteins in a wide mass interval using the data presented in Table D4.3. For globular proteins plot the slope of this is 2/3, which is typical for quasi-spherical particles. Exceptions to the rule are: synthetic peptides (unfolded particles), tropomyosin, fibrinogen, myosin (rod-like particles).

law (Eq. D2.1):

$$f_0 = 6\pi\eta_0 R_0$$

The radius of the sphere is related to its mass via its volume, V_0

$$\frac{4}{3}\pi R_0^3 = V_0 = M\bar{v}/N_A \tag{D4.32}$$

and we can write

$$f_0 = 6\pi\eta_0(3M\bar{v}/4\pi N_A)^{1/3} \tag{D4.33}$$

Inserting this result into Eq. (D4.19), we obtain, after rearrangement,

$$s^* = s^0_{20,w}\bar{v}^{1/3}/(1-\bar{v}\rho) = M^{2/3}/6\pi\eta_0 N_A^{2/3}(3/4\pi)^{1/3} \tag{D4.34}$$

The log–log dependence of s^* on M is a straight line of slope 2/3. When $s^0_{20,w}$ is expressed in Svedberg units and all other quantities in cgs units, Eq. (D4.34) becomes

$$s^* = 12.0 \times 10^{-3} M^{2/3} \tag{D4.35}$$

We now explore to what extent globular proteins behave as a homologous series of spherical particles. A log–log plot of s^* as a function of M is shown in Fig. D4.25 for globular proteins with a wide range of molecular masses using the data presented in Table D4.3. A good straight line fit of $\log s^*$ versus $\log M$ plot is obtained according to a single equation:

$$s^* = 9.1 \times 10^{-3} M^{0.65} \qquad\qquad (D4.36)$$

which establishes that globular proteins do in fact form a homologous series. Comparison of Eqs. (D4.35) and (D4.36) shows that though the exponents in these equations coincide, the coefficients are different. Small deviations from perfect spherical shapes and the existence of hydration shells modify the relation between the Stokes radius and the partial specific volume assumed in Eq. (D2.1), without changing the power law. We conclude from the fact that their sedimentation behaviour can be described by a single equation, Eq. (D4.36), that globular proteins are very close to spherical in shape, and hydrated to about the same degree.

Table D4.3 includes three proteins (tropomyosin, fibrinogen, myosin) that do not obey Eq. (D4.27). Their sedimentation coefficients are smaller than that of a globular protein of the same molecular mass (Fig. D4.25). There are two possible causes for such behaviour: either the molecule has an elongated (rod-like) shape or it is unfolded. The shape of a non-globular protein can be derived *directly* from sedimentation measurements if the molecule can be broken up into fragments that themselves form a homologous series. Collagen is a good example of such an analysis (Comment D4.28).

Comment D4.28   Collagen and its fragments

Sonic irradiation of soluble calf-skin collagen causes fragmentation of the macromolecules into shorter pieces that retain the three-stranded, helical structure. The dependence of the sedimentation coefficient on molecular mass is small and can be written as

$$s_{20,w}^{0} = 0.232 M^{0.20} \qquad\qquad (A)$$

The small value of the exponent in Eq. (A) is a solid argument in favour of the fact that the collagen and its fraction form a homologous series of solid rod-like molecules, for which the asymmetry grows 'head to tail'.

Fig. D4.26 A log–log plot showing the dependence of $s_{20,w}^{0}/(1 - \Phi' \rho)$ on n, where n is the number of amino acid residues and Φ' is the apparent partial specific volume for proteins in 6 M guanidine hydrochloride. (After Tanford *et al.*, 1967.)

D4.10.2 Homologous series of random coils: proteins in guanidine hydrochloride

Most proteins are dissociated into single polypeptide chains and unfolded in 6 M guanidine hydrochloride (GuHCl) or 8 M urea containing 0.1 M mercaptoethanol (EtSH). Figure D4.26 shows the dependence of $s_{20,w}^{0}/(1 - \Phi' \rho)$ on n, where n is

Comment D4.29
**'Ideal' and 'good'
solvents in
polymer chemistry**

The term 'ideal' or
'theta' solvent in
polymer chemistry
means a solvent in
which the second virial
coefficient is equal to
zero. The term 'good'
solvent means a
solvent in which the
second virial
coefficient is positive.

the number of amino acid residues in a polypeptide chain and Φ' is the apparent partial specific volume for proteins in 6 M GuHCl. The plot is a good straight line that follows the equation

$$s^0/(1 - \bar{\upsilon}\rho) = 0.286n^{0.47} \qquad (D4.37)$$

The sedimentation coefficient varies with molecular mass in the manner expected for random coil polymers in a solvent close to 'ideal' (Comment D4.29).

D4.10.3 From slightly flexible rod to nearly perfect random coil: DNA

The study of DNA molecules is a good example of how to extract a molecular shape from sedimentation data. It is well known from early studies that the log–log dependence of sedimentation data on molecular weight for DNA cannot be described by *one* straight line in the full molecular weight interval.

Now that homogeneous fragments of double-stranded DNA can be prepared easily and their precise molecular weight calculated from the sequence, we can reconsider in greater detail the sedimentation coefficient dependence on molecular weight (Fig. D4.27). The new set of data obtained confirms that the log–log dependence is not a straight line, but has an inflection near a molecular weight of 3.5 MDa (Fig. D4.27). For fragments ranging from 170 kbp (molecular weight $\sim$112 MDa) to nearly 5.4 kbp (molecular weight $\sim$3.5 MDa) the dependence is described by

$$s = K_s M^{0.440} \qquad (D4.38)$$

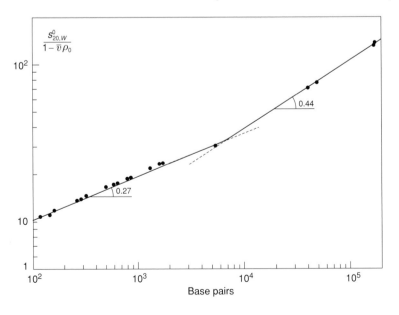

Fig. D4.27 Dependence $s^0_{20,w}$ on molecular mass (in base pairs) corrected for the buoyancy factor $(1 - \bar{\upsilon}\rho_0)$ for DNA fragments. Data are taken from Table D4.4.

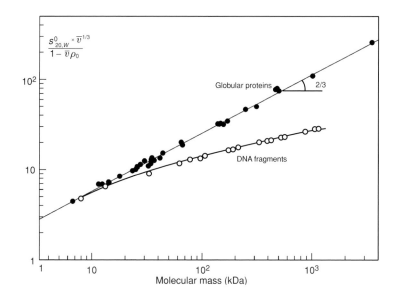

Fig. D4.28 Dependence of log $s^0_{20,w}$ on the logarithm of molecular mass, corrected for the buoyancy factor $(1 - \bar{v}\rho_0)$ for globular proteins (•) and for the PM2 *Hae* III fragments of DNA (○) in the small molecular mass range. Data are taken from Tables D4.2 and D4.3. Sedimentation coefficients for 12 bp and 20 bp DNA fragments are calculated from experimental values for their diffusion coefficients: 13.4 and 10.9 Fick units, respectively, assuming a partial specific volume $\bar{v} = 0.56$ cm^3 g^{-1}.

while in the range 200–5400 bp, it is described by

$$s = K_s M^{0.273} \tag{D4.39}$$

Comparing the new exponent values (Eqs. (D4.38) and (D4.39)) with the early ones reveals a clearer definition of the random coil properties of large molecular weight DNA (a increases from 0.40_5 to 0.44_0), and a more pronounced degree of chain stiffness in the low-molecular-weight range (a decreases from 0.32_5 to 0.27_3).

It is interesting to note that the frictional properties of short DNA fragments (12–20 bp) are similar to those of globular proteins of corresponding molecular weight (Fig. D4.28). At the same time these properties are very different for large molecular weight. This fact reflects the principal difference between these two homologous series. In the globular protein series, asymmetry remains constant with increasing molecular weight, whereas in the DNA fragment series, asymmetry grows series.

D4.10.4 Ribosomal RNAs, ribosomal particles and RNP complexes

Figure D4.29 shows log–log plot dependence of $s^0_{20,w}/(1 - \bar{v}\rho)$ for ribosomal RNAs and their fragments (Comment D4.30) in buffer without Mg^{2+} ions using

Fig. D4.29 Dependence of the logarithm of the sedimentation coefficient corrected for the buoyancy factor, $(1 - \bar{v}\rho_0)$ versus the logarithm of the molecular mass for different ribosomal RNAs (•) and their fragments (○) using the data presented in Table D4.8.

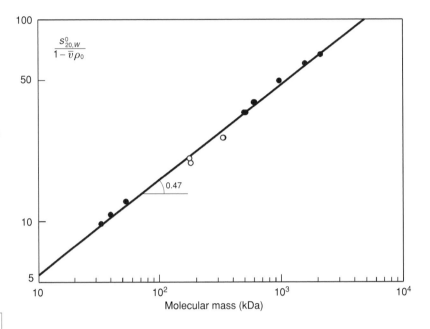

Comment D4.30

Fragments of 16S rRNA

Fragments of the 16 S rRNA of *T. thermophilus* representing the 3′ domain (nucleotides 890–1515), 5′ domain (nucleotides 1–539), and the central domain (nucleotides 547–895) were prepared by transcription in vitro. Incubation of those fragments with a total of 30 S ribosomal proteins of *T. thermophilus* resulted in the formation of the specific ribonucleoprotein complexes (Agalarov *et al.*, 1999).

Fig. D4.30 (a) Three major domains of 16 S RNA of the small ribosomal subunit. The 5′ domain (nucleotides 1–563) corresponds to the central body of the 30 S subunit, the central domain (nucleotides 564–915) is involved in the formation of the side bulge or platform, whereas the 3′ major domain (nucleotides 919–1396) is localised in the head region of the subunit; there is also an additional 3′ minor domain (residues 1397). (Ramakrishnan and Moore, 2001.) (b) The 'beheaded' 30 S ribosomal subunit correspond to the 5′ end and central domains of the 16 S RNA from *T. thermophilus*. The procedure is based on oligodeoxyribonucleotide-directed cleavage of protein-deficient ribonucleoprotein derivatives of the 30 S ribosomal subunits with ribonuclease H. (Serdyuk, *et al.*, 1999.)

Table D4.8. *Ribosomal RNAs and their fragments in buffer containing 100 mM KCl without Mg ions* $(\bar{v} = 0.54 \ cm^3 \ g^{-1})$

Particle	Molecular mass (in bases)	Sedimentation coefficient $s_{20,w}^0$ (Svedberg units)	References
5 S RNA of chloroplast	100	4.5	1
5 S RNA of *E. coli*	120	5.0	1
5 S RNA of rat	160	5.8	1
3'-fragment (1000–1542) of 16 S RNA *E. coli*	542	9.0	2
5'-fragment (1–530) of 16 S RNA *E. coli*	530	9.5	2
3'-fragment (535–1542) of 16 S RNA *E.coli*	1007	12.0	2
5'-fragment (1–996) of 16 S RNA *E.coli*	996	12.0	2
16 S RNA of *E.coli*	1542	16.0	1
16 S RNA of Chloroplast	1490	16.0	1
16 S RNA of *T. thermophilus*	1515	16.0	1
18 S of yeast	1789	18.0	1
18 S RNA of yeast	1825	18.0	1
23 S RNA of *E.coli*	2904	23.0	1
28 S RNA of rat	4750	28.0	1
RNA of TMV	6363	31.0	1

Reference 1: data from Spirin (1999).
Reference 2: data from Serdyuk (unpublished).

the data presented in Table D4.8. The straight line follows the equation

$$s_{20,w}^0/(1 - \bar{v}\rho) = K''M^{0.47} \qquad (D4.40)$$

with a slope equal to 0.47, which is close to the slope of the straight line for proteins in 6 M GuHCl (Eq. (D4.37)). Such a slope is typical for Gaussian coils in solvent close to 'ideal' (Section D2.6).

Figure D4.29 also shows data for ribosomal particles and their subparticles using the data presented in Table D4.9. The straight line follows the equation

$$s_{20,w}^0/(1 - \bar{v}\rho) = K'M^{0.66} \qquad (D4.41)$$

with a slope equal to 0.66, which is typical for compact quasispherical particles (Section D2.6). Using the same approach as above (Section D4.9.1), we note that the exponents in Eqs. (D4.40) and (D4.41) coincide, but the coefficients are different. We conclude that globular proteins and ribosomal particles have similar compact shapes but different hydration.

Table D4.9. *Ribosomal particles, subparticles and RNP complexes*

Particle	Molecular mass (kDa)	Sedimentation coefficient $s_{20,w}^0$ (Svedberg units)	Partial specific volume $\bar{v}$ (cm³ g⁻¹)	$\dfrac{s_{20,w}^o}{(1-\bar{v}\rho)}$	References
T4 RNP complex of *E.coli*	70	5.8	0.63	15.7	1
T3 RNP complex of *E. coli*	93	6.1	0.64	16.9	1
Central domain of 30S *T. th.* (547–895)	195	12.0	0.61	30.8	2
5′-domain of *T. th.* (1–539)	242	13.0	0.58	31.0	3
3′ domain of *T. th.* (890–1373)	280	15.4	0.61	39.5	4
'Beheaded' 30 S of *T. th.*	427	22S	0.59	53.7	5
30 S *E. coli*	910	30.8	0.612	79.4	5
50 S *E. coli*	1550	50.0	0.597	124	5
70 S *E. coli*	2650	70.0	0.610	179	5
80 S from yeast	3700	80.0	0.630	216	5
80 S from *Artemia salina*	3800	81.0	0.630	219	6
80 S from *Eugena*	4400	86.0	0.630	232	6

Data are taken from:
1 Agalarov and Williamson (2000).
2 Agalarov *et al.*(1999).
3 Agalarov *et al.* (1998).
4 Selivanova *et al.* (unpublished result).
5 Ulitin *et al.* (1997).
6 Serdyuk *et al.* (1999).

A comparison of the sedimentation characteristics of RNP complexes, modelling the three major domains (Fig. D4.30), with those of the original ribosomal particles indicates that the complexes have the same compactness as the original particles. This follows from the fact that all RNP complexes lie on the continuation of the straight line formed by ribosomal units and subunits in the log–log plot of sedimentation versus molecular weight (Fig. D4.31). The result demonstrates clearly that each of three major parts (domains) of the 30 S subunit from *T. thermophilus* are capable of self-assembly in vitro relatively independently of each other and that the 22S RNP complex can be considered as a compact 'beheaded' derivative of the 30 S ribosomal subunit.

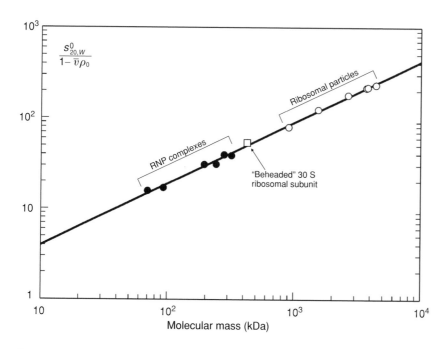

Fig. D4.31 Dependence of the logarithm of the sedimentation coefficient corrected for the buoyancy factor $(1 - \bar{v}\rho_0)$ versus the logarithm of the molecular mass for different ribosomal particles and their subparticles (○), and RNP particles (●) using the data presented in Table D4.9.

D4.11 Checklist of key ideas

- The introduction in the 1990s of the new Beckman Instruments Optima XL ultracentrifuge opened a new epoch in AUC.
- Two complementary methods can be applied in an analytical ultracentrifuge: sedimentation velocity and sedimentation equilibrium.
- Sedimentation velocity provides information about the size, shape and interactions between molecules; a high rotor speed and a long solution column are used to maximise the resolution.
- Sedimentation equilibrium provides information about the molecular mass, association constants and stoichiometries; low rotor speeds and short solution columns are used to minimise the time needed to reach equilibrium.
- The ratio of the measured velocity of the particle to its centrifugal acceleration is called the sedimentation coefficient, s. (1 S (Svedberg unit) $= 10^{-13}$ s.)
- The Lamm equation describes the behaviour of a macromoleculecular solution in the ultracentrifuge field.
- The exact solution of the Lamm equation is an infinite series in terms of integrals, which can be computed only by numerical integration.
- The application of modern computer facilities enables numerical solutions for the Lamm equation and approximate analytical solutions to be proposed under given boundary conditions.
- The partial specific volume of a protein can be calculated very accurately from its amino acid composition, according to the Cohn and Edsall method.

- The molecular mass of a macromolecule can be determined by a combination of sedimentation and diffusion coefficients, according to Svedberg's equation, or directly by the sedimentation equilibrium method.
- The slope of the dependence of the sedimentation constant on molecular mass, on a double logarithmic scale, is related to shape for a homologous series of macromolecules.
- Globular proteins in aqueous solvent form a homologous series of quasi-spherical particles with exponent a equal to $2/3$ in the dependence of s on M.
- Proteins in GuHCl form a homologous series of random coils with exponent a equal to 0.47.
- Collagen molecules and products of their sonic fragmentation form a homologous series of rod-like particles with exponent a equal to 0.2.
- The dependence of s on M for DNA molecules is described by two exponents: $a_1 = 0.273$ in the molecular weight range $1.3 \times 10^4 < M < 3.5 \times 10^6$, and $a_2 = 0.440$ for large molecular weights (beyond 3.5×10^6).
- Ribosome particles and subparticles, the 'beheaded' 30 S subunit, and RNP complexes, modelling the three major domains of 30 S subunits, form a homologous series of quasi-spherical particles with exponent a equal to $2/3$.
- Ribosomal RNAs and their fragments in buffer without Mg ions form a homologous series of random coils with exponent a equal to 0.47.
- Density-gradient equilibrium ultracentrifugation is a powerful technique for separating macromolecules according to their buoyant densities.
- The technique of differential sedimentation allows the measurement of very small differences in sedimentation coefficients.

Suggestions for further reading

Historical review

Schachman, H. K. (1989). Analytical ultracentrifugation reborn. *Nature*, **341**, 259–260.

Schachman, H. K. (1992). Is there a future for the ultracentrifuge? In *Analytical Ultracentrifugation in Biochemistry and Polymer Science*, eds. S. E.Harding, A. J. Rowe and J. C. Horton. Cambridge: Royal Society of Chemistry.

Instrumentation and innovative techniques

Schachman, H. K. (1959). *Ultracentrifugation in Biochemistry*. New York: Academic Press.

Ralston, G. (1993). *Introduction to Analytical Ultracentrifugation.*Fullerton, CA: Beckman Instruments.

Geibeler, R. (1992). The Optima XL-A: A new analytical ultracentrifuge with a novel precision absorption optical system. In *Analytical Ultracentrifugation in Biochemistry and Polymer Science*, eds. S. E. Harding, A. J. Rowe and J. C. Horton. Cambridge: Royal Society of Chemistry,

Van Holde, K. E., Johnson, W. C., and Ho, S. P. (1998). *Principles of Physical Biochemistry*. New Jersey: Prentice Hall.

Velocity sedimentation

Van Holde, K. E. (1975). Sedimentation analysis of proteins. In *The Proteins*, third edition., eds. H. Neurath and R. I. Hill, Volume VI, pp. 225–291. New York: Academic Press.

Perkins, S. J. (1986). Protein volumes and hydration effects: the calculation of partial specific volumes, neutron scattering matchpoints and 280 nm absorption coefficients for proteins and glycoproteins from amino acid sequences. *Eur. J. Biochem.*, **157**, 169–180.

Harding, S. E., Rowe, A. J., and Horton, J. C. (eds.) (1992). *Analytical Ultracentrifugation in Biochemistry and Polymer Science*. Cambridge: The Royal Society of Chemistry.

Stafford, W. F., III (1994). Boundary analysis in sedimentation velocity experiments. *Methods Enzymol.*, **240**, 478–501.

Harding, S. E. (1994). Determination of macromolecular homogeneity, shape, and interactions using sedimentation velocity analytical centrifugation. In *Methods in Molecular Biology*, eds. C. Jones, B. Mulloy and S. Thomas, Volume 22. *Microscopy, Optical Spectroscopy, and Macroscopic Techniques*. Totowa, NJ: Humana Press.

Hansen, J. C., Lebowitz, J., and Demeler, B. (1994). Analytical utracentrifugation of complex macromolecular systems. *Biochemistry*, **33**, 13155–13163.

Hensley, P. (1996). Defining the structure and stability of macromolecular assemblies in solution: the re-emergence of analytical ultracentrifugation as a practical tool. *Structure*, **4**, 367–373.

Laue, T. M., and Stafford, W. F., III (1999). Modern application of analytical ultracentrifugation. *Annu. Rev. Biophys. Biomol. Struct.*, **28**, 75–100.

Carruthers, L. M., Schirf, V. R., Demeller, B., and Hansen, J. C. (2000). Sedimentation velocity analysis of macromolecular assemblies. *Methods Enzymol.*, **321**, 67

Shuck, P. (2000). Size distribution analysis of macromolecules by sedimentation velocity ultracentrifugation and Lamm equation modelling. *Biophys. J.*, **78**, 1606–1619.

Sedimentation equilibrium

Cantor, C., and Schimmel, P. (1980). *Biophysical Chemistry*. Part II. *Technique for the Study of Biological Structure and Function*. San Francisco, CA: W. H. Freeman and Company.

Oberfelder, R. W., Consler, T. G., and Lee, J. C. (1985). Measurement of changes of hydrodynamic properties by sedimentation. *Meth. Enzymol.*, **117**, 27–40.

Schuster, T. M., and Laue, T. M. (1994). *Modern Analytical Ultracentrifugation*. Boston, MA: Birkhauser.

Laue, T. M. (1995). Sedimentation equilibrium as thermodynamic tool. In *Methods in Enzymology*, eds. G. K. Ackers and M. L. Jonson V, pp. 427–452. New York: Academic Press.

Chapter D5
Electrophoresis

D5.1 Historical review

1920s

M. Smoluchowski and **E. Hückel** were the first to elaborate the theory of electrophoresis for thin and thick double layers. **D. C. Henry** provided a theory for the electrophoresis of spherical polyions that was valid for double layers of arbitrary thickness. In the Henry model, charge distribution within the sphere is assumed to be spherically symmetric, and ion relaxation is ignored. In the **1950s** a number of investigators (**J. Th. G. Oberbeek, F. Booth, P. H. Wiersma**) studied the effect of ion relaxation on the electrophoretic mobility of spheres containing a centrosymmetric charge distribution.

1930

A. Tiselius presented a new technique for studying the electrophoretic properties of proteins. He showed that sharp electrophoretic moving boundaries of ionised molecules could be obtained in U-shaped quartz glass tubes and that the protein boundary could be detected with UV light. **A. Tiselius** described moving boundaries corresponding to albumin, α-, β-, and γ-globulin in serum.

1940s

T. Wieland and **E. Fischer** proposed the use of filter paper as a supporting medium for electrophoresis. Simplicity of construction, relative cheapness and compactness, and the requirement of much less than 1mg of protein for analytical or diagnostic purposes popularised paper electrophoresis.

1950s

O. Smithies obtained good resolution of proteins using starch gels. Electrophoresis with molecular sieving of a gel gave a much higher resolution than electrophoresis in free solution. In this decade, many materials (like strips of cellulose acetate, agarose) were introduced as supports for electrophoresis. In **1959** **S. Raymond** and **L. Weintraub** reported the first results of electrophoresis done with polyacrylamide gels (PAGE).

1960s

A. L. Shapiro, **E. Vinuela** and **J. V. Maizel** found a proportional relationship between electrophoretic migration of proteins and their molecular mass in homogeneous PAGE in the presence of sodium dodecyl sulphate SDS. Electrophoresis in PAGE with SDS became widely used for the assessment of purity and the determination of molecular mass of proteins. The ability to separate nucleic acids and their nucleotides was extended to DNA and even chromosomes in agarose gel via pulsed field electrophoresis.

1975

P. O'Farrell and **J. Klose** developed two-dimensional electrophoresis. Proteins were separated by isoelectric focusing in one direction and by molecular mass in another. In the initial experiments, 1100 proteins from *E.coli* were observed. At that time, proteins from gels were identified by comparison of the electrophoresis of known purified proteins or by immunostaining. In **1977 N. G. Andersen** and **N. L. Andersen** pioneered the automatic evaluation of two-dimensional electrophoresis gels and as a result different commercial systems are now available. Protein identification was revolutionised by the development of sequencing-stable membranes, from which proteins were directly N-terminally sequenced. The identification limit was shifted down to 1 pmol. The resolution of two-dimensional electrophoresis was increased to resolve up to 10 000 protein spots on one gel by increasing the gel size and using silver staining for detection.

1981

J. W Jorgensen and **K. D. Lukas** performed an electrophoretic amino acid separation in a capillary with a very high efficiency that previously was obtainable only in the gas phase. The capillary format enables the generation of electric fields at least 50-fold more powerful than those of traditional slab gels, which leads to rapid separation – in the order of minutes. Capillary electrophoresis can be viewed as a fully automated tool for rapid highly sensitive and quantitative analysis of nanolitre amounts of complex samples. Many advanced instruments have been marketed and are increasingly used in connection with sequencing of proteins and synthesis of peptides. The same instruments can be used in conjunction with mass spectrometers.

1990

B. L. Karger's group reported a linear polyacrylamide matrix that could be used for high-resolution DNA separations in a capillary. In the same year **H. Sverdlow** and **R. Gesteland** reported that DNA sequencing in a single 75-μm internal diameter fused-silica capillary was faster and achieved a higher separation efficiency than a single lane in a slab-gel sequencer. In **1993 H. Kambara** and **S. Takahashi** constructed capillary arrays with scanning detection in a confocal microscope geometry. The improvement of electrophoretic separation and subsequent visualisation of nucleic acids as well as nucleotides made possible the efficient mapping of genes and resulted in the biggest concerted endeavour in the history of science, the mapping of the human genome.

2000 to now

Electrophoretic methods are among the most widely used high-resolution techniques for analytical and preparative separations. They have had an unprecedented impact on life science, providing a basis for unique advances in biochemistry, molecular biology, genetics, gene technology and medicine.

D5.2 Introduction to biological problems

Electrophoresis is the migration of charged electrical species when dissolved, or suspended, in an electrolyte through which an electric current is passed.

In 1791 M. Faraday formulated the two fundamental laws of electrolysis that bear his name. Faraday coined the words electrolyte, electrolysis, anode and cathode (for the positive and negative electrodes), and anion and cation (for the ions migrating towards to these respective electrodes). In 1897 F. Kollraush formulated the first theory explaining the behaviour of substances moving in an electric field.

In 1926 T. Svedberg encouraged A. Tiselius to work on electrophoresis. The migration of protein boundaries was observed and recorded in moving boundary electrophoresis and ultracentifugation, but neither method was able to separate different molecules completely. In 1930 A. Tiselius presented his doctoral thesis on moving boundary electrophoresis, which opened the way for biologists to use electrophoresis.

After Tiselius, work on electrophoresis followed two paths. In the first, electrophoretic methods in which there was no gel support medium, so-called free-solution electrophoresis, were created. Up to now these methods have not been widely used as *structural probes*, because the complex interactions between charged molecules, solvent, mobile salt and an external electric field have made it difficult to interpret electrophoretic mobility.

In the second, various electrophoretic methods which employed of a gel support medium were developed. These methods have proven to be of enormous value as *separation techniques*. Thus, two-dimensional polyacrylamide gel electrophoresis provides the highest resolution for protein analysis and, therefore, plays a central role in 'proteomics'. Capillary electrophoresis (CE) continues to grow rapidly as an analytical technique in a wide range of application areas. Attractive features of the technique are a short analysis time, very high resolution and compatibility with on-line detection and automated sample loading with minimal (microlitre) volume requirements. CE in array format brought the completion of the Human Genome Project forward by 3–5 years.

In 2003 M. L. Plenert and J. B. Shear adapted an unusual capillary geometry, the hourglass, to enhance the electric field strength locally at the separation region, and added ultrafast optical-based sample injection to enable separations on the *microsecond* time scale. The application of this approach to the study of short-lived molecules should be possible.

A new technique, called capillary electrochromatography (CEC), which is a hybrid of high-pressure liquid chromatography (HPLC) and CE, has emerged. CEC offers the combination of the best features of each technique. Like HPLC, it can be used in the separation of uncharged species and, like CE, it can provide highly efficient separations of microvolumes of sample solution without the need for a high-pressure pumping system.

In this chapter, we shall first consider free-solution electrophoresis as an analytical technique. Recent developments raise hope that this method could be a viable experimental method for studying comparatively large and highly charged biological macromolecules. We shall then examine the use of electrophoresis as a separation technique (with particular emphasis on CE) and on the application of this technique to various analytical problems.

D5.3 Electrophoretic experiments

There are three ways of extracting structural information from a particle placed in an electric field (Table D5.1). The first is based on measuring the particle's transport properties in constant electric field. The absolute values of electrophoretic mobilities in solution can be measured by *moving boundary electrophoresis* or *free electrophoresis*. The process is similar to high-velocity sedimentation.

In the second, a thin *layer* or *zone* of the macromolecular solution can be electrophoresed through some kind of matrix. The matrix acts as a molecular sieve separating the molecules on the basis of their size and provides stability against convection. The process is similar to the action of the sucrose gradient in zonal centrifugation (Section D4.9).

The third way is to measure the orientation that is produced when molecules with an asymmetric charge distribution are subjected to a variable electric field. A torque tends to orient the molecules in the direction of the field. The orienting of charged particles under the influence of an electric field is called the *Kerr effect* (Chapter D6). This process is similar to flow birefringence in hydrodynamics (Chapter D7).

Table D5.1 presents a summary of electrophoretic approaches used in scientific research. The table shows the method, the experimental conditions, further parameters that can be calculated on that basis, and analogous phenomena in hydrodynamics.

D5.4 Macromolecules in an electric field

D5.4.1 Charge and electrophoretic mobility

Charge is a fundamental property of a macromolecule that is linked to its structure, solubility, stability and interactions.

Table D5.1. *A summary of electrophoretic approaches used in research*

Experimental method	Experimental conditions	Calculated parameter	Analogues in hydrodynamics
Moving boundary electrophoresis	Free electrophoresis	Electrophoretic mobility	Velocity sedimentation
One-dimensional capillary electrophoresis	Free electrophoresis	Electrophoretic mobility	Velocity sedimentation
Steady-state electrophoresis	Membrane confined analytical electrophoresis in zero-field and non-zero field	Diffusion coefficient, charge	Diffusion coefficient using ultracentrifugal boundary-forming cell
Isoelectric focusing in gel	Gradient of pH	Isoelectric point, where mobility is zero	Equilibrium centrifugation in gradient density
Zonal one-dimensional gel electrophoresis	SDS	Separation according to friction properties/mass	Equilibrium centrifugation in gradient density
Zonal two-dimensional gel electrophoresis	Gradient of pH (first dimension) + SDS (second dimension)	Separation according to: charge (first dimension), friction properties/mass (second dimension)	Equilibrium centrifugation in gradient density
Electric birefringence (Kerr effect) (Chapter D6)	Electrical torque	Spectrum of relaxation times	Flow birefringence (Maxwell effect) (Chapter D7)

The most direct means of discerning the charge of a macromolecule is by its response to an applied electric field. If a particle has a charge Q, then application of an electric field $\mathbf{E}$ results in an applied force:

$$\mathbf{F} = Q\mathbf{E} \qquad (D5.1)$$

Shortly after the application of the electric field, the particle reaches a steady-state velocity u, with the particle moving toward one of the electrodes. At this velocity, frictional forces are equal and opposite to the applied force, so

$$u = QE/f \qquad (D5.2)$$

where f is translational friction coefficient of the macromolecule.

The transport of charged particles under the influence of an electric field is called *electrophoretic mobility* (Comment D5.1). The electrophoretic mobility μ can be defined as the velocity per unit field

$$\mu = u/E \qquad (D5.3)$$

Comment D5.1
Units of eletrophoretic mobility

The electrophoretic mobility μ has units in the cgs system of $cm^2\ s^{-1}\ V^{-1}$.

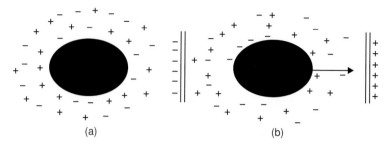

Fig. D5.1 Macromolecules in electric field. (a) In the absence of an electric field, the small ions form an ion atmosphere around the macromolecules in which ions of opposite charge to that of the macromolecules predominate. (b) In the presence of an electric field, the ion atmosphere is distorted by the field and by the motion of the macromolecules. (After Cantor and Shimmel, 1980.)

and hence

$$Q = \mu f \tag{D5.4}$$

Such a simple description of electric mobility does not correspond to reality and may be considered an ideal case. Indeed, an immediate difficulty arises over the definition of the *net charge*.

In any aqueous solution, there are counterions. Some may be associated rather tightly with the macromolecules, others more loosely (Fig. D5.1). Usually, a large amount of supporting electrolyte is introduced in the medium to weaken the effective character of the ion pairing. As the electrolyte forms an ion atmosphere around the macromolecule, Eq. (D5.2) can be written in the form

$$u = Q_{\mathrm{eff}} E / f \tag{D5.5}$$

where Q_{eff} is the effective charge of 'dressed' macromolecules (i.e. macroions + counterions).

Despite its importance, it is difficult to measure the charge in solution. To date, there are no experimental methods that determine the effective charge independently of other macroion properties. Innovations in steady-state and capillary electrophoresis have shown, however, that the effective charge of some molecules, like DNA, can be determined (see Sections D5.5.2 and D5.7.2).

D5.4.2 Modelling of electrophoretic mobility

The most commonly quoted expression for electrophoretic mobility, μ, is the Huckel equation:

$$\mu = Ze / f \tag{D5.6}$$

where the Z is the number of charges, e is the magnitude of the electron charge and f is the frictional coefficient of the charged macromolecule. As with the

---

Comment D5.2   Charge effects influence sedimentation and diffusion constants

In general, charge effects influence sedimentation and diffusion constants as well as intrinsic viscosities even when external electric fields are not present. The best way to understand this is to consider the example of a sedimenting, highly charged polyion such as a DNA fragment. The streaming flow of a solvent past the fragment distorts the ion atmosphere and gives rise to an additional viscous drag, the so-called 'electrolyte friction'. The electrolyte friction reduces sedimentation and diffusion constants. Similarly, ion relaxation increases the intrinsic viscosity of a dilute suspension of rigid polyions placed in a fluid shear field and this phenomenon is called the 'primary electroviscous effect'. For a review of these issues, the reader should refer to the specialist literature (Schmitz, 1993). The effects of charge on sedimentation and diffusion are expected to be small and perhaps negligible for a weakly charged polyion in high salt concentrations, but the effect becomes more important as the polyion charge increases or the ambient salt concentration decreases.

---

sedimentation coefficient, μ is the ratio of the particle's velocity to the strength of the driving field (compare Eq. (D4.16)). If the particle happens to be spherical, Stokes' law applies and we can write

$$\mu = Ze/6\pi \eta_0 R_0 \qquad (D5.7)$$

where R_0 is the particle's radius and η_0 is the viscosity of the solvent.

In principle, Eq. (D5.7) could be generalised to take into account hydration and deviations from a spherical shape (as in the case of sedimentation and diffusion). However, for a charged particle undergoing steady-state motion in a constant electric field, it is also necessary to account for the distortion of the ion atmosphere from its equilibrium state (Comment D5.2). This 'ion relaxation' effect is important unless the macromolecule is weakly charged. The transport of a charged particle is a substantially more difficult problem than the transport of an uncharged particle. Until recently, the classical theory was applied to only the simplest of model structures.

Models used in classical electrophoresis theory are similar to the continuum hydrodynamic models discussed in Chapter D2. They take account of the charge distribution of the particles and the ion atmosphere. Developments in modelling electrostatic and hydrodynamics in complex systems (see Chapter D2) have allowed a reexamination of the theory of free-solution electrophoresis. The problem for polyions of arbitrary shape and charge distribution has been solved by using the boundary element technique (Section D2). An example of the partitioning of the hydrodynamic surface for a 20 bp DNA fragment into 352 triangular platelets is shown in Fig. D2.15(a). It is also illustrated by the wireframe triangles (shell numbers 5–9) corresponding to a single platelet. The assumption is made that the field quantities are constant over a particular platelet but that these field quantities can vary from one platelet element to the next.

D5.5 Electrophoretic methods in the absence of a support medium: free electrophoresis

The various electrophoretic methods have not been widely used as *structural probes*. The reason is simple: the complex interaction between macroion, solvent, salt, gel (if present) and external electric field complicates the interpretation of electrophoretic mobility.

However, in the absence of a gel support medium, i.e. in *free-solution* electrophoresis, the interpretation is more straightforward. Developments in membrane-confined analytical electrophoresis and CE prove that free-solution electrophoresis is a viable experimental method for the study of comparatively large and highly charged biological macromolecules.

Continuous free-solution electrophoresis is also an effective separation technique (Comment D5.3).

D5.5.1 Moving-boundary method

In real electrophoretic experiments, four forces act on the particle (Fig. D5.2). The first is the electrostatic force $F_1 = QE$.

The second is the hydrodynamic frictional force $F_2 = f_t u$, where f_t is the Stokes frictional coefficient and u is the constant particle velocity.

The third force, F_3, results from the fact that a macroion tends to be surrounded by counterions that flow in the opposite direction to its motion. From the perspective of the macroion, it is always moving 'upstream' against this flow, so that its velocity relative to its immediate surroundings is higher than that relative to the laboratory reference frame.

The fourth force, F_4, arises from the distortion in the ionic atmosphere (the field asymmetry effect). This force is attributed to the dipole moment that is established with the redistribution of the surrounding counterions by the external electric field.

As a result, the net charge Q of the macromolecule can be determined from the molecule's mobility under the influence of these four forces:

$$F_2 + F_3 + F_4 = F_1 = fu = Q/E \qquad (D5.8)$$

and the mobility is $\mu = u/E = Q/f_{\mathrm{eff}}$, where f_{eff} is an frictional coefficient produced by forces F_2, F_3 and F_4.

The moving-boundary method has been applied to determine the net charge on the DNA olygonucleotide $(pd(A)_{20}\ pd(T)_{20})$. The boundary velocity was monitored using an apparatus capable of on-line analytical electrophoresis and

Comment D5.3
Free-flow electrophoresis

In so-called free-flow electrophoresis (FFE), a curtain of carrier buffer flows between two narrowly spaced vertical plates. An electric field is applied perpendicular to the carrier flow while a sample solution is injected continuously into the carrier buffer as a narrow band. Sample components are then separated laterally in the electric field according to the differences in their electrophoretic mobilities, and the separated components are collected continuously at the outlets. Because FFE does not use organic solvents, high salt concentrations or supporting media, it is gentle enough for use with proteins and cells.

Fig. D5.2 Four forces are applied in electrophoretic mobility experiments. The forces are described in the text. (Laue *et al.*, 1996.)

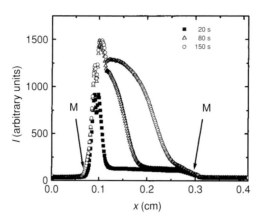

diffusion measurements. The intensity profile of the cell shows the boundary of DNA(pd(A)$_{20}$pd(T)$_{20}$) (Fig. D5.3). The data presented in Figure D5.3 yield an estimate of the electrophoretic mobility of DNA(pd(A)$_{20}$ pd(T)$_{20}$), $\mu = 2.99 \times 10^{-4}$ cm^2 V^{-1} s^{-1} at 20 °C.

D5.5.2 Steady-state electrophoresis (SSE)

In contrast to free boundary motion in mobility experiments, macroions in SSE are trapped in a small chamber whose top and bottom are sealed with semipermeable membranes. An electric field is applied along the chamber so that the macroions crowd up against one of membranes. Diffusion produces a macroion flux in the direction opposite to that due to the electric field. When the steady state is reached, these two fluxes are matched at each point and a stable concentration gradient is observed.

Thus, in SSE the fluxes, as well as the forces, are balanced (Fig. D5.4). At any point x in the cell, the flux due to electrophoresis, J_{eff}, is $C \times u'$, where C is the concentration of macroions at x and u' is their velocity. The flux J_{eff} is the effective flux resulting from all of the forces F_1, F_2, F_3 and F_4. Since

$$f_{\text{eff}}/u' = QE \tag{D5.9}$$

and hence

$$u' = QE/f_{\text{eff}} \tag{D5.10}$$

we have

$$J_{\text{eff}} = (QE/f_{\text{eff}})C \tag{D5.11}$$

We recall that a according to Fick's first law (Eq. (D3.7)) concentration gradients produce flux due to diffusion, $J_{\text{D}} = -D \, dC/dx$. At steady state $J_{\text{eff}} + J_{\text{D}} = 0$, hence

Fig. D5.4 Steady-state measurement in which fluxes J_{eff} and J_{D} are balanced.

$$(QE/f_{\text{eff}})C = D \, dC/dx \tag{D5.12}$$

The solution of the Eq. (D5.12) is

$$C = C_0 \exp[\sigma(x - x_0)] \qquad (D5.13)$$

where C_0 is the concentration at a reference point x_0 and the exponent $\sigma \equiv QE/f_{eff}D_{eff}$ includes D_{eff}, the diffusion coefficient in the presence of the electric field. Now $Q = \sigma f_{eff} D_{eff}/E$. Assuming that the relation $f_{eff} = kT/D_{eff}$ is valid we have

$$Q = \sigma kT/E \qquad (D5.14)$$

Equation (D5.4) provides a simple way of determining the effective charge of the molecule directly from experimental measurements of σ, E, and T, unlike mobility measurements, which also require knowledge of f_{eff} or D_{eff}.

A steady-state concentration gradient for DNA(pd(A)$_{20}$pd(T)$_{20}$) is shown in Fig. D5.5. The diffusion coefficient in the absence of an electric field was obtained from the relaxation of the steady-state concentration gradient and is presented in Fig D5.6. For the data presented here, D is estimated to be $(9.6 \pm 0.2) \times 10^{-7}$ cm^2 s^{-1}.

Assuming that the Einstein–Smoluchovski relation $fD = kT$ remains approximately valid for 'dressed' macroions in an electric field, SSE provides an estimate of the effective electric charge of a macromolecule. The value of Q obtained from the simple SSE theory (Eq. (D5.14)) gives the ratio of the effective charge to the formal charge, $Q/40$ (for 20-bp double-stranded DNA), close to the previous prediction of 0.12 for single-valence salts, neglecting the correction for finite length macro ions.

D5.6 Electrophoretic methods in the presence of a support medium: zonal electrophoresis

The first work of A. Tiselius was performed in free solution. However, it was soon recognised that many problems associated with this approach could be eliminated by stabilising the medium. This was achieved by carrying out electrophoresis on a porous mechanical support, which was wetted by an electrophoresis buffer. The support medium reduces convection currents and diffusion so that the separated components remain as sharp zones.

The earliest supports used were filter paper and cellulose acetate strips. For many years these support media were used in the separation of small molecules such as individual amino acids, peptides and carbohydrates. However, the introduction of gels such as agarose or polyacrylamide gel, as the support medium in electrophoretic methods, greatly improved the analysis of macromolecules (Comment D5.4).

Fig. D5.5 Dependence of the optical absorbance A_{260} on the distance x for (pd(A)$_{20}$pd(T)$_{20}$). The gradient in 100 mM KCl, 20 mM Tris pH 8.0 established at field strength of 0.12 V cm^{-1} is shown decaying after the field was shut off. The initial steady-state gradient (dotted line) was formed by application of the field for > 24 hours. (After Laue et al., 1996.)

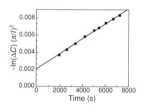

Fig. D5.6 Determination of the diffusion coefficient from the decay of the gradient shown in Fig. D5.5. The slope of a graph of $\ln(A_{2/3} - A_{1/3})$ as a function of time is proportional to $\pi^2 D/L^2$. (After Laue et al., 1996.)

Comment D5.4   Physicist's box: Agarose gels

Agarose is a natural linear polysaccharide isolated from certain agar-bearing seaweeds. Agarose varieties differ in their physical and chemical properties, such as gelling temperature, porosity and gel strength. Agarose gel is formed by suspending dry agarose in aqueous buffer, then boiling the mixture until a clear solution forms. This is poured on to a flat surface and allowed to cool to room temperature to form a rigid gel. Agarose gels are used for electrophoresis of both proteins and nucleic acids.

Polyacrylamide gels (PAGE)

Cross-linked PAGE are formed from the polymerisation of acrylamide monomers in the presence of small amounts of comonomer (bisacrylamide). The cross-linkers are bifunctional acrylic agents that covalently link adjacent linear polyacrylamide chains. Acrylamide monomers are polymerised in a head-to-tail fashion into long chains and occasionally a bisacrylamide molecule is built into a growing chain, thus introducing a second site for chain extension. Acrylamide gels are defined in terms of the total percentage of acrylamide present, and the pore size in the gel can be varied by changing the concentration of both the acrylamide and the bisacrylamide.

Sodium dodecyl sulphate (SDS)

SDS (CH_3-$(CH_2)_{10}$-$CH_2OSO_3^-$ Na^+) is an anionic detergent.

**Comment D5.5
SDS and charge**

Some proteins bind more or less SDS than the average amount, which makes them run anomalously on the gel. If the protein itself has a very large positive or negative charge, this charge may not be negligible compared with the charge produced by the bound SDS. A specific example is histone H1 which has a molecular mass of about 21 kDa, but behaves in gel electrophoresis like a molecule of 30 kDa, because of its strong positive charge.

D5.6.1 Separation according to frictional properties: SDS gel electrophoresis

In the method called SDS gel electrophoresis, the protein to be studied is first heated in a dilute solution of sodium dodecyl sulphate (SDS). This procedure breaks down all native secondary, tertiary and quaternary structures in the protein. A special agent such as β-mercaptoethanol is added to break any disulphide bonds. This treatment fully denatures the protein which forms a rod-shaped structure with a series of negatively charged SDS molecules along the polypeptide chain (Fig D5.7).

On average, one SDS molecule binds to every two amino acid residues. The original native charge on the protein is therefore suppressed by the negatively charged SDS molecules (Comment D5.5). In this way, SDS converts all proteins into similar structures that differ only in molecular mass. Unfolded SDS–polypeptides move through gels in a predictable manner: lower-molecular-mass

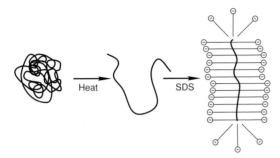

Fig. D5.7 Schematic presentation of a protein in SDS. The protein is denatured and coated with a molecule with a uniform charge per unit length. The intrinsic charge of the protein is neutralised. Addition of β-mercaptoethanol reduces any dusulphide bonds.

molecules migrate faster than larger ones. Figure D5.8 shows the typical relative mobility of various SDS-polypeptide chains in 10% gels as a function of molecular mass.

To a first approximation, migration rates of SDS complexes are inversely proportional to the logarithm of their molecular masses. The absolute molecular mass of a protein can be estimated from the relative mobilities of its polypeptide chains in the SDS-PAGE gel by comparing these to the mobilities of known protein markers (Comment D5.6).

Comment D5.6   Gel pore size and protein molecular mass

Typically, the separating gel used is a 15% polyacrylamide gel. This gives a gel of a certain pore size in which proteins of relative molecular mass 10 kDa move through the gel relatively freely, whereas proteins of 100 kDa can only just enter the pores of the gel. This means that gels of 15% polyacrylamide are suited for separating proteins in the range 10–100 kDa. For proteins with a molecular mass larger than 100 kDa a larger-pored gel (e.g 10% or even 7.5%) would be used.

Fig. D5.8 Relative mobility of 37 different polypeptide chains in 10% gel as a function of molecular mass. (Weber and Osborn, 1969.)

D5.6.2 Separation according to the charge: isoelectric focusing (IEF)

IEF is an electrophoretic method for separating amphoteric molecules in pH gradients. The net charge on an amphoteric molecule is determined by the pH of its local environment (Comment D5.7). Proteins, the best-known examples of amphoteric molecules, carry positive, negative or zero net electrical charge, depending on the pH of their surroundings. For each protein there is a specific pH, called pI at which it carries no charge. Hence the pI is a characteristic property of the protein.

In a pH gradient a protein is moved by an electric field until it reaches a point at which it is isoelectric (Fig. D5.9). Its net charge at that point is zero and the electrical field there no longer influences it. The situation is similar to

Comment D5.7

Amphoteric molecules

Amphoteric molecules are molecules capable of positive, negative or zero net charge, depending on pH of their surrounding.

Fig. D5.9 IEF: a protein is placed in a pH gradient in an electric field. The pH gradient formed by ampholyte molecules under the influence of an electric field is indicated. The gradient increases from acidic (pH 3) at the anode to basic (pH 10) at the cathode. The hypothetical protein in the figure bears a net charge of +2, 0 or −2, at the three positions in the pH gradient shown. At the pI, the net charge on the protein is zero, so that it does not move in the field and becomes focused. (After Garfin, 1995.)

the sedimentation equilibrium in a density gradient experiment, when a particle moves to the point at which its density becomes equal to the density of the solvent and then stops (see Table D5.1 and Section D4.9).

IEF is the ideal method for separating proteins according to charge. The method has high resolution, being able to separate proteins that differ in their isoelectric points by as little as 0.01 pH unit. The most widely used system for IEF utilises horizontal gels on a glass plate. Separation is achieved by applying a potential difference across the gel, which contains a pH gradient.

Having a stable, linear pH gradient is the key to successful IEF. Establishment of such gradient is accomplished by the introduction into the gel of compounds known as ampholites (Comment D5.8). Since IEF requires the protein to move freely according to its charge under the electric field, it is carried out in low percentage gels to avoid any sieving effects within the gel (see Comment D5.9). A typical isoelectric focusing gel is presented in Fig. D5.10.

Fig. D5.10 SDS-PAGE analysis of proteins of ribosomes and ribosomal subunits from *E. coli* and *T. thermophilus*. M, the molecular mass marker, is in kilodaltons. The 15% SDS-PAGE shows that protein S1 from *E. Coli* is present in the 70 S ribosomes and the 30 S subunit of *T. thermophilus* but is absent in the 50 S subunit. Note, that the content of S1 protein in the samples of *T. thermophilus* ribosomes is typically lower than that of protein standard preparation *E. Coli* ribosomes. (After Shiryaev *et al.*, 2002.)

D5.6.3 Separation according to the charge and frictional properties simultaneously: two-dimensional gel electrophoresis

Two-dimensional electrophoresis combines the method of IEF with the size separation method of SDS-PAGE. This combination gives two-dimensional polyacrylamide gel electrophoresis (2-D PAGE), which is at present the most sophisticated analytical method for separating proteins. Figure D5.11 shows a typical two-dimensional gel.

In the initial application of this method, 1100 proteins from *E.Coli* were observed, and it was predicted that up to 5000 proteins could be displayed by this method. Despite these outstanding properties and widespread application, almost 20 years went by before 2-D PAGE gels became an integral part of *proteomics* (see Section B2.2).

(a)   (b)

Fig. D5.11 Electrophoretic separation of proteins of the 30 S ribosomal subunits from *E. Coli* (a) and *T. thermophilus* (b). The electrophoresis in the first dimension was performed at acidic pH (5.5), while that in the second dimension was done in 15% SDS-PAGE. Comparison of the results of the electrophoretic separation of 30 S subunit proteins from *E. Coli* and *T. thermophilus* in this system demonstrates that the position of the spot (marked by •) in (b) corresponds to that of protein S1 from *E.Coli* in (a). (Shiryaev *et al.*, 2002.)

The ability to separate large numbers of proteins by 2-D PAGE and then to identify individual proteins by partial sequence analysis and database searching for this sequence is resulting in a rapid increase in our knowledge of the protein component of cells. Moreover, we now have the ability to follow changes in the expression of all proteins in the cell and to compare information from 'normal' and 'abnormal' cells (see examples in Section B2.6.1).

D5.7 Capillary electrophoresis (CE)

Despite substantial progress in methodology, the slab-gel technique typically suffers from laborious, time-consuming and difficult-to-automate procedures. Capillary electrophoresis (CE) has now emerged as an alternative to slab-gel techniques and has all the advantages of modern automated technology. This technology could be adapted to two-dimensional electrophoresis.

D5.7.1 One-dimensional electrophoresis

As the name suggests, CE involves electrophoresis of a sample in very narrow bore tube (typically 20–100 μm in internal diameter). The fused silica separation capillary is dipped into background electrolyte reservoirs equipped with electrodes for connecting the high-voltage power supply. Electrodes made of an inert material such as platinum are also inserted into the electrolyte reservoirs to complete the electrical circuit. A small volume of sample is injected into one end of the capillary. A schematic diagram of a typical instrument for CE is shown in Fig. D5.12. The separated molecules are observed by a detector that is located at the end of the tube that is opposite to where the sample was introduced. The term, electropherogram, describes the plot of the detector response.

Theoretical consideration of CE generates two equations. One of them defines the electrophoretic mobility, μ, as

$$\mu = L^2/Vt \qquad (D5.15)$$

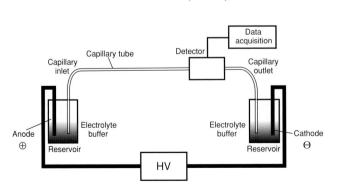

Fig. D5.12 Schematic representation of a typical CE apparatus. (After Wilson, 2000.)

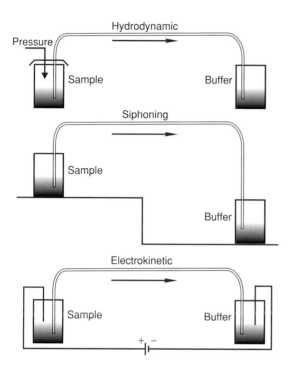

Fig. D5.13 The most
frequently used injection
mode is to dip the
capillary into the sample
solution vial. The vial is
then pressurised causing
a volume of solution to be
forced into the capillary. A
less popular, alternative
sample injection
procedure is to dip the
capillary and electrode
into the sample solution
vial and to apply a
voltage. If the sample is
ionised and the
appropriate voltage
polarity is used, then
sample ions migrate into
the capillary. This type of
injection is known as
electrokinetic sampling.

where t is the time required for the sample to travel to the detector (in seconds), L is the tube length (in centimetres), and V is the applied voltage. The second equation defines the separation efficiency in terms of the total number of theoretical plates, N, as

$$N = \mu V / 2D \qquad (D5.16)$$

where D is the diffusion coefficient of the molecule (Comment D5.10).

Equation (D5.15) shows that the ideal situation is to apply the highest voltage possible to the shortest capillary possible. However, there are practical limits to this approach. As the capillary length is reduced, the amount of heat that must be dissipated increases owing to the decreasing electrical resistance of the capillary. Also the higher the voltage that is applied, the greater the current, and therefore the greater the heat generated. In practical terms a compromise between the voltage used and the capillary length is required. Voltages of 20–50 kV with capillary lengths of 40–100 cm are possible choices. Separations are rapid, typically taking a few minutes. Injection procedures are shown in Fig. D5.13.

The use of fused silica as the capillary material permits direct detection of separated components in the tubes. Part of the polyamide coating is removed at one end of the capillary, which is positioned in the light path of an

Comment D5.10
Detection level in
CE

When considering a
nanolitre amount of
injected volume, the
actual mass detection
limits are in the
attomole (10^{-18} mole)
or zeptomole (10^{-21}
mole) ranges. Indeed,
at the 1-zm level, 600
molecules are
detected. Detection
levels are at times
described in terms of
number of molecules
analysed rather than
the number of moles.

optical detector. On-column monitoring of UV absorbance is the most common detection mode. The amount of sample injected (usually pressure injection) is in the nanolitre range. The very small volumetric flow rates of under 1 μL min^{-1} from electrophoresis capillaries make it feasible to couple the effluent from the capillary of an electrophoretic device directly to the ionisation source of a mass spectrometer.

An important development in CE is the utilisation of lasers as excitation sources for on-column laser-induced fluorescence detection. This approach, particularly with dye-labelled molecules, provides a means for the selective detection of molecules at concentrations of 10^{-11} M or lower (Comment D5.10). Besides having high sensitivity, fluorescence detection is also quite selective, requiring the presence of certain fluorescence moieties within the detected molecule. The situation is quite straightforward in DNA analysis, in which the use of fluorescently labelled nucleotides is a routine practice.

An important consideration in CE, in contrast to gel electrophoresis, is the magnitude of electroosmotic flow (EOF). EOF results in a flow of the solution along the capillary usually towards the detector (Fig. D5.14). This flow can significantly reduce analysis times and can overcome an ion's tendency to migrate towards the electrode to which it is attracted by its charge (Comment D5.11). This flow effectively pumps solute ions along the capillary, generally towards the detector, and could be considered as an 'electrically-driven pump'. To eliminate the EOF, all capillaries are coated with neutral hydrophilic polymers. Coating the capillary walls increases the viscosity of the double layers near the wall, eliminating the EOF.

Figure D5.15 shows an example of the high-resolution capabilities of CE. The separation of a protein mixture was performed in an open-tube format with a

Fig. D5.14 Electroosmotic flow. EOF refers to a flow of liquid in a separation chamber resulting from interactions at charged groups on the chamber walls.

> **Comment D5.11   Flow of ions and water molecules along the capillary**
>
> The walls of a fused silica capillary contain silanols, which ionise in contact with a high pH electrolyte solution. This dissociation produces a negatively charged wall. A layer of metal ions is then established at the wall to preserve electroneutrality. When a voltage is applied these metal ions and their associated solvating water molecules migrate towards the cathode. This movement of ions and their associated water molecules results in a flow of solution towards the detector

Fig. D5.15 Open-tube CE separation of basic proteins at pH 4.5. The capillary, 50 cm × 75 μm is coated with linear polyacrylamide. The electric field strength is 530 V cm^{-1}. 1, lysozyme; 2, cytochrome c; 3, myoglobin; 4, trypsinogen; 5, α-chymotrypsinogen A. (Karger *et al.*, 1995.)

coated capillary to prevent sample adsorption. The separation was fast, and the peak widths are quite narrow. When protein adsorption on the capillary wall is minimised, separation efficiencies exceeding 10^6 theoretical plates per metre (a measure of peak broadening, see Eq. (D5.16)) are possible in CE. This factor is at least 10–100 times better than the efficiency of HPLC.

D5.7.2 Ultrafast capillary electrophoresis

From Eqs. (D5.15) and (D5.16), it appears feasible to achieve faster separation simply by applying the voltage across ever shorter separation channels. Ultimately, however, Joule heating places a practical limit on the strength of the electric field that can be used, and variances caused by the injection-plug length and spatial extent of the detection region limit separation efficiency.

An unusual capillary geometry, the hourglass, has been used to enhance the electric field strength locally at the separation region. A combination of such a geometry with ultrafast sample injection has enabled separations on the

(a)

(b)

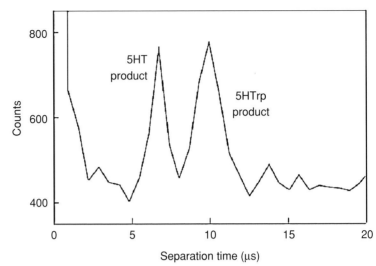

Fig. D5.17 Electrophoretic resolution of 5HT and 5HTrp photoproducts in 10 μs. A field, which was estimated to be 0.15 MV cm^{-1} (35 kV), was used to fractionate components over a separation distance of 9 μm. (After Plenert and Shear, 2003.)

Fig. D5.16 (a) Schematic of the photoreaction and probe regions in an optically gated microsecond separation. Femtolitre reaction plugs are created in flowing reagent streams by a gate focus that is switched to high intensity for ≈ 1–2 μs; transient reaction products migrate according to charge-to-drag ratios, providing the capacity to analyse the reaction mixture with microseconds. (b) A typical hourglass structure created in the central portion of a short fused-silica capillary (29-μm internal diameter and 320-μm outer diameter in unpulled regions. The scale bar represents 250 μm. The inset shows the central (≈ 60 μm) region of the hourglass where separations are performed. (Plenert and Shear, 2003.)

microsecond time scale. Figure D5.16(a) shows a schematic of a tightly focused laser beam used for the detection of electrophoretically resolved components. By reducing the cross-sectional area by a factor of 30–40, fields in excess of 0.1 MV cm^{-1} can be generated in short capillary stretches by using an applied potential of 20 kV. The sample was transported electrokinetically between inlet and outlet vials.

The capability of this system is demonstrated in Fig. D5.17. Here a field of 0.15 MV cm^{-1} was used to fractionate 5-hydroxytryptamine (5HT) and hydroxy-tryptophan (5HTrp) in a microsecond time scale.

The future of this new methodology is very promising, especially to probe short-lived reaction intermediates, which traditionally is the privilege of spectro-scopic methods.

D5.7.3 Electrophoretic mobility of DNA

The electrophoretic mobility observed during CE is the difference between the electrophoretic mobility of the macromolecule and the electrophoretic mobility of the solvent due to EOF (see above).

Figure D5.18 shows the application of the CE technique to the measurement of the free solution mobility of DNA in tris-acetate-EDTA (TAE) buffer. The

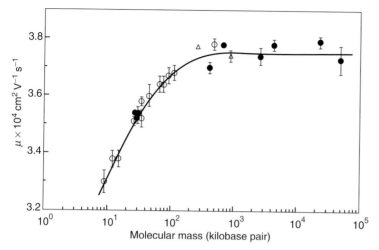

Fig. D5.18 Dependence of the electrophoretic mobility on the logarithm of molecular weight for monodisperse DNA fragments in buffer. (After Stellwagen *et al.*, 1997.)

capillaries were coated with acrylamide monomer, which is stable at basic pH and effectively eliminates EOF due to the capillary walls.

The free solution mobility of DNA in TAE buffer was found to be $(3.75 \pm 0.04) \times 10^{-4}$ cm^2 V^{-1} s^{-1} at 25 °C, independent of the DNA concentration, electric field strength and capillary coating. The free solution mobility of DNA was independent of its molecular mass (from ∼400 base pairs to 48.5 kilobase pairs). It decreases for smaller fragments monotonically with decreasing molecular mass of the smaller fragments. The origin of the molecular mass dependence of the electrophoretic mobility of small DNA fragments is not clear.

Fig. D5.19 Conventional and capillary formats. In a conventional two-dimensional electrophoresis system (a), a gel strip is used for the separation in the first dimension, and (b) the gel strip is connected to a gel slab for electrophoresis in the second dimension. In the capillary format (c), the gel strip is replaced by a capillary, and (d) the gel slab is replaced by multiple parallel capillaries. (e) This diagram suggests the simplest design that connects the capillary used for the first separation to the array of capillaries used in the second separation. (After Chen *et al.*, 2002.)

D5.7.4 Two-dimensional capillary electrophoresis

In standard two-dimensional gel electrophoresis, separation in the first dimension (usually by IEF) is done in a strip of gel. Additional separation according to another property (for instance the size of the protein) is carried out in a slab of gel by electrophoresis in a second, orthogonal direction.

An apparatus prototype for carrying out two-dimensional gel electrophoresis in a capillary format has been developed. In this method, separation in the first dimension is carried out in a one-dimensional capillary, with this system physically isolated from the capillaries that provide the separation in the second dimension (Fig. D5.19). After completion of the first separation, the first-dimension channel is physically connected to the second-dimension capillaries, and a second separation is carried out in an orthogonal set of a parallel capillaries. The ability of polydimethylsiloxane to support the fabrication of three-dimensional microfluidic systems makes it possible to produce membranes that both enclose the gel used in the first separation in a capillary and provide passages for the proteins to migrate into the array of the orthogonal capillaries.

The feasibility of this system was demonstrated using a protein mixture containing fluorescein-conjugated carbonic anhydrase, fluorescein-conjugated BSA, and Texas-Red-conjugated ovalbumin.

D5.7.5 Separation of chiral molecules

Many of the most widely used drugs contain at least one chiral centre, and about 80% are marketed as racemates. In effect, each drug dose may be contaminated with an equal amount of an isomer that does not have therapeutic value and could potentially cause unsuspected side effects. One of the most frequently prescribed cardiac β-adrenergic blocking agents is racemic propanol. The D enantiomer of propanol is 100 times more potent and is more slowly metabolised than the L enantiomer.

The main method for obtaining chiral separations in CE is to add cyclodextrins to the separation electrolyte. Cyclodextrins are non-ionic, cyclic, chiral carbohydrates composed of glucopyranose units bound through α-(1,4) linkages as shown in Fig. D5.20. The structure of a cyclodextrin is that of a truncated cone with a relatively hydrophobic interior and hydrophilic exterior. The larger opening of the cavity is lined with secondary hydroxyls, and the smaller opening is lined with primary hydroxyls.

Cyclodextrins are ideal as chiral selectors on CE because of their ability to form host–guest complexes, also called inclusion complexes, with a wide variety of compounds. As the two drug enantiomers migrate along the capillary one of the enantiomers will interacts more strongly than the other enantiomer (see Fig. D5.21). Therefore the more strongly interacting enantiomer reaches the detector later than the less retained enantiomer. Typically β-cyclodextrin and chemically derivatised cyclodextrins such as dimethylated- or hydroxypropylated-

Fig. D5.20 Chemical structure of β-cyclodextrin. (After Ward, 1994.)

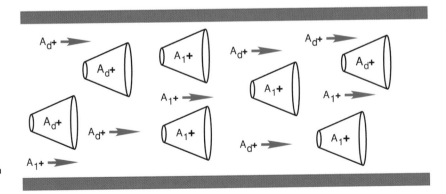

Fig. D5.21 Schematic diagram of interaction between two drug enantiomers and cyclodextrin in CE. (From www.ceandcec.com.)

cyclodextrins are added to the electrolyte at 1–100 mM concentrations. The mechanism for chiral recognition of solutes with cyclodextrins, both bonded and in free solution, was well described by Armstrong and coworkers in 1985 and was later verified by X-ray crystallographic studies. Figure D5.22 shows the separation and enantiomeric resolution of five dansyl-DL-amino acids using γ-cyclodextrins and SDS.

D5.8 Capillary electrochromatography (CEC)

Advances in capillary liquid chromatography with advances in capillary electrophoresis propelled the development of a new technique – CEC. In conventional liquid chromatography the flow of eluent is generated by the application of a pressure gradient, which is usually provided by means of a high-pressure mechanical pump. However, under certain conditions, the mobile phase can be forced to migrate through the chromatographic medium by the application of an electric field. This phenomenon of an electrically induced fluid flow is called EOF. Liquid chromatography based on this effect, in which the pump is substituted by a high-voltage power supply, has come to be known as electrochromatography.

In essence, CEC is a hybrid of HPLC and CE, and can be performed in open tubes as well as in capillaries packed with conventional packing materials. As in CE, an electric field is applied across the separation capillary, generating a solvent flow caused by electroosmosis (Section D5.7). In CEC electroosmosis is used as a 'pumping' mechanism to drive the solvent through a separation column as an alternative to pressure-driven flow.

Two components are associated with the separation process in CEC. One involves solute retention based on differential interactions between the mobile and the stationary phase (i.e. the distribution ratio), as in HPLC. The other involves differential migration, which is characterised by the solute's electrophoretic velocity, which is similar to CE. This process simultaneously involves chromatography and electrophoresis

The stationary phase can be fixed on particulate material, commonly silica based, packed inside the capillary column or fixed on the walls of the open tube (Fig. D5.23). The mobile phase, typically an organic-aqueous solution, contains an electrolyte and is transported through the capillary column by EOF, which is generated at the solid–liquid interface on the silica material. The silica surface is negatively charged because of the deprotonation of the surface sianol groups. Consequently, the solution at the interface bears a net positive charge (caused by ions in solution) to balance the fixed negative charges on the silica surface, forming an electrical double layer. The positive character of the solution near the surface extends into a mobile diffuse region of the double layer. Under the influence of the electric field, solvated cationic species beyond a shear layer in the diffusion region of the double layer are electrically driven toward the negative electrode, dragging the solvent molecules with them. Thus, a flow is generated

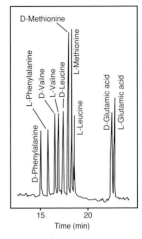

Fig. D5.22 Separation and enantiomeric resolution of five dansyl-DL-amino acids using γ-cyclodextrins and SDS. A 57 cm × 75 μm fused-silica capillary was used. (After Ward, 1994.)

Fig. D5.23 Schematic
representation of CEC
using a packed capillary.
(Adapted from Colon
et al., 1997.)

near the solid–liquid interface of the silica particles, as well as on the walls of the capillary.

Macromolecules possessing a charge can also be influenced by the electric field, leading to the differential migration caused by electrophoresis, as in CE. Therefore, both charged and uncharged species can be separated according to

Fig. D5.24 Separation of five small peptides on a CEC capillary column. Column, 100 μm internal diameter, 25 cm separation length; voltage 25 kV; background electrolyte, 40 mM l^{-1} NH$_4$Ac–AcOH (pH 4.5); acetonitrile (1:1 v/v); detection 214 nm. T, triptorelin; D, desmopressin; O, oxylocin; A, peptide A; U, uracil; C, EOF marker. (After Unger *et al.*, 2002.)

their differential migration through the column, based on the solute's interaction between two phases and the inherent electrophoretic mobilities of the solute. As an example Fig. D5.24 shows the separation of five small peptides on a CEC capillary column.

D5.9 Checklist of key ideas

- The motion of charged particles in an electric field is called electrophoresis; the motion is brought about by the Coulombic forces acting between the particles and field.
- The electrophoretic steady-state velocity of a charged particle per unit of electric field is called the *electrophoretic mobility, $\mu = u/E = Q/f$,* where Q is the charge and f is the frictional coefficient. It has units in the SI system of $m^2\ s^{-1}\ V^{-1}$.
- The absolute values of electrophoretic mobility in solution can be measured by *moving boundary electrophoresis* or *free electrophoresis*; the process can be equated to high-velocity sedimentation.
- The effective charge of the particle can be calculated from measurements of the electrophoretic mobility and the diffusion coefficient, using the techniques of membrane-confined analytical electrophoresis.
- Developments in the theory of free-solution electrophoresis solved the problem of frictional properties for polyions of arbitrary shape and charge distribution by using the boundary element technique.
- The absolute molecular mass of a protein can be estimated from the relative mobilities of its polypeptide chains in SDS-PAGE gel by comparing the mobilities of test proteins to the mobilities of known protein markers.
- IEF is a method of separating proteins according to charge; the method has a high resolution, being able to separate proteins that differ in their isoelectric points (pI) by as little as 0.01 of a pH unit.
- Combination of the IEF method with the size separation method in SDS-PAGE gives two-dimensional polyacryalamide gel electrophoresis (2-D PAGE) in a slab, the most sophisticated analytical method available for separating proteins
- One-dimensional CE yields high-speed, high-resolution separation on exceptionally small sample volumes (0.1–10 nL in contrast to slab electrophoresis, which requires samples in the microlitre range).
- Microsecond electrophoresis is based on an unusual capillary geometry, the hour-glass, which locally enhances the electric field strength at the separation region; microsecond electrophoresis promises a new approach to the study of a range of short-lived molecules.
- Two-dimensional CE can be considered as an alternative to slab gel electrophoresis with all the advantages of modern automated technology.
- Cyclodextrins are ideal as chiral selectors in CE because of their ability to form host--guest complexes with a wide variety of compounds
- CEC is a hybrid of HPLC and CE; it offers several advantages over traditional competitive techniques used for separating charged macromolecules.

Suggestions for further reading

Historical review and introduction to biological problems

Vesterberg, O. (1993). A short history of electrophoretic methods. *Electrophoresis*, **14**, 1243–1249.

Cantor, C., and Schimmel, P. (1980). *Biophysical Chemistry*. Part II. *Technique for the Study of Biological Structure and Function*. San Francisco. A. W. H. Freeman and Co.

Electrophoretic experiments

Van Holde K. E., Johnson, W. C., and Ho, S. P. (1998). *Principles of Physical Biochemistry*. New Jersey: Prentice Hall.

Skoog, D. A., Holler, F. J., and Neiman, T. A. (1998). *Principles of Instrumental Analysis*, fifth edition. Philadelphia: Sounders College Publishing.

Walker J. M. (2000). Electrophoretic techniques. In *Principles and Techniques of Practical Biochemistry*, Ch. 12, eds. K. Wilson and J. Walker. Cambridge: Cambridge University Press.

Garfin, D. E. (1995). Electrophoretic methods. In: *Introduction to Biophysical Methods for Protein and Nucleic Acid Research*, eds. J. A. Glaser and M. P. Deutscher. San Diego: Academic Press.

Macroions in electric field

Manning, G. S. (1978). The molecular theory polyelectrolyte solutions with applications to the electrostatic properties of polynucleotides. *Q. Rev. Biophys.*, **11**, 179–246.

Schmitz, K. S. (1993). *Macroions in Solution and Colloidal Suspension*. New York: VCH Publishers.

Allison, S. A. (2001). Boundary element modelling of biomolecular transport. *Biophys. Chem.*, **93**, 197–213.

Electrophoretic methods in the absence of support medium: free electrophoresis

Laue, T. M., Ridgeway, T. M., Wool, J. O., and Shepard, H. K. (1996). Insights into a new analytical electrophoresis apparatus. *J. Pharm. Sci.*, **85**, 1331–1335.

Stellwagen, N. C., Gelfi, C., and Righetti, P. G. (1997). The free solution mobility of DNA. *Biopolymers*, **42**, 687–703.

Electrophoretic methods in the presence of support medium: zonal electrophoresis

Weber, K., and Osborn, M. (1969). The reliability of molecular weight determinations by dodecyl sulfate-polyacrylamide gel electrophoresis. *J. Biol. Chem.*, **244**, 4406–4412.

O'Farrell, P. (1975). High-resolution two-dimensional electrophoresis of protein. *J. Biol. Chem.*, **250**, 4007–4021.

Celis, J. E., and Gromov, P. (1999). 2D protein electrophoresis: can it be perfected?. *Curr. Opin. Biotech.*, **10**, 16–21.

Capillary electrophoresis

Grossman, P. D. (1992). Free solution capillary electrophoresis. In *Capillary Electrophoresis: Theory and Practice*, eds. P. D. Grossman and J. C. Colburn. San Diego, CA: Academic Press.

Karger, B. L., Chu, Y.-H., and Foret, F. (1995). Capillary electrophoresis of proteins and nucleic acids. *Annu. Rev. Biophys Biomol. Struct.*, **24**, 579–610.

Zubritsky, E. (2002). How analytical chemists saved the human genome project. *Anal. Chem.*, **74**, 23A–26A.

Chen, X., Wu, H., Mao, C. and Whitesides, G. M. (2002). A prototype two-dimensional capillary electrophoresis system fabricated in poly(dimethylsiloxane). *Anal. Chem.*, **74**, 1772–1778.

Plenert, M. L., and Shear, J. B. (2003). Microsecond electrophoresis. *Proc. Natl. Acad. Sci. USA*, **100**, 3853–3857.

Capillary electrochromatography

Colon, L. A., Guo, Y., and Fermier, A. (1997). Capillary electrochromatography. *Anal. Chem. News*, **69**, 461A–467A.

Unger, K. K., Huber, M., Walhagen, K., Hennessy T. P., and Hearn, M. T. W. (2002). A critical appraisal of capillary electrochromatography. *Anal. Chem.*, **74**, 200A–207A.

Chapter D6
Electric birefringence

D6.1 Historical review and introduction to biological problems

1875–1880

J. Kerr first observed the transient electric birefringence effect (TEB) and established that the birefringence is proportional to the square of the field strength. Later, the law was named after him.

1880–1920

During this time high d.c. voltage generation was used in the study of TEB. It seriously hampered the application of TEB to the study of aqueous solutions because of their conductivity. In **1940–1950** a few groups **(W. Kaye** and **R. Devaney, N. A. Tolstoy** and **P. P. Feofilov, C. T. O'Konski** and **B. H. Zimm,** and **H. Benoit)** independently initiated the measurements of the electric birefringence of colloidal systems using a short rectangular electrical pulse. This greatly reduces the heating and polarisation effects, which are most critical in studies biological macromolecules.

1910

L. Langevin proposed the first theory for electric birefringence. In **1918 M. Born** generalised the theory. It was shown that the orientation of optically anisotropic molecules may be caused by a permanent dipole momentum or its electrical anisotropy, or both. In **1931, A. Peterlin** and **H. A. Stuart** extended the theory to suspensions of rigid particles.

1951

H. Benoit developed a theory for transient phenomena by solving the diffusion equation, which describes the rotational diffusion of axially symmetric rigid macromolecules following the establishment and disappearance of an electric field. He showed that the rotational diffusion coefficients can be obtained from the decay curve and the mechanism of electrical orientation can be elucidated from the build-up curve. He also applied the theory to studies of aqueous solutions of TMV and DNA.

1959

O'Konski proposed a method for separately determining the dipole moment, the anisotropy of electrical polarisability and the optical anisotropy factor from the field strength dependence of the birefringence in the case of axially symmetric rigid macromolecules. This method was applied to TMV, synthetic polypeptides and collagen. In **1963 I. Tinoco Jr** extended the birefringence theory to the most general molecular model, and applied it to an interpretation of the data on fibrinogen. In **1973 E. Fredericq** and **C. Houssier's** book '*Electric Dichroism and Electric Birefringence*' was published; this is one of the best books describing the Kerr phenomenon and its application.

1979

W. A. Wegener solved the problem of birefringence decay for a general rigid body of arbitrary shape. He showed that in this case the time dependences of TEB are multiexponential functions containing up to five components. In **1981 B. Zimm** and **P. J. Hagerman** presented a quantitative numerical relationship between the rotational diffusion coefficient, as measured by the decay of optical anisotropy following an electric field pulse, and the flexibility (persistence length) of short, worm-like chains. In the **1990s P. J. Hagerman** and his group made an essential contribution to the theory and application of TEB for the study of the macromolecular structure of DNA.

1990–2000

Two important methodological advances in TEB instrumentation were introduced. First, instrument response times were dramatically reduced (approaching 1 ns), thus enabling one to study the electrooptic–hydrodynamic properties of proteins and nucleic acids that have a longest dimension of less than 10 nm. Second, a new electrically insulated ceramic with excellent thermal conductivity, aluminium nitride, was introduced into TEB experiments. This allows TEB measurements in solutions that have much higher salt concentrations, without significant Joule heating of the samples during the pulse and, therefore, made a higher pulse repetition frequency possible. All these innovations led to greater signal stability, more accurate resolution of the component exponentials in TEB measurements and opened up a new way to study the global conformation and flexibility of RNA and DNA.

D6.2 Macromolecules in an electric field

D6.2.1 Dipole moments of biological macromolecules

The most important dipole is the electric dipole, which consists of two charges of opposite sign ($+ q$ and $- q$), separated by distance L (Fig. D6.1). Many molecular species, such as ionised groups or those with an asymmetric charge distribution

Fig. D6.1 A dipole in an electric field. There is no transport in the electric field, because the net charge is zero. However, there is a torque tending to orientate the molecule in the direction of the field.

Comment D6.1
Dipole moment

Values of the dipole moment, μ, are expressed in the cgs system of units, since this system is universally used by workers in the field. The numerical values are in debye units, D, $(1\ D = 10^{-18}$ electrostatic units of charge $\times$ centimetres). The conversion factor to the SI system is $1\ D = 3.33564\times 10^{-30}$ C m.

in a polar bond, can be treated as dipoles. In these cases the molecule is said to possess a *permanent dipole moment* (Comment D6.1). This charge asymmetry in the molecule is given a magnitude and direction by calculating the dipole moment μ

$$\mu = qL \tag{D6.1}$$

A dipole moment occurs only in asymmetric molecules. For example, the H_2O molecule is bent and it has a dipole moment of 1.85 D. The CO_2 molecule is symmetric and it has no dipole moment.

When a molecule that has no net permanent dipole moment is placed in an electrical field, the field causes the displacement of the electron clouds (negatively charged) away from the nuclei (positively charged), thus inducing a dipole. In these cases the molecule is said to possess an *induced dipole moment.* The induced dipole exists as long as the field remains.

The interaction of the molecules with an external field depends on their polarisability or on the ability of a neutral molecule to become a dipole in an external electric field (Comment D6.2). The magnitude of the dipole moment μ induced by the external field is proportional the field intensity. The proportionality constant is called the 'polarisability':

$$\mu = \alpha E \tag{D6.2}$$

Comment D6.2
Polarisability

Polarisability is a scalar quantity for isotropic substances but it becomes a tensor quantity for anisotropic substances. The dimension of polarisability is cm^3 in the cgs system.

Very complex molecules can be represented by a single vector sum of all the moments of individual dipoles within the molecules. A synthetic polypeptide in the α-helical conformation has a very large dipole moment because all of the small dipole moments of the amino acid residues along the chain add together, whereas the antiparallel double helix of DNA has a small permanent moment but may have a large induced moment.

D6.2.2 Electric birefringence and dichroism

Some crystalline substances transform incident light into two perpendicularly polarised beams which are propagated at different velocities in the medium. This property is called *birefringence* or *double refraction*. The birefringence of the medium is defined as the difference between the refractive indices of the two beams:

$$\Delta n = n_x - n_y \tag{D6.3}$$

and is related to the optical retardation δ (also called the phase shift) by

$$\delta = 2\pi l \frac{\Delta n}{\lambda} \tag{D6.4}$$

where λ is the wavelength of the incident light beam in vacuo, and l is the thickness of the sample.

When the two orthogonal vibrations transmitted by the sample are differently absorbed, the substance is said to exhibit *dichroism,* which is defined as the difference in absorbance between the two beams:

$$\Delta A = A_x - A_y. \qquad (D6.5)$$

These two phenomena constitute *optical anisotropy*. They are usually called linear birefringence and linear dichroism to distinguish them from the circular birefringence and circular dichroism that characterise optical activity (Chapter E3). Like refraction and absorption, birefringence (if present) may be observed at any wavelength while dichroism (if present) necessarily appears in the absorption bands only.

Optical anisotropy may be induced in isotropic solutions by external forces such as hydrodynamic forces (flow birefringence), electric (electric birefringence) and magnetic fields (magnetic birefringence). These forces produce an orientation of the particles in the solution: in such cases, the x and y orthogonal polarisations are chosen parallel and perpendicular to the orientation force, respectively, and the birefringence and dichroism are defined by

$$\Delta n = n_{||} - n_{\perp} \qquad (D6.6)$$

and

$$\Delta A = A_{||} - A_{\perp} \qquad (D6.7)$$

respectively. Depending on the relative magnitudes of $n_{||}$ ($A_{||}$) and $n_{\perp}$ ($A_{\perp}$) the birefringence (dichroism) can be positive or negative.

In principle, TEB and transient electric dichroism (TED) yield the same information regarding macromolecular structure. Both should yield an identical set of field-free decay times and the signs and amplitudes of the birefringence and dichroism signals should provide information regarding the orientation of chromophoric groups within molecules. However, in practice, TEB gives more precise information about the decay constant and is therefore more appropriate for studies of the external shapes of macromolecules, while TED is a more precise means of determining amplitudes and is therefore more suitable for studies involving the determination of internal structure.

In this chapter we deal exclusively with TEB. We will not discuss magnetic birefringence, which results from magnetically and optically anisotropic particles in a strong magnetic field. Interested readers should refer should to specialist literature.

D6.3 Theoretical background of TEB measurements

Two main types of pulses are used in TEB measurements (Fig. D6.2). In the first, a rectangular electric pulse is applied to a solution of a particles (Fig. D6.2(a)). After the pulse the birefringence produces changes with time as shown in

Fig. D6.2 (a) Rectangular
electrical impulse
E versus time.
(b) Birefringence Δn
versus time, showing the
attainment of a
steady-state value Δn_0.
(c) Reversal of the
electrical impulse. The
voltage in the signal is
changed in mid-pulse.
(d) Transient of a
birefringence signal upon
reversal of the field.

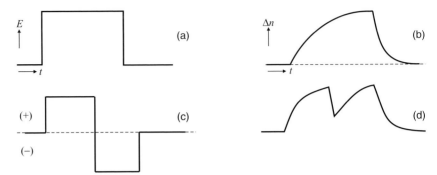

Fig. D6.2 (a) Rectangular electrical impulse E versus time. (b) Birefringence Δn versus time, showing the attainment of a steady-state value Δn_0. (c) Reversal of the electrical impulse. The voltage in the signal is changed in mid-pulse. (d) Transient of a birefringence signal upon reversal of the field.

Fig. D6.2(b). This pulse technique allows the build-up and decay of the birefringence, as well as its steady state, to be studied. In the second, a rectangular pulse is also applied, and, after the steady-state birefringence is achieved, the field is rapidly reversed in sign as shown in Fig D6.2(c). After such a pulse the birefringence produced changes with time as shown in Fig. D6.2(d). This pulse technique is more useful for investigating the mechanism of electrical orientation.

Below we shall briefly summarise the main theoretical results for dilute solutions of rigid particles. For the sake of simplicity, we assume that the macromolecule has a common axis of symmetry for its electric, optical, and hydrodynamic properties and the permanent dipole moment is along this axis. It is further assumed that the solution is diluted so that there are no interactions between the macromolecules.

D6.3.1 Steady-state birefringence

The relation between the optical anisotropy and the electric birefringence of a macromolecular solution is

$$\Delta n = \frac{2\pi \Omega}{n}(g_1 - g_2)\Phi \qquad (D6.8)$$

where Ω is the volume fraction of the solute macromolecules, n is the refractive index of the solution in the absence of an electric field. The birefringence is proportional to a product of an optical anisotropy factor, $g_1 - g_2$, and an orientation distribution function, Φ.

Upon the application of the electric field during a sufficiently long time, the orientation distribution function Φ takes its steady-state form and is given by

$$\Phi = \frac{1}{2}(3\langle \cos^2 \theta \rangle_{\text{av}} - 1) \qquad (D6.9)$$

where θ is angle between the symmetry axis of a macromolecule and the field direction, and $\langle \ldots \rangle$ denotes an ensemble average.

When an orienting effect due to the electric field is just balanced by a disorienting effect due to the rotational Brownian motion,

$$\Delta n = \frac{2\pi\Omega}{n}(g_1 - g_2)\frac{1}{15}\left(\frac{\mu^2}{k^2T^2} + \frac{\alpha_1 - \alpha_2}{kT}\right)E^2 \qquad (D6.10)$$

where α_1 and α_2 are the excess electrical polarisabilities along the symmetry and transverse axes, respectively. Thus the birefringence is proportional to the square of the field strength for limiting low fields. This is Kerr's law. Equation (D6.10) shows that in a TEB experiment the molecules tend to orient in the field through interactions of induced ($\alpha_1 - \alpha_2$), and/or permanent μ dipoles with the applied field.

The proportionality constant K is called the Kerr constant. The Kerr constant is given by

$$K = \frac{2\pi\Omega}{n^2}(g_1 - g_2)\frac{1}{15}\left(\frac{\mu^2}{k^2T^2} + \frac{\alpha_1 - \alpha_2}{kT}\right) \qquad (D6.11)$$

Because the Kerr constant is a product of terms involving the optical ($g_1 - g_2$) and electrical ($\mu^2/k^2T^2 + (\alpha_1 - \alpha_2)/kT$) properties, one cannot determine the individual contributions of these terms in the Kerr law regime. Another problem arises when the solution exhibits form birefringence (Comment D6.3).

D6.3.2 Decay process

If the electric field is removed, rotational Brownian motion disorients the macromolecules, leading to a decay of the birefringence. In this case the decay of the birefringence provides a direct measure of the relaxation time, τ, as the decay is of the form

$$\Delta n = \Delta n_0 \exp(-t/\tau) \qquad (D6.12)$$

where Δn is the birefringence at any time after the start of decay and Δn_0 is the birefringence at time $t = 0$.

We recall that relaxation time t is related to the rotational diffusion constant Θ (Section D2.3.3). Thus the decay of birefringence provides a more direct method for determining the rotational diffusion constant than other methods such as flow birefringence (Chapter D.7) and depolarisation of fluorescence (Chapter D8). The rotational diffusion constant Θ is related to the dimensions and shape of the macromolecules as we discussed in Section D2.3.4.

D6.3.3 Rise process

For a sufficiently low field strength, the birefringence Δn at time t is given by

$$\frac{\Delta n}{\Delta n_0} = 1 - \frac{3A}{2(A+1)}\exp(-2\Theta t) + \frac{A-2}{2(A+1)}\exp(-6\Theta t) \qquad (D6.13)$$

Comment D6.3
Form birefringence

If the molecules have a mean refractive index different from that of the solvent, the solution exhibits *form* birefringence. If the molecules are optically anisotropic, the solution may also exhibit *intrinsic* birefringence. The best way to estimate the contribution of form birefringence is to measure birefringence in solvents of varying refractive index. The form birefringence then disappears when the refractive index of the solvent is equal to that of the solute. Solutions of glycerol and sucrose in water are used for this purpose.

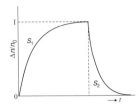

Fig. D6.3 Normalised build-up and decay curves $\Delta n / \Delta n_0$ versus time.

where Δn_0 is the value of Δn_0 as $t \to \infty$ and is the same as the steady-state value given by Eq. (D6.10). The parameter A is the ratio of the permanent dipole contribution to the induced dipole contribution in the expression for the low-field steady-state birefringence:

$$A = \frac{\mu^2}{(\alpha_1 - \alpha_2)kT} \qquad (D6.14)$$

When the orientation is entirely due to the anisotropy of the electrical polarisability (i.e. $A = 0$), Eq. (D6.13) reduces to

$$\frac{\Delta n}{\Delta n_0} = 1 - \exp(-6\Theta t) \qquad (D6.15)$$

In this case the build-up curve and the decay curve are symmetrical. On the other hand, when the orientation is entirely due to the permanent dipole moment (i.e. $A = \infty$), Eq. (D6.13) reduces to

$$\frac{\Delta n}{\Delta n_0} = 1 - \frac{3}{2}\exp(-2\Theta t) + \frac{1}{2}\exp(-6\Theta t) \qquad (D6.16)$$

In this case the build up is much slower than the decay. It is possible to obtain the parameter A from the shape of the build-up curve.

Another method for determining the parameter A is the area method. The ratio of the area S_1 bounded by the build-up curve and the area S_2 bounded by the decay curve in Fig. D6.3 is given by the expression

$$\frac{S_1}{S_2} = \frac{4A + 1}{A + 1} \qquad (D6.17)$$

for low field strength, if the decay begins after the steady state is reached. Note that the ratio S_1/S_2 is equal to unity for $A = 0$ and equal to 4 for $A = \infty$.

D6.3.4 Birefringence in a reversing field

The equivalent of Benoit's rise expression for the reversal case in a low field is

$$\frac{\Delta n}{\Delta n_0} = 1 - \frac{3A}{A + 1}[\exp(-6\Theta t) - \exp(-2\Theta t)] \qquad (D6.18)$$

The most salient feature of the reversal curve is the minimum in orientation or birefringence, which occurs at time t_{min} with a depth d_{min} (Fig. D6.2(d)). According to the low-field theory, both quantities are independent of field strength and, furthermore, t_{min} does not depend on the type of dipole.

D6.4 Measurement of TEB

D6.4.1 Experimental schemes

A schematic a TEB experiment is shown in Fig. D6.4. Light from a laser source, propagating along the z-axis, passes successively through a polariser, the Kerr cell, and an analyser. A beam is polarised by a polariser, which is set typically

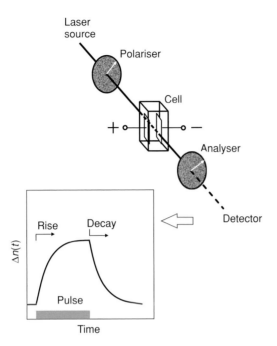

Fig. D6.4 Outline of a basic TEB experiment. Upon application of an electric impulse, the molecules in the cell become partially oriented, rendering the solution birefringent. In the standard optical arrangement, the electric field (and hence the principal axis of orientational asymmetry) lies along the x-axis that is set at 45° with respect to the polariser and analyser, which are disposed at 90° to one another. (Adapted from Hagerman (2000)). A complete apparatus for electric birefringence measurements is not commercially available at the time of writing, and anyone who plans to enter this field of research must be prepared to design and construct the necessary optical and electric trains (see, for example, Hagerman, 2000).

at 45° with respect to the electric field across a Kerr cell. When the solution becomes birefringent with the application of a pulse, the light emerging from the solution is elliptically polarised. The light then passes through an analyser, which is set at 90° with respect to the polariser, and falls on a detector.

The standard optical arrangement includes a quarter-wave plate, which is interposed between the Kerr cell and the analyser, with its fast axis in the direction of the axis of the polariser (Comment D6.4). The analyser is rotated by an angle α

Comment D6.4   Simplest optical arrangement in TEB

In the simplest optical arrangement without a quarter-wave device the change in light intensity at the detector produced by the electric field ΔI_δ is given by

$$\Delta I_\delta = I_0 \sin^2(\delta/2)$$

where δ is the optical retardation due to the electric field, and I_0 is the light intensity, not including the stray light, which would have reached the detector if the polariser and the analyser were parallel and δ were zero. For $\delta \langle\,\rangle 1$ we obtain the approximate formula

$$\Delta I_\delta = I_0 \delta^2/4$$

This scheme has the disadvantage that the sensitivity is low for small values of δ and the sign of δ remains unknown.

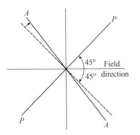

Fig. D6.5 Direction of the analyser axis *A* with respect to the polariser axis *P*.

from the crossed position, as shown in Fig. D6.5. The elliptically polarised light emerging from the birefringent solution is reconverted into linearly polarised light, after passing through the quarter-wave plate, whose angle of polarisation depends on the magnitude of δ, the optical retardation due to the electric field. In this optical system the relation between ΔI_δ and δ for $\delta \ll 1$ is approximated by

$$\Delta I_\delta = I_0 \sin 2\alpha(\delta/2) \tag{D6.19}$$

Consequently ΔI_δ is proportional to δ, instead of δ^2. The sensitivity achieved in this way is greater than any sensitivity achievable without the quarter-wave plate. Furthermore, the light signal increases if δ is positive and decreases if δ is negative. This permits us to determine the sign of the birefringence.

Birefringence decay curves are usually displayed on a semilogarithmic scale as indicated in Fig. D6.6. If the macromolecules under observation are solid and monodisperse, a plot $\ln\Delta n(t)/\Delta n(0)$ versus t is a straight line with a negative slope of $1/\tau$.

Fig. D6.6 Example of birefringence decay. Semilogarithmic plot of the decay curve for a 98-bp DNA restriction fragment displaying single-exponential decay over two orders of magnitude of birefringence signal intensity. Measurement conditions: 2.0 mM Tris-HCl, pH 8.0, 4 °C; 2 μs pulse width; 5 kV cm^{-1} pulse amplitude. Inset: same curve plotted on a linear ordinate scale. (Hagerman, 1985.)

D6.4.2 Interpretation of the birefringence decay profile

For rigid particles of arbitrary shape, the decay of birefringence can be described in terms of up to five decay times depending upon the symmetry of the particle with respect to both its charge distribution (i.e. how it orients in the field) and the directions and magnitudes of the various optical transitions within the particle:

$$\Delta n(t) = \sum_{k=1}^{5} C_k \exp \frac{-t}{\tau_k} \tag{D6.20}$$

The extraction of the five relaxation times τ_k and their corresponding amplitudes C_k from an experimentally measured $\Delta n(t)$ is impossible owing to the well-known ill-conditioned nature of multiexponential fitting.

For the case a tri-axial body, when the main axes of rotational diffusion (Θ), electric polarisability and electric alignment coincide only two relaxation terms are involved in the birefringence decays:

$$\frac{\Delta n(t)}{\Delta n(0)} = a_1 \exp\left(\frac{-t}{\tau_1}\right) + a_5 \exp\left(\frac{-t}{\tau_5}\right) \tag{D6.21}$$

where a_1 and a_5 are the *longest* and *shortest* relaxation times.

In 1996 Garcia de la Torre and coworkers proposed characterising decay in terms of other relaxation times that can be more precisely determined. One of them is the initial relaxation time τ_{ini}, defined by

$$\tau_{\text{ini}}^{-1} = [d\Delta n(t)/dt]_{t=0} \tag{D6.22}$$

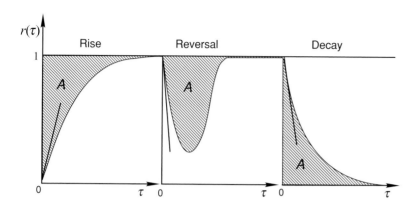

Fig. D6.7 Geometric interpretation of the initial and mean relaxation time for the three electrooptic processes.

so that τ_{ini}^{-1} is the initial slope of the $\Delta n(t)$ curve. A second is the mean relaxation time τ_{mean}

$$\tau_{mean} = \int_0^\infty \Delta n'(t)dt \qquad (D6.23)$$

where $\Delta n'(t) = \Delta n(t)$ for the decay, while $n(t) = 1\Delta n(t)$ for the rise and reversal processes, so that $\Delta n(0) = 1$ and $\Delta n(\infty) = 0$ in all cases. Note that τ_{mean} is the area under the $\Delta n(t)$ decay. The geometric meanings of τ_{ini} and τ_{mean} are illustrated in Fig. D6.7. For a sphere with radius R_0, all the τ_k relaxation times (including τ_{ini} and τ_{mean}) reduce to a single value, denoted by τ_0, and given by

$$\tau_0 = \frac{1}{6\Theta} \qquad (D6.24)$$

D6.5 Global structure and flexibility of proteins

Equation (D6.11) shows that with TEB it is possible to determine two characteristic birefringence constants, permanent and induced dipole moments, electric polarisability and optical anisotropy of macromolecules. Below we give examples of the high efficiency of TEB in studies of molecular motor domains.

D6.5.1 Internal structure of 'motor' domains from TEB

The 'motor' domains, namely the ncd domain of *Drosophila* and the human kinesin domain, facilitate the transport of vesicules and organelles along microtubules (see Chapter F4). The truncated heavy-chain motor domain of human kinesin, kinesin(349), has a molecular mass of 39.200 Da and, contains the first 349 amino acids of human kinesin; the heavy-chain motor domain of *Drosophila* Ncd (Ncd/(335–700)) has a molecular mass of 41.308 Da and contains 366 amino acids. Although the molecular masses of the two molecular motor domains differ by only 5% and possess extensive sequence similarity, they manifest different optical, electric and hydrodynamic properties. The slightly smaller

Fig. D6.8 Comparison of the TEB signals from solutions at 4°X of (a) kinesin (349) and (b) Ncd (335–700) with added ADP and Na_2HPO_4 after the application of 520-ns pulses. The protein concentrations are each 1.5 mg ml^{-1} (After Eden et al., 1995.)

kinesin domain (349 amino acids) displays a rotational decay time that is significantly longer (and field-strength-dependent) than the decay time for the ncd domain (366 amino acids). It is most intriguing that the two domains have opposite signs for the TEB signal. As shown in Fig. D6.8(a) the sign of the birefringence of kinesin is positive, in contrast to that of the ncd motor (Fig. D6.8(b)). The differences in sign and magnitude of the Kerr constants of Ncd (335–700) and kinesine (349) suggest large differences in their electrical properties. The alignment of both the ncd and kinesin motors is primarily via a permanent dipole moment, because the rise of the birefringence is much slower than the decay (Fig. D6.8). Interestingly, the ncd and kinesin motor domains translocate with opposite polarity, although the relationship between this observation and the sign of the TEB signal is not known.

D6.5.2 Hydrodynamic dimensions of globular proteins from TEB

Calculations according Perrin's formula (Eq. (D2.31)) show that a typical spherical molecule with radius 25 Å has a rotational relaxation time in water at 20 °C of about 17 ns. Such a small value requires the apparatus to have a very high resolution time (2–5 ns). Table D6.1 contains rotational diffusion times of a few globular proteins obtained mainly using the high-resolution TEB technique. We also include data for the dimer of gramicidine, the rotational diffusion coefficient of which was obtained using dynamic light scattering. The dependence of rotational relaxation times on molecular mass for proteins on a double logarithmic scale is shown in Fig. D6.9. The slope of this line is close to 1.0. It is clearly seen that relaxation times are proportional to the protein volume (see further discussion on this topic in Chapter D9).

Table D6.1. *Rotational relaxation times of some proteins*

Protein	M (kDa)	τ_{20w} (ns)
Gramicidin (dimer)[a]	2.5	2.7
Kinesin (349)[b]	39.2	41
Ncd (335–700)[b]	41.3	43
BSA[c]	68.0	78

[a] Michielsen and Pecova (1981).
[b] Eden et al. (1995).
[c] Munro et al. (1979).

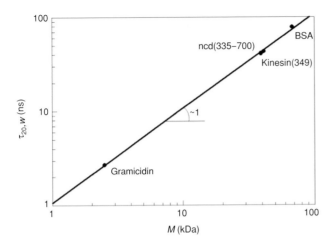

D6.6 Global structure and flexibility of nucleic acids

Early studies of the conformational properties of nucleic acids were confounded by the need to work with relatively long, size-fractionated polymers. Uncertainties in the polymer length limited the precision of the estimates of flexibility from those measurements. The TEB approach is particularly useful as a method for describing the global structure and flexibility of nucleic acids in solution for two reasons: first, because of the ease with which one can produce DNA and RNA molecules of virtually any sequence and length; second, recent advances in TEB instrumentation have resulted in great signal stability and as a consequence more accurate resolution of the component exponentials.

D6.6.1 Flexibility of DNA in B- and A-forms

Native B-form DNA under all the experimental conditions employed, always shows negative dichroism in its UV absorption band ($\lambda = 259$ nm) and negative birefringence in the visible region. This observation together with the negative birefringence and dichroism observed when the DNA molecule is oriented by the hydrodynamic force (Chapter D7) lead to two conclusions: the macromolecule is orientated by the electric field with its long axis in the direction of the field, and the plane of the base pairs is roughly at right angles to the length of the molecule (Fig. D6.10). However, the values of the reduced dichroism, which are measured and then extrapolated to infinite field strength (complete orientation), are quite low. This is mainly due to the flexibility of the DNA chain.

A particularly useful quantitative characterisation of chain flexibility in DNA is the persistence length, introduced in 1948 by Kratky and Porod in their pioneering development of the worm-like-coil model (see Fig. D2.17). In this model the chain is characterised by two parameters: its contain length c and a persistence length

Fig. D6.10 Molecular structure of B-form DNA: ten residues per turn, translation per turn 3.46 Å. (Cantor and Schimmel, 1980.)

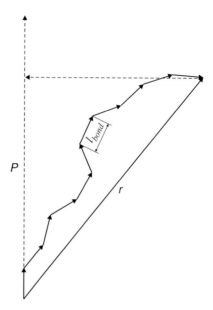

Fig. D6.11 The persistence length P is determined as the average projection of vector **r** connecting the ends of the chain to the direction of the first bond in the limit of infinite chain length. If the bond length l_{bond} became smaller and smaller, the chain would transform into a filament with a smooth curvature variation. This abstract model is called the Kratky–Porod model. The end-to-end distance, $\langle h^2 \rangle$ of a worm-like-coil DNA molecule can be calculated from its contour length, L_c, if the persistence length, P, is known:

$$\langle h^2 \rangle = 2PL_c(1 - P/L_c) + P/L_c[\exp(-L_c/p)] \tag{A}$$

In the limit of large L_c (coil conformation)

$$(\langle h^2 \rangle / L_c)_{L_c \to \infty} = 2P \tag{B}$$

For the small values of L_c Eq. (1) may be expanded for $L_c/P < 1$ to yield

$$\langle h^2 \rangle = L_c^2[1 - \frac{1}{3}(L_c/P) + \frac{1}{12}(L_c/P)^2] \tag{C}$$

For $L_c \ll P$, Eq. (C) states that $\langle h^2 \rangle = L_c^2$, i.e. the end-to-end distance is equal to the contour length. This is exactly the behaviour of a rod-like model. As L_c increases, $\langle h^2 \rangle = L_c^2$ as the effect of curvature becomes manifest. At very high values of L_c, $\langle h^2 \rangle L_c^2 = 2P/L_c$ (Eq. (B)), i.e. the deviation of $\langle h^2 \rangle$ from the rod is determined by the ratio of the persistence length to the contour length.

P. The value of P is a measure of how far a flexible macromolecule extends along the chain (Fig. D6.11).

Equilibrium persistence lengths P can be measured by a wide variety of methods that are sensitive to the equilibrium curvature including static light-scattering, magnetic birefringence, flow birefringence, sedimentation, intrinsic viscosity. Possible problems with interpretation of some of these data and contradictory results regarding the magnitude of the salt dependence of P have been discussed many times in the literature. Despite the problems, there is now a consensus that the equilibrium persistence length of DNA in the B-form in a buffer of moderate salt concentration is about 500 Å.

The availability of fragments of precisely defined length can eliminate many previous problems, particularly the problem of length polydispersity. Table D6.2 lists rotational correlation times $\tau_{20,w}$ of B-from DNA starting with a short DNA fragment (8 bp) and finishing with native DNA from the insect *Drosophila melanogaster* (119.71 Mbp). The rotational diffusion coefficients of synthetic short DNA fragments (8, 12, 20 bp) were measured using depolarised dynamic light scattering (Chapter D10). DNA fragments in the middle range of molecular mass (64–5243 bp) were obtained using a restriction enzyme and their correlation times were measured by TEB technique. Correlation times of native DNAs from different sources were calculated from viscoelastic relaxation measurements.

Figure D6.12 shows the dependence of rotational correlation times $\tau_{20,w}$ on the molecular mass of DNA on a double logarithmic scale. The data are taken from different references, but sheared, sonicated, or otherwise abused calf thymus DNAs are not included. The figure clearly demonstrates three characteristic features (ranges) of DNA behaviour in solution. First, in the range 20–200 bp the dependence of the rotational correlation times on the molecular mass of DNA on a double logarithmic scale can described by a straight line with a slope of 2.47, which is typical for a homologous series of *rod particles of finite diameter* (Section D2.6). In the region of small duplexes (8 and 12 bp) the contribution of the end effect is evident. A hairpin molecule consisting of 13 bases (tridecamer) rotates faster than the B-form duplexes because of its compact, almost spherical shape. Second, relaxation times begin to depart from strictly rod-like behaviour after approximately 180–200 bp. The decay of the birefringence depends on two

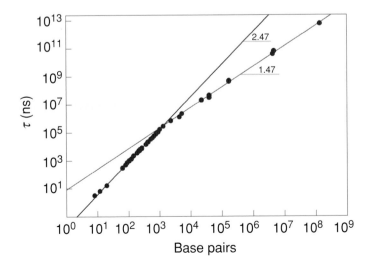

Fig. D6.12 Dependence of rotational relaxation times on the molecular mass of DNA (in bp) on a double logarithmic scale over a interval of molecular mass starting with short DNA fragment (8 bp) and finishing with native DNA from insect *Drosophila melanogaster* (120 Mbp). All measurements were made in 1–2 mM NaCl, pH 6–7. The relaxation times reduce to that of water at 20 °C.

Table D6.2. *Relaxation times of DNA molecules: from oligonucleotides (8 bp) to* Drosophila melongaster *(119 710 000 bp)*

Number of base pairs	Relaxation time	Source of DNA[a]
Tridecamer	1.9 ns	S
8	3.2 ns	S
12	6.4 ns	S
20	16.2 ns	S
64	294 ns	R
80	485 ns	R
89	703 ns	R
104	0.97 μs	R
124	1.34 μs	R
145	2.18 μs	R
188	3.44 μs	R
192	3.8 μs	R
213	4.8 μs	R
213	5.01 μs	R
213	5.01 μs	R
234	5.50 μs	R
234	6.2 μs	R
267	8.0 μs	R
267	7.15 μs	R
267	7.15 μs	R
410	20.65 μs	R
434	22.0 μs	R
520	33.8 μs	R
587	42.0 μs	R
587	42.5 μs	R
659	57.0 μs	R
782	90.0 μs	R
910	110 μs	R
1314	280 μs	R
4361	1.28 ms	R
5243	2.20 ms	R
39 937	32 ms	T7
168 903	450 ms	T4
168 903	520 ms	T4
4 266 814	40 s	*B. subtilis*
4 639 221	65 s	*E. coli*
119 710 000	6012 s	*Drosophila melongaster*

[a]Sources of DNA: S, fragments obtained by synthesis; R, fragments obtained using restriction enzyme; T7, T4, DNA obtained from T7 and T4 bacteriophages, respectively.

exponentials (fast and slow components) for fragments of more than 200 bp. Only the slow component is taken into account in the figure. The 200–8000 bp region is the *transition region from a rod-like shape to a coil shape.* Finally, DNA molecules of more than 8 kbp can be considered to be *typical Gaussian coils.* In the region of 8–4700 kbp the dependence of the rotational correlation times of DNA molecules on molecular mass is a good straight line with a slope of 1.47, which is typical for extended Gaussian coils (Section D2.6).

Figure D6.13 shows the molecular structure of A-form DNA. It has 11 residues per turn, and the diameter is a few ångströms larger than that of B-form DNA. Figure D6.14 shows the dependence of rotational correlation times $\tau_{20,w}$ on the molecular mass (in bp) of A-form DNA for four fragments, 373, 460, 676 and 917 bp, on a double logarithmic scale. The persistence length is about 1500 Å, or about three times longer than for B-form DNA.

The measurements discussed above pertain to equilibrium or static bending of DNA. None of these measurements is dynamic in the sense that it monitors the flexing of DNA per se as a function of time. The dynamic bending rigidity can be obtained from initial rapid transients in the decay of TEB. In such experiments DNA fragments are aligned by a very high electric field pulse of up to 60 kV cm^{-1}. The original data are analysed using a double-exponential decay

$$A(t) = A_1 \exp(-6\Theta t) + A_2 \exp(-t/\tau) \tag{D6.25}$$

where Θ is the rotational diffusion coefficient for the uniform end-over-end tumbling mode, and τ is tentatively assigned to a bending mode. In a weak electric field the uniform mode is observed to dominate the decay ($A_1/A_2 \rangle 1.0$) and to relax much more slowly than the bending mode ($6\Theta \langle (1/\tau)$. The relative amplitude of the fast component (A_2/A_1) rises with increasing field strength to a plateau value. The enhancement of the relative amplitude of the fast component in large electric fields is attributed to steady-state bending of initially perpendicular DNAs by the

Fig. D6.13 The molecular structure of A-form DNA: eleven residues per turn, translation per turn 2.55 Å. (Adapted from Cantor and Schimmel, 1980.)

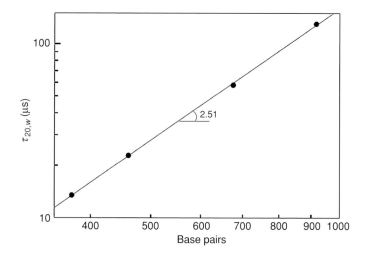

Fig. D6.14 Dependence of rotational correlation times τ on molecular mass (in bp) of A-form DNA in for four fragments, 373, 460, 676, and 917 bp, on a double logarithmic scale. The slope of this line is 2.51, which is typical for rod-like particle. Note that, relaxation times for B-form DNA begin to depart from strictly rod like behaviour after approximately 180–200 bp (Fig. D6.12).

(a)   (b)

Fig. D6.15 Side and top view of space-filling models of double helices formed by Watson–Crick base pairing between complementary DNA (a) and RNA (b); ribose sugars are depicted in yellow and deoxyriboses in green. A ribbon is drawn through the P atoms in the sugar--phosphate backbone. Base atoms exposed in the major groove of the helix are grey; other atoms in the nucleobase are white. A-, B-helices have 20 bp. (Adapted from Lescrinier *et al.*, 2003.)

field. The dynamic bending rigidity is four times larger than the apparent static bending rigidity inferred from equilibrium persistence length measurements on the same fragments.

D6.6.2 Flexibility of dsRNA

The 2′-hydroxyl groups in RNA stabilise the double helix in the A form, which differs from the canonical B-form helix in that the A form has a smaller rise per base pair and a substantial displacement of the base pairs from the helix axis (Fig. D6.15). The dependence of the rotational correlation time $\tau_{20,w}$ of 10 dsRNAs ranging in length diffusion times of 10 dsRNAs ranging in length from 97 to 317 base pairs is shown in Fig. D6.16 using the data of Table D6.3. A hydrodynamic analysis of the birefringence decay data yields a helix persistence length of 720 ± 70 Å, based on a helix with an effective hydrodynamic diameter of 26 Å and a rise per base of 2.7 Å in the presence of Mg^{2+} ions. In contrast to B-form DNA the dependence of rotational correlation times on molecular mass on a double logarithmic scale for A-form DNA can be described by a straight line over the whole range of molecular mass studied with a slope of 2.47, which is typical for a homologous series of *rod-like particles of finite diameter* (Section D2.6). The central conclusion is that RNA is indeed stiffer than DNA.

Table D6.3. *Relaxation times of RNA molecules*

RNA number of bp	Relaxation time μ (in µs)
92	0.44
109	0.62
116	0.69
119	0.75
132	1.00
136	1.08
180	1.90
186	2.02
203	2.48
226	3.46
243	4.03
288	6.63
317	7.31

Data are taken from Kebbekus *et al.* (1995).

Fig. D6.16 Dependence of rotational correlation times $\tau_{20,w}$ on molecular mass (in bp) of RNA for ten fragments on a double logarithmic scale (After Kebbekus *et al.*, 1995).

D6.6.3 Conformational of non-helical elements in RNA

Birefringence decay times are usually interpreted in terms of a hydrodynamic representation of the molecule under study, with various parameters of the model (e.g. helix rise, intrinsic flexibility) being varied until the computed rotational diffusion times are in agreement with experimental results. The interpretation is thus subject to uncertainties in both the model itself and the underlying hydrodynamic theory. In the early 2000s a new variant of TEB (called the τ ratio approach) was proposed for studying the conformational properties of single-stranded RNA in solution. Such studies are greatly facilitated by the relative constancy of the local helix structure; this allows the properties of molecules with non-helix elements to be compared with full duplex control. Figure D6.17 shows the principles of the τ ratio approach used to study the conformational properties of a single-stranded RNA in solution. For a non-helix element of interest (e.g. a bulge), full-duplex and bulge-containing heteroduplex molecules are constructed (Fig. D6.17(a)).

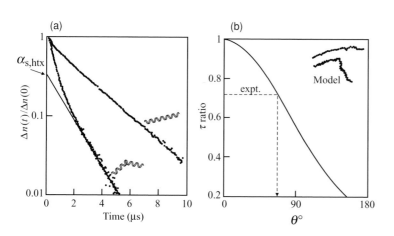

Fig. D6.17 (a) Principle of the τ ratio approach to studying the conformational properties of DNA and RNA. Terminal decay times are determined for bead-containing (e.g. heteroduplex, τ_{htx}) and fully duplex control helices (τ_{dplx}). Each experimental profile comprises two exponential components, $\alpha_{slow(s)}$ and $\alpha_{fast(f)}$; $\alpha_{s,htx}$ is illustrated. (b) The experimental and theoretical results for the ratio of the two decay times (τ ratio $\equiv \tau_{htx}/\tau_{dplx}$) are compared to estimate the angle of the intrinsic (or induced) bond. (From Hagerman, 2000.)

Comment D6.5
Length of RNA in the τ ratio approach

The choice of length for the RNA helix depends on a few general factors. First, helix should be short enough to present a reasonably rigid framework for studying the central bend, and to ensuring decay profiles that are essentially single exponential in the absence of a bend. This gives RNA helices that are comparable to or less than one persistence length (200–250 bp). Second, the total length of the helix arms should be at least 4–5 times the appropriate linear span of the non-helix element being studied. In view of these factors helix arms should be approximately 70 bp per arm. (Hagerman, 2000.)

The two molecules are designed to have identical sequences and lengths except for the presence of the non-helix element in the heteroduplex (Comment D6.5). The decay profiles are determined for both molecules, yielding the (experimental) ratio of the terminal decay times. Determination of the τ ratio for this pair of molecules yields a quantitative estimate for the angle of the intrinsic (or induced) bend (Fig D6.17(b)) (Comment D6.5).

In a series of investigations, the TEB method was used to characterise a number of simple non-helix elements (e.g. bulges, internal loops, etc.) commonly found in RNA. Thus, it was found that for bulges possessing either extra A or U residues, the bulge-induced bend angle increases in a monotonic fashion, from $10–15°$ for single residues in the presence of Mg^{2+} ions to $80–90°$ for six residues in the absence of Mg^{2+} ions. Upon addition of Mg^{2+}, the entire set of U_n bulges experience an approximately two-fold reduction in the bend angle, whereas the angles for A_n bulges remain nearly unchanged. This observation is important because it represents an elementary example of a structural rearrangement in RNA that is both sequence-specific and cation-valence specific.

D6.7 Checklist of key ideas

* If a molecule can be treated as a dipole it is said to possess a *permanent dipole* moment; the existence of a permanent dipole moment implies the molecule is asymmetric.
* If a molecule does not have a net permanent dipole moment but acquires one in an electric field, it is said to possess an *induced dipole moment*; the induced dipole exists as long as the field remains active.
* The magnitude of the dipole moment induced by an external field is proportional to the field intensity. The proportionality constant is called the 'polarisability.'
* Because of their dipolar nature, biological macromolecules in solution are orientated by an applied electric field to form optically birefringent systems.
* TEB is a response of the system to an electric field that is usually in the form of a square-wave pulse.
* The birefringence is proportional to the square of the field strength for limiting low fields. This is Kerr's law. The proportionality constant is called the Kerr constant.
* TEB and TED yield the same information regarding macromolecular structure; however, in practice, TEB is more appropriate for studies of the external shape of macromolecules.
* The rate of decay of the birefringence following the removal of the electric field provides a direct measure of the relaxation time of macromolecules.
* For a general rigid body of arbitrary shape, the time dependence of TEB is represented by a multiexponential function containing up to five components.
* The relaxation time of a spherical macromolecule is proportional to its volume.
* For a rigid rod-like particle the relaxation time is a measure of its length: it is unnecessary for the molecular mass or other dimensions to be known with a high degree of accuracy.
* The *static* persistence length of the B-form of DNA in a buffer of moderate salt concentration is about 500 Å. The analogous value for A-form DNA is about 1500 Å.

- The DNA molecules from 8 bp (oligonucleotides) to 119.71 Mbp (*Drosophila melanogaster*) form a unique homologous series, in which rotational correlation times range from a few nanoseconds to thousands of seconds (12 orders of magnitude).
- In the low-molecular-weight (8–200 bp) region, the worm-like chain of DNA approaches the limit of behaving like a rigid rod while in the high-molecular-weight (more than 8 kbp) region it approaches the limit of flexible coil behaviour.

Suggestions for further reading

Historical review and introduction to biological problems

Fredericq, E., and Houssier, C. (1973). *Electric Dichroism and Electric Birefringence.* Oxford: Clarendon Press.

Hagerman, P. J. (1996). Sometimes a great motion: application of transient electric birefringence to study of macromolecular structure. *Curr. Opin. Struct. Biol.*, **6**, 643–649.

Macromolecules in electric field

Belmonte, P. A., Martinez, C. L., and Garcia de la Torre, J. (1991). Time course of the orientation and birefringence of axially symmetric macromolecules upon reversal of an electric field of arbitrary strength. *J. Phys. Chem.*, **95**, 5661–5664.

Theoretical background of TEB

Charney, E. (1988). Electric linear dichroism and birefringence of biological polyelectrolites. *Quart. Rev. Biophys.*, **21**, 1–60.

Experimental measurement of TEB

Hagerman, P. J. (1985). Application of transient electric birefringence to the study of biopolymer structure. *Methods in Enzymol.*, **117**, 199–215.

Global structure and flexibility of proteins

Kooijman, M., Bloemendal, M., Van Amerongen, H., Traub, P., and Van Grondelle, R. (1994). Characterization of multiple oligomeric vimentin intermediate filament units by transient electric birefringence measurements. *J. Mol. Biol.*, **236**, 1241–1249.

Global structure and flexibility of nucleic acids

Klotz, L. C., and Zimm, B. H. (1972). Size of DNA determined by viscoelastic measurements: results on bacteriophage. *Bacilus Subtilus and Escherichia Coli*, **72**, 779–800.

Hagerman, P. J., and Zimm, B. H. (1981). Monte Carlo approach to the analysis of the rotational diffusion of wormlike chains. *Biopolymers*, **20**, 1481–1502.

Diekmann, S., Hillen, W., Morgeneyer Wells, R. D., and Porschke, D. (1982). Orientation relaxation of DNA restriction fragments and the internal mobility of the double helix. *Biophys. Chem.*, **15**, 263–270.

Pecora, R. (1991) DNA: a model compound for solution studies of macromolecules, *Science*, **251**, 893–898.

Kebbekus, P., Draper, D. E., and Hagerman, P. J. (1995). Persistence length of RNA. *Biochemistry*, **34**, 4354–4357.

Hagerman, P. J. (2000). Transient electric birefringence for determining global conformations of nonhelix elements and protein-induced bends in RNA. *Meth. Enzymol.*, **317**, 440–453.

Lu, Y., Weers, B., Stellwagen, N. C. (2001–2002). DNA persistence length revisited. *Biopolymers*, **61**, 261–275.

Hagerman, P. J. (1981). Investigation of the flexibility of DNA using transient electric birefringence. *Biopolymers*, **20**, 1503–1535.

Stellwagen, N. C. (1996). Electric birefringence of kilobase-sized DNA molecules. *Biophys. Chem.*, **58**, 117–124.

Cantor, C., and Schimmel, P. (1980). *Biophysical Chemistry*. Part III. *The Behavior of Biological Macromolecules*. San Francisco, CA: W. H. Freeman and Company.

Chapter D7
Flow birefringence

D7.1 Historical review

1870

J. Clerk Maxwell was the first to describe the phenomenon of double bire-fringence in flow (DBF), using Canada balsam as the liquid for study. DBF was produced by subjecting the Canada balsam to shear between two concentric cylinders, one of which rotated while the other was held fixed.

1932

P. Boeder presented the first theory of DBF. He formulated the main problem of the theory as finding the relation of the extinction angle and the magnitude of double refraction to the molecular dimensions of colloidal particles. Between **1938** and **1941**, **A. Peterlin** and **H. A. Stuart**, and **O. Snellman** and **Y. Bjornstahl** proposed that this relation could be expressed as a function of a parameter that is equal to the ratio of the velocity gradient to the rotational diffusion coefficient. The crux of the problem was the determination of the rotational diffusion coefficient from experimental measurements of the extinction angle and the magnitude of double refraction.

1950s

H. A. Sheraga, J. T. Edsall and **J. O. Gadd** evaluated the extinction angle and birefringence numerically as a function of velocity gradient over a wide velocity gradient to rotational diffusion coefficients ratio range. This opened the way for the calculation of molecular dimensions of ellipsoidal particles from flow birefringence.

Late 1960s

G. B. Thurston proposed measuring birefringence in a hydrodynamic oscillating field. At the same time **N. Davidson** created a method for determining relaxation times from the decay of flow dichroism measurements.

1970s

R. E. Harrington developed the instrumentation and associated techniques for the measurement of flow birefringence and extinction angles of high-molecular-weight polymer systems at extremely low velocity gradients. This opened the way to studying very dilute high-molecular-weight DNA solutions. **W Kun, H. Kun, R. Serf,** and **V. N. Tsvetkov** made an essential contribution to the theory and practice of flow birefringence of deformable macromolecules, primarily of linear polymers.

1980–2000

In the early 1980s the birefringence flow method started to lose popularity among biophysicists. The reasons for this were twofold: first, because of the development of transparent electric birefringence, which required much a smaller quantity of material; second, because the range of applicability of flow birefringence is limited for very elongated particles, such as rod-like viruses and DNA molecules. Now, however, flow birefrigence is undergoing a renaissance as the recognised method for studying real-time conformational dynamics of individual large DNA molecules in shear flow. Such dynamics can be directly observed by the use of video fluorescence microscopy.

D7.2 Introduction to biological problems

Double refraction is produced in a liquid containing large asymmetrical particles when a velocity gradient is set up in the liquid. This is most readily achieved by subjecting it to shear between two concentric cylinders, one of which rotates while the other is held fixed. This procedure was employed in 1870 by Maxwell, who was the first to describe the phenomenon of double refraction, using Canada balsam as the object of study.

If the external field is applied to an initially isotropic solution, the system becomes birefringent. In the case of a hydrodynamic field, flow birefringence is produced by the shearing force to which the system is subjected. If the solute macromolecules are geometrically and therefore optically asymmetric, the double refraction of the system appears as a result of the equilibrium between the orienting effect of the applied field and the disordering effect of the Brownian motion. Even when the molecules are spherical and optically isotropic in the absence of the field, optical anisotropy can appear as a result of *deformation* of the molecules by shearing force. Therefore flow birefringence is a very useful method of obtaining information about the geometrical, optical and flexible properties of biological macromolecules in solution. Unfortunately the range of application of the birefringence in flow method for studies of biological macromolecules is relatively small in spite of its evident potential. Rotation rates of globular proteins are so great that little orientation occurs. The most interesting results have been obtained for large flexible biological macromolecules, especially DNA and RNA. Interest has focused on single macromolecular dynamics of DNA molecules, which can be observed directly in real time using video fluorescence microscopy.

D7.3 Steady-state flow birefringence

Steady-state orientation in flow is most readily achieved by subjecting the liquid to shear between two concentric cylinders, one of which rotates while the other is held fixed (Fig. D7.1). When both cylinders are at rest as shown in Fig. D7.1(a) the suspended particles have random orientation. When one of the cylinders is set in motion, laminar flow is produced in the liquid and as a result a velocity gradient is set up in the gap between the cylinders. The resulting shearing forces produce an orientation of the particles, as shown schematically in Fig. D7.1(b).

The field between the cylinders is observed by passing light directed parallel to the axis of the cylinder through crossed Nicole prisms (polariser and analyser). Suppose, as in Fig. D7.2, we have a number of molecules oriented so that their axes make an angle θ with the y-axis. Polarised light, with the electric field vibrating in the y-direction, which is incident on the molecules is split into components $E_{\parallel}$ and $E_{\perp}$ because the refractive index is different for vibrations in these two directions. We say that the oriented molecules are birefringent. If the incident electric field is E_0, then $E_{\parallel} = E_0 \cos \theta$ and $E_{\perp} = E_0 \sin \theta$. The light that passes through oriented molecules now must pass through the analyser, which is set to transmit vibrations parallel to the z-axis. If θ is zero or $90°$, the incident polarised light is not split into components, the field incident on the analyser is parallel to y, and no light will be transmitted. For all other orientation angles, some light will be transmitted. As a result application of the external field produces four dark regions in the shape of a cross on an otherwise bright field (Fig. D7.3). In the dark regions, the plane of vibration of the light transmitted by either the polariser or analyser is parallel to the optic axis of the solution.

One of the most important limitations of flow birefringence as a quantitative tool for studying the hydrodynamic properties and optical behaviour of macro-molecules in solution is that it is not sufficiently sensitive to measure precisely very small values of birefringence and extinction angles near $45°$, where the

(a)

(b)

Fig. D7.1 Particles in a fluid placed between concentric cylinders: (a) particles, each schematically represented by a line indicating its optic axis, at rest; (b) orientation of particles caused by rotation of the external cylinder.

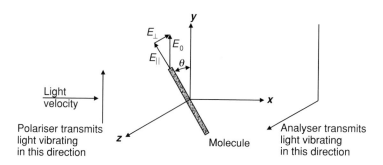

Light velocity

Polariser transmits light vibrating in this direction

Molecule

Analyser transmits light vibrating in this direction

Fig. D7.2 The relative orientation of polariser, molecular axes and analyser. The molecules split the electric field of the incident light into components $E_{\parallel}$ and $E_{\perp}$.

Fig. D7.3 Schematic
illustration of flow
birefringence when all
particles have the same
orientation angle χ with
respect to the flow lines.
The diagram shows the
cross of isocline, which
indicates the four
locations where the optic
axes of the particles are
exactly parallel to the
analyser or polariser
plane. (Adapted from
Tenford, 1965.)

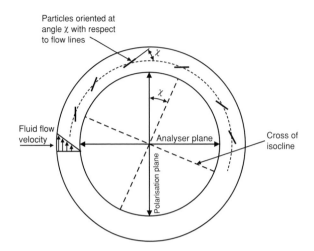

distribution of macromolecular orientations is relatively broad. The standard
methods for measuring the flow birefringence of macromolecules utilise a device
in which the streaming solutions are viewed between crossed polarisers and the
birefringence is determined visually using optical compensation devices (see
above). Unfortunately, an acceptable level of accuracy cannot be attained by
using visual detecting methods if the total birefringence is less than approximately 10^{-8}.

The best approach to the problem of sensitivity in flow birefringence measurements has involved the use of photoelectric detection. Several different types
of photoelectric devices have been developed. One of them allows very high-
molecular-weight macromolecules in solution and at low concentrations to be
studied to shear rates below 0.01 s^{-1} where the Maxwell effect is very small.

D7.3.1 Extinction angle and rotational friction coefficient

Flow birefringence experiments measure the direction of the principal axis of the
elongated particle of refractive index n, or those of the optical axes n_1 and n_2, of
the solution and the difference between the refractive indexes in these directions
(Fig. D7.4). The angle χ between the direction of optical axis of the solution n_1

Fig. D7.4 The velocity
gradient field in a solution
of particles between two
cylindrical walls, one of
which rotates while the
other is held fixed.

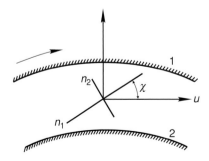

and the direction of the flow u is called the *orientation* or *extinction* angle. The angle χ is one of two basic quantities determined experimentally. The other is the difference of the refractive indexes $n_1 - n_2$. In this subsection we focus our attention mainly on the angle χ because it is directly related to the hydrodynamic properties of elongated particles.

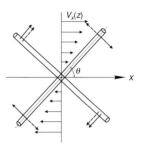

Fig. D7.5 Forces acting on a rigid rod in a velocity gradient. The tangential component of velocity gradient is shown as →, The Brownian motion forces are shown as ---▸.

We consider first rigid rod-like particles, and for simplicity examine the two-dimensional situation depicted in Fig. D7.5. The hydrodynamic forces tend to rotate the particle clockwise, and it is evident that the torque is maximal at an angle $\theta = 90°$ with respect to the flow lines, and minimal at $\theta = 0°$. The molecule moves most quickly through the former angle, and is slow at the latter, thus biasing the distribution of particle orientations. This bias distribution is counteracted by Brownian motion, which tends to randomise the distribution. A steady-state distribution is achieved when these two forces balance.

The balance point is characterised by the parameter $\alpha = G/\Theta$, where G is the velocity gradient and Θ is rotational diffusion coefficient. When $\alpha \ll 1$, the most probable orientation angle is $45°$, where the hydrodynamic and Brownian motion forces are equal and opposite. The least probable orientation angle is $135°$, where the forces add (Fig. D7.5). On the other hand, when $\alpha \gg 1$ the hydrodynamic force greatly suppresses the Brownian motion randomising force. In this case all the particle axes are displaced near an orientation close to zero, and the minimum in the angular distribution is at an orientation angle of $90°$.

Plots of the extinction angle χ versus α for different values of the axial ratio p of the ellipsoid of revolution are shown in Fig. D7.6. This figure demonstrates that for rigid particles χ always approaches $45°$ as the orienting force, G, approaches zero. It is also important that the limiting slope does not depend on the shape of the particle.

For small values the extinction angle is given by

$$\chi = 45° - \frac{1}{12}\frac{G}{\Theta} + k\left(\frac{G}{\Theta}\right)^3 \cdots \qquad (D7.1)$$

where k is a coefficient depending on the shape of the macromolecules. Equation (D7.1) shows that the relation between χ and G should be linear for small values

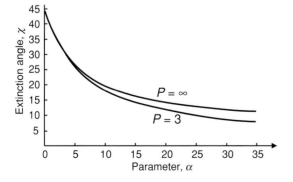

Fig. D7.6 Extinction angle χ versus the parameter α for different values of the axial ratio, P, of the ellipsoid of revolution. (Sheraga *et al.*, 1951.)

of G, with the slope equal to $-G/12\Theta$. Thus the rotational friction coefficient may be determined directly. However, one of the most severe limitations in the use of flow birefringence is its low sensitivity when measuring extinction angles near 45° and very low gradients.

D7.4 Decay of flow dichroism

If molecules have an intrinsic optical anisotropy (Comment D7.1), the orientation can be observed as flow birefringence and/or flow dichroism. If the flow is abruptly stopped, the time course of the relaxation of the molecular orientation back to its equilibrium distribution in the absence of the flow field can be observed. Under these conditions the behaviour of macromolecules may be described as follows

$$\frac{\Delta A(t)}{\Delta A(0)} = \exp(-6\Theta t) \qquad (D7.2)$$

where $\Delta A(0)$ is the absorbance at zero time and $\Delta A(t)$ denotes changes in absorbance with time (see also Section D6.4.2). Typical plots of this type are presented in Fig. D7.7.

Comment D7.1　'Form' and 'intrinsic' birefringence

Most biological macromolecules have a refractive index n that differs from that of the usual aqueous solutions. For example, the refractive index increment dn/dc of DNA is about 0.170. The refractive index increment of protein is about 0.17.

Thus when biological macromolecules in solution are oriented in a velocity gradient, the solution becomes birefringent. That is, it has different refractive indices parallel and perpendicular to the flow lines. This is known as 'form' birefringence, and arises in all solutions of non-spherical molecules whose refractive index differs from that of the solvent. However, even if the refractive index of the particles equals that of the solvent, birefringence arises when the molecule itself is optically anisotropic, with different refractive indices parallel and perpendicular to its long axis. This is called 'intrinsic' birefringence. The reader can find a quantitative treatment of the phenomena of intrinsic and form birefringence in text-books (e.g. Tsvetkov *et al.*, (1971)) (see also Comment D6.3).

At low shear gradients, which corresponds to small deformations of the chain, for non-draining Gaussian coils, the longest relaxation time τ_1 of the chain is

$$\tau_1 = 0.423 \frac{M\,[\eta]\,\eta_0}{RT} \qquad (D7.3)$$

where η is the intrinsic viscosity of the molecule and η_0 is the intrinsic viscosity of the solvent.

Fig. D7.7 Semilogarithmic plots of the decay of the dichroism signals for DNAs of several different molecular weights. (Callis and Davidson, 1969.)

D7.5 Oscillatory flow birefringence

Oscillatory flow birefringence is measured using circularly polarised light which is passed through a narrow fluid-field gap between an oscillating plane surface and a rigid surface. The resulting change in the light, due to flow-induced birefringence in the fluid, is measured to determine both the magnitude of the birefringence and its phase with the respect to the velocity gradient in the gap.

The frequency response for two samples of DNA with molecular masses of 6×10^5 and 5×10^6 Da is shown in Fig. D7.8(a) and (b). These data clearly

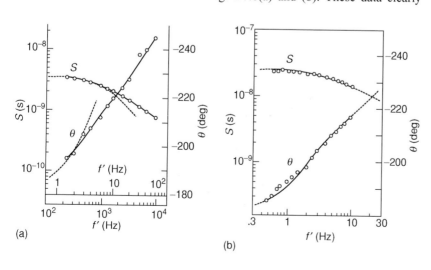

(a)

(b)

Fig. D7.8 (a) The frequency response of oscillatory flow birefringence of DNA ($M = 6 \times 10^5$ Da) in 89% glycerol solution. The dashed lines are the theoretical curves for single-relaxation rigid-particle theory. The frequency scale, f, is the frequency of measurement while the scale f' is the equivalent frequency expected for this same DNA in an aqueous solution. (b) The frequency response of oscillatory flow birefringence of DNA ($M = 5 \times 10^6$ Da) in aqueous solution. (After Wilkinson and Thurston, 1976.)

demonstrate that a DNA molecule with $M = 6 \times 10^5$ Da (~900 bp) can be considered as a rod-like particle with partial flexibility, whereas a DNA molecule with $M = 5 \times 10^6$ Da (~7500 bp) can be considered as a Gaussian coil (see Sections D3.7.4, D4.10.3, D6.7.1, D9.5.5).

D7.6 Orientation of macromolecules in flow as a dynamic phenomenon

Birefringence in flow is a dynamic phenomenon. It may be seen from Fig. D7.9 that the chain is extended when its ends are in the first and third quadrants, and compressed when they are in the second and the fourth quadrants. If the macromolecule is flexible, or soft, it is deformed in phase with its rotational motion (see below).

A very interesting distinctive feature of birefringence in flow is the deformation of macromolecule in the shearing field. The simplest set-up is based on a simple

Fig. D7.10 (a) Simple shear flow as a superposition of purely elongational flow and purely rotational flow. (b) A schematic illustration of a flexible macromolecule stretching in a shear flow, as viewed in the polymer's centre-of-mass frame. As the macromolecule stretches, many different dynamical scenarios may occur. When the chain becomes aligned, the hydrodynamic force decreases and the chain may begin to relax. However, Brownian motion can perturb a stretched chain so that it either continues to stretch or becomes unstable and tumbles end-over-end. (c) A schematic diagram of the measuring apparatus. A uniform shear flow is created in a gap of ~50 μm between two parallel plates by translating one glass plate parallel to the other at a constant field. Isolated molecules of fluorescently labelled λ-DNA (48 502 bp), represented here by dots, were imaged by fluorescence microscopy with an intensified video camera (ICCD). (Smith *et al.*, 1999.)

Fig. D7.11 Images of single lambda bacteriophage DNA chains tagged with fluorescent markers in a viscous sugar solution: (a), (b) and (c) represent different scenarios for the same type of molecule. These DNA molecules had a contour length L of about 22 μm. They contain roughly 440 persistence lengths. Each row of images is a series in time proceeding from left to right. The vertical bar at the left of the top row indicates 5 μm. (Smith *et al.*, 1999.)

shear between two parallel plates, one of which slides in its own plane. This shear flow may be represented as a superposition of a purely rotational flow and a purely elongated flow (Fig. D7.10(a)).

In a pure elongation flow one expects a large deformation of a flexible molecule, whereas in a pure rotational flow one expects only rotation and not deformation. When the magnitudes of the elongational and rotational components are equal the molecule does not attain a stable, strongly stretched state. The conformation of the molecule fluctuates because the stretched state is destabilised by the rotational component of the shear flow (Fig D7.10(b)).

Figure D7.11 shows three example images of individual, isolated molecules of fluorescently labelled lambda bacteriophage DNA (λ-DNA) undergoing conformational changes in steady shear with Wiessenberg's number $W_i = 19$ (Comment D7.2). The first example is a molecule that went from being coiled, to stretched, back to coiled and then stretched again (Fig. D7.11(a)). The second example demonstrates a lump of mass density that propagated down the contour of a partially stretched chain (Fig. D7.11(b)). And finally, Fig. D7.11(c) shows a molecule that became folded into an upward U-shape, unravelled like a rope sliding over a pulley, then tumbled end-over-end and became folded in an inverted U-shape.

D7.7 Checklist of key ideas

* Orientation of macromolecules by hydrodynamic flow is called birefringence in flow, or flow birefringence, or the Maxwell effect.
* Rotation rates of globular proteins are so great that little orientation in flow occurs.
* In flow an asymmetrical particle is continually rotating, but tends to speed up or slow down at various angles with respect to the flow lines, thus biasing the distribution of particle orientations.

Comment D7.2 Weissenberg's number

The strength of an applied shear flow is specified by the dimensionless number $W_i = G/\tau$ (called the Weissenberg number) which is the ratio of the applied shear rate, G, to the relaxation rate of macromolecule $1/\tau_1$. As W_i is increased, the molecule fluctuates more rapidly (relative to its relaxation rate) and reaches larger extensions.

- The quantities which can be measured in flow birefringence experiments are orientation (or extinction) angle χ and the quantity of birefringence $n_1 - n_2$.
- For a solid asymmetrical particle the orientation angle χ is determined only by its hydrodynamic properties and not by optical properties.
- The rotational diffusion coefficient of a solid asymmetrical particle can be directly determined from studies of χ at low values of gradient velocity or from studies of the relaxation of birefringence after suddenly stopping flow.
- The value of birefringence $n_1 - n_2$ depends on both the hydrodynamic and optical properties of the solution of macromolecules.
- 'Form' birefringence is a property of solution of oriented non-spherical molecules in which the refractive index of the solute differs from that of the solvent.
- 'Intrinsic' birefringence is a property of optically anisotropic molecules that have different refractive indices parallel and perpendicular to its long axis.
- For flexible particles the relative roles of the orientation and deformation effects in observed birefringence greatly depend on their geometric, hydrodynamic and optical properties.
- The real-time conformational dynamics of individual large DNA molecules in shear flow can be observed directly using video fluorescence microscopy.

Suggestions for further reading

Historical review and introduction to biological problems

De Gennes, P. G. (1997). Molecular individualism. *Science*, **276**, 1999–1992.

Steady-state flow birefringence

Sheraga, H. A., Edsall, J. T., and Gadd, J. O. (1951). Double refraction of flow: numerical evaluation of extinction angle and birefringence as a function of velocity gradient. *J. Chem. Phys.* **19**, 1101–1108.

Harrington, R. E. (1970). Measurement of flow birefringence at very low velocity gradients. *Biopolymers*, **9**, 141–157.

Tsvetkov, V. N., Eskin, V. E., and Frenkel, S. Ya. (1971). *Structure of Macromolecules in Solution* (translated from Russian), Volume 1, Chapter 7. Boston, UK: National Lending Library for Science and Technology.

Decay of flow dichroism

Callis, P. R., and Davidson, N. (1969). Hydrodynamic relaxation times of DNA from decay of flow dichroism measurements. *Biopolymers*, **8**, 379–390.

Cairney, K. L., and Harrington, R. E. (1982). Flow birefringence of T7 Phage DNA: dependence on salt concentration. *Biopolymers*, **21**, 923–924.

Oscillatory flow birefringence

Schrag, J. L., Guess, J. F., and Thurston, G. B. (1965). Shear-wave interference observed by optical birefringence induced in a viscoelastic liquid. *J. Appl. Phys.*, **36**, 1996–2000.

Wilkinson, S. R., and Thurston, G. B. (1976). The optical birefringence of DNA solutions induced by oscillatory electric and hydrodynamic fields. *Biopolymers*, **15**, 1555–1572.

Orientation of macromolecules in flow as a dynamic phenomenon

Smith, D. E., Babcock, H. P., and Chu, S. (1999). Single-polymer dynamics in steady shear flow. *Science*, **283**, 1724–1727.

Chapter D8
Fluorescence depolarisation

D8.1 Historical review

1926

E. Gaviola created the first instrument for fluorescence measurement, which he called a fluorometer.

1934

F. Perrin formulated the relation between fluorescence depolarisation and Brownian rotary motion for hydrodynamic spherical molecules. In 1936 he extended his treatment to symmetric top molecules.

1952

G. Weber first applied Perrin theory to biological macromolecules. Numerous fluorescence polarisation measurements were carried out under steady-state conditions, utilising constant illumination. Most of them were performed on dye-conjugated macromolecules to determine the harmonic mean rotational relaxation time. Weber's classical review of the application of polarised fluorescence, particularly to proteins, appeared in the 1950s.

1961

A. Jublonski proposed following the rotational relaxation process directly by measuring the decay of the fluorescence polarisation as a function of time. He stressed that time-dependent anisotropy can be interpreted more directly and definitely than its time-average value observed with constant illumination.

1966

P. Wahl, and two years later **L. Stryer**, experimentally measured the time decay of fluorescence polarisation on dye-conjugated globular proteins.

1969

T. Tao derived relations between the time-dependent fluorescence polarisation anisotropy and the Brownian rotational diffusion coefficients of macromolecules. It was shown that in the most general case of a completely asymmetric body, five exponentials appear in anisotropy. In **1972** analogous results were obtained by other groups (**G. G. Belford, R. L. Belford** and **G. Weber**; **T. J. Chuang** and **K. B. Eisenthal**).

Mid-1990s

A few groups reported the synthesis and spectral properties of long-lifetime, highly luminescent, metal–ligand complexes (**B. A. De Graff** and **J. N. Demas, J. R. Lacowicz's group**). These very photostable probes with microsecond decay times allowed measurements of rotational correlation times as long as 2 µs and detection of high-molecular-mass analytes in fluorescence polarisation immunoassays.

1980–1995

J. R. Lakowicz's group made an essential contribution to the theory and practice of depolarised fluorescence spectroscopy. In 1987 the textbook *Principles of Fluorescence Spectroscopy* by **J. R. Lakowicz** appeared.

2000 to now

Fluorescence depolarised spectroscopy is very useful for investigating the rotational motion of macromolecules in aqueous solution in the nanosecond and microsecond time ranges. It has opened the new possibility of measuring interdomain protein movement, lateral diffusion in membranes, large protein and protein–nucleic acid complexes.

D8.2 Introduction to biological problems

The theoretical foundations of fluorescence spectroscopy were established in the first half of the twentieth century by pioneers including E. Gaviola, Jean and Francis Perrin (father and son), P. Pringsheim, S. Vavilov, T. Förster, and, more recently, G. Weber. In 1961 A. Jublonski proposed following the rotational relaxation process directly by measuring the decay of the fluorescence polarisation as a function of time. He stressed that time-dependent anisotropy can be interpreted more directly and definitely than its time-average value observed with constant illumination.

For many years, the most common measurements were of fluorescent intensity only, but since the 1980s, access to instrumentation for measuring fluorescence decays has improved markedly, largely as a result of the more widespread availability of lasers and improvements in electronics. An investigator may now choose to use either time-correlated single-photon counting or multi-frequency phase fluorimetry and be confident of getting the same answers.

Since the 1980s, essential experimental and theoretical contributions to the time-resolved fluorescence spectroscopy were made by J. Lacowicz, D. Millar and R. Cherry and their groups. Time-resolved measurements can now be used in two ways. First, measurement of the decay of the total fluorescence intensity following pulsed excitation may be used as a signal of an *event* (protein conformational changes, ligand binding, protein–nucleic acids interactions, and so on). Second, the values of fluorescent parameters may be used to understand details of *the structure and dynamics of the fluorophore and its environment*. However, our incomplete understanding of the details of photophysics of most fluorophores, such as tryptophan and tyrosine, is a principal reason why interpretation of fluorescence in terms of protein structure and dynamics is still so difficult.

The other type of time-resolved fluorescence experiment involves measurement of the polarisation anisotropy decay to characterise the *rotation* of an entire macromolecule or its separate parts. For many years, the most common measurements were done in the nanosecond time scale. Now fluorescence is capable of detecting microsecond dynamics. This change in time scale was made possible by the development of new metal–ligand complexes, which display decay times ranging from 10 ns to 10 μs. It has opened the possibilities of measuring protein domain movement, lateral diffusion in membranes and microsecond rotational correlation times.

This chapter focuses on the use of fluorescence anisotropy decay methods which provide hydrodynamic data describing the rotational diffusion of biological macromolecules. In Section F3.5 we will discuss the energy transfer Förster effect.

D8.3 Theory of fluorescence depolarisation

In this section we briefly discuss the theory of fluorescence spectroscopy as applied to the study of biological macromolecules in order to understand how rotational friction coefficients are determined from depolarised fluorescent measurements.

D8.3.1 Fluorescence as a physical phenomenon

When an electron is excited to an orbital with a high electronic energy, the excess energy is stored. In most cases this energy is transferred as heat. This is called *non-radiative* energy transfer. In some molecules the electronic energy states are changed so that the energy can be reradiated as the electron drops from the excited state back to the ground state. Non-radiative energy transfer is illustrated in Fig. D8.1(a). Because of the relative configurations of the excited and ground states, the electron can easily make a transition from a vibrational state in the excited system to a nearly degenerate level in the ground state, and it can continue to undergo radiationless energy loss until it reaches the ground state. *Radiative* energy transfer occurs when the electronic configurations of the ground end excited states are arranged as shown in Fig. D8.1(b). Again absorption of a photon leads to the transition of the electron to an excited state with transfer of some energy to the solvent as heat. However, the lowest vibrational level of the excited state is too far above any vibrational level of the ground state for

(a)

(b)

Fig. D8.1 (a) Non-radiative and (b) radiative energy transfer.

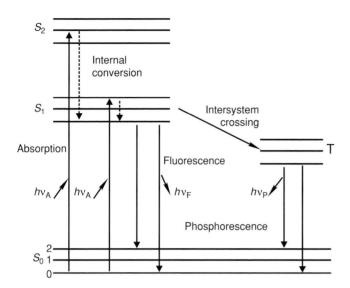

Fig. D8.2 Diagram of the absorption and emission of light (the Jablonski diagram). The ground, first and second electronic states are depicted by S_0, S_1 and S_2, respectively. At each of these electronic energy levels the fluorophores can exist in a number of vibrational energy levels, depicted by 0,1,2, etc. (Lakowicz, 1999.)

energy to be relinquished except with radiation of a photon. The absorption and emission of light in radiative energy transfer is illustrated by the energy-level diagram suggested by Jablonski (Fig. D8.2).

The absorption is quite rapid, taking about 10^{-15} s. Following light absorption, several processes usually occur. A fluorophore is usually excited to some higher vibrational singlet level S_2. Usually, molecules in condensed phases rapidly relax to the lowest vibrational singlet level of S_1 (Comment D8.1). This process is called internal conversion and generally occurs in 10^{-12} s. Since fluorescence lifetimes are typically near 10^{-8} s, internal conversion is generally complete prior to

---

Comment D8.1   Singlet and triplet states

The Pauli exclusion principle states that two electrons in an atom cannot have the same set of four quantum numbers. This restriction requires that no more than two electrons can occupy an orbital and, furthermore, those two must have opposed spin states. The spins are then said to be paired. Because of spin pairing, most molecules exhibit no net magnetic field and are thus said to be diamagnetic – i.e. they are neither attracted nor repelled by static magnetic fields. A molecular electronic state in which all electron spins are paired is called a *singlet* state, and no splitting of electronic energy levels occurs. When one of a pair of electrons in a molecule is excited to a higher energy level, either a singlet or a *triplet* state is formed. In the excited singlet state, the spin of the promoted electron is still paired with the ground-state electron. In the triplet state, however, the spins of two electrons have become unpaired and are thus parallel.

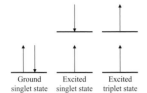

Ground singlet state   Excited singlet state   Excited triplet state

Fig. D8.3 Three different states of electron. Note that the excited triplet state is less energetic than the corresponding excited singlet state.

emission. Hence, fluorescence emission generally results from the thermodynamically equilibrated excited state.

Molecules in the S_1 state can also undergo conversion to the first triplet state T_1. Emission from T_1 is called *phosphorescence*, and generally is shifted to longer wavelength, (low energy). Conversion of S_1 to T_1 is called intersystem crossing. Transition from T_1 to the ground state is forbidden according to quantum mechanical rules, and as a result the rate constant for such an emission is several orders of magnitude smaller than those of fluorescence. Different molecular states are shown in Fig. D8.3, where the arrows represent the direction of spin.

D8.3.2 Lifetime of fluorophore and rotational correlation time

The time interval between excitation and emission for many small molecules in water solution is of the order of 10^{-8} s, while the rotational relaxation time is considerably shorter. It may be expected that by the time the emission takes place, random orientation of the excited molecules will have been regained and the fluorescence will be unpolarised. A large molecule such as a protein, however, will have undergone on the average only a relatively small rotation due to its Brownian movement during a time interval of the order of 10^{-8} s. Consequently the fluorescent light that it emits is partially polarised (Fig. D8.4).

In practice, very few proteins are naturally fluorescent. Most proteins must be converted into fluorescent derivatives by attaching to them suitable fluorescent groups, which have a lifetime about a few nanoseconds. Comment D8.2 gives some typical intrinsic (native) probes, their synthetic analogues and extrinsic fluorescent probes.

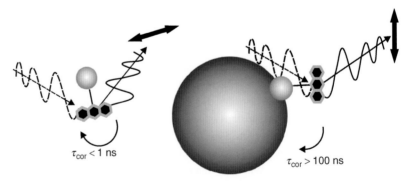

$\tau_{cor} < 1$ ns $\tau_{cor} > 100$ ns

Fig. D8.4 Fluorescence anisotropy and molecular binding. A free ligand labelled with a fluorescein-like fluorophore (left) rotates on the subnanosecond time scale (with rotational correlation time τ_{cor}) and emits depolarised fluorescence. The same ligand when bound to a large macromolecule (right) rotates much more slowly and hence maintains anisotropy. (After Sportsman (2003), Fig. 1.)

Comment D8.2 Fluorescence probes

Fluorescence probes can be broadly placed into three categories: intrinsic probes, analogues of intrinsic probes and extrinsic probes.

Intrinsic probes are any naturally occurring molecules that exhibit sufficient fluorescence to be of use in practice. Examples include the aromatic amino acids, especially tryptophan and tyrosine, NADH, FAD, some porphyrins, some modified nucleic acids (such as the Y base in some tRNAs) and chlorophylls.

In analogues of intrinsic probes, residues in an intrinsic probe are replaced with analogues that have unique spectroscopic properties: examples include 5-hydroxyltryptophan and 7-azatryptophan.

Extrinsic probes are probes that can be introduced into the target system to form a complex, either covalent, or non-covalent. Thousands of probes are now available. Examples of common non-covalent probes are dimethylaminonaphthalene sulphonyl chloride (dansyl chloride), 8-anilino-1-naphthalene sulphonats (ANS), 1,6-diphenyl-1,3,5-hexatriene (DPH), etidium bromide, sulphonyl chloride, isothiocyanate, succinimidyl ester, iodacetamide and maleimide. These and many others are available to target a wide variety of biological macromolecules.

Fig. D8.5 The experimental set-up for measuring the depolarisation of fluorescent light. The angles ϕ and θ define the orientation of μ with respect to the fixed space axes. The polarisation is independent of the angle φ and varies with θ, becoming zero for radiation emitted along the z-direction.

D8.3.3 Steady-state fluorescence depolarisation

The rotational properties of fluorescent molecules may be determined by a method due to Perrin. Suppose that that the solution is illuminated with a beam of plane-polarised light (Fig. D8.5). The exciting light travels along the x-direction, polarised with the electric vector along the z-axis. Two components of the emitted light are measured: $I_{\|}$ is polarised along the z-axis, and $I_{\perp}$ is polarised along the x-axis. μ is the transition dipole of the fluorescent molecule. The anisotropy or polarisation, denoted by A and P, respectively, are then calculated as the difference between these values divided by a sum:

$$A = \frac{I_{\|} - I_{\perp}}{I_{\|} + 2I\perp} \qquad (D8.1)$$

$$P = \frac{I_{\|} - I_{\perp}}{I_{\|} + I_{\perp}} \qquad (D8.2)$$

The denominator in Eq. (D8.1) is the total light that would be observed if no polarisation were used. Referring to Fig. D8.6 and considering the polarisation we can see that $I_{\|}$ and $I_{\perp}$ are emitted along the x-axis as well as the y-axis, but $2I_{\perp}$ is emitted along the z-axis. The total intensity I is proportional to the sum of the emitted light along the three mutually orthogonal Cartesian axes, or to $I = I_{\|} + 2I_{\perp}$.

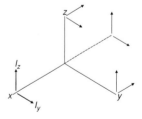

Fig. D8.6 Definition of anisotropy A: non-polarised light incident along x can be resolved into y- and z-polarised components. The total emission can be found by adding the emission along the three Cartesian axes.

To calculate fluorescence polarisation, we need to calculate the probability of exciting molecules with particular orientations, and then to compute the probability that these molecules will emit light polarised in certain directions. First, consider a rigid, isotropic molecule. The molecule must be large enough that no appreciable molecular rotations occur on the fluorescence time scale (typically <100 ns). Calculation shows that if the emitting and absorbing transition dipoles are parallel (i.e. the same electronic transition both absorbs and emits then $I_{||} = 3/5$ and $I_{\perp} = 1/5$. For a totally rigid molecule $P = \frac{1}{2}$ and $A = 2/5$, as one can see by substituting the results of the calculations into Eqs. (D8.1) and (D8.2). When the emission transition dipole of a molecule is perpendicular to the absorption transition dipole, $P = \frac{1}{2}$ and $A = 2/5$. In general, the polarisation and anisotropy for rigid molecule are given by

$$P_0 = \frac{3\cos^2 \zeta - 1}{\cos^2 \zeta + 3} \tag{D8.3}$$

and

$$A_0 = \frac{3\cos^2 \zeta - 1}{5} \tag{D8.4}$$

where ζ is the angle between the absorption and emission transition dipoles.

In the other extreme, the dipole can rotate fast enough to randomise its orientation during the lifetime of the excited state. In this case, by the time emission occurs, all memory of the original photoselection is lost. $I_{||} = I_{\perp}$, and the polarisation and anisotropy are both zero. Most macromolecules of biological interest fall between the extreme cases just described: their rotational motions are not negligible on the fluorescence time scale, but they cannot tumble fast enough to achieve random orientation.

D8.3.4 Time-resolved fluorescence depolarisation

There are two types of experiments that can be performed using time-resolved fluorescence techniques. The first involves measurement of the time for the *decay of the emission* following a brief excitation pulse. This is used to determine the fluorescence lifetime of the fluorophore, which reflects the average time that the molecule remains in the excited singlet state. The fluorescence lifetime is strongly dependent on the molecular properties of the fluorophore environment and can vary from a few picoseconds to tens of nanoseconds.

The other type of time-resolved fluorescence experiment is based on measurement of the fluorescence *anisotropy decay*. The anisotropy decay monitors reorientation of the emission dipole during the excited-state lifetime and gives information on local fluorophore motion, segmental motion and the overall rotational diffusion of macromolecules.

Suppose that a fluorescent probe is rigidly attached to the macromolecules (Fig. D8.4). It is evident that the observed polarisation is some intermediate value between Eq. (D8.3) and zero. To compute this value, we must analyse the relative rates of emission and macromolecular rotational motion. The observed polarisation measurements are a function of both the lifetime and the relaxation time of the protein. This link between the fluorescence lifetime and the observable motion can be seen from the Perrin equation:

$$A_0 = \frac{A_F}{1 + \tau_F/\tau_c} \qquad (D8.5)$$

In this expression A_0 is the steady-state anisotropy and A_F is the anisotropy of the fluorophore in the absence of rotational motion. If the lifetime τ_F is much longer than the rotational correlation time τ_{cor}, the anisotropy A_0 can be taken to be zero. If the lifetime is much shorter than the correlation time, the anisotropy is A_F. The anisotropy A_0 is sensitive to the rate of rotational diffusion only if τ_{cor} is comparable to the lifetime of fluorophore τ_F (Comment D8.3).

D8.4 Instrumentation

The two most popular methods used to record time-resolved fluorescence data are the harmonic-response and impulse-response methods. In the harmonic-response method, time-resolved fluorescence parameters are deduced by analysing the response of the emission to sinusoidally modulated excitation. The light modulators and signal processing circuitry required for these measurements can be incorporated into the design of standard spectrofluorimeters.

The impulse-response method involves direct observation of the time decay of emission following a short excitation pulse. The excitation pulses are typically derived from mode-locked dye lasers or titanium–sapphire lasers, which produce picosecond or subpicosecond pulses and are widely tunable in the visible and UV regions of the spectrum. Emission decay profiles are recorded by the time-correlated, single-photon-counting technique using fast photomultiplier tubes. The use of such a device allows the recording of the fluorescence decay profiles with a high degree of precision and a time resolution of the order of a picosecond or less. Fluorescence anisotropy is technically a rather simple measurement. Two methods are commonly used. These are the L-format and T-format methods. In the L-format method, a single emission channel is used (Fig. D8.7). Fluorescence

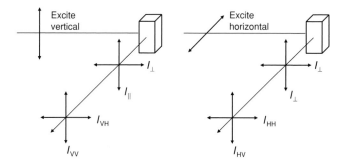

Fig. D8.7 Schematic diagram for the L-format method for the measurements of fluorescence anisotropy: vertical and horizontal excitation.

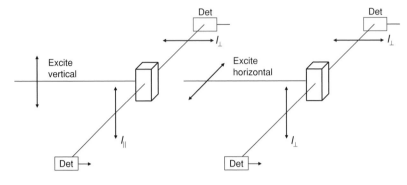

emitted from a sample is detected at right angles to the excitation beam, and these two beams define the scattering plane. For rapid fluorescence anisotropy measurements, two photomultiplier tubes are usually employed in a T-shaped configuration to measure simultaneously the emitted fluorescence light under the two respective relative polarisations, parallel and perpendicular to the polarisation of the incoming beam (Fig. D8.8). In practice, a correction factor G is introduced into Eqs. (D8.1) and (D8.2) to take into account the differences in sensitivity of the detection system in the two polarising directions (Comment D8.4).

A simplified detection system uses only one photomultiplier tube without a polariser. Here, the polarisation of the excitation light is altered between vertical (V) and horizontal (H) by means of a photoelastic modulator (Comment D8.5). As shown in Fig. D8.9(a), when the excitation light is vertically polarised, the sum of the emitted intensities I_{VH} and I_{VV} associated with the μ_x- and μ_z-components, respectively, is detected. Hence, this measurement yields

$$A_V = I_{VH} + I_{VV} \tag{D8.6}$$

The situation for horizontally polarised excitation is depicted in Fig. D8.9(b). In this case the signal detected for horizontally polarised excitation yields:

$$A_H = 2 \times I_{VH} \tag{D8.7}$$

Comment D8.4
Correction factor G

In real measurements a correction factor G is introduced into Eqs. (D8.2) and (D8.3):

$$A = \frac{I_{\|} - GI_{\perp}}{I_{\|} + 2GI_{\perp}}$$

$$P = \frac{I_{\|} - GI_{\perp}}{I_{\|} + GI_{\perp}}$$

The alignment of the optical system can be checked and adjusted using scattering from dilute suspensions of a non-fluorescent substance (e.g., glycogen or colloidal silica in water). The scattered light is 100% polarised, hence the measurements should yield an anisotropy of $A = 1.00$. The alignment is adequate when the measured value is 0.98 or larger (Lakowicz, 1999).

Comment D8.5 Photoelasticity

The phenomenon of photoelasticity is the basis of the photoelastic modulator (PEM). If a sample of transparent solid material is stressed by compression or stretching, the material becomes birefringent, i.e. different linear polarisations of light have slightly different speeds of light when passing through the material.

The PEM may be used in either of two basic modes: as a modulator, to produce polarisation modulation of a light beam, or as an analyser, to determine the polarisation state of a light beam.

(a)

Vertical excitation

(b)

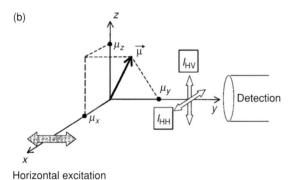

Horizontal excitation

Fig. D8.9 Schematic view of the experimental procedure for measuring fluorescence anisotropy: (a) the excitation polarisation is vertical, and detection without a polariser gives $I_{VH} + I_{VV}$; (b) the excitation polarisation is horizontal, and detection gives $I_{HH} + T_{HV}$. The latter is equal to $2 \times I_{VH}$, if it is assumed that the sample is randomly oriented. Without any additional correction A was found to be 0.95. (After Canet et al., 2001.)

and the anisotropy is calculated from I_{VH} and I_{VV} using the resulting expression:

$$A = \frac{A_V - A_H}{A_V + 0.5 \times A_H} \tag{D8.8}$$

This method does not require the G factor correction (see Comment D8.4).

D8.5 Depolarised fluorescence and Brownian motion

D8.5.1 Steady-state or static polarisation

If constant illumination is employed the measurement is called steady-state or static polarisation. In this case the expressions for average static polarisation and anisotropy are:

$$\bar{P} = \frac{3}{1 + 10(1 + \tau_F/\tau_{cor})(3\cos^2 \zeta - 1)^{-1}} \tag{D8.9}$$

$$\bar{A} = \frac{3\cos^2 \zeta - 1}{5(1 + \tau_F/\tau_{cor})} \tag{D8.10}$$

If measurements are made at such high values of η/T that the macromolecules are unable to rotate during the lifetime of the excited state, then $\tau_F/\tau_{cor} = 0$, and the polarisation and anisotropy become

$$\bar{P}_0 = \frac{3\cos^2\zeta - 1}{\cos^2\zeta + 3} \tag{D8.11}$$

$$\bar{A}_0 = \frac{3\cos^2\zeta - 1}{5} \tag{D8.12}$$

These values, called the limiting polarisation and anisotropy, are identical to those derived for a rigid system (Eqs. (D8.3) and (8.4)).

Using Eqs. (D8.1) and (D8.2) for $\bar{P}_0$ and $\bar{A}_0$, we can rewrite Eqs. (D8.9) and (D8.10) in a convenient form, called the Perrin equations:

$$\frac{1}{\bar{P}} - \frac{1}{3} = \left(\frac{1}{\bar{P}_0} - \frac{1}{3}\right)\left(1 + \frac{\tau_F}{\tau_{cor}}\right) = \left(\frac{1}{\bar{P}_0} - \frac{1}{3}\right)\left(1 + \frac{\tau_F kT}{V_h \eta}\right) \tag{D8.13}$$

$$\frac{1}{\bar{A}} = \frac{1}{\bar{A}_0}\left(1 + \frac{\tau_F}{\tau_{cor}}\right) = \frac{1}{\bar{A}_0}\left(1 + \frac{\tau_F kT}{V_h \eta}\right) \tag{D8.14}$$

On the right-hand side of the Perrin equations, the rotational correlation time τ_{cor} has been expressed using Eq. (D2.31).

Equation (D8.13) and (D8.14) can be tested for consistency by plotting $1/\bar{P}_0$ or $1/\bar{A}_0$ against T/η, when a straight line should be obtained. The slope yields the hydrodynamic volume (V_h), provided that the fluorescence decay time (τ_F) is known. The intercept at $T/\eta \to 0$ yields the limiting anisotropy or polarisation. Figure D8.10 shows an example of such a Perrin plot for protein bovine serum albumin.

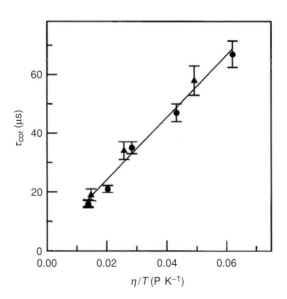

Fig. D8.10 Rotational correlation time τ_{cor} of the erythrosine–BSA complex in different mixtures of glycerol/buffer as a function of η/T, at 273 K constant temperature (•) or at 92% w/w glycerol constant composition (▲). (Ferrer *et al.*, 2001.) The value of τ_{cor} extrapolated to 1 cP, 273 K is 40 ± 2 ns.

D8.5.2 Fluorescence anisotropy decay time

Consider what happens if the sample is excited by a pulse of polarised light, and the time dependence of $I_{||}$ and $I_{\perp}$ is measured. If the emission and absorption dipoles are parallel, the earliest photons to be emitted are highly likely to be z-polarised because the molecules have not had time to reorient. The last protons to be emitted should have random polarisation because by then the system has experienced considerable rotational motion. Therefore, if the polarisation or anisotropy is measured as a function of the time of fluorescence emission, it decays from the initial values of P_0 or A_0 to final values of zero. It is clear, that the rate of decay is a measure of the rate of rotational motion.

As we saw in Section D2.3.4, the rotational behaviour of a sphere can be described by one parameter, the rotational correlation time τ_{cor}. If we define the rotational correlation time τ_{cor} as

$$\tau_{cor} = \frac{1}{6\Theta} = \frac{V_h \eta_0}{kT} \tag{D8.15}$$

then

$$A(t) = (2/5)e^{-t/\tau_{cor}} \frac{3\cos^2 \zeta - 1}{2} \tag{D8.16}$$

Thus, the decay of the fluorescence anisotropy of a spherical molecule is a single exponential.

For a rigid ellipsoid the results become much complicated. Explicit equations have been derived only for rigid ellipsoids of revolution, for which $A(t)$ decays as the sum of three exponential terms. The decay of anisotropy of the fluorescence is given by

$$\frac{A(t)}{A_0} = \alpha_1 \exp\left(-\frac{t}{\tau_{1cor}}\right) + \alpha_2 \exp\left(-\frac{t}{\tau_{2cor}}\right) + \alpha_3 \exp\left(-\frac{t}{\tau_{3cor}}\right) \tag{D8.17}$$

where the pre-exponentials α_i depend only on the relative orientation of the electronic transition moments of the emitting label (dye). If a protein is randomly labelled with an emitting label:

$$[\alpha_1 = \alpha_2 = \alpha_3 = 2/5]$$

the rotational correlation times τ_{cor} are related to the rotational diffusion coefficients of the ellipsoid $D_{||}$ and $D_{\perp}$ by.

$$\tau_{cor} = (4D_{||} + 2D_{\perp})^{-1}, \tau_{2cor} = (D_{||} + 5D_{\perp})^{-1}, \tau_{3cor} = (6D_{\perp r})^{-1}$$

The dependence of the diffusion coefficients on molecular shape, temperature and solvent viscosity can be expressed in compact form by the Stokes–Einstein–Debye equation:

$$D_i = \frac{kT}{6\eta V_h g_i} \quad (i \rightarrow ||, \perp) \tag{D8.18}$$

where V_h is the hydrodynamic volume and η is the solvent viscosity. The Perrin factors g_i depend only on the axial ratio of the ellipsoid $p = a/b$ (where a is the symmetry axis) and are given by

$$g_{\parallel} = \frac{2(p^2 - 1)}{3p(p - S)} \tag{D8.19}$$

$$g_{\perp} = \frac{2(p^4 - 1)}{3p(2p^2 - 1)S - p} \tag{D8.20}$$

$$S_{prol} = (p^2 - 1)^{-1/2} \ln[p + (p^2 - 1)^{-1/2}. \tag{D8.21}$$

$$S_{obl} = (1 - p^2)^{-1/2} \arctan[p^{-1}(1 - p^2)^{-1/2}. \tag{D8.22}$$

In practice, it is rare that the accuracy of the data justifies fitting to more than two exponentials. In this case, the best one can do is to construct various plausible models for the structure and see how they fit the data. One conclusion is evident: an asymmetric molecule has a larger τ_{cor} than a sphere.

For rigid prolate ellipsoids with an axial ratio ≤ 5 and for any oblate ellipsoids $A(t)$ is very close to a single exponential function. Under these conditions the observed decay can be an approximate function with an average correlation time

$$\bar{\tau}_{cor} = \sum_1^3 \alpha_i \tau_{icor} \tag{D8.23}$$

This average $\bar{\tau}_{cor}$ is different from the harmonic mean value that is obtained from the slope of the $A(t)$ function at $t = 0$ (Section D2.3.3).

D8.5.3 Rotational correlation time of globular proteins

Most proteins studied using polarisation techniques exhibit a single rotational correlation time. Figure D8.11 shows an example of anisotropy decay data for anthraniloyl chymotrypsin. The measured rotational correlation time is about 15 ns. A measurement of the decay constant permits the hydrodynamic volume V_h to be calculated if the viscosity of the solvent is known. Equation (D8.15) shows that the larger the molecule, the slower decay of the fluorescence anisotropy.

Simple calculations show that τ_{cor} can be estimated from Eq. (D8.15):

$$\tau_{cor} = \frac{M}{2.55} \times 10^{-12} \text{ s} \tag{D8.24}$$

roughly, τ_{cor} is 1 ns for each 2500 Da of protein molecular mass, if the protein has an approximately spherical shape with a hydration level of 0.3–0.4 g g^{-1} (Chapter D1).

Rotational correlation times $\bar{\tau}_{cor}$ of proteins determined using depolarised fluorescence are presented in Table D8.1. The table also contains protein correlation times obtained by NMR (Chapter J1) and depolarised dynamic light scattering (Chapter D10).

(a)

(b)

Fig. D8.11 (a) Fluorescence decay data of antraniloyl-Ser^{193}-α-chymotrypsin for individual polarised components. (b) Anisotropy on a logarithmic scale after removing the effects of the exciting pulse (G correction). (Stryer, 1968.)

Table D8.1. *Rotational correlation times* $\bar{\tau}_{cor}$ *(20 °C) of proteins determined by different methods*

Protein	Molecular mass (kDa)	Rotational correlation time $\bar{\tau}_{cor}(20\,°C)$ (ns)	Method	Reference
Gramicidin	2.5	2.8	DLS	f
Xfin-Zinc finger	2.93	2.4	NMR	a
BPTI	6.16	4.4	NMR	a
Eglin c	8.15	6.2	NMR	a
Calbindin-D9k apo	8.43	5.1	NMR	a
Calbindin-D9k apo	8.43	4.9	NMR	a
Ubiquitin	8.54	5.4	NMR	a
Cytochrome b_5	9.61	6.1	NMR	a
Barstar c/40/82A	10.14	7.4	NMR	a
Trp-repressor	11.89	23.1	NMR	a
Ribonuclease	13.68	8.3	NMR	a
Azurin	13.95	4.8	ns DPF	b
Lysozyme	14.32	8.3	NMR	a
Lysozyme	14.30	10	DLS	c
Staphylococcal nuclease	15.51	13.3	NMR	a
Interleukin-1β	17.4	12.4	NMR	a
β-Lactoglobulin (monomer)	18.40	9.4	ns DPF	h
β-Lactoglobulin (monomer)	18.40	8.6	μs DPF	k
Leukemia inh. factor	19.1	14.9	NMR	a
S. aurease nuclease B	20.00	9.9	trypt. DPF	h
p21^{ras}	21.00	16	ns DPF	e
HIV-1 protease	21.58	13.2	NMR	a
Tripsin	25.00	14.3	ns DPF	d
Chymotrypsin	25.00	D8.7	ns DPF	d
Savinase	26.70	12.4	NMR	a
Carbonic anhydrase	30.0	12.4	ns DPF	d
β-Lactoglobulin (dimer)	36.00	22.5	ns DPF	d
Apoperoxidase	40.00	28.0	ns DPF	d
Ovalbumin	43.5	30.2	μs DPF	h
F_{AB} fragment Ig G	47.50	33	ns DPF	d
Serum albumin	66.50	40	μs DPF	g
β-Lactoglobulin	18.40	8.6	μs DPF	h
Sucrase-isomaltase	221.0	112	μs DPF	h

Abbreviations: DLS, dynamic light scattering; NMR, nuclear magnetic resonance; ns DPF, depolarised fluorescence in the nanosecond range; μs DPF, depolarised fluorescence in the microsecond range; trypt. DPF, depolarised fluorescence using tryptophan residue.

Data are taken from:

a Garcia de la Torre (2001).   d Yguerabide *et al.* (1970).   g Ferrer *et al.* (2001).
b Kroes *et al.* (1998).   e Hazlett *et al.* (1993).   h Cherry and Schneider (1976).
c Dubin *et al.* (1971).   f Michielsen and Pecora (1981).

Fig. D8.12 Dependence of the rotational correlation times of 32 different proteins on molecular mass. Data are taken from Table D8.1. The dashed straight line corresponds to data predicted by Eq. (D8.24).

Comment D8.6

Rotational relaxation τ_{relax} and correlation τ_{cor} times

Please note that workers in magnetic resonance and electric birefringence sometimes use the rotational relaxation time τ_{relax} and rotational correlation time τ_{cor}. The rotational correlation time τ_{cor} and rotational relaxation time τ_{relax} are related by $\tau_{cor} = \tau_{relax}/3$, $\tau_{cor} = (6D)^{-1}$ and $\tau_{relax} = (2D)^{-1}$. The simplest case is that of a rigid sphere, for which $A(t)$ decays exponentially, $A(t) = A_0 e^{-6\Theta t} = A_0 e^{-t/\tau_{cor}}$, where Θ is the rotational diffusion coefficient.

The dependence of rotational correlation times on molecular mass of 32 different proteins is plotted in Fig. D8.12 (Comment D8.6). The dashed straight line corresponds to the data predicted by Eq. (D8.24). Figure D8.12 shows that practically all proteins (with the exception of carboanhydrase) have a rotational correlation time sufficiently large to correspond to the motion of spherical particles. The difference between predicted and experimental rates arises because the protein is not spherical. Molecules with elongated shapes rotate more slowly. The slope of the dashed straight line, drawn through the experimental points,

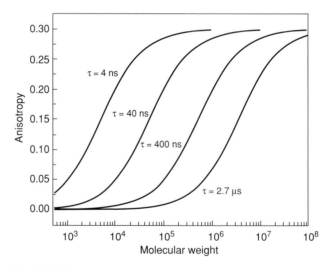

Fig. D8.13 Dependence of the anisotropy on molecular mass for probes with decay times from 4 to 2700 ns. For these calculations Eq. (D8.5) was used. The volume V of a protein is calculated from its molecular weight with the assumption of 20% hydration. (After Lakovicz et al., 2000.)

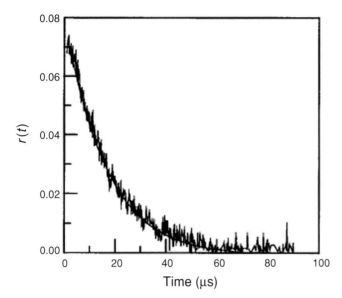

Fig. D8.14 The decay of the phosphorescence anisotropy of the erythrosine–BSA complex and the fitted monoexponential function with rotational correlation time τ = 16 ± 1.5 μs (92% w/w glycerol/buffer, 18 °C). (Ferrer *et al.*, 2001.) Erythrosin is 2′,4′,5′,7′- tetraiodofluorescein.

(a)

(b)

80 Å

shows that the rotational correlation time in the limit of experimental error is proportional to molecular mass in the agreement with Eq. (D8.24).

Most of the fluorescent data presented in Fig. D8.12 were obtained in the nanosecond range. Such conditions put a serious limitation on the range of molecular mass of biological macromolecules available for such measurements. From Eq. (D8.7) it follows that macromolecules with molecular masses in the range 10–50 kDa have relaxation times in the range 4–20 ns, a range that corresponds to the singlet lifetimes of most organic dyes. Figure D8.13 shows the dependence of anisotropy on the molecular mass according to Eq. (D8.5). The figure demonstrates that for a protein with a molecular mass of 100 kDa labelled with a 4-ns decay time probe the emission is expected to be completely polarised. This explains why for a protein the upper molecular mass limit for measuring rotational diffusion with a nanosecond lifetime probe is near 40–50 kDa.

It is now possible to use fluorescence in the microsecond time scale. This change in time scale has been made possible by the development of metal–ligand complexes that display decay time ranging from 10 ns to more than 10 μs, a range that corresponds to triplet lifetimes. Figure D8.14 shows the decay of the phosphorescence anisotropy of the erythrosin–BSA complex on the microsecond time scale in a buffer containing glycerol. These measurements are consistent with the absence of independent motion of large protein segments in solution, in the time range from nanoseconds to fractions of a millisecond, and give a

Fig. D8.15 (a) The elongated ellipsoid (140 Å × 40 Å) used as the standard model for conformation in a solution of human and bovine serum albumin in all textbooks. (b) The heart-shaped structure of human serum albumin at the same scale, adapted from the Brookhaven Protein Data Bank X-ray coordinates, file 1bm0. (After Ferrer *et al.*, 2001.)

(a)

(b)

Fig. D8.16 (a) The rough shell triangular model for BSA with dimensions of 80 Å × 80 Å × 89 Å × 30 Å. (b) The rough shell model derived from the atomic-level structure of human serum albumin (After Ferrer *et al.*, 2001.)

single rotational correlation time τ_{cor} for BSA (1cP, 20 °C) = 40 ± 2 ns (see also Fig. D8.10).

Combining the values of the rotational correlation time (40 ± 2 ns) and intrinsic viscosity (4.1 cm^3 g^{-1}) allows a choice to be made between the heart-shaped structure of human serum albumin observed in single crystals and the standard, textbook hydrodynamic model of BSA: a cigar-shaped ellipsoid with dimensions of 140 Å × 40 Å (Fig. D8.15). A detailed analysis of the BSA hydrodynamics was based on two bead modelling methods. In the first, BSA was modelled as a triangular prismatic shell with optimised dimensions of 84 Å × 84 Å × 84 Å × 31.5 Å, whereas in the second, the atomic-level structure of human serum albumin obtained from crystallographic data was used to build a shell model (Fig. D8.16). In both cases, the predicted and experimental rotational diffusion coefficients were in good agreement. These data show that the elongated ellipsoid model used for decades to interpret the protein solution hydrodynamics at neutral pH is inappropriate.

D8.6 Depolarised fluorescence and molecular interactions

Depolarised fluorescence can be used to analyse most molecular interactions, including protein–protein, DNA–protein and antigen–antibody binding. Depolarised fluorescence is unique among the methods used to analyse molecular binding because it gives a direct, nearly instantaneous measure of a tracer's bound/free ratio, even in the presence of free tracer. Most such measurements can be taken in real time, allowing the kinetic analysis of association and dissociation reactions. Below we will give a few examples of the application of depolarised fluorescence to studies of molecular protein–protein interactions.

Fluorescence polarisation equilibrium binding studies usually involve titrating a small amount of fluorescently labelled protein with a range of concentrations of a second protein. As the fraction of bound tracer increases, the polarisation

Fig. D8.17 Effect of thtrombin activation on factor XIII A and B subunit association. Aliquots of rA$_2$ or thrombin-activated rA$'_2$ were added to a starting solution of B$_2$. The horizontal dashed line shows the half maximum anisotropy change. (After Checovich *et al.*, 1995.)

Fig. D8.18 Time course of fluorescein-labelled casein digested with trypsin. The reaction was started by the addition of 5 pmol fluorescein-labelled casein with: 200 ng trypsin (bottom plot); 100 ng trypsin (middle plot): no trypsin (top plot). Mp, millipolarisation units. (After Checovich *et al.*, 1995.)

> ### Comment D8.7 Physicist's box: Blood coagulation factor XIII
>
> Blood coagulation factor XIII or the fibrin stabilising factor is a plasma zymogen of the enzyme responsible for strengthening the clot network and for endowing it with a higher resistance to lysis. A zymogen or proenzyme is an inactive enzyme precursor that requires a biochemical change, such as hydrolysis, in order to become active in catalysis. Factor XIII is made up of two different subunits in an AB protomeric structure thought to associate into an A_2B_2 tetramer. Only the A subunits are modified by thrombin, realising an N-terminal activation peptide (AP). The hydrolytically cleaved A subunit is denoted A'.

increases. Equilibrium binding constants are calculated from the resulting binding curve (i.e. polarisation versus unlabelled protein concentration). Figure D8.17 shows the effect of thrombin activation on factor XIII A and B subunit association (Comment D8.7). Thrombin cleavage of rA_2 to rA'_2 results in a dramatic decrease in the affinity of rA_2 for labelled B2 (294 nM for rA'_2 versus 19 nM for rA_2) thereby explaining A_2B_2 complex dissociation.

Another example concerns the action of degradation enzymes, such as proteases, DNases and RNase. The action of such enzymes usually results in a change in molecular size, and can therefore be followed using depolarised fluorescence. Figure D8.18 shows a time course of trypsin activity using fluorescein-labelled casein. Almost any substrate can be fluorescently labelled and used to measure the activity of a specific enzyme. The standard curve for the casein-based assay is linear over at least a 100-fold concentration range, and the detection limit is 10 times more sensitive than is possible with the more common acid-precipitation assays.

D8.7 Checklist of key ideas

- Luminescence is the phenomenon of light emission from an electronically excited state of a molecule.
- Emission from a *singlet* state to the ground state is termed *fluorescence*; in fluorescence the interval between excitation and emission is of the order of 10^{-8} s.
- Emission from a *triplet* state to the ground state is termed *phosphorescence*; in phosphorescence the interval between excitation and emission ranges from milliseconds to seconds.
- In practice, very few proteins are naturally fluorescent; most proteins and practically all nucleic acids must be converted into fluorescent derivatives by attaching to them suitable fluorescent probes.
- Fluorescent probes can be placed into three categories: intrinsic probes (aromatic groups in proteins), chemical analogues of intrinsic probes (e.g. 5-hydroxyltryptophan) and extrinsic probes for proteins and nucleic acids, of which there are thousands.

- Most biological macromolecules have rotational motion on a time scale which is longer than the fluorescence time scale, so they cannot tumble fast enough to achieve random orientation.
- A rotational correlation time τ_{cor} can be measured from the decay of the fluorescence or phosphorescence polarisation as a function of time only, if the rate of macromolecular rotational diffusion is comparable to the fluorophore lifetime.
- Globular proteins with molecular masses in the range 10–50 kDa have relaxation times in the range 4–20 ns, a range that corresponds to the singlet lifetimes of most organic dyes.
- The rotational correlation times of larger proteins can be measured with microsecond probes. For a phosphorescence decay time of 2.7 μs it is possible to measure the correlation time of macromolecules with molecular masses of a few million daltons
- Fluorescence anisotropy of a spherical molecule exhibits single-exponential decay; roughly, τ_{cor} is 1 ns for each 2500 Da of protein molecular mass, if the protein is approximately spherical.
- For rigid non-spherical molecules, the experimental rotational correlation time is always greater than that predicted for a sphere, because elongated molecules rotate more slowly.
- The decay of the fluorescence anisotropy of a general rigid body follows a multiexponential function (up to five exponentials); however in practice, it is rare that the accuracy of the experimental data justifies fitting to more than two exponentials.

Suggestions for further reading

Historical review and introduction to biological problems

Tao, T. (1969). Time dependent fluorescence depolarization and Brownian rotational diffusion coefficients of macromolecules. *Biopolymers*, **8**, 609–632.

Prendergast, F. G. (1991). Time-resolved fluorescence techniques: methods and application in biology. *Curr. Opin. Struct. Biol.*, **1**, 1054–1059.

Checovich, W. J., Bolger, R. E., and Burke, T. (1995). Fluorescence polarization – a new tool for cell and molecular biology. *Nature*, **375**, 254–256.

Theory of fluorescence depolarisation

Weber, G. (1953). Rotational Brownian motion and polarization of the fluorescence of solutions. *Adv. Protein Chem.*, **8**, 415–459.

Small, E. W., and Isenberg, I. (1977). Hydrodynamic properties of a rigid molecule; rotational and linear diffusion and fluorescence anisotropy. *Biopolymers*, **16**, 1907–1928.

Instrumentation

Cherry, R. J., and Schneider, G. (1976). A spectroscopic technique for measuring slow rotational diffusion of macromolecules. 2: Determination of rotational correlation times of proteins in solution. *Biochemistry*, **24**, 3657–3661.

Lakowicz, J. R., Gryczynski, I., *et al.* (2000). Microsecond dynamics of biological macromolecules. *Meth. Enzymol.*, **323**, 473–509.

Sportsman, J. R. (2003). Fluorescence anisotropy in pharmacologic screening. *Meth. Enzymol.*, **361**, 505–529.

Lakowicz, J. R., Gryczynski, I., *et al.* (2000). Microsecond dynamics of biological macromolecules. *Meth. Enzymol.*, **323**, 473–509.

Kroes, S. J., Canters, G. W., Giardi, G., Van Hoek, A., and Visser, A. J. W. G. (1998). Time-resolved fluorescence study of azurin variants: conformational heterogeneity and tryptophan mobility. *Biophys. J.*, **75**, 2441–2450.

Depolarised fluorescence and Brownian motion

Lakovicz, J. R. (ed.). (1999) *Principles of Fluorescence Spectroscopy*, second edition. New York: Kluwer Academic/Plenum Publ.

Cherry, R. J., and Schneider, G. (1976). A spectroscopic technique for measuring slow rotational diffusion of macromolecules. 2: Determination of rotational correlation times of protein in solution. *Biochemistry*, **15**, 3657–3661.

Ferrer, M. L., Duchowicz, R., Carrasco, B., Garcia de la Torre, J., and Acuna, A. U. (2001). The conformation of serum albumin in solution: a combined phosphorescence depolarization-hydrodynamic modeling study. *Biophysical J.*, **80**, 2422–2430.

Depolarised fluorescence and molecular interactions

Millar, D. P. (2000). Time-resolved fluorescence methods for analysis of DNA–protein interactions. *Meth. Enzymol.*, **323**, 442–459.

Hazlett, T. L., Moore, K. J. M., Lowe, P. N., Jameson, D. M., and Eccleston, J. F. (1993). Solution of p21[ras] proteins bound with fluorescent nucleotides: a time-resolved fluorescence study. *Biochemistry*, **32**, 13575–13583.

Canet, D., Doering, K., Dobson, C. M., and Dupont, Y. (2001). High-sensitivity fluorescence anisotropy detection of protein-folding events: Application to α-lactalbumin. *Biophys. J.*, **80**, 1996–2003.

Chapter D9
Viscosity

D9.1 Historical review

1846

J. L. Poiseuille produced a theory of liquid flow in a capillary. Based on this theory, **W. Ostwald (1933)** and **L. Ubbelohde (1936)** invented viscometers and introduced them into physical and chemical practice. Later, the instruments were named after them.

1906

A. Einstein was the first to treat the viscosity of suspensions of rigid spherical particles that are large relative to the size of the solvent molecules. He showed that the specific viscosity of the solution is proportional to the volume fraction of the solution occupied by the particles and does not depend on the absolute size of the spheres.

1932

H. Staudinger proposed using the intrinsic viscosity to determine the molecular mass of polymers. He found that the dependence of the intrinsic viscosity on the molecular mass for a homologous series of polymers can be expressed by a simple formula of the type $[\eta] \approx M^{\alpha}$. It became understood later (**H. Mark, R. Houwink, W. Kuhn** and **H. Kuhn**) that the constant α is related to the molecular conformation of the macromolecules in solution.

1962

B. H. Zimm and **D. M. Crothers** proposed an original design for a rotational viscometer that operates at low velocity gradients and which has been very useful for measurements on asymmetric structures such as DNA, fibrous proteins and rod-like viruses.

1985

M. A. Haney invented a 'differential viscometer' which measures the relative viscosity directly. The advantage of using the differential viscometer instead of a conventional glass capillary viscometer is its increased sensitivity, precision, speed and operational ease which allow measurements of the solution viscosity of low sample concentrations ($\sim$1 mg cm^{-3} for globular proteins).

2000 to now

Intrinsic viscosity is very sensitive to the conformation of macromolecules. It can be obtained precisely and easily using different types of viscometers. However, to date the viscosity method has not proved popular mainly owing to the relatively large quantities of material required.

D9.2 Application to biological problems

Intrinsic viscosity is one of the oldest molecular parameters. Einstein considered the intrinsic viscosity of a suspension of spherical particles in 1906. It was understood later that intrinsic viscosity is very sensitive to the conformation of biological macromolecules and can be measured with high precision. Yang's classical review of its measurement and application, particularly to proteins, was published in 1961. In 1997 Harding published a very useful text, which describes the considerable progress that has been made in the measurement of intrinsic viscosity and its interpretation in terms of molecular conformation, hydration and flexibility of biological macromolecules and, primarily, glycopolymers in solution.

Traditionally, the viscosity of a solution of biological macromolecules was measured by timing either the flow of the liquid through a capillary tube or the rotation speed of one cylinder relative to another. Although the instrumentation for both these approaches has become either automatic or semiautomatic with on-line computer data, the basic design has remained the same since Yang's and Zimm's articles were published in the early 1960s. Neither approach is attractive in protein biochemistry, because of the relatively large quantities of material required – normally greater than 1 ml at concentrations of 5–6 mg ml^{-1}.

An important development was the appearance of an instrument based on a quite different principle – the so-called 'differential' viscometer. With this instrument measurements can be performed at a lower concentration (about 1 mg cm^{-3}) and on smaller volumes, making the intrinsic viscosity attractive for use in protein biochemistry once more. However, in spite of this innovation viscosity is not now popular as a method for studying the structure of biological macromolecules.

D9.3 Viscosity measurement

D9.3.1 Capillary viscometer

The capillary or '*Ostwald*' viscometer is still the most commonly used viscometer (Fig. D9.1(a)) and involves essentially just a piece of glassware suspended in a

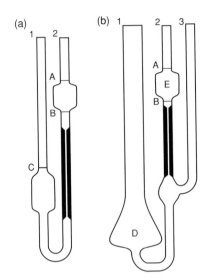

Fig. D9.1 Capillary viscometers: (a) Ostwald and (b) Ubbelohde. The Ostwald
instrument is used in the following way. The solution is added at opening 1 until the
liquid level at rest is at scratch C. Suction is then applied at opening 2 until the liquid
level is above scratch A. The suction is removed and the liquid falls owing to the
difference in height between the two arms. The time t required for the meniscus to
move between scratches A and B is measured. Because of the change in relative
liquid heights, the flow rate is not constant.

  The Ubbelohde instrument is used in the following way. The solution is added at
opening 1 until the liquid level at rest is at scratch C. Opening 3 is closed and suction
is applied at opening 2, until the liquid has been drawn above A. Opening 3 is then
opened with opening 2 closed, which allows bulb D to drain. Opening 2 is then
opened and the times for the meniscus to pass scratches A and B are determined. To
vary the concentration, the liquid in bulb E can be diluted because the amount of
liquid in X does not determine the volume of liquid contained between A and the
bottom of the capillary.

Comment D9.1

Kinematic viscosity

The relative viscosity
without the density
correction is known as
the 'kinematic' (as
opposed to 'dynamic')
relative viscosity $\eta_{rel} =$
(t/t_0); derived from its
parameters: $\eta'_{sp}, \eta'_{red},$
$[\eta]'$, are the
corresponding
kinematic quantities.
To a reasonable
approximation, for
concentrations <1 mg
ml^{-1} $[\eta]' \sim [\eta]$.

constant temperature environment. The Ostwald viscometer consists of a capillary
tube of radius r and length L through which a volume V passes. The viscosity η
and the average velocity gradient G are given by

$$\eta = \pi hg\rho r^4 t/8LV \quad \text{and} \quad G = 8V/3\pi r^2 t \tag{D9.1}$$

in which h is the average liquid height, g is the gravitational constant and ρ is the
density. A precise evaluation of h, r and L is avoided by measuring

$$\eta_{rel} = (t_\rho \rho/t_0 \, \rho_0) \tag{D9.2}$$

where t and t_0 are the flow times for the solution at a particular concentration C
and ρ and ρ_0 are the corresponding solution and solvent densities. The relative
viscosity without the density correction is known as the 'kinematic' viscosity
(Comment D9.1).

To facilitate measurements at a series of concentrations where the dilutions can be performed in situ, a modified form of viscometer, called an *Ubbelohde* viscometer, can be used (Fig. D9.1(b)). This viscometer is designed so that the head of liquid when the flow time is being measured is independent of the amount of solution in the viscometer: progressive dilutions can then be made directly in the viscometer. However, if a macromolecule degrades or denatures appreciably during a series of measurements, this type of viscometer should not be used (Comment D9.2).

D9.3.2 Rotor viscometer

Formally viscosity depends on the orientation of the particles. If they are oriented parallel to the streamlines, as they would be for large velocity gradients, a smaller viscosity increment is anticipated than if the particles are oriented at random, so that some of them lie perpendicular to the streamlines. In other words, particles oriented in parallel show non-Newtonian behaviour even though the solvent is Newtonian.

Non-Newtonion behaviour is rarely important at the velocity gradient normally used in a capillary viscometer for solutions of macromolecules with a small degree of asymmetry. Also, symmetrical non-deformable molecules (e.g. globular proteins) do not show non-Newtonian behaviour at all and hence it is not necessary to measure the viscosity at different velocity gradients. However, for very asymmetrical molecules (such as collagen) or very deformable molecules (such as DNA) extrapolation to zero gradient is necessary.

The Couette viscometer, which was designed for use at relatively low shear gradients, consists of two concentric cylinders separated by a narrow annulus, which is filled with the sample (Fig. D9.2). One cylinder is fixed and the other rotates. This instrument operates at a low, controllable and easily measured value of G.

The Zimm–Crothers floating-rotor viscometer is a modification of the Couette viscometer. This modification allows high-precision work at a very low shear gradient (Fig. D9.3). It differs in principle from the Couette-type in that, with a Couette viscometer, the shear gradient is fixed and the shear stress is measured, whereas, with this viscometer, the shear stress is fixed and the shear gradient is measured. The solution is placed between a stationary outer cylinder and a potentially rotating inner cylinder, which floats in the liquid. The inner cylinder contains a steel pellet. External to the outer cylinder are rotating magnets, which cause the steel pellet, and therefore the inner cylinder, to rotate with a constant angular velocity ω. The relative viscosities of any two solutions can be calculated to be simply $\eta_1/\eta_2 = \omega_2/\omega_1$. The gradient can be altered by varying the strength of the applied magnetic field.

This instrument simply measures the viscosity relative to a standard, such as water, by determining the relative amounts of time required for the inner cylinder to make a given number of rotations.

Comment D9.2
Progressive dilutions in the Ostwald viscometer

The best results can be obtained using the Ostwald viscometer in which the progressive dilutions can be performed directly. In this case the flow times of the solvent depend on the solvent volume. The viscometer can be calibrated for different volumes of liquid.

Rotating axis

Rotating outer cylinder

Side view

Fig. D9.2 A Couette viscometer: the outer cylinder usually rotates; the inner cylinder is stationary.

Fig. D9.3 Cross-section of the Zimm–Crothers viscometer. See text. (After Zimm and Crothers, 1962.)

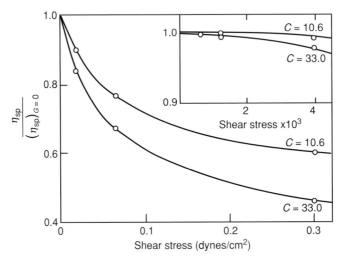

Fig. D9.4 Dependence on shear of the specific viscosity of T2 DNA, expressed as the percentage of the value at zero shear τ_{sp} (τ_{sp}) $G = 0$. Curves for two concentrations ($\mu g/cm^{-3}$) of DNA solution are shown. The viscosity becomes essentially G-independent at very low shear (inset). Note that the standard value of G in a capillary viscometer is about 50 times more than in a Zimm–Crothers viscometer. (After Crother and Zimm, 1965.)

Very asymmetric molecules or easily deformable molecules tend to be oriented or deformed by high velocity gradients. Such a behaviour leads to a reduction in the disturbance of the flow pattern at high velocity gradients and hence to viscosities below those characteristic of unoriented or undeformed molecules. This effect is an important one for the molecules with a very high intrinsic viscosity, and for such molecules it is essential that experimental determinations be made at several different values of velocity gradient and that the intrinsic viscosities be extrapolated to zero velocity gradient. Thus Eq. (D2.40) should be rewritten

$$[\eta] = \lim_{\substack{c \to 0 \\ G \to 0}} (\eta - \eta_0)/\eta_0 c \tag{D9.3}$$

An example of the effect produced is shown by data for a DNA sample in Fig. D9.4. The effect is very large. Thus, it is imperative to make measurements at very low velocity gradients, and preferably to extrapolate to zero velocity gradients.

D9.3.3 Pressure imbalance differential viscometer

The pressure imbalance differential viscometer uses a fluid analogue of a Wheatstone bridge electrical circuit. This is referred to as a 'differential viscometer' since it measures relative viscosity directly. It is also highly sensitive, permitting

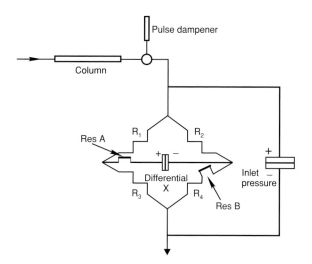

Pulse dampener

Column

Res A

R_1 R_2

$+$ $-$

Differential

R_3 X R_4

Inlet $-$
pressure

$+$

Res B

Fig. **D9.5** Principles of the pressure imbalance differential viscometer. At 'baseline conditions' (i.e. solvent only) the differential pressure across the bridge is zero because there is solvent in all four capillaries, R_1–R_4. When the molecular solution enters the bridge from the column it fills capillaries R_1, R_2 and R_3, while solvent from a delay reservoir (Res B) remains in capillary R_4 and prevents the entry of the macromolecular solution. The transducer X measures the differential pressure ΔP and Y measures the inlet pressure P_i. The reservoir Res A located out of the flow stream acts to compensate the volume so that any flow rate fluctuations cause equal pressure changes on each side of the differential pressure transducer Y. The difference in viscosity causes a pressure imbalance in the bridge that is proportional to the specific viscosity of the sample solution. From knowledge of the concentration (measured using an off- or on-line refractive index detector), the reduced specific viscosity can be obtained. (Adapted from Dutta *et al.*, 1991.)

the accurate measurement of low relative viscosities and hence measurements at low concentration ($\sim$1 mg ml^{-1} for globular proteins). At baseline conditions (i.e. solvent alone) the differential pressure across the bridge is zero because there is solvent in all four capillaries. When the sample enters the bridge (Fig. D9.5) it fills capillaries R_1, R_2 and R_3 while solvent from a delay reservoir remains in capillary R_4. The difference in viscosity between the solvent in R_4 and solution in R_3 causes a pressure imbalance ΔP in the bridge, which from Poiseuille's law can be related to the relative viscosity or the specific viscosity of the solution:

$$\eta_{sp} = 4\Delta P/(P_i - 2\Delta P) \tag{D9.4}$$

This viscometer has great sensitivity at high dilution and rapidity of measurement, and in addition solution can be injected continuously via a flow cell; it can thus be fitted on-line to a concentration detector (refractive index or UV absorbance based) to convert the relative viscosity to the reduced specific viscosities.

D9.3.4 The effect of concentration

The viscosity of a solution of particles is a complex function of concentration because, at high concentrations, distortion of the velocity pattern of the liquid by one particle can affect the shear at a neighbouring particle. Thus, in general, one can expect the viscosity of a solution to be expressed as a power series relative to the viscosity of pure solvent:

$$\eta = \eta_0(1 = k_1 C + k_2 C^2 + \cdots) \tag{D9.5}$$

In such a virial expansion, the coefficient k_1 can be associated with the contribution of individual molecules, k_2 with pairwise effects of clusters of two molecules,

Fig. D9.6 Plot of the relative viscosity η_{rel} versus the concentration C for five globular proteins: (1) human haemoglobin; (2) ovalbumin; (3) bovine albumin; (4) human albumin; and (5) human γ-globulin. (After Monkos and Turczynski, 1991.)

and so on. Figure D9.6 shows a plot of the relative viscosity versus concentration for several globular proteins. As can be seen, this effect is clearly revealed at protein concentrations greater than 150 mg cm^{-3}.

For a dilute solution the well-known Huggins equation can be written as

$$\eta_{sp}/C = [\eta] + K_H[\eta]^2 C \tag{D9.6}$$

where K_H is the (dimensionless) Huggins constant, which is related to the conformation of macromolecules in solution.

An alternative equation for the concentration dependence is

$$\eta_{sp}/C = [\eta](1 + k_\eta C) \tag{D9.7}$$

where

$$k_\eta = [\eta]K \tag{D9.8}$$

The concentration dependence parameter k_η has the same units (ml g^{-1}) as the equivalent parameters from sedimentation velocity, k_s (Eq. (A) of Comment D4.18), and translational diffusion, k_D (Eq. (A) of Comment D3.6). The values of k_η and k_s are related:

$$k_\eta = \frac{k_s V_s}{\bar{v}} \tag{D9.9}$$

where $V_s/\bar{v}$ is the swelling factor of the macromolecule in solution. The factor $V_s/\bar{v}$, a quantity notoriously difficult to estimate, is found very simply from the ratio k_η/k_s, without any assumptions as to particle conformation.

D9.4 Comparison of theory with experiment

In spite of the empirical character of Eq. (D2.58), calculation of the intrinsic viscosity via polarisability gives good practical results for globular proteins. Table D9.1 compares values of intrinsic viscosity predicted from atomic

Table D9.1. *Comparison of protein intrinsic viscosities predicted from atomic coordinates with experimental ones*

Protein	PDB entry	M (kDa)	Predicted intrinsic viscosity ($cm^3 g^{-1}$)	Measured intrinsic viscosity ($cm^3 g^{-1}$)
RNase A	7rsa	13.7	3.26	3.30
Lysozyme	6lys	14.3	2.99	2.98–3.00
Myoglobin	1mbo	17.2	3.14	3.15
Chymotrypsinogen	2cg	25.66	2.93	2.5–3.13

The intrinsic viscosities of proteins were calculated via capacitances assuming that the water of hydration forms a hydration shell of a uniform thickness of 0.09 nm (Zhou, 1995).

coordinates with experimental ones. These data clearly show that inclusion of 0.9-Å-thick hydration shell (universal for all globular proteins) leads to good agreement between theory and experiments (see also Table D3.2 for diffusion coefficients).

D9.5 Determination of the shape of macromolecules from intrinsic viscosities: homologous series

D9.5.1 Synthetic polypeptides

Synthetic polypeptides provide an ideal model to illustrate the high sensitivity of viscosity to the conformation state of macromolecules. It was found by Doty and coworkers as early as 1956 that synthetic polypeptide poly-γ-benzyl-L-glutamates can exist in the form of Pauling and Corey's α-helix in a solvent with low hydrogen–hydrogen interactions. However, in solvents with unusually strong hydrogen–hydrogen interactions the intramolecular hydrogen bonds of the helices are no longer stable and the conformation of the polypeptides is that of a random coil. Figure D9.7 shows the dependence of intrinsic viscosity on molecular weight on double logarithmic scale for poly-γ-benzyl-L-glutamates in two types of solvent. In dichloracetic acid (open circles) the polypeptides exist in the form of a Gaussian coil. In this solvent $[\eta] = 2.78 \times 10^{-3} M^{0.87}$, i.e. the exponent falls in the range expected for a Gaussian coil in a 'good' solvent. In structure-forming solvents such as dimethyl formamide or chloroform saturated with formamide (filled circles) $[\eta] = 2.78 \times 10^{-3} M^{1.7}$. The power of 1.7 is in a perfect accord with Simha's theoretical value (see Eq. (D2.52)).

It should be pointed out that the two curves in the Fig. D9.7 cross and the intrinsic viscosity at the intersection point (M is about 70 kDa) remains unchanged

Fig. D9.7 Dependence of
intrinsic viscosity on
molecular mass for poly-
γ-benzyl-L-glutamates.
The open circles
represent the randomly
coiled form, filled circles
the α-helical form. (After
Doty *et al.*, 1956.)

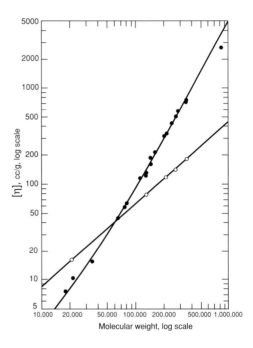

in spite of the coil being transformed into a helix (and vice versa). On the right-hand side of the crossing point a shift from the helix to a coil is accompanied by a decrease in intrinsic viscosity and the reverse is true on the left-hand side of the Fig. D9.7.

D9.5.2 Proteins in globular conformations

Experimental values of the intrinsic viscosity of a number of proteins are listed in Table D9.2. It is clearly seen that there are two groups of proteins. One group is characterised by values of $[\eta]$ between 3.0 and 4.0. These proteins are the globular proteins, the shape of which is close to spherical. It is evident that determining themolecular mass of a globular proteins from measuring of its intrinsic viscosity is impossible. It is interesting to note that the spherical bushy stunt virus with a molecular mass of 10.7 MDa has the same intrinsic viscosity as ribonuclease with a molecular mass of 13.7 kDa. This observation is the best proof of Eq. (D2.48), according to which the volume fraction of the solution occupied by spheres, but not the radius, is related to their specific viscosity.

The second group of proteins, by contrast, have much larger values of $[\eta]$. These proteins are the elongated proteins, the shape of which is far from spherical. The intrinsic viscosities of such proteins, primarily, depend on the asymmetry of the particles. Thus, the intrinsic viscosity of collagen (rod-shaped with an axial ratio of about 100) is 30 times more than the intrinsic viscosity of the tobacco mosaic virus (rod-shaped with an axial ratio of 18).

Table D9.2. *Intrinsic viscosity of a number of proteins and viruses in aqueous solutions*[a]

Protein	Molecular mass (Da)	Intrinsic viscosity ($cm^3\ g^{-1}$)
Ribonuclease	13 683	3.30
Lysozyme	14 211	2.7
Myoglobin	17 836	3.15
Chymotrypsinogen	25 660	2.5–3.15
β-Lactoglobulin	36 800	3.4
Serum albumin	66 296	3.7
Haemoglobin	68 000	3.6
Catalase	250 000	3.9
Bushy stunt virus	10 700 000	3.4
Tropomyosin	70 000	52
Fibrinogen	330 000	27
Collagen	345 000	1150
Myosin	490 000	217
Tobacco mosaic virus	40 000 000	37

[a] For a more comprehensive table, see Harding (1997).

D9.5.3 Proteins in coil conformation

A polypeptide chain in the random coil conformation (lacking secondary or tertiary structure) occupies a much larger volume than it does in a globular conformation, and consequently exhibits an increased intrinsic viscosity. Figure D9.8

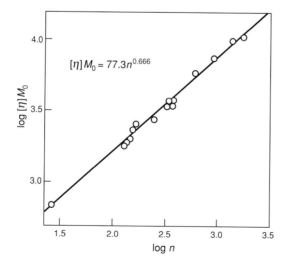

$$[\eta]\,M_0 = 77.3 n^{0.666}$$

Fig. D9.8 Dependence of the intrinsic viscosity on the chain length for polypeptide chains in 5–7.5 M guanidine–HCl at 25 °C. Disulphide bonds were prevented from forming by the presence of excess β-mercaptoethanol (0.15–0.2 mM), or by protection by alkylation, following reduction of all existing bonds. (After Tanford, 1968.)

Table D9.3. *Proteins in random coil conformation (6 M GuHCl + 0.15 mM β-mercaptoethanol)*

Protein	M (Da)	M_0 (Da)	Amino acids (number)	Intrinsic viscosity (cm³ g⁻¹)
Insulin	5 700	113	26	6.1
Ribonuclease	13 683	110	124	16.6
Lysozyme	14 211	111	129	17.1
Haemoglobin	15 500	108	144	18.9
Myoglobin	17 200	112	153	20.9
β-Lactoglobulin	18 400	113	162	22.8
Chymotrypsinogen	25 660	105	245	26.8
Phosphoribosyl transferase	35 000	109	320	31.9
Glyceraldehyde-3-phosphate dehydrogenase	36 300	110	331	34.5
Tropomyosin	38 000	114	333	33.0
Aldolase	40 000	107	376	35.5
Pepsinogen	40 000	108	370	31.5
Serum albumin	66 296	114	605	52.2
Paramyosin	100 000	114	880	65.6
Thyroglobulin	165 000	121	1364	82
Myosin	200 000	115	1739	92.6

shows a plot of log $[\eta]M_0$ versus log n for 16 proteins in guanidine–HCl, where M_0 and n represent the mean residue mass and the number of residues in the protein, respectively. The least squares line fits the equation given in the figure. Some numerical data for $[\eta]$ for denatured proteins in the random coil conformation are given in a Table D9.3.

D9.5.4 DNA

Table D9.4 contains some results for the viscosity of large molecular weight DNAs. Measurements were made at a shear stress, at which no shear dependence

Table D9.4. *Intrinsic viscosity of DNA from some bacteriophages*

Source	Molecular mass (MDa)	Intrinsic viscosity (cm³ g⁻¹)
T2	115	31 600
λ	32	14 000
1/4 T2	27.2	11 500
T7	26.35	11 100
T7 fraction	3.7	2 590

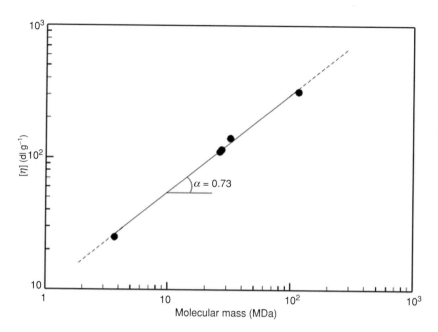

Fig. D9.9 Dependence of $[\eta]$ on the molecular mass of DNA from bacteriophages on a double logarithmic scale. The slope of this straight line is 0.73, suggesting a somewhat extended Gaussian coil.

of the viscosity could be detected. The data are plotted on a double logarithmic scale in Fig. D9.9. It can clearly be seen that data in the range of molecular mass 4–115 MDa can be described by a very good straight line with a slope of $\alpha = 0.73$, which is typical for an extended Gaussian coil (Comment D9.3).

Comment D9.3 Relation between exponents α and a for DNA in the molecular mass range 4–115 MDa

We recall that for a homologous series of Gaussian coils in a good solvent the following *KMX* relations are valid (see Comment D4.27)

$$[\eta] = K_{[\eta]} M^{0.8}$$
$$s = K_s M^{0.4}$$

The experimental values of exponents for DNA in the molecular mass range 4–115 MDa are very close to expected ones. Thus, $\alpha = 0.73$ (Fig. D9.9), $a = 0.44$ (Fig. D4.28). Interestingly that relation between exponents α and a, described by Eq. (E) in Comment D4.27 is fulfilled in the limit of experimental error.

D9.6 Monitoring of conformational change in proteins and nucleic acids

If a globular particle is denatured its intrinsic viscosity increases dramatically. The sensitivity of intrinsic viscosity measurements for monitoring change of

Fig. D9.10 (a)
Dependence of [η] on the
temperature (thermal
denaturation) for
ribonuclease (pH 2) (Van
Holde, 1985). (b)
Dependence of η_{sp} on
temperature (unfolding
RNA) for 16 S ribosomal
16S RNA (Spirin, 1963).

conformation is illustrated in Figure D9.10(a) for protein (ribonuclease H) and in Fig. D9.10(b) for nucleic acid (ribosomal 16S RNA).

D9.7 Checklist of key ideas

- When a fluid flows slowly along a tube or in the space between two parallel planes, the flow pattern observed experimentally is *laminar* flow; in this case η_0 is independent of the velocity gradient and the liquid is said to be Newtonian.
- At sufficiently high velocity the flow pattern is disturbed. The disturbed flow is called *turbulent* flow; in this case η_0 depends on the velocity gradient and it is said to be non-Newtonian.
- The intrinsic viscosity [η] is defined as the fractional increment in the viscosity of the solution due to the addition of 1 g of macromolecule; it can be obtained experimentally as $[\eta] = \lim \eta_{sp}/c$ and has units of cubic centimetre per gram or decilitre per gram as $C \to 0$.
- Very asymmetric molecules or easily deformable molecules tend to be oriented or deformed by flow velocity gradients; for such molecules, intrinsic viscosities should be extrapolated to zero velocity gradient.
- The specific viscosity η_{sp} of a solution of spheres is proportional to the volume fraction occupied by the spheres; η_{sp} is independent of the absolute size of the spheres.
- The intrinsic viscosity can be predicted with good accuracy for a protein from its atomic coordinates, provided hydration contributions are taken into account appropriately.
- The intrinsic viscosity for a homologous series of globular proteins and spherical viruses does not depend on their molecular mass.
- The dependence of the experimental values of intrinsic viscosities on the molecular mass for different proteins in guanidine-HCl plotted on a double logarithmic scale is a straight line of slope 0.67; this slope is typical for Gaussian coils in a 'good' solvent.
- The intrinsic viscosity for a homologous series of rod-like particles is proportional to their molecular mass to the power of about 1.7–1.8.

- Double-stranded DNA macromolecules of high molecular weight (M 3×10^6) forms a homologous series of extended Gaussian coils; for this homologous series the intrinsic viscosity is proportional to molecular mass to the power 0.73.

Suggestions for further reading

Historical review and application to biological problems

Harding, S. E. (1997). The intrinsic viscosity of biological macromolecules. Progress in measurement, interpretation and application to structure in dilute solution. *Prog. Biophys. Mol. Biol.*, **68**, 207–262.

Comparison of theory and experiment

Zhou, H.-X. (1995). Calculation of translational friction and intrinsic viscosity. I. General formulation for arbitrary shaped particles. *Biophysical J.*, **69**, 2286–2297.

Zhou, H.-X. (1995). Calculation of translational friction and intrinsic viscosity. II. Application to globular proteins. *Biophysical J.*, **69**, 2298–2303.

Brenner, H. (1974). Rheology of a dilute suspension of axisymmetric Brownian particles. *Int. I. Multiphase Flow*, **1**, 195–341.

Allison, S. A. (1999). Low Reynolds number transport properties of axisymmetric particles employing stick and slip boundary conditions. *Macromolecules*, **32**, 5304–5312.

Measurement of viscosity

Cantor, C., and Schimmel, P. (1980). *Biophysical Chemistry*. Part II. *Technique for the study of Biological Structure and Function*. San Francisco: W.H. Freeman and Co.

Dutta, R. K., Hammons, K., Willibey, B., and Haney, M. A. (1991). Analysis of protein denaturation by high-performance continuous differential viscometry. *J. Chromatogr.*, **536**, 113–121.

Shapes of macromolecules from their intrinsic viscosities: homologous series

Tsvetkov, V. N., Eskin, V. E., and Frenkel, S. Ya. (1971). *Structure of Macromolecules in Solution* (translated from Russian), Volume 1. Chapter 7. Boston, UK: National Lending Library for Science and Technology.

Tanford, C., Kawahara, K., and Lapanije, S. (1967). Proteins as a random coils. Intrinsic viscosity and sedimentation coefficients in concentrated guanidine hydrochloride. *J. Amer. Chem. Soc.*, **89**, 729–736.

Rowe, A. J. (1977). The concentration dependence of transport processes: a general description applicable to the sedimentation, translation diffusion, and viscosity coefficients of macromolecular solutes. *Biopolymers*, **16**, 2595–2611.

Crothers, D. M., and Zimm, B. H. (1965). Viscosity and sedimentation of the DNA from bacteriophages T2 and T7 and the relation to molecular weight. *J. Mol. Biol.*, **12**, 527–536.

Doty P., Bradbury, J. H., and Holtzer, A. M. (1956). Polypeptides. IV. The molecular weight, configruation and association of poly-γ-glutamate in various solvents. *J. Am. Chem. Soc.*, **78**, 947–954.

Monitoring of conformational change in proteins and nucleic acids

Yangi, J. T. (1961). The viscosity of macromolecules in relation to molecular conformation. *Adv. Protein. Chem.*, **16**, 323–400.

Tanford, C. (1968). Protein denaturation. *Adv. Protein Chem.*, **23**, 121–282.

Van Holde, K. (1985). *Physical Biochemistry*. Englewood Cliffs, NJ: Prentice-Hall.

Chapter D10
Dynamic light scattering

D10.1 Historical review

1869

J. Tyndall performed the first experimental studies on light scattering from aerosols. He explained the blue colour of the sky by the presence of dust in the atmosphere.

1871

Lord Rayleigh presented the theory of scattering from assemblies of non-interacting particles that were sufficiently small compared with the wavelength of light. According to Rayleigh, scattering by a gas occurs because of the fluctuation of the molecules around a position of equilibrium. Rayleigh obtained the formulae that explained the blue colour of the sky as being due to preferential scattering of blue light in comparison with red light by molecules in atmosphere.

1906

L. Mandelshtam raised the question on the nature of scattered light once again. He pointed out that Raylegh's arguments do not fully explain the scattering phenomenon. According to Mandelshtam light scattering is also due to the random fluctuations of molecules near the position of equilibrium. As a result of random fluctuations of order λ^3, where λ is the scattering wavelength, the number of macromolecules will vary with time. It follows from the Mandelshtam theory that the translational and rotational diffusion coefficients of macromolecules could be obtained from the spectrum of the scattered light.

1924

L. Mandelshtam described theoretically and experimentally so-called *combination* light scattering, which was later called Raman scattering. As early as **1926** he recognised that the translational diffusion coefficient of macromolecules can be obtained from the spectrum of the light they scatter. Lack of spatial coherence and monochromaticity in conventional light sources rendered such experiments impossible, however.

1914

L. Brillouin predicted a doublet in the frequency distribution of scattered light due to scattering from thermal sound waves in a solid. This doublet was later named after him. In **1932 E. Gross** performed a series of light-scattering experiments on liquids. He observed not only the Brillouin doublet but also a central peak, the position of which was unshifted. In **1933 L. Landau** and **G. Placzek** gave a theoretical explanation of the central line from a thermodynamic point of view and calculated the ratio of the integrated intensity of the central line to that of the doublet.

1947

L. Gorelik conceived the idea of *optical mixing spectroscopy*. He called this method demodulated analysis of scattered light. In **1955 A. Ferrester, R. Gudmunsen**, and **P. Johnson** realised this idea in practice. The technique was called optical-beating and was used as an alternative to conventional high-resolution spectroscopy. The signal-to-noise ratio was very poor and the technique was not widely adopted for conventional spectroscopy at that time.

1964

R. Pecora showed that the frequency profile of scattered light was broadened by translational diffusion of macromolecules. He postulated that the half-width at half-height of the central peak is a direct measure of translational diffusion coefficient. In **1964 H. Z. Cammins, N. Knable** and **Y. Yeh** used an optical-mixing technique to resolve light-scattering from diluted suspensions of polystyrene latex spheres. They used a He–Ne laser as the light source and what is now commonly called the 'heterodine' optical-beating technique. In **1967 S. B. Dubin, S. J. Lunacek** and **G. B. Benedek** reported Rayleigh linewidth measurements for a series of biological macromolecules. These pioneering theoretical and experimental works have led to the current rich and diverse field. In **1976 B. J. Berne** and **R. Pecora** published their famous book *Dynamic Light Scattering with Application to Chemistry, Biology and Physics*.

1972

B. R. Ware and **W. H. Flygare** reported dynamic light-scattering studies of bovine serum albumin in the presence of a static electric field. They showed that the effect of the electric field is to superimpose a drift velocity that is proportional to the electrophoretic mobility of the species on the random Brownian motion of the charged particles. This technique is referred to as electrophoretic light scattering (ELS).

1976

S. B. Dubin, N. A. Clark and **G. B. Benedek** measured the rotational diffusion coefficient of lysozyme using a high-resolution spherical Fabry–Perot interferometer. Combining this result with the translational diffusion coefficient showed that lysozyme in solution has very nearly the same molecular dimensions as crystalline lysozyme. In 1991 Pecora extended this approach to short oligonucleotides.

1982

Z. Kam and **R. Rigler** used cross-correlation for the separation of translational diffusion from the other contributions to the dynamic fluctuations of laser light-scattering. Rotational diffusion of asymmetric particles, conformational relaxation of random coils and association–dissociation dynamics were determined by cross-correlating scattering intensity fluctuations at different angles.

1997

R. Bar-Ziv with collaborators proposed the technique of localised dynamic light-scattering (LDLS) for the study of dynamical properties of a single object at the nanometre level. This technique was applied to the measurement of biomolecular force constants and to probe the viscoelastic properties of complex media.

2000 to now

DLS is routinely used for fast (a few minutes) and accurate (1–2%) measurements of translational diffusion coefficients of macromolecules in solution and rotational diffusion coefficients of small globular proteins and oligonucleotides, and for anemometric measurements of linear and turbulent flow.

D10.2 Introduction to biological problems

In mid-1960s molecular biology benefitted from the development of a new physical method which has had several names but and which we, like R. Pecora, will call *dynamic light scattering* (DLS). Successive developments over 15 years showed that the method adds the *time dimension* to the classic probe of incoherent, total-intensity scattering of light. Intensive work has been devoted to the analysis of the dynamics of different processes that contribute to fluctuations in intensity of the scattered light. Experiments established the effect of diffusion on broadening and the effect of directed motion on the Doppler shift of the spectrum of light scattered from a wide range of particles of different sizes, shapes and flexibility. Today, DLS is the most powerful technique for accurate measurements of translational diffusion coefficients of macromolecules in solution.

In the last decade new directions have emerged in DLS, which mainly are connected with the move from studies of the collective properties of macromolecules in a large volume to studies of their individual properties in a very small volume. A new technique that unifies the results of light and fluorescent microscopy is called fluorescent correlation spectroscopy (FCS).

In this chapter we describe two types of dynamic light-scattering experiments. In the first, intensity fluctuates as the *molecule distribution* (orientation and position) in a microscopic volume ($\sim \lambda^3$) changes with time. We will call this type of experiment DLS under Gaussian statistics (i.e. for a large number of particles). In the second, intensity fluctuates due to fluctuations in the number of molecules in microscopic volume. We will call this type of experiments DLS under non-Gaussian statistics (i.e. for a small number of particles).

FCS is a logical development of dynamic light scattering. Unlike DLS, which is a *coherent* phenomenon, FCS is an *incoherent* phenomenon, and cannot be observed in an ensemble of molecules. Ideally FCS would be performed with just a single molecule in the observed volume. We will describe FCS in Chapter D11.

Today there are many variations of the original light-scattering technique that was introduced in 1964. The following terms are used to describe the various techniques as distinguished by either the method of detection or the process being examined:

> *Dynamic light scattering*: this is a generic term encompassing all of the light-scattering methods that provide information on molecular dynamics.
>
> *Quasi-elastic light scattering*: this name draws attention to the fact that the process of scattering gives a new wave, the scattered wave, whose central frequency is that of the incident light beam but the amplitude and phase of this scattered wave are frequency modulated.
>
> *Photon correlation spectroscopy*: this is an experimental technique that employs photon counting in the computation of the autocorrelation function, as opposed to using an analogue signal as in the carrier correlators.
>
> *Optical mixing spectroscopy*: this term is used to underline that a technique of very high resolution might be used for the detection of dynamic processes that describe the mobility of biological macromolecules in solution. There are many other variations (intensity fluctuation spectroscopy, Rayleigh broadening, etc.). We will consider all these name variants as synonyms.

In Comment D10.1 we present some information on the interaction of light with matter that is necessary for further discussion.

D10.3 Dynamic light scattering as a spectroscopy of very high resolution

It is fundamental to statistical physics that a system of many particles never exhibits perfect uniformity. The thermodynamic variables describing the system constantly fluctuate around their most probable values. Since these fluctuations occur very slowly compared with the frequency of a light wave, a very high-resolution spectrometer is required. The resolution needed is so high that

---

Comment D10.1 Light and matter

Light is a non-perturbative probe that can be used to obtain information about the structure and dynamics of molecules. Maxwell's equations form the basis of the description of all electromagnetic phenomena. These equations identify light as a transverse electromagnetic wave that oscillates in both space and time, i.e. the direction of oscillation is perpendicular to its direction of propagation.

The mode of interaction of light with matter depends upon the electronic structure of the material as determined by its quantum mechanical properties. If the energy of the incident photon ($h\nu$, where h is Planck's constant and ν is the frequency of light) is equal to the difference in energy between two states in the system, then the photon may be adsorbed by the system. As an oscillating electric field, light also distorts the distribution of the charges in the system that, as accelerated charges, emit radiation in the form of scattered light. If there is no exchange of energy between the photon and the system, then the frequency of the scattered light, ν_s, is equal to ν_0, and the process is referred to as *elastic light-scattering*. If energy is exchanged between the photon and the system, then ν_s differs from ν_0, and the process is referred to as *inelastic light-scattering*.

The response of a system to an external electric field is called the *polarisation* of the system. The magnitude of the polarisation depends on the amplitude of the applied electric field and the ability of the charge distribution to be 'deformed' by the external stimulus. The capacity of the system to be distorted is referred to as the *polarisability* of the system. In general, the polarisibility of a system is not homogenous throughout and the *scattering of light is a result of fluctuations in the polarisability of the medium.*

If light is polarised in the direction perpendicular to the plane defined by the source, the scattered cell and the detector, then the beam is said to have vertical polarisation. If the polarisation of both the incident and the scattered light is in the vertical direction, then the process is referred to as *polarised light-scattering*, denoted by I_{VV}. *Depolarised light scattering*, in which the horizontal component of the scattered light is monitored, is denoted by I_{VH}. *Unpololarised light* can be thought of as combination of several polarised beams with random orientation about the axis of propagation.

---

conventional optical spectroscopy is inadequate and we need to use the technique of optical mixing spectroscopy.

The laser beam incident on the scattering medium can be considered as *'carrier wave'* oscillating at an optical frequency of about 5×10^{14} Hz. The process of scattering gives a new wave, the scattered wave, whose central frequency is that of the incident light beam but whose amplitude and phase are modulated in synchrony with fluctuations of the medium. To obtain the information contained in the modulation, one must demodulate the scattered field. The

---

Comment D10.2 'Homodyne' and 'heterodyne' radio detection

In order to see how to resolve such narrow lines we must remember that similar problems arose with radio frequency. In that case, the resolution problem was solved by the development of 'homodyne' and 'heterodyne' receivers. These devices first shift the central frequency of the carrier wave down to a lower frequency by a process of non-linear mixing. Then at the lower frequency where narrow band filters are available the spectrum is analysed by sweeping the narrow filter across the spectrum which is now centred at a low frequency. Thus, whereas the optical spectrometer filters at the optical frequency, the radio frequency is filtered only after the carrier wave has been shifted down, by means of a mixing element, to a convenient low frequency. Optical mixing spectroscopy is the application of the radio frequency technique to the optical frequency region, where the required mixing element is a photomultiplier.

---

demodulation of *fast* processes is conventionally accomplished with a spectrometer such as a diffraction grating for processes with a time scale of $10^{-11}-10^{-10}$ s or a Fabry–Perot interferometer for processes with a time scale of $10^{-10}-10^{-7}$ s. Such a device operates as an optical filter and therefore cannot resolve spectral components narrower than ~10 MHz.

Dynamic processes describing the mobility of biological macromolecules in solution are referred to as *slow* processes and can produce spectral line widths as narrow as 10 Hz, six orders of magnitude narrower than that detectable optically. Hence we must use a technique of very high resolution. This technique is called *optical mixing spectroscopy* (Comment D10.2).

D10.3.1 Fluctuations and time-correlation functions

In order to obtain dynamic information from a fluctuating system we can adopt one of two different but entirely equivalent approaches. We may analyse the fluctuating system in terms of various frequency components of differing magnitude, thus yielding a *frequency spectrum* for fluctuations of property A, $\langle \Delta A^2(\nu) \rangle$, where ν is the frequency. However, an often more precise method can be used in which one measures the *time correlation function* of the fluctuations $\langle A(t), A(t+\tau) \rangle$ directly. Experiments involving analysis by the time correlation technique are currently more common and it is these we shall discuss.

Construction of time-dependent correlation functions

In the laboratory, the transformation of the fluctuating signal into a correlation function is performed using a device called a correlator. The signal must usually be presented to the correlator in a digital form and the experiments designed accordingly. In light scattering a light signal is detected as a photon count rate and the input to the correlator is a series of pulses.

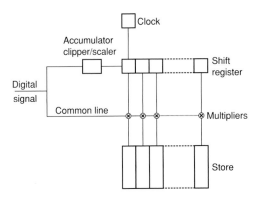

Fig. D10.1 Schematic diagram of a correlator for the analysis of fluctuating processes. The digitised noise signal is fed to two channels: an accumulator and a common line in the correlator. The accumulator counts the pulses in the train for a length of time Δt governed by a crystal clock. At the end of this time the contents of the accumulator are moved to the first stage of a shift register. Simultaneously the accumulator is reset to zero and begins to count pulses for a second identical time interval. This process continues throughout the experiment so that the shift register contains a continuously updated past history of the pulse train extending back in time by $n \times \Delta t$, where n is the number of stages in the shift register and is called the sampling time which can be chosen typically in the range 2 ns to 1s. A plot of counter contents against time interval is proportional to the noise correlation function.

A schematic diagram of a correlator is shown in Fig. D10.1. Note that commercially available digital correlator cards in a personal computer can now perform autocorrelation with submicrosecond time resolution in virtually real time.

Field and intensity correlation functions

Since light scattering is an interference phenomenon, the scattered field at the detector depends on the local inhomogeneity in the system. A completely homogeneous system gives no net scattered field.

For a single particle the induced dipole **p**, which is the source of the scattering, is determined by the incident field **E** and the particle polarasibility α

$$\mathbf{p} = \alpha \cdot \mathbf{E} \qquad (D10.1)$$

The light radiated by all the individual induced dipoles is collected at the detector where the net electric field depends on the relative positions and orientations of all particles in the scattering volume. A typical light-scattering geometry is shown in Fig. D10.2. The correlation function of the scattered electric field,

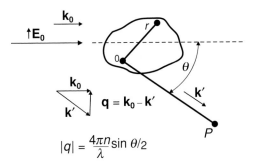

Fig. D10.2 A typical light-scattering geometry. $\mathbf{E_0}$ is the incident laser field; $\mathbf{k_0}$ and $\mathbf{k_1}$ are the incident and scattered wave vectors; $\mathbf{q} = \mathbf{k_0} - \mathbf{k_1}$ is the wave vector of the scattering density fluctuation as defined by the scattering angle θ and the wavelength of incident light λ; n is the refractive index of the medium.

$G^{(1)}(\tau)$, sometimes called the first-order correlation function which carries the undistorted information about this motion, can be defined by

$$G^{(1)}(\tau) = \langle \mathbf{E}(t)\mathbf{E}(t+\tau)\rangle \tag{D10.2}$$

In practice, of course, it is the intensity fluctuations rather than the field fluctuations that are measured. For the intensity $(\mathbf{I} \sim \langle \mathbf{E}^2\rangle)$ a normalised correlation function, $G^{(2)}(\tau)$, sometimes called the second-order correlation function, is given by

$$G^{(2)}(\tau) = \langle \mathbf{I}(t)\mathbf{I}(t+\tau)\rangle \tag{D10.3}$$

For systems containing a large number of independent scatterers, the Gaussian approximation holds and

$$G^{(2)}(\tau) = 1 + |G^{(1)}(\tau)|^2 \tag{D10.4}$$

Correlation functions and frequency spectra of fluctuations

As mentioned before, the frequency spectrum and the correlation function are entirely equivalent descriptions of the dynamical behaviour of a property of an equilibrium system, being related by the Fourier transformation:

$$I(v) = (2\pi)^{-1} \int G(\tau)\exp(-2\pi i v t)d\tau \tag{D10.5}$$

where v is the Hertzian frequency. Thus, if $I(v)$ has a Lorentzian form around zero frequency, then $G(\tau)$ shows a simple exponential decay with a time constant equal to the reciprocal of the Lorentzian half-width. This simple form occurs for all elementary stochastic processes and accounts for the behaviour of most of the systems that we shall discuss. In order to show the link between correlation functions and frequency spectra some simple examples are shown in Fig. D10.3.

Fig. D10.3 Correlation functions and frequency spectra for various models of fluctuation processes: (a) sinusoidal time dependence; (b) invariant $A(t)$; (c) simplest correlated noise. (See also Chapter A3.)

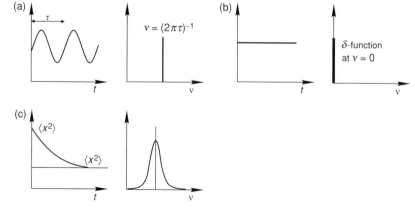

D10.3.2 Measurements of the dynamic part of scattered light

Figure D10.4 Schematically illustrates various techniques used in light experiments: (a) filter; (b) homodyne; and (c) heterodyne. We briefly describe these three techniques below.

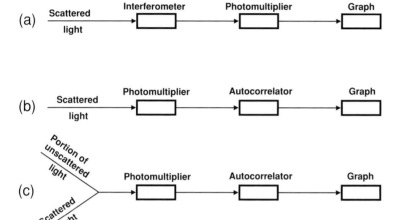

Fig. D10.4 Schematic illustration of the various techniques in light experiments: (a) filter method; (b) homodyne technique; (c) heterodyne technique.

Filter technique

The demodulation of *fast* processes is conventionally accomplished with a spectrometer, such as a diffraction grating for processes with a time scale of $10^{-11}-10^{-10}$ s or a Fabry–Perot interferometer for processes with a time scale $10^{-10}-10^{-7}$ s. (Comment D10.3). Such devices operate as optical filters and

---

Comment D10.3 Fabry–Perot interferometer

A Fabry–Perot interferometer consists of two plane dielectric mirrors held parallel to each other. The inner surfaces of the mirrors are highly reflecting (99%). In light-scattering experiments, light usually enters the interferometer normal to the mirrors and is then reflected back and forth in the interferometer cavity. The condition for survival in the cavity of a wave of wavelength λ is that an integral number of half wavelengths must fit in the cavity, i.e. $m(\lambda/2) = d$, where m is an integer and d is the spacing between the mirrors. At this wavelength there is a maximum in the intensity of the transmitted light. If $d = 0.1$ cm and $\lambda = 5000$ Å, then $m = 2 \times 10^3$. The difference between successive wavelengths, $\Delta\lambda$, that can pass through the interferometer for this value of d is $\Delta\lambda \sim 2.5$ Å. The corresponding frequency range is $\Delta\nu \sim 3 \times 10^{11}$ Hz. Thus a proper setting of d in the interferometer selects only one frequency component (wavelength) of the light-scattering spectrum. Some workers who want especially high resolution use spherical mirrors rather than the flat mirrors described above. Spherical mirrors increase the light-gathering power and also allow easier scanning of a spectrum.

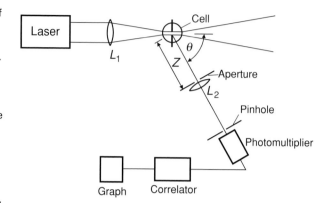

therefore cannot resolve spectral components narrower than $\sim$10 MHz. Because the rotational periods of small globular proteins lie between 10 MHz and 1 MHz (see Table D2.2) a high-resolution spherical Fabry–Perot interferometer can be successfully used to measure them (see below).

Photon correlation spectroscopy

The apparatus necessary to study the self-beat spectrum (see below) is not complicated, although some care in its design is required if optimum signal-to-noise ratio is to be obtained. A diagram of a typical experimental set-up is shown in Fig. D10.5. The lens L_1 focuses the laser beam in the centre of the scattering cell. The aperture lens L_2, a pinhole, and the photomultiplier, which is arranged on an arm rotating about an axis through the centre of the cell, establish a scattering angle of 2θ. The position of the elements on the arm is adjusted so that an image of the scattering volume is focused onto the pinhole. The electronic signal correlator is used to process the signal.

Figure D10.6 shows the general features of the $G^{(2)}(\tau)$ correlation function. The scattered intensity always consists of time-dependent and time-independent parts. The time-independent term, as measured by a dc voltmeter, is the Rayleigh scattering commonly used in the study of macromolecules, and is proportional to the number of particles in the scattering volume.

Note that in photon correlation spectroscopy no 'filter' is inserted between the scattering medium and the photomultiplier.

Fig. D10.6 The correlation function expected in an experiment.

Homodyne and heterodyne techniques

In the most of the applications discussed below, a 'self-beating' technique (or homodyne) is employed in which the scattered light correlates with itself. Another experimental variant of correlation spectroscopy uses the heterodyne technique. In this the scattered light beats against a reference signal, usually the incident laser beam. This gives a somewhat simpler correlation function but the experiment is rather more difficult to construct. In some experiments, such as electrophoretic light scattering and Doppler velocimetry, only the heterodyne technique is used (see below).

D10.3.3 Diffusion coefficients from DLS

Mutual diffusion coefficient D_m

There are many different experimental techniques for determining the diffusion coefficients of macromolecules (Chapter D.3), amongst which is DLS. The best results for DLS experiments are obtained with dilute and semi-dilute solutions of macromolecules. We call the diffusion coefficient determined by DLS D_m to underline that DLS is fundamentally a coherence phenomenon. D_m cannot be determined from DLS experiments on a single level.

Concentration dependence of D_m

As the concentration of the solute molecules is increased we expect to find changes in the observed correlation function reflecting two separate effects. First, the diffusion constant changes due to the thermodynamic and hydrodynamic influence responsible for virial coefficients and modified frictional coefficients, respectively. Perhaps of more interest in terms of the information to be gained about the macromolecules are the changes due to the association of the solute molecules. In this case the correlation function is governed by the diffusion constants of the monomers, dimers and larger aggregates. It is no longer a single exponential, but rather a sum of exponentials, each term representing a different species of molecule.

Multiple decay analysis of the correlation function

The distribution of internal relaxation times in the correlation function leads to the problem of solving Fredholm integral equations of the first kind in the form

$$G^{(1)}(q,t) = \int_0^\infty G(\Gamma)\exp(-Gt)\mathrm{d}\Gamma \qquad\qquad (\text{D10.6})$$

where $G^{(1)}(q, t)$ is the normalised time and scattering vector-dependent first-order or electric-field autocorreletion function and $G(\Gamma)$ is the distribution function or frequencies to be determined from $G^{(1)}(q, t)$ (Comment D10.4).

The analysis of data for a single time constant is quite staightforward. Any of the weighted least-squares fitting programs commonly available is satisfactory and may be expected to converge rapidly. The analysis of the multiexponential signals is, however, of greater difficulty or impossible, particularly if the time constants differ by a factor of 2 or less. The resolution of the method can be improved by combining DLS with multiple angle laser light scattering (Comment D10.5; see also Section G2.6).

Comment D10.4   Solution of a Fredholm integral equation

Experiments in which the DLS method is used serve as examples of physical experiments in which an effect function is measured with the aim of determining the cause function. A mathematical model of such an experiment enables us, by using a Fredholm integral equation of the first kind (Eq. (D10.6)), to relate the effect function (the known correlation function $G^{(1)}(q, t)$ to the cause-function (the unknown distribution of exponential curves $G(\Gamma)$ in it).

It is common knowledge that the solution of Eq. (D10.6) is unstable with respect to minor changes in $G^{(1)}(q, t)$ and, strictly speaking, does not exist while conducting measurements of $G^{(1)}(q, t)$ in real noise experiments. This is attributable to the fact that under all conditions (non-existing, non-uniqueness, instability) the solution is an ill-posed problem.

A number of methods have been developed for solving the ill-posed problem by means of substituting a similar well-posed problem for it. These approaches are called the regularisation technique.

Comment D10.5   MALLS and DLS

In the late 1980s, multiple angle laser light scattering (MALLS) instruments that can measure simultaneously at several angles the *classical* or *static* light scattering signal of a solution of macromolecules became available commercially. Combined with chromatographic separation techniques such as high-performance size exclusion chromatography (HPSEC, also called gel permeation chromatography) and field flow fractionation (flow FFF), which sort a complex sample solution into fractions of similar molecular weight and size particles, and 'user-friendly' analysis software taking advantage of the progress in the power of personal computers, these instruments provided a readily accessible means for the determination of macromolecular mass and size distributions on an absolute scale.

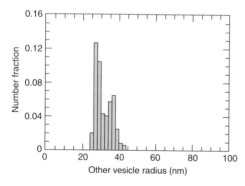

Fig. Comment 10.5(a) Number-weighted size distribution of the mixture of vesicles with two original samples of diameters 27 nm and 36 nm. (Korgel *et al.*, 1998.)

To illustrate the power of this technique a mixture of vesicles of different average size was analysed by flow FFF/MALLS (Figure A). The peaks of the two models of the distribution occur at 27 nm and 36 nm, very close to the peaks in the histogram of the two original vesicle samples. The resolution of this bimodal size distribution illustrates the possible superiority of flow FFF/MALLS for resolving a heterogeneous population. In comparison, DLS alone cannot resolve peaks in such a bimodal distribution. If DLS could be coupled effectively to a flow FFF device, improved resolution might be achievable. However, the explicit time dependence of DLS makes it more amenable to batch as opposed to flow measurements.

D10.4 Dynamic light scattering under Gaussian statistics

In the interpretation of $G^{(1)}(t)$ or $G^{(2)}(t)$ a long list of assumptions is normally made. Rotations and internal movements (if present) are assumed to be independent of both each other and translational motion. Furthermore for most solutions it is supposed that the solutions are sufficiently dilute that the translational motions of different scatterers are independent. Finally all the relaxations are usually assumed to have exponential decays (Brownian translational and rotational motions). In these simplifying circumstances, for a single solute

$$G^{(1)}(q,t) = A \exp(-D_t q^2 t) \left[1 + \sum_j B_j \exp(-t/\tau_j) \right] [1 + C \exp(-t/\tau_c)] \quad \text{(D10.7)}$$

The first exponential term describes the translational diffusion of molecules with diffusion coefficient D_t, the second is concerned with rotation about the jth molecular axis (B_j are the relevant polarisability anisotropy components). The final exponential describes decay of the correlation function due to the internal motion of the solute species with a characteristic relaxation time τ_c. A, B and C are constants giving the magnitude of the respective fluctuations.

A fundamental problem in analysing DLS data from complex systems is to *decompose* Eq. (D10.7) into the individual exponential decay components, given a measured data set $G^{(1)}(t)$ or $G^{(2)}(t)$. This problem will be discussed below.

D10.4.1 Particles small compared with the wavelength of the light

As a first example we shall discuss the application of light scattering to particles small compared with the wavelength of the light. In this case the coefficients B_j and C in Eq. (D10.7) are zero and only the first term remains, so that

$$G^{(1)}(\tau) = a_1 \exp(-D_t q^2 \tau) \quad \text{(D10.8)}$$

and

$$G^{(2)}(\tau) = 1 + a_1 \exp(-2 D_t q^2 \tau) \quad \text{(D10.9)}$$

From the single exponential decay of the measured $G^{(2)}(t)$ autocorrelation function, one obtains the translational diffusion coefficients under study (Comment D10.6).

When the system becomes more complex, for example, when the sample contains a mixture of N species with diffusion coefficients D_i, $G^{(1)}(t)$ and $G^{(2)}(t)$ become sums of single exponential decays with rate constants $q^2 D_i$:

$$G^{(1)}(\tau) = \sum_{i=1}^{N} a_i \exp(-D_i q^2 \tau) \quad \text{(D10.10)}$$

$$G^{(2)}(\tau) = \text{constant} + \left[\sum_{i=1}^{N} a_i \exp(-D_i q^2 \tau) \right]^2 \quad \text{(D10.11)}$$

Comment D10.6

We recall that according to the Einstein–Stokes relation (Eq. (D3.24)) $D_t = kT/(6\pi\eta/_0 R_0)$. Since water has a viscosity $\eta_0 \sim 0.01$ P, the coefficient of translational diffusion D_t at room temperature is $D_t \sim 2 \times 10^{-13}/R_0$. The q value in light-scattering experiments is $q = (4\pi\eta/\lambda)\sin(\theta/2)$ (Section D10.3). For visible light and a scattering angle $\theta \sim 90°$, $q \sim 10^5$ cm^{-1}. In typical light-scattering experiments it follows that $q^2 D_t \sim 10^{-3}/R_0$ and that the correlation time τ_{cor} is of the order $\tau_{cor} \sim R_0 \times 10^3$ so that the measured autocorrelation function $G^{(2)}(t)$ decays more slowly the larger the particle. For spheres of radius of 0.01 μm, $\tau \sim 10^{-3}$ s.

D10.4.2 Rigid particles of dimension comparable to the wavelength of light

If the intensity of light scattered by a molecule depends upon its orientation with respect to the laboratory, it is possible to obtain information concerning the rotational diffusion coefficient D_R of the molecule from DLS. Two separate experimental approaches lead to the required orientation dependence. The first approach is based on using polarised light, whereas the second one operates with depolarised light.

Polarised component of scattered light

If a molecule is non-spherical in shape and has one dimension comparable to the wavelength of light or larger, the scattering varies with orientation because of interference effects between light scattered from various parts of the molecule. For example, if we consider light scattered in the backward direction from a long rod-shaped molecule of length L, we find that when the molecule is perpendicular to the incident light beam, it scatters more light than when it is parallel because in the former case all scattered waves are in phase, whereas in the latter there is a certain amount of destructive interference.

The theory shows that the result for the simpler case when the rotational and translational motions are uncoupled is

$$G^{(1)}(\tau) = \sum_{l\,even} B_l \exp(-D_t q^2 + l(l+1)D_R)\tau \qquad (D10.12)$$

where l takes on the values 0, 2, 4 . . . and the B_l depends upon qL as shown in Fig D10.7. It is seen that for small values of qL, as would be obtained even for light scattered from large molecules if a small scattering angle is used, B_0 is the only important coefficient and only the translational diffusion constant may be obtained. For large qL, however, B_2 becomes important and a second exponential time constant enters $G^{(1)}(\tau)$ from which D_R may be determined. The usual procedure is to extrapolate the small-angle single exponential to large angles. The difference is analysed in terms of an exponential with time constant $D_t q^2 + 6D_R$.

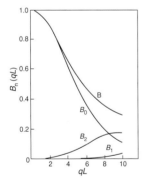

Fig. D10.7 The relative amplitudes of the components of the correlation function given in Eq. (D10.12) that result when light is scattered from a rod-shaped molecule of length L. The curve labeled B gives the sum of all B_n and hence represents the total scattered light. (Pecora, 1968.)

Depolarised component of scattered light

The second approach arises when the polarisability of the scattering molecule depends on the relative orientations of the electric field of the incident light and the axis of the molecule. In this case the polarisability is said to be anisotropic and it is found that the scattered light is no longer polarised as it would be for a molecule with isotropic polarisability but has a component (usually small) polarised in a direction perpendicular to that of the normal scattering. If the self-beat spectrum of the depolarised light is obtained it can be shown that $G^{(1)}(\tau)$ contains a single exponential

$$G^{(1)}(\tau) = \exp(-6D_R)\tau \qquad (D10.13)$$

provided the molecule has cylindrical symmetry. This is a remarkable result as it allows D_R to be determined from the analysis of a single exponential measuring scattering rather than the more complicated analysis required by Eq. (D10.12).

Unfortunately depolarised scattering is usually very weak (Comment 10.7). In spite of that the rotational diffusion coefficient of the tobacco mosaic virus has been obtained using this technique and found to be 350 s^{-1} in good agreement with results obtained by other techniques. In the case of lysozyme, results for D_R and D_t, combined with Perrin's expression for the rotational and translational diffusion coefficients of ellipsoids, showed that lysozyme in solution has very nearly the same molecular dimensions as crystalline lysozyme. Note that these results were obtained in the late 1960s using a simple wave analyser to give a spectrum of scattered light.

In the early 1990s R. Pecora and colleagues applied this technique to measurements of the rotational diffusion coefficients of a homologous series small oligonucleotides. The scattered light at $90°$ in the VH geometry, where V and H correspond to directions that are vertical and horizontal with respect to the scattering plane, was frequency analysed by passing it through a piezoelectrically driven Fabry–Perot interferometer. The interferometer was equipped with a set of 750 MHz free spectral range confocal mirrors to allow the analysis of slow relaxation processes of the order of nanoseconds. Figure D10.8 shows the concentration dependence of three B-duplex oligonucleotides 8, 12, and 20 base pairs in length (see also Table D6.2).

The combination of depolarised and polarised DLS provides a powerful method for obtaining the hydrodynamic dimensions and asymmetry of short rod-like particles (Eq. (D2.34)).

D10.4.3 Flexible macromolecules: DNA

In addition to the rigid body motions already discussed, when the molecule under observation can flex on a scale comparable to the wavelength of light, the relaxation times of the flexing modes can be obtained by careful analysis of the scattered light. Once again the phenomenon responsible for an additional contribution to the field correalation function is the change in light scattering intensity as the molecule assumes a new conformation due to interference effects between the light waves scattered from different segments of the same molecule.

Many efforts have been made during last 30 years to extract internal relaxation times for flexible molecules, and particularly for DNA from DLS measurements. Generally, three strategies have been followed to obtain this information. In the first, a double-exponential fitting was used, assuming that for sufficiently long molecules (more than 200 nm) the correlation function at long times is dominated by the single time constant

$$G^{(1)}(\tau) = D_t q^2 + 2/\tau_1 \qquad \text{(D10.14)}$$

Comment D10.7
Depolarisation ratio for the tobacco mosaic virus (TVM)

Taking TVM as an example, the depolarisation ratio is defined as $\rho_v = H_v/V_v$, where H_v and V_v denote the scattering intensities of horizontal and vertical polarisation in keeping the polarisation of incident light in the vertical sense; this has been measured many times ($\sim$0.0030). This value shows that the extent of the amplitude modulation of the scattered light by the rotational Brownian motion is about 1/300 of the total intensity.

Fig. D10.8 Rotational diffusion coefficients versus concentration corrected to 20 °C for the three oligonucleotides. Plot symbols: o = 20-mer, ● = 12-mer, and □ = 8-mer. (Eimer and Pecora, 1991.)

Comment D10.8   Separation of internal relaxation modes from the translational motion

It is not generally possible to separate internal relaxation modes from the translational motion of semiflexible macromolecules. Such a separation is possible for a monodisperse sample with an optimal molecular weight but only in a *limited q* range. This fact is not surprising, since the frequencies for the translation and the internal modes become narrower for high q. In the low-q limit, on the other hand, the internal modes have a fairly small amplitude; this also makes it difficult to extract them from the correlation function. For the linear pUC19-DNA (2760 bp), for instance, the frequencies at a scattering angle of 45° are a factor of 4 apart, and internal amplitude amounts to 10% of the correlation function, whereas the frequencies at 120° differ only by factor of 1.7 with a contribution to the internal amplitude of 75%. Such narrow frequencies could never be separated by any fitting procedure. It follows that for pUC19-DNA (molecular mass of 1.8×10^6 Da) at an ionic strength of 0.15 M NaCl ($R_g \sim$150 nm) the internal modes can be separated from the translational one only in a very limited range of scattering angles, close to 90° (Seils and Dorfmuller 1991).

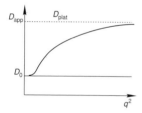

Fig. D10.9 Behavior of the apparent diffusion coefficient of a flexible polymer in a DLS experiment as a function of the square of the scattering vector q^2. (Adapted from Langovski *et al.*, 1992.)

where τ_1 is the relaxation time of the first normal mode of oscillation of the molecule. But this is only correct for a very limited range of q that is not known in advance (Comment D10.8).

In the second strategy, D_t is determined in the low-q range. Using the measured D_t value, the q dependence of the internal modes has been determined. Extrapolation of the internal relaxation times for $q \rightarrow 0$ yields the longest internal relaxation time τ_1.

In the third strategy, one can obtain an apparent diffusion coefficient D_{app} by approximation of the autocorrelation function through a forced single-exponential fit. In this case, an increase of D_{app} with increasing q is observed (Fig. D10.9). The low-q part of the curve approaches the translational diffusion coefficient D_t, the middle-q part corresponds to translational and rotational contributions, while at the high-q end a plateau value is reached which corresponds to the motion of the smallest independent subunits.

D10.4.4 Macromolecules in uniform motion: electrophoretic light scattering (ELS)

Doppler shift and correlation function

The light scattered from moving particles is Doppler shifted in frequency:

$$\Delta v = \frac{1}{2\pi} \mathbf{q} \cdot \mathbf{u} \tag{D10.15}$$

By measuring $\Delta\nu$ one can measure the velocity of the particles, because the scattering vector $\mathbf{q}$ is known from the angle of scattering (Comment D10.9). ELS is a technique for the rapid measurements of electrophoretic drift velocities via the Doppler shifts of scattered light (Comment D10.10). The principal advantage of the technique is its ability to perform the electrophoretic characterisation of many particles simultaneously.

Comment D10.10   How fast is ELS?

The scattering of light by a particle and hence the frequency of Doppler shift caused by a moving particle upon the light's frequency are extremely rapid. However, the magnitudes of typical ELS Doppler shifts (100 Hz) place limits on the required duration of a single measurement. For instance, if a particle is exposed to a sufficiently strong electric field so that the detected scattered light from the particle is Doppler shifted by 100 Hz with respect to the frequency of the incident light, and a 0.5% precision (i.e. 0.5 Hz) is desired in the measurement of the shifted frequency, then that particle must be exposed to the electric field for at least 0.5 Hz^{-1} = 2.0 s, and the consequent scattered light intensity from that particle must be recorded for the same period of time.

When a solution of charged molecules is subject to an electric field, the particles drift towards the electrode of opposite polarity with a characteristic velocity. The effect of the electric field is that an additional term must be included in the correlation function, assuming the absence of reorientations and chemical reactions:

$$G(\tau) = A \exp(-D_t q^2 \tau) \exp[iquE \cos(\theta/2)] \qquad (D10.16)$$

where E is the magnitude of the applied electrical field and u is the electrophoretic mobility. Equation (D10.16) shows that the peak width caused by diffusion exhibits an angular dependence proportional to the square of the amplitude of the scattering vector $\mathbf{q}$, whereas the Doppler-induced peak shift is only linearly proportional to the absolute value of $\mathbf{q}$. Thus the analytical resolution of the technique may be improved, when contribution diffusion to the linewidth is decreased, by working at a low scattering angle.

Figure D10.10 shows the theoretical shapes of the correlation function and the resulting spectrum. In an electric field the autocorrelation function is expected to be a damped cosine whose frequency is related to the electrophoretic drift velocity of the particles and whose damping constant is related to their diffusion coefficient. In a zero field the spectral function is Lorentzian (see Chapter A3); in the presence of the electric field the spectral function is a well-shifted asymmetrical Lorentzian of width similar to the zero-frequency-centred heterodyne peak.

**Comment D10.9
Doppler shift**

By analogy with sound waves from a source moving towards the detector at speed u, the Doppler shift of the frequency for light is given by

$$\Delta\nu = \frac{u}{c}\nu_0$$

where ν_0 is the original frequency in the source's inertial frame of reference and c is the characteristic speed of the wave (i.e. the speed of light for light scattering). If the source is moving at some angle θ with respect to the detector, the magnitude of the Doppler shift is diminished by a factor of $\sin\theta$.

Fig. D10.10 Schematic diagram of spectral shapes and correlation functions observed in ELS.

Experimental geometry

ELS can be considered a special embodiment of both laser Doppler velocimetry (Comment D10.11) and DLS. However, the application of an external electric field and the relatively slow particle velocities encountered in ELS necessitate the use of special light-scattering configurations and sample chambers distinct from those normally employed in DLS experiments (Fig. D10.11). Light from the main laser beam scattered by particles through some angle θ falls on a photodetector simultaneously with unscattered light from a reference beam derived from the same laser source. The scattered light is shifted in frequency relative to the reference beam because of the Doppler effect by an amount given by

$$\Delta \nu = \frac{1}{2\pi} \mathbf{q} \cdot \mathbf{u} = \frac{2nU}{\lambda_0} \sin(\theta/2) \qquad (D10.17)$$

It should be noted that when $\mathbf{u}$ is perpendicular to the scattering plane there is no frequency shift (Doppler shift), but when $\mathbf{u}$ is parallel to $\mathbf{q}$ there is a maximum Doppler shift.

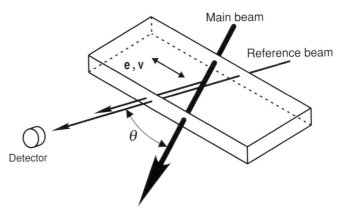

Fig. D10.11 Scattering geometry in an ELS instrument. **E** and **u** are the electric field and particle velocity, respectively. **q** is the scattering vector. (Adapted from Langley, 1992.)

Fig. D10.12 Frequency spectra in ELS: (a) free lysozyme: diffusion broadening smears out the peak except at the two smaller scattering angles; (b) multimers of bovine serum albumin: three frequency spectra were obtained simultaneously at three different scattering angles in one experimental run, and the results scaled are plotted on the same mobility axis. (Adapted from Langley, 1992.)

The range of frequency difference typically encountered in ELS is only about 100 Hz. This shift is so small that no optical filters of sufficient resolution are available. However, it can be easily measured because the detector is sensitive to the frequency difference between the scattered light and the reference beam (heterodyne detection).

Applications

In principle ELS can be applied to the study of the electrokinetic properties of any charged species in solution, but in practice the application of the technique, and particularly its efficacy *vis-à-vis* other electrophoretic techniques, must be considered separately for different particle-size domains. ELS is very difficult to apply to small globular proteins, such as lysozyme, largely because of the substantial contribution of diffusion to the ELS line width (Fig. D10.12(a)). The resolution of ELS essentially increases with increasing molecular mass. The electrophoretic mobility spectrum of a preparation of BSA is shown in Fig. D10.12(b). At least five species of different mobility are resolved.

Note that using a gravitational field is a possible alternative to using an electric field. Both sedimentation and diffusion coefficients can be measured simultaneously from the spectrum of scattered light, and from these the molecular weight can be calculated using the Svedberg equation (Eq. (D4.20)).

D10.5 DLS under non-Gaussian statistics

D10.5.1 Scattering of a small number of particles (number fluctuations)

Let us imagine that we are performing an experiment to monitor the concentration of a solute in a large volume of particles. Generally speaking, in this case the larger the number of particles being studied the more easily are the fluctuations detected by DLS. The total intensity of scattered light received by a detector is the sum of contributions from scattering events at many particles. Changes in intensity due to *number fluctuations* are usually very small (Comment D10.12).

Now let us imagine that we are performing an experiment to monitor the concentration of a solute in a small volume within a large sample of solution and that

Comment D10.12   Why do we not readily notice number fluctuations in a macroscopic sample?

Straightforward application of thermodynamic fluctuation theory gives the magnitude of these number fluctuations in terms of the mean square deviation $\langle \Delta N^2 \rangle$ from the average value $\langle N \rangle$. Angle brackets will be used henceforth to indicate an equilibrium ensemble time average. In this case, it can be shown that

$$\langle \Delta N^2 \rangle = \langle N \rangle \qquad \text{(A)}$$

or

$$\langle \Delta N^2 \rangle^{1/2}/\langle \Delta N \rangle = \langle \Delta N \rangle^{1/2} \qquad \text{(B)}$$

so that the relative magnitude of the fluctuation is inversely proportional to the square root of the number of particles. This is why we do not readily notice the fluctuations in a macroscopic sample where $\langle N \rangle = 10^{20}$.

this defined volume has permeable walls. Experiments of this type were first carried out at the beginning of the nineteenth century using the microscope. The object of such studies was to determine the mean number $\langle N \rangle$ of particles (in this case colloidal) per unit volume. If, however, we were to obtain a number of instantaneous 'snapshots' of the volume as a function of time, it would be found that the number of particles counted would vary with time as a result of particle diffusion in and out of the defined volume.

For the case to be considered here two contributions to the time dependence are important. One is caused by the time fluctuations due to particle diffusion in and out of the scattering volume. The other is caused by the diffusion of particles within the scattering volume over distances of the order of the wavelength of light, which thus causes temporal fluctuations. If the two time fluctuating terms are assumed to be uncorrelated with each other, then, for homodyne detection, the normalised correlation function is given by

$$G^{(2)}(\tau) = A[N_0 \exp(-\gamma t) + B N_0^2 \exp(-2D_t q^2 t)] \qquad \text{(D10.18)}$$

The first term on the right-hand side results from fluctuations in the total number of particles within the scattering volume and the second from a diffusional interference term. N_0 is the average number of particles in the scattering volume; γ is the characteristic time required for a particle to diffuse across the scattering volume and is of the order $L^2/24N_0 D_t$ (where L is a characteristic dimension of the scattering volume); A and B are the constants.

At high concentrations, the second term dominates and the experimentally observed decay constant yields the diffusion constant. At sufficiently low concentrations, however, N_0 is comparable to BN_0^2, and both components contribute to the observed correlation function. Figure D10.13 shows results for an avian

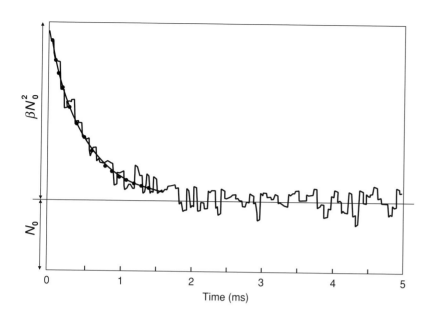

myelblastosis virus sample in which number fluctuation noticeably contributes to the correlation function.

However, the scattering at such concentrations (~2000 particles in the scattering volume) and at such a molecular mass ($= 4 \times 10^8$) is so weak that the virus diffusion coefficient cannot be determined precisely from either the interference fluctuation (the second term in Eq. (D10.18)) or the number fluctuations (the first term in Eq. (D10.11)). The effect of the first term can be fully suppressed if the process being studied is *incoherent* (as in fluorescence). This has been exploited in studies using the fluorescence correlation technique (Chapter D11).

D10.5.2 Cross-correlation (method of two detectors)

Number fluctuations can be observed from DLS, provided the coherent part of light scattering usually prevailing can be essentially reduced. This can be achieved by violation of spatial coherency (Eq. (D10.9)) with appropriate optical arrangements as well as by cross-correlating the intensity fluctuation of the same volume using two detectors. Figure D10.14(a) shows the cross-correlation function due to rotational motion of the tobacco mosaic virus around its short axis of symmetry at various scattering angles. It is evident that the cross-correlation function does not depend on scattering angle. Unlike the correlation due to rotational diffusion relaxation of rigid molecules, the measured cross-correlation function for *E. coli* plasmid DNA due to internal degrees of freedom depends on the scattering vector (Fig. D10.14(b)).

Fig. D10.14 (a) Cross-correlation measured from TMV scattering. The fitted correlation time is 240 ± μs, microseconds, independent of scattering angle. The correlations at 90° and 60° are shifted along the ordinate by 0.25 and 0.5 units, respectively.
(b) Cross-correlation of fluctuation scattering from *E. coli* plasmide DNA molecules. The fitted correlation times are 400, 120, and 80 μs at scattering angles of 60°, 90°, and 120°, respectively. The correlations at 90° and 60° are shifted along the ordinate by 0.25 and 0.5 units, respectively (After Kam and Riger, 1982.)

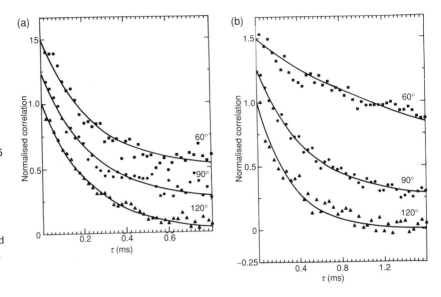

D10.5.3 Scattering of single particle

The most simple illustrative case of non-Gaussian statistics is the single particle. In this case phase correlations are *absent*, diffusion broadening is completely eliminated, and studies of pure form-factor fluctuations become possible. However, in practice, probing single-particle dynamics of submicron dimensions is impossible because of the small signal-to-noise ratio. For a single particle of micron dimensions such studies become possible using the method of localised dynamic light scattering (LDLS). The basic idea of this technique is to perform a DLS experiment with a single scatter under the microscope in a highly inhomogeneous field of light. A probe, based on optical fibers, enables one to collect light under the microscope at different scattering angles. The light is detected by a multiplier or photomultiplier coupled to the optical fibre probe, and the signal is fed into a digital correlator that calculates the intensity autocorrelation function. The dynamic spectrum of a fluctuating object over many decades in time down to the nanosecond regime is calculated. The applicability of this approach has been demonstrated by measuring the force constants of a single bead and a pair of beads trapped by laser tweezers (Section F5.2).

D10.6 Checklist of key ideas

- DLS adds the *time dimension* to the classic probe of incoherent, total-intensity scattering of light.
- In a solution of macromolecules, the local concentration continuously fluctuates due to Brownian motion of the particles; since these fluctuations are *very slow* compared with the frequency of the light wave, a very high-resolution spectrometer is required.

- The resolution needed to determine the translational diffusion coefficients of macro-molecules is so *high* that conventional optical spectroscopy is inadequate and the technique of optical mixing spectroscopy is required; however this resolution is good enough to determine the rotational diffusion coefficients of small globular proteins.
- Time-dependent correlation functions provide a concise method of expressing the degree to which two dynamical properties are *correlated* over a period of time; in the laboratory, the transformation of the fluctuating signal into a correlation function is performed using a device called a correlator.
- The correlation function and the frequency spectrum are entirely equivalent descriptions of the dynamical behaviour of a property of an equilibrium system, being related by the Fourier transformation.
- Two different experimental techniques are employed in DLS; in the 'self-beating' (or homodyne) technique the scattered light correlates with itself, whereas in the heterodyne technique the scattered light beats against a reference signal, usually the incident laser beam.
- For small molecules, the translational diffusion contribution is by far the largest component in the fluctuation of scattering intensity.
- For molecules of a size approaching the wavelength of light, rotational diffusion and internal flexibility (if present) contribute a sizable amplitude to the fluctuation.
- A fundamental problem in analysing DLS data from complex systems is the *decomposition* of the measured correlation function into the individual exponential decay components.
- Cross-correlating measurements at different angles are used to separate translational diffusion from the other contributions such as rotational diffusion of asymmetric particles and conformational relaxation of random coils.
- In standard DLS experiments, the fluctuation of molecules in a large (*macroscopic*) volume is analysed; in this case the larger the number of particles being studied the more easily are the temporal fluctuations detected.
- ELS is a technique for rapid measurement of electrophoretic drift velocities via the Doppler shift of scattered light.
- If the scattering volume contains a relatively small number of particles (*about a few thousand*) the fluctuating intensity possesses *two* commensurable contributions: the first contribution results from fluctuations in the total number of particles within the scattering volume and the second results from a diffusional interference term.
- If the scattering volume contains just a few particles the fluctuating intensity has only one contribution which results from fluctuations in the total *number* of particles within the scattering volume.

Suggestions for further reading

Historical review and introduction to biological problems

Berne, B. J., and Pecora, R. (1976). *Dynamic Light Scattering*. New York: Wiley.

DLS as a spectroscopy of very high resolution

Benedek, G. B. (1969). Optical mixing spectroscopy in physics, chemistry, biology and engineering. In *Polarization: Matière and Rayonnement*, pp. 4–84. Paris: Les Presses Universitaires de France.

Dubin, S. B., Lunacek, J. H., and Benedek, G. B. (1967). Observation of the spectrum of light scattered by solutions of biological macromolecules. *Proc. Natl. Acad. Sci. USA*, **57**, 1164–1171.

Doherty, J. V., and Clarke, H. R. (1980). Noisy solutions: a source of valuable kinetic information, *Sci. Prog. Oxf.*, **66**, 385–419.

DLS under Gaussian statistics

Chu, B. (1991). *Laser Light Scattering. Basic Principles and Practice*. San Diego: Academic Press.

Wada, A., Suda, N., Tsuda, T., and Soda, K. (1968). Rotary-diffusing broadening of Rayleigh lines scattered from optically anisotropic macromolecules in solution. *J. Chem. Phys.*, **50**, 31–35.

Dubin, S. B., Clark, N. A., and Benedek, G. B. (1971). Measurment of the rotational diffusion coefficient of lysozyme by depolarized light scattering: configuration of lysozyme in solution. *J. Chem. Phys.*, **54**, 5158–5164.

Seils, J., and Dorfmuller, T. H. (1991). Internal dynamics of linear and superhelical DNA as studied by photon correlation spectroscopy. *Biopolymers*, **31**, 813–825.

Langowski, J., Kremer, W. and Kapp, U. (1992). Dynamic light scattering for study of solution conformation and dynamics of superhelical DNA. *Meth. Enzymol.*, **211**, 431–448.

Eimer, W., and Pecora, R. (1991). Rotational and translational diffusion of short rodlike molecule in solution: Oligonucleotides. *J. Chem Phys.*, **94**, 2324–2329.

Ware, B. R., and Haas, D. D. (1983). *Electrophoretic Light Scattering in Fast Methods in Physical Biochemistry and Cell Biology*, eds. R. I. Sha'afi and S. M. Fernandez, Elsevier Sci. Publishers.

Langley, K. H., (1992). Developments in electrophoretic laser light scattering and some biochemical application. In *Laser Scattering in Biochemistry*, eds. S. E. Harding, D. B. Sattelle and V. A. Bloomfield. Cambridge: Royal Society of Chemistry.

DLS under non-Gaussian statistics

Bar-Ziv, R., Meller, A., *et al.* (1997). Localized dynamic light scattering: probing single particle dynamics at the nanoscale. *Phys. Rev. Lett.*, **78**, 154–157.

Meller, A., Bar-Ziv R., *et al.* (1998). Localized dynamic light scattering: a new approach to dynamic measurements in optical microscopy. *Biophys. J.*, **74**, 1541–1548.

Weissman, M., Schindler, H., and Feher, G. (1976). Determination of molecular weights by fluctuation spectroscopy: application to DNA. *PNAS*, **73**, 2776–2780.

Kam, Z., and Rigler, R. (1982). Cross-correlation laser scattering. *Biophys. J.*, **39**, 7–13.

Chapter D11
Fluorescence correlation spectroscopy

D11.1 Historical review

1916

M. von Smoluchowski gave the first theoretical description of the amplitude and the temporal decay of number fluctuations in a diffusion system.

1972–1974

D. Magde, E. L. Elson and **W. W. Webb** published a rigorous formalism of fluorescence correlation spectroscopy (FCS) with its various modes of possible application, highlighting the large potential of this variant of the relaxation method.

1990

R. Rigler and coworkers made the final breakthrough for the FCS method. They reached the single-molecule detection limit by combining FCS with a confocal set-up, thus increasing the signal-to-noise ratio dramatically. By tightly focusing a laser beam and inserting a pinhole into the image plane, maximum lateral and axial resolution were achieved.

1994

M. Eigen and **R. Rigler** triggered an important further development by proposing the application of dual-colour cross-correlation for diagnostic purposes. The underlying idea was to separate single-labelled reaction educts from dual-labelled reaction products to discriminate against an excess of free single-labelled species and thus enhance the specificity of detection. In **1997 P. Schwille** and coworkers successfully monitored a hybridisation reaction of two differently labelled oligonucleotides by the dual-colour cross-correlation technique. In **2000 K. G. Heinze** and coworkers, reported the application of dual-colour two-photon cross-correlation to determine enzymatic cleavage of a DNA substrate by endonuclease.

2000 to now

FCS has evolved into a whole family of related methods sharing the basic principle of fluorescence fluctuation analysis.

D11.2 Introduction to biological problems

Although the first theoretical description of the amplitude and the temporal decay of number fluctuations in a diffusion system is as old as Smoluchowski's concept of probability, it has taken a long time to reach a sensitivity adequate for detecting minute fluctuation events.

The analytical strategy of FCS was introduced in the 1970s to measure chemical kinetics and the associated modulation of molecular diffusion by analysis of concentration fluctuations about equilibrium of a small ensemble ($\sim$1000) of particles. The power of the technique was demonstrated by measuring diffusion and the binding kinetics of the small fluorescent drug ethidium to DNA, and by the determination of the molecular weight of giant phage DNA (Comment D11.1). This early research formulated the mathematical framework of the general FCS problem in terms of the eigenvalues and eigenvectors of the coupled diffusion and chemical kinetic rate equations, so that it still provides the foundation for analysis.

Recently the FCS strategy has been enjoying a remarkable renaissance. Experimentally one can now realise small open volumes, often of less then a femtolitre, by sharp focusing of the illumination in a double cone with a waist <1 μm in diameter plus the selection of the fluorescence emitted only from this region using a pinhole in the image plane. A concentration of 1 nM, which corresponds to an average of *less than one molecule* in the volume, is generated in the microscope. Thus, the FCS technique is particularly attractive for researchers striving for a quantitative assessment of interactions of a small molecular quantity in biologically relevant systems.

D11.3 General principles of FCS

FCS can be considered as a branch of DLS under non-Gaussian conditions because fluorescence is fundamentally an *incoherent* phenomenon. FCS is a method in which the fluorescence intensity arising from a very small volume containing fluorescent molecules is analysed (correlated) to obtain information about the processes that give rise to fluctuations in the fluorescence. One important process is Brownian motion, which leads to the stochastic appearance and disappearance of fluorescent molecules in the small observation volume. The relative fluctuation amplitude of a signal is inversely proportional to the number of molecules simultaneously measured (Eq. (A) of Comment D11.1). Thus, the presence of a large number of molecules, as typically encountered in a macroscopic medium, suppresses the effect of fluctuations and only the ensemble average is observed. This simple argument illustrates that FCS measurements require signals from a single molecule or a very small number of molecules in order not to mask the signal fluctuation.

Comment D11.1   Molecular weight of DNA by FCS

In the mid-1970s the FCS technique was successfully applied to the determination of the molecular weight of DNA molecules from different sources. These experiments were done in solutions containing about 1000 molecules to provide a reasonable signal-to-noise ratio. Recalling Eq. (B) of Comment D10.12 we can write

$$\left\langle \left(\frac{\Delta N}{\langle N \rangle}\right)^2 \right\rangle = \left\langle \left(\frac{\Delta C}{\langle C \rangle}\right)^2 \right\rangle = \frac{1}{N} \qquad (A)$$

where C is the (wt/vol) concentration of the molecules. The fewer the number of molecules, the larger are the fractional fluctuations. Therefore, for given concentration $\langle C \rangle$, the size of the fluctuations increases with molecular weight. By measuring these fluctuations (via any parameter that is sensitive to concentration), one can determine from Eq. (A) the number of molecules N in a given volume V within the measured fluctuations. Then knowing the average concentration $\langle C \rangle$, the molecular weight is determined from

$$M = \langle C \rangle \left(\frac{\Delta C}{\langle C \rangle}\right)^2 V N_A \qquad (B)$$

The molecular weights of DNA molecules from different sources were: for T2 phage DNA, 1.14×10^8; and replicating *E. coli* DNA, 3.9×10^9. The molecular weight of nuclei and individual chromosomal DNA molecules of *Drosophila melangaster* were obtained as 3×10^{11} and 4×10^{10}, respectively (Adapted from Weissman *et al.*, 1976.)

D11.3.1 Basics and applications

The general principles of FCS are illustrated in Fig. D11.1. The fluctuations in the fluorescence signal are recorded (horizontal arrow) from which the autocorrelation function is computed (vertical arrow). The fluctuations are caused by diffusion of fluorescent molecules through the cavity or by changes in fluorescence over time caused by chemical kinetics. From the autocorrelation function the averaged dwell time (τ_{av}) and number of molecules in the cavity (N) can be derived (see below and Comment D11.2). Together with the known size of the light cavity, this can be used to determine the diffusion coefficient D of the fluorescence probe. D can then be used to evaluate the formation of molecular complexes as the diffusion of molecular complexes (red end green structures in Fig. D11.1) is slower than that of free protein molecules (green structure), resulting in a smaller value of D.

Three- and two-dimensional diffusion
When a laser beam with a Gaussian beam profile is used for excitation and the fluorescence is collected through a confocal pinhole, the probe volume generated

Fig. D11.1 General
principles of FCS (see
text). (After Bastiaens and
Pepperkok, 2000.)

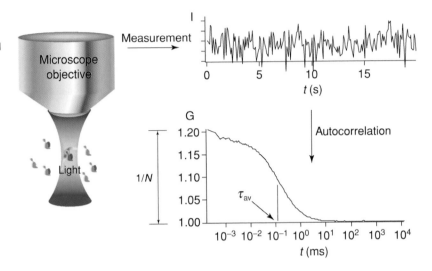

Comment D11.2   Autocorrelation functions in FCS

In most applications of FCS one of the following definitions of the autocorrelation function is used:

$$G^{\delta F}(\tau) = \frac{\langle (\delta F(t)\delta F(t+\tau)\rangle}{\langle F(t)\rangle^2}$$

or

$$G^{F}(\tau) = \frac{\langle F(t)F(t+\tau)\rangle}{\langle F(t)\rangle^2}$$

where $\langle\rangle$ denotes the time average

$$\delta F(t) = F(t) - \langle F(t)\rangle$$

denotes the fluctuation around the mean intensity and for a long time average of F (no bleaching),

$$G^{\delta F} = G^{F} - 1$$

---

can be described to a good approximation by a three-dimensional Gaussian function. If r and l are the characteristic radial and axial dimensions of the volume (i.e. where the Gaussian function drops to $1/e^2$ of its volume at maximum) the relation between the autocorrelation function $G(\tau)$ and the diffusion time τ_{av} is:

$$G(\tau) = \frac{1}{N}\left(\frac{1}{1+\dfrac{\tau}{\tau_{av}}}\right)\left(\frac{1}{1+\left(\dfrac{r}{l}\right)^2\dfrac{\tau}{\tau_{av}}}\right)^{1/2} \tag{D11.1}$$

The diffusion time τ_{av} characterises the average time it takes for a molecule to diffuse through the radial part of the observation volume of the microscope.

However, the diffusion time is not a constant and changes with the size of the observation volume, which depends on the laser wavelength and the optics of the instrument. The relationship between the diffusion time and the diffusion coefficient is

$$\tau_{av} = \frac{r^2}{4D} \tag{D11.2}$$

Full three-dimensional sampling of diffusion can be performed by creating defined light cavities using two-photon excitation (Section F3.3.2). In this case the diffusion time is reduced by a factor of 2: $r^2/8D$.

For large structural parameters r/l or two-dimensional diffusion, as expected in planar membranes, Eq. (D11.1) simplifies to

$$G(\tau) = \frac{1}{N}\left(\frac{1}{1 + \tau/\tau_{av}}\right)^{1/2} \tag{D11.3}$$

Different mobility modes

In a complex biological environment, many different modes of mobility exist and no conventional diffusion constant can be specified (Section D2.3). Model autocorrelation curves showing different mobility modes are shown in Fig. D11.2.

The figure shows that the autocorrelation curve for active transport displays the sharpest decay, whereas the one for anomalous diffusion decreases rather slowly. If one treats the autocorrelation curve as a kind of distribution function of residence times within the focal volume, this behaviour becomes clearer. Particles

1- 3-dimensional diffusion
2- 2-dimensional diffusion
3- active transport
4- anomalous diffusion

Fig. D11.2 Model autocorrelation curves for different particle mobilities: (1) three- and (2) two-dimensional diffusion, (3) active transport and (4) anomalous diffusion (see text). (Adapted from Hausten and Schwille, 2003.)

diffusing within a biological membrane may exhibit a large variety of different mobilities, depending on the time-dependent local environment of each individual chromophore. This results in curves that are typically smeared out across several orders of magnitude in time. Exactly the opposite situation is physically realised in plug flow experiments. Here, all the particles are expected to have a velocity. Only the ellipsoidal shape of the focal volume causes a small variation in the residence times, so that the resulting decay of the autocorrelation curve is very abrupt. The same principles may be applied to two- and three-dimensional diffusion.

Dynamic time range

FCS commonly accesses dynamics in the time range 1μs to 100 ms. For convenient plotting of this very wide dynamic range, a logarithmic time scale is generally used (Fig. D11.3 and D11.4). Depending on the surrounding medium, the molecular mobility changes by several orders of magnitude. The corresponding diffusion coefficients range from 3×10^{-6} cm^2 s^{-1} for free dye in an aqueous buffer solution to 10^{-10} cm^2 s^{-1} for a large receptor in the cell membrane.

Since the FCS temporal resolution is below 100 ns, limitations in FCS for in vivo applications appear in the slow time range. When analysing slowly moving molecules, for example, those interacting with the cytoskeleton, nuclear DNA, or membranes, one has to ensure that the molecules under study are not photobleached during their transit time through the focal volume.

Multicomponent systems

In a multicomponent system, the correlation curve is evaluated, assuming a model of two or more diffusing components, by a standard Marquardt non-linear

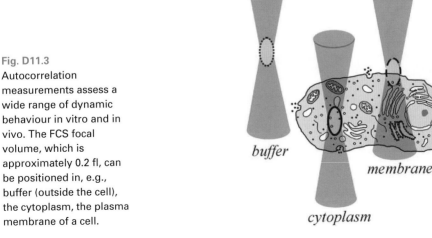

Fig. D11.3
Autocorrelation measurements assess a wide range of dynamic behaviour in vitro and in vivo. The FCS focal volume, which is approximately 0.2 fl, can be positioned in, e.g., buffer (outside the cell), the cytoplasm, the plasma membrane of a cell.

buffer

membrane

cytoplasm

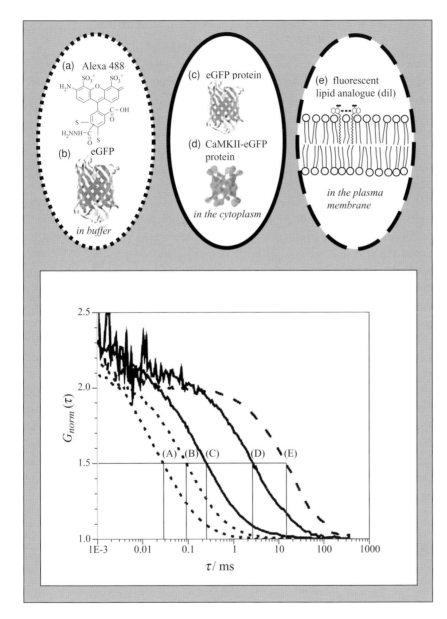

Fig. D11.4 Depending on the probe and the environment, diffusion coefficients of different orders of magnitude are obtained: (a) small dye molecule (Alexa 488, M ~ 0.6 kDa) in water, $D = 3 \times 10^{-6}$ cm^2 s^{-1}; (b) green fluorescent protein (eGFP) in aqueous solution (M ~ 27 kDa); (c) the same protein in the cytoplasm of a HEL cell, $D = 3 \times 10^{-7}$ cm^2 s^{-1}; (d) large protein complex (multimeric complex of calmodulin-dependent kinase II-eGFP fusion protein) in the cytoplasm of a HEK cell, $D = 3 \times 10^{-8}$ cm^{2}s^{-1}; (e) two-dimensional diffusion of a fluorescent lipid analogue (long-chain carbocyanine dye 'dif C_{18}') in the plasma membrane of a HEK cell, $D = 6 \times 10^{-9}$ cm s^{-1}. (Adapted from Bacia and Schwille, 2003.)

least-squares fitting routine. Although this has been successfully carried out in some cases, extensive calibration measurements of all species must be carefully made to fix, for example, the diffusion coefficients or related parameters. The dual-colour cross-correlation scheme described below eliminates these preliminary calibration steps by introducing two spectroscopically separable fluorescence labels that allow simultaneous measurements of two reaction partners and their products.

D11.4 Dual-colour fluorescence cross-correlation spectroscopy

Dual-colour fluorescence cross-correlation spectroscopy uses separate lasers to excite two different colour probe molecules, and independent red and green detector channels are used for fluorescence detection (Fig. D11.5). The correlation between the two channels is then measured, as opposed to the autocorrelation of a single channel in standard FCS measurements.

The dye system is designed to have a green species (G) and a red species (R), as well as an increasing fraction of green-and-red substance (GR) due to the reaction of both partners. Whereas pure G and R should be recorded by only one detector, GR is detected in both of them (Fig. D11.5). Cross-correlation of the detector signals is therefore a means of measuring the reaction product GR independently of fluorescent educts. In contrast to a single-colour autocorrelation analysis, the method in principle yields a *yes-or-no* decision about the presence of a doubly labelled reaction product, without the necessity of mathematical evaluation of the correlation curve. The most prominent example of cross-correlation is the dual-colour mode schematically depicted in Fig. D11.6. Whereas autocorrelation measurements are sensitive only to signal variations within one channel, cross-correlation analysis is used to compare the signals arising from two differently labelled fluorescent species. Only temporally coordinated fluctuations in both channels, i.e. colours, give rise to cross-correlation.

A number of experiments applying this technique to different kinds of reactions have been carried out to probe interactions between different molecular species. One of them was the monitoring of a hybridisation reaction of two differently labelled DNA molecules. In this experiment, two complementary ssDNA molecules were each been fluorescently tagged at one end, one strand with rhodamine green and the other with CY-5 (Fig. D11.7). The hybridisation reaction

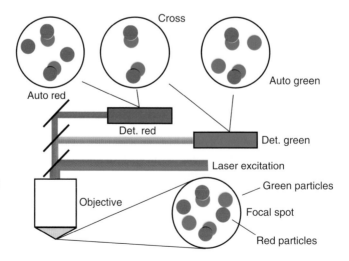

Fig. D11.5 Schematic of the measurement principle: green and red lasers illuminate the sample, which contains diffusing species R, G and GR. Photodetector Det. Red detects R and GR; photodetector Det. Green detects G and GR.

Fig. D11.6 The amplitude of the autocorrelation curve is a direct measure of the number of the doubly labelled particles. Only coordinated movements of the two different chromophores result in non-zero cross-correlation amplitudes. For an enzymatic cleavage reaction of a doubly labelled DNA strand, the maximum amplitude is measured before starting the reaction. The cleavage reaction can be monitored on-line via the gradually decreasing cross-correlation amplitude that directly reflects the decreasing number of doubly labelled particles. (Adapted from Haustein and Schwille, 2003.)

kinetics was followed using cross-correlation FCS analysis. The main advantage of the dual-colour FCS approach is that there is no correlation between the two channels unless individual molecules produce both red and green fluorescence signals. In other words, only DNA molecules that have associated with their complementary strand produce a correlated signal. The amplitude of the cross-correlation signal is thus directly proportional to the concentration of hybridised DNA, and the hybridisation kinetics can be interpreted directly as changes in the

Fig. D11.7 Emission spectra of rhodamine green (1) and Cy-5 (2), as well as the transmission characteristic of second dichroic mirror (3). (After Schwille et al., 1997.)

Fig. D11.8 (a) Time course of the cross-correlation function in the renaturation (hybridisation) reaction. Light dots: 8-min incubation; dots: 15 min; short dashes: 25 min; dashed line: 60 min; solid line: 120 min. The fraction of GR molecules represented by the amplitude increases with time. The diffusion time (curve decay time) remains the same. (b) Plot of cross-correlation amplitude during the reaction (points). By fitting the irreversible second-order reaction (short dashes), we get $k = 8 \times 10^5$ M^{-1} s^{-1}. (After Schwille *et al.*, 1997.)

time zero amplitude of the measured cross-correlation signal. This is in contrast to one-colour FCS measurement in which the ssDNA and dsDNA molecules both contribute to the amplitude of the correlation signal, requiring some analysis to discern the equivalent information. Figure D11.8(a) shows the time course of the cross-correlation function over 120 min. Figure D11.8(b) shows a plot of the cross-correlation amplitude values versus time. The kinetic rate parameter may be evaluated directly from these data.

D11.5 Checklist of key ideas

- FCS is a method in which the stochastic appearance and disappearance of fluorescent molecules in a small (*microscopic*) volume is analysed.
- FCS is an *incoherent phenomenon*, and cannot be observed in an ensemble of molecules. FCS is best performed in the presence of a single molecule in the observed volume.

- Dual-colour fluorescence cross-correlation spectroscopy uses separate lasers to excite two different colour probe molecules, and independent red and green detector channels are used for fluorescence detection.

Suggestions for further reading

Historical review and introduction to biological problems

Bastianes, P. I. H., and Pepperkok, R. (2000). Observing proteins in their natural habitat: the living cell. *TIBS*, **25**, 631–637.

Elson, E. L., and Magde, D. (1974). Fluorescence correlation spectroscopy. I. Conceptual basis and theory. *Biopolymers*, **13**, 1–27.

Haustein, E., and Schwille, P. (2003). Ultrasensitive investigations of biological systems by fluorescence correlation spectroscopy. *Methods*, **29**, 153–166.

General principles of FCS

Magde, D., Elson, E. L., and Webb, W. W. (1972). Thermodynamic fluctuations in a reacting system. Measurement by fluorescence correlation spectroscopy. *Phys. Rev. Lett.*, **29**, 705–708.

Eigen, M., and Rigler, R. (1994). Sorting single molecules: application to diagnostics and evolutionary biothechnology. *Proc. Natl. Acad Sci. USA*, **91**, 5740–5747.

Bacia, K., and Schwille, P. (2003). A dynamic view of cellular processes by in vivo fluorescence auto- and cross-correlation spectroscopy. *Methods*, **29** 74–85.

Schwille, P., and Kettling, U. (2001). Analyzing single protein molecules using optical methods. *Curr. Opin. Biotech.*, **12**, 382–386.

Dual-colour fluorescence cross-correlation spectroscopy

Schwille, P., Meyer-Almes, F. J., and Rigler, R. (1997). Dual-colour fluorescence cross-correlation spectroscopy for multicomponent diffusional analysis in solution. *Biophys. J.*, **72**, 1878–1886.

Koltermann, A., Kettling, U., Stephan, J., Winkler, T., and Eigen, M. (2001). Dual-color confocal flourescence spectroscopy and its application in biotechnology. In *Fluorescence Correlation Spectroscopy – Theory and Application*, eds. R. Rigler and E. Elson. Heidelberg: Springer-Verlag.

Stephan, J., Dorre, K., *et al.* (2001). Towards a general procedure for sequencing single DNA molecules. *J. Biotechnol.*, **86**, 255–267.

Part E
Optical spectroscopy

Chapter E1
Visible and IR absorption spectroscopy

E1.1 Brief historical review and biological applications

1704

In *Opticks* **Isaac Newton** dealt with the formation of a *spectrum* by a prism, and the composition of white light and its dispersion. The Latin word, *spectrum*, means an appearance; a spectrum is obtained when radiation is broken up into its colour or wavelength distribution.

1800

The astronomer, **William Herschel**, discovered infrared (IR) radiation.

1801

The physicist **Johann Wilhelm Ritter** discovered ultraviolet (UV) radiation.

1814

Joseph von Frauenhofer showed that the Sun's spectrum contained dark lines (later named Frauenhofer lines), indicating that light of the corresponding colour was missing because of absorption.

1850–1900

August Beer stated the empirical law, which was named after him, that there is an exponential dependence between the transmision of light through a substance, the concentration of the substance and the path length of the beam through it. The law is also known as the Beer–Lambert law or the Beer–Lambert–Bouguer law, in recognition of the work of **Pierre Bouguer (1729)** and **Johann Heinrich Lambert (1760)**. **Gustav Kirchhoff's** discovery that each pure substance has a characteristic spectrum provided the basis for analytical spectroscopy. **Gustav Kirchhoff** and **Robert Bunsen** identified the chemical elements in the Sun by analysing its spectrum. **Johann Jacob Balmer** identified a numerical series in the spectrum of hydrogen. **Joseph John Thomson** discovered the electron. **Max Planck** introduced the concept of quanta in the treatment of heat radiation and laid the foundation of quantum theory. He was awarded the Nobel prize in **1918**.

1911–1913

Ernest Rutherford proved that atoms comprise positively charged nuclei surrounded by electrons. **Niels Bohr** made the first theoretical calculations of the discrete energy states in the hydrogen atom and the related wavelengths of the emitted radiation lines.

Quantum and wave mechanics, which was developed in the **1920s** by **Erwin Schroedinger, Werner Karl Heisenberg, Wolfgang Pauli** and **Paul Adrien Maurice Dirac**, now provides a firm basis for the theoretical understanding and application of spectroscopic methods to study matter.

The first mid IR spectrometer was constructed less than 35 years after the discovery of IR radiation, and within 90 years IR spectroscopy was finding applications in astronomy and organic and atmospheric chemistry.

1949

E. K. Blout and **R. C. Mellors** demonstrated that IR spectra could provide information concerning the molecular structure of human and animal tissue. This avenue of exploration was not vigorously pursued, however, because of instrumentation limitations and the complexities of the systems under investigation and of the spectra produced.

1950

A. Elliot and **E. J. Ambrose** demonstrated that the IR absorption spectrum of a protein is sensitive to conformation. Since then fundamental theoretical calculations on the subject have been extensively developed by **T. Miyazawa** and coworkers and **S. Krimm** and his colleagues. Nine characteristic amide vibrational bands or group frequencies were identified in the **1980s**. Among them, the amide I band, which is due almost entirely to the $C-O$ stretch vibration, turned out to be the most useful probe for determining protein secondary structure in solution. However, the low sensitivity of conventional dispersive IR spectrometers and the strong water (H_2O) absorption at wavelengths in the amide I region severely limited early experiments to the solid state or heavy water (D_2O) solutions.

1961

H. T. Miles investigated the tautomeric forms of the DNA bases, making use of the power of IR spectroscopy to identify chemical groups. He discovered that the correct tautomer for cytosine is the amine-carbonyl form that is taken for granted today.

1994

After the introduction of Fourier transform IR (FTIR) spectroscopy in **1980**, **B. R. Singh** and coworkers demonstrated the potential usefulness of amide III spectra for the determination of the type of secondary structure of proteins in aqueous solution.

Proteins and nucleic acids have specific absorption signatures in the UV and IR spectral ranges. Biological macromolecules are themselves colourless but they are often associated with prosthetic groups, such as retinal in the vision proteins or the haem group in myoglobin, haemoglobin and the cytochromes, which absorb visible light. The absorption spectra of these chromophores (from the Greek *chroma*, colour, and *phoros*, bearing) depend on their local environment and can be used to probe conformational changes and dynamics in the macromolecule during its biological active cycle. Enzyme activity can be measured with great sensitivity through the observation of the specific optical absorption patterns of substrates, products and cofactors.

Spectrophotometers in the UV–visible spectral range have become indispensable tools in biochemistry and biophysics laboratories for the systematic analysis and characterisation of macromolecular solutions. The determination of the absorption of UV and visible light might well be the most-performed measurement on biological macromolecules. Experiments using UV–visible absorption range from being aimed at the initial detection of material and concentration measurements to sophisticated temperature-dependence approaches used to measure macromolecular dynamics.

Linear IR spectroscopy is one of the classical tools for the study of the structures and interactions of small molecules. At first, it appeared to be too ambitious to apply this technique to biological macromolecules, because of their enormous number of vibrational modes. It was thought that the resulting overlap in absorption bands would not allow the extraction of detailed structural information from the IR spectra. Biological macromolecules, however, exhibit an intrinsic order of repeating units: the peptide bond in the protein backbone, the phosphate ester bond between the 3′-hydroxyl group on the sugar residue of one nucleotide and the 5′-phosphate group of the next nucleotide in nucleic acids and the two-dimensional array of a limited class of molecules in lipid membranes. IR spectra of biological macromolecules are simpler than at first expected, therefore, and detailed band analysis provides useful information on macromolecular structure and interactions. In the case of proteins, linear IR spectroscopy provides insights into secondary structure, complementary to the information obtained from circular dichroism (see Chapter E4). For nucleic acids, information can be obtained on the overall structure and interactions with small molecules such as intercalating drugs or metal ions.

IR spectroscopy has powerful applications in elucidating the molecular mechanisms of protein activity, by providing information on the interaction of participating molecules. The IR *difference* spectra formed between different states of an enzyme, for example, only contain bands of those groups that undergo changes during the transition from one state to the other. In parallel experiments, time-resolved IR spectroscopy provides information on the reaction kinetics. The difference spectroscopy approach, in fact, has greatly increased the power of the FTIR method by allowing the assignment of peaks in often formidably complex spectra, by comparing spectra of the same protein containing amino acids that have been modified by mutation or chemical modification.

Multidimensional IR spectroscopy has been proposed as a means to determine the structure and dynamics of biological assemblies. It provides a wealth of new information beyond that obtained from linear spectroscopy (see Chapter E2).

E1.2 Brief theoretical outline

The full quantum chemistry theory of UV–visible–IR absorption spectroscopy is in terms of the interactions between the electromagnetic wave and the electrons of the bound atoms in sample molecules. Such a treatment is beyond the scope of this book and here we give only a *qualitative* outline of the theory. We emphasise, however, that the usefulness of absorption spectroscopy in the laboratory is based on the *quantitative* interpretation of the absorption curves of biological macromolecules in terms of accurately determined coefficients. Band assignments have been made reliably from the study of model compounds. For example, the absorption of the peptide bond at 190 nm was established from studies of simple amides, amino acids and polypeptides in solid films and aqueous solutions.

In an absorption spectroscopy experiment, electromagnetic radiation of a given wavelength is observed after it passes through a sample to see how much of it has been absorbed. The power of the method is based on the fact that the molecules in the sample and the light beam exchange only specific amounts of energy, depending on their particular molecular states and dynamics (see Chapter A3). As described by quantum mechanics, radiation energy is absorbed in order to allow transitions between these well-defined molecular states. In a classical mechanics description, the interaction of an electromagnetic wave with matter is treated in terms of induced dipoles. An electromagnetic wave is constituted of mutually perpendicular oscillating electric and magnetic fields (see Fig. A3.7). The fields are *transverse* with respect to the wave; i.e. they are also perpendicular to the direction of propagation. As the wave passes through a molecule, depending on its frequency and on how strongly the charges are maintained in place (the *polarisability*), it causes positive and negative charges to oscillate with opposite phase in the electric field plane. This results in the induction of dipoles with an oscillating *dipole moment* (charge multiplied by separation distance). The electromagnetic wave is represented by discrete energy quanta, $h\nu$, where h is Planck's constant and ν is the frequency of the wave (see Chapter A3). Similarly, the electronic charges in the molecule can only form oscillating dipoles of certain defined energies and in certain defined directions with respect to the beam, because of the properties of the atoms and bonding patterns involved.

Absorption spectroscopy has been applied to biological macromolecules in the full range of the electromagnetic spectrum (Table E1.1). In this chapter we discuss

Table E1.1. *Absorption spectroscopies*

λ (m)	ν (Hz s^{-1})	E (kJ mole^{-1})	Method
~ 3	$\sim 10^8$	$\sim 10^{-4}$	Radio-frequency NMR
$\sim 0.1–0.01$	$\sim 10^9–10^{10}$	$\sim 10^{-1}–10^{-2}$	Microwave rotational and EPR spectroscopy
$\sim 10^{-4}–10^{-5}$	$\sim 10^{12}–10^{13}$	$\sim 1–10$	IR vibrational spectroscopy
$\sim 8 \times 10^{-7}–4 \times 10^{-7}$	$\sim 5 \times 10^{14}–10^{15}$	~ 200	Visible electronic spectroscopy
$\sim 10^{-7}$	$\sim 4 \times 10^{15}$	$\sim 10^3$	UV electronic spectroscopy
$\sim 10^{-10}$	$\sim 10^{18}$	$\sim 10^6$	X-ray absorption electronic spectroscopy
$\sim 10^{-13}$	$\sim 10^{21}$	$\sim 10^9$	γ-ray Mössbauer spectroscopy

the absorption of near-UV, visible and IR radiation. Near-UV, with wavelengths of about 300 nm and energies of about 4 eV or 400 kJ mole^{-1}, and visible radiation, with wavelengths between 400 and 800 nm and energies of about 2 eV or 200 kJ mole^{-1}, correspond to transitions between electronic orbital states, and can be used to probe the different types of bonds in molecules. The spectral range of IR radiation, with wavelengths between 10 μm and 100 μm and energies between about 100 and 10 meV (10 kJ mole^{-1} to 1 kJ mole^{-1}) includes thermal energy at ambient temperature (about 25 meV) and can be used to study vibrational states and molecular dynamics.

E1.2.1 The extinction coefficient and absorbance

Consider the simple set-up in Fig. E1.1. The intensity of light absorbed, $-\mathrm{d}I$, by the molecules in the thin sample slab of thickness, $\mathrm{d}l$, and unit area is proportional to the number of moles in the slab, $C\mathrm{d}l$, and to the incident intensity, I:

$$-\mathrm{d}I = C\varepsilon'(\lambda) I \, \mathrm{d}l \qquad (E1.1)$$

where ε', the proportionality constant, depends on the wavelength and on the particular molecular type. It is called the *molar extinction coefficient* and has usual units of (mole litre^{-1})$^{-1}$ cm^{-1}, or M^{-1} cm^{-1}.

Integrating Eq. (E1.1) over the sample path length, l, we obtain

$$\ln(I_0/I_f) = C\varepsilon'(\lambda)l \qquad (E1.2)$$

where I_f is the final intensity transmitted by the sample (Fig. E1.1).

Equation (E1.2) is usually expressed in terms of the base 10 logarithm:

$$\log(I_0/I_f) = C\varepsilon(\lambda)l \equiv A(\lambda) \qquad (E1.3)$$

Fig. E1.1 Schematic of an experiment to measure absorbance. A beam of wavelength, λ, and intensity, I_0 is shone on a sample cell of path lenth, l, containing a solution of concentration, C (moles litre^{-1}). The final transmitted intensity, I_f, is measured by the detector, D.

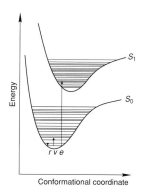

Fig. E1.2 Potential energy diagram of the lowest (S_0) and first excited (S_1) electronic states in a molecule. r, v, and e are rotational, vibrational and electronic transitions, respectively. (After Cantor and Schimmel, 1980.)

Fig. E1.3 Absorption spectrum of a molecule with one type of electronic transition: (a) theoretical, at very low temperature; (b) theoretical at ambient temperature; (c) observed. (After Cantor and Schimmel, 1980.)

Equation (E1.3) is the Beer–Lambert law. The dimensionless (measured) quantity, $A(\lambda)$, is called the *absorbance* or *optical density* (OD) of the sample. OD can be measured accurately by modern spectrophotometers in the range of about 0.1 to ≤ 3. Note that for a value of 3, a fraction of only $1/1000$ of the intensity is transmitted. The molar extinction coefficients, ε' and ε, are related by $\varepsilon = \varepsilon' / e$, where $e = 2.303$. Published values are usually given for ε.

E1.3 The UV–visible spectral range

The UV–visible spectral range corresponds to transition energies of the order of $100\text{–}1000$ kJ mole^{-1} between states, which matches the difference between the ground (lowest-energy) and first excited electronic states (S_0 and S_1, respectively) in small molecules. These energies are much greater than thermal energy at ambient temperature (2.5 kJ mole^{-1} at 300 K).

Each electronic state is itself split into finer energy levels arising from rotational and vibrational energies of the molecule itself (Fig. E1.2). For small molecules, the energy differences between vibrational states are of the order of 40 kJ mole^{-1}, also larger than ambient thermal energy, and it results from statistical quantum mechanics that essentially all the molecules can be considered as occupying the lowest vibrational energy levels in the electronic ground state.

Energy states corresponding to molecular rotations are separated by less than 5 kJ mole^{-1}, so that many rotational energy levels could be occupied at ambient temperature.

Electronic transitions, such as the one depicted in Fig. E1.2, become possible when light of the appropriate wavelength is shone on the molecule. If there were no rotational or vibrational states, the absorption spectrum would be characterised by a sharp line for each type of transition in the molecule (Fig. E1.3(a)). Transitions are possible, however, between the lowest S_0 level and the different vibrational and rotational levels in S_1, resulting in a set of closely spaced lines in the absorption spectrum (Fig. E1.3(b)), whose intensity (*absorbance*) is related to the probability of the transition. In practice, the lines are not only very close together but also each is broadened by various effects, e.g. due to an inhomogeneous local molecular environment or coupling with solvent if the molecule is in solution, so that only a smooth envelope is observed (Fig. E1.3(c)).

For most applications, UV–visible absorption spectra are modelled in terms of single electronic bands, without taking line shape into account. But the line broadening and shape contain information on vibrational and rotational states and an approach has been developed to study protein molecular dynamics from line behaviour as a function of temperature (Comment E1.1).

Biological macromolecules are studied mainly in aqueous solution. Measurements have to be limited to above 170 nm, because water itself absorbs so strongly below that wavelength that even a micron-thick layer appears opaque.

Electronic bands in the macromolecule are also perturbed through coupling with the polar water molecules. These interactions depend on distance and mutual orientation and, since the macromolecular solution is essentially disordered, they are different for each individual molecule resulting in further broadening of the absorption line.

E1.3.1 UV–visible spectrophotometers and measurement strategies

A *spectrophotometer* is an instrument that measures the intensity of a light beam as a function of its *colour*, or wavelength. In absorption experiments, a spectrophotometer provides an accurate quantitative measurement of the fraction of light that passes through a defined path length of solution. Common laboratory spectrophotometers operate in the UV–visible spectral range and sometimes also in the near-IR spectral region. Instruments operating in the middle-IR spectral region have to satisfy different technical requirements, in particular with respect to the photosensor device (the light detector in the instrument) and sensitivity, since all substances emit thermal radiation as IR light (see Section E1.4). Absorption spectrophotometers are either double-beam or single-beam instruments. A double-beam spectrophotometer measures the ratio of light intensity on two different optical paths (e.g. through a sample solution cell and through a cell containing solvent alone). The light intensity with respect to a fixed reference is measured on single-beam instruments. In a single-beam instrument sample solution and solvent absorbance are measured sequentially; the solvent $A(\lambda)$ curve is stored on the instrument's computer and automatically subtracted from the sample solution curve. The optical path in a type of single-beam spectrophotometer is shown in Fig. E1.4.

> **Comment E1.1**
> **Molecular dynamics from band shapes**
>
> Lorenzo Cordone and his collaborators have developed a method to characterise the molecular dynamics of myoglobin from a detailed shape analysis of the experimental visible absorption spectrum of the haem group measured as a function of temperature (Leone *et al.*, 1994).

Fig. E1.4 The optical path in a single/split beam spectrophotometer. The tungsten and deuterium lamps provide light in the visible and UV ranges, respectively. The rotating concave holographic grating geometrically spreads out the beam into its wavelength components, so that the entire wavelength range is measured simultaneously by a photodiode array (detector). A beam splitter diverts part of the beam to a reference detector for continuous calibration.

Modern spectrophotometers are computer-driven and incorporate diverse software packages for specific applications such as time-based kinetics or temperature scans, and for quite a refined analysis of the measured absorbance curves.

Kinetics

For a two-state reaction or conversion, in which each state has a specific absorption spectrum, the spectral change, as the reaction proceeds, displays an *isobestic* point (the point at which the absorbances of the two states are equal). An isobestic point is also observed in cases where the equilibrium between two absorption spectral states is shifted by changing solvent conditions, temperature or pressure, for example (Fig. E1.5). Note, however, that an isobestic point is a necessary but not a sufficient condition for a two-state transition, since an *intermediate* state may well show isobestic points with the initial and final states at the same wavelength.

Difference spectra

Difference spectra are very useful in the study of processes characterised by spectral changes. Figure E1.6 shows difference absorption spectra as a function of time after photoexcitation for retinal in bacteriorhodopsin (see Comment A3.6). Note how the isobestic point (where the difference spectrum crosses zero) varies for the shorter delay times (Fig. E1.6(a)). This indicates the progression of the reaction through *intermediates* with different spectral characteristics. An apparent

Fig. E1.6 Difference absorption spectra of bacteriorhodopsin in purple membrane as a function of delay time after photoexcitation. Numbers between 1 and 16 correspond to delay times between 100 ns and 600 ms, in non-equal intervals (parts (a), (b) are for times shorter and longer than 1 ms, respectively). See also Section E1.3.3. (Váró and Lanyi, 1991.) (Figure reproduced with permission from the American Chemical Society.)

single isobestic point after longer times (Fig. E1.6(b)) suggests the later reaction steps can be interpreted in terms of a two-state transition.

Derivative spectroscopy

The detection sensitivity of small spectral changes can be amplified by *derivative spectroscopy*. Small shoulders in an absorbance curve, for example, yield large oscillatory features when expressed as higher derivatives. Good examples of the effect are given in Section E1.4.

The derivative approach is used mainly to identify and separate small effects in the aromatic region of protein spectra, close to 280 nm (see below).

E1.3.2 UV absorption spectra of proteins

The absorption of UV light by proteins has been analysed in detail and proposed as a structural probe from the very early days of molecular biology (Comment E1.2). The absorption of proteins in this spectral range arises mainly from electronic bands in the peptide group (at 170–220 nm), aromatic amino acid side-chains close to 280 nm and prosthetic groups, cofactors, enzyme substrates or inhibitors (in the full UV–visible range depending on the group). Colour in a protein is always due to a prosthetic group that absorbs in the visible region of the spectrum. The visible absorption spectra of protein-associated molecular groups are treated separately below.

Spectra of amide groups in crystals show electronic coupling, suggesting that the peptide ordering in secondary structures is reflected in the absorption spectra.

Comment E1.2
UV spectra of proteins and amino acids

The treatment given in this chapter is based on a review by D. B. Wetlaufer published in *Advances in Protein Chemistry* in 1962 (Wetlaufer, 1962).

Fig. E1.7 UV absorption
spectra of poly-L-lysine
hydrochloride in aqueous
solution: random coil at
pH 6.0, 25 °C; helix at pH
10.8, 25 °C; β-form at pH
10.8, 52 °C. The x-axis is
the wavelength in nano-
metres (the 'milli-micron'
is an old fashioned unit,
now replaced by 'nano-
metre'). (From Rosenheck
and Doty, 1961, reviewed
by Wetlaufer, 1962.)

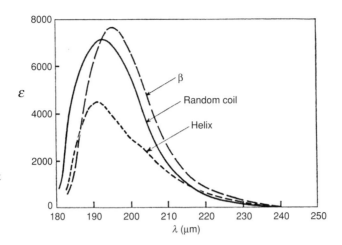

Fig. E1.7 UV absorption spectra of poly-L-lysine hydrochloride in aqueous solution: random coil at pH 6.0, 25 °C; helix at pH 10.8, 25 °C; β-form at pH 10.8, 52 °C. The x-axis is the wavelength in nanometres (the 'milli-micron' is an old fashioned unit, now replaced by 'nanometre'). (From Rosenheck and Doty, 1961, reviewed by Wetlaufer, 1962.)

The molar extinction coefficient of a lysine polypeptide, expressed per amino acid residue, is shown Fig. E1.7 as a function of wavelength in the region of peptide group absorption.

The conformation dependence of the spectra in Fig. E1.7 is obvious, and attempts have been made to estimate the α-helix content in proteins from measured molar extinction values at 190 nm by interpolating between the random coil and helix values. Such measurements are not entirely satisfactory, however, because of the low reliability of the assumptions involved with respect to the effects of other than helix and random coil conformations, as well as with respect to corrections for amino acid side-chain contributions (Fig. E1.8). Secondary structures estimates in proteins are more reliably performed by IR absorption and circular dichroïsm (CD) measurements (see Chapter E4). Histidine, tryptophan, tyrosine and phenylalanine side-chains present extinction coefficients greater than 5000 in the peptide absorption region (190–210 nm). The chemical formulae of the amino acid side-chains are given in Chapter A2.

The near-UV is more easily accessible than the far-UV for accurate measurement of absorption profiles. Above 230 nm, the strongest individual amino acid contribution to absorption in proteins is from tryptophan, followed by tyrosine, then phenylalanine and cystine. Trp absorption has an ε maximum of 5600 at 279.8 nm; Tyr has an ε maximum of 1429 at 274.6 nm – both measured in aqueous solution at pH 6 (Fig. E1.9). Also shown in the figure is the absorption curve of the phenylalanine side-chain with an ε maximum of 197 at 257.4 nm, comparable to the ε values for cystine of 360 at 250 nm and 280 at 260 nm (curve not shown). These four amino acids have to be considered in practice in order to account for the near-UV absorption of a protein. Absorption contributions are additive, so that each amino acid type contributes to the observed absorbance in proportion to how many units of that type there are in the protein. The contribution of Phe

Fig. E1.8 The far-UV absorption spectra of amino acids in aqueous solution at pH 5, except for cystine (pH 3.0). The chloride ion contribution was subtracted from the basic amino acids measured as hydrochlorides. TRY is an old-fashioned abbreviation for tryptophan. CYS (-SH) is cysteine. CYS is cystine. Other abbreviations are the usual ones. (From Sussman and Gratzer, 1962, reviewed by Wetlaufer, 1962.) (Figure reproduced with permission from Elsevier.)

and Cys absorption can be quite significant, therefore, in a protein that contains very few Trp and Tyr residues.

The molar absorption coefficients in Fig. E1.9 are plotted on a log scale, which emphasises qualitative similarities. Fine structure can be seen in the curve maxima (it is especially clear in the spectrum for Phe) resulting from electronic transitions to different vibrational states. These features become sharper at low temperatures, at which broadening due to coupling with rotational modes is reduced.

The UV absorption curve of Tyr depends on its various ionisation states and is especially sensitive to pH. Between pH 6 and pH 13, the peak maximum in the near-UV, at 274.6 nm, shifts to higher wavelengths by about 20 nm and increases in intensity by almost a factor of 2 (Fig. E1.10). This sensitivity has been used

Fig. E1.9 The near-UV
absorption spectra of the
aromatic amino acids in
aqueous solution at pH 6.
TRY is TRP. The *x*-axis is
wavelength in ångströms.
(From Malik, 1961,
reviewed by Wetlaufer,
1962.) (Figure reproduced
with permission from
Elsevier.)

Fig. E1.10 Absorption
spectra and difference
spectra of Tyr measured
for various pH values. The
x-axis is wavelength in
ångströms. The *y*-axis
scale on the right hand
side of the bottom panel
refers to the range above
2430 Å. (From Malik,
1961, reviewed by
Wetlaufer, 1962.)

Comment E1.3   Protein concentration from absorbance measurements (see also Comment A1.2)

In order to measure the absorbance peak close to 280 nm correctly it is essential to display the full background-corrected absorbance curve between about 320 nm and 250 nm. Such a curve is shown schematically below.

(a) The figure illustrates the absorbance of the ribosomal protein S13, for which there is significant light scattering from the solution, perhaps because of protein aggregation. In such a case it is difficult to estimate a correct absolute concentration value from the reading at 280 nm, because the background is sloping.

(b) The reading at 280 nm can be used for the concentration measurement of the ribosomal protein S6, because the background above about 320 nm is flat and at zero. Furthermore, the low absorbance at 260 nm indicates there is no contamination by ribosomal RNA.

Examples of extinction coefficients for proteins

The wavelength (λ, nm), observed ε ($M^{-1}cm^{-1}$) and molecular weight of ribonuclease (A), carboxypeptidase A (B) and bovine serum albumin (C) are:

Protein	λ	ε	M
A	277.5	9 800	13 680
B	278	66 700	34 400
C	280	43 600	66 000

The corresponding absorbance values for 1 mg ml^{-1} solutions in 1 cm optical path length cells are 0.72 for ribonuclease, 1.94 for carboxypeptidase A, and 0.66 for bovine serum albumin. Note that absorbance values for 1 mg ml^{-1} solutions, 1 cm optical path are of the order of 1. The higher value for carboxypeptidase A is because this protein has six Trp residues. Ribonuclease has no Trp residues and bovine serum albumin has 2.

(From the review by Wetlaufer (1962) which gives the original references.)

in the accurate titration of Tyr residues in proteins as well as in the separate determination of Tyr and Trp contributions to an observed absorption spectrum (the absorbance of Trp is much less sensitive to pH than that of Tyr).

Protein concentration measurements

Protein concentration measurements are usually performed by measuring the absorbance of the solution at wavelengths close to 280 nm. Such a determination provides only a *relative* concentration value, however, unless the molar extinction coefficient of the specific protein has been determined independently. We recall that biophysical methods involving determinations of molar mass or molar scattering power require accurate and *absolute* determinations of concentration (see Comment A1.2). Molar extinction coefficients at 280 nm, calculated from amino acid compositions (in particular, Tyr, Trp and Phe contents), have been found to agree reasonably well with experimental absorbance measurements on unfolded proteins in guanidine hydrochloride, but values for native proteins may differ considerably because of the environmental effects on the absorbance of the amino acids discussed above. In early work, molar absorption coefficients were found by determining concentration on an absolute scale via a dry weight measurement (at 105 °C in vacuo, corrected for ash) or a nitrogen determination. The modern approach is based on quantitative amino acid analysis. The molar extinction coefficients of a few different proteins are given in Comment E1.3.

E1.3.3 Visible absorption spectra of protein-associated groups

Chromophore prosthetic groups with strong absorption bands in the UV–visible spectral range have greatly facilitated spectroscopic studies of protein activity. Usually, the chromophores are themselves part of the active site, their electronic configuration, reflected in the observed absorption bands and extinction coefficients, providing a sensitive measure of the progression of a reaction. The light absorption properties of a protein-associated chromophore may be of direct biological relevance as is obviously the case for retinal in vision, or chlorophyll in photosynthesis. Chromophores can be as simple as a single metal ion that changes its colour according to its oxidation state, such as copper in the blue protein azurin with absorption bands at 781 and 625 nm, or an organic molecule, such as retinal in rhodopsin (see below), whose main absorption band shifts in a wide range between 400 and 600 nm, depending on its local environment. Protein-linked chromophores can also be quite complex, such as the metal clusters in ceruloplasmin (absorption band at 793 nm), iron–sulphur clusters in ferredoxins, porphyrin associated with iron in the haem groups of myoglobin, haemoglobin and the cytochromes (different bands between 300 and 800 nm giving the red colour to the proteins), porphyrin associated with magnesium in chlorophyll (three sets

Fig. E1.11 The dimer of the cytochrome b6f complex from *Chlamydomonas reinhardtii*. The stromal side is on the top, the luminal side is on the bottom. The protein subunits are shown as light colour ribbons. The light orange rectangle indicates the location of the membrane. The complex binds a number of chromophore prosthetic groups: haem (red), beta carotene (orange), chlorophyll (green); the iron atoms of iron–sulphur groups are shown as red dots. The blue molecule is tridecylstigmatellin, an analogue of quinone, which is a substrate of the complex. We are grateful to Daniel Picot for the figure. The cytochrome b6f complex structure is described in Stroebel *et al.* (2003).

of bands, respectively, at 360–400, 580–590, close to 780 nm, giving a green colour of course), or the flavin (vitamin B2) and flavin derivative groups that are involved in electron transfer reactions in many enzymes (bands close to 450 nm).

Proteins that have to ensure complex, highly regulated, electron transfer reactions, such as photoreaction centres in photosynthesis or different cytochromes in respiration chains, may include many different chromophore prosthetic groups within the same structure (Fig. E1.11).

Reactions and associated changes in absorption properties of a chromophore can be initiated and followed very rapidly by using laser flash triggers. Below we briefly describe two examples of such studies that have had significant impacts on molecular biophysics: the rebinding of CO to the haem group in myoglobin following flash photolysis, and the light-activated proton pump photocycle in bacteriorhodopsin.

Flash photolysis of CO–myoglobin

Results from flash photolysis experiments on CO–myoglobin constituted one of the bases for the development of current ideas on protein dynamics and its relation to biological function, especially with respect to the conformational substates and dynamical transition models (see Chapter A3). The absorption spectra of

Fig. E1.12 Absorption
spectra of haemoglobin
and oxygen-bound
haemoglobin.

haemoglobin and oxygen-bound haemoglobin are in Fig. E1.12. The absorption spectrum of the protein-bound haem contains three main groups of bands, called the Soret bands, between 325 and 500 nm, the a, b, bands between 500 and 600 nm, and the 'charge transfer' band or band III, in the near-IR, above 650 nm. CO binds much more strongly than oxygen to the iron atom of the haem group (this is why CO is a poisonous gas). A picture of CO bound to myoglobin is given in Fig. A3.32(b). CO binding to myoglobin leads to a shift in the Soret band, a merging of the a, b, bands and a disappearance of band III. In flash photolysis experiments, the Fe–CO bond is broken by a laser flash (photodissociation) and the rebinding kinetics is then followed, with nanosecond time resolution, through the changes in the absorption spectrum (Fig. E1.13).

The time dependence kinetics of the Soret band and band III shifts are very similar (Fig. E1.14). Information was derived from the non-exponential kinetics on the important role played by protein dynamics in the competition between ligand rebinding (geminate recombination, meaning 'repeated', from the Latin, *geminus*, twin) and ligand escape from the haem pocket through the protein structure.

A relation between protein relaxation and ligand rebinding was first formulated for NO geminate recombination, because they occur on similar time scales. In contrast, the recombination rate for CO is several orders of magnitude slower than protein relaxation rates under normal temperatures and solvent conditions. Experiments to examine the effects of protein relaxation on CO rebinding kinetics (see Comment A3.3) have been performed in viscous, glycerol-containing, solvents, and at low temperature to slow down protein relaxation rates.

Fig. E1.13 Time-resolved band III spectra of the recovery of CO–myoglobin after flash photolysis. The equilibrium unbound, deoxy-myoglobin spectrum maximum is given by the dashed line. The first measurement was after 60 ns (top), the last curve (bottom) was taken after 1 ms. The experiment was performed in 75% glycerol-containing solvent at 250 K. The transient spectra are offset vertically for clarity. (After Franzen and Boxer, 1997.)

The absorption band changes result from the influence of the porphyrin and protein environment on the electronic orbitals of the iron atom. Interpretations of the spectral changes in structural terms have been attempted from correlations with crystallography data on different protein states (see Chapter A3) and from studies of different mutants (Fig. E1.15). Two important histidines are associated

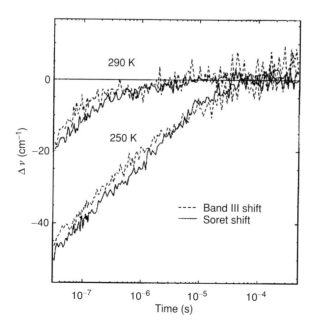

Fig. E1.14 The time dependence kinetics of the Soret band and band III shifts are very similar. Experimental conditions were as for Fig. E1.13. (Franzen and Boxer, 1997.)

Fig. E1.15 Time
dependence of Soret
band shifts following
photodissociation of
CO–myoglobin for the
wild-type protein (WT)
and different distal
hsitidine mutants.
Experimental conditions
were as for Fig. E1.13.
Note the six orders of
magnitude time scale on
the x-axis. (After Franzen
and Boxer, 1997.)

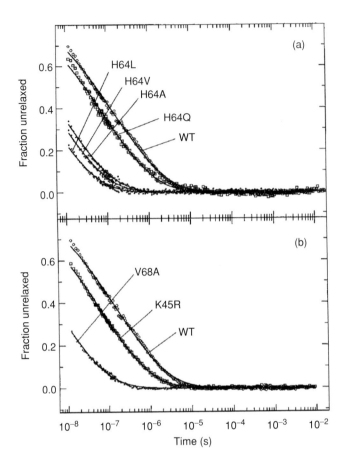

with the haem group in myoglobin: H93, which is bound by a covalent bond
to the iron, is called the *proximal histidine*; H64, the *distal histidine*, is on the
other side of the haem plane, pointing towards the ligand–Fe binding site. The
figure shows the large differences in protein relaxation rates and CO rebinding, as
monitored by the Soret band shift following CO flash dissociation, for different
distal histidine mutants.

The photocycle of bacteriorhodopsin

Bacteriorhodopsin, the light activated proton pump in the purple membrane of
Halobacterium salinarum was described in Comment A3.6. Absorption of green
light, with a maximum wavelength of 569 nm, initiates a photocycle of observable
colour changes in the retinal (Fig. E1.16). The photocycle was established by
careful measurements of absorbance change kinetics following photoactivation
in wild-type bacteriorhodopsin and various bacteriorhodopsin mutants (e.g. see
Fig. E1.6). By using a small number of reasonable assumptions, it was possible to
calculate the absorption spectra of the different states making up the photocycle
(Fig. E1.17).

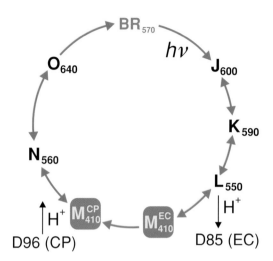

Fig. E1.16 The photocycle of bacteriorhodopsin (BR). The capital letters denote each spectral state of the retinal. The subscripts are the maximum absorption wavelengths in nanometres. Between L and M the Schiff base is accessible to the extracellular (EC) side. A proton is released from amino acid residue D85 to the EC side of the membrane. Within M there is an irreversible transition of the Schiff base that becomes accessible to the cytoplasmic (CP) side. Between M and N, a proton is bound to D96 on the CP side. See also Comment E1.4. (Haupts et al., 1997.)

The analysis led to the discovery of two 'M' states in the photocycle with similar absorbance curves but different structures. Interpretation models were proposed in which the M states constitute the 'switch' (Comment E1.4), by relating absorbance in the visible with other results mainly from IR spectroscopy and X-ray crystallography on different states trapped at low temperatures.

Fig. E1.17 Absorbance (top) and absorbance differences (bottom) of the different states in the bacteriorhodopsin photocycle. (Váró and Lanyi, 1991.) (Figure reproduced with permission from the American Chemical Society.)

Comment E1.4   Light and bacteriorhodopsin (see also Comment A3.6)

In absence of light, bacteriorhodopsin is in the *dark-adapted* state, and the retinal configuration is a mixture of all-*trans* and 13-*cis*. Upon illumination, all-*trans* retinal accumulates in the *light-adapted* state, and the photocycle is activated from that state. Retinal remains bound to bacteriorhodopsin via a Schiff base linkage during the entire photocycle. The orientation and protonation state of the Schiff base vary, however, in the different intermediates. Different models persist for the primary reaction and the electronic origin of the J_{625} intermediate formed about 500 fs after absorption of a photon by the retinal. The K_{590} intermediate, which can be trapped below 150 K and in which retinal is in the 13-*cis* configuration, is formed 5 ps later. After a lifetime of 1 µs, K leads to L_{550}, which decays to M_{410} after 100 µs. During the L to M transition D85 becomes protonated, from the Schiff base and a proton is released on the EC side.

The Schiff base is deprotonated only in the blue-shifted M_{410} state of the photocycle, which is also the longest-lived intermediate with a lifetime in the order of milliseconds. The switch or valve function of the proton pump (see Chapter A3) occurs during the M state, in which the accessibility of the Schiff base changes from the EC to the CP side (Fig. E1.16), associated with a large protein conformational change. The Schiff base is reprotonated between M and N from D96, which is itself reprotonated from the CP side. The retinal isomerises back to all-*trans* between N_{560} and O_{640}, which decays in milliseconds to the ground state with the reprotonation of D85. (Haupts *et al.*, 1997, 1999.)

E1.3.4 UV absorption spectra of nucleic acids

The electronic transitions of the sugar and phosphate moieties of a nucleotide are well beyond 200 nm in the UV and the aromatic bases dominate nucleic acid absorption at more easily observable wavelengths. Chemical derivatives of purine and pyrimidine bases exhibit strong absorption in the 260–290 nm spectral range, through the electronic states of their conjugated double-bond structures (see Chapter A2 for nucleotide chemical structures). The purine and pyrimidine bases in aqueous solution have distinct absorption spectra, which vary characteristically with pH (Fig. E1.18, Table E1.2).

Despite the differences between the spectra of the individual bases, the maxima all occur close to 260 nm, and the minima close to 230 nm. This leads to UV absorbance curves of nucleic acids that are similarly smooth, with a peak at 260 nm and a minimum close to 230 nm, from which it is not possible to extract base composition information.

The molar extinction coefficients of nucleotides at 260 nm are of the order of 10^4 M^{-1} cm^{-1}. The UV absorption spectrum of nucleic acids can be measured with concentrations as low as a few micrograms per millilitre. It provides

Table E1.2. *pH dependence of purine and pyrimidine absorbance maxima*[a]

Base	pH	λ_{max}(nm)
Adenine	1	262.5
Adenine	7	260.5
Adenine	12	269
Cytosine	1	276
Cytosine	7	267
Cytosine	14	282
Guanine	1	248, 276
Guanine	7	246, 276
Guanine	11	274
Thymine	4	264.5
Thymine	7	264.5
Thymine	12	291
Uracyl	4	259.5
Uracil	7	259.5
Uracil	12	284

[a]Data taken from Sober (1997).

a sensitive microdetection method for a variety of applications, ranging from chromatography to AUC (see Chapter D4). Nucleic acid extinction values are much larger than those of amino acid side-chains close to 280 nm, except for Trp (5600 M^{-1} cm^{-1}) and Tyr (1460 M^{-1} cm^{-1}) (see above). There are relatively few aromatic residues in a protein, however, while in any nucleic acid all the nucleotides display strong UV absorption, so that the UV absorption spectrum

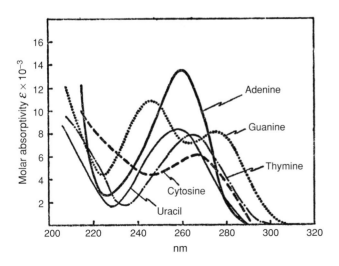

Fig. E1.18 UV absorption curves of purine and pyrimidine bases at pH 7. (Davidson and Secrest, 1972.) (Figure reproduced with permission from Elsevier.)

Fig. E1.19 Increase in the absorbance at 260 nm of T7 DNA as it melts with temperature. T_m is the melting temperature. (After Cantor and Schimmel, 1980.)

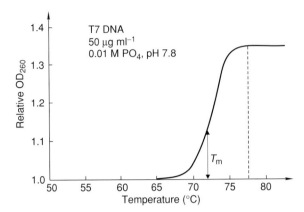

of a nucleic acid–protein complex is strongly dominated by the nucleic acid component. The absorbance ratio at 280 nm/260 nm in UV spectrophotometry is, therefore, a sensitive test for nucleic acid or nucleotide (e.g. many enzyme cofactors are nucleotide derivatives) contamination in a protein solution.

The extinction coefficient (per nucleotide) of a nucleic acid is smaller than that of the free nucleotides so that it increases significantly upon denaturation, degradation or hydrolysis (Fig. E1.19). Note the similar spectra for the denatured (with no tertiary structure) and enzymatic digest (essentially made up of mononucleotides) samples. The decrease in extinction coefficient (*hypochromicity*, from the Greek meaning under colour) upon tertiary structure formation is due to strong electronic interactions between the stacked bases. The increase in extinction coefficient is called the *hyperchromic effect* (from the Greek, over coloured). Hypo- and hyper-chromicity are useful properties that allow nucleic acid stability to be easily assessed by measuring melting temperature as a function of temperature by UV spectrophotometry (see Chapter C2).

E1.4 IR absorption spectroscopy

The energy of IR radiation corresponds to the small energy differences between vibrational and rotational states in molecules (see Fig. E1.2). As we discussed above, radiation absorption by a molecule is related to electric dipole changes, so that in the case of IR radiation these changes must be associated with vibrational or rotational motions.

E1.4.1 IR spectrometers

The IR region of the electromagnetic spectrum encompasses the wavelength range from 0.78 to 1000 μm. In IR spectroscopy, the frequency of a band is usually expressed in terms of wave numbers in units of cm^{-1} (wave number is the inverse of wavelength; recall that velocity = frequency × wavelength, so

Table E1.3. *IR spectral regions*

Region	Wavelength range, μm	Wave number range, cm^{-1}	Frequency range, Hz
Near	0.78–2.5	12 800–4000	$3.8 \times 10^{14} - 1.2 \times 10^{12}$
Middle	2.5–50	4000–200	$1.2 \times 10^{14} - 6.0 \times 10^{12}$
Far	50–1000	200–10	$6.0 \times 10^{12} - 3.0 \times 10^{11}$
Most used	2.5–15	4000–670	$1.2 \times 10^{14} - 2.0 \times 10^{13}$

that wave number = frequency/velocity). The wave number range of IR radiation is from about 12 800 to 10 cm^{-1}. From considerations of the applications as well as the instrumentation, the IR spectrum is conveniently divided into near-, mid-, and far-IR regions; the rough limits of each are shown in Table E1.3. The applications and methodology associated with each of the three IR spectral regions are considerably different. Measurements in the *near*-IR region are made with a spectrophotometer similar in design and components to the instruments described for UV–visible spectrometry (see above).

Until the early 1980s, instruments for the *mid*-IR region were largely of the dispersive type, based on diffraction gratings. Since that time, however, there has been a dramatic change in mid-IR instrumentation. Current instruments are of a Fourier transform (see Chapter A3) type and the method is called Fourier transform IR (FTIR) spectroscopy.

In the past, the *far*-IR region of the spectrum has had limited use because of significant experimental difficulties. The few available sources of this type of radiation are notoriously weak and are further attenuated by the need for order-sorting filters that must be used to prevent radiation of higher grating orders from reaching the detector. FTIR spectrometers, with their much higher throughput, have largely overcome this problem and made the far-IR spectral region much more accessible to chemists and biologists.

Fourier transform infrared (FTIR) spectrometers

All IR spectrometers include a broad-band source (usually an incandescent ceramic material), optics for collimating the IR light and directing it through a sample, an IR detector, and some means for analysing the wavelength dependence of transmitted light. In conventional *dispersive* spectrometers, this last function is accomplished by spatially spreading the light with a diffraction grating, and selecting individual spectral elements with a narrow slit. Only a narrow wavelength range reaches the sample and detector at any time. The collection of a spectrum covering a broad wavelength range requires sequential scanning over different spectral elements.

Attenuated total reflection

IR transmission measurements of biological molecules are often impeded by the intense absorption of water near 1650 and 3300 cm^{-1} bands. For this reason most transmission measurements must be made on very thin samples (10 µm) in D$_2$O, or under partially dehydrated conditions. Attenuated total reflection (ATR) avoids many problems associated with transmission measurements by limiting the effective sample thickness to a thin layer near the surface of internal reflection element (IRE). At each point of internal reflection there is an evanescent electromagnetic wave that penetrates through surface into the medium of lower refractive index (see Section F3.3.3 for details). In the case of a 3 mm thick by 50 mm long IRE crystal, it is routine to obtain 10–20 reflections. The actual penetration depth is described by Eq. (F3.4). For the case of crystalline germanium surrounded by an aqueous medium, this equation gives a penetration depth of ~ 5 µm at 1000 cm^{-1}. Thus ATR measurements solve the problem of thickness of the sample and create appropriate conditions for the transmission measurement of biological samples. These conditions are very suitable for biomembrane studies: a multilayer membrane sample can be kept at the interface between the internal reflection element and well-defined buffer solution without the bulk water contributing significantly to the IR absorption spectrum.

A more efficient approach is to detect the signals from different wavelengths simultaneously. In the case of visible and near-UV dispersive spectrometers, this is accomplished by using multichannel area detectors. Such detectors, however, are not available for the IR spectral region (400–4000 cm^{-1}). The alternative approach employed in FTIR spectrometers is to pass the broad-band IR beam through an interferometer before sending it through the sample and then to a single very sensitive detector (Fig. E1.20). In this case, the intensity of each wavelength component varies as a cosine function of the optical path length difference, ΔL, from a beam splitter to fixed and moving mirrors; the frequency

Fig. E1.20 Simplified layout of an FTIR spectrometer. Steady illumination from a broad-band IR source is modulated by a Michelson interferometer, i.e. a beam splitter that divides the collimated light along different paths to two mirrors. The light reflected from the fixed and moving mirrors is recombined at the beam splitter and is directed through the sample and into the IR detector. The path length difference is measured very precisely with a He–Ne laser directed into the Michelson interferometer (not shown). A computer stores the digitised interferogram and converts it via Fourier transformation to an IR spectrum. (Braiman and Rothschild, 1988.)

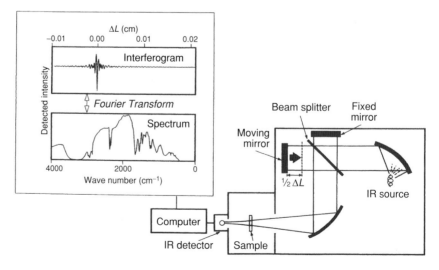

of each cosine function is the reciprocal of the associated IR wavelength (i.e. the wave number $\bar{\nu}$). The single-beam spectrum (the detected IR intensity as a function of $\bar{\nu}$) is obtained from a Fourier transform of the *interferogram*, which is the detected intensity as a function of ΔL. Most modern IR spectrometers are Fourier transform instruments.

E1.4.2 Molecular vibrations

The relative positions of atoms in a molecule fluctuate continuously as a consequence of a multitude of different types of vibrations and rotations about their bonds (see Section A3.2.2). For a simple diatomic or triatomic molecule, it is easy to define the number and nature of such excitations and to relate these to the energies of absorbed radiation. Such an analysis becomes difficult if not impossible for molecules made up of more atoms. Not only do they contain a large number of vibrating centres, but also interactions can occur amongst them and must be taken into account.

Vibrations fall into one of two basic categories: *stretching* and *bending vibrations*. A stretching vibration involves a continuous change in interatomic distance along the axis of the bond between two atoms. Bending vibrations are characterised by a change in the angle between two bonds and are of four types: *scissoring, rocking, wagging and twisting* (see Fig. E1.21). All of the vibration types shown in Fig. E1.21 are possible in a molecule containing more than two atoms. In addition, interaction or coupling between vibrations can occur if they involve a bond to a single central atom. Coupling results in a change in the characteristics of the vibrations involved.

Vibration and polarised light
The various types of vibration of a hypothetical non-linear chemical group made up of three nuclei are illustrated in Fig. E1.22. In-plane stretching can be either

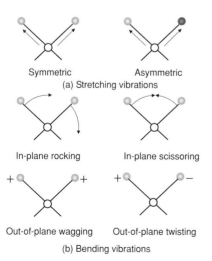

Symmetric   Asymmetric
(a) Stretching vibrations

In-plane rocking   In-plane scissoring

Out-of-plane wagging   Out-of-plane twisting
(b) Bending vibrations

Fig. E1.21 Types of molecular vibration. Note that + indicates motion out of the page toward the reader; – indicates motion away from the reader.

Stretching

In-plane bending

Out-of-plane bending

Fig. E1.22 In-phase and out-of-phase vibrations for three non-linear atoms.

in-phase or *out-of-phase*. There are also in-phase and out-of-phase in-plane bending modes. Finally, out-of-plane bending can be in-phase or out-of-phase. We recall that electromagnetic radiation interacts with vibrating dipoles in the molecule. In order for the oscillations of the IR light to excite the in-phase stretching or in-plane bending modes, the beam needs to be *polarised along* the symmetry axis of the group. For the out-of-phase stretching the light needs to be polarised in-plane and across the symmetry axis. For the in-phase out-of-plane bending mode, the light needs to be polarised perpendicularly to the plane of the group. Out-of-phase and out-of-plane bending modes do not have dipole motions that can be driven by radiation absorption (see Section E1.4.3), and thus will not be excited.

E1.4.3 IR-active and IR-inactive modes

A permanent dipole moment is associated with a *heteronuclear* diatomic molecule (or its chemical bond), because the *electronegative* values of the bonded atoms are different so that there is a separation between the centres of negative and positive charge. If the bond vibrates with the normal mode frequency, v_m (see Chapter A3) then the dipole moment oscillates at the same frequency. According to classical electromagnetic theory, an oscillating dipole absorbs electromagnetic radiation of the same frequency as that of its oscillation. A normal mode of vibration that gives rise to an oscillating dipole in the IR spectral range is said to be IR-*active*. Conversely, a normal mode of vibration that does not give rise to an oscillating dipole moment (e.g. the stretching of a *homonuclear* diatomic molecule) cannot lead to IR absorption and is said to be IR-*inactive*.

As an example, we consider the carbon dioxide molecule (O=C=O), which has four normal modes of vibration (see Section E1.4.5). Symmetry considerations dictate that two of these are bond stretching modes, and two are valence angle bending modes. One stretching mode is symmetrical, involving the simultaneous extension or the simultaneous contraction of each oxygen–carbon double bond. This mode is IR-inactive, because no molecular dipole moment change accompanies the in-phase displacements of the two oxygen atoms with respect to the central carbon. The other stretching mode is antisymmetrical; the carbon–oxygen bonds vibrate out of phase, which is essentially equivalent to linearly displacing the central carbon atom between two fixed oxygen atoms. This mode is IR-active, because it is associated with a large dipole moment fluctuation.

E1.4.4 Quantum mechanical treatment of vibrations

The harmonic approximation

Molecular vibrations can be described in terms of quantum mechanical simple harmonic oscillators and normal modes (see Chapter A3). We recall that in

---

Comment E1.5 The force constants of chemical bonds

IR absorption measurements allow the evaluation of the force constants for various types of chemical bond by using Eq. (E1.4). Generally, k has been found to lie in the range between 3×10^2 and 8×10^2 N m^{-1} for most single bonds, with 5×10^2 N m^{-1} serving as a reasonable average value. Double and triple bonds have associated force constants of about two and three times this value (1×10^3 and 1.5×10^3 N m^{-1}, respectively). By inserting these average experimental values, Eq. (E1.6) can be used to estimate the wave number of the fundamental absorption peak.

---

contrast to classical mechanics, in which vibrators can assume any potential energy, quantum vibrators can take only certain discrete energy values, E. In the case of two masses, m_1, m_2, connected by a Hooke's law spring, E is given by Eq. (E1.4) (see also Eq. (A3.43)):

$$E = \left(n + \frac{1}{2}\right) \frac{h}{2\pi} \sqrt{\frac{k}{m_{1,2}}} \qquad (E1.4)$$

where n is the vibrational quantum number, which can take only positive integer values including zero, h is Planck's constant, $m_{1,2} = m_1 m_2 / (m_1 + m_2)$, and k is the spring force constant (Comment E1.5).

Transitions between vibrational energy levels can be brought about by the absorption of radiation, only if the energy of the radiation exactly matches the difference in energy between levels and certain *selection rules* are obeyed (Comment E1.6). The energy difference, ΔE, between the vibrational quantum states is given by

$$\Delta E = h\nu_m = \frac{h}{2\pi} \sqrt{\frac{k}{m_{1,2}}} \qquad (E1.5)$$

where ν_m can be seen as a fundamental frequency associated with the set of vibrational modes. In IR spectroscopy, the frequency of a band is usually expressed in terms of wave numbers

$$\bar{\nu} = \frac{1}{2\pi c} \sqrt{\frac{k}{m_{1,2}}} = 5.3 \times 10^{-12} \sqrt{\frac{k}{m_{1,2}}} \qquad (E1.6)$$

where $\bar{\nu}$ is the wave number of an absorption peak in cm^{-1}, k is the force constant for the bond in newtons per metre, c is velocity of light in centimetres per second, and the reduced mass, $m_{1,2}$, has units of kilogram (Comment E1.5).

Anharmonicity

Even from a qualitative standpoint, the description of a bond between two atoms in terms of a harmonic spring cannot be fully satisfactory. As the two atoms approach one another, for example, Coulombic repulsion between the two nuclei produces

Comment E1.6
Selection rules

Quantum theory indicates that transitions can take place only between levels for which the vibrational quantum number differs by unity; i.e. there is a *selection rule* that $\Delta n = \pm 1$. Since the vibrational levels are equally spaced (Eq. (E1.4)), only a single absorption peak is observed for a given molecular vibration.

a force that acts in the same direction as the bond restoring force; thus the potential energy can be expected to rise more rapidly than predicted by the harmonic approximation. The corresponding potential energy curve has a shape similar to the one in Fig. A3.24. Such curves depart from harmonic behaviour to varying degrees, depending upon the nature of the bond and the atoms involved. Anharmonicity leads to ΔE not being constant but becoming smaller at higher energies (curve 2 in Fig. A3.24). The selection rule is no longer rigorously followed; as a result, transitions of $\Delta n = \pm 2$ or $\Delta n = \pm 3$ are observed. Such transitions are responsible for the appearance of *overtone lines*, frequencies approximately two or three times that of the fundamental line. Overtone absorption lines are usually of low intensity, however, and may not be observed.

E1.4.5 Vibrational modes of polyatomic molecules

Molecular normal modes of vibration are discussed in Chapter A3. It has been shown rigorously that only the normal modes of vibration of a molecule are capable of the interaction with electromagnetic radiation that gives rise to IR absorption. In the case of biological macromolecules, most of the $3N - 6$ normal modes are highly localised – meaning that, to a good approximation, each involves only a small group of atoms moving as if they were isolated from the vibrations of other molecular groups. For example, if the N atoms of the protein are distributed among K sets of identical groups each of n atoms, and if each such group exhibits the same or nearly the same conformation and environment in the macromolecule, then the collection of $3nk - 6$ hypothetical spectral bands is reduced to $3n - 6$ (see also Section A3.2.2). Consider the case of the amide I mode of an α-helical polypeptide chain of k residues. To a first approximation, each peptide–CONH group in the α-helix will have the same vibrational frequency, resulting in the appearance of a single amide I band, rather than k distinct bands.

The vibrational frequencies of polyatomic molecules are determined, in principle, by the masses and geometrical arrangement of their constituent atoms and the interatomic forces resulting from the distortion of the equilibrium configuration. The frequencies and atomic displacements can be calculated by using empirical, classical mechanics, potential energy functions (see Chapter A3) incorporating simplifying assumptions according to the complexity of the molecule and its environment. Fairly rigorous analysis is now possible for molecules containing 10–20 atoms. Normal mode calculations (by definition in the harmonic approximation) of much larger biological macromolecules are also possible (see Chapter A3).

Amide bands of polypeptides and proteins

The IR spectra of polypeptides and proteins exhibit several relatively strong absorption bands, which are approximately constant in frequency and intensity

Table E1.4. *Frequency range for characteristic absorption bands of secondary amides in the crystalline state (based on model compounds)*

In-plane modes	
Amide A	~3300 cm^{-1}
Amide B	~3100 cm^{-1}
Amide I	1597–1672 cm^{-1}
Amide II	1480-1575 cm^{-1}
Amide III	1229–1301 cm^{-1}
Amide IV	625–767 cm^{-1}
Out-of-plane modes	
Amide V	640–800 cm^{-1}
Amide VI	537–606 cm^{-1}
Amide VII	~200 cm^{-1}

Fig. E1.23 Characteristic amide bands as exhibited by a capillary film of N-methylacetamide. The amide VII band is not shown.

from one sample to another. These bands are associated with the vibrations of the CONH group, the structural unit common to these molecules.

Characteristic bands of the CONH group of polypeptide chains are similar to absorption bands exhibited by secondary amides, in general. This has led to the commonly accepted procedure of naming the prominent absorption bands in proteins and polypeptides the 'amide I band', 'amide II band', etc. There is a total of nine such bands, usually called amide A, amide B, and amide I--VII, in order of decreasing frequency. An isolated planar CONH group would give rise to five in-plane (C=O stretching, C–H stretching, N–H stretching, OCN bending, CNH bending) modes and one out-of-plane (C–N torsion) normal mode. The frequency ranges of the different amide bands in model compounds are given in Table E1.4. Such bands are shown in Fig. E1.23 for N-methylacetamide, a model for the trans peptide group in proteins, which has 30 normal modes of vibration (Fig. E1.24). Neglecting the six methyl hydrogen

N-methylacetamide

Fig. E1.24 N-methylacetamide is an analogue for the *trans.* peptide group in proteins.

Amide A

Amide I

Amide II

Amide III

Fig. E1.25 Atomic displacement vectors for the amide A, I, II, and III modes as calculated for *N*-methylacetamide.

atoms, there are 12 normal modes arising from the remaining six skeletal atoms (CCONHC).

There are four main characteristic IR bands from secondary amides such as N-methylacetamide (Fig. E1.25):

1. In dilute solution (no hydrogen bonding) a sharp band observed at 3400–3460 cm^{-1} (amide A) is easily assigned to an essentially pure N–H stretching motion by comparison with related molecules and by shifts caused by deuterium labelling of the H atom. Upon hydrogen bonding the band shifts to 3120–3320 cm^{-1} and another, usually weaker, band appears close to 3100 cm^{-1} (amide B). The relatively high frequency of the N–H stretching vibration (3100–3400 cm^{-1} range) isolates this type of motion from other molecular vibrations in N-methylacetamide (Fig. E1.23).

2. All secondary amides, polypeptides and proteins show a strong band in the 1600–1700 cm^{-1} region (amide I), which shifts to higher frequencies in dilute solution. By simple qualitative considerations, based on comparison with related molecules, and by the frequency shift upon dilution (related to the breaking of hydrogen bonds), the band has been assigned to a mode essentially involving the stretching vibration of the C=O bond.

3. The amide II band absorbs between 1510 and 1570 cm^{-1}. Unlike the bands associated primarily with C=O stretching and N–H stretching, the amide II band is not easily interpreted by qualitative considerations.

4. Bands in the amide III spectral region (1350–1570 cm^{-1}) are predominantly due to the in-phase combination of N–H in-plane bending and C–N stretching vibrations and are highly sensitive to the secondary structure.

E1.4.6 Resolution enhanced FTIR spectra

The analysis of IR spectra in terms of protein structure is not straightforward and presents serious conceptual and practical problems, despite the well-recognised conformational sensitivity of the IR-active bonds. Bands in both the amide I and amide III regions are broad, not resolved into individual components corresponding to different secondary structure elements.

Resolution enhancement or band-narrowing methods are applied to resolve broad overlapped bands into individual bands. FTIR spectroscopy presents several advantages over conventional dispersive techniques for this type of analysis through the application of *second derivative* and *Fourier self-deconvolution* techniques, including higher resolution, sensitivity, signal-to-noise ratio, and frequency accuracy.

Second derivative spectra

Because FTIR spectra are digitally encoded with one data point n every $\Delta k/(2^m) = \Delta W$ frequency units, where Δk is the nominal instrumental resolution (in cm^{-1}) and m is the number of times the interferogram is zero-filled prior to Fourier transformation, the second derivative of the spectrum may be calculated by a straightforward analytical method. In particular, at data point n the value of the second derivative A_n'' (in units of absorbance /(wave number)^2) is given by

$$A_n'' = (A_{n+1} - 2A_n + A_{n-1})/(\Delta W)^2 \tag{E1.7}$$

where A_n is the absorbance at data point n in the measured spectrum. (Note that this function gives the second derivative spectrum without any smoothing.)

The intrinsic shape of a single IR absorption line of an isolated molecule may be approximated by a Lorentzian function (the Fourier transform of an exponential decay function, see Chapter A3)

$$A = \frac{\sigma}{\pi(\sigma^2 + k'^2)} \tag{E1.8a}$$

$$= \frac{\sigma}{\sigma\pi(1 + \beta k'^2)} \tag{E1.8b}$$

where A is the absorbance, 2σ is the width at half height, $k' = k - k_0$ is the frequency referred to the band centre at k_0, and $B = 1/\sigma^2$. The second derivative of Eq. (E1.8b) is

$$A'' = \frac{-2B(1 - 3Bk'^2)}{\sigma\pi(1 + Bk'^2)^3} \tag{E1.9}$$

The peak frequency for the second derivative is identical to the frequency of the original band centre, k_0. The half-width of the second derivative, σ^{II}, is related to the half-width of the original line by

$$\sigma^{II} = \sigma/2.7 \tag{E1.10}$$

and the peak intensity of the second derivative is

$$A_0'' = -2A_0/\sigma^2 \tag{E1.11}$$

Thus, the peak height of the second derivative is proportional to the original peak height (with opposite sign) and inversely proportional to the square of the original half-width.

The main advantage of using second derivative spectra lies in the ease with which the peak frequencies of unresolved components can be identified. Measured and second derivative spectra of bovine α-chymotrypsin in D_2O solution are shown in Fig. E1.26. The strong amide I band, which is unresolved in the original spectrum, is resolved into seven components by the second derivative technique. The interpretation of second derivative spectra is more complex in the case of overlapping bands that deviate from a Lorentzian shape. The above formulae, nevertheless, still provide a reasonable approximation in such cases.

Fig. E1.26 Original, deconvoluted and second derivative FTIR spectrum of bovine α-chymotrypsin in D_2O solution from 1250 to 1800 cm^{-1}. (Susi and Byler, 1986.)

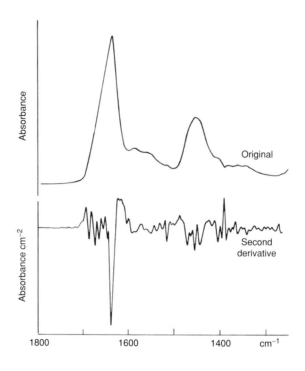

Comment E1.7
Spectral decomposition

Decomposition requires two constants as computer input: the estimated half-width at half-height of the unresolved bands, σ, and the resolution 'efficiency' factor, K, which reflects the *effective increase in resolution*. A choice of σ of about 6–10 and of K of 2–2.5 was found to be satisfactory in most cases. It must be strongly emphasised that an improper selection of σ and/or K can lead to serious errors and artefacts. The spectra are subsequently fitted with Gaussian components (see Chapter A3) by means of a computer program.

Fitting deconvoluted spectra

Observed IR spectra result from the *convolution* (see Chapter A3) of the sample absorption spectrum with an instrumental resolution function (Comment E1.7). An important analytical approach in IR spectroscopy involves curve fitting of deconvoluted amide bands by iterative adjustment of component band positions, heights and relative weights (fraction of the total peak area). Component bands are assigned to different types of secondary structure (α-helices, β-sheets, turns, and irregular loops), and their fractional areas to the fractions of the corresponding secondary structure in the sample protein.

Figure E1.27 demonstrates that the amide I spectra of α-chymotrypsin can be fitted by components with easily measurable areas. Authors call the process deconvolution, but in actual fact it is a decomposition into component spectra (see Section C2.4.6). The bands close to 1627, 1637 and 1674 cm^{-1} are associated with β-strands, the band near 1653 cm^{-1} with the α-helix and the bands close to 1687, 1681, and 1665 cm^{-1} with turns.

The amide III spectra of the same protein is shown in Fig. E1.28(a). The spectral analysis revealed that the 1323, 1311 and 1304 cm^{-1} bands are associated with the α-helix, the bands near 1248, 1234 and 1222 cm^{-1} with β-sheets, and the 1290, 1279, and 1260 cm^{-1} bands with other structures in native α-chymotrypsin.

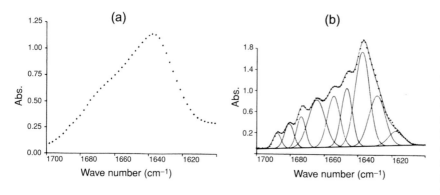

Fig. E1.27 Amide I bands of α-chymotrypsin measured at 5% w/v in D$_2$O, pathlength 0.0075 cm: (a) digitized original FTIR spectrum (crosses); (b) deconvoluted spectrum (crosses) and individual Gaussian components (solid lines) (deconvolution constants: σ = 6.5 cm^{-1}, K = 2.4). The solid overall curve is obtained by summation of the components. (Susi and Byler, 1986.)

The usefulness of FTIR spectroscopy as a tool for protein denaturation studies is illustrated in Fig. E1.28(b). Treatment of α-chymotrypsin with guanidine hydrochloride induced dramatic changes in the spectrum. The bands above 1300 cm^{-1} (most likely corresponding to α-helix) moved to lower wave numbers and their intensities decreased, whereas the intensity of bands between 1248 and 1270 cm^{-1} (most likely corresponding to random coils) increased sharply. The band at 1248–1249 cm^{-1} increased in strength from 20% to 38%. This indicates that this band represented a borderline assignment between a coil and a β-sheet. The native protein bands at 1290 and 1279 cm^{-1} decreased from 20% to 11%, after treatment with 6 M guanidine hydrochloride and the 1279 band shifted to 1282 cm^{-1}. The spectral range was, therefore, tentatively assigned to a β-turn.

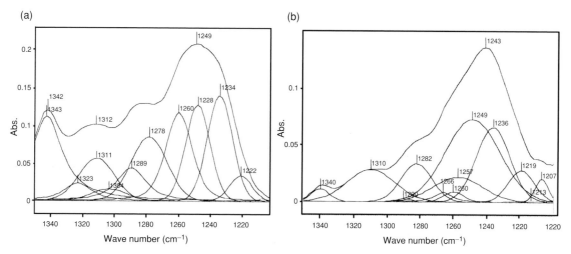

Fig. E1.28 Curve fitting analysis of FTIR spectra of α-chymotrypsin in the amide III region: (a) native state, concentration 1 mg ml^{-1} in H$_2$O buffer; (b) treated with 6 M guanidine hydrochloride. Deconvolution constants: σ = 20 cm^{-1}, K = 2.0. (After Cai and Singh, 1999.)

Comment E1.8
Water absorption

Unfortunately, water
absorbs IR light just
where the most
interesting groups
found in biological
macromolecules also
absorb. To some
extent, this difficulty
can be overcome by
using dried films of the
macromolecule or by
working in D_2O
solution, in which
solvent absorption is
shifted to less
interesting regions
around 1200 and 2500
cm^{-1}. The absorption
of water can also be
brought within
tolerable limits with
the high accuracy of
modern FTIR
instruments by using
cells of 6 μm path
length.

Comment E1.9
Readers can find a
critical review of the
methods evaluating the
amide I band in the
literature (Surewicz
et al., 1993).

E1.4.7 From amide bands to protein secondary structure

Amide I, amide II, and amide III

The amide I mode ($1600–1700$ cm^{-1}) of the peptide group is the most widely used band in studies of protein secondary structure. Due to the strong absorption of water (H_2O) between $1640–1650$ cm^{-1}, most structure determinations by amide I are performed in D_2O solution (Comment E1.8). However, uncertainty in the NH/ND exchange process may cause a certain degree of ambiguity. Also the serious overlap of the random coil and the α-helix band in the amide I region makes it difficult to accurately predict the α-helix contents in a protein (Comment E1.9).

Even though the intensity in the amide II region (1510 and 1570 cm^{-1}) is relatively strong, it is not very sensitive to secondary structure changes in proteins. Furthermore, the amide II bands strongly overlap with bands originating from amino acid side-chain vibrations.

Bands in the amide III spectral region ($1350–1200$ cm^{-1}) are predominantly due to the in-phase combination of N–H in-plane bending and C–N stretching vibrations and are highly sensitive to the secondary structure. In contrast to amide I band, there is no H_2O interference in this region. And even though the signals of these bands are $\sim(5–10)$-fold weaker than those of amide I, they are easily resolved and better defined making the amide III spectral region quite suitable for quantitative analysis of protein secondary structure. The assignment of spectral bands in the amide III region is as follows: $1330–1295$ cm^{-1}, α-helix; $1295–1270$ cm^{-1}, β-turns; $1270–1250$ cm^{-1}, random coils; and $1250–1220$ cm^{-1}, β-sheets.

Comparison with known structure

A number of closely related methods were proposed that avoid spectral deconvolution, and which are based on the use of a calibration matrix of IR spectra of proteins of known (i.e. determined by X-ray crystallography or NMR) secondary structure. This type of data treatment is conceptually very similar to that used in the analysis of circular dichroism spectra of proteins (Chapter E2).

Table E1.5 compares secondary structure estimates from FTIR spectroscopy using the amide I and amide III bands, from the analysis of CD spectra and from X-ray crystallography. The agreement between the FTIR results and the *structural* values derived from X-ray crystallography is quite good (Comment E1.10).

Finally we emphasise that the methodologies currently used for quantitative estimation of protein secondary structure content from IR spectra are not without serious shortcomings. Limitations in the quantitative assessment of protein secondary structure appear to be common to all low-resolution spectroscopic techniques, including CD. Since the potential sources of error in the CD and spectroscopic methods are different, these two techniques are highly complementary and should be used in conjunction.

Table E1.5. *Comparison of secondary structure determinations (%) by FTIR, using the amide I and amide III spectral regions, from the analysis of CD spectra and from X-ray crystallography data*

Protein	α-Helix	β-Sheet	β-Turn	Random coil	Method
α-Chymotrypsin	15	47	20	18	Amide III
	8	50	27	15	X-ray
	9	47	30	14	Amide I
	8–15	10–53	2–12	38–70	CD
Concanavalin A	18	49	25	8	Amide III
	3	60	22	15	X-ray
	8	58	26	8	Amide I
	3–25	41–49	15–27	9–36	CD
Cytochrome c	43	15	26	15	Amide III
	48	10	17	25	X-ray
	42	21	25	12	Amide I
	27–46	0–9	15–28	28–41	CD
Haemoglobin	65	4	11	20	Amide III
	87	0	7	6	X-ray
	78	12	10	–	Amide I
	68–75	1–4	15–20	9–16	CD
IgG	10	60	17	13	Amide III
	3	67	18	12	X-ray
	3	64	28	5	Amide I
Lysozyme	43	16	18	23	Amide III
	45	19	23	13	X-ray
	40	19	27	14	Amide I
Myoglobin	61	11	7	21	Amide III
	85	0	8	7	X-ray
	85	0	8	7	Amide I
	67–86	0–13	0–6	11–30	CD
Ribonuclease A	21	41	25	12	Amide III
	23	46	21	10	X-ray
	15	40	36	9	Amide I
	13–30	21–44	11–22	19–50	CD
Trypsin	14	49	23	13	Amide III
	9	56	24	11	X-ray
	9	44	38	9	Amide I

Partially from Bandecar (1992).

Comment E1.10
Percentage of helix or extended structure in a protein

One cannot emphasise too strongly that values for 'percentage helix' or 'percentage extended structure' are always somewhat subjective even when based on accurate bond lengths and angles. Ambiguity arises from the uncertainty in the choice of the exact point along the peptide chain where one segment of secondary structure begins and another ends. This choice depends not only on the manner in which an ideal helix or sheet has been defined by various investigators but also on just how regular any of these regions of protein substructure really are. With these uncertainties in mind, the agreement between the reported X-ray values, FTIR and CD results is particularly encouraging.

E1.4.8 IR difference spectroscopy

IR spectra are very sensitive to structural alterations as small as those involving a single atomic coordinate. A change in hydrogen bonding distance of only 0.002 Å, for example, shifts the frequency of the amide A (N–H stretch) vibration of a peptide group by approximately 1 cm^{-1}, well within the detection limits of most FTIR spectrometers. In contrast, X-ray crystallography studies of proteins rarely have a resolution below 1 Å. However, it is very difficult, in practice, to detect small, localised structural changes in a biological macromolecule, by IR spectroscopy. All groups in the molecule essentially have IR-active vibrations and, therefore, tend to give rise to a multitude of overlapping spectral bands. For example, bacteriorhodopsin, a 248-amino-acid (26 kDa) membrane protein containing an all-*trans*-retinal prosthetic group (see Chapter A3), is made up of about 1900 heavy (non-hydrogen) atoms that display about 5700 different skeletal vibrational modes.

A simple way around this problem is based on the principle of difference spectroscopy. Figure E1.29 illustrates bacteriorhodopsin undergoing a structural change from state A to state B. The IR spectrum of the protein in state A (top) consists of the sum of contributions from all of the skeletal vibrations. In contrast, the absorbance difference spectrum, computed by subtracting the spectrum recorded in state A from that in state B, shows signals only from those regions of the protein that have undergone a structural change. Since the protein undergoes only a very small structural change as shown, the intensity of peaks in the difference spectrum is quite small.

Genetic engineering is a powerful approach to the assignment of IR bands from a protein. Mutants should show simple difference spectra with peaks corresponding to the changed residues. As always with such studies, however, the interpretation should be done carefully because mutations can have structural effects beyond the target amino acid residue.

Fig. E1.29 The basic principles of FTIR difference spectroscopy applied to membrane proteins. Right: Simplified structure of the active site of a membrane protein (in this case, bacteriorhodopsin), in two different states involved in its physiological reaction (in this case, light-induced proton pumping. Left: The spectrum of the IR absorbance in state A (top), the spectrum of the difference in IR absorbance between states A and B (middle), and the difference spectrum on an enlarged scale (bottom). (Braiman and Rothschild, 1988.)

E1.4.9 Time-resolved IR spectroscopy

Vibrational spectroscopy experiments have an extremely fast intrinsic time resolution, and, in general, changes in the state of chemical bonds can be detected in a period as short as their characteristic vibrational periods, i.e. 10–100 fs.

Photobiological systems

Photobiological systems such as chromoproteins (from the Greek *chroma*, colour) constitute excellent samples for FTIR difference spectroscopy studies, because light is an ideal reaction trigger. In the simplest approach, difference spectra are formed between an initial state before illumination and a final state after illumination. In most cases, however, the photoreaction passes through a number of intermediates.

Studies of the yellow-light photoreceptor PYP (Fig. E1.30) provide excellent examples of how the IR technique can contribute to understanding the processes involved in phototransduction. Upon absorption of a photon by its covalently attached chromophore, PYP undergoes a cyclic sequence of reactions (Fig. E1.31). The ground state P (λ_{max} of 446 nm) is converted into intermediate I_1 (λ_{max} of 465 nm) in ~3 ns. I_1 is then converted into the long-lived blue-shifted intermediate I_2 (λ_{max} of 350 nm), which returns to the ground state. Global fit analysis of the absorbance changes yields four apparent time constants, $\tau_1 = 113$ μs, $\tau_2 = 1.5$ ms, $\tau_1 = 189$ ms, $\tau_4 = 583$ ms.

A three-dimensional representation of the light-induced IR absorbance changes during the photocycle of PYP, measured simultaneously from 1900 to 1000 cm^{-1} is shown in Fig. E1.32. The 30-ns time resolution permitted the monitoring of the light cycle reaction starting from intermediate I_1 and proceeding via the intermediate I_2 back to the ground state P.

Fig. E1.30 Overall structure of the photo-active yellow protein (PYP). PYP is a relatively small protein (14 kDA) containing *para*-hydroxy-cinnamate (pCA) as a prosthetic group bound to Cys 69 via a thioester linkage. The secondary structure is represented as a ribbon (α-helices in red and β-sheets in green) and the molecular surface of PYP is in the background. The chromophore is shown in orange. (Adapted from Heberle and Gensch, 2001.)

E1.4.10 DNA conformation

Nucleic acid vibrations arise from different parts of the macromolecule, resulting in IR absorption bands in several spectral regions. The main vibrations are observed between 1800 and 700 cm^{-1}. Four domains can be distinguished in this region: absorption bands due to the stretching vibrations of double bonds in the base planes are between 1800 and 1500 cm^{-1}; between 1500 and 1250 cm^{-1} we find the bands that are particularly dependent on the glycosidic torsion angle in the base sugars; further strong sugar and phosphate absorption bands occur between 1250 and 1000 cm^{-1}; below 1000 cm^{-1}, there are bands due to the vibrations of the phosphodiester main chain coupled to vibrations of the sugar groups.

A DNA FTIR spectrum recorded in H_2O solution is presented in Fig. E1.33(a). The enlarged area shows the spectral domain corresponding to the vibrations of

Fig. E1.31 Photocycle of PYP. The chemical structure of the chromophore is depicted in the centre. (After Heberle and Gensch, 2001.)

Fig. E1.32
Three-dimensional representation of the IR absorbance changes in PYP. The absorbance changes are shown as a function of wave number (1900–1000 cm^{-1}) and time on a logarithmic scale (50 ns–5 s). The first difference spectrum, recorded 50 ns after the laser flash excitation, is highlighted in black and represents an $I_1 - P$ difference spectrum. The subsequent difference spectra show the reaction from I_1 via I_2 back to the ground state P. Vibrations that characterise changes of specific functional groups during the photocycle were selected for kinetic analysis. Thus, the C=O stretching vibrations of Glu 46 were followed at 1759 cm^{-1} and 1739 cm^{-1}. The coupled C–C/C=C stretching vibration at 1575 cm^{-1} monitors the protonated *cis*-chromophore, etcetera. (Adapted from Brudler *et al.*, 2001.)

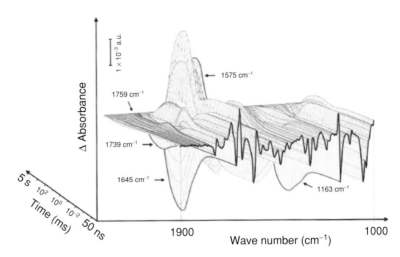

the phosphate groups. We observe an important contribution of the solvent H_2O, which makes it difficult to obtain good data in the region around 1600 cm^{-1} and below 1000 cm^{-1}. To avoid this problem, the spectrum of the same molecule is recorded in D_2O solution (Fig. E1.33(b)). Using H_2O/D_2O solvent combinations makes it possible to examine the complete spectral range necessary for nucleic acid studies.

Different types of information on DNA structure can be obtained by IR spectroscopy (Comment E1.11). For example, it is possible to characterise the molecular geometry by using what is usually called 'IR marker bands', which are conformation sensitive. Table E1.6 lists the most prominent IR marker bands of DNA double-helical structures (A, B and Z conformations). These bands demonstrate the existence of a conformational transition when factors such as temperature, hydration, concentration, counterion type and concentration are varied (B → A, B → Z, helix → coil transitions etc). In the case of conformation mixtures,

---

Comment E1.11

For detailed discussions of nucleic acid spectra–structure correlations, established on the basis of known three-dimensional X-ray structures of oligonucleotide single crystals, interested readers are referred to the review of Thomas and Tsuboi (1993). Another type of information which can be obtained from the FTIR spectra concerns recognition of DNA sequences by a wide variety of molecules, such as oligonucleotides (triple-helical structures), drugs and proteins (Taillandier and Liquier, 1992).

Table E1.6. *Main IR marker bands (in cm^{-1}) of A-, B- and Z-DNA^a*

Conformation			Assignment
A	B	Z	
1705	1715	1695	In-plane base double stretching
		1433	A-T bases
1418	1425	1408	Deoxyribose
1375	1375		dA•dG anti
		1355	dA•dG syn
1335	1344		dT
1335	1328		dA
		1320	dG
1275	1281		T
		1265	G
1240	1225	1215	Antisymmetric phosphate stretching
1188			Deoxyribose
		1065	Deoxyribose

^{a}Data from Taillandier and Liquier (1992).

computer simulations based on the observed marker bands and 'pure form' spectra are used to evaluate quantitatively the amount of each geometric form in the sample. FTIR specra also contain useful information in the recognition of DNA sequences by a wide variety of molecules, such as oligonucleotides (triple-helical structures), drugs and protein samples.

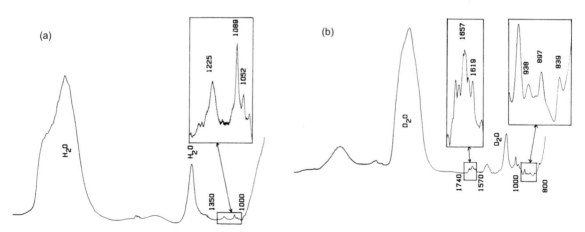

Fig. E1.33 FTIR spectra of DNA in solution: (a) in H$_2$O solvent; (b) in D$_2$O solvent. Enlarged parts of the spectra show the absorptions due mainly to the vibrations of the phosphate groups (top), the double bonds of the bases in their plane (bottom left), and the sugar-phosphate backbone (bottom right). (After Taillandier and Liquier, 1992.)

E1.5 Checklist of key ideas

- In an absorption spectroscopy experiment, a beam of electromagnetic radiation is observed after it passes through a sample to see how much of it has been absorbed at each wavelength.
- The measurement is made with a spectrophotometer.
- Radiation energy is absorbed in order to allow transitions between quantum mechanical electronic states of well-defined energies in the sample molecules. These can be seen as oscillating electric dipoles, induced by the oscillating electric field of the wave.
- Near-UV and visible radiation energies correspond to transitions between electronic orbital states, and can be used to probe the different types of bonds in molecules.
- The range of lower energies of IR radiation includes thermal energy at ambient temperature and can be used to study vibrational states and molecular dynamics.
- Absorption as a function of wavelength is characterised by a dimensionless *absorbance* or *optical density* value, or by a *molar extinction coefficient* in units of $(\text{mole litre}^{-1})^{-1}$ cm^{-1}, or $M^{-1}\,\text{cm}^{-1}$.
- UV–visible absorption spectra are usually modelled in terms of single electronic bands, without taking line shape into account. But there is significant line broadening due to lower-energy transitions between vibrational and rotational states, an inhomogeneous environment and coupling with solvent molecules.
- For a two-state reaction, in which each state has a specific absorption spectrum, the kinetic plot of the absorption spectra displays an *isobestic* point, at which the different spectra cross.
- Difference spectra are very useful in the study of processes characterised by spectral changes.
- The detection sensitivity of small spectral changes can be amplified by *derivative spectroscopy*.
- The absorption of proteins in the UV arises mainly from electronic bands in the peptide group (at 170–220 nm), aromatic amino acid side-chains close to 280 nm, and prosthetic group cofactors, enzyme substrates or inhibitors.
- Amino acid residues do not absorb in the visible and colour in a protein is always due to a prosthetic group, such as the haem which gives haemoglobin its red colour.
- Protein concentration measurements are usually performed by measuring the absorbance of the solution at wavelengths close to 280 nm. Such a determination provides only a *relative* concentration value, however, unless the molar extinction coefficient of the specific protein has been determined independently.
- Protein-associated chromophores can be as simple as a single metal ion that changes its colour according to its oxidation state, or a small organic molecule, such as retinal in rhodopsin, or quite complex, such as metal clusters, iron–sulphur clusters, porphyrin associated to iron in haem, or to magnesium in chlorophyll, or flavin and flavin derivative groups.
- In flash photolysis experiments on myoglobin, the Fe–CO bond is photodissociated by a laser flash and the rebinding kinetics is then followed by observing absorption changes with nanosecond time resolution.

- Upon absorption of a photon, the retinal binding membrane protein bacteriorhodopsin displays a millisecond photocycle of absorption changes associated with proton pump activity. One proton is pumped out of the cell for each photon absorbed.
- Nucleic acids can be identified by their strong absorption close to 260 nm with molar extinction coefficients of the order of 10^4 M^{-1} cm^{-1}.
- The relative positions of atoms in a molecule fluctuate continuously as a consequence of a multitude of different types of *vibrations and rotations* about their bonds.
- A *stretching* vibration involves changes in the length of an interatomic bond along its axis. *Bending* vibrations are characterised by a change in the angle between two bonds.
- For a simple diatomic or triatomic molecule, it is easy to define the number and nature of stretching and bending vibrations and to relate these to absorption energies; such an analysis becomes prohibitively difficult for larger molecules.
- A stretching vibration deviates from harmonic behaviour by varying degrees, depending upon the nature of the bond and the atoms involved.
- In order to absorb IR radiation, a molecule must undergo a net change in dipole moment as a consequence of its vibrational or rotational motion.
- The IR spectra of polypeptides and proteins exhibit several relatively strong absorption bands associated with the CONH (amide) group, which change little in frequency and intensity from one molecule to another.
- Of all the amide modes of the peptide group, the single most widely used one in studies of protein secondary structure is the *amide I* mode (1600–1700 cm^{-1}, stretching vibration of the C=O bond); however, due to the strong absorption of water between 1640–1650 cm^{-1}, most structure determinations based on the amide I mode are performed in D_2O solution.
- The intensity of bands in the *amide II* region (1480–1575 cm^{-1}) is relatively strong, but it is not very sensitive to protein secondary structure changes; furthermore, the amide II bands are strongly overlapped by bands originating from amino acid side-chain vibrations.
- The *amide III* bands (1200–1350 cm^{-1}) are predominantly due to the in-phase combination of N–H in-plane bending and C–N stretching vibrations and highly sensitive to protein secondary structure; in contrast to the amide I band, there is no H_2O interference in this spectral region.
- Despite a well-recognised conformational sensitivity of protein IR bands, the analysis of the spectra in the terms of protein structure is not straightforward and presents serious conceptual and practical problems.
- The estimation of secondary structural elements using amide I and amide III modes correlates quite well with secondary structure estimations from X-ray crystallography.
- Methods currently used to extract information on protein secondary structure from IR spectra are based on empirical correlation between the frequencies of certain vibrational modes and types of secondary structure of polypeptide chains such as α-helix, β-sheet, β-turn and random coil.
- IR difference spectra of different enzyme states contain bands only of those groups which undergo changes during the transition from one state to the other. Difference

spectra are, therefore, significantly simpler to interpret than spectra corresponding to the enzyme states. IR difference spectra of protein mutants can be used to assign bands to particular amino acid residues.

• Time-resolved IR spectroscopy is the principal technique for the study of reaction dynamics.

Suggestions for further reading

Historical review

Seibert, F. B. (1995). Infrared spectroscopy applied to biochemical and biological problems. *Meth. Enzymol.*, **246**, 501–526.

Chalmers, J. M., and Griffiths, P. R. (eds.) (2002). *Handbook of Vibrational Spectroscopy*, 5 volumes. Chichester: John Wiley and Sons Ltd.

Gauglitz, G., and Vo-Dinh, T. (eds.) (2003). *Handbook of Spectroscopy*, 2 volumes. Weinheim: Wiley-VCH Verlag GmbH KgaA.

Jakson, M., Sowa, M. G., and Mantsch, H. H. (1997). Infrared spectroscopy: a new frontier in medicine. *Biophys. Chem.*, **68**, 109–125.

Tinoco, I., Jr. (1995). Optical spectroscopy: General principles and overview. *Meth. Enzymol.*, **246**, 13–18.

Brief theoretical outline

Taillandier, E., Firon, M., and Liquier, J. (1991). In *Spectroscopy of Biological Macromolecules*, eds. R. E. Hester and R. B. Girling. Cambridge: Royal Society of Chemistry.

Mathies, R. (1995). Biomolecular vibrational spectroscopy. *Meth. Enzymol.*, **246**, 377–389.

Sauer K. (1995). Why spectroscopy? Which spectroscopy? *Meth. Enzymol.*, **246**, 1–10.

Tinoco, I. Jr, Sauer, K., and Wang, J. C. (1998). Molecular structure and interactions: physical chemistry. In *Principles and Applications in Biological Science*. Ch. 10. Molecular structures and interactions: Spectroscopy. Upper Saddle River, NJ: Prentice Hall.

The UV–visible spectral range

Miura, T., and Thomas, G. J. Jr, (1995). Optical and vibrational spectroscopic methods. In *Introduction to Biophysical Methods for Protein and Nucleic Acid Research*, eds. J. A. Glasel and M. P. Deutcher. London: Academic Press.

Miyazawa, T., and Blout, E. R. (1961). The infrared spectra of polypeptides in various conformations: Amide I and II bands. *JACS*, **83**, 712–719.

Heberle, J., and Gensch, Th. (2001). When FT-IR spectroscopy meets X-ray crystallography. *Nature Struct. Biol.*, **8**, 195–197.

Surevicz, W. K., Mantsch, H. H., and Chapman, D. (1993). Determination of protein secondary structure by Fourier transform infrared spectroscopy: a critical assessment. *Biochemistry*, **32**, 389–394.

Susi, H., and Byler, D. M. (1986). Resolution-enhanced Fourier transform infrared spectroscopy of enzymes. *Meth. Enzymol.* **130**, 290–311.

Braiman, M. S., and Rotschild, K. J. (1988). Fourier transform infrared techniques for probing membrane protein structure. *Ann. Rev. Biophys. Chem.*, **17**, 541–570.

Bandecar, J. (1992). Amide modes and protein conformation. *Biochim. Biophys. Acta.* **1120**, 123–143.

Arrondo, J. L. R., Muga, A., Castresana, J., and Goni, F. M. (1993). Quantitative studies of the structure of proteins in solution by Fourier-transform infrared spectroscopy. *Progr. Biophys. Mol. Biol.*, **59**, 23–56.

Mantele, W. (1993). Reaction-induced infrared difference spectroscopy for the study of protein function and reaction mechanisms. *TIBS*, **18**, 197–202.

Martin, J-L., and Vos, M. H. (1992). Femtosecond biology. *Annu. Rev. Biophys. Biomol. Struct.*, **21**, 199–222.

Gevert, K. (1993). Molecular reaction mechanisms of proteins as monitored by time-resolved FTIR spectroscopy. *Curr. Opin. Struct. Biol.*, **3**, 769–773.

Taillandier E., and Liquier J. (1992). Infrared spectroscopy of DNA. *Meth. Enzymol.*, **211**, 307–352.

Chapter E2
Two-dimensional IR spectroscopy

The present chapter is included in Part E with the other optical spectroscopy methods; however, the development of two-dimensional IR (2D-IR) spectroscopy is strongly based on two-dimensional NMR, and it is easier to understand after reading the relevant sections in Part J, which the reader is strongly encouraged to do first.

E2.1 Historical review and introduction to biological problems

1950

O. Hann proposed coherent spectroscopy – the use of radiation fields with well-defined phase properties – to extract information about atoms and molecules. The 'spin echo' experiment in nuclear magnetic resonance was the first demonstration of the possibilities of coherent spectroscopy.

1957

R. P. Feynman, **F. L. Vernon Jr.** and **R. W. Hellwarth** published a landmark paper pointing out that if coherent light fields were ever created, it would be possible to use these same methods on optical transitions. The invention of the laser in **1960** was followed quickly by a demonstration of the 'photon echo' – the optical version of the 'spin echo'.

1998

R. M. Hochstrasser and collaborators proposed 2D IR spectroscopy, in analogy with two-dimensional NMR, for the determination of time-evolving structures. The spins associated with the different nuclei in NMR are replaced in the IR experiments by a network of vibrational modes whose coupling can be used to determine molecular structure and dynamics. Structures of dipeptides, tripeptides and pentapeptides were determined by 2D IR spectroscopy. The most exciting aspect of 2D IR spectroscopy, however, is the combination of its sensitivity to structure with time resolution.

Conventional FTIR spectroscopy can monitor protein secondary structure features on the nanosecond–second timescale. Pulsed IR methods can probe vibrations on time scales down to the femtosecond.

E2.2 Linear and multidimensional spectroscopy

Linear spectroscopy, such as visible or IR absorption, Raman scattering and one-dimensional NMR, produces one-dimensional projections of electronic and nuclear interactions onto a single frequency (or time) axis. For simple molecules, direct information can be obtained about energy levels and absorption cross-sections. The situation is very different for complex molecules, with strongly overlapping levels. Here, the microscopic information is highly averaged, and is often totally buried under broad, featureless line shapes, whose precise interpretation often remains a mystery. Multiple pulse techniques have the capacity to induce electronic and vibrational degrees of freedom into non-equilibrium states, and to monitor their subsequent evolution, yielding ultrafast snapshots of dynamical events such as energy or charge transfer pathways, photoisomerisation and structural fluctuations.

E2.3 Principles of 2D-IR spectroscopy

E2.3.1 Pump probe experiments

Heterodyne two-pulse IR correlation spectroscopy (IR COSY) has the same measurement principles as NMR COSY (Chapter J2). In both techniques there is a preparation and a coherence transfer step (Fig. E2.1). In 2D IR spectroscopy, the first pulse (pump) creates a vibrational coherence analogous to the spin coherence created by the first radio-frequency pulse in NMR, but unlike the $\pi/2$ pulse of NMR, this interaction is weak. After some delay, a pair of weak IR pulses (probe pulses) interrupts the initial coherence evolution to generate a free induction decay (FID) from the transferred coherence. The FID electric field is the measured signal in both techniques. When the third IR pulse follows some time after, the pulse sequence is analogous to NMR NOESY or a stimulated spin echo. Like R. Hochstrasser, we refer to this technique as THIRSTY (three IR pulse stimulated echo spectroscopy). The variable delay between the second and third pulses allows time for vibrational coherence and population transfer to occur, analogous to spin transfer in NMR. The signals in the 2D IR experiments are created by the responses of the network of vibrational modes excited by the pulses (Comment E2.1). They can be spatially separated from the transmitted fields and independently measured by a reasonable choice of input beam directions (Fig. E2.1). Such spatial selection of phase is not possible in NMR because of the long wavelengths of the radio frequency fields.

The vibrational FID is measured as a function of the three time intervals and yields a three-dimensional time grid whose Fourier transforms give the IR

Comment E2.1 Pulsed laser techniques

Pulsed laser techniques have made giant strides forward in recent years with the result that complete control of optical and IR electric fields is now possible: the shapes, frequencies, timing and phases of laser pulses can now be manipulated in ways that were thought to be possible only for radio waves. The feasibility of impulsive excitations, which require laser pulses shorter than the corresponding nuclear motion periods, has opened up new possibilities for conducting multiple-pulse coherent measurements to disentangle complex electronic and nuclear motions.

Fig. E2.1 Typical pulse sequences used in (a) NMR spectroscopy and (b) heterodyned 2D-IR spectroscopy. The three pulses in both techniques are separated by time delays t_1 and t_2, and the FID is measured versus t_3. In heterodyned 2D-IR, the three IR laser beams (in colour) and the emitted FID signal beam all travel in different directions, and thus are spatially separated from one another. The electric field of the signal beam cannot be measured directly, as it is in NMR, so it is overlapped with a fourth local-oscillator beam that heterodynes the signal (not shown). When $t_2 = 0$, the 2D-IR spectrum is analogous to a COSY experiment, otherwise it is similar to NOESY. (After Zanni and Hochstrasser, 2001.)

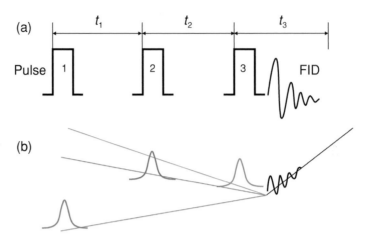

COSY and THIRSTY spectra. The diagonal peaks in these spectra correspond to the vibrational frequencies in the structure distribution, but the cross-peaks show up only if the modes are coupled. To obtain well-resolved 2D-IR spectra, the FID periods need only to be measured for a few tens of picoseconds, thus the method can be used to monitor the structures of intermediates in kinetic experiments.

The first 2D-IR experiments concerned the amide I mode of peptides, which are mainly C=O vibrators (see Chapter E1). In such cases, all of the relevant frequencies of an interacting ensemble of modes can be bracketed by the spectral bandwidth of 120-fs IR laser pulses. The two-dimensional correlation spectra yield the relevant structural and dynamical information (see below).

2D-IR spectroscopy experiments have been developed incorporating two pulses of different frequency (*dual-frequency 2D-IR spectroscopy*), one in the C=O bond (amide I) and the other in the N–H (amide II bond for peptide) region. They have been applied to measure the coupling and angular relation between C=O and N–H modes (Fig. E2.2).

In a dual-pulse experiment on N-methylacetatamide, for example, a *difference frequency generator* produces the first IR pulse of frequency $\omega_1 = 1675$ cm^{-1} propagating towards the sample in direction $\mathbf{k}_1$, and the *local oscillator* pulse, along $\mathbf{k}_{LO}$, required for the heterodyne technique (see Chapter D10). The output of the second difference frequency generator is split into two equal parts to generate two beams of frequency $\omega_2 = 1540$ cm^{-1} and vector directions $\mathbf{k}_2$ and $\mathbf{k}_3$ (Figs. E2.2 and E2.3).

E2.3.2 Selection rules for two-dimensional spectroscopy

A vibrational motion must cause a change in the permanent dipole moment for it to be IR active and revealed in a linear spectrum. This condition forms a selection

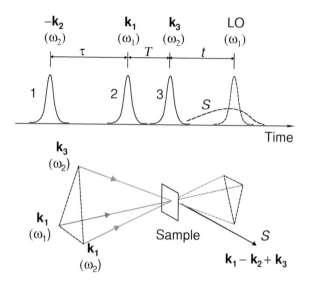

Fig. E2.2 The top panel shows the time sequence in a dual-frequency three-pulse photon-echo experiment. S (dashed line) is a typical signal envelope. LO denotes the local oscillator pulse. Three phase-locked IR pulses, labelled 1, 2 and 3, with wave vectors k_1, k_2 and k_3, respectively, are incident on the sample at time intervals τ, between 1 and 2, and T, between 2 and 3. The bottom panel shows the spatial arrangement of the three beams interacting with the sample. The signal envelope is observed in the $k_1 - k_2 + k_3$ direction. The IR beams were tuned such that one is centred on the amide I band (k_1, ω_1), and the other is close to the amide II band (k_2, and k_3, ω_2). (After Rubtsov et al., 2003.)

rule for linear spectroscopy. In order to observe a signal in a two-dimensional spectrum, a further consideration should be applied. It is well understood that the non-linear optical response from a set of harmonic oscillators vanishes due to destructive interference involving adjacent vibrational levels. Thus observing a non-linear signal requires a deviation from harmonic behaviour. The following conditions must be fulfilled: (1) anharmonicity in the ground state potential; (2) non-linear dependence of the transition dipole on vibrational coordinates; (3) energy-level-dependent dynamics. These conditions form the set of selection

Fig. E2.3 The linear spectrum on an optical density (OD) scale of NMA in DMSO. The laser spectra centered near ω_1 and ω_2 are shown also. LO, is the local oscillator pulse. (After Rubtsov et al., 2003.)

rules for 2D IR spectroscopy, and affect the positions, amplitudes and shapes of the various peaks in a two-dimensional spectrum.

E2.3.3 NMR and 2D-IR spectroscopy: similarity and difference

One of the major achievements of NMR is its capacity to disentangle hopelessly complicated spectra by spreading them over several frequencies or time axes. Carefully designed sequences of radio-frequency pulses induce coherence transfers that make it possible to expose coupling between the spins on different nuclei or to eliminate dipolar interactions which dominate the line widths in solid sate NMR (Part J). Magnetic resonance is the elder brother of coherent laser spectroscopy, and the two share many basic concepts. The ability to shape and control radio-frequency pulses predated similar advances in short pulse laser techniques by many decades, thus many of the optical and IR non-linear techniques have their analogues in early NMR developments.

The time-window and information content of electronic and vibrational multidimensional techniques are very different from NMR. For example, a sequence of IR pulses transfers coherence amongst the components of a network of vibrators. This network has much faster and different dynamical properties than the networks of spins that are interrogated by a sequence of radio-frequency pulses in NMR. The assumptions and theoretical techniques necessary to convert optical data into the relevant structural and dynamical information are fundamentally different from those invoked in NMR. The new peaks representing coupling between excitations, the distributions of structures in the equilibrium ensemble and the intensity of the signals and their line shapes present direct signatures of molecular structure through distance and angular relations between chromophores or vibrations. They also expose the dynamics of electronic or nuclear excitations through the spectral density of their local environments. Amongst systems already studied are aggregates of chromophores, hydrogen bonded complexes and liquids, the secondary and tertiary structure of polypeptides, and protein conformational dynamics. As with NMR, isotope labelling very usefully provides an extra dimension for peak assignment in the 2D IR spectra.

E2.4 From amide bands to protein tertiary structure

We recall that IR spectra of the amide transition provide information about secondary structural motifs in proteins and peptides (see Chapter E1). The amide I band, which consists mainly of the stretching motion of the peptide backbone C=O groups, is a strong IR absorber that is spectrally isolated from other vibrational modes, such as those of the amino acid side groups. Although the group spectrum varies slightly for different positions in a protein structure, the observed spectra are not resolved for proteins and the peaks together constitute a single amide I band.

E2.4.1 Simulations of 2D-IR spectroscopy

The sensitivity of 2D-IR spectroscopy to peptide structure is readily demonstrated by simulations, such as those shown in Fig. E2.4, which examine the coupling between the amide I vibration modes of a β-heptapeptide. The folded β-heptapeptide exhibits a characteristic 2D-IR spectrum of diagonal peaks and cross-peaks, whereas the spectrum of the partially unfolded peptide has a different distribution of couplings and frequencies (Comment E2.2).

E2.4.2 Determination of peptide structures

The first detailed experimental application of IR-COSY involved the acetylproline–NH$_2$ dipeptide and led to the determination of its structure. The peaks generated from the FIDs are shown in Fig. E2.5 for the peptide in

(a)   (b)

Ω_2 [cm^{-1}]

1660

1640

−1640   −1660   −1640   −1660 Ω_3 [cm^{-1}]

Fig. E2.4 Simulated 2D-IR spectra of a β-heptapeptide. The spectra were calculated from two instantaneous conformations during a molecular dynamic simulation when the peptide is: (a) folded and (b) partially unfolded. The simulations treated the peptide as a coupled set of vibrations and vibrational dynamics were deliberately omitted in order to produce narrow lines. There are clear signatures in the spectra for the structural changes between the folded and unfolded states. (Adapted from Zanni and Hochstrasser, 2001.)

Comment E2.2   2D-IR spectra and polypeptide structure

The distribution of polypeptide structures and their nuclear motions are reflected in the shapes and frequencies of the diagonal peaks and cross-peaks in a 2D-IR spectrum. The 2D-IR spectrum of a polypeptide depends on the interpeptide potential energy function and, in the simulation shown in Fig. E2.4 the potential was chosen as a dipole–dipole interaction. When a complete, more realistic distribution of structures is used, the sharp peaks in Fig. E2.4 become spread along the diagonal to form the elliptical profiles seen in the spectra. The cross-peaks are also broadened into shapes that signal an interaction between the peptide units. The minor axes of these ellipses do not depend on the structure distribution, however, but are determined by the dephasing of the vibrational modes.

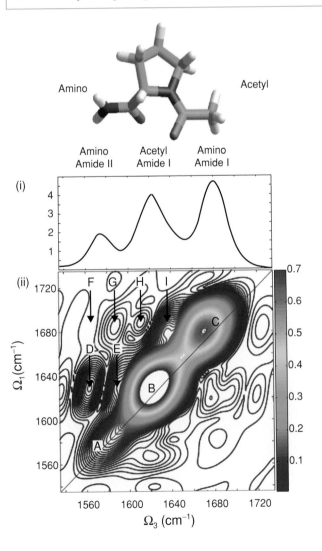

Fig. E2.5 The FTIR and heterodyne 2D-IR (IR-COSY) spectra of acetylproline–NH$_2$ (top) in chloroform. The three peaks in the linear spectrum (i) correspond to the three diagonal features A, B, C in the absolute magnitude spectra (ii). Peaks A and C are from amide I and II vibrational modes that are located on the amino end of the molecule, and peak B is from the amide I mode on the acetyl end. The cross-peaks are labeled D–I in (ii). (After Zanni and Hochstrasser, 2001.)

(a)

(b)

Fig. E2.6 The chemical
formula and (b) the
three-dimensional
structure of the cyclic
pentapeptide
(*cyclo*-Mamb–Abu–Arg–
Gly–Asp). The peptide,
which is stabilised by a
hydrogen bond between
the Mamb-1–Abu-2
peptide bond and the
Arg-3–Gly-4 peptide bond
was specifically designed
to form a single,
well-defined
conformation in solution
with an almost ideal
type II β-turn centred at
the Abu–Arg residues.
(Adapted from Hamm
et al., 1999.)

chloroform. Each vibrational mode produced a pair of peaks along the diagonal (A, B, C) that are separated by the degree of anharmonicity. Each pair of modes results in two pairs of cross-peaks. There are two cross-peaks between each of the acetyl bands and the amino band (cross-peaks F and G between peaks A and C, and cross-peaks H and I between peaks B and C), which indicates that acetylproline–NH_2 must adopt two conformations in chloroform.

A *de novo* designed peptide was the first system whose 2D-IR spectrum was understood quantitatively. The peptide is a cyclic pentapeptide (*cyclo*-Mamb–Abu–Arg–Gly–Asp) for which both the NMR and X-ray structures are known. Its chemical formula and three-dimensional structure are shown in Fig. E2.6. The structure was specifically designed to form a single, well-defined conformation in solution with an almost ideal type II β-turn. The linear absorption spectrum of the peptide in the region between 1570 cm^{-1} and 1680 cm^{-1} exhibits five partially resolved bands (Fig. E2.7(a)). A multiline Lorentzian fit (see Chapter A3) yields peak positions at 1673 cm^{-1}, 1648 cm^{-1}, 1620 cm^{-1}, 1610 cm^{-1}, 1584 cm^{-1}. There are no other strong IR bands between 1500 cm^{-1} and 2000 cm^{-1}. The 2D-IR spectra of the peptide are shown in Fig. E2.7(b) and (c). The time delay between pulses was set such that the broad frequency bandwidth probe pulse (second laser pulse) followed *c.* 800 fs after the fast-rising edge of the pump pulse (first laser pulse). Further information is obtained from the use of laser pulses with different polarization. The spectra for perpendicular and parallel pump and probe polarizations are shown in Fig. E3.7(b) and (c), respectively. The fits to the experimental data are shown in Fig. E3.7(d)–(f).

Fig. E2.7 (a) Absorption spectrum of the pentapeptide shown in Fig.E2.6 in D_2O. The dashed line shows a representative spectrum of the pump pulse used. (b) and (c) Two-dimensional pump-probe spectra of the peptide measured with the polarization of the probe pulse perpendicular and parallel to the polarization of the pump pulse, respectively. (d)–(f) Least-square fits of the experimental data. (After Hamm *et al.*, 1999.)

E2.5 Checklist of key ideas

- Two-pulse IR COSY has the same measurement principles as NMR COSY; however, the time-window and information content of electronic and vibrational multidimensional techniques are very different from NMR.
- The signals in the 2D-IR experiments are created by the responses of the network of vibrational modes excited by the two pulses and can be spatially separated from the transmitted fields and independently measured by a reasonable choice of input beam directions.
- In 2D-IR spectroscopy, the first pulse creates a vibrational coherence analogous to the spin coherence created by the first radio-frequency pulse in NMR, but unlike the $\pi/2$ pulse of NMR, this interaction is weak; after some delay, a pair of weak IR pulses (probe pulses) interrupts the initial coherence evolution to generate an FID from the transferred coherence.
- The three pulses in the 2D-IR and 2D-NMR techniques are separated by time delays t_1 and t_2, and the FID is measured versus t_3; in heterodyned 2D-IR spectroscopy, the three IR laser beams and the emitted FID signal beam all travel in different directions, and thus are spatially separated from one another.

- The electric field of the signal beam cannot be measured directly, as it is in NMR, so it is overlapped with a fourth local-oscillator beam that heterodynes the signal; when $t_2 = 0$, the 2D IR spectrum is analogous to a COSY experiment, otherwise it is simillar to NOESY.

- Multiple-pulse techniques have the capacity to induce electronic and vibrational degrees of freedom into non-equilibrium states, and to monitor their subsequent evolution, yielding ultrafast snapshots of dynamical events such as energy or charge transfer pathways, photoisomerisation and structural fluctuations.

- To date structures of dipeptides, tripeptides and pentapeptides have been determined by 2D-IR spectroscopy; the most exciting aspect of 2D IR spectroscopy, however, is the combination of its sensitivity to structure with time resolution.

- If conventional FTIR spectroscopy can monitor protein secondary structure features on the nanosecond to second time scale, then pulsed IR methods can probe vibrations on time scales down to the femtosecond.

- As with NMR, isotope labelling very usefully provides an extra dimension for peak assignment in the 2D IR spectra.

Suggestions for further reading

Historical review and introduction to biological problems

Zanni, M. T., and Hochstrasser, R. M. (2001). Two-dimensional infrared spectroscopy: a promising new method for the time resolution of structure. *Curr. Opin. Str. Biol.*, **11**, 516–522.

Linear and multidimensional spectroscopy

Multidimensional spectroscopies. *Chem. Phys.* (2001), **266**, 137–351. The complete issue of this journal was devoted to advances being made in optical and IR multidimensional spectroscopy.
Mukalmel, S. (1995). *Principles of Nonlinear Spectroscopy*. New York: Oxford University Press.

Principles of 2D-IR spectroscopy

Asplund, M. C., Zanni, M. T., and Hochstrasser, R. M. (2000). Two-dimensional infrared spectroscopy of peptides by phase controlled femtosecond vibrational photon echoes. *Proc. Natl Acad, Sci. USA*, **97**, 8219–8224.

From amide bands to protein tertiary structure

Hamm, P., Lim, M., DeGrado, W. F., and Hochstrasser R. M. (1999). The two-dimensional IR nonlinear spectroscopy of a cyclic penta-peptide in relation of its three-dimensional structure. *Proc. Natl. Acad. Sci. USA*, **96**, 2036–2041.

Zanni, M. T., Gnakaran, S., Stenger, J., and Hochstrasser, R. M. (2001). Two dimensional infrared spectroscopy of solvent dependent conformations of acetyleproline-NH$_2$. *J. Phys. Chem.*, **105**, 6520–6535.

Rubtsov, I. V., Wang, J., and Hochstrasser, R. M. (2003). Dual-frequency 2D-IR specctroscopy heterodyned photon echo of the peptide bond. *Proc. Natl. Acad. Sci. USA*, **100**, 5601–5606.

NMR and 2D-IR spectroscopy: similarity and difference

Keusters, D., Tan, H. S., and Warren, W. S. (1999). Role of pulse phase and direction in two dimensional optical spectroscopy. *J. Phys. Chem. A*, **103**, 10369–10380.

Chapter E3
Raman scattering spectroscopy

E3.1 Historical review and introduction to biological problems

1928

L. Mandelshtam and **C. V. Raman** independently discovered that a wavelength shift in a small fraction of scattered visible radiation depends on the chemical structure of the molecule responsible for the scattering. **Raman** was awarded the 1931 Nobel Prize in physics for the discovery of the phenomenon, which became known as Raman scattering or the Raman effect (Comment E3.1).

Raman scattering is *inelastic* light scattering that results from the same type of quantised vibrational transitions as those associated with IR absorption and occurs in the same spectral region. The Raman scattering and IR absorption spectra for a given molecule are often similar. Differences between the properties that make a chemical group IR or Raman active, however, make the techniques strongly complementary rather than competitive. Raman spectroscopy has the advantages of minimal or no damage to the sample, and relatively little interference from the water signal in aqueous solution or in vivo. The possibility of observing Raman spectra from crystals as well as from solutions in vitro or in vivo provides a very useful approach to relating molecular structures solved by X-ray crystallography and conformations in solution or in living cells.

Raman spectroscopy was not widely used by molecular biophysicists until lasers became available in the **1960s**. Interference from fluorescence excited in the molecule studied or in sample impurities was also a deterrent to the general use of the method. This problem has now been largely overcome by the use of near-IR laser sources.

> **Comment E3.1**
> Mandelshtam–Raman phenomenon
>
> In Russian scientific literature the effect is called the Mandelshtam–Raman phenomenon. We follow the more general usage and call it the Raman effect.

E3.2 Classical Raman spectroscopy

Classical or *non-resonance* Raman spectroscopy probes radiation scattering by vibrational modes, which have energies in the IR spectral region. The method has proven very useful in molecular biophysics for the determination of protein, nucleic acid and membrane conformational states.

Resonance Raman spectroscopy is discussed in Section E3.3. The resonance effect is between the scattered radiation and *electronic transitions* in the sample molecules. Since these transitions occur in the visible spectral region, which is also readily accessible by the conventional lasers commonly used for Raman spectroscopy, the method has provided considerable information on visual pigments, haem and other metalloproteins. The most studied chromophores are non-fluorescent so that the relatively weak Raman signals can be observed without contamination from broad fluorescence emission.

E3.2.1 Raman spectra

Figure E3.1 shows a portion of a Raman spectrum obtained by irradiating a sample of carbon tetrachloride with an intense beam from an argon ion laser (488.0 nm wavelength, corresponding to 20 492 cm^{-1} wave numbers). The emitted radiation is of three types: Stokes' scattering, anti-Stokes' scattering and Rayleigh scattering. The last, which corresponds to *elastic* scattering of wavelength exactly equal to that of the excitation source, is significantly more intense. The Stokes' and anti-Stokes' lines are *inelastic* scattering corresponding to quantum energy given to or taken away from the sample by the light beam, respectively. The *x*-axis in Fig. E3.1 is the wave number shift $\Delta\bar{v}$, defined as the difference in wave numbers (cm^{-1}) between the observed and excitation radiation. Note the three Raman peaks in the spectrum are found on both sides of the Rayleigh peak, with identical shift magnitudes. Stokes lines are at 218, 314 and 459 cm^{-1}, wave numbers below the Rayleigh peak, while anti-Stokes' lines occur at 218, 314 and 459 cm^{-1} above it. Quite generally, anti-Stokes' lines are appreciably less intense than the corresponding Stokes lines, indicating a higher probability for the sample

Fig. E3.1 Raman spectra for CCl$_4$ excited by laser radiation of $\lambda_0 = 488$ nm and $\lambda_0 = 20\,492$ cm^{-1}. The number above the peaks is the Raman shift, $\Delta\bar{v}$ cm^{-1}. Raman shift magnitudes are independent of the excitation wavelength. Identical shift patterns would be observed for CCl$_4$ regardless of whether excitation was carried out with a krypton ion laser (488.0 nm), a helium–neon laser (632.8 nm), or a Nd:YAG laser (1064 nm). (Adapted from Skoog, *et al.*, 1995.)

to be excited by the beam rather than the other way around. In practice, only the more intense, Stokes portion of a spectrum is analysed.

The appearance of Raman spectral lines of lower energy (longer wavelengths) than the excitation beam is analogous to the Stokes' shifts found in a fluorescence experiment (Chapter D8); for this reason, Raman shifts to longer wavelengths were called *Stokes' shifts*. We see in Section E3.2.2, however, that Raman and fluorescence spectra arise from fundamentally different processes, so that the application of the same terminology to both may be misleading.

E3.2.2 Frequency, intensity and polarisation

Raman spectroscopy is a form of vibrational spectroscopy. Like IR spectroscopy it is related to transitions between vibrational energy levels in a molecule. It differs from IR spectroscopy in that information is derived from *light scattering* whereas in IR spectroscopy information is obtained from *absorption* processes.

Let us assume a light beam of frequency v_{exp} is incident upon a solution of molecules. The electric field E associated with the radiation is described by the wave equation (see Chapter E1)

$$E = E_0 \cos(2\pi v_{exp} t) \tag{E3.1}$$

where E_0 is the amplitude of wave. As the field propagates through the sample the electron charges respond to it. The dipole moment, m, induced in the electron cloud of a molecular bond by the oscillating electric field is given by

$$m = \alpha E = \alpha E_0 \cos(2\pi v_0 t) \tag{E3.2}$$

where α is the polarisability of the bond.

In order to be Raman active (i.e. to be active in inelastic light scattering), the vibration of the atoms joined by the bond must be reflected favourably by polarisability changes in the bond electrons,

$$\alpha = \alpha_0 + (r - r_{eq})\left(\frac{d\alpha}{dr}\right) \tag{E3.3}$$

where α_0 is the polarisibility of the bond at the equilibrium internuclear distance r_{eq} and r is the internuclear separation at any instant. The change in internuclear separation varies with the frequency of the vibration v_v as given by

$$r - r_{eq} = r_m \cos(2\pi v_v t) \tag{E3.4}$$

where r_m is the maximum internuclear separation relative to the equilibrium position. Substituting Eq. (E3.4) into Eq. (E3.3) gives

$$\alpha = \alpha_0 + \left(\frac{d\alpha}{dr}\right) r_m \cos(2\pi v_v t) \tag{E3.5}$$

We can then obtain an expression for the induced dipole moment m by substituting Eq. (E3.5) into Eq. (E3.2):

$$m = \alpha_0 E_0 \cos\left(2\pi \nu_0 t\right) + E_0 r_{\mathrm{m}} \left(\frac{\mathrm{d}\alpha}{\mathrm{d}r}\right) \cos\left(2\pi \nu_{\mathrm{v}} t\right) \cos\left(2\pi \nu_0 t\right) \tag{E3.6}$$

Recall from trigonometry that

$$2 \cos x \cos y = [\cos(x+y) + \cos(x-y)]$$

Applying this identity to Eq. (E3.6) gives

$$m = \alpha_0 E_0 \cos\left(2\pi \nu_0 t\right) + \frac{E_0}{2} r_{\mathrm{m}} \left(\frac{\mathrm{d}\alpha}{\mathrm{d}r}\right) \cos\left[2\pi(\nu_0 - \nu_{\mathrm{v}})t\right]$$

$$+ \frac{E_0}{2} r_{\mathrm{m}} \left(\frac{\mathrm{d}\alpha}{\mathrm{d}r}\right) \cos\left[2\pi(\nu_0 + \nu_{\mathrm{v}})t\right] \tag{E3.7}$$

The first term on the right-hand side of Eq. (E3.7) represents Rayleigh scattering at the excitation frequency ν_0. The second and third terms correspond, respectively, to the Stokes' and anti-Stokes' frequencies of $\nu_0 - \nu_{\mathrm{v}}$ and $\nu_0 + \nu_{\mathrm{v}}$. Here, the excitation frequency has been modulated by the vibrational frequency of the bond. It is important to note that Raman scattering requires that the polarisibility of the bond varies as a function of distance: i.e. $\mathrm{d}\alpha/\mathrm{d}r$ in Eq. (E3.7) must be greater than zero for Raman lines to appear.

The energy shifts observed for a given bond in Raman experiments should be identical to the energies of its IR absorption band, provided that the vibrational modes involved are active toward both IR absorption and Raman scattering. Figure E3.2 illustrates the similarity of the two types of spectra. Several peaks with identical $\bar{\nu}_0$ and $\Delta\bar{\nu}$ values exist for indene. However, the relative sizes of the corresponding peaks are frequently different, and certain peaks that occur in one spectrum are absent in the other.

The differences between a Raman and an IR spectrum are not surprising when it is considered that the basic mechanisms, although dependent upon the same vibrational modes, arise from processes that are mechanistically different. IR

Fig. E3.2 Comparison of Raman and IR spectra for indene. (Adapted from Skoog *et al.*, 1995.)

absorption is observed for vibrational modes which change the *dipole moment* of the molecule, while Raman scattering is associated with modes that produce a change in the *polarisability* of the molecule. Because of this fundamental difference in mechanism, the Raman activity of a given vibrational mode may differ markedly from its IR activity. For example, a homonuclear molecule such as nitrogen has no dipole moment either in its equilibrium position or when a stretching vibration causes a change in the distance between the two nuclei. Thus, absorption of radiation corresponding to the vibration frequency cannot occur. On the other hand, the polarisability of the bond between the two atoms varies periodically in phase with the stretching vibrations, reaching a maximum at the greatest separation and a minimum at the closest approach. A Raman shift is is therefore observed at the frequency of the vibrational mode.

It is interesting to compare the IR and Raman activities of coupled vibrational modes such as those described for the carbon dioxide molecule (Section E3.3.1). In the symmetric mode, no change in the dipole moment occurs as the two oxygen atoms move away from or toward the central carbon atom; thus, this mode is IR-inactive (Fig. E3.3). The polarisability, however, fluctuates in phase with the vibration since distortion of the bonds becomes easier as they lengthen and more difficult as they shorten. Raman activity is associated with this mode. In contrast, the dipole moment of carbon dioxide fluctuates in phase with the antisymmetric vibrational mode. Thus, an IR absorption peak arises from this mode. On the other hand, as the polarisability of one of the bonds increases as it lengthens, the

Vibrational motion in O=C=O molecule

Fig. E3.3 Selection rules for IR and Raman spectra using the carbon dioxide molecule as an example. For vibration to be IR-active, the dipole moment of the molecule must change. Therefore, the symmetric stretch in carbon dioxide is not IR-active because there is no change in the dipole moment. The asymmetric stretch is IR-active due to a change in dipole moment. For vibration to be Raman active, the polarisability of the molecule must change with the vibration motion (Altose *et al.*, 2001.)

polarisibility of the other decreases resulting in no net change in the polarisibility. Thus, the asymmetric stretching vibration is Raman-inactive.

As in the example above, parts of Raman and IR spectra can be complementary, each associated with a different set of vibrational modes within a molecule. Other vibrational modes may be both Raman- and IR-active. For example, all of the vibrational modes of sulphur dioxide yield both Raman and IR peaks. Peak heights differ, however, because the probability for the transitions is different for the two mechanisms.

E3.2.3 Raman spectrometers and Raman microscopes

In order to obtain the Raman spectrum of a protein or nucleic acid, one places the sample at the focal point of a focused laser beam. The scattered light is collected by a lens system and directed through a suitable monochromator to a photon detector. The Raman spectrum consists of a plot of the scattered intensity as a function of its frequency difference with the incident laser frequency. A band at a particular frequency in the Raman spectrum corresponds to a vibrational mode of that frequency. It is a fundamental principle of the Raman effect that the spectrum is independent of the incident laser frequency.

Three types of Raman spectrometer are in general use for the characterisation of protein or nucleic acid structure: (1) the classic laser Raman spectrometer, (2) the laser Raman microscope, and (3) the UV resonance spectrometer (Comment E3.2). The classic spectrometer uses a wavelength in the visible spectral region, and has the advantage that it puts no constraints on the sample, which may be a protein or nucleic acid powder, crystal, fibre or solution. Because these samples are transparent to visible light, there is little or no heating of the sample by the laser beam. Classical Raman spectra of protein or nucleic acid solutions have been obtained from 5–50 μl samples in a glass capillary mounted in a circulating water temperature controlled copper block, with 'tunnels' for the incident and scattered beams. The scattered light is collected at right angles to the incident beam. The concentration of the solution is typically 2.0–20 mg ml^{-1}, or about 20–100 mM in nucleotide or peptide groups.

A powerful advantage of Raman microscopy is that it allows a single protein crystal to be studied in situ in a hanging drop (Fig. E3.4). The Raman spectrum recorded from the crystal is caused only by the focal volume of the laser beam within the crystal, which is typically ~20 μm in diameter and 50 μm in depth. In this volume, the protein concentration is about 50 mM, thus providing very high quality data. Any Raman scattering collected from the surrounding mother liquor makes a negligible contribution because protein concentrations there are less than 1 mM. Thus, Raman microscopy can be used to screen the properties of protein crystals in a hanging drop.

Comment E3.2

Resonance Raman in the UV

In the resonance Raman effect (Section E3.3), the frequency of the incident light falls within the absorption band of an excited electronic state. In the UV region this is usually accompanied by photochemical decomposition, which can be minimised by using a circulating sample. This requires large amounts of material and makes measurements on single crystals and living cells impossible.

Fig. E3.4 Schematic diagram (not to scale) of a Raman microscope. Both video (1) and spectrographic data sets (2) can be displayed on the computer screen. Inset: The arrangement for observing Raman data from a single protein crystal in a hanging drop. (Adapted from Altose et al., 2001.)

In general, Raman spectroscopy and Raman microscopy provide direct comparison of the properties of a molecule in solution and in a single crystal.

E3.2.4 Protein secondary structure from Raman spectra

The different conformations of polypeptide chains have remarkable differences in their Raman spectra. Table E3.1 presents a summary of some of the Raman active modes that are useful for obtaining information about the structure and dynamics of proteins. Of the modes listed in Table E3.1, one of those whose Raman spectrum is sensitive to conformation is the amide I mode. Thus, the amide I vibration changes from 1650 ± 5 cm^{-1} in the α-helical forms in H_2O to 1670 ± 3 cm^{-1} in the β-sheet form in H_2O, indicating that the carbonyl group is more strongly polarised, i.e. has a higher dipole moment in the α-helix than in the β-sheet. Hydrogen–deuterium exchange causes the frequency to decrease because of the increased mass of the deuteron. A frequency of 1632 ± 5 cm^{-1} was found for the α-helical form in D_2O, whereas a frequency of 1659 ± 3 cm^{-1} was found for the β-sheet form in D_2O.

In the estimation of protein secondary structure content in terms of percentage helix, β-strand and turn, the amide I band is analysed as a linear combination of the spectra of the references proteins whose structures are known. The Raman spectra of globular proteins in the crystal and in solution are almost identical, reflecting the compact nature of the macromolecules. Thus one may use the fraction of each type of secondary structure determined in the crystalline state by the X-ray of diffraction studies to study protein in solution.

Table E3.1. *Main marker bands of proteins*[a]

Origin and frequency ($\bar{v}$ cm^{-1})	Assignment	Structural information
	Amide bands	
Amide I	Amide C=O stretch coupled to N–H wagging	Strong band; hydrogen bonding lowers amide I frequencies
1655 ±5	–	α-helix (H$_2$O)
1632 ± 5	–	α-helix (D$_2$O)
1670 ± 3	–	Antiparallel β-related sheet
1661 ± 3	–	α-helix (D$_2$O)
1655 ± 3	–	Disordered structure (H$_2$O)
1658 ± 2	–	Disordered structure (D$_2$O)
Amide III	N–H in-plane bend, C–N stretch	Strong hydrogen bonding raise amide III frequencies
>1275	–	α-helix
1235 ± 5	–	Antiparallel β-related sheet
983 ± 3	–	Disordered structure (D$_2$O)
1245 ± 4	–	Disordered structure
	Amino acid chains	
Tyrosine doublet 850/830	Resonance between ring fundamental and overtone	State of tyrosine
Tryptophan 880/1361	Indole ring	Ring environment
Phenylalanine 1006	Ring breathing	Conformation-insensitive
Histidine 1409	N-deuteroimidazole	Possible probe of ionisation state
S–S		
510	S–S stretch	Indication of conformational heterogeneity among disulphides
525	S–S stretch	*Gauche–gauche–trans*
540	S–S stretch	*Trans–gauche–trans*
C–S		
630–670		*Gauche*
700–745		*Trans*
S–H		
2560–2580	S–H stretch	Environment, deuteration rate

[a] Table adapted from Peticolas (1995).

In addition to the structural information on proteins that is obtained from the vibrations of the polypeptide backbone, the vibrations of the side-chains can also give valuable information on the environment of the side-chain and sometimes on the accessibility to protons and deuterons.

The state of tyrosine in proteins is often of great importance. The ratio of the intensity of the Raman bands at 850–830 cm^{-1} is sensitive to the environment of the phenolic –OH. This doublet is due to resonance between a ring-breathing vibration and an overtone of an out-of-plane ring-breathing vibration. A strong hydrogen bond from the phenolic hydrogen to a negative acceptor results in a low value of I_{850}/I_{830} (about 3:10), whereas hydrogen bonding to the phenolic oxygen from an acidic proton donor yields a higher value (about 10:30) for this ratio.

An example of the use of this technique was in the investigation of the buried tyrosines in fd filamentous phage coat protein (Fig. E3.5, Comment E3.3). The two tyrosines at positions 21 and 24 of the fd coat protein provide a potentially useful probe of the phage structure. Several peaks arising from the tyrosines are clearly evident in the Raman spectrum of the virus. The measured intensity ratio I_{850}/I_{830} is about 10:3. According to the above interpretation, such a ratio indicates that the tyrosine OH groups hydrogen bond with a very strong acidic proton donor. Further titration experiments showed that the tyrosines and their acidic donors in the filamentous phage fd are inaccessible to the solvent. One possible explanation of this fact is that the tyrosine OH groups are the recipients of hydrogen-bonded protons arising from fairly acidic donors, yet these acidic donors do not become titrated over the range pH 7–12.

Comment E3.3

Physicist's box: fd filamentous phage

The filamentous phage fd is a long thin structure of about 895 nm by about 6 nm. The phage particle contains about 2700 copies of a largely α-helical coat protein. This phage is structurally similar to Pf1 phage which is discussed in Comment J2.5.

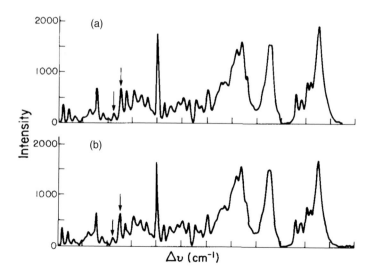

Fig. E3.5 Laser Raman spectra of the fd phage pellets at (a) pH 8.1 and (b) pH 12.0. The positions of the two tyrosine peaks are shown by arrows. Lack of a change in the I_{850}/I_{830} ratio is evident from comparison of the peak intensities (see text). (Dunkerk et al., 1979.)

Fig. E3.6 T_6 crystal
structure of insulin
(Protein Data Bank (PDB)
identifier 4INS).

E3.2.5 Protein conformational dynamics in solution and in crystals

Insulin provides an attractive model for protein conformational dynamics because of its wealth of prior structural characterisation. Insulin is a small globular protein containing two chains, A (21 residues) and B (30 residues). Stored in the pancreatic β cell as Zn^{2+}-stabilised hexamers, the hormone functions as a Zn^{2+}-free monomer. Classical crystallographic studies have focused on Zn^{2+}-stabilised hexamers (Fig. E3.6). Use of engineered insulin monomers and dimers enables

Raman shift cm^{-1}

Fig. E3.7 (a) Amide I region of the Raman spectra (1550–1750 cm^{-1}) of insulin in solution in the hexameric T_6, dimeric T_2 and monomeric T_1 states. The arrow indicates the amide I maximum: spectra are colour-coded as follows: continuous line, experimental data (black); broken line, fitted sum (light purple); dash-dotted line individual components (Gaussian lines) at 1657 cm^{-1} (red), near sum at 1657 cm^{-1}. (b)-(d) Amide I profile deconvolution 1611 cm^{-1} (blue), at 1687 cm^{-1} (green), and at 1585 cm^{-1} (brown); and dotted line, residual (black). The peak at 1657 cm^{-1} in the amide I profiles is due to contributions from α-helices, whereas the shoulder near 1680 cm^{-1} is due to β-sheet and β-strands. Residues in disordered regions are expected to contribute weakly at 1648 cm^{-1} and 1678 cm^{-1}. The unresolved doublet near 1610 cm^{-1} (labelled x and y) and the shoulder at 1585 cm^{-1} are due to ring modes of phenylalanine and tyrosine. (Adapted from Dong et al., 2003.)

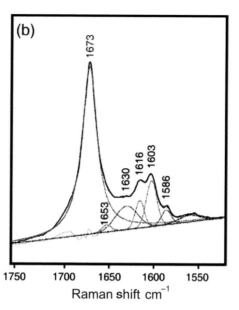

Fig. E3.8 Structure of insulin fibrils formed by gentle agitation of the protein in 60% ethanol and 0.25 M NaCl at 35 °C: (a) electron micrograph of insulin fibrils; (b) Raman spectrum of insulin fibrils obtained using a Raman microscope. The amide I region can be fitted by a line centred at 1673 cm^{-1} and to features centred at 1653 cm^{-1} and 1630 cm^{-1}, respectively. The peak at 1673 cm^{-1} is due to β-sheet formation. (Adapted from Dong et al., 2003.)

direct comparison of the structure and dynamics of a protein in different states of self-assembly.

Solution Raman spectra (1550–1750 cm^{-1} region) of the T_6 hexamer, the engineered dimer designated T_2 and the engineered monomer, herein designated T_1, are shown in Fig E3.7. Strikingly, amide I profiles broaden and change in relative peak heights on disassembly of the protein. Linewidths of Raman bands associated with the polypeptide backbone exhibit progressive narrowing with successive self-assembly. A major increase in width at half-height (WHH) occurs upon dissociation of the aqueous T_6 hexamer (42.9 cm^{-1}) to the T_2 dimer (48.7 cm^{-1}); a further increase is seen in the T_1 monomer (55.1 cm^{-1}).

A further narrowing of linewidths of Raman bands associated with the polypeptide backbone occurs under fibril formation. Figure E3.8 shows the amide profile of the insulin fibrils. The peak at 1673 cm^{-1} has a narrow WHH of 19.7 cm^{-1}. These results provide physical evidence for the flexibility of the isolated monomer and damping of fluctuations in native and pathological assemblies.

Additional information on conformational dynamics for insulin can be obtained from studies in a single crystal in situ in a hanging drop. The partial Raman spectrum of a single insulin crystal in its native conformation is shown in Fig. E3.9(a) and (c). Raman data for a single T_6 insulin crystal in the amide I region (Fig. E3.9(a)) are very close to the data in solution (Fig. E3.7(a)).

Dramatic changes occur when the reducing agent tris(2-carboxyethyl) phosphine is added to the mother liquor containing the crystal. The α-helical 1657 cm^{-1} band (Fig. E3.9(a)) disappears and is replaced by a single β-sheet marker band at 1669 cm^{-1} (Fig. E3.9(b)). The WHH of this band of 24 cm^{-1} compares quite closely to that found for the fibrils. These findings show that

Fig. E3.9 Partial Raman spectra of an insulin single crystal obtained by a Raman microscope: top: native crystal; bottom: β-transformed crystal after the reducing agent tris(2-carboxyethyl) phosphine has been added to the mother liquor. The peaks at 1615 and 643 cm^{-1} are due to tyrosine modes, and the peaks at 1604 and 621 cm^{-1} are due to phenylalanine ring modes. The ring modes are expected to only undergo small changes in intensity on secondary structure change and, thus, act as integral intensity standards. (After Zheng *et al.*, 2004.)

Fig. E3.10 Raman spectra of aqueous solutions of r(CGCGCG) · r(CGCGCG) in the A-form (top), d(CGCGCG) · d(CGCGCG) in the B-form (middle), and d(CGCGCG) d(CGCGCG) in the Z-form (bottom). The top and middle spectra were of samples dissolved in 0.5 M NaCl solution, whereas the bottom spectrum is from a sample dissolved in 6 M NaCl. The marker bands for the A-, B-, and Z-forms are shaded black. All of the spectra were taken with the argon laser line at 514.5 nm. (Adapted from Peticolas and Evertsz, 1992.)

native proteins in the crystalline phase can also undergo large-scale conversion to β-sheets as for solution state described above (Comment E3.4).

E3.2.6 Conformation of DNA

Some of the vibrational normal modes of DNA are conformationally sensitive. This leads to changes in frequencies and/or intensities of the Raman bands with changes in conformation. From a comparison of the Raman spectra taken from different crystals of DNA of known structure, a correlation between the frequencies and intensities of Raman bands and the conformation of the DNA has been established.

Figure E3.10 shows the Raman spectra of hexameric oligonucleotides containing only cytidine (C) and quanine (G). The spectra are of r(CGCGCG) in the A-form (top spectrum), d(CGCGCG) in the B-form, and d(CGCGCG) in the Z-form (bottom spectrum). The bands that change with conformation are highlighted in black. These are the marker bands for guanine, cytosine, and the sugarphosphate backbone as the DNA goes from the A- to the B- to the Z-form.

E3.3 Resonance Raman spectroscopy (RRS)

The central advantage of RRS is that when the wavelength of the exciting laser line is adjusted to coincide with that of an allowed electronic transition in a molecule, the intensities of certain Raman bands are enhanced relative to the off-resonance values. This technique is very selective, because only those vibrational modes that are related to the chromophoric part of the scattering molecule can be enhanced at resonance. This means that the vibrations of the chromophoric group can be

> **Comment E3.4**
> **Amyloidogenic diseases**
>
> Protein unfolding and misfolding events that eventually lead to extensive β-sheet structure, are central to amyloidogenic diseases. Thus, native soluble proteins partially unfold and aggregate, possibly using a residual 'β-sheet' domain as a nucleation motif. This leads to amyloid fibrilogenesis and the formation of highly insoluble fibrils (Booth *et al.*, 1997).

Fig. E3.11 Illustration of the selective enhancement of vibrational modes in a resonance Raman spectrum. The UV–visible absorption spectrum of a *P. aeruginosa* azurin is shown together with two different Raman spectra that derive from laser excitation within the S(Cys) → Cu(II) charge transfer absorption band at 615 nm (647.1 nm, right) and away from the absorption (488.0 nm, left). (After Spiro and Czernuszewicz, 1995.)

monitored in biological macromolecules. If the chromophore is itself a site of functional activity, then RRS can provide structural information associated with reactivity.

This is illustrated in Fig. E3.11 which shows a dramatic increase in the intensity of the Cu-S(Cys) vibrational modes near 400 cm^{-1} as the excitation wavelength (647.1 nm) approaches that of the S → Cu charge transfer electronic transition (∼600 nm) in the blue cooper protein azurin.

Thus the correct identification of the vibrational modes showing RRS enhancement aids in the assignment of the resonant electronic transition and vice versa. Because vibrational frequencies are sensitive to the bond strength, the number of atoms, the geometry and the coordination environment, this technique provides information which is complementary to that obtained by X-ray crystallography and X-ray absorption spectroscopy.

E3.4 Surface enhanced Raman spectroscopy (SERS)

Raman scattering is a very weak effect, with a cross-section between 10^{-30} cm^2 and 10^{-25} cm^2 per molecule. Such small Raman cross-sections require a large number of molecules to achieve adequate conversion rates from exciting laser photons to Raman photons.

SERS involves obtaining Raman spectra in the usual way on samples that are adsorbed on the surface of colloidal metal particles. The same phenomenon can be observed in the visible region (Comment E3.5). For reasons that are not fully

---

Comment E3.5   Plasmon resonant particles (PRP) for biological detection

The light scattering by nanometre-sized colloidal metal particles is dominated by the collective oscillation of the conduction electrons induced by the incident light. The specific colour (i.e. frequency band) of the scattered light is a function of the size, shape, and material properties of the particles. Silver or gold particles with diameters of 30–120 nm efficiently scatter light in the visible spectrum (Fig. (a)). The peak scattering wavelength of a particle is generally referred to as the PR peak (Fig. (b)). Under the same excitation conditions a PRP with diameter of 100 nm can have a brightness that is equivalent to >100 000 fluorescent molecules, and does not undergo photobleaching.

Comment Fig. E3.5 (a) A colour photograph of three PRPs illuminated with white light. The particles were chosen so that their PR peak wavelength would be in the red, green, and blue, respectively. (b) Spectral curves for the three particles shown in (a). (Shultz, 2003).

understood, the Raman lines on the adsorbed molecules are often enhanced by a factor of 10^3–10^6. When surface enhancement is combined with the resonance enhancement technique discussed in the previous section, the net increase in signal intensity is roughly the product of the intensity produced by each of the techniques. In some cases enhancement factors are much larger than the ensemble-averaged values derived from conventional measurements, in the order of 10^{14}–10^{15}. Such large enhancement factors provide in single-molecule SERS effective Raman cross-sections which are of order of 10^{-16} cm^2 per molecule. These cross-sections are comparable to the effective fluorescence cross-sections of common laser dyes and as we will see in Chapter F4 are sufficient for single-molecule detection. A highly sensitive Raman imaging system with submicron spatial resolution can be constructed by combining atomic force microscopy with near-field optics (Chapter F2).

E3.5 Vibrational Raman optical activity

Phenomena that are sensitive to molecular chirality include optical rotation of the plane polarisation of a linearly polarised light beam on passing through a solution of chiral molecules (with equal and opposite optical rotation angles for mirror image enantiomers) and UV circular dihroism (UV CD) where left and right circularly polarised UV light is absorbed slightly differently. Such 'chiroptical' techniques have a special sensitivity to the three-dimensional structure of chiral molecules and were discussed in Chapter E2. The importance of the newer vibrational optical activity methods is that they are sensitive to chirality associated with all the $3N - 6$ fundamental molecular vibrational transitions, where N is the number of atoms, and therefore have the potential to provide more stereochemical information than UV CD, which measures optical activity associated with electronic transitions and so requires an appropriate chromophore.

E3.5.1 Vibrational circular dichroism (VCD)

Proteins with different folds give characteristic VCD shapes, which vary most for the amide I mode (C=O stretch), but are also easily detectable for the broader amide II and III (very weak) modes (N—H deformation and C—N stretch). Examples are shown in Fig. E3.12.

Surveys of protein spectra show that secondary structure determines the dominant contributions to the VCD shape. Analyses of amide I (proteins in D_2O, N–D exchanged), plus amide III and III VCD and ECD (circular dichroism of electronic transition in UV) or FTIR data all yield secondary structure at some level. The important aspects were that VCD sensed sheet and other structural elements (including turns) differently than did ECD, which in turn was superior for helix determination. Combining them gave better determination of all components, especially minimising the impact of outliers on the prediction scheme.

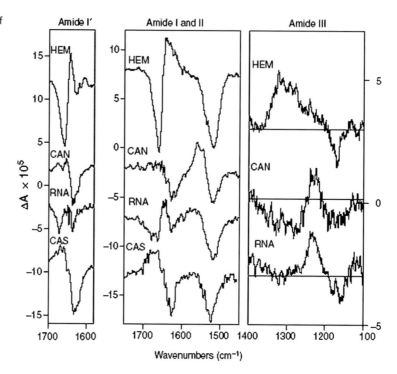

Fig. E3.12 Comparison of typical amide I′ (N-deuterated, left, amide I and II (middle) and amide III (right) VCD spectra for proteins in solution that have different dominant secondary structures. Spectra are for haemoglobin (HEM, top, highly helical); concanavalin A (CAN, highly β-sheet, no helix); ribonuclease A (RNA, second from bottom, sheet and helix mixed); and casein (CAS, bottom, a 'random coil' protein with no extensive secondary structure). (After Keiderling, 2002.)

E3.5.2 Raman optical activity (ROA)

In the ROA technique a small difference in the intensity of Raman scattering is measured using right and left circularly polarised incident light. Interference between the waves scattered via the polarisability and optical activity tensors of the molecule yields a dependence of the scattered intensity on the degree of circular polarisation of the incident light and on a circular component in the scattered light.

ROA is described by the dimensionless circular intensity difference (CID),

$$\Delta = (I^{R} - I^{L})/(I^{R} + I^{L}) \tag{E3.8}$$

as an appropriate experimental quantity, where I^{R} and I^{L} are the scattered intensities in right and left circularly polarised incident light (Comment E3.6).

Polypeptides and proteins

Vibrations of the backbone in polypeptides and proteins are associated with three main regions of the Raman spectrum. These are: the backbone skeletal stretch region $\sim$870–1150 cm^{-1} originating in mainly C_{α}–C, C_{α}–C_{β} and C_{α}–N stretch coordinates; the amide III region $\sim$1230–1310 cm^{-1}, which is often thought to involve mainly the in-phase combination of largely N–H in-plane deformation with the C_{α}–N stretch; and the amide I region $\sim$1630–1700 cm^{-1}, which arises mostly from the C=O stretch. The extended amide III region is particularly

---

Comment E3.6   Circular intensity difference

The theory of ROA shows that unlike the conventional Raman intensity, which is the same in the forward and backward directions, the ROA intensity is maximised in backscattering and is zero in forward scattering. The CID for right-angle scattering takes intermediate values. A given ROA signal-to-noise ratio can be achieved at least an order of magnitude faster in backscattering than in the (easier to implement) right-angle scattering geometry. These considerations led to the important conclusion that backscattering boosts the ROA signal relative to the background biomolecules in aqueous solution (Barron *et al.*, 2000).

---

important for ROA studies because the coupling between N–H and C_α–H deformations is very sensitive to geometry and generates a rich and informative ROA band structure. Bands in the amide II region $\sim$1510–1570 cm^{-1}, which originate in the out-of-phase combination of largely the N–H in-plane deformation with a small amount of the C_α–N stretch, are not usually observed in the conventional Raman spectra of polypeptides and proteins but sometimes make weak contributions to the ROA spectra in H$_2$O solution that are generally enhanced in D$_2$O.

Poly(L-lysin) at alkaline pH and poly(L-glutamic acid) at acid pH adopt well-defined conformations under certain conditions and are able to support a α-helical conformation stabilised by both internal hydrogen bonds and hydrogen bonds to the solvent. Backscattered Raman and ROA spectra of these samples are shown in Fig. E3.13. There are many similarities between the ROA spectra of the α-helical conformations of the two polypeptides.

A good example of the ROA spectrum of a highly α-helical protein is that of human serum albumin shown in Fig. E3.14. The amide I ROA couplet centred at $\sim$1650 cm^{-1}, which is negative at low wave number and positive at high, appears to be a good signature of α-helix in proteins and agrees with the wave number range of $\sim$1645–1655 cm^{-1} for α-helix bands in conventional Raman spectra.

The two polypeptides in α-helical conformations show similar amide I ROA couplets (Fig. E3.13), but shifted to lower wave number, which may be due to the

Fig. E3.13 The backscattered Raman $(I^R + I^L)$ and ROA $(I^R - I^L)$ spectra of poly(L-lysine) and poly(L-glutamic acid) in α-helical conformation in aqueous solution. (Adapted from Barron *et al.*, 2000.)

Fig. E3.14 The
backscattered Raman and
ROA spectra of the
α-helical protein human
serum albumin in H_2O
acetate buffer solutions at
pH 5.4, together with a
MOLSCRIPT diagram of
the X-ray crystal structure
(PDB code 1ao6). The
X-ray crystal structure of
the protein specifies
~70% α-helix, with the
rest made up of turns and
long loops. (Adapted
from Barron et al., 2000.)

particular side-chains or the absence of disordered structure which gives an amide I ROA couplet at high wave number. A positive ROA intensity in the range ~870–950 cm^{-1} also appears to be a signature of an α-helix. Again the two α-helical polypeptides show positive ROA bands in this region. The conservative ROA couplet centred at ~ 1103 cm^{-1}, which is negative at low wave number and positive at high, may also be due to an α-helix since a similar feature appears in the ROA spectra of the two α-helical polypeptides. In addition to a positive ROA band near ~1340–1345 cm^{-1}, an α-helix also gives rise to positive ROA bands in the range ~1297–1312 cm^{-1}.

As well as providing the signature of an α-helix, the ROA spectra of the two polypeptides also provide signatures of β-structure. For example, jack bean concavalin A in H_2O solution (Fig. E3.15) shows a negative ROA band at ~1238 cm^{-1} that is almost unchanged in D_2O which strongly suggests that it originates in the β-strands in the parallel β-sheet within the hydrophobic core.

Fig. E3.15 The
backscattered Raman and
ROA spectra of the
β-barrel protein jack bean
concanavalin A in H_2O
acetate buffer solutions at
pH 5.4, together with a
MOLSCRIPT diagram of
the crystal structure (PDB
code 2cna). The X-ray
crystal structure of the
protein specifies ~40–
50% antiparallel β-sheet.
(Adapted from Barron
et al., 2000.)

Fig. E3.16 The backscattered Raman and ROA spectra of the α-cyclodextrin in H_2O. (Adapted from Barron *et al.*, 2000.)

Carbohydrates

Carbohydrates in aqueous solution are highly favourable samples for ROA studies, giving a rich and informative band structure over a wide range of the vibrational spectrum. The cyclodextrins (Section D5.4) are particularly interesting samples for ROA studies because they exhibit an enormous glycosidic ROA couplet centred at ~918 cm^{-1} as compared to that observed in the corresponding α(1–4)-linked disaccharide D-maltose. An example the ROA spectrum of α-cyclodextrin is shown in Fig. E3.16.

Nucleic acids

Although ROA studies of nucleic acids are still at an early stage, the results obtained so far are most encouraging with ROA being able to probe three distinct sources of chirality: the chiral arrangement of adjacent intrinsically achiral base rings, the chiral disposition of the base and sugar rings with respect to the C–N glycosidic link, and the inherent chirality associated with the asymmetric centres of the sugar rings.

The ROA spectra of polyribonucleotides are generally more intense and informative than those of the corresponding nucleosides since there are fewer degrees of conformational freedom in the polymers which reduces the tendency for cancellation of ROA signals from different conformers. The polyribonucleotide ROA spectra can be conveniently subdivided into three distinct spectral regions, each of which encompasses a well-defined structural subunit. The extent and location of these regions are highlighted in the ROA spectrum of poly(rA).poly(rU), which adopts an A-type double helix, presented in Fig. E3.17. The base stacking region ~1550–1750 cm^{-1} contains ROA patterns from base modes which are characteristic of the particular bases involved as well as being sensitive to the base stacking arrangement. The sugar–base region ~1200–1500 cm^{-1} is dominated by normal modes in which the vibrational coordinates of the sugar and base rings are mixed. The ROA band patterns in this region appear to reflect the mutual orientation of the two rings perhaps together with sugar ring conformation. ROA bands in the sugar–phosphate region ~900–1150 cm^{-1} originate in vibrations of the sugar rings and phosphate backbone. Since the C3′-endo furanose sugar pucker predominates in A-type double helices, the negative-positive-negative ROA triplet (~994, 1049, 1086 cm^{-1}) observed in this region provides a clear marker for this particular conformation. Unfortunately the ROA spectra of naturally occurring DNAs and RNAs are more complicated than those of the polyribonucleotides discussed above.

E3.6 Differential Raman spectroscopy

In general, the protein structure and conformation should change upon ligand binding. These alterations are evident in the Raman difference spectrum, and signals from the protein should be present in the protein ligand-minus-protein difference spectrum. Figure E3.18 shows the Raman spectra of lactate dehydrogenase (LDH) with bound NADH (i.e. reduced nicotinamide adenine dinucleotide) (spectrum (a)) and LDH itself (spectrum (b)). Spectrum (d) demonstrates the difference spectrum between the LDH/NADH binary complex spectrum and that of LDH. The Raman bands in the difference spectrum are mostly from bound NADH, although some protein bands show up owing to the effect of NADH on the protein upon binding (see Fig. E3.18 caption for assignments). A completely dominant protein-ligand-minus-protein difference spectrum can be observed by the isotope editing technique (Comment E3.7).

As a rule of thumb, the Raman spectrum of a protein minus that of the perturbed protein (perturbed by bound ligand or mutated residue) is about 1% that of the protein itself. It requires a protein Raman signal with a signal-to-noise ratio in excess of 300:1. This requirement is consistent with the difference FTIR spectroscopic studies of proteins (Section E1.4.8).

Fig. E3.18 Raman spectrum of (a) lactate dehydrogenase (LDH) with bound NADH and (b) LDH, at 4 °C in 0.1 M phosphate buffer, pH 7.2. (c) The result of the control measurement obtained by calculating the difference between (b) and (b) showing the real signal-to-noise ratio. (d) The result obtained by calculating the difference between (a) and (b). The scale in (c) and (d) is five times that in (a) and (b). The molecular structure of the nicotinamide group of NADH is given in spectrum (a) for references. Assignments of the peaks in (d): A, adenin; N, nicotinamide; P, phosphate; Pr, protein; PP. pyrophosphate; R(ΔR), ribose; S (After Callender and Deng, 1994.)

Comment E3.7 Isotope editing

In many cases, the signal from a bound ligand is similar in size to, or smaller than, that arising from changes in protein structure. Moreover, some proteins are not stable without bound ligands. The strategy of isotope editing of the difference spectrum obviates these problems; here, isotopically labelled ligands are prepared, so that the frequencies of those modes that involve motions of the isotope are shifted. Thus, a subtraction of the labelled ligand from the unlabelled one yields a difference spectrum of positive and negative peaks of just those modes, and all others cancel (Manor *et al.*, 1991).

E3.7 Time-resolved resonance Raman spectroscopy

The inherent scattering processes (10^{-14} s) carries with it the implication that it is theoretically feasible to monitor events on the subpicosecond time scale.

Advances in laser technology (ultrashort pulses) and development of highly sensitive detectors have now made it possible to approach these theoretical limits. Time-resolved resonance Raman methods provide the capability to probe the precise structure of fleeting intermediates which evolve and decay, even on sub-picosecond time scales, throughout the course of a given biological process.

In describing the time-resolved Raman methods, we organise the treatment according to two distinctly different approaches for initiating the reaction. The first and more common approach applies to biological systems that are naturally responsive to light. In these cases, the reaction or process of interest can be initiated by a short pulse of light (called the pump pulse). Other (obviously most) biochemical systems are unresponsive to light pulses, and the reaction must be initiated by rapid mixing of the reactants.

E3.7.1 Light-initiated methods

Molecular systems that are susceptible to the light-initiated approach are those that undergo a sequence of structural changes on exposure to photolysis. There are a number of ways that the photolysis step can be accomplished. By far the most common method is to use pulsed lasers; however, several methods have been devised using continuous wave lasers, and these too are useful if it is not necessary to probe at very short times. Although these latter methods provide time resolution on the microsecond time scale, the approach using a pulsed laser permits investigation at the nanosecond, picosecond, and even subpicosecond levels, the temporal resolution being determined by the widths of the laser pulses.

Fig. E3.19 The biological active centre in haemoglobin and myoglobin molecules. The filled black circles represent the haem skeletal atoms, and the black circles with white dots represent the peripheral groups of the haem. The cross-hatched circles represent the atoms of the proximal and distal histidine and the white circles with black dots a bound CO ligand. (Kincaid, 1995.)

Haem proteins

One of the most important and impressive applications of transient Raman studies deals with the investigation of the structural dynamics and mechanism of
action of haem proteins (Comment E3.8) and, in first place, haemoglobin (Hb)
and myoglobin (Mb) molecules (Comment E3.8). Hb and Mb, two of the first
proteins to be structurally determined by X-ray crystallography, both contain five-
coordinate, ferrous haems at the active sites as shown in Fig. E3.19. In both cases
the haem is attached to the protein through a coordinate linkage to an imidazole
side-chain of a histidine residue (the so-colled proximal histidine). The distal side
of the haem pocket, at which exogenous ligands bind, contains non-polar amino
acid residues as well as an imidazole fragment of a histidine (the so-called distal
histidine), which is known to be capable of forming a hydrogen bond with certain
bound ligands, including dioxygen (Comment E3.9).

Figures E3.20 and E3.21 show the spectra of the deoxyHb photoproduct on
the 7 ns time scale acquired with an excitation line of 416 nm and 532 nm,
respectively. As was explained in the previous Chapter E1 under Soret band
excitation (i.e. 416 nm), the totally symmetric lines (v_2, v_3, v_4) are strongly
enhanced, whereas excitation at 532 nm (Q-band) provides strong enhancement
of non-totally symmetric modes (v_{10}, v_{11}, v_{19}). In Figs. E3.20 and E3.21 the
spectrum of the equilibrium deoxy species is presented along with that of the
7 ns photoproduct. It obvious that all of the features associated with the haem
macrocycle appear at low frequencies in the photoproduct spectra relative to
their positions in the spectra of deoxyHb at equilibrium. Careful comparison
of the frequencies of these core-size markers (Comment E3.10) with a series

---

Comment E3.8   Haem proteins

Haem proteins are a class of biomolecules which possess a haem group embedded
in a single polypeptide chain. The reactivity of the haem group varies greatly from
protein to protein depending on the nature of the active site environment provided by
the polypeptide side-chains, and this variability gives rise to the remarkable
functional diversity that characterises this class of proteins. For example, myoglobin
(Mb) and haemoglobin (Hb) bind dioxygen reversibly and thus serve as O_2 storage
(Mb) and transport (Hb) proteins. Various proxidases and catalases have different
active site environments that impart a different reactivity to the haem group, namely,
the ability to react to H_2O_2 to form highly oxidising intermediates in which the
oxidation equivalents are localised on the haem (and in some cases on surrounding
protein resides) and which are capable of oxidising various substrates. Still other
haem proteins such as cytochrome c and cytochrome b have active site environments
which render the haem unreactive to exogenous ligands, and these serve only as
electron transfer proteins. Finally, one of the most important and frequently studied
haem proteins is cytochrone-c oxidase, whose function is to catalyse the
four-electron reduction of dioxygen to water.

Fig. E3.20 Soret band excited time-resolved Raman spectra of deoxy Hb and Hb* (photoproduct) with 416 nm, 7 ns pulses. Concentration: 0.1 mM; laser energy, 0.5 mJ per pulse; spectral slit width, 8 cm^{-1}; collection time 10 min; and repetition rate 10 Hz. (Kincaid, 1995). Inset: made ν_4 at ten times reduced scale. (Dasgupta and Spiro, 1986.)

Fig. E3.21 Q-band excited resonance Raman spectra of deoxyHb and Hb* with 532 nm, 7 ns pulses. Concentration, 0.5 mM; laser energy, 1 mJ per pulse; spectral slit width, 5 cm^{-1}; collection time, 30 min; and repetition rate 10 Hz. (Adapted from Kincaid, 1995.)

Comment E3.9   Cooperativity of O_2 binding

Whereas the oxygen storage protein, Mb, consists of a single polypeptide chain possessing a proteome at its active site, the oxygen transport, protein, Hb, is actually a tetramer of four (two α and two β) subunits, each of which contains a haem group at its active site. The tetrameric nature of Hb permits interactions between the subunits which give rise to the phenomenon known as cooperativity. Thus the binding curve assumes a sigmoidal shape, leading to more efficient oxygen transport in the sense that it binds O_2 less readily at low O_2 levels (in tissues) but has a much higher affinity at high O_2 levels (in the lungs). This behaviour is consistent with the two-state theory first proposed by Monod in which cooperative behaviour is the result of the existence of a low affinity (the so-called T) state and a higher affinity (R) state; ligand binding to the T state induces a protein structural transition to the more reactive R state. The subsequent crystallographic confirmation of two distinct structures for the ligated (R) and non-ligated (T) forms provided a structural basis for the validity of the thermodynamic arguments. Though the existence of several other states is required to explain the precise details of the binding curves under some conditions, the two-state model remains a valid good approximation under most conditions of ligand binding, indicating that T–R transition is the most important factor for regulating Hb activity. (See also Comment A2.1.)

of structurally well-defined model compounds shows that the photoproduct has a slightly expanded core relative to the deoxy structure and implies that the probable cause for this expansion is restriction of the out-of-plane displacement of the iron brought about by protein restraint on the proximal histidine. Although the preceding examples of time-resolved Raman spectroscopy studies of Hb structural dynamics provide an introduction to these issues, the detailed structural interpretation and precise temporal behaviour of the various modes are quite complex and obviously beyond the scope of this book.

E3.7.2 Rapid-mixing methods

The great majority of biological systems of interest cannot be photo-initiated and for these cases it is necessary to rely on rapid mixing methods, in which the temporal resolution is dictated by the speed with which the reaction mixture can become homogeneous with respect to reactant concentrations. This mixing time is referred to as the 'dead time' of the mixing apparatus, and various designs have been developed to maximise the mixing efficiency of such devices by minimising the dead time.

  All the rapid-mixing time-resolved Raman spectroscopy studies so far reported have employed conventional mixing devices that were developed for use with optical (absorption or fluorescence) detection, in which the observation chamber must necessarily be optically homogeneous. However, given the fact that Raman spectroscopy is based on a scattering phenomenon, novel approaches for rapid mixing have been proposed. The use of one such device is depicted in Fig. E3.22. The essential task to be faced in rapidly and efficiently mixing two solutions is to force them to combine turbulently in a *very small* volume. In the device shown in Fig. E3.22 the two reactant solutions are fed through concentric tubes to a tip which has a platinum sphere positioned very close (10 μm) to the (∼100 μm) orifice of the outer tube. In this way both solutions are forced to flow through the 10 μm circular passage formed by the sphere and the outer tube orifice. The mixing of the two solutions is extremely efficient (dead times as low as 20–40 μs). The progress of the reaction can be conveniently monitored merely by varying the distance between the tip of the outer tube and the probe laser. This approach

Fig. E3.22 Platinum sphere mixing device. (Adapted from Kincaid, 1995.)

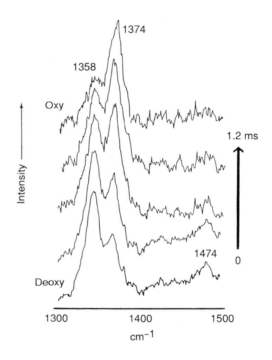

Fig. E3.23 Time-resolved Raman spectra of the reaction of deoxyhaemoglobin (0.25 mM in haem) with an oxygen-saturated buffer, pH 7.0, obtained using the mixing device shown in Fig. E3.22. The excitation line was 457 mm. (Adapted from Kincaid, 1995.)

is very useful in the study of haem proteins. For example, as shown in Fig. E3.23, the monitoring of the reaction of O_2 with deoxyhaemoglobin is followed over the first 1.2 ms by monitoring modes which appear at 1358 cm^{-1} (Hb) and 1374 cm^{-1} (Hb*O_2).

E3.8 Checklist of key ideas

- Raman spectroscopy is a form of vibrational spectroscopy and is related to transitions between the different vibrational energy levels in a biological macromolecule; in this respect Raman spectroscopy is like IR spectroscopy.
- Raman spectroscopy differs from IR spectroscopy in that information is derived from *light scattering* whereas in IR spectroscopy information is obtained from an *absorption* process.
- Raman scattering is associated with normal modes that produce a change in the *polarisability* of the molecule, while IR absorptions are observed for vibrational modes that change the *dipole moment* of the molecule.
- Raman spectroscopy of biological macromolecules is conventionally divided into two types, *resonance and non-resonance*, on the basis of the proximity of the excitation radiation frequency to electronic transitions of the components of the sample.
- Non-resonance or classical Raman spectroscopy is very useful in the determination of protein, nucleic acid and membrane conformational states.
- Resonance Raman spectroscopy has provided considerable information on haem proteins and other metalloproteins and visual pigments.

- The central advantage of resonance Raman spectroscopy is that when the wavelength of the exciting laser line is adjusted to coincide with that of an allowed electronic transition in a molecule, the intensities of certain Raman bands are enhanced relative to the off-resonance values.
- When a molecule is placed near the surface of a particle of metal, such as silver or gold, the Raman signal intensity is enormously enhanced; this effect is called surface enhanced Raman scattering.
- Raman spectroscopy and Raman microscopy provide direct comparison of the properties of a molecule in solution and in a single crystal.
- In the estimation of protein secondary structure content in terms of percentage helix, β-strand and turn, the amide I band is analysed as a linear combination of the spectra of references proteins whose structures are known.
- From a comparison of the Raman spectra taken from crystals of DNA of known structure, a correlation between the frequencies and intensities of Raman bands and the conformation of the DNA has been established.
- Proteins with different folds give characteristic VCD shapes, which vary most for the amide I mode (C=O stretch) but are also easily detectable for the broader amide II and III (very weak) modes (N—H deformation and C—N stretch).
- In the ROA technique a small difference in the intensity of Raman scattering is measured using right and left circularly polarised incident light.
- As well as providing a signature of the α-helix, some of the ROA spectra also provide signatures of β-structure.
- The inherent scattering processes (10^{-14} s) carries with it the implication that it is theoretically feasible to monitor events on the subpicosecond time scale; advances in laser technology (ultrashort pulses) and development of highly sensitive detectors have now made it possible to approach these theoretical limits.

Suggestions for further reading

Historical review and introduction to biological problems

Cary, P. R. (1982). *Biochemical Applications of Raman and Resonance Raman Spectroscopy.* London: Academic Press.

Asher, S. A. (1993). UV resonance Raman spectroscopy for analytical, physical, and biophysical chemistry. Part I. *Analyt. Chem.*, **65**, 59A–66A.

Asher, S. A. (1993). UV resonance Raman spectroscopy for analytical, physical, and biophysical chemistry. Part II. *Analyt. Chem.*, **65**, 201A–210A.

Non-resonance Raman spectroscopy

Tomas, G. J., Jr (1999). Raman spectroscopy of protein and nucleic acid assemblies. *Annu. Rev. Biophys. Biomol. Struct.*, **28**, 1–27.

Skoog, D. A., Holler, F. J., and Nieman, T. A. (1995). *Principle of Instrumental Analysis,* Philadelphia; PA: Saunders College Publishing.

Peticolas, W. L. (1995). Raman spectroscopy of DNA and proteins. *Meth. Enzymol.*, **246**, 389–415.

Callender, R., and Deng, H. (1994). Nonresonance Raman difference spectroscopy: a general probe of protein structure, ligand binding, enzymatic catalysis, and the structures of other biomacromolecules. *Annu. Rev. Biophys. Biomol. Struct.*, **23**, 215–245.

Williams, R. W. (1986). Protein secondary structure analysis using Raman Amide I and Amide III spectra. *Meth. Enzymol.*, **130**, 311–331.

Resonance Raman spectroscopy

Hudson, B., and Mayne, L. (1986). Ultraviolet resonance Raman spectroscopy in biopolymers. *Meth. Enzymol.*, **130**, 331–350.

Spiro, T. G., and Chernuszevich, R. S. (1995). Resonance Raman spectroscopy of metalloprotein. *Meth. Enzymol.*, **246**, 416–459.

Vibrational Raman optical activity

Barron, L. D., Hecht, L., Blanch, E. W., and Bell, A. F. (2000). Solution structure and dynamics of molecules from Raman optical activity. *Prog. Biophys. Mol. Biol.*, **73**, 1–49.

Differential Raman spectroscopy

Zheng, R., Zheng, X., Dong, J., and Carey, P. R. (2004). Proteins can convert to β-sheet in single crystals. *Protein Sci.*, **13**, 1288–1294.

Time-resolved resonance Raman spectroscopy

Kincaid, J. (1995). Structure and dynamics of transient species using time-resolved Resonance Raman spectroscopy. *Meth. Enzymol.*, **246**, 461–501.

Chapter E4
Optical activity

E4.1 Historical review and introduction to biological problems

Optical activity is the property of certain materials to change the angle of polarization of a beam of light that is shone through them as a function of its wavelength (*optical rotation dispersion,* ORD), or to absorb light differently according to its wavelength and polarization (*circular dichroism,* CD, from the Greek related to two colours).

1812–1838

Jean-Baptiste Biot and **Auguste Fresnel** independently made careful studies of *optical activity*. **Biot's** *polarimeter* (an optical instrument used to measure the angle of rotation of polarised light) used sunlight, plane-polarised by reflection off glass, as a light source. **Biot** defined [α] the angle of *specific rotation* of the axis of plane-polarised light by an optically active solution in terms of the concentration in grams per cubic centimetre and the optical path in *decimetres*, a definition still in use. In **1824**, **Fresnel** predicted that helical structures would be optically active.

1848

Louis Pasteur observed that *optically inactive* crystals of sodium ammonium tartrate were in fact mixtures of two classes. When separated, both classes turned out to be optically active with respective specific rotation angles of equal value but opposite in sign. The molecules making up the crystals were themselves in one of two asymmetric forms, *enantiomorphs* (from the Greek for opposite shapes), each a mirror image of the other. Since a two-dimensional molecule and its mirror image can be made to coincide and considered to be identical, the implication was that molecules are three-dimensional (Comment E4.1). The form that turns the axis of polarised light to the right when viewed towards the light source in a *polarimeter* is called *dextro-rotatory* or the *D-form*; the form that turns the axis to the left is called the *L-form* or *laevo-rotatory* (from the Latin for right and left). The same terms apply to *circularly polarised* light (see Chapter A3); the

---

Comment E4.1   Left hand, right hand

Look at your hands. One is a mirror image of the other. If they were two-dimensional, i.e. paper-thin without the palm on one side and the back on the other, then they would superpose exactly and could be considered to be identical. The reason they do not superpose is because they are three-dimensional and asymmetrical. You could make the palms point in the same direction, for example, but then the thumb of one hand would superpose with the little finger of the other.

The word *chiral* (from the Greek, meaning handed) was introduced in 1904 by Lord Kelvin, Professor of Natural Philosophy at the University of Glasgow, with the following definition: 'I call any geometrical figure, or group of points, *chiral*, and say it has *chirality*, if its image in a plane mirror, ideally realized, cannot be brought to coincide with itself.'

---

electric vector of right-circularly polarised light turns in the clockwise direction when viewed towards the light source, the vector of left-circularly polarised light turns anticlockwise. Above a certain temperature (26 °C for sodium ammonium tartrate) the D- and L-forms mix to make up homogeneous crystals. They are said to form a *racemate* or *racemic mixture* (from the Latin, *racemes*, a bunch of grapes (optically inactive tartaric acid was obtained from grape juice)).

1870s

Jacobus Henricus van't Hoff and **Joseph-Achille Le Bel** independently proposed the three-dimensional tetrahedral model for the carbon atom. **Van't Hoff** coined the phrase 'chemistry in space'.

1895

Aimé Cotton accounted for CD and anomalous ORD (in which $[\alpha]$ does not simply increase with decreasing wavelength, but may show sign changes or curve maxima or minima) as due to optically active absorption bands. CD and its associated ORD are termed the *Cotton effect*.

1935

The book *Optical Rotatory Power* by **Thomas Martin Lowry** was published.

1951

Linus Pauling and **L. B. Corey** discovered α-helices and β-sheets and suggested they were components of protein structure. These are optically active structures and the hypothesis strongly stimulated the renaissance of ORD and later the development of CD to identify them in synthetic polypeptides and proteins in solution, and to study the helix–coil transition in polymers.

1954

Jen Tsi Yang and **Joseph F. Foster**, who studied the optical rotation of proteins, showed bovine serum albumin expanded reversibly in acid solution with a marked increase in laevo-rotation at the sodium D-line (negative $[\alpha]_D$) (Comment E4.2).

1955

Carolyn Cohen postulated that right-handed helices would be dextro-rotatory even if made up of left-handed amino acid constituents. In **1957**, **Yang** and **Paul Doty** published the anomalous ORD of polypeptides in helix-promoting solvents, and in the **early 1960s**, the presence of right-handed α-helices in proteins was confirmed by X-ray crystallography.

1956

William Moffit, Yang, D. D. Fitts and **John G. Kirkwood**, first independently, then together, developed the theory of ORD of the α-helix.

1960s and the following decades

High-performance commercial instruments became available that allowed the direct and precise experimental observation of CD bands in the UV spectral region, and CD, the more powerful technique, essentially replaced ORD as a sensitive measure of protein conformation in solution. The method was applied to nucleic acids and protein–nucleic acid interactions. Automation permitted the development of kinetics measurements and stopped-flow equipment was adapted to the spectropolarimeters. The vacuum-UV wavelength limit was pushed down to approach 170 nm.

1980s and the following decades

Synchrotron radiation sources became available, opening the way for an extension of CD applications from the vacuum-UV to the extreme-UV and X-ray ranges (wavelengths ≤ 100 nm). At the other end of the spectrum, ROA studies (see Section E3.5.2) are based on the observation of inelastic scattering of left- and right-handed polarised light, and VCD (see Section E3.5.1) is based on CD in the IR spectral region.

E4.2 Brief theoretical outline

E4.2.1 Plane, circularly and elliptically polarised light

The relationships between plane, circularly and elliptically polarised light have been discussed in Chapter A3 (Figs. E4.1, E4.2). Recall that plane-polarised light can be resolved into equal left and right circularly polarised components. Elliptically polarised light is formed when the right and left components are not equal: in CD this happens when one of the components is absorbed more than the other; the *ellipticity* is defined as the angle θ, which is usually small so that it is

> **Comment E4.2**
> **Optical rotation and protein denaturation**
>
> Secondary structures in a native protein, such as right-handed α-helices, display right-handed optical rotation. Upon loss of secondary structure, a denatured polypeptide shows an increase in left-handed optical rotation, because of its L-amino acid constituents.

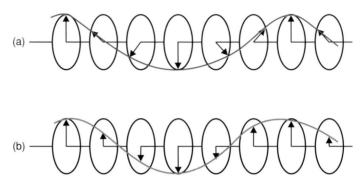

Fig. E4.1 (a) The path traced out by the tip of the electric vector (the arrow) during the propagation of right circularly polarised light traces a right-handed helix. The vector in left circularly polarised light traces a left-handed helix. Note that in both cases the length of the vector remains constant. (b) In linearly polarised light the electric vector stays in the same plane but its length is modulated by the wavelength. (Johnson, 1985.) (Figure reproduced with permission from John Wiley & Sons.)

Fig. E4.2 The projection of the electric vector during light propagation as it would be seen by an observer looking into a polarimeter: (a) right circularly polarised light shown as made up of two plane-polarised components, V and H, of equal magnitude out of phase by 90°; (b) plane-polarised light shown as made up of left (L) and right (R) circularly polarised light of equal magnitude; (c) elliptically polarised light shown as made up of left (L) and right (R) circularly polarised light of unequal magnitude; (d) Elliptically polarised light shown as made up of two plane-polarised components, V and H, of unequal magnitude out of phase by 90°. (Johnson 1985) (Figure produced with permission from John Wiley & Sons.)

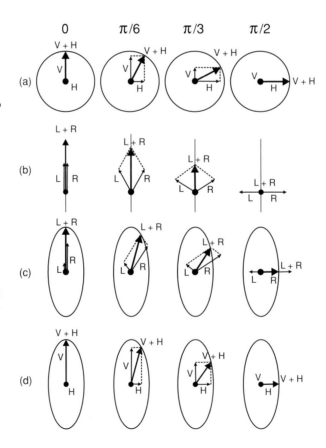

given simply in radians as the value of its tangent, the ratio of the minor to major axes; when the major axis of the ellipse (or axis of the original plane-polarised beam) is rotated by going though an optically active material, the *optical rotation* is defined by the angle α (Fig. E4.3).

E4.2.2 CD, ellipticity and ORD

The absorbance of light of a certain polarization, p, by a material is expressed in terms of an extinction coefficient, ε_p:

$$A_p = \log_{10}\left(I_p^0/I_p\right) = \varepsilon_p C l \tag{E4.1}$$

Fig. E4.3 The ellipse, with major axis *a* and minor axis *b*, formed by the electric vector projection in elliptically polarised light, showing the optical rotation angle α and the ellipticity θ. (Snatzke, 1994) (Figure reproduced with permission from John Wiley & Sons.)

where I_p^0, I_p, are, respectively, the incident and transmitted light intensities of polarization, p, C is the molar concentration and l is the optical path length in the sample. The parameter, ε_p, is called a *decadic* molar extinction coefficient, because it is defined with respect to the base ten logarithm (Eq. (E4.1)).

CD is defined as the difference, ΔA, between the absorption of left circularly and right circularly polarised light, A_l, A_r, respectively, for a *chiral* (from the Greek *cheir*, hand) material:

$$\Delta A = A_l - A_r = \Delta \varepsilon C l \tag{E4.2}$$

where $\Delta \varepsilon$ is called the decadic molar CD.

CD can also be defined in terms of the *molar ellipticity*. We recall that in cases where there is a difference between A_l and A_r the right and left circularly polarised components of the transmitted beam will not be equal, resulting in an elliptically polarised transmitted beam (Fig. E4.3). It can be shown that the ellipticity, θ, and CD, ΔA, are related:

$$\theta = 180 \, (\ln 10)\Delta A/4\pi = 32.98 \Delta A$$
$$[\theta] = 100\theta/Cl \tag{E4.3}$$
$$[\theta] = 3298 \Delta \varepsilon$$

where θ is in now in *degrees* and the ln10 term occurs in the equation because the CD was defined with respect to the base ten logarithm. $[\theta]$ is called the *molar ellipticity*. CD spectra are often expressed in terms of $[\theta]$ rather than $\Delta \varepsilon$ but in either case care must be exercised to specify the basis used to define the molar concentration, which could be, for example, residue concentration, protein concentration or even subunit concentration in the case of oligomeric structures (haem molarity, equivalent to subunit molarity, is usually taken for tetrameric haemogoblin).

Since CD has essentially completely supplanted ORD in molecular biophysics, we shall not discuss ORD spectra in detail. ORD and CD are related to each other, by a mathematical transformation called the Kronig–Kramers transformation (Comment E4.3). The ORD can be calculated when the CD spectrum is known and vice versa. Similarly to a Fourier transformation (see Chapter A3), a CD or optical rotation value at a given wavelength is defined by the entire ORD

Comment E4.3   CD and ORD

The decadic molar CD, $\Delta\varepsilon$, and molar rotation, $[\phi] = 100\alpha / Cl$ (where α is the optical rotation in degrees) are related by the Kronig–Kramers transformation:

$$[\phi] = (9000 \ln 10/\pi^2)\mathbf{P} \int_0^\infty \Delta\varepsilon\,(\lambda)\lambda'\mathrm{d}\lambda'/(\lambda^2 - \lambda'^2)$$

$$\Delta\varepsilon = -(2250 \ln 10\lambda)^{-1}\mathbf{P} \int_0^\infty [\phi\,(\lambda)]\lambda'^2\mathrm{d}\lambda'/(\lambda^2 - \lambda'^2)$$

where $\mathbf{P}$ indicates the 'principal value' of the integral is to be taken, avoiding the point $\lambda = \lambda'$ by limiting processes (Moscowitz, 1962).

---

or CD spectrum, respectively. Because of this interrelationship and the greater wavelength resolving power of CD, interest in ORD decreased significantly when CD instruments became available.

E4.2.3 Electronic transitions, dipole and rotational strengths

The bands observed in absorption and CD spectra of macromolecules in solution correspond mainly to electronic transitions between the lowest vibrational levels in the ground-state potential energy well and various vibrational levels in the first excited-state potential energy well (see Chapter E1).

The integrated area under an absorption band for a transition from state a to state b is proportional to effective values called the *dipole strength, D_{ab}*, and *oscillator strength, f_{ab}* (Comment E4.4):

$$D_{ab} = 9.181 \times 10^{-3} \int (\varepsilon/v)\,\mathrm{d}v\ \text{(debye)}^2$$
$$f_{ab} = 4.315 \times 10^{-9} \int (\varepsilon/v)\,\mathrm{d}v\ \text{(dimensionless)}$$

(E4.4)

Comment E4.4   Dipole strength, transition dipole moment, Debye units and oscillator strength

The dipole strength is usually given in cgs units (erg cm^3) or in units of (debye)^2. The debye (10^{-18} esu cm, where esu is the electrostatic unit of charge), is a measure of dipole moment. The dipole strength is equal to the square of the transition dipole moment. The transition dipole associated with an intense absorption band has a magnitude in the order of debyes (see also Comment D6.1).

The oscillator strength is proportional to the dipole strength. It is dimensionless and in a classical physics interpretation f has the meaning of the number of electrons undergoing the transition as three-dimensional harmonic oscillators (see Chapter A3).

Note that the integration of the extinction coefficient (in units of M^{-1} cm^{-1}) in the band is performed as a function of frequency, ν. A more complex equation is obtained as a function of wavelength. The numerical constant in the expression for D contains Planck's constant, the speed of light and Avogadro's number; in the expression for f, it contains Planck's constant and the mass and charge of the electron.

In an analogous way to the definitions of dipole and oscillator strength for an absorption band, the integrated intensity of the CD band associated with the transition from state a to state b is defined in terms of the *rotational strength*, R_{ab}.

$$R_{ab} = 2.295 \times 10^{-39} \int (\Delta\varepsilon/\lambda)\, d\lambda \,(\text{erg cm}^3) \qquad \text{(E4.5)}$$

The dipole strength is theoretically equal to the square of the transition dipole moment ($\mu_{ab} \cdot \mu_{ab}$), the scalar product of the transition dipole moment vector with itself. The rotational strength depends on both the electric dipole and magnetic dipole transition moments:

$$\begin{aligned} D_{ab} &= \mu_{ab} \cdot \mu_{ab} \\ R_{ab} &= \text{Im}(\mu_{ab} \cdot \mathbf{m}_{ba}) \end{aligned} \qquad \text{(E4.6)}$$

The rotational strength is theoretically equal to the imaginary part of the scalar product of the electric and magnetic transition dipole moment vectors. This means that for a molecule to be optically active the transition magnetic moment must have a component parallel to the transition electric dipole moment (Figure E4.4).

Recall that a light wave consists of perpendicular oscillating and magnetic fields. Pure absorption is dominated by electric effects, which are much stronger than magnetic effects. The electric dipole transition moment can be pictured as a *linear displacement* of charge, induced by the absorption of a photon matching the energy difference of the transition from the ground state a to state b. The magnetic transition moment can be pictured as due to the *circular motion* of charge induced in the transition. The combination of the two results in a helical displacement of charge around the common dipole axis (Fig. E4.4). The implication is that an optically active molecule must be chiral (asymmetric), otherwise it would not display a preferred helical direction. Also, the helical circulation of charge is facilitated in macromolecules with an inherent helical structural organisation making them strongly optically active.

E4.2.4 Rotational strength and structural organisation

The rotational strengths of chiral molecules and structures can be calculated, in principle by using quantum chemistry, in terms of electric and magnetic transitions. The systems studied are extremely complex, however, and the calculations difficult and practically impossible to perform analytically, so that empirical

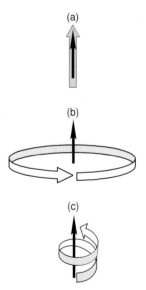

Fig. E4.4 (a) Charge displacement (in grey) due to the transition electric dipole moment vector (black arrow). (b) Charge displacement due to the transition magnetic moment vector. (c) Charge displacement due to both. (After Cantor and Schimmel, 1980.)

approaches are often used. In this section, we give only a simplified qualitative description of the relations linking rotational strength and structural organisation in macromolecules.

A chromophore is optically active if it is itself chiral, if it is perturbed by a chiral field due to the structural environment, or both. In macromolecular structure determination studies, it is the optical activity induced by the chromophore *organisation* that is mainly of interest. *Electric–magnetic* and *exciton* (or *coupled-oscillator*) coupling (Comment E4.5) between the electronic states of a chromophore pair constitute structural contributions to rotational strength, which add to the individual chromophore (*one-electron*) contributions. One-electron, electric–magnetic and exciton contributions to rotational strength can be computed analytically by quantum mechanics for a pair of identical chromophores (a chromophore dimer), each with a single absorption band. The calculation becomes prohibitively complex, however, in the case of a large heterogeneous assembly. The calculation may be simplified by using symmetry arguments and periodicity, in the case of regular repetitive structures such as a polypeptide

---

Comment E4.5   Two-chromophore coupling

The coupling of an electrically allowed transition dipole (linear movement of charge) on one chromophore with a magnetically allowed transition dipole (circular motion of charge) on the other is called *electric–magnetic* or μ-m coupling. The contribution to rotational strength of μ-m coupling is usually small.

The rotational strength of a chromophore dimer is dominated by a term called the *coupled oscillator* or *exciton* term, which depends on the geometry of the molecular arrangement. This term may still be large even in the case of optically inactive individual chromophores. It arises in the dynamic quantum mechanical treatment of the system as a pair of coupled oscillators, in which electrically 'allowed' and magnetically 'forbidden' transition dipoles are coupled by electrostatic interactions. A magnetically forbidden transition, for example, corresponds to a linear movement of charge with respect to an axis through the chromophore. Note, however, that the same motion corresponds to tangential motion along a circle centred on an external axis, which would create a magnetic moment.

The third term contributing to the dimer rotational strength is called the *one-electron* term. It arises from the individual chromophore rotational strengths as they are modified by the static field due to their organisation. This term also is usually weak compared to the coupled oscillator term.

α-helix (note, however, that an α-helix in a protein structure is not perfectly periodic along its length, due to the diversity of the amino acid side-chains, and the local environment).

E4.3 Instruments

The CD signal is the difference in absorption of left and right circularly polarised light, which is of the order of 10^{-3} for usual samples, so that highly sensitive instrumentation is required for its accurate measurement. Since automatic recording CD instrumentation became commercially available in the 1960s, several generations of instrument have been marketed.

E4.4 CD of proteins

If one takes into account the amount of sample required and the ease of the experimental procedure as well as measurement accuracy, CD is the most appropriate method for assessing the secondary structure of a protein in solution. The signal is strong in the far-UV, where the peptide bond absorbs, and fairly dilute solutions can be used, of the order of a tenth of a milligram per millilitre. CD measurements are extremely powerful for monitoring structural differences and conformational changes. Because of the complexity of the protein structure and the sensitivity of the spectrum to precise local conditions within the molecule, however, the interpretation in terms of structure is necessarily based on empirical criteria (see Section E4.2.4). It is known from analysis of the protein structure database that the most abundant secondary structures are the α-helix and β-sheets, and proteins have been classified according to these and their topological organisation (see Chapter A2). CD due to aromatic groups, such as tryptophan and tyrosine, in the near-UV (where these residues absorb, see Chapter E1) is sensitive to tertiary structure. CD also plays an important role in protein folding studies, because the CD signal of an unstructured polypeptide is distinctively different from the signals arising from secondary structures.

E4.4.1 CD of protein secondary structures

CD spectra observed for α proteins, β proteins, $\alpha+\beta$ proteins, α/β proteins and disordered polypeptides are shown in Fig. E4.5. The molar ellipticity is expressed in units per decimole of amino acid residue (see also Table E1.4).

The spectra for α proteins (Fig. E4.5(a)) display negative bands between about 203 and 240 nm with minima at about 209 and 222 nm; the positive bands below 203 nm have a strong maximum close to 192 nm. These bands are characteristic of the α-helix, and are readily observed in model polypeptides in helix-forming solving solvents, such as poly-lysine above pH 10. The differences in band

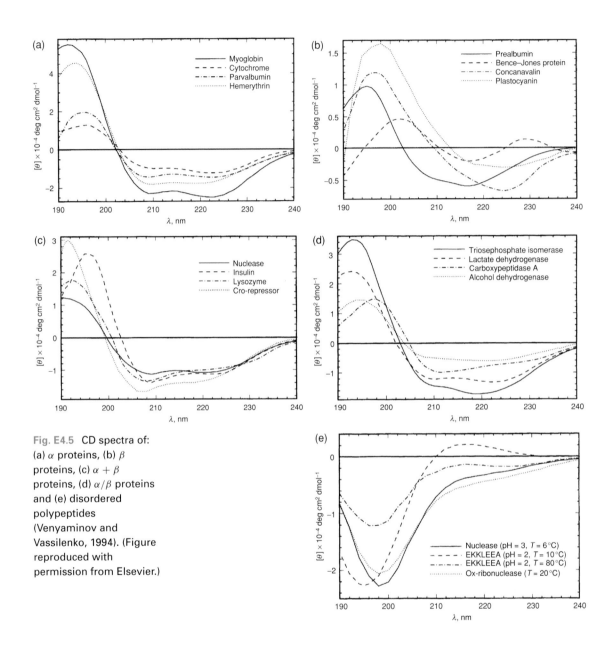

Fig. E4.5 CD spectra of:
(a) α proteins, (b) β
proteins, (c) $\alpha + \beta$
proteins, (d) α/β proteins
and (e) disordered
polypeptides
(Venyaminov and
Vassilenko, 1994). (Figure
reproduced with
permission from Elsevier.)

intensity between the various protein spectra reflect the their different α-helical content (Comment E4.6).

The CD spectra of β proteins (Fig. E4.5(b)) show a greater variety and less intense bands than those for α proteins, and β proteins have been classified in different groups according to their spectrum and β structure type (see below). Regular β proteins have a spectrum similar to that of the β-sheet form of poly-lysine, with a minimum between 210 and 225 nm and a maximum between 190 and 200 nm.

The CD spectra of $\alpha+\beta$ proteins and α/β proteins can be considered as mixtures of α-helix and β structure spectra, in which the stronger α-helix band intensities dominate (Fig. E4.5(c), (d)).

The disordered polypeptides include denatured proteins and oligopeptides and polypeptides with disulphide bonds or associated to prosthetic groups (Fig. E4.5(e)). A strong negative band near 200 nm, and weak positive or negative bands above 210 nm are characteristic of disordered structures.

Fitting spectra with basic sets

If CD coupling between different secondary structures in a protein were negligible to a first approximation, the spectrum could be considered as a simple sum of component spectral bands corresponding to specific secondary structures. The application of such an empirical approach, however, is not straightforward. CD band intensities and wavelengths due to a 'perfect', long α-helix may be modelled well on an absolute per residue scale by the spectrum of α-helical polypeptides. But α-helices in a protein structure deviate from 'perfection', and, in any case, they are certainly not infinitely long and structural distortions at their edges affect the spectrum. We have seen above that β secondary structures are quite varied depending on the structure type and on the protein. In fact, the CD spectra of β-rich proteins could resemble those of either model polypeptide β-sheets or disordered polypeptides (Comment E4.7).

When coordinates from X-ray crystallography and NMR for a large number of structures started to become available in protein data banks, the opportunity arose not only to confirm secondary structure predictions made from CD spectra, but also to set up databases of CD spectra corresponding to different protein secondary structure families, which could be used as basis sets for the fitting of spectra of 'unknown' proteins. The problems to be solved included careful analysis of the structures themselves in order to identify and define limits for the different secondary structures and to interpret how to deal with distortions from 'perfection' within each type of secondary structure. Various databases are proposed on the web for the interpretation of CD spectra.

CD was considered a more convenient experimental approach than FTIR (see Chapter E1) for the determination of protein secondary structures. It appears

---

Comment E4.6
α-helix CD

The CD spectrum of α-helical poly-lysine in solution has negative band ellipticity minima of about $-40\,000$ deg cm^2 dmole^{-1}, and a positive band maximum greater than 70 000 in the same units. The relative band intensities of the proteins in Fig. E4.5(a) reflect their relative α-helix content. Thus myoglobin is more helical than the other proteins, with more than 50% of its residues in α-helices.

Comment E4.7 β structure CD

β-rich proteins are classified according to their CD spectra as β_I or β_{II}. The β_I spectrum is similar to those of model polypeptide β-sheets (Fig. E4.5(b)). The β_{II} spectrum resembles that of a disordered polypeptide. It has been shown that it is compatible with the presence of a conformation called P$_2$ in protein structures. P$_2$ is the type II conformation identified for poly-proline, poly(Pro)II. It is a left-handed helix with three residues per turn (Sreerama and Woody, 2003).

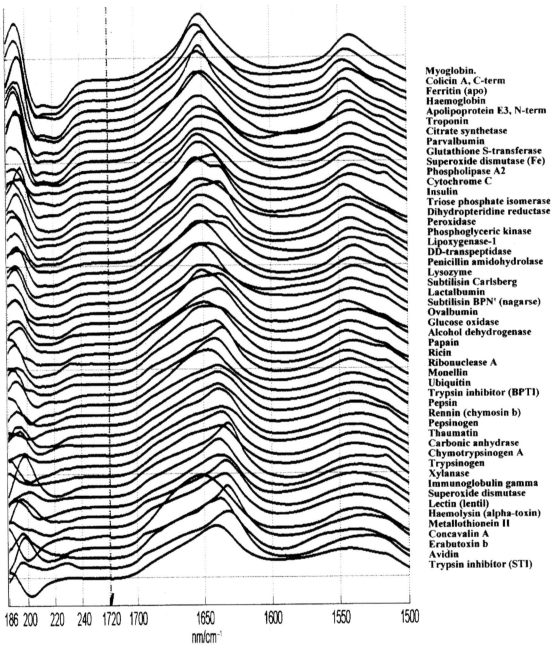

Myoglobin.
Colicin A, C-term
Ferritin (apo)
Haemoglobin
Apolipoprotein E3, N-term
Troponin
Citrate synthetase
Parvalbumin
Glutathione S-transferase
Superoxide dismutase (Fe)
Phospholipase A2
Cytochrome C
Insulin
Triose phosphate isomerase
Dihydropteridine reductase
Peroxidase
Phosphoglyceric kinase
Lipoxygenase-1
DD-transpeptidase
Penicillin amidohydrolase
Lysozyme
Subtilisin Carlsberg
Lactalbumin
Subtilisin BPN' (nagarse)
Ovalbumin
Glucose oxidase
Alcohol dehydrogenase
Papain
Ricin
Ribonuclease A
Monellin
Ubiquitin
Trypsin inhibitor (BPTI)
Pepsin
Rennin (chymosin b)
Pepsinogen
Thaumatin
Carbonic anhydrase
Chymotrypsinogen A
Trypsinogen
Xylanase
Immunoglobulin gamma
Superoxide dismutase
Lectin (lentil)
Haemolysin (alpha-toxin)
Metallothionein II
Concavalin A
Erabutoxin b
Avidin
Trypsin inhibitor (STI)

186 200   220   240   1720 1700   1650   1600   1550   1500

nm/cm^{-1}

Fig. E4.6 Concatenated CD spectra (186–260 nm) and FTIR spectra (1720–1500 cm^{-1}, showing the amide I bands at around 1650 cm^{-1} and amide II bands at around 1550 cm^{-1}) of 50 reference proteins, chosen as a database by Oberg *et al.* (2004). Note that the wavelength range is not continuous over the two methods (1720 cm^{-1}, for example, corresponds to a wavelength of 5800 nm). The spectra have been rescaled and offset for better readability. The proteins are listed according to their α-helix content. Note the typical α-helix CD spectrum of myoglobin at the top. (Oberg *et al.*, 2004.) (Figure reproduced with permission from Blackwell.)

Comment E4.8 The optimisation of protein secondary structure determination by FTIR and CD

The CD and FTIR spectra of a specially designed set of 50 proteins were treated by multivariate statistical analysis in order to optimise secondary structure determination in solution. The proteins in the database were selected for their structural diversity. The main two conclusions of the study were that:

(1) it was not satisfactory to use subsets of the database so that a large database of proteins selected with stringent criteria must be used;
(2) the independent analysis of either FTIR or CD alone identified 'wrong' secondary structure estimates, so that both complementary methods are required.

A further conclusion of the study was that the amide II band in FTIR has high information content and can be used alone (rather than amide I) for secondary structure prediction (Oberg *et al.*, 2004).

---

from a careful analysis, however, that only the comparison of independent CD and FTIR information is able to identify 'wrong' secondary structure estimates in the absence of a known structure (Comment E4.8, Fig. E4.6, and Table E1.4).

Membrane proteins

Various methods have been proposed for the interpretation of membrane protein CD spectra in terms of secondary structures, based on empirical as well as theoretical approaches (Comment E4.9). The CD study of proteins in situ in biological membrane fragments presents a number of difficulties not encountered in soluble protein studies. The membranes, in fact, contribute to differential light absorption and light-scattering effects resulting in optical artefact distortions in the shapes, intensities and/or positions of the CD bands. The example of bacteriorhodopsin in purple membranes (see Chapter A3) is shown in Fig. E4.7. Light scattering by

---

Comment E4.9 Membrane protein CD

Fasman (1996) compares the convex constraint analysis (CCA) deconvolution algorithm, developed in his laboratory, with the fixed reference method of Chang *et al.* (1978), and the variable reference methods of Manavalan and Johnson (1987). The CCA algorithm extracts P common components from a database, where P is defined as an input parameter. The approach offers considerable flexibility to deal with different databases. The value of P is chosen to match the number of significant components. For example, for a database of proteins containing α-helix and random conformations but no β secondary structure, a value of $P = 3$ would be the most appropriate (for transmembrane, peripheral α-helix, and random structures, respectively). Larger values of P would force the algorithm to extract additional poorly defined components (Fasman, 1996).

Fig. E4.7 CD spectra of bacteriorhodopsin and rhodopsin: bacteriorhodopsin in purple membranes (diamonds); bovine retina rhodopsin solubilised in the detergent laurylmaltoside (squares); light adapted bovine rhodopsin in the same detergent (crosses). (Fasman, 1996.)

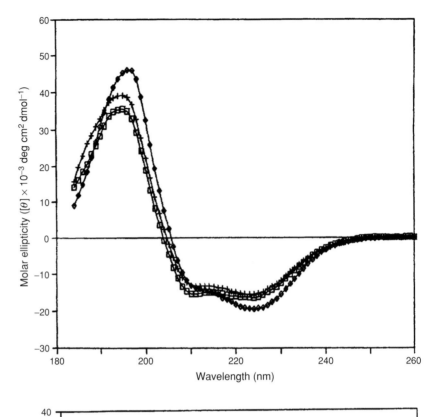

Fig. E4.8 CD spectra of porins (predominantly β-conformation proteins) solubilised in various detergents: maltoporin (triangles); *E. coli* porin (squares); *E. coli* porin in slightly different solvent conditions (crosses); phosphoporin (X); *Rhodobacter capsulatus* porin (diamonds). (Fasman, 1996.)

the membrane patches led to a distortion of the minima at 208 and 222 nm, as well as to a generally more intense spectrum.

Detergent solubilised membrane protein preparations are relatively free of such artefacts (Fig. E4.7). Membrane protein CD experiments should still be designed carefully, however, because different detergents may lead to subtle conformational differences, which will be seen in the spectra.

Typical spectra of membrane proteins made up predominantly of α-helices are shown in Fig. E4.7. The approaches discussed in Comment E4.9 are capable of separating peripheral and transmembrane α-helix components in membrane protein CD spectra. The CD spectra of porins, which have distinctive, transmembrane β-barrel structures, are shown in Fig. E4.8. The differences between the spectra in Fig. E4.8 reflect different minor proportions of α-helix content.

E4.4.2 Near-UV CD and protein tertiary structure

We saw in Chapter E1 how the aromatic amino acids absorb strongly in the near-UV close to 280 nm. The bands also display a strong CD signal (Fig. E4.9). Bovine ribonuclease A (RNase A) does not contain Trp; it has six Tyr, three Phe and four disulphides. Analysis of the absorption spectrum close to 280 nm indicated the Tyr residues reside in different local environments with respect to solvent exposure. We see from Fig. E4.9 that the near-UV CD spectral components of two classes of Tyr residue are distinctive; the 288 nm band corresponds to the buried residues and the 283 nm band to the solvent-exposed residues.

The near-UV CD spectrum of a protein is a sensitive measure of tertiary structure (Fig. E4.10 and see the unfolding studies discussed below).

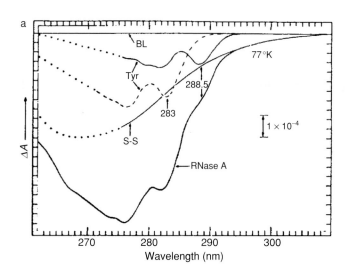

Fig. E4.9 CD spectra of bovine ribonuclease A (RNase A) in the near-UV. The dark lines are observed spectra; the lighter lines and dashed lines are Gaussian components, BL is the background level. The two Tyr lines correspond to two classes for these residues, a solvent exposed and a buried class, respectively. (Horwitz *et al.*, 1970) (Figure reproduced with permission from the American Chemical Society.)

Fig. E4.10 (a) Far- and (b) near-UV CD spectra of the membrane inserting fragment of the peptide colicin A. In each panel the more intense curve corresponds to the acid conditions in which the fragment apparently loses tertiary structure (but maintains secondary structure) in order to insert into the membrane. (van der Groot *et al.*, 1991.) (Figure reproduced with permission from Macmillan Publishers.)

Upon binding to sensitive bacterial cells, the antibiotic colicin A is cleaved to yield a fragment that depolarises the cell membrane (i.e. makes it permeable to ions). In vitro studies indicate that the fragment inserts into lipid vesicles at low pH. The analysis of the far- and near-UV CD data (Fig. E4.10) showed how, while maintaining a predominantly α-helical conformation, the peptide loses its tertiary structure at acid pH to facilitate membrane insertion.

E4.4.3 Protein folding

CD spectra are good empirical indicators of the folding state of a protein. The aromatic region of a spectrum in the near-UV (close to 280 nm) contains information on tertiary conformation, while, as we have seen in some detail above, the peptide bond region in the far-UV CD (close to 222 nm) is sensitive to the presence of α-helix, β-sheet and random chain conformations. Combined with stopped-flow devices, kinetic CD experiments have been used to follow protein unfolding and refolding processes (Fig. E4.11). The time axis is plotted on a logarithmic scale in Fig. E4.11(c), which clearly shows that secondary structure recovery (the 222-nm signal) occurs significantly faster than tertiary recovery (the 289-nm signal). The curves in (a) and (b) were fitted by two-exponential functions, indicating the existence of intermediates in the folding pathway.

Molten globules

The molten globule hypothesis has been formulated for the refolding pathway of globular proteins following the CD analysis of intermediate states (Comment E4.10). After first experiments on α-lactoglobulin, which led to the hypothesis, intermediates with similar properties were also found for other globular proteins. Recall from Chapter C2 that a native, correctly folded, globular protein is characterised by a tertiary structure with a tightly packed core, in which amino

Fig. E4.11 CD stopped-flow refolding kinetics for oxidised cytochrome *c*. The reaction was triggered by a guanidine hydrochloride concentration jump from 4.3 M to 0.7 M. The time courses in the far- and near-UV are in (a) and (b), respectively. The two folding kinetics are compared in (c), after normalisation. The curve fits are for a two-exponential function. Note the slower recovery of the folded state condition for the near-UV compared to the far-UV data. (Elöve *et al.*, 1992.) (Figure reproduced with permission from the American Chemical Society.)

acids fit very favourably through van der Waals interactions. The CD experiments on α-lactoglobulin indicated that during unfolding the far- and near-UV signals followed different time courses (similar to the data shown in Fig. E4.11). An intermediate state appeared in which the secondary structure of the native state was maintained but the tertiary structure was lost. The radius of gyration measured by small angle X-ray scattering (see Chapter G2) and other parameters established that the intermediate had a compact structure, resembling an expanded form of the tertiary structure, in which the core had lost its tight packing; this was named a molten globule.

Molten globule structures are defined by their far- and near-UV CD signatures (secondary structure similar to the native state, loss of tertiary structure), their radius of gyration (only slightly larger than the native value) and their effect on ANS (8-anilino-1-naphthalene sulphonate) fluorescence, which is sensitive to the local environment of the probe; they can penetrate the loosely packed hydrophobic core of a molten globule, but not the tightly packed core of a native, folded, globular protein. A degree of controversy still remains regarding the existence of the molten globule as a folding intermediate for small globular monodomain proteins, however, since no sign of it has been found in calorimetry experiments (see Chapter C2).

E4.5 Nucleic acids and protein–nucleic acid interactions

We recall that the nucleotide, the fundamental unit in nucleic acids, consists of phosphate, a sugar ring (ribose in RNA and deoxyribose in DNA) and an aromatic base (one of adenine, A, guanine, G, cytosine, C, and either thymine, T,

Comment E4.10
The molten globule hypothesis

Ptitsyn and his coworkers analysed the denaturation of α-lactoglobulin and noted the existence of a partially unfolded intermediate that maintained the far-UV secondary structure CD signature of the folded state, but which had lost the near-UV tertiary structure signature. The observation formed the basis of the molten globule hypothesis (Dolgikh *et al.*, 1981).

Comment E4.11
CD spectra of
nucleic acids of
known structure

A compilation of CD
spectra of nucleic
acids in various
conformations has
been published in
Landolt-Börnstein
(Johnson, 1990).

for DNA or uracil, U, for RNA). The bases are planar so that they do not have any intrinsic CD (see Comment E4.1). The CD signal of the bases in nucleic acid comes to a small extent from the asymmetry of the sugar, but is dominated by the hydrophobic stacking of the base-pairs in the helical structures. This technique, which is particularly sensitive to the secondary structure induced by different solvent environments and base compositions, became the basis for the investigation of nucleic acids of unknown conformation by comparing their CD to established spectra for nucleic acids of known structure (Comment E4.11).

E4.5.1 RNA

We show the CD of RNA PK5 as an example in Fig. E4.12. The molecule is a 26-nucleotide strand that displays various structures as a function of rising temperature. CD and normal UV absorption spectra were measured for different temperatures and compared to the basis sets from 58 RNA molecules of known structure. The analysis yielded a pseudoknot structure at the lowest temperature, and a hairpin structure at 30 °C with the PK5 RNA in the A-form (RNA secondary structures are discussed in Chapter A2). The strand was unfolded (melted) at 70 °C. Note how the spectra are sensitive to the different types of stacking.

Fig. E4.12 CD spectra of RNA PK5 at different temperatures: 0 °C (line); 30 °C (dots); 70 °C (dashes) (Johnson and Gray, 1992.)

Fig. E4.13 CD spectra of poly d(GC) · poly d(GC): in the B-form in 0,1M NaCl (full line); in the Z-form in 4M NaCl or in 60% ethanol (dashed line). (Pohl and Jovin, 1972.) (Figure reproduced with permission from Elsevier.)

E4.5.2 DNA

It was well known that double-helical DNA can take up a variety of right-handed structures depending on the solvent conditions and hydration level (see Chapter A2). Then CD observations led to the striking discovery of Z-DNA, a left-handed, double-helical structure that is adopted by poly d(GC) · poly d(GC) in high salt concentrations or in the presence of alcohol (Fig. E4.13). The CD spectra of B- and Z-DNA are characteristically different. The B-form has a positive band close to 280 nm and a negative band close to 255 nm, while the Z-form displays a negative band at 290 nm, and a positive band at 265 nm. Extending the spectra to shorter wavelengths (not shown in Fig. E4.13) revealed a further positive band close to 190 nm for B-DNA and a negative band close to 200 nm and a positive band close to 180 nm for Z-DNA. Its CD signature guided the search for the Z-form in natural DNA (Fig. E4.14). The experiment was based on the complex supercoiling properties (when the double helix twists around itself) of circular DNA in plasmids. The CD of the relaxed plasmid corresponds to the B-form. When the two complementary strands in the plasmid were made to associate under a certain topological constraint, however, the CD spectrum

Fig. E4.14 CD spectra of a relaxed (dashed line) plasmid and the same plasmid under a certain topological constraint. The dashed line is similar to a B-form spectrum, while the full line has several Z-form features. (Johnson, 1985). (Figure reproduced with permission from John Wiley & Sons.)

showed features characteristic of the Z-form (Fig. E4.14), indicating that left-handed regions were used to respond to the imposed supercoiling constraint. The spectrum corresponded to a Z-form content of about 40%. It is interesting to note that CD experiments established also that RNA strands with alternating purine–pyrimidine bases, such as poly r (GC) · poly r(GC) could assume the Z-form if solvent conditions were favourable.

E4.5.3 Protein–nucleic acid interactions

The CD spectrum of a protein–nucleic acid complex above 250 nm is dominated by the secondary structure of the nucleic acid component, and subtle changes in base stacking, for example, can be examined in this spectral region. The interpretation of measurements below 250 nm, where the protein secondary structure contribution becomes dominant, is more difficult, and should be approached with the necessary caution with respect to the spectral width examined, corrections for overlapping nucleic acid bands and the protein database chosen. Different types of complex may lend themselves more readily to a CD analysis. For example, it is easier to correct for the CD spectra of RNA, which are less variable than those of DNA structures. The CD spectra relating to the interaction between protein and RNA in the capsid of a bacteriophage are shown in Fig. E4.15.

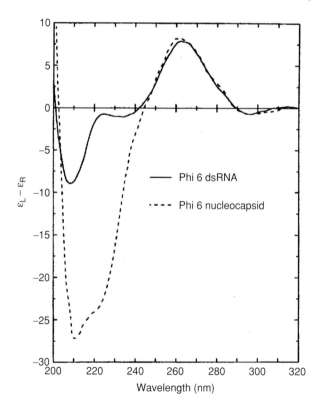

Fig. E4.15 CD spectra, expressed per mole of nucleotide of the double-stranded RNA (full line) and the protein–nucleic acid capsid (dashed line) of bacteriophage ϕ6 (quoted in Gray, Fig. 9 in Fasman (ed.), 1996). (After Steely *et al.*, 1986.)

Bacteriophage $\phi6$ contains three double-stranded (ds) RNA molecules associated with protein within the same nucleocapsid. In the intact virus, the capsid is itself surrounded by a lipid bilayer. Analysis of the isolated ds RNA and nucleocapsid spectra shows clearly how the RNA component dominates above 240 nm (Fig. E4.15). In this spectral region the encapsidated RNA displays a structure intermediate between an A-form and a dehydrated A-form, showing the effect of condensation. A subtraction of the ds RNA spectrum from the nucleocapsid spectrum above 240 nm in order to examine the protein secondary structure contribution was considered to be reliable from inspection of the spectra. The analysis yielded 50% α-helix content in the protein.

E4.6 Carbohydrates

CD analysis of carbohydrates can contribute to elucidating their structures, which are extremely complex in composition as well as in conformation (see Chapter A3). As with other polymeric structures, CD distinguishes between random coil and helix organisation, between rigid and flexible structures and, in favourable cases, the type of linkage between the monomer subunits (which, unlike proteins and nucleic acids, can be extremely varied). Experiments are difficult, however, because the relevant information from saccharides occurs at wavelengths below 200 nm (unless there are substituted aromatic groups), where solvent contributions dominate. The wavelength range is accessible with good laboratory instrumentation and synchrotron sources (see below), and solvent contributions can be minimised by using desolvated *film* samples.

The linkage contribution to a CD spectrum can be estimated empirically from the difference between the spectra of the oligosaccharide and the monomer subunits and modelled by using NMR, X-ray crystallography results and molecular dynamics calculations. The example of trehalose, a dimer of specifically linked glucose subunits (see Chapter A3) is shown in Fig. E4.16. The spectra indicate a positive linkage contribution at 160 nm, which compensates for the negative

Fig. E4.16 CD spectra of (a) methyl α-D-glucopyranoside and (b) α,α-trehalose (quoted in Stevens, in Fasman (ed.), 1996).

Fig. E4.17 CD spectra of: (a) methyl α-D-glucopyranoside; (b) maltose; and (c) amylose (quoted in Stevens, in Fasman (ed.), 1996).

peak in curve (a) and leads to a weak trehalose CD spectrum. The effect of chain length is seen in Fig. E4.17, in which are shown the spectra of a glucose monomer, maltose (a dimer) and amylose (a homopolymer with the same linkage as maltose). There is a large positive linkage contribution in maltose below 175 nm. The difference between the maltose and amylose curves is due to the effects of 'kinks' or 'bends' in the longer polymer (Fig. E4.18). The long helical polymer

Fig. E4.18 Perspective drawing of a segment of amylose chain chosen as representative in a Monte Carlo molecular dynamics calculation. The axes represent the conformational coordinates of the chain. The scales are in ångström units. Circles represent the glycosidic oxygens and lines are virtual bonds spanning the subunits (quoted in Stevens in Fasman (ed.), 1996).

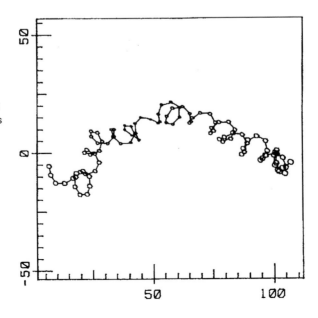

is, in fact, made up of finite lengths of straight parts, as shown in the figure. The *persistence length* of the polymer is a measure of the straight stretches.

E4.7 CD from IR radiation to X-rays

Modern instrumentation and powerful computer-based analysis techniques have extended the experimental usefulness of CD measurements to the vibrational spectral region in the IR, on the one hand, and to the far-UV, also called the *vacuum-UV* (by using intense synchrotron radiation), on the other. An example of the sensitivity of IR CD (VCD) to protein secondary structure is given in Fig. E4.19.

References on developments in VCD and synchrotron radiation CD are given for the interested reader in the book cited at the end of this chapter.

 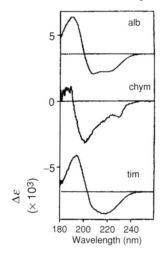

Fig. E4.19 VCD of the amide I band in D_2O solvent and electronic CD for the α protein albumin (alb), the β protein chymotrypsin (chym) and the $\alpha\beta$ protein triosephosphateisomerase (tim) (quoted in Keiderling in Fasman (ed.), 1996).

E4.8 Checklist of key ideas

* Plane-polarised light can be resolved into equal left and right circularly polarised components.
* Elliptically polarised light is formed if the right and left components are not equal – in CD this happens when one of the components is absorbed more than the other.
* In an optically active material, the major axis of the ellipse is rotated by an angle called the *optical rotation* angle with respect to the direction of the incident plane-polarised beam.
* The bands observed in absorption and CD spectra of macromolecules in solution correspond mainly to electronic transitions between the lowest vibrational level in the ground-state potential energy well and various vibrational levels in the first excited state potential energy well.
* The integrated intensity of the CD band associated with the transition is defined in terms of the *rotational strength*.

- A chromophore is optically active if it is itself chiral, if it is perturbed by a chiral field due to the structural environment, or both. It is the optical activity induced by chromophore *organisation* that is mainly of interest in conformational studies of macromolecules by CD.

- α-helical proteins, β-sheet proteins and unfolded polypeptides display characteristic CD bands. β proteins show a greater CD spectral variety and less intense bands than α proteins, and have been classified in different groups according to their spectrum and β structure type. The CD spectra of α/β proteins can considered as mixtures of α-helix and β-structure spectra, in which the stronger α-helix band intensities dominate.

- Databases of empirical CD spectra corresponding to 'known' (from crystallography or NMR) protein secondary structure families (including membrane proteins), with which the spectra of 'unknown' proteins could be fitted, have been set up.

- The near-UV CD spectrum of a protein is dominated by aromatic residues and is a sensitive measure of tertiary structure.

- CD spectra are good empirical indicators of the folding state of a protein; the far-UV spectral region informs on the presence or absence of secondary structure, while the near-UV informs on tertiary structure.

- Combined with stopped-flow devices, kinetic CD experiments have been used to follow protein unfolding and refolding processes.

- The molten globule hypothesis has been formulated for the refolding pathway of globular proteins following the CD analysis of intermediate states.

- The CD signal of nucleic acids is dominated by the hydrophobic stacking of the base-pairs in the helical structures, and is, therefore, sensitive to structure. CD observations led to the striking discovery of Z-DNA, a left-handed double-helical structure, adopted by poly d(GC) · poly d(GC) in high salt concentrations or in the presence of alcohol.

- The CD spectrum of a protein–nucleic acid complex above 250 nm is dominated by the secondary structure of the nucleic acid component, and subtle changes in base stacking, for example, can be examined in this spectral region. The interpretation of measurements below 250 nm, where the protein secondary structure contribution becomes dominant, is more difficult.

- CD analysis of carbohydrates distinguishes between random coil and helix organisation, between rigid and flexible structures and, in favourable cases, the type of linkage between the monomer subunits (which, unlike in proteins and nucleic acids, can be extremely varied). Experiments are difficult, however, because the relevant information from saccharides occurs at wavelengths below 200 nm (unless there are substituted aromatic groups), where solvent contributions dominate. Solvent contributions have been minimised, however, by using desolvated film samples.

Suggestion for further reading

Fasman, G. (ed) (1996). *Circular Dichroism*. New York: Plenum Press.

Part F
Optical microscopy

Chapter F1
Light microscopy

F1.1 Historical review

F1.1.1 Light microscope

Mid-fifteenth century

The simple one-lens microscope was available at this time, with low-power magnifiers being used for the examination of insects. From these early beginnings, glass lenses of increased power were developed. **A. Leeuwenhoek (1632–1723)** produced remarkable microscopes. One of them had a resolving power of 1.35 µm, which was enough for basic cytology. About the same time the basic idea for the compound microscope, in which two or more lenses are arranged in such way to form an enlarged image of an object, occurred independently to several people in the Netherlands (**H. Jansen**, his son **Zacharias** and **H. Lippershey**). Subsequently, compound microscopes became widely available.

1830

J. Lister published the first work in which the construction and design of achromatic microscope lenses was placed on theoretical bases. He showed how two separate achromatic lenses could be combined to act as a single lens so that the image rendered would be completely free from chromatic aberration and from distortion caused by curvature of the lenses. This discovery opened the way to the construction of high-power microscope lenses over the next 50 years. The first achromatic microscope lenses appeared in the Netherlands at about the end of the eighteenth century.

The latter part of the nineteenth century

E. Abbe was the outstanding figure in development of the microscope. Abbe placed the theory of microscopic image formation on a sound basis by emphasising the importance of the light diffracted by the object in resolving its fine details. He introduced the concept of numerical aperture and developed oil-immersion lenses. Very soon this type of lens, with its much improved resolving power and high-quality images, was an essential part of the microscopist's equipment. Abbe showed that an increase of resolving power of the microscope was possible only by increasing in the numerical aperture of the optical system or by using a shorter wavelength. The diffraction limit of resolution power of a microscope according Abbe is about $\frac{1}{2}$ of the wavelength for illuminating light.

Twentieth century

Developments of the microscope have included the introduction of larger stands with built-in illumination systems and higher standards of accuracy of construction. Many different types of microscopes, such as the dark-field microscope, the polarization microscope, fluorescence microscopy and wide-field epifluorescence microscopy have been proposed. Especially notable was the introduction in 1934 of phase-contrast microscopy, which made it possible to examine living cells and tissue specimens without staining them.

F1.1.2 Confocal microscopy

1951

J. Young and **F. Roberts** were the first to examine the idea of confocal scanning optical microscopy. In the **1950s M. Minsky** described in detail the principal features of this type of imaging. He built a microscope that enabled him to view deeper layers in a specimen with astonishing clarity, without first having to cut the specimen into thin sections.

1985

M. Petran and **M. Hadravsky** proposed a different type of confocal scanning optical microscope, the so-called tandem scanning reflected-light microscope, which operated in real time and produced an image that could be directly observed with the naked eye. They used a Nipkow disc which contains many thousands of pinholes distributed in spiral pattern (Fig. F1.10). In the first experiment the microscope was transported to the top of a high mountain to obtain adequate illumination. In 1988 stimulated by Petran's work **G. Kino** developed a modern version of the tandem scanning microscope that was called the real-time scanning optical microscope.

Late 1980s

T. Wilson and **S. J. Hewlett** proposed using a laser beam reflected off vibrating mirrors to generate a three-dimensional image. In laser scanning confocal microscopy, a cathode ray tube recreates the image. The detector signal modulates the cathode ray tube intensity and the detector scan rate is synchronised to the laser scan rate. A computer stores all confocal images. By collecting additional scans after moving the plane of focus up and down, subsequent image processing can reconstruct a three-dimensional image.

F1.1.3 Near-field scanning microscopy

1928

E. Synge was one of the first to bring to the physics community the idea of the near-field scanning microscope (NSOM). As originally conceived an illuminated subwavelength-diameter aperture acts as a light source of subwavelength dimensions that can be scanned close to the object to generate a superresolution image.

1972

E. Ash and **G. Nicholls** used the idea of near-field scanning for imaging purposes. Using 3-cm microwave radiation, they achieved a resolution of $\lambda/60$ in one dimension and $\lambda/20$ for two-dimensional objects. In the **early 1980s A. Lewis's** group demonstrated that subwavelength apertures down to 15 nm gave readily detectable throughputs of transmitted or fluorescent light. Independently **D. Pohl's** group transmitted light through a subwavelength aperture at the tip of an etched quartz rod. In **1995 Lewis's** group demonstrated a new design for the NSOM probe based on a bent optical fibre. Using this optical element a unique near-field optical head has been designed that can convert any far-field optical microscope into a multifunctional near-field optical, far-field optical and scanned probe imaging system.

2000 to now

Twenty years of development of the near-field technique have led to an instrument that can be simply added to the stage of any conventional optical microscope to enable near-field scanning optical microscopy imaging. Today near-field optics produce the highest optical resolution that has ever been achieved in light microscopy and are crashing through the barriers that have been traditionally accepted for this field of science.

F1.2 Light microscopy inside the classical limit

One of the fundamental properties of light is that a beam of light can convey information from one place to another. This information concerns both the source of light and also any object that has partly absorbed or reflected or refracted the light before it reaches the observer. More information reaches the human brain through the eyes than through any other sense organ. Even so, the visual system extracts only a minute fraction of the information that is imprinted on the light that enters the eye. The main goal of an optical instrument is to extract much more information from visual surroundings.

The light microscope is an instrument for producing enlarged images of objects that are too small to be seen unaided. Such images may be viewed directly with a viewing screen or photographic apparatus or special electronic device. The value

> **Comment F1.1   Geometrical and physical approaches in optics**
>
> Under certain conditions the wave nature of light is dominant, and light can be treated as a ray that travels in a straight path. These rays can be bent at the interface between two transparent objects. This forms the basis of geometric optics. Alternatively, we can emphasise the electromagnetic nature of light, in which diffraction and interference are recognised. This underlines the ideas of physical optics. From a practical point of view, these approaches are simply two sides of the same coin: the geometric approach is useful in understanding the ideas of image focus and aberration, while the physical approach leads more easily to an explanation of image contrast and the limits of resolution.

of a microscope lies in its ability not only to magnify objects but also to make their fine details visible (Comment F1.1). These two features depend for the most part on the quality of the lens that forms the primary image. This lens is the critical element in every microscope. The objective lens determines magnification, field of view and resolution; its quality determines light transmission, the contrast and aberrations of the image. Below we briefly describe the main principles of light microscopy.

F1.2.1 The standard light microscope

One-lens microscope

Let us start with an ideal simple lens having two convex curvatures, each with a radius of curvature and focal point (called F and F'). An image is formed by a point-by-point translation of the object, which is said to be in the object plane, into an image that exists in the image plane. Figure F1.1 illustrates this translation. Two light rays coming from the point of the arrow are represented so that one

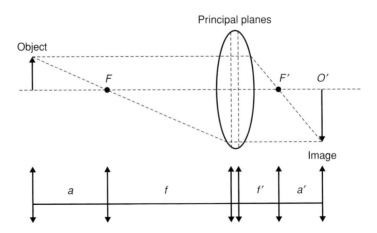

Fig. F1.1 Optical geometry of a simple lens.

> ## Comment F1.2   Aberrations
>
> Aberration commonly comes in several forms: chromatic, point-imaging and astigmatic. Chromatic aberration occurs because the index of refraction is dependent on wavelength. Light of different wavelengths (white light) coming from a single point in the image plane is focused into separate image planes. Correctly engineered, the refractive index of the overall system can be made independent of wavelength. This lens is termed *anachromatic*.
>
> Point-imaging aberrations occur because monochromatic rays from a single point in the object plane do not necessarily pass through the same image plane. Spherical aberration occurs when this effect is caused by the refraction of a single object point by different parts of the lens, with the consequence that the point in the image plane is not in focus. Lenses constructed to eliminate point-imaging aberrations are called *aplanatic*.
>
> Astigmatism is another symmetry-destroying aberration in which the arrangement of object points in Cartesian space is warped so that the image does not maintain a linear point-to-point relationship with the object. Lenses that are corrected for astigmatism are called *anastigmatic*.

travels parallel to the lens axis and the other passes through the focal point. Both pass into the lens. After passing through the lens, the light rays emerge, and the first is refracted through the focal point F, while the second now continues parallel to the lens axis. The point of intersection of the two rays is at the point of the arrow in the image plane. The geometry of the light rays obeys the relation $aa' = ff$, with the magnification given by $-f/a$. The meaning of the negative sign is that the image is inverted in the image plane with respect to the object plain.

An ideal lens brings all light rays from a single point in the object plane to a sharp focus in the image plane. However, real lenses fail to focus all the points in a single image plane and are said to have aberration (Comment F1.2).

Compound microscope

In practice, a single-lens microscope requires a lens with a very short focal length, and the obsever's eye has to be placed extremely close to the lens (Comment F1.3). To overcome this practical restriction, a compound microscope can be used: this consists of an objective lens and a secondary-lens system, the eyepiece (Fig. F1.2). This arrangement allows an object to be placed just beyond the focal point of the objective. The objective-magnified image is further magnified by the eyepiece and then focused onto the retina of the eye.

Diffraction limit of resolution

Because of diffraction effects, the image of a point source of light can never be perfect, even assuming that all the imperfections in the lens are corrected.

> ## Comment F1.3
> ## Reading glass
>
> A simple microscope is often known as a magnifier, or loupe. Supplied in a mount furnished with a handle, such a lens is commonly called a reading glass. The apparent size of an object viewed directly by the eye depends upon its distance from the eye. As an object is brought nearer to the eye, the angle which it subtends at the eye increases, and so the retinal-image size (and hence the apparent object size) also increases.

(a)

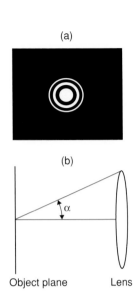

(b)

Object plane   Lens

Fig. F1.2 Schematic illustration of a compound microscope.

Fig. F1.3 (a) The Airy disc is the diffraction image of each point in the image field. Resolution depends on being able to separate these diffraction patterns from one another. (b) The geometry used for determining the resolution of two points by a lens. A commonly used criterion for determining the resolution of an imaging system is the Rayleigh distance, the minimum distance by which two points must be separated to be distinguished when the illumination is incoherent. This Rayleigh resolution limit is theoretically half the wavelength of the imaging of light. Although this suggests that, for a wavelength of 500 nm, 250-nm resolution in possible conventional microscopy never achieves such definition.

In practice, a bright spot of light in the object plane appears as a circle of light surrounded by a series of concentric rings – the result of interference. This pattern is called the Airy disc (Fig. F1.3). The radius of the first dark ring, r, encircling the central bright spot is

$$r = \frac{0.6\lambda}{n \sin \alpha} \tag{F1.1}$$

where α is the angle made by drawing a line from the central axis of the lens to the edge of the lens, λ is the illuminating wavelength of light, and n is the refractive index of the object side of the lens. If two object points are in proximity, the ability of an observer to resolve them into two clearly distinct points (i.e. the resolution) depends not on the separation of the two points but on the dimensions of their Airy discs. Equation (F1.1) shows that resolution can be improved by shortening the wavelength of the illuminating light, increasing the index of refraction on the objective lens side (for example, placing the object under oil), and increasing α. The angle α can be increased either by shortening the distance between the lens and the object or by increasing the diameter of the lens.

Because the state of each point in the object state space is given by the complete set of these many orders of the diffraction, the only way to obtain complete information about the object's state space (image) is to collect all of the observable data carried in the totality of the diffracted orders. This ideal requires the largest possible collection angle between the object and the lens aperture (i.e. the largest α). This is called the angular aperture. However, the detail resolved is proportional not to the angular aperture but to the numerical aperture, NA. In the case of a dry lens, i.e. one in which the cone of light rays from the object passes through air, the numerical aperture is simply $\sin \alpha$. In an immersion system, in which oil or sometimes water or glycerin is placed between the front lens of the objective and the specimen, the numerical aperture is increased and the value of $\sin \alpha$ has then to be multiplied by the value of the refractive index of the immersion medium. Thus, the numerical aperture of a lens is a direct measure of the resolving power of the lens.

In any lens system, be it electron, acoustic or optical, the diffraction effect tends to spread the image of a point object. The intensity in the resulting image varies with distance from the central point. The point spread function (PSF) which is simply the response of the imaging system to a point function describes this variation. The imaging characteristics of a microscope can be derived by convolution of its PSF with a function characterising the detecting of the sample (see also Comments F1.4 and F3.2).

F1.2.2 The problem of contrast

The ability to see the details of the system under microscopic observation requires contrast, i.e. an essential difference in optical properties between point elements in the object's state space. Biological macromolecules are generally transparent to light, and their direct visualisation under bright field illumination is often impossible. Numerous contrast-enhancing techniques have been developed that increase the contrast of cells, including dark-field, phase-contrast, polarization and interference microscopy. All these techniques are briefly discussed below.

Dark-field microscopy

In bright-field studies, light coming from the condenser system illuminates the object plane. In the absence of a specimen in the object plane, light uniformly enters the objective lens, giving an evenly illuminated field of view. If an opaque disc is inserted into the condenser system, an annular or hollow cone of light will fall on the objective plane. The circle of light can be adjusted so that the light surrounds the objective, leaving the field dark (hence, dark field). Though the field is dark, some light reaches this region, and any object placed in the dark field diffracts some of this light. The resolution of the object is quite poor because the information carried in the low orders of diffraction is lost, but the technique illustrates how diffraction can be used to enhance contrast dramatically (Fig. F1.4(a) and (b)).

Comment F1.4
PSF in confocal microscopy

In any lens system, be it electron, acoustic or optical, the diffraction effect tends to spread the image of a point object. The intensity in the resulting image varies with distance from the central point. The PSF which is simply the response of the imaging system to a point function describes this variation. The imaging characteristics of a microscope can be derived by convolving its PSF with a function characterising the detecting of the sample.

Fig. F1.4 Examples of: (a) bright- and (b) dark-field illumination. (Bergethon, 1995.)

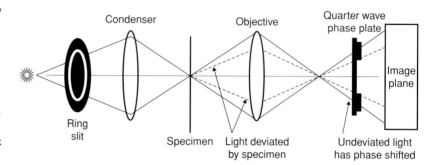

Fig. F1.5 Schematic view
of the optical train in a
phase-contrast
microscope. The hollow
cone of light passes
through the object plane
and falls as a ring of light
on the objective lens. The
image of the annular ring
is thus formed in the back
focal plane of the
objective lens (eyepiece
side). In order to induce
the necessary phase shift,
a quarter wave phase
plate is introduced at the
back focal plane. This
plate is an annular disc of
material which slows the
light so that the light
travelling directly from
the condenser annulus is
retarded one-quarter
wavelength. Any light
that passes through a
specimen placed in the
objective plane and is
diffracted (deviated) is not
focused by the objective
onto the phase plate and
does not have its phase
altered. The deviated and
the phase-enhanced-
undeviated light
recombine at the image
plane where substantial
interference now occurs.
(Adapted from Bergethon,
1995.)

Phase-contrast microscopy

A common technique for visualising cells is to convert the phase difference produced when light passes through cells of different refractive indices into intensity difference. The light passing through each point of a specimen in the object plane is diffracted twice, once by the specimen and once by the objective lens. Each point of the viewed image is the result of the combined interference of each point of the object and each point of objective lens. Though the image contains the information related to the varying diffractive indices of the specimen, the phase differences are generally too small to generate sufficient interference to be perceived as intensity differences.

If the lens system is built to alter the phase of light sufficiently prior to recombination in the image plane, enough destructive interference can be generated to create contrast enhancement. In the phase-contrast microscope (Fig. F1.5), an annular light source is used just as in dark-field microscopy. The interference pattern is seen by the observer as intensity differences and contrast is increased. The phase-contrast method allows cells to be observed in substantial detail.

Polarization microscopy

A polarization microscope contains two polarisers, one in the condenser and one between the objective and the eyepiece, called the analyser (Fig. F1.6). Certain objects possess the property of form birefringence and pass only plane-polarised

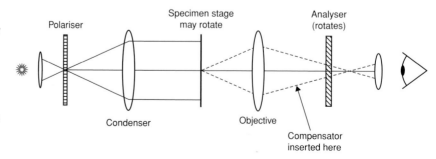

Fig. F1.6 Schematic view
of the optical train of a
polarization microscope.
A polarized light source
can be produced by
placing a polariser in the
condensing system of the
microscope. (Adapted
from Bergethon 1995.)

light when the light is parallel to the long axis of the particles in the object (see Comment D7.1).

When polarised light is passed through such an object at an angle of 45°, the light is resolved into parallel and perpendicular components. If two polarisers are turned at 90° to each other, no light passes, and a dark field is observed. If an object is placed at 45° to both of the polarisers, it passes some light and appears bright against the dark field. An object composed of stacked discs or long molecules in a parallel array appears bright at 45° because both of these structures are birefringent. In the process, use of a compensator provides the information necessary to indicate the positive or negative birefringence of material under investigation.

F1.3 Subwavelength resolution within the restrictions of geometrical optics

F1.3.1 Confocal microscopy

In conventional wide-view microscopic imaging, not only is a sharp image generated from an in-focus area, signals above and below the focal plane are also acquired as out-of-focus blurs that distort the contrast and sharpness of the final image (Fig. F1.7(a)). Confocal microscopy is a technique for increasing the contrast of microscope images, particularly in thick samples. By restricting the observed volume, the technique keeps nearby scatterers from contributing to the detected signal (Fig F1.7(b)).

Figure F1.8 shows how it is possible to view only the focal volume. A point light source is imaged at the object plane, so that the illuminated point and the source are confocal. Then the observation optics forms an image of the illuminated point on a pinhole. Thus there are three points all mutually confocal – hence the name. By moving the laser spot across the specimen with scanning mirrors, a sharp image from the confocal plane can be generated with a sensitive photodetector.

Fig. F1.7 (a) Demonstration of how a lens forms an image of two points of a thick sample, one at the focal point and one away from that point. (b) A pinhole conjugate to the focal point passes all the light from the focal point, and very little of light from the out-of-focus point.

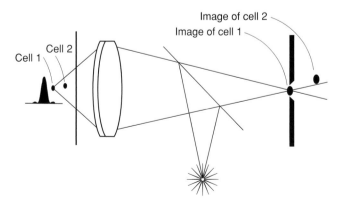

Image of cell 2
Image of cell 1
Cell 2
Cell 1

Fig. F1.8 Light from a point source is imaged at a single point of the object and that, in turn, is imaged on a small pinhole, making all these points confocal. (Webb, 1996.)

Fig. F1.9 Comparison of rhodamine-123-loaded hepatocytes viewed by conventional and confocal fluorescence microscopy: (a) conventional micrograph of rhodamine-123-labelled mitochondria in a cultured hepatocyte; the image was collected with a video camera using a 1.25 oil-immersion lens; (b) a confocal image of rhodamine-123-labelled hepatocyte loaded under similar conditions. The magnification of the two images is approximately the same. (Lemasters et al., 1993.)

Fig. F1.10 The Nipkow disc is a disc with many holes. The holes cover about 1% of the image-plane space at any instant, and rotation of the disc maps out all of required pixels.

Three-dimensional image and resolution

In biological microscopy it is usually more important to image the interior of cells, rather than the cell surface. In contrast to the regular light microscope, the confocal microscope employs a point-like illumination and detection arrangement (Fig. F1.8). Therefore the PSF of the confocal fluorescence microscope is determined by the product of two almost identical intensity PSFs (Comment F1.4). Hence, as a good approximation the effective PSF of a confocal microscope is given by the square of the focal illumination intensity distribution in the objective lens, so that, in a simplistic photon picture, only photons from the closest vicinity of the diffraction-limited spot contribute to the signal. In a confocal microscope, the effective focus acts as a three-dimensional probe that can be scanned through a transparent specimen. Therefore, confocal fluorescence microscopy readily produces three-dimensional images of transparent specimens such as cells.

The improvement of resolution and clarity provided by confocal microscopy is readily apparent when a photomicrograph obtained with a conventional microscope is compared with the corresponding confocal image. Figure F1.9 illustrates cultured rat hepatocytes stained with rhodamine 123, a cationic fluorophore that accumulates electrophoretically into mitochondria in response to the mitochondrial membrane potential. Figure F1.9(a) is an image obtained using conventional epiillumination fluorescence microscopy. Individual mitochondria are easily identified at the periphery where the hepatocyte is thinnest, but in the thicker central portion of the cell, superimposition and out-of-focus fluorescence obscure mitochondrial outlines. Figure F1.9(b) is a confocal image of a hepatocyte labelled under similar conditions. In the confocal image mitochondria are sharply defined spheres and filaments.

Scanning process

The price for restricting the observed volume is that in the confocal microscope we can observe only one point at a time. In the scanning regime the image is built up, pixel by pixel, like a television picture. One of three approaches is generally used to scan the image. The sample or the optics can be mechanically scanned. Another, somewhat faster approach is to scan the beam using a galvanometer mirror or Bragg cells. The third approach is to use a Nipkow disc, which is a disc in which a large number of pinholes have been drilled to allow the entire image to be scanned (Fig. F1.10). This method is capable of producing real-time images at hundreds of frames per second.

F1.4 Lensless microscopy

The optical microscope has one serious drawback – its resolution – which results from the fundamental physics of the lens. This limit of resolution is theoretically half the wavelength of the imaging radiation (Section F1.2.1). This fact was one of the driving forces leading cell biologists to look for new high-resolution imaging tools to probe cellular structure and function. Near-field scanning optical microscopy (NSOM, pronounced 'ensom') is an example of the application of high-resolution imaging in cell biology. NSOM carries optical imaging to resolutions normally associated with electron microscopy. It links the world of conventional optical microscopy with scanned probe imaging techniques, the most common of which is atomic force microscopy.

F1.4.1 Near-field scanning microscopy

The concept of NSOM

The near-field optical interaction between a sharp probe and a sample of interest can be exploited to image surfaces. The easiest such probe to conceptualise consists of a subwavelength-diameter aperture in a optically opaque screen (Fig. F1.11). Light incident upon one side of the screen is transmitted through the aperture and, if the sample is within the near field, illuminates only one small region at any one time. The technique of imaging is based on the use of an 'optical stethoscope', which places a small near-field light source a very short distance from the sample (less than the wavelength of light). Images are obtained by scanning the light source over the sample, and the resolution is limited only by the diameter of the light source. All mechanisms used to obtain contrast in conventional microscopy can be applied to NSOM.

Main component of the near-field microscope

The single most important element in any form of scanning probe microscopy is the probe itself. The ideal aperture is a transparent hole in a thin, perfectly

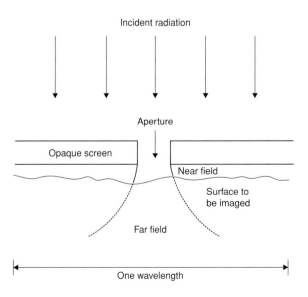

Fig. F1.11 Concept of near-field scanning optical microscopy or 'optical stethoscope'. An illuminated aperture acts as a light source of subwavelength dimensions that can be scanned close to an object to generate a superresolution image.

Fig. F1.12 Magnified view of: (a) a standard cantilevered fibre-optic probe, and (b) a probe that has been etched in hydrofluoric acid buffered with ammonium fluoride. The standard NSOM probe has a diameter of 125 μm and a measured spring constant of ∼400–600 N m^{-1}. For the probe shown in (b), the diameter has been etched down to 10–15 μm, reducing the spring constant to less than 5 N m^{-1}. (Shiku and Dunn, 1999.)

Fig. F1.13 Bent glass tip. Such a structure can be used in any of the available designs for force microscopy. (Lewis *et al.*, 1995.)

conducting metal film. Real metals, however, are fairly poor conductors at optical frequencies. Therefore, the screen around the hole must have a thickness at least an order of magnitude larger than the skin depth to prevent excessive stray radiation. This requirement puts a lower limit on the aperture diameter which cannot be substantially smaller than the film thickness.

A typical probe is a fibre-optic element, typically 10 nm in diameter, that yields a resolution down to 12 nm (< 1/40) with visible light. Figure F1.12 shows examples of such optical elements. Another probe is the submicroscopic aperture at the tip of a glass pipette (Fig. F1.13). Such a bent glass structure can be easily combined with any of the available designs for force microscopy. The result is a microscope with exceptional qualities for both near-field optical imaging and normal force imaging. This microscope allows gold colloids with 80 and 40 nm diameters to be distinguished. This result is unique and cannot be achieved by any far-field optical system.

NSOM in liquids

One major problem in the application of NSOM is the ability to image in liquids. Control of the distance between the sample and the near-field probe is more difficult because of damping of oscillations due both to the presence of the liquid and, in buffer, to electrostatic interactions between the ions on the surface of the near-field probe and ions in solution. One of the approaches to solving this problem is combining a near-field optical microscope and an atomic force microscope. We will discuss this approach in Chapter F3, which is devoted to atomic force microscopy.

F1.5 Checklist of key ideas

- The light microscope is an instrument for producing enlarged images of objects that are too small to be seen unaided; such images may be viewed directly with a viewing screen or photographic apparatus or special electronic device.
- Because biological macromolecules are generally transparent to light numerous contrast-enhancing techniques have been developed including dark-field, phase-contrast, polarization and interference microscopy.
- The diffraction limit of resolution power of a microscope is about half of the wavelength of the illuminating light; an increase of the resolving power of the microscope is possible only by increasing the numerical aperture of the optical system or by using a shorter wavelength.
- In contrast to the ordinary light microscope, the confocal microscope employs a point-like illumination and detection arrangement; by restricting the observed volume, the technique keeps nearby scatterers from contributing to the detected signal.
- The NSOM technique of imaging is based on the use of an 'optical stethoscope', which places a small near-field light source a very short distance from the sample (less than the wavelength of light).
- Images in NSOM are obtained by scanning the light source over the sample, and the resolution is limited only by the diameter of the light source.
- NSOM links the world of conventional optical microscopy with scanned probe imaging techniques, the most common of which is atomic force microscopy.

Suggestions for further reading

Historical review

Kino, G. S., and Corle, T. R. (1989). Confocal scanning optical microscopy. *Physics Today*, **42**, 55–62.

Lichtman, J. W. (1994). Confocal microscopy. *Sci. Am.*, **271**, 40–45.

Light microscopy inside the classical limit

Bergethon, P. R. (1995). In *The Physical Basis of Biochemistry. The Foundation of Molecular Biophysics*. Ch. 16. New York: Springer.

Subwavelength resolution within the restriction of geometrical optics

Webb, R. H. (1996). Confocal optical microscopy. *Rep. Prog. Phys.*, **59**, 427–471.

Lemasters, J. J., Chacon, E., Zahrebelski, G., Reece, J. M., and Nieminen, A.-L. (1993). Laser scanning confocal microscopy of living cells. In *Optical Microscopy. Emerging Methods and Application*. eds. B. Herman and J. J., Lemasters. San Diego: Academic Press.

Webb, R. H. (1999). Theoretical bases of confocal microscopy. *Meth. Enzymol.*, **307**, 3–20.

Lensless microscopy

Betzig, E., and Trautman, J. K. (1992). Near-field optics: microscopy, spectroscopy, and surface modification beyond the diffraction limit. *Science*, **257**, 189–195.

Shiku, H., and Dunn, R. C. (1999). Near field scanning optical microscopy. *Analytical Chem.*, 23A–29A.

Chapter F2
Atomic force microscopy

F2.1 Historical review

Early 1980s

G. Binning and **H. Rohrer** proposed the scanning tunnelling microscope (for which they were awarded the Nobel Prize). This invention has initiated an exciting series of novel experiments to image the surface of conducting as well as insulating solids with atomic resolution. The first attempts at imaging biological molecules using a scanning tunnelling microscope (STM) date back to 1983. In 1987, individual molecules of phthalocyanine, lipid bilayers and ascorbic acid were reported. One year later, **H. Ohtani** with collaborators imaged benzene, the molecule of Kekulé's blue dream, as three-lobed rings. In **1989** a spectacular view of the double-stranded Z-DNA molecule, the first biological macromolecule studied using a STM, was presented.

1986

G. Binning, C. F. Quate and **C. Gerber** invented the scanning force microscope (SFM). In this microscope a sensor tip carried by a flexible cantilever is used to touch and characterise a surface. This was a significant breakthrough which allows biological macromolecules to be scanned in aqueous solution and gives reproducible imaging of DNA and of membrane protein crystals.

Early 1990s

D. J. Keller, **Q. Zhong** and **C. A. J. Putman** independently proposed the so-called tapping mode of the atomic force microscope (AFM), in which the cantilever is oscillated vertically while it is scanned over the sample. In this mode the image is formed by displaying the reduction of the oscillation amplitude at every point of the sample. It was demonstrated that reliable height data for various biological macromolecule assemblies in solution can be obtained using the tapping mode AFM.

2000 to now

Scanning force microscopy has become a powerful method for studying the structure and dynamics of biological macromolecules in solution. Exciting

Fig. F2.1 (a) Iodine atoms on the surface of platinum crystal. A vacancy pit is shown by the arrow.
(b) A liquid crystal (8CB) on graphite. Benzene rings (Kekulé's dream) are shown by the arrow.
(c) First image of putative strands of Z-DNA lying on graphite.

applications of this technique in biology range from visualisation of transcription in real time to the mechanical manipulation of a single molecule. Essential contributions to the theory and technique of application of atomic force microscopy to the study of macromolecular structure biological macromolecules were made by **C. Bustamante**, **D. J. Keller**, **P. K. Hansma**, **H. G. Hansma**, **A. Engel** and **D. J. Müller**.

F2.2 Introduction to biological problems

For many years researchers have struggled to combine the high-resolution advantages of the electron microscope with the in-water operating capabilities of the optical microscope. The invention of the STM by G. Binning and H. Rohrer in 1981 opened a new approach to achieving this goal. The STM was the first member of a new class of instruments called scanning probe microscopes, which are all based on similar principles.

Atomic-resolution surface topographies recorded with the STM in vacuum (Fig. F2.1) fostered the expectation that the STM would become a tool for sequencing DNA molecules and for imaging protein surfaces at atomic resolution. But it turned out that the data were very difficult to reproduce and such dreams have not became reality.

A significant breakthrough came in 1986 when G. Binning, C. Quate and C. Gerber proposed the AFM. In an AFM, the sample is scanned with a cantilever whose deflections are recorded. An AFM could be compared to a blind person scanning the environment with a stick to explore the path ahead.

AFM, in particular, has attracted the attention of biologists, since it theoretically combines the two most important aspects for studying structure–function relationships of biological objects: high-resolution imaging with high signal-to-noise ratio in the molecular/submolecular range and the ability to operate in *aqueous* environments, allowing the observation of dynamic molecular events in real time and physiological conditions.

However, AFM imaging of biological macromolecules has yet to become routine practice in the laboratory, despite progresses in its instrumentation. The major disadvantage of AFM that limits its general use is the fact that its lateral resolution is intrinsically dependent on the sample, the finite size and shape of the tip and the compression due to the probe force. In practice, for each biological object specific modes have to be adopted. However, the most serious problem to be solved is the limited understanding of tip–sample interactions.

F2.3 General principles

Atomic force microscopy (also scanning force microscopy), *does not use lenses to form an image,* but instead uses a sharply pointed sensor, or tip, at the end of a flexible cantilever to scan and sense the topography of a sample. In the simplest

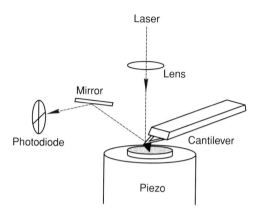

Fig. F2.2 Schematic diagram of a commercial AFM.

mode of operation, the contact mode (see below), a scanning force microscope (SFM) resembles a phonograph; as the sample is scanned underneath the tip, the cantilever deflects up or down, tracking the surface (Fig. F2.2).

Deflections as small as 0.1 nm can be detected. A variety of methods have been devised to detect the cantilever deflection. The most common approach, called the optical lever, is to reflect a laser beam off the back of the cantilever into a four-segment photodetector. The optical lever is essentially a motion amplifier. The detection of the laser spot at the photodetector is proportional to the deflection of the cantilever with a gain factor, which is typically 500–1000. This means that a deflection of 0.1 nm at the cantilever becomes a displacement of 50–100 nm at the detector, which is large enough to generate a measurable signal.

Because it is relatively simple to operate, and imaging does not require an external means of contrast, the AFM has a vast range of imaging applications from single molecules to living cells. In addition, the AFM is a powerful method for single-molecule manipulation, as cantilevers are being used to exert forces on individual molecules to probe their mechanical stability and elasticity. This application of the AFM in the force regime will be discussed in Section F5.3.

F2.3.1 The tip, a key element of an SFM

The AFM is a small instrument; current designs weigh approximately 3 kg, excluding the accompanying computer and controller. The heart of the AFM is a pyramid-shaped stylus mounted on a flexible cantilever (Fig. F2.3). The shape of

Fig. F2.3 The cantilever is the main element of an AFM. For biological applications, cantilevers are made of silicon nitride. The example shown is 200 μm thick and has a force constant of approximately 0.1 N m^{-1}. Thus a measurable deflection is induced by forces of a few piconewtons. (Engel et al., 1999.)

Fig. F2.4 Schematic
diagram showing a tip in
contact with a molecule
deposited on a substrate.
The size and shape of the
tip are defined by its
radius of curvature, R_c
and the opening angle, Φ.
Standard microfabricated
pyramidal tips (silicon
nitride, Si_3N_4) have an R_c
in the range 30–70 nm.

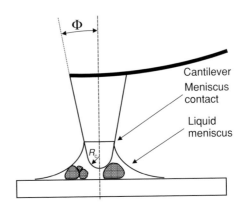

any tip can be described by two parameters: the end radius of curvature, R_c, and opening angle, Φ (Fig. F2.4). Although there is no generally agreed definition of resolution for an AFM, intuitively the smallest discernible features cannot be much smaller than R_c. Similarly, Φ affects how well the AFM visualises steep features, especially around the edges of molecules.

Consider the ideal AFM tip. It should have a high aspect ratio with a cone angle $\Phi = 0°$, a radius that is as small as possible with well-defined and reproducible molecular structure, and be mechanically and chemically robust such that its structure is not altered while imaging in fluid environments. Carbon nanotubules are the only known material that can satisfy all these critical criteria, and thus have the potential to create ideal probes for AFM imaging.

Carbon nanotubes consist of a honeycomb sp^2 hybridised carbon network (called a graphane sheet) that is rolled up into a seamless cylinder (Fig. F2.5(a)), which can be microns in length. Figure F2.5(b) shows a single needle-like carbon microtip grown on a standard pyramidal tip. Because of their flexibility, the tips are resistant to damage from tip crashes, while their slenderness permits imaging of sharp recesses in surface topography.

Fig. F2.5 (a) Fragment of
a stick model of a
single-walled carbon
nanotube. Each vertex in
the honeycomb mesh
corresponds to an
sp^2-bonded carbon atom.
Radii of nanotubes range
from 0.35 to 2.5 nm.
(b) Scanning electron
image of a single-walled
nanotube. The scale bar is
200 nm. (Hafner et al.,
2001.)

(a)   (b)

One of the most exciting results from atomic force microscopy was the discovery that atomic resolution can be achieved when the tip is in contact with the sample while scanning. Conceptually, this 'contact mode' of imaging is like using a stylus *profilometer* to measure the topography of surface atoms. At first, this idea may seem implausible, especially when one considers the macroscopic size of typical AFM tips. The AFM achieves such high resolution by using a very small loading force on the tip – typically 10^{-7}–10^{-11} N – which makes the area of contact between the tip and sample exceedingly small. This small and well-controlled loading force is the essential difference between the contact-mode AFM and earlier stylus profilometers, which typically used loading forces of the order of 10^{-4} N (Comment F2.1).

F2.3.2 Imaging modes

The most common imaging mode is the constant force mode, sometimes referred to as the height mode, which displays the changes in sample height during scanning. This mode gives calibrated height information about the sample surface. Alternatively, one can monitor the small changes in cantilever deflection that signal the feedback loop to make the height adjustments; this imaging mode is called the deflection mode. In addition to raster scanning, the cantilever can oscillate at high frequency ($\sim$300 kHz). When the tip encounters elevation changes on the surface, the amplitude of the oscillation changes. The feedback system then responds to keep the amplitude of cantilever oscillation at a constant level. This mode of operation reduces the lateral forces that can push the sample around. Two types of imaging of this kind are known: non-contact (sometimes called the attractive mode) and tapping, in which the amplitude of the oscillation is larger. In the non-contact mode, forces between the tip and the sample probably affect the oscillation, while in the tapping mode the tip actually touches the surface at the bottom of the oscillation. Below we briefly describe the main imaging modes.

Contact (constant force) mode

In the constant force mode of AFM, the contact force between the tip and the sample, which is preset at a minimum value, is held constant by an electronic feedback loop. Whenever the cantilever begins to deflect upward (or downward), indicating that the tip is encountering an obstacle (or a trough) on the surface, the sample is retracted (or raised) to cancel the deflection of the cantilever, thereby keeping the tip–sample force constant (Fig. F2.6(a)). The surface topography is obtained by displaying the advance and withdrawal of the sample required to prevent cantilever deflection. This operation usually yields stable images, but the compression and shear force generated between the tip and surface may damage the sample. Typical operation forces in the contact mode are 1–10 nN.

Comment F2.1
Spring constant of cantilever

It is perhaps surprising that it is easy to make a cantilever with a smaller spring constant than that of the equivalent spring constant betweens atoms. For example, the vibrational frequencies ω of atoms bound in a molecule or in a crystalline solid are typically 10^{13} Hz or higher. Combining this with the mass of the atoms, of order 10^{-25} kg, gives interatomic spring constants $k = \omega^2 m$, of the order of 10 N m^{-1}. For comparison, the spring constant of a piece of household aluminium foil that is 4 mm long and 1 mm wide is about 1 N m^{-1}.

(a)   Contact AFM

(b)   Tapping AFM

Oscillation (tapping) mode

In the so-called tapping mode, the cantilever (driven by a piezoelectric actuator)
vibrates at its resonance frequency. Upon approaching the sample, the tip briefly
touches the surface at the bottom of each swing, resulting in a decrease in oscillating amplitude (Fig. F2.6(b)). The feedback loop keeps this decrease at a preset
value and a topographic image of the sample surface can be obtained. Because the
restoring cantilever force should be higher than the adhesion force due the water
film present on samples, relatively stiff cantilevers with force constants ranging
from 10 to 100 N m^{-1} and oscillation amplitudes between 50 and 100 nm are
used. In a tapping mode AFM, the destructive influence of the lateral forces, due
to the movement of the tip with respect to the sample, is virtually eliminated,
because the duration of tip–sample contact is short.

The oscillation of a cantilever in a liquid shows important differences compared with oscillation in air or a ultrahigh vacuum. First, the cantilever motion
drags the surrounding liquid, leading to an increase of the effective mass by a
factor of 10–40 and a corresponding decrease of the resonant frequency. Second,
the cantilever oscillation is anharmonic and asymmetric when the quality factor is
low, in contrast with a tapping mode AFM in air, where the cantilever oscillation
is approximately sinusoidal and symmetric.

F2.4 Imaging of biological structures

The mechanism of image formation in the AFM is very different from those of
optical and electron microscopes. The spatial resolution of optical and electron
microscopes is an inherent property of the instrument and depends ultimately on
the design and principles of operation of the microscope. In contrast, the spatial
resolution of the AFM depends as much on the characteristics of the sample as
on the inherent properties of the instrument. Thus, although theories of image
formation for atomic force microscopy exist, there is no general definition of
resolution in force microscopy (Comment F2.2).

Many factors contribute to an AFM image of biological structures in addition
to the topography of the sample surface. These include the size and shape of
the tip, the properties of the feedback loop, and the mechanical and chemical

properties of the sample and imaging environment. Thus, the finite size of the tip limits the lateral resolution, and the shape of the tip contributes significantly to many images. The part of the tip that interacts with structures on the sample surface should not be larger than the structures themselves. Hard and atomically flat surfaces can be imaged to atomic resolution using even relatively blunt tips (i.e. large R_c), because the interaction is very local and is mediated only by the most apical atoms on the tip. On the other hand, soft samples tend to conform to the harder tip surface and tip–sample interactions take place over a larger area. One of the serious problems in AFM imaging of biological macromolecules is control of the tip–sample interaction. The interaction should ideally be highly localised and involve a very low force.

F2.4.1 Imaging of DNA structure

One of the primary interests in AFM in biology is high-resolution imaging of DNA structures. However, even with significant instrumental developments, such as the tapping mode AFM and specially fabricated supersharp microtips, the resolution was not sufficient to resolve even the pitch of the double helix of DNA in aqueous media (Fig. F2.7 and Comment F2.3). Spectacular progress was achieved in high-resolution atomic force microscopy of DNA using cationic lipid bilayers on a freshly cleaved mica surface as the substrate to anchor the DNA with high affinity. Figure F2.8(a) shows an image of a HaeIII restriction fragment of ϕX174. The measured width of this DNA is close to 2 nm, and a periodic modulation is reproducibly detected by the AFM. The measured period is

Comment F2.2
Resolution in AFM

When we ask what is the best resolution that might be achieved in experiments, one needs to look no further than the structure of the tip used for imaging since an AFM image is a convolution of the structure of the macromolecule or macromolecular assembly under investigation and the tip structure. A well-characterised tip is essential for accurately interpreting an image, and moreover, the size of the tip defines the resolution of the image. The situation is somewhat analogous to the need for well-collimated and monoenergetic electron beams in cryoEM and well-defined X-ray sources in X-ray diffraction.

Fig. F2.7 AFM image of plasmid DNA with a ring structure on the ultrasmooth sapphire substrate wet-treated with Na_3PO_4 aqueous solution (image size: 670 nm $\times$ 450 nm). (After Yoshida et al., 1998.)

Fig. F2.8 (a) An AFM image of pBR322 (4.36 kb). Periodic modulations are resolved and were reproduced at different scales and with different specimens. The width of the DNA varied somewhat from AFM tip to tip, but most measurements gave a value below 3 nm. Scale bar: 40 nm. (b) A histogram of measurements of modulation periodicity. The mean value is 3.4 nm with a standard deviation of 0.4 nm, indicating that these modulations are indeed due to the pitch of the DNA double helix. (Adapted from Mou *et al.*, 1995.)

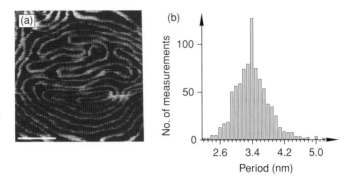

Comment F2.3   Imaging by AFM

Successful imaging by AFM requires that the binding of the sample to the surface be stronger than the contact interaction developed between the tip and the sample during scanning. This may be accomplished either by increasing the strength of attachment to the substrate or by reducing tip–sample forces. Experience has shown that both approaches are necessary. Mica has been used as a deposition substrate since the earliest attempts to image DNA samples. Early attempts at imaging DNA on freshly cleaved mica showed, however, that the molecules were not stably attached to the mica surface, and the scanning tip induced molecular displacement which made the images blurred and difficult to reproduce.

3.4 ± 0.4 nm, in excellent agreement with the known pitch of the double helix in B-form.

The AFM allows direct visualisation of kinked DNA *in situ* (Fig. F2.9). Small DNA circles change from smoothly bent to abruptly kinked shapes. This effect is dependent on the presence of specific divalent cations. The DNA is bent smoothly in the presence of Mg^{2+}, but consists of nearly straight segments connected by kinks in the presence of Zn^{2+}.

Fig. F2.9 Images of 168-bp DNA minicircles in (a) 1 mM $MgCl_2$ and (b) 1 mM $ZnBr_2$, showing a four-fold increase in kink density in Zn^{2+}. Selected molecules are displayed magnified by a factor of 4 to the left of each image. The tip was oscillated at 25 kHz with an amplitude of 5 nm and the image was acquired in five minutes. (Adapted from Han *et al.*, 1997.)

F2.4.2 Imaging protein structure

Individual molecules

Extraction of submolecular information from AFM topography for isolated globular proteins is a very difficult task. On soft biological samples it is not possible
to achieve atomic resolution as with solid-state flat materials or two-dimensional
crystals (see below). Single globular proteins or protein complexes can be
described (or identified) only in a quasi-spherical approximation. The advantages
of nanotube tips are demonstrated in Fig. F2.10(a) for isolated Immunoglobulin G (IgG). The Y-shaped structure was easily resolved. This IgG image is
compared (Fig. F2.10(b)) with the crystal structure to emphasise the very small
ammount of tip-induced broadening in observed images obtained with nanotube
tips.

Fig. F2.10 (a) A nanotube
tip image of
Immunoglobulin G, which
exhibits the characteristic
Y-shaped structure,
compared with (b) the
crystal structure to
emphasise the minimal
tip-induced broadening.
(After Hafner *et al.*, 2001.)

Two-dimensional crystals

The discussion above shows that on soft samples, deformability leads to a pronounced indentation of the sample by the tip. This effect has led to the prediction
that the lateral resolution of such a sample would be limited to several nanometres. However, studies of organic crystals or ordered molecular films show that
they may be imaged at molecular resolution. Below we give two examples of the
application of AFM to studies of biological macromolecules assembled into a
two-dimensional lattice (Comment F2.4).

The first example is a molecular motor of F_0F_1-adenosin triphosphate synthase. It was postulated that rotor of F_0F_1-ATP synthases comprises 12 subunits.
This conclusion was mainly based on cross-linking experiments, genetic engineering and biochemical data that suggest that four protons are required for
the synthesis of one ATP. Furthermore, it was assumed that this stoichiometry
would be constant regardless of the biological species from the ATP synthases
originated.

To prove this structural model AFM was used to image isolated ion driven
rotors from chloroplast and bacterial ATP synthase. In both investigations, the
number of subunits per rotor could be directly counted in the unprocessed AFM
images (Fig. F2.11). Surprisingly, it was found that the rotor from ATP syntase of

**Comment F2.4
Imaging of a
two-dimensional
lattice**

The crucial question in
such studies is whether
this resolution was
obtained by a point
interaction or by the
coherent superposition
of several signals. In
other words, was the
periodicity of the
lattice reproduced, or
were individual
molecules imaged?

Fig. F2.11 Cation driven rotors of F_0F_1-ATP synthases. The individual subunits of the cylindrical rotors represent two transmembrane α-helices connected by a polypeptide loop: (a) the sodium driven rotor of F_0F_1-ATP synthase from *Ilyobacter tartaricus* exhibits 11 subunits; (b) the proton driven rotor of the F_0F_1-ATP synthases from spinach chloroplast is formed by 14 subunits. (Adapted from Muller *et al.*, 2002.)

Ilyobacter tartaricus consisted of 11 subunits (Fig. F2.11(a)), while that from spinach chloroplast ATP synthase exhibited 14 subunits (Fig. F2.11(b)).

Additionally, the X-ray analyses of yeast F_0F_1-ATP synthase yielded a decameric rotor (Sect. F5.3.2). Complementary analysis of a single defective bacterial and chloroplast rotor shows their circular diameter to be independent of the number of subunits missing. This led to the conclusion that the subunits themselves determine the rotor diameter and thereby constrain how many subunits would fit into the rotor.

The second example is the bacteriophage ϕ29 motor. The head-to-tail connector of the *Bacilus subtilus* bacteriophage ϕ29 represents a rotary motor that hydrolyses ATP to power pack double-stranded DNA into a precursor capsid (Section F5.3.3). In contrast to the rotary motor of ATP synthase, the ϕ29 connector converts a mechanical rotation to a translational movement of DNA. In the bacteriophage, the DNA, connector and prohead form a complex of concentric structure with 10-, 12-, and 5-fold symmetry, respectively. These constitute a movable central spindle, intervening ball race and a static outer assembly (stator), which powers the rotor. AFM topographs of native connectors assembled into a two-dimensional lattice clearly show substructures of the connector ends (Fig. F2.12). The 12 subunits of the wide connector end can be directly revealed from unprocessed data.

Fig. F2.12 Rotary rotors of $\dot\phi$29 bacteriophages imaged by contact mode AFM. The two-dimensional crystal exhibits a p4212 symmetry with each thick end of the connector surrounded by the four thin ends of adjacent connectors. (Adapted from Muller *et al.*, 2002.)

Fig. F2.13 Three sequential images (at 0, 12 and 24 minutes) show the digestion of a DNA fragment by BAI 31 restriction nuclease. The DNA fragment was deposited on mica and imaged in buffer using the tapping mode. In the frame taken at 12 min, the pair of arrows enclose the region of initial digestion of the DNA fragment by the endonuclease. (Adapted from Bustamante *et al.*, 1994.)

F2.4.3 Biological macromolecules at work

The ability of the AFM to operate in liquid made the instrument attractive for the study of the dynamics of biological systems. Below we give a few examples of how the development of the tapping mode has made it possible to use the AFM to follow biological processes at the molecular level and get new insights about the variability of activity of individual molecules bound to a surface.

Degradation of DNA by nuclease

Figure F2.13 shows a time-lapse sequence of the digestion of a DNA fragment by BAL 31 nuclease. The first image was taken just before the addition of the nuclease. The nuclease is not seen because it interacts only transiently with the DNA during catalysis. As the DNA is digested, it disappears from the image, leaving an increasingly large gap in the molecule.

RNA polymerase activity

In observing the transcription process a paradoxical problem had to be overcome. To be able to image DNA under fluid with the AFM, not only must the

Fig. F2.14 Complex of RNAP with the 1047 bp dsDNA (a) before and (b) after the addition of 0.5 µM NTPs. These two images were acquired about 6 min apart. (After Kassas *et al.*, 1997.)

DNA molecule be bound sufficiently strongly not to be distributed by the tip, it must also be bound loosely enough for the RNAP to be able to translocate it (see also Section F5.3.1). This compromise is achieved by adjusting the corresponding concentration of Zn^{2+} ions in the reaction chamber since certain divalent cations promote the adhesion of DNA to mica. Figure F2.14 demonstrates the position of RNAP on a 1047-bp DNA template before (Fig. F2.14(a)) and after (Fig. F2.14(b)) the addition of 0.5 μM NTPs. The DNA appears to have been pulled through the RNAP and may well have been transcribed. The transcription rates were observed to be approximately 0.5–2 bases per second at ribonucleoside triphosphate concentrations of approximately 0.5 μM.

Conformational changes in protein

Outer membranes are the protecting surface barriers of Gram-negative bacteria. They are often assembled from regularly packed protein channels. A well-known example is the hexagonally packed intermediate layer of *Dienococus radiodurance*. The pore-forming hexagonally packed intermediate units exhibit a stochastic conformational change that is best described as a switch between an open and a closed state. Figure F2.15 displays a sequence of a scans over one area of the intermediate layer. Repeated imaging of the same surface at 4-min intervals showed that some pores changed their conformation from unplugged to plugged. This conformational change was fully reversible and could be observed over a long time period.

Fig. F2.15 The hexagonally packed intermediate layer of *Dienococus radiodurance* consists of hexameric units that have a central core and connecting arms. At the inner surface, the cores exhibit a central pore that occurs in 'open' (unplugged) and 'closed' (plugged) conformations. The defective pore at the bottom left corner is used as a reference to align the scans. The images were recorded using the contact mode. The different shades correspond to a height range of 6 nm. (Adapted from Engel *et al.*, 1999.)

F2.4.4 AFM probe as nanoscalpel

It is possible to use the AFM tip like a mechanical 'nanoscalpel'. After scanning the area of interest in non-contact mode AFM, chromosome can be cut by the SFM tip with a high force (Fig. F2.16). A cross-section analysis along line AB reveals a cut width of about 100 nm. Minute amounts of material extracted at particular chromosome sites can be processed by the use of a biochemical PCR technique (Comment B2.9).

Fig. F2.16 Topographic SFM micrograph of human chromosome 2 after DNA extraction. For nanoextraction, one line scan with approximately 50 μN contact force was performed. The arrow indicates the dissecting direction of the tip at the beginning of the cut. Next to the arrow shaft, a particle is visible. This particle can be explained as a part of the dissecting tip which broke off when the tip initially snapped onto the surface. (Thalhammer et al., 1997.)

F2.4.5 Study of crystal growth

A number of features of AFM make this a significant development for protein crystallographers. The kind of information found in AFM images is clearly different from that available from other microscopies, such as transmission electron microscopy. First of all, AFM images are obtained from macromolecular crystals *in situ*, i.e. as they exist in their mother liquor, and even while they still growing, without any observable perturbation to the crystals or the growth process.

A second and particularly useful property of AFM images is that unlike transmission electron micrographs, they are not projections of the entire sample onto a single plane, but contain three-dimensional information.

A third feature is that whereas all asymmetric units (generally one macromolecule) are crystallographically equivalent in the interior of a crystal, and therefore have identical chemical and physical environments, this is not true of molecules on the surface of a crystal. Thus an AFM image of a crystal surface, in general, simultaneously presents multiple, symmetrically related perspectives on a single object, the molecule, thus providing three-dimensional visual information.

Figure F2.17 shows as an example the process of two-dimensional crystallization of annexin V on supported planar lipid bilayers. It is clearly seen that AFM enables the process of two-dimensional crystal growths to be visualized in real time.

Fig. F2.17 Growth of crystalline domains of annexin V on PS-containing SPB on mica, imaged by AFM. Successive images are presented in (a)–(b). The time after injection of annexin V in the fluid cell is: (a) 19 min; (b) 23 min; (c) 26 min; (d) 29 min (e) 31 min; (f) 34 min. (a) Circular domains (one domain is labelled with a green asterisk) are homogeneously distributed over the SPB surface. These domains grow and coalesce from (a) to (e). (c). The height profile measured along the green line is shown in the inset. The three crystalline domains are of identical height and the lipid surface is flat. The height measured between the two red arrowheads is 2.6 nm. Scan size, (a)–(c): 5000 nm. (d), (e) Enlarged view of a grain boundary forming between two adjacent domains. Scan size (f) 870 nm (Adapted from Reviakine I. *et al.*, 1998.)

F2.5 Combination of NSOM and AFM

The central component in NSOM is a tapered glass optical element (Section F1.4.1). The cantilevered nature of this optical fiber allows it to work in all the same modes as a conventional AFM probe. It is particularly suitable for the tapping mode. Using standard AFM technology, the point of light at the tip of the cantilever can be brought to within the near-field of a surface that is to be imaged, i.e. closer to it than the dimension of the illuminating light. As the tip scans the surface of the sample, the light emanating from the aperture at the tip interacts with the surface and passes through the sample to measure the absorption or to elicit fluorescence in a region of the surface. After interaction with the surface, the light is collected by the lens of a conventional microscope, detected by a suitable detector and recorded electronically.

An example of imaging using the integrated capability of the NSOM/AFM microscope is shown in a Fig. F2.18. The figure shows an NSOM image of a Giemsa-stained human chromosome and an AFM image of the topography of the same chromosome. A line scan from the NSOM image is displayed to the left of Fig. F2.18(b). In this line scan, the first transition from less staining (less absorption) occurs less than 0.1 μm from the 1-μm line and is completed before the 1-μm line. The alteration from 90% to 10% transmission in such a thin sample occurs over approximately 60 nm, and this provides a measure of the effective resolution in this image. Such optical resolution has not been previously achieved in chromosome imaging.

The development of this technique has led to an instrument that can be simply added to the stage of any conventional optical microscope to enable NSOM

(a)   **(b)**

Fig. F2.18 Simultaneous AFM and NSOM imaging of a Glemsa-stained human metaphase chromosome. AFM (a) measures the topography (i.e. the height variations; brighter means higher), whereas NSOM (b) monitors the light throughput at each position; the light throughout is modulated, in this case, by the binding of the stain to the protein molecules in the chromosome (brighter means more transmission). Bar, 2 μm. (Adapted from Lewis *et al.*, 1999.)

imaging. NSOM, by its very nature, also generates a simultaneous topographic AFM image. With this advance, it became possible to integrate, with overlapping fields of view, images from a single microscope that range from low-resolution optical microscopy to high-resolution NSOM and AFM. The device can also be used to combine these imaging modes with optical-tweezers applications.

F2.6 Checklist of key ideas

- Atomic force microscopy, also scanning force microscopy, *does not use lenses to form an image*, but instead uses a sharply pointed sensor, or tip, at the end of a flexible cantilever to scan and sense the topography of a sample.
- The heart of the AFM is a tip mounted on a flexible cantilever; the ideal AFM tip should have a radius that is as small as possible, a well-defined and reproducible molecular structure and be mechanically and chemically robust such that its structure is not altered while imaging in fluid environments.
- In the simplest mode of operation, the contact mode, the AFM resembles a phonograph. As the sample is scanned underneath the tip, the cantilever deflects up or down, tracking the surface.
- In the tapping mode of operation, the cantilever vibrates at its resonance frequency (typically a few hundred kiloherz); the cantilever could be compared to a blind person scanning the environment with a stick to explore the path ahead.
- Many factors contribute to an AFM image of biological structures in addition to the topography of the sample surface. These include the size and shape of the tip,

the properties of the feedback loop and the mechanical and chemical properties of the sample and imaging environment.

- Extraction of submolecular information from AFM topography for isolated globular proteins is a very difficult task; so far it has not been possible to achieve atomic resolution on soft biological samples.

- High-resolution imaging with high signal-to noise ratio in the molecular/submolecular range can be achieved with two-dimensional crystals of biological macromolecules.

- It is possible to use the AFM tip like a mechanical 'nanoscalpel'; after scanning the area of interest in non-contact mode AFM; biological material (e.g. chromosome bands) can be cut by the AFM tip at a high force.

- AFM enables the process of two-dimensional crystal growth to be visualized in real time.

- Simultaneous AFM and NSOM imaging of a biological material links the world of optical microscopy with scanned imaging techniques: AFM measures the topography, whereas NSOM monitors at each position the light throughput.

Suggestions for further reading

Historical review and introduction to biological problems

Rugar, D., and Hansma, P. (1990). Atomic force microscopy. *Physics Today,* **24**, 23–30.

Engel, A. (1991). Biological applications of scanning probe microscope. *Annu. Rev. Biophys. Biophys. Chem.,* **20**, 79–108.

Bustamante, C., and Keller, D. J. (1995). Scanning force microscopy in biology. *Physics Today,* **48**, 32–38.

Müller, D. J., Janovjak, H., Lehto, T., Kuerschner, L., and Anderson, K. (2002). Observing structure, function and assembly of single proteins by AFM. *Prog. Biophys. Mol. Biol.,* **79**, 1–43.

General principles of AFM

Binning, G., Quate, C. F., and Gerber, C. H. (1986). Atomic force microscope. *Phys. Rev. Lett.,* **56**, 930–933.

Dai, H., Hafner, J. H., Rinzler, A. G., Colbert, D. T., and Smalley, R. E. (1996). Nanotubes as nanoprobes in scanning probe microscopy. *Nature,* **384**, 147–150.

Smith, B. L. (2000). The importance of molecular structure and conformation: learning with scanning probe microscopy. *Prog. Biophys. Mol. Biol.,* **74**, 93–113.

Bustamante, C., Erie, D. A., and Keller, D. (1994). Biochemical and structural applications of scanning force microscopy. *Curr. Opin. Struct. Biol.,* **4**, 750–760.

Imaging of biological structures

Mou, J., Csajkovsky, D. M., Zhang, Y., and Shao, Z. (1995). High-resolution atomic force microscopy of DNA: the pitch of the double helix. *FEBS Letters,* **371**, 279–282.

Bustamante, C., and Rivetti, C. (1996). Vizualizing protein–nucleic acid interactions on a large scale with the scanning force microscopy. *Annu. Rev. Biophys. Biomol. Struct.*, **25**, 395–429.

Kuznetsov, Y. G., Malkin, A. J., *et al.* (1997). Molecular resolution imaging of macromolecular crystals by atomic force microscopy. *Biophys. J.*, **72**, 2357–2364.

Engel, A., Lyubchenko, Y., and Muller, D. (1999). Atomic force microscopy: a powerful tool to observe biomolecules at work. *Trends Cell Biol.*, **9**, 77–80.

Hafner, J. H., Cheung, C.-L., Wooley, A. T., and Lieber, C. M. (2001). Structural and functional imaging with carbon nanotube AFM probes. *Prog. Biophys. Mol. Biol.*, **77**, 73–110.

Combination of NSOM and AFM

Lewis, A., Radko, A., Ami, N. B., Palanker, D., and Lieberman, K. (1999). Near-field scanning optical microscopy in cell biology. *TIBS*, **9**, 70–73.

Chapter F3
Fluorescence microscopy

F3.1 Historical review

1931

M. Goppert-Mayer gave the theoretical background of two-photon excitation in fluorescence. Thirty years later this phenomenon was observed experimentally shortly after the invention of the laser by **W. Kaiser** and **C. Garrett.** In **1990 W. Denk** introduced two-photon excitation in fluorescent microscopy.

1948

T. Forster formulated the principle of fluorescence resonance energy transfer (FRET), a phenomenon that occurs when two different chromophores (donor and acceptor) with overlapping emission/absorption spectra are separated by a suitable orientation and a distance in the range 20–100 Å. In the early 1970s, after a long period of inaction, the ground-breaking work of **L. Stryer** and **R. P Hohland** on FRET revealed the spatial proximity relationships of two fluorescence-labelled sites in biological macromolecules, thereby establishing FRET as a *spectroscopic ruler*. All of this early work used either fluorescent analogues of biomolecules or fluorescent reagents covalently or non-covalently attached to macromolecules as donors or acceptors of FRET. In the 1990s the introduction of the green fluorescent protein (GFP) to FRET-based imaging microscopy gave new life to its use as a sensitive probe of protein–protein interactions and protein conformational changes in vivo.

1964

S. Singh and **L. Bradely** predicted a three-photon absorption mechanism. In 1995 three-photon excited fluorescence spectroscopy and microscopy was demonstrated by a few groups and applied to the imaging of biological specimens and live cells.

Early 1970s

The first lifetime measurements on single points under a microscope were carried out by a few groups. In **1992 J. Lacovich's** and **R. Clegg's** groups proposed independently the principles of a new apparatus that allows lifetime imaging with simultaneous measurements at all positions in the image. In fluorescence lifetime imaging microscopes the nanosecond decay kinetics of the electronic excited-state of chromophores is mapped spatially using a detector capable of high-frequency modulation or fast gating. The spatial resolution of a microstructure on a nanosecond time scale became possible.

1984

D. Axelrod introduced total internal reflection into light microscopy. In total internal reflectance fluorescence microscopy the region near an interface between two media with different refractive indices is illuminated by an evanescent wave. This wave penetrates only a short distance into the medium and selectively excites fluorescence in that region. Microscopic observation further allows one to image probes of structures fluorometrically near the interface with a spatial resolution of about half of the emission wavelength. In **2000 G. Cragg** and **P. So** described a modification of this technique that achieves a lateral resolution of better than one sixth of the emission wavelength.

1985

S. Lanni's group proposed the standing wave fluorescence microscope, in which two coherent plane-wave beams from a laser cross in the specimen volume, where they interfere. In 1993 the same group described a fluorescence microscope in which the axial resolution is increased to better than 100 nm by using the principle of standing-wave excitation of fluorescence.

1992

S. Hell's group proposed so-called 4Pi-confocal microscopy, the primary goal of which is to increase the far-field resolution in the axial direction. In 1998 it was shown by the same group that the combination of two-photon excitation 4Pi-confocal fluorescence microscopy with image restoration leads to a fundamental ($\sim$six-fold) improvement in three-dimensional resolution.

2000

Hell's group demonstrated a novel approach for superresolution imaging. It was accomplished by quenching excited organic molecules at the rim of the focal spot through stimulated emission. Along the optic axis, the spot size was reduced by up to 6 times. A spot volume which can be as low as 0.67 al makes this result relevant for single-molecule spectroscopy.

2000 and to now

Far-field light microscopy, encompassing conventional and confocal fluorescence microscopy, still remains the most widely applied microscopy for biological studies. However, several exciting novel techniques, including standing wave microscopy, 4Pi-confocal microscopy and stimulated emission, have destroyed the standard diffraction barrier of light. Extended resolution microscopy now plays a new role in biological research.

F3.2 Introduction to biological problems

For centuries the light microscope has been used to examine cells and tissues to determine the proximity of structures. Such information is useful to determine the juxtaposition of cellular structures. The limit of resolution of the light microscope is defined by the Rayleigh criterion. According to this criterion if two objects cannot be seen as distinct structures then they may be considered coincident in space. Coincidence may represent a proximity close enough that a molecular association could be possible, although the light microscope does not have sufficient resolution to determine whether coincidence is equivalent to molecular interaction. However, using the FRET technique as a 'molecular ruler' it is possible to determine whether two molecules are within a distance of 20–100 Å of one another, and are thus close enough for molecular interaction to occur.

In the near future, we can expect that most types of biochemical activity in cells will be observed by microscopy. Typical of the reactions that are being or will be studied by microscopy is the catalysis of GTPase and covalent modifications by enzymes such as acetylases. Optical techniques that break the diffraction barrier to increase the spatial resolution of far-field microscopes need to be pushed further to resolve activities within intracellular structures such as individual Golgi stacks, vesicles and tubules, and mitochondrial compartments. Promising progress towards this goal is being made with standing-wave illumination and stimulated depletion microscopy with claims that a 70-nm resolution has been achieved. These far-field high-resolution techniques could also be combined with fluorescence lifetime imaging microscopy to resolve protein reaction states in subcellular compartments. These and other light microscopy techniques will play an important role in the post-genomic era where gene function has to be elucidated in its natural context: the cell.

F3.3 Fluorescence microscopy inside the classical limit

F3.3.1 The standard wide-field fluorescence microscope

Fluorescence techniques are a large part of the life science repertoire. The basic principle of conventional (*one-photon*) fluorescence is that a molecule (fluorophore) absorbs energy from an incident photon of relatively short wavelength (e.g. blue light), and then almost immediately (after a few nanoseconds) reemits a photon of lower energy (longer wavelength; e.g. green light) (Section D8.3). Because of this shift to longer wavelength, filters can be used to block out the excitation light and visualise only the fluorescence. A standard bright-field microscope can, in principle, be modified for fluorescence use simply by positioning

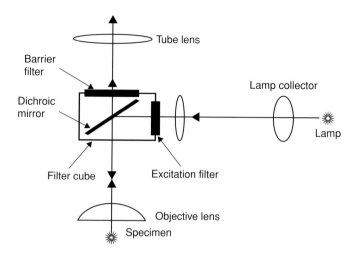

Barrier filter

Dichroic mirror

Tube lens

Lamp collector

Lamp

Filter cube

Excitation filter

Objective lens

Specimen

Fig. F3.1 Diagram of an epifluorescence microscope. A filter cube housing a dichroic mirror, excitation filter and barrier (emission) filter is mounted in the infinity space behind the objective. Excitation light (from a laser or an arc lamp) is focused through the excitation filter and reflected by the dichroic mirror so that an image of the excitation source is formed at the back focal plane of the objective, resulting in even illumination of the specimen. Fluorescent light emitted by the specimen is collected through the objective, passes through the dichroic mirror, and is imaged by the usual arrangement of tube lens and eyepiece. The barrier filter serves to block any excitation light that was transmitted through the dichroic mirror.

a filter to transmit only short wavelengths (the excitation filter) with a second filter (the barrier filter) beyond the objective to block the excitation light while transmitting the emitted fluorescent light. However, this does not work well, because the excitation light is much more intense than the fluorescence light, and is difficult to block completely. Instead, modern fluorescence microscopes utilise epi-illumination, in which the excitation light is introduced into the infinity space of the microscope and illuminates the specimen through the same objective lens that is used to image the fluorescence emission (Fig. F3.1). The key to this arrangement is the use of a dichroic mirror, which reflects the short-wavelength excitation light, while allowing the longer wavelength fluorescent emission to pass through towards the eyepiece. Because dichroic mirrors do not provide a perfect separation between excitation and emission wavelengths, additional excitation and barrier filters are incorporated. All three components are usually integrated within a filter 'cube', and several cubes may be mounted in a turret or slider to allow ready interchange for work with different fluorophores.

F3.3.2 Two-photon excited microscopy

Two-photon excitation arises from the simultaneous absorption of two photons in a single quantitised event. Since the energy of a photon is inversely proportional to its wavelength, the two photons should be about twice the wavelength required for single-photon excitation (Fig. F3.2). For example, a fluorophore that normally absorbs UV light ($\sim$350 nm) can also be excited by two red photons ($\sim$700 nm) if they reach the fluorophore at the same time. In this case, 'the same time' means within about 10^{-18} s. Because two-photon excitation depends on simultaneous absorption, the resulting fluorescence emission depends on the square of the excitation intensity.

Fig. F3.2 Jablonski
diagram of one-photon
versus two-photon
excitation. In one-photon
excitation (left) the
absorption of a single
photon of energy hv
promotes the molecule
from the electronic
ground state into an
excited state. In
two-photon excitation
(right) two photons of half
the energy $hv/2$, are
adsorbed simultaneously
to reach the excited state.
In both cases the
molecule returns via
emission of fluorescence
to its ground state.

To obtain a significant number of two-photon absorption events (where both photons interact with the fluorophore at the same time), the photon density must be approximately a million times what is required to generate the same number of one-photon absorptions. This means that extremely high laser powers are required to generate a lot of two-photon-excited fluorescence (Fig. F3.3). These powers are easily achieved using mode-locked (pulsed) lasers, where the power during the peak of the pulse is high enough to generate significant two-photon excitation, but the average laser power is fairly low. In the case of fluorescence, the resulting two-photon excited state from which emission occurs is the same singlet state that is populated during a conventional fluorescence experiment. Thus, the emission after two-photon excitation is the same as would be generated in a typical biological fluorescence experiment. Figure F3.4 shows photographs of the laser focus for one-(left) and two-photon (right) excitation. In the first case, a double cone is illuminated and spatial resolution has to be enforced by inserting a pinhole into the image plane. Only molecules within a region providing sufficiently high intensities, and thus non-zero probability for the quasi-simultaneous absorption of two photons, are excited.

Three-photon excitation works in much the same way as two-photon excitation except that three photons must interact with the fluorophore at the same time. Because of the quantum-mechanical properties of fluorescence absorption, the photon density required for three-photon excitation is only about ten-fold greater than that needed for two-photon absorption. This makes three-photon excitation attractive for some experiments. For instance, an IR laser ($\sim$1050 nm) can three-photon excite a UV-absorbing fluorophore (350 nm) and simultaneously

Fig. F3.3 Demonstration of photon crowding, which occurs at the focus of a microscope. As the pulsed red laser light passes through a sample containing fluorophores (blue hexagons), there is a nearly zero probability of two photons passing simultaneously within the cross-section of a single fluorophore located outside the focal point (labels 'a' and 'c'). This is true even during the peak of each laser pulse (denoted by the red photon symbols). However, because of the extremely high photon density at the focal point, it is possible for two of them to interact 'simultaneously' with a fluorophore, which becomes excited (stippled blue hexagon, 'b'). (After Piston, 1999.)

Fig. F3.4 Photographs of the laser focus for one-photon (left) and two-photon (right) excitation. (Adapted from Hausten and Schwille, 2003.)

Fig. F3.5 (a) *X–Z* profile of the excitation pattern in a confocal microscope formed by repeatedly scanning a single *X–Y* optical section in a thick film of rhodamine-formvar until fluorescence is completely bleached. The box represents the region from which data are collected by the confocal microscope for this optical section. Nearly uniform bleaching occurs both above and below the focal plane. (b) The same excitation pattern, but for two-photon excitation. Slanted lines denote the path taken by the excitation light to reach the focal plane, but no excitation occurs outside the focal plane. (Piston, 1999.)

two-photon excite a green-absorbing fluorophore ($\sim$525 nm). Three-photon excitation can also be used to extend the region of useful imaging into the far-UV (i.e. use 720-nm light to excite a fluorophore that normally absorbs at 240 nm). This can be useful since UV wavelengths below $\sim$300 nm are very problematic for standard microscopic optics.

The localisation of two- or three-photon excitation to the focal point provides many advantages over confocal microscopy. In a confocal microscope, fluorescence is excited throughout the sample, but only the signal from the focal plane passes through the confocal pinhole, so background-free data can be collected (Section F1.3.1). By contrast, two-photon excitation only generates fluorescence at the focal plane, so there is no background and no pinhole is required. This dramatic difference between confocal and two-photon excitation microscopy is demonstrated by imaging the photobleaching patterns of each method (Fig. F3.5).

Two-photon excitation microscopy excels in thick multicellular systems such as brain slices, intact embryos and other primary culture-tissue preparations. Another demonstration of the power of two-photon excitation microscopy is the imaging of the naturally occurring reduced pyridine nucleotides [NAD(P)H] as an indicator of cellular respiration. [NAD(P)H] has a small absorption cross-section, a low quantum yield and absorbs in the UV – thus, it is difficult to measure and has the potential to cause considerable photodamage. Figure F3.6 shows a typical image of β cell NAD(P)H autofluorescence within an intact islet, displaying a signal from both the cytoplasm and mitochondria. The outlines of single cells are visible, as are the nuclei, both of which appear dark. These imaging experiments simply cannot be performed by confocal microscopy owing to the limitations imposed by photobleaching and UV-induced photodamage.

F3.3.3 Total internal reflectance fluorescence microscopy (TIRFM)

In TIRFM, the region near an interface between two media with different refractive indices is illuminated by an evanescent wave (see Chapter C4). The evanescent wave penetrates only a short distance into the medium of lower refractive index and selectively excites fluorescence in that region (Comment F3.1). As the evanescent wave is set up in medium 2, a standing wave is set up in medium 1. The standing and evanescent waves are depicted schematically in Fig. F3.7.

Several useful experimental situations exist in which the interface is not a simple interface between two media but rather consists of several interfaces between one or more thin layers of materials of refractive indices n_1 and n_2. These intermediate layers may be model membranes, protein layers or membranes of whole cells. The situation that has some particularly interesting consequences for

Comment F3.1 'Evanescent wave'

When light propagating through a medium of high-refractive index encounters an interface with a medium of lower refractive index, it is either reflected or refracted according to Snell's law:

$$n_1 \sin\theta_1 = n_2 \sin\theta_2 \qquad (A)$$

where n_1 and n_2 are the refractive indices of the high- and low-refractive index media, and θ_1 and θ_2 are the angles of incidence relative to the normal to the interface. When $n_2 > n_1$ and θ_1 is greater than the critical angle, θ_c, total internal reflection occurs in medium 1. The critical angle is

$$\theta_c = \sin^{-1}(n_2/n_1) \qquad (B)$$

Although the incident light beam is totally internally reflected under these conditions, an electromagnetic field penetrates a small distance into medium 2. The intensity of this field decays exponentially with the distance z from the interface:

$$I(z) = I_0 \exp(-z/d_p) \qquad (C)$$

with a characteristic penetration depth

$$d_p = \frac{\lambda_0/n_1}{4\pi\sqrt{\sin^2\theta - (n_2/n_1)}} \qquad (D)$$

This field is often called the 'evanescent' wave.

Note that Eq. (D) is valid for IR radiation (see Section E1.4.1).

Fig. F3.6 Optical section of NAD(P)H signal arising from both the cytoplasm and mitochondria, the latter being brighter and somewhat punctate. Cell outlines and nuclei (where there is little or no NAD(P)H) appear dark. Bar, 10 μm. (Piston, 1999.)

Fig. F3.7 Schematic drawing of the evanescent wave produced by a totally reflected laser beam at a quartz–buffer interface. On the right-hand side, the standing wave in a solid medium and the evanescent wave in the liquid medium are shown. When 488-nm laser light is incident at an angle of 72° on a quartz–buffer interface, the penetration depth d_p of the evanescent wave is about 97 nm. (Tamm, 1993).

optical properties near the interface is when proteins are bound to a solid surface. Binding constants and the kinetics of the interaction of protein with different ligands was discussed in Chapter C3.

F3.4 Fluorescence microscopy outside the classical limits

Studies of biological samples require microscopes that combine high resolution with sufficient material contrast. The resolution of far-field instruments is fundamentally limited by the wave nature of light. In the case of fluorescence microscopy, a resolution beyond the common Rayleigh limit can be achieved by different approaches including methods based on a variation of confocal microscopy (4Pi-confocal microscopy) and methods based on non-linear schemes (stimulated emission depletion, standing wave microscopy). Below we briefly describe some of these methods.

F3.4.1 4Pi-confocal microscopy

The primary goal of 4Pi-confocal microscopy is an increase of the far-field resolution in the axial direction. For this purpose, the 4Pi-confocal microscope uses two opposing objective lenses of high numerical aperture with which the same object point is coherently illuminated and observed. The two spherical wave fronts add up to a single wave front so that the total aperture angle is doubled. This is accomplished either for illumination, or detection, or simultaneously for both. With two objective lenses it is not possible to obtain a complete solid angle of 4π. The acronym 4Pi, however, is a reminder of the basic idea behind the concept. Figure F3.8 sketches the 4Pi-confocal microscope for the case of coherent illumination through both objective lenses.

In standard confocal and multiphoton microscopes, a typical resolution is of the order of 150–300 nm in the lateral direction and 500–1000 nm in the axial direction. The 4Pi-confocal microscope improves the axial resolution by a factor of $\sim$3–4 over that of a standard confocal microscope by physical means.

Fig. F3.8 Schematic of the 4Pi-confocal microscopy. The same object point in the sample illuminates two opposing lenses (L1 and L2) of high numerical aperture. This imaging mode is technically the simplest version of 4Pi-confocal microscopy and is referred to as 4Pi-confocal microscopy of type A. (Adapted from Schrader *et al.*, 1998.)

Fig. F3.9 (a) Confocal,
(b) pointed-deconvoluted
4Pi-confocal and
(c) restored 4Pi-confocal
x–y images obtained from
microtubules in a mouse
fibroblast cell. The images
were taken consecutively
at the same site in the cell.
The comparison reveals a
fundamental increase in
the three-dimensional
resolution of far-field
fluorescence microscopy.
(Adapted from Nagorni
and Hell, 1998.)

Comment F3.2   Point spread function

When a specimen point is coherently illuminated through opposing lenses, the two focused, spherical wavefronts add up to form a single wavefront.

For constructive interference of the wavefronts, they produce a PSF, with a pronounced main maximum, as well as an axial lobe above and beneath the focal plane. The 4-times-narrower main maximum resolves better, but, if no other measures are taken, sidelobes introduce ghost images (Nagorni and Hell, 1998).

A further axial resolution can be achieved with standing-wave microscopy (Section F.3.4.3). However, the fundamental improvement in three-dimensional resolution (about 100 nm) of specimens can be obtained by a combination of two-photon excitation 4Pi-confocal fluorescence microscopy and image restoration.

Featuring a three-dimensional resolution in the range 100–150 nm, the 4Pi-confocal (restored) images are intrinsically more detailed than their confocal counterparts. The three images in Fig. F3.9 show the confocal image, 4Pi-confocal axial image and the 4Pi-confocal axial image after image restoration (Comment F3.2). A fundamental increase in three-dimensional resolution is evident.

F3.4.2 Stimulated emission depletion (STED) microscopy

STED microscopy is based on the non-degenerate (different wavelength) pump probe scheme that excites the fluorophores with one beam and quenches them with another. A short (200-fs) visible pulse is focused to a diffraction-limited focal spot and excites all fluorophores in this volume. A second near-IR and longer (40-ps) pulse stimulates the emission from molecules at the outer part of the focus, forcing them into the ground state immediately after they are excited. The longer duration of this second pulse provides full quenching without reabsorption.

Fig. F3.10 Schematic
view of STED microscopy.
(a) Left: A Jablonski
diagram showing the
ground (S_0) and the first
excited singlet (S_1) states
of a fluorophore with the
corresponding vibrational
manifolds. A short pulse
(I_{ex}, 200 fs) excites the
fluorophore into a high
vibrational state of S_1.
The slow STED pulse
(STED, 40 ps) stimulates
emission from the lowest
S_1 vibrational state into a
high vibrational state of
S_0. The stimulated
emission depletes S_1 and
quenches the
fluorescence. Right: The
equivalent wavelength
diagram showing the
absorption curve (dotted
black line), the emission
curve (solid black line),
and the positions of I_{ex},
STED and the filter (F).
The filter blocks the two
lasers and passes the
signal (hatched green).
(b) The combination of
the excitation PSF (blue)
with the STED engineered
intensity distribution (red)
acts as the input to PSFE
by STED. The output PSF
(green) is shrunk. (After
Weiss, 2000.)

Figure F3.10 illustrates the essentials of the technique. Figure F3.10(a) shows
an energy-level diagram (Jablonski diagram, Section D8.2), the absorption and
emission spectra of the fluorophore and the spectral alignment of the excitation
and the STED pulses. Figure F3.10(b) demonstrates how the extent of the PSF is
shrunk in one dimension by using this method (the illustration in one dimension
is for demonstration only; shrinking is effective in all three dimensions).

Figure F3.11 shows the spot probed by a 48-nm bead stained with a fluorophore using standard confocal and STED geometries. Whereas the confocal spot
features an axial full-width-at-half-maximum (FWHM) of 490 nm, that of the
STED-fluorescence spot is only 97 nm. The difference in focal extent becomes
obvious when comparing their $X–Z$ sections shown in the insets. The STED
method allows the diffraction limit to be broken and shrinks the PSF by a factor
of 6 along the optical axis and by a factor of 2 in the radial direction. The spot
volume decreases to 0.67 al.

It is important that the STED method is compatible with live-cell imaging.
Figure F3.12 compares confocal axial images of the live yeast (Fig. F3.12(a))
and *E. coli* (Fig. F3.12(c)) cells with their STED-fluorescence counterparts in

Fig. F3.11 Surface plot of
the $X–Z$ section of a spot
probed by a 48-nm bead
stained with a
fluorophore in (a)
confocal and (b) STED
geometry. (After Klar
et al., 2000.)

(a) x

z

2 µm

(b)

(c) 2 µm

(d)

Fig. F3.12 Resolution improvement in live cells. *X–Z* images of an *S. cerevisiae* yeast cell with labelled vacuolar membrane with (a) standard confocal resolution and (b) axial tesolution improved by STED. Whereas the confocal mode fails in resolving the membranes of small vacuoles, the STED microscopy better reveals their spherical structure. *X–Z* images of membrane-labelled *E. coli* show a three-fold improvement of axial resolution by STED in (d) as compared with their simultaneously recorded confocal counterparts in (c). (Klar *et al.*, 2000.)

Figs. F3.12(b) and (d), respectively. In Fig. F3.12(a), the axial images of the yeast vacuoles are observed as oval-shaped structures. The diffraction-limited spot in the confocal microscope overemphasises the vertical parts of the vacuolar membrane. In contrast, because of the axially narrowed focal spot, the STED-fluorescence image in Fig. F3.12(b) gives a more faithful picture of the spherical shape of the vacuoles of the live yeast cell (see arrow). In particular, it reveals the spherical shape of smaller vacuoles that cannot be recognised as such by the confocal microscope.

F3.4.3 Standing-wave illumination fluorescence microscopy (SWFM)

In SWFM, two coherent plane wave beams from a laser cross in the specimen volume, where they interfere (Fig. F3.13). The nodes and antinodes of this interference field, which are planes parallel to the focal plane, have a spacing $\Delta s = \lambda/(2n\cos\theta)$, where θ is the crossing angle between two fields and n is the specimen refractive index. By controlling the angle θ, the node spacing can be varied down to a minimum value of $\lambda/2n$, when the two beams are counterpropagating along the axis of the microscope. By shifting the phase of one of the beams, the relative position of the field planes within the specimen can be adjusted, at constant node spacing. Because the standing-wave field diameter is comparable to the full field-of-view, images are recorded directly onto an electronic camera, rather than by scanning a focal standing wave field.

Figure F3.14 shows fluorescent beads with a diameter of 100 nm, which is far below the Rayleigh limit for standard fluorescence microscopy and also

Specimen

θ

Piezoelectric drive

Fig. F3.13 Optical configuration for SWFM. Fluorescence is excited in the specimen by the interference pattern of two intersecting coherent beams as described in the text. (Bailey *et al.*, 1993.)

Table F3.1. *Resolving capabilities and probed volume of different fluorescent microscopy methods.*

Method	Resolving power	Probed volume
Conventional microscopy	$\lambda/2$	10 µl
Confocal microscopy	$\lambda/3$	Less than 1fl
4Pi microscopy	$\lambda/3$	About 1fl or less
Standing wave illumination	$\lambda/4$–$\lambda/5$	About 1fl or less
Stimulated emission depletion microscopy	$\lambda/6$	Less than 1 al

Fig. F3.14 Identical area of a sample of 100-nm diameter fluorescent polystyrene beads imaged using different techniques (bar = 1000 nm). (a) The imaged producing using a non-uniform excitation pattern with high spatial frequency components and consists of an extended two-dimensional interference field with closely spaced nodes and antinodes. (b) The image produced using identical lenses with standard illumination. (c) The confocal image. (d) The image recorded with an AFM. See original publication for details. (After Frohn *et al.*, 2000.)

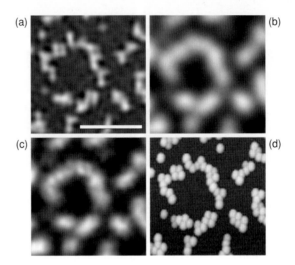

below that of confocal devices. To compare the resolving power of different techniques the same region of water-immersed beads was imaged by using SWFM (Fig. F3.14(a)), standard illumination (Fig. F3.14(b)), and confocal scanning (Fig. F3.14(c)). As a reference for the actual locations of the individual beads an atomic force microscope image is also given (Fig. F3.14(d)) which was acquired in air. The resolution of the standing-wave fluorescence image is clearly superior to that of the standard and also to that of the confocal image.

Table F3.1 summarises the resolving power and probed volume of different fluorescent microscopy techniques. It is evident that the diffraction barrier responsible for a finite focal spot size and limited resolution in far-field fluorescence microscopy is fundamentally broken. The minimal spot volume is now less than 1 al, thus making new approaches relevant to single-molecule spectroscopy.

F3.5 Fluorescence resonance energy transfer (FRET)

FRET is a technique that is more than half a century old, yet, due to advances in the technique, it is undergoing a renaissance. FRET, which relies on the distance-dependent transfer of energy from the donor fluorophore (D) to an acceptor fluorophore (A), is one of the few tools available for measuring nanometre scale distances and changes in distance, both in vitro and in vivo. Advances in the technique have led to qualitative and quantitative improvements, including increased spatial resolution, distance range, and sensitivity. These advances, due largely to genetically encoded dyes, such as green fluorescent protein (GFP), but also to changes in optical methods and instrumentation, have opened up new biological applications.

F3.5.1 FRET as a spectroscopic ruler in static and dynamic regimes

In FRET, a donor (D) fluorophore is excited by incident light, and if an acceptor (A) is in close proximity, the excited-state energy from the donor can be transferred. The process is non-radiative (not mediated by a photon) and is achieved through dipole–dipole interactions. The donor molecule must have an emission spectrum that overlaps the absorption spectrum of the acceptor molecule (Fig. F3.15(a)). FRET is useful as a means of measuring molecule interactions as the efficiency E_T at which Forster-type energy transfer occurs is steeply dependent on the distance R (nm) between the two fluorophores and is given by

$$E_T = \frac{R_0^6}{R_0^6 + R^6} \qquad (F3.1)$$

R_0 (nm), the distance at which 50% energy transfer takes place (Fig F3.15(b)), is expressed as

$$R_0 = [k^2 \times J(\lambda) \times n^{-4} \times Q_D]^{1/6} \times 9.7 \times 100 \qquad (F3.2)$$

where k^2 is the relative orientation of the transition dipoles of the fluorophore, $J(\lambda)$ is the integral of the region of overlap between the donor emission and acceptor absorbance spectra, n is the refractive index of the surrounding medium, and Q_D

(a)

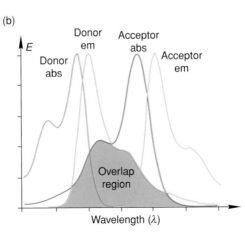

(b)

Fig. F3.15 (a) R_0 is the distance at which 50% energy transfer occurs. (b) Overlap of the donor emission spectrum with the acceptor excitation spectrum. (Adapted from Bastiaens and Pepperkok, 2000.)

Comment F3.3   Calculation of R_0

Depending upon the relative orientations of the donor and the acceptor the k^2 factor can range from 0 to 4. For aligned and parallel transition dipoles $k^2 = 4$, and for oppositely directed and parallel dipoles $k^2 = 1$. Since the sixth root is taken to calculate the distance, variation of k^2 from 1 to 4 results in only a 26% error in R. However, if the dipoles are oriented perpendicular to one another, $k^2 = 0$, which would result in serious errors in the calculated distance. In general, the variation of k^2 does not seem to have resulted in major errors in the calculated distances. It is generally assumed that k^2 is 2/3, which is the value for donors and acceptors that randomise by rotational diffusion prior to energy transfer. This is the value that is generally assumed for calculation of R_0.

---

is the quantum yield of the donor. Typically R_0 is between 20 and 60 Å, which is of the order of protein dimensions (Comment F3.3). Equation (F3.1) could be used as a *spectroscopic ruler*, i.e. by measuring E_T and knowing or calculating R_0, the distance R could be determined.

The dependence on the inverse sixth power of distance was tested in an elegant series of experiments by Stryer and Haugland in 1967. In this study two chromophores were attached covalently to a rigid molecular framework. Then excitation transfer was measured for a series of synthetic species in which the distance between the donor and acceptor molecule differs. The transfer from a dansyl group at one end to a naphthyl group at the other end of oligoprolines with 1–12 monomer units in the rigid chain was examined. The results showed excellent agreement with the R^{-6} dependence for separations of 1.2–4.6 nm, as depicted in Fig. F3.16.

Fig. F3.16 (a) Efficiency of energy transfer as a function of a distance in dansyl-(L-prolyl)$_n$-α-naphthyl, $n = 1, \ldots, 12$. The α-naphthyl and dansyl groups were separated by a defined distance ranging from 12 to 46 Å. Energy transfer is 50% efficient at 34.6 Å. The solid line corresponds to R^{-6} dependence. (b) The dependence of the efficiency of energy transfer on distance is given by slope in this plot of ln $(E^{-1} - 1)$ versus ln$R$. The slope is 5.9, in excellent agreement with the R^{-6} dependence predicted by Förster. (After Stryer and Hougland, 1967.)

(a)

(b)

Fig. F3.17 Schematic illustration of the principle for the detection and characterisation DNA bending that utilises DNA molecules with the helix ends close in space. The difference in the end-to-end distances of the straight 32-bp DNA (a) before and (b) after binding to CAP is small and cannot be detected by FRET. In addition, undesirable dye–protein interaction could have occurred (marked by asterisks). (c) The use of three-bulge DNA molecules with the helix ends close in space circumvents these problems. The dotted lines in the models symbolise FRET; the dotted arrows indicate CAP binding to DNA. (After Hillisch *et al.*, 2001.)

Access to long-range distance information makes FRET a valuable tool for studying nucleic acid molecules (Comment F3.4). Specifically, protein-induced DNA deformation has been characterised using this method. For example, the binding of catabolite activator protein (CAP) to straight DNA leads to a significant bending of the nucleic acid molecule (Fig. F3.17). However, only small differences in the energy transfer efficiencies can be detected for (a) in Fig. F3.17, as the dye-to-dye distance in the complex (b) in Fig. F3.17 is about 90 Å, i.e. considerably larger than R_0 (50 Å for the dye pair fluorescein-TMRh used). This illustrates a general problem. By introducing DNA structural elements into such as bulges, however, longer DNA molecules can be designed in which the helix ends are close to one another in space (in the range of R_0). A molecular model of a two-bulge DNA molecule with close helix ends is shown in Fig. F3.18. Double-helical B-DNA regions were used as substrates for protein binding. Molecules of

---

Comment F3.4 Measuring of distances in protein structures by FRET

FRET is often used to measure distances in protein structures and their assemblies in solution. Proteins contain intrinsic fluorophores (tryptophan and tyrosine). In addition, they can be labelled with dyes at defined sites, e.g. cysteine residues. However, the multiple occurrence of intrinsic fluorophores and tethering sites for dyes reduces the interpretability of FRET data obtained with these chromophores.

Fig. F3.18 Atomic model
of a DNA molecule
containing two DNA
bulges. The molecule
consists of three
double-helical stems
(16, 9 and 17 bp). The
longer fluorescently
labelled strand is
composed of 52 bases
and is coloured green in
the model; the shorter
42-base strand is labelled
with TMRh and is
depicted in orange.
(Adapted from Hillisch
et al., 2001.)

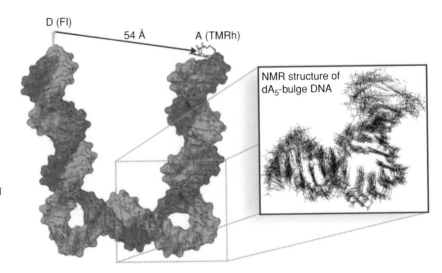

65-bp DNA, containing three dA_5 bulges (with the CAP-binding site inbetween),
were constructed; the 5' end was labelled and used to monitor the binding of
CAP by measuring FRET. A decrease in Forster transfer efficiencies from 21%
with CAP (see (d) in Fig. F3.17) to 8% without CAP (see (c) in Fig. F3.17) was
observed.

Figure F3.18 also shows the NMR structure of a DNA fragment containing
a dA_5 bulge. Combining local (NMR) and global (FRET) structure information
using molecular modelling allows a detailed model for a large DNA molecule in
solution to be deduced (see Chapter J3).

In general FRET is better suited to detecting changes in distance rather than
absolute distance because E_T depends on the orientation of the fluorophores and
the attachment methods cause uncertainty in the probe position with respect to the
macromolecular backbone (Comment F3.5). For an easy quantitative representation, at least one of the dyes should have rotational freedom, so that all dipole
orientations occur with equal probability. In order to achieve this flexibility, the
dyes are chemically attached to the DNA via alkyl chains (two to six carbon
linkers).

Ensemble FRET has long been used as a very effective spectroscopic ruler
for measurement of static distances in the 20–100 Å range. However, dynamic
events such as the relative motion between donor and acceptor molecules cannot
be detected by FRET, usually because of the lack of synchronisation of these
events. FRET measurements on a single pair, in contrast, allow one to look at
dynamic distance changes on a millisecond time scale because only one donor–
acceptor system is observed at a time; hence, time-dependent conformations can
be monitored (Section F4.7).

Comment F3.5   Measuring of distances in DNA structures by FRET

Well-defined sequences of DNA and RNA oligomers are now routinely synthesised and fluorescently labelled at the 3′ and 5′ termini, as well as within the DNA sequence. End labelling has three main advantages (when the shortest possible DNA is used): the dyes can be spatially separated from the binding protein to avoid protein–dye interactions; the DNA end sequence can be specifically designed to provide an identical chemical environment and, thus, the fluorescence properties of the dyes in a series of molecules can be derived; and the exact location of Cy3 on the DNA is known and evidence is accumulated on the position of fluorescein and rhodamine. In most studies, fluorescein and rhodamine are attached to the 5′ ends.

F3.6 Green fluorescent protein (GFP)

Fluorescence imaging of biological reactions in living cells requires techniques to fluorescently label the macromolecules involved. These fluorescent probes need to be specific, interfere as little as possible with the reactions to be visualised, must not perturb the physiological conditions of the cells and must have favourable spectroscopic properties for efficient detection. One of the most exciting advances in fluorescent in vivo labelling techniques was the discovery of the GFP.

Wild-type GFP from the Pacific Northwest jellyfish *Aequoria victoria* is a stable, proteolysis-resistant single chain of 238 residues. The chromophore, resulting from the spontaneous cyclisation and oxidation of the sequence $-\mathrm{Ser^{65}}$ (or $\mathrm{Thr^{65}}$)$-\mathrm{Tyr^{66}}-\mathrm{Gly^{67}}-$, requires the native protein to fold for both formation and fluorescence emission. The crystal structure of GFP with the sequence $-\mathrm{Thr^{65}}-\mathrm{Tyr^{66}}-\mathrm{Gly^{67}}-$ is presented in Fig. F3.19. The chromophore is completely protected from the bulk solvent and is centrally located in the molecule. Mutation of $\mathrm{Ser^{65}}$ to Thr (S65T) simplifies the excitation spectrum to a single peak at 488 nm.

Fig. F3.19 (a) The jellyfish *Aequoria victoria*. (b) A schematic drawing of the backbone of GFP. The chromophore is *p*-hydroxy benzylideneimidazinone formed from the spontaneous cyclisation and oxidation of the sequence-$\mathrm{Ser^{65}}$ (or $\mathrm{Thr^{65}})-\mathrm{Tyr^{66}}-\mathrm{Gly^{67}}-$ and is shown as a ball and stick model. The most distinctive feature of the fold of GFP is an 11-strand barrel wrapped around a single central helix, where each strand consists of approximately 9–13 residues. The barrel forms a nearly perfect cylinder 42 Å long and 24 Å in diameter. The top end of the cylinder is capped by three short, distorted segments, and one short, very distorted helical segment caps the bottom of cylinder (Adapted from Ormö *et al.*, 1996.)

(a)

(b)

Comment F3.6
GFP families

Blue fluorescent
protein (BFP) has the
lowest quantum yield
and is the most
bleachable of all GPF
mutants. To address
these problems, GFP
mutants with longer
wavelengths are now
often used as FRET
partners: CFP (cyan
fluorescent protein)
and YFP (yellow
fluorescent protein).
The CFP–YFP pair
allows FRET to be
detected at larger
distances, as the
calculated R_0 (the
distance at which 50%
energy transfer occurs)
between randomly
oriented chromophores
is 5.2 nm, whereas for
BFP–GFP, it is only 4
nm. Novel YFPs and
red fluorescent
proteins (RFPs) have
been cloned, producing
yet another FRET pair,
GFP–RFP.

GFP has a number of amazing properties that enable it to be used for in vivo imaging. First, GFP can be expressed in a variety of cells, where it becomes spontaneously fluorescent without the aid of a cofactor. Second, GFP can be fused to a host protein to create a fusion protein that usually retains both the fluorescence of the GFP and the biochemical function of the original host. Third, fusion proteins can be targeted to specific organelles, such as the nucleous or endoplasmic reticulum, by adding an appropriate signalling peptide. Finally, and most importantly, mutagenesis of GFP produces many mutants with varying spectral properties that can be used as donors and acceptors of FRET (Comment F3.6).

GFP-based constructs do, however, suffer from limited sensitivity, often precluding single-cell analysis. Moreover, they are relatively large, thereby limiting spatial resolution. They should also be used with caution as it has been shown that GFP can undergo colour changes upon irradiation due to photochemical changes that are independent of FRET. In addition, GFP requires hours to assemble in its final fluorescent form and is thus limited in its ability to monitor kinetic phenomena before final assembly. Nevertheless, GFP is of immense importance, and a new class of red fluorescent proteins indicate that it still has great potential.

F3.6.1 GFP as a conformational sensor

The introduction of GFP to FRET-based imaging microscopy enabled it to be used as a sensitive probe of protein–protein interactions and protein conformational changes in vivo. One pair of fluorescent proteins originally used for FRET is BFP as the donor and GFP as the acceptor. These proteins form FRET pairs with each other or with conventional organic dyes, and can be attached to many proteins of interest, usually at the N- or C-terminus.

Figure F3.20 shows the scheme of an experiment in which the motor protein myosin was labelled at its N- and C-termini with BFP and GFP. When BFP was

Fig. F3.20 To detect the working stroke in the myosin head in operation, fusion proteins were constructed by connecting *Aequorea victoria* GFP and BFP to the amino and carboxyl termini of the motor domain of myosin II, designated S1dC. GPF–BFP pair-based FRET was used to measure conformational changes in myosin upon ATP binding and hydrolysis. (After Suzuki *et al.*, 1998.)

excited by light at 360 nm, intramolecular FRET took place and green light was emitted at 510 nm. The orientation and distance between these probes varied as myosin underwent conformational changes associated with ATP binding and hydrolysis. This result favours what is known as the lever-arm model of muscle contraction (Section F5.3.1).

F3.6.2 GFP as a cellular reporter

Genetically encoded dye, such as GFP, revolutionised the ability to image in vitro and especially in living cells. In practice, the combination of two mutants, cyan-fluorescent protein (CFP) and yellow-fluorescent protein (YFP) is the most suitable for double-colour imaging in living cells (Fig. F3.21).

Another approach is to use two GFPs of different colours, chosen to permit FRET from the shorter to the longer wavelength GFP. Figure F3.22 demonstrates the principal scheme in which GFP as the conformational sensor detects acceptor-sensitised emission alone.

Figure F3.23 demonstrates the imaging by FRET of an induced protein–protein interaction in individual live cells. Cyan-GFP-labelled calmodulin (CFP-CaM) and yellow-GFP-labelled calmodulin binding peptide (YFP-M13) were coexpressed. The left-hand panel shows two cells before stimulation, while the right-hand panel shows the same cells after elevation of cytosolic Ca^{2+} by 0.1 mM histamine. Low Ca^{2+} levels (left) lead to little FRET and mostly blue emission (pseudo-colour green); high Ca^{2+} levels (right) lead to binding and FRET emission of YFP (pseudo-colour red).

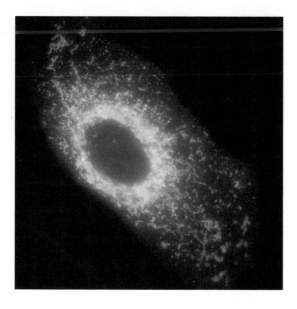

Fig. F3.21 Dual-channel fluorescence micrograph of the cell expressing two GFP fusion proteins that localise to different compartments. A fusion of the peroxisomal membrane protein 24 to CFP (pmp24-CFP) localises to perixomes (blue) and a fusion of LCAT-like lysophospholipase to YFP (LLPL-YEP) localises to mitochondria (yellow). (Adapted from Bastiaens and Pepperkok, 2000.)

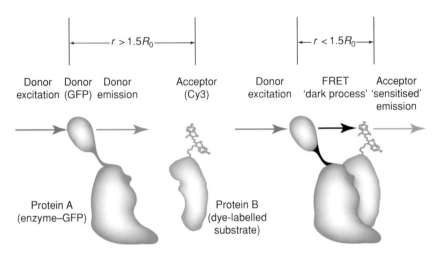

Fig. F3.22 Principal scheme in which GFP as the conformational sensor detects acceptor-sensitised emission alone. The donor GFP-labelled enzyme binds to the dye-labelled substrate. After binding FRET between the donor and the acceptor is detected. Note that FRET is very sensitive to the distance between two phluorophores, r. Acceptor-sensitised emission occurs if only distance $r < R_0$. (After Bastiaens and Pepperkok, 2000.)

Fig. F3.23 FRET constructs for measuring intracellular calcium. CFP-labelled calmodulin (CFP-CaM) and YFP-labelled calmodulin binding peptide (YFP-M13) were coexpressed in HeLa cells. Pseudo-colour hues from blue to magenta indicate increasing ratios of yellow to cyan emissions, resulting either from increased FRET (see overlaid schematic) or excess of YFP over CFP. The presence of Ca^{2+} alters the distance (or orientation) between donor and acceptor, causing the relative emission intensities at the donor and acceptor wavelengths to change. (After Tcien and Miyawaki, 1998.)

F3.7 Fluorescence lifetime imaging (FLI) microscopy: seeing the machinery of live cells

FLI refers to the technique of recording a fluorescence image whereby the fluorescence lifetimes of the fluorophores in a sample are resolved at every location of the image. This can result in either the actual determination of the fluorescence lifetimes (or more often the apparent fluorescence lifetime), or the determination of a spectroscopic parameter that is possible only when the fluorescent signal is lifetime-resolved in the measurement. The spectroscopic information related to the lifetime of the fluorescence decay can then be displayed in an image format, in which every pixel of the image contains the lifetime-resolved information. FLI measurements are analogous to normal intensity fluorescence imaging measurements and are acquired on the same samples, except that the information related to the fluorescence lifetime is directly recorded in addition to the normal measurement of the fluorescence intensity.

Live-cell FRET measurements are more feasible using FLI because lifetimes are independent of probe concentration and light path length. By imaging donor fluorescence lifetime it is possible to calculate FRET efficiencies at each pixel E_T^i of an image:

$$E_T^i = 1 - \left(\frac{\tau_{DA}^i}{\tau_D'} \right) \qquad (F3.3)$$

where τ_{DA}^i is the lifetime of the donor in the presence of the acceptor and τ_D^i is the lifetime of the donor in the absence of acceptor.

Figure F3.24 shows the quantitative determination of protein reaction states by global analysis of FLI microscopy data. Here the relative populations of phosphorilated ErbB1-GFP receptor were determined before (upper row) and after 60 s of epidermal growth factor (EFG) stimulation (lower row). The left-hand column shows the images of ErbB1-GFP, the middle column the images of the antiphosphotyrosine antibody (Cy3-PY72) and the right-hand column the calculated images of the populations of phosphorilated ErbB1-GFP receptor.

By using fluorescence digital imaging microscopy, one can visualise the location of GFPs within a living cell and follow the time course of the changes in

Fig. F3.24 Quantitative determination of protein reaction states by global analysis of FLI microscopy data (see text). (After Bastiaens and Pepperkok, 2000.)

FRET corresponding to cellular events at a millisecond time resolution. The observation of such dynamic molecular events in vivo provides vital insights into the action of biological molecules.

F3.8 Checklist of key ideas

- Two-photon excitation arises from the simultaneous absorption of two photons in a single quantitised event; in this case, 'simultaneous' means within about 10^{-18} s.
- In TIRFM, the region near an interface between two media with different refractive indices is illuminated by an evanescent wave that penetrates only a short distance into the medium (about 250–300 nm).
- The 4 Pi-confocal microscope uses two opposing objective lenses with which the same object point is coherently illuminated and observed. The 4 Pi-confocal microscope improves the axial resolution by a factor of 3–4 over that of a standard confocal microscope.
- In fluorescence microscopy, a resolution beyond the common Rayleigh limit can be achieved by different approaches including methods based on a variation of confocal microscopy and those based on non-linear schemes.
- STED microscopy is based on the different wavelength pump probe scheme that excites the fluorophores with one beam and quenches them with another; it allows the diffraction limit to be broken and shrinks the PSF by a factor of 6 along the optical axis and by a factor of 2 in the radial direction.
- In SWFM, two coherent plane wave beams from a laser cross in the specimen volume, where they interfere.
- In FRET, a donor fluorophore is excited by incident light, and if an acceptor is in close proximity (20–100Å), the excited-state energy from the donor can be transferred; the donor molecule must have an emission spectrum that overlaps the absorption spectrum of the acceptor molecule.
- FRET is useful as a physical phenomenon to measure molecule distances as the efficiency E at which Forster-type energy transfer occurs is steeply dependent on the distance R as R^{-6}.
- FRET is especially suited to the study of protein–nucleic acid complex structures as knowledge of the exact positions of some dyes with respect to specific DNA helix ends enables us to translate the measured distance data into precise structural information.
- Ensemble FRET is used as a very effective *spectroscopic ruler* for measurement of static distances in the 20–100 Å range; however, dynamic events such as the relative motion between donor and acceptor molecules cannot be detected by FRET.
- Single-pair FRET measurements, in contrast, allow one to look at dynamic distance changes on a millisecond time scale because only one donor–acceptor system is observed at a time; hence, time-dependent conformations can be monitored.

- FLI refers to the technique of recording a fluorescence image whereby the fluorescence lifetimes of the fluorophores in a sample are resolved at every location of the image.
- GFPs (27 kDa) enable FRET measurements to be made in living cells; these proteins can be cloned in fusion with proteins of interest and expressed in cell lines.

Suggestions for further reading

Historical review and introduction to biological problems

Lakovicz, J. R. (ed.) (1999). *Principles of Fluorescence Spectroscopy*, second edition. New York: Academic / Plenum Publ.

Fluorescence microscopy inside the classical limit

Herman, B., and Jacobson, K., (eds.) (1990). *Optical Microscopy for Biology*. New York: Wiley-Liss.

Tamm, L. K. (1993). Total internal reflectance fluorescence microscopy. In *Optical Microscopy. Emerging Methods and Applications*. Eds B. Herman and J. J. Lemasters, Academic Press.

Piston, D. W. (1999). Imaging living cells and tissues by two-photon excitation microscopy. *Trends Cell Biol.*, **9**, 66–69.

Fluorescence microscopy outside the classical limits

Gustafsson, M. G. L. (1999). Extended resolution fluorescence microscopy. *Curr. Opin. Struct. Biol.*, **9**, 627–634.

Weiss, S. (2000). Shattering the diffraction limit of light: a revolution in fluorescence microscopy? *Proc. Nat. Acad. Sci. USA.*, **97**, 8747–8749.

Schrader, M., Bahlmann, K., Giese, G., and Hell, S. W. (1998). 4Pi-confocal imaging in fixed biological specimens. *Biophys. J.*, **75**, 1659–1668.

Nagorni, M., and Hell, S. W. (1998). 4Pi-Confocal microscopy provides three-dimensional images of the microtubule network with 100-to-150 nm resolution. *J. Struct. Biol.*, **123**, 236–247.

Cragg, G. E., and So, P. T. C. (2000). Lateral resolution enhancement with standing evanescent waves. *Opt. Lett.*, **25**, 46–48.

Klar, T. A., Jacobs, S., Dyba, M., Egner, A., and Hell, S. W. (2000). Fluorescence microscopy with diffraction resolution barrier broken by stimulated emission. *Proc. Natl. Acad. Sci. USA*, **97**, 8206–8210.

FRET

Stryer, L. (1978). Fluorescence energy transfer as a spectroscopic ruler. *Annu. Rev. Biochem.*, **47**, 819–846.

Selvin, P. R. (2000). The renaissance of fluorescence resonance energy transfer. *Nature Str. Biol.*, **7**, 730–734.

Hillisch, A., Lorenz, M., and Diekmann, S. (2001). Recent advances in FRET: distance determination in protein-DNA complexes. *Curr. Opin. Struct. Biol.*, **11**, 201–207.

GFP

Tcien, R. Y. (1998). The green fluorescent protein. *Annu. Rev. Biochem.*, **67**, 509–544.

Ellenberg, J., Lippincot-Schwartz, J., and Presly, J. F. (1999). Dual-colour imaging with GFP variants. *Trends Cell Biol.*, **9**, 52–60.

FLI microscopy

Bastiaens, P. I. H., and Squire, A. (1999). Fluorescence lifetime imaging microscopy: spatial resolution of biochemical processes in the cell. *Trends Cell Biol.*, **9**, 48–52.

Tcien, R. Y., and Miyawaki, A. (1998). Seeing the machinery of live cells. *Science*, **280**, 1954–1955.

Piston, D. W. (1999). Imaging living cells and tissues by two-photon excitation microscopy. *Trends Cell Biol.*, **9**, 66–69.

Bastiaens, P. I. H., and Pepperkok, R. (2000). Observing proteins in their natural habitat: the living cell. *TIBS*, **25**, 631–636.

Chapter F4
Single-molecule detection

F4.1 Historical review

1924

The history of the study of single molecules in physics starts with the well-known oil-drop experiments of **R. Milliken**, in which the charge-to-mass ratio of a single electron was determined. It may be assumed that the history of studying single molecules in biology started in 1950s when researchers were able to detect the presence of massive molecules such as DNA using conventional electron microscopes in high vacuum. The advent of the scanning tunnelling microscope in the 1980s made it possible to image individual molecules and atoms on the surface. However, the observation of individual molecules within solids or in solution remained an extremely difficult and unsolved problem.

1961

B. Rotman was the first to use fluorescence detection for single-molecule studies in solution. Using a fluorogenic substrate, he measured the presence of a single β-D-galactosidase molecule by detecting the fluorescent product molecules accumulated in a microdroplet through enzymatic amplification. He did not achieve single-molecule sensitivity but clearly demonstrated the great potential of fluorescence detection. In **1976 T. Hirschfeld** reported the use of fluorescence microscopy to detect single antibody molecules tagged with 80–100 fluorescein molecules under evanescent-wave excitation.

1989

W. Moerner's group first used a laser to see single, small organic molecules trapped inside a transparent host crystal. In **1990 M. Orrit**'s group showed that very high sensitivity at the single-molecule level can be achieved by probing 'guest' molecules in solid hosts at cryogenic temperatures. In **1994 F. Guttler** and colleagues described fluorescent experiments with single pentacene molecules under a microscope at a temperature of 1.8 K. However, practical applications of single-molecule detection required measurements in solution at room temperature.

1990

R. Keller's group reported the first efficient detection of individual fluorescent molecules in solution. The observation technique involved exciting fluorescent molecules by passing them in solution through a highly focused laser beam and detecting the subsequently emitted photons. Single-molecule detection was proposed by this group as a tool for rapid base-sequencing of DNA. In **1994** the same group demonstrated a novel method of determining electrophoretic velocities by measuring the time required for individual molecules to travel a fixed distance between two laser beams.

1991

W. Whitten and colleagues solved the main problem of isolating the molecular fluorescence signal from background luminescence and Raman scattering by reducing the excitation volume to the smallest possible size with levitated microdroplets. In **1992 E. Betzig** and **J. Trautman** reduced the optical excitation volume to subwavelength proportions in all three dimensions. This was accomplished using the illumination mode in near-field scanning optical microscopy.

1995

J. Wang and **P. Wolynes** described the mathematical formalism that can help in determinations of the microscopic origin of heterogeneity of behaviour of biological systems. According this formalism the statistics of individual activated events can exhibit intermittency and does not always obey Poisson's law and, therefore, the fluorescence decay function can exhibit a non-exponential character as a result of conformational dynamics. In **1996–1997** the non-exponential characters of two biological processes were described. Thus, **R. Rigler**'s group using confocal fluorescence microscopy monitored conformational transition for single 18-mer deoxyribonucleotide molecule linked to tetramethylrhodamine, and, **B. Cooperman**'s group using lifetimes fluorescent measurements discovered two fluctuating conformational states for t-RNA^{Phe}-probe. These experiments opened a new area of molecular analysis of the biological events that can be understood only on the single-molecule level and could not be extracted from investigation of macroscopic systems.

2000 to now

Advances in ultrasensitive instrumentation enable the detection, identification, and dynamical studies of a single molecule in the condensed phase. These measurements have opened the new avenues for scientists in many scientific regions, such as nanochemistry, structural biology, biosensors and molecular medicine.

F4.2 Introduction to biological problems

Most of the experimental methods in this book represent measurements of ensemble averages by which a multitude of duplicate systems are probed and average responses are recorded. As such, rare events often escape notice because they are buried in the average properties of the system under study. In 1952, Erwin Schrödinger wrote that we would never experiment with just one electron, atom or molecule. Since then, impressive progress has been made that allows the detection of isolated single atoms in traps, single molecules in low-temperature molecular crystals, and the investigation of the more complicated but also more fascinating world of single molecules in solution and on a surface – a world that includes biological macromolecules.

Significant advances in instrumentation involve scanning probe microscopes, of which the atomic force microscope (AFM) is the best known. Scanning probe microscopy holds much promise for molecules bound to a surface, but for molecules embedded in a medium, optical methods involving sensitive detection of emission or absorption appear in general to be more useful. The key concept is that a single molecule can be repetitively cycled between its ground and excited electronic states with a laser beam at a wavelength resonant with this transition, which yields a large number photons for detection. Optical methods that probe chemical chromophores are well suited for non-invasive studies of single molecules in solution.

Optical methods in the far field cannot localise the position of the molecule better than the diffraction limit which is equal to half the wavelength of light. However, it is possible to combine probe microscopy with optical detection. One highly successful example is near-field scanning optical microscopy, in which a tapered single-mode fibre tip is used to provide both optical and graphical information on the same sample area. Another interesting development is a hybrid instrument that combines far-field laser-induced fluorescence and atomic force microscopy via an inverted optical microscope. In this integrated microscope, ultrasensitive optical measurement at the single-molecule level is used to provide molecular information and atomic force imaging is used to resolve individual molecules at subnanometre resolution.

In this chapter we focus our attention on the optical detection of a single molecule. The fundamental principles of scanning probe microscopy, far- and

---

Comment F4.1   Deconstructing DNA for faster sequencing

The international race to sequence the human genome has turned gene sequencing into a high-speed and high-profile endeavour. At its heart are machines that create thousands of copies of DNA fragments as a first step toward decoding the sequencing, one nucleotide base pair at a time. Single-molecule sequencing has the potential to speed up the sequencing process. Currently decoding schemes for DNA are based on relatively short pieces of DNA each about 1000 base pairs long. Researchers wanting to sequence a gene containing, say, 100 000 base pairs must sequence overlapping fragments of the gene and then use complex computer programs to reassemble the pieces in the right order. Single-molecule sequencers, in contrast, hope to sequence DNA pieces as long as 50 000 base pairs, which would simplify and speed up the task of putting the puzzle back together. The key to the technique's long sequencing length is a protein called an exonuclease, which degrades DNA by munching its way through tens of thousands of individual bases one by one. Without tags in the way, exonuclease can clip about 12 bases a second. But the main trick is to detect untagged bases (*Science*. 1999, **283**, 1669).

near-field light microscopies were given in Chapters F1 and F3. We will discuss emerging biological applications of single-molecule detection as well as some possible future directions (Comment F4.1).

Single-molecule detection is only just starting.

F4.3 Fluorescence spectroscopy of single molecules

The central problem of detecting a single absorbing molecule surrounded by quintillions of nominally transparent solvent molecules requires two deceptively simple steps: (1) establishing that only one molecule in the irradiated volume is in resonance with the laser, and (2) ensuring that the signal from the single molecule is larger than any background signal.

Optical detection of single molecules can be achieved by both frequency-modulated absorption and laser-induced fluorescence. Because of the low background and high signal-to-noise ratios, laser-induced fluorescence became the most widely used method. Now it is well established that the ultrasensitive fluorescent technique produces a higher signal-to-noise ratio when the sample background is carefully controlled.

F4.3.1 Laser-induced fluorescence

The fluorescence process begins with absorption of a photon, which drives the molecule from its ground molecular states to its first excited electronic state singlet state (Fig. F4.1). Absorption rates are usually limited only by the laser intensity. Emission of the fluorescence photon occurs in a few nanoseconds for most ionic dyes. The molecule has then returned to the ground electronic state and is ready for another absorption–emission cycle. The key point of the application of laser-induced fluorescence to single-molecule detection is that a single molecule can be repetitively cycled between its ground and excited electronic states with a laser beam at a wavelength resonant with this transition, which yields a large number of photons (Comment F4.2).

The temporal and spectral properties of fluorescence are also advantageous to single-molecule detection (Chapters D3 and D8). Fluorescence emission is typically red shifted significantly with respect to the excitation frequency and is generally a much slower process, occurring in nano- to micro-seconds, than Rayleigh or Raman scattering, which occurs on a femtosecond scale. These properties facilitate separation of a signal from the background in both the time and frequency domains.

There are several important limitations on the magnitude of the signal that can be obtained from a single molecule. First, the fluorescent count rate, defined approximately by the absorption–emission cycle time, is limited at saturation by

Fig. F4.1 Fluorescence cycle of a single molecule. ISC: intersystem crossing from the excited singlet state to the triplet. (Nie and Zare, 1997.)

Comment F4.2 Number of photons in laser-induced fluorescence

By using diffraction-limited laser beams in combination with confocally imaged pinholes or fibre optics, a cylindrically shaped volume element with a radius of 200 nm and a length of 2000 nm can be intensely illuminated. The volume of such a light cavity is about 0.2 fl. One femtolitre is approximately the volume of an *E. coli* cell, and one molecule in such a volume corresponds to a concentration of 10^{-9} M. With a laser intensity of 0.5 mW a photon density of $\sim 1 \times 10^{24}$ photons per square centimetre per second is achieved.

The ability to detect single molecules depends on the density of light quanta emitted. In the ideal case the number of quanta emitted per time interval is solely limited by the transition rate between the excited (singlet) state and the ground state. For example, for rhodamine 6G with an absorption cross section of 1.4×10^{-16} cm^2 per molcule and a photon density of $\sim 1 \times 10^{24}$ photons cm^{-2} s a singlet excitation rate of 1.4×10^8 s^{-1} (per molecule) is reached. With singlet decay rate of 2.5×10^8 s^{-1} in the absence of non-radiative transition (assuming a quantum yield equal to 1) $\sim 25\%$ of the ground state is depleted. Under these conditions it is possible to detect photons at a rate $>100\,000$ per molecule per second. Taking into account the present detection efficiency and the bandwidth of detection, this value corresponds to a rate of 2×10^6 emitted quanta per second. The discrepancy between the maximal possible and the measured emission rate is partly due to existence of the triplet state and partly due to photobleaching during excitation–deexcitation process (Eigen and Rigler, 1994).

---

the finite fluorescence lifetime. Second, dyes also have a finite photochemical lifetime that limits the average number of absorption–emission cycles the molecule may undergo before photochemical destruction or 'photobleaching' takes place. And finally for some combinations of dye and solvent, singlet–triplet intersystem crossing rates (Fig. F4.1) can be significant, resulting in a substantial reduction in fluorescent count rates.

What matters most, however, is not the signal itself but the signal-to-noise ratio and the magnitude of the background signal. Source noise (from Rayleigh and Raman scattering from the solvent and fluorescence from impurities in the solvent) in these types of experiments usually dominates the background. As a rule of thumb, solvent scattering places an upper limit of the probe volume (a few picolitres).

Detection of a single molecule is usually defined by a threshold criterion. In a typical experiment, the threshold level is usually ~ 3 or more times the standard deviation in the non-fluorescent background signal. Most of the 'proof' that a single molecule was detected in liquids is based on the signal-to-noise ratio, supported by small probabilities of more than one molecule occupying the probe volume in a given measurement interval (Comment F4.3).

How accurately can a distinction be made among zero, one, two or more molecules? Just as molecular detection efficiencies are highly dependent on photocount statistics, so is the ability to clearly determine the number of molecules in a given probe volume. In the case of single-chromophore molecules, the similarity between the photocount distributions precludes a clear determination

On average at a
concentration of 3.3×10^{-9} M, only one
target molecule
residues in the probe
volume of 0.5-1.0 fl.
The actual number of
target molecules
fluctuates over time.
Assuming a Poisson
distribution, the
probability of finding
one molecule at this
concentration is 0.368,
that of finding two
molecules is 0.184,
and that of finding
more than two
molecules is 0.0078. In
more dilute solutions,
the detection events
are increasingly
dominated by
single-molecule
events.

of the number of molecules from the fluorescence burst amplitude. If the laser–molecule interaction time is long enough to measure the photobleaching kinetics, it is possible to distinguish between one or more molecules based on the time dependence of the fluorescence count rate versus time.

F4.3.2 Labelling schemes and observable values

Various properties of single fluorescent probes attached to macromolecules can be exploited to provide information on molecular interaction, enzymatic activity and conformational dynamics. 'Native' fluorescence probes such as fluorescence products, and fluorescence enzymes were successfully used to probe enzymatic turnovers of single molecules (see below). Small dye molecules that are covalently and site-specifically attached to macromolecules have also been actively used. And finally we note that another fruitful way to tag single proteins is through fusion with GFP (see Section F3.6).

Several approaches to single-molecule detection can be classified by labelling schemes and observable values. Below we present some of them (Figs. F4.2–F4.7).

A simple, but powerful, use of single-molecule detection localises a single fluorophore with a few tens of nanometre precision (Fig. F4.2). The dimensions of a dye molecule are much smaller than the wavelength of light it emits, and therefore it acts as a point source of light. The response of the optical system to this point source (the PSF) is a spot of light, the centre of which can be localised with great accuracy.

This positioning accuracy can be used for the colocalisation of two different macromolecules (Fig. F4.3). When two macromolecules are labelled with two non-interacting fluorophores that differ in their optical properties (absorption and emission spectra, fluorescence lifetime) they can be colocalised with nanometre accuracy and can report on association, binding and enzymatic-turnover events.

Even higher colocalisation accuracy can be obtained when the two fluorophores interact by FRET (Section F3.5). This technique, which is capable of measuring distances on a scale of 2–80 nm, relies on the distance-dependent energy transfer between a donor fluorophore and acceptor fluorophore (Fig. F4.4, see Section F3.5). This technique can also report on dynamical changes in the

Fig. F4.2 Labelling scheme (left) and observable value (right) for the localisation of a macromolecule labelled with a single fluorophore F. The point-spread function (PSF) can be localised within a few tenths of a nanometre. (Adapted from Weiss, 1999.)

distance between or orientation of the two fluorophores for intramolecular and intermolecular FRET (Fig. F4.5).

The absorption and emission transition dipoles of a single fluorophore can be determined by using polarised excitation light or by analysing the emission polarization, or both. The temporal variation in the dipole orientation of a rigidly attached (Fig. F4.6) or rotationally diffusing tethered probe (Fig. F4.7) can report on the angular motion of the macromolecule or one of its subunits.

Fig. F4.6 The orientation of a single immobilised dipole can be determined by modulating the excitation polarization. The fluorescence emission follows the angle modulation. (Adapted from Weiss, 1999.)

Fig. F4.7 The orientation freedom of motion of a tethered fluorophore can be measured by modulating the excitation polarization and analyzing the emission intensity. The emission dipole electric field vector **E** is defined in space by its angle ϕ in x-y plane and angle θ, relative to the z-axis. The fluorescence emission intensity, I_{xy}, that is detected in the x-y plane as a result of the emission dipole electric field vector **E** is defined as $I_{xy} = I \sin^2\phi = I_x + I_y$ and is also related to the fluorescence polarization angle ϕ as follows: $\cos^2\phi = I_x/(I_x + I_y)$. As defined by this spatial model, observed increases in total fluorescence intensity can be obtained if the emission dipole swings away from the z-axis (i.e. for θ approaching 90°). (Adapted from Weiss, 1999.)

These schemes were used to study ligand–receptor colocalization, to probe equilibrium protein structural fluctuations and enzyme–substrate interactions during catalysis and to identify conformational states and subpopulations of individual diffusing molecules in solution.

With the aid of single-molecule manipulation techniques such as atomic force microscopy and optical and magnetic tweezers, the molecular mechanisms of the functioning of a biological macromolecule can be studied. Examples of the application of these techniques to study overstretching and supercoiling of single DNA molecules, to determine forces and displacements generated during single-molecular motors, and to study mechanically unfolded proteins are given below.

F4.4 Optical detection of single molecules in the solid state

A number of optical methods have been developed to study single molecules. They differ in sampling conditions and the means of delivering excitation energy, but all of them share the need to isolate single molecules for detection. One approach is to isolate individual molecules spectroscopically in low-temperature solids because matrix perturbations cause each molecule to have a slightly different absorption frequency. A more broadly useful approach is to isolate molecules in dilute solution or on a surface, i.e. individual molecules are spatially separated

Fig. F4.8 Hybrid space–frequency image of single molecules of pentacene in a p-terphenyl crystal at 1.8 K, with a 300-MHz laser-frequency range near 506 THz along the horizontal axis and a 40-μm spatial range along the axis into the page. (After Moerner and Orrit, 1999.)

from each other in the volume or area probed by a laser beam. The main principles are discussed below.

Optical detection and spectroscopy of single-chromophore molecules were first achieved in solid matrices at very low temperature (Fig. F4.8). Guest molecules (pentacene) are embedded in a host material (p-terpenyl) at low concentration ($10^{-7}-10^{-9}$ M). Because of the strains and imperfections in the host environment, the optical absorption lines of the embedded molecules are inhomogeneously broadened. These broadened lines consist of a superposition of the homogeneous lines of individual guest molecules and a spread line of imperfections. A laser beam of narrow line width can be tuned to resonance with only one of the guest molecules. Such spectral selection can isolate 1 out of 10 000 molecules in a probe volume which allows the measurement of a single absorber without ensemble averaging.

F4.5 Optical detection of single molecules in the condensed phase

The requirements for successful single measurements in a low-temperature solid host are quite restrictive. Such measurements are limited to solids at temperatures of less than 15 K, which excludes the study of many interesting biological systems. Practical applications requiring single-molecule detection usually require measurements on liquid-phase solutions at room temperature.

F4.5.1 Levitated microdroplets

One of the experimental approaches to detecting single dye molecules in solution is based on using levitated microdroplets. The main concepts are to confine a single molecule in a picolitre-sized droplet and to use an electrodynamic trap to levitate the droplet so that the molecule can be interrogated by a laser beam for an extended period of time. It has been shown that both the spontaneous emission rate and the integrated fluorescence intensity for rhodamine 6G are significantly enhanced in glycerol microdroplets, yielding a signal-to-noise ratio at 10–40 for

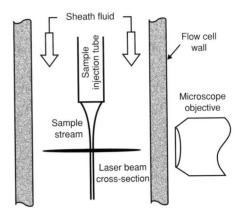

Fig. F4.9 Cross-section of the sheath flow cell (not to scale). The sheath liquid flows through a glass square-bore cell (0.8 mm i.d.) and hydrodynamically focuses the sample stream from the 100-μm i.d. injection tube down to 20 μm. The laser beam travels perpendicular to the plane of the drawing. (Adapted from Castro *et al.*, 1993.)

single rhodamine 6G molecules. Unfortunately, the time required for levitation, fluorescence measurement and droplet size analysis is relatively long (several minutes), which makes this approach impractical for single-molecule counting and sorting applications.

F4.5.2 Liquid streams

The method for detecting single molecules in liquid streams works in much the same way as a flow cytometer counts biological cells. A typical scheme for detecting individual DNA molecules is presented in Fig. F4.9. A dilute liquid sample in a capillary tube flows through a focused laser beam. The flowing sample solution is hydrodynamically focused down to a 20-μm-diameter stream by a rapidly flowing water sheath. Fluorescence signals are collected at 90° with respect to the direction of the laser beam before passing through spatial and spectral filters to the detection system. The signature of the passage of individual molecules is a burst of fluorescence photons that corresponds to single molecules passing across the laser beam.

Figure F4.10 demonstrates the detection of individual λ DNA molecules (48 kbp) stained with TOTO-1 at a base-pair-to-dye ratio of 5:1 (Comment F4.4).

Fig. F4.10 Detection plot for a 3×10^{-15} M solution of DNA–TOTO-1 complex in TE buffer which contains 10 mM Tris, 1 mM EDTA, pH 8.0 (top) and for a similar solution without DNA, i.e. TOTO-1 in TE buffer (bottom). (Adapted from Castro *et al.*, 1993.)

Fig. F4.11 Histogram of the number of molecules versus the fluorescence intensity for a mixture of λ DNA (48 kbp) and λ *Xba* DNA digest (24 kbp) at a concentration 3 × 10⁻¹⁵ M each. (Adapted from Castro *et al.*, 1993.)

Each observed peak is the integrated photon burst emitted by an individual DNA molecule as it traverses the laser beam. Figure F4.11 shows simultaneous detection of λ DNA and λ *Xba* DNA digest (24 kbp). The *Xba* digest of λ DNA yields two fragments of about 24 kbp each. Since the amount of intercalated dye is proportional to the DNA size, one would expect to observe fluorescent molecules of two different brightnesses corresponding to the two DNA fragment sizes. This is indeed the case, as shown by the histogram plot of the number of peaks versus peak amplitude in Fig. F4.11. The distribution centred at about 150 intensity units corresponds to 24-kbp DNA pieces, and the one centred at 300 corresponds to 48-kbp DNA pieces, yielding the expected 2:1 intensity ratio.

F4.5.3 Trapping with capillaries

Even if a single-molecule detection can be achieved with high efficiency in the sheath flow cell, the measurement efficiency may be far less than 1. In this case, the laser beam probes a small portion of the sample stream to reduce the scattering from the solvent. Therefore, the spatial probing efficiency for single-molecule detection in a sheathed flow cell is very low (Comment F4.5).

One simple approach to reducing the probe volume is to confine the molecules in a capillary tube and to irradiate the entire diameter of the capillary with the

---

Comment F4.5   SMD efficiency in flow cell

In a simple 50-μm flow cell, only 4% of the molecules actually flow through a 10-μm laser beam, which means that 96% of the molecules are not detected. The detection efficiency can be significantly improved with a submicrometre-sized detection zone, but the volume throughput is then low for practical use. A compromise may involve the use of a slightly defocused laser beam to completely cover a 2–5-μm detection zone in a 50–100 μm capillary (Nie and Zare, 1997).

Comment F4.4
TOTO is a very efficient fluorescence molecule

Since the fluorescence quantum yield of DNA is very small native DNA molecules are stained with a fluorescent dye, for example TOTO-1. TOTO-1 is a dimer of thiazole orange, which intercalates between the DNA bases and has a binding affinity constant nearly 1000 times larger than that of the most widely used DNA intercalator, ethidium bromide. Also, when bound to DNA, the TOTO-1–DNA complex fluorescence quantum yield increases by a factor of 1100 compared to that of free TOTO-1.

Fig. F4.12 Detection of
single molecules of
lactate dehydrogenase.
The solution (7.6×10^{-16}
M) was incubated in a
20 μm-diameter capillary
tube. Molecules were
electrophoretically driven
at 11 cm/min^{-1} by an
applied voltage of 24 kV.
(After Xue and Yeung,
1995.)

excitation laser beam. A major advantage of this approach is that the capillary
facilitates the analysis of extremely small volumes of solution (about 1 pl). In this
case, every molecule passing through the capillary flow cell should be detected
(Fig. F4.12).

F4.5.4 Electrical trapping

The main goal of electrical trapping is to separate the unbound fluorescent label
from its macromolecular, viral or cellular target with the help of electric forces.
This type of separation is most efficient if the label and the target have opposite
electric charges. The separation can usually be effected for protein–ligand binding. However, application of electrical trapping to nucleic acids presents some
difficulty because both sense and antisense sequences are negatively charged. A
convenient and effective procedure is to use peptide nucleic acid rather than RNA
or DNA strands. (Comment F4.6).

Comment F4.6   Peptide nucleic acid

Peptide nucleic acid is a polymer in which the phosphate sugar backbone of nucleic
acids is replaced by a peptide-like backbone, based on the monomer
2-aminoethylenglycin carrying any of the four nucleobases: A, T, G, C. This
polymer, in contrast to RNA or DNA, is electrically neutral. To positively charge it,
one could add, in addition to fluorescent label, some lysine or any positively charged
residues at one of the termini. The essential point is that this polymer binds
favourably to RNA or DNA (i.e. the RNA or DNA hybrid with peptide nucleic acid
is more stable than the homo-double strand). Below the melting point a (positively
charged) peptide nucleic acid antisense sequence of ∼20 nucleotides binds to its
target RNA or DNA within a suitable equilibration time and remains bound to this
target sequence for days, even if the excess unbound primer is removed. Because the
target RNA or DNA sequence labelled with primer carries an excess negative
charge, it migrates towards the anode, in contrast to the unbound label.

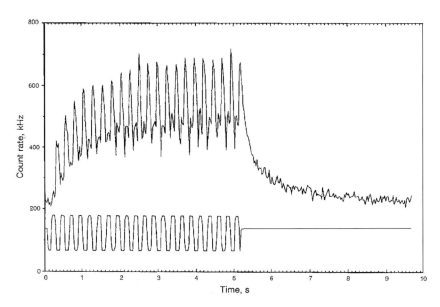

Fig. F4.13 Forced
oscillation of rhodamine
molecules between two
electrodes separated by a
distance $\sim 10^{-3}$ cm in
H_2O. Upper trace: signal
from rhodamine-labeled
dUTP (deoxyuridine
triphosphate); lower
trace: periodic electric
field, 2 V (peak to peak),
4 Hz. The field is
switched on at
$t = 0$ and is switched off at
$t = 5$ s. (Adapted from
Eigen and Rigler, 1994.)

In practice, the method is based on fluorescence correlation spectroscopy, which records spatio-temporal correlations among fluctuating light signals (Chapter D11), coupled with devices for trapping single molecules in an electrical field. In the simplest variant of the method FCS is coupled with capillary electrophoresis (Chapter D5). The equilibrated sample is enclosed in a capillary containing a volume ~ 100 μl. One target molecule in this volume means a concentration of 10^{-20} M. The capillary has an outlet with an opening $\sim 10^{-3}$ cm in diameter at one end, and electrodes are arranged such that negative ions migrate toward this outlet with a suitable speed (typically 0.1 cm s^{-1}). The outlet of the capillary, in particular, poses a high barrier for positively charged ions due to the very large field gradient. The outlet tip is a negative electrode for the static field within the capillary; simultaneously, another field is spanned between this tip and another sharp tip. These two tips are separated by a distance of only 10^{-3} cm and an alternative electric field is applied between them. The light cavity is positioned between these two tips, closer to the sharp second tip with a radius of curvature of only 10^{-4} cm. The alternative field drives a cloud of target molecules through the light cavity. Figure F4.13 shows fluorescence amplitudes from a single molecule when such a cloud of fluorescent molecules oscillates between two electrode tips that are separated by a distance of 10^{-3} cm. A large amplitude of fluorescence is observed when the cloud of labelled targets traverses the light cavity. The fluorescence signal increases to saturation within the light cavity due to the formation of a diffuse double layer between electrodes (Section D5.7). It is obvious that the method is 'electrophoretic' in nature. The critical feature, however, is the trapping of single molecules in the light cavity outside the static field that is superimposed on the separation process.

F4.6 Optical detection of single molecules at a surface

The experimental techniques that have evolved for single-molecule detection in liquids share several common characteristics: they all illuminate and/or image a small volume to minimise background scattering from the solvent, and use high numerical aperture objectives to maximise light collection efficiency. One useful approach is to isolate molecules on a surface, i.e. individual molecules are spatially separated from each other in the area probed by a laser beam in optical microscopy. Aspects of some of these approaches are examined here.

F4.6.1 Near-field scanning optical microscopy (NSOM)

NSOM was developed primarily to break the optical diffraction limit, which restricts the spatial resolution of conventional optical measurements to approximately half of the light wavelength (Section F1.3). In NSOM, a spatial resolution of the order of 50–100 nm is achieved by bringing the sample to within 5–10 nm (the optical near field) of a subwavelength-sized optical aperture so that the distance from the aperture is not great enough for the photons to undergo diffraction. The resolution is limited only by the size of the aperture, which is often a pulled micropipette or a tapered single-mode fibre.

Advances in NSOM have allowed imaging, dynamics spectroscopy and resonance-energy transfer of single molecules (Section F3.5). This capability results from the high photon fluxes delivered by the tapered single-mode fibre, efficient background rejection owing to the small illumination volume and excitation by a strong evanescent wave component near the tip end.

Figure F4.14(a) shows a 4 μm by 4 μm near-field emission image of sulphorhodamine-101 molecules dispersed on a glass cover slip. The bright, subdiffraction-limited spots observed are attributable to emission from single molecules and exhibit a FWHM of 125 nm, limited by the aperture size of the particular near field used (Fig. F4.14(b)).

Fig. F4.14 (a) Near-field fluorescence image of sulphorhodamine 101 molecules dispersed on a glass surface. The bright features in the 4 μm by 4 μm image result from emission from single molecules. (b) Schematic of the near-field scanning optical microscope for detecting single molecules. (Adapted from Xie and Dunn, 1994.)

(a)　　　　　(b)

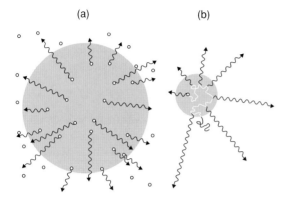

Fig. F4.15 Detecting the fluorescence of excited molecules. (a) Volume of light cavity (V_c) ≫ volume of particle 'territory' (V_T), as required for obtaining large mean intensity. (b) V_c ≪ V_T, as required for producing large fluctuations. The so-called territory is simply the reciprocal of particle concentration. For a 10^{-9} M solution we have 10^{15} particles per litre, and thus a particle territory of 10^{-15} litre or 1 fl. Likewise a 10^{-15} M solution has a particle territory of 1 nl. These numbers clearly show that it suffices to have a light cavity of the order of femtolitres. (After Eigen and Rigler, 1994.)

F4.6.2 Far-field confocal microscopy

The usual way to record a fluorescent signal is to illuminate a sufficiently large volume (Fig. F4.15(a)). This method is called fluorescence correlation spectroscopy, and its concepts were formulated in the early 1970s (Chapter D11). In the beginning of the 1990s the full potential of the method and its extreme sensitivity, thanks to new developments in laser techniques and microscopy, became evident. If the primary goal is to study single molecules, and if subdiffraction spatial resolution is not essential, then the far-field confocal method appears to be a better choice because of its unlimited excitation throughput, its experimental simplicity, and the absence of tip perturbation (Section F1.3).

In confocal microscopy, a laser beam is brought to its diffraction-limited focus inside a sample using an objective with a high numerical aperture ($NA \sim 1.3$). A small pinhole (50–100 μm in diameter) is placed in the image plane to reject light from out-of-focus regions. This arrangement defines a small volume of 0.5-1.0 fl

Fig. F4.16 (a) Schematic diagram of the confocal fluorescence microscopy system. (b) Background from unlabelled molecules. (c) Real-time observation of 14-nm fluorescent microspheres. (Nie *et al.*, 1995.)

Comment F4.7   Diffusion-controlled reactions

Quantitatively, the theory of diffusion-controlled reactions tells us that the encounter time of a particle to enter a sphere-like light cavity of radius R_c is given by

$$\tau_R^{-1} = 4\pi R_c DN$$

where D is the particle diffusion coefficient (in square centimetres per second) and N is the number of particles per cubic centimetre. If R_c is of the order of magnitude of 1 μm and if $D = 10^{-6}$ cm^2 s^{-1}, one obtains for concentration of 10^{-12} M that τ_R is about 1 s. It is apparent that concentrations down to 10^{-15} M (observation times of 15 min) are detectable in this way. By the same approach we can calculate the average time a particle spends inside the light cavity to be

$$\tau_D = R_c^2/3D$$

With above data: $\tau_D \approx 3 \times 10^{-3}$ s. This is more than enough time to yield a burst of several thousand fluorescent quanta during one encounter. Interested readers should refer to the specialist literature for more details (Eigen and Rigler, 1994).

---

in the sample. Figure F4.16(a) shows a schematic diagram of a confocal fluorescent microscope extended to study an individual, fluorescently tagged molecule. The process to be recorded can be described as a diffusion-controlled reaction of an individual fluorescent particle with the intensity illuminated discrete light cavity of 0.1–1 fl (Comment F4.7). This process is illustrated in Fig. F4.15(b). Most of the time the light cavity is free of fluorescent particles; hence only residual noise is recorded (Fig. F4.15(b)). Occasionally, a fluorescent particle diffuses into the light cavity and then emits a burst of fluorescent signal. Figure F4.16(c) shows a real-time observation of 14-nm fluorescent microspheres.

F4.6.3 Wide-field epi-illumination

Wide-field epi-illumination is perhaps the most straightforward way to image single molecules at the diffraction limit. Simple modifications can be made or added to an ordinary epifluorescence microscope to achieve single-molecule detection sensitivity. The optical excitation system generally consists of a laser source, defocusing optics and a high-performance (high-reflection and low-fluorescence) dichroic beamsplitter, and an oil-immersion, low-autofluorescence objective with $NA \sim 1.3$. Highly sensitive detection is achieved by using a cooled, charge-coupled device (CCD) camera. Figure F4.17 shows a schematic diagram of the instrumental setup for single-molecule detection in the thin-layer mode.

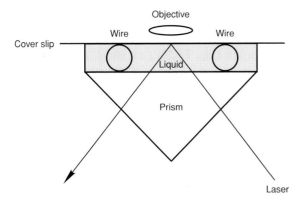

Fig. F4.17 Schematic diagram of the instrumental setup for single-molecule detection in the thin-layer mode. The excitation depth is defined by placement of two 4-μm tungsten wires on the prism. The laser penetrates the solution before being totally reflected at the cover slip surface, providing a lower background than epifluorescence arrangements. The total observation region is of 3.6×10^{-11} litre. Concentrations up to 5 nM can then be used to isolate individual molecules (After Xu and Yeung, 1997.)

Fluorescent dyes or fluorescently labelled macromolecules can be immobilised on a glass surface by covalent attachment, spontaneous non-specific adsorption or solvent evaporation. Figure F4.18 shows imaging of the diffusional path of a single molecule in a phospholipid membrane by using phospholipids carrying one rhodamine dye molecule. For molecules that are not fixed on the surface but can move slowly in a lipid membrane the random two-dimensional motion of single fluorescence-labelled lipid molecules can be measured. Note, that fluorescence microscopy is carried out at a sensitivity such that single fluorescent molecules illuminated for only 5 ms are resolvable at a signal-to-noise ratio of 28.

0 ms

210 ms

35 ms

245 ms

70 ms

280 ms

Fig. F4.18 Nine images of two fluorescence-labelled lipids observed at low surface density in an area $5.4 \times 5.4 \ \mu m^2$ of a lipid membrane; images were taken taken every 35 ms. The two lipid molecules observed in each image undergo motion as is apparent from the lateral shifts of the peaks in the image sequence. The lipid motion is characterised by the lateral diffusion constant $D_{lat} = (1.42 \pm 0.13) \times 10^{-8} \ cm^2 \ c^{-1}$. (Adapted from Schmidt et al., 1996.)

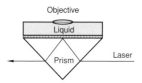

Fig. F4.19 Schematic diagram of the instrumental setup for single-molecule detection in the evanescent mode. The unfocused laser output is directed into a fused-silica right-angle prism located on the microscope stage. The estimated 0.15-μm thickness of the evanescent field defines an observation zone of only 14 al. For fluorophor concentrations lower than 10^{-7} M, the individual molecules are, on average, isolated (Adapted from Xu and Yeung, 1997.)

F4.6.4 Evanescent wave excitation

Evanescent wave excitation is normally achieved by total internal reflection at the glass–liquid or glass–air interface (Fig. F4.19) (see also Chapter C4). At this interface the optical electromagnetic field does not abruptly drop to zero but decays exponentially into the liquid phase (or air). The overall apparatus for evanescent-wave single-molecule imaging is similar to that for epifluorescence imaging (except that the excitation beam is directed to the sample from the other side of the objective). In both techniques, the sensitivity that allows single-molecule detection arises from the small sample volume probed. Although the field of view may be wide, the sample thickness is extremely thin ($\sim$300 nm).

The evanescent method has two advantages over epi-illumination. First, the background signal is generally lower because only a thin layer ($\sim$300 nm) is probed. Second, the dynamics of surface molecular events such as binding can be studied at the single-molecule level because fluorescence signals are detected only when the molecule moves into the evanescent field.

F4.6.5 Integrated optical and atomic force microscopy studies

Although optical methods are sufficiently sensitive to detect single chromophore molecules, it is highly unlikely that they are able to localise a single molecule to a spot comparable to its dimensions. The best spatial resolution achieved in NSOM is approximately 15 nm, which is still far from resolving individual molecules that are often smaller than 20 nm in size. The ultimate measurement in single-molecule studies needs not only to detect a single chromophore but also to resolve the shape and size of a single molecule. When the size and shape of a single molecule can be determined simultaneously, it is immediately clear whether optical detection of a single molecules has been achieved.

One strategy toward meeting this goal is to integrate ultrasensitive fluorescence detection and atomic force microscopy by using an inverted optical microscope. In the integrated microscope, molecular information is obtained by optical imaging and spectroscopy, and atomic force microscopy provides nanometre-resolved topographic information on the same molecule. However, atomic force microscopy normally does not provide chemical or molecular information, and it is difficult to differentiate target molecules from contaminants.

A second strategy is based on the combination of an AFM and optical tweezers. Figure F4.20 shows a schematic view of the experimental setup for the spatial selecting, grafting and manipulation of a single DNA molecule. The cantilever is first aligned in the focal plain. By moving the experimental cell in xy plane, a tethered bead is brought into the visualisation plane of the microscope. The tip of

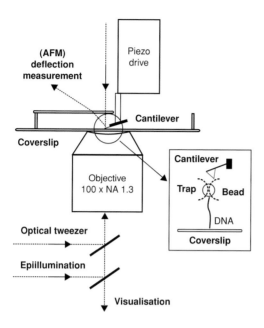

Fig. F4.20 Schematic
view of the setup for
molecular grafting and
manipulation of single
DNA molecules (not to
scale). The main feature
of the experimental setup
is the combination of an
AFM, optical tweezers,
and visualisation in a
unique approach to
manipulate single
molecules. A single DNA
molecule is anchored at
one end to a coverslip
and at the other to a
3.2-μm bead. The
tethered bead is grafted
to the tip of the cantilever
using optical tweezers. In
the microscope field of
view, a large number of
beads are present. Some
are directly bound to the
coverslip, others are
tethered to it by DNA,
with one or many
attachments. Watching
the beads' Brownian
motion, it is easy to
discard those beads glued
to the coverslip, as they
don't move. (After
Shivashankar and
Livchaber, 1997.)

the cantilever is then brought in physical contact with a selected bead, which lies in the optical trap. The ability to select a single DNA polymer and specifically graft it to a localised position on a substrate opens up new possibilities for biosensors and bioelectronic devices (see Chapter C4).

F4.6.6 Scanning near-field Raman spectroscopy

When a molecule is placed near the surface of a metal particle, the Raman signal intensity is enormously enhanced (Chapter E3). This effect is called surface enhanced Raman scattering (SERS). A highly sensitive Raman imaging system with a submicrometre spatial resolution system can be constructed using near-field optics. Figure F4.21 shows a schematic of the experimental setup. The silver-coated AFM tip approaches the near field of a sample. The tip is covered by the SERS active silver surface. A greatly enhanced Raman signal is observed from the area where the tip touches the sample which is only limited by the size of the tip apex (~5–100 nm). Thus, the SERS phenomenon makes it possible to achieve ultrasensitive detection limits allowing Raman spectroscopy down to the single-molecule detection level. We recall that Raman spectroscopy provides more detailed structural information on biological macromolecules than fluorescence techniques through a measurement of the vibrational modes of the molecule.

Table F4.1 summarises the main characteristics of five methods of optical detection.

Table F4.1. *The main characteristic of five methods of single-molecule optical detection*

Method of optical detection	Excitation volume	Signal-to-noise	Spatial resolution	Mode of detection
Near-field optical microscopy	Femtolitres ($\sim 10^{-15}$ l)	High	50–100 nm	Point
Far-field confocal microscopy	Subfemtolitres ($\sim 5 \times 10^{-16}$ l)	High	Diffraction limit	Point
Far-field confocal microscopy in flow stream	Picolitres ($\sim 5 \times 10^{-12}$ l)	High	Diffraction limit	Point
Wide-field epi-illumination	Femtolitres ($\sim 10^{-15}$ l)	Reduced	Diffraction limit	Parallel
Wide-field epi-illumination with total internal reflection excitation	Attolitres (20×10^{-18} l)	High	Diffraction limit	Parallel

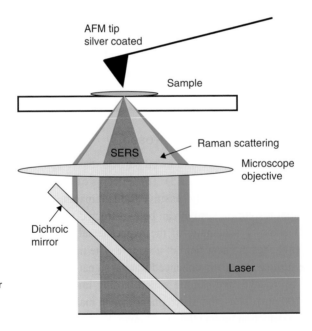

Fig. F4.21 Raman imaging system of the aperture-less near-field optical microscope. The incident laser light illuminates a sample through the thin transparent glass plate. The Raman signal is collected by a lens, using a backscattering geometry, and is separated from the incident laser beam by a dichroic mirror grating and steered to a CCD detector. A special metal probe (a metal-coated AFM tip) is used to scatter the light and efficiently enhance the Raman signal.

F4.7 Single-molecule studies tell more than ensemble measurements

Virtually all our knowledge of chemical reaction kinetics has been deduced from measurements on large populations. This has led us to think of molecular processes as well-ordered sets of events obeying traditional phenomenology based on simple rate laws. For such processes the statistics for small populations should obey Poisson statistics and their activation processes should follow the simple Arrhenius law.

Unfortunately, complex behaviour (e.g., the distribution of a physical entity) does not provide information on whether all molecules share a common distribution or whether each molecule gives its own specific contribution to the distribution seen for many molecules. In other words, phenomena that can be observed for a large number of molecules may not be correctly understood if it is not possible to observe the events on a single-molecule level.

Below we give a few examples of single-molecule studies of biological macromolecules in static and dynamic regimes which show that single-molecule studies indeed reveal more than ensemble measurements.

F4.7.1 Direct observation of enzyme activity

The enzymatic activity of single enzymes can be observed in real time by monitoring their fluorescence properties. Using fluorescence offers the potential to investigate how the enzyme activity varies from molecule to molecule, and for individual molecules as a function of time.

Figure F4.22 shows the variation in the chemical reactivity of individual molecules of lactate dehydrogenase as a function of time, monitored by a fluorescence technique. It was found that the activity of individual electrophoretically pure enzyme molecules can vary by up to factor of 4, and that these activities remain unchanged over a two-hour period. The activity difference may originate from the presence of two stable forms of the enzyme.

Figure F4.23(a) shows the fluorescence image of a single flavoenzyme molecule, cholesterol oxidase (COx), in its oxidised form, and Fig. F4.23(b) shows real-time observation of the enzymatic turnover of a single COx molecule catalysing oxidation of cholesterol molecules (Comment F4.8). The most obvious feature of the turnover trajectory in Fig. F4.23(b) is its stochastic nature. On a single-molecule basis, a chemical reaction event takes place on the

Fig. F4.22 The activity change of individual LDH-1 molecules with time at 21 °C. Two one-hour incubation periods separated by electrophoresis for 3 min were employed in the experiment, creating pairs of peaks. The capillary diameter was 20 μm. The solution concentration of LDH-1 was 4.5×10^{-17} M. (After Xue and Yeung, 1995.)

Fig. F4.23 (a)
Fluorescence image
(8 μm × 8 μm) of single
COx molecules
immobilised in a
10-μm-thick agarose gel.
The emission is from the
fluorescent active site,
which is tightly bound to
the centre of COx. Each
individual peak is
attributed to a single COx
molecule. The intensity
variation between the
molecules is due to
different longitudinal
positions in the light
depths. (b) Real-time
observation of enzymatic
turnovers of a single COx
molecule catalysing the
oxidation of cholesterol
molecules. The figure
shows a portion of an
emission intensity
trajectory recorded in
13.1 ms per channel. The
emission exhibits
stochastic blinking
behaviour as the FAD
toggles between oxidised
(fluorescent) and reduced
(non-fluorescent) states,
each on–off cycle
corresponding to an
enzymatic turnover.
(Adapted from Lu *et al.*,
1998.)

(a)

(b)

Comment F4.8   Physicist's box: Flavoenzymes

"Flavoevzymes are ubiquitous and undergo redox reactions in a reversible manner. Choloesterol oxidase (COx) from *Brevibacterium sp.* is a 53-kDa flavoenzyme that catalyses the oxidation of a cholesterol by oxygen. The active site of the enzyme involves FAD, which is naturally fluorescent in its oxidised form but not in its reduced form. The FAD is first reduced by a cholesterol molecule to FADH$_2$, and is then oxidised by O$_2$, yielding H$_2$O$_2$. The crystal structure of COx shows that the FAD is non-covalently and tightly bound to the centre of the protein and is surrounded by a hydrophobic binding pocket for cholesterol, which is otherwise filled with 14 water molecules.

---

subpicosecond time scale and cannot be time resolved here. However, the time needed for diffusion, thermal activation or both before such an event is usually much longer. The emission on-time and off-time correspond to the 'waiting time' for the flavin adenie dinucleotide (FAD) reduction and oxidation reactions, respectively (Comment F4.8). Statistical analyses of single-molecule trajectories have revealed a significant and slow fluctuation in the rate of cholesterol oxidation by FAD. A molecular memory phenomenon, in which an enzymatic turnover is not independent of its previous turnovers because of a slow fluctuation of protein conformation is evidenced by spontaneous spectral fluctuation by FAD.

These two examples show that single-molecule measurements can reveal the *distribution of molecular properties* in inhomogeneous systems. The distributions, which can be either static, as in the case of lactatedehydrogenase, or dynamic, as in the case cholesterol oxidase, cannot usually be determined by ensemble-averaged measurements.

Fig. F4.24 Simulated structure of the tRNA^{Phe}–TMR complexes. 4-thiouridine (s^{4}U) at position 8 of tRNA^{Phe} was specifically labelled by reacting it with tetramethylrodamine-5-iodacetamide. The dramatic shortening of the lifetime in one of the substrates must involve fluorescence quenching of rhodamine fluorescence. It is well known that the fluorescence of rhodamines is quenched by guanosine through a mechanism that involves the formation of a complex. The only guanosine, G15, is located in the close vicinity of the TMR. Note: TMR yellow, s^{4}U pink, and G15 green. (After Jia *et al.*, 1997.)

F4.7.2 Non-exponential kinetics of a single tRNA molecule

One area in which single-molecule experiments can be particularly valuable in biology is in the determination of the microscopic origins of heterogeneity associated with observables. In kinetic experiments performed on ensembles of molecules, heterogeneity often shows up as non-exponential behaviour. However, it is usually not known how the variations in the properties are distributed over the ensemble. Nevertheless, specific statistical distributions of properties can be constructed, if the experiments are carried out on the level of a single molecule.

For example, the fluorescence decay functions of individual, specifically labelled t-RNA^{Phe} molecules exhibit non-exponential behaviour with the most probable values for the lifetimes being 2.80 and 0.55 ns (Comment F4.7). These two different lifetimes have been attributed to two conformations of the tRNA^{Phe}–TMR complexes. A structural model indicates that the tRNA^{Phe}–TMR complex fluctuates between two states, one of which provides conditions that quench the probe fluorescence (Fig. F4.24).

F4.7.3 RNA catalysis and folding

The *Tetrahymena* ribozyme is a catalytic RNA of about 400 nucleotides (Fig. F4.25). For single-molecule experiments, *Tetrahymena* ribozyme was extended at its 3′-end by annealing it to a complementary oligonucleotide tether with a 5′-biotin, allowing the ribozyme to be immobilised on a surface via a

Fig. F4.25 The secondary structure of the *Tetrahymena* ribozyme. The ribozyme catalyses the cleavage of substrate (S) by guanosine (G). The ribozyme was derived from a self-splicing group I intron. The thin line and the six residues, GGAGGG, represent the L-21 Sca I ribozyme. The thick line represents the extension of the 3′-end of the ribozyme, which is base-paired with complementary DNA oligonucleotide with a 5′-biotin. The residues and duplexes (P1–P.9.2) are numbered on the basis of the *Tetrahymena* self-splicing intron. (After Zhuang *et al.*, 2000.)

streptavidin–biotin interaction. Fluorescent dyes were attached to the ribozyme as shown in Fig. F4.25: a Cy3 dye to the 3′-end of the oligonucleotide substrate (S) bound to the ribozyme and, when required, a Cy5 dye at the 3′-end of the tether. FRET was measured between Cy3 (donor) and Cy5 (acceptor), the positions of which are shown on the medium-resolution model of the *Tetrahymena* ribozyme (Fig. F4.26). The model suggests that the distance between the donor and the acceptor decreases from ∼7 nm in the undocked state to 1–2 nm upon docking state.

Figure F4.27 shows the FRET time trajectory of a single ribozyme molecule. The FRET value fluctuates between two distinct levels with the higher FRET value in the docked state and the lower one in the undocked state. A rarely

Fig. F4.26 The medium-resolution model of the *Tetrahymena* ribozyme. The P1 duplex between S9 (blue) and 3′-end of ribozyme (grey) reversibly docks into the ribozyme (light grey). FRET between Cy3 (green) on S and Cy5 (red) on the tether (black) gives a signal for docking. (Adapted from Zhuang *et al.*, 2000.)

Fig. F4.27 FRET time trace from single ribozyme molecules showing P1 docking and undocking. FRET is defined as $I_A/(I_D + I_A)$, where I_D and I_A are the fluorescence signals from donor and acceptor, respectively. The docked and undocked states give FRET values of ~0.9 and ~0.3, respectively. (After Zhuang *et al.*, 2000.)

populated docked state, not measurable by ensemble methods, was observed. These results established single-molecule fluorescence as a uniquely powerful tool for examining RNA folding.

F4.8 Checklist of key ideas

- Because of the low background and high signal-to-noise ratios, laser-induced fluorescence has become the most widely used method for single-molecule detection.
- Single-molecule detection in solution is achieved using levitated microdroplets, liquid streams and capillaries, which facilitate the analysis of extremely small volumes of solution.
- Single-molecule detection at a surface is accomplished using NSOM, far-field confocal microscopy, wide-field epi-illumination in the thin-layer mode or total internal reflection excitation.
- The combination of an AFM and optical tweezers or ultrasensitive fluorescence detection in an inverted optical microscope provides the ultimate measurement in single-molecule studies.

Suggestions for further reading
Historical review and introduction to biological problems

Weiss, S. (1999). Fluorescence spectroscopy of single biomolecules. *Science*, **283**, 1676–1683.
Gimzewski, J. K., and Joachim, C. (1999). Nanoscale science of single molecules using local probes. *Science*, **283**, 1683–1688.

Fluorescence spectroscopy of single molecules

Nie, S., and Zare, R. N. (1997). Optical detection of single molecule. *Annu. Rev. Biophys. Biomol. Str.*, **26**, 567–596.
Weiss, S. (1999). Fluorescence spectroscopy of single biomolecules. *Science*, **283**, 1676–1683.

Optical detection of single molecules in solid state

Moerener, W. E., and Orrit, M. (1999). Illuminating single molecules in condensed matter. *Science*, **283**, 1673–1700.

Optical detection of single molecules in condensed phase

Nie, S., and Zare, R. N. (1997). Optical detection of single molecule. *Annu. Rev. Biophys. Biomol. Str.*, **26**, 567–596.

Eigen, M., and Rigler, R. (1994). Sorting single molecules: Application to diagnostics and evolutionary biotechnology. *Natl. Acad. Sci. USA*, **91**, 5740–5747.

Optical detection of single molecules on a surface

Betzig, E., and Chichester, R. J. (1993). Single molecules observed by near-field scanning optical microscopy. *Science*, **262**, 1422–1425.

Nie, S., Chlu, D. T., and Zare, R. N. (1994). Probing individual molecules with confocal fluorescence microscopy. *Science*, **266**, 1018–1021.

Single-molecule studies tell us more than ensemble measurements

Lu, H. P., Xun, L., and Xie, X. S. (1998). Single-molecule enzymatic dynamics. *Science*, **282**, 1877–1882.

Jia, Y., Sytnik, A., *et al*. (1997). Nonexponential kinetics of a single tRNA[Phe] molecule under physiological conditions. *Proc. Natl. Acad. Sci. USA*, **94**, 7932–7936.

Zhuang, X., Bartley, L. E., *et al*. (2000). A single-molecule study of RNA catalysis and folding. *Science*, **288**, 2048–2051.

Chapter F5
Single-molecule manipulation

F5.1 Historical review and introduction to biological problems

1986

A. Ashkin and coworkers proposed the idea of an optical trap (tweezers). An optical trap can be produced with a highly focused laser light, and can be used to grab, move and apply measurable forces on micrometre-sized objects, such as dielectric microspheres. A microsphere that is chemically coupled to a molecule of interest provides a means of measuring the molecule's position and the force that it exerts.

1992

K. Bustamante and coworkers pioneered direct mechanical measurements of the elasticity of single DNA molecules using magnetic trapping. In **1994 S. Chu** and coworkers studied the relaxation properties of single DNA molecules using optical trapping and **G. Li** and colleagues made a direct measurement of the force between complementary strands of DNA with atomic force microscopy. Mechanical properties of individual strands and doubled-stranded DNA have been determined. In **1997 G. Shivashankar** and **A. Libchaber** developed a new technique for single DNA molecule grafting and manipulation using combined atomic force microscopy and an optical tweezer. These studies opened up new possibilities in biosensors and bioelectronic devices.

1993

K. Svoboda and coworkers applied optical trapping nanometry (optical tweezers in conjunction with nanometre-precision position detection schemes) to study a single kinesin molecule. Later, a further demonstration of the efficiency of trapping nanometry came from the observation of the forces and displacement produced by a single myosin molecule (**J. Spudlish**, **T. Yanagida** and coworkers, **1994**), direct visualisation of the rotation of the γ-subunit of F_1-ATPase (**H. Noji** and coworkers, **1997**), unfolding individual nucleosomes by stretching single chromatin fibers (**M. Bennik** and coworkers, **2001**). After these experiments it became clear that optical trapping nanometry had fantastic potential for probing molecular mechanisms of functioning biological macromolecules, including molecular linear and rotary protein motors.

Mid-1990s

A few groups demonstrated the possibility of using atomic force microscopy in the force-measuring mode, in which a single molecule is stretched between the microscopic silicon nitride tip of a flexible cantilever and a flat substrate that is mounted on a highly accurate piezoelectric positioner. Direct observation of enzyme activity (**M. Radmacher** and coworkers), the unfolding and folding of a single protein (**M. Rief** and coworkers and, **M. Kellermayer** and his colleagues, **1997**), nucleosomes (**B. D. Brower-Toland** and coworkers, **2002**) and RNA molecules (**J. Liphardt's** group, **2001**) have been demonstrated.

2000 to now

Advances in ultrasensitive instrumentation (optical and magnetic tweezers, atomic force microscopy, etc.) provide extreme precision for position detection (such as $0.1-10$ nm) and in the force regime ($0.1-10\,000$ pN). The mechanical manipulation of single molecules has transformed many fields of biology, such as molecular motor mechanics, protein and RNA folding–unfolding, receptor–ligand interactions. With the aid of single-molecule manipulation techniques such as atomic force microscopy, and optical and magnetic tweezers, overstretching and supercoiling of single DNA molecules have been studied, force and displacements generating during single molecular motor reactions have been observed, and proteins have been mechanically unfolded. Even these examples are just the beginning. There is a lot of molecular machinery in the cell that is now open to study.

F5.2 Nanoscale manipulation techniques

There have been significant advances in the characterisation and manipulation of individual molecules. Optical and magnetic traps, the cantilevers of atomic force microscopy and glass microneedles are widely used for nanoscale manipulation. Below we give a brief description of such devices.

F5.2.1 Optical traps (laser tweezers)

'Optical tweezers' is the name given by A. Ashkin and coworkers to a device they invented that uses light pressure to manipulate tiny objects. By focusing a laser beam through a microscope objective, they found that particles with a high index of refraction, such as glass or plastic particles, or oil droplets, were attracted to an intense region in the beam and could be held permanently at a focal point (Fig. F5.1). Although the theory behind optical tweezers is still being developed, the basic principles are straightforward for objects either much smaller than the wavelength of light or much larger. When a small dielectric object, such as a

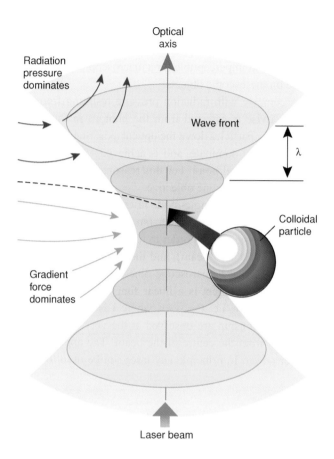

Optical axis

Radiation pressure dominates

Wave front

λ

Colloidal particle

Gradient force dominates

Laser beam

Fig. F5.1 Optical tweezers. A bead with a high index of refraction is attracted to the intense region in the laser beam. Intensity gradients in the converging beam draw small objects towards the focus, whereas the radiation pressure of the beam tends to blow them down the optical axis. Under conditions in which the gradient force dominates, a particle can be trapped near the focal point. (Adapted from Grier, 2003.)

lipid droplet or a polystyrene or silica bead, is placed in the beam, it feels a force pulling it toward to the focus of the beam. The magnitude of the exerted force is a function of number parameters: the size and shape of the object, the difference between the index of refraction of the object and the surrounding medium, and gradient in the intensity of the laser beam (Comments F5.1 and F5.2). Larger

---

Comment F5.1   Optical trap and the dimensions of a bead

Theoretical calculations for objects whose size is small relative to the wavelength of the trap light (Rayleigh particles) suggest a cubic dependence ($F_{max} \sim d^3$), where d is the size of the object. However, for trapped particles of the same approximate size as the light's wavelength (e.g. 500-nm vesicles in an 830-nm trap), this dependence is somewhat weaker, in general somewhere between linear and quadratic. Empirically, for trapped latex beads, the maximum lateral force applied on a 1.02-μm bead is 7 times that on a 0.3-μm bead, and the maximum force on a 2.97-μm bead is 2.47 times more than on the 1.02-μm bead.

Comment F5.2

Optical trap and
refractive index of
a bead

The applied force is a
function of n, the
relative index of
refraction, defined as
$n = n_1/n_2$, where n_1 is
the index of refraction
of the trapped object
and n_2 is the index of
the surrounding
medium. For lateral
forces, the larger n, the
larger the applied
force. Relatively small
corrections are needed
to enable calibrations
to be done in vitro, and
actual measurements
to be done in vivo.
Thus, the applied force
changes by
approximately 20%
when an object (e.g. a
lipid droplet, $n_1 =$
1.52) is in
water ($n_2 = 1.33$)
versus when it is in
cytoplasm ($n_2 = 1.39$).
This is also true for
axial forces.

objects act as lenses, refracting the rays of light and redirecting the momentum of their photons. The resulting recoil draws them toward the higher flux of photons near the focus. This recoil is all but imperceptible for a macroscopic lens but can have a substantial influence on mesoscopic objects.

Optical gradient forces compete with radiation pressure resulting from the momentum absorbed or otherwise transferred from the photons in the beam, which acts like a fire hose to blow particles down the optical axis. Stable trapping requires the axial gradient force to dominate, and is achieved when the beam diverges rapidly enough from the focal point. For this reason, optical tweezers are usually constructed around a microscope objective.

The basic optical trap is quite simple: a single-mode laser with Gaussian beam profile (TM_{00}) is passed through a beam expander (so that the beam fills the back aperture of the microscope objective, in order to use the entire lens, which generates the largest gradient in focused beam), and then steered with a combination of mirrors and lenses into the microscope objective (Fig. F5.2). The force, F, that the trap exerts on an object is a linear function of the object's displacement, x, from the trap centre, i.e. $F = -kx$, where k is a trap stiffness. Deviations from this linear relationship are observed at the very edges of the beam, typically about 200 nm from the centre of the beam. The applied force increases linearly with laser power. In principle any laser can be used to make

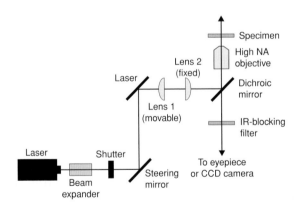

Fig. F5.2 Schematic diagram showing the path of the laser light beam from its origin at the laser to the formation of an optical trap in the specimen. The beam expander is chosen so that the laser beam just fills the back aperture of the microscope objective: if the back aperture is overfilled, laser light is lost; and if underfilled, less of the lens is used, resulting in less of a gradient in a focused beam. The shutter is used to block the laser to minimise the sample's exposure to laser light when an object is not being actively trapped. The steering mirrors are used to control the orientation of the beam, to bring it into the objective. Lenses 1 and 2 allow steering of the trap. After passing through the two lenses, the laser is brought into the microscope's optical path by reflecting it off of a dichroic mirror mounted in line with microscope objective. (After Gross, 2003.)

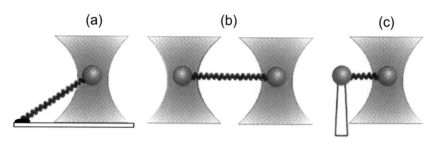

(a) (b) (c)

Fig. F5.3 Different experimental geometries using optical tweezers in which one side of a molecule is attached to the bead. The other side of the molecule is attached to: (a) a coverslip; (b) another bead; (c) a microneedle.

an optical trap; in practice IR lasers are employed because they cause the least optical damage.

For a trap based on a TM_{00} laser beam, the axial force (along the beam path) is weaker than the lateral force (pulling the object in toward the beam centre). However, for both axial and lateral forces, at a given displacement from the trap centre, the force that a trapped object feels is a function of its size: up to about 4 μm, the larger the object, the more force it feels.

Biological macromolecules can be bound to polystyrene or silica beads, which are usually ~1–3 μm in diameter. A trap can be moved to steer a bead into a desired experimental geometry, e.g. to interact with a partner molecule attached to a coverslip (Fig. F5.3(a)). Upon binding between the two molecules, the forces and movements involved can be measured, and the interaction can be perturbed by moving the trap. Similar experiments can be performed with glass microneedles or AFMs, although such probes are typically less compliant than optical ones.

Significant technical difficulties nevertheless exist in measuring movements of single molecules. The motions occur on length scales of ångströms to nanometres and on time scales of milliseconds or less. To obtain the high spatial and temporal sensitivity required, optical tweezers are combined with a dual-beam interferometer to produce an 'optical trapping interferometer'. This instrument provides position detection and trapping functions simultaneously, and can provide controlled, calibratable forces in the piconewton range.

Optical tweezers use light to manipulate small particles, but only one at a time. This is because the manipulating light beam diverges around the first trapped particle, and then cannot be brought back to focus within the short distance to the next particle. So experiments must be done in series, and this is a very time-consuming procedure. The first experiments in which small particles (micrometre-sized) were manipulated in *multiple planes* were demonstrated using a special kind of optical focusing to create a '*Bessel beam*' of light (Fig. F5.4).

Fig. F5.4 A Bessel beam. Light passing through a particular scheme of lenses can form a beam of light with a bright central spot and concentric rings of decreasing intensity; this is known as a Bessel beam, as its varying intensity is described by the zero-order Bessel function. A Bessel beam is used to create optical tweezers capable of manipulating many samples at once. (Adapted from Garces-Chavez *et al.*, 2002.)

F5.2.2 Magnetic traps (magnetic tweezers)

Figure F5.5(a) shows an experimental setup that uses a magnetic bead to manipulate biological macromolecules. Single molecules are bound at multiple sites with one extremity attached to a treated glass coverslip and the other to a magnetic bead.

(a)

(b)

Fig. F5.5 (a) Magnetic tweezers: one end of the macromolecule is bound to a glass coverslip and the other end to a paramagnetic bead. Magnetic tweezers use the gradient of the horizontal magnetic field generated by small magnets to pull vertically on a magnetic bead anchored to a surface by a DNA molecule. A force is applied on the bead with Co–Sm magnets equipped with a polar piece to focus the field in a 2-mm gap located just above the sample. The magnetic device can spin about the optical axis, causing the paramagnetic beads to rotate. (b) Schematic of a bead undergoing transverse and longitudinal fluctuations. (After Strick *et al.*, 1996.)

Comment F5.3   Calibration of stretching force

The stretching force can be determined by analysis of the Brownian fluctuations of the bead. In contrast with other techniques, the stretching force measurement is absolute and does not require a calibration of the sensor. It is based on the analysis of the Brownian fluctuation of the tethered bead, which is completely equivalent to a damped pendulum of length $l = \langle z \rangle$ pulled by a magnetic force F (along the z-axis). Its longitudinal ($\delta z^2 = \langle z^2 \rangle - \langle z \rangle^2$ and transverse δx^2 fluctuations are characterised by effective rigidities $k_{\parallel} = \partial_z F$ and $k_{\perp} = F/l$. By the equipartition theorem they satisfy

$$\delta z^2 = \frac{k_{\mathrm{B}} T}{k_{\parallel}} = \frac{k_{\mathrm{B}} T}{\partial_z F}$$

$$\delta x^2 = \frac{k_{\mathrm{B}} T}{k_{\perp}} = \frac{k_{\mathrm{B}} T l}{F}$$

Thus from the bead's transverse fluctuations one can extract the force pulling on the molecule (the smaller the fluctuations, the greater F) and from its longitudinal one obtains its first derivative.

This method of measurement can be used with magnetic (but not optical) traps because the variation of the trapping gradients occurs on a scale (about 1 mm) that is much larger than the scale on which the elasticity of the molecule changes (about 0.1 μm). In other words, the stiffness of the optical trap is very large compared with F/l.

Supermagnetic beads (typical diameter about a few micrometres) are made of many small ferromagnetic domains. Their magnetic susceptibility is anisotropic, and therefore their response to magnetic fields is similar to that of a compass needle. An external magnetic field can be used to rotate the beads and thus to coil and pull macromolecules. Magnetic tweezers are a useful tool particularly between a few femtonewtons and 100 pN with a relative accuracy of $\sim$10% (Comment F5.3).

A further advantage of the magnetic trap technique is that measurements on DNA at constant force are trivial (the position of the magnets is just kept fixed). With cantilevers or optical tweezers, to work at constant force requires an appropriate feedback to ensure that the displacement of the sensor is kept constant. At the same time, because its stiffness depends on the force, for weak forces ($<$1 pN) the magnetic trap technique has a lower spatial resolution of about 10 nm than the other manipulation methods.

F5.2.3 Cantilever in the force-measuring mode of the AFM

In the force-measuring mode of the AFM (see Chapter F2), single molecules or pairs of interacting molecules are stretched between the silicon nitride tip of a microscopic cantilever and a flat, gold-covered substrate whose position is

Fig. F5.6 AFM instrument design in force-regime mode. An AFM has four basic components: a microcantilever for sensing forces; a means of detecting cantilever deflection; a piezoelectric system for moving the tip in x, y, and z directions at atomic resolution; and a computer control system. In a typical commercial system, the cantilever is microfabricated from silicon nitride or silicon using a semiconductor fabrication method. The photodiode allows the bending motion of the cantilever to be detected to better than 1 Å. (Adapted from Wang et al., 2001.)

controlled by a high-precision piezoelectric positioner (Fig. F5.6). This system allows the suspended molecule(s) to be stretched with subnanometre precision. The forces acting on the molecule as it is extended are transmitted to the cantilever, causing it to bend. The amount of bending is measured by a laser beam reflected off the cantilever tip onto a photodetection system, such that the vertical deflection of the beam registers the angle at which the cantilever is bent. By calibrating the responsiveness of the cantilever, the degree of bending can be translated into the applied force with a precision at the level a few piconewtons.

In a typical experiment, the cantilever tip is pressed into a layer of purified protein adsorbed onto the gold substrate (Fig. F5.7(a)). Protein molecules from the adsorbed layer affix to the cantilever by an unknown mechanism (which could involve the high pressure achieved as the cantilever tip is pressed against the gold substrate) and are stretched as the gold substrate is withdrawn. Suspended

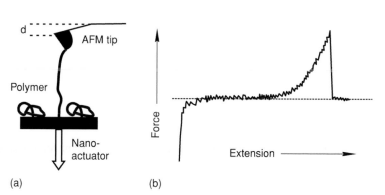

(a)   (b)

Fig. F5.7 A typical experiment in which a single molecule is stretched by an AFM tip. (a) The tip is brought into contact with the sample, which is covered by a layer of protein. Ideally, the attachment is at a known position on the protein, which could be the ends. As is often the case, the ideal situation is not the easiest to attain, and it is possible to study the extension of polymers without knowing the attachment position. (b) If a molecule is bound to the tip it can be stretched and the force is measured via the deflection of the cantilever spring as a function of the extension. When the maximum binding force is exceeded, the molecule breaks away from the tip and the tip is free again.

Comment F5.4

Cantilever and Hooke's law

The cantilever behaves as a Hookian spring with a force constant, k, of the order of 0.01–1 N m^{-1}. The force on the cantilever may be determined directly from the deflection of the cantilever and the force constant of the cantilever using Hooke's law, $F = kx$. The force constant k can be estimated from the dimensions of the cantilever using the Young's modulus of the material, but experimental determination of the force constant is preferable.

molecules resist extension and therefore cause deflection of the cantilever (Comment F5.4). The tension on the cantilever is released either when a force-induced rearrangement increases the distance between the ends of the suspended protein, or when the suspended molecule becomes detached from the cantilever or gold substrate. The resultant data may be expressed as a force extension curve (Fig. F5.7(b)).

To investigate single molecules, the physics and chemistry of the tip should be well defined and tailored for its specific application. The coating of tips with a catalyst to induce reactions locally or with a metal to deposit metallic clusters has been demonstrated. C_{60} molecules attached to tips are geometrically and structurally appealing probes for AFM imaging because they combine molecular dimensions with extreme ruggedness. Chemical functionalisation of carbon nanotubes by coupling basic, hydrophobic, or biochemical terminations of the carboxyl groups to the open tip ends has been reported. Specific chemical binding of pairs of molecules, one sited on the tip, the other on the surface, such as biotin–avidin or conjugated DNA strands, is being commercially developed in combination with magnetic bead technology used for immunoassays.

F5.2.4 Glass microneedles

Using a commercial pipette puller, tapered glass microneedles (typical dimensions: shank diameter 1 mm, tip diameter 1 μm, and tapered length 1 cm) can be prepared (Fig. F5.8). The stiffness varies within a wide range from needle to needle.

Figure F5.8(a) shows a typical video image of a magnetic bead and microneedle under the microscope. The experiments were performed in a liquid buffer, which was retained by a ring glued with paraffin on top of a microscope slide. This assembly formed a 'well' and was placed on an inverted video microscope. The face of the microscope slide in contact with the buffer was coated with an antibody against digoxigenin. The digoxigenin functionalised extremity of the

Fig. F5.8 (a) Typical video image of a magnetic bead and the microneedle under the microscope. (b) Principle of force measurement using a microneedle. (Adapted from Essevaz-Roulet *et al.*, 1997.)

(a)

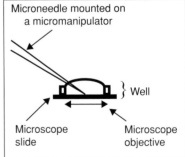

(b)

Table F5.1. *Comparison of atomic force microscopy, the laser trap, the magnetic trap and the glass microneedle*

Method	Atomic force microscopy	Laser trap	Magnetic trap	Glass microneedle
Force generation	Bending of cantilever obeys Hooke's law	Intensity gradient from photons from laser source	Supermagnetic bead in magnetic field	Bending of microneedle obeys Hooke's law
Dynamic range	20 pN–20 nN	0.1–400 pN	A few fN–100 pN	1 pN and as large as desired
Force constant	0.05 N m^{-1} (typically)	0.0002 N m^{-1}	$\sim$10^{-7} N m^{-1}	2 $\times$ 10^{-6} N m^{-1} (typically)

Table adapted from Wang *et al.* (2001).

construction was attached to this face, and the biotinated extremity was anchored to a microscopic bead coated with streptavidin. Figure F5.8(b) demonstrates the principle of force measurement using a microneedle. The bead was attached to a glass microneedle (treated with biotin), which was introduced through the free meniscus and was positioned by an *xyz* micromanipulator. The microneedle served as a force sensor, i.e. the deflection of the tip of the microneedle was monitored under a microscope as a function of the lateral displacement of the surface.

A comparison of the major features of the techniques described above is presented in Table F5.1.

F5.3 Macromolecular mechanics: nanometre steps and piconewton forces

Macromolecular mechanics implies the monitoring of forces at the molecular scale. At the single-molecule level, the characteristic energy is given by the hydrolysis of ATP (20 kT, i.e. 80 pN nm) and the characteristic size by the diameter of a protein (a few nanometres). The resulting forces that physicists must be able to measure and to produce while studying those objects are therefore in the range of hundreds of femtonewtons to hundreds of piconewtons.

The initial applications of macromolecular mechanics involved molecules that produce active movement against a load, e.g. 'molecular motors' such as myosin, which drives muscle contraction. In the 1990s many molecular motors and biological macromolecules have been observed including: stepwise motion of a single kinesin molecule along a microtubule track, displacement of an actin filament by a single myosin molecule, forces and transcriptional pauses associated with RNA polymerase activity, non-linear elasticity of single polymers, reversible

Comment F5.5
Molecular motors in a cell

It is likely that a given cell has 50 or more distinct molecules that operate on actin filaments and microtubules to produce the many types of movement that cells undergo. Taken together with the ATP-driven motor molecules operating along DNA, the movement of ribosomes along messenger RNA, and motors that have yet to be discovered, the number of different molecular motors in a cell is probably closer to 100.

unfolding of single-protein domains by applied force, discrete rotations of a single F_1 subunit of the F_0F_1-adenosin triphosphate (ATP) synthase, DNA packaging in bacteriophage ø29 and chewing of one of DNA's double-helical strands by lambda exonuclease (Comment F5.5).

The parallel development of single-fluorophore detection has allowed these mechanical measurements to be combined with observations of substrate binding, protein position and conformational change. Below we focus mainly on the use of optical trapping technology, including the rival methods that use glass microneedles and AFMs where appropriate.

F5.3.1 Linear molecular motors

The most intensive application of optical tweezers has been focused on linear motor proteins, such as myosin, kinesin, dynein, which convert the energy of ATP hydrolysis into mechanical work. These proteins move along polymer substrates: myosin along actin filaments in muscle and other cells; kinesin and dinein along microtubules. Another example is RNA polymerase (RNAP), which is a highly processive molecular motor, capable of moving through thousands of base pairs without detaching from the DNA template.

Do biological motors move with regular steps? What is the size of the steps? What is the speed and force of such motors? To address these questions many instruments with the spatial and temporal sensitivity to resolve movement on a molecular scale have been used. Figure F5.9 illustrates a typical experimental configuration for manipulating linear molecular motors. The vertical arrows indicate the direction of the light. The force and displacement of the motor are

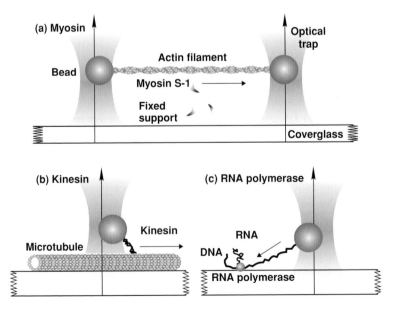

Fig. F5.9 Sketches (not drawn to scale) illustrating typical experimental configurations for manipulating linear molecular motors: (a) myosin; (b) kinesin (or dinein), (c) RNA polymerase. (Adapted from Wang, 1999.)

detected the trapped bead. In Fig. F5.9(a) myosin, an actin filament, suspended by two separate optical traps via two beads attached to its ends, is lowered over a third, fixed bead coated with myosin S-1 fragment (see below). In Fig. F5.9(b) kinesin (or dinein) coated onto a bead moves along a microtubule that is attached to the surface of a microscope coverglass (see below). In Fig. F5.9(c) A RNA polymerase bead is attached to the transcriptional downstream end of DNA so that it becomes tethered to the surface of a microscope coverglass via the RNA polymerase fixed to the coverglass (see below).

To measure the mechanical properties of single motor molecules, it is necessary to manipulate single protein molecules and allow the interaction to be monitored. A laser trap holds a bead by utilising the attractive forces exerted by a focused laser (Fig. F5.9(b) and (c)). By moving the focal point, it is possible to move a bead that attaches to a single protein molecule. A laser trap can also measure the mechanical properties of molecular motors, because the trapped beads behave like a spring. When force is exerted on the protein, the bead becomes displaced and a restoring force is exerted by the trap toward the trap centre to pull the bead back to the original position. The two forces are then balanced.

The laser trap has been used for the myosin–actin system. In contrast to processive molecular motors, conventional myosin readily dissociates from actin. To prevent the actin filament from diffusing away from the myosin during measurements, the actin filament is trapped at both ends (Fig. F5.9(a)).

Kinesin: vesicle transport along microtubules

The most intensive application of optical tweezers has been focused on linear motor kinesin. Most kinesin proteins appear to be dimeric molecules that contain a stalk, a tail-like domain and two globular motor domains (Fig. F5.10). The tail-like domain sometimes includes non-covalently attached light chains and is presumably important for transport specificity or organelle recognition. The globular tail is referred to as the cargo domain to indicate its role in organelle movement. There are many variations in the size and arrangement of the stalks and tails relative to the motor domains.

One of the unusual features of the kinesin superfamily of molecular motors is the existence of two classes of molecule: one, as represented by kinesin, that moves towards the plus-end of the microtubules, and second, as represented by NCD, that moves towards the minus-end. The kinesin super family is radically different from myosin, which only moves in one direction on the actin thin filament. Interestingly the molecular organisation of kinesin and non-claret disjunctional (Ncd) differs at the motor domain residues at the N terminus of kinesin and C terminus of Ncd (Fig. F5.11).

Microtubule-based motors offer special opportunities for understanding how motors work. In part because of its relatively, small size, the mechanism of the action of kinesin is rapidly approaching the same level of understanding as that of myosin, and incorporates three significant new features. First, this motor works

Fig. F5.10 Kinesin domain structure and associated proteins. The top structure represents a protofilament of a microtubule, the 'track' for kinesin motors, containing alternating α- and β-tubulin subunits oriented so that the plus-end points to the right, towards the cell periphery for most microtubules. The blue and red structure in the middle represents a dimeric kinesin heavy chain. Each heavy chain contains a motor domain (head, red and blue) that binds to ATP and microtubules, a neck linker (cyan) whose conformation changes during the ATPase cycle, an α-helical neck and stalk (red). The structure of the globular tail domain is not yet known and is represented as an orange sphere. The tail binds to two light chains (yellow). The light chains contain special protein-interaction motifs, which can dock onto adaptors or cargo receptors, linking kinesin to the cargo to be transported. (Adapted from Mandelkow and Mandelkow, 2002.)

Fig. F5.11 Schematic representation of the architecture of (a) kinesin and (b) NCD. Most kinesins are dimeric and contain two motor domains that are in close proximity and are usually held together by a section of coiled coil. For kinesin it is clear that there is an important interaction between the two heads that confers processivity on this motor protein such that a single kinesin molecule is able to remain in contact with and move along a microtubule for a considerable length of time. This suggests that kinesin moves along the microtubules in a 'hand over hand' fashion: the force generating head does not leave the microtubule until its neighbour is attached. This highly processive nature is lost in a single-headed kinesin, although a single head can still generate movement at a velocity similar to that of the intact molecule. Thus, although the connection between the heads is important, the ability to produce a significant power stroke must reside in a single head. (After Rayment, 1996.)

Kinesin

Neck

*

Myosin

Neck

Fig. F5.12 Comparison of the crystal structures of myosin and kinesin, looking down on the nucleotide (solid black). The structure of NCD (not shown) is almost identical to that of kinesin. The amplifier region adjacent to the motor core is the neck region of kinesin and of myosin. Note: the area between the two arrowheads in the kinesin neck region is not visible in the present crystal structure. The assignment of random coil and helix is based on secondary structure prediction and CD spectral studies; the precise boundary of the neck helix and the location of the neck region relative to the core is not known. The asterisk denotes a kinesin truncation: aa 340 in *Drosophila* kinesin and aa 332 in human kinesin. (Adapted from Vale, 1996.)

processively i.e. unlike myosin, a single molecule moves along the microtubule track for many seconds before dissociating. Second, within the kinesin family there are motors that move in one direction along a microtubule and others that move in the opposite direction. Finally, kinesin motors can be made to move slowly, permitting better time-averaging of position, and they remain attached to the substrate for a substantial fraction of the kinetic cycle, reducing the magnitude of Brownian excursions.

The atomic structures have been solved for the motor domains of two molecules belonging to the kinesin family, kinesin itself and Ncd (the product of the *Drosophila* Ncd gene). These two motors move in opposite directions on microtubule tracks, yet they have highly similar structures. It is startling that these ~45 kDa proteins are also structurally related to the much larger (nearly double the mass) motor domains of myosin (Fig. F5.12). Significant technical difficulties nevertheless exist in measuring movements of single kinesin molecules. The motions occur on length scales of ångströms to nanometres and on time scales of milliseconds or less. To obtain the high spatial and temporal sensitivity required optical tweezers have been combined with a dual-beam interferometer to produce an 'optical trapping interferometer' (see Comment F5.6). This instrument provides position detection and trapping functions simultaneously, and can provide controlled, calibratable forces in the piconewton range.

To characterise its motion at high spatial resolution kinesin is attached to silica beads. A trapped bead is moved near a microtubule that is fixed on a microscope coverslip (Fig. F5.9(b)). Figure F5.13 shows that the kinesin advances in discrete steps of 8 nm (the tubulin repeat unit on the microtubule track), which can be clearly distinguished from Brownian motion. The steps are separated by dwell

Fig. F5.13 Position record of a pulled bead. Kinesin advances in 8-nm increments, which are separated by dwell periods of varable length. (After Mehta *et al.*, 1999.)

Fig. F5.14 Mode of operation for kinesin. Kinesin appears to 'walk' along a microtubule, without detaching both its feet simultaneously and probably using its feet alternately. (Adapted from Kinosita *et al.*, 1998.)

Kinesin walks

Comment F5.6
Kinesin movement

To elucidate the kinetic scheme underlying the 8-nm advances, the distribution of dwell times between steps has been analysed by several investigators. It was concluded that one rate-limiting process, ATP binding, precedes every 8-nm advance. Such measurements eliminate models that postulate two ATP-dependent head movements producing each 8-nm step, suggesting instead that kinesin may move through 16-nm step-by-step movements of the two heads.

---

periods of variable time (Comment F5.6). The kinesin moves the bead away from the trap centre, slows as the resistive force increases, and finally stalls under loads of 5–7 pN. Figure F5.14 shows the mode of operation for kinesin. Kinesin appears to 'walk' along a microtubule, without detaching both its feet simultaneously and probably using its feet alternately.

Dynein: eukaryotic flagella motor

Bacterial flagella differ in every respect but name from eukaryotic flagella (see Comment F5.7). In the eukariotic flagellum, dynein motors powered by ATP generate moving bends by stepping along linear microtubular tracks.

To manipulate the microtubule, optical forces are applied to streptavidine-coated latex beads that are attached as handles to biotinylated single microtubules (Fig. F5.15, see Comment F5.12). Figure F5.16 is an electron micrograph of two singlet microtubules interacting with dynein arms on a doublet microtubule. When the free end of such a microtubule is brought into contact with the surface of a doublet, the microtubule stops exhibiting Brownian motion and, after photolysis of caged ATP, begins to move. If the angle between the singlet microtubule and doublet is less than ∼30°, the singlet microtubule gradually aligns parallel to the doublet and moves along the doublet until finally the bead escapes from the optical trap force (at forces of > 30–40 pN). However, when the microtubule is

---

Comment F5.7   Eukaryotic and bacterial flagella

In the eukaryotic flagellum, dynein motors powered by ATP generate moving bends by stepping along linear microtubular tracks. The eukaryotic flagellum undulates like a snake inside its cellular membrane. In contrast the bacterial flagellum works like a power boat with a rotary motor turning a rigid, helical propeller.

(a)

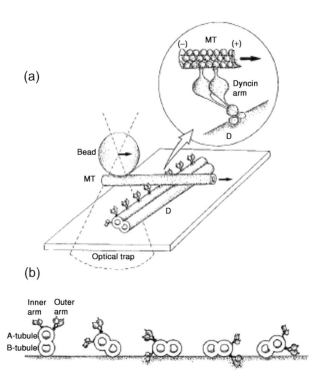

(b)

Fig. F5.15 Interaction of singlet microtubules with dynein arms presented on doublet microtubules. (a) Schematic diagram of the exposed dynein arms, on a doublet microtubule (D), interacting with a singlet microtubule (MT). The singlet microtubule is manipulated by means of an optically trapped streptavidine-coated bead. A two-headed arm pulls the singlet microtubule in the direction of its plus (+) end, causing the bead to move away from the centre of the trap force (black arrows). (b) Possible orientations of a doublet microtubule on the glass surface (bottom) with outer and inner dynein arms. Dynein arms pointing upwards may be capable of interaction with a singlet microtubule (not shown). (After Shingyoji et al., 1998.)

placed obliquely (at 45–90°) on the doublet the bead does not escape and force is recorded repeatedly after each flash of ultraviolet light.

Figure F5.17 shows a typical record of the force generation. Dynein arms generate a peak force of 6 pN and move the singlet microtubule in a processive manner. A remarkable feature of the dynein arm activity is the presence of oscillations (Fig. F5.17(a)). The force oscillates with an amplitude of ~2 pN. The maximum frequency of the oscillation at 0.75 mM ATP is ~70 Hz (Fig. F5.17(b)). Interestingly the oscillation frequency is not far from the frequency of flagellar beating (~350 Hz). This discovery suggests that these oscillation forces of dynein may be the key to rhythmic beating motions of eukaryotic flagella.

Fig. F5.16 Electron micrograph showing two singlet microtubules (MT) interacting with dynein arms (white arrowheads) on a doublet microtubule (D). Scale bar, 100 nm. (After Shingyoji et al., 1998.)

Fig. F5.17 Typical record
of force generation by
dynein arms. (a) The
angle between the singlet
microtubule and the
doublet microtubule to
which the dynein arms
were attached was 89°. A
UV flash was applied at
time zero (arrow).
Displacement refers to
the displacement of the
singlet microtubule over
the doublet by the dynein
arms. (b) Changes in
oscillation frequency and
the velocities of sliding of
untrapped microtubules
along doublets. The line
shows changes in the
estimated ATP
concentration. (After
Shigyoji et al., 1998.)

Fig. F5.17 Typical record of force generation by dynein arms. (a) The angle between the singlet microtubule and the doublet microtubule to which the dynein arms were attached was 89°. A UV flash was applied at time zero (arrow). Displacement refers to the displacement of the singlet microtubule over the doublet by the dynein arms. (b) Changes in oscillation frequency and the velocities of sliding of untrapped microtubules along doublets. The line shows changes in the estimated ATP concentration. (After Shigyoji et al., 1998.)

Myosin: the muscle motor

The term myosin refers to at least 14 classes of proteins, each containing putative or demonstrated actin-based motors and each probably filling different roles. The results described below involve skeletal muscle myosin (myosin II). Figure F5.18 schematically represents the architecture of myosin. Muscle myosin is a highly asymmetric 470-kDa protein, containing two 90-kDa globular N-terminal heads, called subfragment S1, and an α-helical coiled-coil tail that polymerises to form the filament backbone. Myosin heads are also called cross-bridges because they can link the two sets of filaments. In S1, a 'catalytic domain' (CD) contains the ATP and actin-binding sites.

Myosin II shares many structural features with kinesin. Both use ATP to move along their respective tracks, but myosin II is processive – it undergoes (at most) one catalytic cycle per diffusional encounter with its track. Although this means that a single molecule cannot move along its track for large distances, it also means that organised ensembles of molecules can move along the track at higher speeds. Myosin is thought to undergo a conformational change when it binds to actin, resulting in a 'working stroke' (Fig. F5.19).

For the coupling of such transient interactions and movement to a trapped bead, a more complex geometry is needed than that described above for kinesin. Fig. F5.20 shows the experimental geometry for observing a single myosin molecule binding and pulling an actin filament. The filament is attached at either end to a trapped bead. These beads are used to stretch the filament taut and move it near surface-bound silica beads that have been decorated sparsely with myosin molecules. Discrete stepwise movements averaging 11 nm were found in these experiments. The magnitudes of the single forces and displacements are consistent with the predictions of the conventional swinging cross-bridge model of muscle contraction (Comment F5.8).

Fig. F5.18 Schematic
representation of myosin
architecture. (After
Rayment, 1996.)

N terminus   Light chains

                C terminus

Myosin head   Coiled-coil

Fig. F5.19 Schematic view of a myosin molecule. For clarity, only one head (S1, the motor domain of myosin) and a short segment of the α-helical tail are shown. The small curved arrow shows the direction of the conformational change resulting in a stroke of about 10 nm. This stroke of the lever arm of S1 pulls on the myosin tail, which is attached to the myosin thick filament (not shown) and causes a relative movement of the myosin filament and the actin filament. (After Spudlish, 1994.)

> **Comment F5.8   Huxley model**
>
> In 1969 H. Huxley suggested that the sliding force is developed as result of a change in the angle of attachment in the head part of a myosin cross-bridge attached to actin. Based on a mechanical investigation of muscle, Huxley and Simmons proposed a theoretical model, called the 'swinging cross-bridge model'. Conformational changes in the myosin heads coupled with changes in bound ligands were observed in crystals, in solution and in muscle fibres during contraction. The displacement expected from the observed conformational changes in the myosin head is approximately 6 nm.

The most exciting myosin experiment was performed by observing mechanical steps and fluorescent nucleotides simultaneously using total internal reflection microscopy. Figure F5.20 shows the experimental apparatus. A single actin filament with beads attached to both ends is suspended in solution by optical tweezers. The suspended actin filament is brought into contact with a single one-headed myosin molecule in a myosin–rod cofilament bound to the surface of a pedestal formed on a coverslip. The displacement or the force due to actomyosin interactions is determined by measuring bead displacements with nanometre accuracy. Using total internal reflection fluorescence microscopy, individual ATPase reactions are monitored as changes in fluorescent intensity due to association–(hydrolysis–)dissociation events of a fluorescent ATP analogue, Cy3-ATP, with the myosin head. The results show that the myosin head produces a displacement of 15 nm and the force generation does not always coincide with the release of

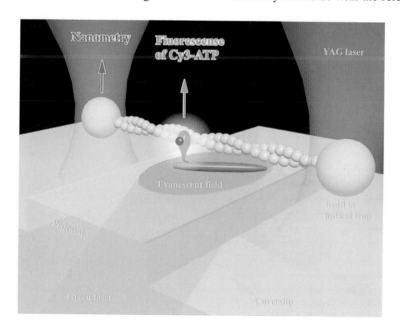

Fig. F5.20 Schematic drawing of the experimental apparatus for single-molecule studies of myosin. (After Ishijima et al., 1998.)

Fig. F5.21 Model of an actomyosin motor:
(a) The lever-arm model: the neck region of myosin acts as a lever arm.
(b) The myosin head is proposed to walk on acting subunits by biased Brownian motion. Multiple steps are produced during a single ATP turnover event, some of which are backward, although backward steps do not occur frequently. (After Yanagida *et al.*, 2000.)

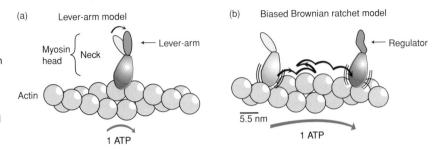

bound nucleotide. Instead the myosin head produces force several hundreds of milliseconds after a bound nucleotide is released. This finding does not support the widely accepted view that force generation is directly coupled to the release of bound ligands. It suggests that myosin has a hysteresis or memory state that stores chemical energy from ATP hydrolysis.

Many studies in crystal and solution have shown that the neck region of the myosin head undergoes conformational changes. The angle of the myosin head changes relative to the neck region. On the basis of these findings, the 'swinging cross-bridge model' of force generation has been refined into the 'lever-arm swinging' model, in which the neck region of the myosin head is postulated to act as a lever arm (Fig. F5.21(a)). The size of displacements reported varies considerably – some investigators have reported a myosin displacement of ~5 nm (per head), fully consistent with the lever-arm model, whereas others have found 10 nm and more. It is crucial to determine the myosin step size unambiguously, taking to account possible different orientations of myosin relative to the actin filament axis. The displacement varies from ~15 nm to 0 nm, with changes in the angle from 5° (near the physiologically 'correct' orientation) to almost 90°. When averaged over all directions, the displacement is ~6 nm. This value is consistent with measurements in which myosin was randomly oriented on the surface. It has been concluded that the small displacements reported might be attributable to the effect of this random orientation of the myosin heads.

The rising phase of the displacements, on an expanded time scale, are shown in Fig. F5.22. The displacements do not take place abruptly, but instead develop

Fig. F5.22 Myosin substeps in the rising phase of displacement recorded on an expanded time scale (After Yanagida *et al.*, 2000.)

Myosin runs (hops)

Fig. F5.23 Mode of operation for myosin. Myosin 'runs' along actin, in that its two feet (the two globular parts of myosin usually referred to as 'heads' are detached from actin for the most time). In fact, myosin can run, skipping many actin monomers in a step, if other myosin molecules pull the actin filament while the first one is detached. If only one myosin molecule interacts with actin, it simply hops and will not move relative to actin while detached (except for random diffusion). (After Kinosita *et al.*, 1988.)

in a step-wise fashion. The size of the substeps is ~5.5 nm. The substeps are stochastic, and the number of steps per displacement varies between one and five, which produces a total displacement of ~5 to ~30 nm.

A single myosin head moves along an actin filament independently of the load with regular 5.5 nm steps that coincide with the distance between adjacent actin subunits in one strand of an actin filament. Multiple steps are produced during a single ATP turnover event, some of which are backward. These results suggest that the myosin head 'runs' on actin subunits using Brownian motion (Fig. F5.23). However, given that backward steps do not occur frequently (only 10% of the total number of observed steps), it seems that this Brownian motion is biased in the forward direction.

RNA polymerase: a processive DNA transcription machine

The enzyme RNA polymerase (RNAP) carries out an essential step of gene expression, the synthesis of an RNA copy of the template DNA (Fig. F5.24). Each RNA molecule is synthesised in its entirety by a single molecule of RNAP

Fig. F5.24 Hypothesised structure of RNAP. Electron microscopy studies of bacterial and yeast RNAPs reveal a groove surrounded by jaw-like protrusions that may clamp around downstream duplex DNA and a tunnel that may function as a channel through which RNA exits from the active site. Chemical experiments show that RNAP protects ~17 bp of DNA on either side of the active site position, keeps the upstream half of this DNA melted, and maintains an ~8-bp RNA:DNA hybrid upstream from the active site. RNAP adds nucleotides to nascent RNA in a biparticle active site that coordinates the RNA 3'-end and the NTP. (After Gelles and Landvick, 1998.)

Fig. F5.25 Different techniques for studying DNA translocation by single molecules of RNAP. RNAP, bound to a glass or mica surface covered by an aqueous solution of NTPs, moves the DNA template in the indicated direction (small arrows) during transcript elongation (not to scale). (a) The tethered particle motion method. A plastic bead or colloidal gold particle undergoing rapid Brownian motion is observed by light microscopy and used to determine how the length of the DNA segment linking the bead to the polymerase changes during transcript elongation. (b) The laser tweezers method. A focused laser beam pulls on a silica or plastic bead attached to the downstream end of the DNA template, keeping the segment of DNA between the bead and the polymerase under tension. Movement of the template by the polymerase pulls the bead away from the focus of the laser beam. This displacement is detected optically. (c) The surface force microscopy method. Template DNA is loosely held against a mica surface by surface-bound cations (circles). The DNA and the position of the RNAP along the DNA contour are measured in the images. (After Gelles and Landvick, 1998.)

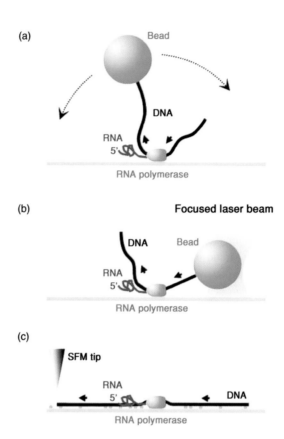

moving processively along the template. Movement is powered by the free energy liberated as appropriate nucleoside triphosphates (NTPs) are joined to the 3′-end of the nascent RNA chain, pyrophosphate (PP) is released and the growing transcript folds.

Figure F5.25 shows different techniques for detecting DNA translocation by single molecules of RNAP. These techniques all isolate a single RNAP molecule by immobilising it on a surface; each then observes DNA translocation by some form of microscopy. Each method has its own advantages: the tethered particle motion approach (Fig. F5.25(a)) is arguably the least perturbational; it neither exerts force on the DNA nor significantly alters its conformation. Since bead diffusion is more rapid than the mean velocity of DNA translocation in a polymerase catalytic cycle, attachment of the bead should not significantly affect the translocation reaction, provided the bead is attached further from the RNAP molecule than the DNA persistence length (∼50 nm or 150 bp). On the other hand, the laser tweezers method (Fig. F5.25(b)) has a high time resolution and thus far is the only method capable of measuring translocation forces as well as displacements. The surface force microscopy technique (Fig. F5.25(c)) is least able to observe features of the RNAP structure (e.g. the bend angle of the DNA at the polymerase) while simultaneously detecting translocation.

Table F5.2. *Comparison of nucleotide-driven motors*

Motor/rail	Step size	Max. force	Max. efficiency	Processivity	Mode
Myosin/actin	Variable	3–6 pN	∼20%	None–poor	Hops
Kinesin/microtubule	8 nm	5 pN	∼50%	Good	Walks
RNA polymerase/DNA	0.34 nm	14–27 pN	∼20%	Excellent	Crawls

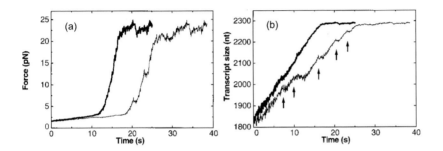

Fig. F5.26 Force and corresponding transcript length at high load. (a) The force was increased until the complex stalled at ∼23 pN. (b) The bottom trace shows movement, which was interrupted occasionally by apparent transcriptional pauses (marked by arrows). The top trace shows the same movement with pauses algorithmically removed, as needed to compute a meaningful transcription velocity. (After Wang *et al.*, 1998.)

The most significant results from the initial round of single-molecule studies are that the enzyme is capable of translocating DNA against large opposing loads and that the deduced efficiency of chemical-to-mechanical energy conversion is high, similar to that of the canonical cytoskeletal motor enzymes. Indeed, the single-molecule stall force is at least three times that measured for myosin and kinesin, showing that RNAP can legitimately be classed with those powerful cytoskeletal motors. The stall force against which RNAP ceased to move was determined as ∼ 21–27 pN (Fig., F5.26(a)). This value exceeds the 5–7 pN that was measured for kinesin, perhaps reflecting the need for RNAP to forcefully disentangle DNA secondary structure. *E. coli* RNAP during transcript elongation progresses along DNA at speeds of about 200 nucleotides per second (Fig. F5.26(b)).

The three motors above are compared in Table F5.2. The lengths of genuine steps of running myosin are expected to be multiples of 5.5 nm, the distance between neighbouring actin monomers. RNA polymerase is expected to step by 0.34 nm, the distance between base-pairs. Kinesin's 8-nm steps were measured. All these step sizes represent intervals of the structural repeats.

F5.3.2 Rotary molecular motors

Rotary protein motors accomplish their primary functions by rotating one group of subunits with respect to the rest. Although they constitute only a small entry in the growing catalogue of cellular molecular motors, they have been studied so intensively that more is known about them than any of the others. The best characterised protein in terms of its atomic structure and biochemistry is ATP synthase (also called F_0F_1 ATPase), which actually consists of two rotary motors connected to a common shaft. The largest, most powerful protein motor is the bacterial flagellar motor, which is found in many species (Comment F5.7).

F_0F_1 ATPase motor

ATP synthase consists of two opposing rotary motors connected in series (Fig. F5.27). The soluble F_1 motor is fuelled by nucleotide hydrolysis and drives the connecting γ-shaft clockwise (viewed from F_1 toward the membrane), whereas the transmembrane F_0 motor is fuelled by the ion motive force and drives the γ-shaft counterclockwise (Comment F5.9). Under normal circumstances, the F_0 motor generates the larger torque and drives the F_1 motor in the synthesis direction. However, in bacteria under anaerobic conditions, when the proton motive force is low, the F_1 motor hydrolyses ATP, driving the F_0 motor in reverse, whereupon it functions as a proton pump. The two motors illustrate two of the principal mechanisms cells use to convert chemical energy into mechanical force. The F_0 motor is made of a transmembrane cylinder of 10–14 subunits, depending on the species (Fig. F5.28). The number of subunits may be related to the ability of the F_0 motor also to act as a proton pump in some species (Section F2.4.2). The stalks coupling F_0 and F_1 are elastic, and thus F_1 does not directly see the steps of F_0. Instead this flexible coupling transmits only the F_0 torque to F_1. Therefore, there is no fixed 'stochiometry' relating the number of subunits in F_0 to the number of subunits in F_1. This elastic coupling also passes energy more efficiently between F_1 and F_0.

Fig. F5.27 Schematic diagram of F_0F_1-ATPase, which is a multidomain complex consisting of two units: a hydrophobic proton channel (F_0) embedded in the mitochondrial membrane and a hydrophilic catalytic unit (F_1) protruding into the mitochondria. Individual F_1-ATPase units are composed of alternating α- and β-subunits surrounding a rotating central γ-subunit. A central rotor of radius ~1 nm, formed by its γ-subunit, turns in a stator barrel of radius ~5 nm, formed by three α- and three β-subunits. When protons flow through F_0, ATP in synthesised in F_1. The synthase is fully reversible in that hydrolysis of ATP in F_1 drives a reverse flow of protons through F_0. Isolated F_1 catalyses only the hydrolysis of ATP, and hence is called the F_1-ATP synthase. (After Kinosita et al., 1998.)

Periplasm

Rotor

Rotor sites

Na^+

Cytoplasm

Outlet channel

Stator

Stator charge

γ-Shaft

Fig. F5.28 The molecular architecture of the sodium-driven F_0 motor of *Propionigenium modestum*. (Adapted from Oster and Wang, 2003.)

Comment F5.9 F_0 motor

Extensive mutational, NMR, cryo-EM and molecular modelling studies have provided a good picture of the general structure for both the proton-driven F_0 motor of mitochondria and *E. coli*, and the sodium-driven F_0 motor of the anaerobic bacterium from *P. modestum*. The difference between the models for sodium- and proton-driven motors lies in the presumed path of the ion as it passes through the motor assembly. In the sodium-motor model, the rotor ion-binding sites are accessible from the cytoplasm when outside the rotor–stator interface; this allows bound sodium ions to dissociate into the low-concentration reservoir. In the proton F_0- motor model these outlet channels are not present. Instead, the stator has a single outlet channel through which protons must exit. It is still unresolved whether or not the sodium F_0 and proton F_0 have the same ion path. The mechanochemical principle is the same in both models (Oster and Wang, 2003).

Rotation of F_1 had been long suspected. On the basis of an extensive kinetic analysis a model for the catalytic mechanism of F_1-ATPase was proposed by Boyer in 1993. Among other properties, this model predicted that the energy was transmitted through rotation of the γ-subunit in the centre of the F_1-ATPase molecule. Boyer's model was almost entirely substantiated by experimental results, with exception of the rotation hypothesis. One year later, the X-ray crystal structure of F_1-ATPase was published. The structure is a ring formation consisting of three α-subunits alternating with three β-subunits that have catalytic sites. The rod-shaped γ-subunit spanned the centre of the ring. This X-ray crystal structure provided a specific and easily understood image that permitted rotational movement of the γ-subunit.

Rotation of F_1 was demonstrated only in 1998 by directly observing the motion of a fluorescent actin filament specifically bound to the rotor element (Fig. F5.29). The His-tagged F_1-ATPase was immobilised, and fluorescent actin filaments were attached to the subcomplex through streptavidin for observation with an

Fig. F5.29 Experimental geometry for observing the rotation of single F$_1$-ATP molecules. The α- and β-subunits were bound by a histidine tag to a coverslip coated with nitrilotriacetic acid, and the opposing face of the γ-subunit was attached to a fluorescent actin filament through a biotin--avidin linkage. Rotation of the filament was observed through a standard fluorescent microscope. (Adapted from Kinosita, 1998.)

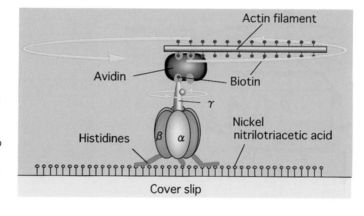

epifluorescence microscope. When ATP was added into the chamber, the actin filaments always rotated counterclockwise.

By observing this rotation at extremely low ATP concentration it was found that the rotation occurs in increments of $120°$, one step per molecule of ATP hydrolysed. In these measurements, the actin filaments attached to the γ-subunit rotated against the viscous resistance of water (Comment F5.10). The amount of rotational torque required for this movement was calculated from the length of the actin filaments and the rate of rotation to provide a mean value of approximately 40 pN nm. For a single step of $120°$ and, hence, a single ATP hydrolysis, this amounts to work of 80 pN nm. This value is in close agreement with the free energy liberated by a single ATP hydrolysis under physiological conditions. From this estimate, it was concluded that F$_1$-ATPase can convert nearly 100% of its ATP-derived energy into mechanical work. In this respect, it appears that nature has far outperformed human engineers, making mechanistic understanding of these extraordinarily efficient motors an important goal (see also Section I1.6.4).

Comment F5.10

The absolute value of torque of F$_1$ motor

Strictly speaking, the long actin filament may have altered the genuine kinetics of the F$_1$ motor, because the filament was subject to a large hydrodynamic friction. The orientation dependence of single-fluorophore-intensity has been exploited in order to videotape the rotation of F1 motor in real time. These results showed that the $120°$ stepping is a genuine property of this molecular motor,

The bacterial flagella as a propeller screw

Many motile bacteria are propelled by flagellar filaments, each of which is turned at its base by a rotary motor driven by a transmembrane ion gradient. These ions are either H$^+$ (protons), or in alkalophilic or marine bacteria, Na$^+$. In *E. coli*, the motor can rotate in either direction, and cells navigate toward regions rich in nutrients by controlling this direction.

The filament of *E. coli* is a long helical polymer of a protein, flagellin, with a molecular mass of about 55 kDa. The filament is connected to the hook by two junctional proteins (Fig. F5.30). Named according to its shape, the flexible hook acts as a universal joint permitting the filament and motor to rotate about different axes. The flagellar filament (propeller)-is a 10-μm-long, thin, rigid, corkscrew-shaped structure, with a helical period of about 2 μm. More so than other motors, the flagellum resembles a machine designed by a human.

Flagellar rotation has traditionally been measured by using the 'tethered cell' assay, in which a single flagellum is attached to the surface, causing the cell body

Fig. F5.30 Schematic of the bacterial flagellum as a mechanical device. The filament, hook and motor constitute the flagellum. The remaining flagellar parts are rings. The L- and P-rings are believed to act as a bushing through which the rotating drive shaft passes The filament or propeller lies outside the cell's inner and outer membranes and is about 10 μm long. In *E. coli*, flagella turning at speeds of 18 000 rpm push cells at 30 μm s^{-1}. It is remarkable that unlike kinesin, NCD, and myosin, flagellar motors can run in both directions. (After Oster and Wang, 2003.)

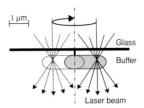

Fig. F5.31 Sketch of a bacterial cell being rotated about its tether by the optical trap (dash–dot line) aligned with the tether (short thick line). The direction of rotation is shown by the circular arrow. Two views of the cell are displayed: at $\theta = 0°$ (solid lines) and $\theta = 180°$ (dotted lines). The laser beam enters from above, through the coverslip, as shown by the crossed rays. The microscope illumination (not shown) enters from below. The tweezers can act directly on the cell or indirectly, via a trapped polystyrene bead. (After Block *et al.*, 1989.)

to counterrotate (Fig. F5.31). Whereas the flagella of swimming *E. coli* rotate at over 100 Hz and are difficult to observe directly, tethered cells rotate at ∼10 Hz and are comparatively easy to monitor. The development of optical tweezers has made it possible to manipulate bacteria with forces that are sufficient to arrest actively swimming bacteria and can overcome torque generated by the flagellar motor of a bacterium tethered to a glass surface by a flagellar filament. It has been shown that a flagella motor generates ∼4500 pN nm of torque. The radius of the rotor is ∼20 nm, so the motor generates at its rim a force of ∼200 pN, far greater than any other molecular motor (see Table F5.3). For

Table F5.3. *Comparison of the main characteristics of rotary motors*

Motor	Rotational speed	Max. torque	Max. efficiency	Processivity
$F_1\alpha\,\beta/F_1\gamma$	∼4 rps	40 pN nm	100%	Perfect
Bacterial flagella	∼300 rps[a] 1700 rps[b]	∼4000 pN nm	Unknown	Perfect
Bacteriophage ϕ29 packaging	100 bp per second[c]	Motor can work against loads of up 57 pN	About 30%	Perfect

[a]Proton-driven motor.
[b]Sodium-driven motor.
[c]The packaging rate decreases as the prohead is filled.

Fig. F5.32 Simplified structure of the ϕ29 virus. The prohead, into which the DNA is packaged, is about 540 Å long and 420 Å wide.

Fig. F5.33 X-ray structure of an isolated connector. The connector structure can be divided into three, approximately cylindrical, regions: the narrow end, the central part and the wide end, having external radii of 33, 47 and 69 Å, respectively. These regions are respectively 25, 28 and 22 Å in height, making the total connector 75 Å long. The internal channel has a diameter of about 36 Å at the narrow end, increasing to 60 Å at the wide end. (After Simpson *et al.*, 2000.)

each revolution of the motor $\sim$1200 protons pass through it, each contributing $\sim$6 $k_B T$.

F5.3.3 The bacteriophage ϕ29 DNA packaging motor

There are other protein motors that are thought to be rotary. One of them is the portal protein in certain viruses that package their DNA into their virial capsid.

The bacteriophage ϕ29 is a 19-kilobase dsDNA virus that infects *Bacillus subtilus* cells. The viral particles are formed by a prolate icosahedral capsid, or head, and tail. Between these two structures there is a connecting region, the so-called connector (Fig. F5.32). This connector, a cone-shaped dodecamer, occupies the pentagonal vertex at the base of the prohead and is the portal for DNA entry during packaging and DNA ejection during infection. The DNA packaging motor lies at a unique portal vertex of the prohead and contains: (1) the connector (the X-ray structure of which is presented in Fig. F5.33); (2) the portion of the prohead shell that surrounds the connector; (3) ϕ29-encoded prohead RNAs, which surround the protruding narrow end of the connector; and (4) viral ATPase which is required for DNA packing.

Bacteriophage ϕ29 packages its 6.6-μm long, dsDNA into a 540 $\times$ 420 Å capsid by means of a portal complex that hydrolyses ATP. This process is remarkable because the entropic, electrostatic and bending energies of the DNA must be overcome to package the DNA to near-crystalline density.

Figure F5.34 shows experimental setup in which force-measuring laser tweezers are used to follow the packaging activity of a single ϕ29 packaging complex in real time. Stalled, partly prepackaged complexes are attached to a polystyrene microsphere by means of the unpackaged end of DNA. This microsphere is captured in the optical trap and brought into contact with a second bead that is held by a microneedle. This bead is coated with antibodies against the phage, so a stable tether is formed between the two beads (Fig. F5.34). In the absence of

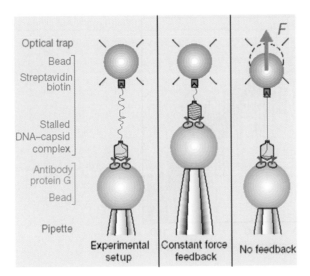

Fig. F5.34 Experimental setup in which laser tweezers are used to measure packaging activity: at the start of a measurement (left); the constant-force feedback mode (middle) and no feedback mode (right). A single φ29 packaging portal complex is tethered between two microspheres. Optical tweezers are used to trap one microsphere and measure the forces acting on it, while the other microsphere is held by a microneedle. To ensure measurement on only one complex, the density of complexes on the microsphere is adjusted so that only about one out of five–ten microspheres yields hook-ups. (Adapted from Smith et al., 2001.)

ATP, the tether displays the elasticity expected for a single DNA molecule. After addition of ATP, the two microspheres move closer together, indicating packaging activity, thus demonstrating that the portal complex is a force-generating motor. In 'constant-force feed back' mode (Fig. F5.34 middle), in which the microsphere distance is adjusted by feedback to maintain DNA tension at a preset value of 5 pN, packaging dynamics is monitored. Figure F5.35 shows that packaging is highly efficient: on average 5.5 min are necessary to package a length of DNA equal to the φ29 genome. Pauses in movement of variable duration (Fig. F5.35, inset) are also clearly evident.

In Fig. F5.36 the packaging rate is plotted against the percentage of genome that has been packaged. These data show a marked reduction in packaging rate

Fig. F5.35 Plots of the tether length versus time for four different DNA constructs with a constant force of ~5 pN, using a length of 1.8 times the φ29 genome. Inset: Increased detail of the regions indicated by the arrows, showing pauses (curves are shifted for clarity). (Smith et al., 2001.)

Fig. F5.36 Packaging rate
versus the amount of
DNA packaged, relative to
the original 19.3-kilobase
ø29 genome. The red line
is an average of eight
measurements. Notably,
the packaging rate
decreases as the prohead
is filled. (Smith *et al.*,
2001.)

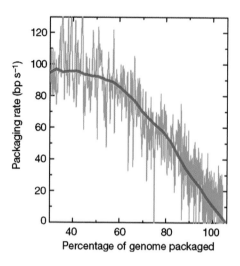

Fig. F5.36 Packaging rate
versus the amount of
DNA packaged, relative to
the original 19.3-kilobase
ø29 genome. The red line
is an average of eight
measurements. Notably,
the packaging rate
decreases as the prohead
is filled. (Smith *et al.*,
2001.)

from when $\sim$50% of the genome is packaged. Initially the rate is $\sim$100 bp s^{-1} and it gradually drops to zero as the capsid fills and the motor stalls.

The force dependence of the motor has been investigated using a 'no feedback' mode in which the positions of the trap and microneedle are fixed (Fig. F5.34). In this mode the tension in the molecule increases as the motor reels in the DNA, and the bead is displaced from the centre of the trap. The measurements show that the motor can work against loads of up to 57 pN on average, making it one of the strongest molecular motors reported to date.

A comparison of main characteristics of rotary motors is given in Table F5.3.

F5.3.4 Molecular motors and Brownian motion

The basic physical principle that governs the operation of all protein motors is quite simple: molecular motors generate mechanical forces by using intermolecular binding energy to capture 'favourable' Brownian motions. They do this in two ways: (1) they may bias against unfavourable Brownian motions by a sequence of small free-energy drops, making backward steps slightly less unlikely than forward ones (Fig. F5.37(a)); or (2) they may rectify a long 'run' of favourable thermal fluctuations by a large free-energy drop, making backward steps extremely unlikely in comparison with forward ones (Fig. F5.37(b)). The former is generally called a 'power stroke' and the latter a 'Brownian ratchet'. According to the power-stroke mechanism, the ligand progresses from its initial weak binding state to its final tight binding state by biasing relatively small Brownian fluctuations of the two binding surfaces (the red curve in the free-energy landscape in Fig. F5.37(a)). This gradual 'zipping' of intermolecular bonds is used to perform mechanical work on an external load.

One power stroke is not sufficient to make a motor, which must operate in a cycle. For this, the catalytic site must realise the tightly bound nucleotide and reset the catalytic site so that a new power stroke can be initiated.

Fig. F5.37 Two extreme mechanisms for using random thermal fluctuations to drive a load. (a) *The power stroke*. A flexible binding site on enzyme slides over the binding surface of a fixed ligand. Its stochastic motion is driven by biased Brownian fluctuations that progressively anneal the short-range attractive forces between the enzyme and substrate 'induced fit'. Thus, the binding energy is converted directly into mechanical work. The red curve is the free-energy landscape seen by the system as the enzyme gradually binds to the substrate. During the binding process, the free-energy decreases encounter only small energy barriers of order $k_B T$. (b) *Brownian ratchet*. A polymer drives an object with a diffusion coefficient D by rectifying its diffusive motion. When the object diffuses against the load F_L by more than the size of a monomer, another monomer can bind to the end of the growing polymer. If binding is successful, the lengthened polymer prevents the object from being driven back to the left by the load forces. The free-energy landscape (the red curve) seen by the object is a tilted staircase. The free energy in each step is much greater than $k_B T$. (Adapted from Oster and Wang, 2003.)

A Brownian ratchet mechanism (Fig. F5.37(b)) involves a filament that polymerises against an object whose diffusion coefficient is D. The load force F_L pushes the object to the left. If an object succeeds in diffusing to the right against the load by further than the size of a monomer, an additional monomer may polymerise onto the end of the filament. This 'rectifies' the object's diffusive motion, blocking it from being pushed by the load force back to the left. Thus, the free-energy landscape resembles a tilted staircase (the red curve in Fig. F5.37(b)). Operation in a cycle is built into the mechanism because the binding of each new monomer reinitiates the cycle.

F5.3.5 Molecular motors and the second law of thermodynamics

The second law of thermodynamics dictates that a motor without an external free-energy supply cannot extract heat from an isothermal environment to do work. Although the notion that molecular motors run on Brownian motion appears at first glance to contravene the second law, we must remember that the second law is a statement about the average behaviour of system (e.g. the macroscopic behaviour). Occasionally, and for short times, an individual molecular motor can 'violate' the second law. For example, a protein-sized object can oscillate to stretch a spring but, without a free-energy supply, such individual violations are not sustainable. When averaged over many objects or over long times for one object, these violations disappear. However, proteins can capture these temporary violations by using an energy supply, such as binding free energy; such behaviour does not violate the second law. So, protein motors can 'extract' heat from their surroundings to do work by using a free-energy supply to bias or rectify thermal fluctuations. This cools down the motor's local environment by taking kinetic energy from the surrounding solvent molecules. Notice that the heat is generally not paid back by the motor itself but by other methabolic processes, such as the manufacture of ATP or the pumping up of the ion gradient.

F5.3.6 DNA mechanics

Native free large DNA (more than 4MDa) adopts a random coil conformation, which maximises its entropy (Chapters D3, D4, D6, D9). Pulling on the molecule reduces this entropy and costs energy. The associated entropic forces result from a reduction of the number of possible configurations of the system consisting of the molecule and its solvent, so that at full extension there is only one configuration left: a straight polymer linking both ends. To reach that configuration work has to be done against entropy, a force has to be applied.

Pioneering measurements of biopolymer elasticity employed optical tweezers and other mechanical probes to stretch DNA. Such experiments pulled macro-molecules beyond their entropically determined regime in which the polymer resists an extension that constrains its range of accessible conformations and into one in which external forces disturb the structure and induce conformational changes. To produce and measure such forces on a DNA molecule, we must use single-molecule manipulation techniques. These techniques have been used to study several regimes of DNA stretching behaviour in detail and have also been extended to study DNA–protein and DNA–drug interactions. High-resolution force measurements have also been used to study the stretching of a single DNA or RNA molecule. Below we give a few examples of the application of the single-molecule technique to studying the stretching of DNA and RNA molecules and their complexes with proteins.

Molecular yoga of DNA

The DNA structure poses some formidable mechanical problems to the cellular machinery which has to read, transcribe and replicate the signals of the genetic code buried inside the double helix. To make the code accessible to the DNA and RNA polymerase enzymes, the molecule has to be unwound and two strands separated. The regulation of the winding and torsional stresses involved in these processes is performed by numerous enzymes known as topoisomerases. To study the function of these molecular motors one has first to understand the mechanical response of the DNA under stress.

The first step in any manipulation experiment is to anchor the DNA (preferentially via its extremities) to an appropriately treated surface (Comment F5.11). Many different methods have been developed to achieve specific DNA binding to surfaces. Most of these methods include biochemical reactions between a (possibly modified) DNA molecule and an appropriately treated surface. For example, the extremity of the molecule can be functionalised with biotin, which can interact specifically with streptavidine bound to a surface (Comment F5.12). Similarly, surfaces coated with oligonucleotides can be used to recognise the complementary extremity of DNA molecules.

Once DNA is bound to a surface by its end(s), a simple way to stretch it is to drain the solution (e.g. by pulling the surface out of it or by letting it evaporate). In Fig. F5.38 the molecule, stained with a fluorescent dye, is stretched by the hydrodynamic force. Because this force is acting locally (at the interface), DNA is stretched uniformly: ~ 1 μm for every 2 kbytes. Fluorescence in situ hybridisation on combed molecules allows one to obtain very accurate (~ 1 kbps) genomic maps. The ordering, orientation and distance between genes and the existence of genomic rearrangements (e.g. deletions) can thus be determined.

One of the first experiments on the overstretching transition of the double helix was demonstrated on λ-phage DNA using the optical trapping technique in combination with a glass micropipette (Fig. F5.39(a)). When the dsDNA molecule is a subject to force of 65 pN or more, it undergoes a highly cooperative transition (~ 2 pN) into a stable form with a 5.8 Å rise per base pair, i.e. 70% longer than the canonical B-form DNA (Fig. F5.39(b)). This new form of DNA is called S-DNA

> **Comment F5.11**
> **Alternative approach for DNA immobilisation on a surface**
>
> An interesting alternative which does not require any modification of the molecules is based on the specific adsorption of DNA by its ends on a hydrophobic surface at a pH ~ 5.5. On many different hydrophobic materials (teflon, polystyrene, silanised glass, etc.) DNA has been observed to adhere strongly and non-specifically at low pH and weakly or not at all at high pH. Between low and high pH there is a narrow range (pH $= 5.5 \pm 0.2$) where DNA binds to the surface by its extremities only.

Comment F5.12   Biotin (vitamin B7 or H) and streptavidine

The interaction between biotin and streptavidine is one of the strongest known non-covalent interactions. The dissociation constant for the streptavidine–biotin complex is $\sim 10^{-15}$ M. Streptavidine, from *Streptomyces avidinii*, is a tetrameric protein containing four high-affinity binding sites for the vitamin biotin. Since streptavidine is multivalent, it is able to serve as a bridge between the biotinylated DNA fragment and the biotin-containing resin. The strong interaction is extremely useful for purification of DNA-binding proteins, because DNA-affinity columns with streptavidine/biotin bridges can be washed under a wide variety of conditions (i.e. 2 M KCl and 1% SDS) without removing either the streptavidine or the biotinylated DNA fragment from the matrix.

Fig. F5.38 Molecular combing of λ-phage DNA stained with YOYO1. The molecules in solution (bottom) are bound at one (A) or both (B) ends. The meniscus extends across the image from the lower left to upper right. The extended molecules left behind the meniscus are visible as a straight segments (C) if bound at one end, or loops (D) if bound at both ends. In molecular combing, the anchored DNA molecules are aligned on the surface by the receding meniscus as algae on the shore are by the receding tide. The force applied on the molecule by the receding air–water interface is large enough to stretch it but not to break its bond(s). (Strick *et al.*, 2000.)

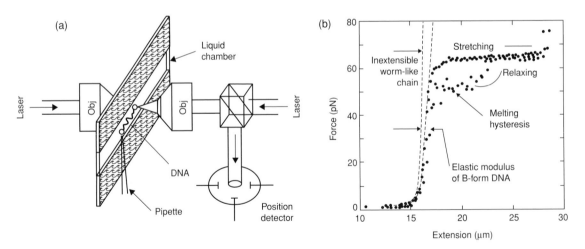

Fig. F5.39 (a) Schematic view of the dual-beam laser-trap instrument. Each end of a single λ-phage dsDNA molecule (48.5 kbp, contour length about 16.4 μm) is attached to a separate microscopic latex bead. One bead is held by suction with a glass micropipette while the other bead is held in the optical trap, which also functions as a force transducer. The DNA molecule is extended by moving the pipette relative to the laser trap. Use of low numerical aperture lenses make it possible to hold beads deep inside a fluid cell, 100 μm away from the glass surface. (b) Stretching of λ-phage DNA in 150 mM NaCl, 10 mM Tris, 1 mM EDTA, pH 8.0. The 'inextensible wormlike chain' curve corresponds to a worm-like chain with a persistence length of 53 nm and contour length of 16.4 μm. (After Smith *et al.*, 1996.)

(Fig. F5.40, right). When the stress is relaxed below 65 pN, the molecules rapidly and reversibly contract to their normal contour lengths. S-form DNA is stable in high salt concentrations up to forces of between approximately 150 pN (for a random sequence) and 300 pN (for poly(dG–dC)). Above these forces S-DNA exhibits the characteristic force–extension behaviour of single-stranded DNA. Single-stranded DNA is more contractile than double-stranded DNA because of its high flexibility but it can be stretched to a greater length because it no longer forms a helix.

In a magnetic tweezers setup (Fig. F5.41) by rotating the magnets placed above the sample, one can twist and supercoil the anchoring DNA molecule (provided it is nick-free). At low stretching forces ($F < 0.3$ pN), the twisting of DNA results in the formation (as in a twisted tube) of plectonemes or supercoils (see below). As the torsional stress increases, it may alter the DNA structure, thus preempting its buckling. When the torque on the DNA reaches about 20 pN nm (in low salt concentrations), the overwound DNA molecule adopts a new inside-out overtwisted structure called P-DNA (Fig. F5.42). Interestingly a P-DNA-like structure has been found in the genome of the Pf1 virus (Comment F5.13). It is very likely that there this unusual structure is stabilised by the proteins of the virus's coat.

Fig. F5.40 Space-filling graphics of relaxed linear dsDNA (left) and one of the models of DNA stretched by factor of 1.7 (right). This elongated DNA model is characterised by a strong base-pair inclination, a narrow minor groove and a diameter roughly 30% less than that of B-DNA. (After Cluzel *et al.*, 1996.)

Comment F5.13 Pauling and Corey model of DNA

Just a few months before J. Watson and F. Crick published their famous model of DNA, L. Pauling and R. Corey proposed another structure for nucleic acids. They believed that the core of the molecule was probably formed of phosphate groups.

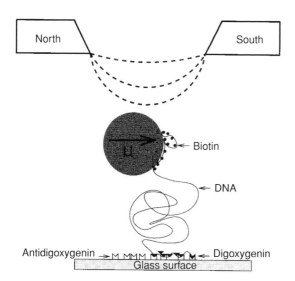

Fig. F5.41 Schematic representation of the magnetic tweezers setup. A digoxigenin (DIG)–biotin end-labelled DNA molecule ($\sim 17\,000$ bp) is attached at one end to a glass surface by DIG--anti-DIG bonds and at the other to a 4.5-μm magnetic bead via streptavidine–biotin links. By varying the distance between the sample and the permanent magnets, the stretching force is controlled, whereas rotating the magnets induces DNA supercoiling. (Adapted from Allemand *et al.*, 1998.)

Fig. F5.42 Structure of P-DNA deduced from molecular modelling. Space-filling models of a $(dG)_{18} \cdot (dC)_{18}$ fragment into (a) linear B-DNA and (b) P-DNA conformations. In P-DNA the phosphodiester backbone winds inside the structure and its bases are exposed outside. The backbones are colored purple, and the bases are coloured blue (guanine) and yellow (cytosine). The anionic oxygens of the phosphate groups are shown in red. P-DNA is a double-helical structure with a pitch of 2.6 bases per turn. (Allemand *et al.*, 1998.)

These results demonstrate that at very low forces and low degrees of super-coiling DNA can locally undergo major structural transitions. These transitions might be relevant to the activity of RNA-polymerase, which is known to exert forces as high as 35 pN and to under(over)wind the molecule up(down)stream.

Topoisomerases and uncoiling of DNA

Topoisomerases are enzymes that control DNA supercoiling and are responsible for disentangling molecules during replication, DNA repair and recombination (Comment F5.14). The uncoiling and decatenation activities of these enzymes can be investigated directly by monitoring the change in extension of a coiled molecule as it is uncoiled by a single topoisomerase.

As we discussed before, an unnicked, double-stranded DNA molecule rigidly anchored at one end to a glass surface and at the other to a small magnetic bead can be stretched with a force F and twisted to a degree of supercoiling σ by translating and rotating small magnets above the sample (see Fig F5.43).

Comment F5.14   Two classes of topoisomerases

There are two classes of topoisomerases, depending on whether one strand (type I) or both strands (type II) of DNA are transiently broken to permit a change in topology. Type I enzymes relax torsion in a molecule in an ATP-independent manner by creating a break in one of the DNA strands, allowing relaxation of torsion about the intact strand. Type II topoisomerases catalyse the transport of one DNA segment through a second one in an ATP-dependent way.

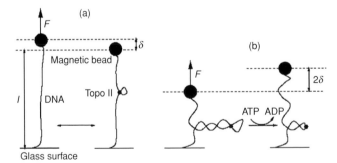

Fig. F5.43 (a) Sketch of a twisted DNA near the bucking instability (point (1) in Fig. F5.44) undergoing topo-II-mediated clamping in the absence of ATP. No supercoiling is relaxed, but stabilisation and destabilisation of a single DNA loop results in a change δ of the system's extension. (b) Supercoil relaxation in the plectonemic regime (point (2) in Fig. F5.44) in the presence of topo II isomerase and ATP. Each enzymatic cycle releases two supercoils, resulting in an increase 2 of δ in the system extension. (Adapted from Strick *et al.*, 2000.)

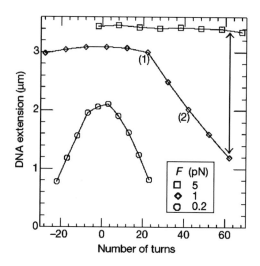

Fig. F5.44 Extension versus supercoiling behaviour for stretched 11-kbp DNA. At $F = 1$ pN, the DNA needs to be overwound by about 25 turns before it reaches the buckling instability (point (1) and forms plectonemes. Increasing the force (arrow) on DNA overwound 60 turns to 5 pN pulls out the plectonemes and causes the extension to increase. At very low forces ($F = 0.2$ pN) the buckling instability is rounded off by thermal fluctuations but the DNA does not denature when unwound. (Adapted from Strick et al., 2000.)

These two mechanical parameters determine the molecule extension l, which is monitored by measuring the three-dimensional position of the tethered bead (Fig. F5.43(a)). As one begins to rotate the magnets, the DNA's extension is unchanged and the torque increases linearly with the twist angle. When a critical torque is reached the molecule buckles (as would a twisted tube), forming a plectoneme and stabilising the torque at its critical value (Fig. F5.44, point (1)). The appearance of these tertiary structures is well known from everyday experience, e.g. the coiling of phone cords. Thereafter it contracts regularly as overwinding generates more supercoils. Addition of topoisomerase II to the system results in an increase in l by an amount σ for every supercoil relaxed (Fig. F5.44, point (2)). In 10 mM ATP, enzyme turnover is slow and relaxation is directly observed as stepwise events of size 2δ (Fig. F5.45).

Prokaryotic type I topoisomerses, 1A, bind preferentially to single-stranded DNA. The type I enzymes relax torsion in a molecule in an ATP-independent manner by creating a break in one of the DNA strands allowing relaxation of torsion about the intact strand.

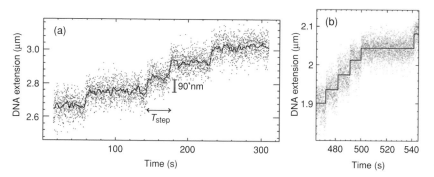

Fig. F5.45 (a) Individual time courses for the topo-II-mediated relaxation of positively supercoiled DNA stretched at $F = 0.7$ pN at 10 mM ATP. (Strick et al., 2000.) (b) Individual time courses for the topo-I-mediated relaxation of DNA stretched at $F = 1.5$ pN in 0.25 mM MgCl$_2$. The DNA contained a symmetric mismatch of 12 nucleotides. (Adapted from Dekker et al., 2002.)

Fig. F5.46 Model for the condensing of a λ-DNA molecule constrained by force-measuring optical tweezers. (a) At a fractional extension y greater than the critical extension y_{crit} the free energy of the stretched DNA chain exceeds the nucleation energy for collapse. (b) At $y = y_{crit}$ thermal motion may create temporary slack, but the collapsed nuclei formed are unstable. (c) At $y < y_{crit}$ the molecular slack generated coalesces into a condensed phase. For the force–extension plot, $y_{crit} = 0.85$. (Adapted from Baumann et al., 2000.)

DNA condensation

In the presence of a critical concentration of multivalent ions, double-stranded DNA condenses and forms very compact structures. Single-molecule stretching methods are used to measure the forces that cause this DNA condensation.

Figure F5.46 shows a model for the transition in a λ-DNA molecule based on force-measuring optical tweezers. If the two ends of a DNA molecule are stretched (more than the critical extension $y_{crit} = 0.85$), they are prevented from condensation (Fig. F5.46(a)). At $y = y_{crit}$ thermal motion may create temporary slack, but the collapsed nuclei formed are unstable (Fig. F5.46(b)). At $y < y_{crit}$ the molecular slack generated coalesces into the condensed phase. The force required to prevent condensation has been shown to be constant, with a magnitude of about 2–3 pN.

Unzipping the DNA double helix

Unzipping the double helix was demonstrated by the mechanical separation of the complementary strands of an individual λ phage dsDNA molecule. A specific molecular construction was designed in which one strand of the DNA to be opened is elongated by a linker arm and two specific functionalisations are introduced to obtain the desired attachments of the DNA construct to the micromechanical device.

The design of the construct used in the experiment appears in Fig. F5.47 and includes two molecules of double-stranded λ-phage DNA, each comprising

Fig. F5.47 The molecular design of the DNA construct used in experiment. The DNA to be opened (DNA-1) and the linker arm (DNA-2) comprise double-stranded λ-phage DNA. Oligonucleotides (thick lines) are used to introduce the biotin and dig attachments and to connect DNA-11 and DNA-2 covalently. (After Essavaz-Roulet, 1997.)

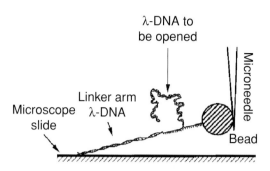

λ-DNA to
be opened

Microscope
slide

Linker arm
λ-DNA

Microneedle

Bead

Fig. F5.48 Principle of the force measurement in which a double-stranded λ-phage DNA is forced open as the surface is displaced to the left. (After Essavaz-Roulet, 1997.)

48.5 kb. The DNA to be opened is DNA-1, and the linker arm is DNA-2. Oligonucleotides (oligo-1, oligo-2 and oligo-3) are used to connect DNA-1 and DNA-2 and to introduce attachment points via a digoxigenin group and a biotin group. The other extremity of DNA-1 is capped with an oligonucleotide forming a hairpin (oligo-4), which prevents the separation of the two strands when the end of the opening process is reached. This allows repeated cycles of opening and closing. The bending of a glass microneedle is used to determine the forces required to pull apart the 3′ and 5′ extremities of the molecule (Fig. F5.48). Strand unzipping occurs abruptly at 10–15 pN and displays a reproducible 'saw-tooth' force variation pattern with an amplitude of ± 0.5 pN along the DNA.

The following picture emerges for the characteristic forces involved for double-stranded DNA: the breaking of the double strand requires ~480 pN ± 20%, the structural transition of DNA overstretching requires 65–70 pN, and strand separation requires about 10–15 pN.

Unfolding of single chromatin fibres and individual nucleosomes

Single-molecule mechanical manipulation techniques offer a direct approach to the investigation of the forces and displacements required for enzymatic access to nucleosome-bound DNA (Comment F5.15). To study these processes single chromatin fibres are assembled directly in the flow cell of an optical tweezers setup. A single λ-phage DNA molecule, suspended between two polystyrene beads, is exposed to a *Xenopus laevis* egg extract, leading to a chromatin assembly with concomitant apparent shortening of the DNA molecule (Fig. F5.49(a)–(c)). Assembly is force-dependent and cannot take place at a force exceeding 10 pN. The assembled single chromatin fibre is subjected to stretching by controlled movement of one of the beads with the force generated in the molecule continuously monitored with the second bead trapped in the optical trap. The force displays discrete, sudden drops upon fibre stretching, reflecting discrete opening events in the fibre structure (Comment F5.16).

Detailed studies of the nucleosome unfolding process were done on very defined individual nucleosome core particles. Nucleosomal arrays for this study were prepared with avian core histones and a 33 684-bp DNA fragment containing

**Comment F5.15
Nucleosomes**

Nucleosomes are the fundamental organisational units of the eukaryotic genome, occuring on average every 200 bp. The foundation of the nucleosome is the nucleosome core particle, consisting of 147-bp DNA wrapped 1.65 times around an octamer of histone proteins.

Fig. F5.49 Schematic representing chromatin assembly on a single λ-DNA molecule. (a) A single λ-DNA molecule suspended between two beads. The arrow indicates the direction of continuous buffer flow. The first bead is held by a glass micropipette using suction. The second bead is maintained downstream from the first one by the drag force. (b) *X. laevis* egg extract containing all core histones and numerous non-histone proteins, but lacking somatic linker histones, is introduced into the flow cell. Histone proteins bind to the single λ-DNA molecule, causing its apparent shortening. (c) Shortening of the single λ-DNA molecule continues and eventually stops. (d) Model of a chromatin fibre with 51–73 bp of linker DNA. Each cylinder represents one histone. (Adapted from Bennink *et al.*, 2001.)

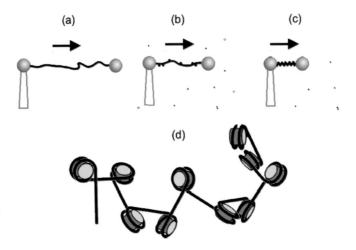

(a)　　　(b)　　　(c)

(d)

Comment F5.16　On the periodicity in the force–extension curves of nucleosomes

These opening events are quantised at increments in fibre length of about ~65 nm and are attributed to unwrapping of the DNA from individual histone octamers. However, the abundance of linker histone-like proteins and other non-histone chromatin-associated proteins in oocyte nuclear extracts used for chromatin assembly in these experiments may in part explain the lack of uniformity and periodicity in the force–extension profiles.

---

17 direct tandem repeats. The experimental setup is shown in Fig. F5.50(a). An example of the force–extention curves obtained for saturated arrays (17 nucleosome) is shown in Fig. F5.50(b). In the high-force range (>15 pN), these data show a sawtooth pattern composed of 17 peaks. At the end of the stretch, the chromatin curve approaches that of the full-length naked DNA (dotted red line), indicating that no histones remained attached to the DNA. The 17 peaks indicate disruption of the 17 positioned nucleosomes. At each sawtooth, DNA remains bound until a peak force is reached, leading to a sudden release of DNA and relaxation to lower tension. Uniform spacing between adjacent peaks (~27 nm) indicates that upon disruption, a relatively constant amount of DNA is released from each nucleosome core particle.

A suggested model for forced disassembly of each individual nucleosome is described by three stages, each involving partial unwrapping of the DNA (Fig. F5.51). The first stage of disruption releases 76 bp of the external DNA. The disruption is gradual, and only low force is required to peel DNA from the protein surface. The second and third stages of the disruption involve the sudden release of the next 80 bp of DNA. The third stage of disruption occurs at even higher loads, releases the remaining 11 bp of DNA and results in a complete

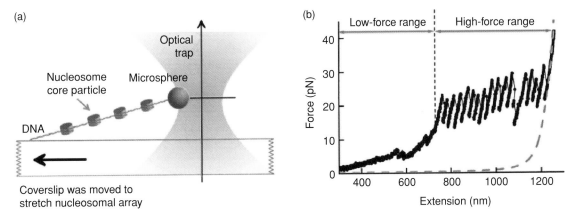

Fig. F5.50 (a) Experimental configuration of nucleosome stretching (not to scale). The DNA is labelled at one end with biotin and at the other end with digoxigenin. Before stretching, one end of each nucleosomal array is attached to the surface of an antidigoxigenin-coated microscope coverslip. A 0.48-μm diameter streptavidin-coated polystyrene microsphere is then attached to the free end of each tethered array. Once a surface-tethered microsphere is optically trapped, the coverslip is moved with a piezoelectric stage to stretch the nucleosomal DNA. (b) Force–extension curve of a fully saturated nucleosomal array. The force–extension characteristic of a full-length naked DNA (red dotted line) is shown for comparison. (Adapted from Brower-Toland et al., 2002.)

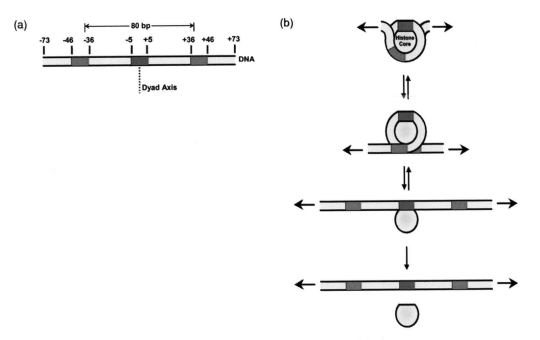

Fig. F5.51 (a) Map of critical DNA–histone interactions within an individual nucleosome core particle. (b) A three-stage model for the mechanical disruption of individual nucleosome. (Adapted from Brower-Toland et al., 2002.)

Table F5.4. *Forces in micromanipulation experiments with dsDNA*

Breaking of the DNA double strands	400–580 pN
Structural transition of uncoiling a double strand upon stretching	60–80 pN
Structural transition of a double strand upon torsional stress	~ 20 pN
Individual nucleosome disruption	20–25 pN
Separation of complementary strands (room temperature, 150 mM NaCl, sequence-specific)	10–15 pN

dissociation of the histones from the DNA. The three-stage model of nucleosome opening suggests the way in which nucleosomes perform their dual function in the eukariotic cell, both to maintain DNA in a condensed state and provide regulated access to the information contained therein.

The following picture emerges for the characteristic forces involved with dsDNA (Table F5.4). Breaking of double strands occurs at a force level of 400–580 pN. The structural transition of uncoiling upon stretching occurs at ~60–80 pN, whereas structural transition upon torsional stress occurs at a force level ~20 pN.

Disruption of individual nucleosomes is associated with a force of 20–25 pN. Strand separation requires 10–15 pN range and is sequence-specific (15 pN for 100% GC sequence and 10 pN for 100% AT sequence).

F5.3.7 RNA mechanics

An RNA structure is generally separated into two levels of organisation – secondary and tertiary structure. The tertiary structure is composed of secondary structural motifs that are brought together to form modules (hairpin loops, symmetric and asymmetric internal loops, junctions), domains and the complete structure.

These structures are very complicated and bulk studies of RNA folding are often frustrated by the presence of multiple species and multiple folding pathways, whereas single-molecule studies can follow folding/unfolding trajectories of individual molecules. Furthermore, in mechanically induced unfolding, the reaction can followed along a well-defined coordinate, the molecular end-to-end distance.

Figure F5.52(a) shows three types of RNA molecules representing major structural units of large RNA assemblies. P5ab is a simple RNA hairpin that typifies the basic unit of RNA structure, an A-form double helix. P5abcΔA has an additional helix and thus a three-helix junction. Finally, the P5abc domain of the *Tetrahymena thermophila* ribozyme is comparatively complex and contains an A-rich bulge, enabling P5abc to pack into a stable tertiary structure in the presence

Fig. F5.52 (a) Sequence and secondary structure of P5ab, P5abcΔA, and P5abc RNAs. The five green dots represent magnesium ions that form bonds (green lines) with groups in the P5c helix and the A-rich bulge. (b) RNA molecules are attached between two 2-μm beads with ∼500-bp RNA/DNA hybrid handles. (Liphardt et al., 2001.)

of Mg^{2+} ions. The individual RNA molecules are attached to polystyrene beads by RNA/DNA hybrid 'handles' (Fig. F5.52(b)). One bead is held in a force-measuring optical trap, and the other bead is linked to a piezoelectric actuator through a micropipette. When the handles alone are pulled, the force increases monotonically with extension (Fig. F5.53(a), red line), but when the handles with P5ab RNA are pulled, the force–extention curve is interrupted at 14.5 pN by an ∼18-nm plateau (black curve), consistent with complete unfolding of the hairpin. The force of 14.5 pN is similar to that required to unzip DNA helices.

Fig. F5.53 (a) Force–extension curves for P5ab RNA when the handles alone are pulled (red line) and when the handles with P5ab RNA are pulled (black line). (b) Stretch (blue line) and relax (green line) force-extension curves for the P5abc domain of the *Tetrahymena thermophila* ribozyme in 10 mM Mg^{2+}. (c) Comparison of P5abc force–extension curves in the presence (blue curve) and the absence of Mg^{2+} (green curve). (Adapted from Liphardt, 2001.)

Fig. F5.54 A model for
the unfolding of P5abc in
the presence of Mg^{2+}, in
which two possible
unfolding paths are
depicted. (After Liphardt,
et al., 2001.)

Figure F5.53(b) shows stretch (blue) and relax (green) force–extension curves
for P5abc domain of the *Tetrahymena thermophila* ribozyme. It is seen that the
tertiary interactions in Mg^{2+} lead to substantial curve hysteresis. Forces of about
19 ± 3 pN are needed before the molecules suddenly unfolds (blue curves). The
blue arrow indicates the typical unfolding force when stretching; the green arrow
shows a refolding transition upon relaxation. The two-step unfolding reveals two
distinct kinetic barriers to mechanical unfolding of P5abc in Mg^{2+}.

Removal of Mg^{2+} removes the kinetic barriers, and folding–unfolding
becomes reversible. Unfolding then begins at 7 pN (Fig. F5.53(c)), showing that
in EDTA the A-rich bulge destabilises P5abc. The refolding curves in Mg^{2+} and
EDTA coincide, except for an offset of 1.5 pN due to charge neutralisation (Fig.
F5.53(c), green curve). In contrast to the all-or-none behaviour of P5ab, refolding
of P5abc both with and without Mg^{2+} has intermediates: the force curve inflects
gradually between 14 and 11 pN (Fig. F5.53(c), black stars) and this inflection is
followed by a fast (< 10 ms) hop without intermediates at 8 pN (green arrows).
The different widths of the transitions and their force separation suggest that the
inflection (Fig. F5.53(c), stars) marks folding of the P5b/c helices, whereas the
hop (Fig. F5.53(c), arrows) marks P5a helix formation.

Figure F5.54 shows a model for the unfolding of P5abc in the presence of
Mg^{2+}, in which two possible unfolding paths are depicted. The blue arrow shows
an unfolding path in which the molecule suddenly unfolds and increases its length
to 26 nm, consistent with data indicated by the blue arrow in Fig. F5.53(b). A
two-step model is shown by the red arrows, in which an intermediate state 13 nm
in length is indicated by the green arrow in Fig. F5.53(b).

F5.3.8 Protein mechanics

Polyproteins

The sensitivity of the AFM has allowed experiments to probe the mechanical
properties of proteins. However, the heterogeneity and complexity of native proteins complicate the interpretation of AFM studies. When a protein containing
multiple different domains is stretched, it is difficult to relate individual unfolding

Fig. F5.55 A schematic diagram of the sequence of events during withdrawal of the gold substrate (grey box) during an AFM experiment. (Adapted from Fisher *et al.*, 2000.)

peaks in the force extension curve to specific domains and therefore to determine the mechanical properties of a specific fold. The solution to this problem was found using a special approach in molecular biology. Thus, by ligating multiple copies of the cDNA encoding a specific domain and expressing the resultant gene in bacteria, it is possible to produce a 'polyprotein' consisting of multiple copies of a single protein fold. An additional benefit of using engineered polyproteins for AFM studies is that they can be constructed from domains with an altered amino acid sequence, thereby allowing dissection of the molecular determinants of mechanical stability.

Figure F5.55 shows a schematic diagram of the sequence of events during withdrawal of the gold substrate (grey box) during an AFM experiment. Prior to the experiment, a layer of proteins is allowed to adsorb onto the gold substrate. Then the AFM cantilever is pressed against the protein layer to allow adsorption onto the cantilever. Upon withdrawal of the gold substrate, the cantiliver is first deflected by interactions with other molecules, such as denatured protein (in green). When these interactions break, the force on a cantilever is released.

The traces in Fig. F5.56 represent force–extension curves obtained from a sample of modular protein composed of 12 identical domains. The final peak in each trace represents the detachment of the final protein molecule(s). These traces demonstrate that even when a sample of pure protein is used, spurious peaks may occur in the force–extension curve because the protein molecule may have been completely or partially denatured due to interactions with the gold substrate or entanglement with other protein molecules. Such interactions, which typically occur when the cantilever is within the gold substrate, may yield a force–extension relationship displaying a single force peak, or displaying several peaks that are irregular in amplitude and spacing (Fig. F5.56(a)).

Fig. F5.56 Extension of modular proteins with the AFM. A series of force–extension curves obtained from a pure sample of protein consisting of 12 identical domains. Force–extension curves can yield unfolding force peaks equal to the number of domains in the protein (as in a trace (d)), but more frequently will yield fewer peaks or no regular peaks at all (as in traces (a) (b) and (c)). (After Fisher *et al.*, 2000.)

The presence of the regularly spaced force peaks seen in Fig. F5.56(b)–(d) is the unmistakable fingerprint of a modular protein. These peaks correspond to the consecutive unfolding of each of the protein domains in a single protein molecule. The force–extension curves are strings of successive enthalpic and entropic portions, reflecting the unfolding of individual domains in the multidomain polypeptide chain, followed by stretching of the unfolded domain (Fig. F5.57). As such proteins are elongated as a result of the initial application of force, they undergo a typical entropic stretching at the beginning. At a certain force, one of the folded domains unfolds, adding significant length to the chain and relaxing the stress on the cantilever, which returns to its non-defected state. The denatured portion of the polypeptide chain can now undergo entropic stretching, behaving like a typical polymer chain. Further extension creates forces high enough to unfold a second domain, which is then stretched entropically, etc. The unfolding and stretching of each individual domain creates an individual peak in the force curve, leading to the characteristic *sawtooth* pattern, illustrated in Figs. F5.56 and F5.57.

Models of elasticity

Sawtooth patterns from an engineered polyprotein may be analysed quantitatively using models that describe the physics of polymer elasticity. Polymer chains that are free in solution exist in a coiled state since this maximises their conformational freedom and therefore entropy. Extension of the molecule generates an

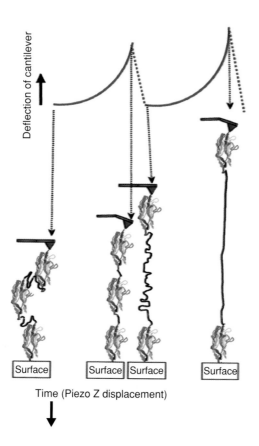

Deflection of cantilever

Surface   Surface Surface   Surface

Time (Piezo Z displacement)

Fig. F5.57 Schematic
representation the
structural transitions in
multidomain proteins
giving multipeak force
curves (saw-tooth
pattern). (After Zlatanova
et al., 2000.)

**Comment F5.17
Freely-jointed
model of elasticity**

The freely-jointed
chain (FJC) model is
also used with varying
degrees of success to
explain the force–
extension behaviour
of single
macromolecules. The
FJC model is based on
polymer chains with a
linear string of rigid
rods rotating freely at
the joints and with no
interactions between
the rods. The elastic
parameters are the
Kuhn segment length
L_k and the total
polymer length L_c. The
Kuhn segment length
L_k in the FJC model is
analogous to the
persistence length P in
the WLC model. Note
that the Kuhn segment
length is double the
persistence length for a
polymer chain of
infinite length.

opposing force due to the reduction in entropy, as the freedom of movement of the
molecule is restricted. The behaviour of polymers under stress may be predicted
using a worm-like model of entropic polymer elasticity (the WLC model). The
entropic elasticity of proteins is described by a worm-like chain equation, which
expresses the relationship between force (F) and extension (x) of a protein using
its persistence length (P) and its contour length (L_c)

$$F(x) = \frac{kT}{P}\left[\frac{1}{4}\left(1 - \frac{x}{L_c}\right)^2 - \frac{1}{4} + \frac{x}{L_c}\right] \qquad (F5.1)$$

The WLC model describes a polymer as continuous string of a given total (or
contour) length. Bending of the polymer at any point influences the angle of
the polymer for a distance, referred to as the persistence length, that reflects the
polymer flexibility. The smaller the persistence lenght the greater the entropy of
the polymer and the greater the resistance to extension. The persistence length
and the contour length comprise the adjustable parameter of the WLC model
(Comment F5.17). WLC model with a single parameter P cannot describe the
polypeptide elasticity equally well over the complete force range (0–300 pN). For
force up to 50 pN, $P = 0.8$ nm describes the polypeptide elasticity well. However,

Comment F5.18
Stiffness, bending and extension

It is useful to note that, in this context, stiffness refers to resistance to bending, not to extension, as is generally used in muscle mechanics. The elasticity parameters L_c and L_k denote the bending configurational freedom for the filamentous chain. In the original WLC and FJC models, the total length of the chain backbone does not change during stretching.

a persistence length of 0.4 nm describes the polypeptide elasticity better. This is due to the fact that additional contributions from bond angle deformations become important at forces above 50 pN (Comment F5.18).

Protein folds and mechanical stability

We now discuss the mechanical properties of four different proteins. Two of them are considered to have an 'all beta' structure and a so-called β-sandwich topology. One of these is from a typical mechanical protein (an immunoglobulin domain from titin) and the other is from a protein involved in secretion (the C2A domain from synaptotagmin I). The remaining two proteins are considered to be 'all alpha' structures. One is calmodulin, a ubiquitous regulatory protein, and the other is spectrin, a major component of the membrane-associated skeleton in erythrocytes.

The 127 module of human cardiac titin, which is 89 amino acids long, has a typical Ig I topology composed of seven β-strands (strands A–G), which fold into face-to-face β-sheets through backbone hydrogen bonds and hydrophobic core interactions (Fig. F5.58(c)). Since the hydrogen bonds in this patch are perpendicular to the direction of the applied force, unfolding requires the simultaneous rupture of this cluster of hydrogen bonds.

Stretching 127_{12} with AFM results in a force–extension curve with peaks that vary randomly in amplitude about a value of 204 ± 26 pN (Fig. F5.58(b)). Fits of the WLC model to the force–extension curves of 127_{12} give a persistence length P of about 4 nm and a variable contour length $L_c = 25$–496 nm (blue lines) with

Fig. F5.58 Mechanical properties of single human cardiac titin immunoglobulin domains: (a) the measured molecular mass of the polyprotein ($'\sim$150 kDa) in a good agreement with predicted molecular mass of a 127_{12} concatemer; (b) force–extension curve for 127_{12} with AFM; (c) top: three-dimensional structure of titin 127 domain; bottom the dashed lines depict the topology of the critical hydrogen-bonds between β-strands under an applied force. (After Carrion-Vazquez et al., 2000.)

(a)

~150 kDa ➤

$[127]_{12}$

(b)

200 pN

100 nm

(c)

(a)

~130 kDa→

[C₂A]₉

(b)

100 pN

100 nm

(c)

Fig. F5.59 Mechanical properties of a single C2 domain. (a) Coomassie blue staining of the purified [C2A]₉ protein. The measured molecular mass of the polyprotein (~130 kDa) is in a good agreement with predicted molecular mass of a [C2A]₉ protein. (b) Force–extension curve for [C2A]₉ protein with AFM. (c) left: three-dimensional structure of [C2A]₉ protein. Right: The dashed lines depict the topology of the critical hydrogen bonds between β-strands under an applied force. (After Carrion-Vazquez et al., 2000.)

a contour length increment $\Delta L_c = 28$ nm. This persistence length is the size of a single amino acid (0.4 nm).

The mechanical function of the C2 domain protein is not known. The C2 domain is a conserved 'all beta' module present in more than 40 different proteins, many of which are involved in membrane interaction and signal transduction. The first C2 domain of synaptotagmin I (C2A) is believed to be the calcium sensor that initiates membrane fusion during neurotransmitter release.

The three-dimensional structure of C2A was found to be a β-sandwich composed of 127 amino acids arranged into eight antiparallel strands with the N- and C-terminal strands pointing in the same direction (Fig. F5.59(c)). In contrast to the I27 domain, in which the hydrogen bonds have a 'shear' topology, those in the C2A domain are in a 'zipper' configuration (i.e. parallel to the direction of the applied force.

In AFM experiments the force–extension curve of C2A polyproteins show a sawtooth pattern with force peaks of ~60 pN separated by a distance of ~38 nm.

Calmodulin is a highly conserved antiparallel 'all alpha' protein that acts as a primary calcium-dependent regulator of many intracellular processes. The three-dimensional structure of calmodulin (148 residues) is dumb-bell-shaped and consists of seven α-helixes distributed in a helical central region capped by two globular regions, each containing two helix–loop–helix motifs that are responsible for Ca^{2+}-binding (Fig.F5.60(c), left). In contrast to the β-sandwich topology, where there is a non-homogenous distribution of the interstrand hydrogen bonds, the α-helix has a homogeneous distribution of intrahelix

Fig. F5.60 Mechanical
properties of a single
polycalmodulin molecule.
(a) Coomassie blue
staining of the purified
[CaM]$_4$ protein. The
measured molecular
mass of the polyprotein is
~80 kDa. (b) Stretching
single calmodulin
polyproteins gives force-
-extension curves with no
evident force peaks. The
force curve is well
described by the WLS
model (continuous lines)
using a contour length of
212 nm and a persistence
length of 0.32 nm. (c) Left:
three-dimensional
structure of rat CaM
showing its α-helical
structure. CaM is made of
seven antiparallel
α-strands and two short
antiparallel β-sheet
hairpins with a zipper
topology that provide a
structural link between
the two Ca^{2+} motifs of
each globular region.
Right: putative
mechanical topology of a
calmodulin domain. (After
Carrion-Vazquez *et al.*,
2000.)

hydrogen bonds. Figure F5.60(b) shows that the stretching of rat polycalmodulin (CaM$_4$) does not yield any force peaks, indicating that the unfolding forces must be below the AFM noise level in the experiments (~20 pN) (Comment F5.19). The origin of this is that two small β-hairpins of calmodulin (Fig. F5.60(c), right) are in a 'zipper' conformation, and therefore should offer little resistance to mechanical unfolding (see also Section J3.5.1).

The only other α-helical structure that reveals force peaks is spectrin, a major component of the membrane-associated skeleton in erythrocytes. As such, it cross-links filamentous actin and contributes to the mechanical properties of the cell. Spectrin consists of two subunits, which form laterally associated

---

Comment F5.19   Mechanical and thermodynamic stability of calmodulin

One of the most striking features of calmodulin is its thermodynamic stability, particularly in the presence of Ca^{2+}. The protein may be exposed to 95 °C with retention of biological activity. The Ca^{2+}-free form of CaM has a melting temperature of ~55 °C, while the Ca^{2+}-bound form denatures only at temperatures exceeding 90 °C. However, a difference in mechanical stability in the presence or absence of Ca^{2+} has not been detected.

(a)

ΔL = 31.7 nm

Force (pN)

Extension (nm)

(b)

Fig. F5.61 (a) Force-extension curve. The continuous lines superimposed on the first curve are WLC fits ($p = 0.8$Å). The gain in length upon each unfolding event is 31.7 nm, which corresponds to the 106 amino acid residues folded in each spectrin repeat. (b) 3-D structure of spectrin. Left: Side view of the dimer. One polypeptide in the chain is shown in red hues, the other in green hues. Right: Side view of one repeat. The B, C loop (white) was inserted by model building. Some of the side-chains that pack to maintain the spacing between the α-helices are shown with carbon in yellow, nitrogen in blue, and oxygen in red. (Adapted from Rief *et al.*, 1999, and Yan *et al.*, 1993.)

heterodimers (Fig. F5.61(b)). The two dimers interact and form a head-to-head dimer. The main part of both chains consists of homologous repeats. Each of these repeats ($\sim$106 amino acid residues) forms triple helical, antiparallel coiled coils. The force required to mechanically unfold these repeats is 25–35 pN (Fig. F5.61(a)). The unfolding forces of the α-helical spectrin domains are 5–10 times lower than those found in domains with β-folds, like cardiac titin immunoglobulin domains. This shows that the forces stabilising the coiled coil lead to a mechanically much weaker structure than multiply hydrogen bonded β-sheets. On the other hand, the melting temperatures of titin domains (50–70 $^{\circ}$C) and spectrin (53 $^{\circ}$C) are rather similar. Melting temperatures are correlated with the free energy of activation for unfolding process (Chapter C2). Obviously a comparison of free energies alone cannot explain the huge difference in mechanical stability.

All these experiments show that different proteins, even those with related structures, display a broad range of mechanical stability. The mechanical phenotypes of proteins may arise from differences in their topology, possibly as a result of variations in the number and position of hydrogen bonds among strands and sheets, and relative to the direction in which the force is applied.

The main results of mechanically unfolding proteins using the AFM can be summarised as follows.

(1) Folds of parallel terminal β-strands in which hydrogen bonds have a 'shear' topology (for example, the 127 module of human cardiac titin) have the maximal mechanical stability ($\sim$200 pN).

(2) β-strands in which hydrogen bonds have a 'zipper' configuration (e.g. the C2 domain of synaptotagmin) have folds with relatively low mechanical stability ($\sim$60 pN).

(3) The α-helical proteins in which the helices form bundles, rather than single α-helices (e.g. spectrin), are proteins with low force peaks ($\sim$30 pN). The forces stabilising the coiled coil lead to a mechanically much weaker structure than multiply hydrogen-bonded sheets.

(4) α-helical proteins in which the α-helices have a homogeneous distribution of intrahelix hydrogen bonds (e.g. calmodulin) do not yield any force peaks.

(5) Protein mechanical stability is not correlated with thermodynamic stability: the melting temperatures of spectrin and titin Ig domains are rather similar, whereas there is a difference in their unfolding forces of up to factor of 10.

Finally we note that the unfolding pathway depends on the pulling geometry and is associated with unfolding forces that can differ by an order of magnitude. Thus the mechanical resistance of a protein is not dictated solely by the amino acid sequence, topology or unfolding rate constant, but depends critically on the direction of applied extension.

F5.3.9 Deformation of polysaccharides

Many of the advances made in polysaccharide characterisation have been possible because of increasingly powerful mass spectrometry (Section B2.10) and nuclear magnetic resonance (Section K3.2.3). Here we briefly describe the main results for the elasticity of single polysaccharide molecules obtained by atomic force microscopy.

Polysaccharides whose glycosidic linkages are attached equatorially to the piranose ring (e.g. cellulose) are found to follow the FJC model of polymer elasticity (Fig. F5.62(c)). However, polysaccharides with axial linkage, such as amilose and pectin, were found to undergo abrupt force-induced length transitions (Fig. F5.62(a), (b)).

Figure F5.63(a) shows several such force curves recorded for various single molecules of the polysaccharide dextran. All of the curves exhibit the same characteristic elastic behaviour. At around 700 pN the curves deviate from a simple shape and shows a kink. Using a molecular dynamics simulation it has been shown that this kink is due to a conformational transition within each dextran monomer, where the C5–C6 bond of the sugar ring flips into a new conformation, thus elongating the monomer by 0.65 Å ($\sim$10% of its length (Fig. F5.63(b)).

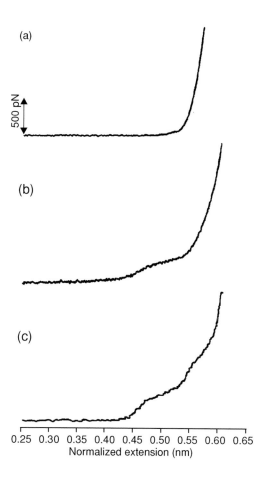

(a)

500 pN

(b)

(c)

0.25 0.30 0.35 0.40 0.45 0.50 0.55 0.60 0.65
Normalized extension (nm)

Fig. F5.62 Force–extension curves for a single polysaccharide structure: (a) cellulose, (b) amylose and (c) pectin. (Brant, 1999.)

(a)

Dextran

Force (pN)

2000
1500
1000
500
0

0   100   200   300   400
Extension (nm)

(b)

1600 pN

800 pN

0 pN

O6 C5

C6   O1

F_{ext}

4.5 Å

Fig. F5.63 (a) Characteristic shapes of the force–extension traces for dextran strands of different lengths. All curves show a kink at around 700 pN where a bond angle within each monomer flips into a new position. The first trace shows the extension and the second trace the relaxation of the same dextran strand. No hysteresis can be observed between the cycles. Also, the force at which the transitions occur is not speed-dependent. This means that the bond flips occur on a faster time scale than the experiment, and therefore stretching is an equilibrium process. For clarity the traces are offset from each other (Rief *et al.*, 1998a). (b) Ring conformation of dextran at different extension forces. (Adapted from Gimzewski and Joachim, 1999.)

The AFM results contradict the view that sugars are inelastic and locked into a stable conformation, raising the tantalising possibility that force-driven sugar conformations play important roles in biological signalling as well as in the elasticity of polysaccharides.

Finally, we mention that the rupture force of single covalent bonds under an external load could be studied by atomic force microscopy. For example, single polysaccharide molecules have been covalently anchored between a surface and an AFM tip and then stretched until they became detached. By using different surface chemistries for the attachment, it has been found that the silicon–carbon bond ruptured at 2.0 ± 0.3 nN, whereas the sulphur–gold anchor ruptured at 1.4 ± 0.3 nN at force-loading rates of 10 nN per second.

F5.4 Checklist of key ideas

- Optical and magnetic traps, the cantilevers of atomic force microscopy and glass microneedles are widely used for nanoscale manipulation. The force range of these facilities is from to 1pN to 10 nN.
- Kinesin (a microtubule-based motor) generates a peak force of 6 pN and appears to 'walk' along a microtubule in discrete steps of 8 nm, without detaching both its feet simultaneously and probably using its feet in an alternate fashion.
- Dinein (eukariotic flagella motor) generates a peak force of 6 pN and moves the singlet microtubule in a processive manner; a remarkable feature of the dinein arm activity is the presence of oscillations with an amplitude of ~ 2 pN and a maximum frequency of ~ 70 Hz.
- A single myosin head (part of the muscle motor myosin) moves along an actin filament in regular 5.5-nm steps that coincide with the distance between adjacent actin subunits in one strand of an actin filament: myosin molecules generate a peak force of 3–6 pN.
- RNA polymerase (DNA transcription machine) progresses along DNA at speeds about 200 nucleotides per second and generates a peak force of 21–27 pN, perhaps reflecting the need for RNAP to forcefully disentangle the DNA secondary structure.
- In *E. coli*, the flagella motor can rotate in either direction, and cells navigate toward regions rich in nutrients by controlling this direction; the flagella motor has rotational speed of 300–1700 rps and generates a torque of 4500 pN nm.
- F_1-adenosine triphosphatase (part of the F_0F_1-ATP synthase complex) has a rotational speed of 4 rps (under filament load) and generates a torque of ~ 40 pN nm; the rotation occurs in increments of 120°.
- When the dsDNA molecule is subject to a force of 65 pN or more, it undergoes a highly cooperative transition (~ 2 pN) into a stable form with 5.8 Å rise per base pair; this stable form is called S-DNA.
- At low stretching forces ($F < 0.3$ pN), the twisting of dsDNA results in the formation of plectonemes or supercoils; when the torque on the dsDNA reaches about 20 pN nm the overwound DNA molecule adopts a new stable form called P-DNA.
- Breaking of the dsDNA strands occurs at a force level of about 500 pN.

- Unzipping of dsDNA occurs abruptly at 10–15 pN and displays a reproducible 'sawtooth' force variation pattern with an amplitude of ± 0.5 pN along the DNA; force is sequence-dependent (~15 pN for the G–C base-pairs and 10 pN for the A–T base-pairs).
- Opening events in a chromatin assembly are quantised at increments in fibre length of about ~27 nm and are attributed to unwrapping of the DNA from individual histone octamers; the forces measured for individual nucleosome disruptions are in the range 20–40 pN.
- The force range for ribozyme unfolding depends on the type of secondary structures inside ribozyme; a force of 14.5 pN is needed before the P5ab RNA molecule (simple hairpin) unfolds, whereas a force of 22 pN is required for the P5abc RNA molecule (an A-rich bulge and 5pc three helix junction).
- Type I topoisomerase relaxes torsion in steps of 40 nm (one turn) at a time, whereas type II topoisomerase relaxes the torsion in a supercoiled DNA molecule in 80-nm steps, implying the relaxation of two turns per enzymatic cycle.
- The sawtooth patterns from an engineered polyprotein may be analysed quantitatively using worm-like or free-joint entropic polymer elasticity.
- The mechanical phenotypes of proteins may arise from differences in their topology as a result of variations in the number and position of hydrogen bonds among strands and sheets, and relative to the direction in which the force is applied.
- β-strand proteins in which hydrogen bonds have a 'shear' topology, as in 127 module of human cardiac titin, have maximal mechanical stability (~200 pN).
- β-strand proteins in which hydrogen bonds have a 'zipper' configuration, as in the C2 domain of synaptotagmin, have relatively low mechanical stability (~60 pN).
- α-helical proteins in which α-helices form bundles, as in spectrin, have low force peaks (~30 pN); the forces stabilising the coiled coil lead to a mechanically much weaker structure than multiple hydrogen-bonded β-sheets.
- α-helical proteins in which the α-helices have a homogeneous distribution of intrahelix hydrogen bonds, as in calmodulin, do not yield any force peaks.
- Protein mechanical stability is not correlated with thermodynamic stability: the melting temperatures of spectrin and titin Ig domains are rather similar, whereas there is a difference in their unfolding forces of up to factor of 10.
- The force–extension curves for dextran strands of different lengths show a kink at around 700 pN; this kink is due to a conformational transition within each dextran monomer where the C5–C6 bond of the sugar ring flips into a new conformation.
- A silicon–carbon bond ruptures at 2.0 ± 0.3 nN, whereas the sulphur–gold anchor ruptures at 1.4 ± 0.3 nN at force-loading rates of 10 nN per second.

Suggestions for further reading

Historical review and introduction to biological problems

Service, R. F. (1997). Chemists explore the power of one. *Science*, **276**, 1027–1029.

Mehta, A. D., Rief, M., Spudlich, J. A., Smith, D. A., and Simmons, R. M. (1999). Single-molecule biomechanics with optical methods. *Science*, **283**, 1689–1695.

Nanoscale manipulation techniques

Wang, M., (1999). Manipulation of single molecules in biology. *Curr. Opini. Biotechnol.*, **10**, 81–86.

Ficher, T. E., Oberhauser, A. F., Carrion-Vazquez, M., Marszalek, P. E., and Fernandez, J. M. (1999). The study of protein mechanics with the atomic force microscope. *TIBS*, **24**, 379–384.

Hegner, M. (2002). The light fantastic. *Nature*, **419**, 125.

Grier, D. (2003). A revolution in optical manipulation. *Nature*, **424**, 810–816.

Gross, S. D. (2003). Application of optical traps in vivo. *Meth. Enzymol.*, **361**, 162–174.

Macromolecular mechanics: nanometer steps and piconewton forces

Ishijima, A., and Yanagida, T. (2001). Single molecule nanobioscience. *Trends Biochem. Sci.*, **26**, 438–444.

Molecular motors

Finer, J. T., Simmons, R. M., and Spudlich, J. A. (1994). Single myosin molecule mechanics: piconewton forces and nanometer steps. *Nature*, **368**, 113–119.

Spudlich, J. A. (1994). How molecular motors work. *Nature*, **372**, 515–518.

Block, S. M. (1995). Nanometers and piconewtons: the macromolecular mechanics of kinesin. *Trends Cell Biol.*, **5**, 169–175.

Kitamura, K., Tokunaga, M., Iwane, A. H., and Yanagida, T. (1999). A single myosin head moves along an actin filament with regular steps of 5.3 nanometers. *Nature*, **397**, 129–134.

Adachi, K., Ysuda, R., *et al.* (2000). Stepping rotation of F_1-Atpase visualized through angle-resolved single-fluorophore imaging. *PNAS*, **97**, 7243–7247.

Oster, G., and Wang, H. (2003). Rotary protein motor. *Trends Cell Biol.*, **13**, 114–121.

Vale, R. D. (1996). Switches, latches and amplifiers: Common themes of G proteins and molecular motors. *J. Cell. Biol.*, **135**, 291–302.

DNA and RNA mechanics

Bustamante, C., Smith, S. B., Liphardt, J., and Smith, D. (2000). Single-molecule studies of DNA mechanics. *Curr. Opin. Struct. Biol.*, **10**, 279–285.

Strick, T., Allemand, J.-F., Croquete, V., and Bensimon, D. (2000). Twisting and stretching single DNA molecules. *Prog. Biophys. Mol. Biol.*, **74**, 115–140.

Liphgardt, J., Onoa, B., Smith, S. B., Tinoco, I. Jr. and Bustamante, C. (2001). Reversible unfolding of single RNA molecules by mechanical force. *Science*, **292**, 733–737.

Brower-Toland, B. R., *et al.* (2002). Mechanical disruption of individual nucleosomes reveals a reversible multistage release of DNA. *Proc. Natl. Acad. Sci. USA*, **99**, 1960–1966.

Protein mechanics

Vazques, M. C., Oberhauser, A. F., *et al.* (2000). Mechanical design of proteins studied by single-molecule force spectroscopy and protein engineering. *Prog. Biophys. Mol. Biol.*, **74**, 63–91.

Fisher, T. E., Marszalek, P. E., and Fernandez, J. M. (2000). Stretching single molecules into novel conformations using the atomic force microscopy. *Nature Struct. Biol.*, **7**, 719–724.

Zlatanova, J., Lindsay, S. M., and Leuba, A. H. (2000). Single molecule force spectroscopy in biology using the atomic force microscope. *Prog. Biophys. Mol. Biol.*, **74**, 37–61.

How strong is a covalent bond?

Rief, M., Oesterhelt, F., Heymann, B., and Gaub, H. E. (1997). Single molecule force spectroscopy on polysaccharides by atomic force microscopy. *Science*, **275**, 1295–1297.

Grandbois, M., Beyer, M., Rief, M., Clausen-Schaumann, H., and Caub, H. E. (1999). How strong is covalent bond? *Science*, **283**, 1727–1730.

Florin, E.-L., Moy, V. T., and Gaub, H. E. (1994). Adhesion forces between individual ligand-receptor pairs. *Science*, **264**, 415–417.

Gimzewski, J. K. and Joachim, C. (1999). Nanoscale science of single molecules using molecular probes. *Science*, **283**, 1683–1688.

Part G
X-ray and neutron diffraction

Chapter G1
The macromolecule as a radiation scattering particle

G1.1 Historical review and introduction to biological applications

The wave nature of light on which diffraction phenomena are based was first suggested by Huygens more than 300 years ago. About 100 years later, Haüy wrote an essay on the regularity of crystal forms that is considered to be the beginning of crystallography.

1690

In his *Treatise on Light* **C. Huygens** wrote that light 'spreads by spherical waves, like the movement of Sound', and explained reflection and refraction by wave constructions.

1784

R.-J. Haüy a mineralogist, published his theory on crystal structure, following observations that calcite cleaved along straight planes meeting at constant angles.

1895

J. J. Thomson discovered electrons during an investigation of cathode rays. He initially called them corpuscles.

1895

W. C. Röntgen discovered X-rays. While experimenting with electric current flow in a partially evacuated glass tube, he noted that a radiation was emitted that affected photographic plates and caused a fluorescent substance across the room to emit light.

1912

P. P. Ewald's doctoral thesis on the passage of light waves through a crystal of scattering atoms led **M. von Laue** to ask what would happen if the wavelength of the light were similar to the atomic spacing, and this led to the first observations of X-ray crystal diffraction by **W. Friedrich, P. Knipping** and **von Laue**. Because of their short wavelengths, X-rays provide a 'ruler' with which to measure distances between atoms.

1912–15
W. H. Bragg and **W. L. Bragg** interpreted diffraction in terms of reflection from crystal planes. They solved the crystal structures of NaCl and KCl and introduced Fourier analysis of the X-ray measurements.

1917
P. P. Ewald introduced the 'reciprocal lattice' construction, a graphical method of expressing the geometrical conditions for crystal diffraction.

1924
W. L. Bragg and collaborators developed the use of absolute intensities in crystal analysis leading to the solution of structures more complex than the monovalent salts.

1924
L.-V. de Broglie proposed the relation between the wave and the particle nature of matter, thus paving the way for the interpretation of scattering of particle beams (such as electrons, and later neutrons) in terms of wave diffraction.

1932
J. Chadwick discovered the neutron, one of the main constituents of atomic nuclei (neutrons and protons have about the same mass; together they make up 99.9% of an atom's mass). Neutrons are emitted spontaneously by certain radioactive nuclei and various elements undergo fission when bombarded by neutrons emitting additional neutrons. Because they are electrically neutral, neutron beams penetrate deeply into matter. Its properties made the neutron a particularly useful probe for investigating structure and dynamics at the molecular level.

1936
W. M. Elsasser, H. v. Halban & **P. Preiswerk** and **D. P. Mitchell** & **P. N. Powers** demonstrated diffraction of neutrons from a radium–beryllium source, and thus their wave nature according to **de Broglie's** relation.

Late 1930s
A. Guinier developed his theory to show that X-ray scattering at small angles, around the direct beam direction, by non-crystalline solids and solutions contained information on particle size and shape.

1945 and the following years

Neutron beams from pile reactors became available for diffraction experiments and crystallography. **C. Shull** performed the first neutron diffraction experiments to investigate material structures. Diffractometers were built at Argonne National Laboratory (USA), followed by Oak Ridge National Laboratory (USA), Chalk River (Canada), and Harwell (UK). In the early **1950s, B. N. Brockhouse** invented the triple-axis spectrometer and measured vibrations in solids by neutron scattering. **Shull** and **Brockhouse** were awarded the Nobel Prize for Physics in 1994.

1953

F. Crick, **J. Watson**, **R. Franklin** and **M. Wilkins** published the double-helix structure of DNA calculated from X-ray fibre diffraction and chemical model building.

Late 1960s

Research groups led by **J. C. Kendrew** and **M. Perutz** published the first ångström resolution structures of proteins (myoglobin and haemoglobin, respectively) from X-ray crystallography.

1960s and 1970s

Medium- and high-flux reactors, and later spallation sources, were built with neutron beams dedicated to the study of matter. Biophysical studies using neutrons provided information on the structure and dynamics of biological membranes and macromolecules that cannot be obtained by other methods.

1980s

The crystallization and first X-ray crystal structures of membrane proteins were obtained by using detergents. Because they are soluble only in complex solvents, the biochemical and structural study of membrane proteins lagged far behind that of water-soluble proteins.

1980s to present

Beam lines at synchrotron facilities that provide very brilliant X-ray sources for macromolecular crystallography and diffraction studies became available. Efficient protein modification, crystallization, data collection and analysis approaches were developed for macromolecular crystallography. Extremely fast data-collection times made it possible to study kinetic intermediates in myoglobin using time-resolved crystallography. High-resolution ribosome structures were obtained from crystallography and electron microscopy.

X-rays provided the foundation on which structural biology has been built and is developing. In the **1920s**, X-ray diffraction was already being observed from complex organic crystals (long-chain carbohydrates, hexamethyl benzene, anthracene, urea), and the first polymer structural studies were performed on rubber, hair and wool fibres. Protein and viruses were shown to form crystals and diffraction diagrams of nerve fibres under different humidity conditions were published in the **1930s**, which led to a model for biological membrane organisation.

The publication of the double-helix structure in **1953** heralded the birth of molecular biology. It was followed by the structures of myoglobin, haemoglobin, lysozyme in the **1960s** and of transfer RNA in the **1970s**, providing a wealth of information about structure–function relationships. Neutron radiation has the special property that it can distinguish between hydrogen and its isotope deuterium. Neutron diffraction studies, using deuterium labelling, of biological membranes, fibres, macromolecules and their complexes in crystals and by small-angle scattering in solution contributed strongly to the understanding of biological structure.

In the last decades, the development of novel crystallographic methods plus faster computers that could implement them and the availability of intense synchrotron sources have brought about a revolution in macromolecular crystallography by greatly increasing the rate at which structures could be solved. The distinction between fundamental and applied science in current X-ray and neutron diffraction experiments on biological systems is increasingly blurred because of the use that can be made of the results in medicine, biotechnology, or food science, for example.

G1.2 Radiation and matter

In a diffraction experiment, waves of radiation scattered by different objects interfere to give rise to an observable pattern, from which the relative arrangement (or structure) of the objects can be deduced. The interference pattern arises when the wavelength of the radiation is similar to or smaller than the distances separating the objects. Radio waves with wavelengths of several metres, for example, are diffracted by buildings in a town. Atomic bond lengths are close to 1 Å unit (10^{-10} m). In practice, three types of radiation are used in diffraction experiments: X-rays of wavelength about 1 Å, electrons of wavelength about 0.01 Å, and neutrons of wavelength about 0.5–10 Å. Electron diffraction is treated with electron microscopy in Chapter H2. X-rays and neutrons are treated together in the current chapter.

A wavelength similar to or smaller than the structural scale examined is not the only criterion to define a useful radiation. It should also present appropriate properties of interaction with matter; it should not be absorbed too strongly and it should be scattered with reasonable efficiency; and, of course, radiation sources of appropriate intensity should be available. X-rays and neutrons broadly satisfy these criteria. There are significant differences in the details of their interaction with matter, however, that make them strongly complementary for diffraction studies of biological molecular structure.

G1.2.1 X-ray and neutron scattering

We start with the neutron case, which is simpler to describe because the scattering centres can be considered as points.

Neutrons

Neutrons are scattered by atomic nuclei in a complex process. Because the neutron wavelength in diffraction experiments ($\lambda \sim 1$ Å $=10^{-10}$ m) is so much larger than the nuclear dimensions ($\sim 10^{-15}$ m), the nuclei act as point scatterers. A point scatters a wave isotropically, i.e. equally in all directions (Comment G1.1).

Heavier elements do not dominate neutron scattering and the scattering power of different isotopes of the same element can be very different. The case of hydrogen (^{1}H) and deuterium (^{2}H or D) is of particular interest in structural biology. The neutron scattering powers of H and D are sufficiently different from each other to allow very useful labelling experiments to observe hydrogen atoms and water molecules in biological samples.

Neutrons and protons (whether they are in atomic nuclei or in beams) are quantum mechanical particles of spin one-half. The neutron scattering power of a nucleus thus also depends on the relative orientation of nuclear and neutron beam spins. Neutron and nuclear spins can be polarised (oriented) by a magnetic field. However, unpolarised beams and samples are used in most diffraction experiments, so that the same type of nucleus in a sample scatters neutrons with different power according to the spin–spin orientations. The distribution of neutron–proton spin–spin orientations is random in the sample resulting in a strong *incoherent* contribution to the scattering (see Comment G1.8). The effect is largest for the proton (the hydrogen-nucleus); it results in hydrogen-containing samples giving a high background signal in neutron diffraction experiments, which does not contain structural information. The analysis of incoherent neutron scattering, however, contains information on sample molecular dynamics (see Chapter I2).

X-rays

X-rays are scattered by electrons. Atomic electron clouds are on the same length scale as the radiation wavelength ($\sim$Å) so that, in the scattering process, atoms appear as extended objects and cannot be considered as points. The scattering decreases with increasing angle. The shape of the atom 'seen' by the X-ray beam is taken into account by describing the angular distribution of its scattering power in terms of a 'form factor' (see Section G1.3).

Comment G1.1
Scattering by a point atom

An object and the pattern of waves it scatters are related by Fourier transformation (see Chapter A3). A point scatters a wave with equal amplitude in all directions.

In other words, the Fourier transform of a Dirac delta function (the point) is a constant amplitude independent of scattering direction.

The X-ray scattering power of an atom increases simply with its number of electrons. There is no isotope effect since isotopes of the same element have the same number of electrons. Heavy atoms dominate the diffraction pattern, allowing them to be used as labels in crystallography. Hydrogen atoms, with only one electron each, can only be 'seen' by X-rays when they are organised with a very high degree of crystallographic order and when the high intensity beams now available with synchrotron radiation are used.

G1.2.2 Absorption

Absorption can severely limit the penetration of radiation in a diffraction experiment to the surface layers of a sample.

X-ray absorption is due to photons exciting electrons to higher energy levels. It is, therefore, energy- (and wavelength-) dependent; absorption increases with increasing wavelength. In practice, X-ray absorption leads to non-negligible radiation damage in the sample that should be corrected for in diffraction experiments. Absorption is severe for wavelengths above 2.5 Å, where even air in the beam path absorbs significantly.

Neutron absorption is due to recombination with the nucleus (recall the neutron is itself a nuclear particle), resulting, for example, in nuclear fission. At the wavelengths used for diffraction studies, however, neutron absorption is very low for most nuclei, even for wavelengths of 10 Å or larger. Fortunately, isotopes of certain nuclei, such as cadmium, boron and lithium, are notable exceptions so that they can be used for shielding and detection.

G1.2.3 Energy momentum and wavelength

Energy–wavelength relations for neutrons and the electromagnetic spectrum are given in Table G1.1. Neutron and X-ray photon properties are given in Comment G1.2.

Table G1.1. *Wavelengths and energies*

Wavelength λ (order of magnitude)	Electromagnetic radiation energy	Neutron λ, energy (temperature)
10 fm (0.1 Å)	Hard X-rays 124 keV	Hot neutrons 0.7 Å, 172 meV (2100 K)
100 fm (1 Å)	X-rays 12.4 keV	Thermal neutrons 1.8 Å, 24.5 meV (300 K)
1 nm (10 Å)	Soft X-rays 1.24 keV	Cold neutrons 7 Å, 1.67 meV (19 K)
10 nm (100 Å)	UV 124 eV	
100 nm (1000 Å)	Visible 12.4 eV	

Comment G1.2   Neutron properties and conversion factors

As particles:

> mass, $m = 1.674928 \times 10^{-27}$ kg
>
> momentum $= m\mathbf{v} = \mathbf{p}$
>
> energy $= E = \frac{1}{2} mv^2$

As waves:

> de Broglie associated wavelength: $\lambda = h/mv$
>
> (h is Planck's constant: 6.6327×10^{-37} kg m^2 s^{-1})
>
> momentum $= \mathbf{p} = h/\lambda$ (in the direction of propagation)
>
> wave vector $= 2\pi\lambda$ (in the direction of propagation) $= 2\pi h\mathbf{p}$
>
> energy, $E = h^2/2m\lambda^2$

$$E[\text{meV}] = 81.81/(\lambda^2[\text{Å}^2])$$
$$= 0.0861737\, T[\text{K}]$$

X-ray properties and conversion factors

As particles:

> Photons of energy $h\nu$ (where ν is frequency) and momentum $h/c\nu$ (in the direction of propagation, where c is the speed of light 2.99792×10^8 m s^{-1})

As waves:

> Wavelength λ, energy hc/λ, momentum h/λ

$$E[\text{keV}] = 12.4/\lambda[\text{Å}]$$
$$= 4.13 \times 10^{-18}\lambda[\text{s}^{-1}]$$

Comment G1.3
Huygens candle

In *Treatise on Light*, Huygens wrote 'Thus in a flame of a candle, having distinguished

the points A, B, C, concentric circles described about each of these points represent the waves which come from them. And one must imagine, the same about every point of the surface and of the part within the flame' (Huygens, 1962).

X-rays are electromagnetic radiation of much higher energy than visible light. Neutrons are particles of mass similar to that of a proton. Their associated wavelength is inversely proportional to their momentum by de Broglie's relation. The velocity of neutrons of wavelengths close to 1 Å is about 4000 m s^{-1}, close to the speed of a bullet leaving the barrel of a gun (and very far from the speed of light!). One-ångström neutrons are therefore not relativistic particles; they behave like billiard balls and their momentum is simply mv, and kinetic energy is simply $\frac{1}{2} mv^2$. Neutron wavelength is, therefore, inversely proportional to the square root of the energy. The temperature of a neutron 'gas' of a given wavelength is also given in Table G1.1.

G1.3 Scattering by a single atom (the geometric view)

Consider an atom as being constituted of a set of points that scatter radiation. When a plane wave of monochromatic radiation is incident upon the atom, each point acts as a source of spherical waves of the same wavelength, similarly to Huygens' historic construction (Comment G1.3).

Comment G1.4   Scattering by a single electron, the Thomson factor and polarization

J. J. Thomson first derived the expression for X-ray scattering by a single electron. The scattering amplitude of an electron is called the Thomson factor:

$$f_{el} = e^2/mc^2 = 2.8 \times 10^{-15} \text{ m}$$

where e and m are respectively the charge and mass of the electron and c is the speed of light; f_{el} is equal to the classical radius of an electron.

The oscillating fields of the incident electromagnetic wave set the electron to oscillate, giving off radiation of intensity I_e, at a distance r from the electron,

$$I_e = (I_0 \cdot f_{el}^2 \cdot 1/r^2) \cdot [1 + (\cos^2 2\theta)/2]$$

I_0 is the incident flux, $1/r^2$ is a measure of solid angle and 2θ is the scattering angle.

The cosine term in the equation is the polarization factor. It takes into account the fact that the scattering process introduces partial polarization of the scattered beam even when the incident beam is unpolarised. Note that the polarization factor is close to 1 for small scattering angles.

Neutrons are scattered by nuclei. As we have seen above, it is a very good approximation to interpret neutron scattering by an atom in terms of scattering by a single point, and to describe it in terms of a single parameter called the *scattering amplitude.*

X-rays are scattered by the atomic electrons. The scattering process can be understood as the oscillating fields of the electromagnetic radiation creating oscillating dipoles in the electron cloud, which, in turn, emit radiation (see Chapter E1, Comment G1.4). In a geometrical interpretation, the atomic electron cloud behaves as an extended object in space; spherical waves scattered from each point in that object interfere to result in a scattered intensity distribution that depends on scattering angle (the *form factor*).

G1.3.1 Point scattering. Scattering length

First, we consider scattering by a point. In the scattering event, the point acts as the source of an isotropic spherical wave. The scattering amplitude (or length), b, of the point is defined as the amplitude of the wave observed per unit *incident flux*, in unit solid angle (in any direction, since the wave is spherical). We recall that the intensity of a wave is equal to the square of its amplitude. The scattered intensity is thus given by

$$I = I_0 b^2 \tag{G1.1}$$

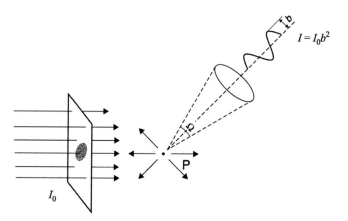

Fig. G1.1 Scattering by a point atom. The incident flux I_0 is defined as the number of radiation particles passing through a 1 m^2 window per second. The point P scatters isotropically. Its scattering length, b, is defined as the amplitude of the wave scattered in unit solid angle Ω for unit incident flux. The intensity of the wave scattered in Ω per incident flux I_0 is $I_0 b^2$. The shaded area in the incident flux window is the area of the effective cross-section (see text).

where I is the scattered intensity in unit solid angle (neutrons s^{-1} or photons s^{-1}) and I_0 is the incident flux, in neutrons s^{-1} m^{-2} or photons s^{-1} m^{-2} (Fig. G1.1).

The total intensity scattered by the point is given by the sum over all directions (4π solid angle):

$$I_{\text{total}} = I_0 4\pi b^2 \qquad (G1.2)$$

The *scattering cross-section*, σ, of the point is then defined as $\sigma = 4\pi b^2$. It has units of [m^2]. It can be understood as an effective projected area perpendicular to the incident beam (Fig. G1.1); each particle that hits the area of cross-section is scattered. The ratio of the area of cross-section to the total area in the definition of incident flux (1 m^2 in this case) can also be understood as the scattering *probability* associated with the point.

In general, the value of the scattering amplitude, b, is a complex number. We recall (Chapter A3) that a phase shift between two waves can be expressed in terms of a complex amplitude. A phase shift of π corresponds to a change in sign of the amplitude. By convention, a positive value of b denotes a phase shift of π between the incident and scattered waves, which is the case for X-ray and neutron scattering by most atoms (Fig G1.2). *Anomalous scattering* refers to a phase shift other than π, resulting in a complex contribution to the scattering amplitude, usually written $f' + if''$ for X-rays (see Chapter G3).

Fig. G1.2 Incident (black) and scattered (red) waves showing: (a) a phase difference of π, equivalent to reflection of an incident wave off a non-absorbing surface, which is the case for most atoms in X-ray and neutron scattering; (b) no phase difference (negative amplitude) which is the case for neutron scattering by ^{1}H.

X-ray scattering amplitudes

Each electron scatters X-rays with the same amplitude, known as the Thomson factor f_{el}, for unit incident flux and in unit solid angle (Comment G1.4). Even for a point-like electron, scattering is not isotropic because of the angle dependence of a polarization term (Comment G1.4). The polarization term, however, is equal to 1 in the forward direction (zero scattering angle), and the waves from all the electrons in the atom interfere constructively, so that the amplitude of the wave scattered at small angles by an atom of n electrons is simply nf_{el}.

Comment G1.5
X-ray and neutron scattering amplitudes

X-ray and neutron form factors, scattering amplitudes and absorption cross-sections are tabulated on the web. *X-ray Data Booklet.* Berkeley, CA: Lawrence Berkeley National Laboratory, University of California (up-dated versions can be found at http://xdb.lbl.gov). *Neuton Data Booklet.* Grenoble: Institut Laue Langevin (http://www.ill.fr).

Comment G1.6
The barn unit

The name *'barn'* was chosen for this very small unit of area with humorous irony since the side of a barn is usually representative of a very large area. Recall the saying for someone who does not know how to shoot that 'he could not hit the side of a barn'.

Due to the spatial extent of the electron cloud, the amplitude of X-rays scattered by the atom decreases with scattering angle, in a similar fashion to the intensity scattered by an assembly of atoms (see Section G1.5). The dependence of amplitude on scattering angle is called the *form factor* of the atom.

Atomic X-ray scattering amplitudes and form factors are usually tabulated simply in terms of the effective number of electrons (Comment G1.5). They are represented by real numbers, which are independent of wavelength provided the incident wavelength (energy) is far from the absorption edge of the scattering atom (see Chapter G3). Close to the absorption edge the X-ray scattering amplitude is represented by a complex number to indicate the phase shift of the scattered wave with respect to the incident wave. The effect, which is usefully exploited in protein crystallography is called *anomalous scattering*.

Neutron scattering amplitudes

Neutron scattering amplitudes do not increase in proportion to the size or mass of the nucleus; they are all of a similar scale between 1 and 10 fm (in scattering methods an amplitude of 1 fm has been defined as 1 Fermi unit) (Comment G1.5). This is an advantage because light elements are as 'visible' as heavy ones in neutron crystallography. The isotope effect, already discussed above, is also particularly useful because of the labelling possibilities that it opens up. The hydrogen nucleus has a negative neutron scattering b value (-3.74 fm), indicating a phase shift of π for neutrons scattered by the proton relative to those scattered by, for example, the oxygen nucleus ($b = 5.85$ fm) (Fig. G1.2). Neutrons also display *anomalous scattering* for certain nuclei. Nuclei with a large neutron absorption have an imaginary component in their b values representing a phase shift. For example, the b value of the boron isotope ^{10}B, a strong neutron absorber used for shielding, is ($-0.1 - 1.066i$) fm.

G1.3.2 Cross-sections and sample size

X-ray and neutron atomic scattering amplitudes are of the order of 10^{-14} m (10 Fermi units); scattering cross-sections are usually given in units of 1 *barn* (10^{-24} cm^2) (Comment G1.6). The cross-section represents the effective scattering area the atom offers the incident flux (Fig. G1.1).

We can appreciate how small atomic scattering cross-sections actually are by calculating how long it would take to detect one scattered neutron from a single carbon atom placed in one of the most intense neutron beams currently available (Comment G1.7). The answer is slightly less than 600 million years. Independently of other considerations, therefore, samples for X-ray or neutron diffraction must contain a very large number of atoms.

The cross-section concept is also applicable to different forms of scattering (coherent or incoherent scattering, see below) and to absorption (Comment G1.5).

Comment G1.7   Scattering by a single atom

What is the length of time we would have to wait to have a good probability of observing one scattered neutron from a single carbon atom?

Answer: From Eq. (G1.2):

I_{total} [neutrons s^{-1}] $= I_0 4\pi b^2$

I_0 is 10^{11} neutrons m^{-2} s^{-1}, and b for carbon is 6.65×10^{-15} m

I_{total} is calculated to be 5.56×10^{-17} neutrons s^{-1}

We should have to wait for $1/(5.56 \times 10^{-17})$ s or almost 600 million years to have a good probability of seeing one neutron.

How many atoms should there be in a sample to scatter 100 neutrons s^{-1}?

Answer: $\sim 10^{19}$ atoms or 10^{16} particles with 1000 atoms in each (e.g. a small protein).

In fact these numbers can be achieved quite easily. For example, 10^{19} atoms of carbon represent a mass of only 0.2 mg.

G1.4 Scattering vector and resolution

Consider the phase difference of a wave scattered by point P and by point O (Fig. G1.3).

A wave front W can be defined perpendicular to the wave propagation as joining points of equal phase. PM is the wave front of the incident wave when it touches P; PN is the wave front of the wave scattered in the direction 2θ. The path difference, Δ, between the waves scattered by P and those scattered by O is $ON - OM$. We now write this in vector notation,

$$OM - ON = \Delta = \mathbf{r} \cdot \mathbf{u}_1 - \mathbf{r} \cdot \mathbf{u}_0 = \mathbf{r} \cdot (\mathbf{u}_1 - \mathbf{u}_0) \qquad (G1.3)$$

where $\mathbf{r}$ is the vector OP and $\mathbf{u}_1$ and $\mathbf{u}_0$ are unit vectors parallel to the incident and scattered waves, respectively.

The phase difference, δ (in radians), between the two waves is given by

$$\delta = \frac{\Delta}{\lambda} 2\pi \qquad (G1.4)$$

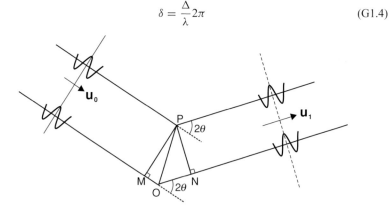

Fig. G1.3 Diffraction from two points.

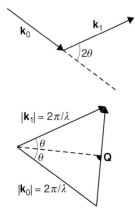

Fig. G1.4 Definition of the scattering vector.

where λ is the wavelength of the incident beam. When we place an atom of scattering amplitude f at P, the equation for the scattered wave relative to a wave from O is written (see Chapter A3):

$$A = f \exp(\mathrm{i}\delta) = f \exp\left(\mathrm{i}\frac{2\pi}{\lambda}(\mathbf{u}_1 - \mathbf{u}_0) \cdot \mathbf{r}\right) \tag{G1.5}$$

We introduce wave vectors $\mathbf{k}_0$ and $\mathbf{k}_1$ of magnitude $2\pi/\lambda$ in the directions of the incident and scattered waves, respectively, and a scattering vector $\mathbf{Q}$, which is the difference between them:

$$\mathbf{k}_0 = \frac{2\pi}{\lambda}\mathbf{u}_0$$

$$\mathbf{k}_1 = \frac{2\pi}{\lambda}\mathbf{u}_1 \tag{G1.6}$$

$$\mathbf{Q} = \mathbf{k}_1 - \mathbf{k}_0$$

The wave from P can now be written simply,

$$A = f \exp(\mathrm{i}\mathbf{Q} \cdot \mathbf{r}) \tag{G1.7}$$

The magnitude of the scattering vector $\mathbf{Q}$ can be calculated from the geometric diagram in Fig G1.4:

$$Q = (4\pi \sin\theta)/\lambda \tag{G1.8}$$

The scattering vector contains in its magnitude not only the angle, 2θ, with respect to the incident wave direction in which the scattered wave is observed, but also the wavelength of the radiation, λ. In fact, expressing scattered intensity as a function of $\mathbf{Q}$ fully defines the interplay of angle and wavelength dependence so that we do not need to know either of these values separately.

The scattering vector is a very useful quantity, which essentially defines the magnification that can be achieved in the diffraction experiment. Waves from atoms separated by a vector $\mathbf{r}$ are out of phase by one cycle (2π phase angle or λ path length), leading to constructive interference when the scattering vector $\mathbf{Q}$ is parallel to $\mathbf{r}$ and for Q equal to $2\pi/r$. If the wavelength and r are on the same length scale, there is increased intensity due to the constructive interference between the two waves at an observable angle (e.g. measured in terms of centimetres on a photographic plate or detector, leading to an effective magnification of 10^8, when r and λ are in the ångström range).

The reciprocal relationship between Q and r arises from the wave construction in Fig. G1.3. It represents a fundamental property of diffraction theory. For smaller values of r, we need to go to larger values of Q to obtain the same path difference. The geometrical space in which Q is depicted is called 'reciprocal space' with respect to the space of r, which is called 'real' or 'direct' space (see Chapter A3). In order to 'resolve' shorter distances in a diffraction experiment it is necessary to go to larger scattering vector values (achieved by increasing the observation angle and/or reducing the wavelength), in order to increase the effective path difference between the scattered waves. The *resolution* of an experiment is the minimum distance between points that can be observed separately. It is given by

$$\text{resolution} \sim 2\pi/Q_{\max} \tag{G1.9}$$

where $Q_{\max}$ is the maximum value of Q for which the scattered intensity is observed.

G1.5 Scattering by an assembly of atoms

The two-atom case treated above provides the basis for building a picture of scattering by an assembly of atoms. We can consider the resulting wave in a direction defined by a given scattering vector as resulting from interference of waves from all possible atom pairs in the assembly.

G1.5.1 Coherent and incoherent scattering

Coherent scattering is defined as the case in which the scattered waves interfere to give a single resultant wave in a given direction, as in Fig. G1.3. The amplitudes of the waves, a_j, from each atom j are added by taking into account their phase relationships. The intensity when the waves are in phase is $(\Sigma a_j)^2$. If the scattering is *incoherent*, the resultant intensity is $\Sigma(a_j^2)$, the sum of intensities scattered individually by each of the atoms (as if the other atoms were not there!). The intermediate case, however, is the one most often observed, in which the scattered waves contain a coherent and an incoherent component.

We recall (Section G1.3.2) that we need to have very large numbers of atoms in a sample because scattering by a single atom is far too weak to observe. The same requirement holds true for a 'structure' made up of two atoms separated by a given vector, for example. The sample must contain a very large number of pairs, all of which are made up of the same two atom types separated by the same vector. In other words, the sample itself presents a coherent structure. If there is disorder in the sample, due to variations in either atom type or the vector separating the atoms, it contributes to incoherence.

For scattering from an assembly of atoms to be coherent, both the incident beam and sample properties must be coherent. Plane wave fronts in perfect phase must cross the entire sample during the scattering process (i.e. the *coherence length* of the source should be longer than the sample), and each equivalent atom in the sample must maintain the same scattering length (i.e. a constant amplitude and phase relationship with the incident wave) and be found in the same spatial environment (i.e. in a constant structure). If equivalent positions in a crystal structure are not all occupied by the same atom type (in X-ray scattering) or by the same isotope in the same spin state (in neutron scattering) the scattered radiation has an incoherent component. Incoherent scattering in a diffraction

---

Comment G1.8   Spin incoherence in neutron scattering

During a neutron scattering event, a nucleus of spin I combines with a neutron of spin $\frac{1}{2}$ to form one of the two intermediate states $I + \frac{1}{2}$ or $I - \frac{1}{2}$, with relative weights, w_+, w_-, respectively. Different scattering lengths, b_+ and b_-, respectively, are associated with each of these states, leading to a total scattering cross-section:

$$\sigma = S + s = 4\pi(w_+ b_+^2 + w_- b_-^2)$$

where S is the coherent and s the incoherent part:

$$S = 4\pi(w_+ b_+ + w_- b_-)^2$$
$$s = \sigma - S$$

---

experiment does not contain structural information and contributes to the background noise.

Compton scattering, which is inelastic, is an important source of incoherent X-ray scattering. Incoherent neutron scattering arises mainly from spin incoherence (Comment G1.8). In practice, the background in a neutron diffraction experiment on biological material is predominantly due to the strong incoherent scattering from hydrogen nuclei. In neutron inelastic scattering experiments the incoherent scattering is analysed to provide information on sample dynamics (see Chapter I2).

G1.5.2 Elastic and inelastic scattering

Elastic scattering is when the scattered beam has the same energy as the incident beam. In an inelastic scattering process, the sample either loses energy to the radiation or gains energy from it (compare Stokes and anti-Stokes lines in light scattering, see Section E3.2). Standard diffraction experiments that provide information on the spatial arrangement of atoms are based on coherent elastic scattering. Inelastic scattering can be either coherent or incoherent. When the energy exchange is with sample excitations it contains information on dynamics (see Chapter I2). Energy analysis of coherent inelastic neutron scattering has, for example, shown how thermal energy propagates in waves of atomic vibration across molecules, while incoherent neutron scattering experiments provide information on individual atomic motions in protein dynamics.

Compton scattering of X-rays is inelastic and related to absorption (Comment G1.9); it is incoherent, in that there is no fixed phase relationship between the incident and scattered waves, and contributes to the background observed in an X-ray diffraction experiment.

**Comment G1.9
Compton
scattering**

A. H. Compton postulated in 1923 that electrons recoil by absorbing some of the energy of the incident X-ray beam, and the scattered beam is therefore of different wavelength. Compton scattering is inelastic. It is also incoherent, contributing to background noise, as there is a random phase relationship between the waves. Compton scattering is larger for low atomic number elements, for which the electron binding energy is lower and the probability of recoil higher.

G1.5.3 Summing waves, Fourier transformation and reciprocal space

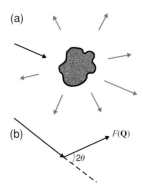

The two-atom example is readily extended to a particle made up of a larger assembly of atoms with a fixed spatial relationship (Fig. G1.5). Recall from Eq. (G1.7), the scattering from an atom at a point P with respect to an arbitrary origin O:

$$A = f \exp(i\mathbf{Q} \cdot \mathbf{r})$$

where f is the scattering amplitude of the atom, $\mathbf{Q}$ is the scattering vector and $\mathbf{r}$ is the vector between P and O. In the case of an assembly of atoms, the scattered wave is given simply by the sum of waves individually scattered, with respect to an arbitrary origin:

$$F(\mathbf{Q}) = \sum_j f_j \exp(i\mathbf{Q} \cdot \mathbf{r}_j) \qquad (G1.10)$$

Fig. G1.5 (a) Scattering by an assembly of atoms in a particle. The black arrow symbolises the incident beam, the red arrows represent the scattered waves of different amplitudes and phases in different directions, corresponding to different scattering angles and $\mathbf{Q}$ vectors. (b) The wave $F(\mathbf{Q})$ for a given vector scattering vector $\mathbf{Q}$.

where f_j, $\mathbf{r}_j$ are respectively the scattering amplitude and position vector with respect to the origin, of atom j. The f_j values are Q-dependent for X-rays (they are the form factors of the atoms) but constant with Q for neutrons. The choice of origin for $\mathbf{r}_j$ is arbitrary; the phase of each scattered wave in Eq. (G1.10) is calculated relative to that of a virtual wave scattered by a point at the origin. The structural information contained in the scattered wave, however, arises from the phase differences between the scattered waves – phase differences that are independent of the origin.

Note that $F(\mathbf{Q})$ is a complex number. It defines the scattered wave for scattering vector $\mathbf{Q}$ in terms of two quantities: its amplitude and its phase. Phase information is lost when the intensity, $| F(\mathbf{Q}) |^2$, of the wave is measured.

Equation (G1.10) establishes in mathematical terms that $F(\mathbf{Q})$ is the Fourier transform of the f_j, $\mathbf{r}_j$ distribution, $f(\mathbf{r})$, which in practice describes the positions of the atoms in the particle. An important property of the Fourier transform is that if A is the transform of B then B is the transform of A (see Chapter A3). The structure of the particle, $f(\mathbf{r})$, can therefore be calculated from the observed scattering amplitude $F(\mathbf{Q})$.

Since the particle scattered amplitude is continuous as a function of $\mathbf{Q}$, it is useful to express the back Fourier transform of Eq. (G1.10) as an integral:

$$f(\mathbf{r}) = \int F(\mathbf{Q}) \exp(-i\mathbf{Q} \cdot \mathbf{r}_j) \, dV_\mathbf{Q} \qquad (G1.11)$$

where $dV_\mathbf{Q}$ is a volume element in $\mathbf{Q}$ space (also called reciprocal space, Comment G1.10) . Similarly, Eq. (G1.10) can be expressed as an integral over the particle volume:

$$F(\mathbf{Q}) = \int f(\mathbf{r}) \exp(i\mathbf{Q} \cdot \mathbf{r}_j) \, dV_\mathbf{r} \qquad (G1.12)$$

Comment G1.10   Real space and reciprocal space

A structure is described in terms of atomic coordinates (vectors **r**) in *real space*. The wave scattered by that structure is described in terms of a phased amplitude F as a function of wave vector **Q**. Because the units of Q are reciprocal to the units of r, the mathematical space in which $F(\mathbf{Q})$ is described is called *reciprocal space*.

Note that the coordinates in real space are far too small to be 'observed' directly in our laboratory frame (in the absence of magnification by a lens system). We observe and measure scattered waves as a function of the scattering angle and the reciprocal of the radiation wavelength in a frame of reference that corresponds to reciprocal space!

There is a one-to-one relationship between the $f(\mathbf{r})$ and $F(\mathbf{Q})$ distributions. Each fully defines and is fully defined by the Fourier integral of the other (see Chapter A3).

A particle is 'observed' by analysing the waves it scatters, $F(\mathbf{Q})$, in as large a Q range as possible. Its structure (atomic arrangement) can then be calculated unequivocally with a resolution of $2\pi/Q_{\mathrm{max}}$ (Section G1.3.2) by using Eq. (G1.11). In order to do so, however, the amplitude *and* phase of each scattered wave, $F(\mathbf{Q})$, should be known.

G1.5.4 The phase problem

In *microscopy*, a magnified image of a particle is obtained from $F(\mathbf{Q})$ by using a lens to recombine the scattered waves while respecting the correct phase relationships. The lens essentially behaves as a Fourier transformation device.

While lenses for light and electron microscopy exist, there are none for X-rays or neutrons in the ångström resolution range. There is a phase problem, therefore, in X-ray and neutron diffraction, because only the intensity, $|F(\mathbf{Q})|^2$, of the scattered waves can be measured and phase information is lost. Solving a structure in the absence of this phase information is the main challenge faced by X-ray and neutron diffraction and crystallography.

G1.6 Solutions and crystals

Atomic scattering cross-sections for X-rays and neutrons are extremely small, and, even with the most intense sources currently available, a very large number of macromolecular particles are required in order to observe an interpretable signal.

The level of resolution obtained from a diffraction experiment depends on how well ordered the particles are with respect to each other in the sample. Thus, for there to be intensity at a **Q** vector corresponding to 2-Å resolution, a significant

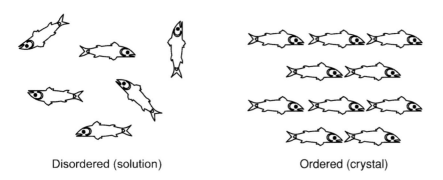

Disordered (solution)   Ordered (crystal)

number of atoms in all particles of the sample should be organised with a position accuracy of 2 Å along the vector parallel to **Q**.

There are two extremes in organisational order for a sample made up of identical particles (such a sample is called *monodisperse*) (Fig. G1.6). In a dilute solution, the particles are located randomly in both position and orientation (see Chapter G2). The only order is that of the distances between atoms within each particle. These distances (not vectors since their orientations are random) represent the most information that can be derived from a diffraction pattern of a solution sample. At the other extreme, in crystals we have particles ordered in three dimensions. Currently, the only way to obtain high-resolution structural information from diffraction experiments is by using crystallographic methods. Between the two extreme cases, we find concentrated solutions, for which interparticle diffraction can arise, and membranes and fibres, which display various degrees of one- and two-dimensional order.

G1.6.1 One-dimensional crystals

A crystal is a periodic arrangement in space of a repeated motif, which is called the *unit cell*. The unit cell may contain one or several macromolecules organised in a symmetrical fashion. The principles involved in calculating the diffraction from a crystal can be explained more simply for a one-dimensional periodic array.

We recall from Chapter A3 that a one-dimensional periodic array can be described as the convolution of a motif with a lattice of Dirac delta functions (Fig. G1.7). If the periodicity of the crystal is d and it contains a very large number of unit cells, crystal diffraction occurs only at the positions of the lattice diffraction with a periodicity proportional to $1/d$. In practice, this means that the waves diffracted in all other directions cancel themselves out by *destructive interference* (see Chapter A3). The peak amplitudes depend on the value of the diffraction of the single unit cell at that position in Q. It is said that the single unit cell diffraction pattern is *sampled* by the lattice diffraction pattern.

Fig. G1.7 (a) Real space:
a one-dimensional crystal
of N fish macromolecules
seen as a convolution of a
single fish with a lattice of
Dirac functions of
periodicity d. (b) The
corresponding Fourier
transforms in reciprocal
space for the case in
which N is a very large
number. If N is not a large
number, the Fourier
transform of the lattice
does not correspond to a
set of delta functions but
to a distribution of
broader peaks.

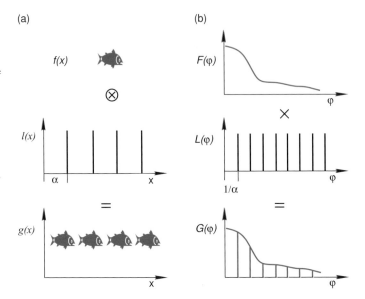

Fig. G1.7 (a) Real space: a one-dimensional crystal of N fish macromolecules seen as a convolution of a single fish with a lattice of Dirac functions of periodicity d. (b) The corresponding Fourier transforms in reciprocal space for the case in which N is a very large number. If N is not a large number, the Fourier transform of the lattice does not correspond to a set of delta functions but to a distribution of broader peaks.

The lattice can be described by a vector $\mathbf{R}$, of length d, displaced n times. The diffraction pattern $G(\mathbf{Q})$ produced by this arrangement is,

$$G(\mathbf{Q}) = F(\mathbf{Q}) \times \sum_{n} \exp(in\mathbf{Q} \cdot \mathbf{R}) \tag{G1.13}$$

where $F(\mathbf{Q})$ is the diffraction pattern of the single unit cell (Eqs. (G1.10) and (G1.12)). Each term in the sum introduces the *phase difference* (see Chapter A3) associated with successive lattice points. This is exactly equivalent to the sampling shown in Fig. G1.7.

G1.6.2 Two-and three-dimensional crystals

It is straightforward to extend Eq. (G1.13) to two dimensions (as in the case of membranes with in-plane organisation) and to three-dimensional crystals by using appropriate vectors $\mathbf{R}$ and $\mathbf{Q}$.

For a three-dimensional crystal, $\mathbf{R} = \mathbf{a} + \mathbf{b} + \mathbf{c}$, where $\mathbf{a}$, $\mathbf{b}$, $\mathbf{c}$ are vectors along the three axes (Fig. G1.8). By substituting for $\mathbf{R}$ in Eq. (G1.13), we write

$$G(\mathbf{Q}) = F(\mathbf{Q}) \times \sum_{t} \exp(it\,\mathbf{Q} \cdot \mathbf{a}) \times \sum_{u} \exp(iu\mathbf{Q} \cdot \mathbf{b}) \times \sum_{v} \exp(iv\mathbf{Q} \cdot \mathbf{c}) \tag{G1.14}$$

where t, u, and v are integral numbers.

Similarly to the one-dimensional case, when the crystal contains a very large number of unit cells, all phases except the ones corresponding to the lattice diffraction positions produce destructive interference so that diffraction is observed only at

$$\mathbf{Q} \cdot \mathbf{a} = 2\pi\, h \qquad \mathbf{Q} \cdot \mathbf{b} = 2\pi\, k \qquad \mathbf{Q} \cdot \mathbf{c} = 2\pi\, l \tag{G1.15}$$

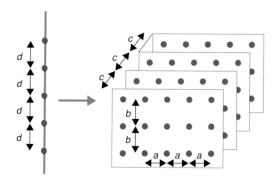

Fig. G1.8 Schematic diagram of one- and three-dimensional crystals. The red dots represent the origins of the scattered waves. In the one-dimensional lattice the origins are separated by a distance *d*, while in the three-dimensional lattice, the origins are displaced by distances, *a*, *b*, and *c* along the respective axes.

It is usual in crystallography to use a scattering vector $\mathbf{S}$ instead of $\mathbf{Q}$, where $\mathbf{Q} = 2\pi.\mathbf{S}$. Equation (G1.15) then becomes

$$\mathbf{S} \cdot \mathbf{a} = h \qquad \mathbf{S} \cdot \mathbf{b} = k \qquad \mathbf{S} \cdot \mathbf{c} = l$$

h, k and l are integral numbers and it is conventional to describe the diffraction peak by its indices (h, k, l); these are called the *Miller indices*.

In summary, the diffraction pattern from a crystal contains two levels of information: the spatial arrangement of the unit cells can be derived from the regular spacing of the diffraction peak pattern (denoted by the *hkl* indices); the amplitudes of the diffraction pattern contain information on the arrangement of atoms within each unit cell.

G1.6.3 Disordered systems

The construction in Fig. G1.7 can be used quite generally and not only in the case of well-ordered lattices. Figure G1.9 illustrates the example of a concentrated

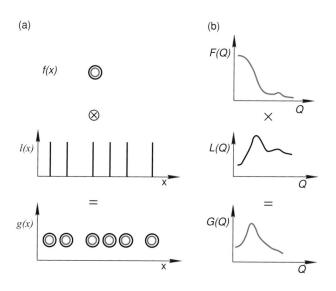

Fig. G1.9 (a) Real space: a liquid of large particles seen as a convolution of a single particle with a disordered lattice of Dirac functions. (b) The corresponding Fourier transforms in reciprocal space.

solution (in one dimension for simplicity but the argument also holds for three dimensions). Note that $L(Q)$, the Fourier transform of the disordered lattice, is a continuous function with a broad peak corresponding to the most frequent interparticle spacing in the liquid. $G(Q) = F(Q)L(Q)$ is also a continuous function. The peak of $G(Q)$, however, results from both the particle's internal structure and its spacing in the solution (see Chapter G2).

G1.7 Resolution and contrast

Biological macromolecules and structures are generally in an aqueous environment. This is obvious for protein solutions. However, protein crystals, DNA fibres, membrane samples… also contain significant volumes of solvent. A diffraction experiment on a perfectly ordered (coherent) sample should provide information on the positions of all atoms, be they in the solvent or macromolecules. In practice, however, resolution may be limited to the dimensions of a volume the size of many atoms. This is particularly true for the disordered solvent regions in a sample (Comment G1.11). In such cases, it is useful to define a *scattering density*,

---

Comment G1.11   Resolution and solvent homogeneity

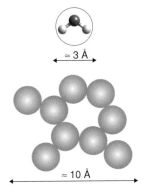

≈ 3 Å

≈ 10 Å

Clearly water is not homogeneous at a resolution better than 3 Å. At 10 Å (maximum $Q \sim 0.6$ Å^{-1}), it can be considered as homogeneous, of constant scattering density,

$$\rho° = (\Sigma f)/V$$

where V is the volume containing atoms with total scattering amplitude Σf.

    The volume per molecule of water in the liquid is calculated to be 30 Å^3, from the density of 1 g cm^{-3}. Since there are 10 electrons in an H_2O molecule

$$\rho° = 0.33 \text{ electrons Å}^{-3} \text{ for X-rays}$$

The corresponding amplitude for neutrons is 0.56 fm Å^{-3}.

i.e. scattering amplitude per unit volume. The concept of *contrast* between two parts of a sample refers to the difference between their scattering densities.

Equation (G1.10) describes the resulting wave scattered by a system of atoms in a direction corresponding to a given scattering vector,

$$F(\mathbf{Q}) = \sum_j f_j \exp(i\mathbf{Q} \cdot \mathbf{r}_j)$$

Fig. G1.10 A particle in solution. The particle is drawn as a fish to emphasise that biological macromolecules fold and attain their quaternary structures through interactions with aqueous solvent.

Consider now the single macromolecule in solution shown schematically in Fig. G1.10. In order to calculate how the system scatters radiation, the sum in Eq. (G1.10) should be over all atoms in the solution, those within the macromolecule as well as those in the solvent. The particle is fully described by the scattering amplitudes f_j at positions $\mathbf{r}_j$, of its atoms, and it is surrounded by an infinite homogeneous solvent of scattering density ρ°.

We now divide the scattering system into the macromolecule on the one hand, and the solvent on the other (Fig. G1.11). The first part, (a), is the particle (including perturbed solvent). The second part, (b), is the homogeneous infinite solvent. Note, however, that the sum of (a) and (b) includes an extra volume of solvent, (c), when compared to the initial scattering system – the volume corresponding to that occupied by the particle. The particle in solution scattering system, therefore, is equal to a + b – c, which is similar to Archimedes' principle for the buoyancy of a particle in a fluid (see Part D and Chapter G2). The sum of waves from each part is given in the right-hand panel of Fig. G1.11. Waves from the homogeneous solvent appear in a very narrow range close to $Q = 0$, so that (b) is

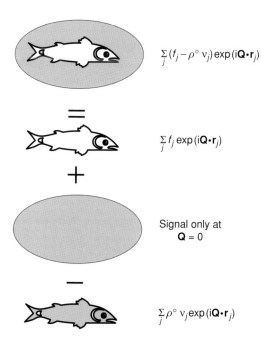

$$\sum_j (f_j - \rho^\circ v_j) \exp(i\mathbf{Q} \cdot \mathbf{r}_j)$$

$$\sum_j f_j \exp(i\mathbf{Q} \cdot \mathbf{r}_j)$$

Signal only at
$\mathbf{Q} = 0$

$$\sum_j \rho^\circ v_j \exp(i\mathbf{Q} \cdot \mathbf{r}_j)$$

Fig. G1.11 Scattering from a particle in solution.

not observed in practice (Comment G1.12). The scattering amplitude from the particle in solution is hence given by

$$F(\mathbf{Q}) = \sum_j (f_j - \rho^\circ v_j) \exp(i\mathbf{Q} \cdot \mathbf{r}_j) \qquad (G1.16)$$

where v_j is the volume of atom j. The $f_j - \rho^\circ v_j$ term is the *contrast amplitude* of atom j with respect to the solvent. The mean contrast amplitude of the particle is given by $\sum (f_j - \rho^\circ v_j)$ or $\sum f_j - \rho^\circ V$, where V is the total volume of the particle. The *mean scattering density contrast* of the particle is $\rho - \rho^\circ$, where $\rho = \Sigma(f_j) / V$.

The contrast considerations are valid for small-angle scattering experiments in solution (Chapter G2) and crystallography, fibre and membrane diffraction at low resolution only (Chapter G3).

G1.8 The practice of X-ray and neutron diffraction

G1.8.1 Complementarity

The very high intensity and narrow divergence of X-ray beams currently available from synchrotron sources make them the radiation of choice in structural biology. Neutron sources are much weaker in that respect but the special properties of the neutron, especially with respect to the scattering powers of hydrogen and deuterium, have allowed the analysis of specific problems that were difficult or impossible to address by using X-rays. Hydrogen and water molecules play essential roles in biological structures; they can be 'seen' in neutron crystallography even when not highly ordered (see Chapter G3). Neutron-specific methods based on hydrogen–deuterium labelling and contrast variation have provided essential information on the organisation of protein–protein, protein–nucleic acid complexes and membranes (see Chapter G2). Neutrons also provide a uniquely suited radiation for the study of molecular dynamics. Because the energy associated with wavelengths of ~1 Å is of the order of thermal energy, they allow the simultaneous measurement of the amplitudes and frequencies of motions in a sample (see Chapter I2).

G1.8.2 Sources and instruments

X-rays

X-ray photons are produced by an electron beam from a heated filament hitting a metal target, or in the form of synchrotron radiation from charges (electrons or positrons) accelerated by magnetic devices in a ring. X-ray sources have evolved from a sealed tube to a rotating anode to several generations of synchrotrons, providing an increase of more than ten orders of magnitude in beam brightness (Comment G1.13). This has made diffraction studies on smaller and smaller samples possible. Protein microcrystals ($\sim 10^{-5}$ mm^3) are routinely studied at third generation synchrotrons. Because of the inherent difficulty in obtaining 'large' crystals of biological macromolecules, the possibility of working with such minute samples is an important advantage of the method, making it applicable to a wide range of structural problems (see Chapter G3).

In the case of X-ray generators with metal anodes, the emitted radiation corresponds to electronic transitions of specific energy. The main wavelength from copper is 1.54 Å while that from a molybdenum target is 0.71 Å. Synchrotron sources are particularly advantageous because they provide a broad spectrum from which specific wavelengths can be chosen using monochromators.

Neutrons

Neutron sources are considerably less intense than X-ray sources. Their relative weakness is compensated for to some extent by very low absorption (the absence of radiation damage compared to X-rays), associated with the possibility of using large beam cross-sections and long wavelengths, which are prohibitive for X-rays because of the high absorption. These possibilities combined with specific contrast variation techniques make neutron sources strongly competitive for low-resolution crystallography and small-angle scattering experiments, for example (Chapters G2 and G3).

A continuous flux of neutrons is produced in reactors by the fission chain reaction of ^{235}U. In spallation sources, pulsed neutron beams are produced by an accelerated proton beam impinging onto a metal target. The high-energy protons tear through the target nuclei breaking them apart with a significant yield of fast neutrons.

The neutrons behave like a gas and are slowed down by thermal equilibration through collisions in a material called the moderator (usually light or heavy water). Neutrons of energy close to that associated with ambient temperature (300 K) are called *thermal neutrons* (energy 24.5 meV, wavelength 1.8 Å) (Table G1.1). 'Cold' and 'hot' sources may also be included in a neutron source to produce neutrons at different wavelengths. The neutron spectrum from a cold source peaks at a wavelength of a few ångströms. Long-wavelength neutrons are not strongly absorbed by matter (unlike long-wavelength X-rays) and are particularly useful for small-angle scattering studies (see Chapter G2). Hot neutrons peak at a wavelength of a fraction of an ångström, and are particularly useful for high-accuracy chemical crystallography. Cold sources may also be included on a spallation source, but hot sources are unnecessary because the spallation flux already contains an appreciable fraction of energetic neutrons.

Large-scale facilities

Because laboratory sources provide only extremely weak beams, neutron scattering experiments are performed at large-scale facilities. Several research institutes with reactor or spallation sources dedicated to providing neutron beams and instrumentation for diffraction and spectrometry experiments in structural

Comment G1.13
Flux and brightness

Flux is defined as number of photons or neutrons crossing a unit area per second (e.g. the usual flux units for neutrons are neutrons cm^{-2} s^{-1}). Neutron beams have quite a large divergence, and angular resolution in a diffraction experiment can be achieved only by using slits or similar devices that reduce intensity.

With synchrotron radiation, it is useful to define an intensity unit that also takes into account the small divergence of the beam (which allows very good angular resolution while maintaining the full intensity in the beam). Brightness is defined as number of photons crossing unit area per second per unit solid angle of beam divergence (e.g. the usual brightness units for a synchrotron beam are photons s^{-1} mm^{-2} $mrad^{-2}$).

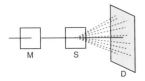

Fig. G1.12 Schematic diagram of a diffractometer. M represents the monochromator area where the wavelength (red) is selected from the source beam spectrum (blue). S is the sample area which may contain a goniometer to align the sample, temperature control devices or other sample environment controls. The detector D converts the scattered intensity (dashed red lines) as a function of scattering angle into a signal that can be read by the instrument computer.

biology now exist in Europe, the USA and Japan, and are open for use by research scientists on the basis of proposal systems. With the advent of synchrotron radiation, X-ray experiments in structural biology are also now largely performed at large-scale facilities.

Diffractometers

X-ray and neutron diffractometers are instruments that measure the scattered intensity from a sample as a function of incident wavelength and scattering angle. They vary in design according to the wavelength range, sample type and angular resolution, which should be optimised for each type of experiment. Spectroscopic experimental setups are also available, in which the energy transfer between sample and radiation is measured by analysing the wavelength (and therefore energy) of the scattered beam by the crystal reflection or time of flight (for neutrons only). The pulsed nature of the neutron beam from spallation sources allows the use of broad wavelength bands for both diffraction and spectroscopy experiments (thus considerably increasing the effective flux); under these conditions both the scattering vector and energy transfer can be analysed by time-of-flight methods (see Chapter I2).

An X-ray or neutron diffractometer consists of three main parts: a monochromator, a sample area and a detector (Fig. G1.12). The incident beam is usually made monochromatic by reflection off a crystal. In the neutron case only, a velocity selector device can also be used; neutron velocities are inversely proportional to wavelength (Comment G1.2), and they are sufficiently slow for a monochromatic beam to be selected by slits in a spinning drum rotating in the 10^3 rpm range. The sample area may contain a *goniometer* (from the Greek *gonia*, angle or corner) to align the sample precisely in the beam, and sample environment controls (temperature, pressure, humidity . . .).

X-ray and neutron detectors transform the energy of the scattered radiation into a signal, which can be processed computationally. They must cope with high data collection rates and accurately and rapidly record the intensity and angular position of the scattering. The oldest type of X-ray detector is photographic film. This detection process is cumbersome, however, as the film can only be used once and must subsequently be developed and scanned into a computer. Modern detectors are based on the measurement of electric currents produced directly or indirectly by the radiation in gas or solid supports.

Heavy gases, like xenon, are ionised directly by the electromagnetic properties of the X-ray photon. X-rays can also excite certain heavy metal ions to fluoresce in the visible spectral range. An image intensifier then converts the visible light into an electronic signal. *Image plate* X-ray detectors have been built based on the properties of europium ions. Eu^{2+} is excited by X-rays to Eu^{3+}, which emits violet light ($\lambda = 390$ nm) when illuminated by a red He–Ne laser ($\lambda = 693$ nm). A photomultiplier is used to detect the violet light and the plate is then 'erased and reset' with white light.

Comment G1.14   Nuclear reaction for neutron detection

$n(^3He, p)^3H$	0.77 MeV
$n(^6Li, \alpha)^3H$	4.79 MeV
$n(^{10}B, \alpha)^7Li + 2.3\ MeV + \gamma\,(0.48\ MeV)$	93%
$n(^{10}B, \alpha)^7Li + 2.79\ MeV$	7%
$n(^{157}Gd, Gd)e^-$	0.182 MeV

In this notation, which is used in nuclear physics, the first line, for example, is equivalent to

$$n + {}^3He \rightarrow p + {}^3H + 0.77\,MeV$$

Since the neutron does not carry an electric charge, its detection is based on nuclear reactions that emit either ionising radiation or charged particles (Comment G1.14). The simplest neutron detector is the *ionisation chamber*, in which a sensitive gas, usually ^3He or $^{10}BF_3$, is contained between two charged plates. Position-sensitive multiwire detectors are based on the same detection principle. Scintillation detectors are based on the light emission of ^6Li when it absorbs a neutron. Image plate detectors have also been developed for neutrons. The matrix contains Gd_2O_3, which emits an electron of sufficient energy to excite the Eu ions (as in the X-ray case) when a neutron is absorbed.

G1.9 Checklist of key ideas

- In a diffraction experiment, waves of radiation scattered by different objects interfere to give rise to an observable pattern from which the relative arrangement (or structure) of the objects can be deduced.
- Three types of radiation are used in diffraction experiments on biological macromolecules: X-rays of wavelength about 1 Å, electrons of wavelength about 0.01 Å, and neutrons of wavelength about 0.5–10 Å
- X-rays are electromagnetic radiation of energy ~ 10 keV, much higher than that of visible light; neutrons are particles of mass similar to that of a proton, with associated wavelengths ~ 1 Å for velocities ~ 4 km /s^{-1} and energies ~ 10 meV.
- Neutrons are scattered by atomic nuclei, which behave as points; X-rays are scattered by atomic electron clouds, which gives rise to a *form factor*, because their spatial extent is similar to the wavelength of the radiation; the X-ray scattering power of an atom increases with its electron number.
- Neutrons are sensitive to different isotopes, and the different scattering powers of hydrogen and deuterium have been especially useful in biophysical studies.

- X-ray absorption leads to non-negligible radiation damage in the sample. It increases significantly with increasing wavelength; neutron absorption is very low for most nuclei, even for wavelengths of 10 Å or more.
- The scattering *amplitude* (or *scattering length*), b, of a point atom is defined as the amplitude of the scattered wave observed per unit *incident flux*, in unit solid angle; the scattering *cross-section* σ is the ratio of the total intensity scattered by the atom in unit time per unit incident flux; σ has units of area and can be understood as an effective projected area perpendicular to the incident beam; there is an analogous definition for the absorption cross-section.
- Scattering is described as a function of the *scattering vector*, a parameter that expresses the angular dependence for a given incident wavelength; in a reciprocal relation, the intensity observed at a given scattering vector magnitude Q contains information on atomic spacings down to a minimum distance equal to $2\pi/Q$.
- The *resolution* of a diffraction experiment is the minimum distance separating points that can be observed separately; it is equal to $2\pi/Q_{max}$.
- *Coherent* scattering is defined as the case in which waves scattered by a set of atoms interfere to give a single wave in a given direction, resulting from the sum of individual wave amplitudes with appropriate phases; if the scattering is *incoherent*, each atom scatters independently of the others, and the resultant intensity is the sum of individually scattered intensities.
- Incoherent scattering does not contain structural information and contributes to the background in diffraction experiments; Compton scattering is an important source of incoherent X-ray scattering; incoherent neutron scattering arises mainly from *spin incoherence*.
- *Elastic* scattering is when the scattered beam has the same energy as the incident beam. In an *inelastic* scattering process, the sample either loses energy to the radiation or gains energy from it.
- The scattering amplitude as a function of the scattering vector (in *reciprocal space*) is calculated from the sum of waves coherently scattered by the atoms in a particle; it is related by Fourier transformation to the particle structure (as a function of the *real-space* vector).
- There is a *phase problem* in X-ray and neutron diffraction, because only the intensity and not the phase of the scattered waves can be measured.
- Atomic scattering cross-sections for X-rays and neutrons are very small, and, even with the most intense sources currently available, a very large number of macromolecular particles are required in order to observe an interpretable signal. The particles can be organised in two extreme ways: with perfect disorder as in a dilute solution or with perfect order as in crystal; the level of resolution obtained from a diffraction experiment depends on how well ordered the particles are with respect to each other in the sample.
- A crystal of particles can be represented as the *convolution* of a lattice function and the particle structure.

- At low resolution, atoms can be grouped to define the *scattering density* in a volume; the *contrast* of part of a particle with respect to another is proportional to the difference between their scattering densities.
- X-rays and neutrons have different properties that make them strongly complementary in biophysical studies; the very high intensity and narrow divergence of X-ray beams currently available from synchrotron sources make them the radiation of choice in structural biology. The different scattering amplitudes of hydrogen and deuterium have allowed the analysis of specific problems by neutron scattering that were difficult or impossible to address by using X-rays. Hydrogen and water molecules play essential roles in biological structures, and they can be 'seen' in neutron crystallography even when not highly ordered. Neutron specific methods based on hydrogen–deuterium labelling and contrast variation have provided essential information on the organisation of protein–protein, protein–nucleic acid complexes and membranes. Neutrons also provide a uniquely suited radiation for the study of molecular dynamics, because the energy associated with a wavelength of ~ 1 Å is of the order of thermal energy, thus allowing the simultaneous measurement of the amplitudes and frequencies of motions in a sample.

Chapter G2
Small-angle scattering

G2.1 Theory of small-angle scattering from particles in solution

Small-angle scattering (SAS) is a very useful method in biochemistry, providing information on molecular masses, shapes and interactions in solution (see Comment G2.1).

G2.1.1 Dilute solutions of identical particles

In order to have an observable signal in X-ray or neutron SAS a solution of the order of 100 μl containing a few milligrams per millilitre of macromolecule is required – corresponding to about 10^{15} particles for a 50 kDa protein at 1 mg ml^{-1} (we note that 1 mg ml^{-1}, the usual unit in biochemistry is in fact equal to 1 g l^{-1}, the unit more conveniently used in equations).

We first consider solution conditions such that the particles do not influence each other, i.e. the position and orientation of each particle is totally independent of that of the others. This is the *infinite dilution* condition, for which we say there is no *interparticle interference*. In practice, it is achieved at different, low, concentrations for different macromolecules and solvents. For example, the condition may well be satisfied for a given protein at a few milligrams per millilitre, in a neutral pH buffer, but tRNA molecules, which are highly charged at pH 7 in low-salt buffer, interact with each other even at these concentrations, and it might not be possible to reach the infinite dilution condition without increasing the solvent salt concentration. In order to make sure that the infinite dilution condition is fulfilled, data should be collected as a function of concentration and extrapolated to zero concentration.

The infinite dilution condition implies that the chance of a wave being scattered by two different particles is practically nil; interference between waves scattered by atoms in different particles can be neglected and does not influence the scattering pattern.

Comment G2.1 Biologist's box: SAS without equations

A good understanding of SAS analysis involves mathematical approaches that may be difficult for the biologist. SAS is, nevertheless, a powerful technique in biochemistry. It is rapid and easy to apply on suitable samples and allows a broad characterisation of macromolecular structures in solution and their interactions. Below, we summarise in non-mathematical terms the information that can be obtained from a SAS experiment, and the practical requirements for setting it up.

Questions addressed:

(1) What is the effective association state (is it a monomer, a dimer . . .) of a protein or other macromolecule in solution or within a membrane environment?

(2) What is its shape or conformation (is it compact and globular or ellipsoidal, long and narrow, or flat and broad, star-shaped or branched . . . is its structure in solution similar to its crystal structure . . .)?

(3) Do different macromolecules in solution interact to form a complex or not?

(4) How do the answers to points 1, 2, 3, vary as a function of solvent conditions (pH, salt, ligand, temperature . . .)? Are there modifications in association state or conformational changes?

(5) The method of *contrast variation* in small angle neutron scattering (SANS) allows one to render *visible* only one component within a complex structure. It is then possible to address questions 1–4 for individual components within, for example, a macromolecular machine made up of various proteins, a protein nucleic acid complex interaction, or a membrane protein in a lipid or detergent environment.

An important point is that the answers to the different questions above are obtained independently of each other. For example, SAS will inform correctly on a monomeric particle (question 1) that is very elongated (question 2), whereas the conclusion of a gel filtration experiment (which essentially measures a diffusion coefficient), for example, might be that because it has a long dimension the particle is a multimer as described by Klein *et al.* (1982).

Practical requirements for a SAS experiment are:

(1) On the order of 100 μl of solution containing a few milligrams per millilitre of macromolecule.

(2) An accurate measurement of the macromolecular concentration (in milligrams per millilitre).

(3) Access to an X-ray or neutron small-angle camera.

SAS applications are illustrated in Section G2.6 by studies of protein–nucleic acid interactions of conformational changes during the working cycle of a chaperone macromolecular machine and a study of the association state of a membrane protein in lipid vesicles.

Fig. G2.1 Schematic diagram of real space structures (left) and corresponding wave intensities in reciprocal space (right): (a) a solution of non-interacting macromolecules described by scattering amplitude density distributions $\rho_1(\mathbf{r})$, $\rho_2(\mathbf{r})$, $\rho_3(\mathbf{r})$, etc.; (b) a solution of identical non-interacting macromolecules.

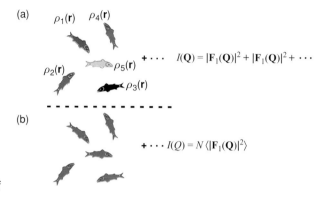

We recall the expression for a wave scattered from a single particle as a function of scattering vector, $\mathbf{Q}$ (Eq. (G1.10)):

$$F(\mathbf{Q}) = \sum_j f_j \exp(i\mathbf{Q} \cdot \mathbf{r}_j)$$

where f_j, $\mathbf{r}_j$ are the scattering amplitude and position, respectively, of atom j. The intensity scattered by the particle is the square of the wave amplitude

$$I(\mathbf{Q}) = |\mathbf{F}(\mathbf{Q})|^2$$

If there is no interparticle interference, the scattered intensity from unit volume of solution containing N particles is given simply by summing the intensities of waves scattered by different particles (Fig. G2.1)

$$I_N(\mathbf{Q}) = \sum_{n=1}^{N} |\mathbf{F}_n(\mathbf{Q})|^2 \qquad (G2.1)$$

In the identical particle (*monodisperse)* case, the particles still differ from each other with respect to their scattering properties because they take up different orientations, and we write (Fig. G2.1(b))

$$I_N(Q) = N \left\langle |\mathbf{F}(\mathbf{Q})|^2 \right\rangle \qquad (G2.2)$$

where the $\langle \cdots \rangle$ brackets denote rotational averaging in space (the different particles take on all different orientations) and time (each particle takes on different orientations).

Note that, apart from the number N, this is equivalent to assuming that one particle moves during the experiment and takes up all orientations. Particle motion can be described fully by a translation and a rotation. Translation is equivalent to a change of origin and does not change the amplitude calculation, which depends on the positions of the atoms relative to each other and not relative to the origin. In practice, therefore, Eq. (G2.1) accounts for the N particles in the solution and for the fact that each particle is moving. Note that due to rotational averaging Q is not a vector in $I_N(Q)$.

The rotational average in Eq. (G2.1) was calculated by Debye in 1915, and is known as the Debye formula (Comment G2.2):

$$\langle |\mathbf{F}(\mathbf{Q})|^2 \rangle = \sum_j \sum_k (f_j - \rho^\circ v_j)(f_k - \rho^\circ v_k)\frac{\sin Qr_{jk}}{Qr_{jk}} \tag{G2.3}$$

where r_{jk} is the distance between atoms j and k, and $(f_j - \rho^\circ v_j)$ are contrast amplitudes when the particle is in a solvent of scattering density ρ° (see Chapter G1).

The scattered intensity from the solution in conditions of infinite dilution and identical particles (Comment G2.3) is given by

$$I_N(Q) = N \sum_j \sum_k m_j m_k \frac{\sin Qr_{jk}}{Qr_{jk}} \tag{G2.4}$$

where we have written $m_j = (f_j - \rho^\circ v_j)$.

If we write $m_j = (\rho(\mathbf{r}_j) - \rho^\circ)dv_j$, where $\rho(\mathbf{r}_j)$ is the local scattering density of the volume dv_j at position $\mathbf{r}_j$ in the particle, we obtain the Debye equation in its integral form:

$$I_N(Q) = N \iint (\rho(\mathbf{r}_j) - \rho^\circ)(\rho(\mathbf{r}_k) - \rho^\circ)\frac{\sin Qr_{jk}}{Qr_{jk}}dv_j dv_k \tag{G2.4a}$$

The double integration is over the volume of the particle. Note that the volume of the particle should include the hydration layer or part of the solvent that is 'disturbed' by the presence of the particle (see Chapter A3). *From the scattering point of view, the particle volume is defined all the way to the boundary beyond which the solvent behaves as free bulk solvent.*

It is useful to normalise $I(Q)$ by the particle concentration C in grams per litre:

$$\frac{I_N(Q)}{C} = \frac{N_A}{M} \sum_j \sum_k m_j m_k \frac{\sin Qr_{jk}}{Qr_{jk}} \tag{G2.5}$$

where N_A (mole^{-1}) is Avogadro's number and M is the particle molar mass (grams per mole).

G2.1.2 The scattering curve at small Q values. The Guinier approximation, the forward scattered intensity and radius of gyration

Expanding $(\sin Qr_{jk})/Qr_{jk}$ in the Debye equation in terms of a series in powers of Qr_{jk} (Eq. (G2.6)), we obtain (for a single particle)

$$I(Q) = \sum_{jk} \frac{m_j m_k}{Qr_{jk}}\left(Qr_{jk} - \frac{Q^3 r_{jk}^3}{3!} + \cdots\right) \tag{G2.6}$$

where we can write

$$I(Q) = \left(\sum_j m_j\right)^2 \left(1 - \frac{1}{3}R_G^2 Q^2 + \cdots\right)$$

Comment G2.2
The Debye formula

In order to calculate the rotational average over $|\mathbf{F}(\mathbf{Q})|^2$ Debye used the classic result

$$\langle \exp(-i\mathbf{Q} \cdot \mathbf{r})\rangle = \frac{\sin Qr}{Qr}$$

Comment G2.3
The two assumptions required for the interpretation of the scattering curve in terms of a single particle structure

1 The solution is monodisperse. All the particles in the solution are identical.

2 The solution is infinitely dilute. There is no correlation between the positions or orientations of the particles in the solution.

and

$$I(0) = \left(\sum_{jk} m_j m_k\right) = \left(\sum_j m_j\right)^2 \text{ and } R_G^2 = \frac{1}{2} \sum m_j m_k r_{jk}^2 \bigg/ \left(\sum_j m_j\right)^2 \quad \text{(G2.7)}$$

Note that in a mechanical analogy R_G is the radius of gyration of the m_j distribution. It can also be written

$$R_G^2 = \sum m_j r_j^2 \bigg/ \sum m_j \quad \text{(G2.8)}$$

where r_j^2 is measured from the centre of mass of the m_j distribution.

Comment G2.4   Contrast amplitudes and Archimedes' principle

In general, $m_j = (f_j - \rho^\circ\, v_j)$ values may be positive or negative, which may lead to a negative value of the radius of gyration squared (R_G^2) through Eq. (G2.10). There is a mechanical analogy to this if we consider a particle immersed in a fluid environment. The effective mass of its components contains the buoyancy term from Archimedes' principle

$$m_{\text{eff}} = (m_j - \rho^\circ v_j)$$

where the m_j terms now correspond to inertial masses and ρ° to solvent mass density. A negative value of effective mass is obtained, for example, in the case of cork in water, which does not sink but rises under the influence of the buoyancy term.

It is interesting to note that the centre of mass of a particle made up of positive and negative masses may lie *outside* the particle. Consider the example in one dimension below:

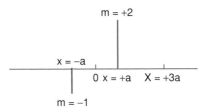

The centre of mass of a positive mass of 2 at a and a negative mass of -1 at $-a$ is calculated to be *outside the particle* at $X = 3$. From

$$\sum m_j (x_j - X) = 0$$

the radius of gyration squared of the particle is

$$R_G^2 = \frac{\sum m_j (x - X)^2}{\sum m_j}$$
$$= \frac{2(4a^2) + (-1)(16a^2)}{2 - 1} = -8a^2$$

It is a negative value resulting from a perfectly plausible physical structure such as, for example, a stick under water with a lead ball at one end and a cork at the other, or a particle with parts of positive and negative contrast amplitude.

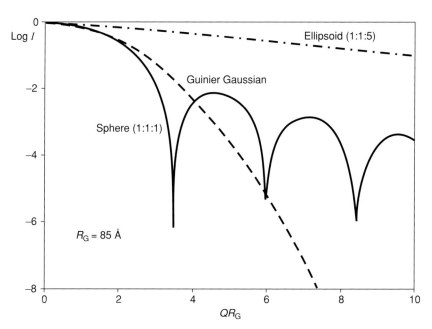

Fig. G2.2 Scattering
curves of a sphere and an
ellipsoid of revolution as
functions of QR_G; the
curve of the Guinier
Gaussian is also shown.
R_G is the radius of
radiation in each case.
Note how the curves
coincide with the Guinier
approximation at small
QR_G.

Note that the radius of gyration always appears as its square (R_G^2) in the equations. This is important. Since the contrast amplitudes, m_j, may be positive or negative, we may obtain physically meaningful negative R_G^2 values (Comment G2.4).

In the 1930s, Guinier reported that a Gaussian function was a much better approximation of the Debye formula at low Q values than just truncating at the Q^2 terms of the series in Eq. (G2.7):

$$I(Q) = I(0)\exp[-(1/3)R_G^2 Q^2]$$
$$\ln I(Q) = \ln I(0) - (1/3)R_G^2 Q^2$$

(G2.9)

Equation (G2.9) is known as the Guinier approximation. It is generally valid for the scattering curve of a particle of any shape, provided QR_G is smaller than or equal to about 1.

Scattering curves for a sphere and a long ellipsoid of axial ratio 1:1:5 are shown in Fig. G2.2. Also shown is the Gaussian function corresponding to the Guinier approximation. In order to observe the influence of particle shape, independently of absolute size, the curves are drawn as a function of QR_G.

The Guinier approximation is valid to higher or lower QR_G values depending on the shape of the particle. For ellipsoids of axial ratios about 1:1:1.7 or 1:1:0.6, the approximation is good significantly beyond the QR_G value of 1. For more compact shapes (e.g. the sphere, which is the most *compact shape*), the scattering curve deviates *below* the Guinier approximation, while for more asymmetric shapes (such as prolate or oblate ellipsoids with axial ratios 1:1:>1.7 or 1:1:<0.6,

Fig. G2.3 A straight line fit: ln $I(Q)$ versus Q^2 in the Guinier approximation. The example shows the small-angle region of neutron scattering data collected by one of the authors on a complex of the protein, seryl-tRNA synthetase, with seryl-tRNA.

The fit is for 0.012 Å^{-1} < Q < 0.043 Å^{-1}. The radius of gyration found from the slope was 28.7 ± 0.2 Å^{-1}, so that the fit corresponded to a range 0.33 < QR_G < 1.23. The $I(0)$ value calculated from the intercept of the straight line, together with the protein and tRNA concentrations, corresponded to the molar mass expected for a one-to-one complex.

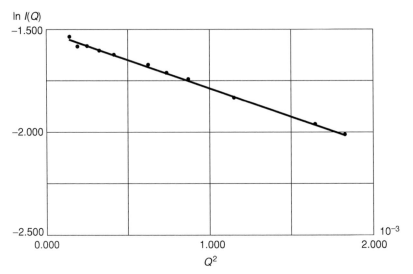

respectively, e.g. the 1:1:5 plot in the figure) the scattering curve deviates *above* the Guinier approximation.

During an experiment it is useful to plot the scattering curve as ln $I(Q)$ versus Q^2 (Fig. G2.3). A straight line can then be fitted to the region where the Guinier approximation is valid. Two *experimental parameters* are obtained from the straight line, its intercept at $Q = 0$ (by extrapolation) and its slope, from which we can calculate (Eq. (G2.9))

$$I(0) = \exp(\text{intercept})$$

and

$$R_G^2 = 3 \times \text{slope}$$

When the assumptions outlined in Comment G2.3 are satisfied (i.e. the solution is effectively dilute and monodisperse), the scattering amplitude contrast of a particle, $\sum_j m_j$, and its radius of gyration can be calculated from the intercept and slope, respectively, of the Guinier approximation fit to the scattering curve and the particle mass concentration. The relation between scattering amplitude contrast and molar mass is easily established from the chemical compositions of the particle and solvent (a calculation is given as an example in Comment G2.5).

The method allows one to determine independently and simultaneously the molar mass of the particle in solution (through the $I(0)$ value and concentration, in order to establish, for example, its oligomeric state in given solvent conditions), and its spatial extent (through the R_G^2 value). This is not trivial information. Analytical centrifugation determines either molar mass (in a sedimentation equilibrium experiment) or a parameter related to molar mass and the translational diffusion coefficient (the sedimentation value in a sedimentation velocity experiment) (see Part D). Gel filtration methods determine molar mass

> **Comment G2.5 Molecular mass calculation**
>
> For a particle of known composition we calculate its contrast amplitude per unit molar mass
>
> $$\frac{1}{M} \sum_j m_j$$
>
> where $\sum_j m_j = \sum_j f_j - \rho^\circ V$ and f_j are atomic scattering amplitudes, ρ° is the solvent scattering density and V is the particle volume. $V[\text{cm}^3]$ can be calculated from $V = M\bar{v}/N_A$ where M is in grams per mole and $\bar{v}$ is the partial specific volume in cubic centimetres per gram. It is usually a good approximation to assume a value between 0.73 and 0.75 for the partial specific volume of any protein (see Chapter A1).
>
> Now M is not necessarily the molar mass of the particle in the solution. For example, if the particle is a monodisperse dimer, its molar mass in solution is $2M$. We, therefore, write M' for the mass in solution. Combining Eqs. (G2.5) and (G2.7) and substituting we obtain
>
> $$\frac{I_N(0)}{C} = \frac{N_A}{M'} \left(M' \frac{\sum_j m_j}{M} \right)^2$$
>
> $$= N_A M' \left(\frac{\sum_j m_j}{M} \right)^2$$
>
> so that for a monodisperse solution of non-interacting particles of known composition and partial specific volume the forward scattered intensity divided by the concentration is proportional to the molar mass.
>
> Note that both $I_N(0)$ and C should be known in absolute units (see Comment A1.2). In the case of SANS absolute calibration by using the scattering of water has been shown to be effective (Jacrot and Zaccai, 1981).

indirectly via the measurement of what is essentially a diffusion coefficient and calibration with known particles (see Part D). If the shape of the calibrating particles is different from that of the unknown particle then the measured molar mass value will be incorrect.

G2.1.3 Asymptotic behaviour of the scattering curve at large Q values. The Porod relation

In the early 1950s, Porod introduced a particle characteristic function from which the asymptotic behaviour of the scattering curve at large Q values can be calculated (Section G2.1.4) . The condition to be fulfilled is that the particle has a well-defined envelope beyond which there is homogeneous solvent. Then, at high Q values, $I(Q)$ oscillates around an asymptotic line corresponding to

$$\lim_{Q \to \infty} \frac{I(Q)Q^4}{2\pi I(0)} = S \tag{G2.10}$$

S is the external surface area of the particle envelope. At high Q values the scattering curve decreases sharply, as Q^{-4}. Formally, a precise measurement of $I(Q)$ in this scattering vector range allows a determination of S. In practice, however, Eq. (G2.10) is used in order to adjust the experimental background level (e.g. due to scattering from the solvent alone) so that after background subtraction from the solution scattering, $I(Q)$ follows a Q^{-4} dependence. This is because the particle scattering intensity is very weak at large Q values (2–3 orders of magnitude weaker than $I(0)$ and very sensitive to the experimental error in the background measurement).

G2.1.4 The full scattering curve. The distance distribution function

The scattering curve is measured in reciprocal space as a function of Q. What information does it contain on the particle structure in real space? We recall from Chapters G1 and A3 that reciprocal and real space are related mathematically by Fourier transformation.

In the SAS case, in which $I(Q)$ results from a rotational average over $|\mathbf{F}(\mathbf{Q})|^2$, the inverse Fourier transform of $I(Q)$ is a function that depends on the distance r between parts of the structure:

$$V\gamma(r) = \frac{1}{2\pi^2} \int_0^\infty Q^2 I(Q) \frac{\sin Qr}{Qr} \, \mathrm{d}Q \qquad (G2.11)$$

This is similar to the Patterson function in crystallography, which is the inverse Fourier transform of *diffracted intensity*. In the equation, $I(Q)$ refers to a single particle of volume V (note that we are still within the assumptions of Comment G2.3). The function $\gamma(r)$ is a *correlation function*, defined as the average of the product of two scattering density contrast values separated by a distance r:

$$\gamma(r) = \langle (\rho(\mathbf{r}_j) - \rho^\circ)(\rho(\mathbf{r}_k) - \rho^\circ) \rangle \qquad (G2.12)$$

where $r = |\mathbf{r}_j - \mathbf{r}_k|$.

The function $V\gamma(r)$ is equivalent to a radial Patterson function. The integral from 0 to ∞ in Eq. (G2.11) is in fact equal to an integral from 0 to D_{max}, the *maximum distance* in the particle, since the scattering density contrast is zero for r greater than that value.

In the case of a particle of homogeneous scattering density ($\rho(\mathbf{r}) = \rho$), we can write

$$\gamma(r) = (\rho - \rho^\circ)^2 \gamma_0(r) \qquad (G2.13)$$

where $\gamma_0(r)$ is known as Porod's characteristic function, which depends only on the particle geometry.

The same product of scattering density contrast values appears in Eq. (G2.12) and in the integral form of the Debye equation (Eq. (G2.4a)), which can now be written (as the Fourier transform of $V\gamma(\mathbf{r})$):

$$I(Q) = V \int 4\pi r^2 \gamma(r) \frac{\sin Qr}{Qr} \, dr \qquad \text{(G2.14)}$$

or

$$I(Q) = \int p(r) \frac{\sin Qr}{Qr} \, dr \qquad \text{(G2.15)}$$

Comment G2.6 The distance distribution function

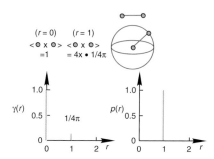

A simple particle of unit volume is made up of two small volumes of unit scattering density contrast ($\rho = 1$) separated by unit distance ($r = 1$). The number of points a distance r away from each volume is proportional to the spherical surface area $4\pi r^2$. Applying Eq. (G2.12) to derive $\gamma(r)$, the mean value of the product of ρ values separated by the distance 1 in the example is equal to $1/4\pi$. The corresponding $p(r)$ is also shown (Eq. (G2.16)). Note that $p(r)$ is in fact a *distance distribution function*. It tells us that there is one distance in the structure of value 1.

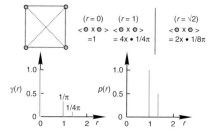

This figure illustrates the same calculations as above for a particle made up of four unit scattering density contrast volumes arranged to form a square of unit side. The pair distribution function now tells us the particle contains four distances of one (the sides of the square) and two distances of $\sqrt{2}$ (the diagonals).

where

$$p(r) = 4\pi r^2 V \gamma(r) \tag{G2.16}$$

The function $p(r)$ is called the *pair-distance distribution function* of the particle. It represents the number distribution of 'distances' joining volume element pairs in the particle, weighted by the product of their scattering density contrast (Comment G2.6).

The pair-distance distribution function is the inverse Fourier transform of $I(Q)$ (combining Eqs. (G2.11) and (G2.16))

$$p(r) = \frac{2}{\pi} r^2 \int\limits_{0}^{\infty} Q^2 I(Q) \frac{\sin Qr}{Qr} dQ \tag{G2.17}$$

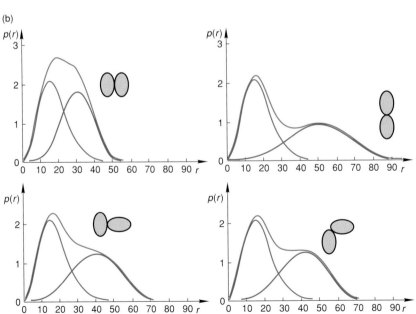

Fig. G2.4 (a) Scattering intensity and pair-distance distribution functions of a sphere (black), an oblate (flat) ellipsoid (green) and a prolate (elongated) ellipsoid (blue) of the same volume.
(b) Pair-distance distribution functions of dimers. The blue line is the monomer function, the red line, the dimer and the green line, the difference between them (from O. Glatter).

Provided, therefore, that the scattering curve can be integrated from zero to infinity, it is possible to calculate the $p(r)$ function of the particle.

The measurement of $I(Q)$ to $Q = 0$ is, of course, not possible, and it is also impossible to measure the scattering curve to infinitely large Q. The evaluation of the integral can be performed if we can fill in the missing parts of the curve by extrapolation in the context of the single-particle assumptions (Comment G2.3): for example, the curve can be extrapolated to $Q = 0$, by using the Guinier approximation, and to $Q \to \infty$ at the high-Q end by using the Porod relation.

Pair-distance distribution functions for various particle shapes are drawn in Fig. G2.4.

G2.1.5 The information content in $p(r)$ and $I(Q)$ for a monodisperse solution of a particle with a well-defined envelope

The $p(r)$ (or $\gamma(r)$) and $I(Q)$ functions are different ways of expressing the same information. Ideally, each fully defines the other since they are related by Fourier transformation. Either function can be used, for example, to compare experimental results to a model calculation, although there are important practical reasons why one may be preferred instead of the other. Since $I(Q)$ is not known with infinite precision from $Q = 0-\infty$, there are inherent assumptions in the calculation of $p(r)$: for example, the assumption that the data correspond to a particle with a well-defined solvent boundary, so that the Porod relation concerning the Q^{-4} asymptotic dependence can be applied at high Q values. The information contained in $p(r)$ or $I(Q)$ sets constraints on the structure of the particle but, of course, does not unambiguously define this structure in terms of spatial coordinates. This can be understood intuitively. The structure of a particle (in real-space coordinates) can be calculated from the observation of waves (in reciprocal space) (see Chapter G1). Note that:

(1) coordinates in real space and reciprocal space are vectors (the particle has a fixed orientation);
(2) the observed waves are defined by their amplitudes and phases (by using a complex number, for example).

In a SAS experiment information has been lost concerning these two points. Because of rotational averaging, the measurement is made as a function of Q and not of the vector $\mathbf{Q}$, and, of course, the phase information is lost, since only the intensity of the wave is observed.

The information content in the scattering curve is obtained by analysing the properties of $p(r)$ and $I(Q)$. The integral of $p(r)$ over the particle is equal to the square of its scattering amplitude contrast, which we recall from the Debye equation (Eq. (G2.4)) is equal to the forward scattered intensity $I(0)$:

$$\int_0^{D_{max}} p(r)r^2 dr = \left| \int (\rho(\mathbf{r}) - \rho^0) dv \right|^2 = \left| \sum m_j \right|^2 = I(0) \tag{G2.18}$$

where D_{max} is the maximum distance (maximum value of r) in the particle.

The radius of gyration of scattering amplitude contrast is related to the second moment of the $p(r)$ distribution (Comment G2.7):

$$R_G^2 = \frac{\dfrac{1}{2} \displaystyle\int_0^{D_{max}} p(r)r^2 dr}{\displaystyle\int_0^{D_{max}} p(r) dr} \tag{G2.19}$$

The second moment of the $I(Q)$ distribution is the Porod invariant:

$$C = \int_0^\infty I(Q)Q^2 dQ \tag{G2.20}$$

Because of the integration to ∞, the Porod invariant is sensitive to the large Q tail of $I(Q)$, which usually has a large error.

The volume V and specific surface S/V of the particle (S is given in Eq. (G2.10)) are given by Eqs. (G2.21) and (G2.22), respectively:

$$V = \frac{2\pi^2 I(0)}{C} \tag{G2.21}$$

$$\frac{S}{V} = \lim_{Q \to \infty} \frac{I(Q)Q^4 \pi}{C} \tag{G2.22}$$

The scattering curve, therefore, is consistent with all particle structures that present the corresponding values of $(\Sigma m)^2$, R_G^2, D_{max}, V and S.

G2.1.6 Polydisperse solutions

Of the two assumptions discussed in Comment G2.3, we relax the assumption that the solution is monodisperse (the solution now contains particles that are not identical to each other), while maintaining the assumption about infinite dilution (there are no interactions between the particles).

Two experimentally useful cases can be distinguished. First, when the particles have similar scattering amplitude contrast radii of gyration, the scattering curve in the Guinier approximation is still fitted reasonably well by a straight line in a given Q range, where QR_G is smaller than or equal to about 1 for all particles (Fig. G2.5(a)). The measured $I(0)$ and R_G^2 values should now be considered as experimental parameters. They correspond to weighted averages over the scattering amplitude contrast values and radii of gyration of the particles in the

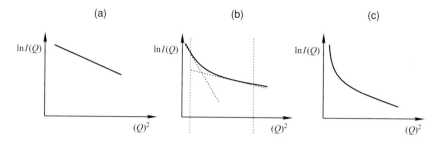

(a) (b) (c)

solution

$$I(0) = \sum_n N_n \left(\sum_j m_j \right)_n^2 \tag{G2.23}$$

$$R_G^2 = \frac{\sum_n^n N_n \left(\sum_j m_j \right)_n^2 R_{Gn}^2}{\sum_n N \left(\sum_j m_j \right)_n^2} \tag{G2.24}$$

where there are N_n particles of scattering amplitude $(\sum_j m_j)_n$ and radius gyration R_{Gn}. The two equations present the basis for a set of very useful SAS experiments to study complex formation and particle association in solution under different conditions (Comment G2.8).

We recall that the Guinier equation is a good approximation to a scattering curve at Q values for which QR_G 1. The second experimentally useful case is when the particles have very different radii of gyration, then the Guinier approximation is still satisfied for each, but in a different Q range. This gives a broken line in the logarithm of I versus Q^2 plot if the particles fall into two sets, one of large particles and one of small ones (as when there is some aggregation in the solution), or a curved line when the particle size varies more smoothly (Fig. G2.5(b), (c)). SAS experiments to study each set of particles separately could be envisaged in a reasonable approximation if the size distribution is such that the Q ranges for Guinier analysis are distinct as in Fig. G2.5(b). On the other hand, practically no information can be deduced from a curved scattering curve other than that the solution is polydisperse in a given size range. An effort should then be made in the biochemical part of the experiment to reduce the polydispersity.

G2.1.7 Interactions between the particles

We now maintain the first assumption of Comment G2.3 concerning monodispersity, but relax the second assumption; the solution is no longer considered as infinitely dilute and each particle feels the presence of the others.

The mean distance between particles in a macromolecular solution as a function of concentration is calculated in Comment G2.9. In vivo protein concentrations are so high that in vitro SAS experiments under similar conditions would

Fig. G2.5 (a) Guinier plot of a polydisperse solution of particles of about the same size. (b) Guinier plot of a solution containing small and large particles. The dotted black lines are straight-line fits to the two parts of the curve. The vertical red dotted line is where $QR_G = 1$ for the large particles. The vertical green dotted line is where $QR_G = 1$ for the smaller particles. In the example the slopes of the data lines are related by about a factor of 10 so that the radius of gyration of the smaller particle is $\sqrt{10} \approx 3$ times smaller than that of the larger particle. (c) Guinier plot of a solution containing particles of smoothly varying size.

**Comment G2.8
Aminoacyl tRNA synthetases and their interactions with tRNA**

An early SANS study of the interaction between methionine tRNA synthetase (MetRS) and tRNA^{Met} is a very good illustration of the rich potential of the application of SAS to an interacting system (Fig. G2.17, Dessen et al., 1978).

Comment G2.9   Molecular crowding

There are N particles in a litre of solution of macromolecules of molar mass M g mole^{-1} and concentration C mg ml^{-1}:

$$N = N_A C/M$$

where N_A is Avogadro's number.

We estimate a mean distance, d, between them, by assuming they are distributed on a grid in a cube of side 10 cm:

$$d = [(10 \times 10^8)/N]^{1/3} \text{Å}$$

As an example we calculate these values for haemoglobin in red blood cells. The protein concentration is about 300 g l^{-1}, a value that is similar to the concentration in cell cytoplasm, in general. Haemoglobin has a molar mass of about 64 kDa. Then

$$N = 3 \times 1021$$

and

$$d = 66 \text{Å}$$

If approximated by a sphere, haemoglobin has a radius of about 30 Å. There is less than 10 Å or two layers of water molecules between particles in the 300 g l^{-1} solution.

---

be very far from satisfying the infinite dilution conditions. The position and orientation of each particle in a concentrated solution is strongly influenced by the others, even when there are no long-range interactions between particles, simply because there is not enough room around it for it to translate or rotate freely. For particles with approximately spherical symmetry (so that all orientations are equivalent) the influence of high concentrations on the SAS scattering curve can be calculated by using a liquid-like potential function, the interparticle interaction and packing being assimilated to that of molecules in a liquid. The problem is much more difficult to solve for asymmetric particles, unless the asymmetry is extreme (very long or very flat particles) in which case they align to form fibres or stacks.

The potential function describing molecules in a liquid, the corresponding particle distribution and the scattering factor (wave amplitude in reciprocal space as a function of Q) for point-like molecules are shown in Fig. G2.6.

We can now predict the scattering curve of a set of particles with spherical symmetry that is organised according to a given interparticle potential by applying the convolution–multiplication relation between real and reciprocal space. In the case of a repulsive potential, the scattering factor rises from Q equal to zero whereas the form factor usually (but not always, see Section G2.3 on contrast

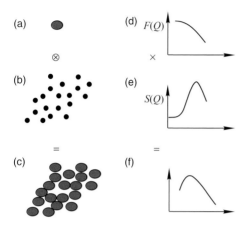

Fig. G2.6 (a) A
quasi-spherical particle
described by a scattering
density distribution.
(b) The distribution of
particle centres of mass
(points) in a liquid. (c) The
quasi-spherical particles
in a liquid described as a
convolution of (a) and (b).
(d) The 'form' factor or
scattered intensity of the
particle in reciprocal
space. (e) The 'structure'
factor of the liquid of
point-like molecules.
(f) The scattered intensity
of the quasi-spherical
liquid molecules resulting
from the multiplication of
(d) and (e). Recall that the
Fourier transform of
convolution is
multiplication (see
Chapter A3).

variation) decreases. The result is a flattening of the scattering curve in the Guinier approximation region, leading to apparent radius of gyration and $I(0)$ values that are smaller than the ones corresponding to the particle in dilute solution. The scattering factor tends to 1 for Q values that are not much larger than those in the Guinier approximation region, depending on the proximity of the particles, so that the effect of interparticle interference on the scattering curve becomes negligible, and high macromolecular concentrations (where available) can be used for these 'higher Q' measurements.

G2.2 Models and simulations

G2.2.1 From structure to scattering curve

The X-ray or SANS curve of a known particle structure can be calculated accurately by using the Debye formula in its sum or integral form (Eqs. (G2.4)).

Two point atoms and two spherical atoms

The two-atom model is of particular interest for the label triangulation method (see below), in which a complex particle is 'simplified' by contrast labelling. The scattering curve of two point atoms separated by a distance r_{12} has been calculated from the Debye formula and is given in Fig. G2.7. The scattering curve of two spherical atoms is also given in the figure, calculated by using the convolution multiplication Fourier transform relation (Chapter A3). In real space, the two-spherical-atom structure can be viewed as the convolution of two delta functions with a sphere. The scattering amplitude (in reciprocal space) of the structure is given by the product of the corresponding Fourier transforms.

Spheres and ellipsoids

As in hydrodynamics, where friction coefficients were initially calculated for simple geometrical shapes (see Part D), a first approach to modelling

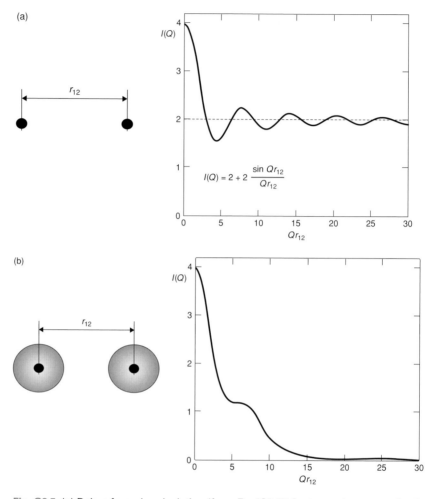

Fig. G2.7 (a) Debye formula calculation (from Eq. (G2.3)) for two point atoms of unit scattering amplitude (delta functions) separated by r_{12}:

$$I(Q) = \sum_j \sum_k \frac{\sin Qr_{jk}}{Qr_{jk}}$$

$$I(Q) = 1 + 1 + 2\frac{\sin Qr_{12}}{Qr_{12}}$$

(b) Scattering curve of two spherical atoms of radius $a = r_{12}/4$ separated by r_{12}.

$$I(Q) = \left(2 + 2\frac{\sin Qr_{12}}{Qr_{12}}\right)\left(3\frac{\sin Qa - Qa\cos Qa}{(Qa)^3}\right)^2$$

scattering curves is in terms of ellipsoids. Provided the comparison is limited to very low resolution ($QR_G \leq 3$), the scattering curve of any globular particle can be simulated successfully as an ellipsoid with a given axial ratio.

The spherical particle is, of course, the extreme case. It is of particular interest because it is unique in that its internal structure expressed as the scattering density distribution as a function of radius, $\rho(r)$, directly corresponds to the rotational average. For the same reason, a sphere is also the only shape whose scattering amplitude, $F(Q)$, can be calculated directly from the square root of the scattering intensity, $I(Q)$. The scattering amplitude corresponds to the Fourier transform of the spherical distribution, which can be expressed in an analytical form by using a result first calculated by Rayleigh in 1911. The example of a homogeneous sphere is illustrated in Fig. G2.8, but the Fourier transform approach to obtaining a radial scattering density can be applied to any spherical distribution, including hollow spheres, for example, and has been used succesfully to solve the internal structure of spherical viruses at low resolution (Comment G2.10).

In a range comprising larger QR_G values where the first subsidiary maximum appears, the information content is higher and more complex shapes are required to simulate the curve successfully (see, for example, Fig. G2.4). Tabulated values of $I(Q)$ are available for oblate and prolate ellipsoids of various axial ratios as well as for a variety of other geometrical solids (Comment G2.10).

Rods and sheets

SAS from two extreme types of asymmetric particle are of interest in structural biology, because of applications to DNA and filamentous complexes (such as certain viruses, for example), on the one hand, and membranes, on the other. These are particles where one dimension is much larger (by a factor of about 10) than the other two (rods) and particles where one dimension is much smaller (by a factor of about 10) than the other two (sheets). We can guess, intuitively, that fluctuations in the large dimensions affect the scattering curve in regions of small Q, and vice versa. When the ratio between the dimensions is large, the corresponding Q regions are well separated. We can apply this consideration to the Debye formula and derive expressions for the scattered intensity from very long or very flat particles that are equivalent to the Guinier approximation for globular particles.

In the case of a rod-like particle whose length $L \gg r$, where r is its cross-sectional dimension (Fig. G2.9(a)), we define

$$I_c(Q) = I(Q)Q/\pi L$$

and a cross-sectional radius of gyration of scattering amplitude R_c:

$$R_c^2 = \frac{\sum_i m_i r_i^2}{\sum_i m_i}$$

where r_i are measured from the centre of scattering mass of the cross-section. Then we have

$$I_c(Q) = \frac{\sum m_i}{L} \exp\left(-\frac{1}{2}R_c^2 Q^2\right) \tag{G2.25}$$

which is equivalent to the Guinier approximation and valid for $QR_c \leq 1$, and where L is the length corresponding to $M = \Sigma m_i$.

Comment G2.10 Spheres and ellipsoids

The special case of spherical particles for which (at low resolution) the spherical average structure is in fact equal to the structure has been exploited efficiently to solve the internal structure of spherical viruses by SANS and contrast variation (Jacrot *et al.*, 1977).

Analytical integrals have been calculated for the scattering functions of ellipsoids based on the fact that for each orientation an ellipsoid presents to the beam a projection equal to that of a sphere. The intensity is then the sum of intensities from spheres of different radii. The result is a smoothing out of the spherical subsidiary maxima.

Formulae and numerical tables for the scattering intensity functions of three-axis ellipsoids are given in Mittelbach and Porod (1962).

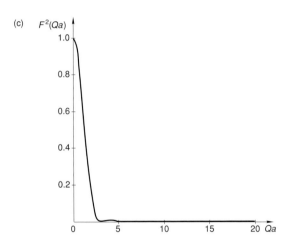

Fig. G2.8 (a) Scattering density contrast, $\rho(r)$; (b) scattering amplitude, $F(Qa)$; and (c) scattering intensity, $[F(Qa)]^2$, of a homogeneous sphere of radius a. The corresponding formulae are

(a)
$$\rho(r) = \begin{cases} 1 \text{ for } r \leq a \\ 0 \text{ for } r > a \end{cases}$$

(b)
$$F(Q) = \rho V \left(3\frac{\sin Qa - Qa\cos Qa}{(Qa)^3} \right)$$

(c)
$$I(Q) = F(Q)^2$$

where V is the volume of the sphere, chosen as equal to 1 in the figure.

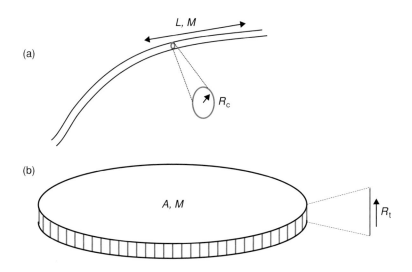

Fig. G2.9 (a) Rod-like particles of scattering mass per length M/L and cross-sectional radius of gyration R_c. (b) Sheet-like particles of scattering mass per area of M/A and transverse radius of gyration R_t.

For a sheet-like particle for which $A^{1/2} \gg t$, where A is the sheet area and t its thickness (Fig. G2.9(b)), we define

$$I_t(Q) = I(Q)Q^2/2\pi A$$

and a transverse radius of gyration of scattering amplitude R_t:

$$R_t^2 = \frac{\sum_i m_i t_i^2}{\sum_i m_i}$$

where t_i is measured from the centre of scattering mass of the thickness. Then we have

$$I_t(Q) = \frac{\sum m_i}{A} \exp(-R_t^2 Q^2) \tag{G2.26}$$

where $QR_t \leq q$ and A is the area corresponding to $M = M = \sum m_i$.

Atomic coordinates

A scattering curve can be calculated unambiguously from the atomic coordinates of a macromolecular structure. The calculation in vacuum is a straightforward application of the Debye formula, putting in the atomic coordinates and scattering amplitudes $\mathbf{r}_j, f_j$, respectively. In solution, however, a volume, v_j, has to be attributed to each atom in order to calculate the contrast amplitude of each atom, $m_j = f_j - \rho^\circ v_j$, and to account for the solvent contribution. Different programs are now available to perform such calculations. Note that because the neutron scattering density, ρ°, of H_2O is small and negative (see Fig. G2.10), the calculation is not very sensitive to the volume terms in $(m_j = f_j - \rho^\circ v_j)$ and the neutron scattering curve of a macromolecule in H_2O solvent may be treated to a good approximation as if the macromolecule were in vacuum.

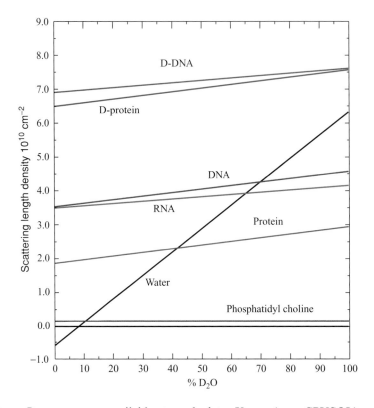

Programs are available to calculate X-ray (e.g. CRYSOL) or neutron
(e.g. CRYSON) scattering curves from crystal structure atomic coordinates
(Comment G2.11). The calculation of neutron scattering curves requires hydrogen atom positions, which are usually not included in atomic coordinate files and
have to be placed in the structure by using chemical arguments. These programs
evaluate the expression

$$I(Q)\langle|A_a(Q) - \rho^\circ A_s(Q) + \delta\rho_G A_b(Q)|^2\rangle \qquad (G2.27)$$

where $A_a(Q)$ is the scattering amplitude of the particle in vacuo calculated from
the atomic positions, $A_s(Q)$ is the scattering amplitude of the solvent-excluded
volume and $A_b(Q)$ is the scattering amplitude of the hydration layer around
the particle. The averaging is over all particle orientations. $A_a(Q)$ is calculated
from atomic form factors and positions. The amplitude $A_s(Q)$ is evaluated by
placing Gaussian spheres (dummy atoms) at the atomic positions and the particle
envelope is represented by connecting the centre of mass of the particle with
the most distant atom in each direction. The hydration shell is modelled with a
thickness and location corresponding to one layer of water (Section G2.3.1). The
excluded particle volume and scattering density in the hydration shell cannot be
obtained accurately from the crystal structure. If experimental data are available,

these two parameters can be varied to minimise the discrepancy between the observed and calculated scattering curves.

Comment G2.12
Spherical harmonics

The approach is discussed in Svergun, *et al.* (1996).

G2.2.2 From the scattering curve to a set of structures

The *inverse scattering problem* arises because it is not possible to reconstruct a unique structure from SAS data. Because of loss of phase information and rotational averaging, there is no 'inverse Debye formula', and, in principle, a very large number of structures may be compatible with the experimental scattering curve. A number of different methods, however, are being developed that successfully reduce the ambiguity and allow the calculation of sets of shapes and even low-resolution structures from measured scattering curves. The fit may be performed on the measured $I(Q)$ itself, or on its Fourier transform, the $p(r)$ function. The experimental invariants of the procedure, with which the model structure should be consistent, are $\sum m_i$, R_G^2, D_{max}, V and S (Section G2.1.5).

Spherical harmonics

Ab initio shape determinations from SAS data have been proposed that express the particle envelope in terms of spherical harmonics of progressively higher order (Comment G2.12). Non-linear minimisation techniques are applied to find the best fit between the calculated structure and observed scattering curve. The method works well with roughly globular shapes, which can be approximated by a low number of spherical harmonics but particles of more complex shape with a re-entrant surface, for example, cannot be modelled successfully by this method.

Monte Carlo and simulated annealing algorithms

Monte Carlo and simulated annealing methods have been proposed for *ab initio* low-resolution structure determinations from SAS data by several authors (Comment G2.13). They are based on more or less random searches of a predefined *configurational space* for structures that best fit the data, and require rapid calculations of scattering curves from large numbers of models and the definition of some goodness-of-fit parameter or 'energy' function to be minimised. In one approach, the starting structure is a sphere of diameter equal to the maximum particle dimension (calculated from the scattering curve, via the $p(r)$) filled with much smaller spherical *dummy* atoms. It is straightforward to calculate the scattering curve corresponding to the model and to compare it with the experimental curve. Further constraints on the model (expressed, for example, as 'penalties' in the 'energy' function) can be applied, for example, with respect to the compactness of the model or the continuity of the occupied volume with no disconnected parts. The method searches for a configuration of the dummy

---

Comment G2.13   *Monte Carlo* and *genetics* methods in SAS: references

Svergun, D. I. (1999). Restoring low resolution structure of biological macromolecules from solution scattering using simulated annealing. *Biophys. J.*, **76**, 2879–2886.

Walther, D., Cohen, F. E., and Doniach S. (2000). Reconstruction of low-resolution three dimensional density maps from one dimensional small-angle X-ray solution scattering data for biomolecules. *J. Appl. Crystallog.*, **33**, 350–363.

Svergun, D. I. (2000). Advanced solution scattering data analysis methods and their applications. *J. Appl. Crystallog.*, **33**, 530–534.

Svergun, D. I., Malfois, M., Koch, M. H. J., Wigneshweraraj, S. R., and Buck, M. (2000). Low-resolution structure of the sigma54 transcription factor revealed by X-ray solution scattering. *J. Biol. Chem.*, **275**, 4210–4214.

The program DAMMIN (Svergun, 2000; Svergun *et al.*, 2000), which takes a simulated annealing approach, also allows the input of information about the point symmetry of the particle.

Chacón, P., Moran, F., Díaz, J. F., Pantos, E., and Andreu, J. M. (1998). Low-resolution structures of proteins in solution retrieved from X-ray scattering with a genetic algorithm. *Biophys. J.*, **74**, 2760–2775.

Goldberg, D. E. (1989). *Genetics Algorithms in Search, Optimisation and Machine Learning*. San Mateo, CA: Addison-Wesley.

Chacón, P., Díaz, J. F., Morán, F., and Andreu, J. M. (2000). Reconstruction of protein form with X-ray solution scattering and a genetic algorithm. *J. Mol. Biol.*, **299**, 1289–1302.

The genetics approach is implemented in the program DALALGA (Chacón *et al.*, 2000).

---

atom model that minimises a function expressing the discrepancy between the calculated and experimental scattering curves.

Genetic algorithms

In an alternative approach, a *genetic algorithm* has been developed to execute the search for the 'best' models to fit an experimental scattering curve (Comment G2.13). Starting models are generated with spherical beads (whose radius is chosen to be small compared with the resolution of the experiment) placed randomly on a grid. The grid represents a restricted search space, which is reasonably compatible with the available information on maximum dimension, volume and radius of gyration. Typically the starting model set consists of a few hundred structures named 'chromosomes' obtained by filling the search space with different numbers of beads. After an 'evaluation and fitness' step the population is allowed to 'reproduce'. The part of the chromosome population that provides scattering curves that fit closest to the data is kept to become part of the next generation (usually corresponding to an arbitrary fraction of 50% of the total population), while the rest of the chromosomes are eliminated to make room for new ones. Applying two genetics operators to randomly chosen chromosomes in the 'best-fit' set generates a new population set. The 'cross-over' operator exchanges information in two parent chromosomes according to predefined rules to yield two offspring. The 'mutation' operator creates new chromosomes by copying members of the 'best-fit' set with a certain (small) error rate. The 'evaluation and fitness' step is then applied to the new population and the cycle is repeated until the 'maximum fitness' no longer improves and convergence is obtained.

G2.3 General contrast variation. Particles in different solvents 'seen' by X-rays and 'seen' by neutrons

The concept of *contrast* and its application to low-resolution diffraction were discussed in Chapter G1. Contrast can be modified, in general, by changing the atomic amplitudes, f_i, and/or solvent scattering density, ρ°. Atomic amplitudes are modified by isotope labelling (for neutron scattering only) or by changing the incident radiation (since amplitudes are different for X-rays and neutrons). The solvent density can be changed by isotope labelling (for neutron scattering only), by changing the incident radiation, or by changing the composition of the solvent (the electron density of a salt or sugar solution, for example, is significantly different from that of water). The fundamental assumption of a contrast variation experiment is that *the particle structure is not modified by changing the contrast conditions*. It is important, therefore, to verify that this is the case when isotope labelling or changes in solvent composition are applied. Isotope labelling for a neutron scattering experiment, for example, does not affect electron density contrast, so that an X-ray scattering curve provides a good control that the structure has not been affected by the labelling.

Mean neutron scattering amplitude densities of biological macromolecules and water as functions of solvent water isotopic content (heavy water, D_2O, percentage) are given in Fig. G2.10. Because atomic compositions are fairly similar within each macromolecular family, it is usually quite reliable to assume the appropriate mean value for a protein or nucleic acid of unknown composition.

For comparison, the electron density of H_2O or D_2O is $0.333e/\text{Å}^3$, corresponding to 9.3×10^{10} cm^{-2} in X-ray scattering amplitude density (see Chapter G1). The mean X-ray scattering amplitude density of protein is close to 12.3×10^{10} cm^{-2}, while that of nucleic acid is slightly higher. Contrast in any solvent condition is proportional to the difference between the macromolecule and solvent scattering densities. The macromolecule scattering density contrast depends on the radiation used, as well as on the solvent composition. Recall that the scattering particle includes the hydration shell, which may itself have a non-negligible contrast with respect to the solvent (see also Chapter A3). The same particle, therefore, presents different scattering curves in X-ray SAS in SANS from H_2O solution and in SANS from D_2O solution, to yield very useful information for solving complex structures.

G2.3.1 Two-component particles and the parallel axes theorem

The generalised contrast variation approach was applied to ribosomes in the early 1970s. The analysis developed is valid for any two-component particle in which the scattering densities of the components are sufficiently different and for a resolution at which each component can be considered as homogeneous (see Chapter G1). Ribosomes are nucleoprotein complexes made up of several proteins and large RNA molecules. The two-component assumption and resolution condition are fulfilled for Q values in the Guinier approximation; $Q \sim 1 / R_G$, where R_G is the radius of gyration of the particle.

Applying Guinier analysis to the two-component particle, the forward scattered intensity and radius of gyration squared for one particle can be written, respectively, as

$$I(0) = (m_1 + m_2)^2 \qquad \text{(G2.28)}$$

where $m_1 = f_1 - \rho^\circ V_1, m_2 = f_2 - \rho^\circ V_2$, and

$$R_G^2 = \frac{m_1 R_1^2 + m_2 R_2^2}{m_1 + m_2} \qquad \text{(G2.29)}$$

where the subscript 1 refers to the RNA component and the subscript 2 to the protein component, and R_1 and R_2 are the radii of gyration about the centre of scattering mass of the particle.

The square root of the scattered intensity should vary linearly with solvent scattering density; this is an important check on the experiment. For example, if sample polydispersity varies with the solvent, the square root of $I(0)$ calculated from the experimental data does not correspond to that of a single particle and a plot of it versus solvent density plot is not a straight line.

Formally, the radius of gyration in SAS follows the same rules as the radius of gyration in mechanics. The parallel axes theorem states that the square of the radius of gyration of a body around an axis is equal to the square of its radius of gyration about a parallel axis through the centre of mass of the body plus the square of the distance between the two axes (Fig. G2.11(a)).

Applying the theorem to an n-component body, we obtain

$$(m_1 + m_2 + \cdots m_n) R_G^2 = m_1 \left(R_{G1}^2 + D_1^2 \right) + m_2 \left(R_{G2}^2 + D_2^2 \right) + \cdots + m_n \left(R_{Gn}^2 + D_n^2 \right) \qquad \text{(G2.30)}$$

where R_{Gi} is the radius of gyration of compound i around its own centre of scattering mass and distances D_i are between the centre of scattering mass of each component and that of the complex. Combining Eq. (G2.30) and the equation defining the centre of mass of the complex ($\Sigma m(\mathbf{r} - \mathbf{r}_{\text{c of m}}) = 0$), we obtain, for a two-component particle

$$R_G^2 = \frac{R_{G1}^2 m_1}{(m_1 + m_2)} + \frac{R_{G2}^2 m_2}{(m_1 + m_2)} + \frac{L_{12}^2 m_1 m_2}{(m_1 + m_2)^2} \qquad \text{(G2.31)}$$

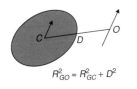

$$R_{GO}^2 = R_{GC}^2 + D^2$$

(a)

(b)

Fig. G2.11 (a) The parallel axes theorem. (b) The radius of gyration of a two-component body, with centres of mass at C_1, C_2, respectively, is given by Eq. (G2.31).

where L_{12} is the distance between the centres of mass of 1 and 2. Equation (G2.31) has the advantage of not requiring knowledge of the centre of mass coordinates (Fig. G2.11(b)).

By defining the respective *scattering fraction* for each component, respectively

$$x = \frac{m_1}{m_1 + m_2}, \qquad (1 - x) = \frac{m_2}{m_1 + m_2}$$

Eq. (G2.31) reduces to

$$R_G^2 = x R_{G1}^2 + (1 - x) R_{G2}^2 + x(1 - x) L_{12}^2 \qquad (G2.32)$$

The radius of gyration of each component within the complex can be calculated by extrapolation or interpolation from measurements in different solvent conditions (Fig. G2.12).

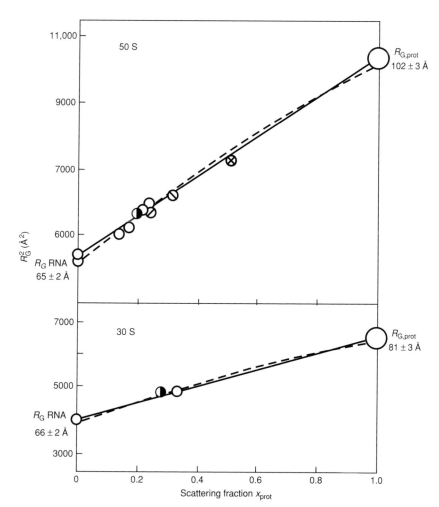

Fig. G2.12 Contrast variation radius of gyration plots of the 50 S and 30 S particles of the *E. coli* ribosome according to Eq. (G2.32). The particles are made up of two components: protein and RNA. The different data points are for measurements by X-ray, light, and neutron scattering in H_2O and D_2O buffers. The radius of gyration of the protein is obtained at the scattering fraction $x_{prot} = 1$; the radius of gyration of the RNA is obtained at $x_{prot} = 0$. (Serdyuk *et al.*, 1979.) (Figure reproduced with permission from Elsevier.)

At $x = 0$ (zero contrast for component 1) the radius of gyration of component 2 is measured, and vice versa at $x = 1$. The curve is parabolic, so that at least three data points that are well spread out in the solvent scattering density range (assuming small errors!) are required to determine the parameters R_{G1}^2, R_{G2}^2, and the distance between the centres of mass of the components 1 and 2. The figure shows the example of a ribosome study (Comment G2.14), in which data from SAS 'views' of the particles by X-rays, (visible) light and neutrons in H_2O and D_2O are plotted together to yield the radii of gyration of the RNA and protein components, respectively.

Contrast variation to see the hydration shell

SAS experiments resolved a controversy concerning the density of the hydration shell around proteins (see Chapter A3). X-ray, and neutron data from H_2O and D_2O were analysed together for a number of proteins, in the context of a two-component model made up in each case of the protein, which was modelled from its atomic coordinates, and the hydration shell (Fig. G2.13). As seen by X-rays, the protein component electron density is higher than that of a dense hydration shell, which is itself higher than that of the bulk solvent. For neutrons in H_2O, only the protein component is observed. The hydration shell scattering and bulk solvent scattering density are close to zero. For neutrons in D_2O, the hydration shell density is higher than that of the solvent (positive contrast), while that of the protein is lower (negative contrast).

The result plotted for lysozyme in Fig. G2.13 established the existence of a dense hydration shell. Recall that the scattering curve of a larger particle falls more steeply with Q. The particle seen by X-rays is largest, because both protein and hydration shell contribute positive contrast; next in size is the particle seen by neutrons in H_2O, for which the hydration shell is essentially invisible; the 'smallest' is the particle seen by neutrons in D_2O, for which the positive contrast of the hydration shell acts against the negative contrast of the protein part.

G2.3.2 The Stuhrmann analysis of contrast variation

The Stuhrmann analysis of contrast variation interprets the contrast dependence of the full scattering curve in terms of density fluctuations in the particle. A general set of equations for contrast variation has been derived by analysing the effect of changing solvent scattering density on the scattering curve of a particle of non-homogeneous scattering density.

The particle is described, in an integral notation, by a scattering density distribution $\rho(\mathbf{r})$, which is divided into two parts: the mean scattering density, ρ, and fluctuations about the mean, $\rho_F(\mathbf{r})$, such that (Fig. G2.14)

$$\int_V \rho_F(\mathbf{r})d\mathbf{r} = 0$$

(a)

Scattering length density, 10^{10} cm^{-2}

(1)
(2)
(3)
(4)

SAXS   SANS in H$_2$O   SANS in D$_2$O

(b)

ln I, relative

∇ (1)

$\bullet$ (2)

$\circ$ (3)

0.0   1.0   2.0   3.0

S (nm^{-1})

Fig. G2.13 (a) Scattering length densities of protein, hydration shell and solvent for X-rays (SAXS) and neutrons (SANS) in H$_2$O and D$_2$O: (1) solvent; (2) a hydration shell 20% denser than bulk solvent; (3) disordered protein side chains entering the hydration shell; (4) protein (the higher density of the protein in D$_2$O for neutrons is due to the exchange of labile hydrogen atoms). (b) Contrast variation scattering curves of lysozyme: (1) is from X-rays; (2) from neutrons in H$_2$O; and (3) from neutrons in D$_2$O. The curves have been normalised to the same scattering intensity at zero angle. The S on the x-axis corresponds to our parameter Q. Note that 1.0 nm^{-1} = 0.1 Å^{-1} (Svergun et al. 1998). (Figure reproduced with permission from the National Academy of Sciences, USA.)

Fig. G2.14 Mean contrast and density fluctuations in Stuhrmann analyis. (Koch and Stuhrmann, 1979.)

Then the contrast at $\mathbf{r}$ in the particle can be written

$$\rho(\mathbf{r}) - \rho^\circ = \rho + \rho_F(\mathbf{r}) - \rho^\circ = \bar{\rho} + \rho_F(\mathbf{r})$$

where $\bar{\rho}$ is the mean contrast density of the particle.

Putting this into the integral form of the Debye formula:

$$I(Q) = \iint (\bar{\rho} + \rho_F(\mathbf{r}_i))(\bar{\rho} + \rho_F(\mathbf{r}_j)) \frac{\sin Q\mathbf{r}_{ij}}{Q\mathbf{r}_{ij}} dV_i dV_j \qquad (G2.33)$$

we obtain

$$I(Q) = \bar{\rho}^2 I_V(Q) + \bar{\rho} I_{VF}(Q) + I_F(Q) \qquad (G2.34)$$

$I_V(Q)$, $I_{VF}(Q)$ and $I_F(Q)$ are the three characteristic functions of the particle:

> $I_V(Q)$ is the scattering curve at infinite contrast; it represents the scattering of a particle of the same shape as the one under study, but of homogeneous scattering density.
>
> $I_F(Q)$ is the scattering curve at zero contrast; it represents the scattering from density fluctuations inside the particle.
>
> $I_{VF}(Q)$ is a cross-term.

Note that the splitting of the intensity into the three characteristic functions can still be performed in the case of non-homogeneous exchange of particle labile hydrogen atoms in a SANS H_2O/D_2O contrast variation experiment. The interpretation of the functions, however, is not as straight forward as described above.

At small values of Q, the forward scattered intensity is obtained from the Guinier approximation, and is equal to

$$I(0) = (\bar{\rho}V)^2 = \{(\rho - \rho^\circ)V\}^2 \qquad (G2.35)$$

The square root of the forward scattered intensity is linear with solvent scattering density, ρ°, and it crosses zero at the particle *contrast match point*, where $\rho^\circ = \rho$ (Fig. G2.15(a)).

The contrast dependence of the radius of gyration is given by

$$R_G^2 = R_V^2 + \frac{1}{\bar{\rho}V}\int \rho_F(\mathbf{r})\mathbf{r}^2 dV + \frac{1}{\bar{\rho}^2 V^2}\left\{\int \rho_F(\mathbf{r})\mathbf{r}dV\right\}^2 \qquad (G2.36)$$

where we write

$$R_G^2 = R_V^2 + \alpha/\bar{\rho} + \beta/\bar{\rho}^2 \qquad (G2.37)$$

R_V^2 is the radius of gyration of a particle of the same shape as the particle under study but of homogeneous density. α is the second moment of the density fluctuations in the particle. A positive value of α denotes that the higher density is towards the periphery of the particle, and vice versa. $\beta/\bar{\rho}^2$ is the separation between the centre of scattering mass of the density fluctuations in the particle and the centre of mass of the particle shape, when the mean contrast conditions correspond to $\bar{\rho}$. Plots of R_G^2 versus $1/\bar{\rho}$ are shown in Fig. G2.15(b):

(a)

(b)

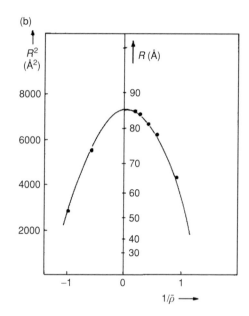

Fig. G2.15 The Stuhrman
analysis parameters:
(a) square root of the
forward scatter of 50 S
ribosomal subunits as a
function of D_2O volume
fraction in the solvent.
The three preparations
were derived from cells
grown on H_2O and
different proportions of
D_2O resulting in different
contrast match points
indicating levels of
deuteration;
(b) radius of gyration
squared as a function of
inverse contrast for 70 S
ribosomes reconstituted
from protonated 50 S and
deuterated 30 S
subunits – the strongly
parabolic shape indicates
the large separation
between the centres of
mass of the dense and
less dense scattering
regions (Koch and
Stuhrmann, 1979). (Figure
reproduced with
permission from Elsevier.)

α positive: higher scattering density in the particle lies further away from its centre
on average than lower scattering density;

α negative: lower scattering density in the particle lies further away from its centre
on average than higher scattering density;

$\beta = 0$, the centres of scattering mass and density fluctuations coincide.

Recalling that three structural parameters are required to describe a two-component particle (the two radii of gyration about their respective scattering mass centres and the distance separating them), it is evident from the above equations that, in a model approach, a maximum of two particle components can be 'resolved' in a contrast variation experiment from the three independent experimental parameters obtained from the scattering curve as a function of contrast at small Q values.

G2.3.3 Deuterium labelling and triangulation

A varied methodology has been developed to apply contrast variation concepts in SANS, by using deuterium labelling (Comment G2.15). We recall the simplicity of the scattering curve of a 'two-atom' structure, and how it can be interpreted to define the distance between the atoms (Section G2.2). The principle of the label triangulation method is to deuterate a complex structure selectively in such a way that the scattering pattern is dominated by the equivalent of a pair of atoms in order to measure the distance between them. Depending on the resolution and size of the complex, the 'atoms' can be individual proteins, as in the case of the ribosome, for example. The complex structure is then 'built up' from its

---

Comment G2.15   Deuterium labelling contrast variation

Specific labelling of biological macromolecules and complexes can be achieved by biosynthesis approaches and reconstitution. *E. coli* strains are now available that grow well in a fully deuterated medium for the expression of chosen proteins. Label triangulation using specific deuteration has been applied to the ribosome (Capel *et al.*, 1988) and the results on protein positions have been used later to solve the crystal structure (see Comment G2.14). The *triple isotopic substitution method* is based on contrast variation from different deuteration levels in the sample. It has the advantages that it can be used on concentrated solutions because it cancels out interparticle effects and in H_2O solvents only, if D_2O has adverse effects on the sample (Serdyuk *et al.*, 1994).

---

components, pair distance by pair distance by using the surveying triangulation method.

G2.4 The thermodynamics approach in SAS

G2.4.1 Fluctuations in hydrodynamics and scattering

So far in this chapter, we have presented the particle approach to SAS. It assumes that radiation is scattered by particles, which have well-defined boundaries separating them from the surrounding solvent. The assumption is intrinsic, for example, in the derivation of Porod's law or the Debye equation. Historically, Rayleigh used a particle approach when he developed his theory on light scattering from the atmosphere. The particle approach, however, is not useful to account for generally more complicated scattering problems, and at the beginning of the twentieth century Smoluchowski and Einstein developed alternative theories based on the analysis of density fluctuations in the scattering system. We could look at density fluctuations as 'transient particles', which scatter radiation in exactly the same way as a classical particle. The contrast in the case of density fluctuations need not arise from a different chemical composition between particle and solvent as it results from the difference in density itself. In this view, even in a pure liquid, thermal density fluctuations lead to scattering. A liquid has a finite compressibility so that its molecules cannot approach beyond certain limits; at the other extreme, the molecules cannot move too far away from each other as they would in a gas. Under the influence of thermal energy, molecules in the liquid state form dynamic clusters by moving towards and away from each other under attractive and repulsive forces, leading to the density fluctuations.

Fluctuation theories have been developed to interpret SAS from interacting macromolecules in solution, in terms of particle–solvent and particle–particle interactions. In 1981, Eisenberg published a review discussing the application of time-averaged fluctuation theory to scattering phenomena, clearly establishing a formal link between sedimentation equilibrium (see Part D) and the forward scattering of light, X-rays and neutrons (Comment G2.16). The thermodynamic basis of the theory is the relation between the observed scattering and directly measurable (in principle anyway) bulk parameters of the solution. For example, the number of particles in the solution is not a measurable quantity since we are unable to count the particles as we introduce them into the sample container. On the other hand, the concentration of the solute is a measurable quantity because we can weigh so many grams of material and add them to a given volume or mass of solvent.

The hydrostatic osmotic pressure (a measurable bulk property) due to a solution of N particles in a volume V is given by an equation similar to the perfect gas equation (see Chapter A1):

$$\Pi V = NRT \qquad \text{(G2.38)}$$

Osmotic pressure constitutes a colligative property of the solution, because its measurement essentially 'counts' the number of particles. Now, if we know how much mass of solute we have added to make up the solution, we can define a molar mass for the particles:

$$M = CV/N \qquad \text{(G2.39)}$$

where c is solute concentration. If C is in grams per millilitre, M is in grams per mole. Then

$$\Pi = (C/M)RT \qquad \text{(G2.40)}$$

Note that the measurement is not an absolute determination of molar mass, because it determines the ratio C/M in units of moles per millilitre, and not either of these quantities separately. The units of M are commensurate to those of C. For example, if C for a DNA sample is given in terms of 'phosphate groups per millilitre', the units of M are 'phosphate groups per mole'.

We consider a solution made up of three components: water, a macromolecule (which is too large to diffuse across a dialysis membrane) and a small solute, e.g. salt (which diffuses across the membrane). In order to have well-defined solvent conditions, the solution is dialysed against a large volume of solvent. It is useful to indicate the various solution components by suffixes: 1 for water, 2 for the macromolecule, 3 for the small solute.

The application of fluctuation theory to sedimentation equilibrium and forward scattering of light, X-rays and neutrons in the solution leads to interestingly analogous equations, in terms of the derivative of osmotic pressure with respect to macromolecule concentration C_2 and $(\partial \rho / \partial C_2)_\mu$, $(\partial n / \partial C_2)_\mu$, $(\partial \rho_{el} / \partial C_2)_\mu$,

Comment G2.16
Time-average
Fluctuation theory

Time-averaged fluctuation theory as applied to scattering phenomena was discussed in detail by Eisenberg (1981).

The invariant particle hypothesis was developed by Luzzati and Tardieu (1980).

$(\partial \rho_n / \partial C_2)_\mu$, which are the mass density, refractive index, electron density and neutron scattering density increments, respectively, at constant chemical potential μ_1, μ_2 of diffusible solutes (see Chapter A1). The derivative of osmotic pressure with respect to C_2 can itself be written in terms of the molar mass M_2.

At vanishing C_2, the equations reduce to

$$\frac{I}{RT} \frac{d\Pi}{dC_2} = M_2^{-1} + A_2 C_2 + A_3 C_2^2 + \cdots \tag{G2.41}$$

where A_i are the virial coefficients, expressing the interparticle interactions in non-dilute solutions (from the Latin *vires* meaning forces).

Sedimentation equilibrium

$$d \ln C_2 / dr^2 = (\omega^2 / 2RT)(\partial \rho / \partial C_2)_\mu M_2 \tag{G2.42}$$

where r is distance to the centre of the rotor, ω is the angular velocity and $(\partial \rho / \partial C_2)_\mu$ is the mass density increment at constant chemical potential of diffusible solvent components (see below).

Light scattering

$$\Delta R(0) = K(\partial n / \partial C_2)_\mu^2 C_2 M_2 \tag{G2.43}$$

where $\Delta R(0)$ is the light scattering in the forward direction (zero angle) in excess of solvent scattering and K is an optical constant.

X-ray scattering

$$I_{el}(0) = K_{el}(\partial \rho_{el} / \partial C_2)_\mu^2 C_2 M_2 / N_A \tag{G2.44}$$

where $I_{el}(0)$ is forward scattering of X-rays, K_{el} is a calibration constant and N_A is Avogadro's number.

Neutron scattering

$$I_n(0) = K_n(\partial \rho_n / \partial C_2)_\mu^2 C_2 M_2 / N_A \tag{G2.45}$$

where $I_n(0)$ is forward scattering of neutrons, K_n is a calibration constant.

The density increments are bulk properties of the solution. We treat the mass density increment as an example (Fig. G2.16). The presence of a macromolecule in a solution perturbs the solvent around it. If, for example, it binds salt and water in a ratio that is different from their ratio in the solvent, then this results in salt or water flowing in or out of the dialysis bag to compensate and maintain a constant chemical potential of diffusible components across the membrane. The mass density increment expresses the increase in mass density of the solution per unit macromolecular concentration. Not only does it account for the presence of the macromolecule itself but also for its interactions with diffusible solvent

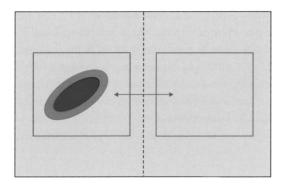

Fig. G2.16 Mass density
increment. In a dialyis
experiment of a particle
(red ellipse) that cannot
cross the membrane
(dashed line), water and
salt flow (double arrow)
to re-establish
equilibrium of diffusible
components. The mass
density increment at
constant chemical
potential of diffusible
solutes is the difference
between the density of a
volume of the solution on
the left hand side (red
rectangle) divided by
particle concentration and
the density of the solution
on the right-hand side
(green rectangle). See
also Chapter A1.

components. It can be written

$$(\partial\rho/\partial C_2)_\mu = (1 + \xi_1) - \rho^\circ(\bar{\partial}_2 + \xi_1\bar{\partial}_1) \tag{G2.46}$$

where ξ_1 is an interaction parameter in grams of water per gram of macro-molecule, and v_x is partial specific volume (in millilitres per gram) of component x. The parameter ξ_1 does not represent water 'bound' to the macromolecules. It represents the water that flows into the dialysis bag to compensate for the change in solvent composition caused by the association (or repulsion) of both water and small solute components with the macromolecule. The mass density increment is essentially equivalent to the buoyancy term arising from Archimedes' principle.

A similar equation is derived in terms of ξ_3, representing the number of grams of the small solute per gram of macromolecule; ξ_1 and ξ_3 are related by $\xi_1 = \xi_3/w_3$, where w_3 is the molality of the solvent in grams of component 3 per gram of water. The parameter ξ_1 (or ξ_3) represents the solvent interactions of the macromolecule in the given solution conditions.

The mass density and refractive index increments are directly measurable in independent experiments on samples of the dialysed solution, even if these measurements are not easy and require high sensitivity. Electron and scattering density increments, on the other hand, are not directly measurable in auxiliary experiments. Neutron scattering density increments are, in principle, measurable by neutron interferometry, but these also are not easy experiments.

The scattering density increments for the different experimental methods are related to solvent interactions and macromolecule properties by

$$(\partial\rho_{el}/\partial C_2)_\mu = (l_2 + \xi_1 l_1) - \rho_{el}^\circ(\bar{\partial}_2 + \xi_1\bar{\partial}_1) \tag{G2.47}$$

$$(\partial\rho_{n}/\partial C_2)_\mu = (b_2 + \xi_1 b_1) - \rho_{n}^\circ(\bar{\partial}_2 + \xi_1\bar{\partial}_1) \tag{G2.48}$$

where l_x, b_x, are electron and neutron scatteringram amplitudes, respectively, per gram of component x. They can be calculated readily from chemical compositions.

In the same way as for the mass density increment, equations for X-ray and neutron scattering are equivalent to buoyancy terms where mass is replaced by scattering amplitudes. The equations can be used jointly to solve for unknown parameters, for example, ξ_1 and ν_2. Combining the mass and X-ray equations is not very useful, because mass and electron density are close to being proportional to each other. On the other hand, atomic neutron scattering amplitudes (which can be positive or negative) are completely independent of mass or electron content, so that combining the neutron equation with mass and/or X-ray results is extremely useful.

G2.4.2 Relating the thermodynamics and particle approaches

The 'buoyancy' equations for the density increments are analogous to the 'contrast' in the particle approach. However, the interaction parameter ξ_1 varies with solvent composition so that plotting the measured density increment, e.g. $(\partial \rho_n / \partial C_2)_\mu$, versus solvent density, e.g. ρ_n°, does not yield a straight line. Eisenberg analysed the 'meaning' of the interaction parameter in terms of B_1 grams of water and B_3 grams of salt bound by the particle:

$$\xi_1 = B_1 - B_3/w_3 \tag{G2.49}$$

When B_1 and B_3 are constant, ξ_1 versus $1/w_3$ is a straight line. The 'bound' water and salt values may be positive or negative (when the component is excluded from the solvation shell). Applying the model to the interaction parameters, we obtain for the buoyancy equations:

$$(\partial \rho / \partial C_2)_\mu = (1 + B_1 + B_3) - \rho^\circ(\bar{\partial}_2 + B_1\bar{\partial}_1 + B_3\bar{\partial}_3) \tag{G2.50}$$

$$(\partial \rho_{el} / \partial C_2)_\mu = (l_2 + l_1 B_1 + l_3 B_3) - \rho^\circ(\bar{\partial}_2 + B_1\bar{\partial}_1 + B_3\bar{\partial}_3) \tag{G2.51}$$

$$(\partial \rho_n / \partial C_2)_\mu = (b_2 + b_1 B_1 + b_3 B_3) - \rho^\circ(\bar{\partial}_2 + B_1\bar{\partial}_1 + B_3\bar{\partial}_3) \tag{G2.52}$$

where the suffix 3 refers to salt. Tardieu and collaborators pointed out that whereas a straight-line dependence of ξ_1 versus $1/w_3$ is consistent with a particle that is invariant in composition (binding constant amounts of water and salt), a straight-line dependence of density increment with solvent density is consistent with a particle in solution that is invariant in both composition and volume (Comment G2.16).

Equations (G2.50)–(G2.52) are the contrast equations in the case of a three-component solution. Note that where the particle composition does not depend on the solvent (this is not the case for SANS experiments on solutions containing heavy water, because of the exchange of particle labile hydrogen atoms) the slope of the density increment versus solvent density is identical for the three methods and equal to the volume of the solvated particle.

G2.5 Interactions, molecular machines and membrane proteins

G2.5.1 Aminoacyl tRNA interactions with tRNA

Aminoacyl tRNA synthetases occupy a key position in protein synthesis (see Comment G2.17). *E. coli* methionyl-tRNA synthetase, MetRS, is a dimer that binds two tRNA molecules in an *anticooperative* fashion (i.e. the binding of the first tRNA molecule to one of the subunits induces a lowering in affinity for the binding of a second tRNA to the other subunit. The forward scattered intensity and radius of gyration values have been followed as a solution of enzyme was titrated by progressive additions of tRNA, and interpreted by using Eqs. (G2.23), (G2.24). Experiments have been performed in H_2O solvents in which the contrasts (see Fig. G2.10) of both protein and tRNA are significant, and in a 77% D_2O 23% H_2O buffer, in which tRNA contrast is close to zero so that only the protein part of the complexes was observed. The contrast variation technique permitted, in a unique way in the analysis of the measured $I(0)$ and R_G values, the separation and identification of *tRNA–protein binding stoichiometries, protein conformational changes and even some protein dimer dissociation associated with the binding events*. The experiments established that the interaction was highly dynamic. They suggested a structural interpretation for the anticooperative binding behaviour and its biological significance in favouring the alternated release of the substrate molecules after the catalytic event (Comment G2.18, Fig. G2.17).

Comment G2.17 Physicist's box: Amino acyl tRNA synthetases

The ribosome is the main site of protein synthesis in the cell. It is where the genetic information encoded in messenger RNA (mRNA) is translated into a polypeptide chain (see Chapter A2). Each amino acid is encoded by a nucleotide triplet (a codon) on mRNA. Transfer RNA acts as an adaptor molecule; one of its ends binds specifically to the codon, while the corresponding amino acid, bound to the other end, is positioned for incorporation into the growing polypeptide. There is a specific tRNA molecule for each codon and amino acid. The correct binding of amino acid AA to $tRNA^{AA}$ is catalysed by an AA specific enzyme: the aminoacyl tRNA synthetase.

Comment G2.18 References

The MetRS study is described in Dessen *et al.* (1978); experiments on yeast valyl-tRNA synthetase and asp-tRNA synthetase are described in Zaccai *et al.* (1979) and Giegé *et al.* (1982), respectively.

Transfer RNA binding stoichiometries for aminoacyl tRNA synthetases were difficult to determine by standard biochemical approaches, because the charged nucleic acid often led to non-specific aggregates of protein forming around it. SANS experiments, because they could easily distinguish protein–protein from protein–nucleic acid complexes by contrast variation, played an important role in establishing these stoichiometries for different enzymes in the family and under different solvent conditions, paving the way for the successful crystallization

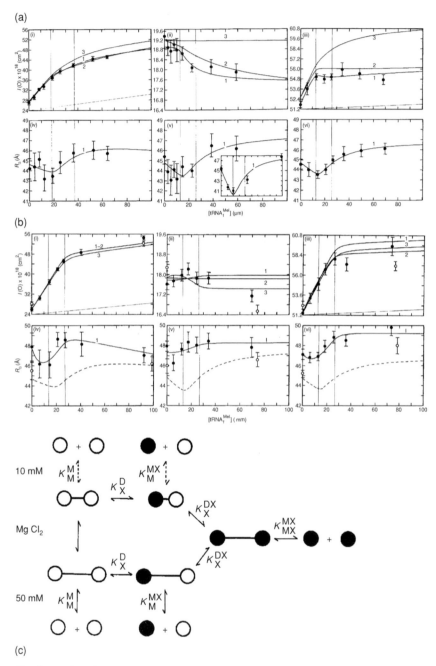

Fig. G2.17 SANS study of the interaction of tRNA specific for methionine with *E. coli* methionyl-tRNA synthetase: (a) Forward scattered intensity (top three panels) and radius of gyration (bottom three panels) changes while tRNA is added to a solution of the dimeric enzyme containing 10 mM $MgCl_2$. The vertical lines just below 20 and 40 μM on the x-axes correspond to tRNA: Protein stoichiometries of 1 and 2, respectively. The left-hand side panels are for the H_2O solution, in which both of enzyme–tRNA complexes. One such experiment was based on changes in forward scattered intensity upon addition of tRNA^{asp} to a solution of yeast AspRS in H_2O solvent. The curve was analysed quantitatively using Eq. (G2.23), with calculated values of m_j for protein and tRNA, respectively, according to their chemical compositions, partial specific volumes and the solvent composition. A clear stoichiometry of 2 tRNA per AspRS dimer was found, not only from the break at this ratio, but also according to the absolute value of the increase in $I(0)$ as tRNA was added to the enzyme solution. Measurements at 77% D_2O, close to the contrast match of the tRNA, confirmed that there was no protein aggregation upon addition of tRNA.

G2.5.2 ATP, solvent- and temperature-induced structural changes of the thermosome

The thermosome is a macromolecular machine in thermophilic archaea that has been described as a group II chaperonin (see Comment G2.19). Its X-ray crystal structure has been solved and shows a rather globular conformation, which is quite different from the cylindrical structure found by cryo-electron microscopy. A SANS study of the thermosome under a variety of conditions established the relation between the two conformations and their places in the reaction cycle (Fig. G2.17).

The distance distribution function $(p(r))$ of the thermosome in solution in a standard buffer clearly indicates an 'open' structure similar to the one found by electron microscopy (Fig. G2.17). Solvent (ATP or ADP binding) and temperature conditions were then found for which the experimental $p(r)$ corresponded to either the one calculated from the crystal structure (the 'closed' conformation) or

←——————————————————————————————————

tRNA and protein are 'visible' with the tRNA contrast being larger than the protein contrast, the middle ones are for 77% D_2O, in which the tRNA is 'invisible', and the right-hand panels are for D_2O, in which the tRNA contrast is smaller than the protein contrast. The data points are shown in each panel, as well as lines corresponding to different interaction models. Note the break in the intensity rise beyond a 1:1 stoichiometry in H_2O solution, indicating anticooperativity, and the fall in intensity and radius of gyration in 77% D_2O, indicating dissociation of the protein dimer. (b) The same as (a) for a solution containing 50 mM $MgCl_2$. The dashed lines in the bottom panels show the radius of gyration dependence in 10 mM $Mgcl_2$. Note the almost straight line increase in intensity in the H_2O solution up to a stoichiometry of 2 indicating essentially independent binding sites under these conditions, and the larger radius of gyration (compared to the 10 mM $MgCl_2$ condition) indicating an 'opening up' of the protein dimer. (c) The interaction model that best fitted the data. Open circles represent protein monomers, filled circles are protein monomers with bound tRNA. The dumb-bell structures are dimers, with the length of the line indicating a large or small radius of gyration. (Dessen et al., 1978.) (Figure reproduced with permission from Elsevier.)

Comment G2.19   Physicist's box: Chaperonins

Chaperonins are cellular macromolecular complexes that assist protein folding by
providing a protected compartment where it can take place. They are multisubunit
macromolecular machines powered by ATP. Bacterial chaperonins, represented by
E. coli GroEL, act together with a cochaperonin (GroES) and have been classified as
group I chaperonins. GroES acts as a cap that opens and closes access to the internal
compartment of GroEL. Extensive studies of GroEL–GroES have led to the
description of a series of structural rearrangements during the activity cycle of the
complex. Archaeal and eukaryotic (see Fig. A2.1) chaperonins act independently of
a cochaperonin, protrusions on the main structure appearing to act as the 'lid' over
the entrance to the folding compartment. These chaperonins have been classified as
group II. Before the study presented in the text, their conformational changes during
the reaction cycle were poorly understood.
   GroEL–GroES structure and activity are reviewed by Sigler *et al.* (1998).

---

the one calculated from electron microscopy (the 'open' conformation), leading
to the model of conformational changes during the reaction cycle in Fig. G2.17(b).

G2.5.3 Membrane proteins

Membrane proteins ensure a large number of vital cellular functions. Their structural study, however, remains a major challenge because of the difficulties in
maintaining their integrity during fractionation out of their complex local environment and in preparing crystals. An integral membrane protein crosses the
lipid membrane; its body is surrounded by the apolar lipid environment with
which it may have specific or non-specific binding interactions; at the lipid bilayer
surface, the protein may interact with the lipid polar headgroups before protruding into an aqueous phase, which, in general, will have a different pH and ionic
composition on each side of the membrane. The low-resolution structural organisation of a membrane protein in its physiological environment is itself often
difficult to solve: is it a monomer? A dimer? The knowledge is fundamental to
the understanding of functional mechanisms, SANS approaches using contrast
variation have been developed specifically for the study of membrane proteins
in detergent and lipid environments. Natural abundance detergent micelles can
be contrast matched between 10 and 15% D_2O. Lipid vesicles, however, cannot
be contrast matched simply because they are inhomogeneous over distances that
are comparable to the size of the proteins to be studied. The scattering density
of the polar headgroups corresponds to that of about 20% D_2O, while that of
the CH, CH_2 and CH_3 groups in the chains is close to the scattering density
of H_2O. In order to bypass this problem, homogeneous scattering lipid vesicles
were prepared by specific deuteration in order to study the association state of the

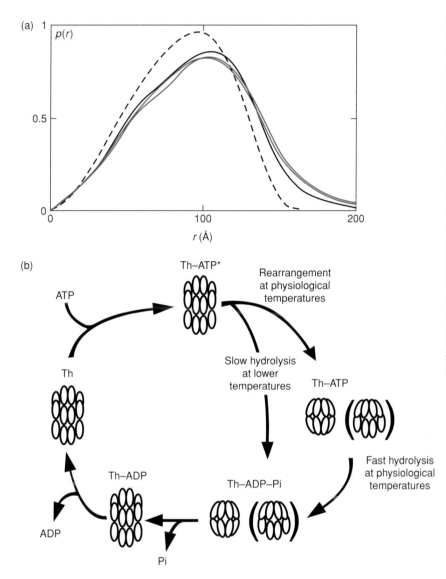

Fig. G2.17
(a) Experimental and calculated pair distribution functions of native ($\alpha\beta$) and recombinant (α) thermosomes. Red, blue lines: experimental data for ($\alpha\beta$), (α), respectively, in a standard buffer at 25 °C; black line: calculated from the electron microscopy structure; dashed line: calculated from the crystal structure.
(b) Model for open (cylindrical) and closed (globular) structures during ATPase cycling, established from the SAS data. (Gutsche et al., 2001.) (Figure reproduced with permission from Elsevier.)

membrane protein, bacteriorhodopsin (Fig. G2.18). BR is the protein that functions as a light driven proton pump in the purple membrane (PM) of *Halobacterium salinarum* (See Comment A3.8). In PM, BR is organised as a trimer; what is its association state in lipid vesicles? Vesicles of dimyristoyl phosphatidyl choline (DMPC) have been prepared with two levels of deuteration (d_{63} and d_{67}) matching, respectively, in 94% D_2O (Fig. G2.15), and 99% D_2O. The Guinier curves of the protein-loaded vesicles at lipid contrast match were parallel, indicating the same radius of gyration in both cases: 16 Å, corresponding to the protein monomer, in accordance with the molar mass of 26 000 g /mole^{-1} calculated

Fig. G2.18 (a) Schematic
of a membrane protein
(blue) in a deuterated
lipid vesicle (red) in H_2O
buffer (light green, left
panel) and D_2O contrast
match buffer for the lipid
(right panel). (b) Guinier
curves for BR in contrast
matched vesicles of
different deuteration
levels. The same radius of
gyration is observed in
both cases,
corresponding to the
protein monomer. (c) Plot
of the scattered intensity
divided by concentration
as a function of the
percentage D_2O in the
solvent, revealing the
contrast match point of
94% D_2O in the lipid
vesicles. (Hunt et al.,
1997.) (Figure reproduced
with permission from
Elsevier.)

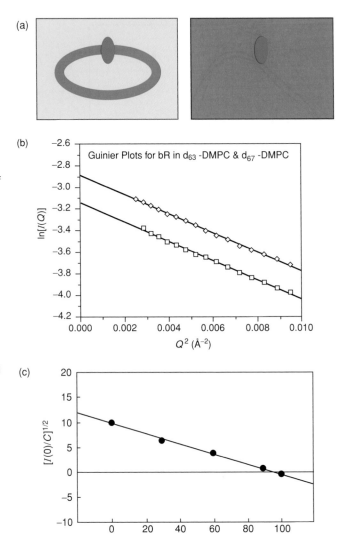

Fig. G2.18 (a) Schematic of a membrane protein (blue) in a deuterated lipid vesicle (red) in H_2O buffer (light green, left panel) and D_2O contrast match buffer for the lipid (right panel). (b) Guinier curves for BR in contrast matched vesicles of different deuteration levels. The same radius of gyration is observed in both cases, corresponding to the protein monomer. (c) Plot of the scattered intensity divided by concentration as a function of the percentage D_2O in the solvent, revealing the contrast match point of 94% D_2O in the lipid vesicles. (Hunt et al., 1997.) (Figure reproduced with permission from Elsevier.)

from the forward scattered intensity and concentration. The use of deuterated lipid for contrast match measurements in D_2O significantly reduces the incoherent background in the measurements, which arises mainly from H nuclei (see Section G1.2.1), while maintaining high contrast for natural abundance protein, leading to a very favourable signal-to-noise ratio for the scattered intensity.

G2.6 Multiangle laser light scattering (MALLS)

The theory of classical or static light scattering by particles in solution was developed early in the twentieth century. As described in Section G2.2.2 for small angle X-ray scattering and SANS, the angular dependence of the absolute scattered intensity contains information on the particle mass and size. Light scattering was further and strongly developed for the study of polymers, but its application to biological macromolecules suffered from a number of technical and sample characterisation difficulties, and the method was used by only a few specialised biophysics groups. Then, in the late 1980s, laser light instruments that simultaneously measured the scattering signal at several angles became available commercially. Combined with chromatographic separation techniques, which sort a complex sample solution into fractions of similar molecular weight and size particles for the MALLS experiment, and 'user-friendly' analysis software taking advantage of the tremendous progress in personal computers, these instruments provided a powerful means for the absolute scale determination of macromolecular mass and size distributions.

The following equation forms the basis of the MALLS analysis:

$$\frac{K^*C}{R(\theta)} = \frac{1}{M_w P(\theta)} + 2A_2C \tag{G2.53}$$

$R(\theta)$ is the light scattered by the sample per unit solid angle at an angle θ in excess of the light scattered by the solvent alone, M_W is the weight-average molecular weight and is a *form factor* (similar to the Guinier approximation, see Section G2.1.2), which contains shape information such as a weight-average radius of gyration, R_G, for example, Because of the wavelength values of visible light, only R_G values larger than about 100 Å can be measured by MALLS. A_2 is the second virial coefficient (Section G2.1.7), C is the particle concentration in mass per volume of solution, and K^* is a constant containing the specific refractive index of the solvent and the refractive index increment due to the dissolved macromolecules, n_0, dn/dC, respectively, and the vacuum wavelength of the light λ_0:

$$K^* = 4P^2 n_0^2 (dn/dC)^2 / \lambda_0^4 N_A \tag{G2.54}$$

where N_A is Avogadro's number. Note that for a protein, the concentration, C, can be determined in parallel, either from a measurement of the UV absorbance if the extinction coefficient is known or from a measurement of the refractive index change with concentration, since the refractive index increment of proteins is fairly constant. Coupling MALLS with high-performance size exclusion chromatography (HPSEC) and other approaches for macromolecular separation, such as flow, electrical, or thermal field-flow fractionation (FFF) techniques (see Comment D10.5), has been of particular significance in the success of the methodology for a wide range of applications.

The study described in Comment G2.20 is an excellent illustration of the application of MALLS in conjunction with small angle X-ray scattering, NMR and CD for the determination of protein association states in solution.

Comment G2.20   MALLS combined with CD, small-angle X-ray scattering and NMR: conversion of phospholamban into a soluble pentameric helical bundle

In an exploration of membrane protein folding and solubility, surface, usually lipid-exposed, amino acid residues of the transmembrane domain of phospholamban (PLB) were mutated and replaced with charged and polar residues. Wild-type PLB forms a stable helical homopentamer within the sarcoplasmic reticulum membrane in eukaryotic cells. The experiments were designed to test if the packing inside the membrane protein could maintain a similar fold even with the lipid-exposed surface redesigned for solubility in an aqueous environment. The CD spectra (see Chapter E4) indicated that the full-length soluble PLB is highly α-helical, similar to the wild-type protein. For the MALLS investigation, samples were subjected first to size exclusion chromatography. Peaks were detected as they eluted from the column with a UV detector at 280 nm (where the protein has a maximum absorbance), a light scattering detector at 690 nm and a refractive index detector. The exact protein concentration was calculated from the change in the refractive index with respect to protein concentration. The weight-average molecular weight of the protein in each elution volume for different concentrations and pH values was then determined by the application of Eq. (G2.53). Small-angle X-ray scattering experiments on selected conditions confirmed the molecular mass of 120 kDa (exactly five times the protein monomer) found by MALLS, and determined the radius of gyration of the particle to be about 50 Å. However, NMR experiments (see Part J) suggested that the redesigned protein exhibits molten globule-like properties, indicating some alteration in native contacts at the core of the protein by the surface mutations. The experiments nevertheless established that the interior of a membrane protein contains at least some of the determinants necessary to dictate folding in an aqueous environment (Li *et al.*, 2001).

G2.7 Checklist of key ideas

- Information on particle shapes and interactions in solution can be obtained from SAS of X-rays and neutrons.
- The two assumptions required for the interpretation of the scattering curve in terms of a single-particle structure are: (1) the solution is monodisperse, i.e. all the particles in the solution are identical; and (2) the solution is infinitely dilute, i.e. there is no correlation between the positions or orientations of the particles in the solution.
- At small scattering vector (Q) values the scattering curve can be interpreted in the Guinier approximation, from which the radius of gyration of the particle and its molar mass are derived independently and in a model-independent manner.
- At large scattering vector values the intensity scattered by a particle with a well-defined border with solvent follows Porod's Q^{-4} law.

- The distance distribution function in the particle $p(r)$ is calculated from the Fourier transform of the scattered intensity $I(Q)$.
- Scattering from polydisperse solutions (mixtures of different particles) can be interpreted in the Guinier approximation in terms of the different radii of gyration and molar masses.
- A 'structure factor' related to the organisation of particles can be derived from the scattering curve of a concentrated solution.
- Scattering curves can be calculated analytically or numerically for particles of different shapes or atomic assemblies as in a crystal structure, for example, for comparison with experimental observations.
- The spherical particle is unique in that its internal structure expressed as the scattering density distribution as a function of radius, $\rho(r)$, directly corresponds to the rotational average; it is the only shape whose scattering amplitude, $F(Q)$, can be calculated directly from the square root of the scattering intensity, $I(Q)$; $\rho(r)$ is the Fourier transform of $F(Q)$.
- By using appropriate plots of the scattering curve the mass per unit length and cross-sectional radius of gyration of rod-like particles (fibre structures) and the mass per unit area and transverse radius of gyration of flat sheet-like particles (membrane structures) can be determined.
- Model structures consistent with scattering curves can be determined by using different approaches including fitting with *spherical harmonics*, *Monte Carlo* and *simulated annealing* algorithms and *genetics* algorithms.
- Contrast variation methods allow one to focus on different parts of a complex particle; contrast can be varied by changing the radiation (because X-ray and neutron scattering amplitudes are different) and/or by deuterium labelling of solvent and/or particle for neutron scattering.
- Different methodological approaches have been developed to plan and interpret contrast variation experiments in terms of the structures of different components within a particle.
- The forward-scattered intensities in neutron and X-ray SAS and light scattering are related to the thermodynamic and hydrodynamic parameters of macromolecule–solvent interactions in solution, and can be used to study these interactions.
- Application examples illustrate the power of SAS as a tool for the study of structural aspects of complex formation and interactions in solution, conformational changes in the reaction cycles of macromolecular machines and the low-resolution structure of membrane proteins in lipid vesicles (by using contrast matched lipids).

Suggestions for further reading

Svergun, D. I. (2000). Advanced solution scattering data analysis methods and applications. *J. Appl. Crystallogr.*, **33**, 530–534.

Zaccai, G., and Jacrot, B. (1983). Small angle neutron scattering. *Ann. Rev. Biophys. Bioeng.*, **12**, 139–157.

Glatter, O., and Kratky, O. (eds.) (1982). *Small Angle Scattering*. London: Academic Press.

Chapter G3
X-ray and neutron macromolecular crystallography

G3.1 Historical review

The discoveries of X-rays, electrons and neutrons as major achievements in the field of biological crystallography have been honoured by Nobel prizes in physics, chemistry and physiology or medicine, spread over the more than 100 years since their inception in **1901**.

The very first Nobel prize in physics was awarded to **W. C. Röntgen** 'in recognition of the extraordinary services he has rendered by the discovery of the remarkable rays subsequently named after him'. **J. J. Thomson** obtained the Nobel physics prize in **1906** 'in recognition of the great merits of his theoretical and experimental investigations on the conduction of electricity by gases'. **M. von Laue** was awarded the physics prize in **1914** 'for his discovery of the diffraction of X-rays by crystals', and **W. H. Bragg** and **W. L. Bragg**, in **1915**, 'for their services in the analysis of crystal structure by means of X-rays'. **J. Chadwick** was awarded the Nobel prize in physics in **1935** 'for the discovery of the neutron'.

The birth of molecular biology was honoured appropriately with the award of the **1962** Nobel prize in physiology or medicine to **J. Watson, M. Wilkins** and **F. Crick** 'for their discoveries concerning the molecular structure of nucleic acids and its significance for information transfer in living material', and, in the same year, the award of the chemistry prize to **M. F. Perutz** and **J. C. Kendrew** 'for their studies of the structures of globular proteins'. **D. C. Hodgkin**, who had been recommended for the prize in the same year as Perutz and Kendrew, was awarded the **1964** prize in chemistry 'for her determinations by X-ray techniques of the structures of important biochemical substances'. The **1982** Nobel prize in chemistry was awarded to **A. Klug** 'for his development of crystallographic electron microscopy and his structural elucidation of biologically important nucleic acid-protein complexes', and the **1985** chemistry prize was given to **H. A. Hauptman** and **J. Karle** 'for their outstanding achievements in the development of direct methods for the determination of crystal structures'. The first high-resolution X-ray crystallography structure of a membrane protein complex was rewarded by the **1988** chemistry prize to **J. Deisenhofer, R. Huber** and **H. Michel** 'for the determination of the three-dimensional structure of a photosynthetic reaction centre'. The physics prize for **1992** was awarded to **G. Charpak** 'for his invention and development of particle detectors, in particular the multiwire proportional chamber'; these detectors also paved the way for the area detectors developed for X-ray and neutron crystallography, which replaced film. The Nobel prize for chemistry in **1997** was shared by the crystallographers **J. E. Walker** and **P. D. Boyer** 'for their elucidation of the enzymatic mechanism underlying the synthesis of adenosine triphosphate (ATP)' and by **J. C. Skou** 'for the first discovery of an ion-transporting enzyme, Na^+, K^+ -ATPase'.

Nobel prizes, however, although they are symbolic of the importance given to the fields of physical, chemical and biological crystallography by the world scientific community, in fact were awarded to only a small number of the scientists, whose contributions were absolutely fundamental. **J. D. Bernal** and **D. Crowfoot** observed the first X-ray diffraction pattern from a protein (pepsin) in **1934**, by using a sample mounting procedure to keep the crystal 'wet' that is still in use today. Dorothy Crowfoot, later known as **D. C. Hodgkin**, with **C. Bunn** solved the structure of penicillin in **1949**. With a molecular weight just under 300, this was the largest structure solved by X-ray crystallography at the time. **D. M. Green**, **V. M. Ingram** and **Perutz** opened the way to solving the phase problem with their publication on *isomorphous replacement* in **1954**; the rigorous application of the method was developed by **D. M. Blow** and **Crick** in a **1959** publication. **Blow** with **M. Rossman** also pioneered the application of *molecular replacement* in the **1960s**. *Anomalous dispersion*, another approach to solving the phase problem was introduced in the study of lysozyme, the first enzyme to be solved by protein crystallography, by **C. C. F. Blake, D. F. Koenig, G. A. Mair, A. C. T. North, D. C. Philips** and **V. R. Sarma** in **1965**. **A. Brunger** and coworkers in **1987** successfully solved the structure of *crambin* by using an NMR structure in a molecular replacement approach. With the advent of high-brilliance synchrotron sources, the use of cryo-protection techniques, developed by **H. Hope** and others for the freezing of crystals became essential, because they greatly reduce X-ray-induced radiation damage. Cryo-techniques such as *flash cooling* also led to the development of *kinetic crystallography* for the study of 'trapped' intermediates in an enzymatic reaction pathway for example. Progress in protein crystallography in the last decades of the twentieth century was astounding, with huge complex structures with internal symmetry such as viruses (**M. Rossman, D. Stuart** and their collaborators), and later cellular molecular machines without internal symmetry such as the ribosome (**A. Yonath, T. Steitz, V. Ramakrishna** and their collaborators) being crystallised and solved to high resolution.

In the **1970s**, the special properties of neutrons, especially with respect to locating hydrogen atoms in molecular structures, were successfully exploited in biological crystallography. Structures of the amino acids were solved by **M. S. Lehmann, T. F. Koetzle and W. C. Hamilton; M. Ramanadham, R. Chidambaram**. **B. Schoenborn** and collaborators initiated the neutron crystallography study of myoglobin. Contrast variation methods were developed for neutron scattering (**H. B. Stuhrmann**); it was shown that, in nucleosomes, DNA is wrapped around the histone core (**J. F. Pardon, D. L. Worcester** and their collaborators); the organisation of nucleic acid and protein in spherical viruses was described (**B. Jacrot** and collaborators); the protein positions in ribosomal subunits were solved (**D. M. Engelman, P. B. Moore** and collaborators) by deuterium labelling and label triangulation (**W. Hoppe**).

G3.2 From crystal to model

The molecular structure of a protein is the basis of its function. A detailed understanding of this relationship is one of the fundamental aims of modern biology. Through the efforts of X-ray crystallography, 2000–3000 biological structures are determined yearly. This chapter details how it is possible to obtain an atomic structure from the X-ray diffraction pattern of a crystal (Comment G3.1). We describe: (1) how proteins and other biological molecules can be coaxed into forming crystals; (2) the steps involved in collecting diffraction data from these crystals; (3) the evaluation of the data by using crystal symmetry; (4) the different methods to determine the structure factor phases in order to calculate an electron density distribution; (5) how to build, refine and assess reliability of a macromolecule structure model that fits the electron density.

Comment G3.1   Biologist's box: Crystal structure and resolution

A picture is worth a thousand words, but how reliable is the information in the structure of a protein obtained from crystallography? The *resolution* in ångström units to which a structure has been solved is a key value in order to judge this reliability. It usually appears in the title of the paper, or at least in the abstract – the lower the number, the 'higher' the resolution.

   In low-resolution structures, worse than 3.5 Å, it is difficult to see individual amino acid side chains. On the other hand, secondary structure elements, especially alpha helices, but sometimes even beta sheets, can be identified, as well as their organisation in domains. At medium resolution (between 2 and 3 Å), most of the individual atoms should be visible and we start to see solvent molecules (water and ions). As the resolution approaches 2 Å, alternative conformations become clear, if the same amino acid side chain moves between different positions. At 1.6 Å resolution, the electron density of an aromatic ring is so well defined that we can see the 'hole' in the middle. At 1.1 Å resolution, hydrogen atoms can be positioned in X-ray crystallography, despite the fact the X-rays have only been scattered by that atom's sole electron. In neutron crystallography, however, hydrogen scattering is similar to other atoms, and they can be positioned in a structure at much lower resolution.

G3.2.1 Reciprocal lattice, Ewald sphere and structure factors

We recall from Chapter G1 that the structure of a macromolecule can be calculated from the Fourier transform of its diffraction waves. (Eq. (G1.11):

$$f(\mathbf{r}) = \int F(\mathbf{Q}) \exp(-i\mathbf{Q} \cdot \mathbf{r}_j) \, dV_Q$$

In the case of a crystal, $F(\mathbf{Q})$ is replaced by $G(\mathbf{Q})$, the crystal diffraction, where from Eq. (G1.14),

$$G(\mathbf{Q}) = F(\mathbf{Q}) \times \sum_t \exp(it\mathbf{Q} \cdot \mathbf{a}) \times \sum_u \exp(iu\mathbf{Q} \cdot \mathbf{b}) \times \sum_v \exp(iv\mathbf{Q} \cdot \mathbf{c})$$

Following crystallographic convention, we replace $\mathbf{Q}$ by $\mathbf{S}$ where $\mathbf{Q} = 2\pi \mathbf{S}$. Recall that $S = 2\sin\theta/\lambda$, and that peaks only appear for values of $\mathbf{S}$ given by the hkl indices (Eq. (G1.15)),

$$\mathbf{S} \cdot \mathbf{a} = h \quad \mathbf{S} \cdot \mathbf{b} = k \quad \mathbf{S} \cdot \mathbf{c} = l$$

Similarly to the one-dimensional *reciprocal lattice* defined in Chapter G1, $\mathbf{S}$ defines a reciprocal lattice in three dimensions. The *Ewald construction* is a geometric representation of the conditions in Eq. (G1.15). The *Ewald sphere* of radius $1/\lambda$ is centred on the crystal. Each point of the reciprocal lattice has hkl coordinates that satisfy Eq. (G1.15). The origin ($h = 0, k = 0, l = 0$) is placed at the point where the incident beam intersects the Ewald sphere (Fig. G3.1).

The hkl lattice is rotated about the origin (O) by rotating the crystal. Diffraction is observed when a lattice point intersects the Ewald sphere (Comment G3.2).

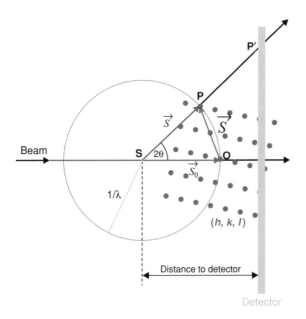

Fig. G3.1 The Ewald construction: diffraction conditions are satisfied at point P, and the diffraction spot appears on the detector at point P′. The angle OSP = 2θ.

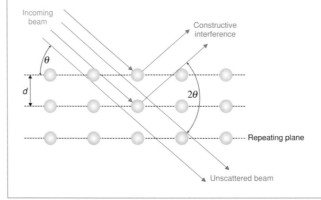

We rewrite Eq. (G1.14) in terms of **S**

$$G(\mathbf{S}) = F(\mathbf{S}) \times \sum_{n} \exp(\mathrm{i}2\pi n\mathbf{S} \cdot \mathbf{R}) \tag{G3.1}$$

$F(\mathbf{S})$ is the Fourier transform of the unit cell contents. In the case of a crystal this
is called a *structure factor*. The unit cell contains atoms of scattering amplitude
f_j at positions $\mathbf{r}_j$. We write

$$\mathbf{r}_j = x_j\mathbf{a} + y_j\mathbf{b} + z_j\mathbf{c} \tag{G3.2}$$

where x, y, z are fractional coordinates in the unit cell. Then,

$$F(\mathbf{S}) = \sum_{j} f_j \exp(\mathrm{i}2\pi \mathbf{S} \cdot \mathbf{r}_j) = \sum_{j} f_j \exp[\mathrm{i}2\pi \mathbf{S} \cdot (x_j\mathbf{a} + y_j\mathbf{b} + z_j\mathbf{c})] \tag{G3.3}$$

Since diffraction only occurs for the special values given by the Miller indices
Eq. (G3.3) becomes

$$F(\mathbf{S}) = F(hkl) = \sum_{j} f_j \exp[\mathrm{i}2\pi(hx_j + ky_j + lz_j)] \tag{G3.4}$$

The structure factor wave can be expressed as $F(hkl) = |F(hkl)| \exp[\mathrm{i}\alpha(hkl)]$
(see Chapter A3) since only its absolute value can be measured from the square
root of the intensity. $\alpha(hkl)$ is the relative phase of the wave.

Table G3.1.

Crystal system	Symbols of conventional unit cells	Unit cell parameters
Triclinic	P	$a \neq b \neq c; \alpha \neq \beta \neq \gamma$
Monoclinic	P, C	$a \neq b \neq c; \alpha = \gamma = 90, \beta > 90°$
Orthorhombic	P, C, I, F	$a \neq b \neq c; \alpha = \beta = \gamma = 90°$
Tetragonal	P, I	$a = b \neq c; \alpha = \beta = \gamma = 90°$
Cubic	P, I, F	$a = b = c; \alpha = \beta = \gamma = 90°$
Hexagonal	P	$a = b \neq c; \alpha = \beta = 90°, \gamma = 120°$
Trigonal	R	$a = b = c; \alpha = \beta = \gamma \neq 90°, < 120°$

G3.2.2 Space group symmetry

Recall that the proportion of X-rays scattered by matter is low and consequently the scattering from a single atom or molecule is too weak to be detected (see Chapter G1). However, in a crystal, the molecules are periodically repeated throughout space. Low amplitudes scattered from sets of individual molecules add to form a significant scattered wave that can be observed. The symmetry of the crystal, therefore, is reflected in its diffraction pattern.

A crystal lattice can be described in terms of one of 230 symmetry relationships called *space groups* (Comment G3.3). The *unit cell* is the basic repeating motif which contains one or more *asymmetric units*, depending on the space group

Table G3.2.

Symmetry elements	Affected reflection	Systematic absences when	
2-fold screw (2_A)	Along *a*	*h*00 where	$h = 2n + 1$
4-fold screw (4_2)	Along *b*	0*k*0	$k = 2n + 1$
6-fold screw (6_3)	Along *c*	00*l*	$l = 2n + 1$
3-fold screw ($3_1, 3_2$)	Along *c*	00*l*	$l = 3n + 1, 3n + 2$
6-fold screw ($6_2, 6_4$)			
4-fold screw ($4_1, 4_3$)			
	Along *a*	*h*00	$h = 4n + 1, 2,$ or 3
	Along *b*	0*k*0	$k = 4n + 1, 2,$ or 3
	Along *c*	00*l*	$l = 4n + 1, 2,$ or 3
C-centred (*C*)			$h + k = 2n + 1$
F-centred (*F*)			$h + k = 2n + 1$
			$k + l = 2n + 1$
			$h + l = 2n + 1$
Body-centred lattice (*I*)			$h + k + l = 2n + 1$

(Table G3.1). The asymmetric unit may contain one or several molecules and even have its own internal symmetry, called *non-crystallographic symmetry* (e.g. two-fold symmetry in a protein dimer).

In protein crystals, certain space groups are not allowed due to the *chirality or handedness* of biological structures (see Chapter A2). Proteins are created from L-amino acids and secondary structure helices are preferentially right-handed in order to avoid steric hindrance in the side-chain conformation. Although there are 230 space groups, only 65 are possible for these chiral objects.

The symmetry results in certain space groups exhibiting missing reflections, called *systematic absences* (Table G3.2). Furthermore, *symmetry related* reflections should have equivalent intensities in the same diffraction pattern, so that measured values can be averaged (Comment G3.4).

G3.2.3 Electron density

In the case of X-ray crystallography it is convenient to define the atomic distribution in the unit cell by an electron density $\rho(x, y, z)$. Following the arguments developed in Chapter A3 and G1, the electron density can be calculated from the Fourier transform of the structure factors:

$$\rho(x, y, z) = \frac{1}{V} \sum_{h=-\infty}^{+\infty} \sum_{h=-\infty}^{+\infty} \sum_{l=-\infty}^{+\infty} |F(h, k, l)| \exp[-2\pi i(hx + ky + lz) + i\alpha(h, k, l)]$$

$$(G3.5)$$

where V is the volume of the unit cell. Note that both the structure factor amplitude and its phase are needed to define electron density in the unit cell.

The theoretical electron density calculation of Eq. (G3.5) has to be modified in order to take into account experimental limitations. Protein crystals are imperfect and of limited size. Consequently, they produce diffraction patterns of limited resolution so that only a limited set of (*hkl*) reflections is measured. Furthermore, phase information is lost since only the intensities of the diffracted waves are recorded (see Chapter G1). The electron density calculated from the experimental data, $|F_{obs}(hkl)|$, is given by

$$\rho_{calc}(x, y, z) = \frac{1}{V} \sum_h \sum_k \sum_l |F_{obs}(h, k, l)| \cos[-2\pi i(hx + ky + lz) + i\alpha_{estim}(h, k, l)]$$

$$(G3.6)$$

The sum is over the measured (*h*, *k*, *l*) indices only and the phases have to be estimated from other information. Several methods have been developed to solve the phase problem (see Section G3.6).

G3.2.4 Technical challenges and the crystallographic model

Macromolecular crystallography involves overcoming a number of technical challenges. Crystals diffract weakly because they have a high solvent content.

Comment G3.5   Atomic coordinates and temperature factors

In principle, each atom in the unit cell is defined by four parameters, the three spatial coordinates x, y, z of its average position, and a *temperature factor B*, which describes how the atom moves. A 10 kDa protein contains about 1000 atoms, which means there are 4000 parameters to be determined.

However, in a crystal structure determination, atomic coordinates are constrained by chemical information, e.g., the length of a carbon–carbon bond. In practice this reduces the number of parameters to be determined.

The structure factor phases must be estimated. The number of useful observations is low compared with the number of parameters required to define an atomic model of the macromolecule in the unit cell (Comment G3.5). An X-ray structure, however, provides a wealth of biological information, which amply justifies and rewards the invested effort.

Structures of biological molecules are currently deposited in the Protein Data Bank (PDB) database. Each atom of the model is described by its position in the unit cell (x, y, z coordinates in ångströms), its temperature factor B and its occupancy (between 0 and 100%, according to the probability of finding it at that position). Experimental details and structure quality information are also contained in the PDB entry.

G3.3 Crystal growth: general principles involved in the transfer of a macromolecule from solution to a crystal form

The initial step in crystallography is to transfer the soluble macromolecule to a solution in which it will form crystals (Comment G3.6). It is striking that such flexible, complex, highly hydrated molecules can arrange themselves in an ordered fashion. A protein crystal suitable for X-ray crystallography contains on average 10^{13}–10^{15} individual molecules.

Protein crystals are very fragile (when poked with a hair whisker they have the consistency of a soft French cheese). They are maintained by weak forces, of the level of a single hydrogen bond. Macromolecular crystallization is an entropy-driven process. The local increase in order gained by organising the macromolecule on a crystal lattice is counterbalanced by the gain in freedom of other species in the solution. The crystallization process is dependent on the concentrations of different solutes, and specific physical chemical parameters, such as pH, which affect macromolecular surface charge. Crystallization trials aim to identify favourable conditions for crystal growth. The main parameters varied during these trials are the ionic conditions, the pH and the concentration of

Comment G3.6
Order and disorder in protein crystals

Bernal realised that protein crystals would only diffract if they were not dried out. This implied that an 'ordered' protein crystal would contain disordered solvent. The fraction of the crystal volume occupied by solvent has been observed to be between 20 and 80%, but is usually between 40 and 60% (Bernal and Crowfoot, 1934).

Fig. G3.2 Phase diagram for the crystallization of a protein showing soluble crystal growth and amorphous (non-crystalline) precipitate regions as a function of precipitant and protein concentrations.

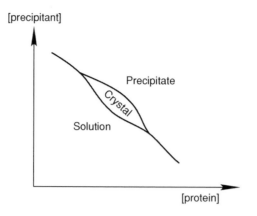

so-called *precipitants*, which include certain salts, polymers and organic solvents (Fig. G3.2).

G3.3.1 Purity and homogeneity

It is imperative that the macromolecule solution is as pure and homogeneous (*monodisperse*) as possible. The presence of glycosilation is often a hindrance to crystallization as the floppy nature of the saccharide chains may impede the creation of stable contacts between proteins. In practice it is best to remove or shorten the sugar moieties before crystallization trials or to use recombinant protein that is not glycosylated.

G3.3.2 Crystallization screens

Since each protein behaves differently, growing crystals is an empirical process. Multifactorial experiments are set up to find the optimum conditions for crystallization (Comment G3.7). The physical parameters usually varied in the first instance are the pH, the ion concentration and the precipitating agent concentrations. There are three categories of precipitant: salts (e.g. ammonium sulphate), polymers (e.g. polyethylene glycol, PEG) and organic solvents (e.g. ethanol). The sample concentration and temperature (e.g. 37 °C, 20 °C and 4 °C) can also be varied. The crystallization process can also be modified by altering the sample volume (which changes the kinetics) or by small concentrations of additives such as certain metal ions, detergents, urea.

After a protein has been crystallized, it is usually necessary to obtain crystals of chemically modified or ligand-bound forms. crystallization screens should still be as exhaustive as possible since the macromolecular surface characteristics may now be very different from those of the native protein.

Comment G3.7
Crystallization kits

Several kits are available commercially for rapid and effective screening of potential crystallization conditions. A screen developed by Jancarik and Kim, for example, has already facilitated the determination of crystallization conditions for more than 500 proteins, peptides, oligonucleotides and small molecules (Jancarik and Kim, 1991).

G3.3.3 Crystallization methods

Several methods have been developed to grow protein crystals by moving the protein into an environment (solution) that is appropriate for crystal nucleation and growth.

Vapour diffusion

The vapour diffusion method is the preferred method for many crystallographers. Not only is it relatively straightforward to set up, but also the resultant crystals can be harvested with ease for X-ray data collection. A drop containing the protein is equilibrated against a larger reservoir of solution (the mother liquor). Volatile species, like water and certain ions and small solutes, diffuse between the two solutions until equilibrium is reached, i.e. the vapour pressure in the drop is equal to that in the reservoir (Comment G3.8). The hope is that then the conditions in the drop are such that the protein will crystallize.

The drop is usually part reservoir solution and part protein-containing weak buffer. The typical protocol is to experiment with several types of reservoir solution in order to obtain the best crystallization conditions. As the drop equilibrates, the protein concentration within it usually increases and in the optimum case brings it within the crystallization phase. Drop volumes typically vary from 100 nl to 2 µl.

Dialysis

Crystals can also be obtained by dialysing the protein solution against a crystallization solution. In this approach, the protein concentration is kept approximately constant.

Seeding

When the protein crystals are too small for X-ray crystallography, *seeding* offers the opportunity to increase the crystal size. It involves taking a crystal and adding it to a new drop containing protein. The crystal can then act as a nucleus from which a larger crystal can grow. Seeding can be done with all crystal sizes. Microcrystals can be seeded by a hair whisker touching the crystals and then being drawn across a new drop. Larger crystals are often 'etched' by passing through water, prior to adding them to a new drop. Crystals of one protein can also help initiate the growth of a crystal of a different but similar protein (Comment G3.9).

Comment G3.8
Vapour diffusion

In vapour diffusion, the drop size usually decreases during equilibration, increasing the constituents' concentrations. Although proteins usually crystallize upon concentration, for some proteins solubility decreases when their solution is diluted by water (reverse diffusion). In such cases, the reservoir is initially more dilute than the drop, which grows in size as it equilibrates. In each case, due to the changing conditions inside the drop, the protein either crystallizes or precipitates out (Jeruzalmi and Steitz, 1997; Richard *et al.*, 2000).

Comment G3.9

Major histocompatability complex (MHC) protein molecules bind and display specific peptides to the immune system. An MHC molecule displaying one type of peptide has been used to grow crystals of the same MHC molecule presenting a different peptide (Bjorkman *et al.*, 1987).

Membrane proteins

Compared with soluble proteins, very few structures of integral membrane protein have been determined by X-ray crystallography (Comment G3.10). Several difficulties are associated with the purification and crystallization of these proteins, because of their insolubility in the usual buffers of biochemistry. The usual approach is to extract the integral membrane protein in a solubilised form with detergents, and to proceed with crystallization trials as in the case of soluble proteins. The choice of detergent and its neutral or ionic character are crucial and depend on the specific characteristics of each protein. Amphiphilic polymer molecules, called *amphipols*, have also been developed for the solubilization of membrane proteins.

Comment G3.10   Examples of membrane protein crystallization

Several difficulties are associated with working with integral membrane protein. In particular, since these proteins are expressed embedded in cellular lipid bilayers, they have to be solubilized (often with detergents) and purified prior to crystallization.

The crystallization step can then be attempted by incorporating in the usual trials either hydrophobic detergents or amphiphilic polymers, named amphipols, which bind to the transmembrane surface of the protein in a non-covalent manner. Another method, showing much promise, is to crystallize the protein from lipidic cubic phases. An integral membrane protein is confined to the three-dimensional network of the curved lipid bilayers, and protein crystals grow within the bulk cubic phase.

Interestingly, in several cases, endogenous membrane lipids cocrystallized with the protein and were observed in the structure (Navarro and Landau, 2002; Popot *et al.*, 2003).

Comment G3.11
Salt crystals

Phosphate, borate and carbonate interact with divalent cations like Mg^{2+}, Zn^{2+} and Ca^{2+} to form crystals. These can usually be identified easily because they are harder, larger, often dichromatic and more beautiful than protein crystals.

Another approach has been to rely upon lipid cubic phases for the crystallization of membrane proteins. These phases are formed by lipids, such as monoolein, under certain conditions of temperature and hydration (see Chapter A2). They present lipid–water interfaces of varying curvature. The membrane proteins are thought to diffuse to patches of lower curvature where they incorporate into a lamellar organization that associates to form highly ordered three-dimensional crystals.

G3.3.4 Identifying crystals and precipitates – crystal shapes and sizes

Under certain conditions, salt crystals may appear. They are usually easy to identify as not being due to protein (Comment G3.11).

The presence of a clear drop indicates that either the drop has not finished equilibrating or that the sample concentration is too low. The presence of precipitate suggests that the sample concentration is too high. If more than 75% of the trials contain protein precipitant, the screen should be repeated with the starting sample concentration halved. However, in propitious circumstances, protein crystals may even grow in a drop displaying precipitant.

Crystals have to be grown sufficiently large for X-ray data collection. The advent of synchrotrons made possible the use of crystals approaching micron size, grown in 0.2 µl drops.

G3.3.5 Cryo-crystallography and cryo-protectants

Synchrotron beam lines are preferred in X-ray crystallography because they permit the rapid collection of high-resolution diffraction data. The high intensity is problematic, however, as X-rays deposit energy into the crystal causing heating and other radiation damage. Direct damage occurs when atoms in the crystal absorb X-ray photons, while indirect damage is due to diffusing reactive radicals produced by the radiation.

In order to reduce radiation damage, the crystal is kept at a low temperature (~ 100 K) such that the indirect damage is greatly reduced because of the lower diffusion rates, allowing sufficient time for data collection. The crystal has to be cooled very rapidly to cryo-temperature (*flash-cooled*) in order to avoid ice formation. Crystalline ice distorts both the lattice of the protein crystal and the structure of the protein itself. The key to the success of the method is the formation of a vitreous water phase. Low-freezing-point chemicals are often added to enable the solvent within the crystal to remain disordered, and favour the formation of an amorphous glass upon freezing. Commonly used cryo-protectants include glycerol, 2-methyl-2,4-pentanediol (MPD), low molecular weight PEG and oils. Further additions may not be necessary if the crystal growth solution itself is already cryo-protected.

It is always important to check that the cryo-protectant solution can be flash-cooled as an amorphous glass on its own. The protein crystal is then drawn through, or equilibrated against the cryo-protectant solution and subsequently flash-cooled.

G3.3.6 Crystal mounting

Goniometer

The crystal is attached to a goniometer which enables it to be rotated freely around an axis in order to examine a large volume of reciprocal space (Fig. G3.3). During data collection, the crystal is usually oscillated over a short angular range. The data corresponding to each angular position is called the *oscillation image*.

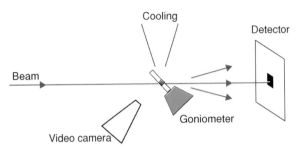

Fig. G3.3 Layout of an X-ray crystallography experiment. The beam (red) has been collimated and made monochromatic. The crystal (blue) in a capillary or a cryo-loop is mounted on a gionometer (green) and centred in the beam with the help of a video camera. The crystal is continuously cooled by a stream of cold dry nitrogen.

Fig. G3.4 Crystals mounted in: (a) a glass capillary and (b) in a cryo-loop.

Room temperature mounting in a capillary tube

In the early days of crystallography, before the advent of cryo-techniques, crystals were exposed to X-rays at room temperature in sealed glass capillary tubes (Fig. G3.4(a)). A drop of mother liquor is included in the tube near the crystal to prevent drying. Data collection is still conducted at room temperature if no suitable cooling conditions have been found.

Cryo-loops

The most widely used method for freezing is to scoop the crystal with a small fibre loop (Fig. G3.4(b)). A thin film of solution spreads across the loop and supports the crystal. An advantage of the method is that it minimises absorption due to glass. Rapid cooling is achieved by plunging the loop into a cryogen at low temperature (e.g. a stream of dry nitrogen at $\sim$100 K). The crystal is then kept continuously frozen during data collection (Fig. G3.3).

G3.3.7 Labelling

Heavy atom derivatives

Binding heavy atoms to a protein can play a substantial role in obtaining structure factor phases (see Sections G3.6.4 and G3.6.5). Heavy atom ions are

Table G3.3. *Most commonly cited heavy-atom derivatising reagents as compiled from macromolecular structures for 1991–1994 (adapted from Rould (1997))*

Reagent	Reagent
K_2PtCl_4	$K_2Pt(CN)_4$
$KAu(CN)_2$	PIP
$Hg(CH_3COO)_2$	$Pb(CH_3COO)_2$
$Pt(NH_3)_2Cl_2$	K_2HgI
$UO_2(CH_3COO)_2$	Mersalyl
$HgCl_2$	*p*-Chloromercuribenzoate
$K_3UO_2F_3$	$CH_3Hg(CH_3COO)$
Ethyl mercurithiosalicylate	TAMM
$(K/Na)AuCl_4$	$SmCl_3$
$(Na/K)_3IrCl_2$	K_2OsO_4
$CH_3CH_2HgPO_4$	$(K/Na)_2OsCl$
K_2PtCl_6	UO_2SO_4
$UO_2(NO_3)_2$	Baker's dimercurial
$K_2Pt(NO_2)_4$	2-Chloromercuri-4-nitrophenol
$(CH_3)_3Pb(CH_3COO)$	$AgNO_3$
CH_3HgCl	CH_3CH_2HgCl
p-Chloromercuribenzene sulphonate	*p*-Hydroxymercuribenzoate

usually soaked directly into the crystal in the hope that they will bind to the protein without destroying the crystal. Protein with bound heavy atoms is called a *heavy-atom derivative*. Table G3.3 lists a selection of commonly used derivatising reagents.

Biosynthetic labelling for anomalous dispersion

Sulphur, which is contained naturally in methionine and cysteine, and phosphorous, which is contained naturally in nucleic acid and lipids, could in principle be used for phasing by anomalous dispersion (see Section G3.6.5). Their absorption edges, however, are at X-ray wavelengths at which absorption is high for most elements, and weak anomalous dispersion signals make experiments very difficult and even impossible. Selenium produces a strong anomalous signal at the X-ray wavelengths more usual for data collection. It has become a standard technique to express proteins containing the modified amino acid selenomethionine to make use of the anomalous dispersion phasing method. The use of selenocysteine is also being developed.

Fig. G3.5 Example of an X-ray diffraction pattern from a myoglobin crystal (Miele *et al.*, 2003).

G3.4 From intensity data to structure factor amplitudes

Oscillation photography is the preferred method for the measurement of the diffraction patterns in protein crystallography. In classical crystallography, the axes of the crystal are carefully aligned to the beam in a series of preliminary exposures. This takes time, however, and, in the case of proteins, it is important to minimise exposure to X-rays because of radiation damage. The *American Method* developed by Rossman is the 'shoot first, think later' technique, in which data are collected from the crystal in whatever orientation, then analysed and interpreted by a battery of powerful computer programs (Comment G3.12).

Spots measured on the detector during data collection (Figs. G3.3, G3.5) have to be interpreted and converted to *hkl* intensity data, from which structure factor amplitudes can be calculated.

G3.4.1 Data collection and processing

The measured intensity has to be corrected for background and other effects. The Lorentz correction accounts for the different rates with which reciprocal lattice points cross the Ewald sphere. There are corrections also for X-ray polarization, extinction and absorption effects and radiation damage (Comment G3.13).

In classical crystallography Wilson statistics allows an absolute scale C factor to be derived from a set of corrected diffraction intensities. The Wilson equation relates the mean intensity to the sum of squared scattering amplitudes and introduces an overall temperature factor B. It is a standard structure factor equation, assuming one atom per unit cell with a scattering intensity equal to the sum of atomic scattering intensities in the unit cell and a single isotropic temperature factor (see Chapters A3 and G1),

$$\langle I(h, k, l) \rangle = C \sum_{n} (f_n)^2 \exp\left(-2B \frac{\sin^2 \theta}{\lambda^2}\right)$$

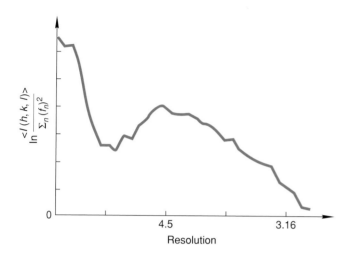

Fig. G3.6 Wilson plot for a protein crystal. The logarithm of average intensity is plotted against $[(\sin\theta)/\lambda]^2$. The corresponding resolution $(1/S)$ is given on the x-axis.

or

$$\ln \frac{\langle I(h, k, l)\rangle}{\sum\limits_{n} (f_n)^2} = \ln C - 2B \left(\frac{\sin^2 \theta}{\lambda^2} \right) \qquad (G3.7)$$

A plot of $\ln \langle I(h, k, l)\rangle / \sum_{n} (f_n)^2$ against $(\sin^2 \theta)/\lambda^2$ is known as a Wilson plot. It should give a straight line from which the experimental values of B and C can be derived. This is not the case for protein crystals for which the Wilson plot looks like the one shown in Fig. G3.6.

The Wilson plot of a protein is characterised by a dip at around 6 Å resolution and a maximum at 4.5 Å resolution. This is due to the fact that because of the presence of α-helices and β-strands, interatomic distances are not distributed evenly. Recall that the protein crystal also contains solvent. At low resolution, the diffraction intensity is not due to the protein's density, but to the contrast in density between the protein and the surrounding solvent (see Chapter G1). As a result, the Wilson plot behaves linearly only at high resolution.

G3.4.2 Indexing Bragg reflections

A diffraction spot is *indexed* by attribution of its *hkl* values. Initial estimates of the unit cell dimensions are made from the positions of the spots in the detector image and from the physical parameters of the experiment (e.g. sample–detector distance). The parameters and indices are refined in a cyclic procedure.

In a theoretically perfect crystal, each diffraction spot's width is infinitesimally small, but due to mosaic spread, this is not the case (Comment G3.13), and the diffraction spot intensity may not be recorded fully. This is corrected for in the scaling step (see Section G3.4.3).

Indexing programs usually work on one oscillation image at a time. They convert raw diffraction data into a file, which contains *hkl* indices, the background, the corrected intensities of the spots on the image and an estimate of their error.

G3.4.3 Scaling the reflection intensities

The images obtained from several orientations of the crystal then have to be incorporated into an overall dataset. Numerous reflections are measured in more than one image and these partial data must be scaled and combined.

The scaling and merging of the data are important stages in the treatment of diffraction data as they allow a separate refinement of the orientation of each image, but with the same unit cell for the whole data set. The global refinement of the crystal parameters should produce precise unit cell values. The integrated data set therefore contains a list of measured *hkl* indices and their intensities.

G3.4.4 Twinning

Crystal growth anomalies may produce the phenomenon of *twinning*. Twinning refers to a specific case of disorder in which there is partial or complete coincidence between the lattices of distinct crystal domains. Each domain diffracts X-rays independently from each other, so that their diffraction is not in phase; so that we have to add intensities rather than structure factors (see Chapters G1 and G2). In the presence of twinning, each measured reflection is due to two or more Bragg spots, which have separate indices. Several procedures have been developed to *de-twin* diffraction data.

In the case of two subcrystal twins (A and B), the measured overall intensity is given by,

$$I_{\text{overall}} = \alpha I_A + (1 - \alpha)I_B$$

where α is fraction of subcrystal A. Perfect twinning occurs when $\alpha = 0.5$, if not the twining is said to be partial.

In the case of *merohedral twinning* (from the Greek *meros*, part, and *hedron*, face), two or more lattices coincide exactly in three dimensions. This is not immediately obvious from the diffraction pattern. However, the presence of twinning can still be identified from the average intensity, $\langle I \rangle$, in each resolution range (Fig. G3.7):

$$\langle I^2 \rangle / \langle I \rangle^2 = 2 \text{ untwinned}$$
$$\langle I^2 \rangle / \langle I \rangle^2 = 1.5 \text{ twinned}$$

Similarly, for the structure factors:

$$\langle F \rangle^2 / \langle F^2 \rangle = 0.885 \text{ untwinned}$$
$$\langle F \rangle^2 / \langle F^2 \rangle = 0.785 \text{ twinned}$$

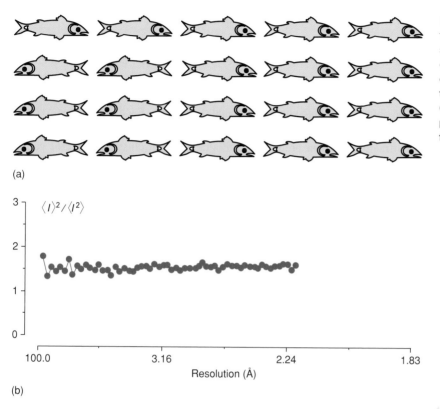

(a)

(b)

G3.4.5 Radiation damage

The issue of radiation damage is treated extensively in this chapter on electron microscopy. In fact, X-rays are much more damaging than electrons. The high intensities obtained from synchrotrons have led to the development of cryo-cooling procedures (see above), which limit indirect radiation damage (Comment G3.14).

Comment G3.14   X-ray radiation damage

Radiation damage is an inherent problem in X-ray crystallography. Experiments with 1-Å wavelength synchrotron radiation suggest that ten absorbed photons are sufficient to 'kill' a 100-Å unit cell and therefore correspondingly on average such a unit cell could contribute only one photon to total Bragg diffraction.

X-ray radiation can cause highly specific damage. For example, disulphide bridges can break, and, in the case of *Torpedo californica* acetylcholinesterase and hen egg white lysozyme, exposed carboxyls including those in these enzymes' active sites, appear more susceptible than other residues to radiation damage (Sliz, Harrison and Rosenbaum, 2003; Weik *et al.*, 2000).

G3.4.6 Determination of the unit cell dimensions

Initial estimates of the unit cell size and orientation are obtained from data collected over a small oscillation of the crystal. Vectors are drawn between all the diffraction spots in the image. The two shortest non-linear vectors define the lattice of diffraction spots. These are the unit vectors from which all other vectors can be obtained through combination and summation. Each of these two vectors defines the orientation and the length of one unit cell axis.

Diffraction spots in protein crystallography often cluster in patterns called *lunes*. A lune (from the Latin *luna*, moon) is the surface bounded by two intersecting arcs. The length and orientation of the third axis are obtained from the shape and distance between the lunes. Recall that the two-dimensional detector records in fact the angular position of the diffraction spots on the Ewald sphere. Based on these starting values, the diffraction pattern at other rotation angles is predicted and spurious spots that do not stem from the diffraction data, such as the *zingers* due to cosmic rays, are identified and omitted.

G3.4.7 Determination of the space group

The space group to which the crystal belongs can be identified from the systematic absences in the diffraction pattern. The presence of symmetry also implies that certain intensities should have similar values. It is therefore not necessary to rotate the crystal fully in order to collect a complete data set.

G3.4.8 Redundancy and statistics

The signal-to-noise ratio (I/σ, where I is the intensity and σ the standard deviation) is a good criterion for assessing the resolution to which meaningful data have been collected. The minimum value of the signal-to-noise ratio is often set to 2 for an individual image. This can be understood from the following statistical relationship:

$$\sigma(\langle I \rangle) = \frac{\sigma(I)}{\sqrt{N}}$$

where N is the number of times each reflection is measured. We can assume that each individual intensity is similar to the average. If a reflection with $I/\sigma(I) \approx 2$ were measured 4 times during the data collection, the redundancy would yield $(\langle I \rangle)/\sigma(\langle I \rangle) \approx 4$. It is accepted practice nevertheless to include as many data as possible.

Several programs have been written to assist in determining the resolution of the diffraction data (Comment G3.1), since the high-resolution limit of the diffraction can vary from image to image. A source of this anisotropy could for example be due to the needle-shaped nature of the crystal.

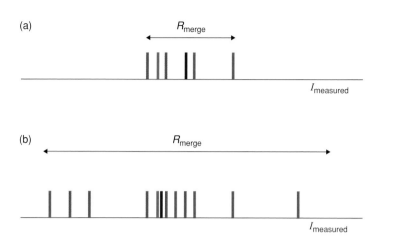

Fig. G3.8 Redundancy and R_{merge}: (a) a reflection is measured four times (red bars) to give the values shown by the red bars; (b) the same reflection is measured a further six times to give a total of ten measurements.

The estimated intensity value of the reflection (blue bar) is the average of the measured intensities (red bars). This average approaches the 'real' intensity value (green bar) as more and more reflections are measured. However, the larger spread in the measured values (see (b)) results in a larger R_{merge} value.

The weighted ratio χ^2 of the difference between the observed and the average value of I, is another useful statistical parameter (this term is often labelled Chi2 in computer printouts):

$$\chi^2 = \sum_{h,k,l} \frac{(I - \langle I \rangle)^2}{\omega^2 N/(N-1)} \tag{G3.8}$$

N is the number of reflections taken into account. The above ratio is corrected by a factor ω which is an explicit declaration of the expected error in the measurement. The aim of the likelihood analysis is to bring χ^2 as close to unity as possible, i.e. to make ω equal to σ, the standard deviation of the repeated measurements. In practice, a χ^2 under 2 before an image is integrated with the other images of the data set is generally acceptable.

The R_{merge} parameter provides an estimate of the precision of individual measurements. It is expressed according to the following formula:

$$R_{merge} = \frac{\sum_{hkl} \sum_{i=1}^{M} |I - \langle I \rangle|}{M \times \sum_{hkl} I} \tag{G3.9}$$

The parameter M takes into account the number of times a given reflection is measured. In contrast to the χ^2 term, the R_{merge} term increases with the redundancy of the data, since the more times a reflection is measured, the more accurate its value should become. R_{merge} is consequently not as informative about the quality of the data, as either the χ^2 term or the signal-to-noise ratio, and may be misleading (Fig. G3.8).

G3.4.9 Molecular packing in the unit cell and the Patterson function

The diffraction pattern intensities from which the unit cell dimensions and crystal symmetry can be determined contain information about the molecular packing within the unit cell.

Solvent content

The fraction of the crystal volume occupied by solvent can be calculated from the crystal density and the partial specific volumes, respectively, of the macromolecule and the solvent. The partial specific volume of all proteins is close to $0.74\ cm^3\ g^{-1}$ or $1.23\ \text{Å}^3\ Da^{-1}$ (see Chapter A1). *Matthew's coefficient* is calculated as the ratio of protein volume to solvent volume in the crystal. The volume of the asymmetric unit is estimated from the unit cell dimensions and the space group. The solvent content of protein crystals has been observed to have values between 20 and 80%, but is usually between 40 to 60% (Matthew's coefficient between 1.5 and 0.66). In many cases, this range is sufficiently restrictive to be able to estimate reliably the total number of molecules per asymmetric unit from the molecular mass and the volume of the crystallographic asymmetric unit.

Patterson function

The orientations of the unit cell's components can be deduced from the Patterson map, which is directly calculated from the intensities measured in the diffraction pattern. The Patterson map is a vector map, with peaks at the positions of vectors between atoms in the unit cell (Figure G3.9).

The Patterson function $P(uvw)$ does not rely upon phase information. It is the Fourier transform of the intensity of structure factors, with their phases set to zero:

$$P(uvw) = \frac{1}{V} \sum_h \sum_k \sum_l \exp\left[-2n(hu + kv + lw)\right] \tag{G3.10}$$

where u, v, w, are relative coordinates in the unit cell, and V is the cell's volume. This nomenclature is used instead of x, y, z in order to avoid confusion even though their dimensions are similar. This function can then be mathematically developed into the following,

$$P(uvw) = \int_x \int_y \int_z \rho(xyz)\rho(x + u, y + v, z + w)\partial_x\partial_y\partial_z \tag{G3.11}$$

The Patterson map's origin therefore contains the vector of each atom with itself, and the peaks throughout the map represent interatomic vectors of the crystal. Following upon this, the intermolecular vectors within the crystal will also be found in the Patterson map.

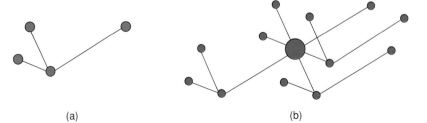

Fig. G3.9 (a) Structure of a four-atom molecule and (b) the resultant Patterson map.

(a)   (b)

These characteristics enable the calculation of relative orientation between similar subunits of the crystal. The Patterson map provides moreover a starting point for many phasing methods (Section G3.6).

Non-crystallographic symmetry

The unit cell consists of a number of asymmetric units that are related through symmetry operators. *Non-crystallographic symmetry* (NCS) arises when there are two or more similar molecules (monomers) present in the asymmetric unit.

NCS can be determined from the diffraction intensity pattern. A *self-rotation function* searches for the direction and angle of rotation of the individual NCS operations, while a *cross-rotation function* searches for the relationship of a structure in one unit cell with similar structures in another cell (Fig. G3.10).

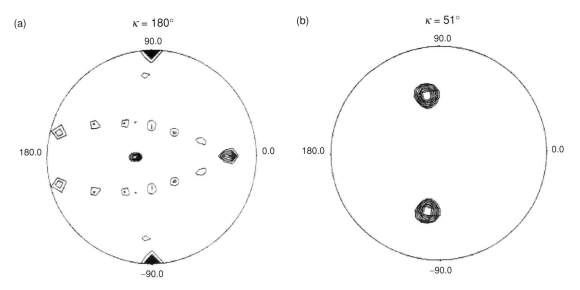

Fig. G3.10 Preliminary crystallographic analysis of the heptamer *Mycobacterium tuberculosis* chaperonin 10.

The crystals belong to the monoclinic space group $P2_1$. Self-rotation functions were calculated, using diffraction data with a resolution between 10 and 5.1Å and a 30Å integration radius, and are displayed in (a) and (b) as stereographic projections. The reciprocal space rotation is described in polar coordinates (ϕ, ω, κ).

In (a), at $\kappa = 180°$, the peaks corresponding to the crystallographic 2_1-axis can be seen at the perimeter of the plot $(\omega = 90°)$ and occur at φ values of 90 and $-90°$. Two strings of seven unique NCS two-fold axes are generated from each double heptamer in the unit cell. These appear as arcs above and below the equator and reflect the tilt in the plane of each heptamer with respect to the *xz* plane owing to their positioning about the 2_1-axis.

In (b), at $\kappa = 51°$, the displacement of each peak from the perimeter at $\varphi = 90°$ and $-90°$ represents the 35° tilt of the NCS seven-fold axis of each double heptamer related by 2_1 symmetry from the *y*-axis (Roberts *et al.*, 1999).

G3.5 Finding a model to fit the data

G3.5.1 The model

The ultimate aim of protein crystallography is to build a model of the crystal contents in terms of atomic positions and temperature factors. The steps involved are: (1) phasing the structure factors; (2) calculating an electron density distribution; (3) fitting an atomic model to the electron density; (4) refining the model with respect to the initial experimental observations and structural chemistry (and of course step; (5) is the deposition of the structure in the data bank and its publication). It is important to emphasize, however, that the experimental observations are the diffracted intensities. The reliability of the model is assessed with respect to: (1) the experimental observations and (2) the chemistry of the structure.

G3.5.2 Assessing agreement between the model and the data

The agreement between the calculated model and the observed data can be expressed in several ways. Structure factor amplitudes calculated from the model, $|F_{calc}|$, can be compared with the observations, $|F_{obs}|$. Recall (Eq. (G3.6)), the electron density is calculated from

$$\rho_{calc}(x, y, z) = \frac{1}{V} \sum_h \sum_k \sum_l |F_{obs}(h, k, l)| \cos[-2\pi i(hx + ky + lz) + i\alpha_{estima}(h, k, l)]$$

A standard linear correlation coefficient (CC) between the observed and calculated structure factor amplitudes is determined from the following.

$$CC = \frac{\sum_{hkl}(|F_{obs}(h, k, l)| - \langle|F_{obs}(h, k, l)|\rangle) \times (|F_{calc}(h, k, l)| - \langle|F_{calc}(h, k, l)|\rangle)}{\left[\sum_{hkl}(|F_{obs}(h, k, l)| - \langle|F_{obs}(h, k, l)|\rangle)^2 \times \sum_{hkl}(|F_{calc}(h, k, l)| - \langle|F_{calc}(h, k, l)|\rangle)^2\right]^{1/2}}$$

$$(G3.12)$$

where $\langle|F_{obs}(h, k, l)|\rangle$ and $\langle|F_{calc}(h, k, l)|\rangle$ are the average structure factor amplitudes for the observed and calculated data, respectively. The values of the correlation coefficient range between 0 and 1. The larger value occurs when the calculated and observed structures factors are identical.

The R_{factor} is traditionally used to indicate the 'correctness' of a model structure. It is defined by

$$R_{factor} = \frac{\sum_{hkl} ||F_{obs}(hkl)| - K|F_{calc}(hkl)||}{\sum_{hkl} |F_{obs}(hkl)|} \qquad (G3.13)$$

where K is a scaling factor.

A distribution of atoms placed at random in the unit cell would have an R_{factor} of 59% for *acentric* (non-centro-symmetric) reflections; the figure is 83% for

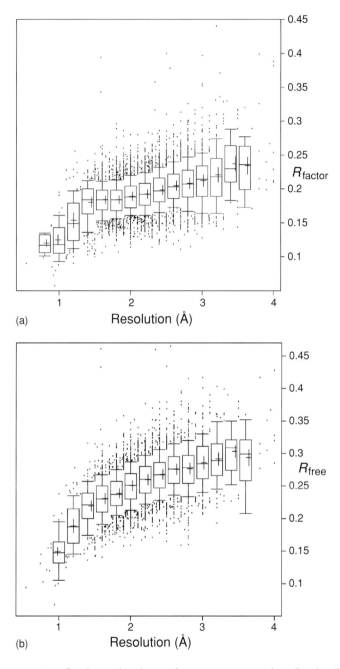

(a)

(b)

Fig. G3.11 (a) The distribution of R_{factor} values versus resolution. (b) The distribution of R_{free} values versus resolution.

The data detail 10 888 macromolecular structures deposited in the PDB between 1991 and 2000. The 'whiskers' indicate the tenth and ninetieth percentile of the data in the bin, whereas the upper and lower boundaries of the box indicate the twenty-fifth and seventy-fifth percentile. The horizontal line inside the box indicates the median (i.e., the fiftieth percentile), and the cross-hair inside the box indicates the average value (in both directions). Finally, in order to show the distribution of the 'outliers,' all individual data points outside the whiskers are shown (Kleywegt and Jones, 2002). (Figures reproduced with permission from Elsevier.)

centric reflections (the phase of a centro-symmetric reflection is 0 or π). During protein structure refinement (see below), the R_{factor} can typically be lowered to values in the 10–25% range (Figure G3.11(a)).

It is possible to *overfit* the model which results in an artificial lowering of the R_{factor}. The term overfitting refers to the case in which too many parameters are refined for the number of observations present. In order to guard against overfitting, an R_{free} term was introduced. It is calculated from structure factor amplitudes that are not included in the refinement (Fig. G3.11(b)). The R_{free} therefore validates the extent to which the model explains the diffraction data.

$$R_{\text{free}} = \frac{\sum\limits_{hkl \subset T} ||F_{\text{obs}}(hkl)| - K|F_{\text{calc}}(hkl)||}{\sum\limits_{hkl \subset T} |F_{\text{obs}}(hkl)|} \tag{G3.14}$$

The scaling factor K used is the same as for the R factor. The subset T contains the reflections set aside from the refinement. An estimated 500 reflections are necessary for the R_{free} to be statistically significant.

G3.5.3 Assessing agreement between the model and chemistry

Chemistry imposes certain rules on the conformation of molecules. Standard bond lengths and bond angles are typically obtained from crystallographic studies of small molecules. Information from high-resolution structures of proteins is also frequently incorporated when defining the stereochemical parameters of atomic models.

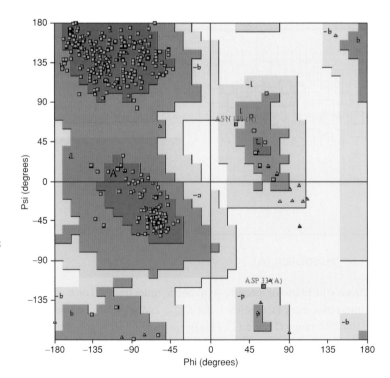

Fig. G3.12 The Ramachandran plot for protein CD1b (PDB accession code 1GZQ). The structure was solved at 2.26Å resolution and has two residues (Asn128 and Asp33) in the generously allowed regions.

Triangles represent glycines and squares the other residues in the protein structure (Gadola *et al.*, 2002).

The main chain of a protein is only allowed certain conformations due to steric clashes between the amino acid side-chains. Recall that the conformation of the polypeptide chain is described by the φ and ψ rotation angles around the C_α carbon bonds and that there are areas for pairs of these values on a Ramachandran plot corresponding to secondary structures such as α-helices and β-sheets and other 'acceptable' conformations (Fig. G3.12 and Chapter A2). The plot contains forbidden areas and if an amino acid residue falls there, its conformation is obviously a mistake in the structure and has to be changed.

An analysis based on the notion that residues have preferred positions in a protein structure can also be used to assess the plausibility of the model. For example, apolar residues are more likely to be in a hydrophobic environment (Comment G3.15).

G3.6 From the data to the electron density distribution – initial phase estimate

Phase information has to be obtained in order to calculate an electron density map. Several experimental procedures are used to phase observed structure factor amplitudes.

G3.6.1 Argand diagram

Recall from Chapter G1 that each structure factor can be expressed as $\mathbf{F} = |\mathbf{F}|.\exp i\alpha$, which can be plotted on an Argand diagram, and where α is a phase angle (see Chapter A3). Each reflection has its own Argand diagram (Fig. G3.13).

The Argand diagram is a fast and easy way of visualising structure factors and phases. A generic Argand diagram (for reflection (hkl)) will be extensively referred to in order to help clarify the different methods used to solve the phase problem.

> **Comment G3.15**
> **VERIFY3D**
>
> The program VERIFY3D checks the plausibility of a model. The incorrect, but first published structures of Rubisco small subunit p21 ras and HIV protease would have been flagged by VERIFY3D (Bowie *et al.*, 1991; Luthy *et al.*, 1992).

Fig. G3.13 Argand diagrams. Each structure factor can be represented on its own diagram. Examples are shown for reflections (100), (200), (615) and for a generic reflection (*hkl*).

G3.6.2 Molecular replacement

A starting model, which is sufficiently similar to the molecule(s) in the crystal, can be used to provide an initial estimate for the phases (Comment G3.16).

Comment G3.16   Molecular replacement programs

Several computational procedures to optimise the orientation superimposition have been developed and these procedures often have differing target functions. They include in the CCP4 package, the programs Molrep and AMoRe and in the CNS package, the cross-rotation.inp script (Vagin and Teplyakov, 2000; Navaza and Saludjian; 1997; DeLano and Brunger, 1995; Tong and Rossmann, 1997).

Fig. G3.14 Argand diagram for molecular replacement. The measured structure factor amplitude of the protein is given the phase of the one calculated from the model.

Reasonable homology can be defined as a structure with at least 30% sequence identity, which typically has been shown to imply an overall root mean square deviation between the coordinates of less than 1.5 Å. Moreover, superposed X-ray structures of homologous proteins lead to better results than a single search model as these ensembles highlight conserved structural features.

The aim of molecular replacement is to choose judiciously the position and orientation of the search model, and hence the phase origin, in order for the structure factors of the search model and of the unknown structure to be similar. The phases calculated for the search model can then be transcribed to the crystal's structure factors (Fig. G3.14). In the case of one molecule per asymmetric unit, six parameters have to be determined in the molecule replacement approach. It is, however, simpler, and less demanding computationally, to separate the three orientation parameters of the model from the translation coordinates. For example, in real space, after the superposition of the centres of mass of two objects, a rotation function can be used to maximise the overlap between them in space. The translation function corresponds to the initial distance between the centres of mass.

The rotation step in molecular replacement is usually calculated from the overlap between the Patterson functions (see Section G3.4.9) calculated, respectively, from the observed data and from the model. The *rotation function* is defined as the overlap between the two Patterson functions. It is maximum when the two overlap optimally. The Patterson function of the model is preferably calculated from the molecule placed in a very large unit cell with no symmetry in order to avoid the contribution of intermolecular vectors.

A translation search is subsequently undertaken with the oriented model in the crystal space group and unit cell. The aim is to obtain the best overlap between the calculated and observed data now including intermolecular vectors.

When there is more than one molecule in the asymmetric unit, a similar (but more complicated) procedure is applied. If a molecule is missing from the molecular replacement solution, the crystal packing may have holes and the omitted molecule should appear as missing density (see Section G3.7.1).

In the presence of multicomponent structures, such as antibodies, slight movement of protein domains will not give clear of molecular replacement solutions. It is hence advised either to do the search domain by domain or to attempt an overall rotation function followed by independent rotations for each domain, prior to an overall translation function.

G3.6.3 Direct methods

Ab initio phasing methods were essential when the first protein structures were determined and must be still be used for new protein structures for which there is no known suitable homologous structure to serve as a search model. Furthermore, molecular replacement suffers from *phase bias*; the resultant structure may resemble the search model even though it is the wrong solution.

Phases derived from Patterson maps

It is possible to interpret the phases directly from the Patterson map. If the individual Patterson peaks are fully resolved, the interatomic vectors can be determined accurately and are therefore sufficient to construct a model structure. In practice, however, only very small molecules are capable of providing such data.

Phase relationships between structure factors

A certain amount of phase information is contained in the reflection intensities. The diffracting crystal is a real object composed of atoms, whose electron density is positive everywhere in the unit cell. These assumptions of atomicity and positivity limit the number of possible phases but more importantly create phase relationships between the different reflections. For example, the phase of a reflection must be such that the respective real-space wave maxima overlap positions of high electron density in the crystal. The phases of three waves intersecting at a position of high electron density (for instance an atom) therefore have a phase relationship, in order for their maxima to overlap in the crystal. In the case of the indices of three reflections which sum to zero, the *triplet relationship* between their respective phases φ is

$$\varphi_h - \varphi_k - \varphi_{h-k} \cong 0[2\pi] \qquad (G3.15)$$

where $[2\pi]$ is the modulus. It is obvious then that the phase for reflection $\mathbf{h}$ can be determined if phases for reflections $\mathbf{k}$ and $\mathbf{h-k}$ are known, or

$$\varphi_h = \varphi_k + \varphi_{h-k}$$

or less succinctly,

$$\varphi_h = \varphi_{k1} + \varphi_{h-k1}, \quad \varphi_h = \varphi_{k2} + \varphi_{h-k2}, \quad \varphi_h = \varphi_{k3} + \varphi_{h-k3} \quad \text{etc.}$$

The *tangent formula* lets us refine the phase of reflections **h** if the phases of reflections **h** and **h−k** are known approximately,

$$\tan(\varphi_h) = \frac{\sum_k |E_k E_{h-k}| \cos(\varphi_k + \varphi_{h-k})}{\sum_k |E_k E_{h-k}| \sin(\varphi_k + \varphi_{h-k})} \qquad \text{(G3.16)}$$

where E_h is the normalised structure factor:

$$E_h^2 = \frac{F_h^2}{\left\langle F_h^2 / \varepsilon \right\rangle}$$

where ε is determined directly from the space group and takes the multiplicity of the reflection into account.

Direct methods are based on these relationships. However, few structures of proteins have been solved *ab initio* in such a way because direct methods require very good data (<1 Å). This methodology is nonetheless often used in conjunction with other phasing techniques. Provided there are a small number of starting phases, a complete data set can be iteratively deduced.

G3.6.4 Single and multiple isomorphous replacement (SIR, MIR)

The addition of heavy atoms to a crystal produces a different diffraction pattern. If the protein is not distorted, i.e. it is *isomorphous*, the structure factors and phases of the protein component of the crystal are unchanged, although the new observed intensities take into account the presence of the extra atoms. The influence of the heavy atoms can then be separated from that of the rest of the molecule as observed in an Argand diagram (Fig. G3.15).

If we recall the Patterson function (Eq. (G3.11))

$$P(uvw) = \iiint_{x\ y\ z} \rho(xyz)\rho(x+u, y+v, z+w)\partial x \partial y \partial z$$

a peak at position **r** in the Patterson map is proportional to the sum of the product of the electron densities of atoms separated by the vector **r**. Since heavy atoms have greater density than the atoms usually found in proteins (hydrogen, carbon, oxygen, nitrogen), peaks at interatomic vectors between heavy atoms appear clearly in the map obtained by subtracting the Patterson diagram of the native protein dataset from that of the protein with heavy atoms.

For example, the extra scattering intensity of one mercury atom added to protein containing 1000 atoms, would be proportional to $1 \times 80^2 = 6400$. Since the mean scattering intensity of the protein on its own is $1000 \times 7^2 = 49\,000$ (as there are on average seven electrons per protein atom), the single mercury atom added represents 1.3% of the mean scattering intensity, which is observable. The

Fig. G3.15 Argand diagram for the presence of a heavy atom. F_P, F_Q, are respectively the structure factors of the protein and heavy atom alone. F_{PQ} is the structure factor of the protein structure containing the heavy atom.

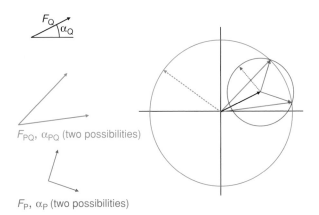

effect of a derivative on the intensities is quantified by

$$R_{\text{derivative}} = \frac{\sum\limits_{hkl} |I_{\text{derivative}} - I_{\text{native}}|}{\sum\limits_{hkl} I_{\text{native}}} \qquad (G3.17)$$

where the *hkl* reflections can be limited to a small oscillation angle ($\sim 10°$).

Due to the small number of heavy atoms per asymmetric unit it is possible to estimate their coordinates from standard methods used in small-molecule crystallography, and therefore phase the structure $\mathbf{F_Q}$ corresponding to the heavy atoms alone. For each *hkl* value, the Argand vector of the derivative ($\mathbf{F_{PQ}}$) is the sum of the Argand vectors of the protein on its own ($\mathbf{F_P}$) and of the heavy atoms ($\mathbf{F_Q}$). As the phases for $\mathbf{F_Q}$ are known, only two possibilities exist where the structure factor amplitudes of the protein and its derivative are compatible (Fig. G3.16). In order to choose the correct phase angle, it is therefore preferable to use data from two or more different heavy atom crystals in order to produce a single solution for the phase of the protein crystal. Using statistical approaches, including error estimates for the intensities, it is also possible to estimate likely phases from a single isomorphous replacement derivative.

G3.6.5 Single and multiple anomalous dispersion (SAD, MAD)

All atoms have the property that their electrons can be excited from a lower to a higher energy by incident photons of a given wavelength. A secondary photon is emitted as the electron returns to its shell. An anomalous signal refers to a phase change in scattering due to the absorption contribution, which results in a complex scattering amplitude written as,

$$f' + \mathrm{i} f'$$

Fig. G3.17 Typical
absorption edge chart of
an atom.

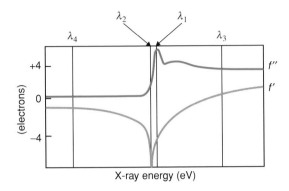

It depends on the wavelength (λ) of the incident X-ray and does not diminish with the diffraction angle (θ), because in practice the electron acts as point scatterer.

The X-ray absorption spectra of atoms are well known and tabulated. The example in Fig. G3.17 displays its key characteristics. The strength of the signal is expressed in electrons.

The total scattering amplitude of the anomalous scatterer is written as

$$ f = f^0(\theta) + f'(\lambda) + \mathrm{i}f''(\lambda) $$

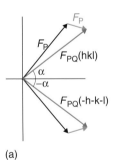

Fig. G3.18 Anomalous
scattering amplitudes.

where $f^0(\theta)$ is the normal form factor (Fig. G3.18). The structure factor hkl due to the anomalous atom alone can now be written

$$ \mathbf{F}(hkl) = \mathbf{F}^0(\theta) + \mathbf{F}'(\lambda) + \mathbf{F}''(\lambda) $$

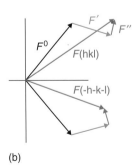

(a)

where $\mathbf{F}^0(\theta)$ and $\mathbf{F}'(\lambda)$ are parallel and perpendicular to $\mathbf{F}''(\theta)$ (Fig. G3.17).

In the case of normal scattering, *Friedel's law* states that the structure factors for (hkl) and ($-h\ -k\ -l$) (*Bijvoet pairs*) have equivalent amplitude but opposite phases (Fig. G3.19). The presence of an anomalous signal would add extra vectors $\mathbf{F}'$ and $\mathbf{F}''$ to the vector and result in the breaking of the symmetry in the Argand diagram (Fig. G3.19(b)).

An *anomalous Patterson map* is calculated from the intensity differences between Bijvoet pairs. The anomalous scattering is very weak, usually equivalent to the strength of a few electrons. Highly redundant data enables the precise determination of each reflection's structure factor amplitude, so as to reduce to a minimum the noise of the Patterson map. Vectors between anomalous scatters are nevertheless present and when identified, provide phasing information.

(b)

Fig. G3.19 (a) Friedel's
law; (b) the effect of
anaomalous dispersion
on $\mathbf{F}(hkl)$ and $\mathbf{F}(-h-k-l)$.

During data collection, the X-ray wavelengths must be chosen with care. The first choice is to choose the wavelength that produces the largest anomalous signal (i.e. where $|f''|$ is largest). The wavelength is then varied to collect at maximal f'. Third and fourth data sets are usually collected at remote points of the absorption spectrum. As in the case of MIR, it is better to have as many anomalous derivatives as possible, each with a different anomalous scatterer.

The anomalous signal produced by heavy atoms, soaked into the crystal, is also often used in combination with MIR. The anomalous Patterson map is correlated with the difference Patterson map so as to position the anomalous scatter in the crystal with less ambiguity. Ligands with a number of anomalous scatters bound together are hence particularly valuable. Since the distances between each atom in the ligand are known, they can be readily identified in the Patterson maps and consequently help phase the rest of the structure (Comment G3.17).

Comment G3.17 Phasing the ribosome

The *T. thermophilus* 30 S ribosomal subunit at 5.5 Å resolution was solved by using the anomalous scattering from osmium and lutetium. The *Haloarcula marismortui* 50 S ribosomal subunit at 5 Å resolution was phased with the help of heavy atoms like tungsten and tantalum cluster compounds, and the anomalous signal of osmium. The whole ribosome *T. thermophilus* (70 S) at 7.8 Å resolution was phased with the help of the anomalous signal from iridium (Yonath *et al.*, 1980; Gluehmann *et al.*, 2001; Wimberly *et al.*, 2000; Ban *et al.*, 1999; Cate *et al.*, 1999).

G3.7 From the electron density to the atomic model – refinement of the model – phase improvement

Because of uncertainties in the data, the structure factor Argand diagrams are composed of vectors whose amplitudes are not known exactly (Fig. G3.20). The

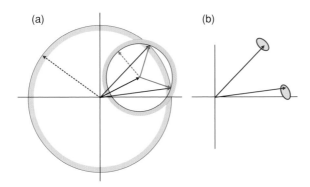

(a) (b)

Fig. G3.20 Argand diagram with errors.
If the measured structure factor amplitudes are not exactly known (a), this results in a level of uncertainty for both the amplitude and phase of the protein's structure factors (b).

Comment G3.18

Programs to
visualise and
calculate electron
density

The program O is often
used. The electron
density maps can be
calculated with the
CNS and CCP4 suite
of programs using the
X-ray data (Jones
et al., 1991).

ambiguity in each measured vector results in further error in the phase angle
estimate. Several procedures have been developed to refine the phases in order
to create a plausible structural model.

G3.7.1 Fitting to the electron density by eye and by hand

The interpretation of the electron density map of the crystal is done with the aid of
interactive molecular graphics programs (Comment G3.18). In order to view the
electron density accounted for by the model, it is usual to calculate as so-called
$2F_o$-F_c map,

$$\rho(x, y, z) = \frac{1}{V} \sum_{hkl} (2|F_{obs}| - |F_{calculated}|) \exp[-2\pi i(hx + ky + lz) + i\alpha_{calculated}]$$

(G3.18)

The structure factor $F_{calculated}$ is determined from the model's electron density,
and the phase angles $\alpha_{calculated}$ of the model are used. In such a map, F_{obs} is said
to be weighted twice with respect to $F_{calculated}$.

In order to visualise the location of the electron density not accounted for by
the model, a F_o-F_c map is calculated instead:

$$\Delta\rho(x, y, z) = \frac{1}{V} \sum_{hkl} (|F_{obs}| - |F_{calculated}|) \exp[-2\pi i(hx + ky + lz) + i\alpha_{calculated}]$$

(G3.19)

There are other weighting schemes for viewing the electron density and some
crystallographers prefer to view their data in $3F_o$-$2F_c$, or $2F_o$-F_c type maps as they
believe that it is less model-biased. These are 'native' maps ($F_{obs} \exp(i\alpha_{calculated})$)
into which difference Fourier terms are introduced.

When the positions of certain atoms are problematic, *simulated annealing
OMIT maps* can be created. Basically, the occupancy of the atoms in the trou-
blesome area in the model is set to zero. This keeps chemical interactions intact
although the atom no longer contributes to the electron density. A simulated
annealing can then be undertaken to rid the model of bias (see Section G3.7.4)
and the new maps hopefully show where the missing density should be.

G3.7.2 Minimisation of a target function
(maximum likelihood)

Phases have a very strong influence on the appearance of the electron density.
In molecular replacement, the phases of the search model are used with the
diffraction data in order to calculate the starting electron density of the structure.
The initial solution of the phase problem fits the information gleaned from the
diffraction data but the resultant macromolecular structure may not make sense
physically, e.g. there may be steric clashes within the unit cell. The purpose of

the refinement procedure is to optimise the agreement between an atomic model and both observed diffraction data and chemical restraints.

An entirely general refinement method would treat the chemical restraint information like the observed diffraction information. The squared residuals arising from the calculated and ideal chemical restraints and from the suitably weighted observed and calculated data can then be added together and minimised. The refinement can hence be conceptualised in terms of finding a minimum for the target function E as parameters are improved in cycles of refinement,

$$E = E_{chem} + W_{X\text{-ray}} \cdot E_{X\text{-ray}} \qquad (G3.20)$$

The term E_{chem} describes empirical information about the chemical interactions in the crystal. $W_{X\text{-ray}}$ is a weight that controls the relative contributions of the E_{chem} and $E_{X\text{-ray}}$ terms. The $E_{X\text{-ray}}$ term describes the difference between the observed and calculated diffraction data and can be understood as the experimental potential energy. It was initially calculated from

$$E_{X\text{-ray}} = \sum_{hkl} \left(|F_{obs}| - k|F_{calculated}| \right)^2 \qquad (G3.21)$$

where k is a relative scale factor.

The $E_{X\text{-ray}}$ term has been subsequently modified into a *maximum likelihood* target function:

$$E_{X\text{-ray}} = \sum_{hkl} \frac{1}{\sigma_{ML^{cv}}^2} (|F_{obs}| - \langle |F_{obs}| \rangle^{cv})^2$$

where the expected value, $\langle |F_{obs}| \rangle^{cv}$, is derived from the calculated structure factors taking into account, approximately, the errors in the atomic model, and where the variance $\sigma_{ML^{cv}}^2$ embodies both data and model errors. The $|F_{obs}|^2$ intensities, for instance, should follow a Wilson distribution.

The model then becomes more and more complete during the course of the refinement. Missing side-chains, ions and water molecules can be added as soon as the noise levels in the electron density maps decline sufficiently. Water molecules, for example, can be identified if they are at the correct distance and orientation from donor and acceptor atoms in the model. Ions and small molecules are usually distinguished by their position with respect to the protein's surface and appear in the F_o-F_c maps, as missing density that the electron density of a water molecule in that location cannot fill (Comment G3.19).

G3.7.3 Crystallographic refinement restraints

Bulk solvent correction
As detailed in the previous discussion on the Wilson plot, the effect of the solvent on low-resolution diffraction data must be accounted for and a bulk correction must be applied.

Comment G3.19
Automatic model building

Automatic building and refining procedures are being developed. An example of a successful program is ARP/wARP which can build structures from initial phase estimates obtained from high-resolution data, or improve on partial models at resolution better than 3 Å (Perrakis *et al.*, 1997).

Chemical restraints

The E_{chem} energetic term describes the physical restrictions that the crystallographic model has to obey. In order for the refinement to proceed, it is therefore necessary to explicitly define the chemical characteristics of the model, such as the presence of covalent bonds and atomic charges.

E_{chem} is a function of the atom positions (bonds, angles, dihedrals, chiral volume, non-bonded contacts . . .). It was initially designed for macromolecular simulations but subsequently modified for structure refinement (see Chapter I1). Weighting factors for certain of the stereochemical restraints had to be increased, for instance in order to keep aromatic rings planar. Similarly, the charges of certain residues have to be set to zero; the charges on arginine, lysine and glutamic and aspartic acids are often screened by the bulk solvent. The covalent bond length distortions are also kept smaller than 0.02 Å and bond angle distortions to less than $3.0°$.

The stereochemical constraints within molecules can be established from comparison with other molecules previously characterised by X-ray crystallography. Perusal of the Chemical Database Service, which contains structures of several hundred thousand small molecules, provides reference to the conformation geometries available to each molecule. As more macromolecules are determined, statistical analysis will further underpin information about a protein structure's general properties. This geometric information is then often codified into dictionaries for use in different refinement programs.

The alteration of the weight $W_{X\text{-ray}}$ in the target function E decides which term, the model's chemistry or the observed X-ray data, should have the greatest influence during the refinement. The optimal weight $W_{X\text{-ray}}$ can be estimated from the gradients between the two energy terms after an unrestrained molecular dynamics simulation and, as the X-ray refinement progresses, it must be adjusted as the estimated positions of the atoms approach their true value in the crystal structure.

Molecular averaging

When more than one molecule is present inside the asymmetric unit cell, the molecules can be related by NCS. The electron densities of the related molecules are essentially equal, although differences may exist in the contact regions between neighbouring molecules. The NCS operators can therefore be imposed to constrain or restrain the molecules to be similar during the refinement. Constraints confine the molecule's features to specific values, while restraints restrict each feature to a realistic range of possibilities.

G3.7.4 Refinement procedures

Phase extension

As discussed in Section G3.6.3, phase relationships exist between different structure factors. The phase extension enables phases for higher-resolution structure factors to be estimated from the phases of lower-resolution data.

Solvent flattening

In solvent regions of the crystal, atoms are disordered so they are not represented by discrete positions in the electron density map. It is therefore possible to set an overall density in those regions. For example, a molecular envelope, or mask, can be created around protein and any density outside that mask is set to a uniform level. This procedure increases the quality of the electron density of the model.

Gradient descent refinement methods

The refinement procedure tries to find the minimum for the function E. In *rigid-body refinement*, the atoms of the model are kept in rigid positions and the whole model is moved in order to optimise correlation with the observed data. In the refinement of multicomponent proteins, each domain can be treated as a rigid body.

The subsequent refinement of each of the model atoms' coordinates increases the agreement between the observed X-ray data and the chemical parameters describing the model (Fig. G3.21).

Fig. G3.21 Energy landscape showing the path followed in a gradient descent (red) and in simulated annealing (blue).

B factor refinement

The scattering power of an atom is strongly dependent on its temperature factor B. In theory, this term reflects the atom's thermal motion, but may also indicate local differences in its conformation. Recall that for atom j, its B factor is

$$B_j = 8\pi^2 \langle u^2 \rangle$$

where $\langle u^2 \rangle$ is the atom's mean square displacement relative to the average position. The structure factor amplitude contribution of each atom is therefore modified:

$$f_j = f_{0,j} \exp\left[-B_j \left(\frac{\sin^2 \theta}{\lambda^2}\right)\right] \tag{G3.22}$$

where $f_{0,j}$ is the scattering factor for atom j calculated at 0 K.

Temperature factor refinement enables a better fit of the calculated structure factor to the observed data but introduces additional parameters into the refinement. It may therefore be necessary to limit the refinement to the overall B factor of the model in the case of low parameter/observation ratios. At very high resolution (1 Å), an anisotropic description of the B factor for each atom is possible, as an atom's thermal motions parallel and perpendicular to its bonds are different. At high resolution, ($\sim$2 Å), an isotropic B factor has to suffice. Nearer 3-Å resolution, it is suggested to have two B factors per residue; one for the main chain and the other for side-chain atoms, and at very low resolution, only one B factor per domain is appropriate.

Simulated annealing

In *gradient descent* methods, the refinement may become trapped in one of several local minima. *Simulated annealing* can overcome such problems as it allows large areas of space to be sampled by the model's atoms because the refinement can proceed against the target energy gradient and therefore escape from local minima. It is therefore particularly useful after molecular replacement as it lowers the amount of bias towards the search model at the beginning of the refinement.

Simulated annealing is based on molecular dynamics simulation protocols (see Chapter I1). At its origin, this procedure involves solving the Langevin equation of motion, which describes the force acting on each atom in terms of the bond force, a frictional force due to solvent effects and a random force due to temperature. Kinetic energy is initially added by increasing the temperature (to 1000 K for example) in order to permit the model to sample many alternative conformations. The model is then slowly cooled until the atoms end up in a conformation more energetically compatible with the X-ray diffraction data. The simulated annealing temperature term has no physical meaning and is, in fact, unreasonable from the biological point of view. It can, however, be understood as the likelihood of overcoming the energetic barriers present in the target function E. The diagram in Fig. G3.21 shows an energy landscape and the simulated annealing procedure.

G3.7.5 Final assessment of the structure

Information and resolution

The quality of a model depends on the quality of the data used in its creation. It is usually assumed that the positional error is roughly 10% of the quoted resolution.

High resolution enables one to place atoms accurately in the model. It may also uncover residues with alternative conformations. High resolution also enables more complete models to be built since water and other molecules can be identified. As the resolution is lowered, it becomes more and more difficult to place first atoms, then individual side-chains in the structure. The presence of NCS and independent phasing increases the mount of information. For example, although data only extend to 4 Å, the structure of the highly symmetric Blue Tongue Virus is very informative. The 60-fold averaging procedure creates a model from which side-chains can be discerned. Similarly, the origins of anomalous diffraction can be identified with certainty even in a low-resolution map. In fact, in the initial maps derived from MAD phasing, the approximate positions of the residues in a protein are deduced from the locations of the selenomethionines identified from the anomalous signal.

Publishing a structure

Scientific journals require that a model's coordinates and B factors are available prior to publication. Structures currently deposited in the PDB list further information such as relevant publications and details about the protein (sequence, production method . . .). The crystallographic details principally include the crystallization conditions, the X-ray source, the computer programs used during the structure determination and refinement and importantly the statistics associated with the data processing and the final model (Comment G3.20). Current trends

---

Comment G3.20
Biologist's box: Crystallographic statistics reported in publications

A crystallographic paper usually has two tables. The first describes the diffraction data and usually includes: space group and unit cell, wavelength, resolution, number of unique reflections, completeness, redundancy, signal-to-noise ratio and R_{merge}. The second describes the refinement and details R_{factor}, R_{free}, the number of non-hydrogen atoms, the number of water molecules, the root mean square deviation (RMSD) from standard values in bond lengths and in bond angles, average B values and Ramachandran plot statistics.

are forcing structure factors to be deposited also as interested scientists can then check the quality of the electron density maps and decide on the validity of the published interpretation.

G3.7.6 Structural genomics

Structure determination of a protein by X-ray crystallography has progressed to the point that several steps in the process can be automated. For example robots have been designed for cloning and expression trials, and software can index diffraction intensities, refine and build structural models which fit the X-ray data.

In conjunction with the short data collection times associated with synchrotron radiation, it has become possible to study hundreds of proteins in parallel. This has several obvious advantages. Numerous protein constructs can be screened for solubility and crystallization. The elucidation of all the proteins of an organism (*structural genomics*) and the comparison of a protein's homologues expressed in different species (*comparative genomics*) offer very interesting biological insights. Finally, the determination of every possible protein fold provides crucial empirical knowledge for structure-from-sequence modelling approaches.

G3.8 Kinetic crystallography

All chemical and biological reactions involve atomic motion, embodied in dynamic structural changes. The aim of kinetic crystallography is to observe a number of structures, associated with reaction intermediates. The lifetime of these states varies between reactions; it can be extremely brief (in the femtosecond range), or last for hours.

G3.8.1 Trapping of intermediate states

During the course of a reaction, a molecule goes through a series of intermediate states. In an enzymatic reaction, these are the steps leading from substrates to products. The associated structures can be determined by physical or chemical trapping techniques, designed both to increase the lifetime of the molecules in a desired state and to trap a structurally homogeneous species.

An obvious trapping mechanism is to initiate a reaction (e.g. with a laser pulse), wait for the reaction to proceed sufficiently and then flash cool the crystal. It is often preferred, however, to use substrate analogues, which block the reaction from going any further. For example, hydrolysis of ATP to ADP, and subsequently to AMP, provides energy to drive many enzymatic reactions. The use of non-hydrolysable forms of ATP and of ADP can inhibit the reaction at specific stages, from which structural mechanisms involved in the reaction can be identified.

G3.8.2 Laue diffraction and time-resolved crystallography

In contrast to trapping the crystal in specific states, time-resolved crystallography follows the evolution of the space-averaged structure in real time. The biochemical reaction is initiated in the crystal and followed. However, in order for this to succeed the crystal must not become disordered during the reaction. As *time* is of the essence, data collection has to be as rapid as possible. The advent of high-brilliance synchrotrons allows collection of meaningful data over very short exposure times. A special diffraction setup called a *Weissenberg camera* is often used as it measures more data in one exposure than a flat detector. A Laue diffraction approach significantly increases the number of data that can be collected concurrently. This method differs from the usual crystallographic studies as it uses polychromatic radiation, If we recall Comment G3.3, in order to satisfy Bragg's law, $n\lambda = 2d \sin\theta$, a crystal only diffracts at particular wavelengths for specific values of d and θ. The use of a broad spectrum of radiation consequently increases the number of crystal planes diffracting constructively for one crystal orientation. Figure G3.22 details the relevant Ewald construct. If wavelengths ranging from λ_{min} to λ_{max} fall on a crystal, all lattice points lying between the Ewald spheres of radii λ_{max}^{-1} and λ_{min}^{-1} are detected. A large number of reflections is therefore observed in a Laue experiment. In a typical experiment, for a crystal diffracting at 2 Å resolution and $\lambda_{max} - \lambda_{min} = 2$ Å, 10–60 times more reflections can be collected in one exposure than for a monochromatic experiment. Although spatial overlap of spots may become a problem, this usually does not affect more than 17% of spots in realistic experimental conditions.

In time-resolved Laue diffraction studies, the structure can be determined directly from the diffraction pattern, although it is much simpler to determine the starting and final conditions from static structures solved with monochromatic resolution. As the reaction proceeds, the time-dependent average structure in the crystal varies due to the time-dependent rise and fall of the concentrations of the intermediate states of the molecule (Fig. G3.23). The structures themselves do not vary with time, but since their concentrations do, the diffraction varies over time. If the crystal contains a set of transient structures $(1, 2, 3, \ldots, i)$ with fractional occupancies $(C_1, C_2, C_3, \ldots, i)$ and structure factors $\mathbf{F}_i(\mathbf{k})$, the total structure factor of the crystal is

$$\mathbf{F}(\mathbf{k}, t) = \sum_i C_i(t)\mathbf{F}_i(\mathbf{k}) \tag{G3.23}$$

where $\sum_i C_i(t) = 1$ at all time t. The data can then be analysed in order to identify the individual transient structures in the crystal.

A typical time-resolved experiment involves *kick-starting* the biochemical reaction in the crystal with a burst of laser light. A subsequent train of short X-ray pulses is then used for data collection with temporal resolution as short as 150 ps. The photolysis of the ligand, protein relaxation and rebinding of CO

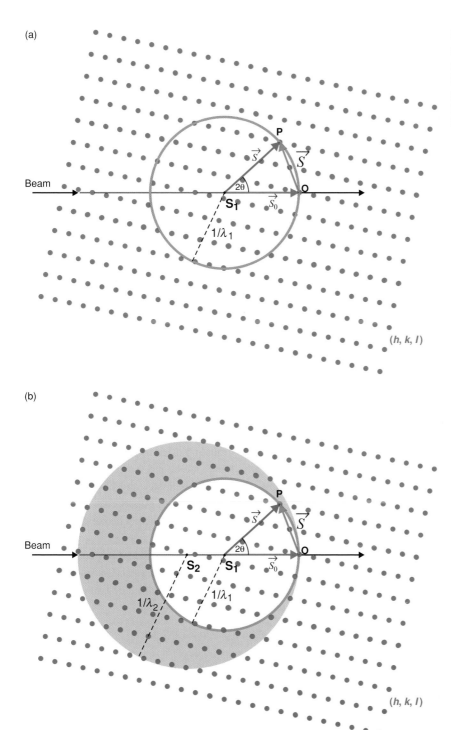

Fig. G3.22 Ewald constructs for: (a) monochromatic radiation (wavelength λ); and (b) polychromatic radiation (wavelengths ranging between λ_1 and λ_2).

Fig. G3.23 Example of a
simple irreversible
sequential chemical
kinetic mechanism with
four intermediates.

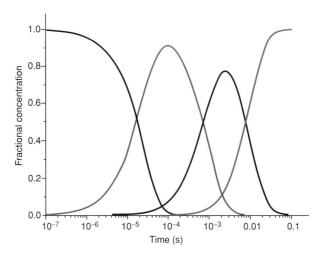

Fig. G3.23 Example of a simple irreversible sequential chemical kinetic mechanism with four intermediates.

to myoglobin, for example, has been determined to nanosecond time resolution (see Chapter A3).

G3.9 Neutron crystallography

Neutron diffraction is similar in theory and practice to X-ray diffraction. However, it provides a different information set due to the differences in atomic scattering and absorption between these two types of radiation (see Chapter G1). Significantly, neutron crystallography has the ability to locate hydrogen atoms in large molecules experimentally. X-rays, on the other hand, require high-resolution diffraction (<1 Å) and hydrogen atoms must be often inferred from the stereochemistry of the heavier atoms to which they are bound. Since typically half the atoms in a protein are hydrogens, a protein's chemistry and physical structure can be precisely detailed by neutron crystallography; in particular in structures where water and protein hydrogens have unusual, but biochemically relevant, conformations.

Neutrons scatter off the atom nuclei of the structure, which act like point sources (see Chapter G1). In neutron diffraction from a crystal (Comment G3.21), we can rewrite Eq. (G3.4), and replace the scattering amplitude f_j of atom j of the structure by the nuclear coherent scattering amplitude b_j:

$$F(hkl) = \sum_j b_j \exp[i2\pi(hx_j + ky_j + lz_j)] \qquad \text{(G3.24)}$$

where atom j is at position (x_j, y_j, z_j) in the unit cell. The interaction of the neutron's magnetic moment with the crystal has not been included, as its effect is negligible for most biological material (Comment G3.22). We recall from Chapter G1 the neutron scattering amplitude is usually a real number (either positive or negative) and does not vary much in magnitude between atom types. Hydrogen and larger atoms like carbon, nitrogen and oxygen are consequently all discernible in 2-Å

> **Comment G3.21   Incoherent neutron scattering from crystals and deuterium labelling**
>
> There are no interference effects due to the incoherent component of the scattering processes, although these contribute to background noise. In order to reduce the substantial incoherent scattering of neutrons by hydrogen atoms, crystals are therefore routinely soaked in D_2O buffer prior to data collection. The labile protons of the crystal are replaced by deuterium atoms, which have much lower incoherent scattering than hydrogen atoms. These atoms are consequently modelled as deuterium or hydrogen in the neutron structure, depending on whether or not they are accessible to solvent exchange. In order to make more efficient use of the neutron fluxes available, crystals are prepared with fully deuterated protein (obtained, for example, by recombinant expression in *E. coli*, grown on fully deuterated cultures), which significantly reduces the incoherent background and increases the coherent scattering signal (Kossiakoff, 1983).

> **Comment G3.22 Magnetic scattering amplitude**
>
> The magnetic scattering amplitude is due to the interaction of the neutron spin with the spins of unpaired electrons. The magnetic scattering amplitude of certain metals, like Co, Ni and Fe can approach the magnitude of their nuclear coherent scattering amplitude.

maps. The atomic structure can then be determined from the neutron diffraction in a similar manner to X-ray crystallography. However, phases are usually obtained from a previously determined X-ray model of the crystal.

The most serious limitation of neutron crystallography is due to the low neutron fluxes available, so a neutron structure is never determined before its X-ray counterpart. Large crystals (in the cubic millimetre range) and long collection times (weeks . . .) are therefore required. Nonetheless, the development of Weissenberg camera-type detectors for neutrons and the use of Laue diffraction methods with neutrons of different energies has expanded the scope of neutron crystallography.

G3.10 Checklist of key ideas

- The molecular structure of a protein is the basis of its function. A detailed understanding of this relationship is one of the fundamental aims of modern biology.
- Phase information is lost during a diffraction expriment since only the diffraction intensities are recorded. The aim of crystallography is to solve this phase problem in order to create a model of the crystal structure.
- The *Ewald construction* is a geometric representation of the diffraction conditions.
- In protein crystals, certain space groups are not allowed due to the handedness of biological structures. The crystal symmetry results in certain space groups exhibiting missing reflections, called *systematic absences*.
- Structures of biological molecules are deposited in the PDB database. Each atom of the model is described by its position in the unit cell (x, y, z coordinates in ångstroms), its temperature factor B and its occupancy (between 0 and 100%, according to the probability of finding it at that position).

- Protein crystals are very fragile. They are maintained by weak forces, of the level of a single hydrogen bond.
- Crystallization trials aim to identify favourable conditions for crystal growth. The main parameters varied are the ionic conditions, pH and *precipitant* concentration.
- The vapour diffusion method is the method preferred by many crystallographers. In this method, a drop containing the protein is equilibrated against a larger reservoir of solution.
- In order to reduce radiation damage, a crystal is often flash frozen.
- During data collection, the intensities measured on the detector have to be interpreted and converted to *hkl* intensity data, from which structure factor amplitudes can be calculated.
- Crystal growth anomalies may produce the phenomenon of *twinning*. Twinning refers to a specific case of disorder in which there is partial or complete coincidence between the lattices of distinct crystal domains.
- Initial estimates of the unit cell size and orientation can be obtained from diffraction data collected over a small oscillation of the crystal. The space group to which the crystal belongs can be identified from the systematic absences in the diffraction pattern.
- The signal-to-noise ratio and the χ^2 value are good criteria for assessing the quality of the collected data. The R_{merge} value can be misleading if the data are highly redundant.
- The solvent content of protein crystals (denoted by *Matthew's coefficient*) has been observed to have values between 20 and 80%, but is usually between 40 and 60%. In many cases, this range is sufficiently restrictive to enable reliable estimation of the total number of molecules per asymmetric unit from the protein's molecular weight.
- *Non-crystallographic symmetry* arises when there are two or more similar molecules present in the asymmetric unit. It can be determined directly from the diffraction intensities.
- The ultimate aim of protein crystallography is to build a model of the crystal contents in terms of atomic positions and temperature factors.
- The R_{factor} is traditionally used to indicate the 'correctness' of a model structure. In order to guard against overfitting, R_{free} is used. This is calculated from structure factor amplitudes that are not included in the refinement.
- The *Argand diagram* is a fast and easy way of visualising structure factors and phases.
- In molecular replacement, a model that is similar to the molecule(s) in the crystal can be used to provide an initial estimate of the phases.
- *Ab initio* phasing methods were essential when the first protein structures were determined and must be still be used for new protein structures for which there are no known suitable homologous structures to serve as search models.
- High-resolution diffraction data contain phase information.
- In SIR and MIR, the addition of heavy atoms to a crystal produces a different diffraction pattern, from which phase information can be obtained.
- In SAD and MAD, the presence of an anomalous signal in crystal diffraction can provide phase information.

- A model of the crystal is built that agrees with both the diffraction data and chemical knowledge of the structure. Several refinement procedures have been developed to refine the model against the data.
- Kinetic crystallography aims to interpret crystallographic data in terms of a small number of time-independent structures, each associated with a reaction intermediate. Intermediate states can be trapped or followed in real time.
- Neutron crystallography is particularly useful for the experimental determination of hydrogen atom positions in medium-resolution structures. The technique requires large crystals, however, because of the low neutron fluxes available.

Suggestions for further reading

Drenth, J. (2002). *Principles of Protein X-ray Crystallography*, second edition. Berlin: Springer-Verlag.

Rossmann, M. G., and Arnold, E. (eds.) (2001). *International Tables for Crystallography*. Volume F: *Crystallography of Biological Macromolecules*. Dordrecht: Kluwer Academic.

Part H
Electron diffraction

Chapter H1
Electron microscopy

H1.1 Historical review

J.-B. Biot and **F. Savart** collaborated on the theory of magnetism. Their epony-
mous law (formulated in **1820**) describes the motion of a charge in a magnetic
field.

Ernst Abbe in **1872** formulated his wave theory of microscopic imaging. In
1878, he was able to correlate resolution and wavelength of light mathematically.
A few years later, through the company Zeiss, he commercialised the first ever
lenses to have been designed based on sound optical theory and the laws of
physics.

In the **second half of the nineteenth century**, **J. C. Maxwell** established the
laws of electromagnetism. His equations first appeared in fully developed form
in his book *Electricity and Magnetism* (1873).

In **1897 J. J. Thomson** discovered electrons by demonstrating cathode rays
were actually units of electrical current made up of negatively charged particles
of subatomic size. For these investigations he won the Nobel Prize for physics in
1906.

In his doctoral thesis of **1924**, **de Broglie** theorised that matter has the properties of both particles and waves. He received the 1929 Nobel Prize for physics.
The particle–wave duality was confirmed experimentally, for the electron, a few
years later by **C. J. Davisson** and **L. H. Germer** and by **G. P. Thomson**. Thomson
and Davisson jointly shared the Nobel Prize in physics in 1937.

In **1926 H. Busch** described a lens for electrons. **H. Ruska** subsequently built
the first working electron microscope in **1930** (Nobel Prize in 1986). Ruska's 1931
paper contains for the first time the term 'electron microscopy'. The principle
behind this type of microscope has not varied significantly since.

In **1932**, **Fritz Zernike** invented the phase-contrast light microscope, which
uses a quarter-wave plate. He was awarded the 1953 Nobel Prize for physics
for this work, which opened up the study of transparent biological materials.
The feasibility of a scanning electron microscope was demonstrated by **Knoll**, in
1935, and the first prototype was built by **M. Van Ardenne** three years later.

By the **late 1930s**, electron micrographs of recognisable value to biologists were being published. In **1942**, **S. E. Luria** obtained the first high-quality electron micrograph of a bacteriophage.

The negative staining technique came into extensive use in the **1940s**, although it was only in **1955** that **Hall** recognised, and deliberately used, phosphotungstic acid as a negative stain.

In **1954**, **V. E. Cosslet** and **M. E. Haine** suggested the use of a field emission gun (FEG) for electron microscopy. Due to technical difficulties associated with the extremely high vacuum required, a manageable FEG microscope was not available until **1996** when one was built by **Crewe**.

In **1959**, **S. J. Singer** used ferritin-coupled antibodies to detect cellular molecules.

In **1968**, **A. Klug** and **D. J. DeRosier** introduced three-dimensional image reconstruction techniques of electron micrographs and opened the way for the interpretation of images obtained with very low radiation doses and without the use of heavy metal stains.

In **1984**, the power of this approach was greatly increased by **J. Dubochet** and coworkers, who developed methods to rapidly freeze biological specimens to achieve a near native frozen-hydrated state. The use of extremely low temperatures reduces the extent of radiation damage by as much as an order of magnitude more than at room temperature.

In **1975**, **R. Henderson** and **N. Unwin** solved the first structure of an unstained integral membrane protein (bacteriorhodopsin) by combining electron micrographs and electron diffraction. The structure was extended to atomic resolution by the same technique and published by **Henderson** and coworkers in 1990.

In the **1990s**, the single particle reconstructions of the ribosome by the groups of **J. Franck** and of **M. van Heel** breached nanometer resolution.

H1.2 Introduction to biological problems

H1.2.1 The electron microscope image

Electron microscopy (or EM) offers structural information, which is complementary and competitive with crystallography, SAS and NMR. Its resolution for biological specimens is usually of the order of 10 Å but has reached ~1 Å.

EM is based on the wave-like behaviour of electrons. It can be understood as operating in a similar manner to light microscopy with electrons, instead of photons, focused by electromagnetic lenses. The features in the EM image are the result of the interaction of electrons with the electrostatic potential distribution of the atoms in the sample.

H1.2.2 Applications of EM

Advances in electron microscopy have opened up a wide variety of macromolecular assemblies to structural and dynamical analysis. The method enables the description of biological macromolecules at nanometre resolution and better. Furthermore it can deal with a wide range of specimen geometries, ranging from asymmetric particles, through helical filaments (one-dimensional crystals), to two-dimensional crystals (and even three-dimensional crystals at low resolution).

EM is hampered primarily by radiation damage leading to a poor signal-to-noise ratio in the data, which limits both resolution and the size of the studied complex. Currently, it is typical to obtain reconstructions with resolutions in the nanometre range, although two-dimensional crystals have yielded atomic resolution reconstructions. In theory, it should be possible to determine the structures of biological molecules of mass down to 100 kDa. However, the current lower limit lies between 300 and 500 kDa for unstained specimens.

H1.2.3 Techniques covered

This chapter focuses on the methods for obtaining EM images of biological macromolecules.

Chapter H2 details how these images are processed in order to determine the three-dimensional structure of the specimen. It will cover helical reconstruction, two-dimensional crystallography and single-particle reconstruction methods.

H1.3 Principles of electron diffraction and imaging

H1.3.1 Properties of electrons

Wave-like properties of electrons
Electrons, like photons, have the dual property of behaving both as a wave and as a particle. EM is based on their wave-like behaviour. The de Broglie relationship is thus applicable to electrons and relates the momentum and wavelength of the particle by Planck's constant h,

$$h = \lambda_e p \tag{H1.1}$$

where λ_e is the wavelength of the electron and p its momentum (Comment H1.1).

Effect of the electron charge
Electrons can also be considered as charged particles; consequently their movement is influenced by electromagnetic fields. A particle of charge q and mass m placed in an electric potential difference of Φ (in volts) accelerates until the particle's kinetic energy is balanced:

$$q\Phi = \frac{1}{2}\frac{p^2}{m} \tag{H1.2}$$

where p is the equilibrium momentum. Since an electron has charge $(-e)$, its

---

Comment H1.1
Effect of an electric field on a charged particle

A charge q placed in an electric field ($\mathbf{E}$) experiences a force $q\mathbf{E}$. The field can be expressed as the gradient of a potential (Φ) such that in one dimension we can write $E = d\Phi/dx$

momentum can be written as

$$p = \sqrt{2m_e e \Phi} \qquad (H1.3)$$

where the mass of an electron m_e is 9.1×10^{-28} g.

H1.3.2 Electromagnetic lens

In electron microscopes, suitably shaped magnetic (or electrostatic) fields are used to control the direction of moving electrons. As in the case of a glass lens that focuses a light beam, these electromagnetic lenses focus electron beams. The electron microscope allows the study of a large range of length scales which overlaps with that of light microscopy (Fig. H1.1). The energy (or wavelength) of an electron is a feature specific to the accelerating voltage. It imposes limits on the electron microscope's resolution. The wavelength is given by Eqs. (H1.3) and (H1.1) as

$$\lambda_e = \frac{h}{\sqrt{2m_e e \Phi}} \approx 12 \Phi^{-1/2} \qquad (H1.4)$$

An electron microscope of 100 keV energy therefore produces electrons of 0.037-Å wavelength. Electron wavelengths are five orders of magnitude smaller than those of visible light.

Recall from Chapter G1 that resolution, wavelength and maximum scattering angle are related. The angular aperture of a lens (2α), i.e. the angle between the most divergent rays that can pass through a lens to form the image of an object, therefore controls the lens's maximum resolution, d:

$$d = \frac{\lambda_e}{\sin \alpha} \qquad (H1.5)$$

Since λ_e is very small, it is often convenient to assume that the EM data are not diffraction-limited. However, in the presence of spherical aberration, the resolution limit is lowered, as rays passing at larger angles through the lens are focused closer to the lens than the central ray. It is given in terms of geometrical optics by

$$d = \left(C_S \lambda_e^3 \right)^{1/4} \qquad (H1.6)$$

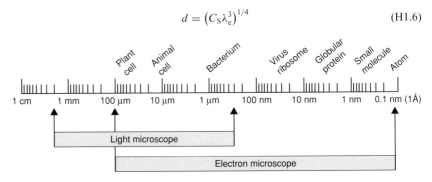

Fig. H1.1 Resolution ranges of optical and electron microscopes.

where C_S is the spherical aberration coefficient. With $C_S=1$ mm, the maximum resolution of a 100 keV microscope is 5 Å. Note also that the resolution of an electron microscope depends on $\lambda_e^{3/4}$, and therefore on its voltage.

H1.3.3 The image recorded by an electron microscope

Signal-to-noise ratio

The principal problem in EM is the radiation damage to the imaged specimen, which limits the amount of information that can be collected. The *signal-to-noise ratio* of the image expresses the confidence with which a structure can be distinguished from the noise in the image. The signal is usually defined as resulting from elastic scattering of the electrons from the sample, and the noise is due to inelastic scattering events, but also to statistical fluctuations in the number of electrons detected in different parts of the image.

The *Rose model* describes the signal-to-noise ratio (S/σ_{Rose}) necessary for the detection at a certain contrast (C) of a uniform object of area A in a uniform background:

$$S/\sigma_{Rose} = C\sqrt{An_{backgrouund}} \qquad (H1.7)$$

where the contrast is given by $|(n_{object} - n_{background})/n_{background}|$, with the mean number of electrons per area due to the background and to the object being respectively $n_{background}$ and n_{object}, and A is the area of the object. Signal-to-noise ratios in the range of 5–7 are adequate to identify features in the image. $S/\sigma=5$ represents an 83% confidence level, while $S/\sigma=1$ represents only a 50% confidence level that a structural feature can be recognised from the noise.

Noise and radiation damage

The image produced by an electron microscope contains a lot of noise. As written above, this is due to the statistical fluctuations in the number of electrons between different parts of the image and to the inelastic scattering of electrons as they travel through the sample.

In *elastic scattering*, the incident electrons do not lose energy when scattered by the sample. In *inelastic scattering*, energy from the incident electrons is deposited into the sample. Electrons from the sample can be expelled with energies 5–10 times greater than valence bond energies. The resulting radiation damage severely limits the amount of structural information that can be obtained. It is, in fact, the main resolution-limiting feature.

In EM of biological objects, due to the energies of the incident electrons, the number of inelastic scattering events always outnumbers the number of scattering events by a factor of 3–4. Energy filters, although not routinely used in biological EM, can be used to eliminate certain electron energies in order to reduce the number of inelastic events. Current trends, however, favour an increase in accelerating voltage since this reduces inelastic scattering.

H1.3.4 Transfer of information from sample to image

Comparison with optical lenses

Magnetic lenses suffer from the same defects as optical lenses (Comment H1.2); in particular, *spherical aberration* and *astigmatism*. *Chromatic defects* happen when certain electron frequencies are affected, but not others.

Comment H1.2   Magnetic lenses

The magnetic lens influences the helical path of the electron. In old microscopes, the image rotated when the defocus was changed.

Light lenses are optically almost perfect. Magnetic lenses are not. Their ever-present spherical aberration results in variable phase shifts across each lens. Their astigmatism produces stretched images, which are clearly noticeable in the image's diffraction pattern; however, the effects of astigmatism can be corrected for prior to recording the image, and subsequently during image processing.

Effect of the microscope on the path of the electron

As the electron wave moves through the microscope and the sample to the image recording plane, its amplitude is modified at different points of its journey. By convention, the electron wave travels along the z-axis.

The incoming electron wave, τ_{in}, is modified as it travels through the sample. On leaving the sample, the scattered electrons form a wave τ_{out} that contains information of the projected specimen potential $\phi_p(x, y)$, along the z-axis:

$$\phi_p(x, y) \propto \int_{-t/2}^{t/2} \phi(x, y, z) \mathrm{d}z \tag{H1.8}$$

where $\phi_p(x, y)$ is the Coulomb potential distribution within the object and t is the width of the object. For simplicity, constants have been cast aside and electron wavelength and mass have been corrected for relativistic effects (the electron velocity is close to the velocity of light). This wave can be depicted as a phase shift induced in the incoming wave, where

$$\tau_{out}(x, y) = \tau_{in} \exp[i\phi_p(x, y)] \tag{H1.9}$$

In EM of biological samples, the specimen can be viewed as a weak phase object, since very few electrons are scattered by the specimen ($\phi_p(x, y)$ is very small), and the electron beam is coherent. This results in a further approximation of $\tau_{out}(x, y)$ (see Chapter A3 and Comment H1.3):

$$\tau_{out}(x, y) = \tau_{in}[1 + i\phi_p(x, y)] \tag{H1.10a}$$

Comment H1.3
Electron beam coherence and transfer of structural information

It is very important to have a coherent beam of electrons as its waves keep in phase as they travel through space. An incoherent beam has no phase correlation and is therefore not able to transfer information to an electron micrograph from weak-phase objects.

However, since some absorption occurs even for thin biological samples, an amplitude component should be included:

$$\tau_{out}(x, y) = \tau_{in}[1 + i\phi_p(x, y) + \mu(x, y)] \qquad \text{(H1.10b)}$$

In the back focal plane of the microscope, the Fourier transform of $\tau_{out}(x, y)$, $T_{out}(u, v)$, is observed:

$$T_{out}(u, v) = \tau_{in}\lfloor \delta(u, v) + i\Phi_p(u, v) + M(u, v)\rfloor \times P(u, v) \qquad \text{(H1.11)}$$

where $\delta(u, v)$ is a Dirac function, $\Phi_p(u, v)$ and $M(u, v))$ denote the Fourier transform of $\phi_p(x, y)$ and $\mu(x, y)$ respectively, and the $P(u,v)$ term is the transfer function of the microscope. $P(u, v)$, which is discussed in further detail in Chapter H2, arises essentially from the wave aberrations of the objective lenses. The extra phase shift due to the lens can be described by

$$T_{out}(u, v) = \tau_{in}[\delta(u, v) + i\Phi_p(u, v) + M(u, v)] \times \exp\left(i\frac{2\pi}{\lambda}\gamma(u, v)\right) \qquad \text{(H1.12)}$$

assuming a rotationally symmetric astigmatism, with

$$\gamma(u, v) = \gamma(\theta) = \frac{-\Delta z\theta^2}{2} + \frac{C_S\theta^4}{4} \qquad \text{(H1.13)}$$

where θ is the scattering angle, Δz represents the objective lens defocus and C_S the spherical aberration constant.

As in light microscopy, the (real-space) image on the electron micrograph is obtained by a further Fourier transform.

Contrast transfer function and defocus

To understand the effect of the transfer function on the image, it is useful to recall that the image intensities can also be given by the product $\tau_{out}(x, y) \times \tau_{out}^*(x, y)$. The Fourier transform of image intensity $I(u,v)$ is then the convolution product:

$$I(u, v) = T_{out}(u, v) \otimes T_{out}(-u, -v)$$

Omitting the constant τ_{in} for simplicity, and since $\Phi_p(u, v)$ and $M(u, v)$ are both very small, the quadratic terms can be neglected, and

$$I(u, v) = \delta(u, v) + 2\left\{\Phi_p \sin\left[\frac{2\pi}{\lambda}\gamma(u, v)\right] - M(u, v)\cos\left[\frac{2\pi}{\lambda}\gamma(u, v)\right]\right\} \qquad \text{(H1.14a)}$$

Experimentally, the amplitude component can be of the order of 35% for negatively stained samples and 7% for unstained samples in ice (see Sections H1.5.2 and H1.5.3). If we define $W(u, v)$ as the ratio of amplitude to phase component, Eq. (H1.14a) simplifies further to

$$I(u, v) = \delta(u, v) + 2\Phi_P \left\{ \sin\left[\frac{2\pi}{\lambda}\gamma(u, v)\right] - W(u, v)\cos\left[\frac{2\pi}{\lambda}\gamma(u, v)\right]\right\} \quad \text{(H1.14b)}$$

or,

$$I(u, v) = \delta(u, v) + \Phi_P \times \text{CTF} \quad \text{(H1.15)}$$

where we have found it useful to define a *contrast transfer function*,

$$\text{CTF} = 2\left\{ \sin\left[\frac{2\pi}{\lambda}\gamma(u, v)\right] - W(u, v)\cos\left[\frac{2\pi}{\lambda}\gamma(u, v)\right]\right\} \quad \text{(H1.16)}$$

where we recall

$$\gamma(u, v) = \gamma(\theta) = \frac{-\Delta z\theta^2}{2} + \frac{C_S\theta^4}{4}$$

For each microscope operating at a given high voltage, C_S and λ are fixed. The image defocus, Δz, is the experimental variable. In order to account for the gaps in the transfer of amplitude and phase information that occur because of the oscillating CTF, exposures are recorded over a broad defocus range (e.g. between -1000 Å and -5 μm defocus).

The Fourier transform of the intensities of each image therefore displays a series of ripples, where the observed amplitudes drop to zero at a certain spatial frequencies (denoted by θ) (Fig. H1.2). At each of these nodes, the phases of the image amplitudes reverse. CTF correction procedures will be described in Chapter H2.

Fig. H1.2 Example of variation of average amplitude according to resolution.

The image as a projection

If we recall the Fourier transform of the image intensities, and reintroduce τ_{in} (the incoming wave amplitudes),

$$I(u, v) = \{\delta(u, v) + \Phi_P \times \text{CTF}\}\, \tau_{in}^2 \qquad \text{(H1.17)}$$

The reverse Fourier transform gives the image intensities ($i(x, y)$),

$$i(x, y) = \{1 + \phi_P \otimes ctf\}\, \tau_{in}^2 = \tau_{in}^2 + \tau_{in}^2 \phi_P \otimes ctf \qquad \text{(H1.18)}$$

where ctf is the Fourier transform of the CTF.

In the presence of a perfect lens ($CTF = 1$) or, more realistically, a corrected CTF, this further reduces to

$$i(x, y) = \tau_{in}^2 + \tau_{in}^2 \phi_P = A + C \int_{-t/2}^{t/2} \phi(x, y, z)\mathrm{d}z \qquad \text{(H1.19)}$$

using Eq. (H1.8) and where C is a constant. The constant A represents the intensities (τ_{in}^2) that would be recorded in the absence of sample. If we omit this background, the image intensities are a linear description of the sample's potential projection along the path of the electron beam, or expressed mathematically

$$i(x, y) \propto \int_{-t/2}^{t/2} \phi(x, y, z)\mathrm{d}z \qquad \text{(H1.20)}$$

H1.4 Electron microscopes

H1.4.1 Electron beam generation

Two main methods have been implemented in electron microscopes to generate electron beams (Fig. H1.3). A thermionic electron gun contains a loop of filament (the cathode), heated by an electric current. A negatively charged Wehnelt cylinder surrounds the loop, and a more positively charged anode then attracts the electrons. Due to the configuration of the gun, the electrons accelerate through the anode and on towards the sample. In the case of a tungsten thermionic gun, the filament has a diameter of $\sim$100 μm and is heated to $\sim$2700 K. The accelerating voltage applied to the Wehnelt cylinder is generally between -500 V and $-50\,000$ V.

Field emission guns are now preferred as they produce much brighter electron beams. The increased spatial and temporal coherency in the electron beams produced by field emission guns appear to also give a real advantage in terms of achievable resolution. The tip of the pointed cathode has a very small radius and is perpendicular to the surface of a flat anode. A voltage of several thousand volts is applied between the anode and the cathode. Due to its shape, there is a very strong electric field gradient close to the cathode, forcing electrons to be emitted

Fig. H1.3 Schematic
diagrams of (a)
a thermionic electron gun
and (b) a field emission
gun.

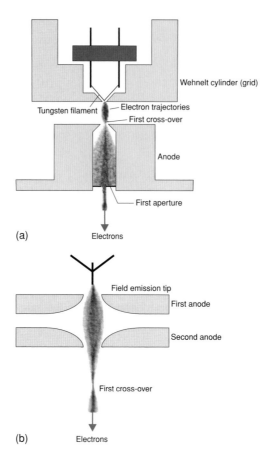

from the tip. These travel through a small hole in the anode and on towards the
sample. The cathode tip is often a tungsten crystal with a diameter between 100
and 1000 Å.

H1.4.2 Transmission and scanning electron microscopes

The transmission electron microscope (TEM) has a layout similar to the phase
contrast light microscope (see Chapter F1 and Fig. H1.4). The image in the TEM
is formed by the interference between the electrons which have passed through
(and interacted with) the sample and with those electrons which were not affected
by the sample, giving a phase contrast image.

In the SEM, a beam of electrons is scanned across the sample, resulting in the
release of secondary electrons, which contribute to the signal. The image is then
formed by correlating the position of the scan and the resulting signal. SEMs are
often coupled with X-ray detectors, since not only electrons but also X-rays are
emitted, with spectra characteristic of the elements present in the scanned portion
of the sample.

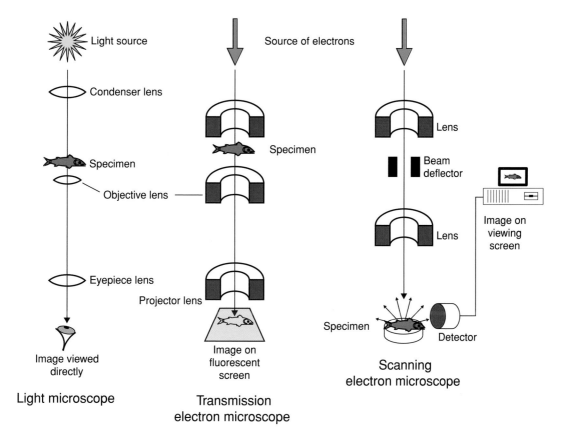

Light microscope — Light source, Condenser lens, Specimen, Objective lens, Eyepiece lens, Image viewed directly

Transmission electron microscope — Source of electrons, Specimen, Objective lens, Projector lens, Image on fluorescent screen

Scanning electron microscope — Lens, Beam deflector, Lens, Specimen, Detector, Image on viewing screen

Fig. H1.4 Schematic diagrams of a light microscope, a TEM and an SEM.

H1.4.3 Electron images

Features of an electron microscope image

Electrons interact with a sample by being repelled by negative charges and attracted to positive charges. The electron microscope records the resulting interaction in an *electron micrograph*, which contains information regarding the electrostatic potential distribution within the sample. The *contrast* in the image is approximately proportional to the atomic number of the atoms imaged.

As we would expect, the information contained in the image decreases with increasing resolution. This fall-off is due to a number of factors, including beam-induced damage, sample movement (drift and temperature factors), loss of coherence of the electron beam at higher resolutions etc.

Imaging phase objects

High-resolution EM is closely analogous to optical phase contrast microscopy, in which ultrathin low contrast samples are imaged. Amplitude contrast and phase contrast are treated in Chapter F1. In optics, transparent objects do not absorb but just scatter photons. They can be conceptualised as pure phase objects. In

Fig. H1.5 Position of
image plane: in focus,
underfocus and
overfocus.

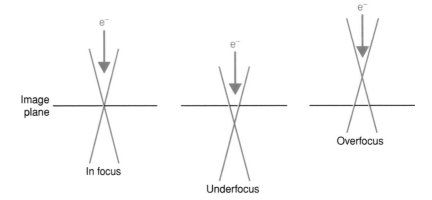

order to visualise their details, the insertion of a quarter-wave plate in the light
microscope transforms a phase difference induced by the imaged object into an
amplitude difference. However, a practical quarter-wave plate for electrons does
not exist. In order to obtain useful phase contrast from which structural details
can be discerned, the samples are imaged at a slight defocus (Fig. H1.5).

Effect of focus

Varying the focus of the electron microscope alters the features of the image
(Comment H1.2). At focus, the image has very low contrast. Amplitude differences are very weak. They result from the loss of electrons scattered at high angles
and removed by the objective lens. In fact, a good indication that an image was
recorded at focus is its featurelessness and smooth background. At underfocus,
the image has enhanced contrast, with a white ring around the imaged object.
The presence of a mass is observed as an increase in intensity. The phase shift is
in the same direction as the intensity contrast and thus they both accentuate the
contrast of the image. At overfocus, the imaged object appears white, with a dark
rim around it. The phase shift and amplitude contrast are in opposite directions.

H1.4.4 Image recording system

Optical density

Fig. H1.6 Response of film
to electron dose: optical
density electron exposure
relationship.

Electron microscope images are usually recorded on photographic films, which
are extremely efficient recording media for electrons. The recorded optical density
is proportional to electron exposure up to a certain electron dose, then the film
saturates. The intensity measured on the film must therefore be rescaled to reflect
the actual dose received (Figure H1.6).

   Electronic area CCD detectors are currently used to record electron diffraction
data. They provide rapid feedback in terms of digitised intensity distribution.

Pixelisation

A photographic image has to be digitised in order for the data to be processed
computationally. This process is never perfect so information is always lost.

It should, however, be emphasised that the image should be sampled so that the information of interest is not lost. The *Shannon sampling theorem* explains that the sampling size should be at least half the final desired resolution. In practice, a finer sampling is used ($1/3 - 1/4$ resolution per pixel), as it reduces interpolation during the data processing. The grey level of each pixel must also be quantified. For example, 256 grey levels are sufficient for an EM reconstruction as the resulting file is still manageable with currently available computer power.

H1.5 Techniques in specimen preparation

H1.5.1 Specimen support

During imaging, the sample has to be kept free from drift. It is usually held on a carbon film that is placed on a grid (Comment H1.4). The grid has numerous holes large enough for the electron beam to travel through but small enough for the sample to form a fine film over the surface, covering the grid's holes. During the imaging process, a goniometer-mounted stage allows the specimen to be tilted with respect to the electron beam.

H1.5.2 Negative staining

Principles of negative staining

Negative staining has been applied widely to the study of biological samples. It is a simple and rapid method for determining the overall morphology of a macromolecule and leads to high-contrast images whose best resolution, from three-dimensional reconstructions, is near 15 Å.

The main constituents of biological molecules (oxygen, carbon and nitrogen) are not very electron dense and so do not interact strongly with the electron beam. The sample is therefore *negatively stained* by embedding it in an electrodense material, like an uranium or lead salt (Fig. H1.7, Comment H1.5). In the electron micrograph, the sample appears as a light region surrounded by a dark background originating from the stains. The particle itself is not observed but its surrounding stain is. The sample structure is hence inferred by the distribution of the heavy stain (and therefore has limited achievable resolution). In the ideal case, the particle is not distorted by the stain; for example, negative staining vividly displays the symmetry of viruses (Fig. H1.7(b)).

The main drawbacks to negative staining are that the stain may distort the particle, and it may cause protein aggregation or dissociation.

Method for negative staining

The grids are prepared by laying a thin film of plastic (usually formvar) over their dull surface and covering it with a further layer of carbon (the plastic could in fact be omitted). The amount of carbon that has been deposited can be estimated by eye from the darkness of the grid. The grids are then glow discharged to render

Fig. H1.7 (a) Diagram to
show how the electron
dense material covers the
biological sample.
(b) Example of a
negatively stained virus.

them hydrophilic. A small volume of sample is pipetted onto the carbon-coated side of the grid and left to spread out. The grid is then rapidly dipped several times in deionised water to wash off excess sample, and subsequently dipped for some time (seconds or longer if necessary) in an electrodense solution containing the stain (e.g. uranyl acetate). Filter paper is then used to remove any excess fluid, and the grid air-dried prior to imaging.

H1.5.3 Freezing of the sample

Reasons for electron cryo-microscopy

In electron cryo-microscopy, the sample is rapidly cooled and observed at liquid nitrogen (and even liquid helium) temperatures. The method improves the preservation of the native structure and significantly reduces radiation damage. The sample can be subjected to more electrons before it is destroyed, or at least before high-resolution information lost. Radiation damage is discussed in greater detail in Chapter G3.

Flash cooling

The aim of the rapid cooling is to avoid ice crystal formation, which would damage the specimen. A small volume (about a microlitre) of the sample is pipetted onto a non-glow discharged holey-carbon film on mesh copper grid. The grid is blotted to remove excess sample fluid and then rapidly cooled. This procedure can be carried out in a humidified atmosphere to avoid water evaporation from the sample and a consequent increase in salt concentration.

The flash cooling is usually performed by plunging the grid into liquid ethane or propane (temperature of 11 K). These liquids have high heat conductance and the water on the grid is transformed into vitreous ice (as it does not have sufficient time to form ice crystals) and thus remains in an amorphous transparent state. The procedure is described schematically in Fig. H1.8. The grid is then transferred to liquid nitrogen (110 K) for storage before loading into the electron microscope.

Fig. H1.8 Schematic diagram of a plunger.

H1.6 Data collection

H1.6.1 Factors to consider during data collection

The aim is to collect sufficient information from the electron images in order to achieve a reconstruction at the desired resolution of the sample structure. The amount of data must not be too large, however, as this would affect the speed and efficiency of the computer calculations.

Magnification

Electron microscopes provide a choice of magnification, from 1000 times to 200 000 times. The choice is dependent on a number of factors. These include the targeted resolution, the required number of imaged particles, the amount of electron radiation the sample can survive, the resolution limits of the recording medium (film or CDD) and image digitiser. Under specified illumination conditions, the image intensity decreases with increasing magnification (M) and the dose per unit area decreases with $1/M^2$. Magnifications in the 40 000–60 000 range are often used for reconstructions of biological samples to 10–20 Å resolution.

Sample concentration

Each micrograph should contain a sufficient number of particles from which to choose, but not be overcrowded. The balance is between the need to reduce the time required for digitisation of the micrograph and particle aggregation. The optimum concentration depends on the sample studied. Protein concentrations in the region of 1 mg ml^{-1} have been often used in electron cryo-microscopy. In the case of smaller molecular mass molecules (500 kDa or less), it is preferable to have a less concentrated sample (0.1 mg ml^{-1}) as it is then easier to identify the molecules in the electron micrographs.

Electron dose

A typical biological sample is very sensitive to the electron beam, and can only tolerate a dose of 10–15 electrons Å^{-2}, before radiation-based structural damage renders the micrograph's data unusable. High-resolution data are generally more sensitive to electron beam damage than lower resolution data.

Stained samples can often be imaged at higher electron doses because it is the stain envelop around the particle that is observed, rather than the particle itself. The imaging of unstained specimens at very low temperatures helps to reduce the radiation damage. Since inelastic events create highly charged and reactive radicals in the sample, low temperatures can limit the progress of the diffusion of these damaging molecules and ions through the sample.

H1.6.2 Data from single particles

The data collection usually uses the SEARCH/FOCUS/EXPOSURE mode. An initial low-dose/low-magnification search identifies areas of interest on the grid. Each area is then assigned a specific defocus, and a high-dose/high-magnification exposure is then recorded.

For example, the following protocol can be used. Pictures are initially taken under low-dose conditions in order to identify the locations on the grids that should be imaged at high resolution. The image is then focused at high magnification ($\times 230\,000$) at a site adjacent (5 μm) to the area of interest. The presence of drift is checked at even higher magnification ($\times 490\,000$). An image of the area of interest is then taken at a magnification of 60 000 with a chosen unique defocus. This procedure is repeated at different places on the grid, sampling a range of defocus conditions.

H1.6.3 Imaging crystals and helical molecules

The presence of symmetry in a sample facilitates its three-dimensional reconstruction. This is discussed in detail in Chapter H2. Several biological molecules have inherent symmetry. For example, biological filaments, like microtubules and tubulin, assemble into helical shaped objects (Fig. H1.9(a)); and many viruses have icosohedral symmetry. The power of using an ordered sample is that symmetry related parts of the image can be averaged, thus allowing the use of very low electron doses.

The integral membrane protein bacteriorhodopsin spontaneously assembles into two-dimensional crystals in the so-called purple membrane (Comment A3.7). Other proteins can be attached to a lipid layer, where their movement is restricted to the lipid plane. Integral membrane proteins embed themselves directly into the lipids, while soluble protein can, for example, bind to specially functionalised lipids. Under the right biochemical conditions (lipid-to-protein ratio and temperature), they self-assemble in an ordered two-dimensional lattice

(a)

(b)

Fig. H1.9 (a) Helical fragment of Ca-ATPase and (b) Two-dimensional crystal of streptavidin. (Togoshima *et al.*, 1993; Frey *et al.*, 1996) (Figures reproduced with permission from Macmillan Publishers Ltd. and National Academy of Sciences USA, respectively.)

(Fig. H1.9(b)). This crystallisation process is very similar to three-dimensional crystallization, discussed in Chapter G3.

A similar procedure to single-particle imaging is implemented in the imaging of two-dimensional crystals and helical molecules. A rapid scan at low resolution over the grid identifies potential areas of interest. Crystals or helices do not, in fact, have to cover the whole field of view of the microscope, since each ordered region is processed individually before being combined during data processing. A Fourier transformation of the image is then computed to provide a diffraction pattern of the ordered sample, with the added advantage over X-ray crystallography that the phase information associated with each reciprocal lattice point is also obtained. The extent of the diffraction pattern highlights the maximum resolution which could be theoretically achieved (see Chapter G1). The EM reconstruction of bacteriorhodopsin crystals for instance was determined to 3.5 Å resolution (see Chapter H2). The direct collection of diffraction patterns from two-dimensional crystals offers several advantages, in particular that specimen drift and defocus do not affect the data. In order to have sufficient signal-to-noise at high resolution, crystals must be large (at least several micrometres across).

H1.6.4 Tomography

In tomography (from the Greek *tomos* slice) data are collected by imaging the same sample at different tilt angles. This technique is particularly advantageous when dealing with non-uniform samples in which the varying shapes make it impossible to average them for reconstruction. Nevertheless, the orientation of each image relative to the others of the tomographic series is known. Typically, the specimen is rotated about a fixed axis in $1°$ to $2°$ angular increments over $\pm 70°$. By using such an approach it has been possible to identify 26 S proteasomes and actin filaments in the cellular environment from their distinctive shapes (50–60 Å resolution) (Fig. H1.10). The cellular regions examined were between 300 and 600 nm in thickness.

Fig. H1.10 EM tomography of a frozen *Dictyostelium discoideum* cell. Colours were subjectively attributed to linear elements to mark in red the actin filaments, in green the other macromolecular complexes (mostly ribosomes), and in blue membranes (Medalia et al., 2002). (Figure reproduced with permission from AAAS.)

Fig. H1.11 Electron micrograph of the herpes simplex virus (HSV-1) after staining with an antibody specific for the portal protein pUL6 followed by an anti-antibody conjugated to gold beads. Note that the gold clusters are found at a single capsid vertex. Bar, 100 nm. (Newcomb *et al.*, 2001.)

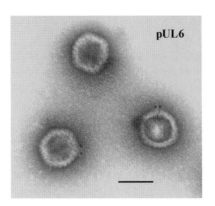

In general, a tomography experiment aims to collect results at as many orientations as possible in order to produce an accurate reconstruction; however, the cumulative electron dose limits the number of exposures because of radiation damage. The resolution of a tomography experiment therefore is usually worse than the resolution achieved from single-particle reconstruction. A current aspiration is to achieve a 20-Å resolution tomographic reconstruction for high-molecular-mass samples.

H1.7 Immunochemistry

A scanning electron microscope can easily identify gold clusters, due to the strong emission of secondary electrons and backscattered electrons. Labelling antibodies with gold has therefore provided a tool by which components in complex biological samples can be localised. The gold particles may be conjugated to primary antibodies for a one-step identification of antigens, but are more usually employed as secondary antibody/protein labels (Fig. H1.11). The location of the antibody is therefore determined from the presence of the gold. Gold has the further advantage that it is available in different sizes (1–40 nm) so the binding pattern of different antibodies can be identified in a single micrograph.

H1.8 Checklist of key ideas

- EM is a powerful tool used in the study of biological specimens.
- EM reconstructions typically attain resolutions in the nanometre range, although two-dimensional crystals have yielded atomic resolution reconstructions.
- An electron microscope is similar to a light microscope but instead of photons its uses electrons.
- In single-particle reconstruction, the data from differently oriented molecules can be combined to produce an accurate three-dimensional reconstruction of the molecule.
- In tomography, the same object is imaged at different orientations, and the images are then combined to give the three-dimensional structure of that object.

- EM is hampered primarily by the poor signal-to-noise ratio of the data, which places a lower limit on both resolution and the size of the studied complex.
- An electron has the dual property of behaving both as a wave and as a particle.
- Like a light lens focuses photons, electromagnetic lenses can focus electron beams.
- The EM map is the result of the interaction of the incident electrons with the electrostatic distribution of the electrons and nuclei in the sample.
- High-resolution EM is closely analogous to optical phase contrast microscopy, in the analysis of weak phase objects.
- EM is usually recorded on a photographic film.
- The Shannon sampling theorem says that the sampling size should be at least half the final desired resolution.
- In negative staining, it is not the particle itself that is observed but the surrounding electrodense material in which it is embedded.
- The presence of symmetry in a sample eases its three-dimensional reconstruction.
- Certain proteins can be coaxed to form two-dimensional crystals.
- In electron cryo-microscopy, the sample is frozen rapidly for imaging. The low temperature reduces damage to the sample on exposure to electrons.
- Electron microscopes provide a choice of magnifications, from $\times 1000$ to $\times 200\,000$ times.
- Biological samples are very sensitive to an electron beam. Typically they can only tolerate a dose of 10–15 electrons Å^{-2}.
- The technique of tomography is particularly advantageous when dealing with non-uniform samples in which the varying shapes make it impossible to average them for reconstruction.
- In immunochemistry, labelling antibodies with gold has provided a tool by which components in complex biological samples can be localised by EM.

Suggestion for further reading

Henderson, R. (1995). The potential and limitations of neutrons, electrons and X-rays for atomic resolution microscopy of unstained biological molecules. *Q. Rev. Biophys*, **28**, 171–93.

Chapter H2
Three-dimensional reconstruction from two-dimensional images

H2.1 EM in biology

H2.1.1 Structural biology with EM

Electron microscopes can provide vivid details of biological macromolecules at the nanometre scale.

Recall from Chapter H1, an electron image is a two-dimensional projection of the object along the electron beam. A three-dimensional reconstruction of the object can be obtained by combining data measured at different angular projections. In *single-particle reconstruction*, a population of molecules is imaged. If these are identical (monodisperse) and present in sufficient quantities, the data from differently oriented molecules can be combined to produce an accurate three-dimensional reconstruction of the structure. The presence of symmetry is taken advantage of in *helical reconstruction* and *two-dimensional crystallography*. In *tomography*, the same object is imaged at different orientations. The different images are then combined together to recreate its structure.

H2.1.2 Examples of electron cryo-microscopy reconstructions

In EM, the image is the result of the interaction of the incident electrons with the electrostatic distribution due to the atomic structure of the sample. The reconstruction of the structure results from the analysis of many images. Three-dimensional reconstructions at different resolutions depict the object at different levels of detail (Figure H2.1).

At atomic resolution, it is a good approximation to understand this structural density as being equivalent to the electron density obtained by X-ray diffraction, but with a larger contribution from hydrogen atoms in the electron diffraction case. It should furthermore be possible to distinguish the charge state of amino acids.

At medium resolution (better than 10 Å), the scattering depends on protein-specific features such as secondary structures. Alpha-helices produce a stronger signal at around 10 Å resolution and β-sheets at about 4.5 Å. Comparable features

(a)

(b)

(c)

(d)

Fig. H2.1 EM reconstructions at different resolutions. (a), (b) Semliki virus: reconstruction at: (a) 22 Å (Venien-Bryan and Fuller, 1994) and (b) at 9 Å (Mancini et al., 2000). The color scheme reflects the radial distance from the centre, increasing from blue to red, as shown in the inserted color scale. Scale bar, 100 Å. (c) Pyruvate kinase: density with superimposed coordinate model (Rosenthal and Henderson, 2003). (d) Bacteriorhodopsin: weighted density maps for wild-type bacteriorhodopsin (3.5 Å, purple) and the D96G, F171C, F219L triple mutant (3.2 Å, yellow). The atomic model of the wild-type is included (Subramaniam and Henderson, 2000). (Figures reproduced with permission from Elsevier ((a), (b), (c)) and Macmillan Publishers Ltd. (d).)

are also observed in X-ray diffraction experiments and the reader should refer to the chapters on X-ray crystallography (Chapter G3) and solution scattering (Chapter G2) for more details.

At low resolution, the electron scattering is determined by the shape of the macromolecule. Reconstructions based on either negative staining or unstained samples therefore yield roughly the same information at 20-Å resolution.

H2.2 EM data preparation

H2.2.1 Preliminary analysis of the image

Prior to any data processing, the quality of the electron micrograph must be checked. The preliminary analysis can be based on a direct examination of the micrograph by eye as particles should be identifiable. A small optical lens can also be used. The particles should be monodisperse and in sufficient number. Aggregated particles are not useful for reconstructions.

Fig. H2.2 Diffraction patterns with different defects such as astigmatism and drift. Drift leads to the Thon ring fading in a direction perpendicular to the direction of the drift.

Perfect lens   Astigmatism   Drift

Optical diffraction is used to evaluate the putative resolution, the degree of underfocus, lack of astigmatism and drift in the electron micrographs (see Chapter H1). Examples of these defects are given in Figure H2.2. In two-dimensional crystallography, the contrast is often so low that optical diffraction is essential to check that the crystal has been imaged.

H2.2.2 Particle selection

The particles are identified in the digitised image file. A reference image can be used to help this process if the shape of the particle is known roughly. Automatic particle selection software is also available, although it is still not fool-proof.

Individual particles are selected and stored in smaller files. Each contains the image of just one particle and its surrounding background. The image can be *band pass filtered* to remove certain bits of spurious high-/low-resolution information to improve the signal-to-noise ratio. The data manipulation for the reconstruction is done on these smaller files.

H2.2.3 Correction for the Contrast Transfer Function

Recall that a micrograph is taken at a specific defocus because of the CTF (see Chapter H1). Expressed as a Fourier transform (see Chapter A3), the CTF gives the average amplitudes of the spatial frequencies to oscillate with a resolution (Fig. H2.3(a)). The lack of information at certain resolution ranges can be overcome by combining data from micrographs with different CTFs. This enables uniform coverage of the resolution range to the highest possible resolution. The CTF can also be corrected by mathematically modifying the signal in certain resolution shells of the data (Fig. H2.3). The simplest correction to the CTF is *phase flipping* (Fig. H2.3(b)). Phases in the resolution ranges with negative amplitudes are flipped, i.e. $180°$ is added to the phase of structure factors with negative CTF. Hence the amplitudes are not modified.

The CTF can be also be corrected by either multiplying or dividing the data by the CTF itself (Fig. H2.3(c)). The phase gains the correct sign while the amplitudes at higher resolutions are increased. When dividing by the CTF, a

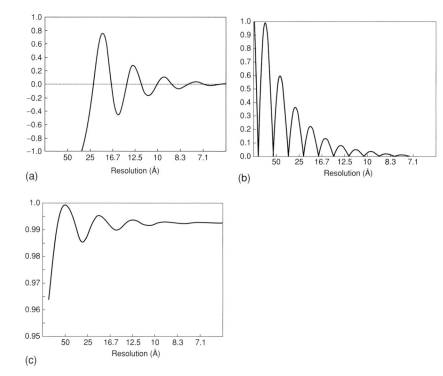

Fig. H2.3 The effect of the CTF on a cross-section of the Fourier transform of an image, and the effect of different correction procedures: (a) CTF with decay: (b) CTF correction by phase slopping; (c) Weiner correction of the CTF.

problem arises at the CTF zeros, so a Weiner-like filter of the form $1/(CTF + s)$ is used, where s is a noise-suppression term. In the case of images of tilted specimens, the variation in defocus across the image can be accounted for by a *tilt transfer function*.

Finally, the decrease in amplitude with resolution can also be corrected for. Weighting schemes are described in further detail in Section H2.7.

H2.3 Single-particle reconstruction procedures

H2.3.1 Coordinate system

An EM reconstruction merges the data from images of differently oriented particles in order to recreate a three-dimensional reconstruction of the particle. The basis of the reconstruction is to identify the spatial relationship between the different images.

A particle is described in terms of its position in the plane of the image (x, y) and a rotation. The description of the z-axis is not possible as the particle is projected in that direction to produce the electron image. Depending on the reconstruction program, different angular conventions are used (Comment H2.1).

Comment H2.1

Coordinate system

The Euler angles (α, β, γ) can describe rotations or relative orientations of orthogonal coordinate systems. However, their definition is not unique. In the IMAGIC software, a γ-rotation around the z-axis is followed by a β-rotation around the y-axis, which is subsequently followed by an α-rotation again around the z-axis.

Three angles define the orientation of the view in the Cartesian coordinates system: θ is the rotation in the xz-plane, positive from z towards x; ϕ rotation is in the xy-plane, positive from x towards y; ω is the rotation of the object about the θ, ϕ view direction.

H2.3.2 Reconstruction from projections

From the arguments detailed in Chapter H1, once the CTF has been corrected, we can assume the electron image, $i_{corrected}$, is a projection of the particle onto the xy-plane, along the z-axis,

$$i_{corrected}(x, y) = \int_{-t/2}^{t/2} \phi(x, y, z)dz \qquad (H2.1)$$

where $i_{corrected}(x, y)$ is the two-dimensional CTF-corrected image and $\phi(x, y, z)$ is the three-dimensional particle potential.

The Fourier transform of an object $\phi(x, y, z)$ can be expressed as

$$\Phi(u, v, w) = \int_{-\infty}^{+\infty} \int_{-\infty}^{+\infty} \int_{-\infty}^{+\infty} \phi(x, y, z) \exp[-i2\pi(ux + vy - wz)]\partial x \partial y \partial z \qquad (H2.2)$$

If we restrict $\Phi(u, v, w)$ to the line defined by $w = 0$,

$$\Phi(u, v, 0) = \int_{-\infty}^{+\infty} \int_{-\infty}^{+\infty} \int_{-\infty}^{+\infty} \phi(x, y, z) \exp[-i2\pi(ux + vy)]\partial x \partial y \partial z \qquad (H2.2a)$$

and therefore the integral can be separated into two:

$$\Phi(u, v, 0) = \int_{-\infty}^{+\infty} \int_{-\infty}^{+\infty} \left\{ \int_{-\infty}^{+\infty} \phi(x, y, z)\partial z \right\} \exp[-i2\pi(ux + vy)]\partial x \partial y \qquad (H2.2b)$$

The term in braces is the projection of the object along the z-axis, which, on referring back to Eq. (H2.1) is the corrected image, or

$$\Phi(u, v, 0) = \int_{-\infty}^{+\infty} \int_{-\infty}^{+\infty} i_{corrected}(x, y) \exp[-i2\pi(ux + vy)]\partial x \partial y \qquad (H2.3)$$

In words, the two-dimensional Fourier transform of the image is equivalent to a section of the three-dimensional Fourier transform of the particle along the direction in which the projection was taken.

Moreover, by the nature of the Fourier transform, if an object $\phi(x, y, z)$ is rotated by an angle with respect to one or more axes, the Fourier transform $\Phi(u, v, w)$ is correspondingly rotated by the same angle with respect to the coresponding axes. A sufficient number of differently oriented sections therefore enable the three-dimensional object to be reconstructed by reverse Fourier transformation (Fig. H2.4). The handedness of the particle is, however, still ambiguous as each image is a projection of the object, although *focal pairs* can solve this problem (see Section H2.7.2).

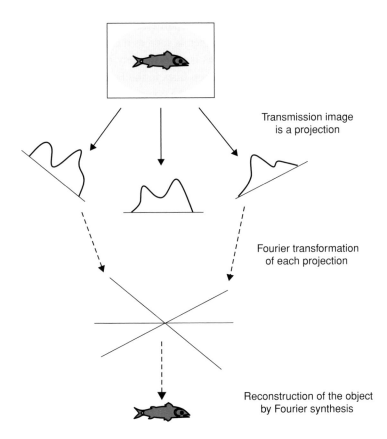

Fig. H2.4 Concept of
common lines: sections
cut in space along a
common line.

Transmission image
is a projection

Fourier transformation
of each projection

Reconstruction of the object
by Fourier synthesis

H2.3.3 Iteration procedure: reprojection method

Due to the noise in each micrograph, the reconstruction procedures have to be iterated until the quality of the final reconstruction no longer improves. The three-dimensional model of the reconstructed particle is often reprojected computationally into different directions to create a new set of reference images with known orientations, against which the raw data set can be compared to determine the orientations more accurately. A new model can then be produced and the procedure continued until the reconstruction solution is stable.

H2.3.4 Common lines reconstruction procedure

If the views of a particle are different, two different sections in Fourier space intersect along a common line. The corresponding planes can each rotate about the line. A third section would also display two other common lines, one for each of the two other sections. The common lines reconstruction procedure is based on the fact that the three planes are now locked in space.

Theoretically images of three differently oriented particles are therefore sufficient to orient each particle with respect to the other as soon as the common lines are identified. In many experimental cases, this is unfortunately not the case since the EM images are too noisy for the common lines in the Fourier transform to be identified accurately. Furthermore, because the common lines extend across Fourier space, they are resolution-dependent. A statistical analysis of the section pairs must therefore be implemented.

When symmetry is present in the particle, two types of common lines appear in the Fourier transform of its image. Application of the symmetry operators leads to the appearance of *self-common lines*. In icosohedral objects, a maximum of 37 self-common lines can be observed, due to the application of five-fold, three-fold and two-fold symmetry and removal of non-unique lines. The *cross-common lines*, on the other hand, are due to the application of the particle's symmetry between pairs of its images. In icosohedra, there are at most 60 pairs of cross-common lines.

H2.3.5 Polar Fourier transform reconstruction

Polar Fourier transform reconstruction relies on the use of a preliminary model (or template), which helps orient the images in order to create a more appropriate model of the imaged sample (Comment H2.2). The analysis starts by constructing a database, which contains differently projected views of a model representing the particle. For icosahedra, these views cover one-half of the asymmetric unit or $1/120$ of the icosohedron (i.e. $\theta = 69–90°$ and $\phi = 0–32°$). In each view, the Cartesian coordinate system is translated into a polar reference frame (r, γ). The radius r varies from 0 (at the chosen origin) to a radius just outside the particle edge and the angle γ indicates the azimuthal direction of the projection. The particle image is also converted to polar coordinates.

In order to implement a polar Fourier transform reconstruction, each particle image must be initially centred and scaled, e.g. by cross-correlation with the circularly averaged projection of the model. The particle images and model projections are then Fourier transformed along the azimuthal axis to produce one-dimensional rotational spectra. The spectrum of every image is then multiplied by the spectrum of each model projection in order to produce a cross-correlation map. A high correlation indicates that the particle in the image is oriented similarly to the projected model. Images with the highest correlation are then used to create a better model. The process is then iterated until the correlation between the image and final model projections does not improve.

This method has the advantage of using all the data in the image. In contrast, the common-lines method makes use of only a small fraction of the Fourier data along the common line.

Comment H2.2
Initial model from negative staining

EM chiefly records the surface of stained particles. The resolution is often limited, but the staining can nevertheless highlight the particles' overall shape and variability. A model based on stained particles can therefore be used in the initial stages of the reconstruction process. In fact, just 5–10 particles are often sufficient.

In a polar Fourier transform reconstruction, the choice of template obviously influences the reconstruction. For scientists' peace of mind, the final reconstruction could be confirmed by using different starting models and comparing the resulting reconstructions. A sufficiently large dataset of images also permits reconstructions based on different subsets of the data. The comparison of the different reconstructions is the basis of the FSC method, which is treated in greater detail below (see Section H2.5.2).

Polar Fourier transform reconstructions are not recommended for non-symmetrical objects since they require the chosen origin to be determined separately and not refined against the average model projections. The use of the centre of mass of the image as an origin could be a way around this problem.

H2.3.6 Real-space reconstruction procedure

Most of the data manipulation and analysis can also be performed in real space. Recall, the particle image is a two-dimensional projection of the particle. The image itself can be projected onto a single line. The rotation of the particle image over 360° creates a sinogram, the collection of all possible line projections (Fig. H2.5).

Sinograms can be used to find common lines between two projections of a particle. The largest correlation between two sinograms corresponds to a pair of shared line projections. This method offers several advantages. It does not depend on defining the particle's image centre. The presence of symmetry is also exhibited in the sinograms, if the sinogram of the image is cross-correlated with itself.

H2.3.7 Focal pairs

Focal pairs can be recorded in order to facilitate image processing. A second image is recorded immediately after the first, but at a large defocus. The resultant strong contrast in the low-resolution features of the second image can then provide estimates of the positions and orientations of the particle in the low-contrast first image.

In a similar fashion, pairs of micrographs can be recorded in which the sample has been rotated for the second image. The hand of the particle can hence be identified in the initial steps of the reconstruction. As the rotation angle is known, the computational process can furthermore be optimised, e.g. by varying the resolution range, reconstruction radii, contrast etc. in order to accurately predict the spatial relationship between the two images of the same particle. The optimised procedure can then be used to deduce efficiently the orientations between all the particles present in the first micrograph. Since it is of poorer quality due

Fig. H2.5 A graphical overview of the relations between two projection images, their 'sinograms', and the sinogram correlation function between two sinograms. The images (a) and (b) are class averages deduced from a large data set of herpes simplex virus type 1 (HSV1) electron cryo-microscopy images. Each line of the sinogram images (c), (d) is generated from the two-dimensional projection image by summing all one-dimensional lines of the two-dimensional images, from top to bottom, after rotation of the image over angles ranging from 0° to 360°. Each point of the sinogram correlation function contains the correlation coefficient of two lines of the two sinograms one is comparing (e). (Van Heel et al., 2000.)

to radiation damage, the data in the second image are not used in the subsequent refinement.

H2.4 Reconstruction procedures for one- and two-dimensional crystals

H2.4.1 Helical reconstruction

Many biological molecules have a helical organisation. The best known is DNA, but key intracellular proteins such as actin and tubulin also assemble into structures with helical symmetry. The common helix can be visualised as the curve formed by the thread of an ordinary screw. Helical symmetry implies that the neighbouring subunits in the helix are related by a displacement and a rotation along the screw axis. For example, seven repeating subunits could be spaced between two helical turns. An image of a helix therefore has the advantage of displaying the repeating subunit in a number of orientations. There is therefore no need to tilt the helix during the data collection.

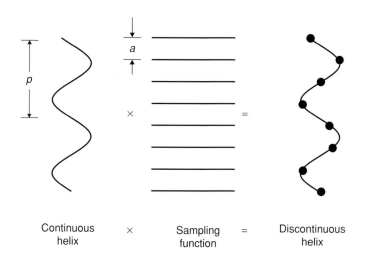

Fig. H2.6 Fourier transforms of a helix with pitch *p* and distance *a* between repeating subunits along its axis.

Continuous helix   ×   Sampling function   =   Discontinuous helix

Fourier transforms

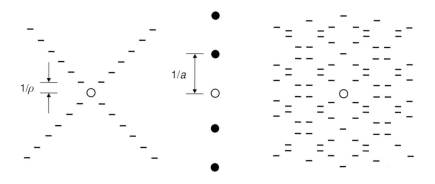

An image reconstruction of the helix makes use of its underlying geometrical relationships. The Fourier transform of a continuous helix is a series of lines fanning out as a cross (see Fig. 1(d) in the Introduction). The presence of a periodic displacement (of length *a*) between the different subunits creates a more complicated pattern due to the repetition of the X-like pattern every $1/a$ (Fig. H2.6). These short lines are called *layer lines*. Each layer-line pair corresponds to a set of helices. In a similar fashion to X-ray crystallography (Chapter G3), they have to be indexed in order to undertake the reconstruction. Since helical symmetry involves just two parameters, only two indices are needed for each layer plane. These indices (*l* and *n*) are derived from the helix's pitch ($l = p^{-1}$) and its rotational frequency *n* (a similar view of the helix appears after rotating the helix by $2\pi/n$ radians).

The advantage of the treatment in reciprocal space is that most of the image noise can be discarded by using only the layer-line amplitudes and phases in the reconstruction. Furthermore, several data sets are often combined in the helical reconstruction. Their size is usually a compromise between keeping the length of helical fragments as short as possible (to eliminate areas with distortions) and generating useful phase and amplitude data for alignment and averaging.

H2.4.2 Electron crystallography

Data from thin two-dimensional crystals can be processed as in X-ray crystallography (Chapter G3). However, in electron crystallography the crystals are usually imaged in real space and the data are computer analysed in Fourier space.

The Fourier transform of the crystal image displays a diffraction lattice from which structure factor intensities and phase information can be estimated. Noise can then be dramatically reduced by using only amplitudes and phases in the diffraction spots (see Fig. H2.7). Often, different orientations within a crystal can be patched together computationally by an *unbending procedure* to give the equivalent of one large crystal. This significantly increases the resolution that can be obtained from the image of a single crystal. Non-crystallographic symmetry, as in X-ray crystallography, can further increase the signal-to-noise ratio (see Chapter G3).

Recall that the electron image is the projection of the sample projections along the electron beam direction. Data from differently oriented crystals must therefore be integrated into a data set from which a three-dimensional structure can be

Fig. H2.7 Example of the electron diffraction pattern from a two-dimensional crystal of bacteriorhodopsin. (Subramaniam *et al.*, 2002.) (Figure reproduced with permission from the Royal Society.)

calculated. For example, merged electron diffraction patterns from 402 crystals have been used to calculate a 3.2-Å structure of the protein bacteriorhodopsin. Since the 2-d crystal lies parallel to the grid, it is necessary to orient the grid at different angles to the electron beam. As angles above a certain orientation are not accessible experimentally due to the grid's thickness (for the bacteriorhodopsin specimen the maximum tilt was 62.5°), a *cone of missing data* occurs in Fourier space resulting in reconstructions with anisotropic resolution, which are therefore elongated perpendicularly to the crystal plane (Comment H2.3).

Data from the integral membrane protein bacteriorhodopsin displayed diffraction spots extending out to 2 Å, although only data from 2.8–3 Å were successfully phased for reconstruction. This resolution, however, was sufficient to allow molecular models to be built integrating both stereochemistry and EM data: the electron diffraction data (intensities transformed into amplitudes) and images (phases) providing the amplitude/phases of the structure factor components. A refinement procedure, similar to that used for X-crystallography (Chapter G3), was then used to match the crystal's structure factors and phases with an atomic model of the structure.

Finally, the structure can be solved by collecting diffraction data and solving the phases by *molecular replacement* (see Chapter G3), thus avoiding the need to record images of tilted crystals, as these, in particular, have different defocuses across the crystal plane.

> **Comment H2.3**
> **Missing data**
>
> The cone of missing information means that data are missing in one direction. The reconstruction of a circle would thus give an ellipse, elongated in that direction. It is therefore essential to check the effect of the missing data in any electron crystallographic reconstruction. This is important as helices often lie perpendicular to the membrane, in the direction of the missing information.

H2.5 Classification procedures

H2.5.1 Statistical analysis

Electron micrographs are unfortunately characterised by a low signal-to-noise ratio and *statistical analysis* is required. The signal-to-nose ratio is proportional to the square root of number of particles averaged.

Images that look similar can be grouped together into *classes*. The average image of a class then has an increased *signal-to-noise* ratio, which in turn facilitates the determination of the angular relationship between the different images.

H2.5.2 Multivariate statistical analysis

Multivariate statistical analysis is implemented in certain reconstruction algorithms. This *eigenvector analysis* of the image reduces the total number of data treated.

An initial step of the procedure is to reassign the data in the two-dimensional image to a different frame of reference, called *hyperspace*. Each pixel of the image corresponds to an axis within this hyperspace, and the pixel's grey level to the coordinate along that axis. For an $n \times n \, (= m)$ pixel image, the corresponding hyperspace has a total of m dimensions. Each image can therefore be reduced

Fig. H2.8 (a) Typical eigenimages. (b) Four characteristic projection views out of 50 class-sum images obtained by multivariate statistical analysis of the H+-ATPase in the ligand-free state. (c) The resultant three-dimensional map. (Rhee *et al.*, 2002.) (Figure reproduced with permission from Macmillan Publishers Ltd.)

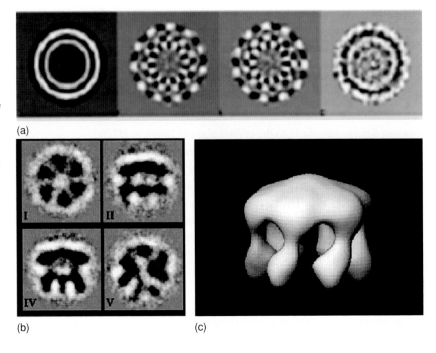

Fig. H2.8 (a) Typical eigenimages. (b) Four characteristic projection views out of 50 class-sum images obtained by multivariate statistical analysis of the H+-ATPase in the ligand-free state. (c) The resultant three-dimensional map. (Rhee *et al.*, 2002.) (Figure reproduced with permission from Macmillan Publishers Ltd.)

(a)

(b)   (c)

to a single point inside the hyperspace with similar-looking images clustering together.

The multivariate statistical analysis aims to collapse the *m*-dimensional hyperspace into fewer dimensions and still represent the data truthfully. New axes in this hyperspace are therefore determined that enable a simplified, but truthful representation of the data. The first axis is oriented according to the data's average. The second axis is oriented in the direction of the next largest variance. The third axis is oriented towards the next largest variance, and so on for each extra axis, until all the data's information has been used. Typically, the first ten axes contain most of the information about the dataset, and are sufficient to summarise the data. Each axis in hyperspace can be monitored by eye as it is equivalent to an *m*-pixel image, called an *eigenimage*. Examples of eigenimages are shown in Fig. H2.8. Each picked particle is described by a linear combination of these new axes. Similar looking particles can then be rapidly identified and clustered together based on these axes. It is optimum to perform this classification on previously aligned images.

H2.6 Determination of the resolution

H2.6.1 Number of images required for a reconstruction

The number of views required to calculate a reconstruction up to a resolution *d*, depends on the *n*-fold symmetry and diameter (*D*) of the particle. The minimum number of views required, N, is

$$N = \frac{\pi D}{nd} \qquad \text{(H2.4)}$$

where d is the target resolution of the reconstruction. The number of images used in the reconstruction is, however, inherently much larger since Eq. (H2.4) is for the perfect case with noise-free images, and an evenly spaced distribution of particle orientations. Hence, more appropriately, the number of images required ($N_{required}$) can be estimated as

$$N_{required} = N_{proj} \left(\frac{\pi D}{nd} \right) \qquad \text{(H2.5)}$$

where N_{proj} is the average number of images in one projection required to reach a certain signal-to-noise threshold. The calculation must also account for the loss of contrast at high resolution due to amplitude decay.

For example, icosohedral viruses, due to their 60-fold symmetry, would require 60 times less views than for the asymmetric ribosome. The very conservative criterion of a signal-to-noise ratio greater than 3σ means 60 000 particles would be required for a 9 Å resolution reconstruction. A resolution criterion related to map interpretability brings this theoretical number down to 2200.

H2.6.2 Definition of resolution

The resolution in EM is difficult to define as there is often no independent way of monitoring it. In X-ray crystallography, the atomic resolution forces the reconstruction to obey certain chemical rules in order to be plausible.

Shannon sampling theorem

The digitisation of the image imposes a maximum resolution. The Shannon sampling theorem states that the measurement (or sampling) frequency must be at least twice the maximum frequency to be measured. The image must therefore be scanned at a step size which is at least twice as fine as the desired resolution: for example, if the pixel size is 4.8 Å, the maximum resolution of the reconstruction is 9.6 Å.

Phase residual

A *phase residual* can be used to underscore the consistency of the phases corresponding to its common-line pairs. After a particle's orientation is estimated, the common lines are calculated. A phase residual of $0°$ average phase along the common lines means there is perfect agreement with the projection of the model, while a phase difference of $90°$ indicates complete lack of agreement.

Fig. H2.9 FSC plotted
against resolution for the
reconstruction of the
yeast RNA polymerase I.
Different thresholds are
shown (1, 0.5 and 0.143)
as well as the 3σ level
(dotted line). (Bischler
et al., 2002.)

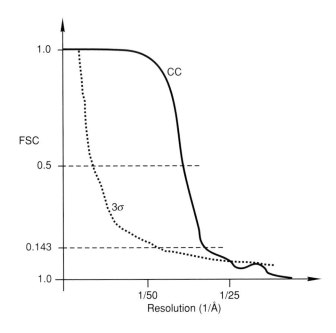

FSC

FSC is often used to qualify the final reconstruction. The data used in the reconstruction are split into two, and two independent reconstructions are calculated. The correlation of each reconstruction in Fourier space is then calculated as

$$\text{FSC} = \frac{\sum F_1 F_2^*}{\sqrt{\sum |F_1|^2 \sum |F_2|^2}} \tag{H2.6}$$

where F_1 and F_2 are the structure factors of two maps calculated from each data set, and the sum runs over a resolution shell. The FSC plotted against resolution is shown in Fig. H2.9. The resolution can be understood as the spatial frequency at which the FSC drops below a certain threshold. The often quoted value of 0.5 is quite pessimistic. However, at the resolution described by a FSC threshold of 0.143, the information contained in such maps should be equivalent to the information gained at that resolution in X-ray crystallography and this value has been adopted by many research groups.

In addition, a resolution cut-off can be determined from the background noise (3σ) of the reconstruction (see Chapter A2). The resolution is found where the FSC is equal to

$$3\sigma \times \sqrt{n} \tag{H2.7}$$

where n is the number of asymmetric units in the reconstruction. The icosahedral symmetry of viruses leads to a resolution with FSC threshold of approximately 0.5.

H2.7 Map enhancement

H2.7.1 Symmetry

The presence of symmetry in the particle is a boon to the reconstruction process. Since symmetry elements can be imposed during the particle's reconstruction, identifying and then applying a symmetry operator on the reconstruction map drastically increases the signal-to-noise ratio of the data (Comment H2.4).

   The task is relatively straightforward for crystalline samples as their translational symmetry can be deduced easily from the Fourier transforms of the images. In a crystal, the symmetry is determined from layout and systematic absences in the diffraction pattern. Furthermore, the extent of the crystallographic symmetry can be evaluated by analysing the phase relationship between symmetry-related reflections in the image's computed Fourier transform, since the apparent symmetry breaks down at high resolution (probably due to low signal-to-noise ratios). The observed symmetry is then applied to the reconstruction.

H2.7.2 Weighting

An amplitude profile can be imposed on the final map in order to down-weight the low resolution data and boost the higher-resolution features. The argument for weighting is that maps are phase-dominated. Provided the phases are reliable to a given resolution, the structure factor amplitudes can be scaled to correct for their fall-off with resolution. For example, the structure factors of the map could be multiplied by

$$\exp(-X/d^2) + \exp(Y/d^2) \qquad (H2.8)$$

where the real and positive values X and Y and the resolution d are in ångströms. By varying the values of X and Y it is possible to highlight features in the map which are only noticeable at certain resolutions. In the case of a two-dimensional crystal, a negative temperature factor is applied, estimated from the Wilson plot of structure factor amplitude decay against resolution (see Chapter G3). It is typically -300 Å^2 to -400 Å^2 for a field emission gun microscope.

   An *envelope decay function* can also be applied to correct for the image degradation. Some variant of the Wiener filter is generally used. The corrections can be estimated from either comparison with focal image pairs or on the basis of X-ray solution scattering data.

   The decay in the EM map's scattering factors can be scaled to follow the decay that would be observed if the structure were determined by X-ray crystallography. The EM map is noise-weighted by incorporating information from the FSC. The map's structure factors are multiplied by

$$\sqrt{\frac{2\mathrm{FSC}}{1 + \mathrm{FSC}}}\exp\left(-B_{\mathrm{restore}}\big/d^2\right) \qquad (H2.9)$$

The exponential is a sharpening term to compensate for the decline in scattering factors with resolution (d) ($B_{restore}$ reflects intrinsic image defects and in experiments is often approximately -1000 Å^2).

H2.7.3 Use of other information

If the atomic structure of a component of the particle has been determined, it is possible to try to fit that model into the EM reconstruction. This is achieved by matching it with the reconstructed density determined by EM. It is important to determine the absolute magnification of the EM structure. Due to experimental limitations, the distance scale of the EM map may vary by up to several percentage points and so has to be adjusted in line with the X-ray crystallographic structure. The relative values of the EM density distribution also have to be scaled to match the densities observed in the X-ray structure. In order to ease the fit, lower-resolution maps of the X-ray structure are often created. The positioning of the X-ray structure within the EM map permits the description of details of the EM reconstruction at a much greater resolution. In the case of a 20-Å EM map of a protein complex, amino acids can be modelled with a 3–5 Å accuracy and thus detail the protein surfaces within the electron map (Comment H2.5). This exercise is akin to protein crystallography (see Chapter G3). Most crystals only diffract X-rays to between 2 and 3 Å, and it is only knowledge of the stereochemistry and the amino acid sequence that enables the description of the protein structure to atomic detail.

H2.8 Applications and examples

H2.8.1 The ribosome

The ribosome is a large RNA–protein complex involved in the synthesis of the protein chain from information encoded in an RNA chain. Single-particle methods of reconstruction of the ribosome have provided much understanding about the structures and dynamics involved in this translation process. Due to its large size ($\sim$2.5 MDa) and lack of symmetry this macromolecular structure has provided an ideal system on which to develop and test EM reconstruction procedures.

Several interesting ideas have emerged from the analysis of numerous EM reconstructions of functional ribosome complexes. For example, the ribosome is able to assume a large number of states, with the different domains being able to act either together or independently (Figure H2.10).

H2.8.2 Icosohedral viruses

The capsids of many viruses have icosohedral symmetry: a geometric construct defined by 12 pentagonal vertices and 20 triangular faces. The presence of such

Fig. H2.10 Details of EM structures of complexes of ribosome 70 S and different ligands. The 50 S subunit is depicted in blue; the 30 S in yellow, EF-G is in red; P-site tRNAs in green; E-site tRNAs in orange. (Valle *et al.*, 2003.) (Figure reproduced with permission from Elsevier).

high symmetry has facilitated EM reconstruction of numerous viruses. For example, a 22-Å resolution reconstruction of Semliki forest virus was based on 50 viral particles. In order to reach 9-Å resolution, 5000 particle images were then used (Figure H2.11(a)). *Radial density functions* are often plotted in the analysis of the reconstructions of viruses (Fig. H2.11(b)). They show how the density varies throughout the virus. For instance, there would be a peak corresponding to the viral capsid. A secondary peak, which represents an inner shell of density, could indicate the presence of packaged DNA or RNA. The radial density function can be used to scale reconstructions of other similar viruses appropriately.

The analysis of epitopes can be carried out by EM. Particles can be incubated with the relevant ligands and imaged. The reconstructions then illustrate the interaction interface between the particles and their ligand (Fig. H2.12). In fact, the presence of bound ligands can often be rapidly appraised from raw micrographs. Native particles usually appear very smooth. In comparison, images of vitrified particle–ligand complexes appear fuzzy because of the presence of bound ligand.

Three-dimensional reconstructions have been made of antibody–particle Fab–particle and receptor–virus complexes (Fig. H2.12(a)). It is often noticed that only the particle and the ligand domains in immediate contact with the particles are well

Fig. H2.11 (a) Example of the Semliki forest virus determined at 9-Å resolution. (b) Radial density function of the Semliki forest virus.

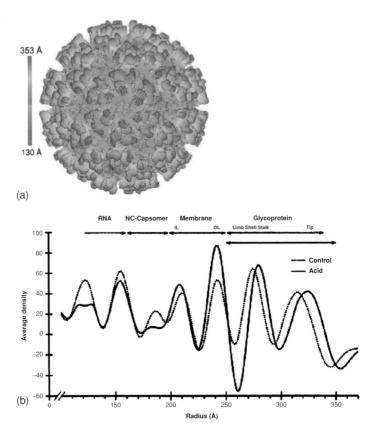

(a)

(b)

Fig. H2.12 Example of receptor–particle complex determined by EM: (a) rhinovirus-14 bound to different antibodies (Che *et al.*, 1998); (b) HBV complexed with a peptide: in red is the difference map between the reconstruction of the virus with peptide and that of the virus on its own (Böttcher *et al.*, 1998). (Figures reproduced with permission from the American Society for Microbiology and from Macmillan Publishers Ltd., respectively.)

defined. The other domains of the ligand are not seen due to their inherent mobility and flexibility, so they are averaged out during the reconstruction procedure.

The site of the ligand binding identifies the ligand's epitope. This footprint can be correlated with mutagenesis data to enable the positioning of key amino acids within the reconstruction. It has been shown that by reconstructing the same virus

(a)

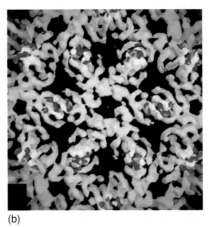

(b)

with and without a bound peptide it is possible to decipher where the peptide is bound to the viral capsid by omitting the peptide-viral map from the map with just the virus (Fig. H2.12(b)).

H2.8.3 Microtubules

Microtubules are essential components of the cytoskeleton of eukaryotic cells. In conjunction with motor proteins like dynein and kinesin, they engage in intracellular transport and in cell division.

Reconstructions from electron micrographs detailed the structural organisation of microtubules (Fig. H2.13). They are 25-nm-diameter hollow tubes with walls made from linear arrangements of tubulin heterodimers, named 'protofilaments'. When followed from protofilament to protofilament around the tube, tubulin describes a 12-nm pitch, left-handed, helical pathway around the microtubule. The number of protofilaments may vary from 9 to 16 resulting in microtubules with different configurations and symmetry. Some of them are truly helical. In the case of the 15-protofilament microtubules, the final map was produced from averaging helical data from 24 data sets, representing over ∼17 000 asymmetric units and layer lines to $1/18$ Å^{-1}.

Docking the atomic X-ray structure of the tubulin $\alpha\beta$-dimer into the lower-resolution EM maps provides an unambiguous orientation of the tubulin dimer within a protofilament. This pseudoatomic resolution microtubule model revealed that the tubulin heterodimer was stacked head-to-tail and also interacted laterally with tubulin from neighbouring protofilaments. Reconstructions of

Fig. H2.13 Surface representations and end-on projections of the three-dimensional reconstruction of an undecorated microtubule (MT), in a complex with monomeric Ncd (MT-M), and in a complex with dimeric Ncd (MT-D). The side-view representations are oriented with the microtubule plus end at the top. The end projections represent the view from the plus end. Scale bars, 100 Å. (Sosa *et al.*, 1997.) (Figure reproduced with permission from Macmillan Publishers Ltd.)

microtubules bound to Ncd, a type of kinesin, subsequently revealed that the binding interface was extensive and primarily constituted of the β-tubulin monomer (Fig. H2.13).

H2.8.4 Integral membrane proteins

Through the efforts of X-ray crystallography and solution NMR, 2000–3000 biological structures are determined yearly; however only one or two concern integral membrane proteins. This is because of experimental difficulties, such as obtaining suitable crystals and collecting interpretable NMR spectra from lipid-or detergent-embedded proteins. Electron cryo-microscopy offers an alternative approach to determining the architecture of macromolecular assemblies. The method has several advantages over other structural techniques. It does not necessitate large amounts of material and even heterogeneous samples can be studied. The two-dimensional crystal, moreover, provides a more native-like environment for the membrane protein than three-dimensional crystals, as it is surrounded by lipid, with little or no detergent, and has the further advantage that it can be characterised by using essentially the same techniques that would be used for the protein in its native membrane (spectroscopy, ligand-binding assays . . .).

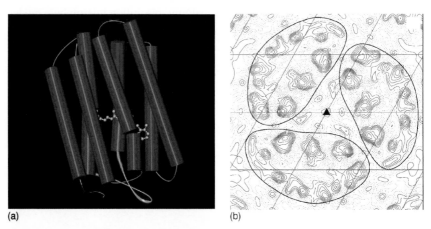

(a)   (b)

Fig. H2.14 The integral membrane protein bacteriorhodopsin and the two-dimensional projection maps identifying its α- helices: (a) Structure of bacteriorhodopsin: the retinal is shown as ball-and-stick. (b) two-dimensional projection Fourier map of trimers of wild-type bacteriorhodopsin at a resolution of 3.5 Å identifying locations of the seven transmembrane helices. The three-fold axis is indicated by the solid triangle. The solid lines around the densities indicate the boundary of each monomer. (Subramaniam et al., 2002; Subramaniam and Henderson, 1999). (Figure reproduced with permission from the Royal Society and Elsevier, respectively).

Work on the structure of the integral membrane protein bacteriorhodopsin has been the basis of several key developments in electron cryo-microscopy. Bacteriorhodopsin, which is a 27-kDa light integral membrane protein found in *Halobacterium salinarum*, uses the energy of an absorbed photon to move a proton (H^+) across the cell membrane (see Comment H3.7). Amazingly, this light-driven proton pump occurs naturally as a two-dimensional crystal in the organism's inner membrane, so this system is particularly amenable to electron crystallographic analysis. The ordered lattice enabled collection of electron diffraction data (3.5 Å in plane resolution), which, combined with phases from the images, enabled the determination of an atomic model of bacteriorhodopsin. The data were, in fact, adequate to resolve unambiguously the majority of the bulkier side-chains (Trp, Tyr, Phe and Leu) The protein consists of seven helices with a lysine covalently bound via a protonated Schiff base linkage to a retinal molecule (Fig. H2.14(a)).

The flash cooling of the sample offered the opportunity to trap the protein at different steps of the pumping mechanism, as the two-dimensional crystals could be frozen after different times after illumination. The EM data established that a single, large protein conformational change occurs within 1 ms after illumination,

Fig. H2.15 Voltage gated opening of an ion channel KvAp. Top: Images of negatively stained (left) and cryo-negatively stained (right) KvAP–33H1 complexes, with selected particles marked with red circles. The scale bar is 50 nm. The Fab fragment 33H1 only binds to KvAp in its open conformation. A Bottom left: three-dimensional reconstruction of the KvAP–33H1 complex at 10.5-Å resolution. The red arrow identifies a putative structural change on channel opening. The red asterisk marks the position of the variable immunoglobulin domain. Bottom right: (FSC) for the three-dimensional map of the KvAP–33H1 complex from two maps calculated using two halves of the data. The 0.5 threshold corresponds to 10.5 Å. (Jiang *et al.*, 2003.) (Figure reproduced with permission from Macmillan Publishers Ltd.)

as the retinal isomerisation from the all-*trans* to the 13-*cis* configuration initiates the release of a proton from the Schiff base to the outside of the cell. An intermediate protein structure follows that enables the base's reprotonation from the cytoplasmic side of the membrane. Retinal reisomerisation brings the protein back into the resting state, and thus ready to start the pumping cycle again. Full three-dimensional reconstructions are not, in fact, necessary, as even from the two-dimensional projection maps it is possible to identify the location of the seven α-helices and any structural rearrangements (Fig. H2.14(b)).

In contrast to X-ray crystallography (Part G) and NMR (Part J), electron cryo-microscopy allows even partially heterogeneous samples to be studied. Single-particle reconstructions of integral membrane protein therefore offer the opportunity to determine the structure of integral membrane proteins. The small size of many of these proteins is still a limiting factor; however, negative staining may be relied upon to facilitate the identification and subsequently the determination of a low-resolution reconstruction. Jiang and coworkers, for example, analysed the opening of a voltage gated ion channel ($\sim$200 kDa) at 10.5-Å resolution using negatively stained samples (Fig. H2.15). In conclusion, the advent of better microscopes, sample preparation and reconstruction procedures is bringing single-particle analysis of integral membrane protein into the realm of the feasible.

H2.9 Checklist of key ideas

- Biological samples are very sensitive to the electron beam. Typically they can only tolerate a dose of 10–15 electrons/Å^{-2}.
- The CTF is due to the imperfections of the electron microscope. Once it has been estimated corrections can be applied.
- The common-lines reconstruction procedure is performed in Fourier space. The Fourier transforms of two different images of the same object intersect along a common line.
- Polar Fourier transform reconstruction is done with Fourier transforms of the EM images described by polar coordinates. The method relies on the use of a template that helps orient the images in order to create a more appropriate model of the imaged sample.
- The back projection method relies on matching EM images and that of a model which has been projected in several directions.
- Focal pairs can ease the initial reconstruction procedures, e.g. by identifying the hand of the object.
- Statistical analysis is used to help in the reconstruction procedures due to the poor signal-to-noise of the EM micrograph.
- In multivariate statistical analysis, the EM images are transformed to reduce the total number of data treated.
- The minimum number of particles needed for a reconstruction at a desired resolution depends on the size of the object, its symmetry and the signal-to-noise ratio deemed acceptable.

- Resolution is not clearly defined in EM.
- The phase residual can be used to underscore the consistency of the phases of common-line pairs to a certain resolution.
- FSC is used to qualify the final reconstruction: it consists of a comparison between two reconstructions independently calculated from separate halves of the dataset.
- An amplitude profile can be imposed on the final map in order to down-weight the low-resolution data and boost the higher-resolution features.
- The lower-resolution EM reconstruction can be combined with atomic models of its different constituents derived from X-ray crystallography to create a 'quasi'-atomic-resolution structure.

Suggestions for further reading

Several excellent reviews have been written on electron cryo-microscopy, including:

Baker, T. S., Olson, N. H., and Fuller, S. D. (2000). Adding the third dimension to virus life cycles: three-dimensional reconstruction of icosahedral viruses from cryo-electron micrographs. *Microbiol. Mol. Biol. Rev.*, **63**, 862–922.

Van Heel, M., Gowen, B., *et al.* (2000). Single-particle electron cryo-microscopy: towards atomic resolution. *Q. Rev. Biophys.*, **33**, 307–369.

Few books on three-dimensional reconstructions have been published but those that have include:

Frank, J. (2005). *Three-Dimensional Electron Microscopy of Macromolecular Assemblies*. New York: Oxford University Press.

Glaeser, K., Dowrling, K., Derosier, D., Baumeister, W., and Frank, J. (2004). *Electron Crystallography of Biological Macromolecules*. Oxford: Oxford University Press.

Part I
Molecular dynamics

Chapter I1
Energy and time calculations

I1.1 Historical review of biological applications

The first molecular dynamics simulation of a biological macromolecule, the small stable protein bovine pancreatic trypsin inhibitor (BPTI) for which an accurate X-ray crystallographic structure was available, was published by **J. A. McCammon, B. R. Gelin** and **M. Karplus** in **1975**. Based on a simple approximation for the molecular mechanics potential, the simulation was performed for the macromolecule in vacuum, and limited to 9.2 ps by the computing power available at the time. The BPTI study was nevertheless effectively instrumental, together with the earlier hydrogen exchange experiments of **K. Linderstrom-Lang** and his collaborators **(1955)**, in establishing the view that proteins are not rigid bodies but dynamic entities whose internal motions must play a role in their biological activity.

In **2003 M. Karplus** wrote that molecular dynamics simulations of biological macromolecules developed from two lines of work: *individual particle trajectory calculations in chemical reactions* and *physical calculations to predict the dynamic behaviour of large particle assemblies*.

The first line of work goes back to the two-body scattering problem for which analytical solutions are available. It is well known in physics, however, that no analytical solutions have been found for the three (or more)-body problem. A prototype trajectory calculation was attempted in **1936** by **J. A. Hirschfelder, H. Eyring** and **B. Topley** for the simple chemical reaction involving the interconversion between hydrogen atoms and molecules, but the calculation could only be completed by using the computers that became available in the **1960s**. Classical chemical trajectory calculations have now evolved to incorporate quantum mechanical effects.

The second line of work has its origins in the work of **J. D. van der Waals** on particle–particle interactions and the statistical mechanics of **L. Boltzmann**. In the **1950s, B. J. Alder** and **T. E. Wainwright** published a study of liquids whose molecules behaved like 'hard spheres'. **A. Rahman** in **1964** published a molecular dynamics simulation of liquid argon, assuming a 'soft' Lennard–Jones potential; the first simulation of liquid water, a much more complex liquid, was published in **1971**, by **F. H. Stillinger** and **A. Rahman**.

The two independent approaches were brought together in the simulation of protein molecular dynamics. In the **late 1960s**, good starting points for the calculations were provided, on the one hand, by the empirical energy functions that had been developed for bonded and non-bonded interactions in small organic molecules in the laboratories of **H. Scheraga** and **S. Lifson** and, on the other hand, by the precise three-dimensional structures that were being solved by X-ray crystallography.

The decades following the publication of the BPTI calculation saw a significant extension of the calculations to larger proteins and longer simulation times, as higher-resolution structures and more powerful computers became available. Calculation methods combining quantum and molecular mechanics (QM/MM) were developed. Macromolecular dynamics simulations are necessarily approximate, due to the complexity and heterogeneity of the systems studied. Parallel developments in experimental approaches provided information on macromolecular internal motions and were essential in order to validate the simulation methods. These included the analysis of Debye–Waller factors in crystallography, fluorescence depolarisation of tryptophan residues, NMR, inelastic neutron scattering, Mössbauer spectroscopy and Fourier transform IR spectroscopy. Symbiotic relationships were established between molecular dynamics calculations and NMR and X-ray crystallography, starting with the incorporation of energy minimisation in structural refinement.

Since **2000** massive progress in computing power has made possible molecular dynamics simulations of systems of more than 100 000 atoms and time scales up to the microsecond. Molecular dynamics simulations are now being applied on the supramolecular and even cellular scale. In the study of molecular machines, like the ATP synthase (the enzyme responsible for most ATP synthesis in living organisms) macromolecular dynamics calculations are building the link between thermodynamics data and the crystallographic structure. Kinetic cryo-crystallography, in which intermediate structures are 'frozen' in and examined, has become one of the powerful molecular dynamics experimental techniques.

I1.2 Dynamics, kinetics, kinematics, and molecular stabilisation forces

The concepts of *dynamics, kinetics, kinematics* and *molecular stabilisation forces* were discussed in Chapter A3.

Our understanding of macromolecular structures is based on an understanding of the forces underlying them – the forces acting to maintain the atoms in their positions. And if, in turn, we gain an understanding of these forces, then we shall also understand how the atoms in the structure move about their mean locations. Expressed in the language of physics, our aim is the understanding of the *force field* or *potential energy function* around each atom in a structure. The forces that maintain biological macromolecular structure are known (see Section C2.4.9). They arise from hydrogen bonds, salt bridges or screened electrostatic interactions, so-called hydrophobic interactions, and van der Waals interactions. Covalent bonds other than the disulphide bond that represents a special case because of its sensitivity to the reducing capacity of the environment, which are very strong with respect to temperature, are not usually broken when a macromolecule unfolds, so that in general they do not contribute to the stabilisation energy of macromolecular tertiary or quaternary structure.

I1.3 Length and time scales in macromolecular dynamics

The amplitudes of atomic motions in macromolecules at ambient temperature (300 K) range from 0.01 Å to >5 Å for time periods from 10^{-15} to 10^3 seconds (1 fs for electronic rearrangements to about 20 min for protein folding or local denaturation) (see Chapter A3).

I1.4 Normal mode analysis

The *normal mode* approach to molecular dynamics has the advantage that it provides an analytical solution for the description of atomic motions and other dynamic parameters of a macromolecular system such as thermodynamic averages of amplitudes, fluctuations and correlations. The approach is based on the assumption that the atoms are coupled by simple harmonic potentials, and their vibration results from a superposition of fundamental modes (see Chapter A3).

The equations of motion of N harmonically coupled atoms in a macromolecule are given in Chapter A3. Their solutions form a set of $3N$ periodic functions. Six of these correspond to the macromolecular translational and rotational degrees of freedom, for which the atoms do not move relative to each other. The system, therefore, has $3N - 6$ normal modes. Atomic motions are a superposition of all the normal modes.

At a given temperature (mean energy) the mean square amplitude of a mode is inversely proportional to its effective force constant, so that the low-frequency modes dominate motion amplitudes (see the example calculation for BPTI in Fig. A3.14).

A normal mode calculation is expected to provide a reasonable description of vibrational protein dynamics at low temperature, below the dynamical transition above which atomic motions sample different conformational substates (see Chapter A3). Normal mode calculations are based on a number of assumptions, however, that are not directly applicable to proteins at ambient temperature. These are: (i) atoms in the macromolecule behave as point particles; (ii) the macromolecular conformation is at an energy minimum; (iii) atoms move in simple harmonic potentials (quadratic functions of the spatial coordinates); (iv) interactions with solvent are negligible. In particular, solvent interactions and other anharmonic effects are far from negligible at room temperature. As we pointed out at the beginning of this section, however, the precise analytical nature of normal mode calculations of protein dynamics yields considerable advantages, and various approaches have been developed to make such calculations applicable even when the assumptions above are not strictly valid. The conformational energy landscape at ambient temperature is anharmonic, made up of the many local minima of the conformational substates (see Chapter A3). It may be approximated by an empirical potential calculated from molecular dynamics simulations (Section I1.5.3) in which solvent effects can be introduced in different ways (Section I1.5.4).

I1.5 Molecular dynamics simulations

In a molecular dynamics simulation, the behaviour of a system is determined as a function of time from an algorithm that includes a starting *structural model* and *force field*. Our knowledge of the determinants involved, however, is very far from complete and the molecular dynamics simulation approach remains firmly anchored in experiment.

The quality of a molecular dynamics result is highly sensitive to the quality of the starting structure (usually a high-resolution structure from X-ray crystallography) and on force field assumptions, which can be tested against experimental spectroscopic data. In fact, the simulation procedure itself is often seen as an *experiment*, in which conditions are varied until good agreement is obtained with data from structural and spectroscopic methods.

Simulation tools fall into two categories: *classical* molecular mechanics approaches to study Newtonian and *stochastic* (statistically random) dynamics (in order to determine atomic motions, conformational changes, ligand binding, solvent effects etc.) and *quantum* mechanics to study electronic distributions (in order to determine enzyme reaction mechanisms etc.). *Ab initio* quantum mechanical calculations are limited by computer power, in 2004, to about 100 atoms during about 10 ps. By using a semi-empirical quantum mechanical approach for a similar number of atoms, the time range is extended to about 1 ns. Molecular mechanics at the atom level deals with the energy of the system and the forces exerted on each of its atoms. Molecular mechanics calculations are relatively fast; they allow the simulation of systems of about 100 000 atoms during several nanoseconds in picosecond steps. The QM/MM approach combines a quantum mechanical calculation (around the atoms in an enzyme active site, for example) with a molecular mechanics calculation (for all the other atoms in the system).

I1.5.1 Force field

The potential energy function in molecular mechanics is given by

$$V_{\text{total}} = V_{\text{cov}} + V_{\text{non-cov}} \tag{I1.1a}$$

$$V_{\text{total}} = V_{\text{bond}} + V_{\text{angle}} + V_{\text{dihedrals}} + V_{\text{elec}} + V_{\text{LJ}} \tag{I1.1b}$$

The first three terms on the right-hand side of Eq. (I1.1b) make up the covalent energy, V_{cov}; they correspond, respectively, to covalent bond stretching, angle and

Comment I1.1   Covalent and non-covalent energy

Covalent bond stretching:

$$V_{\text{bond}} = \sum_{\text{bonds}} \frac{1}{2} k_b (b - b_0)^2$$

Covalent angle energy:

$$V_{\text{angle}} = \sum_{\text{angles}} \frac{1}{2} k_\theta (\theta - \theta_0)^2$$

Covalent dihedral angle energy:

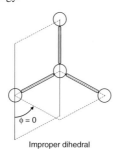

Proper dihedral   Improper dihedral

$$V_{\text{diherdral}} = \sum_{\text{dihedrals}} \frac{1}{2} k_\phi [1 + \cos(n\phi - \delta)]^2$$

Electrostatic energy:

$$V_{\text{elec}} = \frac{1}{4\pi \varepsilon_0 \varepsilon} \sum_{\text{ijpairs}} \frac{q_i q_j}{r_{ij}^2}$$

Lennard–Jones potential

$$V_{\text{LJ}} = \sum_{\text{ijpairs}} 4\varepsilon_{ij} \left\{ \left(\frac{\sigma_{ij}}{r_{ij}} \right)^{12} - \left(\frac{\sigma_{ij}}{r_{ij}} \right)^6 \right\}$$

dihedral angle energy. The last two terms refer to non-covalent bond interactions, $V_{\text{non-cov}}$; they are the electrostatic and Lennard–Jones potentials (which describe the van der Waals interaction) (Comment I1.1). Other energy terms that may be considered include a polarization term due to charge distribution distortion by the internal electric field in the macromolecule and the special case of the hydrogen bond.

The force on atom i along a coordinate vector $\mathbf{r}_i$ is given by the slope of the potential:

Comment I1.2

MD programs

The CHARMM webpage is www.charmm.org. The AMBER web page is www.amber.ucsf.edu.

$$\mathbf{F}_i = -\frac{\partial V_{\text{total}}}{\partial \mathbf{r}_i} \tag{I1.2}$$

I1.5.2 Parameterisation of the force field

The potentials in Eqs. (I1.1) are described by phenomenological functions, expressed in terms of coordinates, coordinate deviations and coefficients (Comment I1.1), which are optimised to reproduce experimental data and quantum calculations. The total energy function is itself empirical. It is expressed as a function of the position of each atom, so that it is important to have as good a starting structure as possible.

The parameters in the potential energy terms have been obtained from experimental structural data (X-ray crystallography or NMR for bond lengths and angles, for example), spectroscopy (IR and Raman, for the force constants) and quantum calculations on model systems for charges and Lennard–Jones potential constants.

Molecular dynamics simulation programs propose different expressions for the force field and include parameter files for amino acids, proteins, nucleic acids and many other organic molecules, especially enzyme substrates and inhibitors (Comment I1.2).

I1.5.3 Potential energy surface and energy minimisation

We recall from Chapter A3 that the potential energy of a protein as a function of the positions of all its atoms (expressed in terms of a conformational coordinate) can be described as a rugged landscape with dips and rises (Fig. I1.1).

The *global minimum* is the lowest energy state. Depending on the available energy (proportional to the temperature), however, the molecule may be trapped in a *local minimum*, one of the conformational substates discussed in Chapter A3. Transitions between local minima pass through *saddle points*.

Before starting a molecular dynamics simulation, it is necessary to allow a structure to reach a locally stable energy state. At a minimum in the energy surface, the first derivative or gradient is zero and the second derivative is posiive ($(\partial V/\partial r = 0$ and $\partial^2 V/\partial r^2 > 0)$, and the set of atomic coordinates that satisfy these equations must be found. This is a very complex problem to solve for

Fig. I1.1 Schematic potential energy surface as a function of conformational coordinate for a protein.

proteins, because of the large number of atoms and interactions involved, and various mathematical approaches have been developed to address it at different levels of approximation. It is, in particular, very difficult to explore large areas of conformational space by crossing saddle point barriers.

One approach used for energy minimisation is based on *simulated annealing*. The temperature of the system is increased (e.g. to 800 K), then rapidly decreased (e.g. at a rate of 50 K ps^{-1}) to 300 K. Simulated annealing tends to bring the system into a relatively stable area of potential energy space, and increases the chances of finding a global minimum in conformational space.

I1.5.4 Modelling the solvent

The importance of the role played by the solvent in protein dynamics was discussed in Chapter A3. In molecular dynamics simulations, the solvent can be accounted for by different approaches.

In the *explicit solvent* approach, the protein is positioned in a box of water molecules. The infinite character of the solvent is taken into account by introducing periodic boundary conditions in which a tractable number of water molecules is placed in an effective unit cell that is repeated a great number of times. Periodic boundary conditions have also been applied to simulate a protein crystal with a unit cell containing protein and water molecules (Comment I1.3). Solvent and protein mutually influence each other, and molecular dynamics simulations, such as the one described in the comment, also address the important point concerning how the hydration layer structure and dynamics is strongly influenced by interactions with the protein surface.

Where they are applicable, *implicit solvent* approaches can allow a substantial gain in computational time. Water molecules are not represented explicitly and solvent effects are included in the simulations in various ways, by attributing continuum dielectric constant values to the solvent and protein volumes, or by introducing a dielectric constant within the macromolecule that depends on the distance from the protein–solvent interface effectively screening the charges on the protein surface.

Solvent effects can also be described by using a friction coefficient and random force on the macromolecular atoms due to water molecule collisions. The general equation of motion (Eq. (I1.3)) used is due to Langevin, and describes what has been called *Langevin dynamics*:

$$m\frac{d^2\mathbf{r}}{dt^2} = -\frac{\partial V}{\partial \mathbf{r}} - m\beta\frac{d\mathbf{r}}{dt} + \mathbf{R}(t) \qquad (I1.3)$$

where m is the mass of the atom under consideration, $\mathbf{r}$ is its positional vector, V is the potential energy function at that position, β is a friction coefficient and $\mathbf{R}(t)$ is a random force as a function of time, t. Simulation of the random force implies simulating the macromolecule in a thermal water bath at constant temperature.

Comment I1.3 Molecular dynamics simulation of an RNase A crystal

The simulation box corresponds to one unit cell of the crystal, containing two protein molecules (light atoms) and 817 water molecules. The box is replicated infinitely in three dimensions by applying periodic boundary conditions. The simulation protocol uses the CHARMM force field (Comment I1.2) and the so-called SPC/E model for water. Two constant temperatures were examined, 150 K and 300 K. Data on single particle and collective water dynamics were derived that compared favourably with X-ray diffraction and inelastic neutron scattering observations (see also Chapters G3, I2) (Tarek and Tobias, 2002; Bon *et al.*, 1999, 2002).

Note that the frictional force due to solvent viscosity reduces the particle energy, whereas the collisions due to the thermal agitation of the water molecules increase the energy.

I1.5.5 Typical molecular mechanics simulation protocol

The starting point for a molecular mechanics simulation of a macromolecule is the best possible available X-ray crystallography structure. A solvent model is then chosen and the potential energy function of the system is defined and minimised.

Each atom, i, is represented by a point at the position described by the vector, $\mathbf{r}_i$, that obeys Newtonian dynamics (or Langevin dynamics if this solvent model was adopted):

$$m_i \frac{d^2 \mathbf{r}_i}{dt^2} = -\frac{\partial V}{\partial \mathbf{r}_i} \qquad (I1.4)$$

where subscript i refers to atom i.

A velocity $\mathbf{v}_i$ associated with each atom is chosen in order to obtain the desired temperature

$$T = \frac{1}{3 N k_B} \sum_i m \langle \mathbf{v}_i^2 \rangle \qquad (I1.5)$$

where N is the number of atoms.

The equations of motion are integrated (by various algorithms) to determine the *trajectory* of each atom, i.e. its position and velocity between time t and time $t + dt$. The trajectories of all the atoms expressed as a chosen number of time-steps are then stored for further analysis.

A calculation time-step of 1 fs is usually used in protein dynamics simulations, so that a 1 ns simulation requires 10^6 force and energy calculations per atom.

I1.5.6 Analysis of results

Temperature
The temperature of the simulated system is defined by Eq. (I1.5).

Radius of gyration
The radius of gyration, R_G, is a useful parameter to define the overall conformation of the system (see Chapter G2). It can be calculated in two steps. The centre-of-mass coordinates, $\mathbf{R}_C$, are obtained first from

$$\sum m_i (\mathbf{r}_i - \mathbf{R}_C) = 0 \qquad (I1.6)$$

The radius of gyration is related to the moment of inertia and defined by

$$R_G^2 = \sum m_i (\mathbf{r}_i - \mathbf{R}_C)^2 / M \qquad (I1.7)$$

where M is the total mass of the system.

The radius of gyration can also be obtained from the system coordinates, without going through the centre-of-mass calculation, from

$$R_G^2 = \sum \sum m_i m_j (\mathbf{r}_i - \mathbf{r}_j)^2 / M^2 \tag{I1.8}$$

Root mean square deviation (RMSD)

The RMSD follows how the structure evolves during the simulation. It is defined by

$$\text{RMSD} = \sqrt{\frac{1}{N} \sum_{i=1}^{N} (\mathbf{r}_i - \mathbf{r}_i^{\text{ref}})^2} \tag{I1.9}$$

where $\mathbf{r}_i^{\text{ref}}$ are the coordinate vectors of the reference structure (usually, the starting structure, used for the simulation). The RMSD plotted during the trajectory of a structure as it is allowed to equilibrate at a constant temperature should reach a plateau value in times of the order of 100 ps.

Mean square fluctuations and Debye–Waller factors

Crystallographic *Debye–Waller factors* describe the fluctuations of atoms about their mean positions in a crystal structure (see Part G). Since crystallographic studies are time-averaged, fluctuation amplitudes include terms due to time-dependent thermal motions and terms due to deviations from atomic mean positions because of disorder in the crystal.

$$\langle u_i^2 \rangle_{\text{total}} = \frac{3 B_i}{8\pi^2} = \langle u_i^2 \rangle_{\text{thermal}} + \langle u_i^2 \rangle_{\text{disorder}} \tag{I1.10}$$

where B_i is the crystallographic B-factor as it is usually defined. In practice it is possible to distinguish between the static and dynamic contributions to the Debye–Waller factor by collecting crystallographic data as a function of temperature. The disorder term is independent of temperature, while the square of the thermal amplitudes for harmonic vibrations varies linearly with absolute temperature (see Chapter A3).

Mean square fluctuations, calculated from a molecular dynamics normal modes calculation, compare very favourably with fluctuations derived from Debye–Waller factors (Fig. I1.2), showing the larger mobility of loops outside secondary structure elements.

Diffusion coefficients

When the atomic displacements are large during the time of the simulation it is useful to express them in terms of a diffusion coefficient (see Part D). The diffusion coefficient of atom i is given by

$$D_i = \frac{1}{6t} \langle (\mathbf{r}_i(t) - \mathbf{r}_i(0))^2 \rangle \tag{I1.11}$$

where t is time.

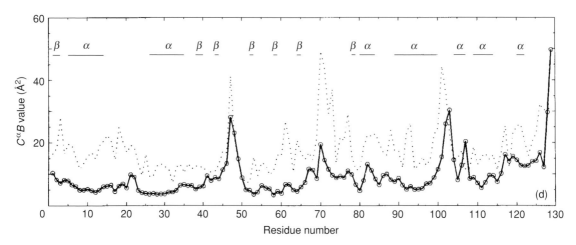

Fig. I1.2 Experimental
B-factors (dotted line) and
B-factors from the full set
of normal modes
calculated for 300 K in
lysozyme (full line), as a
function of residue
number. The *y*-axis
corresponds to $\langle u^2 \rangle 8\pi^2/3$
(see Eq. (I1.10)). The
secondary structure
stretches in the sequence
are shown also. (Levitt
et al., 1985.) (Figure
reproduced with
permission from Elsevier.)

I1.6 Application examples

I1.6.1 BPTI and lysozyme

The bases for molecular dynamics simulations of proteins were set between 1975 and 1977. The first molecular dynamics simulation of a protein was published for BPTI, a small, 58-amino-acid residue protein of known structure, in 1977. BPTI was also the first protein for which normal mode calculations were performed in the early 1980s. As discussed above, normal mode calculations are complementary to molecular dynamics simulations. The normal mode approach has the limitation that a quadratic approximation must be used for the free-energy potential, but the advantage is that it provides a complete description of atomic vibrations in the system, in the fundamental 0.1–10 ps time interval. The normal modes approach provides analytical expressions for thermodynamic averages of amplitudes, fluctuations and correlations. Early normal mode calculations illustrated collective atomic motions in globular proteins, and showed how the lower-frequency modes dominated amplitudes (see Section I1.4). When calculated from the normal mode fluctuations, Debye–Waller factors were found to be in reasonable agreement with the experimental crystallographic values corresponding to thermal motions (Fig. I1.2).

Normal mode calculations revealed similar dynamics in the collective motions of small monodomain proteins, like BPTI, and in enzymes, like lysozyme and ribonuclease A, whose active sites are sandwiched between two domains. It is interesting to note, however, that in the larger molecules they also showed collective domain motions about the active site (Fig. I1.3).

I1.6.2 Protein folding

The first microsecond molecular dynamics simulation of a protein was published in 1998. It described the thermodynamic folding of HP-36, a 36-residue subdomain.

Fig. I1.3 Domain motion about the active site in the low-frequency normal mode of period 11.2 ps, in lysozyme. (Levitt *et al.*, 1985.) (Figure reproduced with permission from Elsevier.)

In vitro HP-36 folds spontaneously between 10 and 100 μs, so that only the very first steps of the process could be simulated. The 1-μs calculation on the 36 residues and about 3000 explicit water molecules occupied 256 processors for 4 months.

I1.6.3 Structure refinement

The objective of structure refinement in high-resolution crystallography and NMR (see Parts G, J) is to reach the best possible agreement between a model structure and experimental data. This is obtained in practice by minimising a function representing differences between observed parameters and corresponding parameters calculated from the model. The function can be expressed as a total energy to be minimised:

$$E = E_{chem} + w_{data} E_{data} \qquad (I1.12)$$

where E_{chem} represents the potential energy calculated for the model structure and E_{data} is related to the difference between measured and model calculated diffraction data; w_{data} is a factor to weight the relative contributions of the two terms (Comment I1.4).

> **Comment I1.4**
> Energy minimisation in structure refinement
>
> See Brunger and Adams (2002).

I1.6.4 ATP synthase, a molecular machine

Spectacular progress in macromolecular crystallography, yielding high-resolution structures of large protein complexes and molecular machines in their various states, has stimulated a parallel effort in molecular dynamics simulations.

Comment I1.5   Mitchell's chemiosmotic hypothesis

Chemical energy in biological systems is stored mainly in the form of adenosine triphosphate (ATP) and it is recovered when a phosphate bond is cleaved by hydrolysis to form adenosine diphosphate (ADP). After it had been established that the phosphorylation of ADP involved electron transfer to molecular oxygen during respiration, the mystery remained as to how the energy derived from these reactions was converted to ATP. A high-energy intermediate that would drive ATP synthesis by phosphate group transfer was postulated but not found. Peter Mitchell proposed a revolutionary solution to the problem, the 'chemiosmotic hypothesis' in which the 'high-energy intermediate' is, in fact, a proton concentration gradient across a membrane between two aqueous compartments. The gradient, established by respiration or photosynthesis, would thus couple electron transfer to ATP synthesis, which would be driven by the return flow of protons (see figure above). At first controversial, the main tenets of the chemiosmotic hypothesis were finally accepted by the scientific community and Mitchell was awarded the Nobel prize in 1978. ATP synthase is the key enzyme which couples the proton gradient with ATP synthesis, the red square in the figure. P. D. Boyer, J. E. Walker and J. C. Skou shared the 1997 Nobel Prize in chemistry for their work on ATPase enzymes.

The enzyme F_1F_0-ATPase is responsible for most of the ATP synthesis in living organisms (Comment I1.5). It is a large membrane-bound molecular machine made up of several protein subunits. The catalysis of ATP synthesis from ADP is driven by proton translocation across the membrane through the F_0 domain – the proton gradient having been created previously by respiration or photosynthesis. The catalytic domain, F_1, is made up of a nine-subunit, $\alpha_3\beta_3\gamma\delta$, globular structure outside the membrane. The $\alpha_3\beta_3$ subunits are arranged in a ring around a central stalk made up of the γ subunit associated to δ and , whose foot makes extensive contact with a ring of c subunits in the F_0 domain. The central stalk in F_1 and the c ring in F_0 are believed to act similarly to the rotor in an electric motor and rotate together relative to the rest of the enzyme, by using the proton translocation energy. The rotation modulates the binding affinities of the β subunits for substrate and product according to the *binding change* mechanism, in which the catalytic subunits cycle between three states, *open* (or *empty*), *loose* and *tight*. During the cycle 120° rotation of the stalk converts an open site (with low affinity for both substrate and product) to a loose site (which binds substrate); a further 120° rotation converts the loose site to the tight site (in which the reaction product is formed) and a further 120° rotation brings the subunit back to the open state to release the product. The three β states have been identified in crystal structures according to their binding properties for ADP and various derivatives. The F_1 domain retains its ability to hydrolyse ATP to ADP (enzyme catalysis can go in both forward and backward reaction directions). Hydrolysis of ATP leads to the rotation of the central stalk, strikingly visualised in a microscope by attaching an actin filament or a bead to its exposed foot (Comment I1.6, see also Chapter F5). Important hints on the mechanisms underlying the rotation were obtained from comparisons of the different subunit structures observed in the crystallographic studies, and MD simulations were performed in order to analyse the dynamics of the changes and their sequence in time (Comment I1.7).

There will certainly be great progress in the coming years in molecular dynamics simulations of ATP synthase mechanisms, in the wake of 'better' crystal structures and improved calculation methods. The first results, however, have already provided fascinating insights into the workings of this molecular motor.

> **Comment I1.6**
> **Direct observation of ATPase rotation by microscopy**
>
> See Noji *et al.* (1997).

Comment I1.7 Targeted and biased molecular dynamics simulations of ATP synthase

The targeted molecular dynamics method used by Karplus and his collaborators for the simulation of ATP synthase applies constraints to a system and looks for a low-energy pathway connecting two well-defined end-states. Targeted molecular dynamics results were supplemented by a biased molecular dynamics simulation, in which a biasing force was applied only to the γ subunit to determine the response of the α and β subunits. The simulations corresponded to one 120° rotation step of the γ subunit (clockwise, when viewed from the membrane, in the direction of ATP synthesis) between two end-states, for which the conformational differences between the α and β subunits were inferred from the crystal structures.

The targeted and biased molecular dynamics methods were implemented in the CHARMM program, using standard molecular mechanics force fields for all atoms in the structures and solvent screening. The targeted molecular dynamics trajectory was calculated after 50 ps equilibration at 300 K, in 2 fs steps for a total duration of 250 ps or 500 ps. The coordinates changed smoothly during the trajectory, suggesting there were no high-energy barriers to be crossed. The duration of the biased molecular dynamics was 1 ns with a 1 fs time-step. Calculation times were much shorter than 250 µs, the time scale of the 120° rotation of γ in the ATP synthase observed in the hydrolysis reaction cycle. The correspondence of the results from simulations of different lengths, however, provided support for identifying the dynamics during the trajectory with the physical motions. It is interesting to note (for future comparisons in this rapidly progressing field) that the 500 ps TMD simulation required 240 cpu hours on a single SGI indigo processor (Ma *et al.*, 2002).

Fig. I1.4 Structures from the molecular dynamics trajectory: (a) is the initial structure, (b), (c), (d), (e), are after 150, 300, 350 and 425 ps, respectively, (f) is the final structure. Subunit colours are red for α, yellow for β, purple for γ, green for δ, and light yellow for ε. (Ma et al., 2002.) (Figure reproduced with permission from Elsevier.)

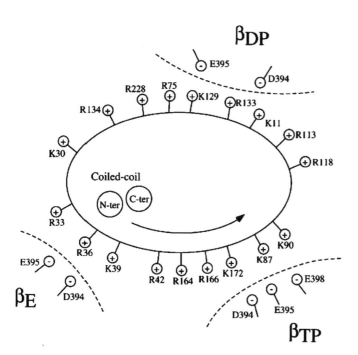

βDP

Fig. I1.5 Schematic view of the ionic track interactions between the γ subunit and β subunits (Ma *et al.*, 2002). (Figure reproduced with permission from Elsevier.)

Structures during the molecular dynamics trajectory are shown in Fig. I1.4. The molecular dynamics simulations showed how the rotation of the γ subunit induces the opening and closing of the catalytic β subunits, with, of particular interest, the demonstration of the existence of an *ionic track* guiding their motion during the transition (Fig. I1.5). (See also Section F5.3.2.)

I1.7 Checklist of key ideas

- Molecular dynamics is the study of forces that underlie molecular structure and motions.
- The folded native structures of biological macromolecules are maintained by forces arising from hydrogen bonds, salt bridges or screened electrostatic interactions, so-called hydrophobic interactions and van der Waals interactions.
- The *force field* is the *potential energy function* around each atom in the macromolecule.
- The amplitudes of atomic motions in macromolecules at ambient temperature range from 0.01 Å to >5 Å for time periods from 10^{-15} to 10^3 s (1 fs for electronic rearrangements to about 20 min for protein folding or local denaturation).
- Our understanding of protein dynamics is based on results from various spectroscopic experiments, and crystallographic studies of time-averaged structures and transient intermediates.
- Normal mode calculations are based on simple harmonic (quadratic) potentials and provide analytical solutions for the atomic motions and thermodynamic averages of amplitudes, fluctuations and correlations.

- A three-dimensional system of N atoms has $3N - 6$ normal modes.
- Low-frequency normal modes dominate vibration amplitudes.
- In a molecular dynamics simulation, the behaviour of a system is determined as a function of time from an algorithm that includes a starting structural model and force field.
- Classical *molecular mechanics* simulations are used to study Newtonian or stochastic dynamics (atomic motions, conformational changes, ligand binding, solvent effects . . .). Such calculations are currently limited by computer power to a maximum of about 100 000 atoms, moving for a few nanoseconds.
- *Quantum mechanics* is used to study electronic distributions (enzyme reaction mechanisms . . .). Current limits are 100 atoms moving for 10 ps.
- The QM/MM approach combines quantum mechanics for the study of a limited part of the structure (for example, the active site of an enzyme) and molecular mechanics for the rest of the molecule.
- Force fields include covalent bond terms and terms referring to non-covalent bond interactions, such as electrostatic and Lennard–Jones potentials.
- The potential energy of a protein as a function of a conformational coordinate is represented by a rugged *energy landscape* with local minima (corresponding to the conformational substates), separated by saddle points, and a global energy minimum.
- In an *energy minimisation* procedure, a structure is allowed to evolve towards a local or global minimum.
- *Simulated annealing* is a method that increases the chances of finding a global energy minimum, by controlling the temperature during a molecular dynamics simulation.
- Solvent and protein mutually influence each other during a molecular dynamics simulation.
- In the *explicit solvent* simulation approach, the protein is positioned in a box of water molecules and periodic boundary conditions are applied.
- In *implicit solvent* approaches, solvent effects are included in the simulations by attributing continuum dielectric constant values to the solvent and protein volumes.
- In *Langevin dynamics* the solvent is taken into account by introducing a friction coefficient and random force on the macromolecular atoms due to water molecule collisions.
- Results of a molecular dynamics simulation include a *trajectory* describing the time evolution of conformational and fluctuation parameters, and values for the temperature and pressure of the system, its radius of gyration, atomic RMSD, and atomic diffusion coefficients.
- RMSDs from molecular dynamics simulations compare favourably with experimental Debye–Waller factors derived values from crystallography.
- Normal mode calculations in two domain enzymes, like lysozyme and ribonuclease A, reveal collective domain motions about the active site.
- The first microsecond molecular dynamics simulation of a protein described the first steps in the thermodynamic folding of HP-36, a 36-residue subdomain, in the presence of 36 water molecules.

- Energy minimisation by molecular dynamics simulation approaches is used in the refinement of high-resolution structures in crystallography and NMR.
- Targeted molecular dynamics simulations are being developed to study the mechanisms in large molecular machines such as ATP synthase or chaperones.

Suggestions for further reading

McCammon, J. A., and Harvey, S. C. (1987). *Dynamics of Proteins and Nucleic Acids.* Cambridge: Cambridge University Press.

Karplus, M. (2003). Molecular dynamics of biological macromolecules: a brief history and perspectives. *Biopolymers*, **68**, 350–358.

Field, M. (2005). *A Practical Introduction to the Simulation of Molecular Systems.* Cambridge: Cambridge University Press.

Chapter I2
Neutron spectroscopy

I2.1 Historical overview and introduction to biological applications

We recall from Chapter A1 that biological events occur on an immensely extended range of time scales – from the femtosecond of electronic rearrangements in the first step of vision, to the 10^9 years of evolution. Thermal energy is expressed as atomic fluctuations on the picosecond–nanosecond time scale that constitute the basis of molecular dynamics. These fluctuations are of particular interest in biophysics because they result from and reflect the forces that structure biological macromolecules and the atomic motions and molecular flexibility associated with biological activity.

Thermal energy propagates through solids in waves of atomic motion, such as the normal modes discussed in Chapters A3 and I1. We can estimate values for the frequencies, wavelengths and amplitudes of thermal excitation waves from an order of magnitude calculation, e.g. by considering the movement of a mass similar to the mass of an atom moving in a simple harmonic potential of energy equal to Boltzmann's constant multiplied by 300 K (ambient temperature). It turns out that the frequencies are of the order of 10^{12} s^{-1}, while the wavelengths and amplitudes are on the ångström scale. We saw in Part E that the energy associated with thermal vibrations corresponds to the IR frequency range in optical spectroscopy. Similarly, *neutron spectroscopy* takes advantage of the fact that the energies of *thermal neutron* beams match vibrational energies in solids and liquids. This is not surprising because *thermal neutrons* are produced by equilibration at ambient temperature, so that by definition, their energy is equal to the corresponding thermal energy. The kinetic energy of a neutron is given by the same equation as for a billiard ball: $E = \frac{1}{2}mv^2$. The quantum mechanical wavelength associated with a neutron beam is h/v, where h is Planck's constant (see Chapter A3). The mass of the neutron is very close to that of a proton and therefore to that of a hydrogen atom, or about one atomic mass unit ($1/N_A = 1.66 \times 10^{-24}$ g), where N_A is Avogadro's number. Together neutrons and protons make up 99.9% of an atom's mass, the mass of the electron being several orders of magnitude smaller. Equating the neutron kinetic energy to thermal energy at ambient temperature, we can calculate the neutron velocity (about 4 km s^{-1}) and associated wavelength (about 1 Å).

We recall that thermal fluctuation amplitudes are of the order of 1 Å, so that not only do thermal neutron energies match thermal fluctuation energies but also thermal neutron wavelengths match thermal fluctuation amplitudes. In contrast, the wavelength of IR radiation is of the order of 1000 Å. There is, therefore, an important difference between neutron and optical spectroscopy, because neutron experiments, as well as yielding information on excitation frequencies also yield fluctuation amplitudes, i.e. they allow the direct determination of the *dispersion relations* of thermal excitations, the relations between wavelength and frequency of the corresponding waves. This unique physical property provided an impetus to develop neutron sources and spectrometers dedicated to the study of excitations in solids and liquids.

The neutron was discovered in **1932** by **J. Chadwick**. Neutrons are emitted spontaneously by certain radioactive nuclei and various elements undergo fission when bombarded by neutrons emitting additional neutrons. Because they are electrically neutral, neutron beams penetrate deeply into matter. These and other properties, as we have seen above, make the neutron a particularly useful probe for investigating structure and dynamics at the molecular level.

In **1935**, **J. R. Dunning** showed neutrons could be sorted according to their velocity by using 'choppers', rotating discs with slits, thus setting the basis for the development of time-of-flight spectroscopy decades later.

Elsasser, H. v. Halban & **P. Preiswerk** and **D. P. Mitchell** & **P. N. Powers** demonstrated diffraction of neutrons from a radium–beryllium source in **1936**, and thus their wave nature according to **de Broglie's** relation, paving the way for the use of neutrons in crystallography.

Following the developments in nuclear science stimulated by the Second World War tragedy, neutron beams from pile reactors became available for diffraction experiments and crystallography. **C. Shull** performed the first neutron diffraction experiments to investigate material structures. The first inelastic neutron scattering experiments were performed by **B. Jacrot** and published in **1955**. He used the time-of-flight method to measure the energy distribution of neutrons scattered by a copper sample. **B. N. Brockhouse** invented the triple-axis spectrometer and measured vibrations in solids by neutron scattering. **Shull** and **Brockhouse** shared the Nobel Prize for physics in **1994**.

In the **late 1960s** and **early 1970s,** high-flux neutron beam reactors were dedicated to condensed matter physics and chemistry and biophysics. Neutron spectrometers were built with increasing energy resolution, based on time of flight, back-scattering and later the spin echo technique. Experiments on protein dynamics became possible and contributed strongly to our present understanding (see Chapter A3). The high-flux reactor of the international Institut Laue Langevin (ILL) in Grenoble, France, delivered its first neutron beams in **1973**; the ILL remains the foremost neutron scattering centre in the world due to the high flux of its neutron beams and its variety of instrumentation.

Comment I2.1

Neutron momentum and energy (recalled from Chapter G1)

Momentum of a neutron:

$$\mathbf{p} = m\mathbf{v} = \hbar\mathbf{k}$$

Wavelength:

$$\lambda = \frac{h}{p} = \frac{2\pi}{k}$$

Energy:

$$E = \hbar\omega$$

Dispersion relation:

$$E = \frac{\hbar^2 k^2}{2m} = \frac{p^2}{2m}$$

Since **2000** neutron spectroscopy experiments on biological macromolecules have been being performed at several neutron scattering centres around in the world. We are now waiting for the second generation neutron spallation sources that should become operational in the next decade and which should significantly extend the possibilities of the method for characterising macromolecular dynamics.

I2.2 Theory

The treatment of energy-resolved neutron scattering offered in this chapter uses many concepts introduced in Chapters A3 and G1.

I2.2.1 Momentum and energy, distance travelled and time

We recall the expressions for neutron momentum and energy in terms of wavelength, velocity and frequency in Comment I2.1. Scattering neutrons from a sample and observing their energy and momentum changes is, in principle, a direct way to obtain information on the atomic motions in that sample. Because we have chosen their direction and wavelength or velocity carefully, the incident neutron particles have well-defined energy and momentum. Neutron spectroscopy is based on the observation of the momentum and energy of scattered neutrons, by measuring their scattering angle and wavelength or velocity, respectively. The momentum ($\mathbf{p}$) and energy ($\hbar\omega$) taken away from the sample in the scattering process are then calculated simply from conservation laws (Fig. I2.1):

$$\hbar\mathbf{Q} = \Delta\mathbf{p} = \hbar(\mathbf{k}_1 - \mathbf{k}_0) \tag{I2.1a}$$

$$\hbar\omega = \hbar\omega_1 - \hbar\omega_0 \tag{I2.1b}$$

It is usual to drop Planck's constant in Eqs. (I2.1) and write momentum and energy changes respectively as the scattering vector $\mathbf{Q}$ and angular frequency ω.

Fig. I2.1 A neutron beam of wavelength λ_0 and corresponding energy $\hbar\omega_0$ incident on a sample S. Scattered waves of wavelength λ_1 and energy $\hbar\omega_1$ are observed by a detector (the 'eye') at an angle 2θ with respect to the incident beam. The incident and scattered wave vectors $\mathbf{k}_0$, $\mathbf{k}_1$ have amplitudes $2\pi/\lambda_0$, $2\pi/\lambda_1$ respectively.
$\mathbf{Q} = \mathbf{k}_1 - \mathbf{k}_0$, is the scattering vector, and the energy increase of a neutron in the scattering process is $\hbar\omega = \hbar\omega_1 - \hbar\omega_0$.

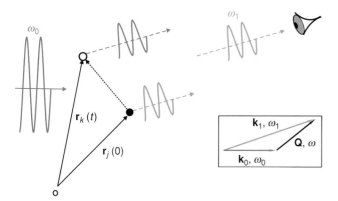

Fig. I2.2 Interference of waves scattered by a moving atoms. The white atom is at the position given by the vector $\mathbf{r}_j$ at time zero; the black atom is at the position given by the vector $\mathbf{r}_k$ at time t. The incident beam of frequency ω_0 is in red. As the neutron wave passes through the sample it is scattered by both atoms. Waves (blue and green) scattered by the atoms interfere to yield the purple wave. Because of the energy exchanged between the neutron and moving atoms the frequency ω_1 of the purple wave is different from ω_0. The inset shows the diffraction diagram with the incident and scattered wave vectors and momentum transfer (see Fig. I2.1).

Note that incident and scattered wave vectors and scattering vector in Fig. I2.1 and Eqs. (I2.1) refer to the same quantities as the ones already discussed in Chapter G1. In diffraction, however, only elastic scattering, for which the energy change is zero and $|\mathbf{k}_1| = |\mathbf{k}_0|$, is taken into account. In spectroscopy we deal with the more general case in which *energy as well as momentum* can be exchanged between the moving atoms and the neutrons.

In a similar way to scattered intensity $I(\mathbf{Q})$ in diffraction being expressed as a function of $\mathbf{Q}$, in spectroscopy we write the scattered intensity as $I(\mathbf{Q}, \omega)$. The information we seek on a system of moving atoms is the vector of each atomic displacement as a function of time (*how far each atom moves in a given direction in a given lapse of time*).

By analogy with scattered wave interference in diffraction methods discussed in Chapter G1, $I(\mathbf{Q}, \omega)$ is the result of the interference of waves scattered by atoms at different positions in space and at different times (Fig. I2.2). The neutron wave speed approximately matches the speed of the atomic motions, so that it can be diffracted by both atoms shown in the figure at different times.

The momentum transfer $\mathbf{Q}$ gives us information on the length scale of fluctuations, and the energy transfer ω gives us information on their time scale. In a similar way to the scattering vector $\mathbf{Q}$ being the reciprocal of a real-space vector $\mathbf{r}$ in the Fourier transform procedure (see Chapter G1), the frequency ω is the reciprocal of time t (see Chapter A3). As in the treatment of diffraction from a static group of atoms, the information in terms of $\mathbf{r}$ and t is calculated from the scattered intensity $I(\mathbf{Q}, \omega)$ by a succession of Fourier transforms.

I2.2.2 The dynamic structure factor, intermediate scattering function and correlation function

The scattered intensity $I(\mathbf{Q}, \omega)$ is expressed in absolute units as a *double differential cross-section*:

$$\frac{\mathrm{d}^2\sigma}{\mathrm{d}\Omega\mathrm{d}\omega} = \frac{I}{I_0} \tag{I2.2}$$

where I is number of scattered neutrons per unit time in solid angle $d\Omega$, with an energy change $d\omega$, and I_0 is the incident flux (neutrons per unit area per unit time). Recall from Chapter G1 that the units of cross-section in diffraction are *area* or $(length)^2$; the units of the double differential cross-section are $(length)^2/energy$.

The scattered intensity is related to the sample dynamic properties via the *dynamic structure factor* $S(\mathbf{Q}, \omega)$:

$$\frac{d^2\sigma}{d\Omega d\omega} = \frac{k_1}{k_0} N b^2 S(\mathbf{Q}, \omega) \tag{12.3}$$

The equation is written for a system containing N atoms of the same type with scattering length b. Equation (I2.3) contains a further complication, when compared with simple diffraction, in the k_1/k_0 term. Because of the energy difference, the velocities of incident and scattered neutrons (proportional to k_0, k_1, respectively) are different and the k_1/k_0 term accounts for the difference between the rates of the incident and scattered waves entering and exiting the sample, respectively.

Following the wave interference arguments of Section I2.2.1, the dynamic structure factor is the double Fourier transform of a *space-time correlation function* (defined below). The transformation is performed in two steps. First, integrating over time

$$S(\mathbf{Q}, \omega) = \frac{1}{2\pi} \int_{-\infty}^{+\infty} (I(\mathbf{Q}, t) \exp(-i\omega t) dt \tag{12.4}$$

where $I(\mathbf{Q}, t)$ is called the *intermediate scattering function*. Second, integrating over space

$$I(\mathbf{Q}, t) = \frac{1}{2\pi} \int_{-\infty}^{+\infty} g(\mathbf{r}, t) \exp\{-i\mathbf{Q} \cdot \mathbf{r}\} d\mathbf{r} \tag{12.5}$$

where $g(\mathbf{r},t)$ is an atomic correlation function in space and time (see Chapter D10):

$$g(\mathbf{r}, t) = \frac{1}{N} \sum_{j,k}^{N} \langle \delta\{\mathbf{r} - \mathbf{r}_{jk}\} \rangle \tag{12.6}$$

where the sum is over all atoms and the angular brackets refer to the thermal average. The vector $\mathbf{r}_{jk}$ is between atom j at time *zero* and atom k at time t. The Dirac delta function is equal to 1 when $\mathbf{r} = \mathbf{r}_{jk}$ and equal to *zero* otherwise. The correlation function counts the fraction of atoms that are separated by the vector $\mathbf{r}$ after a lapse of time t. Recall that a similar formulation is used in the analysis of dynamic light scattering (Chapter D10).

Combining Eqs. (I2.5) and (I2.6), we find

$$I(\mathbf{Q}, t) = \frac{1}{N} \sum_{j,k} \langle \exp[i\mathbf{Q} \cdot r_k(t) \exp[i\mathbf{Q} \cdot r_j(0)] \rangle \tag{12.7}$$

I2.2.3 Coherent and incoherent cross-sections

Coherent and incoherent scattering due to neutron spin was discussed in Chapter G1. We recall the relation between the total cross-section and the coherent and incoherent scattering lengths and cross-sections:

$$b_{inc}^2 = \langle b^2 \rangle - \langle b \rangle^2 \tag{I2.8}$$

where $\langle b^2 \rangle$ is b_{total}^2 and $\langle b \rangle^2$ is b_{coh}^2.

By using Eq. (I2.8), we separate the $b^2 S(\mathbf{Q}, \omega)$ term in Eq. (I2.3) into coherent and incoherent terms:

$$b_{total}^2 S(\mathbf{Q}, \omega) = b_{coh}^2 S_{coh}(\mathbf{Q}, \omega) + b_{inc}^2 S_{inc}(\mathbf{Q}, \omega) \tag{I2.9}$$

Equation (I2.9) introduces the coherent and incoherent dynamical structure factors.

Coherent scattering occurs when waves scattered by *different* atoms interfere, whereas incoherent scattering is scattering from a single atom. When the atom is motionless, it behaves like a point scatterer for neutrons so that the incoherent scattering is constant as a function of $\mathbf{Q}$.

A very interesting situation arises, however, when an atom is moving. The wave scattered by the atom at time zero interferes with the wave scattered by the *same atom* at time t, so that the displacement of the atom in the time lapse is reflected in a $\mathbf{Q}$ dependence of the scattered intensity. In the notation of Fig. I2.2 there is interference of waves from atom k at time zero and position $\mathbf{r}_k(0)$ and the same atom k at time t and position $\mathbf{r}_k(t)$. For incoherent scattering, therefore, the correlation function $g(\mathbf{r}, t)$ only contains *autocorrelation* terms for which $j = k$. The intermediate *incoherent* scattering function is given by

$$I_{inc}(\mathbf{Q}, t) = \frac{1}{N} \sum_k \langle \exp[i\mathbf{Q} \cdot \mathbf{r}_k(t)] \exp[i\mathbf{Q} \cdot \mathbf{r}_k(0)] \rangle \tag{I2.10}$$

Similarly, the incoherent dynamic structure factor reports on the motions of individual atoms.

The coherent dynamic structure factor is sensitive to interference between waves scattered by different atoms according to their positions in space and as they move in time; it is, therefore, sensitive to *collective dynamics* in the sample, as, e.g. when atoms move coherently in a normal mode of vibration.

I2.3 Applications

Most neutron spectroscopy applications to biological macromolecules are based on incoherent scattering for the practical reason that the incoherent cross-section of the hydrogen nucleus is more than an order of magnitude larger than the total scattering cross-sections of the other atoms. The information obtained, therefore, is on the motions of individual hydrogen atoms in the structures. It is nevertheless very useful information because hydrogen atoms are distributed homogeneously in biological macromolecules. In the picosecond–nanosecond time scale of the neutron experiments hydrogen atoms reflect the motions of the larger groups (methyl groups, amino acid side-chains etc.) to which they are bound, so that they are a good gauge of molecular dynamics. The data can also be used to test collective dynamics models through simulations of their effects on individual motions. Furthermore, since the incoherent cross-section of deuterium, ^{2}H, is very much smaller than that of ^{1}H, isotope labelling can be used to reduce the contribution of parts of a complex structure in an approach similar to contrast variation in SAS (see Chapter G2).

I2.3.1 Energy and time resolution

The *energy resolution* of an experiment is the smallest energy change ω_{min} that can be measured. The *energy range* is given by the maximum ω value. In the reciprocal time frame, the energy resolution corresponds to the *longest* lapse of time t_{max} over which the experiment is sensitive. Consider an atom moving over a distance equal to its own diameter in time t. If t is shorter than t_{max}, then there is interference between waves scattered by the atom in its two end positions and the displacement is observed. If t is longer than t_{max}, then the atom appears to be immobile.

The energy resolution of neutron spectrometers is usually given in electron volts (eV). A resolution of 1 μeV corresponds to a maximum time of about 1 ns, 10 μeV to 0.1 ns, etc. (see Section I2.4).

I2.3.2 Space-time window

The elastic peak in a diffraction or spectroscopy experiment is the peak of radiation scattered without energy change, i.e. for $\omega = 0$. In an ideally perfect spectrometer the elastic peak is an infinitely narrow Dirac delta function. In a real spectrometer the width of the elastic peak corresponds to the energy resolution (Fig. I2.3). Quasi-elastic and inelastic scattering which are also shown in the figure are discussed further below.

A neutron spectrometer opens *a window in space and time* defined by the minimum and maximum scattering vector values, Q_{min}, Q_{max}, accessible, and by the energy resolution and range (Fig. I2.4).

Referring to Fig. I2.4, the motion of an atom is observed only if it lies in the light blue part of the diagram. If the atom remains within the yellow circle it appears as an immobile point and its incoherent scattering is a constant as a function of $\mathbf{Q}$. If its motion takes it outside the light blue circle in the time lapse of the experiment its incoherent scattering peaks close to $\mathbf{Q} = 0$, too close to the direct beam to be observed.

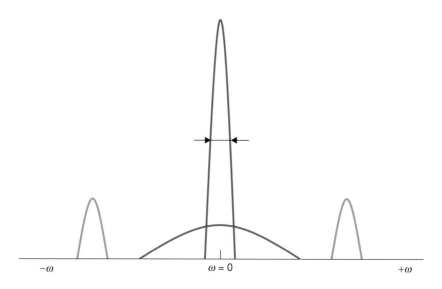

Fig. 12.3 The elastic peak (in red, the energy resolution is the full width at half height shown by the double arrow), quasi-elastic scattering (blue) and inelastic scattering (green) in a neutron spectroscopy experiment (see text). The intensity peaks are observed at a given value of **Q**.

We now recall the properties of Fourier transforms discussed in Chapter A3 to discuss the information content in $S(\mathbf{Q}, \omega)$. At $\mathbf{Q} = 0$, the intensity is sensitive to all distances in the sample (in practice, up to a maximum distance given by the minimum Q value) but cannot distinguish between them. Displacements $2\pi/Q$ are resolved at a scattering vector value Q. Analogously, at $\omega = 0$ the intensity is sensitive to events taking place in the total time lapse up to the maximum time given by the energy resolution (minimum energy transfer observable) but cannot

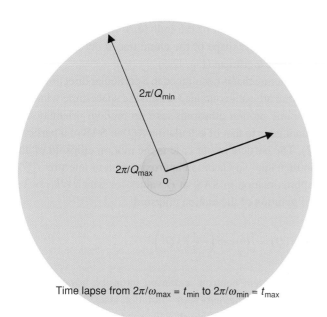

Fig. 12.4 A window in space and time. The displacement **r** of an atom from the origin O in a time lapse between the minimum and maximum time given by the energy range and resolution, respectively, is observed if it lies in the light blue area of the circle which is defined by the scattering vector range.

Comment I2.2
Experiments in H₂O solution and the space-time window

A neutron spectrometer with an energy resolution of about 10 μeV, corresponding to a time lapse of about 100 ps is not sensitive to water diffusion in a protein solution sample. A hydrogen atom bound to the protein fluctuates over about 1 Å, whereas the mean square displacement of a freely diffusing water molecule (calculated from its translational diffusion coefficient) is about 100 Å², well outside the space-time window of the experiment (Tehei *et al.*, 2001, Gabel, 2005).

distinguish between different times within that time frame. An event taking place during a time $2\pi/\omega$ is resolved at energy transfer ω.

The space-time window as a filter

Samples studied by neutron spectroscopy are usually complex, displaying motions on different time and length scales. A pure protein solution, for example, contains: (1) atoms that fluctuate while remaining firmly bound within the macromolecule, and (2) atoms in water molecules, which diffuse freely. Because neutron scattering is strongly dominated by the incoherent scattering of ^{1}H, experiments to study protein dynamics were performed in D_2O solutions, to minimise water scattering. This was not very satisfactory, however, because D_2O itself can modify dynamics, since the hydrogen bonds and deuterium bonds have different properties. Experiments on proteins in H_2O solution became possible when it was realised that the motion of freely diffusing water and hydrogen atoms bound to the protein are so different that they can be separated by choosing spectrometers with appropriate space-time windows (Comment I2.2).

I2.3.3 Q dependence of the elastic intensity

We recall from Part G that the Debye–Waller factors in crystallography define the fluctuations in atomic positions due to thermal energy, from the scattered intensity dependence on $\mathbf{Q}$. Briefly, the motion of each atom in the structure describes a cloud (also called the *thermal ellipsoid*), which is not time resolved in the elastic intensity. Scattering by the cloud has a $\mathbf{Q}$ dependence (a *form factor*) according to its shape. Similarly, the dependence of the elastic incoherent intensity on $\mathbf{Q}$, even though it cannot resolve the motion of the individual atom as a function of time, contains information on the shape of the cloud traced out by the atom within the time window of the experiment (Fig. I2.5).

In the case of a sample structure that is not aligned in a particular direction, and for motion displacements that are well contained within the space-time window (*localised motion*), the atomic motion ellipsoids take up random orientations. The problem is strictly analogous to that of calculating $I(Q)$ in SAS of a particle in solution (Chapter G2). The particle in our case is the motion ellipsoid of an individual atom. We can then apply a Gaussian approximation to interpret $I(Q)$ similar to the Guinier approximation in SAS, at Q values for which $QR_G \sim 1$, where R_G is the radius of gyration of the motion ellipsoid,

$$I(Q) = I(0) \exp\left(-\frac{1}{3}R_G^2 Q^2\right)$$

In the analysis of the incoherent neutron scattering intensity, however, it is conventional to use the mean square fluctuation $\langle u^2 \rangle$ notation instead of the radius of gyration notation, where $\langle u^2 \rangle = 2R_G^2$; note that the mean square fluctuation refers to the full amplitude of the motion; while the radius of gyration, R_G, refers

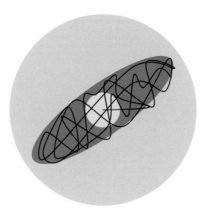

Fig. I2.5 Localised motion of an atom (black line) describing an ellipsoid (red) within the space-time window. A particularly complex diffusive pathway is shown in the example but a more oscillatory path would be expected for an atom anchored within a macromolecule.

to the displacement from the centre of mass of the motion of the ellipsoid:

$$I_{\text{elastic}}(Q) = I(0) \exp\left(-\frac{1}{6}\langle u^2 \rangle Q^2\right) \tag{12.11}$$

The mean square fluctuation is calculated from the slope of the $\ln I_{\text{elastic}}(Q)$ versus Q^2 linear plot.

Mean square fluctuations and effective force constants

In an *elastic temperature scan* the neutron incoherent elastic intensity $I_{\text{elastic}}(Q)$ is measured for different temperatures, and the mean square fluctuation plotted as a function of temperature (Fig. I2.6). Mean square fluctuations against absolute temperature T for myoglobin powders are shown in the figure; the elastic intensity data from which they were calculated were collected for Q values around 1 Å^{-1} with an energy resolution of about 10 μeV, i.e. the mean square fluctuations refer to about 1 Å amplitude motions taking place in 100 ps or less. They correspond to the thermal dynamics of the atoms in the protein.

The plot is discussed in Section A3.3.3. The temperature axis (T) is, in effect, an energy scale ($k_B T$, where k_B is Boltzmann's constant) for the motions. We recall that for a simple harmonic oscillator the square of the amplitude is proportional to the energy, with a proportionality constant related to the spring force constant. The straight-line dependence of $\langle u^2 \rangle$ versus T at low temperatures, therefore, indicates that the mean atomic motions within the protein in this temperature range can be described in terms of a mean simple harmonic oscillator with a mean force constant, which can be calculated from the slope of the line:

$$\langle k \rangle = 2k_B \left(\frac{d\langle u^2 \rangle}{dT}\right)^{-1} \tag{12.12}$$

where the $\langle x^2 \rangle$ value plotted in Fig. I2.6 is equal to $\langle u^2 \rangle/6$. Equation (12.12) is an extremely useful relation because it allows quantification of the forces that maintain an atom within a macromolecular structure.

Above what has been called a dynamical transition at about 180 K atomic motions are no longer harmonic, but an effective force constant $\langle k' \rangle$ can still be

Fig. I2.6 Mean square fluctuations as a function of absolute temperature, in a myoglobin powder, hydrated by heavy water D_2O ($\triangle$) and in a trehalose glass (o). The break in slope at about 180 K is called a dynamical transition. Effective force constants, k and k' were calculated for different parts of the curves as explained in the text. (From Zaccai, 2000, using data from Doster *et al.*, 1989, and Cordone *et al.*, 1999). (Figure reproduced with permission from *Science*.)

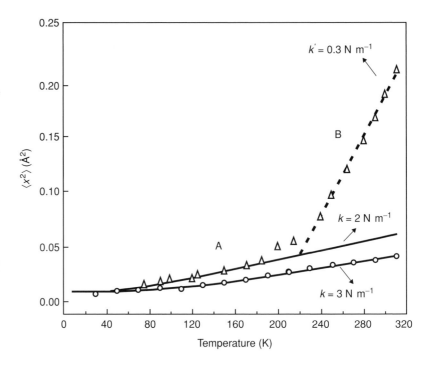

obtained from the slope of the line by applying a quasi-harmonic approximation (Comment I2.3). The force constants associated with different parts of the plot are shown in Fig. I2.6. There is a 'softening' of an order of magnitude between the protein structure in the trehalose glass and above the dynamical transition in the hydrated state, the force constant decreasing from 3 to 0.3 N m^{-1}. It is interesting to note that values in the same range are obtained by direct force measurements in single-molecule manipulation experiments (see Part F).

The neutron elastic temperature scan uniquely allows a direct probe of the mean forces acting on atoms in a macroscopic sample. Such scans have been used effectively to explore stabilisation forces in pure protein samples and even *in situ* in live bacterial cells, illustrating the potential of the method for the characterisation of the crowded internal cell environment (Comment I2.4).

Comment I2.3
Effective force constants

Calculations justifying the quasi-harmonic approximation and the derivation of an effective force constant from the general dependence of the mean square displacement on temperature have been reviewed by Gabel *et al.* (2002).

Comment I2.4   Protein stabilisation forces

Force constants have been measured by neutron spectroscopy for proteins in solution (Tehei *et al.*, 2001) as well as for macromolecules *in situ* in bacterial cells (Tehei *et al.*, 2004). These data allowed correlation between protein dynamics and adaptation to extreme conditions, by quantifying intramolecular forces and showing how stronger force constants stabilised thermophilic proteins and weaker force constants permitted the flexibility of psychrophilic (cold temperature loving) proteins necessary to their activity at low temperatures.

I2.3.4 Quasi-elastic scattering and diffusion

As we start to look outside the elastic peak, we gain information on the time constants of different dynamics processes. The *quasi-elastic scattering* is shown as the blue curve in Fig. I2.3. As its name suggests it is centred on zero energy transfer and appears as a broadening of the elastic peak. Similarly to the case of dynamic light scattering (Chapter D10), quasi-elastic neutron scattering arises from a diffusing particle, with the energy transfer width of the curve inversely proportional to a relaxation time τ. •

In the simplest case of linear diffusion obeying Fick's law (see Section G3.4), the intermediate scattering function calculated from the correlation function is written:

$$I(Q, t) = \exp\left(-\frac{t}{DQ^2}\right) \tag{I2.13}$$

where D is the diffusion coefficient and $DQ^2 = 1/\tau$. We recall that the Fourier transform of an exponential decay function is a Lorentzian function (Chapter A3), so that the dynamic structure factor of the quasi-elastic scattering is given by (from Eq. (I2.4):

$$S(Q, \omega) = \frac{1}{\pi} \frac{DQ^2}{[DQ^2]^2 + \omega^2} \tag{I2.14}$$

which describes a curve of full width at half maximum Γ equal to $2DQ^2$ in frequency units.

If the diffusion model is appropriate, plotting experimental $\Gamma(Q)$ versus Q^2 yields a straight line passing through zero from the slope of which it is possible to calculate the translational diffusion coefficient. In practice, however, a simple diffusion model is more likely to be valid at high Q values, i.e. for small displacements for which it is less likely the diffusion process will be hampered by interactions with the atomic environment. We can see this in Fig. I2.7, in which measured curve width values, assuming a single Lorentzian fit to the quasi-elastic scattering of protein hydration water, are plotted versus Q^2 for different temperatures. Clearly at small Q the data do not lie on straight lines passing through the origin. Diffusion coefficients could nevertheless be calculated from the slopes at high Q, quantifying the slowing down of protein hydration water with respect to bulk water, at three different temperatures.

Considerably more complex models than simple diffusion have been used in the analysis of quasi-elastic scattering data, with respect to the behaviour of water in different biological environments, as well as to the slower motions of macromolecule bound atoms, leading to quite a sophisticated understanding of the phenomena involved. These are discussed in the review cited in the figure caption.

Fig. I2.7 Translational diffusion line width of the protein hydration water (H_2O) in a sample of fully deuterated C-phycocyanin (purified from algae grown in D_2O). The meV energy range of the spectrometer as well as the full deuterium labelling of protein hydrogen atoms ensured that the scattering signal was predominantly from the water. Data are shown for three temperatures. The sample was a powder containing 0.4 g of water per gram of protein. (From Bellissent-Funel *et al.*, 1996; reviewed by Gabel *et al.*, 2002.)

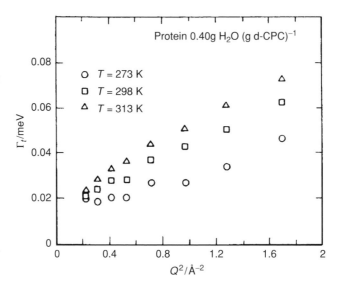

I2.3.5 Inelastic scattering and vibrations

In Fig. I2.3 inelastic neutron scattering is shown in green. Similarly to the Stokes and anti-Stokes lines in light scattering (see Section E3.2), there are peaks for neutron energy gain (at positive values of ω) and energy loss (at negative values of ω). Although the peaks in the figure are shown as symmetrical, this is not the case experimentally, because the probability of scattering with neutron energy gain is higher.

As with the lines observed in light scattering the neutron peaks arise from energy exchange with atomic vibrations in the sample. In contrast to light scattering, however, there are no selection rules for neutrons since the scattering is due to a direct interaction with the moving nuclei, and not with electric dipoles or other such effects. Also, because the wavelength of light is large compared with the fluctuation amplitudes, light scattering experiments are effectively limited to measurements at zero **Q** values. The wavelengths associated with neutrons of comparable energy are similar to the fluctuation amplitudes and the whole **Q** range can be examined.

Coherent inelastic scattering arises from the interaction between the neutron and collective vibrational modes, which can be seen as waves of given frequency and momentum propagating through an ordered sample. These waves are called *phonons*, and the neutron–phonon scattering interaction obeys energy and momentum conservation rules. Inelastic neutron scattering has been extremely powerful for the determination of phonon *dispersion relations*, the functions relating the energy with the momentum of the vibration waves. Phonon propagation has been observed in collagen samples and in oriented membrane samples but, so far, suitably large protein crystals have not been available for phonon dispersion measurements by neutron scattering.

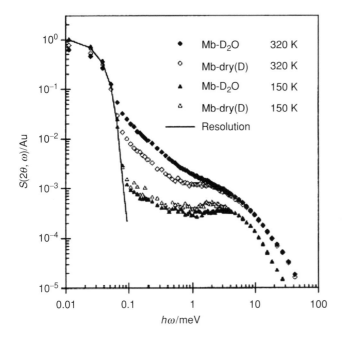

Fig. I2.8 The dynamical structure factor expressed in arbitrary units in terms of scattering angle 2θ instead of scattering vector measured for D_2O hydrated (0.35 g g^{-1}) and dried (from D_2O so that the exchangeable hydrogens are deuterated) at 150 K and 320 K (Diehl et al., 1997, reviewed by Gabel et al., 2002).

In the case of *incoherent inelastic scattering*, the peaks arise from the vibrations of individual atoms (here too, dominated by the hydrogen atoms because of their higher scattering cross-section). Inelastic incoherent spectra have been measured for a wide variety of proteins under different conditions. The results have guided the development of molecular dynamics simulations (Chapter I1) and have been essential in understanding the forces acting in a protein structure and the important role of the hydration environment. It is interesting to note that because of the $\mathbf{Q} \cdot \mathbf{r}$ term in the intermediate scattering function (Eq. (I2.5)) the experimental intensity is dominated by the larger vibration amplitudes and, therefore, by the lower-frequency modes. Recall from Chapters I1 and A3 that the first few normal modes in a protein dominate the atomic displacement amplitudes. Since the effect of a vibration mode on biological activity is likely to increase with the extent of the atomic displacements, we find there is a good match between the sensitivity to amplitude of the neutron scattering measurements and biological relevance.

Neutron incoherent inelastic scattering spectra of wet and dry powder samples of the protein myoglobin at low and high temperature, measured with an energy resolution of 30 μeV, are shown in Fig. I2.8. The elastic peak below 0.1 meV, fitted by the full line, represents the experimental energy resolution. The inelastic scattering is best seen at low temperature with a peak at 2 meV (corresponding to a period of a few picoseconds) for the hydrated sample, which moved to lower frequencies when the protein was dried. This low-frequency peak in the protein vibrations has been simulated approximately by molecular dynamics calculations.

The fact that the hydrated sample at low temperature is stiffer than the dry one has been interpreted in terms of the hydrogen (deuterium) bond network surrounding the protein. At high temperature, the inelastic peak appears at about the same frequency values, and quasi-elastic scattering from diffusional rather than vibrational motions fills in the gap between 0.1 and 1 meV. The higher level of quasi-elastic scattering from the wet sample indicates the contribution of the solvent diffusion. Within the context of the physical model for protein dynamics (see Chapter A3), the quasi-elastic scattering from the protein at higher temperatures has been suggested to arise from the sampling of different conformational substates.

I2.4 Samples and instruments

Relatively large samples (of the order of a few hundred milligrams) are required for neutron spectroscopy, because incident fluxes and scattering cross-sections are low. Samples for the triple-axis spectrometer (discussed below) must be crystalline or show preferred orientation as in the case of membranes or fibres. Ordered samples as well as disordered hydrated powder or solution samples can be used on the other instruments. Powders of hydrated proteins or other macromolecules do not contain free bulk water and are suitable for measurements over a wide temperature range including below the freezing point of water. Experimental data and MD calculations have shown that provided there is sufficient hydration, the internal dynamics of proteins is similar in powder samples and in solution. Powder samples are also suitable for studies of the effects of different hydration levels on macromolecular dynamics. Care must be exercised with macromolecular solution samples, in order to account for the contribution of the solvent to the signal, as well as for effects due to diffusion of the macromolecular particles themselves.

The dynamics of macromolecule and solvent or of different parts of a macromolecule can be measured separately by using H_2O/D_2O exchange and/or specific H-D labelling. Because of important applications in neutron scattering and NMR, specific deuteration methods are in full development for the labelling of amino acids within proteins or domains in complex structures. Neutron spectroscopy has been used successfully to study the dynamics of hydrogen labels in a fully deuterated natural membrane (see Fig. A3.29).

Below we present neutron spectrometers in order of decreasing energy resolution width (increasing time scale): $\sim$1 meV (ps) for triple-axis spectrometry to $\sim$10 μeV (100 ps) for time-of-flight, spectrometry to 1 μeV (1 ns) for backscattering spectrometry to 10 ns for spin echo spectrometry, which measures dynamics directly in the time domain.

The triple-axis spectrometer

B. Brockhouse was awarded the Nobel prize in 1994 for the invention of the triple-axis spectrometer. By measuring scattered neutrons with accurate values of

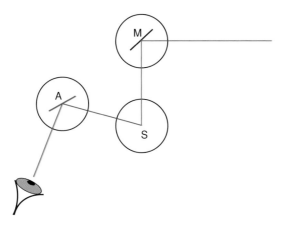

Fig. I2.9 Schematic diagram of a triple-axis spectrometer. M is the monochromator, S the sample, and A the analyser. The 'eye' is the detector. The beam from the neutron source is in red.

momentum and energy, the triple-axis spectrometer allowed fundamental discoveries in solid state physics through the determination of phonon dispersion curves in a wide variety of materials. The principles underlying a triple-axis spectrometer are shown schematically in Fig. I2.9. The basic principle of the instrument is the selection of specific wavelengths by Bragg reflection off *monochromator* and *analyser* crystals. On the first axis, a monochromator crystal M diffracts neutrons of a given wavelength (blue line) from the incident beam (red line). The 'blue line' neutrons are scattered by the sample S, on the second axis, with an energy change (green line). The wavelength of the 'green line' neutrons is measured by Bragg diffraction off an analyser crystal A on the third axis and scattered towards the 'eye' detector. By rotating M, S and A around their respective axes, it is possible to scan the wavelengths, the diffraction angles of neutrons incident and scattered by the sample as well as the orientation of the sample reciprocal lattice (see Chapter G1) and thus map the coherent $S(\mathbf{Q}, \omega)$ distribution in great detail.

Because of the difficulty in producing suitable samples the application of triple-axis spectrometry to biological samples is still in its infancy.

Time of flight

In biophysical experiments, time-of-flight spectrometers are mainly used to measure quasi-elastic and inelastic incoherent scattering from proteins. A time-of-flight spectrometer is shown schematically in Fig. I2.10. Choppers are rotating discs with slots on their rims that select neutrons of a given velocity (wavelength). A pair of choppers selects neutrons of wavelength λ but also wavelengths $\lambda/2$, $\lambda/3$ etc. Several choppers in line are required to form a well-shaped pulse of neutrons with a specific velocity. The energy resolution of time-of-flight spectrometers is typically in the 10 μeV–1 μeV range depending on the incident wavelength chosen, which also determines the Q range of the experiment.

Fig. I2.10 Schematic diagram of a time-of-flight spectrometer. The neutron beam (red line) passes through a series of 'choppers' (see text) that shape a neutron pulse (blue), which is incident on the sample S at a defined time. The scattered beam (green) is detected by a detector bank (eye) located at a distance L from the sample. L being known, the velocity of the scattered neutron is calculated from the time of flight between S and the 'eye', which detects it. The scattering angle is given by the position of the detector on the circumference of the instrument.

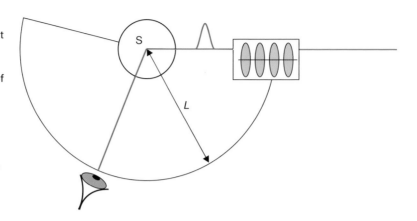

Back-scattering

Back-scattering spectrometers are based on the fact that Bragg's law is extremely wavelength selective for diffraction with scattering angle $2\theta = 180°$ ('back' scattering). Bragg's law (see Chapter G3) is written

$$2d \sin\theta = \lambda \tag{I2.15}$$

where d is the crystal spacing, 2θ is the scattering angle and λ is the wavelength. By taking the derivative of the equation, we can estimate the wavelength spread dependence on scattering angle:

$$2d \cos\theta \cdot \Delta\theta = \Delta\lambda \tag{I2.16}$$

$$\cot\theta \cdot \Delta\theta = \frac{\Delta\lambda}{\lambda}$$

Note that $\cot\theta = 0$ for $\theta = 90°$ leading to zero wavelength spread (excellent wavelength resolution) in the back-scattering position regardless of angular spread. A back-scattering spectrometer is shown schematically in Fig. I2.11. Back-scattering spectrometers are 'reverse geometry' spectrometers. In usual spectrometers, the incident beam wavelength is fixed and the scattered beam wavelength is analysed. In back-scattering spectrometers, the scattered beam wavelength is fixed by the choice of analyser spacing and the incident beam wavelength is varied by scanning the monochromator. The Q value of the scattering is defined by the position of the analyser crystal and the corresponding detector. Back-scattering spectrometers can attain lower than micro-eV resolution and are used to analyse elastic and quasi-elastic scattering from macromolecular samples.

Spin echo

Neutron spin-echo spectrometers are based on the same phenomenon as spin-echo NMR (see Chapter J2). The phenomenon being very sensitive to small energy changes, the neutron spin-echo technique achieves a higher energy resolution than back-scattering. The measurements, however, are performed in the time domain

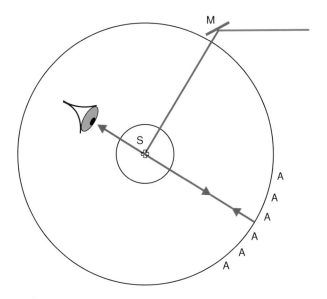

Fig. I2.11 Schematic
diagram of a
back-scattering
spectrometer. The
neutron beam (red line) is
reflected by a
monochromator M, which
selects a wavelength
(blue line) that is varied
during the experiment.
After scattering by the
sample S, only neutrons
of a specified wavelength
(green lines) are back
scattered by analyser
crystals A and detected
by the detector (eye).

(similarly to NMR) and the resolution is expressed accordingly ($\sim$10 ns) and not in energy units. A neutron spin-echo spectrometer is shown schematically in Fig. I2.12. Neutrons are spin-1/2 particles (see Chapter G1) that can be aligned (*polarised*). A polarised neutron beam behaves like a classical magnetic moment and displays Larmor precession when exposed to a magnetic field perpendicular to it (see Section J1.2.2). In a neutron spin-echo experiment, the neutron beam (red line in the figure) passes through a velocity selector V, which reduces the white beam wavelength spread, then through a polariser P, which aligns the spins to produce a *polarised beam*; the spins are rotated by 90° or 180° in spin *flippers* F; magnetic fields along the paths Prec1 and Prec2 (the echo) set the spins precessing in opposite directions, so that in absence of interaction with the sample S, the beam reaches the analyser A in the same state as when it emerged from P, i.e. the total precession angle comes back to zero. The precessing spins are very sensitive to small energy changes. An exchange of energy with the sample results in a total precession angle different from zero, and the analyser direction has to be set accordingly to allow the neutron beam to pass to the detector. It can be shown that the method provides a direct measure of the intermediate scattering function, through the relation of $I(Q, \omega)$ with the analyser direction. The Q value is set

Fig. I2.12 Schematic
diagram of a neutron
spin-echo spectrometer.
V is a velocity selector, P a
polariser, $F_{\pi/2}$ and F_π are
90° and 180° spin flippers,
respectively, Prec1 and
Prec2 are paths in which
the neutron spins are
made to precess by a
magnetic field, S is the
sample, A an analyser and
the 'eye' is the detector.
The beam from the
neutron source is in red.

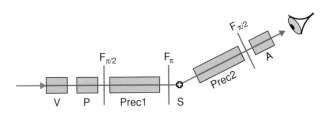

by the angle formed by the arm of the spectrometer downstream of the sample. An important advantage of the neutron spin-echo method is that all neutrons are counted and beyond the velocity selector there is no selection of neutrons by their energy either before or after scattering by the sample.

The NSE time scale is suitable for probing diffusion coefficients of macro-molecules in solution or within cellular environments, or slow domain movements in macromolecular complexes.

I2.5 Checklist of key ideas

- The energies of *thermal neutron* beams match vibrational energies in solids and liquids (with periods in the picosecond to nanosecond time scale), and their wavelengths match thermal fluctuation amplitudes (in the ångström range), so that neutron experiments allow the direct determination of the *dispersion relations* of thermal excitations.
- Neutron spectroscopy is based on the observation of the momentum and energy of scattered neutrons, by measuring their scattering angle and wavelength or velocity, respectively.
- Momentum transfer $\mathbf{Q}$ gives us information on the length scale of fluctuations, and energy transfer ω gives us information on their time scale.
- Scattering intensity is analysed in terms of an intermediate scattering function $I(\mathbf{Q}, t)$ and its Fourier transform, the dynamic structure factor $S(\mathbf{Q}, \omega)$; the intermediate scattering function is itself the Fourier transform of a space-time correlation function, relating the positions of the atoms in the sample as they move in time.
- Coherent scattering occurs when waves scattered by *different* atoms interfere, whereas incoherent scattering is scattering from a single atom; the incoherent dynamic structure factor reports on the motions of individual atoms; the coherent dynamic structure factor is sensitive to *collective dynamics* in the sample, as when atoms move coherently in a normal mode of vibration, for example.
- The incoherent cross-section of hydrogen nuclei strongly dominates the scattering signal; deuterium labelling enables one to modulate the scattering contribution of different parts of a complex sample; in the picosecond–nanosecond time scale hydrogen atoms reflect the motions of the larger groups (methyl groups, amino acid side-chains etc.) to which they are bound, so that they are a good gauge of molecular dynamics.
- The energy resolution of neutron spectrometers is usually given in electron volts (eV). A resolution of 1 μeV corresponds to a maximum time of about 1 ns, 10 μeV to 0.1 ns, etc.
- A neutron spectrometer opens *a window in space and time* defined by the minimum and maximum scattering vector values, Q_{min}, Q_{max}, accessible, and by the energy resolution and range; the space-time window is a filter that permits one to focus on specific motions in the sample.
- The elastic intensity provides information on localised motions occurring during the maximum time given by the instrumental resolution; mean square fluctuations and

effective force constants can be derived from an analysis of the elastic intensity as a function of temperature; the quasi-elastic intensity informs on the correlation times of diffusional motions; the inelastic intensity informs on vibrational modes.

- Relatively large samples (of the order of a few hundred milligrams) are required for neutron spectroscopy, because incident fluxes and scattering cross-sections are low.

- By measuring scattered neutrons with accurate values of momentum and energy, the triple-axis spectrometer allowed fundamental discoveries in solid state physics through the determination of phonon dispersion curves in a wide variety of materials.

- Time-of-flight spectrometers are used to measure quasi-elastic and inelastic incoherent scattering from macromolecules with an energy resolution in the 0.01–1 meV range; back-scattering spectrometers can attain lower than micro-eV resolution and are used to analyse elastic and quasi-elastic scattering from macromolecular samples.

- Neutron spin-echo spectrometers, based on the neutron being a quantum spin-1/2 particle and its interactions with magnetic fields, achieve high energy resolution with a corresponding time domain $\sim$10 ns; they provide a direct determination of the intermediate scattering function $I(\mathbf{Q}, t)$; the time scale is suitable for probing diffusion coefficients of macromolecules in solution or within cellular environments, or slow domain movements in macromolecular complexes.

Suggestions for further reading

Bée, M. (1988). *Quasielastic Neutron Scattering*. Bristol: Adam Hilger.

Smith, J. C. (1991). Protein dynamics: Comparison of simulations with inelastic neutron scattering experiments. *Q. Rev. Biophys.*, **24**, 227–291.

Gabel, F., Bicout, D., *et al.* (2002). Protein dynamics studied by neutron scattering. *Q. Rev. Biophys.*, **35**, 327–367.

Tehei M., Franzetti B., *et al.* (2004). Adaptation to extreme environments: Macromolecular thermal dynamics in psychrophile, mesophile and thermophile bacteria compared, *in-vivo*, by neutron scattering. *EMBO Rep.*, **5**, 66–70.

Part J
Nuclear magnetic resonance

Chapter J1
Frequencies and distances

J1.1 Historical review

1924

W. Pauli proposed the theoretical basis for NMR spectroscopy. He suggested that certain atomic nuclei have properties of spin and magnetic moment and, as a consequence, exposure to a magnetic field leads to splitting of their energy levels. **W. Gerlach** and **O. Stern** observed the splitting in atomic beam experiments, providing proof for the existence of nuclear magnetic moments.

1938

I. I. Rabi and colleagues first observed NMR by applying electromagnetic radiation in atomic beam experiments. Energy was absorbed at a sharply defined frequency, causing a small but measurable deflection of the beam. **Rabi** received the Nobel prize for physics in **1944**.

1946

Research groups led by **F. Bloch** and **E. M. Purcell** reported the observation of proton NMR in liquid water and solid paraffin wax. **Bloch** and **Purcell** shared the **1953** Nobel prize for physics.

1946

F. Bloch suggested a new method of excitation using a short radio-frequency pulse and in **1949 E. L. Hahn** showed that this did indeed produce a free precession signal. **Hahn** also established that pulse sequences could be used to generate additional information in the form of a spin echo. For many years, however, these methods were of little use to chemists because of the complexity of the signal obtained. In **1956, I. J. Lowe** and **R. E. Norberg** pointed out that the time-domain signal and the frequency-domain spectrum are related by Fourier transformation. The first high-resolution multichannel Fourier transform NMR spectrum was measured by **R. R. Ernst** and **W. A. Anderson**.

1950

W. G. Proctor and **F. C. Yu** observed two unexpected ^{14}N resonance frequencies for NH_4NO_3. At about the same time, **W. C. Dickinson** noticed similar effects for ^{19}F in several compounds. In **1951 J. T. Arnold** and colleagues introduced the term *chemical shift* following the observation of several resonance peaks for ^{1}H in ethanol, with the relative intensity in each peak corresponding to the relative number of protons in each chemical environment.

1951

H. S. Gutowsky and **D. W. McCall** suggested that interactions between spins of neighbouring nuclei were responsible for multiple resonance lines. In **1951 N. F. Ramsey** and **E. M. Purcell** proposed the concept of indirect *spin–spin coupling* or *scalar coupling*. It was found that in certain cases spin coupling failed to produce the expected multiplets, leading to the development of the concept of *chemical exchange*.

1953

A. W. Overhauser explored the dynamic polarization of nuclei, in metals where the electron spin resonance had been saturated. The effect he discovered was called the 'Overhauser effect'. The potential of nuclear Overhauser enhancement (NOE) signals for providing information on the conformation of molecules in solution was first demonstrated by **F. A. L. Anet** and **A. J. R. Bourn** in **1965**. In **1970, R. A. Bell** and **J. K. Saunders** reported a direct correlation between NOE and internuclear distances and **R. E. Schrimer** with colleagues demonstrated that relative internuclear distances can be determined quantitatively from NOE measurements on a system containing three or more spins.

By the **mid-1950s** the basic physics of NMR and its potential value in chemistry had been elucidated, and commercial instruments were available. In **1956**, the observation frequency for ^{1}H NMR spectroscopy on the HR-30 Varian Spectrometer with a 0.7 T electromagnet was fixed by a crystal at 30 MHz. In order to improve sensitivity and increase chemical shift dispersion, commercial instrument development focused on increasing the magnetic field strength. The development of persistent superconducting solenoids (cryo-magnets) in the **early 1960s** constituted a major milestone for NMR applications. In the **late 1990s** 18.8 T (corresponding to a resonance frequency of 800 MHz for ^{1}H NMR) spectrometers were installed in many NMR laboratories, with the first 900 MHz (21.1 T) instruments becoming available.

1957

M. Sauders, A. Wishnia and **J. G. Kirkwood** reported the first NMR spectrum of a protein, and a small number of similar studies reports followed in the next decade. Because of technical limitations in sensitivity and spectral resolution, however, these early NMR applications in structural biology did not bear directly on macromolecular three-dimensional structure. The fundamental theory of NMR was published in **1961** in a landmark book, *The Principles of Nulcear Magnetism*, by **A. Abragam**.

1966

Ernst proposed a Fourier transform method, which provided a major leap forward with respect to the amount of information accessible by NMR. The inherent advantages of greater sensitivity, high resolution, and the absence of line-shape distortions contributed to make Fourier spectroscopy the preferred experimental technique in the field.

1971

J. Jeener first suggested the idea of two-dimensional Fourier transform NMR (FT-NMR), based on the Fourier transformation of signals in two independent time domains to yield a plot with respect to two orthogonal frequency axes. In **1975, R. R. Ernst** with colleagues reported the first two-dimensional NMR (2D-NMR) ^{13}C spectrum of hexane. This was followed in **1976** by a seminal publication presenting a comprehensive theoretical treatment of 2D-NMR correlation spectroscopy (COSY). **Ernst** received the **1991** Nobel Prize for chemistry for his many contributions to NMR.[†]

1972

P. C. Lauterbur demonstrated the feasibility of macroscopic imaging by NMR. In the same year, **R. Damadian** used the method for investigations of the human body, in particular for cancer detection, paving the way for the non-invasive imaging of entire biological organisms. In **2003 Lauterbur** and **P. Mansfield** shared the Nobel Prize for medicine or physiology for contributions to magnetic resonance imaging (MRI).

1983

T. A. Cross and **S. J. Opella** showed that high-resolution structural constraints could be obtained from solid-state NMR experiments, and the potential of the approach was rapidly established.

1985

K. J. Wüthrich and coworkers reported the complete three-dimensional structure of a protein, BPTI, in solution based on NOE distance constraints only. There has since been spectacular progress in the development and application

---

[†] J. Jeener originated the idea of two-dimensional NMR spectroscopy in 1971. Unfortunately, his first two-dimensional spectra were never published; the only reference to the work is in a set of lecture notes for a summer school.

of NMR methodology to protein structure determination in solution. Because of the growing number of peaks and larger peak widths with increasing molecular mass, the method was initially limited to macromolecules of a few tens of kilodaltons. Isotopic labelling extended the molecular mass range to 40–50 kDa. In **1998**, **Wüthrich** with coworkers discovered that in very high magnetic fields narrow resonance peaks can result from interference between dipole–dipole coupling and chemical shift anisotropy, and proposed the technique called transverse relaxation optimised spectroscopy (TROSY). TROSY experiments should make possible the determination of three-dimensional structures of proteins close to 100 kDa in molecular mass. **Wüthrich** was the **2002** Nobel laureate in chemistry

Late 1980s

H. Oschkinat, D. Marion, A. Bax and **G. W. Vuister** introduced a third frequency dimension in NMR spectra (3D-NMR), and in the **early 1990s L. E. Kay** and coworkers and **G. M. Clore** and coworkers expanded the technique to 4D-NMR. By using isotopic enrichment 3D- and 4D-NMR have become powerful experimental approaches, widely applicable in structural biology.

2000 to present

NMR has become one of the most powerful spectroscopic techniques in physics, chemistry and biology. Powerful experimental methods have been devised for observing different NMR phenomena in detail. NMR in structural biology maintains all the typical signs of a young, emerging field of research, with fundamental contributions continuing to be made by many scientists, including **A. Bax, S. Grzesiek, A. M. Groenborn, D. Marion**, and others.

The field of NMR proved to be remarkable through the number of revolutionary innovations that have occurred since the first experimental observation of the phenomenon more than 55 years ago. The introduction of the second frequency dimension constituted a critical step for biological applications. A large variety of experimental schemes were developed to extend NMR applications to the characterisation of complex molecules, such as small synthetic polymers, peptides and sugars. Small proteins and oligonucleotides became accessible to study following the introduction of new procedures. The addition of a third frequency dimension (3D- NMR) was the next important development with advances in genetic engineering enabling the overproduction of proteins and their labelling in microorganisms with NMR- stable isotopes. Isotope enrichment, in fact, increased NMR resolution sufficiently that it became possible to add a fourth frequency dimension to the spectra.

NMR now occupies a very special place in the armoury of physical techniques available to biologists. At the end of 2004 the PDB contained more than 4200 NMR-derived structures out of a total of about 24 500. NMR also provides information on protein and nucleic acid dynamics in a time domain spanning from picoseconds to days. Its unique versatility for the study of molecular structure and dynamics makes NMR one of the most powerful tools in modern structural biology.

J1.2 Fundamental concepts

NMR is a field of spectroscopy based on the absorption of electromagnetic radiation in the radio-frequency region, 10 MHz–1 GHz. In contrast to UV, visible and IR absorption spectroscopy, which involve outer-shell atomic electrons, NMR arises from the magnetic properties of atomic nuclei, which, when placed in an intense magnetic field, develop the energy states required for absorption to occur.

Figure. J1.1 shows the electromagnetic spectrum from the radio-frequency region (frequency $\nu = 10^4$ Hz) through microwaves, IR, visible, UV and X-rays to γ-rays ($\nu = 10^{22}$ Hz). NMR utilises the low-frequency end of the spectrum. Other spectroscopic methods are concerned with larger energy level splitting and hence higher frequencies.

In common with optical spectroscopy (Part E), quantum mechanics is essential for a full understanding of NMR in terms of absorption frequencies and nuclear energy states. A classical treatment, however, while limited, may yield a clearer physical picture of the process and of how it is measured.

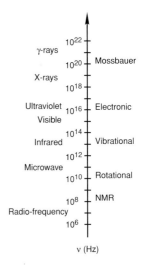

Fig. J1.1 The electromagnetic spectrum from γ-rays ($\nu = 10^{22}$ Hz) to radio-frequency (10^4 Hz) radiation.

J1.2.1 Quantum mechanical description

Magnetic properties of nuclei: nuclear spin

The concept of *spin* in quantum mechanics cannot be explained rigorously in classical terms. However, the properties of certain nuclei can be understood in terms of a model in which they behave as spherical bodies with the nuclear charge distributed uniformly over their surfaces. A non-spinning nucleus does not have a magnetic moment because there is no circulation of charge (Fig J1.2). It is said to have a nuclear spin value equal to zero, and does not give an NMR signal. All nuclei with an even mass number and an even nuclear charge Z have a nuclear spin of zero (Table J1.1 and Comment J1.1). Two nuclei of considerable importance in biology, ^{12}C and ^{16}O, are of this type.

Comment J1.1   Spin quantum number

The spin quantum number of a nucleus is determined by the number of unpaired protons and neutrons it contains. For example, ^{12}C has even numbers of protons and neutrons: each proton pairs with a proton of opposite sign, as does each neutron, giving a net spin angular momentum of zero ($\mathbf{I} = 0$). A nucleus with odd numbers of protons and neutrons (e.g. ^{14}N) generally has an integral non-zero quantum number, because the total number of unpaired nucleons is even, and each contributes $\frac{1}{2}$ to the quantum number. The discussion can be extended to nuclei with even numbers of protons and odd numbers of neutrons, or vice versa, which usually have half-integral quantum numbers due to the odd number of unpaired nucleons.

Fig. J1.2 Schematic view of a spherical non-spinning nucleus.

Table J1.1. *Charge Z, number of neutrons N and nuclear spin quantum number I for nuclei of particular interest in biology*

Nucleus	Z	N	I
^{1}H	1	0	$\frac{1}{2}$
^{2}H	1	1	1
^{12}C	6	6	0
^{13}C	6	7	$\frac{1}{2}$
^{14}N	7	7	1
^{15}N	7	8	$\frac{1}{2}$
^{16}O	8	8	0
^{17}O	8	9	$\frac{5}{2}$
^{19}F	9	10	$\frac{1}{2}$
^{31}P	15	16	$\frac{1}{2}$
^{32}S	16	16	0

Fig. J1.3 Schematic view of a spinning spherical nucleus.

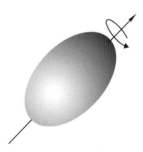

Fig. J1.4 Schematic view of an ellipsoidal (prolate) spinning nucleus.

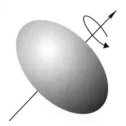

Fig. J1.5 Schematic view of an ellipsoidal (oblate) spinning nucleus.

It is not as unfortunate as it might seem that the principal isotope of carbon has no NMR signal. Most NMR information is obtained from proton resonance peaks and if ^{12}C had a nuclear magnetic moment, the ^{1}H-NMR spectra of most biological macromolecules would be much more complicated than they are, and probably impossible to interpret. Moreover, ^{13}C has a magnetic moment, so that this isotope can be used, either at its low natural abundance concentration or in labelling experiments, for the observation of carbon resonance peaks.

A number of nuclei of particular importance to structural biology (^{1}H, ^{13}C, ^{15}N, ^{19}F, and ^{31}P) have nuclear spin values of $\frac{1}{2}$. In the model, they act as though they were *spinning* spherical bodies of uniform surface charge distribution (Fig. J1.3). A spinning nucleus has circulating charge, and this generates a magnetic field so that a nuclear magnetic moment results. The spherical charge distribution ascribed to nuclei with a spin of $\frac{1}{2}$ means that a probing charge approaching them experiences the same electrostatic field regardless of the direction of approach and, therefore, as with the spherical non-spinning nuclei, the electric quadropole moment is zero.

Most magnetic nuclei, however, act as though they were spinning bodies with non-spherical charge distributions and are assigned spin values of unity or larger integral multiplies of $\frac{1}{2}$. Such nuclei may be considered as approximate ellipsoids spinning about a principal axis (Figs. J1.4 and J1.5). A spinning prolate ellipsoid, with charge uniformly distributed over its surface, presents an anisotropic electrostatic field. The electrostatic work in bringing a unit charge to a given distance is different if the charge approaches along the spin axis or at some angle to it. By convention, the electric quadrupole moment of a nucleus ascribed the shape of a prolate ellipsoid is assigned a value greater than zero. Two nuclei of considerable importance in biology, ^{2}H and ^{14}N, are of this type. Nuclei that behave like charged, oblate ellipsoids also present an anisotropic electric field to a probing charge and by convention are assigned negative electric-quadrupole-moment values. Nuclei of this type include ^{17}O, ^{33}S, ^{35}Cl.

Spin quantum numbers for nuclei of interest in biology are given in Table J1.1.

Magnetic quantum number

An important quantum mechanical property of a spinning nucleus is that the average value of the component of its magnetic moment vector along a defined direction takes up specific values described by a set of magnetic quantum numbers $m = 2I + 1$, in integral steps between $+I$ and $-I$

$$m = I, I-1, I-2, \ldots, -I+1, -I \tag{J1.1}$$

where I is the nuclear spin value. The spin angular momentum vector has magnitude $[I(I+1)]^{1/2}\hbar$ and its z-component is m, where m is given by Eq. (J1.1) (Comment J1.2). Note that the z-axis (the axis of quantization) is arbitrary in the absence of an external magnetic field, so that the spin angular momentum has no preferred direction. The direction of an external magnetic field defines the spin z-axis. The angular momentum component of a spin $\pm\frac{1}{2}$ nucleus (e.g. ^1H, ^{13}C) along the external field has two permitted directions, $I_z = \pm\frac{1}{2}\hbar$, while a nucleus with $I = 1$ has three possible states, $I_z = 0, \pm\hbar$ (Fig. J1.6).

Fig. J1.6 Space quantization of spin $\frac{1}{2}$ (a) and spin 1 nuclei (b).

Comment J1.2   Biologist's box: Spin angular momentum

Bohr's work on the spectrum of the hydrogen atom introduced the postulate that the angular momentum of a system was quantized, i.e. it could only take values which are integer multiples of $h/2\pi$, where h is Planck's constant. It was suggested later by Sommerfeld that the directions of orientation of the electronic angular momentum vector were restricted to certain orientations when the electron was in a closed orbit. In other words, the direction as well as magnitude of the angular momentum vector is quantized. Spin angular momentum $\mathbf{I}$ is a vector quantity, which has magnitude and orientation and should not be confused with the spin quantum number I.

Nuclear magnetization

A charged spinning nucleus creates a magnetic field that is analogous to the field produced when electricity flows through a coil of wire. The resulting magnetic moment μ (a vector quantity) is orientaled along the axis of spin and is directly proportional to the angular momentum vector $\mathbf{I}$ with a proportionality constant, γ, known as the *gyromagnetic* ratio or, less commonly, the *magnetogyric* ratio:

$$\mu = \gamma\mathbf{I} \tag{J1.2}$$

Gyromagnetic ratios of some NMR nuclei are given in Table J1.2. The magnetic moment of a nucleus is parallel (or antiparallel, for nuclei with negative γ) to the spin angular momentum, $\mathbf{I}$.

In the absence of an external magnetic field, all $2I + 1$ orientations of a spin-I nucleus have the same energy. This degeneracy is removed when a magnetic field

Table J1.2. *Gyromagnetic ratios of some NMR nuclei. The SI units for these constants are radian tesla^{-1} second^{-1}*

Atom	$\gamma (10^7$ T^{-1} s^{-1})
^{1}H	26.75
^{13}C	6.73
^{15}N	-2.71
^{19}F	25.18
^{31}P	10.84

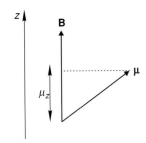

Fig. J1.7 The relationship between the magnetic field $\mathbf{B}_0$, the nuclear magnetic moment μ, and its component along the field direction, μ_z (the scalar product of $\mathbf{B}_0$ and μ).

is applied: the energy of a magnetic moment μ in a magnetic field $\mathbf{B}_0$ is minus the scalar product of the two vectors:

$$E = -\mu \cdot \mathbf{B}_0 \qquad (J1.3)$$

In the presence of a strong field, the quantization axis z is no longer arbitrary, but coincides with the field direction z. Therefore

$$E = -\mu_z B_0 \qquad (J1.4)$$

where μ_z is the z component of μ (the projection of μ onto $\mathbf{B}_0$) and B_0 is the strength of the field (the magnitude of $\mathbf{B}_0$) (Fig. J1.7). From Eqs. (J1.1) and (J1.2) $I_z = m$ and $\mu_z = \gamma I_z$; and so

$$E = \mu \hbar \gamma B_0 \qquad (J1.5)$$

Fig. J1.8 Energy levels for hydrogen ($I = \frac{1}{2}$) nucleus in magnetic field $\mathbf{B}_0$.

Spin $\frac{1}{2}$ nuclei give rise to only two states corresponding to $m = +\frac{1}{2}$ and $-\frac{1}{2}$ (Fig. J1.8). The energy spacing between them is given by

$$\Delta E = \hbar \gamma B_0 \qquad (J1.6)$$

or, expressed in terms of frequency,

$$\nu = \gamma B_0 / 2\pi \text{ Hz} \qquad (J1.7)$$

If the gyromagnetic ratio is positive (e.g. for ^{1}H and ^{13}C), then the $+\frac{1}{2}$ state lies lower in energy, and vice versa for negative γ values (e.g. ^{15}N).[†]

Distribution of nuclei between magnetic quantum states

In the absence of a magnetic field, there is no preference for one or other of the two possible states, for a spin $\frac{1}{2}$ nucleus, so that in a large assemblage of such nuclei there are exactly equal numbers with m equal to $+\frac{1}{2}$ and m equal to $-\frac{1}{2}$.

When an external magnetic field is applied, positive γ nuclei tend to assume the magnetic quantum number $+\frac{1}{2}$, which represents alignment with the field;

[†] The electron has a spin of $\frac{1}{2}$ and a magnetic moment that results in electron spin resonance in a magnetic field.

$m = +\frac{1}{2}$ represents a more favourable energy state than $m = -\frac{1}{2}$. The tendency of the nuclei to align with the field is opposed by thermal agitation. The equilibrium percentage of nuclei in each quantum state can be calculated as a Boltzmann distribution by using values for the nuclear moment, the external field strength and the temperature:

$$N_{\text{upper}}/N_{\text{lower}} = \exp\left(-\Delta E/kT\right) = \exp\left(-h\nu/kT\right) \tag{J1.8}$$

where N_{upper} is the number of nuclei in the higher energy state ($m = -\frac{1}{2}$), N_{lower} is the number in the lower state ($m = +\frac{1}{2}$), ΔE is the energy difference between the states. Since $h\nu$ is very much smaller than kT at temperatures normally used in an NMR experiment there is an small excess of spins in the low state. This can be calculated to be 1 part in 10^9 at normal temperatures. Since this excess is proportional to the signal inducible in the probe, NMR is a very insensitive tehnique compared with other forms of spectroscopy, where the energy difference is very much larger (Comment J1.3). Substituting Eq. (J1.4) into Eq. (J1.6) gives

$$N_{\text{upper}}/N_{\text{lower}} = \exp(-\gamma h B_0/2\pi kT) \tag{J1.9}$$

Comment J1.3   Sensitivity of NMR spectroscopy

In any form of spectroscopy, an electromagnetic field excites molecules or atoms or electrons or nuclei as the case may be, from a lower energy level to an upper one with the same probability as it induces the reverse transition, from excited state to ground state. The net absorption of energy, and hence the intensity of the spectroscopic transition, is therefore dependent on the difference in population between the two levels. In NMR spectroscopy, where the upward transitions outnumber the downward transitions by only one in 10^4–10^6, it is as if one detects only one nucleus in every 10^4–10^6. If we add to this the fact that spectroscopy at higher frequencies is much more sensitive as a rule, because higher-energy photons are easier to detect, it becomes clear why NMR signals are rather weak. It is therefore of crucial importance to optimise signal strengths, e.g. by using strong magnetic fields to maximise ΔE. Similarly, nuclei with a large gyromagnetic ratio and a high natural abundance are favoured (Table J1.2). Hence, ^{1}H is very popular as an NMR nucleus.

The energy-level diagram for a Boltzman distribution of $2n$ nuclei is shown in Fig. J1.9(a). At equilibrium the individual magnetic moment vectors are distributed in two cones about the $+z$- and $-z$-axes (Fig. J1.9(b)). If γ is positive, slightly more than half the nuclei are aligned with the applied field. The excess spin population results in a net magnetization in the $+z$ direction (Fig. J1.9(c)). Expanding the right-hand side of Eq. (J1.9) as a Maclaurin series, and truncating after the second term, we obtain the important result that

$$N_{\text{upper}}/N_{\text{lower}} = 1 - \gamma h B_0/2\pi kT \tag{J1.10}$$

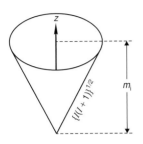

Fig. J1.10 The vector model of angular momentum for a single spin 1/2 nucleus. The angle around the z-axis is indeterminate.

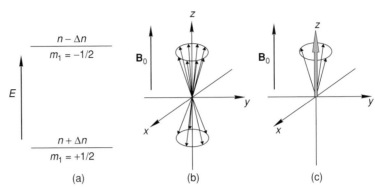

Fig. J1.9 (a) Energy-level diagram for a total of $2n$ nuclei (with positive γ), showing the excess population in the lower-energy state. (b) Schematic presentation showing a greater number of spins aligned with the magnetic field (lower-energy state). (c) The excess spin population aligned with the field results in a net magnetization in the $+z$ direction. The actual excess is much smaller than shown. (Adapted from King and Williams, 1989.)

The ratio of nuclei in the upper- and lower-energy states is linearly related to the magnetic field. Resonant absorption of radio-frequency energy corresponds to a transition between the lower- and upper-energy states. The number of transitions depends on the population ratio between the two states. Equation (J1.8) shows that the intensity of an NMR signal increases linearly with field strength, leading manufacturers to produce increasingly powerful magnets for NMR.

J1.2.2 Classical mechanical description

In order to understand the workings of an NMR experiment, and thus to predict the results, we require a formalism with which to 'visualise' the evolution in time of a spin system. If we were to consider a single spin, a quantum mechanical formalism would be required. This is because atomic phenomena do not behave classically, i.e. they do not obey Newtonian mechanics. For example, if we were to attempt to measure the x or y components of the magnetization of a single proton, we would get one of two answers, $+\frac{1}{2}$ or $-\frac{1}{2}$ (in units of $\hbar$). However, if we repeat the measurement, the result would not always be the same. In other words, a single nucleus does not behave classically. The signal observed in an NMR experiment derives from a large ensemble of spins, so the detected signal (expectation value of the x or y component of the magnetization) behaves in a classical manner. In some respects this must be intuitively obvious, since the NMR sample (as opposed to a single nucleus) is a classical object and we therefore expect it to behave classically.

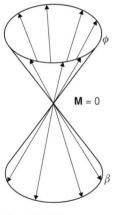

Fig. J1.11 The magnetization of the sample of spin-1/2 nuclei is the resultant of all their magnetic moments. In the absence of an externally applied field, there are equal numbers of spins with different energy at random angles around the z-axis (the field direction) and magnetization is zero.

Now we consider a sample composed of many identical nuclei with spin quantum number $I = \frac{1}{2}$. As we saw in Section J1.2.1, an angular momentum can be represented by a vector of length $[I(I+1)]^{1/2}$ units with a component of length m_i units along the z-axis. Since the uncertainty principle does not allow us to specify the x and y components of the angular momentum, all we know is that the vector lies somewhere on the cone around the z-axis (Fig. J1.10). In the absence of a magnetic field, the sample consists of equal numbers of spins of different energy with their vectors lying at random angles ϕ on the cones (Fig. J1.11). The angles ϕ are unpredictable, and at this stage we consider the spin vectors as stationary. The magnetization $\mathbf{M}$ of the sample, its net nuclear moment, is zero.

Fig. J1.12 Magnetic compass needle.

The effect of the static field

Two changes occur in the magnetization when a magnetic field is present. The first is nuclear precession.

The behaviour of a compass needle is a good starting point for the discussion of this phenomenon. If displaced from alignment with the field and then released, in the absence of friction, the needle fluctuates back and forth indefinitely in a plane about the field axis (Fig. J1.12).

A quite different kind of motion occurs if the needle is spinning rapidly around its north–south axis. Because of the gyroscopic effect, the force applied by the field to the axis of rotation causes movement not in the plane of the force but perpendicular to this plane. The axis of the rotating particle moves in a circular orbit. It is said to *precess* around the magnetic field vector (Fig. J1.13). The precession angular velocity, ω_0, is proportional to the applied field strength B_0

$$\omega_0 = \gamma B_0 \text{ rad/s} \tag{J1.11}$$

The angular velocity can be converted to the precession frequency, ν_0, by dividing by 2π

$$\nu_0 = \gamma B_0/2\pi \text{ Hz} \tag{J1.12}$$

Equation (J1.12) is called the Larmor equation. By considering the nucleus as a spinning magnet, the Larmor equation can be used to describe the fundamental phenomenon of NMR. A comparison of Eq. (J1.5) with Eq. (J1.12) shows that we can equate the Larmor frequency with the resonant frequency derived from quantum mechanical considerations. The proportionality constant, γ, therefore, corresponds to the gyromagnetic ratio of the nucleus in the quantum mechanical description. In the classical description, the spinning nucleus 'resonates' with a precession (Larmor) frequency proportional to the applied field via its gyromagnetic ratio. It is clear, from the lack of any term in Eq. (J1.10) involving the angle of precession, that the energy of the system does not depend on the magnetic moment of the spinning nucleus, μ, but only on the projection of μ onto the magnetic field B_0 axis (conventionally defined as the z-axis).

In a magnetic field strength of 14.1 T, which is typical for current NMR experiments, Eq. (J1.10) predicts a resonance frequency, $\nu = 6 \times 10^8$ Hz or 600 MHz for the ^{1}H nucleus (see Table J1.2 for the γ value). The frequency falls in the radio-frequency region of the electromagnetic spectrum and corresponds to a wavelength of 50 cm. Since the proton is by far the most popular NMR nucleus,

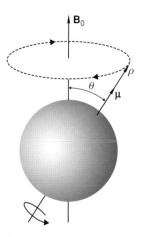

Fig. J1.13 According to classical mechanics, an individual magnetic moment μ precesses about the axis of the applied magnetic field $\mathbf{B_0}$ under angle. This is called Larmor precession. It is analogous to the precession of a spinning gyroscope allowed to topple in the earth's gravitational field.

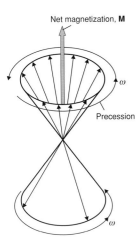

Fig. J1.14 In the presence of an external magnetic field, the spins precess around their cones and there are also changes in the populations in the two spin states. As a result, there is a net magnetization along the z-axis.

Table J1.3. *NMR frequencies (in 14.1 T), and natural abundance of selected nuclei*

Nucleus	ν (MHz)	Natural abundance
^{1}H	600	99.985
^{13}C	150.9	1.108
^{15}N	60.75	0.37
^{19}F	564.75	100.0
^{31}P	243.15	100.0

NMR spectrometers are usually classified by their proton frequencies rather than by the strength of their magnetic fields.

Table J1.3 summarises the NMR properties of several magnetic nuclei in an external magnetic field of 14.1 T. Note that of these nuclei ^{1}H has the largest gyromagnetic ratio.

The second change that occurs in a external magnetic field is that the population in the two spin states alters according to the Boltzmann distribution (Section J1.2.1). Despite a tiny imbalance of populations there is a net magnetization that can be represented by a vector **M** pointing in the z-direction and with a length proportional to the population difference (Fig. J1.14).

Laboratory frame and rotating frame

The classical formalism describes nuclear resonance in terms of the precession of magnetization vectors about the applied static magnetic field $\mathbf{B}_0$ with Larmor precession frequency. In order to visualise the effect of radio-frequency pulses, it is convenient to work with this vector model as it stands. One approach by which this can be achieved is to view the system in terms of the rotating frame of coordinates. In this representation, the normal (laboratory) Cartesian axes x, y, z are replaced by axes x', y', z' which are presumed to rotate at the Larmor frequency of the spins. This representation is known as the rotating frame of coordinates. In this frame the bulk magnetization vector appears stationary (Comment J1.4).

Comment J1.4   Rotating and laboratory coordinate systems

The concept of a rotating coordinate system should be familiar, since we all customarily refer our positions and motion to the earth, a coordinate system that rotates at the rate of $2\pi/24$ rad h^{-1}. A man standing 'at rest' on the equator appears to a distant observer to be moving with a velocity of over 1000 miles h^{-1}. And if our subject repeatedly threw a ball 'vertically' upward and allowed it to fall in the earth's gravitational field, the ball would appear to him to describe a simple vertical straight line path and to be subject to no horizontal forces. To the distant observer, however, the ball would traverse a complicated path composed of parabolic sectors.

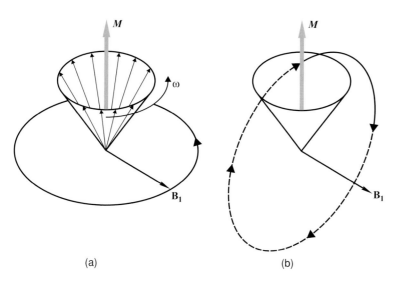

The concept of the rotating frame is enormously useful, since it simplifies the treatment of the otherwise complex gyration of the bulk magnetization vector. It is far easier to understand the workings of complex pulse sequences (see Chapter J2) if the bulk magnetization vector can be considered to be stationary at equilibrium.

The effect of the radio-frequency field

We now consider the effect of the magnetic component of the radio-frequency field in the *xy* plane. Suppose we choose the frequency of the oscillating field to be equal to the Larmor frequency of the spins. The nuclei now experience a steady **B₁** field (Fig. J1.15) because the rotating magnetic field is in step with the precessing spins. Under the influence of this steady field, the magnetization vector begins to precess around its direction. If we apply the **B₁** field in a pulse of certain duration the magnetization precesses into the *xy* plane (Fig. J1.16) at

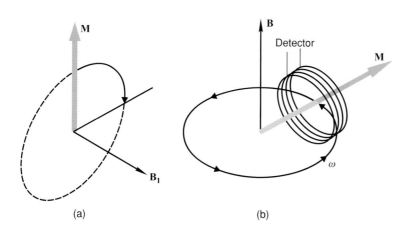

Fig. J1.16 (a) If the radio-frequency field is applied for a certain time, the magnetization vector is rotated into the *xy* plane. (b) To an external observer, the vector is rotating at the Larmor frequency, and can induce a signal in the receiver coil.

the Larmor frequency. The rotating magnetization induces a signal in the coil, which can be amplified and processed.

Relaxation processes

There are two kinds of relaxation process in NMR. The first is related to the establishment of thermal equilibrium in an assemblage of nuclear magnets with different energy. In the absence of a magnetic field the two energy levels available to spin $\frac{1}{2}$ nuclei are of equal energy and are hence equally populated. As we have discussed above, in the presence of a magnetic field the two energy levels and their populations are no longer equal. Provided the Larmor frequencies of nuclei are similar they are in phase and capable of exchanging energy. As the system reverts to thermal equilibrium exponentially, the z component of magnetization approaches its equilibrium value M_0 with a time constant called the *longitudinal relaxation time* T_1. The constant T_1 reflects the efficiency of the coupling between a nuclear spin and its surroundings (lattice) and is also called the *spin-lattice relaxation time*. Spin-lattice relaxation is an energy effect. A shorter T_1 value means that coupling is more efficient and vice versa. Spin-lattice relaxation times lie between 10^{-3} and 10^2 s for liquids, and the range is even larger for solids.

The second kind of relaxation is illustrated in Fig. J1.17. Consider a group of nuclei, precessing in phase about a common magnetic field along the z-axis, like a tied-up bundle of sticks. They produce a resultant rotating magnetic vector with the component in the xy plane. If by any process the nuclei lose their phase coherence, there are as many positive as negative components in xy plane and the resultant vector moves toward the z axis (Fig. J1.17(b)). The randomisation, i.e. the decay of the y or x component of magnetization to zero, occurs exponentially with a time constant called the *transverse relaxation time*, T_2.

The T_2 relaxation time is related to the width of spectral lines as

$$\Delta\nu_{1/2} = \frac{1}{\pi T_2} \tag{J1.13}$$

where $\Delta\nu_{1/2}$ is the line width at half the maximum height. Typical values of T_2

Fig. J1.17 The effect of a radio-frequency (rf) pulse on the magnetic moments of the individual spins in an NMR sample (looking down the z-axis). Starting from the equilibrium state with random phases (a), a pulse along the x-axis in the rotating frame causes the spins to bunch together to some extent (b), producing a net y magnetization in the sample. This phase correlation amongst the spins is known as coherence. After switching the field the randomization occurs with transverse relaxation time T_2 (c).

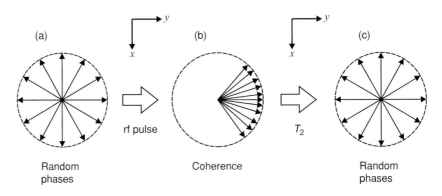

(a) Random phases   (b) Coherence   (c) Random phases

rf pulse T_2

in proton NMR are of the order of seconds corresponding to line widths of about 0.1 Hz.

So far, we have assumed that the equipment, and in particular the magnet, are perfect, and that the differences in Larmor frequencies arise solely from inter-actions within the sample. In practice, the magnet is not perfect, and the field is different at different locations in the sample despite sample spinning. The inhomogeneity dominates the broadening resonance lines. It is usual to express the extent of inhomogenous broadening in terms of an effective transverse relax-ation time T_2^*, similar to Eq. (J1.13)

$$T_2^* = \frac{1}{\pi \, \Delta \nu_{1/2}} \tag{J1.14}$$

where $\Delta \nu_{1/2}$ is the observed width at a half the maximum height. For example, if a line in a ^{1}H-NMR spectrum has a width of 5 Hz, then the effective transverse relaxation time is

$$T_2^* = \frac{1}{\pi \, 5s} = 64 \text{ ms} \tag{J1.15}$$

Experimental NMR schemes for measuring T_1 and T_2 are described in Section J2.3.

J1.2.3 Nuclear environment effects on NMR

NMR would not be a very useful technique if, during an experiment, every nucleus of the same species in a sample were subjected to exactly the same magnetic field defined by the spectrometer magnet. If this were the case, we could measure gyromagnetic ratios with great accuracy but not much else. The Larmor frequency of a given nucleus, however, is strongly affected by its chemical environment. As a consequence, NMR signals from molecules provide a wealth of spectral information that can serve to elucidate their chemical structure.

The spectra of ethanol, shown in Fig. J1.18, illustrate two types of envi-ronmental effect. The curves in Fig. J1.18(a), obtained with a low resolution instrument, show three resonance lines, whose surface areas, in the ratio 1:2:3, correspond to protons with different precession frequencies. It appears logical to attribute the peaks to the hydroxyl, methylene, and methyl protons, respectively

Fig. J1.18 60-MHz NMR spectra of ethanol at different resolutions: (a) low resolution: the areas under peaks stand roughly in the ratio 1:2:3, as would be expected if each peak corresponded to the chemically different OH, CH_2, and CH_3 protons; (b) high resolution: the proton spectra show a considerably greater number of lines – the CH_2 resonance is split into four lines and the CH_3 resonance into three lines. (Adapted from Skoog, et al., 1995.)

(Comment J1.5). The shift in absorption frequency of a nucleus depending on the group to which it is bound is called the *chemical shift*. The higher resolution spectrum in Fig. J1.18(b) reveals that two of the three proton peaks are further split into additional peaks. This secondary environmental effect is called *spin–spin splitting*.

Both the chemical shift and spin–spin splitting are very important in structural analysis. Experimentally, the two effects are easily distinguished. The peak separation resulting from a chemical shift is directly proportional to the field strength, while spin–spin splitting is, in general, independent of the strength of the external magnetic field.

Chemical shift

Chemical shifts arise because the magnetic field $\mathbf{B}$ experienced by a atomic nucleus differs slightly from the external field $\mathbf{B}_0$: $\mathbf{B}$ is slightly smaller than $\mathbf{B}_0$, because of shielding by surrounding electrons. The external field induces the electrons to circulate within their atomic orbitals, much like an electric current passing through a coil of wire. This generates a small magnetic field $\mathbf{B}'$ in the opposite direction to $\mathbf{B}_0$ (Fig. J1.19). $\mathbf{B}'$ is proportional $\mathbf{B}_0$ and typically 10^4–10^5 times smaller. The field at the nucleus may be written

$$\mathbf{B} = \mathbf{B}_0 - \mathbf{B}' = \mathbf{B}_0(1 - \sigma) \qquad (\text{J1.16})$$

where the proportionality constant σ is called the *shielding* or *screening* constant (Comment J1.6). The resonance condition (Eq. J1.5) becomes

$$\nu = \gamma \mathbf{B}_0(1 - \sigma)/2\pi \qquad (\text{J1.17})$$

i.e. the resonance frequency of a nucleus within its atom is slightly lower than that of the same nucleus if it were bare, 'stripped' of all its electrons (Fig. J1.20). The shielding constant is very sensitive to the chemical environment. For protons in a methyl group, it is larger than for methylene protons, and smaller than for the proton in a hydroxyl group. For an isolated hydrogen nucleus, the shielding constant is zero. In order to bring any of the protons in ethanol into resonance at a given excitation frequency ν, therefore, it is necessary to employ an external field

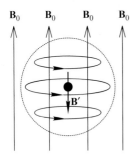

B_0 $\quad$ B_0 $\quad$ B_0 $\quad$ B_0

Fig. J1.19 An applied magnetic field $\mathbf{B}_0$ causes the electrons in an atom to circulate within their orbitals. This motion generates an extra field $\mathbf{B}'$ at the nucleus in opposition to $\mathbf{B}_0$.

B correspondingly greater than $\mathbf{B}_0$, the resonance value for the isolated proton (Eqs. (J1.13) and (J1.14)) Alternatively, if the applied field is held constant, the excitation frequency must be increased in order to bring about the resonance condition.

The shielding constant σ is an inconvenient measure of the chemical shift. Since absolute shifts are rarely needed and are difficult to determine, it is common practice to define the chemical shift in terms of the difference in resonance frequencies between that of the nucleus of interest (v_x) and a reference nucleus (v_{ref}) by means of a dimensionless parameter δ:

$$\delta = (v_x - v_{\text{ref}})/v_0 \qquad (J1.18)$$

where v_0 is the frequency of the spectrometer. The frequency difference $v_x - v_{\text{ref}}$ is divided by v_0 in order to define δ as a molecular property, independent of the external magnetic field. Values of δ are quoted in parts per million (or ppm) (Comment J1.7). A distinct advantage of the definition is that for a given peak, δ is the same regardless of the frequency of the instrument used to measured it; e.g. the same value of δ is obtained with a 60 and a 400 MHz spectrometer (Fig. J1.21). Conventionally, NMR peaks are plotted on a linear δ scale, with the field increasing from left to right. In the example in Fig. J1.21, the TMS reference peak defines the zero value for the δ scale and the value of δ increases from right to left.

The chemical shift of a nucleus depends on many factors, but the surrounding electron density is often the dominant one. A high electron density causes a large shielding effect, which means that the applied magnetic field must be increased to obtain resonance. The up-field shift leads to a lower value of δ. Conversely, a low electron density leads to a down-field shift and an increase in δ value. For

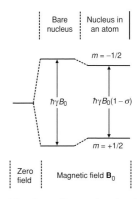

Fig. J1.20 Energy levels of a spin $\tfrac{1}{2}$ nucleus.

Comment J1.7   Reference signals

The reference signal is most conveniently obtained by adding a small amount of a suitable compound to the NMR sample. For ^{1}H and ^{13}C spectra this is usually tetramethylsilane $(CH_3)_4Si$, or TMS. This molecule is inert, soluble in most organic solvents, and gives a single, strong ^{1}H resonance from its 12 identical protons. Moreover, both ^{1}H and ^{13}C nuclei are strongly shielded (large δ values), so that the TMS resonance falls at the low-frequency end of the spectrum. Unfortunately, TMS is not water soluble and in aqueous media the sodium salt of 2,2-dimethyl-2-silapentane -5- sulphonic acid (DSS) is normally used as a reference. The methyl protons of this compound produce a peak at virtually the same place in the spectrum as TMS. However, the methylene protons of DSS give a series of small peaks that may interfere with the measurements. For this reason, most of the DSS now on the market contains deuterated methylene groups, which eliminate these undesirable peaks.

Fig. J1.21 Standard abscissa scales for NMR spectra. (Skoog *et al.* 1995.) (Adapted from Skoog *et al.*, 1995.)

Fig. J1.22 Approximate chemical shifts in ppm relative to a reference for different types of proton. The smaller the value of δ, the greater the chemical shielding and the more up-field the proton signals occur. The reference signal is TMS for non-aqueous solutions or DSS for aqueous samples. (Adapted from Tinoco *et al.*, 1998.)

example, the chemical shifts of the methyl protons (in italics below) in CH_3X relative to TMS become larger as X becomes a better electron-withdrawing group: $\delta(CH_3-CH_3) \approx 1$, $\delta(CH_3-C_6H_6) \approx 2$, $\delta(CH_3-OH) \approx 4$. If the proton is attached directly to an electronegative atom such as in a carboxyl group, which has a very low electron density, chemical shifts can have a very high δ value, $\delta(COOH) \approx 10$. Figure J1.22 shows the approximate values of chemical shifts for various types of proton. Most proton peaks lie in the range $\delta=1$ to $\delta=13$.

For a given compound the appearance of the spectrum is governed by an intramolecular chemical shift difference, i.e. a difference in resonance frequencies for different nuclei of the same molecules. As an example, Fig. J1.23 shows the 60-MHz proton resonance spectrum of 2,2-dimethoxypropane: the methyl and methoxy protons give absorption at different positions in the spectrum. Figure J1.23 illustrates another of the important features of NMR spectra, namely that the intensity of absorption is strictly proportional to the concentration of the nuclei (in the present case protons). This is of great importance for the structure

Fig. J1.23 Proton NMR spectrum of 2, 2-dimethoxypropane, $(CH_3)_2C(OCH_3)_2$, in CCl_4 solution: the methyl and methoxy protons give absorption at different positions in the spectrum.

Fig. J1.24 One-dimensional ^{1}H NMR spectrum of a freshly prepared D_2O solution of the protein, BPTI (protein concentration = 5 mM, pD = 4.5, T = 318 K, ^{1}H NMR frequency = 360 MHz). HDO identifies the solvent water resonance. (Adapted from Wuthrich, 1995.)

Fig. J1.25 Schematic presentation of amino acid side-chains in (a) an extended, random-coil conformation and (b) a folded, globular conformation. (Adapted from Wuthrich, 1995.)

determination. For instance, if the compound used in Fig. J1.23 was of unknown structure, the NMR spectrum would immediately show that it contained two types of proton and that there were equal numbers of each type. This use of NMR could be described as providing a relative proton count.

Figure J1.24 shows the one-dimensional ^{1}H NMR spectrum of BPTI, in which the resonance positions of different hydrogen atoms are given by their chemical shift δ in ppm. It is seen that chemical shift is determined primarily by the chemical structure. The resonance lines of all methyl groups appear on the extreme right at around 1 ppm; those of the labile amide protons are on the extreme left from about 7–11 ppm; in between, we observe the methylene groups at 2–3 ppm, the α-protons at 4–5 ppm, and the protons of the aromatic rings at 6–7.5 ppm.

All amino acid side-chains in an extended, flexible polypeptide chain are exposed to the same solvent environment so that multiple copies of a specified amino acid in the sequence have nearly identical ^{1}H chemical shifts (Fig. J1.25(a)). Therefore, the ^{1}H NMR lines of random coil polypeptides correspond closely to the sum of the resonance peaks of the constituent amino acid residues. Figure J1.26 shows the random-coil spectrum of denatured BPTI computed as such a sum. The increased complexity of the NMR spectrum of native BPTI (Fig. J1.26) results primarily from conformation-dependent ^{1}H chemical-shift dispersion, which is a consequence of a generalised solvent effect: interior peptide segments in globular proteins are shielded from the solvent and are surrounded by other peptide segments (Fig. J1.25(b)). When the interior of a globular protein is highly aperiodic, each amino-acid residue is subjected to a unique microenvironment. In Fig. J1.26 these effects are shown in detail for the threonyl and tyrosyl residues in BPTI. In the unfolded polypeptide chain the methyl groups of the three threonyl residues give rise to a single line corresponding

Fig. J1.26 Computed
random-coil ^{1}H NMR
spectrum of BPTI in ^{2}H$_2$O
solution, where all the
labile protons in N–H and
O–H groups are replaced
by ^{2}H. All resonance lines
in these spectra have
been assigned to distinct
H atoms of BPTI. Stick
diagrams indicate the
positions and intensities
of the γCH$_3$ resonances
the threonyl residues in
the sequence positions
11, 32, and 54, and the
aromatic protons of the
tyrosyl residues 10, 21,
and 35. (Adapted from
Wutrich, 1995.)

in intensity to nine protons, whereas in the folded BPTI the chemical shifts of the
three methyl groups are dispersed so that three separate lines, corresponding to
three protons each, are observed. The NMR lines between 7.5 and 11 ppm in a
globular protein correspond to amide protons that exchange slowly with solvent
^{2}H$_2$O, because of their location in the tertiary structure. They have no counterpart
in the unfolded protein because the spectrum of Fig. J1.26 was calculated with
the assumption that these protons had been completely exchanged with ^{2}H.

The detailed examination of the spectra of compounds containing double or
triple bonds reveals that local effects are not sufficient to explain the position of
certain proton peaks. For example, the δ values change in an irregular fashion for
protons in the following hydrocarbons, arranged in order of increasing acidity of
the groups to which they are bonded:

$$CH_3\text{–}CH_3(\delta = 0.9), \ CH_2\text{–}CH_2(\delta = 5.8), \ HC\equiv CH(\delta = 2.9)$$

The effect of multiple bonds on the chemical shift can be explained by taking into
account the anisotropic magnetic properties of these compounds. For example,
the magnetic properties of crystalline aromatic compounds were found to differ
appreciably, depending upon the orientation of the aromatic ring with respect to
the applied field. The effect is called the *ring-current effect*. The anisotropy can
be understood readily from the model shown in Fig. J1.27(a). When the plane
of the ring is perpendicular to the magnetic field a ring current of π electrons is
induced. The induced field is in the opposite direction to the applied field above
and below the plane of the molecule, and in the same direction on the sides of
the molecule as shown in Fig. J1.27(a). Therefore, a proton near an aromatic
group is shielded (decrease in δ) if it is above or below the centre of the planar

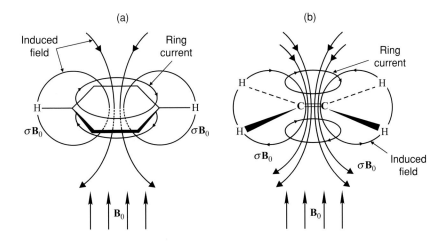

Fig. J1.27 Approximate model of the ring-current effect in NMR. 'Deshielding' of (a) aromatic and (b) ethylene protons. (After Skoog *et al.*, 1995.)

ring, and its δ is increased if it is at the outside of the planar ring. This effect is either absent or self-cancelling in other orientations of the ring. Ring effects are most useful for interpreting local protein conformations near phenylalanine or tyrosine residues, and nucleic acid conformations, in which the purines, adenine and guanine, have the largest ring-current effects. An analogous model is valid for carbonyl double bonds. In this case, we may imagine π electrons circulating in a plane along the bond axis where the molecule is oriented with the field as presented in Fig. J1.27(b). Again, the secondary field produced acts upon the proton to reinforce the applied field.

Finally, it is necessary to point out that the chemical shift range is greater for certain nuclei other than ^{1}H. The chemical shift for ^{13}C in various functional groups typically lies in the range 0–220 ppm. For ^{19}F, the range of chemical shifts may be as large as 800 ppm, while for ^{31}P it is 300 ppm or more.

Spin–spin coupling

As may be seen in Fig. J1.28, the absorption bands for the methyl and methylene protons in ethanol consist of several narrow peaks that can be easily separated with a high-resolution spectrometer (Comment J1.8). Careful examination of these peaks shows that the spacing of the three components of the methyl band is identical to that of the four peaks of the methylene band. This spacing in hertz is called the *coupling constant* for the interaction and is given the symbol J. Moreover, the peak areas within a multiplet are in an integer ratio to one another. The ratio of areas for the methyl triplet is 1:2:1, whereas it is 1:3:3:1, for the methylene quartet.

The results of detailed theoretical calculations are consistent with the concept that coupling take place via interaction between the nuclei and the bound electrons rather than through free space. Let us first consider the effect of the methylene protons in ethanol on the resonance of the methyl protons. Reference to

Fig. J1.28 The high-resolution ^{1}H NMR spectrum of ethanol showing the splitting produced by spin–spin coupling. The bold letters denote the protons giving rise to the resonance peak. The step-like curve is the integrated peak. The CH$_3$ protons form one group of nuclei with $\delta \approx 1$. The two CH$_2$ protons are in a different part of the molecule. They experience a different local magnetic field, and resonate at $\delta \approx 3$. Finally, the OH proton is in another environment, and has a chemical shift of $\delta \approx 4$.

Comment J1.8   Spectrum of ethanol

The methylene protons and the OH proton are separated by only three bonds, so coupling should increase the multiplicity of both OH and methylene peaks. Indeed, the spectrum of highly purified ethanol shows additional splitting of OH (the triplet) and CH$_2$ (the eight methylene peaks). However, if we add a trace of acid or base to a pure sample, the spectrum reverts to the form shown in Fig. J1.28. Both acids and bases and also impurities in ethanol catalyse the exchange of OH protons. It is thus plausible to associate the decoupling observed in the presence of these catalysts to an exchange process. If exchange is rapid, each OH group has several protons associated with it during any brief period; within this interval, all of the OH protons experience the effects of the three spin arrangements of the methylene protons. Thus, the magnetic effects on the ethanol proton are averaged, and a single sharp peak is observed.

Fig. J1.29 shows that the two methylene protons may have any one of four possible magnetic quantum number combinations.

   The spins of the two methylene protons are paired and aligned against or towards the external field, in two of the combinations. There are two other combinations in which the spins oppose one another. The magnetic effect that is transmitted to the methyl protons on the adjacent carbon atom is determined by the instantaneous spin combinations in the methylene group. If the spins are paired and opposed to the external field, the effective applied field on the methyl protons is slightly decreased, and a somewhat higher field is needed to bring them into resonance, resulting in an up-field shift. Spin pairs that are aligned with the

Fig. J1.29 Possible nuclear spin orientations of ethyl and methyl group protons and expected spin–spin splitting patterns.

field result in a down-field shift. Neither of the opposed spin combinations has an effect on the resonance of methyl protons. There results a splitting of the methyl resonance into three peaks, with the unperturbed resonance in the middle. The area under the middle peak is twice that of either of the other two, since two spin combinations are involved.

Let us now consider the effect of the three methyl protons upon the methylene peak (Fig. J1.29). We have eight possible spin combinations. Among these, however, there are two groups containing three combinations that have equivalent magnetic effects. The methylene peak is thus split into four peaks having areas in the ratio 1:3:3:1.

The example of adjacent methyl and methylene groups in ethanol suggests the general rule that the number of peaks in a split band in a first-order spectrum is equal to $n + 1$, where n is the number of magnetically equivalent protons.

The spin–spin splitting expressed as a frequency J is independent of the applied magnetic field (unlike the chemical shift). J values for protons range from 0 to about 20 Hz. If we are measuring proton NMR in a hydrocarbon or carbohydrate, there are no effects from the carbon or oxygen nuclei, because they have no magnetic moment (except from the very small amounts of ^{13}C and ^{17}O). Naturally occurring ^{14}N has a spin of 1 and tends to broaden neighbouring proton lines rather than split them. Consequently, in practice we need to consider only proton–proton splittings.

Three-bond couplings

The most useful spin–spin couplings are those involving nuclei separated by three bonds, e.g. $^3J_{HH}$ in H–C–C–H fragments. The value of this coupling is given by the Karplus equation:

$$J(\theta) = A \cos^2 \theta + B \cos \theta + C \qquad (J1.19)$$

where θ is the dihedral angle between protons (Fig. J1.30). The values A, B, C depend upon the precise system under investigation, and in particular are sensitive

Fig. J1.30 Definition of the dihedral angle θ in the Karplus equation.

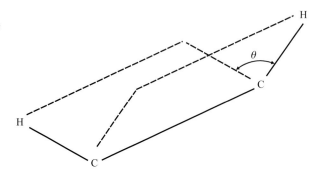

Fig. J1.31 (a) Typical dependence of a three-bond H–C–C–H coupling constant on the dihedral angle θ, calculated using Eq. (J1.19). (b) Part of the backbone of a peptide chain, showing the H–N–C_α–H dihedral angle. R is an amino acid side-chain. (After Hore, 1995.)

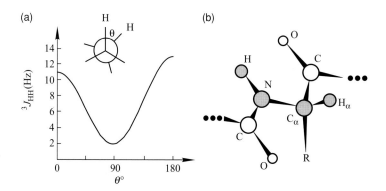

to the electronegativity of the substituents on the carbon. Typical values are $A = 2$ Hz, $B = -1$ Hz, and $C = 10$ Hz, which give a θ variation of the type shown in Fig. J1.31(a). Due to its empirical nature, care should be taken in the generalisation of any parameterisation to an unknown system.

The Karplus equation finds valuable applications in studies of protein structure. For example, the couplings between the amide (NH) and C_α protons in a polypeptide chain provide information on the conformation of the protein backbone (Fig. J1.31(b)). In particular, two major elements of secondary structure in proteins – helices and sheets – have characteristic H–N–C_α–H dihedral angles; $\approx 120°$ and $\approx 180°$, respectively. Thus, amide–C_α proton–proton coupling constants smaller than 6 Hz usually indicate a helix structure, while couplings larger than 7 Hz generally arise from sections of the protein with a β-sheet structure.

Proton–proton coupling constants are generally very small (<1 Hz) when the nuclei are separated by more than three bonds.

Weakly and strongly coupled spins

There are two frequently encountered spin systems (sets of coupled nuclei), which give rise to distinctive multiplets (doublets, triplets, quartets, etc.). Figure J1.32

Fig. J1.32 Spectral simulations of a pair of spin-1/2 nuclei for a range of $\Delta\nu$ values between $16J$ and zero. As the difference in resonance frequencies is reduced, keeping J fixed, the inner component of each doublet steadily increases in amplitude, while the outer components become weaker. The extreme case is that of equivalent nuclei, for which $\Delta\nu = 0$. This case demonstrates that equivalent nuclei do not split each other. (After Hore, 1995.)

Fig. J1.33 Classification of molecules according to their spin system: (a) 1,2-cis-dichlorofluoroethylene is an example of AX proton spin system; (b) chlorbromethylene is an example of AB proton spin system; (c) 1-fluoro-3,4,5-trichlorobenzene: two protons are magnetically equivalent; (d) 1,4-difluoro-2,3-dichlorobenzene: two protons are magnetically non-equivalent.

shows spectral simulations for two coupled spins for a range of $\Delta\nu$ values between $16\,J$ and zero. For the first system it is assumed that all pairs of spins are *weakly coupled*, i.e. that the difference in resonance Larmor frequency of the two nuclei $\Delta\nu$ greatly exceeds their mutual coupling ($\Delta\nu/J \gg 1$). Each spin in such a spin system is labelled with a letters of the Roman alphabet. According to convention these letters are chosen to be well spaced in the alphabet to indicate that the spin system is loosely coupled. Such a two-spin system is referred to as an AX system. This always applies if the nuclei are of different isotopes or elements. Thus *cis-*1,2- dichlorofluoroethylene (Fig. J1.33(b)) is said to possess an AX spin system (it is irrelevant whether A refers to H and X to F or vice versa; the chlorines may be ignored).

In the second system all pairs of spins are *strongly coupled*, i.e. the difference in resonance frequency is close to their mutual coupling ($\Delta\nu/J \sim 1$). This system is referred to as an AB system. The letters are chosen to be close in alphabet to indicate that that the system of interest is strongly coupled. Chlorbromethylene (Fig. J1.33(b)) is example of such system.

A group of n equivalent spins with identical chemical shifts and identical spin–spin couplings to all other spins in the system are denoted A_n. In 1-fluoro-3,4,5-trichlorobenzene (Fig. J1.33(c)) two protons are magnetically equivalent by symmetry. In the case of 1,4-difluoro-2,3-dichlorobenzene (Fig. J1.33(d)) protons are chemically equivalent, but magnetically non-equivalent, because the (H_b, F_d) relationship is not the same as the (H_c, F_d) one – they are *meta* and *ortho* respectively.

In the case of three spins, we can describe an AMX spin system which represents three loosely coupled spins. Similarly we can define an ABX spin system to indicate that protons A and B are weakly coupled to X. Multiplet patterns expected for some of the simple spin systems discussed below are collected together in Fig. J1.34.

$M{\downarrow}X{\downarrow}$ $M{\downarrow}X{\uparrow}$ $M{\uparrow}X{\downarrow}$ $M{\uparrow}X{\uparrow}$

Fig. J1.35 NMR spectrum of nucleus A in an AMX spin system. The four components of the A multiplet, a doublet of doublets, arise from the four combinations of M and X spins, indicated ↑($m = +1/2$) and ↓↑($m = -1/2$).

Fig. J1.34 Multiplet patterns for the A nucleus in various spin systems. M and X are spin $\tfrac{1}{2}$ systems. Weak coupling is assumed throughout. Spectra are drawn for $|J_{AX}| > |J_{AM}|$.

The concept of a spin system can help to show how a multiplet structure arises. Moreover, in some simple cases we can illustrate how the multiplet pattern can be used to determine or verify the structures of molecules, without prior knowledge of the magnitudes of the chemical shifts or coupling constants involved.

The simplest case is the AX system. In this case interaction of nucleus A with X causes the A resonance to split into two equally intense lines centred at the chemical shift of A (a doublet), with spacing equal to the AX coupling constant, J_{AX}.

A step up from the previous case is the AMX spin system, which consists of three nuclei with different chemical shifts and three distinct coupling constants: J_{AM}, J_{AX}, J_{MX}. Four lines are expected because there are four non-degenerate arrangements of the M and X spins ($M{\uparrow} X{\uparrow}$, $M{\uparrow}X{\downarrow}$, $M{\downarrow} X{\uparrow}$, $M{\downarrow} X{\downarrow}$). These peaks are displaced from the chemical shift of A by simple combination of the couplings to spin A (J_{AM} and J_{AX}). The A multiplet should therefore be a doublet of doublets, as shown in Fig. J1.35. The two equivalent spin 1/2 nuclei (AX_2) are a special case of the AMX spin system, with $J_{AM} = J_{AX}$. As seen from Fig. J1.36, the two central lines of the doublet of doublets coincide to give a triplet centred at the chemical shift of A, with a line spacing equal to the coupling constant, and relative intensities 1:2:1.

The multiplet of A in an AX_3 spin system (three identical AX coupling constants) is a four-line quartet (Fig. J1.37). The quartet arises from the eight combinations of the three X spins (Comment J1.9).

Comment J1.9   Resonance lines for AX, AX_2 and AX_3 spectra

The results for AX, AX_2 and AX_3 can be generalised. For n equivalent X (spin 1/2) nuclei, the A resonance is split into $n + 1$ equally spaced lines, with the relative intensity given by the coefficient in the binomial expansion of $(1 + x)^n$.

Fig. J1.36 NMR spectrum of nucleus A in an AX₂ spin system. The triplet arises from the four combinations of the two X spins as indicated.

The high-resolution spectrum of ethanol discussed before (Fig. J1.28) can now be understood. The ethyl protons make up an A_3X_2 spin system: the triplet arises because each of the CH_3 protons couples equally with the two equivalent CH_2 protons interacting identically with each of the CH_3 protons. The rapid internal rotation around the C–C bond averages out the chemical shift differences associated with the different conformations of the molecule, and renders the three methyl protons magnetically equivalent, and similarly the two methylene protons. The absence of splittings from coupling between the CH_2 group

Fig. J1.37 NMR spectrum of nucleus A in an AX_3 spin system. The quartet arises from the eight combinations of the three X spins, as indicated.

> **Comment J1.10 First-order spectra**
>
> The interpretation of spin–spin splitting patterns is relatively straightforward for first-order spectra. First-order spectra are those in which the chemical shift between interacting groups of nuclei is large with respect to their coupling constant J. Rigorous first-order behaviour requires that J be smaller than 0.05δ. The ethanol spectrum shown in Fig. J1.28 is an example of a pure first-order spectrum, in which J is 7 Hz for both the methyl and methylene peaks, while the separation between the centres of the two multiplets is about 140 Hz.
>
> Interpretation of second-order NMR spectra is relatively complex and will not be dealt with in this book. Note, however, that because δ increases with increases of the magnetic field while J does not, spectra obtained with a spectrometer having a high magnetic field are much more readily interpreted than those produced by a spectrometer with a weaker magnet (Skoog *et al.*, 1995)

and the OH proton in the ethanol spectrum (Fig. J1.28) were discussed before (Comment J1.10).

The concept of a spin system can in many cases considerably simplify calculations in macromolecules. For example, important features of many NMR spectra can be calculated by considering only those spins that are scalar coupled. In the case of proteins, each amino acid residue can often be considered essentially an isolated spin system. Likewise, the bases in nucleic acids and monosaccharide residues of oligosaccharides can often be considered in isolation. Complete listings for the aliphatic spin systems and for aromatic spin systems can be found in the literature.

J1.3 Checklist of key ideas

- Certain atomic nuclei can be considered as a having 'spin'; the term spin implies that each nucleus can be considered as a rotating electrical charge and consequently, along with its electrical properties, it also possesses an angular magnetic momentum.
- A number of nuclei of particular importance in structural biology may be assigned the nuclear spin value I of $\frac{1}{2}$ (^{1}H, ^{13}C, ^{15}N, ^{19}F, and ^{31}P). The nuclei of ^{12}C and ^{16}O have an I value equal to zero.
- A nucleus of spin I has $2I + 1$ energy levels, equally spaced with separation $\Delta E = \mu \cdot B_0 / I$, where B_0 is the applied magnetic field, and μ is the nuclear magnetic moment.
- Nuclear magnetic moment μ is given by $\mu = \gamma h I / 2\pi$, where γ is the gyromagnetic ratio, a constant for a given nucleus, and h is Planck's constant; if the gyromagnetic ratio is positive (e.g. ^{1}H and ^{13}C), then the $+\frac{1}{2}$ state lies lower in energy, but the opposite is true for a nucleus with negative γ (e.g. ^{15}N).
- Nuclei with large γ and high natural abundance are favourable for use in practice: hence the ^{1}H as an NMR nucleus is so popular.

- In the absence of a magnetic field the two energy levels available to spin $\frac{1}{2}$ nuclei are of equal energy and are hence equally populated; in the presence of a magnetic field the energy levels are no longer equal in energy.
- The torque exerted on a magnetic moment by a magnetic field inclined at any angle relative to the moment causes the nuclear magnetic moment to precess about the direction of the field with a frequency given by the Larmor equation $v_0 = \gamma B_0 / 2\pi$; such a movement is analogous to the motion of a gyroscope.
- In a single-pulse experiment the flip angle of magnetization α is given by $\alpha = \gamma B_1 t_p$, where t_p is the duration of the pulse. At a constant amplitude of B_1, the pulse length can be varied so as to produce, e.g. a $45°$, $90°$, or $180°$ pulse.
- A typical magnetic field strength B_0 used today for NMR is about 14 T; for hydrogen nuclei, the Larmor equation predicts a resonance frequency of $v = 6 \times 10^8$ Hz or 600 MHz; for ^{13}C the frequency is 150 MHz.
- In a standard one-dimensional NMR spectrum the relative intensities of different resonance lines reflect the number of nuclei manifested by these lines.
- NMR spectroscopy differs fundamentally from optical spectroscopy: in optical spectroscopy, an excited molecule returns to equilibrium by spontaneous emission almost instantaneously, whereas in NMR spectroscopy the probability of spontaneous emission is negligible.
- The nuclear spin system returns to equilibrium with its surroundings (the 'lattice') by a relaxation process characterised by a time T_1, the *spin–lattice* relaxation time or *longitudinal* relaxation time.
- The nuclear spin system returns to internal equilibrium by a relaxation process characterised by a time T_2, the *spin–spin* relaxation time or the *transverse* relaxation time.
- The *chemical shift* δ defines the location of the NMR line along the radio-frequency axis; it is a characteristic measure of nuclear–electron interaction and exquisitely sensitive to local geometry.
- The *chemical shift* is proportional to the applied magnetic field B_0; it is commonly indicated in parts per million (ppm) relative to a reference compounds.
- The *spin–spin splitting*, J, is a measure of the interactions (through-bond) of two or more neighboring nuclei, where the interaction is transmitted by the intervening electrons.
- The spin–spin splitting constant J does not depend on the applied magnetic field and is customarily quoted in hertz.

Suggestions for further reading

Historical review

Jeener, J. (1996). In *Encyclopedia of Nuclear Magnetic Resonance*. Vol. 1. Eds. D. M. Grant and R. K. Harris. Chichester: John Wiley and Sons.

Tinoco, I. Jr., Sauer, K., and Wang, J. C. (1998). *Physical Chemistry. Principles and Applications in Biological Science.* New Jersey: Prentice Hall.

Fundamental concepts

King, R. W., and Williams, K. R. (1989). The Fourier transform in chemistry. Part 1 Nuclear magnetic resonance: introduction. *J. Chem. Education.* **66**, A213–A219.

Skoog, D. A., Holler, F. J., and Nieman, T. A. (1995). *Principle of Instrumental Analysis*, Philadelphia: *Saunders College Publishing.*

Harris, R. (1983). *Nuclear Magnetic Resonance Spectroscopy*, Pitman.

Hore, P. J. (1995). *Nuclear Magnetic Resonance.* Oxford: Oxford University Press.

Chapter J2
Experimental techniques

J2.1 Fourier transform NMR spectroscopy

J2.1.1 Principles

Experiments in the early years of NMR spectroscopy (1945–1970) used so-called continuous wave methods, in which the sample was irradiated with a weak, fixed amplitude, radio-frequency field (Fig. J2.1(a)). Spectra were obtained either by keeping the electromagnetic frequency fixed, while slowly sweeping the magnetic field strength, or vice versa, so as to bring spins with different chemical shifts sequentially into resonance. The 1970s were dominated by the revolutionary development of pulse Fourier spectroscopy (Fig. J2.1(b)), which paved the way for modern NMR and an unprecedented expansion of its applications. The starting point was the design of a multichannel spectrometer, which allowed the simultaneous measurement of many points of a frequency spectrum. It was soon recognised, however, that the instrumental effort became exorbitant as the number of channels increased.

Traditional continuous wave spectrometers have now been almost completely replaced by pulse Fourier instruments. The inherent advantages of greater sensitivity, high resolution and the absence of line-shape distortions contributed to make Fourier spectroscopy the preferred experimental technique in NMR.

In pulse Fourier instruments, data are invariably collected in the time domain; i.e. they are stored in the computer memory as a function of time. However, spectroscopists are interested in the frequency-domain response of a spin system since the energy differences between spin states possess characteristic resonance lines at specific frequencies. In fact, the time domain and the frequency domain are inextricably linked, and we can convert between the two using a procedure known as Fourier transformation (Comment J2.1).

The Fourier transform relates the time-domain data $f(t)$ with the frequency-domain data $f(\nu)$ by the following equation:

$$f(\nu) = \int_{-\infty}^{+\infty} f(t)\exp(i\nu t)\mathrm{d}t \qquad (\text{J2.1})$$

The existence of two related domains allows us to define *Fourier pairs*. Several of these are of particular importance in NMR. For example, the Fourier transform

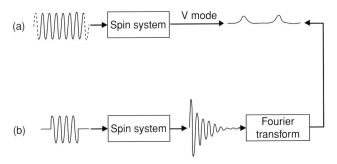

Fig. J2.1 The two basic methods of obtaining an NMR spectrum: (a) by applying a continuous excitation and varying energy; (b) by applying a pulse of energy and Fourier transforming the result.

of a time-domain decaying exponential is a Lorentzian line (see Chapter A3) at zero frequency (Fig. J2.2(a)). This is identical to the well-known relationship between the free induction decay and the NMR spectrum (see Section J2.2).

If the time-domain signal is an exponentially decaying sinusoidal or cosinusoidal oscillation, then again the frequency-domain signal is a Lorentzian line, but offset from zero frequency by the frequency of oscillation of the sinusoidal or cosinusoidal waveform (Fig. J2.2(b)). Recall from Chapter A3 that the Lorentzian is the Fourier transform of an exponential decay function.

A third and equally important example of Fourier pairs is the Fourier transform of a radio-frequency pulse. A basic result of the time–frequency Fourier transform relation is that a short pulse in time can be considered as a multifrequency source, which could allow the simultaneous excitation of different resonance frequencies in an NMR experiment. In order to clarify this point, we examine the magnetization of a sample under the influence of static and radio-frequency fields. In the frame of reference rotating at the angular frequency, ω, of the radio-frequency field, the nuclear magnetization M_j of nuclei of resonant frequency ω_i, precesses about an effective field given by

$$|B_{\mathrm{eff}}| = (1/\gamma)[(\nu_i - \nu)^2 + (\gamma B_1)^2]^{1/2} \tag{J2.2}$$

If B_1 is chosen large enough so that

$$\gamma B_1 \gg 2\pi \Delta \tag{J2.3}$$

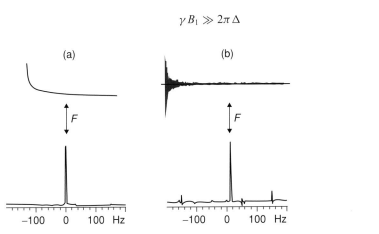

Fig. J2.2 Two 'Fourier pairs': (a) a decaying exponential gives a Lorentzian line at zero frequency after Fourier transformation; (b) an exponentially decaying cosinusoid gives a Lorentzian line offset from zero frequency by an amount equal to the frequency of the oscillation of the cosinusoid.

Fig. J2.3 Schematic spectrum at constant B_0, showing resonance lines covering a range of frequencies Δ'. Reference frequency values are shown within the range (ν') and outside the range (ν).

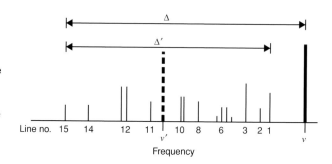

where Δ (in hertz) is the entire range of chemical shifts in the sample, measured with respect to the radio frequency, then for any ν_i within the spectrum the term $(\nu_i - \nu)$ can be neglected, and

$$B_{\text{eff}} \approx B_1 \qquad (J2.4)$$

The magnetization vector precesses about B_1, which is along the x'-axis, (Section J1.2), for *all* nuclei with Larmor frequencies in the range Δ. The width of the time pulse required to cover this frequency range should therefore be $\ll 1/\Delta$.

We call Δ' the frequency range to be examined, as indicated in Fig. J2.3. Because of the heterodyne nature of the detection scheme normally employed, it is not the actual resonance frequencies, ν_i, that are important, but the differences between them and the applied radio-frequency field, $(\nu_i - \nu)$. If ν is chosen within the range Δ', as indicated by the vertical dashed line in Fig. J2.3, then some frequency differences are positive and some are negative. During data acquisition, however, the detector, which measures in the time domain, cannot distinguish positive and negative frequencies (lines 8 and 12 of Fig. J2.3, for example, will appear to be very close together). Since positive and negative frequencies do differ in phase, two phase detectors are used to unravel the spectrum.

Fourier analysis is also essential for the understanding of the effect of the pulse itself. A rectangular pulse in time of monochromatic radiation of frequency, ν_0, can be described in the frequency domain as a band about ν_0 (Fig. J2.4). The monochromatic frequency, ν_0, is produced by a pulse generator and is called

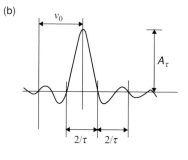

Fig. J2.4 A time pulse of monochromatic radiation of frequency ν_0 with a rectangular envelope, (a), can be described in the frequency domain as a band centred on ν_0, (b).

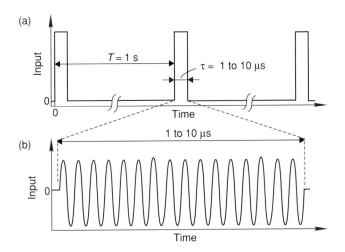

(a)

Input

$T = 1$ s

$\tau = 1$ to $10\ \mu$s

0

0

Time

1 to 10 μs

(b)

Input

0

Time

Fig. J2.5 Typical input signal for pulsed NMR: (a) pulse sequence; (b) expanded view of radio-frequency pulse, typically at a frequency of several hundred megahertz. The time axis is not drawn to scale. It is assumed further that the length of the pulse is short relative to T_1 and T_2, so that no relaxation occurs during the pulse time. (Adapted from Skoog et al., 1995.)

the spectrometer frequency. Following the rules of Fourier transformation, as the pulse length decreases, the width of the frequency band must increase (Fig. J2.4). For a typical pulse length of 10 μs the flat central portion of the frequency band, where the amplitude is within 1% of the peak value, is about 16 kHz wide. For a 7.05 T field ($\nu = 300$ MHz for protons, 75 MHz for ^{13}C) this region easily spans the range of chemical shifts of the commonly observed nuclei (15 ppm or 4.5 kHz for protons, 200 ppm or 15 kHz for ^{13}C) (see Section J1.2.3).

The waveform in Fig. J2.5(a) illustrates a typical pulse train, pulse width and time interval between pulses. The expanded view of one of the pulses it is actually a packet of radio-frequency radiation ($10^2 - 10^3$ MHz). The width of the pulse, τ, is usually less than 10 μs. The interval, T, between pulses is typically one to several seconds. During T, a time domain radio-frequency signal, called the free induction decay (FID) is emitted by the excited nuclei as they relax. FID is detected with a radio receiver coil, perpendicular to the static magnetic field.

J2.1.2 The Fourier transform NMR spectrometer

A schematic lay-out of a typical NMR spectrometer designed for the liquid state is shown in Fig. J2.6. The sensitivity and resolution of the spectrometer depend

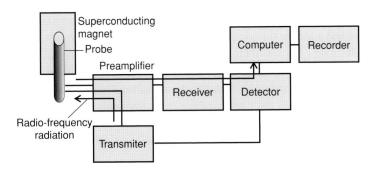

Superconducting magnet

Probe

Preamplifier

Computer — Recorder

Receiver — Detector

Radio-frequency radiation

Transmiter

Fig. J2.6 A schematic presentation of a typical NMR spectrometer.

critically upon the strength and quality of the magnet, which is thus the key component of the instrument. It is advantageous to operate at the highest possible field strength. In addition, the field must be highly homogenous and reproducible. These very stringent specifications are only met by superconducting solenoids. At the time of writing, the highest magnetic field available for NMR is 21 T, corresponding to a proton frequency of 900 MHz (Comment J2.2).

The radio-frequency coil acts as both a transmitter and a detector of the resonance frequency. The measured signal processed by the computer is a low-frequency line resulting from the difference between the transmitted and detected frequencies.

The sample is placed in the centre of the cylindrical magnet to ensure that all the magnetic nuclei experience the same average field. Although a superconducting magnet operates at liquid helium temperature (4 K), the sample itself is normally at room temperature.

In order to perturb the spin system with radio-frequency energy, the spectrometer contains sophisticated pulse programmer and transmitter units, which allow the application of complex pulse sequences to the sample of interest (Section J2.5). The radio-frequency source is a very stable crystal oscillator unit which runs continuously, and all frequencies in the spectrometer are derived from it. The continuous-wave signal derived from the source is controlled in amplitude and phase by the gate unit. Current pulse programmers have better than $1°$ phase control.

The pulse/acquisition/delay sequence is repeated N times until signal-to-noise ratio is satisfactory. The detection of weak signals such as ^{13}C NMR at natural isotopic abundance (1%) is now routine.

J2.2 Single-pulse experiments

J2.2.1 Data acquisition and processing

The benefits of the Fourier transform method have been enhanced by the introduction of techniques using various radio-frequency pulse sequences. In this section we describe single-pulse NMR experiments to obtain proton or ^{13}C spectra. Figure J2.7 shows a typical sequence of events. After an *initial delay*, the pulse is applied to the sample with the radio-frequency coil. The receiver (detector) is turned on, and after a very short *dead time* the response of the spin system is measured, digitised and added to the computer memory. The acquisition time should be long enough for the response of the spin system to have decayed to a negligible level (a few times T_2). A further delay allows the sample spins to equilibrate fully (T_1 relaxation). We recall that in the rotating frame of reference the sample is in an external field $\mathbf{B}_0$ along the z-axis and the radio-frequency pulse creates a non-oscillating field $\mathbf{B}_1$ along the x'-axis. In the absence of the radio-frequency pulse, i.e. in the fixed external magnetic field, sample magnetization is small and

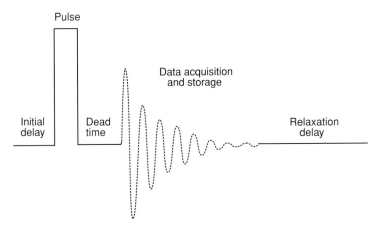

Pulse

Data acquisition
and storage

Initial
delay

Dead
time

Relaxation
delay

Fig. J2.7 Sequence of
events in a single-pulse
experiment.

along $\mathbf{B}_0$, reflecting the imbalance of low- and high-energy spins described by the Botzmann distribution (Fig. J1.9). When the pulse frequency, ν_0, corresponds exactly to the Larmor frequency, the nuclear magnetization vector 'tips' out of the z direction (Fig. J1.15). The pulse is along x', and the tip angle is towards y', in accordance with the gyroscopic effect (Comment J2.3). An insight into why NMR is a 'resonance' technique can be gained from Fig. J2.8. Resonance is a condition in which energy is transferred in such way that a small periodic perturbation produces a large change in some parameter of the system being perturbed. NMR is a resonance technique because the small periodic perturbation $\mathbf{B}_1$ produces a large change in the orientation of the sample magnetization vector $\mathbf{M}$. In most experiments, $\mathbf{B}_1$ is a few orders of magnitude smaller than $\mathbf{B}_0$. The tip angle, α, is given (in radians) by

$$\alpha = \gamma B_1 t_p \qquad (J2.5)$$

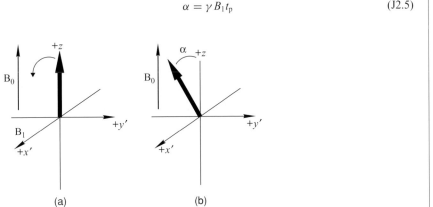

(a) (b)

Fig. J2.8 The effect of a short radio-frequency pulse along x' viewed in the rotating frame. Initially the magnetization vector is parallel to the external field along the z-axis. (a) A pulse creating a field along x' exerts a force on the magnetization vector tipping it towards minus y'. (b) After the $\mathbf{B}_1$ field has been removed, the magnetization vector maintains its tip angle α with the z-axis.

Comment J2.3
Optimum combination of tip angle and $T1$ relaxation time

In order to avoid signal saturation, the sum of the acquisition time and the relaxation delay should be several times T_1, but the resulting long experimental times may make unreasonable demands on the available instrumental resources. For best sensitivity, there is an optimum combination of the tip angle, α, the T_1 relaxation time and the time between pulses, t_{rep}, given by Ernst's equation

$$\cos(\alpha) = \exp(-t_{rep}/T_1)$$

The length of a typical $90°$ pulse is about 10^{-5} s. For the proton this requires a $\mathbf{B}_1$ field of about 6×10^{-4} T. Note that this is much smaller than the external field, $\mathbf{B}_0$.

Comment J2.4

Selective and non-selective pulses

A non-selective pulse is a radio-frequency pulse with a wide frequency bandwidth (short, high-power pulse) that excites all nuclei of a given type (e.g. all protons in the sample). A selective pulse is a radio-frequency pulse with a narrow frequency bandwidth (long, low-power pulse) that excites nuclei in a limited chemical shift range.

A crucial feature of pulsed NMR is the ability to excite nuclei with different chemical shifts uniformly and simultaneously. For example, a typical range of ^{1}H resonance frequencies is 4 kHz (10 ppm $\times$ 400 MHz); a 90° pulse of strength $\gamma B_1/2\pi \gg 4$ kHz therefore rotates the magnetization vectors of all protons irrespective of their resonance frequencies.

The radio-frequency pulse is usually designated by the value of α that it produces and the axis to which it is applied. The term 90°_x or $(\pi/2)_x$ refers to a pulse directed along the $+x'$-axis. At the end of a 90°_x pulse the magnetization vector points along the minus y' direction. Application of the $\mathbf{B}_1$ field for twice as long ($\theta = \pi$) results in inversion of $\mathbf{M}$. Such a pulse is called an 180° or π pulse. Single pulses are used both as a method for perturbing the spin system from equilibrium and as a means for detecting the magnetization (Comment J2.3).

The pulse produces a band of frequencies of almost equal amplitude ($\mathbf{B}_1$ value) (Fig. J2.3), so that the nuclei throughout the entire chemical shift range are all effectively tipped by the same angle (Eq. J2.5). As far as the excitation pulse is concerned, each chemically shifted nucleus may be considered to belong to a frame rotating at the corresponding Larmor frequency. However, in order to understand what happens after excitation it is necessary to assign the frame to a single frequency, which is chosen to be the spectrometer frequency, ν_0, the value at the centre of the frequency band (Comment J2.4).

J2.2.2 Free induction decay (FID)

Figure J2.9 shows the behaviour of the magnetization vector following excitation by a 90°_x pulse. The $\mathbf{B}_0$ field is still present, and the nuclei continue to precess about it. Focusing first on the nuclei whose Larmor frequency corresponds exactly to the frequency of the rotating frame, the magnetization vector remains directed along the $-y'$-axis, in the absence of relaxation effects (Fig. J2.9(a)). However, T_1 and T_2^* processes both act to reduce the $-y'$ component of magnetization (Section J1.2.2). $\mathbf{B}_0$ inhomogenities, the principal source of T_2^* effects, cause groups of nuclei in different parts of the sample to experience slightly different local $\mathbf{B}_0$ fields. Figure J2.9(b) shows only two of the many such microscopic sample sections (called *isochromats* because the field is the same within the section). One group is shown precessing slightly ahead of the frame, and the other group is lagging behind. At the same time, T_1 relaxation processes cause a gradual return of the magnetization towards the z-axis (Fig. J2.9(c)), further decreasing the component along $-y'$. Because field inhomogeneities usually cause T_2^* to be less than T_1, an intermediate situation often occurs as in Fig. J2.9(d), where the $-y'$ component of the magnetization decays to zero before the spin population can achieve Boltzmann equilibrium. At some later time (after at least 5 times T_1) the magnetization has again returned to its equilibrium value (Fig. J2.9(e)). Returning now to the laboratory frame, the $-y'$ component corresponds to a magnetization vector rotating in the xy plane. As the $-y'$ component decreases, the oscillating voltage from the coil decays, and it reaches zero when the condition of the spins corresponds to the situation depicted in Fig. J2.9(d). The record of the receiver voltage in the time domain is called the FID, because the nuclei are allowed to precess 'freely' in the absence of the $\mathbf{B}_1$ field. The FID is the sum

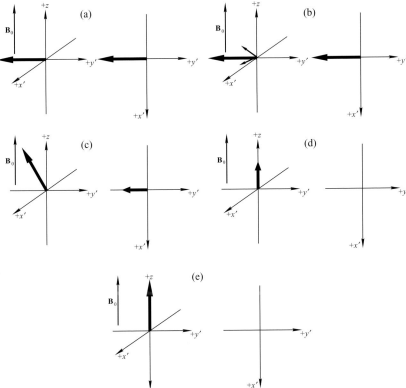

Fig. J2.9 Behaviour of nuclei after a 90°_x pulse. The frame is rotating at the spectrometer frequency, f_s, which is assumed to be equal to the Larmor frequency. (a) Immediately after the pulse the tipped magnetization vector is fixed in the rotating frame. (b) T_2^* effects (mainly $\mathbf{B}_0$ inhomogenity) cause divergent local Larmor frequencies. Only two divergent vectors are shown here. (c) At the same time T_1 relaxation is taking the magnetization back towards the z-axis. (d) T_2^* effects have resulted in complete loss of magnetization in the $x'y'$ plane, but the nuclei have not returned to equilibrium. (e) Complete restoration of equilibrium z-axis magnetization. (Adapted from King and Williams, 1989a.)

of the individual oscillating voltages from the various nuclei in the sample, each with a characteristic offset frequency (i.e. chemical shift and spin–spin couplings), amplitude and T_2. It contains all the information necessary to obtain an NMR spectrum. The FID is the Fourier partner of the NMR frequency spectrum.

The FID of ethanol is shown in Fig. J2.10. The frequency-domain spectrum obtained by Fourier transformation is in Fig. J1.18(b). The FID curve in Fig. J2.10

Time

Fig. J2.10 FID of a sample of ethanol. (King and Williams (1989b).)

is very complex because it is composed of eight components (minimum), each with a characteristic frequency.

J2.3 Multiple-pulse experiments

In a multiple pulse experiment a specific radio-frequency pulse sequence is applied to the sample before the FID is measured. Such experiments have greatly enriched the potential of NMR applications because they provide information that is difficult or impossible to obtain by the single-pulse technique. A pulse sequence is defined by the amplitude and width of each pulse and the time delay in between (Comment J2.5).

There is an enormous variety of multiple-pulse experiments. They fall into three main groups presented in Table J2.1.

The first group (group 1 in Table J2.1) includes pulse sequences for *relaxation time* measurements. A simple two-pulse sequence is the basis of the *inversion-recovery* method for the measurement of the *spin-lattice longitudinal relaxation* time T_1. The spin-echo method for measurements of the *spin–spin transverse relaxation characteristic time T_2* is based on a different pulse sequence, which was originally designed to refocus effects of $\mathbf{B}_0$ inhomogeneity.

The second group (group 2 in Table J2.1) includes heteronuclear polarization transfer techniques for sensitivity enhancement. These techniques greatly contribute to the success of NMR for the observation of rare and low-γ nuclei. The basic trick involves 'borrowing' polarization from rich spin nuclei of high γ-value (see Section J2.3.3). The group includes many very sophisticated pulse sequences with special time delays and sensitive polarization transfer simultaneously for all spins in the sample. They are represented in Table J2.1 by the insensitive nuclei enhanced by polarization transfer (INEPT) method which is conceptually one of the simplest.

Almost all heteronuclear multidimensional NMR experiments use INEPT to transfer magnetization between different spin species, but, unfortunately, the efficiency of INEPT deteriorates with increasing rotational correlation time (molecular tumbling). In contrast, transfer polarization by the cross-relaxation induced polarization technique (CRIPT) is independent of rotational correlation time. The combination of INEPT and CRIPT leads to another highly efficient transfer protocol (CRINEPT) for solution NMR on very high molecular mass samples (up to 100 kDa).

The third group in Table J2.1 uses the nuclear Overhauser effect (NOE) based on polarization transfer via space dipolar coupling.

In this section we discuss the simplest multiple pulse schemes (inversion recovery and spin echo for relaxation measurements and the INEPT polarization technique). NOE will be discussed in Section J2.4 and Crinept in Sect. J2.5.

Comment J2.5
Multiple-pulse sequences and quantum mechanics

In many cases the explanation of how a pulse sequence acts requires quantum mechanical analysis at a level of complexity that is beyond the scope of this book. The interested reader is referred to the various texts and publications on this subject such as Derome (1987), Ernst *et al.* (1987), Brey (1988).

Table J2.1. *Experimental multiple-pulse experiments*

Multiple-pulse scheme	Physical phenomenon	Pulse sequence	Applications
Group 1 Inversion recovery	Relaxation process		Measurements of longitudinal relaxation time T_1
Spin echo	Relaxation process		Measurements of transversal relaxation time T_2^*
Group 2 Insensitive nuclei enhanced by polarization transfer (INEPT)	Polarization transfer via scalar couplings		Enhancement sensitivity of NMR experiments on rare and low gyromagnetic ratio nuclei
Cross-relaxation insensitive nuclei enhanced polarization transfer (CRINEPT)	Highly efficient polarization transfer for ^{15}N, ^1H and other nuclei		Key element of multidimensional NMR (building block to create magnetization and coherence where required)
Group 3 Nuclear Overhauser phenomenon (NOE)	Polarization transfer via space dipolar coupling		Key constraint for three-dimensional structure determination (distance measurements)

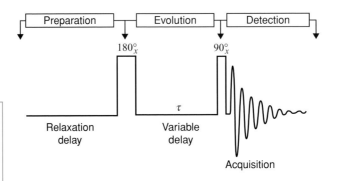

Comment J2.6

Compact representation of a pulse sequence

Although the schematic pulse sequence shown in Fig. J2.12 is a useful pictorial device, it may be represented more compactly as:

equilibration delay–180°_x –τ–90°_x–acquisition

The sequence may be divided into three periods: preparation (equilibrium delay and the 180°_x pulse), evolution (the time delay) and detection (the 90°_x pulse and the data acquisition). These periods are easy to pick out in the simple sequences and are important when analysing the more involved pulse sequences present.

J2.3.1 The inversion recovery method to measure spin-lattice relaxation time T_1

The method is shown schematically in Fig. J2.11. The nuclei are first subjected to a radio-frequency pulse of sufficient width to cause **M** to rotate through 180° to the z-axis. Following the pulse there is a delay period, τ, whose length is chosen appropriately. The delay period, τ, is a key component of most multiple-pulse methods. Typically, eight or ten values of τ are used, ranging from a small value to four or five times T_1, in an inversion-recovery experiment (Comment J2.6). During the τ interval the system returns to equilibrium by the spin-lattice relaxation process. Spin–spin relaxation, which involves magnetization components only in the xy plane, is not involved, because **M** lies along the z-axis. As the individual magnetic moments gradually return to their favoured orientations along the $+z$-axis, the net magnetization vector becomes shorter in the $-z$ direction. Depending on the length of the delay, **M** passes through zero and eventually recovers its full original magnitude along the $+z$-axis.

In order to determine the extent of relaxation, **M** must be converted into observable magnetization in the xy plane. This is done by applying a second pulse of half the length of the first to the $+x$-axis; the 90°_x pulse rotates **M** to lie along the y-axis (Fig. J2.12). If **M** is still negative before the 90° pulse, then the

Fig. J2.12 Vector description of the process for measurement T_1 by *inversion-recovery* method.

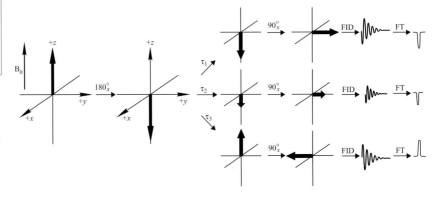

magnetization is rotated to the $+y$ direction, but, when relaxation is complete, the magnetization after the 90° pulse lies along the $-y$-axis. The intensity I of the peak resulting from the Fourier transform of the FID varies with τ from a maximum negative value for $\tau = 0$ to a maximum positive value for $\tau = \infty$ (in practice, 4 or 5 times T_1). The intensities are related exponentially to T_1 by the equation:

$$I = I_\infty[1 - 2\exp(-\tau/T_1)] \qquad (J2.6)$$

A typical plot of peak intensity as a function of τ is shown in Fig. J2.13. The peak intensity goes through zero when $\tau/T_1 = \ln(2)$, i.e. when $\tau = 0.693T_1$.

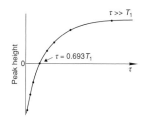

Fig. J2.13 Plot of transformed peak intensity as a function of τ. (After Williams and King, 1990a.)

J2.3.2 The spin-echo effect to measure T_2

We recall that magnetization in the xy plane may be lost by processes that do not affect the z component. The rate constant for this relaxation is the reciprocal of the time constant of the FID, T_2^*, the effective spin–spin relaxation time. T_2^* may be broken down into a $\mathbf{B}_0$ inhomogeneity component (the major cause of line broadening in NMR) and the spin–spin relaxation time, T_2. The value of T_2 is determined by the same factors that are responsible for T_1, plus other processes such as spin and chemical exchange.

A spin echo is a magnetic analogue of an audio echo: a pulse of magnetization is formed, is allowed to spread, is reflected, and is then detected as another pulse a short time later.

The *spin-echo* pulse sequence shown in Fig. J2.14 allows the independent measurement of the $\mathbf{B}_0$ inhomogeneity component (Comment J2.7). The initial 90_x° pulse turns the total magnetization vector to the $-y$-axis. If the axes are rotating at the exact Larmor frequency of the nuclei, the effect of T_2 is a gradual shortening of $\mathbf{M}$ in the $-y$ direction. However, because $\mathbf{B}_0$ is not perfectly homogenous, some of the sample isochromats (Section J2.1) precess faster than the rotating frame and move in a counterclockwise direction; the fastest group is labelled 1 in Fig. J2.15, and the slightly slower one is labelled 2. Similarly, isochromats 3 and 4 precess

Comment J2.7
Abbreviate form of the echo sequence

In customary abbreviated form, the echo sequence is

equilibration delay–90_x°–τ–180_x°–τ–acquisition

Again the three periods corresponding to preparation (equilibrium delay and 90° pulse), evolution (the τ–180_x°–τ segment), and detection are evident in the pulse sequence.

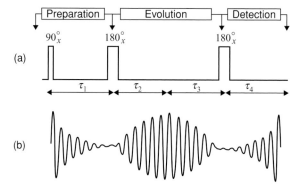

Fig. J2.14 (a) Pulse sequence in a spin-echo experiment. (b) NMR signal showing 'echoes' at the end of τ_2, τ_4. The NMR signal is shown throughout the entire sequence, but in practice the receiver is activated only at the time of appearance of the echo (i.e. after τ_2, τ_4, etc.).

Fig. J2.15 Vector description of a spin-echo experiment. A spin echo is a magnetic analogue of an audio echo: a pulse of magnetization is formed is, allowed to spread, is reflected, and is then detected as another pulse a short time later (After Williams and King, 1990a.)

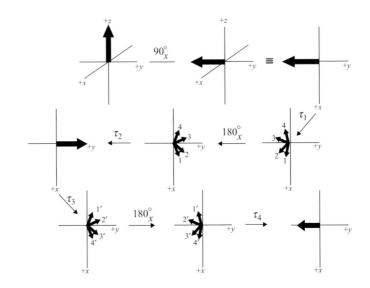

slower than the frame and rotate clockwise. As a result of this disparity the spin isochromat vectors fan out in the xy plane during the delay τ_1, after the 90° pulse. At the end of the first τ delay a 180°_x pulse flips the spins about the x-axis. The isochromats are still precessing in the same directions at the same rates, so, for example, isochromat 1 that moved fastest counterclockwise now has farthest to go to reach the $+y$-axis. After a second τ period of equal length all the isochromats refocus along the $+y$-axis to give a maximum xy magnetization. If the NMR signal is monitored during this $90^\circ_x - \tau - 180^\circ_x - \tau$ sequence, it dies away during the first τ interval (τ_2, τ_4 in Fig J2.15), then it recovers, to another maximum, the spin echo, when the isochromats are refocused after τ_2. Since the magnetization is now along the $+y$-axis, the echo signal is inverted with respect to the first signal observed. After the period shown as τ_3 in Fig. J2.15, the FID again decays to a minimum, because of the same $\mathbf{B}_0$ inhomogeneity effects as before. A second $180^\circ_x - \tau$ sequence results in an echo at the end of τ_4 along the $-y$-axis. The amplitude is reduced by T_2 relaxation and also by any residual spin-lattice effects in the system. The time constant for the decrease in the magnitude of the echo is the 'true' T_2, since the effects of $\mathbf{B}_0$ inhomogeneity were eliminated by the refocusing process (Fig. J2.16).

With small modifications, the spin-echo sequence can provide some very useful structural information. The simplest example is the attached proton test, which allows the determination of whether each carbon in a molecule has an odd or an even number of protons attached to it.

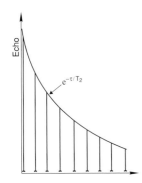

Fig. J2.16 The exponential decay of the spin echoes gives a transverse relaxation time T_2^*

J2.3.3 Polarization transfer

Another important group of multiple-pulse experiments includes the *polarization transfer* methods for sensitivity enhancement. A nucleus such as ^{13}C or ^{15}N suffers

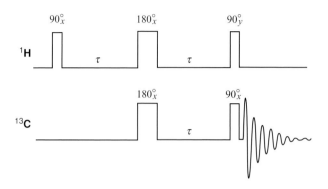

Fig. J2.17 Pulse sequence for INEPT experiments.

from poor sensitivity because of its low natural abundance (Table J1.3) and also because of its small gyromagnetic ratio, γ, which, for a given $\mathbf{B}_0$, determines the energy difference between the two spin states (Eq. (J1.6)). Since the gyromagnetic ratio of the proton is very nearly four times that of ^{13}C, the population difference is in the same proportion (Table J1.2). If some of the polarization of the proton can be transferred by some means to a less sensitive nucleus, then the signal of the latter will be enhanced. One pathway for polarization transfer is the nuclear Overhauser effect (NOE), which operates *via the same through-space dipolar interactions* that are mostly responsible for spin-lattice relaxation in the ^{13}C–^{1}H system. In contrast, polarization transfer pulse sequences were developed to achieve the desired result solely *via spin coupling effects.*

Conceptually one of the simplest experiments of this type uses INEPT, which is shown in Fig. J2.17 (Comment J2.8). The initial 90°_x preparation pulse non-selectively rotates the vectors for all of protons resonances to the $-y$-axis (Comment J2.9). During the first τ interval vectors rotate in the xy plane as a result of chemical shift and spin–spin coupling. However, only the precession due to spin coupling must be considered, because the $\tau-180^\circ-\tau$ segment refocuses

---

Comment J2.9 Abbreviated form for the INEPT sequence

In customary abbreviated form the INEPT sequence is

for ^{1}H nuclei ↓ equilibration delay $- 90^\circ_x-\tau-180^\circ_x-\tau-90^\circ_y$
for ^{13}C nuclei ↓ equilibration delay $- 180^\circ_x-\tau-90^\circ_x-$acquisition

Refocusing of the rotation due to spin coupling is prevented by the application of simultaneous 180° pulses to both coupled nuclei. The spin-echo sequence removes the rotation due to chemical shift, but the exchange of labels means that spin-coupling effects are not refocused and persist in the spectrum. This result is used in many multiple-pulse methods to separate chemical-shift- and spin-coupling-induced precession.

Comment J2.8 Polarization transfer

Polarization transfer uses a suitable pulse combination to impart the greater equilibrium polarization of protons to a coupled nucleus with a smaller magnetogyric ratio. It is an alternative to NOE for sensitivity enhancement.

Selective polarization transfer is polarization transfer experiments in which only one signal in the spectrum is enhanced.

Fig. J2.18 Vector description of the spin system in INEPT experiments.

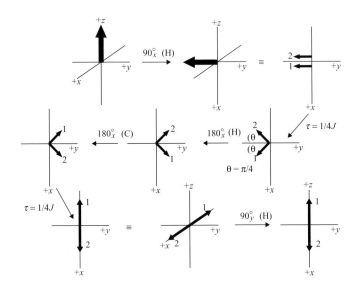

Fig. J2.18 Vector description of the spin system in INEPT experiments.

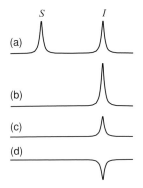

Fig. J2.19 Schematic presentation of NOE. A molecule contains two inequivalent spins, I and S, with no scalar coupling, so that the NMR spectrum consists of a singlet at each of the chemical shifts. (a) Conventional spectrum of two neighbouring spins S and I. (b)–(d) Possible spectra resulting from saturation of S resonance: the I peak gets stronger (b), weaker (c), or inverts (d) depending on the conditions.

dispersion of the spin vectors arising from the chemical shift difference. One essential feature of the experiment is the length of the τ delay, which is made equal to $1/(4J_{C\text{-}H})$. This means that at the end of the first τ interval the two vectors have rotated 90° apart from one another, shown as $+$ and $-45°$ from the $-y$-axis in the Fig. J2.18. The protons are then subjected to a 180°_x pulse, which ensures that chemical shifts effects are refocused. However, because of the second part of the sequence, the usual spin echo at the $+y$-axis is not observed. Simultaneously with the proton 180°_x pulse, a 180°_x pulse applied to the carbons causes the labels of the spin states of the protons to be preserved. Protons that were formerly bonded to carbons in an α spin state now find themselves coupled to ^{13}Cs with β spins, and vice versa. This means that the rotation direction of the vectors is also switched; the counterclockwise-rotating vector becomes the clockwise-rotating one, etc. Instead of meeting on the point the y-axis at the end of the second $1/(4J_{C\text{-}H})$ interval, the vectors arrive 180° out of the phase along $+$ and $-x$-axes. The 90°_x pulse on the protons rotates the antiphase vectors to the z-axis, and the polarization transfer to the coupled carbons is produced. The immediate 90°_x pulse rotates the carbon magnetization to the $-y$-axis, and the FID is obtained in the usual manner.

J2.4 Nuclear Overhauser Enhancement (NOE)

NOE is illustrated for a two-spin system in Fig. J2.19. A saturating radio-frequency field is applied to the high-γ spins S. The resulting population redistribution leads to a polarization enhancement of the I spins, provided the relaxation processes are favourable. This transfer of polarization is called NOE.

In one-dimensional NMR, NOE can be observed in a number of ways, all of which involve the application of a selective radio-frequency pulse at the position of one of the resonance peaks in a system containing two or more. Quantitatively, the NOE is expressed in terms of the relative increase in signal intensity

$$NOE = (I - I_0)/I_0 \qquad (J2.7)$$

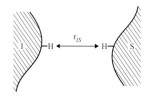

Fig. J2.20 The basis of NOE experiments. r_{IS} is the distance between the protons I and S.

where I and I_0 are the intensities with and without the enhancement due to the Overhauser effect.

The principle of NOE is relatively simple (Fig. J2.20). Consider a system of two protons in a molecule, the resonance peak of one of which is saturated as described above. NOE is a consequence of the modulation of dipole–dipole coupling between the two spins by the molecular Brownian motion. The equation describing NOE has the general form

$$NOE \propto f(r_{IS})f(\tau_c) \qquad (J2.8)$$

where $f(r_{IS})$ is a function of the distance r_{IS} between the protons and $f(\tau_c)$ is a function of the molecular rotational correlation time τ_c (Section D2.3).

First, we discuss the $f(r_{IS})$ term. This term is a cross-relaxation effect due to the dipole–dipole interaction

$$\left\langle \frac{1}{r^3} \times \frac{1}{r^3} \right\rangle = \left\langle \frac{1}{r^6} \right\rangle$$

A relationship between the observed intramolecular NOE and the sixth power of the internuclear distance was confirmed from studies of single proton–proton interactions and of the interaction between a proton and the protons of a methyl group in a series of molecules related to alkaloids,

$$f_I(S) = 1 \Big/ \left(2 + A r_{IS}^6 \right) \qquad (J2.9)$$

We recall that r_{IS} is the separation between proton I (whose resonance peak is observed) and proton S (whose resonance peak has been saturated) and A is a constant.

It follows from Eq. (J2.9) that NOE is applicable only for very close neighbours, typically, nuclei closer than 5 Å to each other. The short range of the effect limits its applications, but provides great specificity for the assignment of proton NMR peaks (Section J2.5).

The maximum value of NOE for the heteronuclear case is given by

$$NOE_{max} = 0.5 \gamma_S/\gamma_I + 1 \qquad (J2.10)$$

where γ_S/γ_I is the ratio of gyromagnetic ratios of the saturated and observed nuclei. The maximum NOE is 4.0 if protons are irradiated while ^{13}C nuclei are observed. For nuclei with negative gyromagnetic ratios such as ^{15}N (Table J1.2), irradiation of protons causes a negative NOE, leading to partial or complete loss of the signal.

Fig. J2.21 Direct cross-relaxation between spins 1 and 2, and 1 and 3 (solid arrows) and spin diffusion pathway from spin 1 via spin 2 to spin 3 (broken arrow).

Fig. J2.22 Dependence of NOE on the product of the Larmor frequency ω_0, and the rotational correlation time τ_c. calculated according Eq. (J2.8) on a double logarithmic scale.

Spin diffusion is the cause of one of the fundamental difficulties that arise when trying to correlate NOE intensities with intramolecular distances. In a network of like-spins contained in a macromolecule tumbling with a correlation time τ_c such that $\omega\tau_c > 1$, spin diffusion by two or several subsequent cross-relaxation steps can greatly influence the observed NOE. In the simple example of three spins shown in Fig. J2.21, a two-step pathway for cross-relaxation, spin 1 to spin 2 followed by spin 2 to spin 3, may under certain experimental conditions be more efficient than direct cross-relaxation between spin 1 and spin 3. The NOE on spin 3 is then no longer a faithful manifestation of the internuclear distance $r_{1,3}$. In the spatial structure of proteins and nucleic acids, the geometric arrangement of hydrogen atoms usually allows for a variety of spin diffusion pathways in addition to direct cross-relaxation between distinct groups of protons. Next, we discuss the rotational correlation time dependence of NOE, i.e. $f(\tau_c)$ in Eq. (J2.8). Theory shows that for two dipolar protons

$$\text{NOE} = \frac{5 + \omega^2\tau_c^2 - 4\omega^4\tau_c^4}{10 + 23\omega^2\tau_c^2 + 4\omega^4\tau_c^4} \tag{J2.11}$$

Equation (J2.11) shows that the NOE phenomenon is intimately related to spin relaxation. Analogously to the spin relaxation times T_1 and T_2, the NOE varies as a function of the product of the Larmor frequency ω_0, and τ_c (Fig. J2.22). Rotational correlation times are of the order of $10^{-10}-10^{-12}$ s for small molecules, while in the range of field strengths used in NMR spectrometers, Larmor frequencies are in the range $3.6 \times 10^8-3.6 \times 10^9$ rad/s. Thus, $\omega_0\tau_c$ is about $0.1-0.01$. The correlation times, however, become longer with increasing molecular mass and solvent viscosity (see Section D2.3.3).

We see from Fig. J2.22 that as long as $\omega\tau_c < 0.1$, the term in $\omega^2\tau_c^2$ contributes very little and the NOE is close to $+ 0.5$. When $\omega\tau_c = 10$ or more, NOE approaches a limit of -1, which corresponds to the disappearance of the I signal. The change from positive to negative NOE occurs at $\omega\tau_c = 1.118$.

An example of changing the sign of NOE in the same spin system by changing the Larmor frequency is shown in Fig. J2.23. The 90 and 250 MHz spectra of valinomycin in dimethylsulphoxide solution are plotted with and without irradiation on the D-Val-NH resonance (S). The effect on the hydroxyvaline α-proton (I) is positive at 90 MHz, and negative at 250 MHz.

The four curves in Fig. J2.24 describe the maximum NOEs for four nuclear spins (^{13}C, ^{31}P, ^{15}N, ^{1}H) interacting with ^{1}H, if the preirradiation is on ^{1}H and the relaxation of the observed spin is entirely by dipole–dipole coupling with the preirradiation proton. For ^{1}H$-^{13}$C and ^{1}H$-^{31}$P the NOE is positive over the entire τ_c range and becomes very small for long τ_c. For ^{15}N$-^1$H NOE is negative throughout because of the negative value of γ (Table J2.1).

Fig. J2.23 Demonstration of positive NOE at 90 MHz and negative NOE at 250 MHz for the D-hydroxyvaline α-proton in valinomycin in deutero, dimethylsulphoxide solution: (a), (d) normal spectra; (b),(e) spectra with irradiation of the D-Val NH resonance; (c), (f) difference spectra. (After Pitner *et al.*, 1976.)

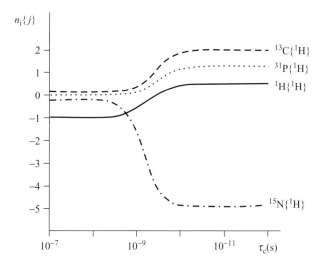

Fig. J2.24 Plots of the maximum NOE (Eq. (J2.11)) versus log τ_c for ^1H, ^{13}C, ^{15}N, and ^{31}P interacting with ^1H: $B_0 = 11.74$ T, preirradiation on 1H; it is assumed that the relaxation is entirely by dipole–dipole coupling with preirradiated proton. (Wutrich, 1986.)

J2.5 Two-dimensional NMR

The term one-dimensional NMR refers to experiments in which the transformed signal is presented as a function of a single frequency. By analogy, in a two-dimensional NMR spectrum the coordinate axes correspond to two frequency domains.

Fig. J2.25 Basic scheme for two-dimensional time-domain spectroscopy, with four distinct intervals leading to a time domain signal $s(\tau_p, t_1, \tau_m, t_2)$. In standard experiments τ_p, τ_m are fixed values and the time-domain signal is a function of only two parameters, t_1, t_2.

Preparation	Evolution	Mixing	Detection
τ_p	t_1	τ_m	t_2

The general scheme of two-dimensional NMR spectroscopy is demonstrated in Fig. J2.25. There are four successive time periods: *preparation, evolution, mixing* and *detection*. The preparation period usually consists of a delay time, during which thermal equilibrium is attained. Following this period, the spin system is prepared to an initial out-of-equilibrium state. Normally this is done by a 90° radio-frequency pulse as shown in Fig. J2.7. Coherence means that an ensemble of like spins in the sample have the same phase in the $x'y'$ plane (see Fig. J1.17).

Unlike one-dimensional NMR, two-dimensional NMR is characterised by the introduction of the *evolution* and *mixing* periods. The former defines a time variable t_1. Signals from the evolution and detection time periods, t_1 and t_2, respectively, are mixed in the time period, τ_m. The data acquisition of NMR signals is performed only during the detection period t_2. By varying the delay time t_1 systematically, while keeping other experimental conditions unaltered, we obtain a matrix type of data set $s(t_1, t_2)$. Double Fourier transformation yields a two-dimensional spectrum frequency $S(\omega_1, \omega_2)$.

The general scheme of two-dimensional NMR is very flexible (Fig. J2.25), and can be adapted to the type of sample, and to the different parameters to be measured. Different classifications have been proposed for two-dimensional NMR experiments. Wuthrich proposed a classification into two groups (Comment J2.10). Below we describe the classification into three groups, according to Ernst:

(1) Experiments designed to *correlate* transitions of coupled spins by transferring transverse magnetization or multiple-quantum coherence from one transition to another in the course of a suitably designed mixing process. This type of experiment is called correlation spectroscopy known under the acronym (COSY).

(2) Experiments designed to *separate* different interactions (e.g. chemical shifts and spin–spin couplings) in orthogonal frequency dimensions, with the purpose of resolving one-dimensional spectra by spreading overlapping resonances in a second dimension. These experiments require conditions such that the spectra in the evolution and detection periods contain different information. The method is called homonuclear or heteronuclear two-dimensional *J*-resolved spectroscopy or spin-echo spectroscopy.

(3) Nuclear Overhauser enhanced spectroscopy (NOESY), which is concerned with the study of *dynamic processes* such as chemical exchange, cross-relaxation, or transient Overhauser effects.

After Ernst proposed this classification, the transverse relaxation-optimised spectroscopy (TROSY) method was developed. TROSY suppresses transverse nuclear

---

Comment J2.10

On the classification of two-dimensional NMR experiments

According to Wuthrich it is useful to classify two-dimensional NMR experiments into two groups:
(1) experiments for delineating *through-bond*, scalar spin–spin connectivity, such as *correlated NMR spectroscopy (COSY)*, and
(2) experiments for delineating through-space, dipolar spin–spin connectivity; such as *nuclear Overhauser spectroscopy (NOESY)*.

spin relaxation, which is the direct cause of the deterioration of NMR spectra of large molecular structures. The combination of TROSY and CRINEPT
(Section J2.3) allows the collection of high-resolution spectra from structures with
molecular masses > 100 kDa, significantly extending the range of macromolecular systems that can be studied by NMR in solution.

J2.5.1 Correlation spectroscopy (COSY)

The COSY technique is one of the simplest, and yet most useful, of the various
two-dimensional NMR techniques. In fact COSY was the first two-dimensional
NMR experiment to be described. The simplest COSY experiment is shown in
Fig. J2.26. The pulse sequence of the type $90°-kt_1-90°$-data acquisition, where
$k = 0, 1, 2, \ldots, 2^n$. The experiment is repeated for each k value. FID signals are
measured and plotted as $s(t_1, t_2)$, where t_1 and t_2 are independent time variables
corresponding to the two sampling times. Digital values of t_1 and t_2 are taken in
order to apply a fast Fourier transformation algorithm to obtain two-dimensional
spectra $S(\omega_1, \omega_2)$ in frequency space.

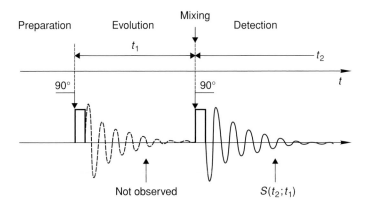

Fig. J2.26 The COSY pulse sequence.

Before describing the theoretical and practical aspects of this important technique in detail, it is worthwhile demonstrating in a purely classical manner how
a two-dimensional spectrum can result from this sequence. Consider two protons A and B, which are scalar coupled. A simple pulse and collect sequence,
followed by Fourier transformation, generates an NMR spectrum as shown dia-
grammatically in Fig. J2.27. We use the terms $\omega_A, \omega_B, J_{AB}$, in obvious notation to

Fig. J2.27 Fourier transformation of the FID resulting from the application of a radio-frequency pulse to a two-spin system results in the NMR spectrum.

Fig. J2.28 Schematic illustration of the contour plot of an ^{1}H–^{1}H COSY spectrum of two coupled spins. The filled circles represent diagonal peaks, whereas the open circles represent off-diagonal peaks or cross-peaks. The dashed line illustrates how cross-peaks correlate diagonal peaks derived from scalar coupled spins. (Adapted from Homans, 1989.)

Comment J2.11

2, 3-dibromo-thiophene

Chemical formula:
$C_4H_2Br_2S$

Structural formula:

---

denote the Larmor frequencies of spins A and B and their scalar (J) coupling. In essence, spin A has precessed with a Larmor frequency of ω_A during t_1, and spin B has precessed with a Larmor frequency of ω_B during t_1, although in fact two frequencies are involved for each spin ($\omega_A \pm \frac{1}{2} J_{AB}$ and $\omega_B \pm \frac{1}{2} J_{AB}$) due to the spin–spin coupling. Now consider what would happen if we were to use the pulse sequence of Fig. J2.28. Let us postulate that we can record data during both the t_1 and t_2 intervals. After 90° pulse and the delay t_1 the situation is analogous to that of Fig. J2.27, i.e. each proton resonates with its characteristic Larmor frequency during t_1. After the second 90° pulse, the situation becomes more complicated. The second pulse may appear to have no effect on the spins, i.e. they continue to resonate at ω_A and ω_B during t_2 just as they did in t_1. Alternatively, a portion of the magnetization associated with spin A during t_1 may transfer to spin B during t_2. This result derives from the quantum mechanical process known as *coherence transfer*. In an analogous manner, a proportion of the magnetization which precessed at ω_B during t_1 now precesses at ω_A during t_2. Now, since we are observing the NMR spectrum with respect of two time periods it follows that we must employ a two-dimensional Fourier transform to observe the frequency components described above. This means that we may display the spectrum in two orthogonal dimensions on a plane. Such a display is shown in Fig. J2.28 for the two-spin case described above.

The real spectrum obtained for 2,3-dibromothiophene (Comment J2.11) with a pulse sequence like the one in Fig. J2.26 is shown in Fig J2.29. Two types of two-dimensional peaks, spread with non-vanishing intensities, can be seen. Eight peaks around the diagonal line (the dashed straight line in Fig. J2.29) on the spectra result from the evolution of the magnetization which has transition frequencies originating from spin flips of the same spins in both transitions. The other eight peaks off the diagonal line correspond to the spin flips belonging to different coupled spins in the two transitions. The latter eight peaks are very important in this two-dimensional NMR method since the appearance of these peaks just indicates that there is scalar coupling between two nuclei.

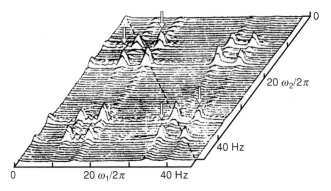

Fig. J2.29 The two-dimensional spectrum of 2,3-dibromothiophene (an *AX* spin system), obtained by applying the pulse sequence in Fig. J2.26.

(a)

HDO

10   8   6   4   2   0 δ(ppm)

(b)

ω_1 (ppm)

—0
—2
—4
—6
—8
—10

10   8   6   4   2   0 ω_2(ppm)

(c)

ω_1 (ppm)

—0
—2
—4
—6
—8
—10

10   8   6   4   2   0 ω_2 (ppm)

a b c d e f g h

Fig. J2.30 ^{1}H COSY spectrum of a D_2O solution of inhibitor K (0.01 M, pD 3.4, 25 °C, 360 MH): (a) the one-dimensional ^{1}H NMR spectrum; (b) stacked plot representation of the COSY spectrum; (c) contour plot of the COSY spectrum. (After Wutrich, 1986.)

These cross-peaks are utilised in protein NMR to elucidate the coupling of various proton species even in a complicate spectrum where many resonance lines overlap.

In a second example, Fig. J2.30(a) illustrates the one-dimensional NMR and the COSY spectra for a small 57-amino-acid residue protein, the protease inhibitor K. The two-dimensional spectrum can be plotted in different ways (Comment J2.12). A *stacked plot* (Fig. J2.30(b)) conveniently shows the peak heights in the third dimension, while a *contour* plot (Fig. J2.30(c)) is more convenient to show peak positions in the two-dimensional plane. In the stacked plot of the ^{1}H COSY spectrum (Fig. J2.30(b)), the complete one-dimensional NMR spectrum (Fig. J2.30(a)) can be recognised on the diagonal from the upper right to the lower left. For example, the highest field methyl resonance at -0.9 ppm has a coordinates ($\omega_1 = -0.9$ ppm, $\omega_2 = -0.9$ ppm), and the lowest field amide-proton line at 10.3 ppm is at ($\omega_1 = 10.3$ ppm, $\omega_2 = 10.3$ ppm). The arrangement of the cross-peaks is best seen in the contour plot (Fig. J2.30(c)). The spectrum is *symmetrical* with respect to the diagonal. Using the empirical rule that COSY cross-peaks are with few exceptions observed only between protons separated by three or less covalent bonds in the amino acid structure allows one to outline the regions a–h (Fig. J2.30(c), below the diagonal) between special protons in amino acids.

J2.5.2 Nuclear Overhauser enhanced spectroscopy (NOESY)

The NOESY pulse sequence is similar to that for proton spin systems in COSY, with an additional third 90° pulse (Fig. J2.31). The resulting sequence is: $90°-t_1-90°-\tau_m-90°-t_2$–acquisition.

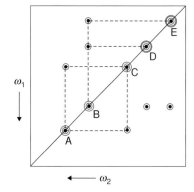

Fig. J2.31 The top trace shows the pulse sequence for two-dimensional NOESY. The bottom trace shows a contour plot of a schematic two-dimensional NOE spectrum. For the five resonance lines A–E, spin proximity is manifested by the NOE cross-peaks between A and C, B and D, B and E. (Adapted from Kumar *et al.*, 1980.)

As in COSY, the equilibrium delay and initial $90°$ pulse prepare the nuclei by turning z-axis magnetization to the xy plane. During the variable t_1 evolution period the nuclei are frequency-labelled by their chemical shift values. The second $90°$ pulse rotates part of the magnetization back to the z-axis. During the τ_m interval (which is usually kept constant) the longitudinal magnetization is allowed to relax. This produces a mixing of the magnetization of nuclei that are related by the dipolar relaxation mechanism, thus correlating their chemical shifts. The third $90°$ pulse is needed to rotate the vectors back to the xy plane, where they can be detected during the t_2 acquisition period.

The general features of a two-dimensional NOESY spectrum are outlined in the lower part of Fig. J2.31. Magnetization components which do not exchange with other components during the mixing time τ_m maintain their frequencies during t_1 and t_2. Hence the corresponding peaks in the two-dimensional spectrum lie on the diagonal. An exchange of magnetization between two components due to dipolar coupling during the mixing period is manifested by cross-peaks. Peak A is dipole–dipole coupled with peak C, and peak B with D and E. The two-dimensional NOESY spectrum of BPTI is plotted in Fig. J2.32. It shows,

Fig. J2.32 Contour plot of a proton two-dimensional NOESY spectrum at 360 MHz for BPTI. The protein concentration was 0.02 M, solvent ^2H_2O, pH 3.8, $T = 18\ °C$. The mixing time was 100 ms. The total accumulation time was 18 h. Cross-relaxation connectivities for selected amino acid residues are indicated by the broken lines (see text). Connected peaks are identified by the one-letter symbol for amino acids (A = alanine, T = threonine, C = cysteine, Q = glutamine, F = phenylalanine, X = tyrosine), the position in the amino acid sequence and the type of protons observed. Bands of intense signals extend from the water line at 4.85 ppm parallel to both the ω_1 and ω_2 axes. (After Kumar et al., 1980.)

for example, that the backbone amide proton of Glu 31 is connected with the α-proton of Cys 30, and the proximity of the 3.5 ring protons of Tyr 23 and the α- and methyl protons of Ala 25. Fundamentally, the two-dimensional NOESY spectrum manifests all possible proton combinations in the protein. A cross-peak indicates that the proton pair is separated by less than about 5 Å, while the absence of a peak is indicative of longer distances. A NOESY spectrum contributes in an important way to macromolecular structure resolution by NMR (Section J3.2).

J2.5.3 Transverse relaxation-optimised spectroscopy (TROSY)

During the past 20 years, the highest magnetic field available for NMR has increased in steps corresponding to proton resonance frequencies of 400, 500, 600, 750, 800 and now 900 MHz. Each of these steps benefitted NMR in structural biology, through improved intrinsic sensitivity and peak separation. For commonly used heteronuclear experiments, however, the advantages of higher magnetic fields are offset partly by field-dependent line broadening due to increased transverse relaxation rates. TROSY was developed to overcome this limitation. The TROSY pulse sequence for two-dimensional $^{15}N-^1H$ correlation spectroscopy is shown in Fig. J2.33.

In this technique, one first creates proton magnetization. This is then transferred to ^{15}N via a polarization transfer element. After 'frequency labelling' of the ^{15}N magnetization during the evolution period, t_1, magnetization is transferred back to ^1H via a reverse polarization transfer element and then detected on ^1H during the acquisition period, t_2. Transverse relaxation (Section J2.3) occurs throughout multidimensional NMR experiments (i.e. during polarization transfer periods as well as during frequency labelling periods. The rate of transverse nuclear spin relaxation increases with molecular mass and has a dominant impact

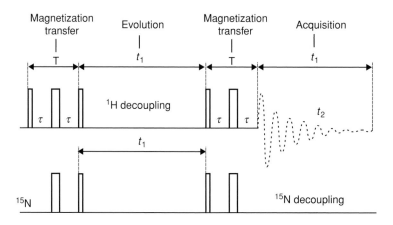

Fig. J2.33 The basic pulse sequence of two-dimensional $^{15}N-^1H$ correlation spectroscopy experiments.

on the upper size limit for macromolecular structures that can be studied by NMR in solution.

In conventional heteronuclear NMR experiments, magnetization is transferred between the different types of nuclei via scalar spin–spin couplings in so-called INEPT transfers (Section J2.3). The time periods T required for the INEPT transfers can be comparable to the mean duration of the evolution period (Fig. J2.33). CRINEPT overcomes this limitation by combining INEPT with cross-correlated relaxation-induced polarization transfer CRIPT, which becomes a highly efficient transfer mechanism for molecular sizes above 200 kDa in aqueous solution at ambient temperature (Section J2.6). Furthermore, in contrast to INEPT, TROSY is active during the CRINEPT transfer periods.

Technically, the TROSY approach has the following basis. In heteronuclear two-spin systems, such as $^{15}N-^1H$ and aromatic $^{13}C-^1H$, the NMR signal of each nucleus is split into two components by the scalar spin–spin coupling. Therefore, in two-dimensional correlation experiments a four-line fine structure is observed (Fig. J2.34(b)) (see also the example in Fig. J2.35). With the advent of modern multidimensional NMR, this four-line pattern has routinely been collapsed into a single, centrally located line using broad-band decoupling (Fig. J2.34(a)) during the evolution and acquisition periods to obtain a simplified spectrum and improved sensitivity. However, at high magnetic fields, chemical shift anisotropy (CSA) of ^1H, ^{15}N and ^{13}C nuclei can be a significant source of transverse relaxation, in addition to the omnipresent relaxation as a result of dipole–dipole coupling. TROSY exploits constructive interference between dipole–dipole coupling and CSA relaxation, and actually uses CSA relaxation at higher fields to cancel field-independent dipolar relaxation. Using the TROSY technique, the multiplet

Fig. J2.34 Contour plots of a tryptophan indole $^{15}N-^1H$ cross-peak from different types of $^{15}N-^1H$ correlation spectra: (a) conventional broad-band decoupled [$^{15}N-^1H$]-COSY spectrum; (b) the same as (a), without decoupling the evolution and detection periods (see Fig. J2.33). (c) [$^{15}N-^1H$]-TROSY spectrum consisting of the sharp component. (After Wider and Wuthrich, 1999.)

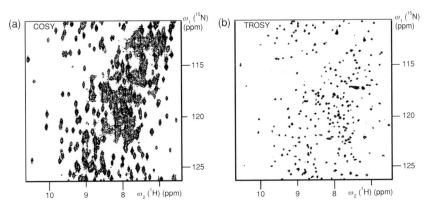

Fig. J2.35 A comparison of the $^{15}N-^1H$ correlation spectra of a protein of molecular mass 45 kDa recorded using: (a) a conventional procedure (COSY) and (b) TROSY. Both spectra were measured at a proton resonance frequency of 750 MHz, using a 0.8 mM sample of uniformly ^{15}N- and ^2H-labelled gyrase-45 from *Staphylococus aureus* in water at 25 °C and pH 8.6. (Adapted from Wider and Wuthrich, 1999.)

structure is not decoupled and only the narrowest, most slowly relaxing line of each multiplet is retained (Fig. J2.34(c)).

A comparison of the ^{15}N–^{1}H correlation spectra of a protein with molecular mass of 45 kDa recorded by using COSY and TROSY, respectively, is presented in Fig. J2.35. The figure demonstrates the essential improvement in quality obtained by the TROSY technique.

Finally we would like to point out that the implementation of [^{15}N–^{1}H]-TROSY in triple resonance experiments results in a several-fold improved sensitivity for ^{2}H/^{13}C/^{15}N-labelled proteins and approximately two-fold sensitivity gain for ^{1}H/^{13}C/^{15}N-labelled proteins. By applying TROSY, the spectra have been obtained of proteins of molecular mass close to 100 kDa.

J2.6 Multi-dimensional, homo- and hetero-nuclear NMR

By analogy with two-dimensional NMR the coordinate axes in three- and four-dimensional NMR spectra correspond to three and four frequency domains, respectively. Figure J2.36 summarises the relationship between two-, three- and four-dimensional NMR pulse sequences.

We recall that two-dimensional experiments all comprise the same basic scheme: namely a preparation pulse (P), followed by an evolution period ($E(t_1)$), during which the spins are labelled according to their chemical shifts, a mixing period (M), during which the spins are correlated with each other; and finally a detection period ($D(t_2)$).

A three-dimensional pulse scheme simply combines two two-dimensional pulse sequences, leaving out the detection period of the first and the preparation pulse of the second (Fig. J2.36). The two evolution periods, t_1 and t_2, are incremented independently, and are followed by a detection period t_3. Further extension to a fourth dimension is easily implemented by combining three two-dimensional schemes, leaving out the preparation pulses of the second and third experiments and the detection period of the first and second (Fig. J2.36).

The first three-dimensional experiments performed on proteins were of the ^{1}H homonuclear variety, in which a scalar correlation pulse scheme was combined with a NOESY sequence. Despite the elegance of the method, the applicability of ^{1}H homonuclear three-dimensional experiments is limited to molecular masses less than or equal to about 10 kDa, line widths becoming too wide for larger proteins. A more useful approach in three- and four-dimensional NMR employs

Fig. J2.36 Relationship between the pulse sequences for recording two-, three-, and four-dimensional NMR spectra. Abbreviations are: P, preparation; E, evolution; M, mixing; and D, detection. (Adapted from Clore and Gronenborn, 1991.)

2D NMR: $P_a \rightarrow E_a(t_1) \rightarrow M_a \rightarrow D_a(t_2)$

3D NMR: $P_a \rightarrow E_a(t_1) \rightarrow M_a \rightarrow E_b(t_2) \rightarrow M_b \rightarrow D_b(t_3)$

4D NMR: $P_a \rightarrow E_a(t_1) \rightarrow M_a \rightarrow E_b(t_2) \rightarrow M_b \rightarrow E_c(t_3) \rightarrow M_c \rightarrow D_c(t_4)$

uniformly (>95%) labelled ^{15}N and/or ^{13}C proteins (see Section J2.8.1). The larger heteronuclear couplings can be resolved more efficiently to allow the study of molecular masses up to about 25 kDa.

As an illustration, Fig. J2.37 shows a three-dimensional ^{15}N-correlated [^{1}H,^{1}H] NOESY spectrum. The third axis corresponds to the NMR frequencies of the ^{15}N. As a result, the same number of NMR peaks that would have been observed in a two-dimensional [^{1}H,^{1}H] NOESY spectrum are now distributed amongst a number (typically 64 or 128) of ^{1}H,^{1}H ones (Fig. J2.38). In addition to the ^{1}H NMR data, the approach provides ^{13}C and ^{15}N NMR information, which may be used to support the structure determination and to provide supplementary data on dynamic features of the molecule studied. Nonetheless, the key purpose of heteronuclear NMR experiments is to obtain the maximum possible number of individually assigned ^{1}H–^{1}H NOE peaks in the system studied.

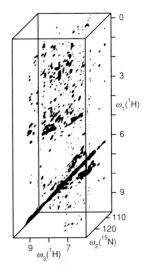

Fig. J2.37 Three-dimensional ^{15}N-correlated [^{1}H,^{1}H] NOESY spectrum of oxidised *E. coli* glutaredoxin (protein concentration = 1.5 mM, solvent = 90% H$_2$O/10% D$_2$O, pH = 7.0, T = 301 K, ^{1}H frequency = 500 MHz, mixing time = 100 ms). The protein was uniformly labelled with ^{15}N to the extent of $\geq$ 95%. (After Sodano et al., 1991.)

Fig. J2.38 Scheme illustrating the improved peak separation in a three-dimensional heteronuclear-resolved [^{1}H,^{1}H] NMR experiment (a), when compared with the corresponding two-dimensional [^{1}H,^{1}H] NMR experiment (b). The two spectra contain the same number of peaks. In the three-dimensional spectrum these are distributed among multiple ω_1(^{1}H)ω_3(^{1}H) planes that are separated along the heteronuclear chemical-shift axis, ω_2(X). (Otting & Wuthrich, 1990.)

J2.7 Sterically induced alignment

Towards the end of the 1990s, a new class of experiment was reported that dramatically changed the range of applicability of NMR structural methods. It is based on anisotropic magnetic interactions that were not normally observable in high-resolution liquid-state NMR spectra. The experiments provide structural long-range constraints that are *orientational*, rather than *distance*-based. They rely on the measurement of residual dipolar couplings, and, in some cases, CSA, under particular conditions. Powerful applications include structural studies of multidomain proteins, macromolecular complexes, and nucleic acids. The global orientational information contained in dipolar couplings ideally complements the distance information contained in NOE peaks. Moreover, dipolar coupling provides a tool for evaluating the quality of NMR structures in an objective manner.

Fig. J2.39 Magnetic dipole–dipole coupling, illustrated for a $^{15}N-^1H$ spin pair. ^{15}N and ^1H magnetic moments are aligned parallel (or antiparallel) to the static magnetic field $\mathbf{B}_0$. The total magnetic field in the $\mathbf{B}_0$ direction at the ^{15}N position can increase or decrease relative to $\mathbf{B}_0$, depending on the orientation of the $^{15}N-^1H$ vector and the spin state of the proton (parallel or antiparallel to $\mathbf{B}_0$). (Bax, 2003.)

J2.7.1 Residual dipolar couplings

Dipolar couplings are potentially large interactions caused by the magnetic flux lines of the nucleus affecting the static magnetic field at the site of another nucleus (Fig. J2.39). Only the component parallel to the external much stronger magnetic field leads to the interaction. Components orthogonal to $\mathbf{B}_0$ have a negligible effect on the total vector sum of the external and dipolar field. The z component of the dipolar field of nucleus I changes the resonance frequency of nucleus S by an amount that depends on the internuclear distance r and on the angle θ between the internuclear vectors $I-S$ and the direction of the magnetic field $\mathbf{B}_0$ (Fig. J2.40).

A dipolar interaction D_{IS} between nuclei I and S is described by

$$D_{IS} = \text{const} \frac{\gamma_I \gamma_S h \mu_0}{4\pi^2 r_{IS}^3} \left\langle \frac{(3\cos 2\theta - 1)}{2} \right\rangle \tag{J2.12}$$

where the γ_i are the gyromagnetic ratios, r_{IS} the internuclear distance and the symbol $\langle \rangle$ denotes a time average. Equation (J2.12) shows that the dipolar interaction, D_{IS}, provides direct information on the angle θ, i.e. on the orientation of the internuclear vector, and that it scales with the inverse of the cubed internuclear distance.

In an isotropic solution, rotational Brownian diffusion rapidly averages out the internuclear dipolar interaction of Eq. (J2.12) to zero. As a result, a solution NMR spectrum shows narrow resonance lines, which can be assigned to individual nuclei in the protein, but which no longer contain valuable orientational information.

The properties of scalar and dipolar couplings are summarised in Table J2.2.

Most biological macromolecules display some degree of anisotropic magnetic susceptibility that causes them to align in a strong magnetic field and leads to anisotropic Brownian motion (tumbling). For example, additional structural parameters were derived from the $^1H-^{15}N$ dipolar coupling measured in

Fig. J2.40 Dipolar coupled $^{15}N-^1H$ spin pair in an amide bond. The bond length, r, is assumed fixed and the primary variable is the angle, θ, between the magnetic field, $\mathbf{B}_0$, and the internuclear vector. (Hansen *et al.*, 1998.)

Table J2.2. *Comparative properties of scalar and dipolar couplings*

Scalar couplings	Dipolar couplings
Through bonds	Through space
Strength ≤100 Hz	Strength depends on distance and nucleus (e.g. for N–H pair ∼ 20 kHz)
Isotropic	$3\cos^2\theta - 1$
Small for >3 bonds	$1/r^3$
Produce splittings	Splittings when motion restricted

Comment J2.13   Bicelles

Bicelles are fragments of lipid bilayers, a well-accepted building block of biological membranes. They were found to form liquid crystal arrays when prepared at 20–30 weight% lipid to aqueous buffer using mixtures of long-chain phospholipids such as dimyristoylphosphatidylcholine (DMPC) and lipids with detergent-like properties such as dihexanoylphosphatidyl-choline (DHPC). The bilayer fragments appear to be small discoidal particles a few hundred ångstroms in diameter, and have become known as bicelles (Sanders *et al.* 1994).

Fig. J2.41 Induced protein orientation by dilute phospholipid bicelles. The protein tumbles rapidly, but anisotropically, in large aqueous interbicelle spaces. (Prestegard, 1998.)

a magnetically oriented paramagnetic protein cyano methmyoglobin. However, the degree of alignment in the systems was relatively low and the largest dipolar coupling was less than 5 Hz.

Tunable degrees of alignment can be achieved in a magnetic field in the presence of an orienting agent such as planar phospholipid bicelles, rod-shape viruses and phages, purple membrane and polymeric strained gels. Methods for creating weakly aligned states also include direct magnetic alignment in electrical fields

The large magnetic susceptibility of lipid bicelles was used to achieve high degrees of alignment (Comment J2.13). At 30–40 °C bicelles orient with the normals to the bilayer surfaces perpendicular to the magnetic field. Figure J2.41 illustrates protein orientation in the presence of a dilute solution of phospholipid bicelles. Proteins and peptides become oriented by the presence of the oriented discoidal surfaces. The protein tumbles rapidly, but anisotropically, in the large aqueous interbicelle space.

The best tunable degree of macromolecular alignment has been achieved in a filamentous phage medium (Comment J2.14). The intrinsic structural properties of these phages make them an ideal system for inducing alignment of nucleic acids and proteins. These properties are: (1) rod-like phages are already fully aligned in 7-T magnetic fields (300-MHz proton frequency) so that ultrahigh magnetic fields are not required for creating aligned systems; (2) the degree of alignment

Comment J2.14   Bacteriophage Pf1

The bacteriophage Pf1 consists of a 7349-nucleotide single-stranded circular DNA genome that is packaged at a ∼1:1 nucleotide/coat protein ratio into a ∼60Å diameter by ∼20 000-Å long particle that has a negatively charged surface. The coat protein forms an α-helical structure that runs roughly parallel to the long axis of the phage. The coat proteins form a repeating network of carbonyl groups that are believed to be the source of the phage's large anisotropic magnetic susceptibility, with the long axis of the phage aligning parallel to the magnetic field. A solution of oriented bacteriophage particles forms a liquid crystalline organisation.

Fig. J2.42 1D ^{2}H spectra of a 10 mM Tris, pH 8.0, 90% H_2O/10% D_2O sample containing 0, 9, 17, 29, 41 and 58 mg ml^{-1} Pf1 phage. The vertical scales of individual spectra were adjusted so that all peaks have the same vertical heights. (After Hansen *et al.*, 1998.)

Fig. J2.43 A portion of the two-dimensional (^{15}N–^{1}H) NMR spectra collected on bovine apo-calmodulin: (a) in absence of phage and (b) in a 25 mg ml^{-1} phage solution. This region of the spectrum shows the effect of phage on the ^{1}H–^{15}N couplings for the G25 and G61 amides. (Hansen *et al.*, 1998.)

of the dissolved macromolecule can be tuned easily, in order to modulate the size of the dipolar couplings, by simply changing the phage concentration; (3) Pf1 phages are extremely stable under usual protein buffer conditions and readily produced, with high purity and in large quantities; (4) for all nucleic acids and proteins studied so far, there appears to be no effect on the rotational correlation time of the macromolecule, which means that standard high-resolution spectra are still obtained; at the concentrations used, the large phage particles lead to high macroscopic viscosity, but do not affect the microscopic tumbling rates of the individual macromolecules (see also Section J2.7.2).

The magnetic alignment of Pf1 phage solutions has been monitored by one-dimensional ^{2}H NMR. Figure J2.42 shows the ^{2}H NMR spectra of a 90% H_2O/10% D_2O solution as a function of phage concentration. The splitting of the HOD signal arises from the large deuterium quadrupole moment, which appears because it is not isotropically averaged for water molecules bound to the aligned phage particles. The observed splitting varies approximately linearly with phage concentration (up to 60 mg ml^{-1} of phage). Figure J2.42 also shows that the deuterium quadrupole splitting of the water resonance is a good indicator of the purity and aggregation state of the phage.

The observation of dipolar coupling in a weakly aligned protein is illustrated in Fig. J2.43. ^{15}N-labelled calmodulin was dissolved in 25 mg ml^{-1} Pf1 phage solution. A comparison of parts of the amide regions of the two-dimensional (^{15}N, ^{1}H) NMR spectra collected in the presence and absence of phage shows

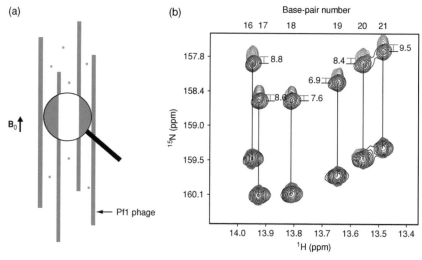

Fig. J2.44 Acquisition of residual dipolar couplings. (a) Sketch showing the steric interaction between Pf1 phage and the DNA, resulting in a slightly anisotropic environment and thus preventing the dipolar couplings from averaging to zero. B_0 is a static magnetic field. (b) Overlaid [^{15}N, ^{1}H] NMR spectra of the DNA duplex (d(GGCAAAAACGG)/d(**CCGTTTTTGCC**)). The spectra were recorded in the absence (black) and presence (red) of Pf1 phage. The bold letters represent nucleotides that are uniformly labelled with ^{13}C and ^{15}N. (After MacDonald and Lu., 2002.)

that the ^1H–^{15}N amide couplings for residues G25 and G61, are -13 ± 2 and $+8 \pm 2$ Hz, respectively.

The Pf1 phage approach appears to align RNA, DNA or protein systems by a steric mechanism, with no evidence of binding between macromolecules and phages. There are no significant changes in chemical shifts or other NMR parameters that would suggest structural changes in the macromolecule in the presence of phage. Alignment appears to be induced by collisions of the asymmetric macromolecule with the magnetically aligned phage.

Figure J2.44(a) schematically illustrates the steric interaction between Pf1 phage and the DNA duplex d(GGCAAAAACGG)/d(**CCGTTTTTGCC**). The bold letters represent nucleotides uniformly labeled with ^{13}C and ^{15}N. The residual dipolar couplings are shown in Fig. J2.44(b).

J2.7.2 Chemical shift anisotropy (CSA)

Residual dipolar coupling is not the only anisotropic spin interaction that can provide useful structural information. Chemical shifts are also anisotropic. Chemical shifts arise because nuclei in various molecular functional groups resonate at different frequencies depending on shielding by the local electronic environment. Electronic environments are seldom isotropic, and hence, shielding is different for different orientations of the functional groups. An angular dependence of the chemical shift occurs that is analogous to the one seen for residual dipole interaction.

(a)   (b)

Fig. J2.45 Segments from a proton coupled, nitrogen decoupled, ^{15}N–^1H NMR spectra of a 0.4 mM solution of a barley lectin fragment in a 5% DMPC/DHPC 3:1 bicelle (doped with a positively charged amphiphile): (a) an isotropic spectrum at 25 °C; (b) an oriented spectrum at 35 °C. The bicelle preparations have a convenient property in that an isotropic phase containing the same lipids, presumably in more symmetric micelles, is obtained by lowering the temperature slightly. Both increases and decreases in couplings are observed. (After Prestegard, 1998.)

An illustration of anisotropic chemical shift effects is given for ^{13}C of the peptide carbonyl group in Fig. J2.44; if the peptide bond has a preferred orientation with the field perpendicular 10 its plane, the carbonyl carbon would resonate at a higher field (or lower frequency) than in the isotropic case.

NMR spectra of a two-domain fragment of a carbohydrate binding protein from barley, which has been ordered in a 5% 3:1 DMPC/DHPC bicelle dispersion are shown in Fig. J2.45. Comparing isotropic (Fig. J2.45(a)) and oriented (Fig. J2.45(b)) spectra, Lys 53 shows a negative 5-Hz residual dipolar contribution and Lys 95 shows a positive 2-Hz contribution. These are indicative of an average N–H vector orientation perpendicular and parallel to the magnetic field, respectively.

The results above show that residual dipolar coupling and CSA in oriented macromolecules no longer average to zero, and are readily measured. The resulting parameters provide valuable information on long-range order for structure determination.

J2.8 Isotope labelling of proteins and nucleic acids

J2.8.1 Labelling strategies for proteins

Proteins used for NMR studies can be labelled by using a number of different protocols to produce molecules with different patterns of ^{2}H, ^{13}C and ^{15}N incorporation.

Uniform or random labelling strategies result in ^{2}H incorporation throughout a protein in a roughly site-independent manner. Amino-acid-specific labels have been used since the earliest ^{1}H-based NMR studies of proteins. Specifically protonated amino acids were introduced in otherwise fully deuterated molecules by growing microorganisms on minimal media with high levels of D_2O, supplemented with fully or partially protonated amino acids. Most of the 20 natural amino acids have been successfully incorporated in this manner, either individually or in different residue. Amino-acid-specific auxotrophic strains have been employed in cases where it is necessary to use smaller amounts of the potentially expensive labelled compounds or to limit undesirable metabolic spreading of the label. Similar approaches have been used in the context of specific ^{15}N and ^{13}C labelling.

A high level of isotopic labelling at specific sites within residues can also be achieved. Uniformly ^{15}N, ^{13}C and ^{2}H-labelled proteins have been produced, in which methyl groups in alanine, valine, leucine and isoleucine were protonated.

An approach that promises to further extend studies to high-molecular-mass, single polypeptide chains is one in which a segmentally isotope-labelled protein is produced through the ligation of labelled and unlabelled polypeptides. The ligation of two protein fragments under very mild reaction conditions has been shown to be possible through a self-catalytic protein splicing process involving naturally occurring domains called inteins. The method has been used to selectively ^{15}N-, ^{13}C-label various segments of maltose-binding protein, as well as the C-terminal domain of the RNA polymerase α subunit.

The majority of isotopically labelled proteins studied by NMR have been obtained by heterologous protein expression in *E. coli*. Although bacterial expression remains the most economical method for producing labelled proteins, it is not always possible to obtain samples in this manner, particularly if there are problems with protein folding or bacterial toxicity. In addition, many eukaryotic proteins undergo post-translational modifications that do not occur in bacterial expression systems. Alternative methods of protein expression that can address these limitations were developed, using the yeast *Pichia pastoris*, Chinese hamster ovary cells and cell-free expression systems.

J2.8.2 Labelling strategies for RNA

The two significant problems encountered in the application of NMR to larger RNA molecules are spectral crowding and large line widths in the ribose region. Uniform ^{13}C and ^{15}N labelling has been applied to RNA structure determination to great advantage. However, introduction of these isotopes results in broader line widths for protons, due to the strong dipolar interaction with directly bonded ^{13}C and ^{15}N nuclei. This effect is particularly problematic for larger RNA molecules.

Deuteration offers the advantage of spectral simplification without causing broader line widths. Since large RNAs are most readily prepared by transcription with T7 RNA polymerase, the main problem is how to prepare deuterated nucleoside triphosphates (NTPs). One method is to harvest total cellular RNA from bacteria that have been grown on deuterated media, enzymatically digest the RNA to nucleoside monophosphates (NMPs), and finally enzymatically convert the NMPs to NTPs. Uniform ^{13}C, ^{15}N, and ^{13}C/^{15}N labelling of NTPs has also been accomplished by this method, by growing *E. coli* or *Methylophilus methylotrophus* on ^{13}C- and/or ^{15}N-labelled media. Chemical approaches and approaches combining chemical and biochemical steps have also been developed to produce isotopically labelled NTP (Fig. J2.46). D,L-ribose-$3',4',5',5'$-d_4

Fig. J2.46 Strategy for synthesis of deuterated nucleotides. Chemical synthesis is used to convert glycerol-d_8 (a) into D-ribose-3,4,5,5-d_4 (D-d_4-ribose, -1) (b), and then four enzymatic reactions are used to convert that ribose into the four d_4-NTPs (**2–5**) (c). (Adapted from Tolbert and Williamson, 1996.)

Fig. J2.47 500-MHz
NOESY spectrum of HIV-2
TAR RNA: (a) unlabelled
(inset is the secondary
structure of HIV-2 TAR
RNA); (b) d_4-labelled TAR
RNA. (After Tolbert and
Williamson, 1996.)

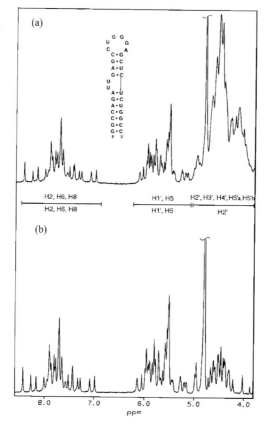

is first synthesised from glycerol-d_8 by chemical methods. Four multienzime reactions are used that mimic purine salvage and pyrimidine biosynthetic pathways to convert D-d_4-ribose (1) into d_4-ATP (2), d_4-GTP (3) and d-$_4$UTP (4). Using this strategy, a 30-nucleotide RNA derived from the HIV-2 TAR RNA has been prepared with protonated and deuterated NTP at specific positions.

Figure J2.47 illustrates the dramatic effect of deuteration on the NMR spectra of RNA. The spectral crowding in the ribose region of the deuterated RNA is greatly reduced, and individual H2′ (see Fig. J2.47(c)) resonance peaks are clearly observed between 4 and 5 ppm.

J2.9 Encapsulated proteins in low-viscosity fluids

Increasing protein size is accompanied by several important limitations for NMR. Increasing size leads to slower tumbling and correspondingly shorter spin–spin relaxation times, T_2, so that line widths become broader. Also, increasing size leads to increasingly complex spectra, simply because of the larger number of nuclei. Spectral overlapping complicates the assignment of resonance peaks in NOE signals. Accordingly, the standard triple-resonance experiment (three-dimensional NMR) becomes unreliable at room temperature for proteins larger than 30 kDa and largely fails for proteins above 35 kDa in the absence of elevated temperature and/or extensive deuteration.

An approach that can be used to increase T_2 directly is based on reducing the rotational correlation time of the protein. The global tumbling correlation time of a spherical molecule, τ_c, is related linearly to the bulk solvent viscosity (see Section D2.3.4). The protein is encapsulated within the water cavity of a reverse micelle (Fig. J2.48), which is itself in a suitable low-viscosity solvent. The basis of the method is that although the reverse micelle containing the protein is larger than the free protein, it still tumbles faster in the low-viscosity solvent than the free protein in water (Comment J2.15).

Fig. J2.48 Chemical structure of bis(2-ethylhexyl) sodium succinate (AOT). AOT is commonly used as a reverse micelle surfactant. The radius of the reverse micelle is about 15 Å. Careful choice of surfactant concentration and relative water content provides for a micelle radius corresponding to the simple sum of the encapsulated protein's effective hydrated radius and the chain length of the surfactant.

Comment J2.15   Dimensions of a reverse micelle

A globular protein of molecular mass 50 kDa has a radius of about 26 Å and a rotational correlation time in water (η 0.85 cP at 300 K) of about 15 ns. The radius of the reverse micelle encapsulating the protein is the sum of the protein radius and thickness of the surfactant shell ($\sim$15 Å for AOT): $\sim$ 41 Å. The corresponding rotational correlation times would be $\sim$ 60 ns in water, $\sim$ 11 ns in butane, $\sim$ 7 ns in propane, and $\sim$ 2.5 ns in ethane. Note that a protein of molecular mass 10 kDa tumbles in water with a correlation time of 2.5 ns (see Section D8.5.3).

The resulting increase in T_2 due to the low-viscosity solvent makes high-resolution solution NMR directly applicable to a 50-kDa protein. Moreover, the relative advantage of the lower-viscosity solution increases with molecular mass, since, for example, encapsulation of a 100-kDa protein (spherical hydrated radius of 33 Å) would have a corresponding reverse micelle radius of $\sim$48 Å, with rotational correlation times of $\sim$95 ns in water, $\sim$18 ns in butane, 11 ns in propane, and $\sim$4 ns in ethane.

The preparation of stable well-behaved solutions of reverse micelles in ethane requires pressure approaching 200 bar (Flynn and Wand, 2001).

A number of significant issues need to be addressed for the encapsulated protein method to be useful in practice. The most obvious is the fact that the most promising low-viscosity fluids require the application of significant pressure in order to be in the liquid phase at room temperature. The method was tested on the protein, ubiquitin, encapsulated in AOT reverse micelles (Fig. J2.48) dissolved in liquid propane at a presure of 12 bar. The NMR spectrum, under these conditions, was closely similar to that obtained in aqueous buffer. These results indicate that proteins can be encapsulated in reverse micelles dissolved in low-viscosity fluids without significant distortion. The approach holds the promise of extending NMR-based structural analysis to soluble proteins of 50 kDa and beyond. The approach is also applicable to other macromolecules such as nucleic acids, carbohydrates, and, perhaps, most important, to membrane proteins.

J2.10 Checklist of key ideas

- FID describes the response of the sample to the pulse. FID is converted to a frequency spectrum by a Fourier transform: an *amplitude versus time* signal is transformed into an *amplitude versus frequency* signal.
- NOE is observed experimentally as the fractional change in the intensity of one NMR line when another resonance is irradiated in a double irradiation experiment.
- NOE is due to dipolar interactions (through space) between different nuclei and is correlated with the inverse *sixth power* of the internuclear distance; as consequence, NOE is usually observed only for protons pairs separated by $\leq$5–6 Å.
- Polarization transfer uses a suitable pulse combination to impart the greater equilibrium polarization of protons to a coupled nucleus with a smaller gyromagnetic ratio (e.g. ^{13}C or ^{15}N); polarization transfer is an alternative to NOE for sensitivity enhancement.
- The term one-dimensional NMR refers to experiments in which the transformed signal is presented as a function of a single frequency. By analogy, in two-, three- and four-dimensional NMR spectra the coordinate axes correspond to two, three and four frequency domains.
- COSY is a multidimensional method in which peaks appear at the coordinates of two nuclei related by a mutual interaction (*J*-coupling, NOE, or chemical exchange).
- COSY identifies pairs of protons that are coupled to each other via scalar spin–spin coupling. NOESY identifies pairs of protons that are related by the NOE.
- TROSY makes use of transverse relaxation optimisation to record high-quality solution NMR spectra of large molecular structures.
- The main source of geometric *short-range* information contained in the experimental NMR restraints is provided by the NOE; additional structural refinements include the use three-bond coupling constants and secondary ^{13}C and ^{1}H shifts.
- The main source of geometric *long-range* information is measurements of residual dipolar couplings and, in some cases, chemical shift anisotropy in a weakly aligned medium.

Suggestions for further reading

Fourier transform NMR spectroscopy

King, R. W., and Williams, K. R. (1989). The Fourier transform in chemistry, Part 2, Nuclear magnetic resonance: the single pulse experiment. *J. Chem. Education*, **66**, A243–A248.

Williams, K. R., and King, R. W. (1990). The Fourier transform in chemistry – NMR. Part 3. Multiple-pulse experiments. *J. Chem. Education*, **67**, A93–A99.

Williams, K. R., and King, R. W. (1990). The Fourier transform in chemistry – NMR. Part 4. Two-dimensional methods. *J. Chem. Education*, **67**, A125–137.

Single-pulse experiments

Skoog, D. A., Holler, F. J., and Nieman, T. A. (1995). *Principle of instrumental analysis*. Philadelphia: Saunders College Publishing.

Brey, W. S. (ed.). *Pulse Methods in 1D and 2D Liquid-Phase NMR* (1988). San-Diego: Academic.

Multiple-pulse experiment

Wuthrich, K. (1986). *NMR of Proteins and Nucleic Acids*. New York: Wiley-Interscience.

Nuclear Overhauser effect

Kumar, A., Ernst, R. R., and Wuthrich, K. (1980). A two-dimensional nuclear Overhauser enhancement (2D NOE) experiment for the elucidation of complete proton–proton cross-relaxation networks in biological macromolecules. *Biochem. Biophys. Res. Commun.*, **95**, 1–6.

Two-dimensional NMR

Derome, A. E. (1987). *Modern NMR Techniques for Chemistry Research*. New York: Pergamon.

Ernst, R. R., Bodenhausen, G., and Wokaun, A. (1987). *Principles of Nuclear Magnetic Resonance in One and Two Dimensions*. Oxford: Oxford University Press.

Multi-dimensional, homo- and hetero-nuclear NMR

Clore, G. M., and Gronenborn, A. M. (1991). Two-, three-, and four-dimensional NMR methods for obtaining more precise three-dimensional structure of proteins in solution. *Ann. Rev. Biophys. Chem.*, **20**, 29–63.

Sterically induced alignment

Prestergard, J. H. (1998). New techniques in structural NMR – anisotropic interactions. *Nature Struct. Biol.*, **5**, 517–522.

Hansen, M. R., Mueller, L., and Pardi, A. (1998). Tunable alignment of macromolecules by filamentous phage yields dipolar coupling interaction. *Nature Struct. Biol.*, **5**, 1065–1074.

Sanders, C. R., Hare, B. J., Howard, K. P., and Prestegard, J. H. (1994). Magnetically oriented phospholipid micelles as a tool for the study of membrane-associated molecules. *Prog. Nucl. Magn. Res. Spectr.*, **26**, 5.

Bax, A. (2003). Weak alignment offers new NMR opportunities to study protein structure and dynamics. *Protein Sci.*, **12**, 1–16.

Isotope labelling of proteins and nucleic acids

Gardner, K. H., and Kay, L. E. (1998). The use ^{2}H, ^{13}C, ^{15}N muldimensional NMR to study the structure and dynamics of protein. *Ann. Rev. Biophys. Biomol. Struct.*, **27**, 357–406.

Goto, N. K., and Kay, L. E. (2000). New developments in isotope labeling strategies for protein solution NMR spectroscopy. *Curr. Opin. Struct. Biol.*, **10**, 585–592.

Tolbert, T. J., and Williamson, J. R. (1996). Preparation of specifically deuterated RNA for NMR studies using a combination of chemical and enzymatic synthesis. *JACS*, **116**, 7929–7940.

Encapsulated proteins in low viscosity

Wand, A. J., Ehrhardt, M. R., and Flynn, P. F. (1998). High-resolution NMR of encapsulated proteins dissolved in low-viscosity fluids. *PNAS*, **95**, 15299–15302.

Flynn, P. F., and Wand, A. J. (2001). High-resolution nuclear magnetic resonance of encapsulated proteins dissolved in low viscosity fluids. *Meth. Enzymol.*, **339**, 54–70.

Chapter J3
Structure and dynamics studies

J3.1 Structure calculation strategies from NMR data

Unlike X-ray crystallography, which generates only one set of coordinates, an ensemble of sometimes more than 50 models is produced from NMR data analysis. This situation is inherent in the method, and arises from the interpretation of the data and also from the dynamic nature of the protein molecule itself. Because of this, the general consensus is that NMR-derived models should be judged by different criteria than those used in the assessment of the more 'rigid' structures from X-ray crystallography, and it has been proposed that for the solution structures a type of statistical description might eventually have to be used (Comment J3.1).

Irrespective of the algorithm used, any structure determination by NMR seeks to find the global minimum region of a target function E_{tot} given by

$$E_{tot} = E_{cov} + E_{vdw} + E_{NMR} \qquad (J3.1)$$

where 'E_{cov},' 'E_{vdw}' and 'E_{NMR}' are terms representing the covalent geometry (bonds, angles, planarity, and chirality), the non-bonded contacts, and the experimental NMR restraints, respectively. The uncertainties associated with the first two terms are relatively small, and the major determinant of accuracy resides in the number and quality of the experimental NMR restraints that enter into the third term, E_{NMR}. Comparable minimisation algorithms are also used in X-ray crystallography (see Chapter G3).

The main source of geometric information contained in the experimental NMR restraints is provided by the NOE. The NOE is proportional to the inverse sixth power of the distance between the protons, so its intensity falls off very rapidly with increasing distance between proton pairs (see Section J2.4). Despite the short-range nature of the observed interactions, the short approximate interproton distance restraints derived from NOE measurements can be highly conformationally restrictive, particularly when they involve residues that are far apart in the sequence but close together in space.

The accuracy of NMR structures is also affected by errors in the interproton distance restraints, arising from three sources: misassignments, errors in distance estimates and incomplete sets of NOE restraints. The best check on the correctness is provided by verifying that all short interproton distances (<3.5 Å) in a structure are in fact observed in the NOE spectra.

Although the interproton distance restraints derived from NOEs provide the mainstay of NMR structure determination, direct refinement against other experimental NMR restraints is both feasible and desirable. These additional experimental restraints that also provide *short-range* structural information are: specifically three-bond coupling constants, secondary ^{13}C chemical shifts, and ^{1}H chemical shifts.

NMR structure determination used to rely exclusively on these restraints whose information a priori is entirely local and restricted to atoms close in space. To date there are two new approaches in which restraints that define *long-range* order a priori can supply invaluable information. The first is based on the dependence of heteronuclear (^{15}N or ^{13}C) longitudinal (T_1) and transverse (T_2) relaxations times, specially T_1/T_2 ratios, on rotational diffusion anisotropy, and the second relies on residual dipolar couplings in oriented macromolecules (Section J2.7). The two methods provide restraints that are related in a simple geometric manner to the orientation of one-bond internuclear vectors (e.g. N–H and C–H) relative to external tensor. In the case of T_1/T_2 ratios, the tensor is the diffusion tensor. In the case of residual dipolar couplings, the tensor may be the *magnetic susceptibility tensor* for molecules aligned in magnetic field, or *the molecular alignment tensor* for molecules aligned by anisotropic media such as liquid crystals. The two methods permit the relative positioning of structural elements that do not have many short interproton distance contacts between them. Examples of such systems are discussed in Section J3.2.

As mentioned above, the addition of dipolar coupling terms to the NMR energy function during simulated annealing results in a highly rippled surface with innumerable sharp local minimum. In the absence of long-range distance information from NOEs, it is usually very difficult to find a good starting model close to the true structure. One simple method for obtaining such a starting model simply breaks the protein of interest into overlapping fragments of 7–10 residues in length. Then, the entire PDB is searched for fragments that are compatible with the experimental dipolar couplings and chemical shifts. Figure J3.1 illustrates the assembly of the backbone of the RecA-inactivating protein DinI by the molecular fragment replacement approach. The approach yields a backbone structure that differs by less than 2 Å from the previously determined NMR structure.

Table J3.1 summarises all experimental NMR constraints used today for macromolecular structure determination.

Table J3.1. *Experimental NMR constraints used in macromolecular structure determination.*

Conformational constraints	Main geometrical parameter	Range order
Interproton distances from NOE measurements (Section J2.4)	Interproton distances	Short, <5 Å
Three-bond coupling constants from *J* correlated spectroscopy (Section J1.2.3)	Torsion angle θ	Short
^{13}C secondary chemical shifts (Section J1.2.3)	Torsion angles θ and ψ	Short
^{1}H chemical shifts (Section J1.2.3)	Distance and orientation of aromatic ring to the proton of interest	Short
Heteronuclear relaxation time measurements (Section J1.2.2)	Orientation of the internuclear vectors relative to molecular diffusion tensor	Long
Residual dipolar coupling (Section J2.7)	Orientation of the internuclear vectors relative to molecular magnetic susceptibility tensor (alignment tensor)	Long

Fig. J3.1 Assembly of a protein backbone by the molecular fragment replacement approach, using seven-residue fragment lengths. The assembly is illustrated for the backbone of the RecA-inactivating protein DinI. The partially overlapping fragments are strung together, while maintaining their correct orientation. (After Bax, 2003.)

J3.2 Three-dimensional structure of biological macromolecules

J3.2.1 Proteins

Solution NMR structural studies are based on high-quality spectra recorded with good sensitivity and resolution. We recall that with increasing molecular mass, these basic requirements are difficult to achieve, because of line broadening and overlap. Advances have been obtained with both novel NMR techniques and new biochemical approaches. In particular, the application of TROSY and CRINEPT (Section J2.5), in combination with suitable isotope-labelling schemes (Section J2.8), allowed the extension of the protein size limit by several-fold. A few examples of NMR studies of proteins follow, classified according to molecular mass range.

Small proteins (6–10 kDa)
In principle, the solution structures of small compact proteins (100 amino acids or less) can be obtained from constraints imposed by NOE signal information alone. In practice, these are commonly supplemented with further constraints based on *J* coupling and chemical bonding. Figure J3.2 shows the first *de novo* atomic resolution NMR structure determination of a globular protein, the bull seminal protease inhibitor (Comment J3.2).

Proteins of medium size (20–60 kDa)
For proteins of molecular mass greater than 10 kDa, the standard sequential assignment procedure may not yield an unambiguous result because of very extensive overlap in the spectra. The overlap problem can be alleviated significantly by employing isotope labelling.

In a double-labelling technique, one type of amino acid is labelled with ^{13}C in the carbonyl position and a second type is labelled with ^{15}N. Dipeptide fragments, in which the ^{13}C label precedes the ^{15}N label, can be identified easily via the *J* coupling peak splitting in the ^{1}H–^{15}N correlation spectrum.

Fig. J3.2 The first protein structure determined by NMR: (a) a representation of the NMR structure of the proteinase inhibitor IIA from bull seminal plasma (BUSI IIA); (b) superposition of the core region (residues 23–42) in the NMR structure of BUSI IIA (green) on the corresponding polypeptide segment in the X-ray crystal structure of the homologous porcine pancreatic secretory trypsin inhibitor (PSTI) (blue). (After Wüthrich, 2001.)

(a)   (b)

Comment J3.2   The first steps of NMR protein structure determination

K. Wuthrich in the article 'The way to NMR structure of proteins' wrote:

> The completion of the first protein NMR structure brought new, unexpected challenges.
> When I presented the structure of BUSI (Fig. J3.2(a)) in some lectures in the spring of
> 1984, the reaction was one of disbelief, and because of the close coincidence
> (Fig. J3.2(a)) with results from an independent crystallographic study of the homologous
> protein PSTI (porcine pancreatic secretory trypsin inhibitor) it was suggested that our
> structure must have been modeled after this crystal structure. In the discussion following
> a seminar in Munich on May 14, 1984, Robert Huber (Nobel Prize in Chemistry, 1988)
> proposed that we settle the matter by independently solving a new protein structure by
> X-ray crystallography and by NMR. For this purpose, each one of us received an sample
> supply of the α-amilase inhibitor tendamistat. Virtually identical three-dimensional
> structures of tendamistat were obtained in our laboratory by NMR in solution and in
> R. Huber's laboratory by X-ray diffraction in single crystals.
> (Wuthrich, 2001)

A new era of NMR had dawned.

The development of heteronuclear triple-resonance NMR experiments (three-dimensional NMR), in conjunction with essentially complete ^{13}C and ^{15}N enrichment dramatically increased the scope of the NMR method, and structure determinations of proteins of 150–200 amino acid residues became a matter of routine. With the exception of a few very favourable cases, however, studies of larger proteins (beyond 200–300 amino acid residues) remained difficult. The sensitivity of heteronuclear triple-resonance NMR experiments was improved by partial random deuteration, which decreased peak overlap by decreasing the number of proton resonances observed. The optimum deuteration level depends on molecular mass. For example, a level of ~50% deuteration gave the best results for the SH3 domain protein from chicken brain alpha spectrin $M = 7.2$ kDa. For proteins with a molecular mass of about 30 kDa, a higher level of deuteration is required. The solution structure of the 259-residue, 30-kDa, N-terminal domain of enzyme I of the *Escherichia coli* phosphoenolpuruvate (EIN) was solved at a deuteration level of 90% (for the non-exchangeable protons). Figure J3.3 shows ribbon diagrams of EIN in solution (red) and in the crystal (blue).

Structural studies of proteins of more than 30 kDa require higher levels of deuterium enrichment. For example, assignment of over 90% of the backbone amide ^{1}H, ^{15}N, ^{13}C$^{\alpha}$ and ^{13}C$^{\beta}$ resonance peaks of a 64-kDa ternary complex consisting of two tandem dimers of trp-repressor was achieved on samples with 95% deuteration, by using site-specific isotope labelling (Section J2.8) and special pulse sequences (Section J2.5).

Fig. J3.3 NMR solution structure (red) of the 259-residue, 30 kDa, N-terminal domain of enzyme I of the *Escherichia coli* (EIN). The structure determination was based on a total of 4251 experimental constraints, and the precision of the coordinates for the final 50 simulated structures is 0.79 ± 0.18 Å for the backbone atoms and 1.06 ± 0.15 Å for all atoms. A comparison with the X-ray structure of EIN (blue) indicates that there are no significant differences between the solution and crystal structures within the errors of the coordinates. (After Garret *et al.*, 1997.)

72K

(a)

472K

(b)

Fig. J3.4 (a) GroES
(72 kDa) is an oligomeric
protein with seven
identical units, one of
which is shown in gold;
(b) a molecular model of
the complex of GroES and
SR1 (472 kDa). The blue
coloured SR1 is not
isotope-labelled and
therefore not visible in
this NMR experiment.
(After Flaux *et al.*, 2002.)

Large and very large macromolecular complexes

Special techniques (TROSY and CRINEPT) were developed for NMR studies of
large and very large macromolecular complexes (Section J2.5). Their resolving
power was tested on isotope-labelled heptameric co-chaperonin GroES (72 kDa),
free in solution, in complex with the homotetra-decameric chaperonin GroEL
(800 kDa), or with the single-ring GroEL variant SR1 (400 kDa) (Fig. J3.4,
Comment J3.3). Most amino acids of GroES showed the same resonance peaks,
whether free in solution or in complex with chaperonin; however, residues 17–32
showed large chemical shift changes upon binding. These amino acids belong to
a mobile loop region that forms contacts with GroEL.

**Comment J3.3   Physicist's box: Molecular chaperones (see also
Comment G2.19)**

Molecular chaperones are required for the correct folding, transport and degradation
of many proteins in vivo. In the *Escherichia coli* chaperonin system, the double
ring-shaped homotetradecamer GroEL (800 kDa), interacts with the dome-shaped
homoheptamer GroES (72 kDa; Fig. J3.4) in the presence of ATP to form a chamber
that is the site of productive polypeptide folding (Xu *et al.*, 1997). The
homo-oligomeric nature and high symmetry of GroEL, SRI, GroES and their
complexes can facilitate NMR spectral analysis because the number of resonances
corresponds to the size of an individual subunit, here 58 kDa for GroEL and 10 kDa
for GroES, even though the large functional assembly is investigated.

Protein–protein interactions

An elegant NMR strategy for the determination of the interface residues in a large
protein–protein complex has been proposed. The approach requires an unlabelled
target protein (containing ^{1}H and ^{14}N spins), and a uniformly ^{2}H-,^{15}N-labelled
ligand protein, in which only the amide proton positions are occupied with ^{1}H.
The relative populations of ^{15}N–^{1}H and ^{15}N–^{2}H groups in the ligand protein
are further adjusted by variation of the solvent ^{1}H$_2$O–^{2}H$_2$O ratio. Figure J3.5
illustrates the resulting situation. The target protein (T) contains a dense population of aliphatic ^{1}H spins, which resonate at chemical shifts, δ, from 0 to 3
ppm in the ^{1}H NMR spectrum (green circles), and aromatic and amide ^{1}H spins
(pink circles), which resonate between 6 and 10 ppm and are hence well separated
from the aliphatics (Fig. J3.6). The ligand protein (L) contains ^{15}N–^{1}H groups
(red circles), but no other hydrogen atom positions are occupied with ^{1}H spins.
The complex is immersed in solvent ^{1}H$_2$O (green circles), which has a ^{1}H resonance line near 5 ppm. (Fig. J3.6). If only the aliphatic proton resonance lines in
a large protein with natural isotope abundance (T in Fig. J3.5) are irradiated with
a radio-frequency field, then its aromatic and amide protons are instantaneously

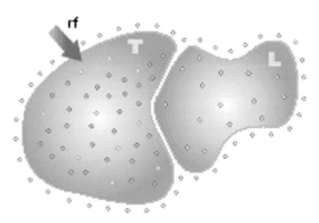

Fig. J3.5 Isotope labelling scheme used for identification of intermolecular contacts between two proteins, T and L, by NMR saturation transfer. T denotes the unlabelled target protein, which contains proton spins in aliphatic $^{12}C-^{1}H$ groups (green circles) and in aromatic $^{12}C-^{1}H$ as well as $^{14}N-^{1}H$ groups (pink circles), which can be saturated but are not observed in the presently discussed experiment. L denotes the uniformly $^{2}H-$, ^{15}N-labeled ligand protein, which contains NMR-observable $^{15}N-^{1}H$ groups (red circles), but is otherwise devoid of ^{1}H spins. The complex is surrounded by solvent water, of which only a first hydration layer is shown (purple circles). The green arrow represents the radio-frequency field, which is applied to saturate the aliphatic $^{12}C-^{1}H$ protons (green circles). (After Wuthrich (2000).)

saturated due to spin diffusion. Spin diffusion is highly effective only for closely spaced ^{1}H spins. Therefore, although the uniformly $^{2}H-,^{15}N$-labelled protein is not directly affected by the radio-frequency field (Fig. J3.6), saturation can be transferred from the unlabelled target molecule to the labelled protein through close $^{1}H-^{1}H$ contacts across the protein–protein interface in the complex (Fig. J3.5).

The new strategy was applied to studies of a 64-kDa immunoglobulin complex with a domain of protein A, a cell wall component of *Staphylococcus aureus* that specifically binds to the Fc portion of immunoglobulin G (IgG). It was shown that helices I (Gln 10–His 19) and II (Glu 25–Asp 37) are primarily responsible for the binding of FB to Fc fragment. The contact residues thus identified overlap with those determined by the X-ray crystallography study (Fig. J3.7(a)). Interestingly, the values of the NMR intensity ratios of the cross-peaks for helicies I and II are smaller for every third or fourth residue (Gln 11, Tyr 15 and Leu 18 in helix 1,

Fig. J3.6 Distribution of NMR signals along the ^{1}H chemical shift axis. The colour code is the same as in Fig. J3.5: green, aliphatic $^{12}C-^{1}H$; pink, aromatic $^{12}C-^{1}H$ and $^{14}N-^{1}H$; red, $^{15}N-^{1}H$; purple, $^{1}H_2O$. The green arrow indicates the radio-frequency irradiation on the spectral region of the aliphatic protons. (Wuthrich, 2000.)

(a)

(b)

Fig. J3.7 Comparison of the binding sites of the FB-Fc complex. (a) The X-ray crystallography study: residues with accessible surface areas that are covered upon binding of the Fc fragment (Phe 6, Gln 10, Gln 11, Asn 12, Phe 14, Tyr 15, Leu 18, His 19, Arg 28, Asn 29, Ile 32, Gln 33 and Lys 36) are coloured red. (b) The cross-saturation NMR experiment: residues Gln 10, Gln 11, Phe 14, Tyr 15, Glu 16, Leu 18, His 19, Arg 28, Asn 29, Ile 32, Gln 33, Leu 35 and Lys 36 are given in a different colour. The colour reflects different intensity ratio of the cross peaks. (Adapted from Takahashi *et al.*, 2000.)

> **Comment J3.4   X-ray and NMR structures in the PDB**
>
> Although in absolute figures the number of deposited protein structures is still much larger, the proportion of DNA and RNA structures in the PDB is increasing steadily. The statistics show a substantial improvement in multidimensional NMR spectroscopy and in the capabilities of advanced isotope-labelling techniques for both RNA and DNA. The absolute values are as follows: in 1997, 137 nucleic acid structures and 784 protein structures were deposited; in 2001, 419 nucleic acid structure and 1775 protein structures were deposited. The numbers for nucleic acids also include the structures of protein–DNA/RNA and drug–DNA/RNA complexes (Zidek *et al.*, 2001).

and Asn 29, Ile 32 and Lis 36 in helix II), suggesting that one side of each of the helices is responsible for binding the Fc fragment (Fig. J3.7(b)).

In identifying the interface of a large protein–protein complex, the method proposed is obviously superior to traditional NMR methods that rely on chemical shift perturbation and H–D exchange. This is because the method described here extracts more direct information on through-space interactions between the two molecules in a complex. The method should be generally applicable to large protein complexes and, with appropriate modifications, to other complexes, such as protein–DNA and protein–lipid complexes.

J3.2.2 Nucleic acids

There has been significant progress in NMR structure determination of nucleic acids, despite the fact that, in principle, it is much more difficult than for proteins (Comment J3.4). The low proton density (0.35 protons per atom versus 0.52 protons per atom for proteins), and the absence of a globular fold result in insufficient NOE constraints (Fig. J3.8). Even the accurate determination of local geometry is difficult because of the low redundancy in the measured NOE. It could be argued, however, that the need for accurate solution structures is even greater for nucleic acids than for proteins, because of their flexible structures whose conformation can be significantly affected by crystallization. Nucleic acid NMR has benefitted from the development of methodology based on selective isotopic labelling (Section J2.8), and molecular alignment in ordered media, such as solutions of disc-shaped phospholipid bicelles and filamentous bacteriophage (Section J2.7).

Figure J3.9 demonstrates the solution structure of the DNA dodecamer d(CGCGAAT-TCGCG)$_2$, in an aqueous liquid crystalline medium containing 5% w/v bicelles. The structure, determined by using dipolar couplings, together with torsion angles derived from *J* couplings and interproton distances obtained from NOE, shows a well-defined B-form helix. Overall agreement with the corresponding X-ray structures is reasonable, although several of the irregular

Fig. J3.8 A comparison of NOE contacts in proteins and nucleic acids. (a) Proteins have NOE contacts that are >5 Å along their primary sequences, but <5 Å apart in their tertiary structure. This allows long-range NOE constraints to be acquired. (b) Nucleic acids lack these long-range NOE constraints between nucleotides distant in their primary structure; this makes their total structure less defined in the absence of residual dipolar couplings. (Adapted from MacDonald and Lu, 2002.)

features seen in the crystalline state are less pronounced or absent in solution (Comment J3.5).

In multidomain molecules, such as higher-order DNA structures, tRNA or RNA, dipolar coupling can be used to determine the relative orientation of the various domains. Figure J3.10 shows the NMR model for tRNA^{Val} superimposed on the X-ray structure. The global orientation of helical domains was determined

Fig. J3.9 Two stereoviews of structures of d(CGCGAAT-TCGCG)$_2$. The NMR structure is shown in blue and the X-ray structure in red. (b) The view in (b) is rotated by 90° around the helix axis relative to (a) and highlights the deviation from cylindrical symmetry, caused by the very different widths of the minor and major grooves. A nearly complete set of one-bond ^{1}H–^{13}C dipolar couplings, supplemented by ^{1}H–^{15}N and ^{1}H–^{1}H dipolar couplings, was obtained in the bicelle-based liquid crystal medium. (After Tjandra et al., 2000.)

**Comment J3.5
X-ray and NMR
data for DNA
dodecamer**

The original X-ray structure exhibited an 18° bend in its helix axes over 12 base pairs, whereas the NMR dipolar structure has significantly less curvature, 7° over 12 base pairs. Most parameters, such as sugar pucker, propeller twist, roll, and the near absence of helical bending, are well determined by the NMR data. In contrast, other parameters such as the base-pair opening and the ^{31}P–^{31}P distance across the minor groove, are less accurately defined by the NMR.

(a)   (b)

Fig. J3.10 Comparison of the model structure of tRNA^{Val} (red) with the global structure determined from residual dipolar couplings: (a) global structure resulting from domain orientation using the X-ray structure-based model for the helical arms (green): (b) global structure resulting from domain orientation using an A-form model for the helical arms (green). The tRNA was partially oriented by addition of 10 mg ml^{-1} of filamentous phage Pf1. (Mollova *et al.*, 2000.)

Comment J3.6
X-ray and NMR data for tRNA

The interarm angles are 86° for the tRNA^{Phe} crystal structure and $99 \pm 2°$ for the tRNA^{Val} structure determined from the dipolar couplings. There is significant variation for the interarm angles in three-dimensional X-ray structures of different tRNAs, ranging from 76° to 96°, and the angle determined here for tRNA^{Val} in solution is near the upper end of the range that been observed by X-ray (Westhof *et al.*, 1988; Basavappa and Sigler, 1991).

precisely from a set of residual dipolar couplings. There is a slight difference in the angle between the helical arms for the two models (Comment J3.6). These results demonstrated that a small number of residual dipolar couplings can be used to determine the global structure of tRNA, if the local structure of the helical stems is known. To test how well the dipolar couplings can be used to determine global structure of an RNA with a known secondary structure, the four stem regions in tRNA^{Val} were replaced with A-form helices. A conformational search was performed to find the orientations of the anticodon and acceptor arms that gives the best fit to the measured dipolar couplings. The resulting structure is given in Fig. J3.10(b). It has an interarm angle of $101 \pm 2°$, in excellent agreement with the angle determined with the helices from the X-ray structure.

Finally we point out that NMR has provided a unique direct observation of hydrogen bonds in nucleic acid base pairs, through the scalar coupling of imino proton donors and acceptor nitrogens.

J3.2.3 Carbohydrates

The structural characterisation of carbohydrates is especially challenging because of their conformational flexibility. There are two kinds of internal motions in oligosaccharides (Fig. J3.12). The first kind consists of modest variations of the glucosidic dihedral angles, of the order of $\pm 15°$. Oligosaccharides exhibiting this type of internal motion are called 'rigid' in the sense that experimental NOE data may agree with those calculated for a model having a single conformation. The second kind features larger variations of glycosidic dihedral angles, between distinct energy minima. Oligosaccharides exhibiting this type of internal motion are called 'flexible' in the sense that experimental NOE data cannot be reconciled

Fig. J3.11 Picture of a sugar showing conventional angles and motions.

Fig. J3.12 Structure of trimannoside, methyl 3,6-di-O-(α-D-mannopyranosyl)-α-D-mannopyranoside. Rings are labelled with roman numerals. This trimannozide forms a common core in N-linked oligosaccharides. Three mannoses are connected together through two glycosides linkage α-(1,3) and α-(1,6). (Adapted from Tian et al., 2001.)

with any single conformation. In this case, molecules can be considered as an ensemble of conformers.

Advances in high-resolution NMR allow direct measurements of dipolar coupling by dissolving the molecules in oriented media such as liquid crystals composed of bicelles or phage (Section J2.7). Methyl 3,6-di-O-(α-D-mannopyranosyl)-α-D-mannopyranoside (trimannoside, Fig. J3.12) has been used to illustrate the application of residual dipolar coupling to the conformational analysis of oligosaccharides. The residual dipolar couplings were calculated as the difference of the measured couplings in aligned media (bicelle and phage samples at 38 °C) and isotropic media (bicelle sample at 20 °C).

Fig. J3.13 (a) The average structure of trimannoside ring I and ring III assembled from bicelle media. (b) The same from phage media. The structure is oriented to show the similarity of conformations about the linkage connecting ring I and ring III. (After Tian et al., 2001.)

Fig. J3.14 (a) The average
structure of trimannoside
ring II and ring III
determined from bicelle
media. (b) The same from
phage media. The
structure is aligned to
superimpose ring III with
that in part (a). Note that
there is a variation in the
apparent orientation of
ring II. (After Tian *et al.*,
2001.)

Figure J3.13(a) shows the average structure of trimannoside ring I and ring III determined from bicelle media, and Fig. J3.13(b) illustrates the average structure of trimannoside ring I and ring III determined from phage media. The structure is oriented to show the similarity of conformations about the linkage connecting ring I and ring III. Figure J3.14(a) shows the average structure of trimannoside ring II and ring III determined from bicelle media and Fig. J3.14(b) illustrates the average structure of trimannoside ring II and ring III determined from phage media. The structure is aligned to superimpose ring III with that in Fig. J3.14(a) to show the similarity of conformations about the linkage connecting ring II and ring III.

Comparison of data presented in Figs. J3.13 and J3.14 indicates that internal motion between rings I and II (at the α-(1,3) linkage) is limited, while significant motion is suggested between rings II and III (at the α-(1,6) linkage). The results clearly demonstrate that trimannoside exists as a set of conformers.

J3.3 Dynamics of biological macromolecules

Molecular dynamics on a broad range of time scales can be monitored using various types of NMR experiments. Nuclear spin relaxation is sensitive to motions occurring in shorter times than the molecular correlation time (10–20 ns for a protein in solution, see Sections D6.6.2 and D8.5.3). Chemical shift exchange between two sites occurs in microseconds to milliseconds. Line broadening and spectral resolution effects take longer, of the order of seconds. These features make NMR a unique and powerful tool for the study of macromolecular dynamics.

J3.3.1 Protein dynamics from relaxation measurements

We assume that the internal atomic motions in a macromolecule are independent of its overall geometric motion arising from tumbling or translational diffusion. In other words, the effective internal correlation time, τ_e, is much shorter than τ_{corr}, the rotational correlation time (see Section D2.3). The three commonly measured NMR relaxation rates, the spin–lattice relaxation rate, $R_1 = 1/T_1$, the spin–spin relaxation rate, $R_2 = 1/T_2$, and the heteronuclear NOE are all sensitive to internal motion on the subnanosecond times cale. In a model-free approach, one obtains two independent parameters: (1) the order parameter S^2, which is a measure of spatial restrictions on the motion, and (2) τ_e, the effective internal correlation time. The analysis does not invoke a specific model for internal motions.

The order parameter S^2 measures the magnitude of the angular fluctuation of a chemical bond vector such as the N–H, C–H, C–C bonds in a protein and thus reflects the flexibility of the polypeptide chain at these sites (Fig. J3.15). S^2 is interpreted as describing the amount of spatial freedom the vector μ has as a result of internal motion. If the bond vector μ diffuses in a cone as depicted in the inset, S^2 decreases rapidly as the cone angle θ_c increases from $0°$ to $75°$. If the motion is completely free (i.e. all orientations of μ are equally probable), then $S = 0$. On the other hand, if the motion is completely restricted, then $S = 1$. Since ^{15}N labelling of proteins became commonly available for proteins with efficient expression systems (Section J2.8), studies of mobility based on the model-free approach have become routine.

The model-free approach is based on a very simple form for the total auto-correlation function that describes the overall and internal motions when it is assumed that they are independent:

$$G(\tau) = \frac{1}{5}G_0(\tau)G_i(\tau) \tag{J3.2}$$

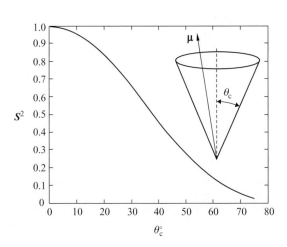

Fig. J3.15 Relationship between internal motion and the model-free parameter S^2. The parameter S^2 provides information about the angular amplitude of the internal motions of bond vectors (N–H, C–H, C–C) at numerous sites in the protein sequences. If the bond vector μ diffuses in a cone as depicted in the inset, S^2 decreases rapidly as the cone angle θ_c increases from $0°$ to $75°$, and remains small for $75° < \theta_c < 180°$. When the bond is rigid $S^2 = 1$, and when internal motion is isotropic $S^2 = 0$. (After Ishima and Torchia, 2000.)

where $G_0(\tau)$ is the correlation function for overall motion (Section D8.5.2):

$$G_0(\tau) = \exp(-\tau/\tau_{corr}) \qquad (J3.3)$$

and $G_i(\tau)$ is the simplest form for internal motions:

$$G_i(\tau) = S^2 + (1 - S^2)\exp(-\tau/\tau_i) \qquad (J3.4)$$

Internal correlation times τ_e, τ_i and τ_{corr} are related as

$$\frac{1}{\tau_i} = \frac{1}{\tau_{corr}} + \frac{1}{\tau_e} \qquad (J3.5)$$

After Fourier transformation, the resulting spectral density $J(\omega)$ is

$$J(\omega) = \frac{2}{5}\left\{\frac{S^2\tau_{corr}}{1 + (\omega\tau_{corr})^2} + \frac{(1 - S^2)\tau_i}{1 + (\omega\tau_i)^2}\right\} \qquad (J3.6)$$

In the limit that S^2 is unity $J(\omega)$ reduces to the familiar Lorentzian distribution for the rotational diffusion of rigid tumbling molecules (see Sections A3.2.3 and D10.3.1). The spectral density function of Eq. J3.6 contains essentially only two fitting parameters, S^2 and τ_e. The overall correlation time τ_{corr} is assumed to be the same for all nuclei of a protein. If τ_{corr} is known, two measurements of relaxation parameters, such as T_1 and T_2 are sufficient to determine these two parameters. The measurement of a third parameter is a check on the validity of the model. In the original experiments the values of S^2 and τ_e were obtained from a pair of relaxation parameters (NOE and T_1) at two fields.

As an example, extensive studies of the relaxation properties were carried out for *Bacillus subtilis* glucose permease IIA domain. The order parameters are represented mapped onto the protein structure in Fig. J3.16. The amide nitrogen

Fig. J3.16 Schematic representation of the solution structure of the *Bacillus subtilis* IIA domain showing the amide nitrogen positions colour-coded according to their calculated order parameters, S^2. Atoms are coloured on a continuous scale from red ($S^2 \leq 0.4$) to indigo ($S^2 = 0.9$). Backbone nitrogenes for which order parameters were not calculated are shown as small white spheres. Every tenth backbone nitrogen is labelled numbered. (After Stone *et al.*, 1992.)

Comment J3.7   HYDRONMR program

Hydrodynamic calculations have been used to interpret the ^{15}N NMR relaxation data of globular proteins using the program HYDRONMR. HYDRONMR uses a shell model (Section D2.2.2), which is constructed from the atomic representation of the protein contained in the PDB file. As described in Section D2.2.4 this procedure avoids the bead overlap problems, but requires computer time. HYDRONMR contains a single adjustable parameter, the atomic element radius (the radius of the bead in the terminology of Section D2.2.2). Comparison shows that there is good agreement between experimental relaxation values (R_1/R_2) and those calculated using HYDRONMR and the three-dimensional structure using an atomic element radius that ranges from 2.8 Å to 3.8 Å with the distribution centred at 3.3 Å. Deviations from the usual range towards larger values are associated with aggregation of proteins. Deviations to lower values may be related to large-scale motions or inappropriate model structures (Bernado *et al.*, 2002).

---

positions are colour-coded according to their calculated order parameters, S^2. Atoms are coloured on a continuous scale from red ($S^2 \leq 0.4$) to indigo ($S^2 = 0.9$). All regions of higher mobility were located in three areas: at the termini, surface loops or regions of irregular secondary structure.

In the simplest analysis, overall motion is assumed to be isotropic and this contribution is described by a single exponential. In the case of rotational diffusion anisotropy the time correlation is a multiexponential function, whose relative weighting depends on the orientation of the different relaxation interactions with respect to the principal components of the diffusion tensor. Thus, the determination of the rotational diffusion tensor is a prerequisite in the analysis of internal dynamics using heteronuclear relaxation. The determination of the overall rotational diffusion tensor can also provide important information concerning the overall form of globular proteins (Comment J3.7, see also Section D2.3.6).

The rotational diffusion tensor is specified by three principal components – the diffusion constants for rotation about the X, Y, Z principle axes – and by the orientation of the principal axes relative to that of the protein. The principal axes, X, Y, Z, are fixed in the protein, and defined as the coordinate axes in which the diffusion tensor assumes a simple, diagonal form.

The rotational diffusion can be characterised accurately using the relaxation rate ratio (R_1/R_2), a parameter which becomes independent of internal dynamics in the fast-motion limit while remaining highly sensitive to overall diffusion. As an example, the analysis of the rotational tensor of the chitin binding of the anti-fungal protein from *Streptomyces tendae* (AFP1) has been performed by measuring ^{15}N relaxation parameters. Relaxation rates R_1 and R_2, as well as ^{15}N–^{1}H heteronuclear NOE values were obtained for a total 65 backbone amides (78% of all possible). AFP1 exhibits a large degree of asymmetry in its overall

Fig. J3.17 Representation of the inertial and diffusion tensors of AFP1. The inertial tensor **I**, with its principal components labeled I_{xx}, I_{yy}, and I_{zz} (where $I_{xx}:I_{yy}:I_{zz} =$ 1.00:0.94:0.61) is shown in black. The principal axes of the two axially symmetric diffusion tensors, corresponding to the oblate or prolate ellipsoid model, are drawn in red and green, respectively. The axes for the fully anisotropic solution are coloured yellow. (After Campos-Olivas *et al.*, 2001.)

shape. In fact, the relative magnitudes of the principal components of inertial tensor **I** are calculated as 1:0.94:0.61 for the average regularised structure. The orientation of the principal components of **I** with respect the molecular frame is shown in **Fig. J3.17**. The figure also shows the orientation of the diffusion tensor **D** obtained for the three non-isotropic solutions and its relation to the inertial tensor **I**. The principal component of the prolate solution, $\mathbf{D}_{\parallel}^{PRO}$, or the direction about which diffusion is maximal (green in Fig. J3.17) is coincident with the direction of $\mathbf{I}_{yy}$ and it is perpendicular to the direction of the maximal inertia $\mathbf{I}_{xx}$. Coincidence of $\mathbf{D}_{\parallel}^{PRO}$ (fastest diffusion) with $\mathbf{I}_{zz}$ (minimal inertia) would be expected if the anisotropy were simply determined by molecular asymmetry. However, for the oblate solution, the direction of the slowest diffusion, $\mathbf{D}_{\parallel}^{OBL}$ (red in Fig. J3.17), is coincident with the direction of maximal inertia, as expected if molecular shape determines the rotational anisotropy. Therefore, the oblate solution, representing the best fit to relaxation data, is in perfect agreement with the asymmetry of the molecule and should be considered as the most appropriate description of the real rotational tumbling behaviour of AFP1 in solution.

Protein function, protein–ligand and protein–protein interactions are often affected by motions on much slower time scales ranging from microseconds to seconds. These interactions often contribute to the rate of transverse spin relaxation, R_2, via the chemical exchange mechanism. In the 'model-free'

Comment J3.8 Two-site chemical exchange in the spin-echo method

The exchange term R_{ex} in the following form approximates the effect of two-site chemical exchange in modified spin-echo method for measuring nuclear relaxation times

$$R_{ex} = \frac{p_1 p_2 (\Delta\omega)^2}{k_{ex}} \times \left[1 - \frac{2}{k_{ex}\tau_{cp}} \tanh\left(\frac{k_{ex}\tau_{cp}}{2}\right) \right]$$

where τ_{cp} is the time between $180°$ pulses, $\Delta\omega$ is the difference in chemical shift of the nuclei between two conformational states; $k_{ex} = k_{-1}/p_1 = k_1/p_2$; $1 \geq p_1 \geq 0.5$, and $p_2 = (1 - p_1)$ are the populations of the conformational states; k_1 is the forward exchange rate constant, and k_{-1} is the reverse exchange rate constant (Luz and Meiboom 1963).

---

formalism the evidence of this class of motions emerges via the introduction of the 'exchange' term R_{ex}, which is the exchange contribution to R_2. The exchange term R_{ex} may or may not be present (Comment J3.8).

As an example, the NMR exchange broadening arising from specific low-affinity protein self-association of the N-terminal domain of the rat T-cell adhesion protein (CD2d1) has been performed by measuring of ^{15}N nuclear relaxation at wide intervals of protein concentration. Figure J3.18 shows exchange contribution R_{ex} along the polypeptide chain for two protein concentrations, of 0.1 mM (black bars) and 1.2 mM (grey bars). The magnitudes of the R_{ex} terms for Thr 37

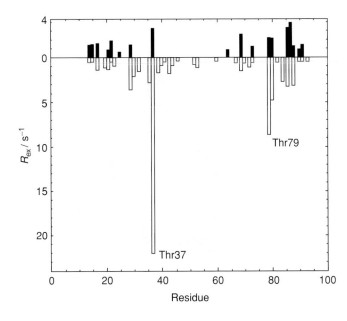

Fig. J3.18 Exchange contribution R_{ex} along the polypeptide chain for rat CD2d1 at 10 mM phosphate buffer, pH = 6.0, $T = 25\,°C$, a proton resonance frequency of 500 MHz and protein concentrations of 0.1 mM (black bars) and 1.2 mM (grey bars). A small number of residues – most notably Thr37 and Thr79 – show very large deviation on the R_{ex} panel. (After Pfuhl et al., 1999.)

Fig. J3.19 (a) Variation of
^{15}N line broadening as a
function of protein
concentration for Thr37 at
10 mM phosphate buffer,
pH 6.0 and $T = 25\,^{\circ}$C.
(b) The same dependence
for variation of ^{15}N
chemical shifts δ in ppm.
(c) The same dependence
for the apparent rotational
correlation time τ_c. (After
Pfuhl *et al.*, 1999.)

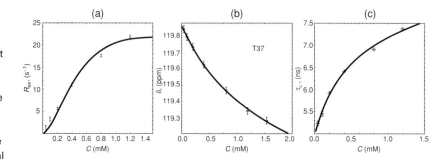

Fig. J3.19 (a) Variation of
^{15}N line broadening as a
function of protein
concentration for Thr37 at
10 mM phosphate buffer,
pH 6.0 and $T = 25\,^{\circ}$C.
(b) The same dependence
for variation of ^{15}N
chemical shifts δ in ppm.
(c) The same dependence
for the apparent rotational
correlation time τ_c. (After
Pfuhl *et al.*, 1999.)

($22\ \mathrm{s}^{-1}$) and Thr 79 ($8\ \mathrm{s}^{-1}$) are exceptionally large and very strongly dependent on the protein concentration. Figure J3.19 shows the variation of experimental parameters as a function of protein concentration for one residue, Thr 37. The value of R_{ex} drops to very small values in more dilute samples (Fig. J3.19(a)). The broadened resonance also displays considerable concentration-dependent chemical shift variation (Fig. J3.19(b)). These findings, together with the observation of concentration-dependent values of the rotational correlation time τ_c (Fig. J3.19(c)) allow low-affinity self-association of CD2d1 to be identified as a major factor in the explanation of its NMR behaviour.

Concentration-dependent line-width effects are observed for the resonances of a specific limited region of the protein surface, which is coincident with the major intermolecular contact in crystals of the complete extracellular portion of the CD2 protein (Fig. J3.20).

These observations clearly demonstrate that very weak protein–protein interactions can be detected by NMR.

Fig. J3.20 CD2d1 dimer
interface as seen in the rat
CD2 crystal structure. For
clarity the two monomers
are shown in different
shades of grey. Residues
with $R_{\mathrm{ex}} > 1.5\ \mathrm{s}^{-1}$ are
marked by a sphere
centred on the position of
the corresponding
backbone nitrogen. The
radius of the sphere is
approximately
proportional to the
magnitude of R_{ex}. (After
Pfuhl *et al.*, 1999.)

J3.3.2 Protein dynamics from amide proton exchange

Measurements of amide proton exchange rates in proteins can provide useful information about global stability and the dynamics of local structural fluctuations. Exchange between amide protons in proteins and bulk solvent can be measured in a variety of ways. Slow exchange lifetimes (from minutes to days) are determined by following the loss of amide proton signal intensity of a protein dissolved in D_2O, and provide information about the relative solvent accessibility of various components of protein secondary and tertiary structure.

Faster exchange lifetimes (5–500 ms) can be monitored by following the exchange of amide proton magnetization with that of water protons. At high pH the base-catalysed proton exchange reaction is very fast ($\sim 10^5$ s^{-1} at pH 10), and these magnetization exchange measurements directly measure the time scale of rate-limiting protein conformational openings that permit solvent access to the exchanging amide sites. Such measurements have been performed on a highly thermostable protein, rubredoxin from *Pyrococcus furiosus*. At 28 °C solvent access to all amide sites in the protein occurs on the millisecond time scale and solvent exchange protection factors for all sites are similar to those observed for mesophilic homologues. This observation suggests that the flexibility of thermophilic rubredoxin is similar to that of the mesofilic homologues, and brings into question the general hypothesis that the high stability of thermophilic protein comes from the rigidity of its native structure.

J3.3.3 Dynamics of slow events: translational diffusion

Traditionally, diffusion coefficients of biological macromolecules are determined by the numerous physical methods presented in Table D3.3. Although effective, these techniques are often carried out at micromolar concentrations that are three orders of magnitude less than the millimolar concentrations required for NMR. This discrepancy is particularly important for the concentration dependence of diffusion processes. In contrast, diffusion measurements of the NMR sample report under conditions identical to those used for structure determination. This provides knowledge of oligomeric state of macromolecules prior to structure calculations using NMR data.

Diffusion coefficients of molecules can be measured using a spin-echo experiment in pulsed field gradient (PFG) NMR (Comment J3.9). The method relies on two gradient pulses surrounding the 180° pulse in the spin echo: the first dephases the transverse magnetization in a spatially dependent manner along the *z*-axis and the second gradient then rephases the magnetization. If the molecules move along the *z*-axis during the time between the two gradients, the magnetization does not refocus completely. Therefore the attenuation of its resonance is large, if the molecule diffuses rapidly; the attenuation is relatively small if

Comment J3.9 Spin echo in pulsed field NMR

The original experimental scheme for measuring the diffusion coefficient using spin-echo experiment in pulsed field gradient NMR was:

$(90° - \tau(\text{PFG}) - 180°$
$-\tau(\text{PFG}) - \text{Acquisition})$

(Stejskal and Tanner, 1965).

the molecule diffuses slowly. The echo measured amplitude $A(t)$ is differentially attenuated in each spectrum due translational diffusion, which is related to D_t by

$$A(t) = A(0)\exp[-(\gamma\delta)^2 D_t(\Delta - \delta/3)G^2] \qquad (J3.7)$$

where γ is the ^{1}H gyromagnetic ratio of a proton, δ is the PFG duration time (s), G is the gradient strength (G/cm), and Δ is the time between PFG pulses (s). Equation (J3.7) shows that correct measurements of D_t require at least 1 order of magnitude signal decay to accurately fit the data to Eq. (J3.7). The limit on measurable D_t values, therefore, depends on the area of the gradient pulses (δG) and the time allowed for diffusion (Δ). Larger values of D_t ($\geq 10^{-5}$ cm^2 s^{-1}) can be reliably measured using this type of NMR experiment. However, as one attempts to measure D_t for proteins (where D_t is about $\sim 10^{-6}$ cm^2 s^{-1}), the time necessary to observe diffusion, Δ, becomes long relative to the transverse relaxation time ($T_2^* \sim 12$–40 ms), and the signal decays from spin–spin interactions, independent of diffusion.

The stimulated spin-echo experiment $(90°-\tau(\text{PFG})-90°-T-90°-\tau(\text{PFG}) - \text{Acq})$ was designed to avoid T_2 relaxation effects by storing the magnetization along the z–axis during time T, so that relaxation depends primarily on T_1, which is usually much longer than T_2 for proteins. Longer diffusion times (Δ) can then be used for larger proteins without significant signal loss from relaxation.

Stimulated spin-echo NMR experiments have been used (see Section J2.2) to determine the diffusion coefficient of lysozyme (Comment J3.10). The diffusion coefficient was obtained by fitting the signal attenuation to Eq. (J3.7), which yields a value of 10.8×10^{-7} cm^2 s^{-1}. This result is in a good agreement with data obtained by others methods (see Table D3.3).

The stimulated spin-echo NMR technique has been used for the characterisation of protein unfolding. Figure J3.21(a) and (c) show stack plots of the

Fig. J3.21 Stimulated spin-echo spectra obtained from a 1.4 mM solution of hen egg white lysozyme in D$_2$O at pH 2.0 and 20 °C. A small amount of 1,4-dioxane was added as the radius standard. The diffusion gradients were varied between 5% and 100% of their maximum strength (60 G cm^{-1}). (a) Attenuation of the aromatic signal from His^{15} taken from the spectrum of lysozyme in 0 M of urea: (b) attenuation of the dioxane signal used as a standard reference molecules; (c) attenuation of the aromatic signal from His 15 taken from the spectrum of lysozyme in 8 M of urea. (After Jones *et al.*, 1997.)

signal attenuation of the aromatic signal from His[15] taken from the spectrum of lysozyme in different concentrations of urea (0 M and 8 M, respectively). Figure J3.21(b) shows the attenuation of the dioxane signal used as the reference molecule in the same solution. The observed diffusion coefficients of both dioxane and lysozyme decreased with increasing urea concentration as a result of the increased viscosity, but the ratio D_{diox}/D_{lys} increased, as shown in Fig. J3.22. The effective hydrodynamic radius increases by 38% on unfolding in urea, a result in a good agreement with a study by small-angle X-ray scattering. Unfortunately, the small diffusion coefficients ($D < 10 \times 10^{-7}$ cm^2 s^{-1}) are difficult to measure by PFG NMR methods using stimulated echoes described above. This limitation results from the rapid longitudinal relaxation of the nuclei (usually protons) that carry the information about the localisation of the molecules during the diffusion interval. The duration of this interval can be increased by about one order of magnitude by storing the information in the form of longitudinal magnetization of heteronuclei, such as ^{15}N, that have much longer spin-lattice relaxation times than protons. This novel method, called 'heteronuclear stimulated echoes', allows one to gain about one order of magnitude in measurement of D, so that the diffusion coefficient of molecules with molecular masses in excess of 100 kDa should be readily measured.

Finally, we point out that stimulated spin-echo NMR has been used to measure translational diffusion coefficients of the three short DNA duplexes (12, 14 and 24 base pairs). The results were in good agreement with data obtained by DLS method (see Table D3.4).

J3.4 Solid-state NMR

In 1983 the first paper that demonstrated that high-resolution structural constraints could be obtained by solid-state NMR was published. The potential of this approach was rapidly established and it evolved through the development of new types of constraints and advances in NMR technology.

In solid-state NMR studies, the sites of interest are immobile on the time scales defined by the relevant nuclear spin interactions, ranging from about 10^3 Hz for chemical shift interactions to 10^6 Hz for nuclear quadrupole interactions. Suitable immobile sites are present in protein microcrystals and in supramolecular structures in solution, such as large nucleoprotein or membrane–protein complexes. In single crystals any arbitrary orientation of the sample can be studied. In uniaxially oriented samples, the direction of orientation must be parallel to that of applied magnetic field.

Obtaining the information contained in solid-state NMR spectra is more difficult than analysing liquid NMR spectra. It is necessary to apply more complicated pulse sequences, use special sample spinning techniques, and apply higher radio-frequency energies (Comment J3.11). All of the experiments benefit from using a high magnetic field.

Fig. J3.22 Diffusion coefficients for 1.4 mM of hen egg white lysozyme as a function of urea concentration. The filled circles and the solid line show data at pH 2.0, while open circles show data at pH 5.5. Diffusion coefficients are presented as a ratio of D_{diox}/D_{lys}, in which the hydrodynamic radius of dioxane was taken as a standard (1.7 Å). It is assumed that for protein and reference molecules in the same solution,

$$R_H^{prot} = \frac{D_{ref}}{D_{prot}} \times R_H^{ref}$$

(After Jones *et al.*, 1997.)

Comment J3.11   The sample spin rate in liquid and solid state NMR

In liquid NMR, the sample spin rate is in the range of $\sim$15–30 Hz. With solids, the spinning is used to average out strong interactions, so the rate is commonly >10 kHz. This significantly changes the engineering of the probe and requires compressed gas and a turbine rotor (Smith and Peerson, 1992).

Assignment methods are also being developed for very fast MAS speeds (20–40 kHz), which have several potential advantages including sensitivity enhancement. However, it is not clear which approach will prove most useful: for example, sensitivity will be increased by very fast MAS because this maximises the centre band intensity at ultrahigh fields, (Fig. J3.24), but it will be decreased because the probes use reduced sample volume.

J3.4.1 Solution and solid-state NMR: comparative analysis.

From an NMR spectroscopist's viewpoint, the main difference between solids and liquids is the mobility of the molecules in the sample. The main restriction for solution NMR, that samples must be tumbling isotropically, does not apply to solid state NMR: for solid-state NMR, the more rigid the samples the better. This reflects the distinction between solid-state NMR (observation of anisotropic systems) and solution NMR (observation of isotropic systems). Consequently, the 'correlation time problem' discussed as a molecular mass limit for solution NMR (Section J2.4) does not apply to solid-state NMR. The molecular mass range that can be studied by solid-state NMR is nearly without bounds, hence there have been studies of silk fibroin, the 16-MDa filamentous virus fd, and colicin E1 P190 in planar lipid bilayers.

The NMR spectra of solids actually contain more information than is available by liquid NMR, but the information is hidden under broad, overlapping peaks with poor resolution. To eliminate or greatly reduce the spectral line-broadening caused by dipolar interactions and CSA (see Section J3.4.2), most solid-state NMR experiments rely on magic angle spinning (MAS) – titling samples to the 'magic angle' of 54.7° relative to the external magnetic field (Fig. J3.23). In MAS, the sample is physically rotated at high speeds in the magnetic field in order average the nuclear spin interactions. Both the chemical shift and the dipolar coupling contribution have terms that contain factors of $(3\cos^2\theta - 1)$, and at the magic angle, these terms vanish (see also Section J2.7.1). For example, the CSA, σ_{izz}, can conveniently be expressed in terms of the angle of rotation of the sample with respect to the applied field:

$$\sigma_{izz} = (3\cos^2\theta - 1)(\text{other terms}) + (3/2\sin^2\theta)\sigma_i \qquad (J3.8)$$

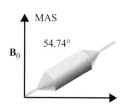

Fig. J3.23 The magic angle in solid-state NMR.

When θ is chosen to be 54.7°, $3\cos^2\theta - 1 = 0$, $\sin^2\theta = 2/3$ and $\sigma_{izz} = \sigma_i$, the isotropic chemical shift, i.e. that which is observed in NMR of liquids. This results in high-resolution ^{13}C NMR spectra of solids.

Figure J3.24 illustrates the distinctive ^{13}C line-shapes for the carboxyl and α-carbon resonances of free glycine. These line-shapes arise from the random orientation of glycine molecules in a polycrystalline sample and are representative of the broad NMR resonances mentioned above. A single molecule or crystallite in the sample contributes a narrow Lorentzian component to the overall line-shape at a frequency that is dependent on its orientation with the respect to the external magnetic field. In Fig. J3.24(a) the line-shapes represent the sum of all of the molecular orientations in the sample and result solely from anisotropy in the chemical-shift interaction because dipolar interactions between ^{13}C and ^1H spins have been eliminated by proton decoupling. In Fig. J3.24(b) and (c), the CSA has been averaged by spinning the sample at the magic angle.

J3.4.2 Solution three-dimensional structure in solid-state NMR

Solid-state NMR has the inherent ability to detect single atomic sites and, hence, the potential to yield high-resolution data. There are many approaches for gaining structural insights into proteins from solid-state NMR. The main ones have orientational, distance, and torsional constraints (Comment J3.12).

Orientation-dependent approaches
The assembly of three-dimensional structures is similar to that in solution NMR where distances and torsional constraints are involved (Section J3.1). However,

Fig. J3.24 Solid-state ^{13}C NMR spectra of glycine illustrating: (a) the broad NMR resonances in static samples; (b) the effect of magic spinning at 2.0 kHz and (c) the effect of magic spinning at 7.2 kHz. The principal values of the chemical-shift tensor are shown for the ^{13}C carboxyl resonance and correspond to the down-field inflection point (σ_{11}), the maximum (σ_{22}), and the up-field inflection point (σ_{33}). MAS collapses the broad line-shapes into sharp centre-bands at the isotropic chemical shifts (σ_{iso}) and the rotational side-bands spaced at the spinning frequency. The frequency scale is centred on the carboxyl centre-band. (After Smith and Peersen, 1992.)

Comment J3.12   Biologist's box: Different constraints in NMR

Orientational constraints are observed in samples that have unique orientation with respect to the magnetic field axis of the NMR spectrometers. By obtaining numerous constraints, all with respect to the same axis, three-dimensional structure can be achieved.

Distance constraints that define the relative separation of two atoms can be obtained through observation of residual dipolar interactions. Both homonuclear and heteronuclear interactions can be observed with considerable precision. As is routine in solution NMR, three-dimensional structures can be solved by distance constraints.

Torsional constraints are defined through the observation of a relative orientation of spin interaction tensors in adjacent sites within the protein. Such constraints lead directly to the definition of the structural torsion angles, the ultimate goal for defining three-dimensional structure.

Fig. J3.25 Representation of the peptide backbone structure of crambine: (a) α-carbon atoms connected by vectors; (b) peptide planes outlined. (Adapted from Opella and Stewart, 1989.)

orientational constraints are unique in this regard. Orientational constraints are absolute and independent, unlike the torsional and distance ones. Moreover, because these constraints are very precise even when there are relatively few constraints, high-resolution structure is achievable.

One solid-state NMR method focuses on the determination of torsion angles in the polypeptide backbone of protein structures solely from orientation constraints. The pertinent features of the protein backbone structure are illustrated with the small 46-residue protein crambin (Fig. J3.25). The backbone of a protein consists of linked planes with quite regular geometry. The α-carbons of adjacent amino acids in a polypeptide chain are joined by the amide C–N bonds. The six atoms that form this peptide linkage lie roughly in a plane, and the relative orientation of adjacent planes is defined by the ϕ and ψ torsion angles (Fig. J3.26). The secondary structure of the peptide backbone can be determined by sequentially establishing the orientation of each peptide plane relative to common axis, the z-axis of the external magnetic field. Measurements of both the dipolar and chemical-shift interactions are necessary to limit the number of possible orientations and define the peptide structure.

Dipole interactions, CSA and quadrupole interactions

The dipole–dipole interactions, CSA and quadrupole interactions in polypeptide backbone sites are highly anisotropic. The dipole–dipole interaction tensor between two spins depends on the distance between the two spins as well as on the orientation of the internuclear vector with respect to the direction of the applied magnetic field; therefore this spin interaction provides spatial as well as angular information. In contrast, the CSA and quadrupole interactions, which reflect local electronic properties, provide only orientational information.

Additional constraints for peptide-plane orientations are derived from the amide ^{15}N and carbonyl ^{13}C chemical shifts tensors. The orientation dependence of the chemical-shift interaction has the form:

$$\sigma = \sigma_{11}\cos^2\alpha\,\sin^2\beta + \sigma_{22}\sin^2\alpha\cos^2\beta + \sigma_{33}\cos^2\beta \qquad (J3.9)$$

Fig. J3.26 Illustration of the peptides planes of a dipeptide and the approximate orientations of the amide ^{15}N and carbonyl ^{13}C chemical-shift tensors with respect to the molecular frame. The rotation axes of the torsion angles ϕ and ψ are defined with the arrows. The angles ϕ and ψ define the orientation of the planar peptide linkages and are determined by measuring the N–C and N–H bond orientations, as well as the ^{15}N and ^{13}C chemical-shift tensors. (After Smith and Peersen, 1992.)

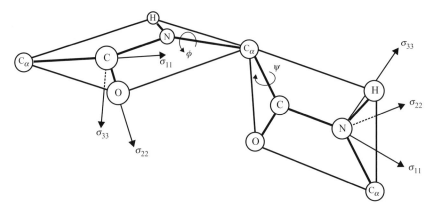

where σ is observed chemical shift, σ_{11}, σ_{22} and σ_{33} are the principal components of the chemical shift tensor and α and β are the Euler angles relating the principal axis system of the chemical-shift tensor to the laboratory frame.

The quadrupole interaction of the spin $S = 1$, ^{14}N amide site is useful for determining angles in solid-state NMR studies and has several important advantages over the other observable interactions. ^{14}N is over 99% naturally abundant, thus no isotopic labelling is required and sensitivity is high; the large quadrupole interaction provides high spectral resolution. The ^{14}N quadrupole tensor has been well characterised for model peptides. Like the chemical-shift tensor, it is non-axially symmetric and thus there is a dependence on two angles, α and β, which describe the orientation of the principal axes of the tensor with respect to the magnetic field vector. Because ^{14}N is an $S = 1$ nucleus there are two fundamental $\Delta m = 1$ transitions and the spectra from oriented samples are doublets centred at the ^{14}N Larmor frequency. The observed splitting between the two frequencies gives angular information described by

$$\Delta \nu_Q = (3/4)\,\varsigma[3\cos^2\beta - 1 + \kappa\sin^2\beta\cos2\alpha] \qquad (J3.10)$$

where ς is quadrupole coupling constant and κ is the asymmetry parameter.

^{2}H quadrupole interactions can also be useful for determining the peptide backbone structure. For a single deuterium bonded to carbon, the quadrupole tensor is axially symmetric with the largest principal component along the C–^{2}H bond. Thus, the observed quadrupole splitting depends on a single angle. By specifically replacing C_α protons with deuterons, it is possible to measure the ^{2}H quadrupole splitting. The angular dependence of the splitting is described by

$$\Delta \nu_Q = (3/4)\,\varsigma[3\cos^2\beta - 1] \qquad (J3.11)$$

where ς is the quadrupole coupling constant, which is approximately 180 kHz, and β is the angle between the C–^{2}H bond and the magnetic field. This angle is the same as would be found from ^{1}H–^{13}C$_\alpha$ dipole–dipole splitting and both measurements have the same angular dependence. Figure J3.27 shows a calculated spectrum that is a doublet with a splitting, $\Delta\nu_D$, corresponding to a heteronuclear ^{15}N–^{1}H dipole-dipole coupling of 17.8 kHz.

Distance-dependent approaches

The second class of methods used to determine protein structure by solid-state NMR relies on an accurate determination of weak dipolar couplings between nuclear spins. A set of distances determined from dipolar couplings serve to constrain the structure, analogous to the use of ^{1}H . . .^{1}H distance constraints from NOE data in solution NMR (Section J2.4). The difficulty in these measurements is that the weak couplings are virtually impossible to observe in the broad NMR spectra, while they average to zero in magic spinning spectra. Two special methods, rotational echo double-resonance (REDOR) and rotational resonance (RR), have been developed to overcome the difficulty.

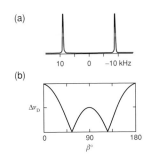

Fig. J3.27 (a) Calculated spectrum for heteronuclear dipole–dipole splitting of 17.8 kHz. (b) Plot of the dipole–dipole splitting, $\delta\nu_D$, versus the angle β. (After Opella and Stewart, 1989.)

Finally we point out that today's solid-state NMR is well suited for studies of membrane proteins immobilised in phospholipid bilayers. Below we demonstrate that the most promising approach to determining three-dimensional structures and extensions to larger membrane proteins is based on the effects of sample orientation in both solution NMR and solid-state NMR.

J3.4.3 Structure of membrane proteins by NMR

Membrane proteins are responsible for many significant biological functions, including those of enzymes, channels and receptors. Unfortunately, membrane proteins present serious obstacles for applications of the most widely used methods in structural biology, such as X-ray crystallography and NMR.

NMR studies of membrane proteins are difficult because of the characteristics of the samples rather than those of the protein themselves. Aside from the fact that membrane proteins are generally insoluble in water, these proteins do not have the unusual features. Membrane proteins are typically monomers in membrane environments and are characteristically organised around long, rigid, hydrophobic helices and shorter amphipatic helices separated by loops, some of which are mobile, and N- and C-terminal segments.

The two well-accepted membrane environments for membrane proteins, (a) detergent micelles and (b) phospholipid bilayers, are shown schematically in Fig. J3.28. Micelles and lipid bilayers provide membrane environments for quite different types of NMR studies of membrane proteins, because of the vast differences in the overall reorientation rates exhibited by polypeptides in these two systems. The reorientation rate of a 25-residue polypeptide (10^{-8} s) in micelles averages the nuclear-spin interaction tensors to their isotropic values, with resonances narrow enough for most multidimensional solution NMR experiments. At the same time, the same polypeptides in bilayers are immobile on the relevant NMR time scales, and require the use of solid-state NMR techniques. Both solution NMR experiments of phospholipid micelle samples and solid-state NMR experiments on oriented phospholipid bilayer samples may provide an approach to measuring orientational and distance parameters for structure determination. A small 5-kDa protein in a micelle behaves like a large 30-kDa protein in water. Although spectral complexity is minimal in this situation, the most serious spectroscopic difficulties of large proteins, broad lines and rapid spin-diffusion, exist

(a)   (b)

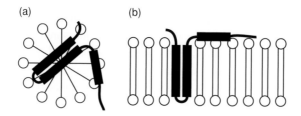

Fig. J3.28 Schematic presentation of a protein in a membrane environment: (a) detergent micelle; (b) phospholipid bilayer.

Fig. J3.29 The overall architecture of the three recombinant constructs, with the hydrophobic helix in blue and both amphipathic helices in red. (After Marassi *et al.*, 1999.)

for the best micelle preparation. This explains why solution NMR is applied only to membrane peptides and relatively short membrane proteins.

Vpu, one of the four accessory proteins encoded in the HIV-1 genome membrane protein has been used to illustrate the high efficiency of the joint use of solution and solid-state NMR approaches. Vpu is an ideal target for NMR because it has a relatively low molecular mass (81 residues), and its individual domains can be obtained by the recombinant technique. NMR spectra of three uniformly ^{15}N-labelled recombinant Vpu constructs have been studied in dihexanoyl phosphatidylcholine micelles. One-dimensional solid-state ^{15}N NMR spectra of the three Vpu constructs were obtained at 0 °C in oriented lipid bilayers. Figure J3.29 summarises the architecture of recombinant Vpu constructs. Vpu has three helical segments, which fold into two distinct domains. Both of the amphipatic helices in the cytoplasmic domain lie in the plane of the bilayer, while, in contrast, the transmembrane hydrophobic helix is perpendicular to the plane of the bilayer. Such structural motifs are highly correlated with specific functions of this protein.

The same approach has been successfully applied to determining the three-dimensional structure of a channel-forming polypeptide corresponding to the M2 segment of the nicotinic acetylcholine receptor and of the gramicidin ion-channel peptide. These results suggest that the determination of three-dimensional structures and extensions to larger membrane proteins rely on the effects of sample orientation in both solution NMR and solid-state NMR experiments.

J3.5 NMR and X-ray crystallography

At the end of 2004 there were more than 4000 solved NMR structures deposited in the PDB, out of a total of more than 24 000 structures. One reason that NMR has not played a larger role in structure determination is its limitation on protein size. Resolution depends inversely on resonance line-widths, and the widths depend in turn on the effective molecular mass. Limits imposed by resolution have been pushed back over the years, by the use of isotope enrichment (^{13}C, ^{15}N, and ^{2}H), by the extension of spectra to three and four dimensions, by measuring residual coupling in oriented samples and more recently by TROSY techniques in combination with high magnetic fields. However, complete structure determination of monomeric proteins still has not pushed beyond a molecular mass limit of 50 kDa, and determinations not requiring deuteration of the protein still seem to be limited to 25–30 kDa. While this is a severe limitation, it is mitigated by the

> **Comment J3.13   Time for data collection**
>
> More severe limitations in the application of NMR are that the time required for data acquisition and analysis is long and sample preparation requires the use of isotopically labelled media (^{15}N-, ^{13}C- and ^{2}H-labelled proteins). The weeks of acquisition and subsequent months-long periods required for assignment and structure determination are still a major obstacle for NMR.
>
> The promise of reducing actual data collection time has come from the introduction of NMR cryoprobes in which the receiver coil and associated electronic components are cooled to a very low temperature (Montelione *et al.*, 2000). Sensitivity improvements of a factor of ~3 can be achieved for protein samples. This saves a factor of 9 in time when substantial signal averaging is required. For example a protein of 180 amino acids required 1.5 days for backbone assignments (Medek *et al.*, 2000).

fact that the average domain size of encoded proteins appears to be about 100 amino acids (Comment J3.13).

Since X-ray diffraction in crystals and NMR in solution can both be used independently to determine the complete three-dimensional structure of proteins, application of the two methods to the same proteins provides a basis for meaningful comparisons of the corresponding structures in single crystals and in non-crystalline states. This is highly relevant, since the solution conditions for NMR studies may coincide closely with the natural physiological environment of the protein, or they may be varied over a wide range for studies of structural transitions with pH, temperature or ionic strength. Extensive similarities between corresponding crystal and solution structures as well as major differences in the conformational features of the two states have now been well documented, and are briefly surveyed in this section.

> **Comment J3.14**
> **Molecular internal motion and flexibility**
>
> The occurrence of internal motion and flexible parts in molecules always presents a challenge in NMR-based structure determination. It is often difficult to distinguish whether the multiple conformers observed in NMR structures reflect real motion, or simply result from insufficient restraints (see Section J3.1).

J3.5.1 Structure of macromolecules in crystal and in solution: comparative studies

A large number of protein structures have been determined by both methods and comparison of these structures allows a further important assessment of the 'reality' of the models. In general, such comparisons provide reassuring confirmation of the similarity of protein structures in the crystalline and solution states, in addition to providing a method for identifying errors in one or other of the techniques (Comment J3.14).

Provided the structures have been correctly determined in both methods, the comparisons show that the fold and conformations both of interior residues and of hydrogen-bonded secondary structures are very similar. However, it has generally been concluded that protein surfaces can have different structures and dynamic

Fig. J3.30 Stereo diagram showing ribbon traces of the superimposed backbone of the four independently determined human interleukin-4 structures. Two different NMR structures (green and yellow) and two different X-ray structures (blue and red) are shown. (After Smith *et al.*, 1994.)

properties in crystals and in solution, and there are several examples in which the differences appear to be real and significant.

One example is human recombinant interleukin-4. The protein has a 129-residue four-helix bundle, for which four independent structures, two by NMR techniques and two by X-ray crystallography, have been compared (Fig. J3.30). The largest differences between the four structures were found in the exposed surface loop regions, which were inadequately defined in all four structures. In the X-ray structures, the diffuse electron density made chain tracing difficult (see Chapter G3).

A second example is the structure of oestrogen receptor DNA-binding domain (84 residues). The domain is monomeric in solution, but two molecules bind cooperatively to specific DNA sequences. The NMR-derived structure is compared with the X-ray crystal structure in Fig. J3.31. Although the two structures are very similar over the regions of ordered residues (C_α RMS deviation $= 1.07$ Å), the NMR-derived structure of the monomer shows that the 15 internal residues (Cys43–Cys59) disordered in solution make contact with both DNA and the corresponding region of the monomer. The results suggest that these residues become ordered during DNA binding, forming the dimer interface and thus contributing to the cooperative interaction between monomers.

A third example is calmodulin. Calmodulin has two globular domains that bind calcium (see also Section F5.3.8). In the protein crystal, a helix connects the two domains, yielding a dumb-bell structure (Fig. J3.32, left panel). In fact, the connecting helix appears to be an artefact of crystallization. When calmodulin is in solution, part of the helical rod melts into a flexible linker, which enables calmodulin to wrap itself around its target (Fig. J3.32, right panel).

Fig. J3.31 Orthogonal views (top and bottom) of the X-ray structure of oestrogen receptor DNA–binding domain–DNA complex: (left) X-ray structure and (right) three-dimensional NMR structure modelled as a dimer with DNA. Two monomers are shown, for clarity, in blue and yellow. (After Schwabe *et al.*, 1993).

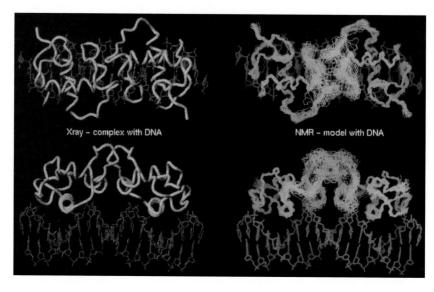

Xray – complex with DNA   NMR – model with DNA

(a)

Fig. J3.32 (a) Calmodulin structure in crystal (pdb file 1osa). (b) Calmodulin (blue) bound to the target helix from calmodulin-dependent protein kinase II (yellow) is shown from two orientations: looking from the side of the target helix (left) and looking down the target helix (right).

(b)

J3.5.2 NMR and structural genomics

The majority of eukaryotic genes do not encode for aqueous single-domain proteins, but for multidomain, membrane and 'unstructurated' proteins. NMR spectroscopy is unique among the techniques of structural biology in its ability to observe and characterise such polypeptide chains in solution. Figure J3.33 shows influence of unstructured regions in proteins on the choice of structure determination method. For proteins with a low percentage of disorder, related X-ray structures are abundant, but for proteins with a higher percentage of disorder (more than 10%), the number of related X-ray structures drops dramatically. NMR-based structures persist to a much greater percentage of disorder. Such observations clearly support the suggestion that NMR-based structure determination is more applicable to proteins with disordered regions, possibly because these proteins may be difficult to crystallise.

Fig. J3.33 The influence
of unstructured regions in
proteins on the choice of
structure determination
approach. The subset
NMR_ONLY means that
only the NMR method
was used for structure
determination, the subset
NMR_XRAY means that
NMR and X-ray methods
were used for structure
determinations. (After
Prestegard *et al.*, 2001.)

Structure determinations have revealed the existence of native, folded proteins that contain long flexible coils attached to well-structured globular domains. A striking example is the prion protein (Fig. J3.34), with a globular domain containing α-helical and β-sheet secondary structure, and an N-terminal domain of nearly equal size that forms a highly mobile extended coil. The length of this extended coil exceeds the diameter of the globular domain by almost ten-fold. A similar structure consisting of a globular domain and a flexibly extended coil has been observed for a yeast heat-shock transcription factor and many others proteins.

Considering the practical difficulties in preparing proteins with long extended coils for structural studies, one is tempted to speculate that this structure type has so far largely escaped detailed characterisation and may be quite common in nature (Comment J3.15).

Whether or not NMR is able to produce structures for disordered proteins, a simple ability to identify these proteins would be a valuable contribution to structural proteomic. Screening of expressed and purified proteins for sample conditions that are apt to promote crystallization or give good NMR samples is therefore an important factor in structural studies. One experiment incorporates a simple test for rapidly exchanging amide protons (Comment J3.16).

Comment J3.15

Extended polypeptide segments in nucleosomes

In this context, it is interesting that the crystal structure of the nucleosome core particle contains numerous extended polypeptide segments in the multimolecular aggregate, indicating that formation of the ollgomeric structure may start with subunits that contain sizable extended coils.

Fig. J3.34 NMR structure
of the recombinant
murine prion protein. In
the intact protein the
segment with residues
126–226 forms a globular
domain, whereas the
segment with residues
23–126 forms an
extended coil. (After Riek
et al., 1997.)

Fig. J3.35 One-dimensional NMR amide proton exchange spectra of BFP at 0.5 mM: (a) with Ca^{2+} and (b) without Ca^{2+}. The spectra were each acquired in about 15 min from on 0.5 mM protein samples using a standard 600-MHz spectrometer. Such experiments can be made quite efficient, and even automated, using NMR flow probes and micromanipulator robots. A special NMR technique that eliminates artefacts due to transfers from α-protons underlying the water resonance was used. (Adapted from Prestegard *et al.*, 2001.)

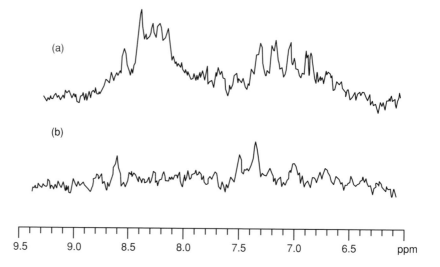

Figure J3.35(b) shows the amide region of a one-dimensional spectrum of BFP in the presence of Ca^{2+}. (See also Section F3.6) The few signals in 8.1--8.9 ppm region are typical of a well-folded protein with a few unstructured or surface-exposed amides. The spectrum suggests that at least in the presence of Ca^{2+} production of quality crystals may be possible. The intensity of the spectrum in Fig. J3.35(b) in the same region is abnormally high and indicative of partial unfolding of the backbone. The data suggest that the protein is difficult to crystallise in the absence of Ca^{2+}, but the protein may be sufficiently folded for an NMR study.

These results show that NMR can be used very efficiently to screen proteins for foldedness and the propensity for crystallization.

J3.6 NMR imaging

At first sight, it may not be obvious why macroscopic imaging should employ just NMR and how an imaging concept could be realised. Consider first the

Fig. J3.36 Illustration of attenuation of different types of radiation by human tissue. All electromagnetic radiation is absorbed except in the X-ray (wavelengths less than 20 Å) and radio-frequency (wavelengths more than 1 m) ranges. Acoustic radiation is strongly absorbed for wavelengths below 1 mm.

attenuation of radiation by human tissue, schematically indicated in Fig. J3.36, taking into account both electromagnetic and acoustic radiation. It is obvious from this figure that nature provides three windows which permit us to look inside the human body.

The X-ray window has been exploited since the basic experiments by Röntgen in 1895 and has completely revolutionised medical diagnosis. In more recent years, X-ray computer tomography has had a further significant impact on medicine, despite the potential dangers of non-negligible radiation doses.

The second window which has been taken advantage of is the low-frequency ultrasonic window. It led to the development of ultrasonic image scanners that permit one to obtain images of acceptable quality in very fast sequence.

The radio-frequency window, on the other hand, was not exploited until 1972. This is not astonishing considering the achievable resolution, which is usually limited by the wavelength of the applied radiation through the uncertainty relation. The maximum radio frequency useful for imaging is about 100 MHz, leading to a resolution of 3 m which is not sufficient even for imaging elephants.

The crucial idea of magnetic resonance imaging MRI is to utilise a magnetic field *gradient* to disperse the NMR resonance frequencies of the various volume elements. The basic principle is shown in Fig. J3.37, where an image of two tubes of water in a linear field gradient is shown.

NMR theory shows that a system of nuclei of a protons precesses with identical Larmor frequencies (ω), if the effective field B_0 experienced by all nuclei is the same (Section J1.2.3). In a high-resolution NMR experiment, we detect the sensitivity of each nucleus to small changes in B_0 dictated by the microenvironment. Since these changes are very small, it is necessary to employ a magnetic field that is spatially homogenous to a very high degree across the sample volume. A linear magnetic field gradient imposes corresponding linear shifts in the Larmor frequencies ($\delta\omega$) of nuclei across the sample volume since each experiences an effective field that is different from B_0 (Comment J3.17).

Fig. J3.37 Simple illustration of magnetic resonance imaging of two tubes of water in linear field gradient. See text for details.

Comment J3.17   Magnetic field gradient and Larmor frequency

Equation (J1.12), the Larmor equation, suggests that these nuclei exhibit difference resonance frequencies, v_1 and v_2 in this instance. Thus, for example, if a magnetic field gradient of 1×10^{-5} T/cm is applied along the bore, or the z-axis, of an MRI magnet, a resonance frequency range of $(2.68 \times 10^8$ radians s^{-1} T$^{-1})$ $(1 \times 10^{-5}$ T/cm)/$(2\omega$ radians) $= 425$ Hz results. In other words, protons 1 cm apart along the field gradient in the subject have resonance frequencies that differ by 425 Hz. Thus, by changing the centre frequency of the NMR probe pulse in increments of 425 Hz, it is possible to probe successive 1-cm positions in the direction of the magnetic field gradient. Each consecutive radio-frequency pulse produces an FID signal that encodes the concentration of protons at each 1-cm position along the direction of the field gradient. When the FIDs are subjected to Fourier transformation, concentration information is produced by the heights of the peaks at the bottom of Fig. J3.37.

Consider two tubes of water placed in a linear gradient (Fig. J3.37). Nuclei at location A experience a static field $B_0 + \delta B_0$ and therefore possess a chemical shift $\omega + \delta\omega$. Similarly, nuclei at location B possess a chemical shift $\omega - \delta\omega$. In addition, since intensity is proportional to the number of spins per unit volume, the NMR spectrum of tubes of water maps a cross-section through the tubes (Fig. J3.37). This section is, however, only a two-dimensional image. To get a complete three-dimensional image, the object or the field gradient must be rotated to produce several cross-sections along different directions. From a series of projections, the object can be reconstituted using image reconstruction. This is known as the projection-reconstruction technique (see Chapter H2).

Magnetic resonance imaging (MRI) is not subject to the same limitations found in optical microscopy. As illustrated Fig. J3.38, MRI can perform

Fig. J3.38 Magnetic resonance image of *Homo sapiens*. The image look approximately 5 min to record. Bar: 10 mm.

Fig. J3.39 Light microscopy (a) and microscopic resonance imaging (side view, (b) and top view, (c) of a living embryo. A single blastomere was injected at the 32-cell stage and the animal was allowed to develop to stage 32 before imaging. The embryo was labelled with bifunctional contrast agent, GRID. GRID contains a paramagnetic metal (Gadolinium III), which is the T_1 agent and super-paramagnetic iron oxide which serves as the T_2 agent. To detoxify the metal ions, they are held within an organic chelator diethylenetriaminepen-taacetic (DPTA). This agent was made visible in fluorescence microscopy by covalently linking a fluorophore to DTPA gadolinium dextran polymer (17 kDa). Bar, 100 μm. (After Jacobs et al., 1999.)

three-dimensional non-invasive imaging of optically opaque specimens. In the vast majority of cases, the magnetic resonance image is based upon the NMR signal from protons of water molecules. The relative amount of signal arising from a particular region is determined by the physical and chemical properties of the region – such as water concentration, temperature, viscosity, magnetic suscep-tibility and the underlying tissue microstructure. These characteristics affect the water diffusion D, spin-lattice relaxation time T_1 and spin–spin relaxation time T_2. One of the principal achievements in MRI was the development of micro-scopic MRI (μMRI) techniques appropriate for imaging embryos. The success of these techniques is evident from MR imaging with a spatial resolution of ≤ 10 μm that have been achieved at field strength ranging from 4.7 T to 14 T using novel MRI contrast agents. Contrast agents have the same function in MRI as fluorescent labels in optical microscopy (see Section F4.3.2). Agents that affect T_1 yield local bright regions and those primarily effecting T_2 yield local dim regions. Figure J3.39 shows a comparison of fluorescence and MRI microscopy of an early frog embryo in which a single cell was injected with rhodamine –DTP A (Gd)-dextran. Notice that the light microscope generates excellent images of the labelled cells near the surface of the embryo but is largely unable to image the labelled cells more central in the animal, such as the labelled clone of cells in the forming brain and spinal chord. These cells are clearly visible in both the side and top views of the (μMRI) images.

In the context of multidimensional Fourier transform NMR, a second technique exists, which is known as a Fourier imaging. In the manner of two-dimensional NMR, the two (or three) frequency coordinates that correspond to a given volume element are measured sequentially in an experiment with one (or two) evolution periods together with a detection period.

J3.7 Checklist of key ideas

- Alignment of biological macromolecules can be achieved in high magnetic fields by including orienting agents such as planar phospholipid bicelles, rod-shaped viruses and purple membrane phages, polymeric strained gels, and so on.
- The NMR spectra of solids actually contain more information than is available from liquid NMR, but the information is hidden under broad, overlapping peaks with poor resolution. To eliminate the spectral line-broadening most solid-state NMR experiments rely on using the magic angle ($54° \, 44'$ relative to the external magnetic field).
- Provided the structures have been correctly determined using NMR and X-ray crystal-lography methods, comparisons show that the fold and conformations of both interior

residues and hydrogen-bonded secondary structures are very similar, but it has been concluded that protein surfaces can have different structures and dynamic properties in crystals and in solution.

- NMR is able to produce three-dimensional structures for partially disordered proteins in solution.
- A prominent NMR application is MRI. Along with X-ray computer tomography and ultrasonic image scanners, three-dimensional NMR imaging today provides significant impact on medicine, especially in cancer research.

Suggestions for further reading

Strategy of structure calculations from NMR data

Wuthrich, K., Wider, G., Wagner, G., and Braun, W. (1982). Sequential resonance assignments as a basis for determination of spatial protein structures by high resolution proton nuclear magnetic resonance. *J. Mol. Biol.*, **155**, 311–319.

Spronk, C. A. E. M., Linge, J. P., Hilbers, C. W., and Vuister, G. W. (2002). Improving the quality of protein structures derived by NMR spectroscopy. *J. Biomolecular NMR*, **22**, 281–289.

Tjandra, N., Garrett, D. S., Gronenborn, A. M., Bax, A., and Clore, G. M. (1997). Defining long range order in NMR structure determination from the dependence of heteronuclear relaxation times on rotational diffusion anisotropy. *Nat. Struct. Biol.*, **4**, 443–449.

Tjandra, N., Omichinski, J. G., Gronenborn, A. M., Clore, G. M., and Bax, A. (1997). Use of dipolar 1H–15N and 1H–13C couplings in the structure determination of magnetically oriented macromolecules in solution. *Nat. Struct. Biol.*, **4**, 732–738.

Three-dimensional structure of biological macromolecules

Wuthrich, K. (1995). NMR – This other method for protein and nucleic acid structure determination. *Acta Cryst.*, **D51**, 249–270.

Wider, G., and Wuthrich, K. (1999). NMR spectroscopy of large molecules and multimolecular assemblies in solution. *Curr. Opin. Struct. Biol.*, **9**, 594–601.

Garret, D. S., Seok, Y.-J., *et al.* (1997). Solution structure of the 30 kDa N-terminal domain of enzyme I of the Escherichia coli phosphoenolpuruvate: sugar phosphotransferase system by multidimensional NMR. *Biochemistry*, **36**, 2517–2530.

Flaux, J., Bertelsen, E. B., Horwich, A. L., and Wuthrich, K. (2002). NMR analysis of a 900K GroEL–GroES complex. *Nature*, **418**, 207–211.

Wuthrich, K. (2000). Protein recognition by NMR. *Nat. Struct. Biol.* **7**, 188–189.

Takahashi, H., Nakanishi, T., Kami, K., Arata, Y., and Shimada, I. (2000). A novel NMR method for determining the interfaces of large protein–protein complexes. *Nat. Struct. Biol.* **7**, 220–223.

Zidek, L., Stefl, R., and Sklenar, V. (2001). NMR methodology for the study of nucleic acids. *Curr. Opin. Struct. Biol.*, **11**, 275–281.

Tjandra, N., Tate, S., Ono, A., Kainosho, M., and Bax, A. (2000). The NMR structure of a DNA dodecamer in an aqueous dilute liquid crystalline phase. *JACS*, **122**, 6190–6200.

Mollova, E. T., Hansen, M. R., and Pardi, A. (2000). Global structure of RNA determined with residual dipolar coupling. *JACS*, **122**, 11561–11562.

Tian, F., Al-Hashimi, H. M., Craighead, J. L., and Prestegard, J. H. (2001). Conformational analysis of a flexible oligosaccharide using residual dipolar coupling. *JACS*, **123**, 485–492.

Dynamics of biological macromolecules

Ishima, R., and Torchia, D. (2000). Protein dynamics from NMR, *Nat. Str. Biol.*, **7**, 740–743.

Stejskal, E. O., and Tanner, J. E. (1965). Spin diffusion measurements: spin echoes in the presence of a time dependent field gradient. *J. Chem. Phys.*, **42**, 288–292.

Jones, J. A., Wilkins, D. K., Smith, L. J., and Dobson, C. M. (1997). Characterization of protein unfolding by NMR diffusion measurements. *J. Biomolecular NMR*, **10**, 199–203.

Solid-state NMR

Smith, S. O., and Peersen, O. B. (1992). Solid-state NMR approaches for studying membrane protein structure. *Ann. Rev. Biophys. Biomol. Str.*, **21**, 25–47.

Opella, S. J., and Stewart, P. L. (1989). Solid-state nuclear magnetic resonance structural studies of proteins. *Meth. Enzymol.*, **176**, 242–275.

Marassi, F. M., Ma, C., *et al.* (1999). Correlation of the structural and functional domains in the membrane protein Vpu from HIV-1. *PANS*, **96**, 14336–14341.

NMR and X-ray crystallography

Smith, L. J., Redfield, C., *et al.* (1994). Comparison of four independently determined structures of human recombinant interleikin-4. *Struct. Biol.*, **1**, 301–310.

Schwabe, J. W. R., Chapman, L., Finch, J. T., Rhodes, D., and Neuhaus, D. (1993). DNA recognition by the oestrogen receptor: From solution to the crystal. *Structure*, **1**, 187–204.

Prestegard, J. H., Valafar, H., Glushka, J., and Tian, F. (2001). Nuclear magnetic resonance in the era of structural Genomics. *Biochemistry*, **40**, 8677–8685.

NMR imaging

Jacobs, R. E., Ahrens, E. T., Meade, T. J., and Fraser, S. E. (1999). Looking deeper into vertebrate development. *TIBS*, **9**, 73–76.

References

Agalarov, S. C., and Williamson, J. R. (2000). A hierarchy of RNA subdomains in assembly of the central domain of the 30S ribosomal subunit. *RNA*, **6,** 402–408.

Agalarov, S. C., Selivanova, O. M., *et al.* (1999). Independent *in vitro* assembly of all three major morphological parts of the 30S ribosomal subunit of *Thermus thermophilus. Eur. J. Biochem.*, **266**, 533–537.

Agalarov, S. C., Sheleznyakova, E. N., *et al.* (1998). In vitro assembly of a ribonucleoprotein particle corresponding to the platform domain of the 30S ribosomal subunit. *Proc. Natl. Acad. Sci. USA*, **95**, 999–1003.

Allemand, J.-F., Bensimon, D., Lavery, R., and Croquette, V. (1998). Stretched and overwound DNA forms a Pouling-like structure with exposed base. *Proc. Natl. Acad. Sci. USA*, **95**, 14152–14157.

Allison, S. A. (1999). Low Reynolds number transport properties of axisymmetric particles employing stick and slip boundary conditions. *Macromolecules*, **32**, 5304–5312.

Allison, S. A. (2001). Boundary element modelling of biomolecular transport. *Biophys. Chem.*, **93,** 197–213.

Altieri, A. S., Hinton, D., and Byrd, R. A. (1995). Association of biomolecular system via pulse field gradient NMR self-diffusion measurements. *JACS*, **117**, 7566–7567.

Altose, M. D., Zheng, Y., Dong, J., Palfey, B. A., and Carey, P. R. (2001). Comparing protein–ligand interactions in solution and single crystals by Raman spectroscopy. *PNAS*, **98**, 3006–3011.

Bacia, K., and Schwille, P. (2003). A dynamic view of cellular processes by in vivo fluorescence auto- and cross-correlation spectroscopy. *Methods*, **29**, 74–85.

Bailey, B., Farkas, D. L., Taylor, D. L., and Lanni, F. (1993). Enhancement of axial resolution in fluorescence microscopy by standing-wave excitation. *Nature*, **366**, 44–48.

Ban, N., Nissen, P., *et al.* (1999). Placement of protein and RNA structures into a 5 Å-resolution map of the 50S ribosomal subunit. *Nature*, **400(6747)**, 841–847.

Banachowicz, E., Gapinski, J., and Patkowski, A. (2000). Solution structure of biopolymers: A new method of constructing a bead model. *Biophys. J.*, **78**, 70–78.

Bandecar, J. (1992). Amide modes and protein conformation. *Biochim. Biophys. Acta*, **1120**, 123–143.

Barnes, W. L., Dereux, A., and Ebbesen, T. W. (2003). Surface plasmon subwavelength optics. *Nature*, **424(6950)**, 824–830.

Barron, L. D., Hecht, L., Blanch, E. W., and Bell, A. F. (2000). Solution structure and dynamics of biomolecules from Raman optical activity. *Prog. Biophys. and Mol. Biol.*, **73**, 1–49.

Basavappa, R., and Sigler, P. B. (1991). *EMBO J.*, **10**, 3105–3111.

Bastiaens P. I. H., and Pepperkok, R. (2000). Observing proteins in their natural habitat: the living cell. *TIBS*, **25**, 631–636.

Baumann, C. G., Bloomfield, V. A., Smith, S. B., Bustamante, C., Wang, M. D., and Block, S. M. (2000). Stretching of single collapsed DNA molecules. *Biophys. J.*, **78**, 1965–1978.

Bax, A. (2003). Weak alignment offers new NMR opportunities to study protein structure and dynamics. *Protein Sci.*, **12**, 1–16.

Belke, J., and Ristau, O. (1997). Analysis of interacting biopolymer systems by analytical centrifugation. *Eur. Biophys. J.*, **25**, 325–332.

Bellissent-Funel, M. C., Zanotti, J. M., *et al.* (1996). Slow dynamics of water molecules on the surface of globular proteins. *Faraday Discuss.*, **103**, 281–294.

Belov, M. E., Gorshkov, M. V., Udeseth, H. R., Anderson, G. A., and Smith, R. D. (2000). Zeptomole-sensititivity electrospray ionization – Fourier transform ion cyclotron resonance mass spectrometry proteins. *Anal. Chem.*, **72**, 2271–2279.

Benner, W. H. (1997). A gated electrostatic ion trap to repetitiously measure the charge and *m/z* of large electrospray ions. *Anal. Chem.*, **69**, 4162–4168.

Bennett, M. J., and Eisenberg, D. (1994). Refined structure of monomeric diphtheria toxin at 2.3 Å resolution. *Protein Sci.*, **3(9)**, 1464–1475.

Bennink, M. L., Leuba, S. H., Leno, G. H., Zlatanova, J., De Grooth, B. G., and Greve J. (2001). Unfolding individual nucleosomes by stretching single chromatin fibers with optical tweezers. *Nature Str. Biol*, **8**, 606–610.

Berg, H. (1983). *Random Walks in Biology*. Princeton: Princeton University Press.

Bergethon, P. R. (1995). *The Physical Basis of Biochemistry. The Foundation of Molecular Biophysics*. New York: Springer.

Bernado, P., Garcia de la Torre, J., and Pons, M. (2002). Interpretation of 15N NMR relaxation data for globular proteins using hydrodynamic calculations with HYDRONMR. *J. Biomol. NMR*, **23**, 139–150.

Bernal, J. D., and Crowfoot, D. (1934). X-ray photographs of crystalline pepsin. *Nature*, **134**, 794–795.

Biemann, K. (1992). Mass spectrometry of peptides and proteins. *Annu. Rev. Biochem.*, **61**, 977–1010.

Bischler, N., Brino, L., *et al.* (2002). Localization of the yeast RNA polymerase I-specific subunits. *Embo. J.*, **21(15)**, 4136–4144.

Bjorkman, P. J., Saper, M. A., *et al.* (1987). Structure of the human class I histocompatibility antigen, HLA-A2. *Nature*, **329(6139)**, 506–512.

Blattner, F. R. (1997). The complete genome sequence of *Eshcherichia coli* K-12. *Science*, **277**, 1453–1474.

Block, S. M., Blair, D. F., and Berg, H. C. (1989). Compliance of bacterial flagella measured wih optical tweezers. *Nature*, **338**, 514–518.

Bon, C., Dianoux, A. J., *et al.* (2002). A model for water motion in crystals of lysozyme based on an incoherent quasielastic neutron-scattering study. *Biophys. J.*, **83(3)**, 1578–1588.

Bon, C., Lehmann, M. S., *et al.* (1990). Quasi Laue neutron-diffraction study of the water arrangement in crystals of triclinic hen egg-white lysozyme. *Acta Crystallogr.* D, **55**, 978–987.

Booth, D. R., Sunde, M., *et al.* (1997). Instability, unfolding and aggregation of human lysozyme variants underlying amyloid fibrillogenesis. *Nature*, **385**, 787–793.

Bottcher, B., Tsuji, N., *et al.* (1998). Peptides that block hepatitis B virus assembly: analysis by cryomicroscopy, mutagenesis and transfection. *Embo. J.*, **17(23)**, 6839–6845.

Bowie, J. U., Luthy, R., and Eisenberg, D. (1991). A method to identify protein sequences that fold into a known three-dimensional structure. *Science*, **253(5016)**, 164–170.

Braiman, M. S., and Rothschild, K. J. (1988). Fourier transform infrared techniques for probing membrane protein structure. *Ann. Rev. Biophys. Biophysical Chem.*, **17**, 541–570.

Brändén, C.-I., and Tooze, J. (1999). *Introduction to Protein Structure*. New York: Garland Pub.

Brant, D. A. (1999). Novel approaches to the analysis of polysaccharide structure. *Curr. Opin. Struct. Biol.*, **9**, 556–562.

Brey, W. S. (ed.) (1988). *Pulse Methods in 1D and 2D Liquid-Phase NMR*. San-Diego: Academic.

Brooks, B. R., Bruccoleri, R. E., *et al.* (1983). CHARMM: A program for macromolecular empirical energy modelling. *J. Comp. Chem.*, **4**, 187–230.

Brower-Toland, B. R., Smith, C. L., Yeh, R. S., Lis, J. T., Peterson, C. L., and Wang, M. D. (2002). Mechanical disruption of individual nucleosomes reveals a reversible multistage release of DNA. *Proc. Natl. Acad. Sci. USA*, **99**, 1960–1966.

Brudler, R., Rammelsberg, R., *et al.* (2001). Structure of the I1 early intermediate of photoactive yellow protein by FTIR spectroscopy. *Nat. Struct. Biol.*, **8(3)**, 265–270.

Brune, D., and Kim, S. (1993). Predicting protein diffusion coefficients. *J. Am. Chem. Soc.*, **90**, 3835–3839.

Brunger, A. T., and Adams, P. D. (2002). Molecular dynamics applied to X-ray structure refinement. *Acc. Chem. Res.*, **35(6)**, 404–412.

Brunger, A. T., Adams, P. D., *et al.* (1998). Crystallography & NMR system: A new software suite for macromolecular structure determination. *Acta Crystallogr. D Biol. Crystallogr.*, **54** (Pt 5), 905–921.

Buchanan, M. V., and Hettich, R. L. (1993). Fourier transform mass spectrometry of high mass molecules. *Anal. Chem.*, **65**, 245A–259A.

Burlingame, A. L., and Carr, S. A. (1996). *Mass Spectrometry in the Biological Sciences*. Totowa: Humana Press.

Burlingame, A. L., Carr, S. A., *et al.* (2000). *Mass Spectrometry in Biology and Medicine*. Totowa: Humana Press.

Burton, A., and Sinsheimer, R. L. (1965). The process of infection with bacteriophage X174. VII. Ultracentrifugal analysis of the replicative form. *J. Mol. Biol.*, **14**, 327–347.

Bustamante, C., Erie, D. A., and Keller, D. (1994). Biochemical and structural applications of scanning force microscopy. *Curr. Opin. Struct. Biol.*, **4**, 750–760.

Byron, O. (1997). Construction of hydrodynamic bead models from high resolution x-ray crystallographic or nuclear magnetic resonance data. *Bioph. J.*, **72**, 406–415.

Caffrey, M. (2000). A lipid's eye view of membrane protein crystallization in mesophases. *Curr. Opin. Struct. Biol.*, **10(4)**, 486–497.

Cai, S., and Singh, B. R. (1999). Identification of beta-turn and random coil amide III infrared bands for secondary structure estimation of proteins. *Biophys. Chem.*, **80(1)**, 7–20.

Callender, R., and Deng, H. (1994). Nonresonance Raman difference spectroscopy: a general probe of protein structure, ligand binding, enzymatic catalysis, and the structures of other biomacromolecules. *Annu. Rev. Biophys. Biomol. Struct.*, **23**, 215–245.

Callis, P. R., and Davidson, N. (1969). Hydrodynamic relaxation times of DNA from decay of flow dichroism measurements. *Biopolymers*, **8**, 379–390.

Campos-Olivas, R., Horr, I., Bormann, C., Jung G., and Gronenborn, A. M. (2001). Solution structure, backbone dynamics and chitin binding of the anti-fungal protein from Streptomyces tendae TU901. *J. Mol. Biol.*, **308**, 765–782.

Canet, D., Doering, K., Dobson, C. M., and Dupont, Y. (2001). High-sensitivity fluorescence anisotropy detection of protein-folding events: application to α-lactalbumin. *Biophys. J.*, **80**, 1996–2003.

Cantor, C., and Schimmel, P. (1980). *Biophysical Chemistry*. Part II. *Technique for the Study of Biological Structure and Function*. San Francisco: W. H. Freeman and Company.

Caprioli, R. M., and Suter, M. J.-F. (1995). Mass spectrometry. In *Introduction to Biophysical Methods for Protein and Nucleic Acid Research*. San Diego: Academic Press.

Carr, S. A., and Burlingame, A. L. (1996). The meaning and usage of the terms monoisotopic mass, average mass, mass resolution, and mass accuracy for measurements of biomolecules. In *Mass Spectrometry in the Biological Sciences*, eds. A. L. Burlingame and S. A. Carr, pp. 546–552. Totowa: Humana Press.

Carra, J. H., Murphy, E. C., *et al.* (1996). Thermodynamic effects of mutations on the denaturation of T4 lysozyme. *Biophys. J.*, **71(4)**, 1994–2001.

Carrasco, B., and Garcia de la Torre, J. (1999). Hydrodynamic properties of rigid particles: comparison of different modelling and computational procedure. *Biophys. J.*, **75**, 3044–30572.

Carrion-Vazquez, M., Overhauser, A. F., *et al.* (2000). Mechanical design of proteins studied by single-molecule force spectroscopy and protein engineering. *Prog. Biophys. Mol. Biol.*, **74**, 63–91.

Castro, A., Fairfield, F. R., and Shera, E. B. (1993). Fluorescence detection and size measurement of single DNA molecules. *Anal. Chem.*, **65**, 849–852.

Cate, J. H., Yusupov, M. M., *et al.* (1999). X-ray crystal structures of 70S ribosome functional complexes. *Science*, **285(5436)**, 2095–2104.

Chacon, P., Moran, F., *et al.* (1998). Low-resolution structures of proteins in solution retrieved from X-ray scattering with a genetic algorithm. *Biophys. J.*, **74(6)**, 2760–75.

Chacon, P., Diaz, J. F., *et al.* (2000). Reconstruction of protein form with X-ray solution scattering and a genetic algorithm. *J. Mol. Biol.*, **299(5)**, 1289–1302.

Charney, E., Chen, H.-H., and Rau, D. (1991). The flexibility of A-form DNA. *J. Biolmol. Struct. Dyn.*, **9**, 353–362.

Che, Z., N. Olson, H., *et al.* (1998). Antibody-mediated neutralization of human rhinovirus 14 explored by means of cryoelectron microscopy and X-ray crystallography of virus–Fab complexes. *J. Virol.*, **72(6)**, 4610–4622.

Checovich, W. J., Bolger, R. E., and Burke, T. (1995). Fluorescence polarization – a new tool for cell and molecular biology. *Nature*, **375**, 254–256.

Chen, X., Wu, H., Mao, C., and Whitesides, G. M. (2002). A prototype two-dimensional capillary electrophoresis system fabricated in poly(dimethylsiloxane). *Anal. Chem.*, **74**, 1772–1778.

Cherry, R. J., and Schneider, G. (1976). A spectroscopic technique for measuring slow rotational diffusion of macromolecules. 2: Determination of rotational correlation times of protein in solution. *Biochemistry*, **15**, 3657–3661.

Chervenka, 1969. *A Manual of Methods for the Analytical Ultracentrifuge*. Palo Alto: Spinco Division, Beckman Instruments.

Chong, B. E., Lubman, D. M., *et al.* (1999). Rapid screening of protein profiles of human breast cancer cell lines using non-porous reversed-phase high performance liquid chromatography separation with matrix-assisted laser desorption/ionization time-of-flight mass spectral analysis. *Rapid. Commun. Mass Spectr.*, **13(18)**, 1808–1812.

Chong, B. E., Lubman, D. M., Rosenspire, A., and Miller, F. (1998). Protein profiles and identification of high performance liquid chromatography isolated proteins of cancer cell lines using matrix-assisted laser desorption/ionization time-of-flight mass spectrometry. *Rapid Commun. Mass Spectr.*, **12**, 1986–1993.

Clore, G. M., and Gronenborn, A. M. (1991). Two-, three-, and four-dimensional NMR methods for obtaining more precise three-dimensional structure of proteins in solution. *Ann. Rev. Biophys. Chem.*, **20**, 29–63.

Cluzel, P., Lebrun, A., Heller, C., Lavery, R., Viovy, J.-L., Chatenay, D., Caron, F. (1996). DNA: an extensible molecule. *Science*, **271**, 792–794.

Cohn, E. J., and Edsall, J. T. (1965). In *Proteins, Amino Acids and Peptides as Ions and Dipolar Ions*, eds. E. J. Cohn and J. T. Edsall, pp. 370–381. New York: Hafner Publ. Co.

Collins, K. D. (1997). Charge density-dependent strength of hydration and biological structure. *Biophys. J.*, **72(1)**, 65–76.

Colon, L. A., Guo, Y., and Fermier, A. (1997). Capillary electrochromatography. *Anal. Chem. News*, **69**, 461A–467A.

Cordone, L., Ferrand, M., *et al.* (1999). Harmonic behavior of trehalose-coated carbon-monoxy-myoglobin at high temperature. *Biophys. J.*, **76**, 1043–1047.

Crichton, R. R., Engelman, D. M., *et al.* (1977). Contrast variation study of specifically deuterated Escherichia coli ribosomal subunits. *Proc. Natl. Acad. Sci. USA*, **74**, 5547–5550.

Crick, F. H. (1968). The origin of the genetic code. *J. Mol. Biol.*, **38**, 367–379.

Crothers, D. M., and Zimm, B. H. (1965) Viscosity and sedimentation of the DNA from bacteriophages T2 and T7 and the relation to molecular weight. *J. Mol. Biol.*, **12**, 527–536.

Dasgupta, S., and Spiro, T. G. (1980). Resonance Raman characterization of the 7-ns photoproduct of (carbonmonoxy) hemoglobin: implications for hemoglobin dynamics. *Biochemistry*, **25**, 5941–5948.

Davidson, I. W., and Secrest, W. L. (1972). Determination of chromium in biological materials by atomic absorption spectrometry using a graphite furnace atomizer. *Anal. Chem.*, **44(13)**, 1808–1813.

De la Torre, G. (2001) Hydration from hydrodynamics. General consideration and applications to bead modelling to globular proteins. *Biophys. Chem.*, **93**, 159–170.

Dekker, N. H., Rybenkov, V. V., *et al.* (2002). The mechanism of type IA topoisomerases. *Proc. Natl. Acad. Sci. USA*, **99**, 12126–12131.

Delano, W. L., and Brunger, A. T. (1995). The direct rotation function: rotational Patterson correlation search applied to molecular replacement. *Acta Cryst. D*, **51**, 740–748.

Derome, A. E. (1987). *Modern NMR Techniques for Chemistry Research*. New York: Pergamon.

Dessen, P., Blanquet, S., Zaccai, G., and Jacrot, B. (1978). Antico-operative binding of initiator transfer RNAMet to methionyl-transfer RNA synthetase from Escherichia coli: neutron scattering studies. *J. Mol. Biol.*, **126**, 293–313.

Dickerson, R., and Geiss, I. (1969). *The Structure and Action of Proteins*. Menlo Park: Benjamin Cummings.

Diehl, M., Doster, W., *et al.* (1997). Water-coupled low-frequency modes of myoglobin and lysozyme observed by inelastic neutron scattering. *Biophys. J.*, **73**, 2726–32.

Dobo, A., and Kaltashov, I. A. (2001). Detection of multiple protein conformational ensembles in solution via deconvolution of charge-state distribution in ESI MS. *Anal. Chem.*, **73**, 4763–4773.

Dolgikh, D. A., Gilmanshin, R. I., *et al.* (1981). Alpha-lactalbumin: compact state with fluctuating tertiary structure? *FEBS Lett.*, **136**, 311–315.

Dong, J., Wan, Z., Popov, M., Carey, P. R., and Weiss, M. A. (2003). Insulin assembly damps conformational fluctuations: Raman analysis of amide I linewidths in native states and fibrils. *JMB*, **330**, 431–442.

Doster, W., Cusack, S., *et al.* (1989). Dynamical transition of myoglobin revealed by inelastic neutron scattering. *Nature*, **337**, 754–756.

Doty, P., Bradbury, J. H., and Holtzer, A. M. (1956). Polypeptides. IV. The molecular weight, configuration and association of poly-γ-glutamate in various solvents. *J. Am. Chem. Soc.*, **78**, 947–954.

Dubin, S. B., Clark, N. A., and Benedek, G. B. (1971). Measurement of the rotational diffusion coefficient of lysozyme by depolarised light scattering: configuration of lysozyme in solution. *J. Chem. Phys.*, **54**, 5158–5164.

Dunkerk, A. K., and Williams, R. W. (1979). Ultraviolet and LASER Raman investigation of the buried tyrosines in fd phage. *J. Biol. Chem.*, **254**, 6446.

Dutta, R. K., Hammons, K., Willibey, B., and Haney, M. A. (1991). Analysis of protein denaturation by high-performance continuous differential viscometry. *J. Cromatog.*, **536**, 113–121.

Dykxhoorn, D. M., Novina, C. D., *et al.* (2003). Killing the messenger: short RNAs that silence gene expression. *Nat. Rev. Mol. Cell Biol.*, **4**, 457–467.

Eastman, J. E., Taguchi, A. K., *et al.* (2000). Characterization of a *Rhodobacter capsulatus* reaction center mutant that enhances the distinction between spectral forms of the initial electron donor. *Biochemistry*, **39**, 14787–14798.

Eden, D., Luu, B. Q., Zapata, D. J., Sablin, E. P., and Kuul, F. J. (1995). Solution structure of two molecular motor domains: nonclaret disjunctional and kinesin. *Biophys. J.*, **68**, 59s–65s.

Eigen, M., and Rigler, R. (1994). Sorting single molecules: applications to diagnostic and evolutionary biotechnology. *Proc. Natl. Acad. Sci. USA*, **91**, 5740–5747.

Eimer, W., and Pecora, R. (1991). Rotational and translational diffusion of short rodlike molecules in solution: oligonucleotides. *J. Chem. Phys.*, **94**, 2324–2329.

Eimer, W., Williamson, J. R., Boxer, S. G., and Pecora, R. (1990). Characterization of the overall and internal dynamics of short oligonucleotides by depolarised dynamic light scattering and NMR relaxation measurements. *Biochemistry*, **29**, 799–811.

Eisenberg, H. (1981). Forward scattering of light, X-rays and neutrons. *Q. Rev. Biophys.*, **14**, 141–172.

Elias, J. G., and Eden, D. (1981). Transient electric birefringence study of the persistence length and electrical polarizability of restriction fragments of DNA. *Macromolecules*, **14**, 410–419.

Elöve, G. A., Chaffotte, A. F., *et al.* (1992). Early steps in cytochrome c folding probed by time-resolved circular dichroism and fluorescence spectroscopy. *Biochemistry*, **31**, 6876–6883.

Engel, A., Lyubchenko, Y., and Muller, D. (1999). Atomic force microscopy: a powerful tool to observe biomolecules at work. *Trends Cell Biol.*, **9**, 77–80.

Ernst, R. R., Bodenhausen, G., and Wokaun, A. (1987). *Principles of Nuclear Magnetic Resonance in One and Two Dimensions*. Oxford: Oxford University Press.

Essavaz-Roulet, B., Bockelman, U., and Heslot, F. (1997). Mechanical separation of the complementary strands of DNA. *PNAS*, **94**, 11935–11940.

Fan, E., Merritt, E. A., *et al.* (2000). AB(5) toxins: structures and inhibitor design. *Curr. Opin. Struct. Biol.*, **10**, 680–686.

Fasman, G. D. (1996). *Circular Dichroism and the Conformational Analysis of Biomolecules*. New York: Plenum Press.

Ferrer, M. L., Duchowicz, R., Carrasco, B., Garcia de la Torre, J., and Acuna, A. U. (2001). The conformation of serum albumin in solution: a combined phosphorescence depolarization–hydrodynamic modelling study. *Biophys. J.*, **80**, 2422–2430.

Feynman, R. P., Leighton, R. B., *et al.* (1963). *The Feynman Lectures on Physics*. Reading, MA: Addison-Wesley Pub. Co.

Fisher, T. E., Marszalek, P. E., and Fernandez, J. M. (2000). Stretching single molecules into novel conformations using the atomic force microscopy. *Nature Stuctr. Biol.*, **7**, 719–724.

Flaux, J., Bertelsen, E. B., Horwich, A. L., and Wuthrich, K. (2002). NMR analysis of a 900 K GroEL–GroES complex. *Nature*, **418**, 207–211.

Flynn, P. F., and Wand, A. J. (2001). High-resolution nuclear magnetic resonance of Encapsulated proteins dissolved in low viscosity fluids, *Methods Enzymol.*, **339**, 54–70.

Franklin, S. E., and Gosling, R. G. (1953). Molecular configuration in sodium thymonucleate. *Nature*, **171**, 740–741.

Franks, F., Gent, M., *et al.* (1963). Solubility of benzene in water. *J. Chem. Soc.*, **8**, 2716–2723.

Franzen, S., and Boxer, S. G. (1997). On the origin of heme absorption band shifts and associated protein structural relaxation in myoglobin following flash photolysis. *J. Biol. Chem.*, **272**, 9655–9660.

Frauenfelder, H., Parak, F., *et al.* (1988). Conformational substates in proteins. *Ann. Rev. Biophys. Chem.*, **17**, 451–479.

Freifelder, D. (1970). Molecular weights of coliphages and coliphage DNA. IV, Molecular weights of DNA from bacteriophages T4, T5, and T7 and general problem of determination of molecular weight. *J. Mol. Biol.*, **54**, 567.

Freire, E. (1995). Thermal denaturation methods in the study of protein folding. *Methods Enzymol.*, **259**, 144–168.

Frey, W., Schief, W. R., Jr., *et al.* (1996). Two-dimensional protein crystallization via metal-ion coordination by naturally occurring surface histidines. *Proc. Natl. Acad. Sci. USA*, **93**, 4937–4941.

Frohn, J. T., Knapp, H. F., and Stemmer, A. (2000). True optical resolution beyond the Rayleigh limit achieved by standing wave illumination. *Proc. Natl. Acad. Sci. USA*, **97**, 7232–7236.

Fuller, S. D., and Argos, P. (1987). Is Sindbis a simple picornavirus with an envelope? *Embo. J.*, **6**, 1099–1105.

Fuller, W., Forsyth, T., *et al.* (2004). Water-DNA interactions as studied by X-ray and neutron fibre diffraction. *Phil. Trans. R. Soc. Lond B Biol. Sci.*, **359**, 1237–1247; discussion 1247–1248.

Gabel, F., Bicout, D., *et al.* (2002). Protein dynamics studied by neutron scattering. *Q. Rev. Biophys.*, **35**, 327–367.

Gabel, F. (2005). Protein dynamics in solution and powder measured by incoherent elastic neutron scattering: the influence of Q-range and energy resolution. *Eur. Biophys. J.*, **34**, 1–12.

Gadola, S. D., Zaccai, N. R., *et al.* (2002). Structure of human CD1b with bound ligands at 2.3 Å, a maze for alkyl chains. *Nat. Immunol.*, **3**, 721–726.

Ganem, B. (1993). Detecting noncovalent interactions: new frontiers for mass spectrometry. *Am. Biotechnol. Lab.*, **11**, 32–34.

Ganem, B., Li, Y.-T., and Henion, J. D. (1991). Observation of noncovalent enzyme. Substrate and enzyme product complexes by ion spray mass spectrometry. *J. Am. Chem. Soc.*, **113**, 7818–7819.

Garces-Chavez, V., McGloin, D., Melville, H., Sibbett, W., and Dholakia, K. (2002). Simultaneous micromanipulation in multiple planes using a self-reconstruction light beam. *Nature*, **419**, 145–147.

Garcia de la Torre, J. (2001). Hydration from hydrodynamics. General consideration and applications of bead modelling to globular proteins. *Biophys. Chem.*, **93**, 159–170.

Garcia de la Torre, J., Martinez, M. C. L., and Tirado, M. M. (1984). Dimensions of short, rodlike macromolecules from translational and rotational diffusion coefficients. Study of the gramicidin dimer. *Biopolymers*, **23**, 611–615.

Garcia de la Torre, J., Navarro, S., and Lopez Martinez, M. C. (1994). Hydrodynamic properties of a double-helical model for DNA. *Biophys. J.*, **66**, 1573–1579.

Garfin, D. E. (1995). Electrophoretic methods. In: *Introduction to Biophysical Methods for Protein and Nucleic Acid Research*, eds. J. A. Glaser and M. P. Deutscher. San Diego: Academic Press.

Garret, D. S., Seok, Y.-J., Liao, D.-I., Peterkofsky, A., Gronenborn, A. M., and Clore, G. M. (1997). Solution structure of the 30 kDa N-terminal domain of enzyme I of the Escherichia coli phosphoenolpuruvate: sugar phosphotransferase system by multidimensional NMR. *Biochemistry*, **36**, 2517–2530.

Gelles, J., and Landick, R. (1998). RNA polymerase as a molecular motor, *Cell*, **93**, 13–16.

Giege, R., Lorber, B., *et al.* (1982). Formation of a catalytically active complex between tRNAAsp and aspartyl-tRNA synthetase from yeast in high concentrations of ammonium sulphate. *Biochimie*, **64**, 357–362.

Gilbet, W. (1986). The RNA world. *Nature*, **319**, 618.

Gimzewski, J. K., and Joachum, C. (1999). Nanoscale science of single molecules using molecular probes. *Science*, **283**, 1683–1688.

Gluehmann, M., Zarivach, R., *et al.* (2001). Ribosomal crystallography: from poorly diffracting microcrystals to high-resolution structures. *Methods*, **25**, 292–302.

Go, N., Noguti, T., *et al.* (1983). Dynamics of a small globular protein in terms of low-frequency vibrational modes. *Proc. Natl. Acad. Sci. USA*, **80**, 3696–3700.

Godovach-Zimmermann, J., and Brown, L. R. (2001). Perspectives for mass spectrometry and functional proteomics. *Mass Spectr. Rev.*, **20**, 1–57.

Goldberg, D. E. (1989). *Genetic Algorithms in Search, Optimization, and Machine Learning.* Reading, MA: Addison-Wesley Pub. Co.

Gomez, J., Hilser, V. J., *et al.* (1995). The heat capacity of proteins. *Proteins*, **22**, 404–412.

Goodsell, D. S., and, Olson, A. J. (1993), Soluble proteins: size, shape and function. *TIBS*, **18**, 65–68.

Gordon, D. B. (2000). Mass spectrometric technique. In *Principles and Techniques of Practical Biochemistry*, 5th edn., eds. K. Wilson and J. Walker. Ch. 11. Cambridge: Cambridge University Press.

Greis, K. D., Hayes, B. K., *et al.* (1996). Selective detection and site-analysis of O-GlcNAc-modified glycopeptides by beta-elimination and tandem electrospray mass spectrometry. *Anal. Biochem.*, **234**, 38–49.

Grier, D. (2003). A revolution in optical manipulation. *Nature,* **424**, 810–816.

Griko, Y. V., Freire, E., *et al.* (1995). The unfolding thermodynamics of c-type lysozymes: a calorimetric study of the heat denaturation of equine lysozyme. *J. Mol. Biol.*, **252**, 447–459.

Griko, Y. V., Makhatadze, G. I., *et al.* (1994). Thermodynamics of barnase unfolding. *Protein Sci.*, **3**, 669–676.

Gross, S. (2003). Application of optical traps in vivo. *Methods in Enzymol.*, V **361**, 162–174.

Grotjahn, L., Frank, R., and Blocker, H. (1982). Ultrafast sequencing of oligodeoxyribonucleotides by FAB-mass spectrometry. *Nucl. Acids Res.*, **10**, 4671–4677.

Gutsche, I., Holzinger, J., *et al.* (2001). ATP-induced structural change of the thermosome is temperature-dependent. *J. Struct. Biol.*, **135**, 139–146.

Haag, L., Garoff, H., Xing, L., Hammar, L., Kan, S. T., Cheng, R. H. (2002). Acid-induced movements in the glycoprotein shell of an alphavirus turn the spikes into membrane fusion mode. *EMBO J.*, **21**, 4402–4410.

Hafner, J. H., Cheung, C.-L., Wooley, A. T., and Lieber, C. M. (2001). Structural and functional imaging with carbon nanotube AFM probes. *Progr. Biophys. Mol. Biol.*, **77**, 73–110.

Hagerman, P. J. (1981). Investigation of the flexibility of DNA using transient electric birefringence. *Biopolymers*, **20**, 1503–1535.

Hagerman, P. J. (1985). Application of transient electric birefringence to the study of biopolymer structure. *Methods Enzymol.*, **117**, 199–215.

Hagerman, P. J. (2000). Transient electric birefringence for determining global conformations of non-helix elements and protein-induced bends in RNA. *Methods Enzymol.*, **317**, 440–453.

Hahn, T. and International Union of Crystallography (2002). *International Tables for Crystallography*. Brief teaching edition of volume A, Space-group symmetry. Dordrecht; Boston, Published for the International Union of Crystallography by Kluwer Academic Publishers.

Hamm, P., Lim, M., and Hochstrasser, R. M. (1999). Structure of the amide I band of peptides measured by femtosecond non-linear-infrared spectroscopy. *PNAS*, **96**, 6123–6128.

Han, W., Lindsay, S. M., Dlakic, M., and Harrington, R. E. (1997). Kinked DNA. *Nature*, **386**, 563.

Hansen, J. C., Lebowitz, J., and Demeler, B. (1994). Analytical ultracentrifugation of complex macromolecular systems. *Biochemistry*, **33**, 13155–13163.

Hansen, M. R., Mueller, L., and Pardi, A. (1998). Tunable alignment of macromolecules by filamentous phage yields dipolar coupling interaction. *Nat. Struct. Biol.*, **5,** 1065–1074.

Harding, S. E. (1980). The combination of the viscosity increment with the harmonic mean rotational relaxation time for determining the conformation of biological macromolecules in solution. *Biochem. J.*, **189**, 359–361.

Harding, S. E. (1981). A compound hydrodynamics shape function derived from viscosity and molecular covolume measurements. *Int. J. Biol. Macromol.*, **3**, 398–399.

Harding, S. E. (1995). On the hydrodynamic analysis of macromolecular conformation. *Biophys. Chem.*, **55**, 69–93.

Harding, S. E., and Rowe, A. (1982). Modelling biological macromolecules in solution: 1. The ellipsoid of revolution. *Int. J. Biol. Macromol.*, **4**, 160–164.

Harding, S. E., Horton, J. C., and Colfen, H. (1997). The *ELLIPS* suite of macromolecular conformation algorithms. *Eur. Biophys, J.*, **25**, 347–359.

Harpaz, Y., Gerstein, M., Chothia, C. (1994). Volume changes on protein folding. *Structure*, **2**, 641–649.

Harris, R. (1983). *Nuclear Magnetic Resonance Spectroscopy*. London: Pitman.

Haupts, U., Tittor, J., *et al.* (1997). General concept for ion translocation by halobacterial retinal proteins: the isomerization/switch/transfer (IST) model. *Biochemistry*, **36**, 2–7.

Haupts, U., Tittor, J., and Oesterhelt, D. (1999). Closing in on bacteriorhodopsin: progress in understanding the molecule. *Ann. Rev. Biophys. Biomol. Struct.*, **28**, 367–399.

Hausten, E., and Schwille, P. (2003). Ultrasensitive investigations of biological systems by fluorescence correlation spectroscopy. *Methods*, **29**, 153–166.

Hazlett, T. L., Moore, K. J. M., Lowe, P. N., Jameson, D. M., and Eccleston, J. F. (1993). Solution of p21 [ras] proteins bound with fluorescent nucleotides: a time-resolved fluorescence study. *Biochemistry*, **32**, 13575–13583.

Heberle, J., and Gensch, T. (2001). When FT-IR spectroscopy meets X-ray crystallography. *Nat. Struct. Biol.*, **8**, 195–197.

Hellweg, T., Eimer, W., Krahn, E., Schneider, K., and Muller, A. (1997). Hydrodynamic properties of nitrogenase – the MoFe protein from *Azotobacter vinelandii* studied by dynamic light-scattering and hydrodynamic modelling. *Biochim. Biophys. Acta*, **1337**, 311–318.

Hensley P. (1996). Defining the structure and stability of macromolecular assemblies in solution: the re-emergence of analytical ultracentrifugation as a practical tool. *Structure,* **4**, 367–373.

Hillisch, A., Lorenz, M., and Diekmann, S. (2001). Recent advances in FRET: distance determination in protein–DNA complexes. *Curr. Opin. Struct. Biol.*, **11**, 201–207.

Hirao, I., and Ellington, A. D. (1995). Re-creating the RNA world. *Curr. Biol.*, **5**, 1017–1022.

Homans, S.W., Edge, C. J., Ferguson, M. A., and Dwek, R. A. (1989). Solution structure of the glycosylphosphatidylinositol membrane anchor glycan of *Trypanosoma bruccei* variant surface glycoprotein. *Biochemistry*, **28**, 2881–2887.

Hore, P. J. (1995). *Nuclear Magnetic Resonance*. Oxford: Oxford University Press.

Horwitz, J., Strickland, E. H., and Billups, C. (1970). Analysis of the vibrational structure in the near-ultraviolet circular dichroism and absorption spectra of tyrosine derivatives and ribonuclease-A at 77 K. *J. Am. Chem. Soc.*, **92**, 2119–2129.

Hu, C.-M., and Zwanzig, R. (1974). Rotational friction coefficients for spheroids with the slipping boundary conditions. *J. Chem. Phys.*, **60**, 4354–4357.

Hunt, J. F., McCrea, P. D., *et al.* (1997). Assessment of the aggregation state of integral membrane proteins in reconstituted phospholipid vesicles using small angle neutron scattering. *J. Mol. Biol.*, **273**, 1004–1019.

Hutchens, J. O. (1970). *Handbook of Chemistry and Selected Data for Molecular Biology*, ed. H. A. Sober Cleveland, OH: Chemical Rubber Co; *International Tables for Crystallography: Spzae Group Symmetry* (2002). Dordrecht: Kluwer.

Huygens, C. (1690). *Treatise on Light*, New York: Dover (1962) of the English translation first published by Macmillan and Co. in 1912.

Ishijima, A., Kojima, H., *et al.* (1998). Simultaneous observation of individual ATPase and mechanical events by a single myosin molecule during interaction with actin. *Cell*, **92**, 161–171.

Ishima, R., and Torchia, D. (2000). Protein dynamics from NMR. *Nat. Struct. Biol.*, **7**, 740–743.

Jacobs, R. E., Ahrens, E. T., Meade, T. J., and Fraser, S. E. (1999). Looking deeper into vertebrate development. *TIBS*, **9**, 73–76.

Jacrot B. (1976). The study of biological structures by neutron scattering from solution. *Rep. Prog. Phys.*, **39**, 911–953.

Jacrot, B., and Zaccai, G. (1981). Determination of molecular weight by neutron scattering. *Biopolymers*, **20**, 2414–2426.

Jacrot, B., Chauvin, C., and Witz, J. (1977). Comparative neutron small-angle scattering study of small spherical RNA viruses. *Nature*, **266**(5601), 417–421.

Jaenicke, R. (2000). Do ultrastable proteins from hyperthermophiles have high or low conformational rigidity? *Proc. Natl. Acad. Sci. USA*, **97**, 2962–2962.

Jancarik, J., and Kim, S.-H. (1991). Sparse matrix sampling: a screening method for crystallization of proteins. *J. Appl. Cryst.*, **24**, 409–411.

Jancarik, J., Scott, W. G., *et al.* (1991). Crystallization and preliminary X-ray diffraction study of the ligand-binding domain of the bacterial chemotaxis-mediating aspartate receptor of Salmonella typhimurium. *J. Mol. Biol.*, **221**, 31–34.

Jeener, J. (1996). In *Encyclopedia of Nuclear Magnetic Resonance*, eds. Grant D. M., Harris R. K., Vol. 1, p. 40, Chichester: John Wiley and Sons.

Jeffrey, P. D., Nichol, L. W., Turner, D. R., and Winzor, D. J. (1977). The combination of molecular covolume and frictional coefficient to determine the shape and axial ratio of a rigid macromolecule. Studies on Ovalbumin. *J. Phys. Chem.*, **81**, 776–781.

Jeruzalmi, D. and Steitz, T. A. (1997). Use of organic cosmotropic solutes to crystallize flexible proteins: application to T7 RNA polymerase and its complex with the inhibitor T7 lysozyme. *J. Mol. Biol.*, **274**, 748–756.

Jia, Y., Sytnic, A., Li, L., Vladimirov, S., Cooperman, B. S., and Hochstrasser, R. M. (1997). Nonexponencial kinetics of a single tRNA Phe molecule under physiological conditions. *Proc. Natl. Acad. Sci. USA*, **94**, 7932–7936.

Jiang, Y., Ruta, V., *et al.* (2003). The principle of gating charge movement in a voltage-dependent K+ channel. *Nature*, **423**, 42–48.

Johnson, W. C. (1985). *Circular Dichroism and Its Empirical Application to Biopolymers*. Methods of Biochemical Analysis, Vol. 31, ed. D. Glick. Chichester: John Wiley and Sons Inc.

Johnson, K. H., and Gray, D. M. (1992). Analysis of an RNA pseudoknot structure by CD spectroscopy. *J. Biomol. Struct. Dyn.*, **9**, 733–745.

Jolly, D., and Eisenberg, H. (1976). Photon correlation spectroscopy, total intensity light with laser radiation, and hydrodynamic studies of a well fractionated DNA sample. *Biopolymers*, **15**, 61–95.

Jones, J. A., Wilkins, D. K., Smith, L. J., and Dobson, C. M. (1997). Characterization of protein unfolding by NMR diffusion measurements. *J. Biomolecular NMR*, **10**, 199–203.

Jones, T. A., Zou, J. Y., *et al.* (1991). Improved methods for building protein models in electron density maps and the location of errors in these models. *Acta Crystallogr. A*, **47** (Pt 2), 110–119.

Kabsch, W. (1988). Evaluation of single-crystal X-ray diffraction data from a position-sensitive detector. *J. Appl. Cryst.*, **21**, 916–924.

Karger, B. L., Chu, Y.-H., and Foret, F. (1995). Capillary electrophoresis of proteins and nucleic acids. *Annu. Rev. Biophys. Biomol. Struct.*, **24**, 579–610.

Kassas, S., Thomson, N. H., *et al.* (1997). *Esherichia coli* RNA polymerase activity observed using atomic force microscopy. *Biochemistry*, **36**, 461–468.

Kebbekus, P., Draper, D. E., and Hagerman, P. (1995). Persistence length of RNA. *Biochemistry*, **34**, 4354–4357.

Keiderling, T. A. (2002). Protein and peptide secondary structure and conformational determination with vibrational circular dichroism. *Curr. Opin. Chem. Biol.*, **6**, 682–688.

Kincaid, J. R. (1995). Structure and dynamics of transient species using time-resolved resonance Raman spectroscopy. *Methods Enzymol.*, **246**, 460–501.

King, R. W., and Williams, K. R. (1989a). The Fourier transform in chemistry. Part 1, Nuclear magnetic resonance: introduction. *J. Chem. Education*, **66**, A213–A219.

King, R. W., and Williams, K. R. (1989b). The Fourier transform in chemistry, Part 2, Nuclear magnetic resonance: the single pulse experiment. *J. Chem. Education*, **66**, A243–A248.

Kinosita, K., Jr., Yasuda, R., Noji, H., Ishiwata, S., and Yoshida, M. (1998). F1-ATPase: a rotary motor made of a single molecule. *Cell*, **93**, 21–24.

Klar, T. A., Jacobs, S., Dyba, M., Egner, A., and Hell, S. W. (2000). Fluorescence microscopy with diffraction resolution barrier broken by stimulated emission. *Proc. Natl. Acad. Sci. USA*, **97**, 8206–8210.

Klein, G., Satre, M., *et al.* (1982). Spontaneous aggregation of the mitochondrial natural ATPase inhibitor in salt solutions as demonstrated by gel filtration and neutron scattering. Application to the concomitant purification of the ATPase inhibitor and F1-ATPase. *Biochim. Biophys. Acta*, **681**, 226–232.

Kleywegt, G. J., and Jones, T. A. (2002). Homo crystallographicus – quo vadis? *Structure (Camb)*, **10**, 465–472.

Klotz, L. C., and Zimm, B. H. (1972). Size of DNA determined by viscoelastic measurements: results on bacteriophage, *Bacillus subtilus* and *Escherichia coli*, **72**, 779–800.

Koch, M. H., and Stuhrmann, H. B. (1979). Neutron-scattering studies of ribosomes. *Methods Enzymol.*, **59**, 670–706.

Konermann, L., and Douglas, D. J. (1998a). Equilibrum unfolding of proteins monitored by electrospray ionization mass spectrometry: distinguishing two-state from multi-state transition. *Rapid Commin. Mass Spectr.*, **12**, 435–442.

Konermann, L., and Douglas, D. J. (1998b). Unfolding of proteins monitored by electrospray ionization mass spectrometry: a comparison of positive and negative ion modes. *J. Am. Soc. Mass Spectrom.*, **9**, 1248–1254.

Koppel, D. E. (1979). Fluorescence redistribution after photobleaching. *Biophys. J.*, **28**, 281–291.

Korgel, B. A., Van Zanten, J. H., and Monbouquette, H. G. (1998). Vesicle size distributions measured by flow field-flow fractionation coupled with multiangle light scattering. *Biophys. J.*, **74**, 3264–3272.

Kossiakoff, A. A. (1983). Neutron protein crystallography: advances in methods and applications. *Ann. Rev. Biophys. Bioeng.*, **12**, 159–182.

Kovacic, R. T., and van Holde, K. E. (1977). Sedimentation of homogeneous double-strand DNA molecules. *Biochemistry*, **16**, 1490–1498.

Kroes, S. J., Canters, G. W., Giardi, G., van Hoek, A., and Visser, A. J. W. G. (1998). Time-resolved fluorescence study of azurin variants: conformational heterogeneity and tryptophan mobility. *Biophys. J.*, **75**, 2441–2450.

Kumar, A., Ernst, R. R., and Wuthrich, K. (1980). A two-dimensional nuclear Overhauser enhancement (2D NOE) experiment for the elucidation of complete proton–proton cross-relaxation networks in biological macromolecules. *Biochem. Biophys. Res. Commun.*, **95**, 1–6.

Kunst, F., Oyasawara, N., *et al.* (1997). The complete genome sequence of the Gram-positive bacterium *Bacillus subtilis*. *Nature*, **390**, 249–256.

Kuntz, I. D., Jr., and Kauzmann, W. (1974). *Hydration of Proteins and Polypeptides*. In *Advances in Protein Chemistry*, eds. C. B. Anfinsen, J. T. Edsall and F. M. Richards, Vol. 28. New York: Academic Press.

Lakowicz, J. R. (ed.) (1999). *Principles of Fluorescence Spectroscopy*, second edn. New York: Kluwer Academic/Plenum Publ.

Lakowicz, J. R., Gryczynski, I., *et al.* (2000). Microsecond dynamics of biological macromolecules. *Methods Enzymol.*, **323**, 473–509.

Langan, P., Nishiyama, Y., *et al.* (1999). A revised structure and hydrogen bonding system in cellulose II from a neutron fibre diffraction analysis. *J. Am. Chem. Soc.*, **121**, 9940–9946.

Langley, K. H. (1992). Developments in electrophoretic laser light scattering and some biochemical applications. In *Laser Scattering in Biochemistry*, eds. S. E. Harding, D. B. Sattelle, and V. A. Bloomfield. Cambridge: Royal Society of Chemistry.

Langowski, J., Kremer, W., and Kapp, U. (1992). Dynamic light scattering for study of solution conformation and dynamics of superhelical DNA. *Methods Enzymol.*, **211**, 431–448.

Laue, T. M., Ridgeway, T. M., Wool, J. O., and Shepard, H. K. (1996). Insights into a new analytical electrophoresis apparatus. *J. Pharm. Sci.*, **85**, 1331–1335.

Lay, J. O. (2001). MALDI-TOF mass spectrometry of bacteria. *Mass Spectr. Rev.*, **20**, 172–194.

Leavitt, S., and Freire, E. (2001). Direct measurement of protein binding energetics by isothermal titration calorimetry. *Curr. Opin. Struct. Biol.*, **11**, 560–566.

Lehnert, U. (2002). Hydration dependence of local thermal motions in the Purple Membrane explored by neutron scattering and isotopic labeling. Ph D Thesis. Université Joseph Fourier, Grenoble.

Lemasters, J. J., Chacon, E., Zahrebelski, G., Reece, J. M., and Nieminen, A.-L. (1993). Laser scanning confocal microscopy of living cells. In: *Optical Microscopy. Emerging Methods and Application*, eds. B. Herman and J. J. Lemasters. San Diego: Academic Press.

Leone, M., Cupane, A., *et al.* (1994). Thermal broadening of the Soret band in heme complexes and in heme-proteins: role of iron dynamics. *Eur. Biophys. J.*, **23**, 349–352.

Lescrinier, E., Froeyen, M., and Herdewijn, P. (2003). Difference in conformational diversity between nucleic acids with six-membered 'sugar' unit and natural 'furanose' nucleic acids. *Nucleic Acids. Res.*, **31**, 2975–2989.

Leslie, A. G. (1999). Integration of macromolecular diffraction data. *Acta Crystallogr. D Biol. Crystallogr.*, **55** (Pt 10), 1696–1702.

Levitt, M., Sander, C., *et al.* (1985). Protein normal-mode dynamics: trypsin inhibitor, crambin, ribonuclease and lysozyme. *J. Mol. Biol.*, **181**, 423–447.

Lewis, A., Lieberman, K., *et al.* (1995). New design and imaging concepts in NSOM, *Ultramicroscopy*, **61**, 215–220.

Lewis, A., Radko, A., Ami, N. B., Palanker, D., and Lieberman, K. (1999). Near-field scanning optical microscopy in cell biology. *TIBS*, **9**, 70–73.

Li, H., Cocco, M. J., Steitz, T., and Engelman, D. M. (2001). Conversion of phospholamban into a soluble pentameric helical bundle. *Biochemistry*, **40**, 6636–6645.

Li, Y., Hunter, R. L., and Mciver, R. T., Jr. (1994). High-resolution mass spectrometer for protein chemistry. *Nature*, **370**, 393–395.

Liphardt, J., Onoa, B., Smith, S. B., Tinoco, I. Jr., and Bustamante, C. (2001). Reversible unfolding of single RNA moleculees by mechanical forces. *Science*, **292**, 733–737.

Longsworth, L. G. (1953). Diffusion measurements, at 25°, of aqueous solution of amino acids, peptides and sugars. *J. Am. Chem. Soc.*, **75**, 5705–5709.

Lu, H. P., Xun, L., and Xie, X. S. (1998). Single-molecule enzymatic dynamics. *Science*, **282**, 1877–1882.

Luthy, R., Bowie, J. U., and Eisenberg, D. (1992). Assessment of protein models with three-dimensional profiles. *Nature*, **356**, 83–85.

Luz, Z., and Meiboom, S. (1963). *J. Chem. Phys.*, **39**, 366–370.

Luzzati, V., and Tarclieu, A. (1980). Recent developments in solution X-ray scattering. *Annu. Rev. Biophys. Bioeng.*, **9**, 1–29.

Ma, J., Flynn, T. C., *et al.* (2002). A dynamic analysis of the rotation mechanism for conformational change in F(1)-ATPase. *Structure (Camb)*, **10**, 921–931.

MacDonald, D., and Lu, P. (2002). Residual dipolar couplings in nucleic acid structure determination. *Curr. Opinion Str. Biol.*, **12**, 337–343.

MacGregor, I. K., Anderson, A. L., and Laue, T. M. (2004). Fluorescence detection for the XLI analytical ultracentrifuge. *Biophys. Chem.*, **108**, 165–185.

Machtle, W. (1999). High-resolution, submicron particle size distribution analysis using gravitational-sweep sedimentation. *Biophys. J.*, **76**, 1080–1091.

Madern, D., Ebel, C., *et al.* (2000). Halophilic adaptation of enzymes. *Extremophiles*, **4**, 91–98.

Makhatadze, G. I., and Privalov, P. L. (1990). Heat capacity of proteins. I. Partial molar heat capacity of individual amino acid residues in aqueous solution: hydration effect. *J. Mol. Biol.*, **213**, 375–384.

Mancini, E. J., Clarke, M., *et al.* (2000). Cryo-electron microscopy reveals the functional organization of an enveloped virus, Semliki Forest virus. *Mol. Cell.* **5**, 255–266.

Mandelkow, E., and Mandelkow, E. M. (2002). Kinesin motor and disease. *TICB*, **12**, 585–591.

Manor, D., Weng, G., Deng, H., Cosloy, S., Chen, C. X. (1991). An isotope edited classical Raman difference spectroscopic study of the interactions of guanine nucleotides with elongation factor Tu and H-*ras* p21. *Biochemistry*, **30**, 10914–10920.

Mark, J. E. (1996). *Physical Properties of Polymers Handbook*. Woodbury, NY: AIP Press.

Marassi, F. M., Ma C., *et al.* (1999). Correlation of the structural and functional domains in the membrane protein Vpu from HIV-1. *PNAS*, **96**, 14336–14341.

Mattei, B., Borch, J., and Roepstorff, P. (2004). Biomolecular interaction analysis and MS. *Anal. Chem.*, **76**, 18A–26A.

McLafferty, F. W. (1993). *Interpretation of Mass Spectra*, fourth edn. Mill Valley, CA: University Science Books.

Medalia, O., Weber, I., *et al.* (2002). Macromolecular architecture in eukaryotic cells visualized by cryoelectron tomography. *Science*, **298**, 1209–1213.

Medek, A., Olejniczak E. T., Meadows, R. P., and Fesik, S. W. (2000). An approach for high-throughput structure determination of proteins by NMR spectroscopy. *J. Biomol. NMR*, **18**, 229–238.

Mehta, A. D., Rief, M., Spudlich, J. A., Smith, D. A., and Simmons, R. M. (1999). Single-molecule biomechanics with optical methods. *Science*, **283**, 1689–1695.

Merritt, E. A., Sarfaty, S., *et al.* (1994). Crystal structure of cholera toxin B-pentamer bound to receptor GM1 pentasaccharide. *Protein Sci*, **3**, 166–175.

Merzel, F., and Smith, J. C. (2002). Is the first hydration shell of lysozyme of higher density than bulk water? *Proc. Natl. Acad. Sci. USA*, **99**, 5378–5383.

Meselson, M., and Stahl, F. W. (1958). The replication of DNA in *Escherichia coli*. *Proc. Natl. Acad. Sci. USA*, **44**, 671–682.

Meselson, M., Stahl, F. W., and Vinograd, J. (1957). Equilibrium sedimentation of macromolecules in density gradient. *Proc. Natl. Acad. Sci. USA*, **43**, 581–588.

Michielsen, S., and Pecora, R. (1981). Solution dimensions of the gramicidin dimer by dynamic light scattering. *Biochemistry*, **20**, 6994–6997.

Miele, A. E., Federici, L., *et al.* (2003). Analysis of the effect of microgravity on protein crystal quality: the case of a myoglobin triple mutant. *Acta Crystallogr. D Biol. Crystallogr.*, **59** (Pt 6), 982–988.

Mitellbach, P., and Porod, G. (1962). Zur Röntgenkleinwinkelstreuung die Berechning der Steukurven von Dreiachsigen Ellipsoiden. *Acta Physica Austriaca*, **15**, 122–147.

Moerner, W. E., and Orrit, M. (1999). Illuminating single molecules in condensed matter. *Science*, **283**, 1670–1676.

Mollova, E. T., Hansen, M. R., and Pardi, A. (2000). Global structure of RNA determined with residual dipolar coupling, *JACS*, **122**, 11561–11562.

Monkos, K., and Turczynski, B. (1991). Determination of the axial ratio of globular proteins in the aqueous solution using viscometric measurements. *Int. J. Biol. Macromol.* **13**, 341–344.

Montelione, G. T., Zheng, D., Huang, Y. J., Gundalus, K. S., and Szyperski, T. (2000). Protein NMR spectroscopy in structural genomics. *Nat. Struct. Biol.*, **7**, 982–985.

Moscowitz, A. (1962). Theoretical aspects of optical activity. Part one: Small molecules. *Adv. Chem. Phys.*, **4**, 67–112.

Mou, J., Csajkovsky, D. M., Zhang, Y., and Shao, Z. (1995). High-resolution atomic force microscopy of DNA: the pitch of the double helix. *FEBS Lett.*, **371**, 279–282.

Muller, D. J., Janovjak, H., Lehto, T., Kuerschner, L., and Anderson, K. (2002). Observing structure, function and assembly of single proteins by AFM. *Progress Biophys. Mol. Biol.*, **79**, 1–43.

Mullett, W. M., Lai, E. P., *et al.* (2000). Surface plasmon resonance-based immunoassays. *Methods*, **22**, 77–91.

Munro, I., Pecht, I., and Stryer, L. (1979). Subnanosecond motions of tryptophan residues in proteins. *Proc. Natl. Acad. Sci. USA*, **76**, 56–60.

Myers, E. W., Sutton, G. G., *et al.* (2000). A whole-genome assembly of *Drosophila*. *Science*, **287**, 2196–2204.

Myszka, D. G., Sweet, R. W., *et al.* (2000). Energetics of the HIV gp120-CD4 binding reaction. *Proc. Natl. Acad. Sci. USA*, **97**, 9026–9031.

Nagorni, M., and Hell, S. W. (1998). 4Pi-Confocal microscopy provides three-dimensional images of the microtubule network with 100- to 150-nm resolution. *J. Struct. Biol.*, **123**, 236–247.

Navarro, S., Lopez Martinez, M. C., Garcia de la Torre, J. (1995). Relaxation times in electric birefringence of flexible polymer chains. *J. Chem. Phys.*, **103**, 7631–7639.

Navaza, J., and Saludjian, P. (1997). AMoRe: an automated molecular replacement program package. *Methods Enzymol.*, **276**, 581–594.

Newcomb, W. W., Juhas, R. M., *et al.* (2001). The UL6 gene product forms the portal for entry of DNA into the herpes simplex virus capsid. *J. Virol.*, **75**, 10923–10932.

Nie, S., and Zare, R. N. (1997). Optical detection of single molecule. *Annu. Rev. Biophys. Biomol. Struct.*, **26**, 567–596.

Nie, S., Chlu, D. T., and Zare, R. N. (1995). Real-time detection of single molecules in solution by confocal fluorescence microscopy. *Anal. Chem.*, **67**, 2849–2857.

Nielsen, M. L., Bennet, K. L., Larsen, B., Monniate, M., and Mann, M. (2002). Peptide end sequencing by orthogonal MALDI tandem mass spectroscopy. *J. Proteome Res.*, **1**, 63–71.

Nir, S., and Stein, W. D. (1971). Two modes of diffusion in liquids. *J. Chem. Phys.*, **55**, 1598–1603.

Nishiyama, Y., Langan, P., *et al.* (2002). Crystal structure and hydrogen-bonding system in cellulose Ibeta from synchrotron X-ray and neutron fiber diffraction. *J. Am. Chem. Soc.*, **124**, 9074–9082.

Nishiyama, Y., Okano, T., *et al.* (1999). High resolution neutron fibre diffraction data on hydrogenated and deuterated cellulose. *Int. J. Biol. Macromol.*, **26**, 279–283.

Noji, H., Yasuda, R. *et al.* (1997). Direct observation of the rotation of F1-ATPase. *Nature*, **386**, 299–302.

Nollert, P., Navarro, J., and Landau, E. M. (2002). Crystallization of membrane proteins in cubo. *Methods in Enzym.*, **343**, 183–99.

Oberg, K. A., Ruysschaert, J. M., *et al.* (2004). The optimization of protein secondary structure determination with infrared and circular dichroism spectra. *Eur. J. Biochem.*, **271**, 2937–2948.

Ogorzalek Loo, R. R., Cavalcoli, J. D., *et al.* (2001). Virtual 2-D gel electrophoresis: visualization and analysis of the *E. coli* proteome by mass spectrometry. *Anal. Chem.*, **73**, 4063–4070.

Opella, S. J., and Stewart, P. L. (1989). Solid-state nuclear magnetic resonance structural studies of proteins. *Methods Enzymol.*, **176**, 242–275.

Orgel, L. E. (1968). Evolution of the genetic apparatus. *J. Mol. Biol.*, **38(3)**, 381–93.

Ormo, M., Cubbit, A. B., Kallio, K., Gross, L. A., and Tsien, R. Y. (1996). Crystal structure of the *Aequorea victoria* green fluorescent protein. *Science*, **273**, 1392–1395.

Oster, G., and Wang, H. (2003). Rotary protein motor. *Trends Cell Biol.*, **13**, 114–121.

Otting, G., and Wuthrich, K. (1990) Heteronuclear filters in two-dimensional [^{1}H, ^{1}H]-NMR spectroscopy: combined use with isotope labelling for studies of macromolecular conformation and intermolecular interactions. *Q. Rev. Biophys.*, **23**, 39–56.

Otwinowski, Z., and Minor, W. (1997). Processing of X-ray diffraction data in oscillation mode. *Methods Enzymol.*, **276**, 307–326.

Pauling L., and Corey, R. (1953). A proposed structure for the nucleic acids. *PNAS*, **39**, 84–97.

Pecora, R. (1968). Spectral distribution of light scattered by monodisperse rigid rods. *J. Chem. Phys.*, **48**, 4126–4128.

Perrakis, A., Sixma, T. K., *et al.* (1997). wARP: improvement and extension of crystallographic phases by weighted averaging of multiple-refined dummy atomic models. *Acta Cryst. D.*, **53**, 448–455.

Peticolas, W. L. (1995). Raman spectroscopy of DNA and proteins. *Methods Enzymol.*, **246**, 389–416.

Peticolas, W. L., and Evertsz, E. (1992). Conformation of DNA in vitro and in vivo from laser Raman scattering. *Methods in Enzymol.*, **211**, 335–352.

Pfuhl, M., Chen, H. A., Kristinsen, S., Driscool, P. C. (1999). NMR exchange broadening arising from specific low affinity protein self-association: Analysis of nitrogen-15 relaxation for rat CD2 domain 1. *J. Biomolecular NMR*, **14**, 307–320.

Piston, D. W. (1999). Imaging living cells and tissues by two-photon excitation microscopy. *Trends Cell Biol.*, **9**, 66–69.

Pitner, T. P., Walter, R., and Glickson, J. D. (1976). Mechanism of the intramolecular ^{1}H nuclear Overhauser effect in peptides and depsipeptides. *Biochem. Biophys. Res. Commun.*, **70**, 746–751.

Plenert, M. L., and Shear, J. B. (2003). Microsecond electrophoresis. *Proc. Natl. Acad. Sci. USA*, **100**, 3853–3857.

Pohl, F. M., and Jovin, T. M. (1972). Salt-induced co-operative conformational change of a synthetic DNA: equilibrium and kinetic studies with poly (dG-dC). *J. Mol. Biol.*, **67**, 375–396.

Popot, J. L., Berry, E. A., *et al.* (2003). Amphipols: polymeric surfactants for membrane biology research. *Cell Mol. Life Sci.*, **60**, 1559–1574.

Prestegard, J. H. (1998). New techniques in structural NMR – anisotropic interactions. *Nature Struct. Biol.*, **5**, 517–522.

Prestegard, J. H., Valafar, H., Glushka, J., and Tian, F. (2001). Nuclear magnetic resonance in the era of structural Genomics. *Biochemistry*, **40**, 8677–8685.

Price, N. C., Dwek, R. A., *et al.* (2001). *Principles and Problems in Physical Chemistry for Biochemists.* Oxford: Oxford University Press.

Price, P. B. (2000). A habitat for psychrophiles in deep Antarctic ice. *Proc. Natl. Acad. Sci. USA*, **97**, 1247–1251.

Privalov, G. P., and Privalov, P. L. (2000). Problems and prospects in microcalorimetry of biological macromolecules. *Methods Enzymol.*, **323**, 31–62.

Privalov, G., Kavina, V., *et al.* (1995). Precise scanning calorimeter for studying thermal properties of biological macromolecules in dilute solution. *Anal. Biochem.*, **232**, 79–85.

Privalov, P. L. (1980). Scanning microcalorimeters for studying macromolecules. *Pure & Appl. Chem.*, **52**, 479–497.

Privalov, P. L. (1982). Stability of proteins. Proteins which do not present a single cooperative system. *Adv. Protein Chem.*, **35**, 1–104.

Privalov, P. L., and Khechinashvili, N. N. (1974). A thermodynamic approach to the problem of stabilization of globular protein structure: a calorimetric study. *J. Mol. Biol.*, **86**, 665–684.

Privalov, P. L., and Makhatadze, G. I. (1990). Heat capacity of proteins. II. Partial molar heat capacity of the unfolded polypeptide chain of proteins: protein unfolding effects. *J. Mol. Biol.*, **213**, 385–391.

Privalov, P. L., and Makhatadze, G. I. (1992). Contribution of hydration and non-covalent interactions to the heat capacity effect on protein unfolding. *J. Mol. Biol.*, **224**, 715–723.

Purcell, E. M. (1977). Life at low Reynolds number. *Am. J. Phys.*, **45**, 311.

Ramakrishnan, V., and Moore, P. B. (2001). Atomic structures at last: the ribosome in 2000. *Curr. Opinion in Str. Biol.*, **11**, 144–154.

Rau, D. C., and Bloomfield, V. A. (1979) Transient electric birefringence of T7 virial DNA. *Biopolymers*, **18**, 2783–2805.

Rayment, I. (1996). Kinesin and myosin: molecular motor with similar engine. *Structure*, **4**, 501–504.

Records, M. T., Jr., Woodbury, C. P., and Inman, R. B. (1975). Characterization of rodlike DNA fragments. *Biopolymers*, **14**, 393–408.

Rhee, K. H., Scarborough, G. A., *et al.* (2002). Domain movements of plasma membrane H(+)-ATPase: 3D structures of two states by electron cryo-microscopy. *Embo. J.*, **21**, 3582–3589.

Richard, S. B., Madern, D., *et al.* (2000). Halophilic adaptation: novel solvent protein interactions observed in the 2.9 and 2.6 A resolution structures of the wild type and a mutant of malate dehydrogenase from *Haloarcula marismortui*. *Biochemistry*, **39**, 992–1000.

Rief, M., Fernandez, J. M., and Gaub, H. E. (1998). Elastically coupled two-level system as a model for biopolymer extensibility, *Phys. Rev. Letters*, **81**, 4764–4767.

Rief, M., Oesterhelt, T. F., Heymann, B., and Gaub, H. E. (1997). Single molecule force spectroscopy on polysaccharides by atomic force microscopy. *Science*, **275**, 1295–1297.

Rief, M., Pascual, J. Saraste, M., and Gaub, H. E. (1999). Single molecule force spectroscopy of spectrin repeats: low unfolding forces in helix bundles. *J. Mol. Biol.*, **286**, 553–561.

Riek, R., Hornemann, S., Wider, G., Glockshuber, R., and Wuthrich, K. (1997). NMR characterization of the full-length recombinant murine prion protein, mPrP(23–231). *FEBS Lett.*, **413**, 282–288.

Roberts, M. M., Coker, A. R., *et al.* (1999). Crystallization, X-ray diffraction and preliminary structure analysis of Mycobacterium tuberculosis chaperonin 10. *Acta Crystallogr. D Biol. Crystallogr.*, **55** (Pt 4), 910–914.

Rosenheck, K., and Doty, P. (1961). The far ultraviolet absorption spectra of polypeptide and protein solutions and their dependence on conformation. *Proc. Natl. Acad. Sci. USA*, **47**, 1775–1785.

Rosenthal, P. B., and Henderson, R. (2003). Optimal determination of particle orientation, absolute hand, and contrast loss in single-particle electron cryomicroscopy. *J. Mol. Biol.*, **333**, 721–745.

Rostom, A. A., and Robinson, C. V. (1999). Disassembly of intact multiprotein complexes in the gas phase. *Curr. Opin. Struct. Biol.*, **9**, 135–141.

Rostom, A., Fucini, P., *et al.* (2000). Detection and selective dissociation of intact ribosomes in a mass spectrometer. *Proc. Natl. Acad. Sci. USA*, **97**, 5185–5190.

Rould, M. A. (1997). Screening for heavy-atom derivatives and obtaining accurate isomorphous differences. *Methods in Enzym*, **276** 461–472.

Rould, M. A., and Carter, C. W., Jr. (2003). Isomorphous difference methods. *Methods in Enzym.*, **374**, 145–63.

Rowe, A. (1977). The concentration dependence of transport processes: a general description applicable to the sedimentation, translational diffusion, and viscosity coefficients of macromolecular solutes. *Biopolymers*, **16**, 2595–2611.

Rubtsov, I. V., Wang, J., and Hochstrasser, R. M. (2003). Dual-frequency 2D-IR spectroscopy heterodyned photon echo of the peptide bond, *PNAS*, 5601–5606.

Salmeen, I., Rimai, L., Liebes, L., Rich, M. A., and McCormick, J. J. (1975). Hydrodynamic diameters of RNA tumor viruses. Studies by laser beat frequency light scattering spectroscopy of avian myeblastosos and Rauscher murine leukemia viruses. *Biochemistry*, **14**, 134–141.

Sanders, C. R., Hare, B. J., Howard, K. P., Prestegard, J. H. (1994). Magnetically oriented phospholipid micelles as a tool for the study of membrane-associated molecules. *Prog. NMR Spectros.*, **26**, 5.

Schachman, H. K. (1989). Analytical ultracentrifugation reborn. *Nature*, **341**, 259–260.

Schmidt, B., and Reisner, D. (1992). A fluorescence detection system for analytical ultracentrifuge and its application to proteins, nucleic acids, viroid and viruses. In *Analytical Ultracentrifugation in Biochemistry and Polymer Science*, eds. S. E. Harding, A. J. Rowe and J. C. Horton. Cambridge: Royal Society of Chemistry.

Schmidt, T., Schultz, G. J., Baumgartner, W., Gruber, H. J., Schindler, H. (1996). Imaging of single molecule diffusion. *PNAS*, **93**, 2926–2929.

Schmitz, K. S. (1993). *An Introduction to Dynamic Light Scattering of Molecules*. Boston: Academic Press.

Schmitz, K. S., and Schurr, J. M. (1973). Rotational relaxation of macromolecules determined by dynamic light scattering. II. Temperature dependence for DNA. *Biopolymers*, **12**, 1543–1564.

Scholtan, W., and Lange, H. (1972). Bestimmung der Teilchegrobenverteilung von Latices mit der Ultracentrifuge. *Kolloid-Z., u. Z. Polimere.*, **250**, 782–796.

Schouten, S., Hopmans, E. C., *et al.* (2000). Widespread occurrence of structurally diverse tetraether membrane lipids: evidence for the ubiquitous presence of low-temperature relatives of hyperthermophiles. *Proc. Natl Acad. Sci. USA*, **97**, 14421–14426.

Schrader, M., Bahlmann, K., Giese, G., and Hell, S. W. (1998). 4Pi-confocal imaging in fixed biological specimens. *Biophys. J.*, **75**, 1659–1668.

Schuck, P. (2004). A model for sedimentation in inhomogeneous media. I. Dynamic density gradients from sedimenting co-solutes. *Biophys. Chem.*, **108**, 187–200.

Schultz, D. A. (2003). Plasmon resonant particles for biological detection. *Curr. Opin. Biotechnol.*, **14**, 13–22.

Schwabe, J. W. R., Chapman, L., Finch, J. T, Rhodes, D., and Neuhaus, D. (1993). DNA recognition by the oestrogen receptor: from solution to the crystal. *Structure*, **1**, 187–204.

Schwille, P., Meyer-Almes, F. J., *et al.* (1997). Dual-colour fluorescence cross-correlation spectroscopy for multicomponent diffusional analysis in solution. *Biophys. J.*, **72**, 1878–1886.

Seelig, J. (2004). Thermodynamics of lipid–peptide interactions. *Biochim. Biophys. Acta.*, **1666**, 40–50.

Seils, J., and Dorfmuller, Th. (1991). Internal dynamics of linear and superhelical DNA as studied by photon correlation spectroscopy. *Biopolymers*, **31**, 813–825.

Serdyuk, I. N., Grenader, A. K., *et al.* (1979). Study of the internal structure of Escherichia coli ribosomes by neutron and x-ray scattering. *J. Mol. Biol.*, **135**, 691–707.

Serdyuk, I. N., Pavlov, M., *et al.* (1994). The triple isotopic substitution method in small angle neutron scattering. Application to the study of the ternary complex EF-Tu.GTP.aminoacyl-tRNA. *Biophys. Chem.*, **53**, 123–130.

Serdyuk, I., Ulitin, A., *et al.* (1999). Structure of a beheaded 30S ribosomal subunit from *Thermus thermophilus. J. Mol. Biol.*, **292**, 633–639.

Sheetz, M. P., Turne,Y. S., Qian, H., and Elson, E. L. (1989). Nanometre level analysis demonstrates that lipid flow does not drive membrane glycoprotein movements. *Nature,* **340**, 284–285.

Sheraga, H. A., and Mandelkern, L. (1953). Consideration of the hydrodynamic properties of proteins. *J. Am. Chem. Soc.*, **75**, 179–184.

Sheraga, H. A., Edsall, J. T., and Gadd, J. O. (1951). Double refraction of flow: numerical evaluation of extinction angle and birefringence as a function of velocity gradient. *J. Chem. Phys.*, **19**, 1101–1108.

Shiku, H., and Dunn, R. C. (1999). Near field scanning optical microscopy. *Anal. Chem.*, **71**, 23A–29A.

Shingyoji, C. Higuchi, H., Yoshimura, M., Katayama, E., and Yanagida, T. (1998). Dynein arms are oscillating force generators. *Nature,* **393**, 711–714.

Shiryaev, V. M., Selivanova, O. M., Hartsch, T., Nazimov, I. V., and Spirin, A. S. (2002). Ribosomal protein S1 from *Thermus thermophilus*: its detection, identification and overproduction. *FEBS Letters*, **525,** 88–92.

Shivashankar, G. V., and Livchaber, A. (1997). Single DNA molecule grafting and manipulation using a combined atomic force microscope and an optical tweezer. *Appl. Phys. Lett.*, **71**, 3727–3729.

Sigler, P. B., Xu, Z., *et al.* (1998). Structure and function in GroEL-mediated protein folding. *Ann. Rev. Biochem.*, **67**, 581–608.

Simpson, A. A., Tao, Y., *et al.* (2000). Structure of the bacteriophage Φ29 DNA packaging motor. *Nature*, **408,** 745–750.

Skoog, D. A., Holler, F. J., and Nieman, T. A. (1995). *Principle of Instrumental Analysis*. Philadelphia: Saunders College Publishing.

Sliz, P., Harrison, S. C., and Rosenbaum, G. (2003). How does radiation damage in protein crystals depend on X-ray dose? *Structure (Camb)*, **11**, 13–19.

Smith, D. E., Babcock, H. P., and Chu, S. (1999). Single-polymer dynamics in steady shear flow. *Science*, **283,** 1724–1727.

Smith, D. E., Tans, S. J., *et al.* (2001). The bacteriophage Φ29 portal motor can package DNA against a large internal force. *Nature*, **413**, 748–752.

Smith, L. J., Redfield, C., *et al*. (1994). Comparison of four independently determined structure of human recombinant interleikin-4. *Struct. Biol.,* **1**, 301–310.

Smith, M. H. (1970). Molecular weight of proteins and some other materials including sedimentation diffusion and frictional coefficients and partial specific volumes. In *Handbook of Biochemistry. Selected Data for Molecular Biology*, ed. H. A. Sober, pp. C3–C47. Cleveland, OH: The Chemical Rubber Company.

Smith, R. D., Bruce, J. E., *et al*. (1996). The role of Fourier transform ion cyclotron resonance mass spectrometry in biological research – new development and applications. In *Mass Spectrometry in the Biological Science*, eds. A. L. Burlingame and S. A. Carr, pp. 25–68. Totowa, NJ: Humana Press.

Smith, S. B., Cui, Y., and Bustamante, C. (1996). Overstretching B-DNA: the elastic response of individual double-stranded and single-stranded DNA molecules. *Science*, **271**, 795–799.

Smith, S. O., and Peersen, O. B. (1992). Solid-state NMR approaches for studying membrane protein structure. *Ann. Rev. Biophys. Biomol. Str.*, **21**, 25–47.

Snatzke, G. (1994). Circular dichroism: an introduction. In *Circular Dichroism: Principles and Applications.*, eds, K. Nakanishi, N. Berova and R. W. Woody. New York, VCH Publishers.

Sodano, P., Chary, K. V., *et al*. (1991). Nuclear magnetic resonance studies of recombinant Escherichia coli glutaredoxin. Sequence-specific assignments and secondary structure determination of the oxidised form. *Eur. J. Biochem*, **200**, 369–377.

Sober, H. A. (ed.) (1970). *Handbook of Biochemistry*, 2nd edn. Cleveland: CRC Press.

Sorlie, S. S., and Pecora, R. (1988). A dynamic light scattering study of a 2311 base pair DNA restriction fragment. *Macromolecules*, **21**, 1437–1441.

Sosa, H., Dias, D. P., *et al.* (1997). A model for the microtubule-Ncd motor protein complex obtained by cryo-electron microscopy and image analysis. *Cell*, **90**, 217–224.

Spirin, A. S. (1963). *Some Problems of Macromolecular Structure of Ribonucleic Acids* (in Russian). Moscow: Academy of Science USSR.

Spirin A. S. (2000). *Ribosomes*. New York: Kluwer Academic/Plenum Publishers.

Spiro, T. G., and Chernuszevich, R. S. (1995). Resonance Raman spectroscopy of metalloprotein. *Methods Enzymol.*, **246**, 416–459.

Spolar, R. S., and Record, M. T., Jr. (1994). Coupling of local folding to site-specific binding of proteins to DNA. *Science*, **263**, 777–784.

Sportsman, J. R. (2003). Fluorescence anisotropy in pharmacologic screening. *Methods Enzymol.*, **361**, 505–529.

Spronk, C. A. E. M., Linge, J. P., Hilbers, C. W., and Vuister, G. W. (2002). Improving the quality of protein structures derived by NMR spectroscopy, *J. Biomolecular NMR*, **22**, 281–289.

Spudlish, J. A. (1994). How molecular motors works. *Nature*, 372, 515–518.

Squire, P. G. (1970). An equation of consistency relating the harmonic mean relaxation time to sedimentation data. *Biochim. Biophys. Acta*, **221**, 425–429.

Srajer, V., Ren, Z., *et al.* (2001). Protein conformational relaxation and ligand migration in myoglobin: a nanosecond to millisecond molecular movie from time-resolved Laue X-ray diffraction. *Biochemistry*, **40**, 13802–13815.

Sreerama, N., and Woody, R. W. (2003). Structural composition of betaI- and betaII-proteins. *Protein Sci*, **12**, 384–388.

Stafford, W. F., and Braswell, E. H. (2004). Sedimentation velocity, multi-speed method for analyzing polydisperse solutions. *Biophys. Chem.*, **108**, 273–279.

Steely, H. T., Jr., Gray, D. M., and Lang, D. (1986a). Study of the circular dichroism of bacteriophage $\phi 6$ and $\phi 6$ nucleocapsid. *Biopolymers* **25**, 171–188.

Steely, H. T., Jr., Gray, D. M., Lang, D., and Maestre, M. F. (1986b). Circular dichroism of double-stranded RNA in the presence of salt and ethanol. Biopolymers **25**, 91–117.

Stejskal, E. O., and Tanner, J. E. (1965). Spin diffusion measurements: Spin echoes in the presence of a time dependent field gradient. *J. Chem. Phys*, **42**, 288–292.

Stellwagen, N. C. (1981). Electric birefringence of restriction enzyme fragments of DNA: Optical factor and electric polarizability as a function of molecular weight. *Biopolymers*, **20**, 399–434.

Stellwagen, N. C. (1996). Electric birefringence of kilobase-sized DNA molecules. *Biophys. Chem.*, **58**, 117–124.

Stellwagen, N. C., Gelfi, C., and Righetti, P. G. (1997). The free solution mobility of DNA. *Biopolymers*, **42**, 687–703.

Stoeckli, M., Chaurand, P., Hallahan, D. E., and Caprioli, R. (2001). Imaging mass spectrometry: A new technology for the analysis of protein expression in mamalian tissues. *Nat. Medicine*, **7**, 493–496.

Stoeckli, M., Chaurand, P., *et al.* (2001). Imaging mass spectrometry: a new technology for the analysis of protein expression in mammalian tissues. *Nat. Med.*, **7**, 493–96.

Stone, M. J., Fairbrother, W. J., *et al.* (1992). Backbone dynamics of the *Bacillus subtilis* glucose permease IIA domain determined from 15N NMR relaxation measurements. *Biochemistry*, **31**, 4394–4406.

Strick, T. R., Allemand, J. -F., Bensimon, D., Bensimon, A., Croquettet, V. (1996). The elasticity of a single supercoiled DNA molecule. *Science*, **271**, 1835–1837.

Strick, T., Allemand, J. -F., Croquete, V., and Bensimon, D. (2000). Twisting and stretching single DNA molecules. *Progr. Biophys. Mol. Biol.*, **74**, 115–140.

Stroebel, D., Choquet, Y., Popot, J. L., and Picot, D. (2003). An atypical haem in the cytochrome b(6)f complex. *Nature*, **426**, 413–418.

Stryer, L. (1968). Fluorescence spectroscopy of proteins. *Science*, **162**, 526–533.

Stryer, L., and Hougland, R. P. (1967). Energy transfer: A spectroscopic ruler. *PNAS*, **58**, 719–726.

Stuhrmann, H. B., and Miller, A. (1978). Small-angle scattering of biological structures. *J. Appl. Cryst.*, **11**, 325–345.

Subramanian, S., and Henderson, R. (1999). Electron crystallography of bacteriorhodopsin with millisecond time resolution. *J. Struct. Biol.*, **128**, 19–25.

Subramanian, S., Hirai, T., and Henderson, R. (2002). From structure to mechanism: electron crystallographic studies of bacteriorhodopsin. *Phil. Trans. A Math. Phys. Eng. Sci.*, **360**, 859–874.

Subramaniam, S., and Henderson, R. (2000). Molecular mechanism of vectorial proton translocation by bacteriorhodopsin. *Nature*, **10**, 653–657.

Subramaniam, S., Lindahl, M., *et al.* (1999). Protein conformational changes in the bacteriorhodopsin photocycle. *J. Mol. Biol.*, **287**, 145–1461.

Surewicz, W. K., Mantsch, H. H., *et al.* (1993). Determination of protein secondary structure by Fourier transform infrared spectroscopy: a critical assessment. *Biochemistry*, **32**, 389–394.

Susi, H., and Byler, D. M. (1986). Resolution-enhanced Fourier transform infrared spectroscopy of enzymes. *Meth. Enzymol.*, **130**, 290–311.

Suzuki, Y., Yasunaga, T., Ohkura, R., Wakabayashi, T., and Sutoh, K. (1998). Swing of the lever arm of a myosin motor at the isomerization and phosphate-release steps. *Nature*, **396**, 380–383.

Svergun, D. I. (1999). Restoring low resolution structure of biological macromolecules from solution scattering using simulated annealing. *Biophys. J.*, **76**, 2879–2886.

Svergun, D. I. (2000). Advanced solution scattering data analysis methods and their applications. *J. Appl. Crystallog.*, **33**, 530–534.

Svergun, D. I., Barberato, C., *et al.* (1995). CRYSOL – a program to evalate X-ray solution scattering of biological macromolecules from atomic coordinates. *J. Appl. Crystallogr.*, **28**, 768–773.

Svergun, D. I., Malfois, M., *et al.* (2000). Low resolution structure of the sigma54 transcription factor revealed by X-ray solution scattering. *J. Biol. Chem.*, **275**, 4210–4214.

Svergun, D. I., Richard, S., *et al.* (1998). Protein hydration in solution: experimental observation by x-ray and neutron scattering. *Proc. Natl. Acad. Sci. USA*, **95**, 2267–2272.

Svergun, D. I., Volkov, V. V., *et al.* (1996). New developments in direct shape determination from small-angle scattering 2. Uniqueness. *Acta. Crystallog.*, **A 52**, 419–426.

Taillandier, E., and Liquier, J. (1992). Infrared spectroscopy of DNA. *Meth. Enzymol.*, **211**, 307–335.

Takahashi, H., Nakanishi, T., Kami, K., Arata, Y., and Shimada, I. (2000). A novel NMR method for determining the interfaces of large protein–protein complexes. *Natur. Struct. Biol.*, **7**, 220–223.

Tamm, L. K. (1993). Total internal reflectance florescence microscopy in optical microscopy. In *Emerging Methods and Applications*, eds. Herman and Lemasters. Academic Press.

Tanford, C. (1968). Protein denaturation. *Adv. Protein Chem.*, **23**, 121–282.

Tanford, C., Kawahara, K., and Lapanje, S. (1967). Proteins as random coils. I. Intrinsic viscosities and sedimentation coefficients in concentrated guanidine hydrochloride. *J. Am. Chem. Soc.*, **89**, 729–736.

Tardieu, A., Vachette, P., *et al.* (1981). Biological macromolecules in solvents of variable density: characterization by sedimentation equilibrium, densimetry, and X-ray forward scattering and an application to the 50S ribosomal subunit from *Escherichia coli*. *Biochemistry*, **20**, 4399–4406.

Tarek, M., and Tobias, D. J. (2002). Single-particle and collective dynamics of protein hydration water: a molecular dynamics study. *Phys. Rev. Lett.*, **89**, 275501.

Tcien, R. Y., and Miyawaki, A. (1998). Seeing the machinary of live cells. *Science*, **280**, 1954–1955.

Tehei, M., Franzetti, B., *et al.* (2004). Adaptation to extreme environments: macromolecular dynamics in bacteria compared in vivo by neutron scattering. *EMBO Rep.*, **5**, 66–70.

Tehei, M., Madern, D., *et al.* (2001). Fast dynamics of halophilic malate dehydrogenase and BSA measured by neutron scattering under various solvent conditions influencing protein stability. *Proc. Natl. Acad. Sci. USA*, **98**, 14356–1461.

Tenford, C. (1965). *Physical Chemistry of Macromolecules.* New York: John Wiley and Sons.

Thalhammer, S., Stark, R. W., Müller, S., Weinberg, J., and Heck, W. M. (1997). *J. Struct. Biol.*, **119**, 232–237.

Thomas, G. J., and Tsuboi, M. (1993). Raman spectroscopy of nucleic acids and their complexes. *Adv. Biophys. Chem.*, **3**, 1–69.

Thompson, D. S., and Gill, S. J. (1967). Polymer relaxation times from birefringence relaxation measurements. *J. Chem. Phys.*, **47**, 5008–5017.

Tian F., Al-Hashimi, H. M., Craighead, J. L., and Prestegard, J. H. (2001). Conformational analysis of a flexible oligosaccharide using residual dipolar coupling, *JACS*, **123**, 485–492.

Tinoco, I., Jr., Sauer, K., and Wang, J. C., *Physical Chemistry. Principles and Applications in Biological Science*. Prentice Hall. New Jersey (1998).

Tirado, M. M., Martinez, C. L., and Garcia de la Torre, J. (1984). Comparison of theories for the translational and rotational diffusion coefficients of rod-like macromolecules. Application to short DNA fragments. *J. Chem. Phys.* **81**, 2047–2051.

Tjandra, N., Tate, S., Ono, A., Kainosho, M., and Bax, A. (2000). The NMR structure of a DNA dodecamer in an aqueous dilute liquid crystalline phase. *JACS*, **122**, 6190–6200.

Tolbert, T. J. and Williamson, J. R. (1996). Preparation of specifically deuterated RNA for NMR studies using a combination of chemical and enzymatic synthesis. *JACS*, **118**, 7929–7940.

Tong, L., and Rossmann, M. G. (1997). Rotation function calculations with GLRF program. *Methods in Enzymology*, **276**, 594–611.

Toyoshima, C., Sasabe, H., and Stokes, D. L. (1993). Three-dimensional cryo-electron microscopy of the calcium ion pump in the sarcoplasmic reticulum membrane. *Nature*, **362**, 469–471.

Tristram-Nagle, S., and Nagle, J. F. (2004). Lipid bilayers: thermodynamics, structure, fluctuations, and interactions. *Chem. Phys. Lipids*, **127(1)**, 3–14.

Tsvetkov, V. N. (1989). *Rigid-chain Polymers. Hydrodynamic and Optical Properties in Solution*. Consultants Bureau, New York and London.

Tsvetkov, V. N., Eskin, V. E., and Frenkel, S. Y. (1971). *Structure of Macromolecules in Solution* (translated from Russian), V. 1. Chapter 7. National Lending Library for Science and Technology, Boston, UK.

Ulitin, A. B., Agalarov, S. C., and Serdyuk, I. N. (1997). Preparation of a "beheaded" derivative of the 30S ribosomal subunit. *Biochimie*, **79**, 523–526.

Unger, K. K., Huber, M., Walhagen, K., Hennessy, T. P., and Hearn, M. T. W. (2002). A critical appraisal of capillary electrochromatography. *Anal. Chem.*, **74**, 200A–207A.

Vagin, A., and Teplyakov, A. (2000). An approach to multi-copy search in molecular replacement. *Acta Crystallogr. D Biol. Crystallogr.*, **56**, 1622–1624.

Vale, R. D. (1996). Switches, latches, and amplifier: common themes of proteins and molecular motors. *J. Cell Biol.*, **135**, 291–302.

Valle, M., Zavialov, A., *et al.* (2003). Locking and unlocking of ribosomal motions. *Cell*, **114**, 123–34.

van der Groot, F. G., Gonzàlez-Mañas, J. M., Lakey, J. H. and Pattus, F. (1991). A 'molten globule' membrane-insertion intermediate of the pore-forming domain of colicin A. *Nature*, **354**, 408–410.

van Heel, M., Gowen, B., *et al.* (2000). Single-particle electron cryo-microscopy: towards atomic resolution. *Q. Rev. Biophys.*, **33**, 307–369.

Van Holde (1985). *Physical Biochemistry*. Englewood Cliffs, NJ: Prentice Hall.

Varki, A. (1999). *Essentials of Glycobiology*. Cold Spring Harbor, NY: Cold Spring Harbor Laboratory Press.

Váró, G., and Lanyi, J. K. (1991). Effects of the crystalline structure of purple membrane on the kinetics and energetics of the bacteriorhodopsin photocycle. *Biochemistry*, **30**, 7165–7171.

Velazquez-Campoy, A., Kiso, Y., *et al.* (2001). The binding energetics of first- and second-generation HIV-1 protease inhibitors: implications for drug design. *Arch. Biochem. Biophys.*, **390**, 169–175.

Velazquez-Campoy, A., Leavitt, S. A., *et al.* (2004). Characterization of protein-protein interactions by isothermal titration calorimetry. *Methods Mol. Biol.*, **261**, 35–54.

Venyaminov, S. Y. and Vassilenko, K. S. (1994). Determination of protein tertiary structure class from circular dichroism spectra. *Anal. Biochem.*, **222**, 176–184.

Vénien-Bryan, C., and Fuller, S. D. (1994). The organization of the spike complex of Semliki Forest virus. *J. Mol. Biol.*, **236**, 572–583.

Vysotski, E. S., Liu, Z. J., Rose, J., Wang, B. C., and Lee, J. (1999). Preparation and preliminary study of crystals of the recombinant calcium regulated photoprotein obelin from the bioluminescent hydroid *Obelia longissima*. *Acta Crystallogr.*, **D55**, 1965–1966.

Wakia, S. (1971). Slow motion in shear flow of a doublet of two spheres in contact. *J. Phys. Soc. Jpn*, **31**, 1581–1587.

Walker, J. M. (2000). Electrophoretic techniques. In *Principles and Techniques of Practical Biochemistry*, eds. K. Wilson and J. Walker. Cambridge: Cambridge University Press.

Walther, D., Cohen, F. E., and Doniak, S. (2000). Reconstruction of low-resolution three dimensional density maps from one dimensional small-angle X-ray solution scattering data for biomolecules. *J. Appl. Crystallognr.*, **33**, 350–363.

Wand, A. J., Ehrhardt, M. R., and Flynn, P. F. (1998). High-resolution NMR of encapsulated proteins dissolved in low-viscosity fluids. *PNAS,* **95**, 15299–15302.

Wang, K., Forbes, J. G., and Jin, A. J. (2001). Single molecule measurements of titin elasticity. *Progr. Biophys. Mol. Biol.*, **77**, 1–44.

Wang, M. (1999). Manipulation of single molecules in biology. *Curr. Opin. Biotech.*, **10**, 81–86.

Wang, M. D., Schnitzer, M. J., *et al.* (1998). Force and velocity measured for single molecules of RNA polymerase. *Science*, **282**, 902–907.

Ward, T. J. (1994). Chiral media for capillary electrophoresis. *Anal. Chem.*, **66**, 633A–640A.

Watson, J. D., and Crick, F. H. (1953). Molecular structure of nucleic acids; a structure for deoxyribose nucleic acid. *Nature*, **171**, 737–738.

Webb, R. H. (1996). Confocal optical microscopy. *Rep. Progr. Phys.*, **59**, 427–471.

Weber, G. (1953). Rotational Brownian motion and polarization of the fluorescence of solutions. *Adv. Protein Chem.*, **8**, 415–459.

Weber, K., and Osborn, M. (1969). The reliability of molecular weight determinations by dodecyl sulfate-polyacrylamide gel electrophoresis. *J. Biol. Chem.*, **244**, 4406–4412.

Weber, P. C., and Salemme, F. R. (2003). Applications of calorimetric methods to drug discovery and the study of protein interactions. *Curr. Opin. Struct. Biol.*, **13**, 115–121.

Weik, M. (2003). Low-temperature behavior of water confined by biological macromolecules and its relation to protein dynamics. *Eur. Phys. J. E. Soft Matter*, **12**, 153–158.

Weik, M., Lehnert, U., and Zaccai, G. (2005). Liquid-like water confined in stacks of biological membranes at 200 K and its relation to protein dynamics. *Biophys. J.*, **89**, 3639–3646.

Weik, M., Ravelli, R. B., *et al.* (2000). Specific chemical and structural damage to proteins produced by synchrotron radiation. *Proc. Natl. Acad. Sci. USA*, **97**, 623–628.

Weiskopf, A. S., Vouros, P., and Harvey, D. (1997). Characterization of oligosaccharide composition and structure by quadropole ion trap mass spectrometry. *Rapid Commun. Mass Spectr.*, **11**, 1493–1504.

Weiss, S. (1999). Fluorescence spectroscopy of single biomolecules. *Science*, **283**, 1676–1683.

Weiss, S. (2000). Shattering the diffraction limit of light: a revolution in fluorescence microscopy? *Proc. Nat. Acad. Sci. USA*, **97**, 8747–8749.

Weissman, M., Schindler, H., and Feher, G. (1976). Determination of molecular weights by fluctuation spectroscopy: application to DNA. *Proc. Nat. Acad. Sci. USA*, **73**, 2776–2780.

Westhof, E., Dumas, P., and Moras, D. (1988). *Acta Cryst.*, A**44**, 122–123.

Wetlaufer, D. B. (1962). Ultraviolet spectra of proteins and nucleic acids. *Adv. Prot. Chem.*, **17**, 303–390.

Wider, G., and Wuthrich, K. (1999). NMR spectroscopy of large molecules and multimolecular assemblies in solution. *Curr. Opin. Struct. Biol.*, **9**, 594–601.

Wilkinson, S. R., and Thurston, G. B. (1976). The optical birefringence of DNA solutions induced by oscillatory electric and hydrodynamic fields. *Biopolymers*, **15**, 1555--1572.

Willcox, B. E., Gao, G. F., *et al.* (1999). TCR binding to peptide-MHC stabilizes a flexible recognition interface. *Immunity*, **10**, 357–65.

Williams, K. R., and King, R. W. (1990a). The Fourier transform in chemistry – NMR. Part 3. Multiple-pulse experiments. *J. Chem. Educ.*, **67**, A93–A99.

Williams, K. R., and King, R. W. (1990b). The Fourier transform in chemistry – NMR. Part 4. Two-dimensional methods. *J. Chem. Educ.*, **67**, A125–A137.

Wilson, K., and Walker, J. M. (2000). *Principles and Techniques of Practical Biochemistry*. Cambridge: Cambridge University Press.

Wimberly, B. T., Brodersen, D. E., *et al.* (2000). Structure of the 30S ribosomal subunit. *Nature*, **407**, 327–339.

Winston, R. L., and Fitzgerald, M. C. (1997). Mass spectrometry as a readout of protein structure and function. *Mass Spectr. Rev.*, **16**, 165–179.

Woese, C. R. (1967). *The Genetic Code; The Molecular Basis for Genetic Expression*. New York: Harper & Row.

Wuthrich, K. (1986). *NMR of Proteins and Nucleic Acids*, New York: Wiley-Interscience.

Wuthrich, K. (1995). NMR – this other method for protein and nucleic acid structure determination. *Acta Cryst.*, D**51**, 249–270.

Wuthrich, K. (2000). Protein recognition by NMR, *Nat. Struct. Biol.*, **7**, 188–189.

Wuthrich, K. (2001). The way to NMR structures of protein. *Nat. Struct. Biol.*, **8**, 923–925.

Wuthrich, K., Wider, G., Wagner, G., and Braun, W. (1982). Sequential resonance assignments as a basis for determination of spatial protein structures by high resolution proton nuclear magnetic resonance. *J. Mol. Biol.*, **155**, 311–319.

Wyer, J. R., Willcox, B. E., *et al.* (1999). T cell receptor and coreceptor CD8 alphaalpha bind peptide-MHC independently and with distinct kinetics. *Immunity*, **10**, 219–225.

Xie, X. S., and Dunn, R. C. (1994). Probing single molecule dynamics. *Science*, **265**, 361–364.

Xu, H.-X., and Yeung, E. S. (1997). Direct measurement of single-molecule diffusion and photodecomposition in free solution. *Science*, **275**, 1106–1109.

Xu, Z., Horwich, A. L., and Sigler, P. B. (1997). *Nature*, **388**, 741–750.

Xue, Q., and Yeung, E. S. (1995). Difference in the chemical reactivity of individual molecules of an enzyme. *Nature*, **373**, 681–683.

Yan, Y., Winograd, E., et al. (1993). Crystal structure of the repetitive segments of spectrin. *Science*, **262**, 2027–2030.

Yanagida, T., Kitamura, K., Tanaka, H., Iwane, A. H., and Esaki, S. (2000). Single molecule analysis of the actomyosin motor. *Curr. Opin. Cell. Biol.*, **12,** 20–25.

Yguerabide, J., Epstein, H. F., and Stryer, L. (1970). Segmental flexibility in an antibody molecule. *J. Mol. Biol.*, **51**, 573–590.

Yonath, A., Miissig, J., et al. (1980). Crystallization of the large ribosomal subunit from B. stearothermophilus. *Biochem. Int.*, **1**, 428.

Yoshida, K., Yoshimoto, M., et al. (1998). Fabrication of new substrate for atomic force microscopic observation of DNA molecules from an ultrasmooth sapphire plate. *Biophys. J.*, **74**, 1654–1657.

Yoshizaki, T., and Yamakawa, H. (1980). Dynamics of spheroid–cylindrical molecules in dilute solution. *J. Chem. Phys.*, **72**, 57–69.

Zaccai, G. (2000). How soft is a protein? A protein dynamics force constant measured by neutron scattering. *Science*, **288**, 1604–1607.

Zaccai, G., Morin, P., et al. (1979). Interactions of yeast valyl-tRNA synthetase with RNAs and conformational changes of the enzyme. *J. Mol. Biol.*, **129**, 483–500.

Zanni, M., and Hochstrasser, R. (2001). Two-dimensional infrared spectroscopy: a promising new method for the time resolution of structure. *Curr. Opin. Struct. Biol.*, **11**, 516–562.

Zheng, R., Zheng, X., Dong, J., and Carey, P. R. (2004). Proteins can convert to β-sheet in single crystals. *Protein Science*, **13**, 1288–1294.

Zhou, H.-X. (1995). Calculation of translational friction and intrinsic viscosity. I. General formulation for arbitrarily shaped particles. *Biophys. J.*, **69**, 2286–2297.

Zhou, H.-X. (2001). A unified picture of protein hydration: prediction of hydrodynamic properties from known structures. *Biophys. Chem.* **93**, 171–179.

Zhuang, X., Bartley, L. E., et al. (2000). A single-molecule study of RNA catalysis and folding. *Science*, **288**, 2048–2051.

Zidek, L., Stefl, R., and Sklenar, V. (2001). NMR methodology for the stydy of nucleic acids. *Curr. Opin. Struct. Biol.*, **11**, 275–281.

Zipper, P., and Durchshlag, H. (2000). Prediction of hydrodynamic and small angle scattering parameters from crystal and electron microscopic structure. *J. Appl. Cryst.*, **33**, 788–792.

Zlatanova, J., Lindsay, S. M., and Leuba, A. H. (2000). Single molecule force spectroscopy in biology using the atomic force microscope. *Progr. Biophys. Mol. Biol.*, **74**, 37–61.

Index of eminent scientists

Subject index